INTERNATIONAL RESIDENTIAL CODE®
FOR ONE- AND TWO-FAMILY DWELLINGS

CODE AND COMMENTARY

VOLUME I

2006

INTERNATIONAL
CODE COUNCIL®

First Printing: September 2006

ISBN -13: 978-1-58001-481-6
ISBN -10: 1-58001-481-X

COPYRIGHT © 2006
by
INTERNATIONAL CODE COUNCIL, INC.

PRINTED IN THE U.S.A.

PREFACE

The principal purpose of the Commentary is to provide a basic volume of knowledge and facts relating to building construction as it pertains to the regulations set forth in the 2006 *International Residential Code*. The person who is serious about effectively designing, constructing and regulating buildings and structures will find the Commentary to be a reliable data source and reference to almost all components of the built environment.

As a follow-up to the *International Residential Code*, we offer a companion document, the *International Residential Code Commentary—Volume I*. Volume I covers Chapters 1 through 11 of the 2006 *International Residential Code*. The basic appeal of the Commentary is that it provides in a small package and at reasonable cost thorough coverage of many issues likely to be dealt with when using the *International Residential Code* — and then supplements that coverage with historical and technical background. Reference lists, information sources and bibliographies are also included.

Throughout all of this, strenuous effort has been made to keep the vast quantity of material accessible and its method of presentation useful. With a comprehensive yet concise summary of each section, the Commentary provides a convenient reference for regulations applicable to the construction of buildings and structures. In the chapters that follow, discussions focus on the full meaning and implications of the code text. Guidelines suggest the most effective method of application, and the consequences of not adhering to the code text. Illustrations are provided to aid understanding; they do not necessarily illustrate the only methods of achieving code compliance.

The format of the Commentary includes the full text of each section, table and figure in the code, followed immediately by the commentary applicable to that text. At the time of printing, the Commentary reflects the most up-to-date text of the 2006 *International Residential Code*. As stated in the preface to the *International Residential Code,* the content of sections in the code which begin with a letter designation (i.e., Section 307) are maintained by another code development committee. Each section's narrative includes a statement of its objective and intent, and usually includes a discussion about why the requirement commands the conditions set forth. Code text and commentary text are easily distinguished from each other. All code text is shown as it appears in the *International Residential Code*, and all commentary is indented below the code text and begins with the symbol ❖. All code figures and tables are reproduced as they appear in the IRC. Commentary figures and tables are identified in the text by the word "Commentary" (as in "see Commentary Figure 704.3"), and each has a full border.

Readers should note that the Commentary is to be used in conjunction with the *International Residential Code* and not as a substitute for the code. **The Commentary is advisory only;** the code official alone possesses the authority and responsibility for interpreting the code.

Comments and recommendations are encouraged, for through your input, we can improve future editions. Please direct your comments to the Codes and Standards Development Department at the Chicago District Office.

TABLE OF CONTENTS

Part I — Administrative

Chapter 1:
Administration

General Comments

Chapter 1 of the code is largely concerned with maintaining due process of law in enforcing the performance criteria contained in the body of the code. Only through careful observation of the administrative provisions can the building official reasonably hope to demonstrate that equal protection under the law has been provided. While it is generally assumed that the administrative and enforcement section of a code is addressed to the building official, this is not entirely true. The provisions also establish the rights and privileges of the design professional, the contractor and the building owner. The position of the building official is to review the proposed and completed work and to determine whether the residential structure conforms to the code requirements. The design professional, if one is used, is responsible for the design of the structure. The contractor is responsible for constructing the building in strict accordance with the code and any approved construction documents.

During the course of the construction of a building, the building official reviews the activity to make certain that the intent and letter of the law are being met and that the structure will provide adequate protection for the health, safety and welfare of the users. As a public servant, the building official enforces the code in an unbiased, professional and honest manner. Every individual is guaranteed equal enforcement of the code. Furthermore, design professionals, contractors and building owners have the right of due process for any requirement in the code.

Section R101 establishes the title, scope and purpose of the document. Section R102 establishes the applicability of the code. Section R103 establishes the Department of Building Safety. Section R104 establishes the duties and powers of the building official. Section R105 addresses the requirements for permits. Section R106 establishes the requirements for construction documents. Section R107 addresses the topic of temporary structures and uses. Section R108 establishes permit fees, payment of fees, building permit valuations, and related fees and refunds. Section R109 establishes the requirements for inspections. Section R110 establishes the requirements for occupancy, the issuance of an occupancy certificate and the revocation of an occupancy certificate. Section R111 regulates the connection and disconnection of utilities. Section R112 establishes the board of appeals and its authority. Section R113 addresses the topic of violations of the code. Section R114 establishes the authority for the building official to stop work.

Purpose

A construction code is intended to be adopted as a legally enforceable document that will safeguard health, safety, property and public welfare. A code cannot be effective without adequate provisions for its administration and enforcement. The building official charged with the administration and enforcement of construction regulations has a great responsibility, and with this responsibility goes authority. No matter how detailed the code may be, the building official must exercise judgement in determining code compliance. He or she is responsible for assuring that the homes in which the citizens of the community reside are designed and constructed to be reasonably free from hazards associated with the building's use. *The International Residential Code®* (IRC®) establishes a minimum acceptable level of safety.

SECTION R101
TITLE, SCOPE AND PURPOSE

R101.1 Title. These provisions shall be known as the *Residential Code for One- and Two-family Dwellings* of [NAME OF JURISDICTION], and shall be cited as such and will be referred to herein as "this code."

❖ This code is formally known as the *International Residential Code® (IRC®) for One- and Two-Family Dwellings*, generally referred to as the *International Residential Code* or IRC for short. Upon adoption by the jurisdiction, it is known as the *Residential Code for One- and Two-Family Dwellings* of the adopting jurisdiction, and in the document is often referred to as "this code." It is offered for adoption as a model document of prescriptive provisions to jurisdictions as a stand-alone residential code that establishes minimum regulations for one- and two-family dwellings and townhouses. The forum under which this code is developed encourages consistency of application of its provisions, and it is offered ready for adoption by all communities, large and small, internationally.

R101.2 Scope. The provisions of the *International Residential Code for One- and Two-family Dwellings* shall apply to the construction, alteration, movement, enlargement, replacement, repair, equipment, use and occupancy, location, removal and demolition of detached one- and two-family dwellings and townhouses not more than three stories above-grade in height with a separate means of egress and their accessory structures.

❖ The provisions of the IRC apply to all aspects of construction for detached one- and two-family dwellings, multiple single-family dwellings defined as townhouses and all structures accessory to the dwellings and townhouses. This section sets a limitation in its scope of application to include only those townhouses and dwellings that are up to and including three stories above grade. Additionally, the provisions require each two-family dwelling or townhouse to have separate egress systems for each of the dwelling units. Where a dwelling or townhouse exceeds the allowed height in stories, does not provide individual egress for each dwelling unit or does not conform to the prescriptive provisions of the code, the structures are then beyond the scope of the IRC, and the provisions of the code cannot be applied. The building must then meet the provisions of the *International Building Code®* (IBC®) or other legally adopted building code of the jurisdiction. The actual limiting height of the building, measured in feet and as applied to the height of each story, is limited by the governing provisions for each specific material as found in Chapter 6 of the code. The user of the code will discover that, depending upon which material is selected for the wall construction, the result may be buildings of different permitted heights. For instance, where the wall system is of insulating concrete form construction as prescribed in Section R611, the building is limited to two stories above grade and each story is limited to 10 feet (3048 mm) in height. If wood stud wall framing is used pursuant to the requirements of Section R602, the allowable story height and the overall building height will greatly exceed those permitted for the insulating concrete form wall construction method. The code does not limit the area of the building.

The provisions address all aspects of constructing, altering, repairing, maintaining, using, occupying, enlarging, locating, removing or demolishing any one-family dwelling, two-family dwelling, townhouse or accessory structure. The code regulates any and all activities that modify the buildings as well as any structures that are of incidental use to the main buildings and that are also located on the same lot. The code regulates all construction, plumbing, mechanical, electrical, equipment, fixture and gas piping installations that are done to the building and its operating systems as well as to other structures incidental to the main building and on the same lot. Even work that is specifically exempted from permits must comply with the requirements of the code.

R101.3 Purpose. The purpose of this code is to provide minimum requirements to safeguard the public safety, health and general welfare through affordability, structural strength, means of egress facilities, stability, sanitation, light and ventilation, energy conservation and safety to life and property from fire and other hazards attributed to the built environment.

❖ With the adoption and establishment of a set of minimum construction standards, a community can impose reasonable standards for construction that will maintain the livability of the community while reducing factors that contribute to substandard and hazardous conditions that risk the public health, safety or welfare. Adoption of a modern construction code such as this one raises the level of safety and quality in the built environment, and it is a necessary instrument used to reduce substandard conditions or construction by establishing minimum levels of acceptable construction practice. A reduction in blighted and slum conditions benefits the general public welfare and contributes toward maintenance of a consistent base for the property tax assessments that local governments typically use to fund their general budgets. By applying minimum structural, health, sanitation, fire-safety and life-safety criteria that must be met through the prescriptive or performance provisions of the code, a standard is set that assures the public and individual building occupants they will not be exposed to construction that has gone unchecked or unregulated. The regulation and inspection of plumbing, electrical and mechanical installations also enhances safety for the public health and welfare. The imposition of construction requirements that are in excess of the minimum standards would, in most cases, be considered unreasonable and would most likely suffer from a lack of support, which in turn could undermine the purpose of construction regulations.

SECTION R102
APPLICABILITY

R102.1 General. Where, in any specific case, different sections of this code specify different materials, methods of construction or other requirements, the most restrictive shall govern. Where there is a conflict between a general requirement and a specific requirement, the specific requirement shall be applicable.

❖ The provisions of this section provide guidance to both building officials and other code users on the application of the IRC when different sections of the code specify different materials, methods of construction, or other requirements. The importance of this section should not be understated. It resolves the question of how to handle conflicts between the general and specific provisions found in the code or those instances where different sections specify different requirements. This section provides a necessary hierarchy for application of code provisions and clarifies code applications that would otherwise leave persistent questions and lead to debate. The code requires that where different sections of the code apply but contain different requirements, the most restrictive provisions govern. The code also resolves conflicts between the general requirements of any particular issue with any specific requirements of the same issue by indicating that the specific requirements take precedence over the general requirements.

The following example illustrates the principle. Section R311.5 applies to all stairway types within the purview of the IRC. Section R311.5.3.1 limits the maximum height of risers to $7^3/_4$ inches (196 mm), thus providing a general requirement for stairway riser height. Section R311.5.8.1 limits risers within a spiral stairway to a maximum height of $9^1/_2$ inches (241 mm). This provision is specific to spiral stairways. At first it may appear that these two sections have requirements that are in conflict with one another. However, Sections R311.5.3.1 and R311.5.8.1 are subordinate requirements of Section R311.5. In this case, the specific requirements of Section R311.5.8.1 take precedence over the general requirements of Section R311.5.3.1 in those applications specific to spiral stairways.

Another example would be in relating the requirements for foam plastics to the requirements for wall and ceiling finishes. The code might be interpreted to state that foam plastic boards meeting the requirements of Section R314.3 with a maximum flame-spread rating of 75 could be used as the final surface finish for walls and ceilings because Section R315 allows a flame-spread classification for wall and ceiling finishes with a rating of up to 200. This, however, would be a mistake. The provisions of Section R314.4 require the foam plastic to be covered by a finish material equivalent to a thermal barrier that limits the average temperature rise of the unexposed surface to no more than 250°F (139°C) after 15 minutes of fire exposure to the ASTM E 119 standard time-temperature curve, or to be covered with minimum $1/_2$-inch (12.7 mm) gypsum wallboard. In this case, the uncovered foam plastic must be covered to meet the requirements of Section R314.4 and have the thermal barrier installed. Additionally, the final surface finish material that is chosen must comply with Section R314.4, and it must also meet the required flame-spread rating of 200 or less, as specified in Section R315.

To summarize, where several code sections apply to the use of a material or a method of construction, the most restrictive requirements apply.

R102.2 Other laws. The provisions of this code shall not be deemed to nullify any provisions of local, state or federal law.

❖ Compliance with the requirements of the IRC does not entail authorization, approval or permission to violate the regulations of other local, state or federal laws. Other laws, ordinances and regulations not regulated or enforced by the building official could be in existence and enforced by another authority having juris-

diction over those provisions. Although the requirements may have similar provisions to those of this code, the work must be in conformance with the other regulations.

R102.3 Application of references. References to chapter or section numbers, or to provisions not specifically identified by number, shall be construed to refer to such chapter, section or provision of this code.

❖ There are many instances in the IRC where a reference is merely a chapter number, section number, or in some cases, a provision not specified by number. In all such situations, these references are to the IRC and not some other code or publication.

R102.4 Referenced codes and standards. The codes and standards referenced in this code shall be considered part of the requirements of this code to the prescribed extent of each such reference. Where differences occur between provisions of this code and referenced codes and standards, the provisions of this code shall apply.

> **Exception:** Where enforcement of a code provision would violate the conditions of the listing of the equipment or appliance, the conditions of the listing and manufacturer's instructions shall apply.

❖ The IRC references numerous standards. Chapter 43 is a full list of these standards, which are extensions of the code and become law to the extent to which they are referenced. A standard is a published technical document that represents an industry consensus on how a material or assembly is to be designed, manufactured, tested or installed in order for a specific level of performance to be obtained. Although the code establishes the minimum quality and performance criteria for a material, installation or method of design, the code relies on the reference standards to provide the criteria to determine whether a material or method is in compliance with the code provisions.

Referenced standards in this code are considered part of the code, which allows them to be as enforceable as the code itself. The reference standards, however, are enforceable only to the extent that the code prescribes. This means that only the specific portions of a standard that are specified apply, and the remaining portions of the standard should be considered only as advisory, voluntary, and in some cases, not applicable.

For example, the requirements in Section R703.2 for water-resistive barriers state that the one layer of No. 15 asphalt felt must be free from holes and breaks and comply with ASTM D 226, or any other approved weather-resistant material must be applied over studs or the sheathing of all exterior walls as required by Table R703.4. The referenced standard specification covers asphalt-saturated felts, with or without perforations. It is clear the code does not allow perforations of any kind, simply because the stated code language is "free from holes and breaks." Therefore, only water-resist barriers complying with ASTM D 226 fulfill the code requirements.

The provisions of the code take precedence over the provisions of a referenced standard where differences exist. However, this principle does not apply when enforcement of the provisions of the code would cause a piece of equipment to be installed in violation of the conditions of the listing for the equipment. The equipment and its installation must meet the conditions of the listing and the manufacturer's instructions.

R102.5 Appendices. Provisions in the appendices shall not apply unless specifically referenced in the adopting ordinance.

❖ Provisions of the appendix do not apply unless the jurisdiction has adopted the appendix by statute or ordinance.

R102.6 Partial invalidity. In the event any part or provision of this code is held to be illegal or void, this shall not have the effect of making void or illegal any of the other parts or provisions.

❖ There may be a situation where one or more specific provisions of the code are found to be void or illegal. This may be because a local, state or federal ordinance, statute or law has precedence over the adopted construction provisions. Under such conditions, only those specific provisions found to be void or illegal are affected; the rest of the code remains in force.

R102.7 Existing structures. The legal occupancy of any structure existing on the date of adoption of this code shall be permitted to continue without change, except as is specifically covered in this code, the *International Property Maintenance Code* or the *International Fire Code*, or as is deemed necessary by the building official for the general safety and welfare of the occupants and the public.

❖ Buildings that exist legally at the time the code is adopted are allowed to have their existing use and occupancy continued if the use or occupancy of the structure was also legally in existence. This means that as long as a structure or building remains in a safe and sanitary condition it need not be upgraded to meet the more current standards. However, any new construction, addition or remodeling will require such work to conform to the requirements of the new code. A change of occupancy of the building also will force the building to conform to the new standards.

The existence of a building prior to the adoption of a new edition of the code does not grant it the status of a legal existence. A building is thought of as being "grandfathered" under prior rules and not needing to be brought up to current requirements when there are records to show that it was constructed to meet the regulations of the jurisdiction in force at the time it was built. The most common way to demonstrate legal compliance with the construction codes of a community is through the public records. Copies of past building permits can be researched at the jurisdictional archives. Upon discovery that a building does not have a legal existence, corrective actions will be needed in order to bring the structure into compliance with the

regulations of the jurisdiction at the time the building was built.

R102.7.1 Additions, alterations or repairs. Additions, alterations or repairs to any structure shall conform to the requirements for a new structure without requiring the existing structure to comply with all of the requirements of this code, unless otherwise stated. Additions, alterations or repairs shall not cause an existing structure to become unsafe or adversely affect the performance of the building.

❖ Buildings and structures that have additions, alterations or repairs made to them are required to meet the requirements of the currently adopted code. The existing building is not required to comply with all of the requirements that are required for new buildings; however, the new construction is not allowed to create an unsafe condition in the existing structure. In this case, the existing building elements that would be negatively affected must be made to comply with the new requirements.

SECTION R103
DEPARTMENT OF BUILDING SAFETY

R103.1 Creation of enforcement agency. The department of building safety is hereby created and the official in charge thereof shall be known as the building official.

❖ This section establishes a building department that provides plan review and inspections for buildings regulated by this code. It also establishes the position of building official, who will be the administrator for the enforcement of the jurisdictional codes. The employees of the department may be given varying degrees of authority by the building official. The Department of Building Safety is charged with the responsibility for enforcing the provisions of the code.

R103.2 Appointment. The building official shall be appointed by the chief appointing authority of the jurisdiction.

❖ The building official is an appointed officer of the jurisdiction, charged with the administrative responsibilities of the Department of Building Safety. The building official is appointed to the position by the chief appointing authority of the jurisdiction. Typically the appointment is made by the mayor, council or commission and carried out through the city manager or other administrative authority.

R103.3 Deputies. In accordance with the prescribed procedures of this jurisdiction and with the concurrence of the appointing authority, the building official shall have the authority to appoint a deputy building official, the related technical officers, inspectors, plan examiners and other employees. Such employees shall have powers as delegated by the building official.

❖ The building official has the authority, acting in conjunction and in agreement with the appointing authority of the jurisdiction, to appoint officers and employees of the department. The building official can delegate certain powers of his or her authority to a deputy building official as well as to all technical officers and employees of the department of building safety. This group of deputies typically includes inspectors and plans examiners.

SECTION R104
DUTIES AND POWERS OF THE
BUILDING OFFICIAL

R104.1 General. The building official is hereby authorized and directed to enforce the provisions of this code. The building official shall have the authority to render interpretations of this code and to adopt policies and procedures in order to clarify the application of its provisions. Such interpretations, policies and procedures shall be in conformance with the intent and purpose of this code. Such policies and procedures shall not have the effect of waiving requirements specifically provided for in this code.

❖ The building official is appointed by the legislative body of the jurisdiction to serve as the employee with the authority and responsibility for the proper administration of the code enforcement agency. The building official establishes policies and procedures that will clarify the applications of the code. The development of those policies and procedures should not be simply for the convenience of the jurisdiction's employees but should be viewed as a way to effectively communicate to all interested parties involved in the construction process how the department will process applications, review construction documents, make inspections, approve projects, and determine and clarify the application of the code provisions. Properly developed, these policies and procedures can make the code enforcement department more predictable for those who are regulated and will also establish improved code compliance and public relations.

When interpretation of the code is needed, the building official is the one individual of the jurisdiction with the legal authority to interpret the code and determine how the provisions should be applied, in both general and specific cases. Some departments formalize the interpretation process and require the person with a question to submit their question in writing. Departments are encouraged to develop policies for both formal (written) and informal (verbal) requests for code interpretations. Any such interpretations must be in conformance with the intent and letter of the code and may not waive any requirements. It may be necessary in some cases for the building official to write these code interpretations into the permit.

R104.2 Applications and permits. The building official shall receive applications, review construction documents and issue permits for the erection and alteration of buildings and structures, inspect the premises for which such permits have been issued and enforce compliance with the provisions of this code.

❖ This section states that the building official must receive applications, review construction documents, issue permits, conduct inspections and enforce the provisions of the code. She or he is to provide the services

required to carry the project from application for the permit to final approval. The building official is to accept all properly completed applications and not refuse the receipt of an application that meets the policy requirements. This same principle holds for the review of the construction documents, issuance of permits, inspections and for the enforcement of the code's provisions. The requirements of the code must be met, and approval will be granted only when compliance is verified.

R104.3 Notices and orders. The building official shall issue all necessary notices or orders to ensure compliance with this code.

❖ Building officials are to communicate in writing the disposition of their findings regarding code compliance. If an inspection shows that the work fails to comply with the code provisions, the building official or technical officer who conducted the inspection must issue a written report noting the corrections that are needed. A copy of the report is to be provided to the permit holders or their agent.

R104.4 Inspections. The building official is authorized to make all of the required inspections, or the building official shall have the authority to accept reports of inspection by approved agencies or individuals. Reports of such inspections shall be in writing and be certified by a responsible officer of such approved agency or by the responsible individual. The building official is authorized to engage such expert opinion as deemed necessary to report upon unusual technical issues that arise, subject to the approval of the appointing authority.

❖ This code gives the building official the authority to conduct all required inspections. The building official also has the authority to accept reports from other inspection agencies or private inspectors who have been granted prior approval by the building official to conduct inspections and provide reports. The reports submitted by the approved inspection agencies or individuals must be in writing and must be certified by only those individuals who have been approved by the building official as being qualified to submit the reports. When unusual technical issues arise during the course of construction, the building official has the authority to hire the services of an expert to report on the conditions and technical issues germane to the subject at hand. Prior to hiring the expert for consultation services, the building official must seek approval from the appointing authority.

R104.5 Identification. The building official shall carry proper identification when inspecting structures or premises in the performance of duties under this code.

❖ When the building official and other employees of the jurisdiction are performing their duties and inspecting structures or premises of construction, they are required to carry and display identification that will identify them as employees of the jurisdiction. Commentary Figure R104.5 is an example of the proper identification.

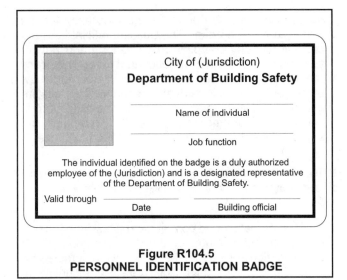

Figure R104.5
PERSONNEL IDENTIFICATION BADGE

R104.6 Right of entry. Where it is necessary to make an inspection to enforce the provisions of this code, or where the building official has reasonable cause to believe that there exists in a structure or upon a premises a condition which is contrary to or in violation of this code which makes the structure or premises unsafe, dangerous or hazardous, the building official or designee is authorized to enter the structure or premises at reasonable times to inspect or to perform the duties imposed by this code, provided that if such structure or premises be occupied that credentials be presented to the occupant and entry requested. If such structure or premises be unoccupied, the building official shall first make a reasonable effort to locate the owner or other person having charge or control of the structure or premises and request entry. If entry is refused, the building official shall have recourse to the remedies provided by law to secure entry.

❖ The building official and employees of the jurisdiction are authorized to conduct inspections in order to enforce the provisions of the jurisdiction's construction codes. Section R109 identifies specific progress points for inspections that must be conducted for both the intermediate phases of the work being done and the final inspection and approval. The officials of the jurisdiction also have the authority to conduct an inspection when the building official has reasonable cause to believe that conditions exist that constitute a violation of the code. The officials of the jurisdiction must take the necessary precautions so the constitutional rights of the owner or tenants are not violated.

The code authorizes the building official to enter and conduct an inspection at reasonable times.

Unless there is an immediate life-threatening condition or an immediate safety concern, the building official must secure an invitation or be requested to conduct a scheduled inspection. Additionally, the building official can conduct an inspection when permission has been granted by the person with immediate control of the premises.

When permission to conduct an inspection is denied, the building official has the authority to seek an inspection warrant.

R104.7 Department records. The building official shall keep official records of applications received, permits and certificates issued, fees collected, reports of inspections, and notices and orders issued. Such records shall be retained in the official records for the period required for the retention of public records.

❖ An important function of the Department of Building Safety is the record keeping associated with the department's functions. It is the responsibility of the building official to maintain adequate and accurate records that can be referenced for a variety of reasons. For example, staffing levels and other budgetary concerns can be better addressed through a historical review of department activity. Also, the department can provide a wealth of information to assist in community development and planning procedures, and a sound record-keeping process best supports legal actions.

Appropriate and complete department records must be maintained for an adequate period of time, based on the value of the information being retained. In all cases, those public records that must be retained by state or local law must be kept for the minimum time period required by the law. Any questions or concerns about the minimum time period for the retention of official records should be directed to the legal representative for the jurisdiction.

R104.8 Liability. The building official, member of the board of appeals or employee charged with the enforcement of this code, while acting for the jurisdiction in good faith and without malice in the discharge of the duties required by this code or other pertinent law or ordinance, shall not thereby be rendered liable personally and is hereby relieved from personal liability for any damage accruing to persons or property as a result of any act or by reason of an act or omission in the discharge of official duties. Any suit instituted against an officer or employee because of an act performed by that officer or employee in the lawful discharge of duties and under the provisions of this code shall be defended by legal representative of the jurisdiction until the final termination of the proceedings. The building official or any subordinate shall not be liable for cost in any action, suit or proceeding that is instituted in pursuance of the provisions of this code.

❖ The building official, members of the appeals board and other employees charged with the duty to enforcing the provisions of the IRC are relieved from personal liability by the jurisdiction when they are acting in good faith during the performance of their official duties. This provision does not provide absolute immunity from all tort liability for employees of the jurisdiction in all cases; an employee acting maliciously is not relieved from his or her personal liability and will most likely not be defended by the jurisdiction. Building officials should not fear lawsuits even if their state does not guarantee them absolute immunity for their actions. Rather, the building official should understand those elements that a plaintiff must show in a lawsuit in order to prevail.

Public officials should familiarize themselves with the laws of their state regarding their exposure to tort liability. Only a few states grant absolute immunity from liability for any actions taken by public officials when those actions are within the scope of their employment.

R104.9 Approved materials and equipment. Materials, equipment and devices approved by the building official shall be constructed and installed in accordance with such approval.

❖ The code is a compilation of criteria with which materials, equipment, devices and systems must comply to be suitable for a particular application. Where the building official grants approval for the use of specific materials, equipment or devices as a part of the construction process, it is important that the approved items be constructed or installed in a manner consistent with the approval. For example, the manufacturer's instructions and recommendations are to be followed if the approval of the material was based even in part on those instructions and recommendations.

The approval authority given the building official is a significant responsibility and is a key to code compliance. The approval process is first technical and then administrative. For example, if data to determine code compliance are required, the data should be in the form of test reports or engineering analysis and not simply taken from a sales brochure.

R104.9.1 Used materials and equipment. Used materials, equipment and devices shall not be reused unless approved by the building official.

❖ In keeping with the authority of the building official to evaluate construction materials based on their equivalency to those specified in the code, the use of used materials and equipment is limited to those approved by the building official. Testing and materials technology has permitted the development of new criteria that old materials may not satisfy. As a result, used materials are often evaluated by the building official in the same way as new materials. It is a common practice to require that used materials and equipment be equivalent to those required by the code if they are to be used in a new installation.

R104.10 Modifications. Wherever there are practical difficulties involved in carrying out the provisions of this code, the building official shall have the authority to grant modifications for individual cases, provided the building official shall first find that special individual reason makes the strict letter of this code impractical and the modification is in compliance with the intent and purpose of this code and that such modification does not lessen health, life and fire safety requirements or structural. The details of action granting modifications shall be recorded and entered in the files of the department of building safety.

❖ The building official has the authority to accept modifications of the code provisions in specific cases. For the building official to allow a modification, he or she must first determine that the strict application of the code is impractical for a specific reason. When the building official grants a modification, it is not a waiver

from the requirements. It should be thought of as fulfilling the requirements to the greatest extent possible but deviating from the requirements slightly to satisfy the intent of the provisions. The modification must not lessen the health, fire-safety, life-safety or structural requirements of the code. All modification actions must be recorded in the files of the building department.

R104.10.1 Areas prone to flooding. The building official shall not grant modifications to any provision related to areas prone to flooding as established by Table R301.2(1) without the granting of a variance to such provisions by the board of appeals.

❖ Section R324 contains provisions for determining flood hazard areas. A modification cannot be granted by the building official for structures located in areas that are prone to flooding without the board of appeals first granting a variance to the provisions. The regulations of the National Flood Insurance Program (NFIP) 44 CFR 60.3 require that proposals meet or exceed the minimum provisions of the program. Requests for modifications to any provision related to flood hazard areas are to be handled as formal variances. The criteria for issuance of variances are given in Section R112.2.2.

R104.11 Alternative materials, design and methods of construction and equipment. The provisions of this code are not intended to prevent the installation of any material or to prohibit any design or method of construction not specifically prescribed by this code, provided that any such alternative has been approved. An alternative material, design or method of construction shall be approved where the building official finds that the proposed design is satisfactory and complies with the intent of the provisions of this code, and that the material, method or work offered is, for the purpose intended, at least the equivalent of that prescribed in this code. Compliance with the specific performance-based provisions of the *International Codes* in lieu of specific requirements of this code shall also be permitted as an alternate.

❖ Although the code reflects current technologies, it is impossible to foresee all potential applications of new materials, construction techniques or design methods. The code encourages the use of new materials and technologies by allowing them to be presented to the building official for approval. The building official must approve a proposed alternative when it is found to be satisfactory and in compliance with the intent of the provisions of the code and is equivalent to that prescribed by the code. Approval may also be granted for the use of any alternative that is in compliance with the performance-based provisions from the other members of the International Code Council® codes family.

R104.11.1 Tests. Whenever there is insufficient evidence of compliance with the provisions of this code, or evidence that a material or method does not conform to the requirements of this code, or in order to substantiate claims for alternative materials or methods, the building official shall have the authority to require tests as evidence of compliance to be made at no expense to the jurisdiction. Test methods shall be as specified in this code or by other recognized test standards. In the absence of recognized and accepted test methods, the building official shall approve the testing procedures. Tests shall be performed by an approved agency. Reports of such tests shall be retained by the building official for the period required for retention of public records.

❖ The building official has the authority to require tests to substantiate the claim that an alternative is equivalent and meets the intent of the code. Any tests must be in compliance with those specified in the code or other recognized test standards approved by the building official. The cost of any tests will be borne by the proponent seeking the approval of the alternative.

SECTION R105
PERMITS

R105.1 Required. Any owner or authorized agent who intends to construct, enlarge, alter, repair, move, demolish or change the occupancy of a building or structure, or to erect, install, enlarge, alter, repair, remove, convert or replace any electrical, gas, mechanical or plumbing system, the installation of which is regulated by this code, or to cause any such work to be done, shall first make application to the building official and obtain the required permit.

❖ This section lists the types of work or installations of equipment or utilities that will require an owner or authorized agent to obtain permits, which are to be obtained before work begins. In general, a permit is required for all activities that are regulated by the code, and these activities cannot begin until the permit is issued.

R105.2 Work exempt from permit. Permits shall not be required for the following. Exemption from permit requirements of this code shall not be deemed to grant authorization for any work to be done in any manner in violation of the provisions of this code or any other laws or ordinances of this jurisdiction.

Building:

1. One-story detached accessory structures used as tool and storage sheds, playhouses and similar uses, provided the floor area does not exceed 120 square feet (11.15 m²).

2. Fences not over 6 feet (1829 mm) high.

3. Retaining walls that are not over 4 feet (1219 mm) in height measured from the bottom of the footing to the top of the wall, unless supporting a surcharge.

4. Water tanks supported directly upon grade if the capacity does not exceed 5,000 gallons (18 927 L) and the ratio of height to diameter or width does not exceed 2 to 1.

5. Sidewalks and driveways.

6. Painting, papering, tiling, carpeting, cabinets, counter tops and similar finish work.

7. Prefabricated swimming pools that are less than 24 inches (610 mm) deep.

8. Swings and other playground equipment.

9. Window awnings supported by an exterior wall which do not project more than 54 inches (1372 mm) from the exterior wall and do not require additional support.

Electrical:

Repairs and maintenance: A permit shall not be required for minor repair work, including the replacement of lamps or the connection of approved portable electrical equipment to approved permanently installed receptacles.

Gas:

1. Portable heating, cooking or clothes drying appliances.

2. Replacement of any minor part that does not alter approval of equipment or make such equipment unsafe.

3. Portable-fuel-cell appliances that are not connected to a fixed piping system and are not interconnected to a power grid.

Mechanical:

1. Portable heating appliances.

2. Portable ventilation appliances.

3. Portable cooling units.

4. Steam, hot or chilled water piping within any heating or cooling equipment regulated by this code.

5. Replacement of any minor part that does not alter approval of equipment or make such equipment unsafe.

6. Portable evaporative coolers.

7. Self-contained refrigeration systems containing 10 pounds (4.54 kg) or less of refrigerant or that are actuated by motors of 1 horsepower (746 W) or less.

8. Portable-fuel-cell appliances that are not connected to a fixed piping system and are not interconnected to a power grid.

The stopping of leaks in drains, water, soil, waste or vent pipe; provided, however, that if any concealed trap, drainpipe, water, soil, waste or vent pipe becomes defective and it becomes necessary to remove and replace the same with new material, such work shall be considered as new work and a permit shall be obtained and inspection made as provided in this code.

The clearing of stoppages or the repairing of leaks in pipes, valves or fixtures, and the removal and reinstallation of water closets, provided such repairs do not involve or require the replacement or rearrangement of valves, pipes or fixtures.

❖ This section of the code lists the types of work in five categories—building, electrical, gas, mechanical and plumbing—that do not require permits. However, all work, even work that does not require a permit, must be done in a manner that will comply with the code requirements.

R105.2.1 Emergency repairs. Where equipment replacements and repairs must be performed in an emergency situation, the permit application shall be submitted within the next working business day to the building official.

❖ Occasionally repairs or replacement work must be done under emergency conditions. The code does not intend that such emergency work be held up until the necessary permits are secured. It is important, however, that the permit application be obtained as quickly as possible once the emergency has been controlled. Any required permit must be applied for within the next business day following the emergency repair or replacement.

R105.2.2 Repairs. Application or notice to the building official is not required for ordinary repairs to structures, replacement of lamps or the connection of approved portable electrical equipment to approved permanently installed receptacles. Such repairs shall not include the cutting away of any wall, partition or portion thereof, the removal or cutting of any structural beam or load-bearing support, or the removal or change of any required means of egress, or rearrangement of parts of a structure affecting the egress requirements; nor shall ordinary repairs include addition to, alteration of, replacement or relocation of any water supply, sewer, drainage, drain leader, gas, soil, waste, vent or similar piping, electric wiring or mechanical or other work affecting public health or general safety.

❖ There is a variety of ordinary repair, replacement or connection work that is exempt from the permit application process. This section identifies a number of general situations in which a permit is not required. The provisions then state the types of repairs for which a permit is required. Repair work done without a permit must still comply with the applicable provisions of the code.

R105.2.3 Public service agencies. A permit shall not be required for the installation, alteration or repair of generation, transmission, distribution, metering or other related equipment that is under the ownership and control of public service agencies by established right.

❖ When the ownership and control of equipment is held by a public service agency, such as a county water district, permits are not required for any work that might be done on that equipment. The scope of this provision includes not only repair activities but also any installation or alteration work. It is clear from this section that public service agencies are self-regulating when it comes to work involving equipment used for generation, transmission, distribution and metering.

R105.3 Application for permit. To obtain a permit, the applicant shall first file an application therefor in writing on a form furnished by the department of building safety for that purpose. Such application shall:

1. Identify and describe the work to be covered by the permit for which application is made.

2. Describe the land on which the proposed work is to be done by legal description, street address or similar description that will readily identify and definitely locate the proposed building or work.

3. Indicate the use and occupancy for which the proposed work is intended.

4. Be accompanied by construction documents and other information as required in Section R106.1.

5. State the valuation of the proposed work.

6. Be signed by the applicant or the applicant's authorized agent.

7. Give such other data and information as required by the building official.

❖ The code lists the minimum information required in an application for a permit. The owner or agent is to fully describe the location of the site, the type and nature of the work to be done and all other pertinent information regarding the job. This provides the jurisdiction with a clear understanding of what will actually be done under the permit. It is a common belief by some owners that all that is required of them is to complete an application form, and then the permit will be issued over the counter. Although issuance of over-the-counter permits may be practical when a water heater is to be installed or for other minor work, this is not the case for more complex work. The general public is often surprised when they discover a complete application includes the requirements to provide construction drawings as required by Section R106, detailed information about the property, engineering data and other information as required by the building department. The department should develop informational fliers and packets that will assist the applicant with understanding the requirements for obtaining a permit. Recognizing that incomplete applications do not serve the applicant's best interest (and create unnecessary delays with processing applications), many building departments have personnel who specialize in providing assistance to customers to ease them through the process.

R105.3.1 Action on application. The building official shall examine or cause to be examined applications for permits and amendments thereto within a reasonable time after filing. If the application or the construction documents do not conform to the requirements of pertinent laws, the building official shall reject such application in writing, stating the reasons therefor. If the building official is satisfied that the proposed work conforms to the requirements of this code and laws and ordinances applicable thereto, the building official shall issue a permit therefor as soon as practicable.

❖ When an application is determined to be incomplete, or when the construction documents do not show compliance with the code provisions or other requirements of the jurisdiction, the building official is to reject them. When the construction documents are rejected, the building official is required to list the reasons for the rejection for the applicant explaining why the plans have been rejected, so the applicant will know what action to take for subsequent approval.

When the application and construction documents are determined to be in compliance with the requirements of the code and other regulations of the jurisdiction, the building official must issue a permit. This section mandates the issuance of a permit when the plans and other data show compliance with all regulations of the jurisdiction, and it mandates that the permit be issued without delay.

R105.3.1.1 Determination of substantially improved or substantially damaged existing buildings in flood hazard areas. For applications for reconstruction, rehabilitation, addition or other improvement of existing buildings or structures located in an area prone to flooding as established by Table R301.2(1), the building official shall examine or cause to be examined the construction documents and shall prepare a finding with regard to the value of the proposed work. For buildings that have sustained damage of any origin, the value of the proposed work shall include the cost to repair the building or structure to its predamage condition. If the building official finds that the value of proposed work equals or exceeds 50 percent of the market value of the building or structure before the damage has occurred or the improvement is started, the finding shall be provided to the board of appeals for a determination of substantial improvement or substantial damage. Applications determined by the board of appeals to constitute substantial improvement or substantial damage shall meet the requirements of Section R324.

❖ The definitions of "substantial damage" and "substantial improvement" are included in federal regulations (44 CFR 59.1 Definitions). The following is from the National Flood Insurance Program regulations:

Substantial damage means damage of any origin sustained by a structure whereby the cost of restoring the structure to its before-damaged condition would equal or exceed 50 percent of the market value of the structure before the damage occurred.

Substantial improvement means any reconstruction, rehabilitation, addition, or other improvement of a structure, the cost of which equals or exceeds 50 percent of the market value of the structure before the "start of construction" of the improvement. This term includes structures [that] have incurred "substantial damage," regardless of the actual repair work performed. The term does not, however, include either:

1. Any project for improvement of a structure to correct existing violations of state or local health, sanitary, or safety code specifications [that] have been identified by the local code enforcement official and [that] are the minimum necessary to assure safe living conditions, or

2. Any alteration of a "historic structure," provided that the alteration will not preclude the structure's continued designation as an "historic structure."

These definitions are consistent with the definitions in Section 1612.2 of the *International Building Code*. "Substantial improvement" is also defined in Section R112.2.1 of the IBC.

Long-term reduction in exposure to flood hazards, including exposure of older buildings, is one of the purposes for regulating floodplain development. Existing buildings or structures located in flood hazard areas are to be brought into compliance with the flood-resistance provisions of Section R324 when the value of improvements or the repair of damage exceeds a certain value.

Section R105.3 requires the applicant to state the valuation of the proposed work as part of the information submitted to obtain a permit. If the proposed work will be performed on existing buildings or structures in flood hazard areas, including restoration of damage from any cause, the building official is to prepare a finding to determine the value of the proposed work, including the value of the property owner's labor, as well as the value of donated labor and materials. For damaged buildings, the value of the proposed work is the value of work necessary to restore the building to its pre-damage condition, even if the applicant is proposing less work.

To make a determination about whether a proposed repair, reconstruction, rehabilitation, addition or improvement of a building or structure will constitute a substantial improvement or correction of substantial damage, the cost of the proposed work is to be compared to the market value of the building or structure before the work is started. To determine market value, the building official may require the applicant to provide such information as allowed under Section R105.3. For additional guidance, refer to FEMA 213: *Answers to Questions about Substantially Damaged Buildings*, and FEMA 311: *Guidance on Estimating Substantial Damage Using the NFIP Residential Substantial Damage Estimator*.

The building official's finding is to be provided to the board of appeals if the finding is that the value of the proposed work equals or exceeds 50 percent of the market value of the building, less land value. The board of appeals is to determine whether the proposed work constitutes a substantial improvement of the existing building or whether a damaged building sustained substantial damage. If the board finds that the work is a substantial improvement, the existing building is to be brought into compliance with the flood-resistance provisions of Section R324.

R105.3.2 Time limitation of application. An application for a permit for any proposed work shall be deemed to have been abandoned 180 days after the date of filing unless such application has been pursued in good faith or a permit has been issued; except that the building official is authorized to grant one or more extensions of time for additional periods not exceeding 180 days each. The extension shall be requested in writing and justifiable cause demonstrated.

❖ Applications for permits are considered valid for 180 days. The permit application and review process must be done in a timely manner within that period. The applicant must be responsive to requests for additional information made by the building department. The 180-day limitation is not intended to penalize an applicant for the lack of action on the part of the jurisdiction. It is merely a measure that is used to void an application when it is no longer reasonable to keep it active because the applicant is delaying the process and is not responding to legitimate requests for information. The building official can extend the time limit of an application in increments of 180 days, provided the applicant can show a valid reason for an extension. The applicant must make this request in writing.

R105.4 Validity of permit. The issuance or granting of a permit shall not be construed to be a permit for, or an approval of, any violation of any of the provisions of this code or of any other ordinance of the jurisdiction. Permits presuming to give authority to violate or cancel the provisions of this code or other ordinances of the jurisdiction shall not be valid. The issuance of a permit based on construction documents and other data shall not prevent the building official from requiring the correction of errors in the construction documents and other data. The building official is also authorized to prevent occupancy or use of a structure where in violation of this code or of any other ordinances of this jurisdiction.

❖ A permit authorizes the permit holder to proceed with work that complies with the code requirements. A permit is not valid if it is issued for work that does not comply with the code requirements. If, after the permit has been issued, errors in the plans or construction are discovered, the building official has the authority to require correction of the plans or the work to comply with the requirements of the code.

R105.5 Expiration. Every permit issued shall become invalid unless the work authorized by such permit is commenced within 180 days after its issuance, or if the work authorized by such permit is suspended or abandoned for a period of 180 days after the time the work is commenced. The building official is authorized to grant, in writing, one or more extensions of time, for periods not more than 180 days each. The extension shall be requested in writing and justifiable cause demonstrated.

❖ Once a permit has been issued, the permit holder has 180 days to begin the construction work; otherwise, the permit will become invalid. If at any time the work stops for a period of 180 days or more, the permit is invalid. Extensions can be granted by the building official in writing in increments of 180 days when the applicant makes a written request.

R105.6 Suspension or revocation. The building official is authorized to suspend or revoke a permit issued under the provisions of this code wherever the permit is issued in error or on the basis of incorrect, inaccurate or incomplete information, or in violation of any ordinance or regulation or any of the provisions of this code.

❖ Any permit that has been issued based on false, misleading or incorrect information can be revoked or suspended by the authority of the building official. The building official also has the authority to suspend or revoke any permit issued when a violation exists with regard to the building code or any other ordinance, code, law or regulation that is legally in effect in the jurisdiction.

R105.7 Placement of permit. The building permit or copy thereof shall be kept on the site of the work until the completion of the project.

❖ The permit must be displayed at the work site until the certificate of occupancy has been issued. Because paperwork at a job site is sometimes lost, the code allows a copy of the permit to be kept at the site and the original to be retained in a more secure place. Keeping a record of permits at the project location satisfies the legal requirements set forth in the code, and any interested party can verify that a valid permit has been obtained.

R105.8 Responsibility. It shall be the duty of every person who performs work for the installation or repair of building, structure, electrical, gas, mechanical or plumbing systems, for which this code is applicable, to comply with this code.

❖ This is one of several code provisions that emphasize the required compliance with the code provisions for every aspect of the project. Although the permit holder is designated as having the primary responsibility for overall code compliance, it is the responsibility of each and every person working on the job to adhere to the requirements of the code.

SECTION R106
CONSTRUCTION DOCUMENTS

R106.1 Submittal documents. Construction documents, special inspection and structural observation programs and other data shall be submitted in one or more sets with each application for a permit. The construction documents shall be prepared by a registered design professional where required by the statutes of the jurisdiction in which the project is to be constructed. Where special conditions exist, the building official is authorized to require additional construction documents to be prepared by a registered design professional.

Exception: The building official is authorized to waive the submission of construction documents and other data not required to be prepared by a registered design professional if it is found that the nature of the work applied for is such that reviewing of construction documents is not necessary to obtain compliance with this code.

❖ This section provides the minimum requirements for construction documents that an applicant must provide along with the permit application form for the application package to be considered complete. Construction documents are not just a set of drawings. Construction documents are the entire set of all submitted forms and information necessary to accurately communicate the scope of the construction. The submittals may include written special inspection and structural observation programs, construction drawings and details, reports, calculations, specifications, shop drawings, manufacturer's installation instructions, site plans and other graphic and written forms that will describe the proposed work in detail. The building official can waive the submission of construction documents for types of work where the review of documents is not necessary to show compliance with the requirements of the code.

This code is prescriptive and makes possible the design of a dwelling or townhouse without the requirement for a licensed design professional. However, the construction documents must be prepared by a licensed design professional when required by the statutes of the state or jurisdiction. Additionally, the building official has the authority to require that plans be prepared by licensed design professionals when not otherwise required by the statutes of the jurisdiction if in the opinion of the building official special or unique conditions exist or if the design of a building does not meet the prescriptive provisions of the code.

R106.1.1 Information on construction documents. Construction documents shall be drawn upon suitable material. Electronic media documents are permitted to be submitted when approved by the building official. Construction documents shall be of sufficient clarity to indicate the location, nature and extent of the work proposed and show in detail that it will conform to the provisions of this code and relevant laws, ordinances, rules and regulations, as determined by the building official.

❖ The emphasis of this section is on the clarity, completeness and accuracy of the construction documents. A wide variety of individuals will be using the construction documents to perform their specific tasks. Therefore, it is critical that there be no confusion about the intent of the designer based on the information in the plans and other documents.

Electronic submittal of construction documents is rapidly gaining popularity. Where the department of public safety has the means to review plans electronically, it may request such submittals. Many departments continue to use hard copy documents for the plan review process, with a request for electronic media copies of documents for archival purposes.

R106.1.2 Manufacturer's installation instructions. Manufacturer's installation instructions, as required by this code, shall be available on the job site at the time of inspection.

❖ Throughout the IRC, the code directs that materials or equipment be installed in accordance with the manufacturer's installation instructions. An example is the installation of modified bitumen roofing as set forth in Section R905.11.3. The code recognizes that the manufacturer can best relate the specific installation requirements applicable to its specific product. Where the code mandates that the manufacturer's installation instructions be followed, those instructions must be available at the job site. In this way, both the installers and the inspector are able to see that the directives of the manufacturer are being followed.

R106.1.3 Information for construction in flood hazard areas. For buildings and structures located in whole or in part in flood hazard areas as established by Table R301.2(1), construction documents shall include:

1. Delineation of flood hazard areas, floodway boundaries and flood zones and the design flood elevation, as appropriate;

2. The elevation of the proposed lowest floor, including basement; in areas of shallow flooding (AO zones), the

height of the proposed lowest floor, including basement, above the highest adjacent grade; and

3. The elevation of the bottom of the lowest horizontal structural member in coastal high hazard areas (V Zone); and

4. If design flood elevations are not included on the community's Flood Insurance Rate Map (FIRM), the building official and the applicant shall obtain and reasonably utilize any design flood elevation and floodway data available from other sources.

❖ This section details the information to be included in an application for a permit to build within a flood hazard area. The site plan is to show sufficient detail and information about the designated flood hazard area, including floodway and flood zones, to allow for a complete review of the proposed activities. Flood Insurance Rate Maps (FIRM) are flood hazard maps prepared by the Federal Emergency Management Agency (FEMA). FIRMs may show specific Base Flood Elevations (BFE). If the community adopts a flood hazard map other than the FIRM, the Design Flood Elevations (DFE) must be at least as high as the BFEs [see Table R301.2(1)].

Construction documents are to include the proposed elevation of the lowest floor and the elevation of the bottom of the lowest horizontal structural member, which is to be at or above the minimum as given in Section R324. In flood hazard areas except coastal high hazard areas, applicants may propose placing fill with the intent of later constructing buildings with excavated basements. When excavated into fill, basements may be subject to damage, especially where waters remain high for more than a few hours. Fill materials can become saturated and provide inadequate support, or water pressure can collapse below-grade walls. Basements below residential buildings are not to be constructed below the DFE, even if excavated into fill that is placed above the DFE.

If elevated on individual fill pads in flood hazard areas, buildings will be surrounded by water during general conditions of flooding. Local emergency personnel responsible for evacuations should be consulted during the subdivision approval process. Many states and communities have provisions that require uninterrupted access to all buildings during a flood. These requirements may be administered by agencies responsible for permitting the construction of public rights-of-way or may be imposed as part of subdivision approval.

Many FIRMs show flood hazard areas without specifying the BFEs, indicating that FEMA has not prepared engineering analyses for those areas. These flood hazard areas are often referred to as unnumbered A zones. An important step in regulating development of these areas is the determination of the DFE. The building official and the applicant are to search for and use data from other sources, which may include the U.S. Army Corps of Engineers, the Natural Resources Conservation Service, the state, a local flood control agency or district, or historical records. If flood elevation information is not available, the building official may require the applicant to develop the DFE in accordance with accepted engineering practices. Local officials unfamiliar with establishing DFEs are encouraged to contact the NFIP State Coordinator or the appropriate FEMA regional office. For additional guidance, refer to FEMA 265: *Managing Floodplain Development in Approximate Zone A Areas: A Guide for Obtaining and Developing Base (100-Year) Flood Elevations.*

R106.2 Site plan. The construction documents submitted with the application for permit shall be accompanied by a site plan showing the size and location of new construction and existing structures on the site and distances from lot lines. In the case of demolition, the site plan shall show construction to be demolished and the location and size of existing structures and construction that are to remain on the site or plot.

❖ One valuable part of the construction documents is the site plan. As a part of reviewing the building's location on the site for conformance with the code, the building official must know the size and location of any other structures on the lot as well as their physical relationship to the new structure. The proximity of the new structure to lot lines and any public ways must also be shown. The distance between a building and the lot lines may trigger a variety of code requirements, the most notable being exterior wall fire-resistance rating and opening prohibition (see Section R302).

The code also requires that a site plan be submitted when a building or structure is to be demolished. The plan must identify the location of the building to be demolished as well as any surrounding structures that will remain in place.

R106.3 Examination of documents. The building official shall examine or cause to be examined construction documents for code compliance.

❖ The building official must review the construction documents or have qualified employees or consultants examine the plans to determine whether they comply with the requirements of the jurisdiction.

R106.3.1 Approval of construction documents. When the building official issues a permit, the construction documents shall be approved, in writing or by a stamp which states "APPROVED PLANS PER IRC SECTION R106.3.1." One set of construction documents so reviewed shall be retained by the building official. The other set shall be returned to the applicant, shall be kept at the site of work and shall be open to inspection by the building official or his or her authorized representative.

❖ Approval of the construction documents is the first in a series of reviews and approvals throughout the design and construction process. The building official or authorized representative must indicate that the construction documents are approved for construction, in writing or by a stamp which specifically states "APPROVED PLANS PER IRC R106.3.1" and including any additional information that is necessary for the project. A set of the documents must be retained by the

building department as its record for the life of the project. It is not uncommon throughout the job for questions to arise that require referencing the approved set of plans in the office. An efficient filing system should be developed to make the retrieval of construction documents a simple process.

In addition to the set of construction documents retained by the department of building safety, at least one set of approved plans is to be returned to the permit holder. These construction documents must be maintained at the job site for reference purposes throughout the project. To avoid confusion, they must duplicate the documents that were approved and stamped. They must be available for review by building department personnel during the numerous inspections that may take place. Additionally, the contractor cannot determine compliance with approved construction documents unless those approved documents are available. Another reason to have the documents available is that these plans will generally indicate any special items or issues identified by the plans examiner that may not be shown on any other construction documents. The approved plans usually must be available at the job site prior to an inspection.

R106.3.2 Previous approvals. This code shall not require changes in the construction documents, construction or designated occupancy of a structure for which a lawful permit has been heretofore issued or otherwise lawfully authorized, and the construction of which has been pursued in good faith within 180 days after the effective date of this code and has not been abandoned.

❖ The code does not require work to be brought up to the requirements of a code that was adopted since the edition of the code under which the existing valid permit was issued. If the work is abandoned for 180 days or more, and an extension of the permit is not granted, the original permit is no longer valid, and a new permit must be sought under the provisions of the latest edition of the code.

R106.3.3 Phased approval. The building official is authorized to issue a permit for the construction of foundations or any other part of a building or structure before the construction documents for the whole building or structure have been submitted, provided that adequate information and detailed statements have been filed complying with pertinent requirements of this code. The holder of such permit for the foundation or other parts of a building or structure shall proceed at the holder's own risk with the building operation and without assurance that a permit for the entire structure will be granted.

❖ Phased approval is needed for projects that use the "fast track" construction method, which allows construction to begin before completion of all of the plans and specifications. Although it is preferable to issue permits for projects in their entirety, the building official has the authority to issue, at his or her discretion, a permit for a portion of the construction. The building official must be satisfied that the information provided shows in satisfactory detail that the partial construction will conform to the requirements of the code. In such a case, the permit holder proceeds at his or her own risk with the construction and has no assurance of the entire permit ever being issued.

R106.4 Amended construction documents. Work shall be installed in accordance with the approved construction documents, and any changes made during construction that are not in compliance with the approved construction documents shall be resubmitted for approval as an amended set of construction documents.

❖ The code requires that all work to be done in accordance with the approved plans and other construction documents. Where the construction will not conform to the approved construction documents, the documents must be revised, and they must be resubmitted to the building official for review and approval. The building official must retain one set of the amended and approved plans. The other set is to be kept at the construction site, ready for use by the jurisdiction's inspection staff.

R106.5 Retention of construction documents. One set of approved construction documents shall be retained by the building official for a period of not less than 180 days from date of completion of the permitted work, or as required by state or local laws.

❖ Construction documents must be retained in case a dispute arises after completion of the project. Unless modified because of state or local statutes, the retention period for the approved construction documents is a minimum of 180 days following the completion of the work, typically the date the certificate of occupancy is issued. Any further retention of plans by the jurisdiction as an archival record of construction activity in the community is not required by the code.

SECTION R107
TEMPORARY STRUCTURES AND USES

R107.1 General. The building official is authorized to issue a permit for temporary structures and temporary uses. Such permits shall be limited as to time of service, but shall not be permitted for more than 180 days. The building official is authorized to grant extensions for demonstrated cause.

❖ The building official can authorize temporary use permits when the applicant has met all regulations governing such use. Permits for temporary structures may also be granted subject to the provisions of this section. Although the permit is to be granted for a time period consistent with the temporary use, 180 days is the maximum period of time for which a temporary use or structure can be valid. This code section often applies to structures that are commonly used for a short duration of time at a specific location, then easily dismantled and removed, often to be reconstructed at a different site. It is common for this type of structure to be used at events like street fairs, carnivals, circuses, parades, sporting events, weddings, concerts or revivals, or they may also be used to house construction offices. These structures range in size from small tents or shade structures, which may

house a vendor at a street fair, to large tents for circuses or religious revivals. Because of the large size, large occupant load and complexity that can result from smaller structures being grouped together, the fire- and life-safety concerns as covered in Section R107.2 must be duly considered before a permit is issued.

R107.2 Conformance. Temporary structures and uses shall conform to the structural strength, fire safety, means of egress, light, ventilation and sanitary requirements of this code as necessary to ensure the public health, safety and general welfare.

❖ This section gives the building official discretion in determining the specific criteria for conformance. The issues of structural strength, fire safety, egress, light, ventilation and sanitation are mentioned in this section as the key areas of concern for temporary structures or uses. The levels of required conformance must be determined by the building official to achieve the jurisdiction's required level of safety, health and welfare. Full compliance with the code is not required for a temporary structure or temporary use, but it is clear that the level of performance set forth by the code must be considered in the development of the requirements.

R107.3 Temporary power. The building official is authorized to give permission to temporarily supply and use power in part of an electric installation before such installation has been fully completed and the final certificate of completion has been issued. The part covered by the temporary certificate shall comply with the requirements specified for temporary lighting, heat or power in the ICC *Electrical Code.*

❖ The *International Code Council Electrical Code Administrative Provisions* is referenced regarding temporary lighting, heat and power. The code allows temporary electrical service to be provided prior to the completion of the entire electrical system if the building official approves such a connection.

R107.4 Termination of approval. The building official is authorized to terminate such permit for a temporary structure or use and to order the temporary structure or use to be discontinued.

❖ Where the use of a temporary structure is not consistent with that approved by the building official, or where an unsafe condition exists, the building official is authorized to terminate the temporary permit.

SECTION R108
FEES

R108.1 Payment of fees. A permit shall not be valid until the fees prescribed by law have been paid. Nor shall an amendment to a permit be released until the additional fee, if any, has been paid.

❖ This section addresses the costs necessary to operate a department of building safety by creating a mechanism for fee collection. Such fees are typically set to provide enough funds to adequately pay for the costs of operating the various department functions, including administration, plans examination and inspection.

All fees, including those for changes to the permit, are to be collected before a valid permit is issued.

R108.2 Schedule of permit fees. On buildings, structures, electrical, gas, mechanical and plumbing systems or alterations requiring a permit, a fee for each permit shall be paid as required, in accordance with the schedule as established by the applicable governing authority.

❖ The code states that fees for activities regulated by the jurisdiction and administered by the building official should be equitably assessed and adequate to fund the administration, inspection and plan review services required by the code. Exorbitant fees are unreasonable and can often be a basis for some citizens' distrust of government agencies, while unrealistically low fees serve little purpose. In either situation, the effectiveness of the department suffers from either distrust and poor relations with the public or under-trained and under-staffed departments that are hopelessly unable to retain the best-qualified employees. Either fee schedule does not serve the best interests of the public, which depends on the jurisdiction to enforce the code provisions effectively to maintain a minimum level of safety in the built environment. An example of a building department fee schedule is shown in Appendix L.

R108.3 Building permit valuations. Building permit valuation shall include total value of the work for which a permit is being issued, such as electrical, gas, mechanical, plumbing equipment and other permanent systems, including materials and labor.

❖ Most jurisdictions develop a fee schedule based on the projected construction cost of the work to be done. Two methods are used to determine this cost valuation: 1) a "per-square-foot" factor based on the use or occupancy of the building and the type of construction involved, or 2) the "bid cost" factor based on the total accepted bid price for doing the work. The valuation is determined by including the value of the construction process, including both materials and labor. It is important that a realistic valuation be determined for every project so that permit fees are applied fairly and accurately.

R108.4 Related fees. The payment of the fee for the construction, alteration, removal or demolition for work done in connection with or concurrently with the work authorized by a building permit shall not relieve the applicant or holder of the permit from the payment of other fees that are prescribed by law.

❖ A building permit and the fees attached to that permit do not necessarily cover all aspects of the work to be performed. All fees of the jurisdiction, including those for additional permits, reinspections, investigations or other departmental functions, must be paid along with those for a building permit.

R108.5 Refunds. The building official is authorized to establish a refund policy.

❖ The building official has the authority to develop policies for refunding permit fees. The refund policy should retain fees for which the jurisdiction has provided services; only that portion of a fee for which no service has been rendered should be refunded.

SECTION R109
INSPECTIONS

R109.1 Types of inspections. For onsite construction, from time to time the building official, upon notification from the permit holder or his agent, shall make or cause to be made any necessary inspections and shall either approve that portion of the construction as completed or shall notify the permit holder or his or her agent wherein the same fails to comply with this code.

❖ Inspections are necessary to verify that the construction conforms to the code requirements, and this section outlines the minimum required inspections. Besides the minimum required inspections that are specifically listed, the building official has the authority to require additional inspections so that compliance with the code can be determined. It is the duty of the permit holder or an authorized agent of the permit holder to notify the building department that some or all of the work covered by the permit is ready and available for inspection. At that point, the appropriate jurisdictional inspector performs the necessary on-site inspection. The inspector then must inform the permit holder or agent that the work has been inspected. This may be by a telephone call, an electronic message, or in many cases, a written record of the inspection posted at the job site.

R109.1.1 Foundation inspection. Inspection of the foundation shall be made after poles or piers are set or trenches or basement areas are excavated and any required forms erected and any required reinforcing steel is in place and supported prior to the placing of concrete. The foundation inspection shall include excavations for thickened slabs intended for the support of bearing walls, partitions, structural supports, or equipment and special requirements for wood foundations.

❖ The foundation inspection is typically the first inspection of the job site by a representative of the building safety department. At that time, the inspector will verify that the footing and/or foundation is located as shown on the approved plans and in accordance with the jurisdictional requirements for building setbacks and easements. Where a footing system is to be used, the footing trenches must be excavated and any required reinforcing steel must be in place. For foundation walls, the wall forms must be completely erected with the appropriate steel reinforcement placed within the forms. Column pads, thickened slabs and other foundation work must also be ready for inspection prior to concrete placement.

R109.1.2 Plumbing, mechanical, gas and electrical systems inspection. Rough inspection of plumbing, mechanical, gas and electrical systems shall be made prior to covering or concealment, before fixtures or appliances are set or installed, and prior to framing inspection.

> **Exception:** Back-filling of ground-source heat pump loop systems tested in accordance with Section M2105.1 prior to inspection shall be permitted.

❖ The various trade rough-in inspections must be completed prior to the framing inspection. The electrical, gas, mechanical and plumbing systems that are to be concealed must be inspected and approved prior to their concealment. In addition, the effect of these systems on the structural integrity of the building must be reviewed after they are completed and ready to be covered. All of the trades inspections must be completed and approvals granted prior to the inspection of the framing system.

R109.1.3 Floodplain inspections. For construction in areas prone to flooding as established by Table R301.2(1), upon placement of the lowest floor, including basement, and prior to further vertical construction, the building official shall require submission of documentation, prepared and sealed by a registered design professional, of the elevation of the lowest floor, including basement, required in Section R324.

❖ For any building constructed in a flood-prone area, the elevation of the lowest floor level must be established immediately after placement of the floor. Documentation of the elevation of the lowest floor is evidence of compliance (see commentary for Section R323.1.9). If the documentation is submitted when the lowest floor level is established and before further vertical construction, errors in elevation may be corrected at the least cost. The elevation of the lowest floor is used by insurance agents to compute flood insurance premium costs. An error in the minimum required elevation may significantly increase the cost of flood insurance, which may be required by mortgage lenders. If the elevation of the lowest floor goes uncorrected, all future owners will incur the additional increase in cost.

R109.1.4 Frame and masonry inspection. Inspection of framing and masonry construction shall be made after the roof, masonry, all framing, firestopping, draftstopping and bracing are in place and after the plumbing, mechanical and electrical rough inspections are approved.

❖ The framing inspection is usually the final opportunity for the inspector to view all of the items that will be concealed within the structure. The inspection includes the structural framework of the building as well as any fireblocking or draftstopping that will be contained within concealed spaces. All of the electrical, mechanical, gas and plumbing inspections must be completed and approved prior to the framing inspection. This allows any framing members to be repaired while they are accessible.

R109.1.5 Other inspections. In addition to the called inspections above, the building official may make or require any other

inspections to ascertain compliance with this code and other laws enforced by the building official.

❖ A variety of other inspections, such as those for gypsum board or insulation, may be mandated by the building department to verify compliance with the code or other city ordinances. These inspections are generally established as a result of local concern regarding a specific portion of the construction process. The procedure for these inspections should be consistent with the other provisions of this section.

R109.1.5.1 Fire-resistance-rated construction inspection. Where fire-resistance-rated construction is required between dwelling units or due to location on property, the building official shall require an inspection of such construction after all lathing and/or wallboard is in place, but before any plaster is applied, or before wallboard joints and fasteners are taped and finished.

❖ There are a limited number of situations in which fire-resistance-rated construction is required by this code. For example, where two or more dwelling units are located within the same structure, they are required by Section R317 to be completely separated from each other by a specified level of fire resistance. This is the case for both two-family dwellings and townhouses. A second example would be the proximity of the building to an adjoining property line. Section R302 requires exterior walls located less than 3 feet (914 mm) from a property line (unless abutting a public way) to be of minimum one-hour fire-resistance-rated construction. If either of these conditions should occur, an inspection of the fire-resistance-rated construction is required.

The inspection for compliance with the fire-resistance requirements of the IRC should be made at a point of construction when the membrane materials are in place, but the fasteners are still exposed. This allows the inspector to verify the appropriate fastener type and location based on the specific fire-resistance listing of the portion building under consideration.

R109.1.6 Final inspection. Final inspection shall be made after the permitted work is complete and prior to occupancy.

❖ The final inspection should occur after all of the work addressed by the code is complete, but prior to occupancy of the building. The issues addressed in the final inspection cover all aspects of construction, including fire safety, life safety and structural safety, as well as electrical, plumbing, gas and mechanical items. The final inspection must be approved before a certificate of occupancy can be issued.

R109.2 Inspection agencies. The building official is authorized to accept reports of approved agencies, provided such agencies satisfy the requirements as to qualifications and reliability.

❖ It is not uncommon for the building official to rely on other agencies for informational or inspection reports regarding various aspects covering methods of the construction process and/or materials. This reliance should be based on the building official's approval of the qualifications and reliability possessed by the third-party inspection or testing service.

R109.3 Inspection requests. It shall be the duty of the permit holder or their agent to notify the building official that such work is ready for inspection. It shall be the duty of the person requesting any inspections required by this code to provide access to and means for inspection of such work.

❖ The individual doing the authorized work has the responsibility for notifying the building department when the work is ready for inspection. Each building department establishes its own procedures on how and when requests should be made. Once an inspection has been scheduled, access to the area ready for inspection must be provided. The individuals performing the work should make the inspection process run as smoothly as possible.

R109.4 Approval required. Work shall not be done beyond the point indicated in each successive inspection without first obtaining the approval of the building official. The building official upon notification, shall make the requested inspections and shall either indicate the portion of the construction that is satisfactory as completed, or shall notify the permit holder or an agent of the permit holder wherein the same fails to comply with this code. Any portions that do not comply shall be corrected and such portion shall not be covered or concealed until authorized by the building official.

❖ Work must not continue past the point of a required inspection until that inspection has been approved by the building department. It is possible that if the work progresses beyond this point and is not in total compliance with the code, some of the work may have to be removed. It is critical that each individual stage of the project be approved prior to continuance of construction.

As indicated in Section R109.1, inspections must be performed when requested, and the inspector must indicate whether the construction is satisfactory or is not compliant. If the work is not approved, it must be corrected, and a reinspection must be requested. No work may be concealed until the building department approves it.

SECTION R110
CERTIFICATE OF OCCUPANCY

R110.1 Use and occupancy. No building or structure shall be used or occupied, and no change in the existing occupancy classification of a building or structure or portion thereof shall be made until the building official has issued a certificate of occupancy therefor as provided herein. Issuance of a certificate of occupancy shall not be construed as an approval of a violation of the provisions of this code or of other ordinances of the jurisdiction. Certificates presuming to give authority to violate or cancel the provisions of this code or other ordinances of the jurisdiction shall not be valid.

Exceptions:

1. Certificates of occupancy are not required for work exempt from permits under Section R105.2.

2. Accessory buildings or structures.

❖ The tool the building official employs to control the uses and occupancies of the various buildings in a jurisdiction is the certificate of occupancy. This section establishes the conditions of a certificate of occupancy, commonly called a C.O., and identifies the information the certificate must contain. The building official must be satisfied that the structure meets the requirements of the code before a certificate of occupancy can be given, and the structure cannot be legally occupied until a certificate has been issued. If the occupancy classification of an existing building has changed, such as changing an old Victorian home to a small office building, a new certificate of occupancy must be issued. The certificate of occupancy is the legal notification from the Department of Building Safety that the building may be occupied for its intended purpose.

The granting of a certificate of occupancy does not necessarily indicate that no violations of the code or other jurisdictional laws exist. The building official should make every effort to determine compliance with all applicable code provisions and other ordinances of the jurisdiction. It is important that violations be corrected. It is possible that the certificate of occupancy will be revoked if it is found that the certificate was issued in error.

R110.2 Change in use. Changes in the character or use of an existing structure shall not be made except as specified in Sections 3406 and 3407 of the *International Building Code.*

❖ When an existing building's character or use is modified, the provisions of Sections 3405 and 3406 of the *International Building Code* dealing with change of occupancy and historic buildings must be met. Because the *International Residential Code* is limited in scope to specific residential occupancies, any change of occupancy to a use beyond the scope of IRC Section R101.2 will be governed by the IBC.

R110.3 Certificate issued. After the building official inspects the building or structure and finds no violations of the provisions of this code or other laws that are enforced by the department of building safety, the building official shall issue a certificate of occupancy which shall contain the following:

1. The building permit number.

2. The address of the structure.

3. The name and address of the owner.

4. A description of that portion of the structure for which the certificate is issued.

5. A statement that the described portion of the structure has been inspected for compliance with the requirements of this code.

6. The name of the building official.

7. The edition of the code under which the permit was issued.

8. If an automatic sprinkler system is provided and whether the sprinkler system is required.

9. Any special stipulations and conditions of the building permit.

❖ Prior to use or occupancy of the building, the building official shall perform a final inspection as addressed in Section R109.1.6. If the official finds no violations of the code and other laws enforced by the department of building safety, the official is required to issue a certificate of occupancy. Commentary Figure R110.3 illustrates the information that must be provided on the certificate of occupancy.

R110.4 Temporary occupancy. The building official is authorized to issue a temporary certificate of occupancy before the completion of the entire work covered by the permit, provided that such portion or portions shall be occupied safely. The building official shall set a time period during which the temporary certificate of occupancy is valid.

❖ Where a portion of a building is intended to be occupied prior to occupancy of the entire structure, the building official may issue a temporary certificate of occupancy. Prior to issuance of a temporary certificate of occupancy, it is critical that the building official determine that the portions to be occupied provide the minimum levels of safety required by the code. In addition, the building official must establish a definitive length of time for the temporary certificate of occupancy to be valid.

R110.5 Revocation. The building official shall, in writing, suspend or revoke a certificate of occupancy issued under the provisions of this code wherever the certificate is issued in error, or on the basis of incorrect information supplied, or where it is determined that the building or structure or portion thereof is in violation of any ordinance or regulation or any of the provisions of this code.

❖ In essence, the certificate of occupancy certifies that the described building or portion of that building complies with the requirements of the code for the intended use. However, any certificate of occupancy may be suspended or revoked by the building official under one of three conditions: 1) when the certificate is issued in error, 2) when incorrect information is supplied to the building official, or 3) when it is determined that the building or a portion of the building is shown to be in violation of the code or any other ordinance or regulation of the jurisdiction.

SECTION R111
SERVICE UTILITIES

R111.1 Connection of service utilities. No person shall make connections from a utility, source of energy, fuel or power to any building or system that is regulated by this code for which a permit is required, until approved by the building official.

❖ This section addresses the connection and disconnection, either permanent or temporary, of any utilities that service a building or structure regulated by this code. The building official is authorized to control the connection for any service utility when the connection is to a building that is regulated by the code and requires a permit. Prior to the connection of a utility, source of energy, fuel or power, all conditions for the connection must be met and verified by required inspections.

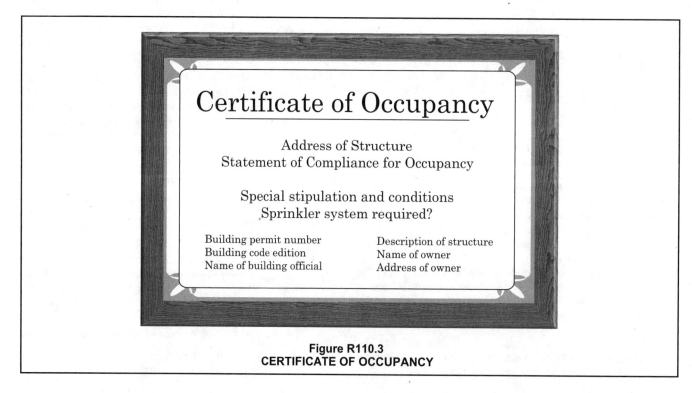

Figure R110.3
CERTIFICATE OF OCCUPANCY

R111.2 Temporary connection. The building official shall have the authority to authorize and approve the temporary connection of the building or system to the utility, source of energy, fuel or power.

❖ Temporary service utility connections, such as for temporary electrical service during the construction process, are permitted when approved by the building official after the necessary inspections have been performed and any additional conditions met.

R111.3 Authority to disconnect service utilities. The building official shall have the authority to authorize disconnection of utility service to the building, structure or system regulated by this code and the referenced codes and standards set forth in Section R102.4 in case of emergency where necessary to eliminate an immediate hazard to life or property or when such utility connection has been made without the approval required by Section R111.1 or R111.2. The building official shall notify the serving utility and whenever possible the owner and occupant of the building, structure or service system of the decision to disconnect prior to taking such action if not notified prior to disconnection. The owner or occupant of the building, structure or service system shall be notified in writing as soon as practical thereafter.

❖ When an immediate hazard to life or property exists, the building official has the authority to order disconnection of the utility services. This can also occur when the utility service has been connected without the necessary approvals required by the code. Whenever possible, the building owner and the building occupant or occupants should be notified prior to the disconnection of the services. Then at the first practical opportunity, the building official is to formally notify the building owner in writing of the disconnection activities. As with all administrative functions, all aspects of due process must be followed.

SECTION R112
BOARD OF APPEALS

R112.1 General. In order to hear and decide appeals of orders, decisions or determinations made by the building official relative to the application and interpretation of this code, there shall be and is hereby created a board of appeals. The building official shall be an ex officio member of said board but shall have no vote on any matter before the board. The board of appeals shall be appointed by the governing body and shall hold office at its pleasure. The board shall adopt rules of procedure for conducting its business, and shall render all decisions and findings in writing to the appellant with a duplicate copy to the building official.

❖ This section holds that any aggrieved party with a material interest in the decision of the building official may appeal such a decision before a board of appeals. This provides a forum other than the court of the jurisdiction in which the building official's action can be reviewed. A board of appeals is to be created by the jurisdiction. The primary function of the board of appeals is hearing and acting on the appeal of orders or decisions the building

official has made on the application or interpretation of the code. The building official is an ex-officio member of the board of appeals, but he or she has no vote on any issue that comes before the board. The appellant is to be provided with a written copy of the board's findings and decisions, with a duplicate copy sent to the building official. Appendix B of the *International Building Code* supplies example rules of procedure.

R112.2 Limitations on authority. An application for appeal shall be based on a claim that the true intent of this code or the rules legally adopted thereunder have been incorrectly interpreted, the provisions of this code do not fully apply, or an equally good or better form of construction is proposed. The board shall have no authority to waive requirements of this code.

❖ The code does not grant authority to the board to hear appeals regarding the administrative provisions of the code, nor does it grant the board the authority to waive any code requirements. An appeal must be based on the claim that the provisions of the code have been misinterpreted, that the provisions do not apply to the appellant's circumstances or that where an alternate method or technique of construction is used it has been shown to be at least equal to the code requirements.

R112.2.1 Determination of substantial improvement in areas prone to flooding. When the building official provides a finding required in Section R105.3.1.1, the board of appeals shall determine whether the value of the proposed work constitutes a substantial improvement. A substantial improvement means any repair, reconstruction, rehabilitation, addition or improvement of a building or structure, the cost of which equals or exceeds 50 percent of the market value of the building or structure before the improvement or repair is started. If the building or structure has sustained substantial damage, all repairs are considered substantial improvement regardless of the actual repair work performed. The term does not include:

1. Improvements of a building or structure required to correct existing health, sanitary or safety code violations identified by the building official and which are the minimum necessary to assure safe living conditions; or

2. Any alteration of an historic building or structure, provided that the alteration will not preclude the continued designation as an historic building or structure. For the purpose of this exclusion, an historic building is:

    2.1. Listed or preliminarily determined to be eligible for listing in the National Register of Historic Places; or

    2.2. Determined by the Secretary of the U.S. Department of Interior as contributing to the historical significance of a registered historic district or a district preliminarily determined to qualify as an historic district; or

    2.3. Designated as historic under a state or local historic preservation program that is approved by the Department of Interior.

❖ "Substantial improvement," found in Sections R105.3.1.1 and R112.2.1, is defined in this section and is consistent with the definition used by the National Flood Insurance Program (NFIP) and Section 1612.2 of the *International Building Code*. One of the long-range objectives of the NFIP is to reduce the exposure of older buildings that were built in flood hazard areas before local jurisdictions adopted flood hazard area maps and regulations. Section R105.3 of the code directs the applicant to state the valuation of the proposed work as part of the application for a permit. To determine whether a proposed alteration, repair, addition or improvement of a building or structure constitutes a substantial improvement, the cost of the proposed work is to be compared to the market value of the building or structure before the work is started. To determine market value, the building official may require the applicant to provide the information listed in Section R105.3. For additional guidance, refer to FEMA 213: *Answers to Questions about Substantially Damaged Buildings*, and FEMA 311: *Guidance on Estimating Substantial Damage Using the NFIP Residential Substantial Damage Estimator*.

When the board of appeals makes its determination, certain items are not included in the valuation of proposed work. Specifically, if certain health, sanitation or safety-code violations have been cited previously, the cost of the minimum repairs required to correct those violations is not included in the determination.

Alteration of a historic structure is not considered a substantial improvement or repair of substantial damage, provided the proposed work does not alter the building to the extent that it would no longer qualify as a historic structure. The building official may require applicants to consult an appropriate historic preservation authority to determine whether a proposed alteration will jeopardize a structure's historic designation. The exception for historic structures does not apply to structures located within designated historic districts unless those structures are individually listed as historic structures.

R112.2.2 Criteria for issuance of a variance for areas prone to flooding. A variance shall only be issued upon:

1. A showing of good and sufficient cause that the unique characteristics of the size, configuration or topography of the site render the elevation standards in Section R324 inappropriate.

2. A determination that failure to grant the variance would result in exceptional hardship by rendering the lot undevelopable.

3. A determination that the granting of a variance will not result in increased flood heights, additional threats to public safety, extraordinary public expense, nor create nuisances, cause fraud on or victimization of the public, or conflict with existing local laws or ordinances.

4. A determination that the variance is the minimum necessary to afford relief, considering the flood hazard.

5. Submission to the applicant of written notice specifying the difference between the design flood elevation and the elevation to which the building is to be built, stating that the cost of flood insurance will be commensurate with the increased risk resulting from the reduced floor elevation, and stating that construction below the design flood elevation increases risks to life and property.

❖ All the criteria set forth in this section must be met in order for the board of appeals to 1) consider granting a variance for construction in flood-prone areas, and 2) provide relief from selected provisions for flood-resistant construction. Granting of the variance must not cause additional public safety concerns beyond those already present.

The board of appeals is empowered to hear requests for variances from the flood hazard provisions of this code. Variances from these provisions may place people and property at significant risk. Therefore, communities are cautioned to carefully evaluate the impacts of issuing a variance, particularly variances to the requirements for elevating buildings to the Base Flood Elevation (BFE). The elements that are to be evaluated include impacts on the site, the applicant and other parties who may be affected such as adjacent property owners and the community as a whole. Flood plain development that is not undertaken in accordance with the flood-resistance provisions of the code will be exposed to increased flood damages. As a consequence, flood insurance premium rates will be significantly higher. Variance decisions made by the board of appeals are to be based solely on technical justifications outlined in this section and not on the personal circumstances of an owner or applicant.

Applicants sometimes request variances to the minimum elevation requirements for the lowest floor of buildings in flood hazard areas to improve access for the disabled and the elderly. Generally, variances of this nature are not to be granted because these are personal circumstances that will change as the property changes ownership. Not only would persons of limited mobility be at risk from flooding, but a building built below the BFE would continue to be exposed to flood damage long after the personal need for a variance ends. More appropriate alternatives are to be considered to serve the needs of disabled or elderly persons, such as varying setbacks to allow construction on less flood-prone portions of sites or the installation of personal elevators.

All variances are to be the minimum necessary to afford relief. The board of appeals will address each listed condition for a variance, especially the requirement that it determine whether the failure to grant the variance would result in exceptional hardship by rendering the lot undevelopable. By itself, this determination may be insufficient to result in an exceptional hardship if other conditions for issuing a variance cannot be met. The determination of hardship is to be based on the unique characteristics of the site and not on the personal circumstances of the applicant.

In guidance materials, FEMA cautions that economic hardship alone is not an exceptional hardship. Building officials and boards of appeals are cautioned that granting a variance does not affect how the building will be rated for the purposes of NFIP flood insurance. Even if circumstances justify granting a variance to build a lowest floor that is below the Design Flood Elevation, the rate used to calculate the cost of a flood insurance policy will be based on the risk to the building. Flood insurance, required by certain mortgage lenders, may be extremely expensive. Although the applicant may not be required to purchase flood insurance, the requirement may be imposed on subsequent owners. The building official is to provide the applicant a written notice to this effect, along with the other cautions listed in this section.

R112.3 Qualifications. The board of appeals shall consist of members who are qualified by experience and training to pass on matters pertaining to building construction and are not employees of the jurisdiction.

❖ The members of the board of appeals are to be selected by the appointing authority of the jurisdiction; the members must be qualified to pass judgment on appeals associated with building construction. The purpose of the board is to provide a review of the appeal independent from that of the building department, so no jurisdictional employee is permitted to be a member. Additional information on suggested qualifications of members of the board of appeals is contained in Appendix B of the *International Building Code*.

R112.4 Administration. The building official shall take immediate action in accordance with the decision of the board.

❖ Decisions made by the board of appeals must be enacted as quickly as possible. The building official is to take whatever action is necessary to see that the orders of the board are carried out.

SECTION R113
VIOLATIONS

R113.1 Unlawful acts. It shall be unlawful for any person, firm or corporation to erect, construct, alter, extend, repair, move, remove, demolish or occupy any building, structure or equipment regulated by this code, or cause same to be done, in conflict with or in violation of any of the provisions of this code.

❖ This section describes the citing, recording and subsequent actions to be taken when code violations are found.

R113.2 Notice of violation. The building official is authorized to serve a notice of violation or order on the person responsible for the erection, construction, alteration, extension, repair, moving, removal, demolition or occupancy of a building or structure in violation of the provisions of this code, or in violation of a detail statement or a plan approved thereunder, or in violation of a permit or certificate issued under the provisions of this code. Such order shall direct the discontinuance of the illegal action or condition and the abatement of the violation.

❖ The building official is required to notify the person responsible for the construction or use of a building found to be in violation of the code. The section of the code that is being violated must be cited so that the responsible party can respond to the notice.

R113.3 Prosecution of violation. If the notice of violation is not complied with in the time prescribed by such notice, the building official is authorized to request the legal counsel of the jurisdiction to institute the appropriate proceeding at law or in equity to restrain, correct or abate such violation, or to require the removal or termination of the unlawful occupancy of the building or structure in violation of the provisions of this code or of the order or direction made pursuant thereto.

❖ When the building owner, the owner's agent or the tenant does not correct the condition causing the violation as directed, the building official must pursue, through the use of legal counsel of the jurisdiction, legal means to correct the violation. This is not optional.

Any extensions that allow the violations to be corrected voluntarily must be for a reasonable, bona fide cause, or the building official may be subject to criticism for "arbitrary and capricious" actions. In general, it is better to have a standard time limitation for correction of violations. Departures from this standard must be for a clear and reasonable purpose, usually stated in writing by the violator.

R113.4 Violation penalties. Any person who violates a provision of this code or fails to comply with any of the requirements thereof or who erects, constructs, alters or repairs a building or structure in violation of the approved construction documents or directive of the building official, or of a permit or certificate issued under the provisions of this code, shall be subject to penalties as prescribed by law.

❖ The jurisdiction must establish penalties for a variety of violations that may occur. Violations specifically addressed by the code include failure to comply with the 1) code, 2) approved plans or 3) directives of the building official.

SECTION R114
STOP WORK ORDER

R114.1 Notice to owner. Upon notice from the building official that work on any building or structure is being prosecuted contrary to the provisions of this code or in an unsafe and dangerous manner, such work shall be immediately stopped. The stop work order shall be in writing and shall be given to the owner of the property involved, or to the owner's agent or to the person doing the work and shall state the conditions under which work will be permitted to resume.

❖ The stop work order is a tool authorized by this code that enables the building official to demand that work on a building or structure be temporarily suspended. Typically used under rare circumstances, this order may be issued where the work being performed is dangerous, unsafe or significantly contrary to the provisions of the code.

The stop work order is to be a written document indicating the reason or reasons for the suspension of work, identifying those conditions where compliance is necessary before work is allowed to resume. All work addressed by the order must cease immediately. The stop work order must be presented to either the owner of the subject property, the agent of the owner or the individual doing the work. Commentary Figure R114.1 is an example of a stop work order.

R114.2 Unlawful continuance. Any person who shall continue any work in or about the structure after having been served with a stop work order, except such work as that person is directed to perform to remove a violation or unsafe condition, shall be subject to penalties as prescribed by law.

❖ The only activity permitted in a building or a portion of a building subject to a stop work order is that work necessary to eliminate the violation or unsafe condition. Otherwise, penalties prescribed by laws of the jurisdiction must be imposed for illegal construction activity in defiance of a stop work order.

Bibliography

The following resource materials are referenced in this chapter or are relevant to the subject matter addressed in this chapter.

FEMA 213, *Answers to Questions About Substantially Damaged Buildings*. Washington, DC: Federal Emergency Management Agency, 1991. (For ordering information, see the Bibliography for Chapter 3).

FEMA 257, *Mitigation of Flood and Erosion Damage to Residential Buildings in Coastal Areas*. Washington, DC: Federal Emergency Management Agency, 1994.

FEMA 265, *Managing Floodplain Development in Approximate Zone A Areas: A Guide for Obtaining and Developing Base (100-Year) Flood Elevations*. Washington, DC: Federal Emergency Management Agency, 1994.

FEMA 311, *Guidance on Estimating Substantial Damage Using the NFIP Residential Substantial Damage Estimator*. Washington, DC: Federal Emergency Management Agency, 1995.

FEMA 348, *Protecting Building Utilities from Flood Damage*. Washington, DC: Federal Emergency Management Agency, 1999.

IBC-2006, *International Building Code*. Falls Church, VA: International Code Council, Inc., 2006.

IFC-2006, *International Fire Code*. Falls Church, VA: International Code Council, Inc., 2006.

IPMC-2006, *International Property Maintenance Code*. Falls Church, VA: International Code Council, Inc., 2006.

Legal Aspects of Code Administration. Country Club Hills, IL: Building Officials and Code Administrators International, Inc.; Whittier, CA: International Conference of Building Officials; Birmingham, AL: Southern Building Code Congress International, 1984.

LEGAL NOTICE

Date _____

WHEREAS, VIOLATIONS OF
{ Article _____ , Section _____ of the Zoning Ordinance
Article _____ , Section _____ of the Building Code
Article _____ , Section _____ of the _____ Code }
have been found on

these premises, IT IS HEREBY ORDERED in accordance with the above Code that all persons cease, desist from, and

STOP WORK

at once pertaining to construction, alterations or repairs on these premises known as _____

All persons acting contrary to this order or removing or mutilating this notice are liable to arrest unless such action is authorized by the Department.

_____
CODE OFFICIAL

Figure R114.1
STOP WORK ORDER

O'Bannon, Robert E. *Building Department Administration*. Whittier, CA: International Conference of Building Officials, 1973.

Readings in Code Administration, Volume 1: History/Philosophy/Law. Country Club Hills, IL: Building Officials and Code Administrators International, Inc., 1974.

Chapter 2:
Definitions

General Comments

The code user should be familiar with the terms in this chapter because 1) the definitions are essential to the correct interpretation of the *International Residential Code®* (IRC®), and 2) the user might not be aware that a particular term encountered in the text has the special definition found herein.

Section R201.1 contains the scope of the chapter. Section R201.2 establishes the interchangeability of the terms in the code. Section R201.3 establishes the use of terms defined in other chapters. Section R201.4 establishes the use of undefined terms, and Section R202 lists terms and their definition according to this code.

Purpose

Codes are technical documents, so literally every word, term and punctuation mark can add to or change the meaning of the provision. Furthermore, the IRC, with its broad scope of applicability, includes terms inherent in a variety of construction disciplines. These terms can often have multiple meanings depending on the context or discipline being used at the time. For these reasons it is necessary to maintain a consensus on the specific meaning of terms contained in the IRC. Chapter 2 performs this function by stating clearly what specific terms mean for the purpose of this code.

SECTION R201
GENERAL

R201.1 Scope. Unless otherwise expressly stated, the following words and terms shall, for the purposes of this code, have the meanings indicated in this chapter.

❖ This section clarifies the terminology used in the IRC. The terms defined in the IRC often have very specific meanings, which can be different from their typical meanings. This section gives guidance to the use of the defined words relative to tense and gender and also provides the means to resolve those terms not defined.

R201.2 Interchangeability. Words used in the present tense include the future; words in the masculine gender include the feminine and neuter; the singular number includes the plural and the plural, the singular.

❖ Although the definitions contained in Chapter 2 are to be taken literally, gender and tense are considered to be interchangeable; thus, any gender and tense inconsistencies within the code text should not hinder the understanding or enforcement of the requirements.

R201.3 Terms defined in other codes. Where terms are not defined in this code such terms shall have meanings ascribed to them as in other code publications of the International Code Council.

❖ When a word or term appears in the code that is not defined in this chapter, other code publications of the *International Code Council®* (ICC®) may be used to find its definition. These code documents include the *International Building Code®* (IBC®), *International Mechanical Code®* (IMC®), *International Plumbing Code®* (IPC®), *International Fuel Gas Code®* (IFGC®), *International Fire Code®* (IFC®) and others. These codes contain additional definitions (some parallel and duplicated) which may be used in the enforcement of this code or in the enforcement of the other ICC codes by reference. When using a definition from another code, keep in mind the admonition from "Purpose": "These terms can often have multiple meanings depending on the context or discipline being used at the time."

R201.4 Terms not defined. Where terms are not defined through the methods authorized by this section, such terms shall have ordinarily accepted meanings such as the context implies.

❖ Another possible source for the definitions of words or terms not defined in this code or in other codes is their "ordinarily accepted meanings." Dictionary definitions may suffice, provided that the definitions are in context.

Sometimes construction terms used throughout the code may not be defined in Chapter 2, in another code, or in a dictionary. In such cases, one would first turn to the definitions contained in the referenced standards (see Chapter 43) and then to textbooks on the subject in question.

SECTION R202
DEFINITIONS

ACCESSIBLE. Signifies access that requires the removal of an access panel or similar removable obstruction.

❖ In general, where immediate access is not required because of the low level of hazard involved, an accessible method is all that is mandated, such as the removal of an access panel. The electrical definitions found in Chapter 34 contain two additional definitions for accessible that apply specifically to electrical wiring methods and electrical equipment. See Section 1102 of the IBC for the meaning of "accessible" as it applies to the requirements for "accessible" dwelling units in Section R326.

ACCESSIBLE, READILY. Signifies access without the necessity for removing a panel or similar obstruction.

❖ Where this term is designated as a requirement by other sections of the code, it is intended that access to the device, controls, shut-off valves or other element be extremely easy. "Readily accessible" means that the device must be reachable directly without a panel, door or equipment needing to be moved to gain access. Chapter 24, Fuel Gas, has a similar definition noted as "Ready access (to)."

ACCESSORY STRUCTURE. A structure not greater than 3,000 square feet (279 m²) in floor area, and not over two stories in height, the use of which is customarily accessory to and incidental to that of the dwelling(s) and which is located on the same lot.

❖ As it applies to the scope in Section R101.2, this term describes structures that are designed for accessory use to one- or two-family dwellings and multiple single-family townhouses. These structures are commonly used as garages, carports, cabanas, storage sheds, tool sheds, playhouses and garden structures. The structures all house uses that are incidental to the primary use, which is the dwelling unit, and the activities that take place in accessory structures occur as a result of the primary building. Their use is secondary or minor in importance to the primary residence.

ADDITION. An extension or increase in floor area or height of a building or structure.

❖ Where a structure is increased in size vertically or horizontally, the increase is considered an addition. The addition will typically contribute additional floor area to the building, although an increase in roof height only is also an addition.

AIR ADMITTANCE VALVE. A one-way valve designed to allow air into the plumbing drainage system when a negative pressure develops in the piping. This device shall close by gravity and seal the terminal under conditions of zero differential pressure (no flow conditions) and under positive internal pressure.

❖ An air admittance valve is an alternative to traditional vent piping configurations. Such valves allow the venting of fixtures where conventional venting may be impractical. The valve is designed to open and admit air into the drainage system when negative pressures occur. The valve closes by gravity and seals the vent terminal when the internal drain pressure is equal to or exceeds atmospheric pressure. The seal prevents sewer gas from entering the building [see Commentary Figures R202(1) and R202(2)].

AIR BREAK (DRAINAGE SYSTEM). An arrangement in which a discharge pipe from a fixture, appliance or device drains indirectly into a receptor below the flood-level rim of the receptor, and above the trap seal.

❖ An air break is an indirect drainage method by which waste discharges to the drainage system through piping that terminates below the flood level rim of an approved receptor. An air break is commonly used to protect mechanical equipment from sewage backup in the event that stoppage occurs. It also protects the drainage system from adverse pressure conditions caused by pumped discharge [see Commentary Figure R202(3)].

AIR CIRCULATION, FORCED. A means of providing space conditioning utilizing movement of air through ducts or plenums by mechanical means.

❖ Forced-air systems use a central heating or cooling unit equipped with a fan to return unconditioned air through return ducts and to deliver conditioned air through supply ducts or plenums to occupied spaces.

AIR-CONDITIONING SYSTEM. A system that consists of heat exchangers, blowers, filters, supply, exhaust and return-air systems, and shall include any apparatus installed in connection therewith.

❖ This definition is limited to the components commonly used in a mechanical air-conditioning system. Additional parts of the air-conditioning system include thermostats, humidistats, dampers and any other controls needed for the system to operate properly.

AIR GAP, DRAINAGE SYSTEM. The unobstructed vertical distance through free atmosphere between the outlet of a waste pipe and the flood-level rim of the fixture or receptor into which it is discharging.

❖ In a drainage system air gap, waste is discharged to the drainage system through piping that terminates at a specified distance above the flood level rim of an approved receptor.

The air gap serves as an impossible barrier for sewage to overcome in the event that stoppage occurs in the receptor drain, because sewage backup would overflow in the receptor drain flood level before it comes in contact with the drain line above [see Commentary Figure R202(3)].

AIR GAP, WATER-DISTRIBUTION SYSTEM. The unobstructed vertical distance through free atmosphere between the lowest opening from a water supply discharge to the flood-level rim of a plumbing fixture.

❖ An air gap is the most reliable and effective means of backflow protection. It is simply the vertical air space between the potable water supply outlet and the possible source of contamination. This air gap prevents possible contamination of the potable water supply by preventing the supply outlet, such as a faucet, from backsiphoning waste water from a basin, for example. Many manufacturers have developed air gap fittings that provide a rigid connection to the drainage system while maintaining the minimum level of protection from contamination.

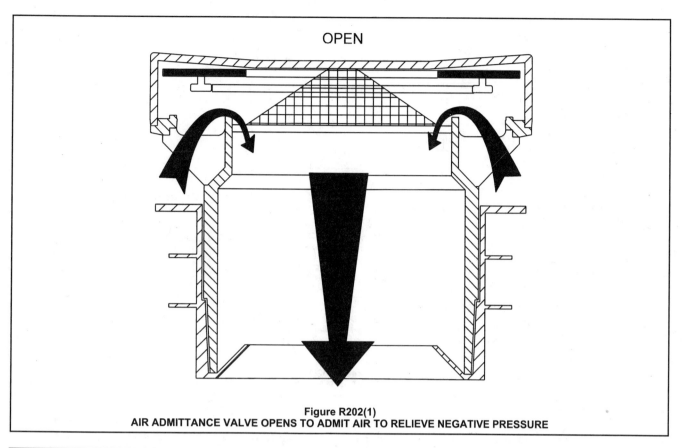

Figure R202(1)
AIR ADMITTANCE VALVE OPENS TO ADMIT AIR TO RELIEVE NEGATIVE PRESSURE

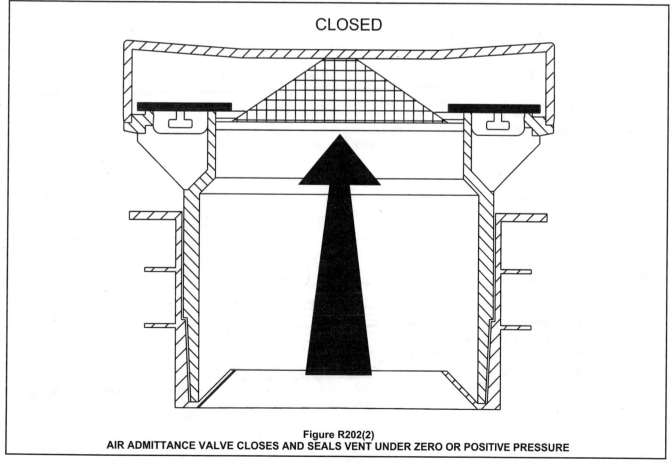

Figure R202(2)
AIR ADMITTANCE VALVE CLOSES AND SEALS VENT UNDER ZERO OR POSITIVE PRESSURE

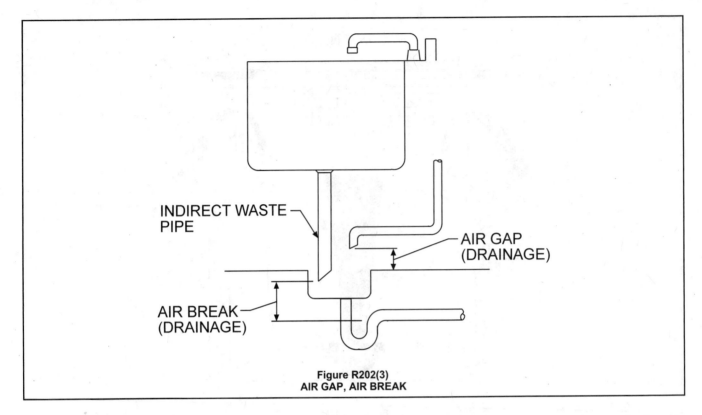

Figure R202(3)
AIR GAP, AIR BREAK

ALTERATION. Any construction or renovation to an existing structure other than repair or addition that requires a permit. Also, a change in a mechanical system that involves an extension, addition or change to the arrangement, type or purpose of the original installation that requires a permit.

❖ The modification of an existing structure without adding any floor area or height to the structure is an alteration. Section R105 of the code specifies that a permit for the alteration work is required before work begins. The term "alteration" also applies to mechanical work where the original installation is altered in a manner requiring a permit. The repairs described in Section R105.2.2 are not alterations because a permit is not required.

ANCHORS. See "Supports."

❖ Anchors are the same types of devices as supports. Any device, such as a hanger used to support or secure piping and fixtures, is an anchor.

ANTISIPHON. A term applied to valves or mechanical devices that eliminate siphonage.

❖ The term refers to the function of certain valves and devices used to break or prevent the siphon effect that can be created in plumbing systems.

APPLIANCE. A device or apparatus that is manufactured and designed to utilize energy and for which this code provides specific requirements.

❖ An appliance is a manufactured component or assembly of components that converts one form of energy into a different form of energy to serve a specific purpose. The term "appliance" generally refers to residential and commercial equipment that is manufactured in standardized sizes or types. The term is generally not associated with industrial equipment. For the application of the code provisions, the terms "appliance" and "equipment" are mutually exclusive.

Examples of appliances include furnaces; boilers; water heaters; room heaters; refrigeration units; cooking equipment; clothes dryers; wood stoves; pool, spa and hot tub heaters; unit heaters ovens; and similar fuel-fired or electrically operated appliances. See the definition of "Equipment."

APPROVED. Acceptable to the building official.

❖ Throughout the code, the term "approved" is used to describe a specific material or method of construction, such as the approved drainage system mentioned in Section R408.5. Where "approved" is used, it means that the design, material or method of construction is acceptable to the building official. It is imperative that the building officials base their decision of approval on the result of investigations, tests or accepted principles or practices.

APPROVED AGENCY. An established and recognized agency regularly engaged in conducting tests or furnishing inspection services, when such agency has been approved by the building official.

❖ The building official will occasionally rely on the expertise of others to assist in plan review and inspection activities, or the owner will employ an individual or company to provide testing services for portions of the construction work. An agency qualified to perform such activities is an "approved agency." The building official will review an agency's qualifications, expertise and reliability to determine whether the agency should be granted approval for a specific activity.

ASPECT RATIO. The ratio of the height to width (h/w) of a shear wall. The shear wall height is the maximum clear height from top of foundation or diaphragm to bottom of diaphragm framing above and the shear wall width is the sheathed dimension in the direction of applied force on the shear wall.

❖ The shear wall or braced wall panel height divided by the width or length of the shear wall or braced wall panel is defined as the aspect ratio. Generally, the code places limits on shear wall aspect ratios to limit shear wall deformation.

ATTIC. The unfinished space between the ceiling joists of the top story and the roof rafters.

❖ Several provisions apply to the attic area of a building, such as those relating to ventilation of the attic space. The code identifies an attic as the unfinished space between the ceiling joists of the top story and the roof rafters. Such a space would be the top story, rather than the attic, if it is finished and occupiable.

BACKFLOW, DRAINAGE. A reversal of flow in the drainage system.

❖ Backflow is a condition found in plumbing systems where the contents of the piping flow in a direction opposite of the intended direction.

BACKFLOW PREVENTER. A device or means to prevent backflow.

❖ Backflow preventers are designed for different applications, depending on the pressures and the degree of hazard posed by the cross connection. There are six basic types of backflow preventers that can be used to correct cross connections: air gaps; barometric loops; vacuum breakers, both atmospheric and pressure type; double check valve with intermediate atmospheric vent; double check-valve assembly and reduced-pressure-principle devices.

BACKFLOW PREVENTER, REDUCED-PRESSURE-ZONE TYPE. A backflow-prevention device consisting of two independently acting check valves, internally force loaded to a normally closed position and separated by an intermediate chamber (or zone) in which there is an automatic relief means of venting to atmosphere internally loaded to a normally open position between two tightly closing shutoff valves and with means for testing for tightness of the checks and opening of relief means.

❖ This device is one of the most reliable mechanical devices (second only to an air gap) for the prevention of backflow. It uses two spring-loaded check valves with a relief valve in between that monitors the system pressure upstream and downstream of the device. The zone between the check valves is maintained at a pressure that is less than the water supply pressure.

BACKFLOW, WATER DISTRIBUTION. The flow of water or other liquids into the potable water-supply piping from any

sources other than its intended source. Backsiphonage is one type of backflow.

❖ There are primarily three types of backflow conditions that could occur in a water distribution system: 1) backpressure; 2) backpressure, low-head; and 3) backsiphonage.

1. Backpressure. Liquid in a pipe flows from a high pressure condition to a low pressure condition, such as when the pressure within the water distribution system is greater than that in the water service. The water will reverse its intended direction of flow and return to the water supply unless a device to prevent backflow has been installed in the line. Elevated pressure can be created through mechanical means, such as a pumping system, or it can be the result of a loss of pressure within the main water supply. Backpressure often occurs in a system or a portion of a system that is closed (not open to the atmosphere).

2. Backpressure, low-head. This is the back-pressure created where the source of backpressure comes from a hose elevated to a level that produces 10 feet (3048 mm) or less of water column at the outlet. Low-head backpressure typically occurs at a hose connection, such as a wall hydrant or sillcock. Certain types of backflow assemblies and devices are designed specifically for either backpressure or low-head backpressure.

3. Backsiphonage. See definition of backsiphonage below.

BACKPRESSURE. Pressure created by any means in the water distribution system, which by being in excess of the pressure in the water supply mains causes a potential backflow condition.

❖ Liquid in a pipe flows from a high-pressure condition to a low-pressure condition, such as when the pressure within the water distribution system is greater than that in the water service. The water will reverse its intended direction of flow and return to the water supply, unless a device to prevent backflow has been installed in the line. Elevated pressures can be created through mechanical means, such as a pumping system, or can be the result of a loss of pressure within the main water supply. Backpressure will often occur in a system or a portion of a system that is closed, i.e., it is not open to the atmosphere.

BACKPRESSURE, LOW HEAD. A pressure less than or equal to 4.33 psi (29.88 kPa) or the pressure exerted by a 10-foot (3048 mm) column of water.

❖ Low-head backpressure is created where the source of backpressure comes from a hose elevated to a level that produces 10 feet (3048 mm) or less of water column at the outlet. A hose connection, such as a wall hydrant or faucet, is where "low-head backpressure" is typically experienced. Also see the commentary to "Backpressure."

BACKSIPHONAGE. The flowing back of used or contaminated water from piping into a potable water-supply pipe due to a negative pressure in such pipe.

❖ Backsiphonage occurs when the pressure within a potable water distribution system drops below atmospheric pressure or negative gauge pressure. It, like backpressure, allows the normal direction of flow to reverse. A siphon can result in an unprotected cross connection, causing contamination or pollution of the potable water supply. The main difference between backsiphonage and backpressure is the pressure in the system. Water distribution systems having a pressure less than zero gauge have the potential to create a siphon, resulting in backsiphonage. Backpressure typically occurs in a closed system, and backsiphonage typically occurs in an open system (open to the atmosphere).

BACKWATER VALVE. A device installed in a drain or pipe to prevent backflow of sewage.

❖ A backwater valve is a type of check valve designed for installation in drainage piping. The valve has a lower invert downstream of the flapper than upstream to help prevent solids from interfering with valve closure.

BALCONY, EXTERIOR. An exterior floor projecting from and supported by a structure without additional independent supports.

❖ This term describes an exterior floor that is supported by the structure without the use of independent supports, such as post or columns. A balcony typically cantilevers from the supporting structure, whereas a deck is supported not only by the main structure but also by independent columns or posts. An exterior balcony lacks the structural redundancy of a deck.

BALL COCK. A valve that is used inside a gravity-type water closet flush tank to control the supply of water into the tank. It may also be called a flush-tank fill valve or water control.

❖ A ball cock is a float-actuated valve used in a traditional gravity-flush water closet flush tank. The valve serves to control the water supply for refilling the flush tank and restoring the trap seal in the water-closet bowl. Ball cock valves are available with or without an antisiphon device, depending on the installation requirements.

BASEMENT. That portion of a building that is partly or completely below grade (see "Story above grade").

❖ Defined as "that portion of a building which is partially or totally below grade," a basement is further identified in the provisions in Section R202 for a "story above grade." Specific provisions, including the requirement for emergency escape and rescue openings in Section R310.1, are applicable for those floor levels meeting the criteria for basements. The presence of occupiable space below grade level causes various concerns that are addressed by the code.

BASEMENT WALL. The opaque portion of a wall that encloses one side of a basement and has an average below grade wall area that is 50 percent or more of the total opaque and non-opaque area of that enclosing side.

❖ A basement wall is defined in respect to the energy efficiency provisions of Chapter 11. A wall that does not qualify as an exterior wall is a basement wall.

BASIC WIND SPEED. Three-second gust speed at 33 feet (10 058 mm) above the ground in Exposure C (see Section R301.2.1) as given in Figure R301.2(4).

❖ Basic wind speed, the basis for a number of structural decisions in the IRC, is the 3-second gust speed measured at a point 33 feet (10 058 mm) above the ground. The assumed exposure used in determining the basic wind speed is that for generally open terrain having scattered obstructions.

BATHROOM GROUP. A group of fixtures, including or excluding a bidet, consisting of a water closet, lavatory, and bathtub or shower. Such fixtures are located together on the same floor level.

❖ Special consideration is given to a bathroom group when sizing water distribution systems and drain, waste and vent piping because of fixture usage. "Bathroom group" has historically referred to a single water closet, a lavatory and a bathtub or shower, all located within a single room. Such an arrangement would normally allow only one occupant in the room; therefore, the likelihood of simultaneous fixture discharge is remote. For the fixture-usage theory of the bathroom group to be valid, the fixtures in the group must be located within the same bathroom. A bidet has been added to the group of fixtures constituting a bathroom group.

BEND. A drainage fitting, designed to provide a change in direction of a drain pipe of less than the angle specified by the amount necessary to establish the desired slope of the line (see "Elbow" and "Sweep").

❖ This term denotes a one-way fitting (often referred to as an elbow) that is typically used in a drainage system to make a change in direction within a single run of pipe. Although bends (or elbows) are used in other plumbing and mechanical piping systems, in this instance the term refers to a drainage fitting that will meet the required radius of turn.

BOILER. A self-contained appliance from which hot water is circulated for heating purposes and then returned to the boiler, and which operates at water pressures not exceeding 160 pounds per square inch gage (psig) (1102 kPa gauge) and at water temperatures not exceeding 250°F (121°C).

❖ Boilers are usually manufactured of steel or cast iron, and are used to transfer heat (from the combustion of a fuel or from an electric-resistance element) to water for supplying steam or hot pressurized water for heating or other process or power purposes.

Boilers are usually installed in closed systems where the heat transfer medium is recirculated and retained within the system. Hot water supply boilers are normally part of open systems where the heated water is supplied and used externally to the boiler. Large domestic (potable) water heating systems often employ hot water supply boilers.

Boilers must be labeled and installed in accordance with the manufacturer's installation instructions and the applicable sections of the code. Boilers are rated in accordance with standards published by the American Society of Mechanical Engineers (ASME), the Hydronics Institute, the Steel Boiler Institute (SBI), the American Gas Association (AGA) and the American Boiler Manufacturers Association (ABMA). Boilers can be classified in accordance with the following: working temperature, working pressure, type of fuel used (or electric boilers), materials of construction and whether or not the heat transfer medium changes phase from a liquid to a vapor.

BOND BEAM. A horizontal grouted element within masonry in which reinforcement is embedded.

❖ Composed of masonry, grout and reinforcement, a bond beam is a horizontal structural member constructed to transfer loads while tying the wall together.

BRACED WALL LINE. A series of braced wall panels in a single story constructed in accordance with Section R602.10 for wood framing or Section R603.7 or R301.1.1 for cold-formed steel framing to resist racking from seismic and wind forces.

❖ Commonly referred to as a shear wall, a braced wall line is used to resist the lateral racking forces created by seismic or wind loading. A number of braced wall panels work together to form a braced wall line. The code has specific requirements for the method of constructing a braced wall line. Where a building does not comply with the bracing requirements of Chapter 6, those portions of the building not meeting the requirements must be designed in accordance with accepted engineering practice.

BRACED WALL PANEL. A section of a braced wall line constructed in accordance with Section R602.10 for wood framing or Section R603.7 or R301.1.1 for cold-formed steel framing, which extend the full height of the wall.

❖ A braced wall line is composed of a series of braced wall panels constructed in accordance with the provisions of Chapter 6. A braced wall panel must extend the full height of the wall.

BRANCH. Any part of the piping system other than a riser, main or stack.

❖ A branch refers to piping that connects to a riser, main or stack in a plumbing system. Branch piping is generalized as horizontal piping, can include vertical offsets and is smaller in size or capacity than the main artery or trunk from which it extends. It is analogous to branches of a tree.

BRANCH, FIXTURE. See "Fixture branch, drainage."

❖ See the commentary for "Fixture branch, drainage".

BRANCH, HORIZONTAL. See "Horizontal branch, drainage."

❖ See the commentary for "Horizontal branch, drainage."

BRANCH INTERVAL. A vertical measurement of distance, 8 feet (2438 mm) or more in developed length, between the connections of horizontal branches to a drainage stack. Measurements are taken down the stack from the highest horizontal branch connection.

❖ A branch interval is the vertical distance between a drainage stack's branch connections. In typical construction practice, the branch drain will connect to a stack near the floor level; therefore, the branch interval will correspond to the height of the story in which it is located. Depending on the design of the building, however, a branch interval may not correspond to the height of the story because stories are not limited in height. Therefore, the code requires a branch interval to be at least 8 feet (2438 mm) in developed length between the connections of the horizontal branches to the drainage stack.

Branch intervals are a design factor in drainage pipe sizing and venting design. Drain, waste and vent system design must consider the nature of waste and airflow in a stack and the effects that branch connections have on that flow.

BRANCH, MAIN. A water-distribution pipe that extends horizontally off a main or riser to convey water to branches or fixture groups.

❖ This applies to the primary water supply system piping that extends horizontally off a main or riser to supply water to branches or groups of fixtures.

BRANCH, VENT. A vent connecting two or more individual vents with a vent stack or stack vent.

❖ See Commentary Figure R202(4).

BTU/H. The listed maximum capacity of an appliance, absorption unit or burner expressed in British thermal units input per hour.

❖ This term stands for British thermal units per hour. Fuel-fired appliances and equipment are rated based on their Btu/h (W) input or output.

BUILDING. Building shall mean any one- and two-family dwelling or portion thereof, including townhouses, that is used, or designed or intended to be used for human habitation, for living, sleeping, cooking or eating purposes, or any combination thereof, and shall include accessory structures thereto.

❖ A building according to the IRC may be a single-family dwelling, a two-family dwelling, a townhouse or an accessory structure to such buildings. The use of a building, excluding an accessory structure, is human habitation, which specifically includes living, sleeping, cooking or eating.

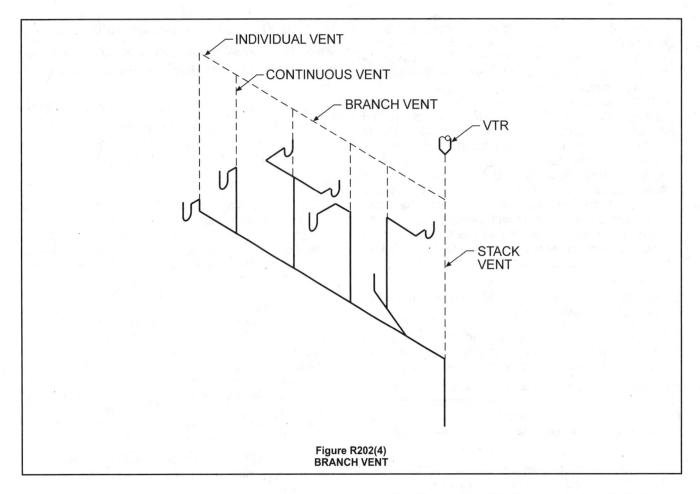

Figure R202(4)
BRANCH VENT

BUILDING DRAIN. The lowest piping that collects the discharge from all other drainage piping inside the house and extends 30 inches (762 mm) in developed length of pipe, beyond the exterior walls and conveys the drainage to the building sewer.

❖ A building drain is usually the main drain of a system within a structure. Building drains are horizontal (including vertical offsets) and are the portion of the drainage system that is at the lowest elevation in the structure. All horizontal drains above the elevation of the building drain are horizontal branches. The building drain terminates at the point where it exits the building [see Commentary Figure R202(5)].

BUILDING, EXISTING. Existing building is a building erected prior to the adoption of this code, or one for which a legal building permit has been issued.

❖ There are two definitions for an existing building. The most general use of the term is for a building that has been constructed prior to the adoption of the 2000 IRC. In other words, any building that is in existence at the time the current code is adopted by the jurisdiction is "existing." A second definition relates to those buildings that have been issued a building permit. Once a building is under construction, it is an "existing build-

ing" and is regulated by the code under which the permit was issued.

BUILDING LINE. The line established by law, beyond which a building shall not extend, except as specifically provided by law.

❖ The building line is not established by the code but rather by the jurisdiction as the point beyond which the construction of a building is not permitted. Often known as a setback line or a building setback, the building line is established to maintain order within the community's building stock. The IRC does not regulate construction with regard to any jurisdictional building line.

BUILDING OFFICIAL. The officer or other designated authority charged with the administration and enforcement of this code.

❖ Regardless of title, the individual designated by the jurisdiction as the person who administers and enforces the IRC is the building official. In addition, the building official may appoint various other individuals to assist in the activities of the Department of Building Safety. In many jurisdictions, the authority of the building official is extended to plans examiners and inspectors to some degree. Section R104 sets forth the duties and responsibilities of the building official.

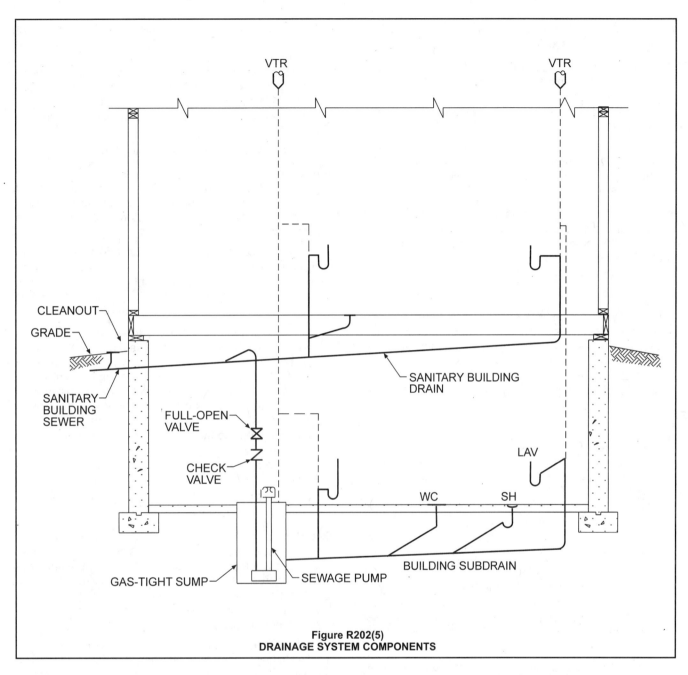

Figure R202(5)
DRAINAGE SYSTEM COMPONENTS

BUILDING SEWER. That part of the drainage system that extends from the end of the building drain and conveys its discharge to a public sewer, private sewer, individual sewage-disposal system or other point of disposal.

❖ The building sewer is the extension of the building drain and is located entirely outside the building exterior walls.

BUILDING THERMAL ENVELOPE. The basement walls, exterior walls, floor, roof and any other building element that enclose conditioned spaces.

❖ The building thermal envelope includes the roof/ceiling assembly, wall assemblies and floor assemblies that surround a conditioned area, which is the space that is being intentionally heated and/or cooled. The building

envelope is the assembly that separates conditioned space from unconditioned space or the outdoors. For example, a wall between a conditioned dwelling and an unconditioned garage is part of the building thermal envelope. Other elements of the building envelope include attic kneewalls, the perimeter joist between two conditioned floors and skylight wells.

BUILT-UP ROOF COVERING. Two or more layers of felt cemented together and surfaced with a cap sheet, mineral aggregate, smooth coating or similar surfacing material.

❖ A common roof covering for buildings with relatively flat roofs, built-up roof covering uses two or more felt layers with a cap-sheet surfacing. A number of materials standards regulate the wide variety of built-up roofing materials that are available.

CEILING HEIGHT. The clear vertical distance from the finished floor to the finished ceiling.

❖ Ceiling height as regulated by Section R305 is measured from the finished floor surface to the finished ceiling. Depending on the space under consideration, the minimum ceiling height required by the IRC can vary greatly.

CHIMNEY. A primary vertical structure containing one or more flues, for the purpose of carrying gaseous products of combustion and air from a fuel-burning appliance to the outside atmosphere.

❖ Chimneys differ from vents in the materials they are made of and the type of appliance they are designed to serve. Chimneys are capable of venting flue gases of much higher temperatures than vents.

CHIMNEY CONNECTOR. A pipe that connects a fuel-burning appliance to a chimney.

❖ Chimney connectors are sections of pipe used to convey combustion products from an appliance flue outlet to a chimney inlet. Factory-built chimneys can connect directly to some appliances without the need for a connector; however, masonry chimneys cannot connect directly to an appliance because of the chimney's weight.

CHIMNEY TYPES

Residential-type appliance. An approved chimney for removing the products of combustion from fuel-burning, residential-type appliances producing combustion gases not in excess of 1,000°F (538°C) under normal operating conditions, but capable of producing combustion gases of 1,400°F (760°C) during intermittent forces firing for periods up to 1 hour. All temperatures shall be measured at the appliance flue outlet. Residential-type appliance chimneys include masonry and factory-built types.

❖ The chimneys addressed in the IRC are generally for use with residential-type appliances. Such chimneys are limited in the temperature of the combustion gases they can exhaust. Both masonry and factory-built chimneys can be residential type.

CIRCUIT VENT. A vent that connects to a horizontal drainage branch and vents two traps to a maximum of eight traps or trapped fixtures connected into a battery.

❖ A circuit vent vents multiple fixtures using only one or two vents. The horizontal drainage branch actually serves as a wet vent for the fixtures located downstream of the circuit vent connection. See Section P3110.

CLADDING. The exterior materials that cover the surface of the building envelope that is directly loaded by the wind.

❖ Sheathing, siding or other materials used on the exterior envelope of a building are "cladding" for the purpose of wind design.

CLEANOUT. An accessible opening in the drainage system used for the removal of possible obstruction.

❖ Cleanouts are broadly defined as any access point into the drainage system that allows the removal of a clog or other obstruction in the drain line. A cleanout provides convenient access to the piping interior without significant disassembly of the plumbing installation. A "P" trap with a removable U-bend is typically acceptable as a cleanout for the fixture drain or branch to which the trap discharges.

CLOSET. A small room or chamber used for storage.

❖ A closet is simply a small storage room. Closets are generally found throughout dwelling units but are not considered habitable spaces. The typical hazard related to closets is the probable storage of combustible materials.

COMBINATION WASTE AND VENT SYSTEM. A specially designed system of waste piping embodying the horizontal wet venting of one or more sinks or floor drains by means of a common waste and vent pipe adequately sized to provide free movement of air above the flow line of the drain.

❖ This special system is employed when no other conventional system of separate waste and vent is practical. The combination waste and vent system is used only for venting floor drains, standpipes, sinks and lavatories. The drainage piping is larger than required for draining purposes only, and the drainage branch and stack should be provided with vent piping. The system is commonly used where floor drains are installed in large open areas that cannot accommodate vertical vent risers from the floor drains. See Section P3111.

COMBUSTIBLE MATERIAL. Any material not defined as noncombustible.

❖ Any material that ASTM E 136 does not qualify as a noncombustible material is considered combustible. The presence of sizable quantities of combustible material increases the potential fire hazard. Buildings containing or constructed of combustible materials are typically more highly regulated that those that are primarily noncombustible.

COMBUSTION AIR. The air provided to fuel-burning equipment including air for fuel combustion, draft hood dilution and ventilation of the equipment enclosure.

❖ The process of combustion requires a specific amount of oxygen to initiate and sustain the combustion reaction. Combustion air includes primary air, secondary air, draft hood dilution air and excess air. Combustion air is the amount of atmospheric air required for complete combustion of a fuel and is related to the molecular composition of the fuel being burned, the design of the fuel-burning equipment and the percentage of oxygen in the combustion air. Too little combustion air will result in incomplete combustion of a fuel and the possible formation of carbon deposits (soot), carbon monoxide, toxic alcohols, ketones, aldehydes, nitrous oxides and other byproducts. The required amount of combustion air is usually stated in terms of cubic feet per minute (m^3/s) or pounds per hour (kg/h).

COMMON VENT. A single pipe venting two trap arms within the same branch interval, either back-to-back or one above the other.

❖ Where two or more fixtures are connected to the same drainage pipe, a common vent could be installed. This common vent is sized and classified as an individual vent. Any two fixtures may be common vented to either a vertical or horizontal drainage pipe.

A typical form of common venting connects two fixtures at the same level. Two fixtures connecting at different levels but within the same story are also common vented. When one of the fixtures connected is at a different level, the vertical drain between the fixtures is oversized because it functions as a drain for the upper fixture and a vent for the lower fixture.

CONDENSATE. The liquid that separates from a gas due to a reduction in temperature, e.g., water that condenses from flue gases and water that condenses from air circulating through the cooling coil in air conditioning equipment.

❖ Condensate forms when the temperature of a vapor is lowered to its dew point temperature. Air conditioning systems produce condensate when an airstream contacts cooling coils. The moisture in the air condenses on the cold surface of the coils and the air is "dehumidified." Condensate also forms within improperly designed chimneys and vents when the products of combustion (which contain water vapor) contact the colder walls of the flue. If the temperature of the products of combustion is lowered to the dew point temperature of the water vapor, condensate will form on the inside walls of the flue. Condensed steam in hydronic systems is also referred to as "condensate."

CONDENSING APPLIANCE. An appliance that condenses water generated by the burning of fuels.

❖ Such units include compressors and condenser heat exchangers; they convert refrigerant vapor to liquid.

CONDITIONED AIR. Air treated to control its temperature, relative humidity or quality.

❖ Conditioned air is air that is heated, cooled, humidified, dehumidified or decontaminated.

CONDITIONED AREA. That area within a building provided with heating and/or cooling systems or appliances capable of maintaining, through design or heat loss/gain, 68°F (20°C) during the heating season and/or 80°F (27°C) during the cooling season, or has a fixed opening directly adjacent to a conditioned area.

❖ The conditioned area is the space within the building that receives conditioned air. The space can be directly conditioned or indirectly conditioned by an adjacent conditioned space.

CONDITIONED FLOOR AREA. The horizontal projection of the floors associated with the conditioned space.

❖ The conditioned floor area is the floor area in square feet (m²) of a conditioned space.

CONDITIONED SPACE. For energy purposes, space within a building that is provided with heating and/or cooling equipment or systems capable of maintaining, through design or heat loss/gain, 50°F (10°C) during the heating season and 85°F (29°C) during the cooling season, or communicates directly with a conditioned space. For mechanical purposes, an area, room or space being heated or cooled by any equipment or appliance.

❖ If a space is heated, cooled or humidified/dehumidified, it is conditioned space. The building must have a system that is capable of maintaining the space at 50°F (10°C) or above for heating and 85°F (29°C) or below for cooling during normal operation. Also, a basement is "conditioned" if the floor/ceiling assembly between the basement and the conditioned first floor is uninsulated and there is enough heat transfer across the floor to maintain the space at 50°F (10°C) or above. In addition, uninsulated duct systems located in an unconditioned basement may raise the temperature of the space to 50°F (10°C) and above during normal operation of the system. The basement would then be conditioned space. If the space is conditioned, the building envelope that surrounds the space must meet the thermal requirements in Table N1102.1.

CONFINED SPACE. A room or space having a volume less than 50 cubic feet per 1,000 Btu/h (4.83 L/W) of the aggregate input rating of all fuel-burning appliances installed in that space.

❖ This term is used in conjunction with Chapter 20 and other sections related to combustion air.

CONSTRUCTION DOCUMENTS. Written, graphic and pictorial documents prepared or assembled for describing the design, location and physical characteristics of the elements of a project necessary for obtaining a building permit. Construction drawings shall be drawn to an appropriate scale.

❖ Construction documents are the necessary drawings, specifications and support materials created for the design and construction of a building. Addressed in Section R106, construction documents are required at the time of permit application unless the work is of such a minor nature that the documents are not necessary to provide compliance with the code. The construction documents must be drawn to an appropriate scale so they may be easily interpreted.

CONTAMINATION. An impairment of the quality of the potable water that creates an actual hazard to the public health through poisoning or through the spread of disease by sewage, industrial fluids or waste.

❖ The Environmental Protection Agency (EPA) sets maximum levels for various chemicals and bacteria in drinking water. When an unacceptable level of one or more contaminants is present, the water is considered to be nonpotable. One of the primary purposes of the code is to protect potable water supplies from contamination. Contamination consists of either hazardous chemicals or raw sewage, and is considered by the code to represent a high hazard when present in the water supply. Refer to Table P2902.2 for the required potable protection requirements related to high-hazard situations.

The code and other related regulations, such as those from the EPA, are intended to reduce, if not totally eliminate, the risk to the general public of poisoning or health impairment through sewage, industrial fluids or waste entering the potable water supply. Note that these contaminants would have a harmful effect on the occupants of a building should the water be consumed. Contamination is quite different from pollution, which is a situation where the water supply is still considered to be nonpotable, but would not kill or harm if ingested. Refer to the definition of "Pollution" for additional information on that term.

CONTINUOUS WASTE. A drain from two or more similar adjacent fixtures connected to a single trap.

❖ Continuous waste is the piping described and regulated in Section P3201.6. The most common occurrences are where multiple compartments of a kitchen sink or double lavatories are connected to a single trap.

CONTROL, LIMIT. An automatic control responsive to changes in liquid flow or level, pressure, or temperature for limiting the operation of an appliance.

❖ Limit controls are safety devices used to protect equipment, appliances, property, and persons. Such controls act at their set point to limit a condition such as temperature or pressure and include high and low limits. This definition applies to controls for all types of fuel: gas, liquid and solid.

CONTROL, PRIMARY SAFETY. A safety control responsive directly to flame properties that senses the presence or absence of flame and, in event of ignition failure or unintentional flame extinguishment, automatically causes shutdown of mechanical equipment.

❖ This is a device designed to shut down mechanical equipment in response to changes in liquid or gas flow rates that could extinguish a flame.

CONVECTOR. A system-incorporating heating element in an enclosure in which air enters an opening below the heating element, is heated and leaves the enclosure through an opening located above the heating element.

❖ See Commentary Figure R202(6) for an illustration of an electric hydronic convector.

CORROSION RESISTANCE. The ability of a material to withstand deterioration of its surface or its properties when exposed to its environment.

❖ Under those conditions where corrosion of materials is detrimental to the integrity of the building's construction and use, the code mandates corrosion-resistant materials. Corrosion resistant materials control or withstand the deterioration of their surfaces and retain their physical properties in their intended environment. Corrosion resistance can be provided by the material or by a process applied to the material.

COURT. A space, open and unobstructed to the sky, located at or above grade level on a lot and bounded on three or more sides by walls or a building.

❖ An exterior area is a court if it is enclosed on at least three sides by exterior walls of the building or other enclosing elements and is open and unobstructed to the sky above. By virtue of being substantially open to the exterior and to the sky, a court may be used to obtain natural light and ventilation for the building.

CRIPPLE WALL. A framed wall extending from the top of the foundation to the underside of the floor framing of the first story above grade plane.

❖ Cripple walls are built on the top of footings or foundation walls. They can typically be found along the top of stepped foundation walls where the grade adjoining the structure changes height. Cripple walls must be properly braced to resist lateral forces. They are often treated the same as a first-story wall. Provisions for the bracing of cripple walls are in Section R602.10.2.

CROSS CONNECTION. Any connection between two otherwise separate piping systems whereby there may be a flow from one system to the other.

❖ Cross connections are the links through which it is possible for contaminated materials to enter a potable water supply. The contaminant enters the potable water supply when the pressure of the polluted source exceeds the pressure of the potable source. The action may be called backsiphonage or backpressure.

Many serious outbreaks of illnesses and disease have been traced to cross connections. The intent of the code is to eliminate cross connections or prevent backflow where cross connections cannot be eliminated. Refer to the definitions and commentary for the terms "Backflow" and "Backsiphonage."

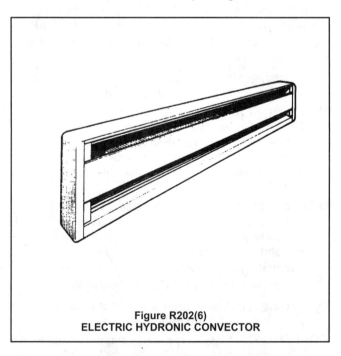

Figure R202(6)
ELECTRIC HYDRONIC CONVECTOR

DALLE GLASS. A decorative composite glazing material made of individual pieces of glass that are embedded in a cast matrix of concrete or epoxy.

❖ Various types of decorative glazing materials are exempt from the safety glazing provisions when installed in side-hinged doors, adjacent to doors, or in large glazed openings. Dalle glass is one such material. Made up of multiple individual pieces of glass, dalle glass is exempt because of its decorative function.

DAMPER, VOLUME. A device that will restrict, retard or direct the flow of air in any duct, or the products of combustion of heat-producing equipment, vent connector, vent or chimney.

❖ A damper is used in an air distribution system as a restrictor to regulate airflow through duct work. When used in flues venting combustion gases, a damper is used to regulate draft.

DEAD END. A branch leading from a DWV system terminating at a developed length of 2 feet (610 mm) or more. Dead ends shall be prohibited except as an approved part of a rough-in for future connection.

❖ "Dead end" refers to horizontal piping that does not conduct waste flow but is connected to piping that does conduct such flow. A dead end will collect solid waste as a result of the normal flow depth in the drain pipe to which it is connected. A dead end will also collect solids as a result of drainage system stoppages.

DEAD LOADS. The weight of all materials of construction incorporated into the building, including but not limited to walls, floors, roofs, ceilings, stairways, built-in partitions, finishes, cladding, and other similarly incorporated architectural and structural items, and fixed service equipment.

❖ Dead loads are considered in the structural design of a building and are used in the span tables for sizing floor joists, ceiling joists and rafters. The weight of the materials of construction joists, sheathing, studs and gypsum board is counted as the dead load, as are stairways, cladding and fixed service equipment. The dead loads make up the fixed load placed in the building and are independent from the live loads.

DECK. An exterior floor system supported on at least two opposing sides by an adjoining structure and/or posts, piers, or other independent supports.

❖ Where an exterior floor system is supported on a minimum of two opposing sides, it is a deck. The support may be totally by posts or piers, or, as often occurs, by the dwelling on one side with posts or piers on the opposite side.

DECORATIVE GLASS. A carved, leaded or Dalle glass or glazing material whose purpose is decorative or artistic, not functional; whose coloring, texture or other design qualities or components cannot be removed without destroying the glazing material; and whose surface, or assembly into which it is incorporated, is divided into segments.

❖ Decorative glass is not designed for functional purposes such as a vision panel, but rather is used for artistic reasons. Some types of decorative glass include

carved, leaded, and dalle glass. Decorative glass is typically made up of multiple pieces of glass arranged in some design or pattern. Fixed-in-place decorative glass is exempt from some of the requirements for glazing in hazardous locations.

DESIGN PROFESSIONAL. See definition of "Registered design professional."

❖ "Design professional" is another term for registered design professional. See the definition of "Registered design professional."

DEVELOPED LENGTH. The length of a pipeline measured along the center line of the pipe and fittings.

❖ This term identifies a concept necessary for computing the actual length of piping in a plumbing system. The developed length is measured along the actual flow path of piping and includes the piping lengths in all offsets and changes in direction. Several code requirements are dependent on the actual length of plumbing piping.

DIAMETER. Unless specifically stated, the term "diameter" is the nominal diameter as designated by the approved material standard.

❖ "Diameter" is the length of a line passing though the center of a pipe from one side to the other. Depending on the pipe standard, this distance can be measured to the inside portions (inside diameter, or ID) or the outside portions (outside diameter, or OD) of the pipe.

DIAPHRAGM. A horizontal or nearly horizontal system acting to transmit lateral forces to the vertical resisting elements. When the term "diaphragm" is used, it includes horizontal bracing systems.

❖ In the load path for forces transmitted through a building or structure, the diaphragm is that horizontal portion of the path where lateral forces are transferred to the vertical resisting elements. Roof diaphragms and floor diaphragms carry lateral loads to the shear walls.

DILUTION AIR. Air that enters a draft hood or draft regulator and mixes with flue gases.

❖ Dilution air is associated with draft-hood-equipped Category I appliances. Fan-assisted Category I appliances do not use dilution air; all of the air brought in by the fan passes directly through the combustion chamber of the appliance. Dilution air causes lowering of the dew point and cooling of flue gases, but to a point that is still above the point when condensation will occur inside the vent.

DIRECT-VENT APPLIANCE. A fuel-burning appliance with a sealed combustion system that draws all air for combustion from the outside atmosphere and discharges all flue gases to the outside atmosphere.

❖ Such appliances are equipped with independent exhaust and intake pipes, or they have concentric pipes that vent through the inner pipe and convey combustion air in the annular space between pipe walls. See the commentary for Sections R304.1 and R804.1.

DRAFT. The pressure difference existing between the appliance or any component part and the atmosphere, that causes a continuous flow of air and products of combustion through the gas passages of the appliance to the atmosphere.

> **Induced draft.** The pressure difference created by the action of a fan, blower or ejector, that is located between the appliance and the chimney or vent termination.

> **Natural draft.** The pressure difference created by a vent or chimney because of its height, and the temperature difference between the flue gases and the atmosphere.

❖ Draft is the negative static pressure measured relative to atmospheric pressure that is developed in chimneys and vents and in the flue-ways of fuel-burning appliances. Draft can be produced by hot flue-gas buoyancy ("stack effect"), mechanically by fans and exhausters or by a combination of both natural and mechanical means.

DRAFT HOOD. A device built into an appliance, or a part of the vent connector from an appliance, which is designed to provide for the ready escape of the flue gases from the appliance in the event of no draft, backdraft or stoppage beyond the draft hood; prevent a backdraft from entering the appliance; and neutralize the effect of stack action of the chimney or gas vent on the operation of the appliance.

❖ Draft hoods are integral to or supplied with natural-draft, atmospheric-burner gas-fired appliances. Appliances equipped with draft hoods are classified by the manufacturer as Category I appliances. Because of new minimum-efficiency standards and the popularity of mechanical and other special proprietary venting systems, draft-hood-equipped appliances are becoming rare in the marketplace.

DRAFT REGULATOR. A device that functions to maintain a desired draft in the appliance by automatically reducing the draft to the desired value.

❖ Excessive draft reduces the efficiency of an appliance because of energy loss in the form of heat loss out the vent. Draft regulators automatically adjust the draft by allowing cooler air to enter the vent at a preset flow rate, reducing the draft.

DRAFT STOP. A material, device or construction installed to restrict the movement of air within open spaces of concealed areas of building components such as crawl spaces, floor-ceiling assemblies, roof-ceiling assemblies and attics.

❖ Draft stops divide a large concealed space into smaller compartments in wood-frame floor construction. They limit the movement of air within the cavity, reducing the potential for rapid fire spread. Draftstopping materials include gypsum board, wood structural panels, particleboard or other substantial materials that are adequately supported.

DRAIN. Any pipe that carries soil and water-borne wastes in a building drainage system.

❖ A drain is any pipe in a plumbing system that carries sanitary waste, clear-water waste or storm water. In the case of wet vents, common vents, waste stack vents, circuit vents and combination drain and vents, the drain may also be conducting airflow.

DRAINAGE FITTING. A pipe fitting designed to provide connections in the drainage system that have provisions for establishing the desired slope in the system. These fittings are made from a variety of both metals and plastics. The methods of coupling provide for required slope in the system (see "Durham fitting").

❖ These fittings will accommodate the necessary slope and sweep required for a drain line. See Section P3005.1.

DUCT SYSTEM. A continuous passageway for the transmission of air which, in addition to ducts, includes duct fittings, dampers, plenums, fans and accessory air-handling equipment and appliances.

❖ Duct systems are part of an air distribution system and include supply, return and relief/exhaust air systems.

DURHAM FITTING. A special type of drainage fitting for use in the durham systems installations in which the joints are made with recessed and tapered threaded fittings, as opposed to bell and spigot lead/oakum or solvent/cemented or soldered joints. The tapping is at an angle (not 90 degrees) to provide for proper slope in otherwise rigid connections.

❖ This is a specific type of fitting used in Durham piping systems, which employ specially designed cast-iron threaded fittings and threaded galvanized steel piping.

DURHAM SYSTEM. A term used to describe soil or waste systems where all piping is of threaded pipe, tube or other such rigid construction using recessed drainage fittings to correspond to the types of piping.

❖ See the definition of "Durham Fittings."

DWELLING. Any building that contains one or two dwelling units used, intended, or designed to be built, used, rented, leased, let or hired out to be occupied, or that are occupied for living purposes.

❖ A dwelling is a building that contains either one or two dwelling units. The purpose of a dwelling is occupation for living purposes, regardless of the manner of ownership. Single-family houses and duplexes fall under the definition of dwelling. See also "Dwelling Unit."

DWELLING UNIT. A single unit providing complete independent living facilities for one or more persons, including permanent provisions for living, sleeping, eating, cooking and sanitation.

❖ The specific purpose of a dwelling unit is to provide the essential amenities necessary for complete and independent facilities. Commonly, dwelling units are thought of as single-family houses or individual living units in duplexes or townhouses.

DWV. Abbreviated term for drain, waste and vent piping as used in common plumbing practice.

❖ This term is used extensively to identify systems and materials suitable for drain, waste and vent installations. It will often be labeled on specific materials approved for such use.

EFFECTIVE OPENING. The minimum cross-sectional area at the point of water-supply discharge, measured or expressed in terms of diameter of a circle and if the opening is not circular, the diameter of a circle of equivalent cross-sectional area. (This is applicable to air gap.)

❖ An effective opening is used to determine the minimum air gap required between a potable water supply outlet or opening and the flood level rim of a receptacle, fixture or other potential source of contamination.

ELBOW. A pressure pipe fitting designed to provide an exact change in direction of a pipe run. An elbow provides a sharp turn in the flow path (see "Bend" and "Sweep").

❖ See definition of "Bend."

EMERGENCY ESCAPE AND RESCUE OPENING. An operable exterior window, door or similar device that provides for a means of escape and access for rescue in the event of an emergency.

❖ In the case of an emergency, particularly a fire, immediate action must be taken. Quick evacuation of the dwelling unit is often required. If occupants are sleeping at the time of the incident, the time for evacuation is extended, often to the point where normal egress is not possible. In such situations, a door or window to the exterior can be used. It is also possible to use such an exterior opening for rescue. See Section R310 for code requirements.

EQUIPMENT. All piping, ducts, vents, control devices and other components of systems other than appliances that are permanently installed and integrated to provide control of environmental conditions for buildings. This definition shall also include other systems specifically regulated in this code.

❖ Throughout the code, the terms "equipment" and "appliance" have been used as necessary to match the terms with the intent and context of the code text. Appliances are not referred to as equipment and vise versa. Traditionally, the term "equipment" meant large machinery and specialized hardware not thought of as an "appliance." See the definition of "Appliance."

EQUIVALENT LENGTH. For determining friction losses in a piping system, the effect of a particular fitting equal to the friction loss through a straight piping length of the same nominal diameter.

❖ A straight run of piping will contain a friction loss [typically figured in friction loss per every 100 feet (30 480 mm)]. Where fittings are included, the result is additional friction loss. Although it is possible to find friction loss through calculations, the prescriptive method of water supply sizing in Chapter 29 uses a multiplier of 1.2 to provide an average friction-loss calculation. Therefore, the equivalent length of a 100-foot (30 480 mm) run of piping with fittings will be 120 feet (36 576 mm) to accommodate the friction loss expected. The equivalent length will be used in applying the sizing tables in Chapter 29. See Section P2903.7, Item 2.

ESSENTIALLY NONTOXIC TRANSFER FLUIDS. Fluids having a Gosselin rating of 1, including propylene glycol; mineral oil; polydimenthyoil oxane; hydrochlorofluorocarbon, chlorofluorocarbon and hydrofluorocarbon refrigerants; and FDA-approved boiler water additives for steam boilers.

❖ Transfer fluids are liquids or gases that transfer heat to or remove heat from another fluid, such as water. The exchange of heat energy takes place through a heat exchange material, which separates the transfer fluid from the fluid being heated or cooled.

ESSENTIALLY TOXIC TRANSFER FLUIDS. Soil, water or gray water and fluids having a Gosselin rating of 2 or more including ethylene glycol, hydrocarbon oils, ammonia refrigerants and hydrazine.

❖ See the commentary for the definition of "Essentially nontoxic transfer fluids."

EVAPORATIVE COOLER. A device used for reducing air temperature by the process of evaporating water into an airstream.

❖ Also known as "swamp coolers," such units are used in arid climates and use water as a refrigerant. Such units substantially increase the humidity of the air being conditioned.

EXCESS AIR. Air that passes through the combustion chamber and the appliance flue in excess of that which is theoretically required for complete combustion.

❖ The nature of natural-draft, fuel-fired appliances is such that the introduction and efficient intermixing of primary and secondary combustion air is not precise and is coupled with the induced air inflow caused by the internal draft. To achieve complete oxidation (combustion), there will always be more air (excess) introduced into the appliances than is theoretically necessary for the complete combustion of the fuel.

EXHAUST HOOD, FULL OPENING. An exhaust hood with an opening at least equal to the diameter of the connecting vent.

❖ See Commentary Figure R202(7) for an illustration of a full opening exhaust hood.

EXISTING INSTALLATIONS. Any plumbing system regulated by this code that was legally installed prior to the effective date of this code, or for which a permit to install has been issued.

❖ "Existing installations" is a term that applies to all plumbing work that has been legally installed prior to the effective date of the code, as well as to all plumbing work for which a permit to install has been issued prior to the effective date of the code. Plumbing that has been illegally installed prior to the effective date of the code is not "existing" and is subject to all of the code requirements for new installations.

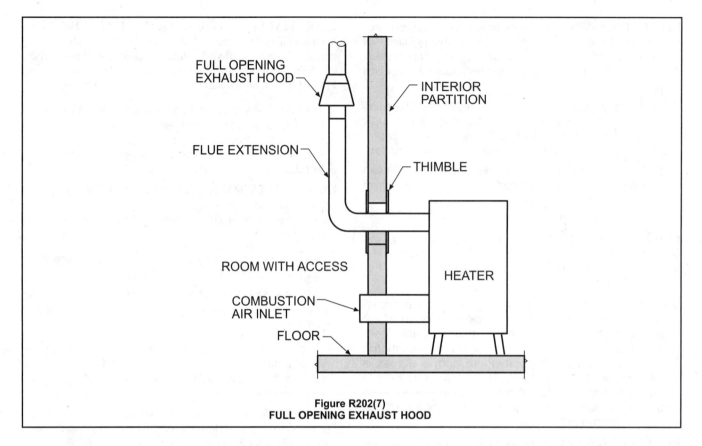

Figure R202(7)
FULL OPENING EXHAUST HOOD

EXTERIOR INSULATION FINISH SYSTEMS (EIFS). Synthetic stucco cladding systems typically consisting of five layers: adhesive, insulation board, base coat into which fiberglass reinforcing mesh is embedded, and a finish coat in the desired color.

❖ A variety of popular exterior finish systems have been developed, with the typical construction consisting of five layers of material. The code has relatively few provisions regulating EIFS; most of the requirements are those of the manufacturer.

EXTERIOR WALL. An above-grade wall that defines the exterior boundaries of a building. Includes between-floor spandrels, peripheral edges of floors, roof and basement knee walls, dormer walls, gable end walls, walls enclosing a mansard roof and basement walls with an average below-grade wall area that is less than 50 percent of the total opaque and nonopaque area of that enclosing side.

❖ An exterior wall is any wall located above grade level enclosing conditioned space. Regulated by this definition for the purpose of energy efficiency, a basement wall is exterior where less than 50 percent of its enclosing surface area is below grade.

FACTORY-BUILT CHIMNEY. A listed and labeled chimney composed of factory-made components assembled in the field in accordance with the manufacturer's instructions and the conditions of the listing.

❖ A factory-built chimney is a manufactured, listed and labeled chimney that has been tested by an approved agency to determine its performance characteristics.

Factory-built chimneys are manufactured in two basic designs: a double-wall insulated design or a triple-wall air-cooled design. Both designs use stainless steel inner liners to resist the corrosive effects of combustion products.

FENESTRATION. Skylights, roof windows, vertical windows (whether fixed or moveable); opaque doors; glazed doors; glass block; and combination opaque/glazed doors.

❖ Exterior windows and doors are fenestrations in the exterior wall. Skylights and other sloped glazing are also included in the definition, as are glass doors, opaque doors and opaque/glazed doors. Chapter 11 regulates fenestrations for energy efficiency.

FIBER CEMENT SIDING. A manufactured, fiber-reinforcing product made with an inorganic hydraulic or calcium silicate binder formed by chemical reaction and reinforced with organic or inorganic non-asbestos fibers, or both. Additives which enhance manufacturing or product performance are permitted. Fiber cement siding products have either smooth or textured faces and are intended for exterior wall and related applications.

❖ Fiber cement siding is produced from flat non-asbestos fiber cement sheets. It is intended for exterior application and consists of fiber-reinforced cement which is formed either with or without pressure and cured either under natural or accelerated conditions. The surface of the sheet to be exposed is either smooth, granular or textured. This code permits its use as either panel siding or horizontal lap siding.

FIREBLOCKING. Building materials installed to resist the free passage of flame to other areas of the building through concealed spaces.

❖ Fireblocking materials include lumber; wood structural panels, particleboard, gypsum board, cement-based millboard, batts or blankets of mineral wool or glass fiber, or any other approved material that will resist the passage of flame from one concealed area to another. Regulated by Section R602.8 for wood-frame construction, fireblocking is often used to isolate vertical cavities such as stud spaces from horizontal concealed areas such as attics or floor-ceiling assemblies.

FIREPLACE. An assembly consisting of a hearth and fire chamber of noncombustible material and provided with a chimney, for use with solid fuels.

❖ Fireplaces burn solid fuels such as wood and coal and are not referred to as appliances in the code. Both masonry and factory-built fireplaces are regulated by the IRC.

Factory-built fireplace. A listed and labeled fireplace and chimney system composed of factory-made components, and assembled in the field in accordance with manufacturer's instructions and the conditions of the listing.

❖ Factory-built fireplaces are solid-fuel burning units having a fire chamber that is either open to the room or, if equipped with doors, operated with the doors either open or closed. The term "fireplace" describes a complete assembly, which includes the hearth, fire chamber and chimney. A factory-built fireplace is composed of factory-built components representative of the prototypes tested, and it is to be installed in accordance with the manufacturer's installation instructions to form a completed fireplace.

Masonry chimney. A field-constructed chimney composed of solid masonry units, bricks, stones or concrete.

❖ Masonry chimneys can have one or more flues within them and are field constructed of brick, stone, concrete and fire-clay materials. A masonry chimney can stand alone or be part of a masonry fireplace.

Masonry fireplace. A field-constructed fireplace composed of solid masonry units, bricks, stones or concrete.

❖ Masonry fireplaces must be constructed in accordance with the requirements found in Section R1003. These specific requirements are based on tradition and field experience and describe the conventional fireplace that has proven to be reliable where properly constructed, used and maintained.

FIREPLACE STOVE. A free-standing, chimney-connected solid-fuel-burning heater designed to be operated with the fire chamber doors in either the open or closed position.

❖ Fireplace stoves are generally of the free-standing type and heat a space by direct radiation. There are various types of fireplace stoves, and their installation must be in compliance with the listing of the stove.

FIREPLACE THROAT. The opening between the top of the firebox and the smoke chamber.

❖ The fireplace throat is the point where the smoke and heat from the fireplace pass into the chimney.

FIRE SEPARATION DISTANCE. The distance measured from the building face to one of the following:

1. To the closest interior lot line; or

2. To the centerline of a street, an alley or public way; or

3. To an imaginary line between two buildings on the lot.

The distance shall be measured at a right angle from the face of the wall.

❖ This is the distance between the exterior surface of a building and one of the following three locations: the nearest interior lot line; the centerline of a street, alley, or public way; or an imaginary line placed between two buildings on the same lot. The measurement is perpendicular to the lot line [see Commentary Figure R202(8)]. The fire separation distance is important in determining exterior wall protection and the potential prohibition of exterior openings based on the proximity to the lot lines.

FIXTURE. See "Plumbing fixture."

FIXTURE BRANCH, DRAINAGE. A drain serving two or more fixtures that discharges into another portion of the drainage system.

❖ See Commentary Figure R202(9).

FIXTURE BRANCH, WATER-SUPPLY. A water-supply pipe between the fixture supply and a main water-distribution pipe or fixture group main.

❖ See the definition of "Branch."

FIXTURE DRAIN. The drain from the trap of a fixture to the junction of that drain with any other drain pipe.

❖ Commonly referred to as a trap arm, a fixture drain is the horizontal section of pipe connecting the outlet weir of a trap to a stack, fixture branch or any other drain [see Commentary Figures R202(9) and (10)].

FIXTURE FITTING

Supply fitting. A fitting that controls the volume and/or directional flow of water and is either attached to or accessible from a fixture or is used with an open or atmospheric discharge.

Waste fitting. A combination of components that conveys the sanitary waste from the outlet of a fixture to the connection of the sanitary drainage system.

❖ This includes the various portions of a faucet from its connecting supply lines to the termination point where water is delivered. When applied to waste components, it includes tubing, fittings and traps used for the connection to the drainage system.

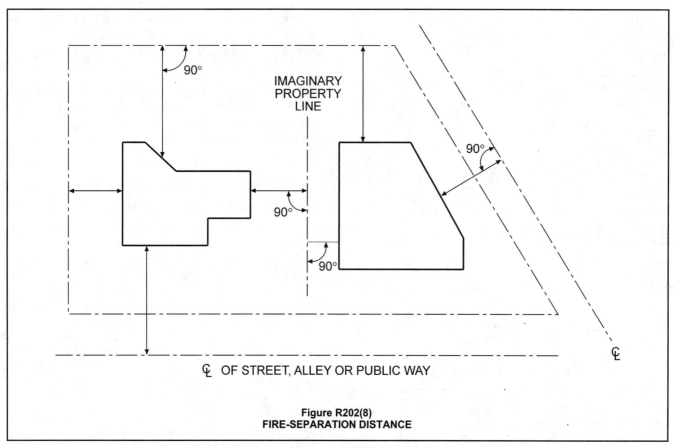

Figure R202(8)
FIRE-SEPARATION DISTANCE

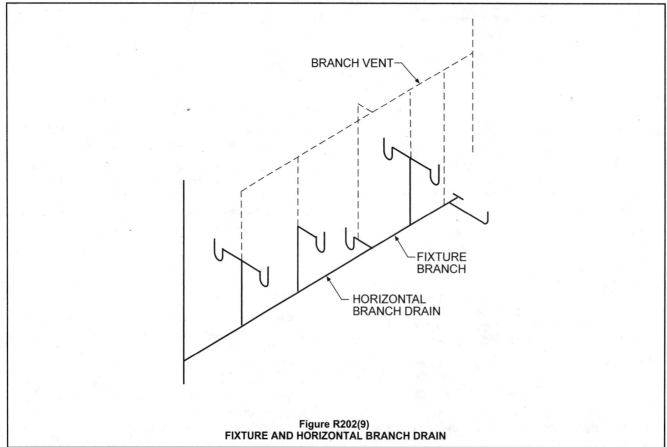

Figure R202(9)
FIXTURE AND HORIZONTAL BRANCH DRAIN

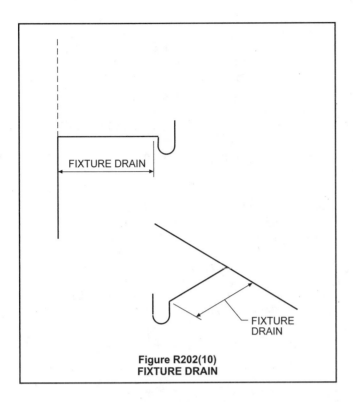

**Figure R202(10)
FIXTURE DRAIN**

FIXTURE GROUP, MAIN. The main water-distribution pipe (or secondary branch) serving a plumbing fixture grouping such as a bath, kitchen or laundry area to which two or more individual fixture branch pipes are connected.

❖ This is the main or branch water-distribution pipe that extends to a group of fixtures (i.e. bathroom, laundry or kitchen).

FIXTURE SUPPLY. The water-supply pipe connecting a fixture or fixture fitting to a fixture branch.

❖ A fixture supply is a water supply pipe serving a single fixture.

FIXTURE UNIT, DRAINAGE (d.f.u.). A measure of probable discharge into the drainage system by various types of plumbing fixtures, used to size DWV piping systems. The drainage fixture-unit value for a particular fixture depends on its volume rate of drainage discharge, on the time duration of a single drainage operation and on the average time between successive operations.

❖ The conventional method of designing a sanitary drainage system is based on drainage fixture unit (d.f.u.) load values. The fixture-unit approach takes into consideration the probability of load on a drainage system. The d.f.u. is an arbitrary loading factor assigned to each fixture relative to its impact on the drainage system. D.f.u. values are determined based on the average rate of water discharge by a fixture, the duration of a single operation, and the frequency of use or interval between each operation.

Because d.f.u. values have a built-in probability factor, they cannot be directly translated into flow rates or discharge rates.

A d.f.u. is not the same as the water supply fixture unit (w.s.f.u.) described in Chapter 29 and below.

FIXTURE UNIT, WATER-SUPPLY (w.s.f.u.). A measure of the probable hydraulic demand on the water supply by various types of plumbing fixtures used to size water-piping systems. The water-supply fixture-unit value for a particular fixture depends on its volume rate of supply, on the time duration of a single supply operation and on the average time between successive operations.

❖ When estimating peak demand for water supply systems, sizing methods use water supply fixture units (w.s.f.u.). This is a numerical factor on an arbitrary scale assigned to intermittently used fixtures to calculate their load-producing effects on the water supply system.

The use of fixture units makes it possible to simplify the difficult task of calculating load-producing characteristics of any fixture to a common basis. The fixture units of different kinds of fixtures can be applied to a single basic probability curve found in various sizing methods.

FLAME SPREAD. The propagation of flame over a surface.

❖ During a fire, flames will travel across a building's interior surfaces in different forms and at different speeds. The movement of the flames along the surface material is the flame spread.

FLAME SPREAD INDEX. The numeric value assigned to a material tested in accordance with ASTM E 84.

❖ To regulate the spread of fire over an interior surface material, limitations are imposed on the characteristics of the material. The standardized test procedure described in ASTM E 84 measures the flame spread under specified conditions, and a numerical value is assigned to the material being tested. This value is the flame spread index

FLOOD-LEVEL RIM. The edge of the receptor or fixture from which water overflows.

❖ The flood-level rim is the highest elevation that liquid can be contained in a receptacle without spilling over the side of the fixture or receptor.

FLOOR DRAIN. A plumbing fixture for recess in the floor having a floor-level strainer intended for the purpose of the collection and disposal of waste water used in cleaning the floor and for the collection and disposal of accidental spillage to the floor.

❖ Floor drains are typically installed as emergency fixtures, preventing the flooding of a room or space.

FLOOR FURNACE. A self-contained furnace suspended from the floor of the space being heated, taking air for combustion from outside such space, and with means for lighting the appliance from such space.

❖ Such units supply heat through a floor grille placed directly over the unit's heat exchanger. Typically floor furnaces are classified as gravity-type furnaces. The circulation of air from a gravity-type floor furnace is accomplished by convection. Floor furnaces may be equipped with factory fans to circulate the air.

FLOW PRESSURE. The static pressure reading in the water-supply pipe near the faucet or water outlet while the faucet or water outlet is open and flowing at capacity.

❖ Each fixture requiring water has a minimum flow pressure necessary for the fixture to operate properly and to help protect the potable water supply from contamination. "Flow pressure" is a factor used in sizing the water supply system. The flow pressure of a plumbing system is less than the static pressure because energy is lost in putting the fluid in motion.

FLUE. See "Vent."

FLUE, APPLIANCE. The passages within an appliance through which combustion products pass from the combustion chamber to the flue collar.

❖ This is the passage through an appliance used for the removal of combustion products.

FLUE COLLAR. The portion of a fuel-burning appliance designed for the attachment of a draft hood, vent connector or venting system.

❖ See the commentary on the definition of "Draft hood."

FLUE GASES. Products of combustion plus excess air in appliance flues or heat exchangers.

❖ The exact composition of flue gases depends on the materials being burned. The primary components of flue gases are nitrogen, carbon dioxide, water vapor, particulates, carbon monoxide and myriad compounds and trace elements that vary with the nature of the fuel.

FLUSH VALVE. A device located at the bottom of a flush tank that is operated to flush water closets.

❖ A flush valve consists of the flapper and the flush-valve seat in a flush-tank type water closet.

FLUSHOMETER TANK. A device integrated within an air accumulator vessel that is designed to discharge a predetermined quantity of water to fixtures for flushing purposes.

❖ This device is a hybrid between a flushometer valve and a gravity flush tank. A flushometer tank employs a compression tank that holds water under a pressure equivalent to the water-supply pressure. The flushing action created by the flushometer tank is similar to that of the flushometer valve, and the amount of water used is limited as it is with a gravity flush tank.

FLUSHOMETER VALVE. A flushometer valve is a device that discharges a predetermined quantity of water to fixtures for flushing purposes and is actuated by direct water pressure.

❖ This flushing device is a type of metering valve that is used primarily in public occupancy applications because of its powerful flushing action.

FOAM BACKER BOARD. Foam plastic used in siding applications where the foam plastic is a component of the siding.

❖ Products that combine an exterior siding material and foam plastic are available in the marketplace. The

foam plastic component used in these products is referred to as a foam backer board. By using a foam layer in addition to the siding the thermal properties can be enhanced.

FOAM PLASTIC INSULATION. A plastic that is intentionally expanded by the use of a foaming agent to produce a reduced-density plastic containing voids consisting of open or closed cells distributed throughout the plastic for thermal insulating or acoustic purposes and that has a density less than 20 pounds per cubic foot (320 kg/m^3) unless it is used as interior trim.

❖ An expanded plastic produced through use of a foaming agent that has a reduced density. This plastic contains voids consisting of open or closed cells distributed throughout the plastic, which provide thermal insulation and/or acoustic control. The density of the material is to be less than 20 pounds per cubic foot (320 kg/m^3) unless it is used as interior trim. Foam plastic used in siding applications where the foam plastic is a component of the siding.

FOAM PLASTIC INTERIOR TRIM. Exposed foam plastic used as picture molds, chair rails, crown moldings, baseboards, handrails, ceiling beams, door trim and window trim and similar decorative or protective materials used in fixed applications.

❖ An expanded plastic used as a decorative trim which is produced through use of a foaming agent. This foam plastic contains voids consisting of open or closed cells distributed throughout the material with a minimum density of at least 20 pounds (320 kg/m^3) per cubic foot (pcf). The foam plastic trim is left exposed in applications such as picture moldings, chair rails, crown moldings, baseboards, handrails, ceiling beams, door trim, window trim and similar decorative or protective applications.

FUEL-PIPING SYSTEM. All piping, tubing, valves and fittings used to connect fuel utilization equipment to the point of fuel delivery.

❖ As used in this code, this term includes the tubing and pipe used to convey fuel gas from the point of delivery to the appliance.

FULLWAY VALVE. A valve that in the full open position has an opening cross-sectional area equal to a minimum of 85 percent of the cross-sectional area of the connecting pipe.

❖ Gate valves and ball valves are typical fullway (full-open) valves. Unlike a globe valve, a fullway valve will provide the most open and straight run for the contents to flow through because the closing portion is perpendicular to the run of pipe.

FURNACE. A vented heating appliance designed or arranged to discharge heated air into a conditioned space or through a duct or ducts.

❖ The single most distinguishing characteristic of furnaces is that they use air as the heat transfer medium. Furnaces can be fueled by gas, oil, solid fuel or electricity and can use fans, blowers and gravity (convection) to circulate the heated air to and from the unit. In the context of the code, the primary use of the term "furnace" refers to heating appliance units that combine a combustion chamber with related components, one or more heat exchangers and an air-handling system.

GLAZING AREA. The interior surface area of all glazed fenestration, including the area of sash, curbing or other framing elements, that enclose conditioned space. Includes the area of glazed fenestration assemblies in walls bounding conditioned basements.

❖ The glazing area includes not only the surface area of the exposed glazing but also the framing elements, including the sash and curbing. The amount of glazing area is regulated for natural light by Section R303; however, this definition applies to the energy efficiency provisions of Chapter 11.

GRADE. The finished ground level adjoining the building at all exterior walls.

❖ This is the point at which the finished exterior ground level intersects the exterior wall of the building. The grade around a building may remain relatively constant, such as on a flat site, or may change dramatically from one point to the next if the site is steeply sloping.

GRADE FLOOR OPENING. A window or other opening located such that the sill height of the opening is not more than 44 inches (1118 mm) above or below the finished ground level adjacent to the opening.

❖ In the requirements for emergency escape and rescue openings found in Section R310, the size of the openings may be reduced if they are grade floor openings. These are windows or other openings that are located within close proximity to the finished ground level. The sill of a grade floor opening may be located either above or below the adjacent ground level, provided it is located no more than 44 inches (1118 mm) vertically from the level of the ground.

GRADE, PIPING. See "Slope."

GRADE PLANE. A reference plane representing the average of the finished ground level adjoining the building at all exterior walls. Where the finished ground level slopes away from the exterior walls, the reference plane shall be established by the lowest points within the area between the building and the lot line or, where the lot line is more than 6 ft (1829 mm) from the building between the structure and a point 6 ft (1829 mm) from the building.

❖ This definition can be important in determining the number of stories within a building as well as its height in feet. In some cases, the finished surface of the ground may be artificially raised with imported fill to create a higher grade plane around a building to decrease the number of stories or height. The definition requires that the lowest elevation within 6 feet (1829 mm) of the exterior wall be used to determine the grade plane.

GRIDDED WATER DISTRIBUTION SYSTEM. A water distribution system where every water distribution pipe is inter-

connected so as to provide two or more paths to each fixture supply pipe.

❖ These systems offer the advantage of a simplistic design, typically smaller sized distribution lines and aid water conservation. In a traditional water distribution system, the water contained in the larger diameter piping is wasted when the line is opened and the user has to wait until the water reaches the desired temperature.

Parallel or gridded water distribution systems differ from branch systems which have individual supply pipes that extend to each fixture or outlet from a central supply point [see Commentary Figure R202(1)]. The central supply point is a multiple-outlet manifold to which the distribution lines connect [see Commentary Figure R202(2)].

GROSS AREA OF EXTERIOR WALLS. The normal projection of all exterior walls, including the area of all windows and doors installed therein.

❖ The calculation for determining the gross area of exterior walls for energy efficiency purposes is based on the total area of the entire exterior surface, including openings such as windows and doors.

GROUND-SOURCE HEAT PUMP LOOP SYSTEM. Piping buried in horizontal or vertical excavations or placed in a body of water for the purpose of transporting heat transfer liquid to and from a heat pump. Included in this definition are closed loop systems in which the liquid is recirculated and open loop systems in which the liquid is drawn from a well or other source.

❖ This definition assists the user with a ready means of distinguishing ground-source heat pump loop systems from other hydronic systems.

GUARD. A building component or a system of building components located near the open sides of elevated walking surfaces that minimizes the possibility of a fall from the walking surface to the lower level.

❖ A guard is a component or system of components whose function is the prevention of falls from an elevated area. Placed adjacent to an elevation change, a guard must be of adequate height, strength and configuration to help prevent people, especially small children, from falling over or through the guard to the area below.

HABITABLE SPACE. A space in a building for living, sleeping, eating or cooking. Bathrooms, toilet rooms, closets, halls, storage or utility spaces and similar areas are not considered habitable spaces.

❖ An area within a building used for living, sleeping, dining or cooking is a habitable space. Those areas not meeting this definition include bathrooms, closets, hallways and utility rooms. Habitable spaces are typically occupied, and as such they are more highly regulated than accessory use areas.

HANDRAIL. A horizontal or sloping rail intended for grasping by the hand for guidance or support.

❖ Typically used in conjunction with a ramp or stairway, a handrail provides support for the user along the travel path. Through its height, size, shape, continuity and structural stability, a handrail assists users at elevation changes inside and outside a building. A handrail may also be used as a guide to direct the user in a specified direction.

HANGERS. See "Supports."

HAZARDOUS LOCATION. Any location considered to be a fire hazard for flammable vapors, dust, combustible fibers or other highly combustible substances.

❖ The environment in which mechanical equipment and appliances operate plays a significant role in the safe performance of the equipment installation. Locations that may contain ignitable or explosive atmospheres are classified as hazardous locations with respect to the installation of mechanical equipment and appliances. For example, repair garages can be classified as hazardous locations because they can contain gasoline vapors from vehicles stored within them as well as other volatile chemicals. Public and private garages can be considered as hazardous locations because of the presence of motor vehicles and because of the owners' propensity for storing paint, varnish, thinners, lawn and home maintenance products and other chemicals within the space.

HEATING DEGREE DAYS (HDD). The sum, on an annual basis, of the difference between 65°F (18°C) and the mean temperature for each day as determined from "NOAA Annual Degree Days to Selected Bases Derived from the 1960-1990 Normals" or other weather data sources acceptable to the code official.

❖ Heating Degree Days are a measure of the heating requirements for buildings in a particular climate. The greater the number of HDD, the colder the climate. HDD is calculated by first determining the average temperature for each day of the year over a 24-hour period. The average temperature for each day is then subtracted from a base temperature of 65°F (18°C) to determine the HDD value for the day. For days on which the average temperature is over 65°F (18°C), the HDD is zero. The HDD for each day are then totaled to determine the HDD value for the particular area. The average temperatures are calculated over a number of years. These values are published in the ASHRAE Handbook of Fundamentals. They can also be found in the NOAA Annual Degree Days to Selected Bases Derived from the 1960-1990 Normals.

Climate Zone numbers listed in Table N1101.2 can also be used to approximate the HDD value for a particular county. Each climate zone is equal to 500 HDD. For example, Climate Zone 1 is equal to 0-499 HDD, and Climate Zone 2 is equal to 500-999 HDD.

HEAT PUMP. An appliance having heating or heating/cooling capability and that uses refrigerants to extract heat from air, liquid or other sources.

❖ Heat pumps are referred to as reverse-cycle refrigeration systems. Special controls and valves allow the system to be used as a comfort heating system, comfort cooling system and potable water heating system. Auxiliary heat is often installed in heat pumps in the form of electric resistance heat or fuel-fired furnaces. In most cases, the design and installation of heat pumps is more critical than for other types of comfort air-conditioning systems because of the heat-transfer process and the amount of air that must be circulated. Heat-pump types include water source (hydronic), ground source (earth loop), water source (wells) and air source.

HEIGHT, BUILDING. The vertical distance from grade plane to the average height of the highest roof surface.

❖ Once the elevation of the grade plane has been calculated, it is possible to determine a building's height, which is measured vertically from the grade plane to the average height of the highest roof surface. Commentary Figure R202(11) contains examples of this measurement.

If the building is stepped or terraced, it is logical that the height is the maximum height of any segment of the building. It may be appropriate under certain circumstances that the number of stories in a building be determined in the same manner. Each case should be judged individually based on the characteristics of the site and construction.

HEIGHT, STORY. The vertical distance from top to top of two successive tiers of beams or finished floor surfaces; and, for the topmost story, from the top of the floor finish to the top of the ceiling joists or, where there is not a ceiling, to the top of the roof rafters.

❖ Within the building itself, the height of an individual story is determined by measuring the vertical distance between two successive finished floor surfaces. In a single-story building or for the uppermost floor in a multistory structure, the measurement is taken from the finished floor surface to the top of the ceiling joists or to the underside of the roof deck (top of the roof rafters) where a ceiling is not provided [see Commentary Figure R202 (12)].

HIGH-TEMPERATURE (H.T.) CHIMNEY. A high temperature chimney complying with the requirements of UL 103. A Type H.T. chimney is identifiable by the markings "Type H.T." on each chimney pipe section.

❖ This is a factory-built chimney used to vent solid-fuel-burning appliances when a masonry chimney is not available. See Commentary Figure R202(13) for an illustration of a Type H.T. chimney.

HORIZONTAL BRANCH, DRAINAGE. A drain pipe extending laterally from a soil or waste stack or building drain, that receives the discharge from one or more fixture drains.

❖ See Commentary Figure R202(9).

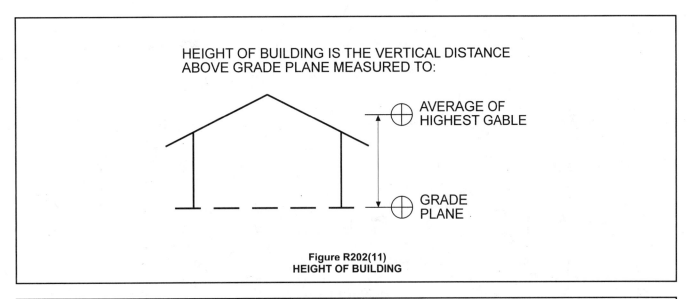

HEIGHT OF BUILDING IS THE VERTICAL DISTANCE
ABOVE GRADE PLANE MEASURED TO:

AVERAGE OF HIGHEST GABLE

GRADE PLANE

**Figure R202(11)
HEIGHT OF BUILDING**

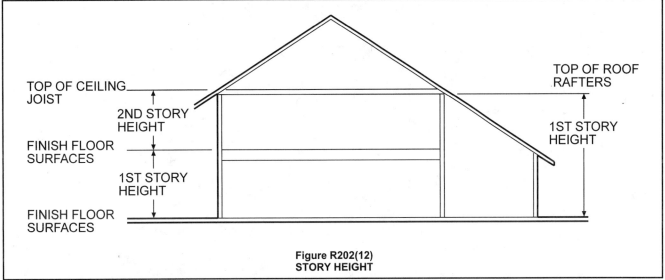

TOP OF CEILING JOIST

TOP OF ROOF RAFTERS

2ND STORY HEIGHT

FINISH FLOOR SURFACES

1ST STORY HEIGHT

1ST STORY HEIGHT

FINISH FLOOR SURFACES

**Figure R202(12)
STORY HEIGHT**

HORIZONTAL PIPE. Any pipe or fitting that makes an angle of less than 45 degrees (0.79 rad) with the horizontal.

❖ This definition is needed for the application of many code provisions that apply only to horizontal piping.

HOT WATER. Water at a temperature greater than or equal to 110°F (43°C).

❖ The code does not specify a maximum temperature for hot water; however, it should be limited to a temperature that will minimize 1) the risk of a user's burn and 2) thermal stress to plumbing system components. The designer, installer or owner has the choice of determining the temperature. Most water heaters are now shipped with a factory-set temperature of 120°F (49°C). The minimum temperature is established to determine what is considered as hot water. Water temperature below the minimum temperature is either tempered or cold. Care must be used when installing a system that has temperature capability of exceeding 120°F (49°C) because of the possibility of a user being scalded.

HURRICANE-PRONE REGIONS. Areas vulnerable to hurricanes, defined as the U.S. Atlantic Ocean and Gulf of Mexico coasts where the basic wind speed is greater than 90 miles per hour (145 km/h), and Hawaii, Puerto Rico, Guam, Virgin Islands, and America Samoa.

❖ Those geographical areas subject to hurricanes are "hurricane-prone regions." The specific areas include those having a minimum basic wind speed exceeding 110 miles per hour (177 km/h or 49 m/s) along the United States coastline of the Atlantic Ocean and the coasts bordering the Gulf of Mexico, as well as the indicated islands.

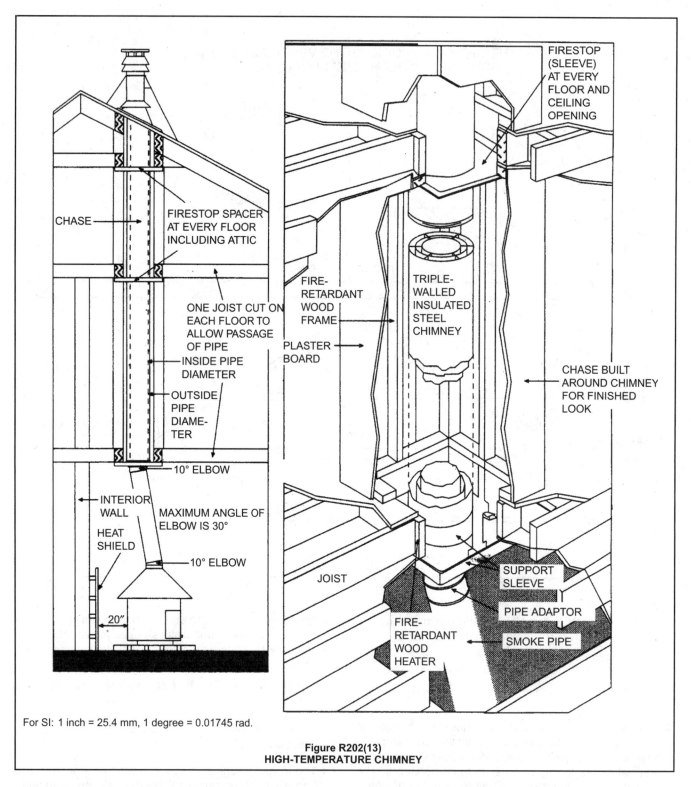

For SI: 1 inch = 25.4 mm, 1 degree = 0.01745 rad.

Figure R202(13)
HIGH-TEMPERATURE CHIMNEY

HYDROGEN GENERATING APPLIANCE. A self-contained package or factory-matched packages of integrated systems for generating gaseous hydrogen. Hydrogen generating appliances utilize electrolysis, reformation, chemical, or other processes to generate hydrogen.

❖ Hydrogen is being generated to fuel a new line of developmental vehicles and to generate electric power in fuel cell power systems. Several different processes are being used to generate hydrogen. One such process removes hydrogen from tap water (H_2O) in an electrically powered appliance.

Installation requirements for these appliances are found in Chapter 7 of the IFGC and ventilation requirements for spaces containing the appliances are found in Section M1307.4 of this code.

IGNITION SOURCE. A flame, spark or hot surface capable of igniting flammable vapors or fumes. Such sources include appliance burners, burner ignitions and electrical switching devices.

❖ Any energized portion of an electrical system that can generate a spark or that can produce significant heat (often referred to as "glow," as in a heating element) is an ignition source should it come in contact with flammable gasses or liquids. Understandably, any open flame (such as a pilot light) is also an ignition source.

INDIRECT WASTE PIPE. A waste pipe that discharges into the drainage system through an air gap into a trap, fixture or receptor.

❖ Often referred to as "open site" or "safe waste," indirect waste pipe installations protect fixtures, equipment and systems from the backflow or backsiphonage of waste.

INDIVIDUAL SEWAGE DISPOSAL SYSTEM. A system for disposal of sewage by means of a septic tank or mechanical treatment, designed for use apart from a public sewer to serve a single establishment or building.

❖ When installed in accordance with the *International Private Sewage Disposal Code*® (IPSDC®) and approved by the administrative authority, these systems are acceptable as a point of disposal for sewage when a public sewer system is not available.

INDIVIDUAL VENT. A pipe installed to vent a single-fixture drain that connects with the vent system above or terminates independently outside the building.

❖ An individual vent is the simplest and most common method of venting a single plumbing fixture.

INDIVIDUAL WATER SUPPLY. A supply other than an approved public water supply that serves one or more families.

❖ This type of water supply includes wells, springs, streams and cisterns. Surface bodies of water and land cisterns can be used when properly treated and approved.

INSULATING CONCRETE FORM (ICF). A concrete forming system using stay-in-place forms of rigid foam plastic insulation, a hybrid of cement and foam insulation, a hybrid of cement and wood chips, or other insulating material for constructing cast-in-place concrete walls.

❖ This is a relatively new construction technique for foundation/basement walls and above-grade walls that uses stay-in-place insulation forms and concrete. A wide variety of systems are available, with several types of insulating materials used for the formwork. Sections R404.4 (foundation walls) and R611 (above-grade walls) address three different systems consisting of insulating concrete forms: flat, waffle grid, and screen grid.

INSULATING SHEATHING. An insulating board having a minimum thermal resistance of R-2 of the core material.

❖ A sheathing board having an insulating core material is insulating sheathing if its thermal resistance is at least R-2.

JURISDICTION. The governmental unit that has adopted this code under due legislative authority.

❖ The governmental entity that adopts and enforces the IRC is the "jurisdiction." A jurisdiction can take on a variety of forms such as a state, city, town, county, fire district, improvement district or federal department. The jurisdiction appoints a building official to administer and enforce the provisions of the code.

KITCHEN. Kitchen shall mean an area used, or designated to be used, for the preparation of food.

❖ This is an area designed to be used for the preparation of food. A kitchen is a habitable space and is subject to a number of requirements.

LABEL. An identification applied on a product by the manufacturer which contains the name of the manufacturer, the function and performance characteristics of the product or material, and the name and identification of an approved agency and that indicates that the representative sample of the product or material has been tested and evaluated by an approved agency. (See also "Manufacturer's designation" and "Mark.")

❖ A label identifies a product or material and provides other information that can be investigated if there is a question about the suitability of the product or material for a specific installation.

The applicable reference standard often states the minimum identifying information required on a label. The information on a label as mandated by the code includes the name of the manufacturer, the product's function or performance characteristics, the name and identification of the approved labeling agency and the approval of the testing agency. The product name, serial number, installation specifications or applicable tests and standards might be additional information that is provided.

LABELED. Devices, equipment or materials to which have been affixed a label, seal, symbol or other identifying mark of a testing laboratory, inspection agency or other organization concerned with product evaluation that maintains periodic inspection of the production of the above labeled items that attests to compliance with a specific standard.

❖ When a product is labeled, the label indicates that the material has been tested for conformance to an applicable standard and that the component is subject to a third-party inspection that verifies that the minimum level of quality required by the appropriate standard is maintained. Labeling provides a readily available source of information that is useful for field inspection of installed products.

The labeling agency performing the third-party inspection must be approved by the building official, and the basis for this approval may include, but is not necessarily limited to, the capacity and capability of the agency to perform the specific testing and inspection.

LIGHT-FRAMED CONSTRUCTION. A type of construction whose vertical and horizontal structural elements are primarily formed by a system of repetitive wood or light gage steel framing members.

❖ Light-framed construction is by far the predominant construction technique used for dwelling units. Consisting primarily of repetitive framing members such as studs, floor joists, ceiling joists and rafters, light-framed construction can use wood or light-gage steel as framing members. The repetitive members work together as a system to provide the structural integrity of the building. The IRC devotes most of its structural provisions to structures of light-framed construction.

LISTED AND LISTING. Terms referring to equipment that is shown in a list published by an approved testing agency qualified and equipped for experimental testing and maintaining an adequate periodic inspection of current productions and whose listing states that the equipment complies with nationally recognized standards when installed in accordance with the manufacturer's installation instructions.

❖ As stated in the definition, the listing states that either the equipment or material meets nationally recognized standards or it has been found suitable for use in a specific manner. The listing becomes part of the documentation the building official can use to approve or disapprove the equipment or appliance.

LIVE LOADS. Those loads produced by the use and occupancy of the building or other structure and do not include construction or environmental loads such as wind load, snow load, rain load, earthquake load, flood load or dead load.

❖ A variety of loads are imposed on a structure that must be considered in the building design. One group is live loads, produced through the use of the building. Different types of uses or occupancies result in different live loads, due primarily to the number of people involved, the type of activity that takes place and the movable furnishings and equipment that are typically a part of the use.

Design loads that are not live loads include environmental loads such as wind loads, snow loads, rain loads, earthquake loads and flood loads, as well as construction loads. Fixed-in-place equipment is a dead load rather than a live load.

LIVING SPACE. Space within a dwelling unit utilized for living, sleeping, eating, cooking, bathing, washing and sanitation purposes.

❖ The living space within a dwelling extends beyond what is defined as habitable space. Living space includes not only those areas used for living, sleeping, eating and cooking but also includes bathing and washing areas such as bathrooms. Storage areas such as closets and garages are not a portion of the living space.

LOT. A portion or parcel of land considered as a unit.

❖ A lot is a piece of property regulated as a single unit. The code regulates the construction of one or more buildings based on their location on a single lot. Owners are expected to control what occurs on their own lots, with no control over any adjacent property. Therefore, several provisions are based upon the proximity of the building to the lot line.

LOT LINE. A line dividing one lot from another, or from a street or any public place.

❖ The lot line defines the exterior perimeter of the lot. It is used in several provisions of the code as the basis for requirements dealing with the proximity of an exterior wall or opening to the lot line.

MACERATING TOILET SYSTEMS. A system comprised of a sump with macerating pump and with connections for a water closet and other plumbing fixtures, that is designed to accept, grind and pump wastes to an approved point of discharge.

❖ See Section P2723.

MAIN. The principal pipe artery to which branches may be connected.

❖ This generic term applies to the primary or principal pipe artery in either water-supply systems or drainage systems to which branches may be connected.

MAIN SEWER. See "Public sewer."

MANIFOLD WATER DISTRIBUTION SYSTEMS. A fabricated piping arrangement in which a large supply main is fitted with multiple branches in close proximity in which water is distributed separately to fixtures from each branch.

❖ A manifold is a multiple-opening header to which one or more branch lines connect. See Section P2903.8.

MANUFACTURED HOME. Manufactured home means a structure, transportable in one or more sections, which in the traveling mode is 8 body feet (2438 body mm) or more in width or 40 body feet (12 192 body mm) or more in length, or, when erected on site, is 320 square feet (30 m²) or more, and which is built on a permanent chassis and designed to be used as a dwelling with or without a permanent foundation when connected to the required utilities, and includes the plumbing, heating, air-conditioning and electrical systems contained therein; except that such term shall include any structure that meets all the requirements of this paragraph except the size requirements and with respect to which the manufacturer voluntarily files a certification required by the secretary (HUD) and complies with the standards established under this title. For mobile homes built prior to June 15, 1976, a label certifying compliance to the Standard for Mobile Homes, NFPA 501, in effect at the time of manufacture is required. For the purpose of these provisions, a mobile home shall be considered a manufactured home.

❖ Appendix E of the IRC contains provisions for the use of manufactured housing as dwellings. Appendix E is applicable only if so specified in the adoption ordinance of the jurisdiction. The provisions of this appendix are limited to three areas: 1) the construction, alteration or repair of a foundation system used to support a manufactured home unit; 2) the construction, installation, addition, alteration, repair or maintenance of building services equipment necessary for connecting the manufactured home to water, fuel, power supplies or sewage systems; and 3) alterations, additions or repairs to existing manufactured homes.

The code describes a manufactured home as being transportable in one or more sections, sets forth minimum dimensions for the home and indicates that a manufactured home is constructed on a permanent chassis. Under the provisions of the IRC, a mobile home is a manufactured home.

MANUFACTURER'S DESIGNATION. An identification applied on a product by the manufacturer indicating that a product or material complies with a specified standard or set of rules. (See also "Mark" and "Label.")

❖ Identification applied on a product by the manufacturer indicating that a product or material complies with a specified standard or set of rules.

MANUFACTURER'S INSTALLATION INSTRUCTIONS. Printed instructions included with equipment as part of the conditions of listing and labeling.

❖ A set of instructions provided by the manufacturer of a product that guides the user or installer through the steps necessary to ensure a complying installation. These instructions help ensure that an installer does not violate the listing of a product.

MARK. An identification applied on a product by the manufacturer indicating the name of the manufacturer and the function of a product or material. (See also "Manufacturer's designation" and "Label.")

❖ Identification applied on a product by the manufacturer indicating the name of the manufacturer and the function of a product or material.

MASONRY CHIMNEY. A field-constructed chimney composed of solid masonry units, bricks, stones or concrete.

❖ Masonry chimneys can have one or more flues within them and are field-constructed of brick, stone, concrete, or fire-clay materials. Masonry chimneys can stand alone or be part of a masonry fireplace.

MASONRY HEATER. A masonry heater is a solid fuel burning heating appliance constructed predominantly of concrete or solid masonry having a mass of at least 1,100 pounds (500 kg), excluding the chimney and foundation. It is designed to absorb and store a substantial portion of heat from a fire built in the firebox by routing exhaust gases through internal heat exchange channels in which the flow path downstream of the firebox includes at least one 180-degree (3.14-rad) change in flow direction before entering the chimney and which deliver heat by radiation through the masonry surface of the heater.

❖ Masonry heaters are appliances designed to absorb and store heat from a relatively small fire and to radiate that heat into the building interior. They are thermally more efficient than traditional fireplaces because of their design. Interior passageways through the heater allow hot exhaust gases from the fire to transfer heat into the masonry, which then radiates into the building.

MASONRY, SOLID. Masonry consisting of solid masonry units laid contiguously with the joints between the units filled with mortar.

❖ In solid masonry construction, walls are made up of solid masonry units. The joints between the units are also solid because they must be filled with mortar. Solid masonry walls may be either single-wythe or multiple- wythe construction. Where multiple wythes are used, the space between the wythes is filled with mortar or grout.

MASONRY UNIT. Brick, tile, stone, glass block or concrete block conforming to the requirements specified in Section 2103 of the *International Building Code*.

❖ The IBC contains the material requirements for masonry units. Various materials are used to construct masonry units, including brick, tile, stone, glass and concrete.

Clay. A building unit larger in size than a brick, composed of burned clay, shale, fire clay or mixtures thereof.

❖ Clay masonry units are composed of burned clay, shale, fire clay or a combination of these materials. A clay masonry unit is somewhat larger than a brick.

Concrete. A building unit or block larger in size than 12 inches by 4 inches by 4 inches (305 mm by 102 mm by 102 mm) made of cement and suitable aggregates.

❖ Masonry units constructed of cement, aggregates, water and suitable admixtures are concrete masonry units. The minimum size for a concrete masonry unit is 4 inches by 4 inches by 12 inches.

Glass. Nonload-bearing masonry composed of glass units bonded by mortar.

❖ Glass masonry units are unique in that they are permitted for use only in nonload-bearing conditions. The units, either solid or hollow, are connected with mortar in the manner described in Section R610.

Hollow. A masonry unit whose net cross-sectional area in any plane parallel to the loadbearing surface is less than 75 percent of its gross cross-sectional area measured in the same plane.

❖ The net cross-sectional area of a hollow masonry unit is less than 75 percent of its gross cross-sectional area measured in the same plane and parallel to the load-bearing surface. A unit having a greater ratio of net area to gross area than a hollow masonry unit is considered a solid masonry unit.

Solid. A masonry unit whose net cross-sectional area in every plane parallel to the loadbearing surface is 75 percent or more of its cross-sectional area measured in the same plane.

❖ The net cross-sectional area of a solid masonry unit is at least 75 percent of its gross cross-sectional area measured in the same plane and parallel to the load-bearing surface. A unit having a ratio of net area to gross area below 75 percent is a hollow masonry unit.

MASS WALL. Masonry or concrete walls having a mass greater than or equal to 30 pounds per square foot (146 kg/m^2), solid wood walls having a mass greater than or equal to 20 pounds per square foot (98 kg/m^2), and any other walls having a heat capacity greater than or equal to 6 Btu/ft^2 · °F [266 J/(m^2 · K)].

❖ A mass wall must have a heat capacity of 6 Btu/ft^2·°F. In general terms, the heat capacity is a measure of how well a material stores heat. The higher the heat capacity the greater amount of heat stored. For example, a 6-inch, heavyweight concrete wall has a heat capacity of 14 Btu/ft^2·°F (286 kJ/m^2 · K) compared to a conventional 2-inch by 4-inch wood framed wall with a heat capacity of approximately 3 Btu/ft^2·°F (63 kJ/m^2 · K). The code defines a mass wall as a masonry or concrete wall having a mass greater than or equal to 30 lb/ft^2, a solid wood wall (e.g., a log house) having a mass of greater than 20 lb/ft^2, or any other walls having a heat capacity of greater than or equal to 6 Btu/ft^2·°F (123 kJ/m^2 · K).

MEAN ROOF HEIGHT. The average of the roof eave height and the height to the highest point on the roof surface, except that eave height shall be used for roof angle of less than or equal to 10 degrees (0.18 rad).

❖ For a relatively flat roof (up to 10 degrees slope), the mean roof height is the height of the eave. For steeper roofs, the mean roof height is measured to the average point between the eave and the highest point on the roof surface, typically the highest ridge.

MECHANICAL DRAFT SYSTEM. A venting system designed to remove flue or vent gases by mechanical means, that consists of an induced draft portion under nonpositive static pressure or a forced draft portion under positive static pressure.

❖ Mechanical draft systems do not depend on draft; they use fans or blowers.

Forced-draft venting system. A portion of a venting system using a fan or other mechanical means to cause the removal of flue or vent gases under positive static pressure.

❖ Power exhausters and some power-burner systems are examples of forced-draft systems. Vents and chimneys must be listed for positive pressure applications where used with forced-draft systems.

Induced draft venting system. A portion of a venting system using a fan or other mechanical means to cause the removal of flue or vent gases under nonpositive static vent pressure.

❖ Induced-draft venting is commonly accomplished with field-installed inducer fans designed to supplement natural-draft chimneys or vents.

Power venting system. A portion of a venting system using a fan or other mechanical means to cause the removal of flue or vent gases under positive static vent pressure.

❖ See the commentary for "Forced-draft venting system" above.

MECHANICAL EXHAUST SYSTEM. A system for removing air from a room or space by mechanical means.

❖ A mechanical exhaust system uses a fan or other air-handling equipment to exhaust air to the outdoors. Mechanical exhaust systems include those used for hazardous exhaust and commercial kitchen exhaust. Mechanical exhaust systems may or may not use ductwork as part of the system and may include air-cleaning or air-filtering equipment and fire-suppression equipment.

MECHANICAL SYSTEM. A system specifically addressed and regulated in this code and composed of components, devices, appliances and equipment.

❖ Mechanical systems include, among others, refrigeration systems, air-conditioning systems, exhaust systems, piping systems, duct systems, venting systems, hydronic systems and ventilation systems. A mechanical system may be any of the above systems or incorporate one or more of the systems. The system includes all of the equipment and appliances required to perform the function for which the system is designed.

METAL ROOF PANEL. An interlocking metal sheet having a minimum installed weather exposure of at least 3 square feet (0.28 m^2) per sheet.

❖ A metal roof panel is specifically defined here in order to identify its difference from a metal roof shingle. Metal roof panels are each at least 3 square feet in surface area and interlock to provide a suitable roof covering. The provisions of Section R905.10 address roof coverings consisting of metal roof panels.

METAL ROOF SHINGLE. An interlocking metal sheet having an installed weather exposure less than 3 square feet (0.28 m^2) per sheet.

❖ Section R905.4 regulates roof coverings consisting of metal roof shingles, which are smaller than 3 square feet (0.28 m^2) per sheet.

MEZZANINE, LOFT. An intermediate level or levels between the floor and ceiling of any story with an aggregate floor area of not more than one-third of the area of the room or space in which the level or levels are located.

❖ Floor levels consisting of relatively small floor areas opening into larger rooms are loft mezzanines. Located at an intermediate level within a high-ceiling space, a loft mezzanine is limited to one-third of the area of the room in which it is located. Because of its openness and relatively small size, a loft mezzanine is not an additional story.

MODIFIED BITUMEN ROOF COVERING. One or more layers of polymer modified asphalt sheets. The sheet materials shall be fully adhered or mechanically attached to the substrate or held in place with an approved ballast layer.

❖ Polymer-modified asphalt sheets, applied in one or more layers, may be used as a roof covering when installed in accordance with Section R905.11. There are several approved methods for attaching the asphalt sheets, including 1) full adhesion to the substrate, 2) mechanical attachment to the substrate, and 3) use of a layer of ballast.

MULTIPLE STATION SMOKE ALARM. Two or more single station alarm devices that are capable of interconnection such that actuation of one causes all integral or separate audible alarms to operate.

❖ A multiple station smoke alarm consists of two or more interconnected single station smoke alarms. When one of the smoke alarms is actuated, all of the audible alarms sound. Most dwelling units will require several smoke alarms throughout the unit; Section R317.1 requires that they be interconnected.

NATURAL DRAFT SYSTEM. A venting system designed to remove flue or vent gases under nonpositive static vent pressure entirely by natural draft.

❖ Natural-draft chimneys and vents do not rely on any mechanical means to convey combustion products to the outdoors. Draft is produced by the temperature difference between the combustion gases (flue gases) and the ambient atmosphere. Hot gases are less dense and more buoyant; therefore, they rise or produce a draft.

NATURALLY DURABLE WOOD. The heartwood of the following species: Decay-resistant redwood, cedars, black locust and black walnut.

Note: Corner sapwood is permitted if 90 percent or more of the width of each side on which it occurs is heartwood.

❖ Because of their natural ability to resist deterioration, the harder portions of some species of wood are considered to be naturally durable. The code specifies that "occasional" sapwood is permitted if heartwood constitutes 90 percent of each side.

Decay resistant. Redwood, cedar, black locust and black walnut.

Redwood, cedar, black locust and black walnut lumber are known to resist deterioration caused by the action of microbes that enter the wood fibers. The code defines these species of lumber as being decay resistant.

Termite resistant. Redwood and eastern red cedar.

Redwood and eastern red cedar are considered to be resistant to infestation by termites and are thus listed as naturally durable. The Formosan Termite, however, is capable of destroying all naturally durable species of wood.

NONCOMBUSTIBLE MATERIAL. Materials that pass the test procedure for defining noncombustibility of elementary materials set forth in ASTM E 136.

❖ A material that successfully passes the ASTM E 136 test is considered to be noncombustible. The test determines whether a building material will act to aid combustion or add appreciable heat to a fire. A material may have a limited combustible content but not contribute appreciably to a fire; thus, it may still qualify as noncombustible.

NONCONDITIONED SPACE. A space that is not a conditioned space by insulated walls, floors or ceilings.

❖ Defined for the purpose of regulating energy efficiency, this is an area of a building that does not qualify as conditioned. A nonconditioned space is not enclosed with the appropriate insulating materials of the kind that are considered in the definition of conditioned space.

OCCUPIED SPACE. The total area of all buildings or structures on any lot or parcel of ground projected on a horizontal plane, excluding permitted projections as allowed by this code.

❖ The entire floor area of a building is generally the occupied space. It encompasses all floor levels, including basements. Any exterior space without exterior walls that is covered by a roof or floor above is occupied space. Only those areas beneath projections permitted by Section R302 do not contribute to the total occupied space.

OFFSET. A combination of fittings that makes two changes in direction bringing one section of the pipe out of line but into a line parallel with the other section.

❖ Offsets are necessary to route piping around or through structural elements such as beams, joists, trusses and columns. An offset always involves at least two changes in direction; this is necessary to keep the piping heading in the same direction as the pipe before the offset. Any other arrangement would be considered a bend.

OWNER. Any person, agent, firm or corporation having a legal or equitable interest in the property.

❖ The code places certain responsibilities on the owner in the construction and maintenance of a building. The owner is considered the individual, firm or corporation who, under applicable laws, has a legal interest in the property.

PELLET FUEL-BURNING APPLIANCE. A closed combustion, vented appliance equipped with a fuel feed mechanism for burning processed pellets of solid fuel of a specified size and composition.

❖ This is an appliance that uses solid fuel in the form of pellets in a closed combustion system. The pellets consist of a blend of wood waste products (sawdust and chips) mixed with resin (binders) and compressed into small pellets. The pellets are fed into the appliance in a regulated manner to provide consistent burning.

PELLET VENT. A vent listed and labeled for use with a listed pellet fuel-burning appliance.

❖ See the commentary for the definition for "Vent."

PERMIT. An official document or certificate issued by the authority having jurisdiction that authorizes performance of a specified activity.

❖ The department of building safety is authorized to require a permit according to Section R105. The permit is the official document issued by the jurisdiction authorizing construction-related activity to take place.

PERSON. An individual, heirs, executors, administrators or assigns, and also includes a firm, partnership or corporation, its or their successors or assigns, or the agent of any of the aforesaid.

❖ As used in the IRC, a person is not only an individual but also any heirs, executors, administrators, firm, partnership or corporation, including its or their successors or assigns. An agent for any person as described is also a "person."

PITCH. See "Slope."

❖ The pitch of a pipe run is the fall of the pipe in relationship to a horizontal plane. It is also known as the slope and is measured as a ratio between change in horizontal units and vertical units along a length of pipe.

PLATFORM CONSTRUCTION. A method of construction by which floor framing bears on load bearing walls that are not continuous through the story levels or floor framing.

❖ In light-frame construction, platform construction is the technique of stacking floor systems on load-bearing stud walls. As opposed to walls in balloon construction, the load bearing walls in platform construction do not continue beyond each individual story level. Platform construction is by far the most popular framing method in use today.

PLENUM. A chamber that forms part of an air-circulation system other than the occupied space being conditioned.

❖ A plenum can also be a room or space that supplies conditioned air to another room or space. Corridors cannot be used as plenums.

PLUMBING. For the purpose of this code, plumbing refers to those installations, repairs, maintenance and alterations regulated by Chapters 25 through 32.

❖ This term refers to the collective system of piping, fixtures, fittings, components, devices or appurtenances that transport potable water and liquid and solid wastes associated with cleaning, washing, bathing, food preparation, drinking and the elimination of bodily wastes. Plumbing also includes appliances, equipment and systems such as water conditioners, water heaters, storm drainage systems, water coolers, water filters and waste treatment systems.

The word "plumbing" comes from the Latin word for the element lead (plumbum), because lead was used extensively in the construction of piping systems. The practice of installing piping systems and the materials used in such systems became known as plumbing.

Piping systems such as fuel gas, fuel oil, heating, cooling, air conditioning, lawn irrigation and fire suppression systems are not plumbing, even though plumbers often perform the work necessary to install these systems. Chapters 25-32 regulate plumbing.

PLUMBING APPLIANCE. An energized household appliance with plumbing connections, such as a dishwasher, food-waste grinder, clothes washer or water heater.

❖ Examples of plumbing appliances include water heaters, hot water dispensers, garbage disposals, dishwashers, clothes washers, water purifiers and water

softeners. Typically, an appliance is dependent on a source of energy for its operation.

PLUMBING APPURTENANCE. A device or assembly that is an adjunct to the basic plumbing system and demands no additional water supply nor adds any discharge load to the system. It is presumed that it performs some useful function in the operation, maintenance, servicing, economy or safety of the plumbing system. Examples include filters, relief valves and aerators.

❖ Examples of plumbing appurtenances include water closet seats, hand-held showers, manifolds, backflow preventers, water-hammer arrestors, strainers and filters. An appurtenance by default becomes part of the entire fixture.

PLUMBING FIXTURE. A receptor or device that requires both a water-supply connection and a discharge to the drainage system, such as water closets, lavatories, bathtubs and sinks. Plumbing appliances as a special class of fixture are further defined.

❖ Plumbing fixtures include water closets, bidets, lavatories, sinks, showers, bathtubs and floor drains.

A separate class of plumbing fixtures defined as plumbing appliances includes clothes washing machines, dishwashers, water heaters, water softeners, hot water dispensers, garbage disposals and water purifiers. See "Plumbing appliance."

PLUMBING SYSTEM. Includes the water supply and distribution pipes, plumbing fixtures, supports and appurtenances; soil, waste and vent pipes; sanitary drains and building sewers to an approved point of disposal.

❖ The plumbing system includes all piping, fixtures and components that transport potable water and convey liquid and liquid-borne solid wastes. See "Plumbing." Devices that treat the water prior to its use are also included.

POLLUTION. An impairment of the quality of the potable water to a degree that does not create a hazard to the public health but that does adversely and unreasonably affect the aesthetic qualities of such potable water for domestic use.

❖ Books have been written trying to define the term "pollution." For use in the code, pollution is anything that reduces the quality of the potable water supply so that it is undesirable for consumption or use. A pollutant in the water supply causes the water to look, smell or taste bad, but drinking the water would not be harmful. This does not change the fact that polluted water is still considered to be nonpotable. Pollutants are thus not considered potentially harmful, whereas a contaminant would be considered potentially harmful. Refer to the definition of the term "Contamination" to find additional information and the differences between pollution and contamination.

The methods for determining potable and nonpotable water supplies are very closely regulated by the EPA. For the purposes of protecting a water supply, the code considers a polluted source to represent a low hazard when selecting a backflow preventer. Refer to Table P2902.2 for requirements related to low-hazard situations in a potable water supply.

PORTABLE FUEL CELL APPLIANCE. A fuel cell generator of electricity, which is not fixed in place. A portable fuel cell appliance utilizes a cord and plug connection to a grid-isolated load and has an integral fuel supply.

❖ These appliances are a smaller version of the stationary fuel cell power plants in Section M1903 of this code. Fuel cells generate electricity from hydrocarbon fuels and, when pure hydrogen is used, release water and heat as the only by-products. Portable fuel cell appliances are generally fueled by gas tanks or some other local source and are connected to the house current or are used only to charge the batteries of electric cars. Connecting the appliance to the local electrical power grid or hard-piping fuel to the appliance would constitute a permanent installation and require permitting in accordance with Section 105.

POSITIVE ROOF DRAINAGE. The drainage condition in which consideration has been made for all loading deflections of the roof deck, and additional slope has been provided to ensure drainage of the roof within 48 hours of precipitation.

❖ Unless a roof is designed to support a specified amount of water, adequate drainage must be provided to approved locations. In considering the appropriate drainage pattern, all of the possible roof deck deflections should be included. The code mandates that the water be removed from the roof within a period of 48 hours from the time of deposit.

POTABLE WATER. Water free from impurities present in amounts sufficient to cause disease or harmful physiological effects and conforming in bacteriological and chemical quality to the requirements of the public health authority having jurisdiction.

❖ In the United States, the EPA Clean Drinking Water Act defines the quality requirements for potable water. "Potable" means fit to drink.

PRECAST CONCRETE. A structural concrete element cast elsewhere than its final position in the structure.

❖ Precast concrete members may be cast on site and subsequently lifted into place (e.g., tilt-up wall construction) or cast off site, usually in a plant out of the outside weather and transported to the construction site when needed. Precast concrete which has been manufactured under plant conditions has closer controlled tolerances than cast-in-place concrete construction. This results in the need for lesser minimum concrete cover for reinforcement versus that needed for cast-in-place concrete (see ACI 318, Sections 7.7.1 and 7.7.2).

PRESSURE-RELIEF VALVE. A pressure-actuated valve held closed by a spring or other means and designed to automatically relieve pressure at the pressure at which it is set.

❖ This valve is designed to be actuated by excessive pressures for the purpose of preventing the rupture or explosion of tanks, vessels or piping. The degree of valve opening is directly proportional to the pressure acting on the valve disk.

PUBLIC SEWER. A common sewer directly controlled by public authority.

❖ A public sewer is the piping system that receives the discharge from building sewers and conveys it to treatment facilities.

PUBLIC WATER MAIN. A water-supply pipe for public use controlled by public authority.

❖ A water main is owned and operated by municipalities, rural water districts, privately owned water purveyors and other such entities.

PUBLIC WAY. Any street, alley or other parcel of land open to the outside air leading to a public street, which has been deeded, dedicated or otherwise permanently appropriated to the public for public use and that has a clear width and height of not less than 10 feet (3048 mm).

❖ A public way is essentially a street, alley or any parcel of land that is permanently appropriated to the public for public use. Therefore, the public's right to use such a parcel of land is guaranteed. A public way is open to the exterior and available for use as an egress path if necessary. To qualify as a public way, the space must be at least 10 feet (3048 mm) wide and must have a clear height of at least 10 feet (3048 mm).

PURGE. To clear of air, gas or other foreign substances.

❖ This term applies to a process used for clearing piping systems of impurities, making the system pure prior to the introduction of a desired substance. Although the practice of flushing a system to remove debris is sometimes referred to as purging the system, true purging involves a specific process of forcing a product into a system until impurities are completely removed. One such method would be the introduction of nitrous gas within a copper system serving a medical gas installation. Such a purge is required to clear the system of impurities that may have been left during installation.

QUICK-CLOSING VALVE. A valve or faucet that closes automatically when released manually or controlled by mechanical means for fast-action closing.

❖ A quick-closing valve is any type of solenoid-actuated valve; spring-loaded, self-closing faucets; or any other device capable of instantaneously reducing water flow from full flow to no flow. This rapid change from a full-flow to a no-flow condition can cause water-hammer-related damage. Although most faucets can be manually closed fast enough to produce water hammer, the code does not designate manually closed valves or faucets as "quick closing."

Typical quick-closing valves include electrically actuated valves such as those found in dishwashing machines, clothes washing machines and boiler makeup-water feeders. Some self-closing valves and faucets are not quick closing. For example, typical metering faucets and typical flushometer valves are slowly, automatically closed by a diaphragm/bleed orifice or pilot valve assembly. Also, a typical float-actuated valve is automatic yet slow in closing.

R-VALUE, THERMAL RESISTANCE. The inverse of the time rate of heat flow through a building thermal envelope element from one of its bounding surfaces to the other for a unit temperature difference between the two surfaces, under steady state conditions, per unit area (h · ft² · °F/Btu).

❖ The *R*-value is a unit of thermal resistance used for comparing insulating values of different materials. The higher the *R*-value of a material, the greater its insulating properties and the slower heat flows through it. *R*-values are used to rate roof/ceiling, wall and floor insulation.

RAMP. A walking surface that has a running slope steeper than 1 unit vertical in 20 units horizontal (5-percent slope).

❖ A pedestrian travel surface that has a slope of 1:20 (5 percent) or less is a walking surface. A surface that is steeper than 1 unit vertical in 20 units horizontal is a ramp. Because steeply sloping walking surfaces present a considerable hazard to users, ramps are regulated for maximum slope, handrails and landings by Section R313.

RECEPTOR. A fixture or device that receives the discharge from indirect waste pipes.

❖ This term applies to a fixture used for collecting drainage from fixtures that are not directly connected to the drainage system. A floor sink is the most common type of receptor. However, a laundry sink that is used for collecting the drainage from a clothes washer could be defined as a receptor.

REFRIGERANT. A substance used to produce refrigeration by its expansion or evaporation.

❖ The refrigerant is the working fluid in refrigeration and air-conditioning systems. In vapor-refrigeration cycles, refrigerants absorb heat from the load side at the evaporator and reject heat at the condenser. The selection of a refrigerant must take into consideration suitable thermodynamic properties, chemical stability, flammability, toxicity and environmental compatibility. Refrigeration is a result of the physical laws of vaporization (evaporation) of liquids. Basically, evaporation of liquid refrigerant is an endothermic process, and condensing of vapors is an exothermic process.

REFRIGERANT COMPRESSOR. A specific machine, with or without accessories, for compressing a given refrigerant vapor.

❖ A compressor is the heart of mechanical refrigeration systems. It is used in a vapor-refrigeration cycle to raise the pressure and enthalpy of the refrigerant into the superheated vapor state, at which point the refrigerant vapor enters the condenser and transfers heat energy to a cooler medium.

REFRIGERATING SYSTEM. A combination of interconnected parts forming a closed circuit in which refrigerant is circulated for the purpose of extracting, then rejecting, heat. A direct refrigerating system is one in which the evaporator or condenser of the refrigerating system is in direct contact with the air or other substances to be cooled or heated. An indirect refrigerating system is one in which a secondary coolant cooled or heated by the refrigerating system is circulated to the air or other substance to be cooled or heated.

❖ Such systems include at minimum a pressure-imposing element or generator, an evaporator, a condenser and interconnecting piping. A single piece of equipment can contain multiple refrigeration systems (circuits).

REGISTERED DESIGN PROFESSIONAL. An individual who is registered or licensed to practice their respective design profession as defined by the statutory requirements of the professional registration laws of the state or jurisdiction in which the project is to be constructed.

❖ All states have imposed some degree of registration or licensing requirements on design professionals, such as architects and engineers, who practice in their state. An individual who meets the statutory requirements in the laws of the state, and possibly even an individual jurisdiction, is a registered design professional.

RELIEF VALVE, VACUUM. A device to prevent excessive buildup of vacuum in a pressure vessel.

❖ This device will automatically open to allow air into a vessel or system in the event of backsiphonage. Such a device is used where a water heater without an anti-siphon dip tube is installed above the fixtures served. Where the water supply to the water heater is closed and the lower fixtures opened, the vacuum relief valve will open, thus breaking the siphoning effect.

REPAIR. The reconstruction or renewal of any part of an existing building for the purpose of its maintenance.

❖ When an existing building undergoes some form of remodel, reconstruction or renewal, the work is defined as a repair. A repair maintains a portion of the building in safe and sound working order, without reducing the intended level of protection or structural safety. Major repair work may be subject to a permit where required by Section R105.

REROOFING. The process of recovering or replacing an existing roof covering. See "Roof recover."

❖ The description of reroofing includes two different processes: 1) recovering an existing roof covering, and 2) replacing an existing roof covering. Recovering involves the installation of an additional roof covering over an existing roof covering without the removal of the existing roof covering. Replacement addresses those cases where the existing roof covering is completely removed prior to installation of the new roof covering.

RETURN AIR. Air removed from an approved conditioned space or location and recirculated or exhausted.

❖ Return air is air that is being returned to the air handler. Only air in excess of the required ventilation air can be recirculated.

RISER. A water pipe that extends vertically one full story or more to convey water to branches or to a group of fixtures.

❖ See the commentary for "Water pipe, riser."

ROOF ASSEMBLY. A system designed to provide weather protection and resistance to design loads. The system consists of a roof covering and roof deck or a single component serving as both the roof covering and the roof deck. A roof assembly includes the roof deck, vapor retarder, substrate or thermal barrier, insulation, vapor retarder, and roof covering.

❖ A roof assembly may be either a single component serving as both the roof covering and the deck or, as typically occurs, a combination of individual roof deck and roof covering components used together to form a complete assembly. A roof assembly may include the roof deck, vapor retarder, substrate, thermal barrier, insulation and roof covering.

ROOF COVERING. The covering applied to the roof deck for weather resistance, fire classification or appearance.

❖ Roof covering provides a building with weather protection, fire retardancy or decoration.

ROOF COVERING SYSTEM. See "Roof assembly."

❖ "Roof covering system" is another term for "Roof assembly."

ROOF DECK. The flat or sloped surface not including its supporting members or vertical supports.

❖ Often composed of wood structural panels, a roof deck is the flat or sloped surface on which roof covering is typically attached. Roof decks assist in both the weather protection and structural integrity of the building. The rafters and associated structural roof members are not a part of the roof deck, nor is the insulation or roof covering that is placed on the deck.

ROOF RECOVER. The process of installing an additional roof covering over a prepared existing roof covering without removing the existing roof covering.

❖ Under the provisions of Section R907, it is acceptable under specific conditions to install new roof covering over an existing roof without first removing the old roof covering. In a roof recover, the existing roof covering must be prepared in an approved manner prior to installation of the new roof covering.

ROOF REPAIR. Reconstruction or renewal of any part of an existing roof for the purposes of its maintenance.

❖ A roof repair occurs when only minor reconstruction or renewal work to an existing roof is necessary. The extent of a roof repair is limited by Section R907.1.

ROOFTOP STRUCTURE. An enclosed structure on or above the roof of any part of a building.

❖ Often used to house mechanical equipment or installed for decorative purposes, a rooftop structure is any enclosed structure on or above the roof. A rooftop

structure must be adequately supported and comply with the construction provisions applicable to the remainder of the building.

ROOM HEATER. A freestanding heating appliance installed in the space being heated and not connected to ducts.

❖ Room heaters are space-heating appliances that are not connected to ducts. They are also called "unvented room heaters."

ROUGH-IN. The installation of all parts of the plumbing system that must be completed prior to the installation of fixtures. This includes DWV, water supply and built-in fixture supports.

❖ Rough-in refers to that stage of construction prior to the installation of any building materials that would conceal plumbing installations in the building structure. This term generally refers to the installation of drain, waste, vent and water supply piping that will be concealed in floor, ceiling and wall cavities or concealed underground or under slab. At this stage of construction, the plumbing systems are tested for leaks and the entire plumbing "rough-in" is inspected for code compliance.

RUNNING BOND. The placement of masonry units such that head joints in successive courses are horizontally offset at least one-quarter the unit length.

❖ A common method of placing masonry units, running bond describes the type of masonry construction where the units are horizontally offset on each successive course. The length of the offset must be at least one-fourth the length of the masonry unit, otherwise it is considered as stack bond. Commentary Figure R202(14) shows an example of running bond.

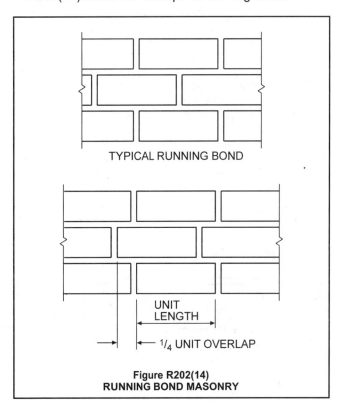

TYPICAL RUNNING BOND

UNIT LENGTH

¹/₄ UNIT OVERLAP

Figure R202(14)
RUNNING BOND MASONRY

SANITARY SEWER. A sewer that carries sewage and excludes storm, surface and groundwater.

❖ A sanitary sewer extends from the sanitary building drain to the public sewer system or to a private sewage disposal system. Most jurisdictions arbitrarily set a distance such as 3 or 5 feet (914 or 1524 mm) outside of the building foundation as the point at which the building drain becomes the building sewer.

SCUPPER. An opening in a wall or parapet that allows water to drain from a roof.

❖ Various design methods are used to drain water from a roof. A common approach is to slope the roof toward one or more exterior walls and provide drainage over the edge of the roof, often into a gutter and downspout system. Another method is the use of roof drains that capture water from numerous low points of the roof and use interior drain lines to remove the water. A third procedure is the use of scuppers, which are openings in the exterior wall surface that allow water to flow off the roof.

Where scuppers are used, the code specifies their size and height above the roof. Scuppers are often included as a portion of the overflow or emergency drain system, a system required by the code and designed to shed water when the primary roof drains are blocked.

SEISMIC DESIGN CATEGORY. A classification assigned to a structure based on its Seismic Group and the severity of the design earthquake ground motion at the site.

❖ The structural provisions of the IRC vary based on several factors, one of which is the probability of an earthquake causing damage to the building. The term used by the code to assign a rating or level of seismic risk is "Seismic Design Category." Classification into one of the categories is based on two factors: 1) the Seismic Group, and 2) the severity of the ground motion at the site. The Seismic Group is the classification assigned based upon use of the structure. Residential construction is considered in Seismic Group I, with Groups II and III assigned to structures whose loss would create a substantial public hazard or those structures considered as essential facilities. The ground-motion criterion used in the IRC is based on Short Period Design Spectral Response Accelerations (0.2 second). Sections 1613 through 1623 of the *IBC* contain further explanation of the seismic considerations.

SEPTIC TANK. A water-tight receptor that receives the discharge of a building sanitary drainage system and is constructed so as to separate solids from the liquid, digest organic matter through a period of detention, and allow the liquids to discharge into the soil outside of the tank through a system of open joint or perforated piping or a seepage pit.

❖ This type of tank is typically used in private sewage disposal systems. See Appendix I.

SEWAGE. Any liquid waste containing animal matter, vegetable matter or other impurity in suspension or solution.

❖ This term refers to the discharge from all plumbing fixtures, which includes human bodily wastes and the wastes associated with cleaning, washing, bathing and food preparation.

SEWAGE PUMP. A permanently installed mechanical device for removing sewage or liquid waste from a sump.

❖ Sewage pumps are designed to transport liquid/solid mixtures without clogging and are typically equipped with large suction and discharging ports.

SHALL. The term, when used in the code, is construed as mandatory.

❖ The term "shall" means that the requirement is a mandatory provision. Where a provision is allowed but not required, the term "is permitted" generally appears in the code.

SHEAR WALL. A general term for walls that are designed and constructed to resist racking from seismic and wind by use of masonry, concrete, cold-formed steel or wood framing in accordance with Chapter 6 of this code and the associated limitations in Section R301.2 of this code.

❖ Shear walls are walls designed to carry lateral loads such as seismic or wind loads. Shear walls may be of a variety of materials; however, they must be designed in the specific manner required by the code. Shear walls are typically used in conjunction with diaphragms to provide a load path for transmitting lateral loads through the structure.

SIDE VENT. A vent connecting to the drain pipe through a fitting at an angle less than 45 degrees (0.79 rad) to the horizontal.

❖ A typical application of a side vent would be where the drain from a water closet flange extends downward vertically with the vent for the water closet, connecting to the vertical drain and extending horizontally, perpendicular to the drain.

SINGLE PLY MEMBRANE. A roofing membrane that is field applied using one layer of membrane material (either homogeneous or composite) rather than multiple layers.

❖ In contrast to a built-up roof, a single-ply membrane roof system consists of one layer of membrane material. Chapter 9 addresses both thermoset and thermoplastic single-ply roofing systems.

SINGLE STATION SMOKE ALARM. An assembly incorporating the detector, control equipment and alarm sounding device in one unit that is operated from a power supply either in the unit or obtained at the point of installation.

❖ A single-station smoke alarm contains all components for operation of the device, including the power supply in battery-operated units. For a hard-wired unit, the power is available directly at the alarm device. Where two or more single-station smoke alarms are interconnected, they are multiple-station alarms.

SKYLIGHT AND SLOPED GLAZING. See Section R308.6.1.

❖ Any glass or other transparent or translucent glazing installed in the building construction at a slope greater than 15 degrees from the vertical is sloped glazing. Skylights are the most common type of sloped glazing, which may also include solariums, sun rooms, roofs and sloped walls.

SKYLIGHT, UNIT. See Section R308.6.1.

SLIP JOINT. A mechanical-type joint used primarily on fixture traps. The joint tightness is obtained by compressing a friction-type washer such as rubber, nylon, neoprene, lead or special packing material against the pipe by the tightening of a (slip) nut.

❖ A slip joint is commonly found in fixture tailpieces and in "P" traps. See the commentary for Section P2704.

SLOPE. The fall (pitch) of a line of pipe in reference to a horizontal plane. In drainage, the slope is expressed as the fall in units vertical per units horizontal (percent) for a length of pipe.

❖ Commonly referred to as pitch, fall or grade, slope is the gradual change in elevation of horizontal piping. Slope is necessary to cause waste flow by gravity.

SMOKE-DEVELOPED RATING. A numerical index indicating the relative density of smoke produced by burning assigned to a material tested in accordance with ASTM E 84.

❖ The ASTM E 84 test standard for flame spread also addresses the potential for smoke development of specific materials. The numerical value assigned to the material being tested, based on the density of the smoke produced during the test, is the smoke-developed rating, which is typically regulated in conjunction with the flame spread index.

SOIL STACK OR PIPE. A pipe that conveys sewage containing fecal material.

❖ In the plumbing trade, "soil pipe" is a common name for cast-iron drainage pipe. The term, however, is not specific to cast-iron pipe and refers to any drainpipe that conveys the discharge of water closets, urinals or any other fixture that receives human waste.

SOLAR HEAT GAIN COEFFICIENT (SHGC). The solar heat gain through a fenestration or glazing assembly relative to the incident solar radiation (Btu/h · ft^2 · °F).

❖ The solar heat gain coefficient (SHGC) measures the ratio of the solar radiation that passes through the window to the total solar radiation that strikes the window. Sunlight, or solar radiation, striking a window contains visible light and heat. As the sunlight strikes the window, a portion of the visible light and heat pass through into the building. A portion of the heat is absorbed by the glass and window sash. Moreover, a portion of the visible light and heat is reflected to the outdoors. The lower the SHGC, the lower the solar radiation that is allowed to pass through the window. For example, a window with an SHGC value of 0.40 allows only 40 percent of the total solar radiation that strikes the window to pass through.

SOLID MASONRY. Load-bearing or nonload-bearing construction using masonry units where the net cross-sectional area of each unit in any plane parallel to the bearing surface is not less than 75 percent of its gross cross-sectional area. Solid masonry units shall conform to ASTM C 55, C 62, C 73, C 145 or C 216.

❖ Solid masonry can consist of either load-bearing or nonload-bearing construction made up of masonry units. Although not required to be totally of solid material, solid masonry must have a net cross-sectional area equal to at least 75 percent of its gross cross-sectional area. The percentage must be maintained in any plane parallel to the bearing surface.

STACK. Any main vertical DWV line, including offsets, that extends one or more stories as directly as possible to its vent terminal.

❖ Traditionally, the term "stack" has been used to describe vertical piping (90 degrees [1.57 rad] from the horizontal) that extends through one or more stories of a building and that constitutes a main to which branch piping connects. Vertical vent branch piping that extends to the open air or through two or more stories is a stack.

STACK BOND. The placement of masonry units in a bond pattern is such that head joints in successive courses are vertically aligned. For the purpose of this code, requirements for stack bond shall apply to all masonry laid in other than running bond.

❖ Where the head joints in successive courses of masonry construction are not offset horizontally by at least one-fourth of the unit length, the method of placement is a stack bond. Vertical joints directly aligned on each course are typical in stack bond masonry.

STACK VENT. The extension of soil or waste stack above the highest horizontal drain connected.

❖ A stack vent is the dry extension of a soil or waste stack. Generally, a stack vent extends to the open air and can serve as a main vent to which branch vents connect.

STACK VENTING. A method of venting a fixture or fixtures through the soil or waste stack without individual fixture vents.

❖ This is a method of wet venting where a uniformly oversized vertical drain serves as the vent for one or more fixtures. The fixtures being vented are connected to the stack independently. The use of this venting method is limited to waste fixtures. It is also referred to as waste stack venting; see Section P3109.

STANDARD TRUSS. Any construction that does not permit the roof/ceiling insulation to achieve the required R-value over the exterior walls.

❖ The term "standard truss" describes typical roof truss construction where the insulation cannot achieve the necessary R-value mandated for the roof-ceiling assembly.

STATIONARY FUEL CELL POWER PLANT. A self-contained package or factory-matched packages which constitute an automatically-operated assembly of integrated systems for generating useful electrical energy and recoverable thermal energy that is permanently connected and fixed in place.

❖ Fuel cells generate electricity from hydrocarbon fuels and, when pure hydrogen is used, release water and heat as the only by-products (see commentary for Section M1903 for a description of the conversion process). Stationary fuel cell appliances cannot exceed 1000 kW of power output, listed and tested in accordance with ANSI Z21.83 and installed in accordance with NFPA 853 and the manufacturer's instructions. These appliances may be independent of or connected to the local electrical power grid and may be fueled by fuel tanks or permanent piping systems.

STORM SEWER, DRAIN. A pipe used for conveying rainwater, surface water, subsurface water and similar liquid waste.

❖ This is the general term used to describe the piping that conducts rainwater or surface-water waste from structures to the point of disposal.

STORY. That portion of a building included between the upper surface of a floor and the upper surface of the floor or roof next above.

❖ A story is that portion of a building from a floor surface to the floor surface or roof above. In the case of the topmost story, the height of the story is measured from the floor surface to the top of the ceiling joists of an attic. Where a ceiling does not create an attic, such as a cathedral ceiling, the story height is measured to the top of the roof rafters. See "Story above grade."

STORY ABOVE GRADE. Any story having its finished floor surface entirely above grade, except that a basement shall be considered as a story above grade where the finished surface of the floor above the basement is:

1. More than 6 feet (1829 mm) above grade plane.

2. More than 6 feet (1829 mm) above the finished ground level for more than 50 percent of the total building perimeter.

3. More than 12 feet (3658 mm) above the finished ground level at any point.

❖ The code defines a story above grade as any story having its finished floor surface entirely above grade. However, the critical part of the definition involves whether or not a basement is a story above grade. A level that is a story above grade may be either an inhabited story or unused under-floor space. Three criteria are important to the determination of whether a given floor level is either a story above grade or a basement:

1. If the finished floor above the level under consideration or above the under-floor space is more than 6 feet (1829 mm) above the grade plane as defined in Section R202, the level under consideration is a story above grade.

2. If the finished floor level above the level under consideration or above the under-floor space is more than 6 feet (1829 mm) above the finished ground level for more than 50 percent of the total

perimeter of the building, the level under consideration is a story above grade.

3. It the finished floor level above the level under consideration or above the under-floor space is more than 12 feet (3658 mm) above the finished ground level at any point, the floor level under consideration or the under-floor space is a story above grade.

Conversely, if the finished floor level above the level under consideration is 6 feet (1829 mm) or less above the grade plane, and is 6 feet (1829 mm) or less above the finished ground level for more than 50 percent of the perimeter and does not exceed 12 feet (3658 mm) at any point, the floor level under consideration is a basement. Or, described a bit differently, a basement is a floor level that does not qualify as a story above grade. Commentary Figure R202(15) illustrates the definition of story above grade.

STRUCTURAL INSULATED PANELS (SIPS). Factory fabricated panels of solid core insulation with structural skins of oriented strand board (OSB) or plywood.

❖ Structural insulated panels are construction elements composed of solid-core insulation panels enclosed within structural wood-panel membranes. These panels are fabricated at the factory, then brought to the job site and installed.

STRUCTURE. That which is built or constructed.

❖ This definition is intentionally broad so as to include within its scope—and therefore the scope of the code (see Section R101.2)—everything that is built as an improvement to real property.

SUMP. A tank or pit that receives sewage or waste, located below the normal grade of the gravity system and that must be emptied by mechanical means.

❖ This is the receiving tank or vessel used to collect and store waste from drainage systems that are incapable of draining by gravity. A sump generally houses an ejector or pump used to evacuate the contents. A sump can refer to a receiver for either waste water or storm water.

SUMP PUMP. A pump installed to empty a sump. These pumps are used for removing storm water only. The pump is selected for the specific head and volume of the load and is usually operated by level controllers.

❖ A typical sump pump [see Commentary Figure R202(16)] receives its primary power from the building primary wiring system; however, some are models equipped with a battery or natural-gas-activated backup system. This backup system is helpful during electrical storms or any condition where there is an unexpected loss of power.

SUNROOM. A one-story structure attached to a dwelling with a glazing area in excess of 40 percent of the gross area of the structure's exterior walls and roof.

❖ Sunrooms are unique attached rooms that create a space which differs in character from that provided by conventional additions. Sunrooms and other highly glazed structures are sometimes called conservatories or solariums. This definition distinguishes a sunroom from other conventional structures because they are limited to one-story in height and the total glazing area needs to be at least 40 percent of the exterior wall and roof area. This definition is important because Section R303.2 permits a sunroom addition to be considered for light and ventilation. Sunroom additions are discussed in Section R303.7.2. See the commentary to "Thermal Isolation".

SUPPLY AIR. Air delivered to a conditioned space through ducts or plenums from the heat exchanger of a heating, cooling or ventilating system.

❖ As opposed to return air, supply air is delivered to the conditioned space by an air handler and may or may not be returned to the air handler. Supply air can include ventilation air.

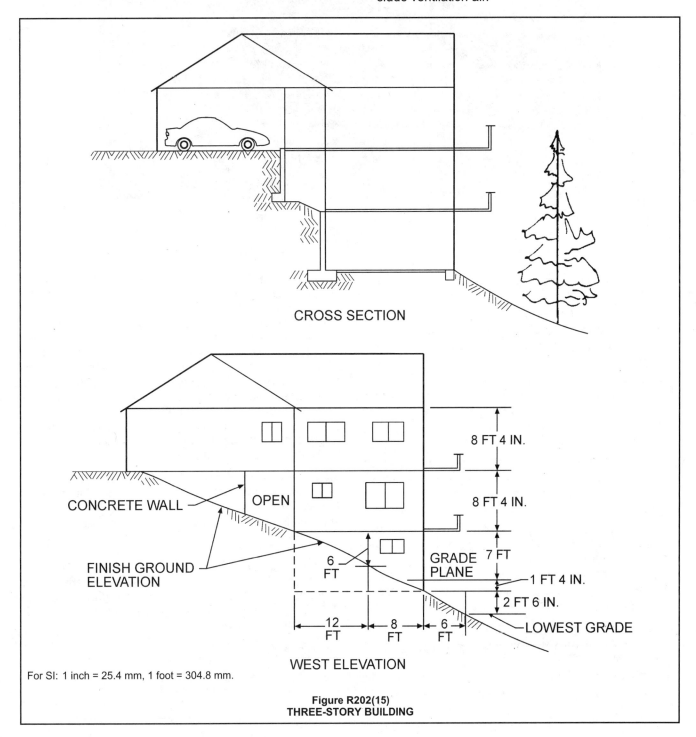

CROSS SECTION

WEST ELEVATION

For SI: 1 inch = 25.4 mm, 1 foot = 304.8 mm.

Figure R202(15)
THREE-STORY BUILDING

Figure R202(16)
SUMP PUMP

SUPPORTS. Devices for supporting, hanging and securing pipes, fixtures and equipment.

❖ Types of supports include hangers, anchors, braces, brackets, strapping and any other material or device used to support plumbing piping and components.

SWEEP. A drainage fitting designed to provide a change in direction of a drain pipe of less than the angle specified by the amount necessary to establish the desired slope of the line. Sweeps provide a longer turning radius than bends and a less turbulent flow pattern (see "Bend" and "Elbow").

❖ See the definition of "Bend."

TEMPERATURE- AND PRESSURE-RELIEF (T AND P) VALVE. A combination relief valve designed to function as both a temperature-relief and pressure-relief valve.

❖ These devices combine the components of pressure-relief and temperature-relief valves and are designed to discharge automatically at excessive pressures or temperatures or both.

TEMPERATURE-RELIEF VALVE. A temperature-actuated valve designed to discharge automatically at the temperature at which it is set.

❖ This valve is designed to discharge at excessive temperatures to prevent dangerously high water temperatures that can cause tank, vessel or pipe explosions.

THERMAL ISOLATION. Physical and space conditioning separation from conditioned space(s). The conditioned space(s) shall be controlled as separate zones for heating and cooling or conditioned by separate equipment.

❖ This term is conceptually somewhat similar to the separation provided by the "building envelope" but instead of being between conditioned space and the exterior or conditioned space and unconditioned space, this separation occurs between two conditioned spaces, the separation between the existing dwelling and the new sunroom addition. Unless a new wall is constructed, the existing "exterior wall" of the dwelling will generally provide the thermal isolation between the dwelling and the sunroom.

THERMAL RESISTANCE, *R*-VALUE. The inverse of the time rate of heat flow through a body from one of its bounding surfaces to the other for a unit temperature difference between the two surfaces, under steady state conditions, per unit area (h · ft^2 · °F/Btu).

❖ *R*-value is a unit of thermal resistance used for comparing insulating values of different materials. The greater the *R*-value of a material, the greater its insulating properties and the slower heat flows through it. *R*-values are used to rate roof/ceiling, wall and floor insulation.

THERMAL TRANSMITTANCE, *U*-FACTOR. The coefficient of heat transmission (air to air) through a building envelope component or assembly, equal to the time rate of heat flow per unit area and unit temperature difference between the warm side and cold side air films (Btu/h · ft² · °F).

❖ *U*-factor is the rate of heat transfer in Btu per hour through a square foot of a surface when the difference between the air temperatures on either side is one Fahrenheit degree. The *U*-factor is the reciprocal of the *R*-value. Vertical windows, skylights and doors are rated with *U*-factors. The lower the *U*-factor, the more thermally efficient the material or assembly.

TOWNHOUSE. A single-family dwelling unit constructed in a group of three or more attached units in which each unit extends from foundation to roof and with open space on at least two sides.

❖ A configuration of three or more single-family dwellings attached together in a single structure constitutes a townhouse if both of the following conditions exist: 1) each unit extends vertically from the foundation to the roof (townhouses cannot be stacked), and 2) each unit is open to the exterior on at least two sides, providing some degree of independence from other units.

For townhouses to be designed and constructed under the IRC, they must also comply with the limitations of Section R101.2. Townhouses are limited to three stories, and each unit must have independent egress to the exterior.

TRAP. A fitting, either separate or built into a fixture, that provides a liquid seal to prevent the emission of sewer gases without materially affecting the flow of sewage or waste water through it.

❖ The sole purpose of a trap is to isolate the interior of occupiable spaces from the sanitary drainage and vent system. It traps liquid waste and retains it to form a seal or barrier through which gases and vapors in the drainage and vent system cannot pass under normal operating conditions. Traps are either separate devices or are integral with fixtures. Other than a grease trap or a trap that is integral with a fixture, the only type

of trap permitted by the code is the "P" type. See Commentary Figure R202(17) and "Fixture drain."

TRAP ARM. That portion of a fixture drain between a trap weir and the vent fitting.

❖ This is the common name for a "Fixture drain."

TRAP PRIMER. A device or system of piping to maintain a water seal in a trap, typically installed where infrequent use of the trap would result in evaporation of the trap seal, such as floor drains.

❖ This is a device or system of piping that maintains a water seal in a trap. See Section P3201.2.

TRAP SEAL. The trap seal is the maximum vertical depth of liquid that a trap will retain, measured between the crown weir and the top of the dip of the trap.

❖ A trap seal is the vertical depth of liquid held in the dip of a trap. Trap seals are required to have a minimum depth.

TRIM. Picture molds, chair rails, baseboards, handrails, door and window frames, and similar decorative or protective materials used in fixed applications.

❖ Where limited amounts of decorative or protective materials are installed as a part of the building construction, they are regulated as trim. Examples include chair rails, door and window frames, handrails, baseboards and cove molding.

TRUSS DESIGN DRAWING. The graphic depiction of an individual truss, which describes the design and physical characteristics of the truss.

❖ If a floor or roof truss system is to be used in a building's construction, truss design drawings must be prepared and submitted to the building official prior to truss installation. The truss design drawings should graphically illustrate each specific type of truss to be installed and include a variety of other information as required in the code.

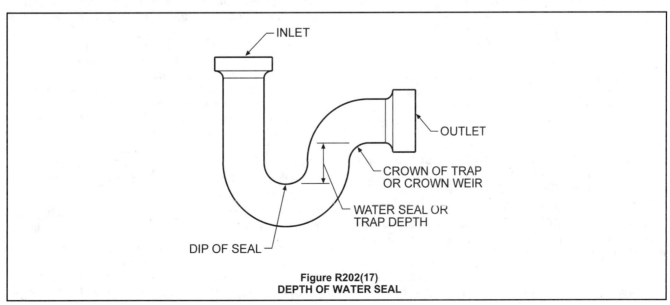

**Figure R202(17)
DEPTH OF WATER SEAL**

TYPE L VENT. A listed and labeled vent conforming to UL 641 for venting oil-burning appliances listed for use with Type L vents or with gas appliances listed for use with Type B vents.

❖ See the commentary for the definition of "Vent."

***U*-FACTOR, THERMAL TRANSMITTANCE.** The coefficient of heat transmission (air to air) through a building envelope component or assembly, equal to the time rate of heat flow per unit area and unit temperature difference between the warm side and cold side air films (Btu/h $\cdot$ ft^2 $\cdot$ °F).

❖ See "Thermal transmittance, *U*-factor."

UNCONFINED SPACE. A space having a volume not less than 50 cubic feet per 1,000 Btu/h (4.8 m^3/kW) of the aggregate input rating of all appliances installed in that space. Rooms communicating directly with the space in which the appliances are installed, through openings not furnished with doors, are considered a part of the unconfined space.

❖ For simplicity of calculation, the volume per Btu/h (W) input rating ratio can be expressed as 1 cubic foot per 20 Btu/h (4.8 m^3/kW).

UNDERLAYMENT. One or more layers of felt, sheathing paper, nonbituminous saturated felt, or other approved material over which a roof covering, with a slope of 2 to 12 (17-percent slope) or greater, is applied.

❖ Underlayment for roof covering is the initial layer of weather protection installed over a roof deck. The underlayment may consist of a single or multiple layers of material and is to be used on roofs having a minimum slope of 2 to 12. Examples of underlayment include felt, sheathing paper and nonbituminous saturated felt.

UNUSUALLY TIGHT CONSTRUCTION. Construction in which:

1. Walls and ceilings comprising the building thermal envelope have a continuous water vapor retarder with a rating of 1 perm (5.7 $\cdot$ 10^{-11} kg/Pa $\cdot$ s $\cdot$ m^2) or less with openings therein gasketed or sealed.

2. Storm windows or weatherstripping is applied around the threshold and jambs of opaque doors and openable windows.

3. Caulking or sealants are applied to areas such as joints around window and door frames between sole plates and floors, between wall-ceiling joints, between wall panels, at penetrations for plumbing, electrical and gas lines, and at other openings.

❖ As a result of energy conservation efforts, buildings are being constructed to limit the infiltration of outdoor air into the building. Buildings of such construction cannot provide combustion air by infiltration alone.

VACUUM BREAKERS. A device which prevents backsiphonage of water by admitting atmospheric pressure through ports to the discharge side of the device.

❖ A vacuum breaker is a device used as a backflow preventer. The device neutralizes (breaks) the nega-

tive pressures (vacuum) in closed systems by allowing atmospheric pressure to enter the system.

VAPOR PERMEABLE MEMBRANE. A material or covering having a permeance rating of 5 perms (2.9 $\cdot$ 10^{-10} kg/Pa $\cdot$ s $\cdot$ m^2) or greater, when tested in accordance with the desiccant method using Procedure A of ASTM E 96. A vapor permeable material permits the passage of moisture vapor.

❖ A vapor permeable membrane can consist of breather paper or other materials that allow the passage of moisture vapor. An acceptable vapor permeable membrane can be any material that has a vapor permeance of no less than 5 perm (2.9 $\cdot$ 10^{-10} kg/Pa $\cdot$ s $\cdot$ m^2) when tested in accordance with the dessicant method using Procedure A of ASTM E 96.

VAPOR RETARDER. A vapor resistant material, membrane or covering such as foil, plastic sheeting, or insulation facing having a permeance rating of 1 perm (5.7 $\cdot$ 10^{-11} kg/Pa $\cdot$ s $\cdot$ m^2) or less, when tested in accordance with the dessicant method using Procedure A of ASTM E 96. Vapor retarders limit the amount of moisture vapor that passes through a material or wall assembly.

❖ An acceptable vapor retarder can be any material that has been tested in accordance with the dessicant method using Procedure A of ASTM E 96 that does not have a permeance rating greater than 1.0. Permeance is a measure of water vapor transmission (wvt) rate in grains per hour per square foot divided by the vapor pressure difference in inches of mercury. Materials tested to the standard are papers, plastic films, asphalt-saturated construction papers and sheet materials. Vapor retarders are installed on the warm-in-winter side of walls, floors and ceilings of the building's thermal insulation.

VEHICULAR ACCESS DOOR. A door that is used primarily for vehicular traffic at entrances of buildings such as garages and parking lots, and that is not generally used for pedestrian traffic.

❖ A typical vehicular access door would be an overhead door that is counterbalanced by a spring mechanism that uses a roller system on a fixed set of tracks that it follows to the fully open position above and beyond the opening. This door system can have a single leaf or multiple leaves. These units are available in a wide range of sizes, and may be wood, steel or aluminum. In addition, some of the access door manufacturers offer a fully insulated product. Section R314.5.6 specifically addresses foam-filled garage doors. Automatic door openers are available for most of these products; however, it is important to consult the manufacturer's installation instructions to ensure that proper safety measures are followed.

VENT. A passageway for conveying flue gases from fuel-fired appliances, or their vent connectors, to the outside atmosphere.

❖ In code terminology, vents are distinguished from chimneys and usually are constructed of factory-made listed and labeled components intended to function as a system. Type B and BW vents are constructed of galvanized steel and aluminum sheet metal and are double-wall and air insulated. Such vents are designed to vent gas-fired appliances and equipment that are equipped with draft hoods or specifically listed (labeled) for use with Type B or BW vents. Type L vents are typically constructed of sheet steel and stainless steel. They are double-wall, air insulated and designed to vent gas and oil-fired appliances and equipment. Mid- to high-efficiency appliances are designed for use with corrosion-resistant vents such as those made of plastic pipe and special alloys of stainless steel. Pellet vents (PL) are specialized vents similar in design and construction to Type L vents. Pellet vents must not be used with any appliances other than pellet-burning appliances.

VENT COLLAR. See "Flue collar."

VENT CONNECTOR. That portion of a venting system which connects the flue collar or draft hood of an appliance to a vent.

❖ In most cases, appliances are not located directly in line with the vertically rising chimney or vent; therefore, a vent connector is necessary to connect the appliance flue outlet with the vent. Vent connectors can be single- or double-wall pipes and are usually made from galvanized steel, stainless steel or aluminum sheet metal. In many installations, the vent connectors must be constructed of the same material as the vent, as is typically done with Type B vent systems.

VENT DAMPER DEVICE, AUTOMATIC. A device intended for installation in the venting system, in the outlet of an individual, automatically operated fuel burning appliance and that is designed to open the venting system automatically when the appliance is in operation and to close off the venting system automatically when the appliance is in a standby or shutdown condition.

❖ The purpose of this device is the conservation of energy by preventing the cooling of the combustion chamber of the appliance. This is accomplished by preventing the flow of air (by means of connection) through the combustion chamber when the appliance is not in operation.

VENT GASES. Products of combustion from fuel-burning appliances, plus excess air and dilution air, in the venting system above the draft hood or draft regulator.

❖ These products of combustion are the result of burning fuel gas.

VENT STACK. A vertical vent pipe installed to provide circulation of air to and from the drainage system and which extends through one or more stories.

❖ This is a dry vent that extends from the base of a waste or soil stack and typically runs parallel and adjacent to the waste or soil stack it serves. The vent stack can connect to the horizontal drain downstream of the drain stack, or it can connect directly to the drainage stack.

VENT SYSTEM. Piping installed to equalize pneumatic pressure in a drainage system to prevent trap seal loss or blow-back due to siphonage or back pressure.

❖ This system is a piping arrangement designed to maintain pressure fluctuation at fixture traps to within plus or minus 2 inches of water column (249 Pa).

VENTILATION. The natural or mechanical process of supplying conditioned or unconditioned air to, or removing such air from, any space.

❖ Ventilation can be used for comfort cooling, the control of air contaminants, equipment cooling and replenishing oxygen levels.

VENTING. Removal of combustion products to the outdoors.

❖ This is the desired result of the venting system.

VENTING SYSTEM. A continuous open passageway from the flue collar of an appliance to the outside atmosphere for the purpose of removing flue or vent gases. A venting system is usually composed of a vent or a chimney and vent connector, if used, assembled to form the open passageway.

❖ Venting systems operate by gravity (natural draft) or mechanical means. Venting systems include both vents and chimneys.

VERTICAL PIPE. Any pipe or fitting that makes an angle of 45 degrees (0.79 rad) or more with the horizontal.

❖ Vertical pipes are angled 45 degrees (0.79 rad) or more above the horizontal.

VINYL SIDING. A shaped material, made principally from rigid polyvinyl chloride (PVC), that is used to cover exterior walls of buildings.

❖ Vinyl siding is a manufactured plastic product whose primary component is polyvinyl chloride (PVC). Vinyl siding is either co-extruded in layers or thermoformed, depending on the style or profile of the product. Additives are used to improve weathering performance. Vinyl siding products are available in a wide range of sizes, shapes, colors and textures and are intended for exterior wall and related applications.

WALL, RETAINING. A wall not laterally supported at the top, that resists lateral soil load and other imposed loads.

❖ Retaining walls are discussed in Section R404.5 of this code. These walls retain soil that would otherwise slump or cave to a more natural slope. The lateral pressure of soil against a retaining wall is greatly influenced by soil moisture and it is this lateral pressure that constitutes the largest part of the load that the wall must withstand.

WALLS. Walls shall be defined as follows:

Load-bearing wall is a wall supporting any vertical load in addition to its own weight.

Nonbearing wall is a wall which does not support vertical loads other than its own weight.

❖ Walls are either load-bearing or nonbearing. A load-bearing wall supports not only its own weight but also an additional vertical load. An example is a wood stud wall supporting roof rafters and ceiling joists. A nonbearing wall does not support any vertical load other than its own weight.

WASTE. Liquid-borne waste that is free of fecal matter.

❖ Waste is any plumbing fixture discharge other than the discharge from a water closet or urinal. This material does not contain human fecal matter.

WASTE PIPE OR STACK. Piping that conveys only liquid sewage not containing fecal material.

❖ A waste pipe or stack is any waste-conducting pipe not falling under the definition of soil pipe. Waste pipes do not convey human fecal matter.

WATER-DISTRIBUTION SYSTEM. Piping which conveys water from the service to the plumbing fixtures, appliances, appurtenances, equipment, devices or other systems served, including fittings and control valves.

❖ This is all the water piping within a building from the point of connection to the water service to the various connections of the fixture supplies or outlets.

WATER HEATER. Any heating appliance or equipment that heats potable water and supplies such water to the potable hot water distribution system.

❖ The types of water heaters include storage (tank), circulating and instantaneous. Point-of-use water heaters typically supply hot water to a single fixture or outlet and are located in close proximity to the fixture or outlet.

WATER MAIN. A water-supply pipe for public use.

❖ Water mains are owned and operated by municipalities, rural water districts, privately owned water purveyors or other such entities.

WATER OUTLET. A valved discharge opening, including a hose bibb, through which water is removed from the potable water system supplying water to a plumbing fixture or plumbing appliance that requires either an air gap or backflow pre-vention device for protection of the supply system.

❖ A water outlet is a discharge opening through which water is supplied to a fixture, into the atmosphere (except into an open tank that is part of the water supply system), to a boiler or heating system or to any device or equipment requiring water to operate but which is not part of the plumbing system.

WATER-RESISTIVE BARRIER. A material behind an exterior wall covering that is intended to resist liquid water that has penetrated behind the exterior covering from further intruding into the exterior wall assembly.

❖ Protection of the building envelope is a primary concern. The ability of the water-resistive barrier to provide weather resistance and maintain the integrity of the building envelope is key to controlling water based issues like mold, decay and deterioration of a structure. Water-resistive barriers are discussed in Section R703.2 and in Section 1404.2 of the IBC.

WATER-SERVICE PIPE. The outside pipe from the water main or other source of potable water supply to the water-distribution system inside the building, terminating at the service valve.

❖ Water-service pipe is any acceptable piping that conveys potable water.

WATER-SUPPLY SYSTEM. The water-service pipe, the water-distributing pipes and the necessary connecting pipes, fittings, control valves and all appurtenances in or adjacent to the building or premises.

❖ The water supply system includes all piping and components that convey potable water from the public main or other source to the points of water usage.

WET VENT. A vent that also receives the discharge of wastes from other fixtures.

❖ Although Section P3108 describes "wet venting" systems, the term "wet vent" applies to any pipe that is used concurrently as both a drain and a vent. This arrangement can be found in the following systems: Common Vent (Section P3107), Wet Venting (Section P3108), Waste Stack Vent (Section P3109), Circuit Venting (Section P3110), and Combination Waste and Vent System (Section P3111).

WIND BORNE DEBRIS REGION. Areas within hurricane-prone regions within one mile of the coastal mean high water line where the basic wind speed is 110 miles per hour (49 m/s) or greater; or where the basic wind speed is equal to or greater than 120 miles per hour (54 m/s); or Hawaii.

❖ The IRC regulates the exterior openings of a building where there is a concern about the extensive damage that can be caused by debris carried through the air by extremely high winds. Wind borne debris regions are those areas where the potential for such damage is high. The code identifies three such areas: 1) hurricane-prone regions within one mile of the coastal mean high water line, provided the wind speed is at least 110 mph; 2) those locations where the wind speed is at least 120 mph; and 3) the islands of Hawaii. See also the definition for "Hurricane-prone Region."

WINDER. A tread with non-parallel edges.

❖ A winder is a tread in a winding stairway. Winder treads are used as components of stairs that change direction, just as straight treads are components in straight stairs. A winder serves as a tread but its shape allows the additional function of a gradual turning of the stairway direction. The tread depth of a winder at the walk line and the minimum tread depth at the narrow end can control the turn made by each winder.

WOOD STRUCTURAL PANEL. A panel manufactured from veneers; or wood strands or wafers; bonded together with waterproof synthetic resins or other suitable bonding systems. Examples of wood structural panels are plywood, OSB or composite panels.

❖ Wood structural panels are sheet products commonly known as plywood, wafer board, oriented strand board (OSB) or chip board manufactured to meet the United

States Department of Commerce (DOC) standards for Construction and Industrial Plywood, PS 1, and Performance Standards for Wood-Based Structural-Use Panels, PS 2.

YARD. An open space, other than a court, unobstructed from the ground to the sky, except where specifically provided by this code, on the lot on which a building is situated.

❖ Used throughout the code to describe an open space at the exterior of a building, a yard must be unobstructed from the ground to the sky and located on the same lot on which the building is situated. The openness of a yard allows it to be used to supply buildings with natural light and ventilation. A court, which is bounded on three or more sides by the exterior walls of the building, is not a yard.

Bibliography

The following resource materials are referenced in this chapter or are relevant to the subject matter addressed in this chapter.

ANSI Z21.83-1998, *Fuel Cell Power Plants.* New York, NY: American National Standards Institute, 1998.

ASTM C 55-03, *Standard Specification for Concrete Brick.* West Conshohocken, PA: ASTM International, 2003.

ASTM C 62-03, *Standard Specification for Building Brick (Solid Masonry Units Made from Clay or Shale).* West Conshohocken, PA: ASTM International, 2003.

ASTM C 73-99a, *Standard Specification for Calcium Silicate Face Brick (Sand Lime Brick).* West Conshohocken, PA: ASTM International, 1999.

ASTM C 90-03, *Standard Specification for Load-Bearing Concrete Masonry Units.* West Conshohocken, PA: ASTM International, 2003.

ASTM C 216-04a, *Standard Specification for Facing Brick (Solid Masonry Units Made from Clay or Shale).* West Conshohocken, PA: ASTM International, 2004.

ASTM E 84-04, *Standard Test Method for Surface Burning Characteristics of Building Materials.* West Conshohocken, PA: ASTM International, 2004.

ASTM E 96-00e04, Standard Test Method for Water Vapor Transmission of Materials. West Conshohocken, PA: ASTM International, 2004.

ASTM E 136-99e01, *Standard Test Method for Behavior of Materials in a Vertical Tube Furnace at 750°C.* West Conshohocken, PA: ASTM International, 2001.

DOC PS 1-95, *Construction and Industrial Plywood.* Gaithersburg, MD: U.S. Department of Commerce, 1995.

DOC PS 2-93, *Performance Standard for Wood-Based Structural-Use Panels.* Gaithersburg, MD: U.S. Department of Commerce, 1993.

IBC-2006, *International Building Code.* Falls Church, VA: International Code Council, Inc., 2006.

IFC-2006, *International Fire Code.* Falls Church, VA: International Code Council, Inc., 2006.

IFGC-2006, *International Fuel Gas Code.* Falls Church, VA: International Code Council, Inc. 2006.

IMC-2006, *International Mechanical Code.* Falls Church, VA: International Code Council, Inc., 2006.

IPC-2006, *International Plumbing Code.* Falls Church, VA: International Code Council, Inc., 2006.

NFPA 501-03, *Standard on Manufactured Housing.* Quincy, MA: National Fire Protection Association, 2003.

NFPA 853-03, *Installation of Stationary Fuel Power Power Plants,* Quincy, MA: National Fire Protection Association, 2003.

UL 103-2001, *Chimneys, Factory-Built, Residential Type and Building Heating Appliance*—with Revision through March 1999. Northbrook, IL: Underwriters Laboratories, 2001.

UL 641-95, *Type L, Low-Temperature Venting Systems*—with Revisions through April 1999. Northbrook, IL: Underwriters Laboratories, 1999.

Chapter 3:
Building Planning

General Comments

Chapter 3 is a compilation of the code requirements specific to the building planning sector of the design and construction process. The provisions address a wide variety of issues important to designing a building that is both safe and usable. The limitations placed on the materials and methods of construction contribute to the development of a structurally sound building. Snow, wind and seismic design and flood-resistant construction are regulated, as are the live and dead loads, in Chapter 3.

Fire-resistance-rated assemblies are necessary under two different conditions: 1) where a building is situated very close to a property line, the code addresses the concern for radiant heat exposure in a fire; 2) where two or more dwelling units are housed in a single structure, the code mandates a minimum level of fire separation between all units. Other concerns related to fires include the limitations on wall and ceiling finishes, the requirement for emergency escape and rescue openings, the required installation of smoke alarms throughout the dwelling unit and limitations on the use of foam plastics and other insulation materials. In addition, the specific construction requirements for the common wall between the house and garage and the ceiling assembly between the garage and habitable space are addressed in Chapter 3.

This chapter sets forth traditional code requirements dealing with light, ventilation, sanitation, room size, ceiling height and environmental comfort. Life-safety provisions include limitations on glazing used in hazardous areas, specifications on the use of guards at elevated surfaces and basic rules for the exiting system. This chapter also contains most of the regulations found in the code that deal with the planning and design of dwelling units.

Section R301 establishes the design criteria, including dead loads, live loads, roof loads, floor loads, snow loads, wind loads and seismic loads. Section R302 identifies the requirements for the building based on its location on the lot. Section R303 establishes the light, ventilation and heating requirements for dwelling units. Section R304 establishes the minimum requirements for rooms in dwelling units. Section R305 establishes the ceiling height requirements for dwelling units. Section R306 contains requirements for sanitation. Section R307 contains requirements for toilets and bath and shower spaces. Section R308 contains requirements for glazing, hazardous locations of glazing, site-built windows and skylights. Section R309 contains provisions for garages and carports. Section R310 contains provisions for emergency escape and rescue openings. Section R311 establishes the means of egress requirements, including provisions for exterior balconies, hallways, doors including their landings, stairways including their width, treads, risers, landings and handrails, special stairways and ramps including their landings and handrails. Section R312 addresses guards. Section R313 contains the requirements for smoke alarms. Section R314 addresses the use of foam plastic. Section R315 addresses flame spread and smoke-density requirements for walls and ceilings. Section R316 contains provisions for insulation. Section R317 addresses the separation of dwelling units. Section R318 addresses moisture vapor retarders. Section R319 contains requirements for decay protection. Section R320 contains requirements for termite protection. Section R321 provides the requirements for premise identification (site address). Section R322 contains the scoping requirements for accessibility. Section R323 addresses elevators and platform lifts. Section R324 establishes flood-resistant construction provisions.

Purpose

Chapter 3 provides guidelines for a minimum level of structural integrity, life safety, fire safety and livability for inhabitants of dwelling units regulated by the code. The chapter sets forth the requirements that affect the most basic planning and design aspects of dwelling construction. It identifies the various structural loads that are imposed on a building, and it establishes criteria that address each of the imposed loads. In the design of residential structures scoped by the *International Residential Code*® (IRC®), there are many climatic and geographical issues that must be considered. This chapter provides guidance in the determination of all appropriate design criteria. In addition, it sets forth the limiting conditions under which a building may be designed and constructed using the IRC.

Fundamental issues of livability and sanitation are satisfied through the regulation of minimum room sizes and ceiling heights, as well as basic requirements for toilet rooms and kitchens. Life-safety concerns are addressed in a number of areas, including provisions regulating emergency escape and rescue openings, glazing in areas subject to human impact and exiting. The chapter establishes minimum specifications for a number of different building components, including stairways, ramps, landings, handrails and guards. It deals with fire-safety issues such as early fire detection by smoke alarms, exterior wall protection for proximity to property lines, separation of dwelling units in multiple-family buildings and control of fire spread across wall and ceiling finishes. Property protection is also a concern, with provisions established for protection against decay and termites.

SECTION R301
DESIGN CRITERIA

R301.1 Application. Buildings and structures, and all parts thereof, shall be constructed to safely support all loads, including dead loads, live loads, roof loads, flood loads, snow loads, wind loads and seismic loads as prescribed by this code. The construction of buildings and structures in accordance with the provisions of this code shall result in a system that provides a complete load path that meets all requirements for the transfer of all loads from their point of origin through the load-resisting elements to the foundation. Buildings and structures constructed as prescribed by this code are deemed to comply with the requirements of this section.

❖ This section specifies the minimum design loads required for structures built in accordance with the provisions of the IRC. In structural design, loads are generally divided into two categories: gravity loads, which act vertically, and lateral loads, which act horizontally. Lateral loads typically result from either wind (see Section R301.2.1), earthquakes (see Section R301.2.2) or flood loads (see Section R301.2.4. Although wind, flood and earthquake design may concern themselves with lateral loads, there are also vertical force components that should be considered.

All structures must be designed to support these loads and provide a complete load path capable of transferring these loads from their point of origin through the appropriate load-resisting elements and foundation and ultimately to the supporting soil. The charging statement specifically states that any building or structure that has been built in strict compliance with this code provides a complete load path that meets all requirements for load transfer from the point of origin to the foundation. A load path that is either incomplete or inadequate will expose the structure to damage just as surely as an undersized structural member will. The concept of a complete load path is a fundamental principle in structural engineering, and the code makes it clear that a complete load path must be provided.

R301.1.1 Alternative provisions. As an alternative to the requirements in Section R301.1 the following standards are permitted subject to the limitations of this code and the limitations therein. Where engineered design is used in conjunction with these standards the design shall comply with the *International Building Code*.

1. American Forest and Paper Association (AF&PA) *Wood Frame Construction Manual* (WFCM).

2. American Iron and Steel Institute (AISI) *Standard for Cold-Formed Steel Framing—Prescriptive Method for One- and Two-Family Dwellings* (COFS/PM) *with Supplement to Standard for Cold-Formed Steel Framing-Prescriptive Method for One- and Two-Family Dwellings.*

❖ This section permits the use of alternate prescriptive framing methods. Wood framing is permitted to comply with the provisions of the American Forest and Paper Association, *Wood Frame Construction Manual for One-and Two-Family Dwellings.* Cold-formed steel framing is permitted to comply with American Iron and Steel Insti-

tute, *Standard for Cold-Formed Steel Framing-Prescriptive Method for One-and Two-Family Dwellings with Supplement to Standard for Cold-Formed Steel Framing-Prescriptive Method for One- and Two- Family Dwellings.* Engineered design in accordance with the *International Building Code* (IBC) is required when a building is beyond (or exceeds) the applicability limits of either standard.

R301.1.2 Construction systems. The requirements of this code are based on platform and balloon-frame construction for light-frame buildings. The requirements for concrete and masonry buildings are based on a balloon framing system. Other framing systems must have equivalent detailing to ensure force transfer, continuity and compatible deformations.

❖ The requirements of the IRC are based on platform or balloon-frame construction for light-frame buildings (see the definitions of platform construction and light-framed construction in Chapter 2) and on a balloon-framing system for concrete and masonry buildings.

R301.1.3 Engineered design. When a building of otherwise conventional construction contains structural elements exceeding the limits of Section R301 or otherwise not conforming to this code, these elements shall be designed in accordance with accepted engineering practice. The extent of such design need only demonstrate compliance of nonconventional elements with other applicable provisions and shall be compatible with the performance of the conventional framed system. Engineered design in accordance with the *International Building Code* is permitted for all buildings and structures, and parts thereof, included in the scope of this code.

❖ Generally, proper application of the IRC requires a clear understanding of and adherence to its prescriptive limitations, which are based on conventional construction. However, a building may contain structural elements that are either unconventional or exceed the prescriptive limitations of the code. This is acceptable, if these elements are designed in accordance with accepted engineering practice by a design professional.

R301.2 Climatic and geographic design criteria. Buildings shall be constructed in accordance with the provisions of this code as limited by the provisions of this section. Additional criteria shall be established by the local jurisdiction and set forth in Table R301.2(1).

❖ This section establishes the design criteria that vary based on location and/or climate. Some of the criteria reflect loading such as earthquake, flood and wind; others reflect susceptibility to damage from hazards such as weather exposure or termites. Additional criteria may be established by local jurisdictions as necessary. These would include, for example, whether a site is within a wind-borne debris region as described in Section R301.2.1.2, of this commentary. Table R301.2(1) lists the criteria that must be established within each jurisdiction for any project constructed under the IRC. The Table must be filled in by the jurisdiction adopting the code for their particular area. Note that some of these criteria (e.g., wind exposure category or flood hazard) can vary within a given jurisdiction and may need to be established on a site-by-site (or project-by-project) basis. The table serves as a useful reminder for code enforcement personnel, builders, designers and owners. Verifying this information up front aids compliance with the IRC.

R301.2.1 Wind limitations. Buildings and portions thereof shall be limited by wind speed, as defined in Table R301.2(1) and construction methods in accordance with this code. Basic wind speeds shall be determined from Figure R301.2(4). Where different construction methods and structural materials are used for various portions of a building, the applicable requirements of this section for each portion shall apply. Where loads for wall coverings, curtain walls, roof coverings, exterior windows, sky-

lights, garage doors and exterior doors are not otherwise specified, the loads listed in Table R301.2(2) adjusted for height and exposure using Table R301.2(3) shall be used to determine design load performance requirements for wall coverings, curtain walls, roof coverings, exterior windows, skylights, garage doors and exterior doors. Asphalt shingles shall be designed for wind speeds in accordance with Section R905.2.6.

❖ As described above, wind loads are a major consideration in designing a structure's lateral force-resisting system. Commentary Figure R301.2.1 is a schematic representation of the lateral component of wind loading on a building. Wind loads affect more than the lateral load system, as evidenced by provisions such as roof tie-downs in Section R802.11.

TABLE R301.2(1)
CLIMATIC AND GEOGRAPHIC DESIGN CRITERIA

GROUND SNOW LOAD	WIND SPEED[d] (mph)	SEISMIC DESIGN CATEGORY[f]	SUBJECT TO DAMAGE FROM			WINTER DESIGN TEMP[e]	ICE BARRIER UNDERLAYMENT REQUIRED[h]	FLOOD HAZARDS[g]	AIR FREEZING INDEX[i]	MEAN ANNUAL TEMP[j]
			Weathering[a]	Frost line depth[b]	Termite[c]					

For SI: 1 pound per square foot = 0.0479 kPa, 1 mile per hour = 0.447 m/s.

a. Weathering may require a higher strength concrete or grade of masonry than necessary to satisfy the structural requirements of this code. The weathering column shall be filled in with the weathering index (i.e., "negligible," "moderate" or "severe") for concrete as determined from the Weathering Probability Map [Figure R301.2(3)]. The grade of masonry units shall be determined from ASTM C 34, C 55, C 62, C 73, C 90, C 129, C 145, C 216 or C 652.

b. The frost line depth may require deeper footings than indicated in Figure R403.1(1). The jurisdiction shall fill in the frost line depth column with the minimum depth of footing below finish grade.

c. The jurisdiction shall fill in this part of the table to indicate the need for protection depending on whether there has been a history of local subterranean termite damage.

d. The jurisdiction shall fill in this part of the table with the wind speed from the basic wind speed map [Figure R301.2(4)]. Wind exposure category shall be determined on a site-specific basis in accordance with Section R301.2.1.4.

e. The outdoor design dry-bulb temperature shall be selected from the columns of $97^1/_2$-percent values for winter from Appendix D of the *International Plumbing Code*. Deviations from the Appendix D temperatures shall be permitted to reflect local climates or local weather experience as determined by the building official.

f. The jurisdiction shall fill in this part of the table with the seismic design category determined from Section R301.2.2.1.

g. The jurisdiction shall fill in this part of the table with (a) the date of the jurisdiction's entry into the National Flood Insurance Program (date of adoption of the first code or ordinance for management of flood hazard areas), (b) the date(s) of the currently effective FIRM and FBFM, or other flood hazard map adopted by the community, as may be amended.

h. In accordance with Sections R905.2.7.1, R905.4.3.1, R905.5.3.1, R905.6.3.1, R905.7.3.1 and R905.8.3.1, where there has been a history of local damage from the effects of ice damming, the jurisdiction shall fill in this part of the table with "YES". Otherwise, the jurisdiction shall fill in this part of the table with "NO".

i. The jurisdiction shall fill in this part of the table with the 100-year return period air freezing index (BF-days) from Figure R403.3(2) or from the 100-year (99%) value on the National Climatic Data Center data table "Air Freezing Index- USA Method (Base 32°Fahrenheit)" at www.ncdc.noaa.gov/fpsf.html.

j. The jurisdiction shall fill in this part of the table with the mean annual temperature from the National Climatic Data Center data table "Air Freezing Index-USA Method (Base 32°Fahrenheit)" at www.ncdc.noaa.gov/fpsf.html.

❖ Table R301.2(1) is designed so that jurisdictions recognize certain climatic and geographic design criteria that vary from location to location. Communities are directed to complete the table with a variety of factors. See the table footnotes for the sources of the information to be determined by the local jurisdiction so that it can complete the table.

The required information includes the date of entry into the NFIP, which is the date of adoption of the first code or ordinance for management of flood hazard areas. With respect to the official map that shows flood hazard areas, the community inserts the date of the currently effective Flood Insurance Rate Map (and Flood Boundary and Floodway Map, if applicable) or the date of other maps that are adopted. Another flood hazard map may be specified if it shows flood hazard areas that are larger than those shown on the FIRM, as may be the case if a community elects to define its floodplains based on higher standards such as the "flood of record," "ultimate development" or "no-rise" rules to define the floodway.

From time to time, FEMA floodplain maps and studies may be revised and republished. When maps are revised and flood hazard areas are changed, FEMA involves the community and provides a formal opportunity to review the documents. Once the revisions are finalized, FEMA requires adoption of the new maps by the community. Communities may be able to minimize having to adopt each revision by referencing the date of the original map and study and all future revisions. This is a method by which subsequent revisions to flood maps and studies may be adopted administratively without requiring legislative action on the part of the community. Communities will need to determine whether this adoption-by-reference approach is allowed under their state's enabling authority and due process requirements. If not allowed, communities are to follow their state's requirements, which normally require public notices, hearings and specific adoption of revised maps by the community's legislative body.

Buildings must be capable of withstanding the wind loads specified in this section. Wind speeds used for design are determined from Figure R301.2(4), which portrays basic wind speeds for a 50-year mean recurrence interval. For the most part, no further calculation is required because the prescriptive tables of the code are based on wind speed and exposure.

Wall coverings, curtain walls, roof coverings, exterior windows, skylights, garage doors and exterior doors must be capable of withstanding the component and cladding wind pressures of Table R301.2(2) adjusted by the height and exposure coefficients given in Table R301.2(3).

Section R905.2.6 addresses the attachment details for asphalt shingles. Roofs with higher slopes or in areas subject to higher wind speeds may require special methods of attachment. Please see the commentary to Section R905.2.6.

Table R301.2(1) requires the local municipality or code user to insert a frost line depth entry for the particular geographical area they are located in. Commentary Figure R301.2(8) provides some guidance to do this more accurately based on the United States Weather Bureau information.

R301.2.1.1 Design criteria. Construction in regions where the basic wind speeds from Figure R301.2(4) equal or exceed 100 miles per hour (45 m/s) in hurricane-prone regions, or 110 miles per hour (49 m/s) elsewhere, shall be designed in accordance with one of the following:

1. American Forest and Paper Association (AF&PA) *Wood Frame Construction Manual for One- and Two-Family Dwellings* (WFCM); or

2. *Southern Building Code Congress International Standard for Hurricane Resistant Residential Construction* (SSTD 10); or

3. *Minimum Design Loads for Buildings and Other Structures* (ASCE-7); or

4. American Iron and Steel Institute (AISI), *Standard for Cold-Formed Steel Framing—Prescriptive Method For One- and Two-Family Dwellings (COFS/PM) with Supplement to Standard for Cold-Formed Steel Framing—Prescriptive Method For One- and Two-Family Dwellings.*

5. Concrete construction shall be designed in accordance with the provisions of this code.

❖ With the exception of concrete construction, the prescriptive provisions of the IRC apply only where the basic wind speed is equal to or exceeds 100 miles per hour (45 m/s) in hurricane prone regions, or 110 miles per hour (49 m/s) elsewhere. Concrete provisions are applicable in regions with wind speeds up to 150 mph (67 m/s), where the site is classified as Exposure B, C or D (see the commentary on Section R611.2). Where the basic wind speed exceeds the limitations given in the code, structures must be designed for wind loads. For this purpose, the IRC expressly allows the use of AF&PA WFCM, *Wood Frame Construction Manual for One- and Two-Family Dwellings*; SBCCI SSTD 10,

Standard for Hurricane Resistant Residential Construction; ASCE 7, *Minimum Design Loads for Buildings and Other Structures;* or AISI, *Standard for Cold-Formed Steel Framing-Prescriptive Method for One- and Two-Family Dwellings with Supplement.*

R301.2.1.2 Protection of openings. Windows in buildings located in windborne debris regions shall have glazed openings protected from windborne debris. Glazed opening protection for windborne debris shall meet the requirements of the Large Missile Test of an approved impact resisting standard or ASTM E 1996 and ASTM E 1886 referenced therein.

Exception: Wood structural panels with a minimum of $^{7}/_{16}$ inch (11 mm) and a maximum span of 8 feet (2438 mm) shall be permitted for opening protection in one- and two-story buildings. Panels shall be precut so that they shall be attached to the framing surrounding the opening containing the product with the glazed opening. Panels shall be secured with the attachment hardware provided. Attachments shall be designed to resist the component and cladding loads determined in accordance with either Table R301.2(2) or Section 1609.6.5 of the *International Building Code.* Attachment in accordance with Table R301.2.1.2 is permitted for buildings with a mean roof height of 33 feet (10 058 mm) or less where wind speeds do not exceed 130 miles per hour (58 m/s).

❖ Windows in buildings located in wind-borne debris regions must have glazed openings that provide protection from wind-borne debris. The opening protection, usually in the form of permanent shutters or laminated glass, must meet the requirements of the Large Missile Test of ASTM E 1996 and ASTM E 1886, listed in Chapter 43. The exception provides a prescriptive and more economical approach for one- and two-story buildings. This method of opening protection uses wood structural panels that are limited to a maximum span of 8 feet (2438 mm), with fastening as specified in Table R301.2.1.2 for buildings with a mean roof height of 33 feet (10 058 mm) or less where wind speeds do not exceed 130 miles per hour (58 m/s). The builder must precut these to fit each glazed opening and provide the necessary attachment hardware.

To determine where this requirement applies, refer to the definition in Chapter 2, which states that wind-borne debris regions include areas in hurricane- prone regions within one mile of the coastal mean high water line where the basic wind speed [from Figure R301.2(4)] is 110 mph (49 m/s) or greater; areas where the basic wind speed is at least 120 mph (54 m/s); or Hawaii. For the definition of a hurricane-prone region, see Section R202.

Table R301.2(2). See next page.

❖ This table lists wind pressures for components or cladding building elements. The term cladding is defined in Section R202. Although the basic wind speed from Figure R301.2(4) is applicable to the structure as a whole, this table applies to elements of the building that typically are not part of the wind-force-resisting system. Wind pressures on these elements vary depending on the tributary area, the location/orientation

TABLE R301.2(2)
COMPONENT AND CLADDING LOADS FOR A BUILDING WITH A MEAN ROOF HEIGHT OF 30 FEET LOCATED IN EXPOSURE B (psf)

	ZONE	EFFECTIVE WIND AREA (feet²)	85	90	100	105	110	120	125	130	140	145	150	170
			BASIC WIND SPEED (mph—3-second gust)											
Roof > 0 to 10 degrees	1	10	10.0 -13.0	10.0 -14.6	10.0 -18.0	10.0 -19.8	10.0 -21.8	10.5 -25.9	11.4 -28.1	12.4 -30.4	14.3 -35.3	15.4 -37.8	16.5 -40.5	21.1 -52.0
	1	20	10.0 -12.7	10.0 -14.2	10.0 -17.5	10.0 -19.3	10.0 -21.2	10.0 -25.2	10.7 -27.4	11.6 -29.6	13.4 -34.4	14.4 -36.9	15.4 -39.4	19.8 -50.7
	1	50	10.0 -12.2	10.0 -13.7	10.0 -16.9	10.0 -18.7	10.0 -20.5	10.0 -24.4	10.0 -26.4	10.6 -28.6	12.3 -33.2	13.1 -35.6	14.1 -38.1	18.1 -48.9
	1	100	10.0 -11.9	10.0 -13.3	10.0 -18.5	10.0 -18.2	10.0 -19.9	10.0 -23.7	10.0 -25.7	10.0 -27.8	11.4 -32.3	12.2 -34.6	13.0 -37.0	16.7 -47.6
	2	10	10.0 -21.8	10.0 -24.4	10.0 -30.2	10.0 -33.3	10.0 -36.5	10.5 -43.5	11.4 -47.2	12.4 -51.0	14.3 -59.2	15.4 -63.5	16.5 -67.9	21.1 -87.2
	2	20	10.0 -19.5	10.0 -21.8	10.0 -27.0	10.0 -29.7	10.0 -32.6	10.0 -38.8	10.7 -42.1	11.6 -45.6	13.4 -52.9	14.4 -56.7	15.4 -60.7	19.8 -78.0
	2	50	10.0 -16.4	10.0 -18.4	10.0 -22.7	10.0 -25.1	10.0 -27.5	10.0 -32.7	10.0 -35.5	10.6 -38.4	12.3 -44.5	13.1 -47.8	14.1 -51.1	18.1 -65.7
	2	100	10.0 -14.1	10.0 -15.8	10.0 -19.5	10.0 -21.5	10.0 -23.6	10.0 -28.1	10.0 -30.5	10.0 -33.0	11.4 -38.2	12.2 -41.0	13.0 -43.9	16.7 -56.4
	3	10	10.0 -32.8	10.0 -36.8	10.0 -45.4	10.0 -50.1	10.0 -55.0	10.5 -65.4	11.4 -71.0	12.4 -76.8	14.3 -89.0	15.4 -95.5	16.5 -102.2	21.1 -131.3
	3	20	10.0 -27.2	10.0 -30.5	10.0 -37.6	10.0 -41.5	10.0 -45.5	10.0 -54.2	10.7 -58.8	11.6 -63.6	13.4 -73.8	14.4 79.1	15.4 -84.7	19.8 -108.7
	3	50	10.0 -19.7	10.0 -22.1	10.0 -27.3	10.0 -30.1	10.0 -33.1	10.0 -39.3	10.0 -42.7	10.6 -46.2	12.3 -53.5	13.1 -57.4	14.1 -61.5	18.1 -78.9
	3	100	10.0 -14.1	10.0 -15.8	10.0 -19.5	10.0 -21.5	10.0 -23.6	10.0 -28.1	10.0 -30.5	10.0 -33.0	11.4 -38.2	12.2 -41.0	13.0 -43.9	16.7 -56.4
Roof > 10 to 30 degrees	1	10	10.0 -11.9	10.0 -13.3	10.4 -16.5	11.4 -18.2	12.5 -19.9	14.9 -23.7	16.2 -25.7	17.5 -27.8	20.3 -32.3	21.8 -34.6	23.3 -37.0	30.0 -47.6
	1	20	10.0 -11.6	10.0 -13.0	10.0 -16.0	10.4 -17.6	11.4 -19.4	13.6 -23.0	14.8 -25.0	16.0 -27.0	18.5 -31.4	19.9 -33.7	21.3 -36.0	27.3 -46.3
	1	50	10.0 -11.1	10.0 -12.5	10.0 -15.4	10.0 -17.0	10.0 -18.6	11.9 -22.2	12.9 -24.1	13.9 -26.0	16.1 -30.2	17.3 -32.4	18.5 -34.6	23.8 -44.5
	1	100	10.0 -10.8	10.0 -12.1	10.0 -14.9	10.0 -16.5	10.0 -18.1	10.5 -21.5	11.4 -23.3	12.4 -25.2	14.3 -29.3	15.4 -31.4	16.5 -33.6	21.1 -43.2
	2	10	10.0 -25.1	10.0 -28.2	10.4 -34.8	11.4 -38.3	12.5 -42.1	14.9 -50.1	16.2 -54.3	17.5 -58.7	20.3 -68.1	21.8 -73.1	23.3 -78.2	30.0 -100.5
	2	20	10.0 -22.8	10.0 -25.6	10.0 -31.5	10.4 -34.8	11.4 -38.2	13.6 -45.4	14.8 -49.3	16.0 -53.3	18.5 -61.8	19.9 -66.3	21.3 -71.0	27.3 -91.2
	2	50	10.0 -19.7	10.0 -22.1	10.0 -27.3	10.0 -30.1	10.0 -33.0	11.9 -39.3	12.9 -42.7	13.9 -46.1	16.1 -53.5	17.3 -57.4	18.5 -61.4	23.8 -78.9
	2	100	10.0 -17.4	10.0 -19.5	10.0 -24.1	10.0 -26.6	10.0 -29.1	10.5 -34.7	11.4 -37.6	12.4 -40.7	14.3 -47.2	15.4 -50.6	16.5 -54.2	21.1 -69.6
	3	10	10.0 -25.1	10.0 -28.2	10.4 -34.8	11.4 -38.3	12.5 -42.1	14.9 -50.1	16.2 -54.3	17.5 -58.7	20.3 -68.1	21.8 -73.1	23.3 -78.2	30.0 -100.5
	3	20	10.0 -22.8	10.0 -25.6	10.0 -31.5	10.4 -34.8	11.4 -38.2	13.6 -45.4	14.8 -49.3	16.0 -53.3	18.5 -61.8	19.9 -66.3	21.3 -71.0	27.3 -91.2
	3	50	10.0 -19.7	10.0 -22.1	10.0 -27.3	10.0 -30.1	10.0 -33.0	11.9 -39.3	12.9 -42.7	13.9 -46.1	16.1 -53.5	17.3 -57.4	18.5 -61.4	23.8 -78.9
	3	100	10.0 -17.4	10.0 -19.5	10.0 -24.1	10.0 -26.6	10.0 -29.1	10.5 -34.7	11.4 -37.6	12.4 -40.7	14.3 -47.2	15.4 -50.6	16.5 -54.2	21.1 -69.6
Roof > 30 to 45 degrees	1	10	11.9 -13.0	13.3 -14.6	16.5 -18.0	18.2 -19.8	19.9 -21.8	23.7 -25.9	25.7 -28.1	27.8 -30.4	32.3 -35.3	34.6 -37.8	37.0 -40.5	47.6 -52.0
	1	20	11.6 -12.3	13.0 -13.8	16.0 -17.1	17.6 -18.8	19.4 -20.7	23.0 -24.6	25.0 -26.7	27.0 -28.9	31.4 -33.5	33.7 -35.9	36.0 -38.4	46.3 -49.3
	1	50	11.1 -11.5	12.5 -12.8	15.4 -15.9	17.0 -17.5	18.6 -19.2	22.2 -22.8	24.1 -24.8	26.0 -25.8	30.2 -31.1	32.4 -33.3	34.6 -35.7	44.5 -45.8
	1	100	10.8 -10.8	12.1 -12.1	14.9 -14.9	16.5 -16.5	18.1 -18.1	21.5 -21.5	23.3 -23.3	25.2 -25.2	29.3 -29.3	31.4 -31.4	33.6 -33.6	43.2 -43.2
	2	10	11.9 -15.2	13.3 -17.0	16.5 -21.0	18.2 -23.2	19.9 -25.5	23.7 -30.3	25.7 -32.9	27.8 -35.6	32.3 -41.2	34.6 -44.2	37.0 -47.3	47.6 -60.8
	2	20	11.6 -14.5	13.0 -16.3	16.0 -20.1	17.6 -22.2	19.4 -24.3	23.0 -29.0	25.0 -31.4	27.0 -34.0	31.4 -39.4	33.7 -42.3	36.0 -45.3	46.3 -58.1
	2	50	11.1 -13.7	12.5 -15.3	15.4 -18.9	17.0 -20.8	18.6 -22.9	22.2 -27.2	24.1 -29.5	26.0 -32.0	30.2 -37.1	32.4 -39.8	34.6 -42.5	44.5 -54.6
	2	100	10.8 -13.0	12.1 -14.6	14.9 -18.0	16.5 -19.8	18.1 -21.8	21.5 -25.9	23.3 -28.1	25.2 -30.4	29.3 -35.3	31.4 -37.8	33.6 -40.5	43.2 -52.0
	3	10	11.9 -15.2	13.3 -17.0	16.5 -21.0	18.2 -23.2	19.9 -25.5	23.7 -30.3	25.7 -32.9	27.8 -35.6	32.3 -41.2	34.6 -44.2	37.0 -47.3	47.6 -60.8
	3	20	11.6 -14.5	13.0 -16.3	16.0 -20.1	17.6 -22.2	19.4 -24.3	23.0 -29.0	25.0 -31.4	27.0 -34.0	31.4 -39.4	33.7 -42.3	36.0 -45.3	46.3 -58.1
	3	50	11.1 -13.7	12.5 -15.3	15.4 -18.9	17.0 -20.8	18.6 -22.9	22.2 -27.2	24.1 -29.5	26.0 -32.0	30.2 -37.1	32.4 -39.8	34.6 -42.5	44.5 -54.5
	3	100	10.8 -13.0	12.1 -14.6	14.9 -18.0	16.5 -19.8	18.1 -21.8	21.5 -25.9	23.3 -28.1	25.2 -30.4	29.3 -35.3	31.4 -37.8	33.6 -40.5	43.2 -52.0
Wall	4	10	13.0 -14.1	14.6 -15.8	18.0 -19.5	19.8 -21.5	21.8 -23.6	25.9 -28.1	28.1 -30.5	30.4 -33.0	35.3 -38.2	37.8 -41.0	40.5 -43.9	52.0 -56.4
	4	20	12.4 -13.5	13.9 -15.1	17.2 -18.7	18.9 -20.6	20.8 -22.6	24.7 -26.9	26.8 -29.2	29.0 -31.6	33.7 -36.7	36.1 -39.3	38.7 -42.1	49.6 -54.1
	4	50	11.6 -12.7	13.0 -14.3	16.1 -17.6	17.8 -19.4	19.5 -21.3	23.2 -25.4	25.2 -27.5	27.2 -29.8	31.6 -34.6	33.9 -37.1	36.2 -39.7	46.6 -51.0
	4	100	11.1 -12.2	12.4 -13.6	15.3 -16.8	16.9 -18.5	18.5 -20.4	22.0 -24.2	23.9 -26.3	25.9 -28.4	30.0 -33.0	32.2 -35.4	34.4 -37.8	44.2 -48.6
	5	10	13.0 -17.4	14.6 -19.5	18.0 -24.1	19.8 -26.6	21.8 -29.1	25.9 -34.7	28.1 -37.6	30.4 -40.7	35.3 -47.2	37.8 -50.6	40.5 -54.2	52.0 -69.6
	5	20	12.4 -16.2	13.9 -18.2	17.2 -22.5	18.9 -24.8	20.8 -27.2	24.7 -32.4	26.8 -35.1	29.0 -38.0	33.7 -44.0	36.1 -47.2	38.7 -50.5	49.6 -64.9
	5	50	11.6 -14.7	13.0 -16.5	16.1 -20.3	17.8 -22.4	19.5 -24.6	23.2 -29.3	25.2 -31.8	27.2 -34.3	31.6 -39.8	33.9 -42.7	36.2 -45.7	46.6 -58.7
	5	100	11.1 -13.5	12.4 -15.1	15.3 -18.7	16.9 -20.6	18.5 -22.6	22.0 -26.9	23.9 -29.2	25.9 -31.6	30.0 -36.7	32.2 -39.3	34.4 -42.1	44.2 -54.1

For SI: 1 foot = 304.8 mm, 1 square foot = 0.0929 m², 1 mile per hour = 0.447 m/s.

NOTES: For effective areas between those given above the load may be interpolated, otherwise use the load associated with the lower effective area.

Table values shall be adjusted for height and exposure by multiplying by the adjustment coefficient in Table R301.2(3).

See Figure R301.2(7) for location of zones.

Plus and minus signs signify pressures acting toward and away from the building surfaces.

(i.e., a wall that is typically vertical versus a roof and its slope), the region or zone, as well as the basic wind speed. "Zones" refer to various portions of a structure as illustrated in Figure R301.2(7). Some zones, such as at roof edges and the ends of a wall, represent discontinuities in the exterior building surfaces. Discontinuities interrupt the flow of wind and result in higher wind pressures at these locations. For each combination of these various factors there are two entries in the table, one positive and one negative. The positive number denotes the pressure acting inward or toward the exterior surface of the building; the negative number denotes the pressure acting outward or away from the building surface. The latter is of interest in designing roof elements and their hold-downs that must resist wind uplift.

Unlike the IBC, the IRC does not elaborate on the distinctions between wind loads on the main wind-force -resisting system versus components and cladding wind loads. Instead, compliance with the components and cladding pressures is necessary only where code provisions explicitly refer to this table. This occurs in the establishment of design wind loads for skylights, exterior windows and doors (Sections R301.2.1 and R613.2) and setting the threshold for requiring rafter ties (Sections R611.9, R802.11, and R804.4).

R301.2.1.3 Wind speed conversion. When referenced documents are based on fastest mile wind speeds, the three-second gust basic wind speeds, V_{3s}, of Figure R301.2(4) shall be converted to fastest mile wind speeds, V_{fm}, using Table R301.2.1.3.

❖ Prior to publication of the 2002 edition of ASCE 7, most wind-related code provisions were based on fastest-mile wind velocity, V_{fm}. Because many of the documents referenced in the code still use this criterion as a basis for wind design rather than the three-second gust wind velocities, V_{3s}, of Figure R301.2(4), Table R301.2.1.3 provides the necessary conversion to fastest-mile wind velocities used in those documents. For example, the AF&PA *Wood Frame Construction Manual* and the SSTD 10 standard that are referenced in Section R301.2.1.1 are based on fastest-mile wind velocity.

R301.2.1.4 Exposure category. For each wind direction considered, an exposure category that adequately reflects the characteristics of ground surface irregularities shall be determined for the site at which the building or structure is to be constructed. For a site located in the transition zone between categories, the category resulting in the largest wind forces shall apply. Account shall be taken of variations in ground surface roughness that arise from natural topography and vegetation as well as from constructed features. For any given wind direction, the exposure in which a specific building or other structure is sited shall be assessed as being one of the following categories:

1. Exposure A. Large city centers with at least 50 percent of the buildings having a height in excess of 70 feet (21 336 mm). Use of this exposure category shall be limited to those areas for which terrain representative of Exposure A prevails in the upwind direction for a distance of at least 0.5 mile (0.8 km) or 10 times the height of the building or other structure, whichever is greater. Possible channeling effects or increased velocity pressures due to the building or structure being located in the wake of adjacent buildings shall be taken into account.

2. Exposure B. Urban and suburban areas, wooded areas, or other terrain with numerous closely spaced obstructions having the size of single-family dwellings or larger. Exposure B shall be assumed unless the site meets the definition of another type exposure.

TABLE R301.2(3)
HEIGHT AND EXPOSURE ADJUSTMENT COEFFICIENTS FOR TABLE R301.2(2)

MEAN ROOF HEIGHT	EXPOSURE		
	B	C	D
15	1.00	1.21	1.47
20	1.00	1.29	1.55
25	1.00	1.35	1.61
30	1.00	1.40	1.66
35	1.05	1.45	1.70
40	1.09	1.49	1.74
45	1.12	1.53	1.78
50	1.16	1.56	1.81
55	1.19	1.59	1.84
60	1.22	1.62	1.87

❖ This table provides adjustment factors for the components and cladding wind pressures of Table R301.2(2), which are based on a mean roof height of 30 feet (9144 mm) and wind exposure category B (see the commentary on Section R301.2.1.4 for a discussion of exposure). For mean roof heights and exposure categories that vary from those assumed in Table R301.2(2), the resulting pressures will change accordingly. Thus, adjustments to the pressures are required using the factors given in this table.

3. Exposure C. Open terrain with scattered obstructions, including surface undulations or other irregularities, having heights generally less than 30 feet (9144 mm) extending more than 1,500 feet (457 m) from the building site in any quadrant. This exposure shall also apply to any building located within Exposure B type terrain where the building is directly adjacent to open areas of Exposure C type terrain in any quadrant for a distance of more than 600 feet (183 m). This category includes flat open country, grasslands and shorelines in hurricane prone regions.

4. Exposure D. Flat, unobstructed areas exposed to wind flowing over open water (excluding shorelines in hurricane prone regions) for a distance of at least 1 mile (1.61 km). Shorelines in Exposure D include inland waterways, the Great Lakes and coastal areas of California, Oregon, Washington and Alaska. This exposure shall apply only to those buildings and other structures exposed to the wind coming from over the water. Exposure D extends inland from the shoreline a distance of 1,500 feet (457 m) or 10 times the height of the building or structure, whichever is greater.

❖ Wind loading on structures is a function of site exposure category, which reflects the characteristics of ground surface irregularities and accounts for variations in ground surface roughness that arise from natural topography and vegetation as well as from constructed features. These categories range from Exposure A for minimal exposure to Exposure D for the most severe exposure category. Exposure B should be used unless the site meets the definition for another category of exposure. Exposure B includes urban and suburban areas, wooded areas and other terrain with numerous, closely spaced obstructions having the size of single-family dwellings or larger. The commentary of ASCE 7 contains a detailed discussion of wind exposure categories.

R301.2.2 Seismic provisions. The seismic provisions of this code shall apply to buildings constructed in Seismic Design Categories C, D_0, D_1 and D_2, as determined in accordance with this section.

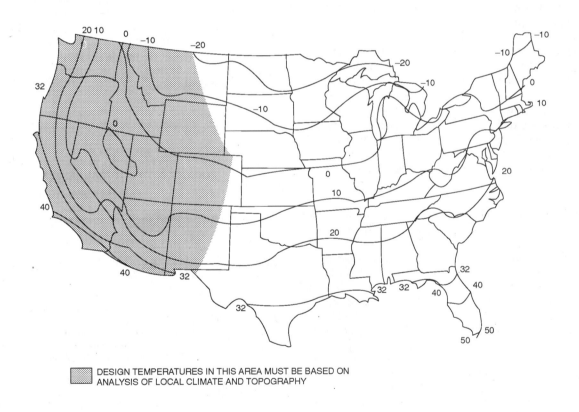

DESIGN TEMPERATURES IN THIS AREA MUST BE BASED ON ANALYSIS OF LOCAL CLIMATE AND TOPOGRAPHY

For SI: °C = [(°F)-32] /1.8.

FIGURE R301.2(1)
ISOLINES OF THE 97$^1/_2$ PERCENT WINTER (DECEMBER, JANUARY AND FEBRUARY) DESIGN TEMPERATURES (°F)

❖ This figure establishes the winter design temperature, which is a criterion for determining the need for dwelling unit heating (Section R303.8) as well as determining the need for freeze protection of piping (Sections M2301.2.5, P2603.6 and P3001.2).

Buildings in Seismic Design Category E shall be designed in accordance with the *International Building Code*, except when the seismic design category is reclassified to a lower seismic design category in accordance with Section R301.2.2.1.

Exception: Detached one- and two-family dwellings located in Seismic Design Category C are exempt from the seismic requirements of this code.

The weight and irregularity limitations of Section R301.2.2.2 shall apply to buildings in all seismic design categories regulated by the seismic provisions of this code. Buildings in Seismic Design Category C shall be constructed in accordance with the additional requirements of Section R301.2.2.3. Buildings in Seismic Design Categories D_0, D_1 and D_2 shall be constructed in accordance with the additional requirements of Section R301.2.2.4

❖ Earthquakes generate internal forces in a structure caused by inertia. This is depicted in Commentary Figure R301.2.2. These forces can cause a building to be distorted and severely damaged. The objective of earthquake-resistant construction is to resist these forces and the resulting distortions.

Providing for the adequate earthquake performance of any structure begins with assessing the degree of hazard at the proposed site. As in the IBC, this is accomplished in the IRC by assigning a seismic design category (see Section R301.2.2.1 and the definition in Chapter 2). These range from Seismic Design Category A, representing the lowest level of seismic hazard, to Seismic Design Category E, representing the highest hazard. Buildings in Seismic Design Category E must be designed in accordance with the IBC, except as allowed in Section R301.2.2.1. Under the IRC, the seismic design category depends on building location according to Figure R301.2(2), unless one of the options permitting a more precise determination is exercised.

The IRC seismic provisions apply to buildings classified as Seismic Design Categories C, D_0, D_1, or D_2. Detached one- and two-family dwellings need to comply with only the seismic requirements of this code when classified as Seismic Design Category D_0, D_1 or D_2. In other words, only townhouses constructed under IRC Section R101.2 need to comply with Seismic Design Category C requirements. This indicates that all buildings classified as Seismic Design Category A or B, as well as detached one- and two-family dwellings classified as Seismic Design Category C, will perform satisfactorily when constructed in accordance with the basic prescriptive requirements of the IRC.

Values for seismic loads on structures are based on a percentage of the dead load, and the IRC prescriptive provisions for structural elements that resist seismic forces are based on an assumed dead load. Generally, the building dead loads as described in Section R301.4 are reflected in the prescriptive design tables throughout the code. The weight and irregularity limitations of Section R301.2.2.2 apply to buildings in the seismic design categories specifically regulated by the code (i.e., SDC C, D_0, D_1, and D_2).

FEMA 232, *Home Builder's Guide to Seismic-Resistant Construction,* is an excellent source of additional information on seismic-resistant construction, particularly as it pertains to residential buildings.

R301.2.2.1 Determination of seismic design category. Buildings shall be assigned a seismic design category in accordance with Figure 301.2(2).

❖ Unlike the IBC, the IRC does not require a calculation to determine the seismic design category of a structure. Instead, the seismic design category is taken directly from Figure R301.2(2). The seismic design categories shown in Figure R301.2(2) are based solely on the Short Period Design Spectral Response Accelerations, S_{DS}, assuming soil Site Class D, as defined in Section 1615.1 of the IBC.

A convenient alternative to the maps of Figure R301.2(2) and the calculations described in Section R301.2.2.1.1 is the Seismic Design Parameters program prepared by USGS and available on CD-ROM from the ICC. Another option is the USGS National Seismic Hazard Mapping Project web site (http://geohazards.cr.usgs.gov/eq/index.html). Either of these allows a computer user to view the seismic hazard map of the IRC in more detail. The Seismic Design Parameters program also provides determination of a site's seismic design category using either the zip code or the longitude and latitude of the site. Because use of longitude and latitude is more precise, this approach is preferable.

TABLE R301.2.1.2
WINDBORNE DEBRIS PROTECTION FASTENING SCHEDULE FOR WOOD STRUCTURAL PANELS[a, b, c, d]

FASTENER TYPE	FASTENER SPACING (inches)		
	Panel span ≤ 4 feet	4 feet < panel span ≤ 6 feet	6 feet < panel span ≤ 8 feet
No. 6 Screws	16″	12″	9″
No. 8 Screws	16″	16″	12″

For SI: 1 inch = 25.4 mm, 1 foot = 304.8 mm, 1 pound = 4.448 N,
1 mile per hour = 0.447 m/s.

a. This table is based on 130 mph wind speeds and a 33-foot mean roof height.
b. Fasteners shall be installed at opposing ends of the wood structural panel. Fasteners shall be located a minimum of 1 inch from the edge of the panel.
c. Fasteners shall be long enough to penetrate through the exterior wall covering and a minimum of $1^1/_4$ inches into wood wall framing and a minimum of $1^1/_4$ inches into concrete block or concrete, and into steel framing a minimum of 3 exposed threads. Fasteners shall be located a minimum of $2^1/_2$ inches from the edge of concrete block or concrete.
d. Where screws are attached to masonry or masonry/stucco, they shall be attached using vibration-resistant anchors having a minimum ultimate withdrawal capacity of 490 pounds.

❖ The fastening schedule shown in this table allows selection of the appropriate fastener size and spacing when the prescriptive alternative is applied to a tested assembly as required by Section R301.2.1.2. Installing the specified fasteners at both ends of the wood structural panel provides the necessary panel support.

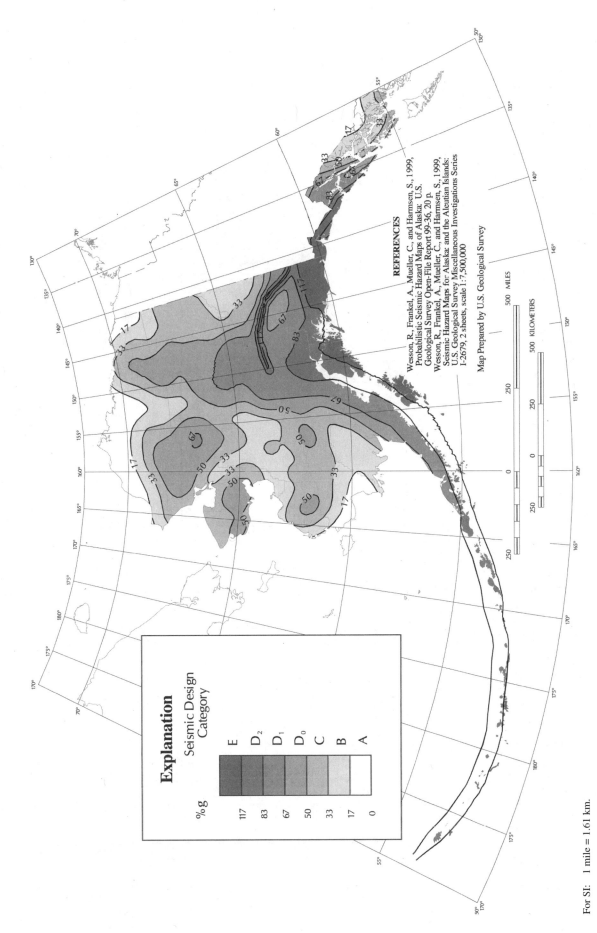

FIGURE R301.2(2)
SEISMIC DESIGN CATEGORIES—SITE CLASS D

(continued)

For SI: 1 mile = 1.61 km.

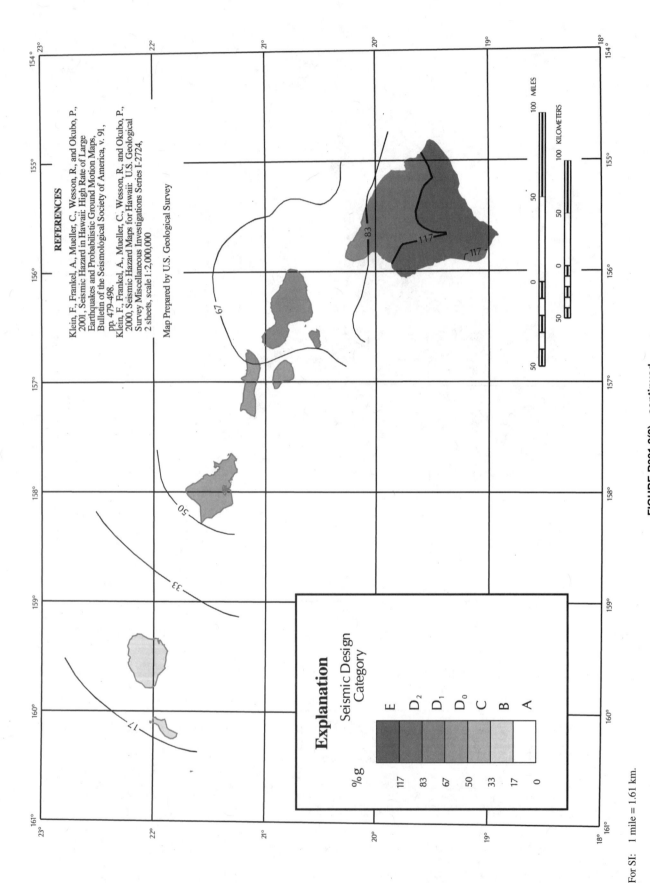

**FIGURE R301.2(2)—continued
SEISMIC DESIGN CATEGORIES—SITE CLASS D**

(continued)

For SI: 1 mile = 1.61 km.

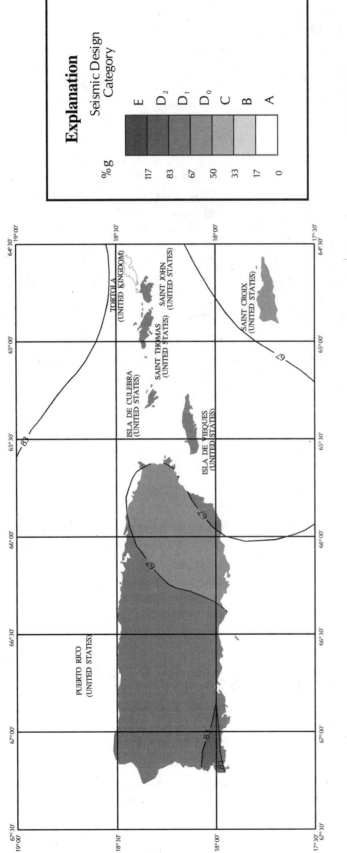

Explanation

Seismic Design Category

%g	
117	E
83	D₂
67	D₁
50	D₀
33	C
17	B
0	A

REFERENCES

Mueller, C., Frankel, A., Petersen, M, and Leyendecker, E., 2003, Documentation for 2003 USGS Seismic Hazard Maps for Puerto Rico and the U.S. Virgin Islands, U.S. Geological Survey Open-File Report 03-379.

Mueller, C., Frankel, A., Petersen, M., and Leyendecker, E., 2004, Seismic-Hazard Maps for Puerto Rico and the U.S. Virgin Island, Sheet 2 - 2% Probability of Exceedance in 50 Years for Peak Horizontal Acceleration and Horizontal Spectral Response Acceleration for 0.2, 0.3, and 1.0 Second Periods U.S. Geological Survey Geologic Investigation Series (in progress).

Map Prepared by U.S. Geological Survey

Scale 1:2,000,000

FIGURE R301.2(2)—continued
SEISMIC DESIGN CATEGORIES—SITE CLASS D

(continued)

For SI: 1 mile = 1.61 km.

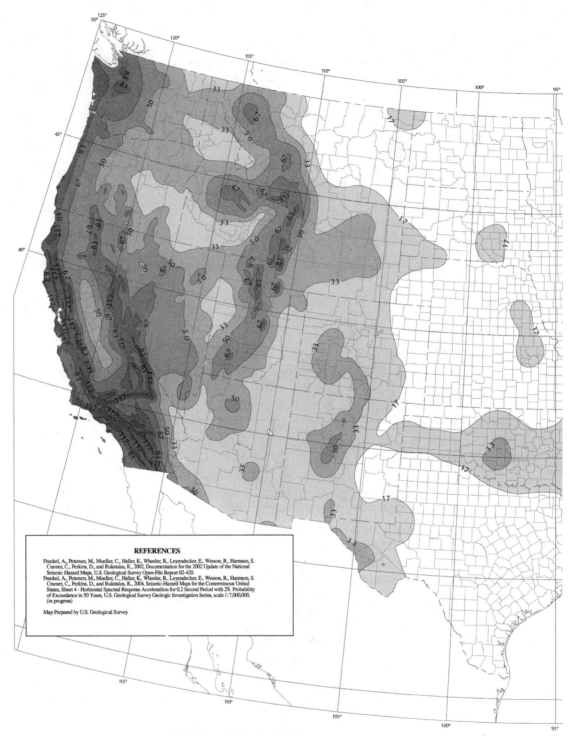

FIGURE R301.2(2)—continued
SEISMIC DESIGN CATEGORIES—SITE CLASS D

(continued)

REFERENCES

Frankel, A., Petersen, M., Mueller, C., Haller, K., Wheeler, R., Leyendecker, E., Wesson, R., Harmsen, S. Cramer, C., Perkins, D., and Rukstales, K., 2002, Documentation for the 2002 Update of the National Seismic Hazard Maps, U.S. Geological Survey Open-File Report 02-420.

Frankel, A., Petersen, M., Mueller, C., Haller, K., Wheeler, R., Leyendecker, E., Wesson, R., Harmsen, S. Cramer, C., Perkins, D., and Rukstales, K., 2004, Seismic-Hazard Maps for the Conterminous United States, Sheet 4 - Horizontal Spectral Response Acceleration for 0.2 Second Period with 2% Probability of Exceedance in 50 Years, U.S. Geological Survey Geologic Investigation Series, scale 1:7,000,000. (in progress)

Map Prepared by U.S. Geological Survey

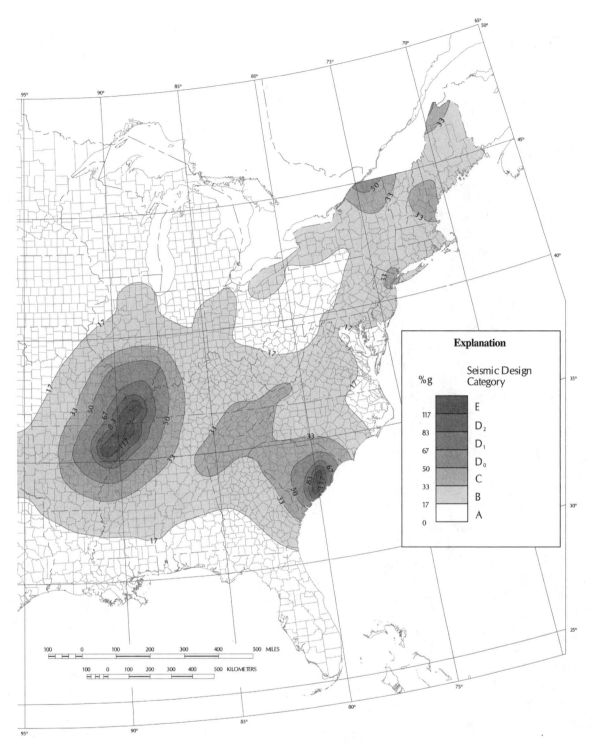

FIGURE R301.2(2)—continued
SEISMIC DESIGN CATEGORIES—SITE CLASS D

❖ This figure establishes the seismic design category for a site as presented in Section R301.2.2.1. It provides the most direct determination of seismic design category in the code. The earthquake-related provisions of the code state requirements as a function of the seismic design category.

For over 50 years, the United States Geological Survey (USGS) has prepared seismic hazard maps that have been used to establish building code provisions for earthquake-resistant construction. The seismic provisions incorporated into the IBC and IRC are largely based on the 1997 National Earthquake Hazards Reduction Program Provisions. The seismic hazard maps from these provisions are the latest USGS ground-motion spectral-response maps. The map incorporated into the IRC is essentially a plot of Short Period Design Spectral Response, S_{DS}, assuming a site class D. In the IRC, the seismic design category is dependent only on the value of S_{DS}. This allows the seismic design category to be read directly from the map in the same manner as the seismic zone was read off a single map in prior building codes.

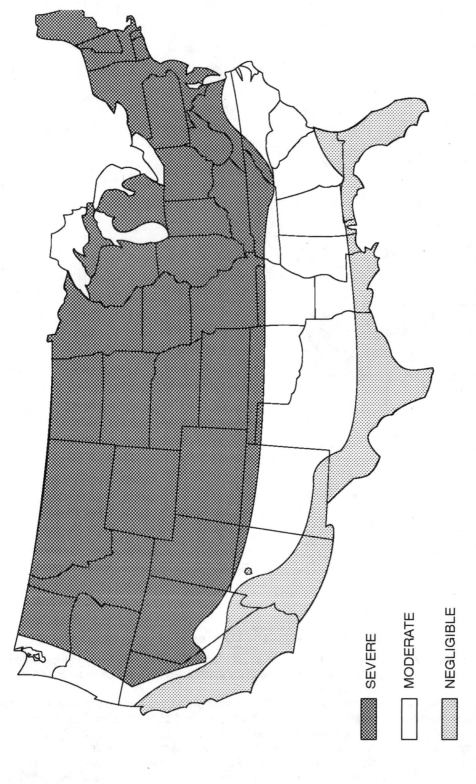

FIGURE R301.2(3)
WEATHERING PROBABILITY MAP FOR CONCRETE

SEVERE

MODERATE

NEGLIGIBLE

a. Alaska and Hawaii are classified as severe and negligible, respectively.
b. Lines defining areas are approximate only. Local conditions may be more or less severe than indicated by region classification. A severe classification is where weather conditions result in significant snowfall combined with extended periods during which there is little or no natural thawing causing deicing salts to be used extensively.

❖ The weathering probability map for concrete in Figure R301.2(3) classifies regions as severe, moderate or negligible. These classifications are based on climatic data, surveys of state transportation departments and field surveys from Portland Cement Association regional offices. The map provides a guideline for the severity of weathering that can be expected for concrete surfaces exposed to weather or de-icing chemicals. As indicated in note "b," the map boundaries are approximate, and local conditions may warrant a classification that is different from that shown on the map.

The weathering index from this map indicates the need to increase the strength of concrete, to add air-entrainment agents to concrete (see Section R402.2) or both. In masonry construction, the index determines the grade of masonry units required in accordance with the ASTM standards listed in Note a to Table R301.2(1).

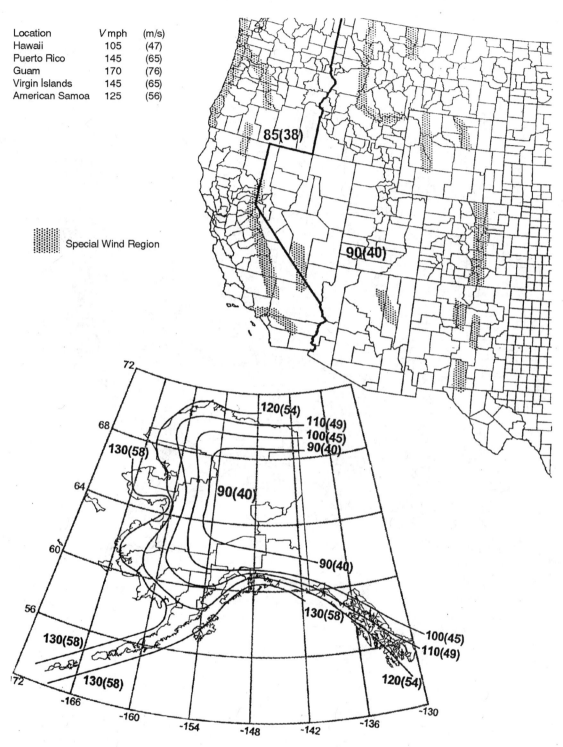

Location	V mph	(m/s)
Hawaii	105	(47)
Puerto Rico	145	(65)
Guam	170	(76)
Virgin Islands	145	(65)
American Samoa	125	(56)

Special Wind Region

For SI: 1 foot = 304.8 mm, 1 mile per hour = 0.447 m/s.

a. Values are nominal design 3-second gust wind speeds in miles per hour at 33 feet above ground for Exposure C category.

b. Linear interpolation between wind contours is permitted.

c. Islands and coastal areas outside the last contour shall use the last wind speed contour of the coastal area.

d. Mountainous terrain, gorges, ocean promontories and special wind regions shall be examined for unusual wind conditions.

e. Enlarged view of Eastern and Southern seaboards are on the following pages.

FIGURE R301.2(4)
BASIC WIND SPEEDS FOR 50-YEAR MEAN RECURRENCE INTERVAL

(continued)

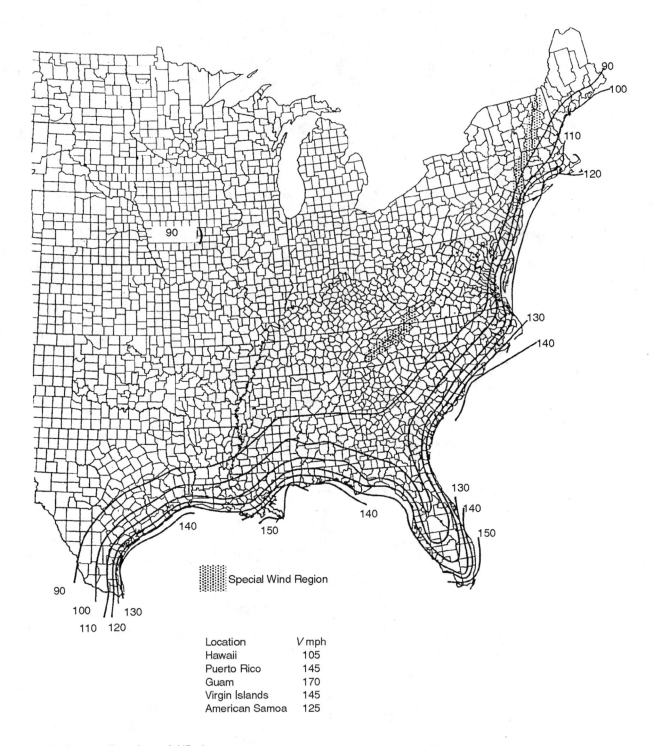

Location	V mph
Hawaii	105
Puerto Rico	145
Guam	170
Virgin Islands	145
American Samoa	125

For SI: 1 foot = 304.8 mm, 1 mile per hour = 0.447 m/s.

a. Values are nominal design 3-second gust wind speeds in miles per hour at 33 feet above ground for Exposure C category.

b. Linear interpolation between wind contours is permitted.

c. Islands and coastal areas outside the last contour shall use the last wind speed contour of the coastal area.

d. Mountainous terrain, gorges, ocean promontories and special wind regions shall be examined for unusual wind conditions.

e. Enlarged view of Eastern and Southern seaboards are on the following pages.

FIGURE R301.2(4)—continued
BASIC WIND SPEEDS FOR 50-YEAR MEAN RECURRENCE INTERVAL

(continued)

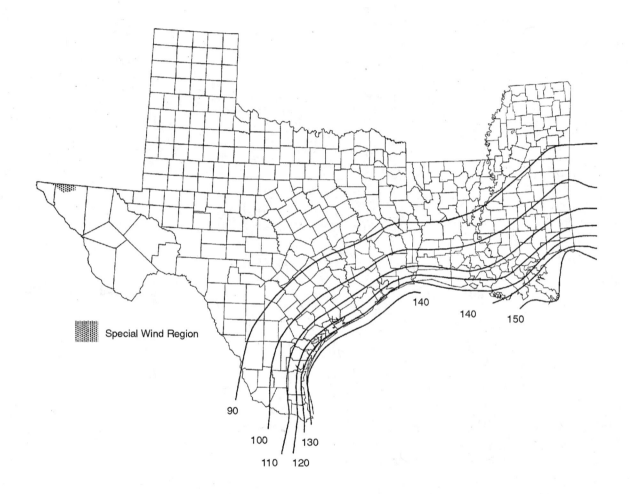

For SI: 1 foot = 304.8 mm, 1 mile per hour = 0.447 m/s.

a. Values are nominal design 3-second gust wind speeds in miles per hour at 33 feet above ground for Exposure C category.

b. Linear interpolation between wind contours is permitted.

c. Islands and coastal areas outside the last contour shall use the last wind speed contour of the coastal area.

d. Mountainous terrain, gorges, ocean promontories and special wind regions shall be examined for unusual wind conditions.

FIGURE R301.2(4)—continued
BASIC WIND SPEEDS FOR 50-YEAR MEAN RECURRENCE INTERVAL

(continued)

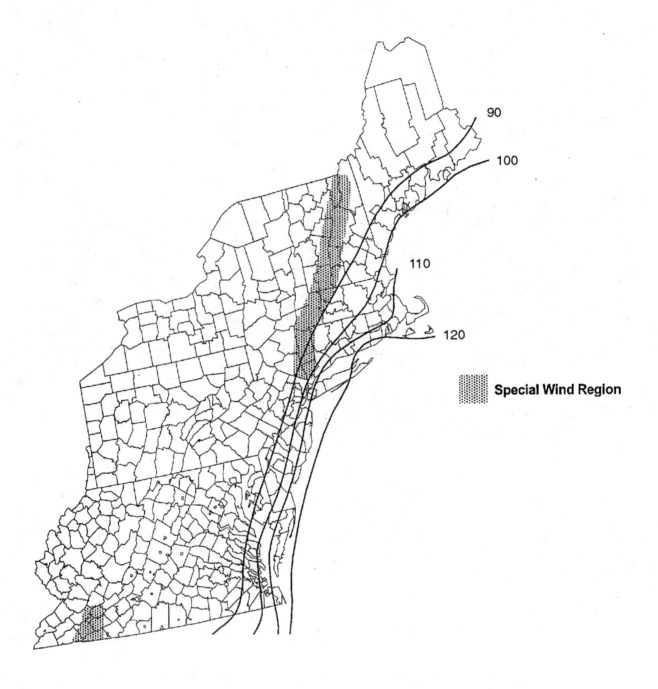

90

100

110

120

Special Wind Region

For SI: 1 foot = 304.8 mm, 1 mile per hour = 0.447 m/s.

a. Values are nominal design 3-second gust wind speeds in miles per hour at 33 feet above ground for Exposure C category.

b. Linear interpolation between wind contours is permitted.

c. Islands and coastal areas outside the last contour shall use the last wind speed contour of the coastal area.

d. Mountainous terrain, gorges, ocean promontories and special wind regions shall be examined for unusual wind conditions.

FIGURE R301.2(4)—continued
BASIC WIND SPEEDS FOR 50-YEAR MEAN RECURRENCE INTERVAL

(continued)

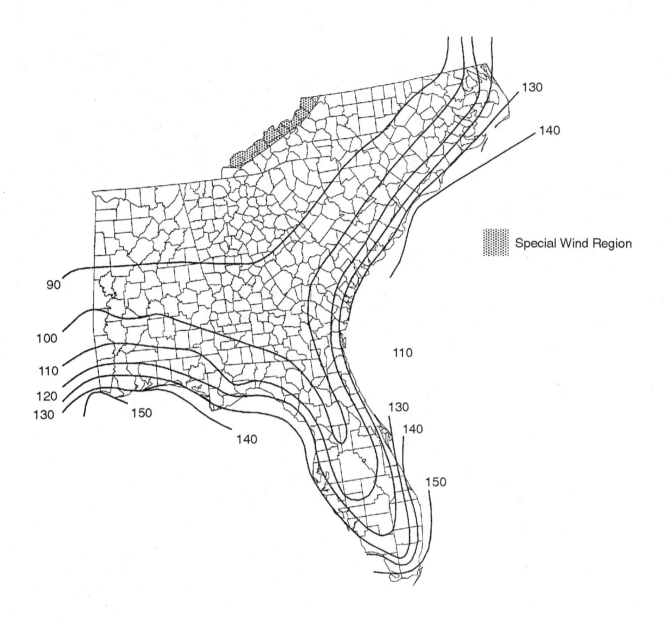

For SI: 1 foot = 304.8 mm, 1 mile per hour = 0.447 m/s.

a. Values are nominal design 3-second gust wind speeds in miles per hour at 33 feet above ground for Exposure C category.

b. Linear interpolation between wind contours is permitted.

c. Islands and coastal areas outside the last contour shall use the last wind speed contour of the coastal area.

d. Mountainous terrain, gorges, ocean promontories and special wind regions shall be examined for unusual wind conditions.

FIGURE R301.2(4)—continued
BASIC WIND SPEEDS FOR 50-YEAR MEAN RECURRENCE INTERVAL

❖ The basic wind-speed map is taken from the ASCE 7 wind provisions and is used to establish the basic wind speed of any particular site. This latest map depicts 3-second gust velocities having a 50-year mean recurrence interval, or a 2-percent annual chance of being exceeded. Because the National Weather Service has phased out the measurement of fastest-mile wind speeds, the wind speed used in design has been revised to a measure of wind speed consistent with current weather data and wind speeds reported by the media.

Although the code does not provide any specific guidance on the topic, one should take note of the areas identified as "special wind regions" as well as other areas described in the note where wind anomalies are observed. As the note explains, these areas must be examined for unusual wind conditions. ASCE 7 discusses the use of regional climatic data in these areas.

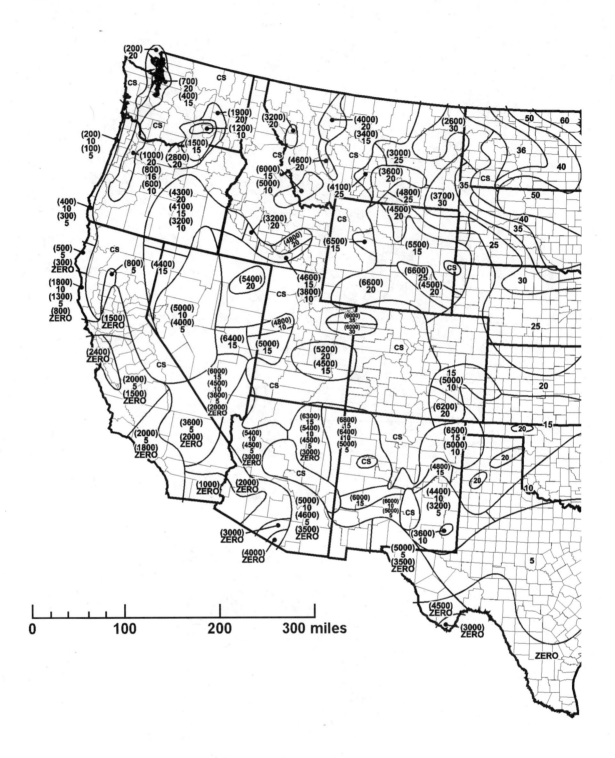

For SI: 1 foot = 304.8 mm, 1 pound per square foot = 0.0479 kPa, 1 mile = 1.61 km.

a. In CS areas, site-specific Case Studies are required to establish ground snow loads. Extreme local variations in ground snow loads in these areas preclude mapping at this scale.

b. Numbers in parentheses represent the upper elevation limits in feet for the ground snow load values presented below. Site-specific cases studies are required to establish ground snow loads at elevations not covered.

FIGURE R301.2(5)
GROUND SNOW LOADS, P_g, FOR THE UNITED STATES (lb/ft^2)

(continued)

For SI: 1 foot = 304.8 mm, 1 pound per square foot = 0.0479 kPa.

FIGURE R301.2(5)—continued
GROUND SNOW LOADS, P_g, FOR THE UNITED STATES (lb/ft²)

❖ The ground snow-load map is taken from the ASCE 7 snow-load provisions and is based on statistical analysis of ground snow data. This map provides ground snow loads for direct use in the prescriptive provisions and tables of the code.

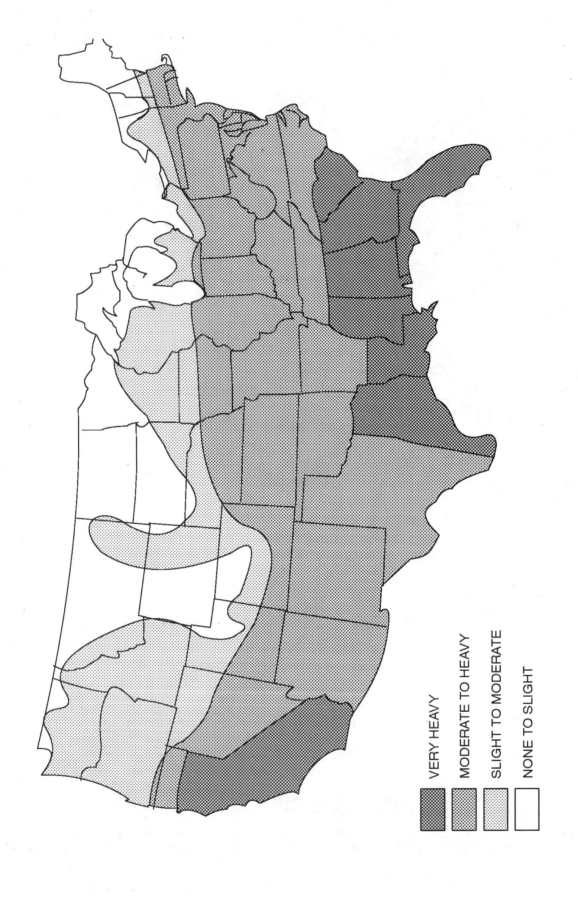

VERY HEAVY

MODERATE TO HEAVY

SLIGHT TO MODERATE

NONE TO SLIGHT

NOTE: Lines defining areas are approximate only, local conditions may be more or less severe than indicated by the region classification.

FIGURE R301.2(6)
TERMITE INFESTATION PROBABILITY MAP

❖ This figure establishes the potential for termite damage in areas of the continental United States. The boundaries for the various classifications are approximate, and local experience should be relied on either to affirm or to modify the classification determined from this map.

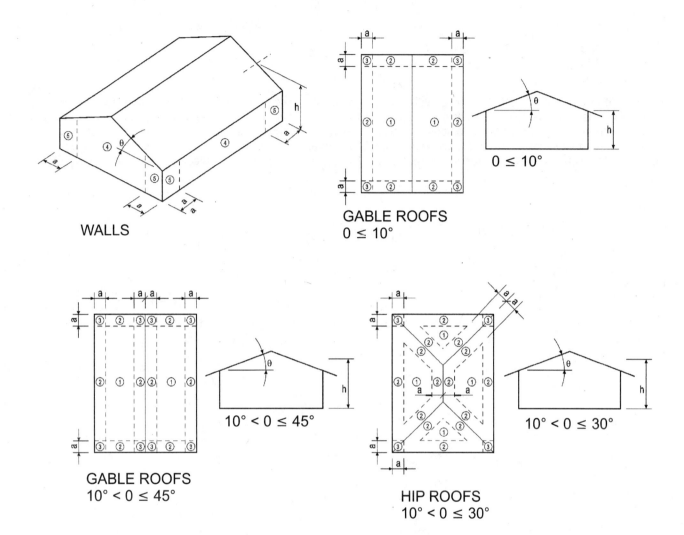

WALLS

GABLE ROOFS
$0 \leq 10°$

$0 \leq 10°$

GABLE ROOFS
$10° < 0 \leq 45°$

$10° < 0 \leq 45°$

HIP ROOFS
$10° < 0 \leq 30°$

$10° < 0 \leq 30°$

For SI: 1 foot = 304.8 mm, 1 degree = 0.0175 rad.
NOTE: a = 4 feet in all cases.

FIGURE R301.2(7)
COMPONENT AND CLADDING PRESSURE ZONES

❖ This figure depicts the zones for the components and cladding wind pressures that are given in Table R301.2(2). See the commentary to this table for a discussion of components and cladding.

TABLE R301.2.1.3
EQUIVALENT BASIC WIND SPEEDS[a]

3-second gust, V_{3s}	85	90	100	105	110	120	125	130	140	145	150	160	170
Fastest mile, V_{fm}	71	76	85	90	95	104	109	114	123	128	133	142	152

For SI: 1 mile per hour = 0.447 m/s.

a. Linear interpolation is permitted.

❖ Prior to publication of the latest edition of ASCE 7, most wind-related code provisions were based on fastest-mile wind velocity, V_{fm}. Because many of the documents referenced in the code still use this criterion as a basis for wind design rather than the three-second gust wind velocities, V_{3s}, of Figure R301.2(4), Table R301.2.1.3 provides the necessary conversion to fastest-mile wind velocities used in those documents. For example, the AF&PA *Wood Frame Construction Manual* and the SSTD 10 standard that are referenced in Section R301.2.1.1 are based on fastest-mile wind velocity.

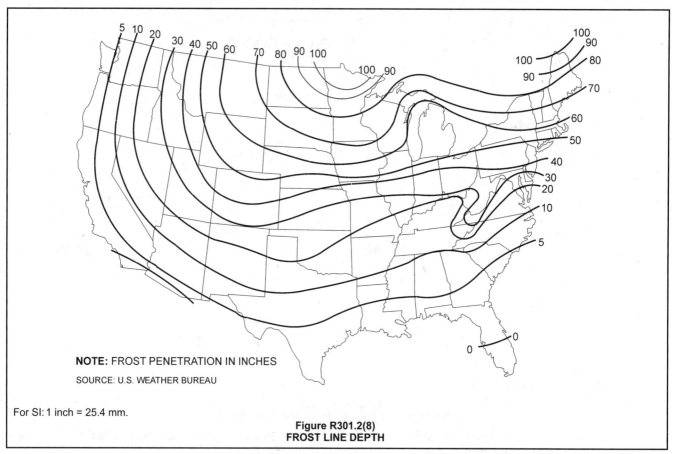

NOTE: FROST PENETRATION IN INCHES

SOURCE: U.S. WEATHER BUREAU

For SI: 1 inch = 25.4 mm.

Figure R301.2(8)
FROST LINE DEPTH

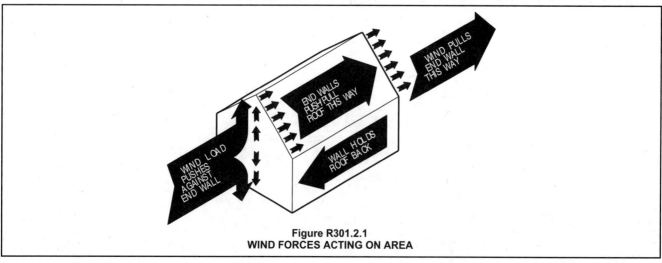

Figure R301.2.1
WIND FORCES ACTING ON AREA

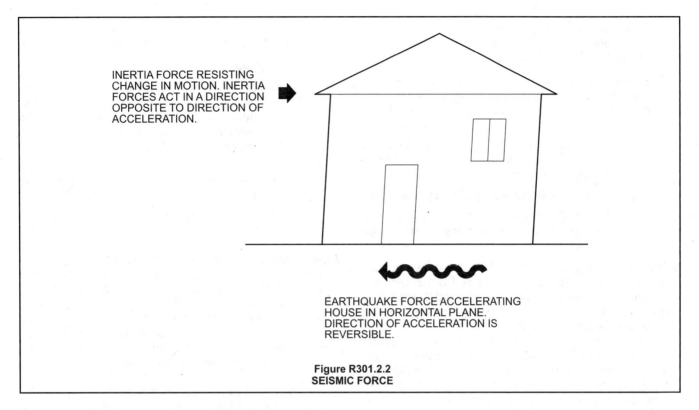

INERTIA FORCE RESISTING CHANGE IN MOTION. INERTIA FORCES ACT IN A DIRECTION OPPOSITE TO DIRECTION OF ACCELERATION.

EARTHQUAKE FORCE ACCELERATING HOUSE IN HORIZONTAL PLANE. DIRECTION OF ACCELERATION IS REVERSIBLE.

Figure R301.2.2
SEISMIC FORCE

R301.2.2.1.1 Alternate determination of seismic design category. The Seismic Design Categories and corresponding Short Period Design Spectral Response Accelerations, S_{DS} shown in Figure R301.2(2) are based on soil Site Class D, as defined in Section 1615.1.1 of the *International Building Code*. If soil conditions are other than Site Class D, the Short Period Design Spectral Response Acceleration, S_{DS}, for a site can be determined according to Section 1615.1 of the *International Building Code*. The value of S_{DS} determined according to Section 1615.1 of the *International Building Code* is permitted to be used to set the seismic design category according to Table R301.2.2.1.1, and to interpolate between values in Tables R602.10.1, R603.7, and other seismic design requirements of this code.

❖ In recognition that there may be an advantage to determining a building's seismic design category using the more detailed procedure of the IBC, the IRC allows this as an option. If soil conditions are other than Site Class D, for instance, the Short Period Design Spectral Response Acceleration (S_{DS}) for a site may be determined using Section 1615.1 of the IBC. This value of S_{DS} can then be used to establish the seismic design category according to Table R301.2.2.1.1. Determining S_{DS} in this manner, Although requiring more work, will provide the benefit of reducing the amount of wall bracing that would otherwise be required by the provisions in Chapter 6, based solely on seismic design category. This is accomplished, in Table R602.10.1 for instance, by interpolating between the tabulated values using the values of S_{DS} determined as described above.

TABLE R301.2.2.1.1
SEISMIC DESIGN CATEGORY DETERMINATION

CALCULATED S_{DS}	SEISMIC DESIGN CATEGORY
$S_{DS} \leq 0.17g$	A
$0.17g < S_{DS} \leq 0.33g$	B
$0.33g < S_{DS} \leq 0.50g$	C
$0.50g < S_{DS} \leq 0.67g$	D_0
$0.67g < S_{DS} \leq 0.83g$	D_1
$0.83g < S_{DS} \leq 1.17g$	D_2
$1.17g < S_{DS}$	E

❖ This table lists the seismic design category for ranges of Short Period Design Spectral Response Acceleration, S_{DS}. These numbers correspond exactly to the legend in the Seismic Design Category Map, Figure R301.2(2). Use of this table is necessary only if the alternative seismic design category determination is used.

R301.2.2.1.2 Alternative determination of Seismic Design Category E. Buildings located in Seismic Design Category E in accordance with Figure R301.2(2) are permitted to be reclassified as being in Seismic Design Category D_2 provided one of the following is done:

1. A more detailed evaluation of the seismic design category is made in accordance with the provisions and maps of the *International Building Code*. Buildings located in Seismic Design Category E per Table R301.2.2.1.1, but

located in Seismic Design Category D per the *International Building Code,* may be designed using the Seismic Design Category D$_2$ requirements of this code.

2. Buildings located in Seismic Design Category E that conform to the following additional restrictions are permitted to be constructed in accordance with the provisions for Seismic Design Category D$_2$ of this code:

   2.1. All exterior shear wall lines or braced wall panels are in one plane vertically from the foundation to the uppermost story.

   2.2. Floors shall not cantilever past the exterior walls.

   2.3. The building is within all of the requirements of Section R301.2.2.2.2 for being considered as regular.

❖ The IRC permits construction of buildings classified as Seismic Design Category E in Figure R301.2(2) in accordance with the provisions for Seismic Design Category D$_2$ if the building has no irregular features as described in Section R301.2.2.2.2. In addition, out-of-plane offsets in the exterior shear-wall lines or braced wall panels are not allowed, and floors are not permitted to cantilever beyond the exterior walls. In other words, this reclassification to a lower seismic design category is limited to very regular, "box-like" structures.

Also, a building classified as Seismic Design Category E using Figure R301.2(2) may possibly be reclassified to Seismic Design Category D$_2$ if a more detailed evaluation in accordance with the provisions of the *IBC* is undertaken. If application of the IBC provisions results in a classification of Seismic Design Category D, then the building may be constructed as Seismic Design Category D$_2$ under the IRC.

R301.2.2.2 Seismic limitations. The following limitations apply to buildings in all Seismic Design Categories regulated by the seismic provisions of this code.

❖ This section establishes the limitations for weights, irregularity and structures in Seismic Design Categories C, D$_0$, D$_1$ and D$_2$.

R301.2.2.2.1 Weights of materials. Average dead loads shall not exceed 15 pounds per square foot (720 Pa) for the combined roof and ceiling assemblies (on a horizontal projection) or 10 pounds per square foot (480 Pa) for floor assemblies, except as further limited by Section R301.2.2. Dead loads for walls above grade shall not exceed:

1. Fifteen pounds per square foot (720 Pa) for exterior light-frame wood walls.

2. Fourteen pounds per square foot (670 Pa) for exterior light-frame cold-formed steel walls.

3. Ten pounds per square foot (480 Pa) for interior light-frame wood walls.

4. Five pounds per square foot (240 Pa) for interior light-frame cold-formed steel walls.

5. Eighty pounds per square foot (3830 Pa) for 8-inch-thick (203 mm) masonry walls.

6. Eighty-five pounds per square foot (4070 Pa) for 6-inch-thick (152 mm) concrete walls.

Exceptions:

1. Roof and ceiling dead loads not exceeding 25 pounds per square foot (1190 Pa) shall be permitted provided the wall bracing amounts in Chapter 6 are increased in accordance with Table R301.2.2.2.1.

2. Light-frame walls with stone or masonry veneer shall be permitted in accordance with the provisions of Sections R702.1 and R703.

3. Fireplaces and chimneys shall be permitted in accordance with Chapter 10.

❖ As stated in Section R301.2.2, the dead loads of structures in seismic design categories regulated by the code must be within the limits listed in this section. Exceeding these limits would necessitate a design of the lateral-force-resisting system as described in Section R301.1.2. In some cases, these dead load limits are more restrictive than the dead load allowances in the prescriptive tables of the code [e.g., Table R802.5(1) for rafters and Table R502.3.1(1) for floor joists].

Exception 1 permits the use of roof/ceiling assemblies that exceed the 15 psf (720 Pa) dead load limit if additional wall bracing is installed as determined using the adjustment factors of Table R301.2.2.2.1. This would permit the use of heavier roof coverings such as tile. The additional wall bracing compensates for the additional seismic force that would be attributed to the added dead load.

Exception 2 directly references Sections R702.1 and R703 specifically addresses interior stone and masonry veneer and Section 703 contains the requirements for the support and anchorage of masonry veneer.

Exception 3 references Chapter 10 for the seismic limitations for chimneys and fireplaces. Seismic reinforcing requirements for masonry fireplaces are found in Section R1001.3 and seismic anchorage of masonry fireplaces is covered in Section R1001.4.

Seismic reinforcing of masonry chimneys is addressed in Section R1003.2.

TABLE R301.2.2.2.1
**WALL BRACING ADJUSTMENT FACTORS BY
ROOF COVERING DEAD LOAD**[a]

WALL SUPPORTING	ROOF/CEILING DEAD LOAD 15 psf or less	ROOF/CEILING DEAD LOAD 25 psf
Roof only	1.0	1.2
Roof plus one story	1.0	1.1

For SI: 1 pound per square foot = 0.049 kPa.
a. Linear interpolation shall be permitted.

❖ The adjustment factors in this table must be used to increase the amount of wall bracing if the exception is used from Section R301.2.2.2.1, which allows the installation of a roof/ceiling assembly exceeding the 15 psf (720 Pa) dead load limit. The resulting additional wall bracing accounts for the additional seismic forces to avoid overloading the lateral force resisting system.

R301.2.2.2.2 Irregular buildings. Prescriptive construction as regulated by this code shall not be used for irregular structures located in Seismic Design Categories C, D₀, D₁ and D₂. Irregular portions of structures shall be designed in accordance with accepted engineering practice to the extent the irregular features affect the performance of the remaining structural system. When the forces associated with the irregularity are resisted by a structural system designed in accordance with accepted engineering practice, design of the remainder of the building shall be permitted using the provisions of this code. A building or portion of a building shall be considered to be irregular when one or more of the following conditions occur:

1. When exterior shear wall lines or braced wall panels are not in one plane vertically from the foundation to the uppermost story in which they are required.

   Exception: For wood light-frame construction, floors with cantilevers or setbacks not exceeding four times the nominal depth of the wood floor joists are permitted to support braced wall panels that are out of plane with braced wall panels below provided that:

   1. Floor joists are nominal 2 inches by 10 inches (51 mm by 254 mm) or larger and spaced not more than 16 inches (406 mm) on center.

   2. The ratio of the back span to the cantilever is at least 2 to 1.

   3. Floor joists at ends of braced wall panels are doubled.

   4. For wood-frame construction, a continuous rim joist is connected to ends of all cantilever joists. When spliced, the rim joists shall be spliced using a galvanized metal tie not less than 0.058 inch (1.5 mm) (16 gage) and 1¹/₂ inches (38 mm) wide fastened with six 16d nails on each side of the splice or a block of the same size as the rim joist of sufficient length to fit securely between the joist space at which the splice occurs fastened with eight 16d nails on each side of the splice; and

   5. Gravity loads carried at the end of cantilevered joists are limited to uniform wall and roof loads and the reactions from headers having a span of 8 feet (2438 mm) or less.

2. When a section of floor or roof is not laterally supported by shear walls or braced wall lines on all edges.

   Exception: Portions of floors that do not support shear walls or braced wall panels above, or roofs, shall be permitted to extend no more than 6 feet (1829 mm) beyond a shear wall or braced wall line.

3. When the end of a braced wall panel occurs over an opening in the wall below and ends at a horizontal distance greater than 1 foot (305 mm) from the edge of the open-

ing. This provision is applicable to shear walls and braced wall panels offset in plane and to braced wall panels offset out of plane as permitted by the exception to Item 1 above.

   Exception: For wood light-frame wall construction, one end of a braced wall panel shall be permitted to extend more than 1 foot (305 mm) over an opening not more than 8 feet (2438 mm) wide in the wall below provided that the opening includes a header in accordance with the following:

   1. The building width, loading condition and framing member species limitations of Table R502.5(1) shall apply and

   2. Not less than one 2×12 or two 2×10 for an opening not more than 4 feet (1219 mm) wide or

   3. Not less than two 2×12 or three 2×10 for an opening not more than 6 feet (1829 mm) wide or

   4. Not less than three 2×12 or four 2×10 for an opening not more than 8 feet (2438 mm) wide and

   5. The entire length of the braced wall panel does not occur over an opening in the wall below.

4. When an opening in a floor or roof exceeds the lesser of 12 feet (3657 mm) or 50 percent of the least floor or roof dimension.

5. When portions of a floor level are vertically offset.

   Exceptions:

   1. Framing supported directly by continuous foundations at the perimeter of the building.

   2. For wood light-frame construction, floors shall be permitted to be vertically offset when the floor framing is lapped or tied together as required by Section R502.6.1.

6. When shear walls and braced wall lines do not occur in two perpendicular directions.

7. When stories above-grade partially or completely braced by wood wall framing in accordance with Section R602 or steel wall framing in accordance with Section R603 include masonry or concrete construction.

   Exception: Fireplaces, chimneys and masonry veneer as permitted by this code.

When this irregularity applies, the entire story shall be designed in accordance with accepted engineering practice.

❖ Conventional light-frame construction typically allows cantilevers, offsets, etc. within certain limits in order to accommodate common design features and options. These features are not well suited to resisting earthquake forces. This becomes more of a concern in areas of higher seismic hazard. Where this is the case, it is preferable that the building, or more importantly the lateral-force-resisting system, be more "box-like" or regular.

Accordingly, this section identifies building features that are irregular and are not permitted under the concrete or conventional light-frame construction provisions in Seismic Design Categories C, D$_0$, D$_1$ and D$_2$. These irregular portions of structures must be designed in accordance with accepted engineering practice. A portion of a building is classified as "irregular" when one of the several conditions described in the code occurs.

Out-of-plane offsets in exterior braced wall lines have been recognized as a factor in buildings damaged by earthquakes. Where an offset occurs, the earthquake forces must be transferred through the floor framing. Thus, Item 1 stipulates that exterior shear-wall lines or braced wall panels that are not in one plane vertically from the foundation to the uppermost story in which they are required constitute an irregularity as illustrated in Commentary Figure R301.2.2.2.2(1). The exception, illustrated in Commentary Figure R301.2.2.2.2(2), allows out-of-plane offsets not exceeding four times the nominal depth of the wood floor joists supporting the braced wall line above. There are five additional limitations that must be met to assure the adequacy of the supporting floor

framing. Use of this exception is limited to light-frame wood construction.

Typically, braced wall lines are located around the perimeter of floors and walls where forces are greatest. As Item 2 indicates, an irregularity occurs where a portion of a floor or roof is not laterally supported by shear walls or braced wall lines on all edges. This is shown in Commentary Figure R301.2.2.2.2(3). The exception permits a floor or roof to extend beyond a braced wall line as illustrated in Commentary Figure R301.2.2.2.2(4).

Where the end of a braced wall panel occurs over an opening in the wall below, an irregularity will be present if the panel extends a horizontal distance greater than 1 foot (305 mm) from the edge of the opening. As Commentary Figure R301.2.2.2.2(5) illustrates, this applies to braced wall panels offset out of plane as well as well as in plane. An exception for wood light-framed construction permits a braced wall panel to extend more than 1 foot (305 mm) over an opening in the wall below if the opening includes a header in accordance with this exception, and the entire length of the braced wall panel does not occur over the opening.

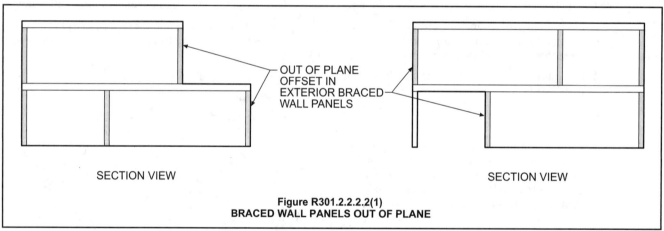

SECTION VIEW   **SECTION VIEW**

Figure R301.2.2.2.2(1)
BRACED WALL PANELS OUT OF PLANE

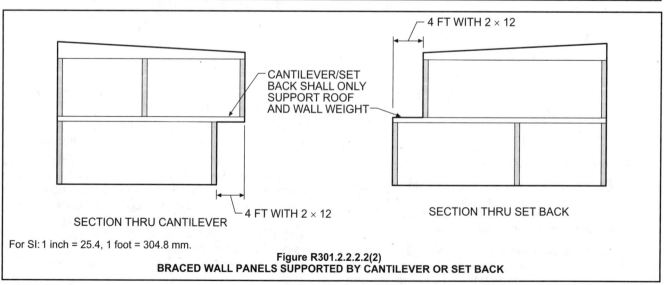

SECTION THRU CANTILEVER   **SECTION THRU SET BACK**

For SI: 1 inch = 25.4, 1 foot = 304.8 mm.

Figure R301.2.2.2.2(2)
BRACED WALL PANELS SUPPORTED BY CANTILEVER OR SET BACK

Openings in floors are required in most buildings to provide for stairways, equipment chases, etc. If the size of the opening is large compared to the size of the floor diaphragm, earthquake forces may overstress portions of the diaphragm. Because this could result in premature failure of the diaphragm and thus constitute an incomplete load path, the code limits the size of such openings allowed under conventional light-frame construction. Commentary Figure R301.2.2.2.2(6) illustrates the irregularity created by an opening in a floor or roof that exceeds the lesser of 12 feet (3657 mm) or 50 percent of the least floor or roof dimension.

Commentary Figure R301.2.2.2.2(7) illustrates Item 5. An irregularity results where portions of a floor level are vertically offset. This limitation does not apply to framing that is supported directly by continuous foundations at the perimeter of the building. Also, in light-frame wood construction, floors may be vertically offset if the floor framing is lapped or tied together as required by Section R502.6.1.

More often than not, the shape of conventional light-frame buildings is regular with wall lines perpendicular to one another. Commentary Figure R301.2.2.2.2(8) depicts a non-orthogonal system. This type of irregularity occurs where shear walls or braced wall lines are not oriented perpendicular to each other.

In addition, an irregularity occurs if shear walls or braced wall lines are constructed of dissimilar bracing systems on any story level above grade. This requirement applies to the underlying framing system material, such as steel-framed shear walls or wood-framed braced wall panels combined with masonry or concrete construction in the same story.

Where this condition occurs, the entire story must be designed in accordance with accepted engineering practice.

Masonry or concrete fireplaces, chimneys and masonry veneer can be combined with steel or wood framing and not be considered to be irregular.

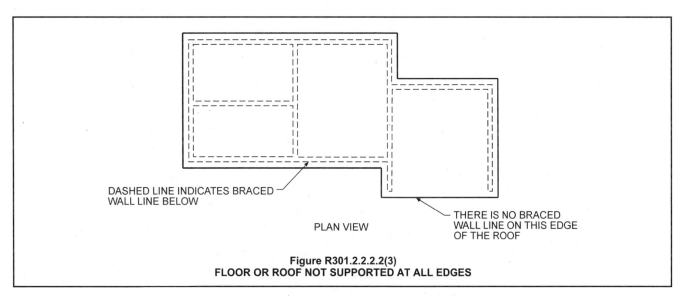

DASHED LINE INDICATES BRACED WALL LINE BELOW

THERE IS NO BRACED WALL LINE ON THIS EDGE OF THE ROOF

PLAN VIEW

Figure R301.2.2.2.2(3)
FLOOR OR ROOF NOT SUPPORTED AT ALL EDGES

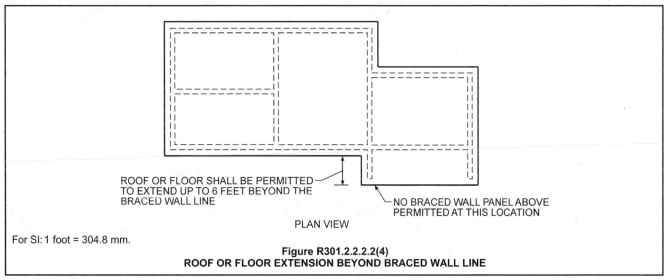

ROOF OR FLOOR SHALL BE PERMITTED TO EXTEND UP TO 6 FEET BEYOND THE BRACED WALL LINE

NO BRACED WALL PANEL ABOVE PERMITTED AT THIS LOCATION

PLAN VIEW

For SI: 1 foot = 304.8 mm.

Figure R301.2.2.2.2(4)
ROOF OR FLOOR EXTENSION BEYOND BRACED WALL LINE

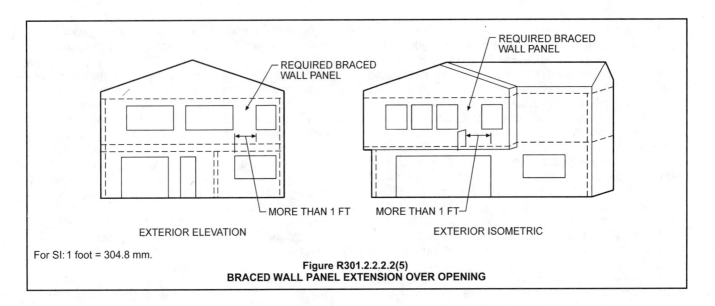

For SI: 1 foot = 304.8 mm.

Figure R301.2.2.2.2(5)
BRACED WALL PANEL EXTENSION OVER OPENING

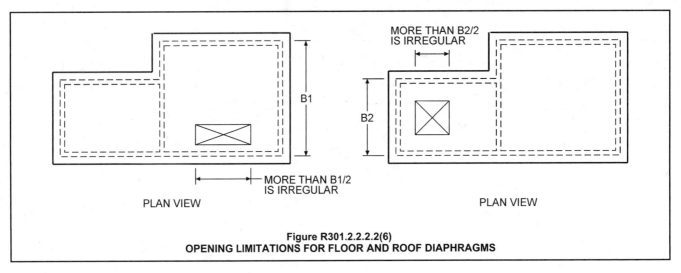

Figure R301.2.2.2.2(6)
OPENING LIMITATIONS FOR FLOOR AND ROOF DIAPHRAGMS

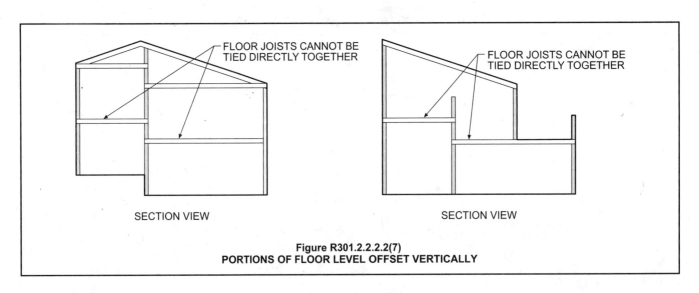

Figure R301.2.2.2.2(7)
PORTIONS OF FLOOR LEVEL OFFSET VERTICALLY

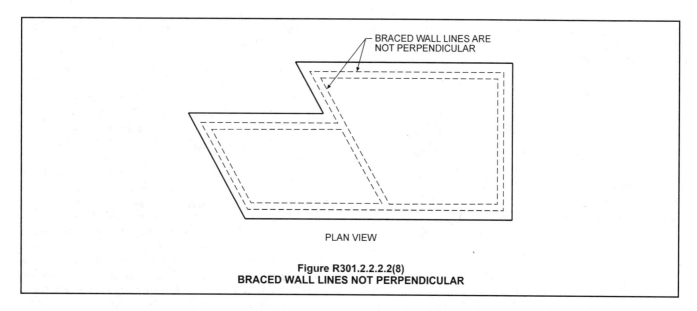

Figure R301.2.2.2.2(8)
BRACED WALL LINES NOT PERPENDICULAR

R301.2.2.3 Seismic Design Category C. Structures assigned to Seismic Design Category C shall conform to the requirements of this section.

❖ This section establishes the limitations for structures in Seismic Design Category C.

R301.2.2.3.1 Stone and masonry veneer. Stone and masonry veneer shall comply with the requirements of Sections R702.1 and R703.

❖ This section limits the height of stone and masonry veneer so the bracing capacity will not be exceeded during an earthquake. This section references Sections R702.1 and R703. Table R703.7(1) and Table R703.7(2) provide specific information on maximum allowed height of veneer and the maximum allowed thickness and weight of veneer.

R301.2.2.3.2 Masonry construction. Masonry construction shall comply with the requirements of Section R606.11.2.

❖ This is a cross reference to the section applicable to masonry seismic requirements.

R301.2.2.3.3 Concrete construction. Concrete construction shall comply with the requirements of Section R611 or R612.

❖ This is a cross reference to the section applicable to concrete seismic requirements.

R301.2.2.4 Seismic Design Categories D_0, D_1 and D_2. Structures assigned to Seismic Design Categories D_0, D_1 and D_2 shall conform to the requirements for Seismic Design Category C and the additional requirements of this section.

❖ Seismic requirements are cumulative as the seismic design category becomes higher. A structure that is assigned to Seismic Design Category D_0, D_1 or D_2 must comply with all the seismic requirements for Seismic Design Category C structures, in addition to those for seismic Design Categories D_0, D_1 and D_2.

R301.2.2.4.1 Height limitations. Wood framed buildings shall be limited to three stories above grade or the limits given in Table R602.10.1. Cold-formed steel framed buildings shall be limited to two stories above grade in accordance with COFS/PM. Mezzanines as defined in Section 202 shall not be considered as stories.

❖ This section limits the building height so the lateral-force-resisting elements will not be overloaded during an earthquake. It provides a cross reference to other sections that limit the height of structures. Wood framed buildings are generally limited to three stories except in Seismic Design Category D_2, where the limit is two stories in accordance with Table R602.10.1. Floor levels meeting the limitations for mezzanines (see the definition in Chapter 2) need not be counted as a story.

R301.2.2.4.2 Stone and masonry veneer. Stone and masonry veneer shall comply with the requirements of Sections R702.1 and R703.

❖ Stone and masonry veneer are addressed in Chapter 7 of this code. Table R703.7(1) lists the stone and masonry veneer limitations based on seismic category.

R301.2.2.4.3 Masonry construction. Masonry construction in Seismic Design Categories D_0 and D_1 shall comply with the requirements of Section R606.11.3. Masonry construction in Seismic Design Category D_2 shall comply with the requirements of Section R606.11.4.

❖ This section provides the cross reference to the masonry sections in this code for the seismic requirements.

R301.2.2.4.4 Concrete construction. Buildings with above-grade concrete walls shall be in accordance with Section R611, R612, or designed in accordance with accepted engineering practice.

❖ This section provides a cross reference to the prescriptive provisions for concrete construction in Seismic Design Categories D_0, D_1 and D_2. As an alternative, concrete construction may be designed in accordance with accepted engineering practice.

R301.2.2.4.5 Cold-formed steel framing in Seismic Design Categories D₀, D₁ and D₂. In Seismic Design Categories D_0, D_1 and D_2 in addition to the requirements of this code, cold-formed steel framing shall comply with the requirements of COFS/PM.

❖ The prescriptive requirements for cold-formed steel framing in Seismic Design Category C is contained in this code. In addition to the seismic requirements found in this code, the cold-formed steel framing must comply with AISI, *Standard for Cold-Formed Steel Framing—Prescriptive Method for One- and Two-family Dwellings* (COFS/PM) with Supplement.

R301.2.3 Snow loads. Wood framed construction, cold-formed steel framed construction and masonry and concrete construction in regions with ground snow loads 70 pounds per square foot (3.35 kPa) or less, shall be in accordance with Chapters 5, 6 and 8. Buildings in regions with ground snow loads greater than 70 pounds per square foot (3.35 kPa) shall be designed in accordance with accepted engineering practice.

❖ This section specifies the maximum ground snow load where the prescriptive provisions in Chapters 5, 6 and 8 apply.

The prescriptive provisions for floors, walls and roofs in Chapters 5, 6 and 8 apply in regions with ground snow loads of 70 psf (3.35 kPa) or less. Structures in regions with ground snow loads exceeding 70 psf (3.35 kPa) are beyond the limitations of these prescriptive provisions and, as stated in Section R301.1, a design would be required for all elements that carry snow loads. The ground snow load is to be determined using Figure R301.2(5) or by site-specific case studies as discussed in Section 7.0 of ASCE 7.

R301.2.4 Floodplain construction. Buildings and structures constructed in whole or in part in flood hazard areas (including A or V Zones) as established in Table R301.2(1) shall be designed and constructed in accordance with Section R324.

Exception: Buildings and structures located in whole or in part in identified floodways as established in Table R301.2(1) shall be designed and constructed as stipulated in the *International Building Code.*

❖ See IBC Section 1612.2 for the definitions of the flood terminology used in this section. The flood hazard areas shown on maps prepared by FEMA are determined using the base flood, which is defined as having a 1-percent chance (one chance in 100) of occurring in any given year. The maps do not show the worst case flood, nor the "flood of record," which usually refers to the most severe flood in the history of the community. Although a 1-percent chance seems fairly remote, larger floods occur regularly throughout the United States. Application of the flood-resistant provisions of the code cannot prevent or eliminate all future flood damage. These provisions represent a reasonable balance of the knowledge and awareness of flood hazards, methods to guide development to less hazard-prone locations, methods of design and construc-

tion intended to resist flood damage and each community's and landowner's reasonable expectations to use the land.

FEMA uses multiple designations for the flood hazard areas shown on each Flood Insurance Rate Map (FIRM), including AO, AH, A1-30, AE, A99, AR, AR/A1-30, AR/AE, AR/AO, AR/AH, AR/A, VO, V1-30, VE, and V. Along many open coasts and lake shores, where wind-driven waves are predicted, the flood hazard area is commonly referred to as the "V Zone." Flood hazard areas that are inland of coastal high-hazard areas subject to high velocity wave action, plus flood hazard areas along rivers and streams, are commonly referred to as "A Zones." Because of waves, the flood loads in areas subject to high velocity wave action differ from those in other flood hazard areas. Users are cautioned that post-flood field investigations and laboratory testing indicated that significant structural damage can occur in areas shown as A/AE zones that are inland of V zones and inland of some shorelines without V Zones, and where breaking wave heights are less than 3-feet (914 mm) but more than 1.5 feet (457 mm). These areas are called "Coastal A Zones". For additional guidance, see ASCE 24, *Flood Resistance Design and Construction* (2005 edition).

Some coastal communities have FIRMs that show areas that are designated as units of the Coastal Barrier Resource System (CBRS) established by the Coastal Barrier Resource Act (CoBRA) of 1982 and subsequent amendments. Within these designated areas, the National Flood Insurance Program (NFIP) is prohibited from offering flood insurance on new or substantially improved buildings. The community is responsible for application of the provisions of the IRC concerning flood resistance in all designated flood hazard areas, whether or not flood insurance is made available.

An exception to Section R301.3.4 states explicitly that buildings in floodways are not to be approved under the IRC and that the provisions of the IBC apply. Because floodways typically are characterized by deeper and faster moving water, it is appropriate that flood loads be computed for foundation design.

R301.3 Story height. Buildings constructed in accordance with these provisions shall be limited to story heights of not more than the following:

1. For wood wall framing, the laterally unsupported bearing wall stud height permitted by Table R602.3(5) plus a height of floor framing not to exceed 16 inches.

   Exception: For wood framed wall buildings with bracing in accordance with Table R602.10.1, the wall stud clear height used to determine the maximum permitted story height may be increased to 12 feet without requiring an engineered design for the building wind and seismic force resisting systems provided that the length of bracing required by Table R602.10.1 is increased by multiplying by a factor of 1.20. Wall studs are still subject to the requirements of this section.

2. For steel wall framing, a stud height of 10 feet, plus a height of floor framing not to exceed 16 inches.

3. For masonry walls, a maximum bearing wall clear height of 12 feet plus a height of floor framing not to exceed 16 inches.

> **Exception:** An additional 8 feet is permitted for gable end walls.

4. For insulating concrete form walls, the maximum bearing wall height per story as permitted by Section 611 tables plus a height of floor framing not to exceed 16 inches.

Individual walls or walls studs shall be permitted to exceed these limits as permitted by Chapter 6 provisions, provided story heights are not exceeded. An engineered design shall be provided for the wall or wall framing members when they exceed the limits of Chapter 6. Where the story height limits are exceeded, an engineered design shall be provided in accordance with the *International Building Code* for the overall wind and seismic force resisting systems.

❖ This section defines the story height and wall height limits. These story heights are the basis for the wind- and seismic-force-resisting system. The wall heights are based on the materials used for wall construction. The wall height limits are prescribed for wood, steel, masonry and concrete. The story height permitted is the wall height plus the height of floor framing not to exceed 16 inches (406 mm). The story height may vary depending on the material used for wall construction. Chapter 6 may permit wall heights to exceed the limits of this section. However, if the story heights of this section are exceeded, an engineered design in accordance with the IBC must be provided for the structure for the overall wind and seismic lateral forces.

R301.4 Dead load. The actual weights of materials and construction shall be used for determining dead load with consideration for the dead load of fixed service equipment.

❖ The actual weights of materials of construction must be used to determine the dead loads. The code requires the weight of fixed service equipment (see the definition in Chapter 2) to be included as well. It is important to verify that the actual dead loads of proposed buildings are within the limits of the dead load allowance in the prescriptive design tables of subsequent chapters. Otherwise, in accordance with Section R301.1, a design is required. Also, the dead load limits of Section R301.2.2.2 must be met where required by a structure's seismic design category as explained in the commentary for Section R301.2.2.

The dead load of a building or other structure is the weight of all permanent construction, such as floors, roofs, permanent partitions, stairways and walls. The actual weights of materials and construction should be considered in building design along with the dead load of fixed service equipment. Commentary Table R301.4 lists the weight of typical residential building components. Sources of information for weights of fixed ser-

vice equipment include manufacturer's literature and trade association publications.

R301.5 Live load. The minimum uniformly distributed live load shall be as provided in Table R301.5.

❖ Table R301.5 lists the minimum uniformly distributed live loads (see the definition in Chapter 2) required for design of various portions of a residence. These loads are the basis for the prescriptive tables for floor systems in Chapter 5.

R301.6 Roof load. The roof shall be designed for the live load indicated in Table R301.6 or the snow load indicated in Table R301.2(1), whichever is greater.

❖ The basic 20-psf (960 Pa) roof live load is a severe enough loading condition on the roof to support live loads created by maintenance workers, including their equipment and materials. The code permits reduction of this basic live load based on the tributary area supported by any structural member of the roof. The rationale for this reduction of roof live load is that it is highly improbable that structural members with large tributary areas would be loaded over the entire area with the full live load. In addition, the reduction of the basic roof live load for roofs is also a function of the slope of the roof because it becomes less probable that the loads on the roof members would be at maximum levels as the roof slope increases.

The code requires that the roof be designed to resist snow loading and does not permit a reduction of snow loads based on the tributary area, as it does for the roof live load. Be aware, however, that snow on the roof rarely accumulates evenly. The design should account for unbalanced snow loading. One case would be the loading of one slope of a gable roof with snow while the other slope is unloaded. Many roofs fail from accumulation of snow at valleys, parapets, roof structures and offsets in roofs. However, the code does not specify criteria for the determination of how these potential accumulations are to be handled. Snow loading provisions in the IBC may be consulted for these loading scenarios.

R301.7 Deflection. The allowable deflection of any structural member under the live load listed in Sections R301.5 and R301.6 shall not exceed the values in Table R301.7.

❖ The allowable deflection of structural members from the design live load must not exceed the values in Table R301.7. These limits are expressed in terms of the span length. Brittle finishes, such as plaster ceilings and exterior stucco walls, are protected by limiting the deflection of those elements.

Commentary Figure R301.7 shows a simply supported beam with dead load deflection before the live load has been applied. It also shows the same beam after the live load has been applied. The vertical distance that the center of the beam has moved from the initial dead load deflection position is called the live load deflection. If a finish material is applied to this beam, the finish material is subject to distortion in proportion to this deflection.

Table R301.4
MINIMUM DESIGN DEAD LOADS FOR TYPICAL RESIDENTIAL COMPONENTS

COMPONENT	LOAD (psf)	COMPONENT	LOAD (psf)
CEILINGS		**FLOORS AND FLOOR FINISHES**	
Acoustical fiber tile	1	Asphalt block (2-inch), 2-inch mortar	30
Gypsum board (per $^1/_8$ inch thickness)	0.55	Cement finish (1-inch) on stone-concrete fill	32
Mechanical duct allowance	4	Ceramic or quarry tile ($^3/_4$-inch) on 2-inch mortar bed	16
Plaster on tile or concrete	5	Ceramic or quarry tile ($^3/_4$-inch) on 1-inch mortar bed	23
Plaster on wood lath	8	Concrete fill finish (per inch thickness)	12
Suspended steel channel system	2	Hardwood flooring, $^7/_8$-inch	4
Suspended metal lath and cement plaster	15	Linoleum or asphalt tile, $^1/_4$-inch	1
Suspended metal lath and gypsum plaster	10	Marble and mortar on stone-concrete fill	33
Wood-furring suspension system	2.5	Slate (per inch thickness)	15
COVERINGS, ROOF AND WALL		Solid flat tile on 1-inch mortar base	23
Asbestos-cement shingles	4	Subflooring, $^3/_4$-inch	3
Asphalt shingles	2	Terrazzo (12-inch) directly on slab	19
Cement tile	16	Terrazzo (1-inch) on stone-concrete fill	32
Clay tile (for mortar add 10 lb)		Terrazzo (1-inch) 2-inch stone concrete	32
Book tile, 2-inch	12	Wood block (3-inch) on mastic, no fill	10
Book tile, 3-inch	20	Wood block (3-inch) on 2-inch mortar base	16
Ludowici	10	**FLOOR, WOOD-JOIST (NO PLASTER) DOUBLE WOOD FLOOR**	
Roman	12		
Spanish	19		

	Joint Sizes (inches)	12-in. spacing (lb/ft^2)	16-in. spacing (lb/ft^2)	24-in spacing (lb/ft^2)
Composition:				
Three-ply ready roofing — 1	2×6	6	5	5
Four-ply felt and gravel — 5.5	2×8	6	6	5
Five-ply felt and gravel — 6				
Copper or tin — 1	2×10	7	6	6
Corrugated asbestos-cement roofing — 4	2×12	8	7	6

COMPONENT	LOAD (psf)	COMPONENT	LOAD (psf)
Deck, metal, 20 gage	2.5	**FRAME PARTITIONS**	
Deck, metal, 18 gage	3	Moveable steel partitions	4
Decking, 2-inch wood (Douglas fir)	5	Wood or steel studs, 2-inch gypsum board each side	8
Decking, 3-inch wood (Douglas fir)	8	Wood studs, 2×4, unplastered	4
Fiberboard, 2-inch	0.75	Wood studs, 2×4, plastered one side	12
Gypsum sheathing, 2-inch	2	Wood studs, 2×4, plastered two sides	20
Insulation, roof boards (per inch thickness)		**FRAME WALLS**	
Cellular glass	0.7	Exterior stud walls:	
Fibrous glass	1.1	2×4 @ 16 inches, $^5/_8$-inch gypsum	
Fiberboard	1.5	insulated, $^3/_8$-inch siding	11
Perlite	0.8	2×6 @ 16 inches, $^5/_8$-inch gypsum	
Polystyrene foam	0.2	insulated, $^3/_8$-inch siding	12
Urethane foam with skin	0.5		
Plywood (per $^1/_8$-inch thickness)	0.4	Exterior stud walls, with brick veneer	48
Rigid insulation, 2-inch	0.75	Windows, glass, frame and sash	8
Skylight, metal frame, $^3/_8$-inch wired glass	8	**MASONRY PARTITIONS AND WALLS**	
Slate, $^3/_{16}$-inch	7	Clay tile:	
Slate, $^1/_4$-inch	10	4-inch	18
		6-inch	24
		8-inch	24
Waterproofing members		Concrete block, heavy aggregate:	
Bituminous, gravel-covered	5.5	4-inch	30
Bituminous, smooth surface	1.5	6-inch	42
Liquid applied	1.0	8-inch	55
Single-ply, sheet	0.7	12-inch	85
Wood sheathing (per 1-inch thickness)	3	Concrete block, light aggregate:	
		4-inch	20
		6-inch	28
Wood shingles	3	8-inch	38
		12-inch	55

For SI: 1 pound per square foot = 0.0479 kPa, 1 inch = 25.4 mm, 1 pound = 0.4536 kg.

NOTE: Weights of masonry include mortar but not plaster. For plaster add 5 lb/ft^2 for each inch of plaster. Values given represent averages. In some cases there is a considerable range of weight for the same construction.

TABLE R301.5
MINIMUM UNIFORMLY DISTRIBUTED LIVE LOADS
(in pounds per square foot)

USE	LIVE LOAD
Attics with limited storage[b, g, h]	20
Attics without storage[b]	10
Decks[e]	40
Exterior balconies	60
Fire escapes	40
Guardrails and handrails[d]	200[i]
Guardrails in–fill components[f]	50[i]
Passenger vehicle garages[a]	50[a]
Rooms other than sleeping rooms	40
Sleeping rooms	30
Stairs	40[c]

For SI: 1 pound per square foot = 0.0479 kPa, 1 square inch = 645 mm², 1 pound = 4.45 N.

a. Elevated garage floors shall be capable of supporting a 2,000-pound load applied over a 20-square-inch area.

b. Attics without storage are those where the maximum clear height between joist and rafter is less than 42 inches, or where there are not two or more adjacent trusses with the same web configuration capable of containing a rectangle 42 inches high by 2 feet wide, or greater, located within the plane of the truss. For attics without storage, this live load need not be assumed to act concurrently with any other live load requirements.

c. Individual stair treads shall be designed for the uniformly distributed live load or a 300-pound concentrated load acting over an area of 4 square inches, whichever produces the greater stresses.

d. A single concentrated load applied in any direction at any point along the top.

e. See Section R502.2.1 for decks attached to exterior walls.

f. Guard in-fill components (all those except the handrail), balusters and panel fillers shall be designed to withstand a horizontally applied normal load of 50 pounds on an area equal to 1 square foot. This load need not be assumed to act concurrently with any other live load requirement.

g. For attics with limited storage and constructed with trusses, this live load need be applied only to those portions of the bottom chord where there are two or more adjacent trusses with the same web configuration capable of containing a rectangle 42 inches high or greater by 2 feet wide or greater, located within the plane of the truss. The rectangle shall fit between the top of the bottom chord and the bottom of any other truss member, provided that each of the following criteria is met:

   1. The attic area is accessible by a pull-down stairway or framed opening in accordance with Section R807.1; and

   2. The truss has a bottom chord pitch less than 2:12.

h. Attic spaces served by a fixed stair shall be designed to support the minimum live load specified for sleeping rooms.

i. Glazing used in handrail assemblies and guards shall be designed with a safety factor of 4. The safety factor shall be applied to each of the concentrated loads applied to the top of the rail, and to the load on the in-fill components. These loads shall be determined independent of one another, and loads are assumed not to occur with any other live load.

❖ The uniform live loads of this table are consistent with Table 1607.1 of the IBC, which in turn is similar to ASCE 7. The table states the minimum loads based on the use of a particular area or portion of a structure that must be considered for the design of corresponding structural elements of any residence constructed under the code. For instance, bedrooms (sleeping rooms) require use of a 30-psf (1.44 kPa) live load, while all other rooms must be designed for 40-psf (1.92 kPa) uniform live load.

Attics with storage must be designed for 20-psf (960 Pa) live loads, while a 10-psf live load (480 Pa) is required without storage. Note b explains that when the maximum clear height between the top of the joist and the rafter is less than 42 inches (1067 mm) there is insufficient clearance or headroom to accumulate a significant amount of storage.

Note h points out that any attic served by a fixed stair needs to be designed to support the same live load as a sleeping room which would be 30 pounds per square foot (1.44 kPa).

TABLE R301.6
MINIMUM ROOF LIVE LOADS IN POUNDS-FORCE
PER SQUARE FOOT OF HORIZONTAL PROJECTION

ROOF SLOPE	TRIBUTARY LOADED AREA IN SQUARE FEET FOR ANY STRUCTURAL MEMBER		
	0 to 200	201 to 600	Over 600
Flat or rise less than 4 inches per foot (1:3)	20	16	12
Rise 4 inches per foot (1:3) to less than 12 inches per foot (1:1)	16	14	12
Rise 12 inches per foot (1:1) and greater	12	12	12

For SI: 1 square foot = 0.0929 m², 1 pound per square foot = 0.0479 kPa, 1 inch per foot = 83.3 mm/m.

❖ See the commentary for Section R301.6.

TABLE R301.7
ALLOWABLE DEFLECTION OF STRUCTURAL MEMBERS[a,b,c]

STRUCTURAL MEMBER	ALLOWABLE DEFLECTION
Rafters having slopes greater than 3/12 with no finished ceiling attached to rafters	L/180
Interior walls and partitions	H/180
Floors and plastered ceilings	L/360
All other structural members	L/240
Exterior walls with plaster or stucco finish	H/360
Exterior walls—wind loads[a] with brittle finishes	L/240
Exterior walls—wind loads[a] with flexible finishes	L/120

Note: L = span length, H = span height.

a. The wind load shall be permitted to be taken as 0.7 times the Component and Cladding loads for the purpose of the determining deflection limits herein.

b. For cantilever members, L shall be taken as twice the length of the cantilever.

c. For aluminum structural members or panels used in roofs or walls of sunroom additions or patio covers, not supporting edge of glass or sandwich panels, the total load deflection shall not exceed L /60. For sandwich panels used in roofs or walls of sunroom additions or patio covers, the total load deflection shall not exceed L/120.

❖ Generally, it is not necessary to calculate the deflections indicated in Table R301.7 because the deflection limit is typically accounted for in the prescriptive design tables of the code. Because this is not the case for elements that require design in accordance with Section R301.1.2, the following example illustrates this requirement.

Example:

What is the maximum allowable deflection of a 20'-0" (6096 mm) span for the following conditions?

a) Rafter with no ceiling load having a slope of 5 units vertical in 12 units horizontal (42% slope)

b) Rafter with no ceiling load having a slope of 2 units vertical in 12 units horizontal (17% slope)

c) A floor joist supporting a finished floor

Solution:

Length = Span = 20'-0" (6096 mm)

a) D MAX = L/180 = 20' x 12"/180 = 1.33" or approx. $1^3/_8$" (35 mm)

b) D MAX = L/240 = 20' x 12"/240 = 1.00" or approx. 1" (25.4 mm)

c) D MAX = L/360 = 20' x 12"/360 = 0.67" or approx. $^5/_8$" (15.9 mm)

R301.8 Nominal sizes. For the purposes of this code, where dimensions of lumber are specified, they shall be deemed to be nominal dimensions unless specifically designated as actual dimensions.

❖ Because solid sawn lumber sizes are normally referred to using the nominal lumber dimensions, this clarification explains that any code references to lumber dimensions are to be taken as nominal dimensions unless explicitly stated otherwise.

SECTION R302
EXTERIOR WALL LOCATION

R302.1 Exterior walls. Construction, projections, openings and penetrations of exterior walls of dwellings and accessory buildings shall comply with Table R302.1. These provisions shall not apply to walls, projections, openings or penetrations in walls that are perpendicular to the line used to determine the fire separation distance. Projections beyond the exterior wall shall not extend more than 12 inches (305 mm) into the areas where openings are prohibited.

Exceptions:

1. Detached tool sheds and storage sheds, playhouses and similar structures exempted from permits are not required to provide wall protection based on location on the lot. Projections beyond the exterior wall shall not extend over the lot line.

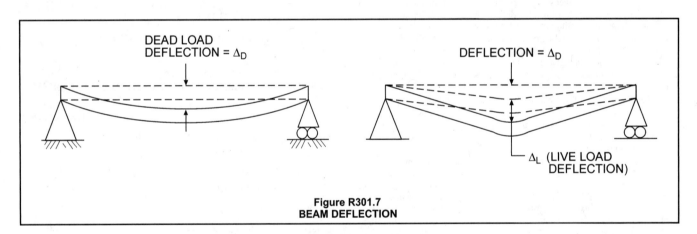

Figure R301.7
BEAM DEFLECTION

TABLE R302.1
EXTERIOR WALLS

EXTERIOR WALL ELEMENT		MINIMUM FIRE-RESISTANCE RATING	MINIMUM FIRE SEPARATION DISTANCE
Walls	(Fire-resistance rated)	1 hour with exposure from both sides	0 feet
	(Not fire-resistance rated)	0 hours	5 feet
Projections	(Fire-resistance rated)	1 hour on the underside	2 feet
	(Not fire-resistance rated)	0 hours	5 feet
Openings	Not allowed	N/A	< 3 feet
	25% Maximum of Wall Area	0 hours	3 feet
	Unlimited	0 hours	5 feet
Penetrations	All	Comply with Section R317.3	< 5 feet
		None required	5 feet

N/A = Not Applicable.

❖ See commentary to Section R302.1.

2. Detached garages accessory to a dwelling located within 2 feet (610 mm) of a lot line are permitted to have roof eave projections not exceeding 4 inches (102 mm).

3. Foundation vents installed in compliance with this code are permitted.

❖ This section provides details for issues related to building location on the property, including the fire rating of exterior walls, permitted openings and projections. Table R302.1 is a tabular overview of the requirements of this section.

Concerning exterior wall protection, the IRC assumes that an owner has no control over an adjoining property. Thus, the location of buildings on the owner's property relative to the property line requires regulation. In addition, Section R309, which lists the separation requirements for garages and carports, specifically requires garages located less than 3 feet (914 mm) from a dwelling unit on the same lot to have not less than $^{1}/_{2}$-inch (12.7 mm) gypsum board applied to the interior side of the walls. Opening protection for these walls is regulated by Section R309.1.

The property line concept is a convenient means of protecting one building from another as far as exposure is concerned. Exposure is the potential for heat to be transmitted from one building to another during a fire in the exposing building. Radiation is the primary means of heat transfer.

Walls less than 5 feet (1525 mm) from the property line must be of 1-hour fire-resistant construction. The fire-resistance rating also requires the rating exposure to be for both sides. Specific provisions are not found in the IRC for the fire-resistance ratings of walls or other elements, so one may use either the *IBC* or another manual of fire-resistant assemblies. Projections may not be closer than 4 feet (1219 mm) to the lot line. Projections should not extend more than 12 inches (305 mm) into an area where walls are required to be of 1-hour fire-resistant construction must be protected on the underside with 1-hour fire-resistant construction [see Commentary Figure R302.1(1)].

Unlike the IBC, the IRC does not set a distance from the property line at which openings must be protected. Openings are not permitted in exterior walls where the

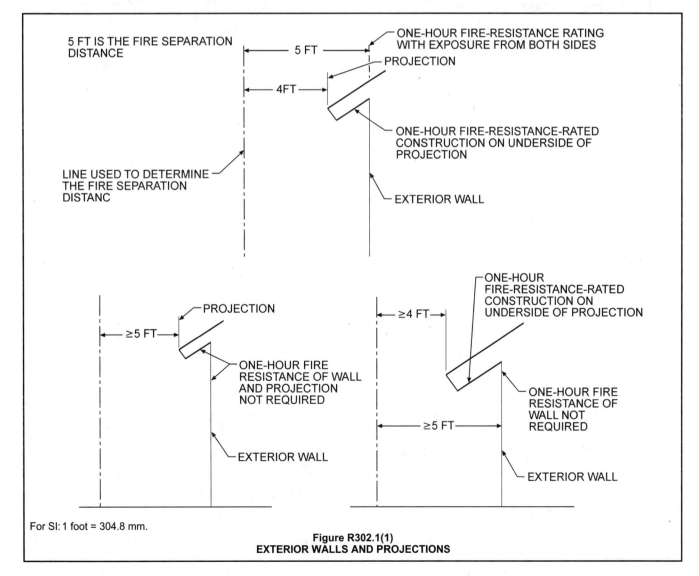

For SI: 1 foot = 304.8 mm.

Figure R302.1(1)
EXTERIOR WALLS AND PROJECTIONS

exterior wall has a fire-separation distance of less than 3 feet (914 mm) from the lot line. Openings in a wall located at a distance in excess of 3 feet (914 mm) from the lot line can not exceed 25 percent of the maximum wall area (see Commentary Figures R302.1(2) and (3)]. Openings, projections or penetrations that are 90 degrees (1.57 rad) (perpendicular) to the line used to determine fire separation distance are not prohibited. Section R317.3 describes through penetrations and membrane penetrations in detail.

See the definition of Fire Separation Distance in Chapter 2. The requirements for protection do not apply to walls that are at a right angle (perpendicular to) the property line (see Commentary Figures R302.1(2) and (3)].

Exception 1 applies to detached tool and storage sheds, playhouses and similar structures that are exempt from permits. Projections from these structures, however, are not permitted to extend over the property line.

Exception 2 will allow a roof eave projection to within 2 feet (610 mm) of a lot line, however the roof eave projection amount is limited to 4 inches (102 mm).

Exception 3 allows foundation vents installed in compliance with the IRC in areas where openings are otherwise prohibited.

SECTION R303
LIGHT, VENTILATION AND HEATING

R303.1 Habitable rooms. All habitable rooms shall have an aggregate glazing area of not less than 8 percent of the floor area of such rooms. Natural ventilation shall be through windows, doors, louvers or other approved openings to the outdoor air. Such openings shall be provided with ready access or shall otherwise be readily controllable by the building occupants. The minimum openable area to the outdoors shall be 4 percent of the floor area being ventilated.

Exceptions:

1. The glazed areas need not be openable where the opening is not required by Section R310 and an approved mechanical ventilation system capable of producing 0.35 air change per hour in the room is installed or a whole-house mechanical ventilation system is installed capable of supplying outdoor ventilation air of 15 cubic feet per minute (cfm) (78 L/s) per occupant computed on the basis of two occupants for the first bedroom and one occupant for each additional bedroom.

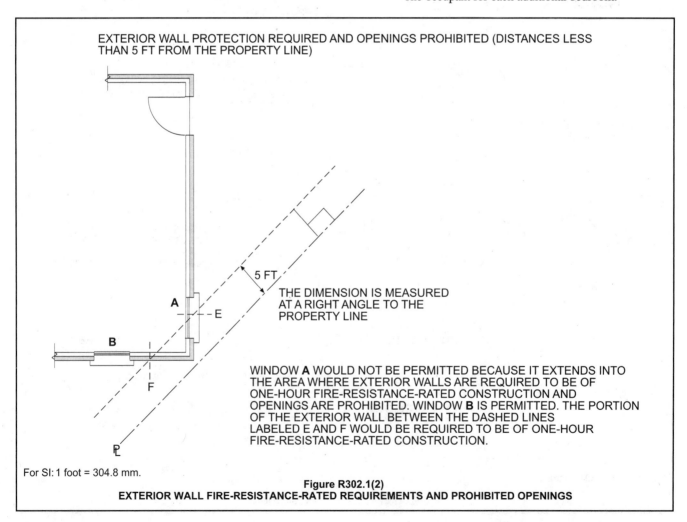

EXTERIOR WALL PROTECTION REQUIRED AND OPENINGS PROHIBITED (DISTANCES LESS THAN 5 FT FROM THE PROPERTY LINE)

5 FT

THE DIMENSION IS MEASURED AT A RIGHT ANGLE TO THE PROPERTY LINE

WINDOW **A** WOULD NOT BE PERMITTED BECAUSE IT EXTENDS INTO THE AREA WHERE EXTERIOR WALLS ARE REQUIRED TO BE OF ONE-HOUR FIRE-RESISTANCE-RATED CONSTRUCTION AND OPENINGS ARE PROHIBITED. WINDOW **B** IS PERMITTED. THE PORTION OF THE EXTERIOR WALL BETWEEN THE DASHED LINES LABELED E AND F WOULD BE REQUIRED TO BE OF ONE-HOUR FIRE-RESISTANCE-RATED CONSTRUCTION.

For SI: 1 foot = 304.8 mm.

Figure R302.1(2)
EXTERIOR WALL FIRE-RESISTANCE-RATED REQUIREMENTS AND PROHIBITED OPENINGS

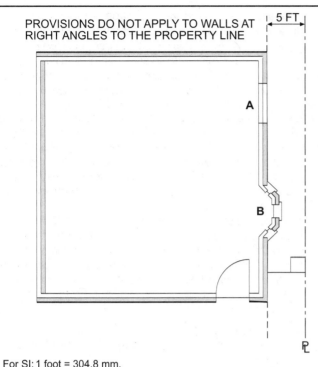

PROVISIONS DO NOT APPLY TO WALLS AT RIGHT ANGLES TO THE PROPERTY LINE

5 FT

A

B

THE WINDOW LABELED **A** IN THE PLAN IS ACCEPTABLE BECAUSE THE EXTERIOR WALL IS 5 FT OR MORE FROM THE PROPERTY LINE. THE BAY WINDOW LABELED **B** WOULD NOT BE ACCEPTABLE BECAUSE IT PROJECTS INTO THE AREA WHERE EXTERIOR WALLS WOULD BE REQUIRED TO HAVE A FIRE-RESISTANCE RATING AND WHERE OPENINGS ARE PROHIBITED. THE EXTERIOR WALLS ADJACENT TO THE PROPERTY LINE WOULD NOT HAVE TO BE FIRE-RESISTANCE RATED EXCEPT FOR THE PORTION THAT FORMS THE BAY WINDOW BECAUSE THE REMAINDER OF THE WALL IS AT LEAST 5 FT FROM THE PROPERTY LINE.

For SI: 1 foot = 304.8 mm.

PL

Figure R302.1(3)
EXTERIOR WALL FIRE-RESISTANCE-RATED REQUIREMENTS AND PROHIBITED OPENINGS

2. The glazed areas need not be installed in rooms where Exception 1 above is satisfied and artificial light is provided capable of producing an average illumination of 6 footcandles (65 lux) over the area of the room at a height of 30 inches (762 mm) above the floor level.

3. Use of sunroom additions and patio covers, as defined in Section R202, shall be permitted for natural ventilation if in excess of 40 percent of the exterior sunroom walls are open, or are enclosed only by insect screening.

❖ All habitable rooms are to be constructed with an aggregate glazing area of 8 percent of the floor area, with a minimal openable area of 4 percent of the floor area being ventilated. The purpose of the glazing and the portion that is openable is to provide both natural light and ventilation for habitable spaces.

Three exceptions modify the requirement of Section R303.1. Exception 1 permits the glazed area to be fixed or "not openable" when an approved mechanical ventilation system or whole-house mechanical ventilation system is installed. The mechanical ventilation system must provide a 35-percent air change per hour in each habitable room. As an example, if a bedroom is 10 feet by 12 feet (3048 mm by 3658 mm) with an 8 foot (2438 mm) ceiling height, the air change per hour would be as follows:

10 x 12 x 8 = 960 cubic feet

960 x 0.35 = 333 cubic feet/hour

or 5.55 cubic feet/minute

The whole-house mechanical ventilation system must provide outdoor ventilation at the rate of 15 cubic feet per minute (7.08 L/s) per occupant, based on two occupants in the first bedroom and one occupant in each additional bedroom. Thus, a three bedroom dwelling unit would require the whole-house ventilation to provide outdoor ventilation at the rate of 60 cfm. This is calculated as follows:

First Bedroom – 2 Occupants

Second Bedroom – 1 Occupant

Third Bedroom – 1 Occupant

Total Occupants – 4 Occupants

4 x 15 cfm = 60 cfm (28.32 L/s)

Exception 2 to Section R301.1 allows glazing to be completely deleted from habitable rooms where artificial light is available that is capable of producing an average illumination of 6 foot candles (6.46 lux) over the room at a height of 30 inches (762 mm) above the floor level and where a mechanical or whole-house mechanical ventilation system complying with Exception 1 to Section R303.1 is installed.

Exception 3 allows for the openings in a Category I sunroom (those in which more than 40 percent of the exterior sunroom walls is open) to provide natural ventilation when the exterior walls of the sunroom are open. Insect screening for these seasonal rooms is also acceptable to make use of this exception.

Calculations should be submitted for review for each of the exceptions. Various handbooks are available from various lighting manufacturers for calculating the required illumination.

R303.2 Adjoining rooms. For the purpose of determining light and ventilation requirements, any room shall be considered as a portion of an adjoining room when at least one-half of the area of the common wall is open and unobstructed and provides an opening of not less than one-tenth of the floor area of the interior room but not less than 25 square feet (2.3 m²).

Exception: Openings required for light and/or ventilation shall be permitted to open into a thermally isolated sunroom addition or patio cover, provided that there is an openable area between the adjoining room and the sunroom addition or patio cover of not less than one-tenth of the floor area of the interior room but not less than 20 square feet (2 m²). The minimum openable area to the outdoors shall be based upon the total floor area being ventilated.

❖ Where rooms do not have access to an exterior wall for light and ventilation, they may be considered as part of an adjoining room if they are open to that room as shown in Commentary Figure R303.2.

The exception deals with a very common circumstance, especially in residential construction. As long as the sunroom addition is large enough and is thermally isolated, the building owner need not move openings for lighting and/or ventilation when installing an addition that meets the definition of Sunroom Addition.

R303.3 Bathrooms. Bathrooms, water closet compartments and other similar rooms shall be provided with aggregate glazing area in windows of not less than 3 square feet (0.3 m²), one-half of which must be openable.

Exception: The glazed areas shall not be required where artificial light and a mechanical ventilation system are provided. The minimum ventilation rates shall be 50 cubic feet per minute (24 L/s) for intermittent ventilation or 20 cubic feet per minute (10 L/s) for continuous ventilation. Ventilation air from the space shall be exhausted directly to the outside.

❖ Like habitable rooms, bathrooms must be provided with natural light and ventilation or adequate artificial light and mechanical ventilation. For natural light, exterior glazing of at least 3 square feet (0.279 m²) must be provided. To meet the natural ventilation requirement, at least 1.5 square feet (0.140 m²) must be openable.

For mechanical ventilation, the ventilation rates must be either 50 cfm (23.6 L/s) for intermittent ventilation or 20 cfm (9.4 L/s) for continuous ventilation. Intermittent ventilation would be controlled by the occupant on an as-needed basis. Continuous ventilation is on at all times. In either case, the ventilation must be to the outside atmosphere.

R303.4 Opening location. Outdoor intake and exhaust openings shall be located in accordance with Sections R303.4.1 and R303.4.2.

❖ This section specifies the locations of outdoor intake and exhaust openings. These locations are intended to prevent the introduction of contaminants into the ventilation air of a building or to avoid the exhausting of contaminants onto areas that may be occupied by people or into other buildings.

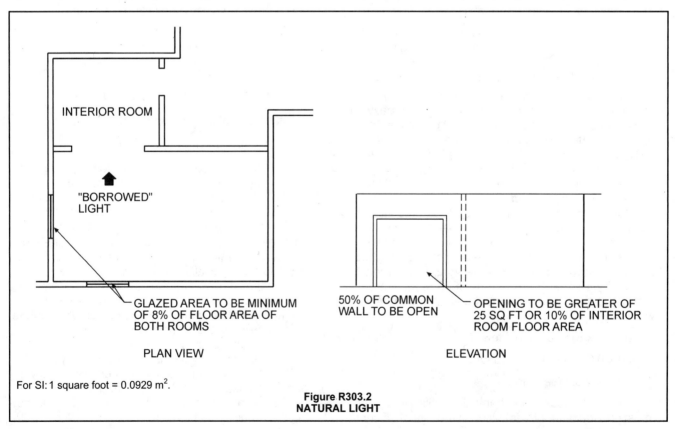

INTERIOR ROOM

"BORROWED" LIGHT

GLAZED AREA TO BE MINIMUM OF 8% OF FLOOR AREA OF BOTH ROOMS

PLAN VIEW

50% OF COMMON WALL TO BE OPEN

OPENING TO BE GREATER OF 25 SQ FT OR 10% OF INTERIOR ROOM FLOOR AREA

ELEVATION

For SI: 1 square foot = 0.0929 m².

Figure R303.2
NATURAL LIGHT

R303.4.1 Intake openings. Mechanical and gravity outdoor air intake openings shall be located a minimum of 10 feet (3048 mm) from any hazardous or noxious contaminant, such as vents, chimneys, plumbing vents, streets, alleys, parking lots and loading docks, except as otherwise specified in this code. Where a source of contaminant is located within 10 feet (3048 mm) of an intake opening, such opening shall be located a minimum of 2 feet (610 mm) below the contaminant source.

For the purpose of this section, the exhaust from dwelling unit toilet rooms, bathrooms and kitchens shall not be considered as hazardous or noxious.

❖ In the context of this section, intake openings include windows, doors, gravity air intakes, soffit vents, combustion air intake openings, outside air intakes for air handlers, makeup air intakes and similar openings that naturally or mechanically draw in air from the building exterior. This section identifies specific locations that are known to generate or emit noxious contaminants, and requires that both mechanical and gravity air intake openings be located a minimum of 10 feet (3048 mm) from such hazards to avoid introducing contaminants into the building. As an alternative, mechanical and gravity air intakes can be located within 10 feet (3048 mm) of such sources of contamination if the intakes are located at least 2 feet (610 mm) below the contaminant source. A 2-foot (610 mm) vertical separation distance will allow the noxious gases and contaminants to disperse into the atmosphere before they can be drawn into an air intake opening. Placing the source of contamination above an air intake takes advantage of the fact that normally encountered sources of contamination are lighter (less dense) than the surrounding air and, therefore, will rise above the vicinity of an air intake located below. Commentary Figure R303.4.1 shows an example of the relative locations for intake air openings for a building where sources of contaminants are present.

Particular types of exhausts may have more specific restrictions on their location that would supercede this section (see Section M1602.2, for example).

The air exhausts discharging from a dwelling unit (clothes dryer, kitchen and bathroom) are not considered to be significantly hazardous or noxious and are of low volume. In these situations, the code official must determine an appropriate distance or location for the relative placement of intake and exhaust openings. In evaluating each installation, consideration should be given to the orientation of the exhaust or intake louver and its spatial relationship to any source of contaminant or adjacent intake opening, as well as to the direction of the prevailing winds at the location.

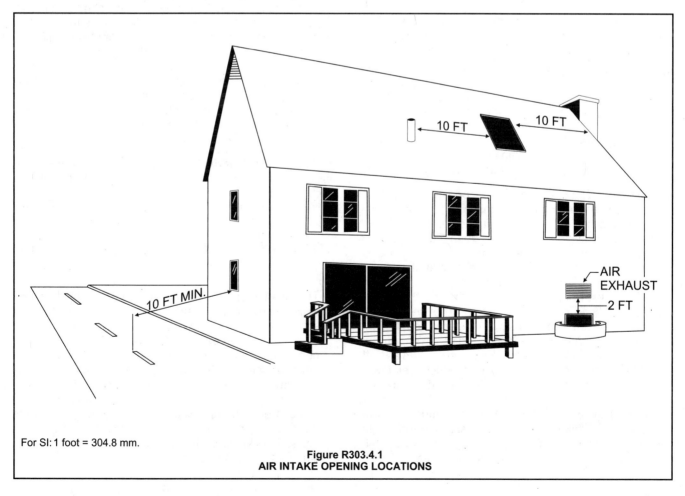

For SI: 1 foot = 304.8 mm.

Figure R303.4.1
AIR INTAKE OPENING LOCATIONS

R303.4.2 Exhaust openings. Outside exhaust openings shall be located so as not to create a nuisance. Exhaust air shall not be directed onto walkways.

❖ Mechanical exhaust air openings must not be directed onto walkways or be located where they create a nuisance. A "Nuisance" is defined in Chapter 2 of the *International Plumbing Code®* (IPC®) as " . . . whatever is dangerous to human life or detrimental to health; whatever structure or premises is not sufficiently ventilated, sewered, drained, cleaned or lighted, with respect to its intended occupancy; and whatever renders the air, or human food, drink or water supply unwholesome." A nuisance is defined as much more than or much worse than simply "bothersome." This definition is rather broad, and allows the code official to decide what may or may not constitute a nuisance. Unfortunately, it is not an easy task to determine whether or not a nuisance will be present because the conditions under which an exhaust system performs vary considerably with the change of seasons, the ambient temperatures and the prevailing winds. The code official should gather as much information regarding the installation as possible to evaluate the hypothetical worst-case scenario. This would include the characteristics and geometry of the installation as well as the local ambient conditions, so that an educated guess may be made to determine the "nuisance effect" of the exhaust outlet.

For obvious health reasons, exhaust air cannot be directed onto walkways in a manner that subjects the users of the walkway to the exhaust airstream.

R303.5 Outside opening protection. Air exhaust and intake openings that terminate outdoors shall be protected with corrosion-resistant screens, louvers or grilles having a minimum opening size of $^1/_4$ inch (6 mm) and a maximum opening size of $^1/_2$ inch (13 mm), in any dimension. Openings shall be protected against local weather conditions. Outdoor air exhaust and intake openings shall meet the provisions for exterior wall opening protectives in accordance with this code.

❖ Outside air exhaust and intake openings must have corrosion-resistant screens, grilles or louvers to prevent foreign objects (such as insects or debris) from entering the system or building. Also, such openings must be protected against the entry of falling or wind-driven water, snow and ice. Exhaust systems sometimes incorporate rotating hoods over the opening to prevent high winds from restricting the flow of exhaust gases out of the system. The hoods align themselves with the direction of the wind to allow the unimpeded, and actually induced, discharge from the exhaust outlet. Rotating turbines are also used as both weather protection and a means of inducing airflow.

The opening sizes for louvers and grilles and the mesh sizes for screens must be within the specified range as indicated. The opening size must be large enough to inhibit blockage by debris, to prevent significant resistance to airflow, and still be small enough to keep out what must be kept out. A screen of such mesh size would restrict the passage of rodents and large insects and would be resistant to blockage by lint, debris and plant fibers.

R303.6 Stairway illumination. All interior and exterior stairways shall be provided with a means to illuminate the stairs, including the landings and treads. Interior stairways shall be provided with an artificial light source located in the immediate vicinity of each landing of the stairway. For interior stairs the artificial light sources shall be capable of illuminating treads and landings to levels not less than 1 foot-candle (11 lux) measured at the center of treads and landings. Exterior stairways shall be provided with an artificial light source located in the immediate vicinity of the top landing of the stairway. Exterior stairways providing access to a basement from the outside grade level shall be provided with an artificial light source located in the immediate vicinity of the bottom landing of the stairway.

Exception: An artificial light source is not required at the top and bottom landing, provided an artificial light source is located directly over each stairway section.

❖ Interior and exterior stairs may be illuminated in two ways. The first option is to install artificial lighting in the vicinity of each landing. This would include top, intermediate and bottom landings. For interior stairs, the artificial light must be capable of illuminating treads and landings to not less than 1 foot candle (11 lux). The measurement of 1 foot candle is to be taken at the center of landings and treads. Exterior stairs require illumination only at the top landing (see Commentary Figure R303.6).

Exterior stairs to a basement must have artificial illumination near the bottom landing.

The exception allows the light source to be installed over each individual stair section, thus eliminating the lighting over the landings.

R303.6.1 Light activation. Where lighting outlets are installed in interior stairways, there shall be a wall switch at each floor level to control the lighting outlet where the stairway has six or more risers. The illumination of exterior stairways shall be controlled from inside the dwelling unit.

Exception: Lights that are continuously illuminated or automatically controlled.

❖ The location of the light activation is based on the need to illuminate an area before it is used. Interior stairway lighting control is required at each floor level of the stair. There is an exception for lights that are continuously illuminated or automatically activated. When manual switches are installed, they should be accessible without the switch operator stepping onto the stair. Exterior stairway light control must be from the interior of the structure. This not only helps provide security for the dwelling but enables the occupants to illuminate the stair for guests without first having to traverse the stair in darkness to do so.

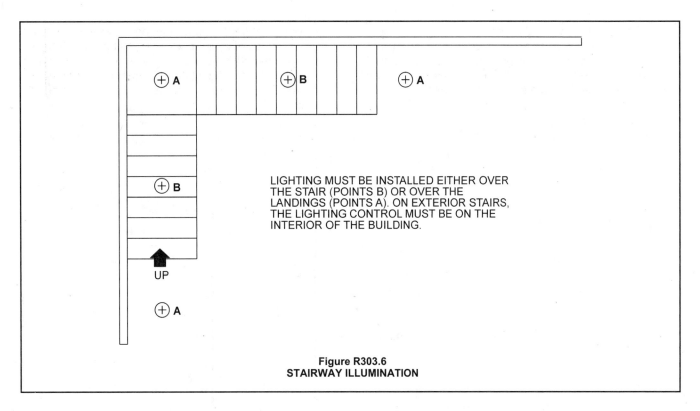

Figure R303.6
STAIRWAY ILLUMINATION

R303.7 Required glazed openings. Required glazed openings shall open directly onto a street or public alley, or a yard or court located on the same lot as the building.

❖ Glazed openings must be placed so that natural light will be available even after future adjacent construction occurs. Where glazing is required by the code, it must open onto a street, public alley, yard or court of the lot on which the building is located. Glazed openings are not permitted in walls that must be of 1-hour fire-resistant construction in accordance with Section R302.1.

R303.7.1 Roofed porches. Required glazed openings may face into a roofed porch where the porch abuts a street, yard or court and the longer side of the porch is at least 65 percent open and unobstructed and the ceiling height is not less than 7 feet (2134 mm).

❖ The porch must be open so that natural light will be available. Glazed openings required by the code are permitted to open onto a porch if it abuts a yard, street or court and the longer side is open at least 65 percent. The porch ceiling height must be at least 7 feet (2134 mm).

R303.7.2 Sunroom additions. Required glazed openings shall be permitted to open into sunroom additions or patio covers that abut a street, yard or court if in excess of 40 percent of the exterior sunroom walls are open, or are enclosed only by insect screening, and the ceiling height of the sunroom is not less than 7 feet (2134 mm).

❖ This section permits openings for lighting to open into a sunroom that meets the criterion of being more than 40 percent open. Both enclosed porches and Category I sunrooms must be a minimum of 7 feet (2134 mm) high and abut a court, yard or public space. Open

areas near the sunroom enable the sunlight to effectively reach the structure. Category I sunrooms are those which are not conditioned and are either open or enclosed by screening or plastic.

R303.8 Required heating. When the winter design temperature in Table R301.2(1) is below 60°F (16°C), every dwelling unit shall be provided with heating facilities capable of maintaining a minimum room temperature of 68°F (20°C) at a point 3 feet (914 mm) above the floor and 2 feet (610 mm) from exterior walls in all habitable rooms at the design temperature. The installation of one or more portable space heaters shall not be used to achieve compliance with this section.

❖ Minimum heating requirements are for health reasons. In areas where the design temperature is based on Table R301.2(1) and is below 60°F (16°C), dwelling units must be provided with heating facilities capable of maintaining a minimum room temperature of 68°F (20°C). The primary need here is that of human comfort. The minimum temperature is measured at 3 feet (914 mm) above the floor and 2 feet (610 mm) from the exterior walls of all habitable rooms (see Commentary Figure R303.8). Portable space heaters cannot be used to achieve compliance with this section.

SECTION R304
MINIMUM ROOM AREAS

R304.1 Minimum area. Every dwelling unit shall have at least one habitable room that shall have not less than 120 square feet (11 m²) of gross floor area.

❖ The interior living environment is affected by a number of issues. Among these are the size of the room, tightness of construction, ceiling height, number of occupants and ventilation. These all interact and impact the interior living conditions including odors, moisture and disease transmission. The IRC regulates room sizes to assist in maintaining a safe and comfortable interior environment. See Commentary Figure R304.1. At least one habitable room must be at least 120 square feet (11.2 m²) of gross floor area. Because the definition of habitable space in Section R202 includes rooms used for living, sleeping, eating or cooking, any one of these rooms can be used to meet the requirement.

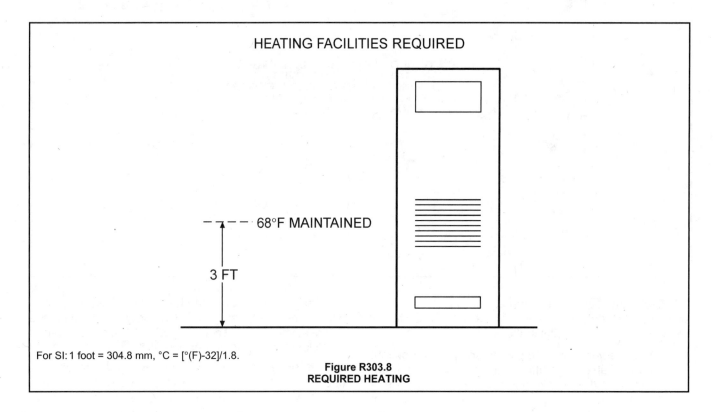

HEATING FACILITIES REQUIRED

68°F MAINTAINED

3 FT

For SI: 1 foot = 304.8 mm, °C = [°(F)-32]/1.8.

**Figure R303.8
REQUIRED HEATING**

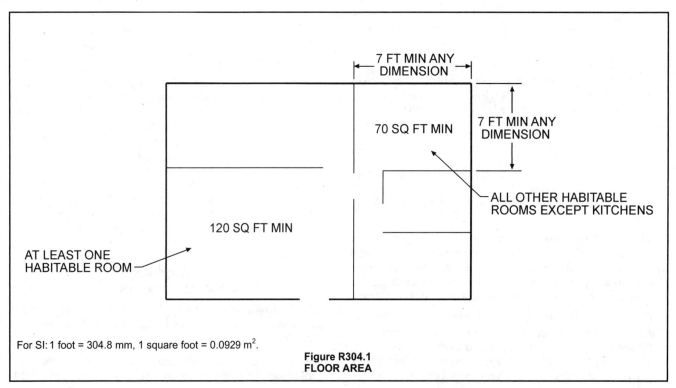

7 FT MIN ANY DIMENSION

7 FT MIN ANY DIMENSION

70 SQ FT MIN

ALL OTHER HABITABLE ROOMS EXCEPT KITCHENS

120 SQ FT MIN

AT LEAST ONE HABITABLE ROOM

For SI: 1 foot = 304.8 mm, 1 square foot = 0.0929 m².

**Figure R304.1
FLOOR AREA**

R304.2 Other rooms. Other habitable rooms shall have a floor area of not less than 70 square feet (6.5 m²).

Exception: Kitchens.

❖ One habitable room must comply with the provisions of Section R304.1. The remainder of the habitable rooms except kitchens are required to have a floor area of 70 square feet (6.5 m²). Kitchens are exempt from the minimum floor area requirement.

R304.3 Minimum dimensions. Habitable rooms shall not be less than 7 feet (2134 mm) in any horizontal dimension.

Exception: Kitchens.

❖ Except for kitchens, all habitable rooms must have a minimum horizontal dimension in any direction of at least 7 feet (2134 mm).

R304.4 Height effect on room area. Portions of a room with a sloping ceiling measuring less than 5 feet (1524 mm) or a furred ceiling measuring less than 7 feet (2134 mm) from the finished floor to the finished ceiling shall not be considered as contributing to the minimum required habitable area for that room.

❖ In a room with a sloping ceiling, any portion of the room with a vertical ceiling height less than 5 feet (1524 mm) from the finished floor does not provide the minimum required floor area for habitation. Likewise, the area under a furred ceiling with a vertical height less than 7 feet (2134 mm) from the finished floor is not part of the habitable area. An example of the first case would be an A-frame structure, which consists of a sloping roof and no or minimal exterior walls. This condition could also exist in any room that has a sloping ceiling. The

low height makes those portions of the room generally unusable for adults (see Commentary Figure R304.4).

SECTION R305
CEILING HEIGHT

R305.1 Minimum height. Habitable rooms, hallways, corridors, bathrooms, toilet rooms, laundry rooms and basements shall have a ceiling height of not less than 7 feet (2134 mm). The required height shall be measured from the finish floor to the lowest projection from the ceiling.

Exceptions:

1. Beams and girders spaced not less than 4 feet (1219 mm) on center may project not more than 6 inches (152 mm) below the required ceiling height.

2. Ceilings in basements without habitable spaces may project to within 6 feet, 8 inches (2032 mm) of the finished floor; and beams, girders, ducts or other obstructions may project to within 6 feet 4 inches (1931 mm) of the finished floor.

3. For rooms with sloped ceilings, at least 50 percent of the required floor area of the room must have a ceiling height of at least 7 feet (2134 mm) and no portion of the required floor area may have a ceiling height of less than 5 feet (1524 mm).

4. Bathrooms shall have a minimum ceiling height of 6 feet 8 inches (2036 mm) over the fixture and at the front clearance area for fixtures as shown in Figure R307.1. A shower or tub equipped with a showerhead shall have a minimum ceiling height of 6 feet 8 inches

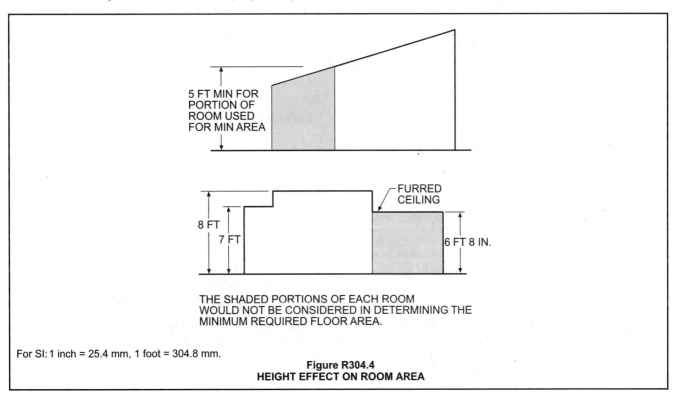

5 FT MIN FOR PORTION OF ROOM USED FOR MIN AREA

FURRED CEILING

8 FT

7 FT

6 FT 8 IN.

THE SHADED PORTIONS OF EACH ROOM WOULD NOT BE CONSIDERED IN DETERMINING THE MINIMUM REQUIRED FLOOR AREA.

For SI: 1 inch = 25.4 mm, 1 foot = 304.8 mm.

Figure R304.4
HEIGHT EFFECT ON ROOM AREA

(2036 mm) above a minimum area 30 inches (762 mm) by 30 inches (762 mm) at the showerhead.

❖ Minimum ceiling heights are required for habitable rooms hallways, corridors, bathrooms, toilet rooms, laundry rooms and basements. The minimum required height of 7 feet (2134 mm) helps maintain a healthy interior environment. The dimension must be measured to the lowest projection from the ceiling.

For ceilings having exposed beams that project down from the ceiling surface, the ceiling beam members may project no more than 6 inches (152 mm) below the required ceiling height, provided the beams or girders are placed at no less than 4 feet (1219 mm) on center. Basements need have a ceiling height of only 6 feet, 8 inches (2033 mm) or more, with at least 6 feet, 4 inches (1932 mm) of clear height under beams, girders, ducts and similar obstructions.

For rooms with sloped ceilings, the code requires only that the prescribed ceiling height be maintained in one-half the area of the room. However, no portion of the room that has a ceiling height of less than 5 feet (1524 mm) must be used in the computations for minimum floor area.

Exception 4 defines the required minimum ceiling height over toilet, bath and shower fixtures. This exception would allow a sloping ceiling over toilet, bath or shower fixtures if the minimum ceiling height of 6 feet, 8 inches (2036 mm) is maintained over the fixture and at the front clearance area.

SECTION R306
SANITATION

R306.1 Toilet facilities. Every dwelling unit shall be provided with a water closet, lavatory, and a bathtub or shower.

❖ Dwelling units have at least one each of the fixtures indicated in the code, and the fixtures must be connected to an approved sanitary sewer or private sewage disposal system. A water closet, lavatory and bathtub or shower are the minimum fixtures needed to maintain the occupant's health and cleanliness.

R306.2 Kitchen. Each dwelling unit shall be provided with a kitchen area and every kitchen area shall be provided with a sink.

❖ Dwelling units must have a kitchen area with a sink for the basic preparation of food.

R306.3 Sewage disposal. All plumbing fixtures shall be connected to a sanitary sewer or to an approved private sewage disposal system.

❖ To maintain sanitary conditions in the dwelling, plumbing fixtures must be attached to either a sanitary sewer or approved private sewage disposal system.

R306.4 Water supply to fixtures. All plumbing fixtures shall be connected to an approved water supply. Kitchen sinks, lava-

tories, bathtubs, showers, bidets, laundry tubs and washing machine outlets shall be provided with hot and cold water.

❖ To provide proper sanitation for occupants of dwelling units, each plumbing fixture must be connected to an approved water supply. Additionally, specific fixtures must have both a hot and a cold water supply.

SECTION R307
TOILET, BATH AND SHOWER SPACES

R307.1 Space required. Fixtures shall be spaced as per Figure R307.1.

❖ Fixtures require certain clearances to be accessible and usable. Figure R307.1 shows the minimum fixture clearances.

R307.2 Bathtub and shower spaces. Bathtub and shower floors and walls above bathtubs with installed shower heads and in shower compartments shall be finished with a nonabsorbent surface. Such wall surfaces shall extend to a height of not less than 6 feet (1829 mm) above the floor.

❖ Wall surfaces subject to water spray by showerheads must be protected with a nonabsorbent surface to a height of at least 6 feet (1829 mm) above the floor of a bathtub or shower.

SECTION R308
GLAZING

R308.1 Identification. Except as indicated in Section R308.1.1 each pane of glazing installed in hazardous locations as defined in Section R308.4 shall be provided with a manufacturer's designation specifying who applied the designation, designating the type of glass and the safety glazing standard with which it complies, which is visible in the final installation. The designation shall be acid etched, sandblasted, ceramic-fired, laser etched, embossed, or be of a type which once applied cannot be removed without being destroyed. A label shall be permitted in lieu of the manufacturer's designation.

Exceptions:

1. For other than tempered glass, manufacturer's designations are not required provided the building official approves the use of a certificate, affidavit or other evidence confirming compliance with this code.

2. Tempered spandrel glass is permitted to be identified by the manufacturer with a removable paper designation.

❖ Once glass is installed in a window frame by a manufacturer, whether that glass is safety glazing or not is not easily determined. In theory, this can be established only by breaking the particular piece of glass, in which case the glass is no longer useable. Thus, the code requires that safety glazing be marked with a manufacturer's designation that is visible during the final building inspection. Except for tempered glass labels, labels may be omitted where approved by the

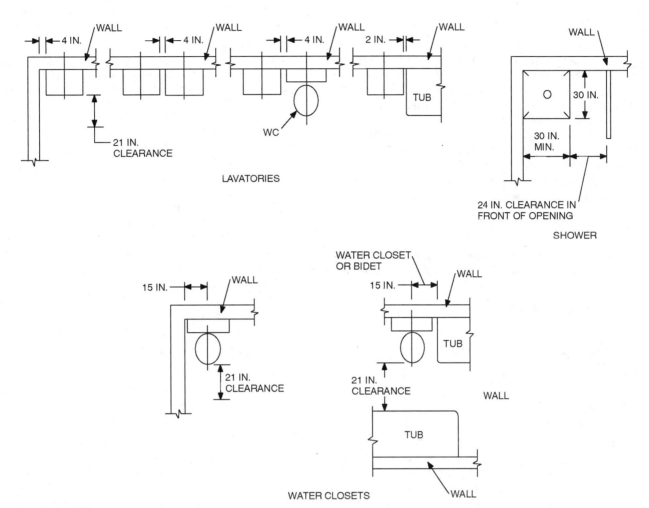

For SI: 1 inch = 25.4 mm.

FIGURE R307.1
MINIMUM FIXTURE CLEARANCES

building official and an affidavit, certificate or other evidence is submitted indicating compliance with the code. A manufacturer can identify safety glazing with a removable paper designation, provided it is destroyed during removal. This ensures that the designation will not be applied to a noncomplying piece of glass.

R308.1.1 Identification of multipane assemblies. Multipane assemblies having individual panes not exceeding 1 square foot (0.09 m²) in exposed area shall have at least one pane in the assembly identified in accordance with Section R308.1. All other panes in the assembly shall be labeled "16 CFR 1201."

❖ Multipane assemblies of glass need identification for the same reasons noted in Section R308.1. This provision allows labeling of only one pane of glass per Section R308.1.1, when the exposed area of each pane is 1 square foot (0.09 m²) or less. All other panes must be labeled "16 CFR 1201."

R308.2 Louvered windows or jalousies. Regular, float, wired or patterned glass in jalousies and louvered windows shall be no thinner than nominal $^3/_{16}$ inch (5 mm) and no longer than 48 inches (1219 mm). Exposed glass edges shall be smooth.

❖ The requirements for louvered windows exist because there is no edge support on the longitudinal edges of these panes. The code requires that the exposed edges be smooth for safety. The minimum thickness and maximum span are specified so that the glass has sufficient resistance to human impact loads.

R308.2.1 Wired glass prohibited. Wired glass with wire exposed on longitudinal edges shall not be used in jalousies or louvered windows.

❖ Wired glass is not permitted if the wire is exposed on the longitudinal edge because it would be a hazard.

R308.3 Human impact loads. Individual glazed areas, including glass mirrors in hazardous locations such as those indicated as defined in Section R308.4, shall pass the test requirements of CPSC 16 CFR, Part 1201. Glazing shall comply with CPSC 16 CFR, Part 1201 criteria for Category I or Category II as indicated in Table R308.3.

Exception: Louvered windows and jalousies shall comply with Section R308.2.

❖ The code requires that glazing in hazardous locations subject to human impact pass the test requirements of CPSC 16 CFR, Part 1201. This is the standard developed by the Consumer Product Safety Commission (CPSC) when it was determined that former standards (such as ANSI Z97.1) were inadequate to properly protect the public from injury caused by accidental impact with glazing. The exception provides for louvered windows and jalousies complying with Section R308.2. See the categories listed in Table R308.3.

R308.4 Hazardous locations. The following shall be considered specific hazardous locations for the purposes of glazing:

1. Glazing in swinging doors except jalousies.

2. Glazing in fixed and sliding panels of sliding door assemblies and panels in sliding and bifold closet door assemblies.

3. Glazing in storm doors.

4. Glazing in all unframed swinging doors.

5. Glazing in doors and enclosures for hot tubs, whirlpools, saunas, steam rooms, bathtubs and showers. Glazing in any part of a building wall enclosing these compartments where the bottom exposed edge of the glazing is less than 60 inches (1524 mm) measured vertically above any standing or walking surface.

6. Glazing, in an individual fixed or operable panel adjacent to a door where the nearest vertical edge is within a 24-inch (610 mm) arc of the door in a closed position and whose bottom edge is less than 60 inches (1524 mm) above the floor or walking surface.

7. Glazing in an individual fixed or operable panel, other than those locations described in Items 5 and 6 above, that meets all of the following conditions:

   7.1. Exposed area of an individual pane larger than 9 square feet (0.836 m²).

   7.2. Bottom edge less than 18 inches (457 mm) above the floor.

   7.3. Top edge more than 36 inches (914 mm) above the floor.

   7.4. One or more walking surfaces within 36 inches (914 mm) horizontally of the glazing.

8. All glazing in railings regardless of an area or height above a walking surface. Included are structural baluster panels and nonstructural infill panels.

9. Glazing in walls and fences enclosing indoor and outdoor swimming pools, hot tubs and spas where the bottom edge of the glazing is less than 60 inches (1524 mm) above a walking surface and within 60 inches (1524 mm) horizontally of the water's edge. This shall apply to single glazing and all panes in multiple glazing.

10. Glazing adjacent to stairways, landings and ramps within 36 inches (914 mm) horizontally of a walking surface when the exposed surface of the glass is less than 60 inches (1524 mm) above the plane of the adjacent walking surface.

11. Glazing adjacent to stairways within 60 inches (1524 mm) horizontally of the bottom tread of a stairway in any direction when the exposed surface of the glass is less than 60 inches (1524 mm) above the nose of the tread.

Exception: The following products, materials and uses are exempt from the above hazardous locations:

1. Openings in doors through which a 3-inch (76 mm) sphere is unable to pass.

2. Decorative glass in Items 1, 6 or 7.

3. Glazing in Section R308.4, Item 6, when there is an intervening wall or other permanent barrier between the door and the glazing.

4. Glazing in Section R308.4, Item 6, in walls perpendicular to the plane of the door in a closed position, other than the wall toward which the door swings when opened, or where access through the door is to a closet or storage area 3 feet (914 mm) or less in depth. Glazing in these applications shall comply with Section R308.4, Item 7.

5. Glazing in Section R308.4, Items 7 and 10, when a protective bar is installed on the accessible side(s) of the glazing 36 inches ± 2 inches (914 mm ± 51 mm) above the floor. The bar shall be capable of withstanding a horizontal load of 50 pounds per linear

TABLE R308.3
MINIMUM CATEGORY CLASSIFICATION OF GLAZING

EXPOSED SURFACE AREA OF ONE SIDE OF ONE LITE	GLAZING IN STORM OR COMBINATION DOORS (Category Class)	GLAZING IN DOORS (Category Class)	GLAZED PANELS REGULATED BY ITEM 7 OF SECTION R308.4 (Category Class)	GLAZED PANELS REGULATED BY ITEM 6 OF SECTION R308.4 (Category Class)	GLAZING IN DOORS AND ENCLOSURES REGULATED BY ITEM 5 OF SECTION R308.4 (Category Class)	SLIDING GLASS DOORS PATIO TYPE (Category Class)
9 sq ft or less	I	I	NR	I	II	II
More than 9 sq ft	II	II	II	II	II	II

For SI: 1 square foot = 0.0929 m².
NR means "No Requirement."

❖ See the commentary for Section 308.3.

foot (730 N/m) without contacting the glass and be a minimum of 1½ inches (38 mm) in height.

6. Outboard panes in insulating glass units and other multiple glazed panels in Section R308.4, Item 7, when the bottom edge of the glass is 25 feet (7620 mm) or more above grade, a roof, walking surfaces, or other horizontal [within 45 degrees (0.79 rad) of horizontal] surface adjacent to the glass exterior.

7. Louvered windows and jalousies complying with the requirements of Section R308.2.

8. Mirrors and other glass panels mounted or hung on a surface that provides a continuous backing support.

9. Safety glazing in Section R308.4, Items 10 and 11, is not required where:

   9.1. The side of a stairway, landing or ramp has a guardrail or handrail, including balusters or in-fill panels, complying with the provisions of Sections 1013 and 1607.7 of the *International Building Code*; and

   9.2. The plane of the glass is more than 18 inches (457 mm) from the railing; or

   9.3. When a solid wall or panel extends from the plane of the adjacent walking surface to 34 inches (863 mm) to 36 inches (914 mm) above the floor and the construction at the top of that

wall or panel is capable of withstanding the same horizontal load as the protective bar.

10. Glass block panels complying with Section R610.

❖ This section lists ten specific hazardous locations where safety glazing is required. Some of these locations are shown in Commentary Figures R308.4(1) through R308.4(9). In addition to the hazardous locations shown in the nine drawings, safety glazing is also required in a number of other locations, including fixed and sliding panels of sliding door assemblies, storm doors and glass railings.

Commentary Figure R308.4(1) illustrates several locations where safety glazing may or may not be required. To facilitate discussion, each glazed panel has been numbered. Panel 1 is not required to have a safety glazing because a protective bar has been installed in compliance with Exception 1 to Item 7, the details of which are illustrated in Commentary Figure R308.4(2). Panels 4 and 7 require safety glazing because they are door sidelights. The exception mentioned above does not apply to panels adjacent to a door, so even though Panel 7 has a protective bar, safety glazing is still required.

Commentary Figures R308.4(3) and (4) illustrate where safety glazing is required for panels adjacent to a door. This requirement applies to both fixed and operable panels. Where there is an intervening wall or permanent barrier, as shown in Commentary Figure R308.4(5),

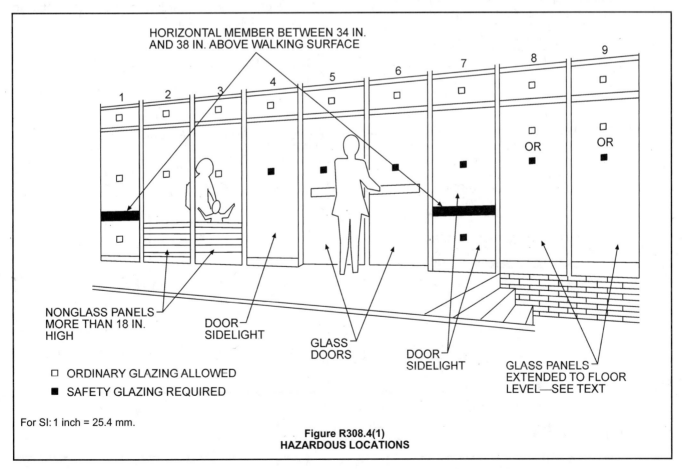

HORIZONTAL MEMBER BETWEEN 34 IN. AND 38 IN. ABOVE WALKING SURFACE

NONGLASS PANELS MORE THAN 18 IN. HIGH

DOOR SIDELIGHT

GLASS DOORS

DOOR SIDELIGHT

GLASS PANELS EXTENDED TO FLOOR LEVEL—SEE TEXT

□ ORDINARY GLAZING ALLOWED
■ SAFETY GLAZING REQUIRED

For SI: 1 inch = 25.4 mm.

Figure R308.4(1)
HAZARDOUS LOCATIONS

safety glazing would not be required. Commentary Figure R308.4(6) illustrates Exception 4, which applies to glazing positioned perpendicular to the plane of the door when it is in the closed position. Only one side is considered to be the hazardous location, the side that the door swings toward. The other side need not be safety glazing. This wall has a much lower risk of problems. When a door swings open to a perpendicular wall with glazing within 24 inches (610 mm), it is possible that if the door were caught by a strong wind it could slam into the wall and break the glass, or the door knob could hit the glass and break it. There is also the possibility that someone could be caught behind the door when it is opened and they could be pushed into/through the glass. Thus, this would be an appropriate area to have the required safety glazing to protect the occupants.

Panels 8 and 9, as well as Panels 2 and 3, fall under Item 6 of Section R308.4. Under this item all four stated conditions must occur before safety glazing is required. These conditions are as follows:

6.1 The area of an individual pane must be more than 9 square feet (0.84 m^2);

6.2 The bottom edge must be less than 18 inches (457 mm) above the floor:

6.3 The top edge must be more than 36 inches (914 mm) above the floor; and

6.4 One or more walking surfaces must be within 36 inches (914 mm), measured horizontally from the glazed panel.

Panels 2 and 3 do not require safety glazing because their bottom edges are not less than 18 inches (457 mm) from the floor.

If Panels 8 and 9 have a walking surface within 36 inches (914 mm) of the interior, safety glazing would be required. From the exterior side, as shown in Commentary Figure R308.4(1), the bottom of the panel appears to be more than 18 inches (457 mm) above the exterior walking surface, so the exterior condition would have no bearing on the determination. Panels 5 and 6 are glass doors, which require safety glazing based on the provisions of Items 1 or 4. Most ingress and egress doors (except jalousies), unframed swinging doors and glazing in storm doors require safety glazing, but there are several exceptions. If openings in a door will not pass a 3-inch diameter (76 mm) sphere, the glazing is exempt, as are assemblies of leaded, faceted or carved glass used for decoration. The latter exception applies to not only doors but also to sidelights and other glazed panels covered by Items 5 and 6.

Commentary Figure R308.4(7) illustrates the condition where a window occurs within a shower enclosure. If this window is less than 60 inches (1524 mm) above a standing surface, safety glazing would be required. This same requirement applies not only to showers but also to windows installed adjacent to hot tubs, whirlpools, saunas, steam rooms and bathtubs. Because of the presence of moisture, all of these locations represent slip hazards and need safety glazing to prevent injury in case of a fall.

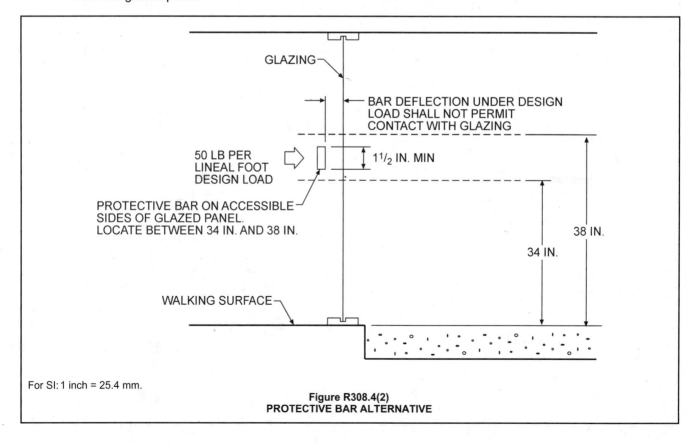

GLAZING

BAR DEFLECTION UNDER DESIGN LOAD SHALL NOT PERMIT CONTACT WITH GLAZING

50 LB PER LINEAL FOOT DESIGN LOAD

1^1/$_2$ IN. MIN

PROTECTIVE BAR ON ACCESSIBLE SIDES OF GLAZED PANEL. LOCATE BETWEEN 34 IN. AND 38 IN.

38 IN.

34 IN.

WALKING SURFACE

For SI: 1 inch = 25.4 mm.

Figure R308.4(2)
PROTECTIVE BAR ALTERNATIVE

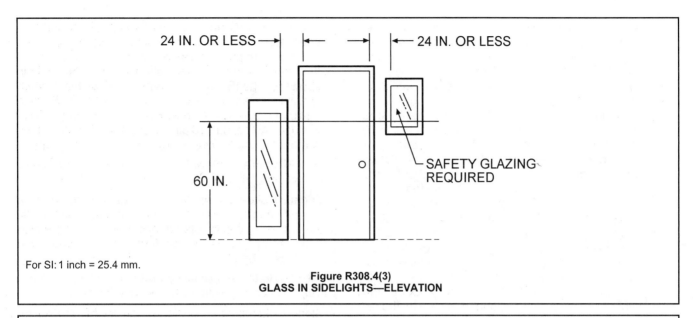

For SI: 1 inch = 25.4 mm.

Figure R308.4(3)
GLASS IN SIDELIGHTS—ELEVATION

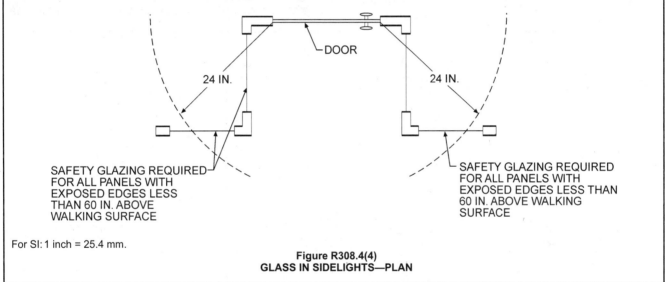

For SI: 1 inch = 25.4 mm.

Figure R308.4(4)
GLASS IN SIDELIGHTS—PLAN

Glass in railings, balusters panels and in-fill panels, regardless of their height above a walking surface, require safety glazing. Because of the high probability that people will strike guards, it is critical that an increased level of protection be provided.

Commentary Figure R308.4(8) illustrates the requirements of Item 9. This provision applies to walls and fences used as barriers for indoor or outdoor swimming pools and spas. Before safety glazing is required, the glazed panels must be within 5 feet (1524 mm) above the decking for a pool or spa. Items 10 and 11 address the hazardous locations to be considered for stairways, landings and ramps. Stairways and ramps present users with a greater risk for injury caused by falling than a flat surface. Not only is the risk of falling greater when using a stair, but the injuries are generally more severe. Unlike falling on a flat surface where the floor will, for the most part, break a person's fall, there is nothing to stop someone from

continuing to fall until he or she reaches the bottom of the stair. The increased risks inherent in stairways, as well as attempting to be consistent with other chapters in the code that mandate more restrictive requirements when addressing safety issues involving stairs and ramps, account for the more restrictive requirements for glazing in and around stairways and ramps. Item 10 includes any glazing within 36 inches (914 mm) horizontally of any walking surface when the exposed surface of that glazing is within 60 inches (1524 mm) of the walking surface. The walking surface in question would be part of a stair or ramp itself, including top, bottom and intermediate landings. It does not include adjacent floors or other walking surfaces [see Commentary Figure R308.4(9)].

In item 11, the concern is any glass that may be located within 60 inches (1524 mm) from the bottom tread in a run of stairs and within 60 inches (1524 mm) vertically of the walking surface of a stair. The code does not distinguish between a bottom tread at the primary floor level or at an intermediate landing. The last tread in a run of steps is the bottom tread. The 60-inch (1524 mm) dimension is from any point on the bottom tread, horizontally in any direction to any surface of any glazing within that range. Commentary Figure R308.4(9) illustrates the applications of Item 11. Safety glazing is not required for Items 10 and 11 where (1) the side of the stairway, landing or ramp has a guardrail or handrail, including the balusters or in-fill panels which comply with Sections 1013 and 1607.7 of the IBC which address the load resistance requirements of handrails and guardrails, (2) the actual plane of the glazing is located at least 18 inches (457 mm) from the glass and (3) a solid wall or panel extends to cover the area between 34 (863 mm) and 36 inches (914 mm) above the landing installing a bar is no longer necessary if the wall can resist the same load as the bar. The wall must be able to resist the same horizontal load at the top as the required protective bar listed in Exception #5 above.

Both the IRC and the IBC exempt glass unit masonry from hazardous locations. Glass block is becoming more prevalent in the design of homes. One of the more common uses is enclosures for walk-in showers. The IBC allows the use of glass block when installed in accordance with Section 2110. Section R610 of the IRC is almost identical to Section 2110 of the UBC. Exception 10 allows glass block installations without having to meet safety glazing requirements.

R308.5 Site built windows. Site built windows shall comply with Section 2404 of the *International Building Code*.

❖ Because site-built windows are not constructed in a manufacturing facility that follows industry standards, they must be constructed in accordance with Section 2404 of the IBC, which sets forth the wind, snow and dead loads on glass.

R308.6 Skylights and sloped glazing. Skylights and sloped glazing shall comply with the following sections.

❖ Sloped glazing and skylights consist of glazing installed in roofs or walls that are on a slope of more than

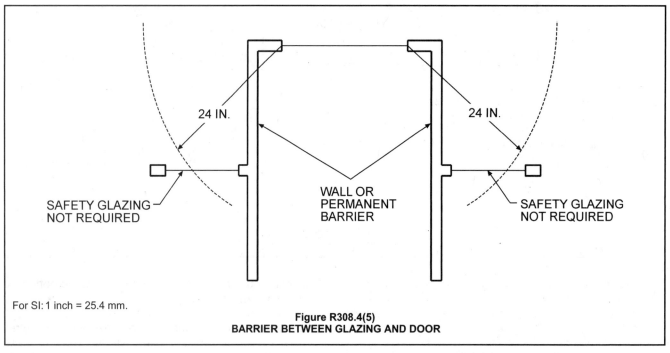

For SI: 1 inch = 25.4 mm.

Figure R308.4(5)
BARRIER BETWEEN GLAZING AND DOOR

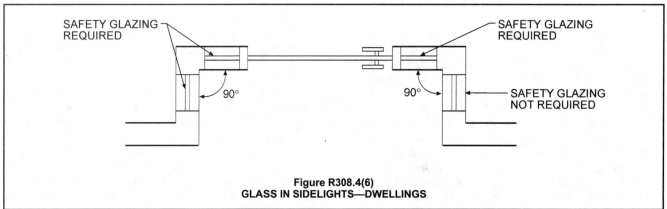

Figure R308.4(6)
GLASS IN SIDELIGHTS—DWELLINGS

15 degrees (0.26 rad) or more from the vertical. The provisions of the IRC address loads normally attributed to roofs. The provisions also enhance the protection of the occupants of a building from the possibility of falling glazing materials.

R308.6.1 Definitions.

SKYLIGHTS AND SLOPED GLAZING. Glass or other transparent or translucent glazing material installed at a slope of 15 degrees (0.26 rad) or more from vertical. Glazing materials in skylights, including unit skylights, solariums, sunrooms, roofs and sloped walls are included in this definition.

❖ The failure of skylights and sloped glazing could result in injury and building damage. This definition establishes the criteria to which the code requirements of Section R308.6 are to apply.

UNIT SKYLIGHT. A factory assembled, glazed fenestration unit, containing one panel of glazing material, that allows for natural daylighting through an opening in the roof assembly while preserving the weather-resistant barrier of the roof.

❖ Unit skylights are a specific type of sloped glazing assembly which is factory assembled. The IBC and IRC contain specific code provisions that are appropriate for this type of building component. Factory-assembled units, as opposed to site-built skylights, can be designed, tested and rated as one component which incorporates both glazing and framing, if applicable. The individual components of site-built glazing must be designed to resist the design loads of the codes individually, and are not usually rated as an assembly.

R308.6.2 Permitted materials. The following types of glazing may be used:

1. Laminated glass with a minimum 0.015-inch (0.38 mm) polyvinyl butyral interlayer for glass panes 16 square feet (1.5 m²) or less in area located such that the highest point of the glass is not more than 12 feet (3658 mm) above a walking surface or other accessible area; for higher or larger sizes, the minimum interlayer thickness shall be 0.030 inch (0.76 mm).

2. Fully tempered glass.

3. Heat-strengthened glass.

4. Wired glass.

5. Approved rigid plastics.

❖ The provisions of this section limit glazing materials in skylights and sloped glazing to those specified, and they outline glazing materials and protective measures for sloped glazing and skylights. The materials and their characteristics and limitations are as follows:

Laminated glass. Laminated glass is usually constructed with an inner layer of polyvinyl butyral, which has a minimum thickness of 30 mil (0.76 mm). Such glass is highly resistant to impact and as a result requires no further protection below. When used within dwelling units, laminated glass is permitted to have a 15-mil (0.38 mm) polyvinyl butyral inner layer if each pane of glass is 16 square feet (1.5 m²) or less in area, and the highest point of the glass is no more than 12 feet (3658 mm) above a walking surface or other accessible area.

For SI: 1 inch = 25.4 mm.

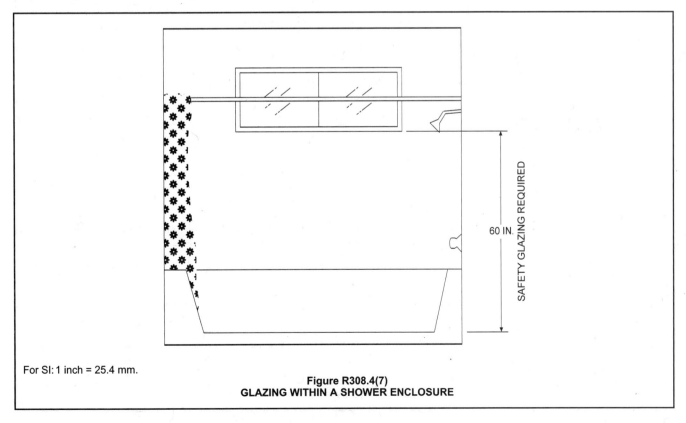

Figure R308.4(7)
GLAZING WITHIN A SHOWER ENCLOSURE

Fully tempered glass. Tempered glass is glass that has been specifically heat-treated or chemically treated to obtain high strength. When broken, the entire piece of glass immediately breaks into numerous small granular pieces. Because of its high strength and manner of breakage, tempered glass has been considered in the past to be a desirable glazing material for skylights that have no protective screens. However, as a result of studies by the industry that show that tempered glass is subject to spontaneous breakage that can result in large chunks of glass falling, the IRC requires screen protection below tempered glass.

Heat-strengthened glass. Heat-strengthened glass is glass that has been reheated to just below its melting point and then cooled. This process forms a compression on the outer surface and increases the strength of the glass. However, heat-strengthened glass requires screen protection below the skylight to protect the occupants from falling shards.

Wired glass. Wired glass is resistant to impact and when used as a single-layer glazing requires no additional protection below.

Approved rigid plastics. Rigid plastics are fairly durable as a glazing material.

Annealed glass. Annealed glass is not allowed because it is subject to breakage by impact and has very low strength. Annealed glass is also unsatisfactory for use as a skylight because it breaks up under impact into large sharp shards, which, when they fall, are hazardous to occupants of a building.

R308.6.3 Screens, general. For fully tempered or heat-strengthened glass, a retaining screen meeting the requirements of Section R308.6.7 shall be installed below the glass, except for fully tempered glass that meets either condition listed in Section R308.6.5.

❖ As a general rule, single-layer glazing of heat-strengthened glass and fully tempered glass must be fitted with screens below the glazing material.

R308.6.4 Screens with multiple glazing. When the inboard pane is fully tempered, heat-strengthened or wired glass, a retaining screen meeting the requirements of Section R308.6.7 shall be installed below the glass, except for either condition listed in Section R308.6.5. All other panes in the multiple glazing may be of any type listed in Section R308.6.2.

❖ As does Section R308.6.3, this section states that screens are required for the inbound plane of glazing when it is fully tempered, heat-strengthened or wired glass. The screen must comply with Section R308.6.7. Screens are not required for either approved laminated glass or approved rigid plastics.

R308.6.5 Screens not required. Screens shall not be required when fully tempered glass is used as single glazing or the

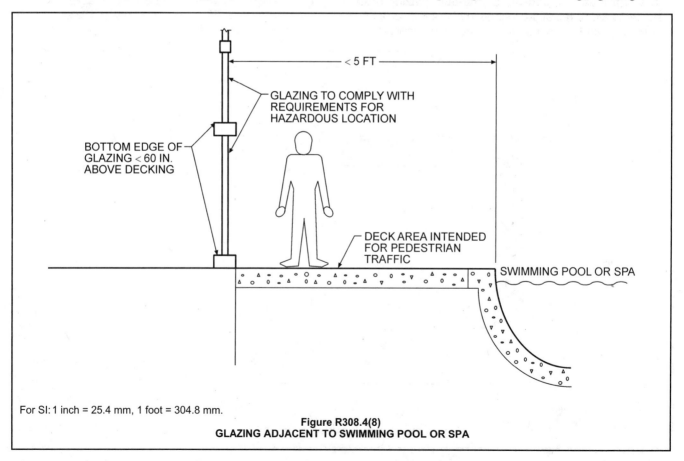

For SI: 1 inch = 25.4 mm, 1 foot = 304.8 mm.

Figure R308.4(8)
GLAZING ADJACENT TO SWIMMING POOL OR SPA

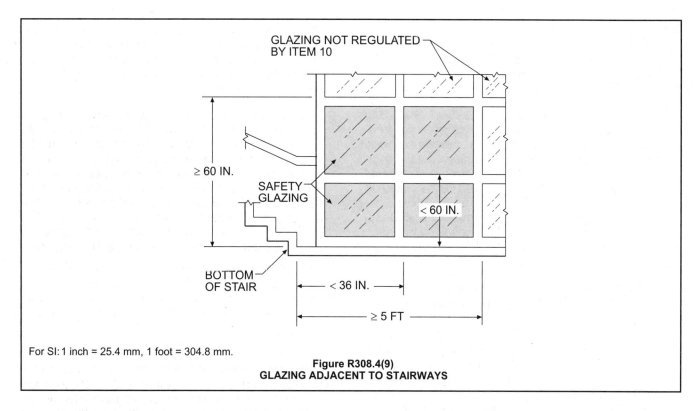

GLAZING NOT REGULATED BY ITEM 10

SAFETY GLAZING

≥ 60 IN.

< 60 IN.

BOTTOM OF STAIR

< 36 IN.

≥ 5 FT

For SI: 1 inch = 25.4 mm, 1 foot = 304.8 mm.

Figure R308.4(9)
GLAZING ADJACENT TO STAIRWAYS

inboard pane in multiple glazing and either of the following conditions are met:

1. Glass area 16 square feet (1.49 m²) or less. Highest point of glass not more than 12 feet (3658 mm) above a walking surface or other accessible area, nominal glass thickness not more than ³/₁₆ inch (4.8 mm), and (for multiple glazing only) the other pane or panes fully tempered, laminated or wired glass.

2. Glass area greater than 16 square feet (1.49 m²). Glass sloped 30 degrees (0.52 rad) or less from vertical, and highest point of glass not more than 10 feet (3048 mm) above a walking surface or other accessible area.

❖ Section R308.6.5 states two exceptions to the provisions of Section R308.6.3 and Section R308.6.4. The first exception applies to glazing that is no larger than 16 square feet (1.49 m²) and is no more than 12 feet (3658 mm) above a walking surface or other accessible area. The second exception applies to sloped glazing with a maximum slope of 30 degrees (0.52 rad) from vertical, is not larger than 16 square feet (1.49 m²) in area, and is no greater than 10 feet (3048 mm) above a walking surface or other accessible area. Generally, installed skylights and sloped glazing will meet one of these exceptions, so screens are not required.

R308.6.6 Glass in greenhouses. Any glazing material is permitted to be installed without screening in the sloped areas of greenhouses, provided the greenhouse height at the ridge does not exceed 20 feet (6096 mm) above grade.

❖ The glazing regulations for greenhouses are less stringent because greenhouses are seldom occupied during storms that might break the glass. These provisions also explain an exception to the screening provisions of Sections R308.6 and R308.6.4, specifically for the sloped glazing areas within greenhouses. Screens are not required for sloped areas of greenhouses if the ridge of the greenhouse is not more than 20 feet (6096 mm) above grade.

R308.6.7 Screen characteristics. The screen and its fastenings shall be capable of supporting twice the weight of the glazing, be firmly and substantially fastened to the framing members, and have a mesh opening of no more than 1 inch by 1 inch (25 mm by 25 mm).

❖ It is critical that screens be installed in a manner that will adequately support the weight of the glass. In using a safety factor of 2, the screen and its fastenings must be capable of supporting twice the weight of the glazing. To accomplish this, the screen is to be fastened firmly to the framing members.

R308.6.8 Curbs for skylights. All unit skylights installed in a roof with a pitch flatter than three units vertical in 12 units horizontal (25-percent slope) shall be mounted on a curb extending at least 4 inches (102 mm) above the plane of the roof unless otherwise specified in the manufacturer's installation instructions.

❖ Skylights installed on low-sloped roofs are more susceptible to leaking than those on higher pitched roofs because water does not drain as quickly on low-sloped roofs.

For these skylights to be properly flashed to prevent leakage, they must be placed on a 4-inch (102 mm) high curb unless the manufacturer's installation instructions indicate otherwise.

R308.6.9 Testing and labeling. Unit skylights shall be tested by an approved independent laboratory, and bear a label identifying manufacturer, performance grade rating and approved inspection agency to indicate compliance with the requirements of AAMA/WDMA/CSA 101/I.S.2/A440.

❖ The referenced standard, AAMA/WDMA 101/I.S.2/A440, *Voluntary Performance Specification for Windows, Skylights and Glass Doors*, includes a separate rating system for positive and negative pressure on skylights which allows the manufacturer to design and fabricate products that are best suited for the climate in which they will be used. Standard 101/I.S.2/NAFS establishes the performance requirements for skylights based on the desired performance grade rating which includes minimum requirements for resistance to air leakage, water infiltration and the design load pressures. The resulting performance grade rating states the design load pressure used to rate the product, but it also includes consideration of these additional performance characteristics. For skylights certified for only one performance grade, the rating is based on the minimum requirements met for both positive and negative design pressure. Skylights certified for two performance grades are rated separately for positive and negative design pressure.

Skylights must be capable of withstanding the component and cladding wind pressures of Table R301.2(2) adjusted by the height and exposure coefficients given in Table R301.2(3).

The most critical load on a skylight is determined by the climate in which it is installed. In a colder climate with heavier snow loads and moderate design wind speeds, the positive load on a skylight from the combined snow and dead load will be more critical than the negative load from wind uplift. The opposite will be the case in warmer, coastal climates with higher design wind speeds, and little or no snow load.

SECTION R309
GARAGES AND CARPORTS

R309.1 Opening protection. Openings from a private garage directly into a room used for sleeping purposes shall not be permitted. Other openings between the garage and residence shall be equipped with solid wood doors not less than $1^3/_8$ inches (35 mm) in thickness, solid or honeycomb core steel doors not less than $1^3/_8$ inches (35 mm) thick, or 20-minute fire-rated doors.

❖ Openings to sleeping rooms from garages are not allowed because a person might not wake up in time if there was a hazard from carbon monoxide fumes or smoke from the garage. Openings from the garage are permitted only into rooms that are not used for sleeping. Those openings must be protected by the installation of a door complying with the provisions of this section. Solid wood doors $1^3/_8$-inch (35 mm) thick, solid or honeycomb steel doors and 20-minute fire-rated doors are the only acceptable doors for use in the opening between the garage and the dwelling unit.

R309.1.1 Duct penetration. Ducts in the garage and ducts penetrating the walls or ceilings separating the dwelling from the garage shall be constructed of a minimum No. 26 gage (0.48 mm) sheet steel or other approved material and shall have no openings into the garage.

❖ Ducts are permitted to penetrate the required separation (see Section R309.2) between the garage and the dwelling unit when the ducts within the garage and the portion of the duct penetrating the wall are of No. 26 gage (0.48 mm) sheet steel or other materials acceptable to the building official. Steel ducts are required to help prevent the passage of an undetected fire within the garage to the dwelling unit (see Commentary Figure R309.1.1). The opening limitation in the garage is to limit the path for smoke to enter the dwelling unit.

R309.1.2 Other penetrations. Penetrations through the separation required in Section R309.2 shall be protected by filling the opening around the penetrating item with approved material to resist the free passage of flame and products of combustion.

❖ This section addresses the annular space which results from a penetration of the common wall by pipes, conduits or ductwork. It is important that the code official verify that these spaces are properly filled and do not compromise the protection offered by the common wall between the residence and garage against the free passage of smoke, fire, noxious gases and odors.

R309.2 Separation required. The garage shall be separated from the residence and its attic area by not less than $^1/_2$-inch (12.7 mm) gypsum board applied to the garage side. Garages beneath habitable rooms shall be separated from all habitable rooms above by not less than $^5/_8$-inch (15.9 mm) Type X gypsum board or equivalent. Where the separation is a floor-ceiling assembly, the structure supporting the separation shall also be protected by not less than $^1/_2$-inch (12.7 mm) gypsum board or equivalent. Garages located less than 3 feet (914 mm) from a dwelling unit on the same lot shall be protected with not less than $^1/_2$-inch (12.7 mm) gypsum board applied to the interior side of exterior walls that are within this area. Openings in these walls shall be regulated by Section R309.1. This provision does not apply to garage walls that are perpendicular to the adjacent dwelling unit wall.

❖ Numerous potential hazards exist within garages because occupants of dwelling units tend to store a variety of hazardous materials there. Along with this and the potential for carbon monoxide build-up within the garage, the IRC requires that the garage be separated from the dwelling unit and the attic with at least $^1/_2$-inch (12.7 mm) gypsum board or other equivalent material. If a habitable room is above the garage, the separation must be at least $^5/_8$-inch (15.9 mm) Type X gypsum board or equivalent. In addition, garages located less than 3 feet (305 mm) from an adjacent structure must be protected with at least $^1/_2$ inch (12.7 mm) gypsum board applied to the interior side of the garage. See Section R302. The close proximity to adjacent structures requires the additional protection.

There are two primary reasons for the enhanced fire endurance of a garage ceiling located beneath a habitable room. First, a fire occurring in a garage may well go undetected for an extended period prior to activation of a detector or other visual alerting.

Second, the inherent fire load and hazardous household activities associated with a garage necessitate this additional level of protection if fire suppression forces are to have a reasonable opportunity to contain a garage fire to the area of origin.

The single layer of $^5/_8$-inch (15.9 mm) Type X gypsum at the garage ceiling increases the fire endurance of the assembly considerably, from 15 minutes for a $^1/_2$-inch (12.7 mm) layer, to at least 40 minutes, or a 167 percent increase in endurance. When added to the rating for floor joists and certain subflooring combinations, the final endurance is close to 1 hour.

Additionally, the exterior walls of the garage are required to have $^1/_2$-inch (12.7 mm) gypsum board on the interior face where they support floors separating all or part of a dwelling unit above the garage. Commentary Figure R309.2 shows two locations of gypsum wallboard; each achieves the protection required by the code.

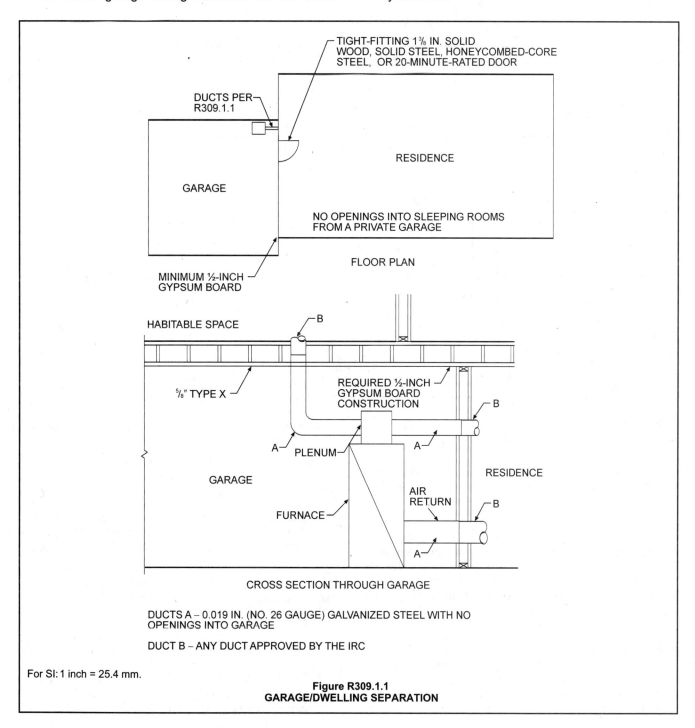

DUCTS A – 0.019 IN. (NO. 26 GAUGE) GALVANIZED STEEL WITH NO OPENINGS INTO GARAGE

DUCT B – ANY DUCT APPROVED BY THE IRC

For SI: 1 inch = 25.4 mm.

Figure R309.1.1
GARAGE/DWELLING SEPARATION

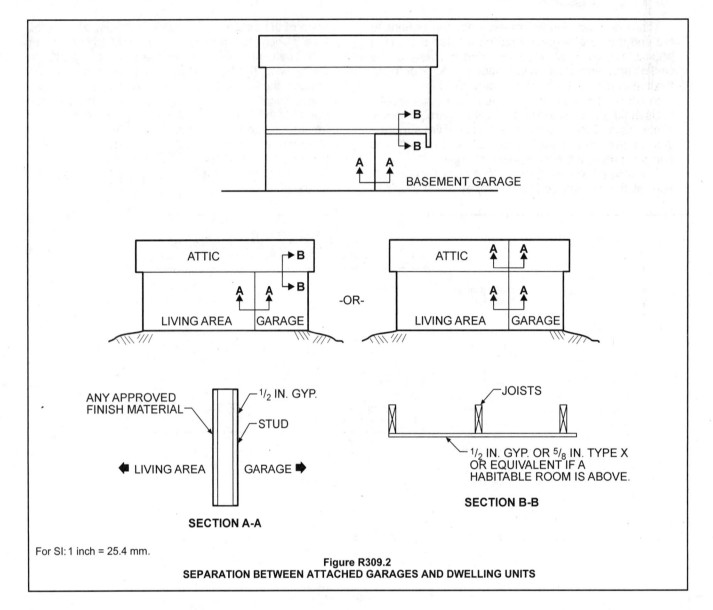

For SI: 1 inch = 25.4 mm.

Figure R309.2
SEPARATION BETWEEN ATTACHED GARAGES AND DWELLING UNITS

R309.3 Floor surface. Garage floor surfaces shall be of approved noncombustible material.

The area of floor used for parking of automobiles or other vehicles shall be sloped to facilitate the movement of liquids to a drain or toward the main vehicle entry doorway.

❖ Garage floor surfaces must be of an approved noncombustible material such as concrete. Additionally, the floor surface must either slope toward the garage door opening or slope to an approved drain. This allows grease, flammable liquids or other hazardous materials that might drain from an automobile to drain from the garage.

R309.4 Carports. Carports shall be open on at least two sides. Carport floor surfaces shall be of approved noncombustible material. Carports not open on at least two sides shall be considered a garage and shall comply with the provisions of this section for garages.

Exception: Asphalt surfaces shall be permitted at ground level in carports.

The area of floor used for parking of automobiles or other vehicles shall be sloped to facilitate the movement of liquids to a drain or toward the main vehicle entry doorway.

❖ Carports must have at least two sides open to outside air. If two sides are not open, the structure is a garage and comes under the provisions of Sections R309.1, R309.2, R309.3 and R309.5.

R309.5 Flood hazard areas. For buildings located in flood hazard areas as established by Table R301.2(1), garage floors shall be:

1. Elevated to or above the design flood elevation as determined in Section R324; or

2. Located below the design flood elevation provided they are at or above grade on all sides, are used solely for parking, building access, or storage, meet the require-

ments of Section R324, and are otherwise constructed in accordance with this code.

❖ Garage floors of buildings in flood hazard areas must meet one of two requirements. The first option is to simply construct the garage floor above the design flood elevation. The second option allows the floor of the garage to be below the design flood elevation if construction is compliant with the applicable provisions of Section R324 for enclosures, including: 1) the floor is at or above finished exterior grade (not a basement); 2) it complies with provisions for enclosed areas below the design flood elevation (Section R324.2.2); 3) materials and finishes below the design flood elevation are flood-resistant (Section R324.1.7); and 4) service equipment and systems comply with Section R324.1.5. If a garage below the design flood elevation does not meet the requirements for enclosures, the garage floor is the "lowest floor," and the building is noncompliant. Noncompliance means that although the living area is elevated, flood insurance will be rated as though the building is not.

R309.6 Automatic garage door openers. Automatic garage door openers, if provided, shall be listed in accordance with UL 325.

❖ This code does not require an automatic garage door opener. However, if one is installed, it must be listed in accordance with UL 325. Federal law requires automatic residential garage door openers to conform to the entrapment protection requirements of UL 325.

SECTION R310
EMERGENCY ESCAPE AND RESCUE OPENINGS

R310.1 Emergency escape and rescue required. Basements and every sleeping room shall have at least one operable emergency escape and rescue opening. Such opening shall open directly into a public street, public alley, yard or court. Where basements contain one or more sleeping rooms, emergency egress and rescue openings shall be required in each sleeping room, but shall not be required in adjoining areas of the basement. Where emergency escape and rescue openings are provided they shall have a sill height of not more than 44 inches (1118 mm) above the floor. Where a door opening having a threshold below the adjacent ground elevation serves as an emergency escape and rescue opening and is provided with a bulkhead enclosure, the bulkhead enclosure shall comply with Section R310.3. The net clear opening dimensions required by this section shall be obtained by the normal operation of the emergency escape and rescue opening from the inside. Emergency escape and rescue openings with a finished sill height below the adjacent ground elevation shall be provided with a window well in accordance with Section R310.2. Emergency escape and rescue openings shall open directly into a public way, or to a yard or court that opens to a public way.

Exception: Basements used only to house mechanical equipment and not exceeding total floor area of 200 square feet (18.58 m²).

❖ Because so many fire deaths occur as the result of occupants of residential buildings being asleep during a fire, the IRC requires that all basements and sleeping rooms have windows or doors that may be used for emergency escape or rescue. These emergency openings must open directly into a public street, public alley, yard or court. The requirement for emergency escape and rescue openings in sleeping rooms exists because a fire will usually have spread before the occupants are aware of the problem, and the normal exit channels may be blocked. The requirement for basements exists because they are so often used as sleeping rooms. For example, a fire in a mechanical room adjacent to a stair could engulf the only means of egress for the basement without the egress window or door.

Openings required for emergency escape or rescue must be located on the exterior of the building so that rescue can be performed from the exterior. Alternatively, occupants may escape through that opening to the exterior of the building without having to travel through the building itself. Therefore, where openings are required, they should open directly into a public street, public alley, yard or court. After the occupants pass through the emergency escape and rescue opening, their continued egress is essential. Where a basement contains sleeping rooms and a habitable space, an emergency escape and rescue opening is required in each sleeping room, but is not required in adjoining areas of the basement.

There is an exception for basements used only to house mechanical equipment with a total floor area not exceeding 200 square feet.

The dimensions prescribed in the code, and as illustrated in Commentary Figure R310.1 for exterior wall openings used for emergency egress and rescue, are based in part on extensive testing by the San Diego Building and Fire Departments to determine the proper relationships of the height and width of window openings to adequately serve for both rescue and escape. The minimum of 20 inches (508 mm) for the width is based on two criteria: the width necessary to place a ladder within the window opening and the width necessary to admit a fire fighter with full rescue equipment including breathing apparatus. The minimum 24-inch (610 mm) height is based on the minimum size necessary to admit a fire fighter with full rescue equipment. By requiring a minimum net clear opening size of the least 5.7 square feet (0.53 m²), the code provides for an opening of adequate dimensions. To be accessible from the interior of the sleeping room or basement, the emergency escape and rescue opening cannot be located more than 44 inches (1118 mm) above the floor. The measurement is to be taken from the floor to the bottom of the clear opening.

The required opening dimensions must be achieved by the normal operation of the window, door or hatch from the inside without the use of keys, tools or special knowledge. The window industry is a highly competitive market. Manufacturers are constantly developing new products that are easier to clean and possess higher thermal protection properties. It is important to keep in mind that no special knowledge for operation of the egress window is a key operational constraint. It is impractical to assume that all occupants can operate a window that requires a special sequence of operations to achieve the required opening size. Although most occupants are familiar with the normal operation to open the window, children and guests are frequently unfamiliar with special procedures necessary to remove the sashes. The time spent comprehending special operations unnecessarily delays egress from the bedroom and could lead to panic and further confusion. Thus, windows that achieve the required opening dimensions only by performing a special sequence of operations such as the removal of sashes or mullions are not permitted. For example, if a specific area of the window has to be depressed or manipulated to allow the sash to be removed or released to achieve the open area requirement of 5.7 square feet (0.53 m^2), the window does not qualify as an egress window.

R310.1.1 Minimum opening area. All emergency escape and rescue openings shall have a minimum net clear opening of 5.7 square feet (0.530 m^2).

Exception: Grade floor openings shall have a minimum net clear opening of 5 square feet (0.465 m^2).

❖ Where an emergency escape and rescue window is located at grade level, the opening size requirement is reduced to be 5 square feet (0.46 m^2). This results from the increased ease of access from the exterior and the probability that a ladder will not be needed.

R310.1.2 Minimum opening height. The minimum net clear opening height shall be 24 inches (610 mm).

❖ The minimum opening height for emergency space and rescue opening is 24 inches (610 mm), based on the minimum dimension of a fire fighter with full rescue equipment.

R310.1.3 Minimum opening width. The minimum net clear opening width shall be 20 inches (508 mm).

❖ This section establishes a minimum width of 20 inches (508 mm) for emergency space and rescue openings,

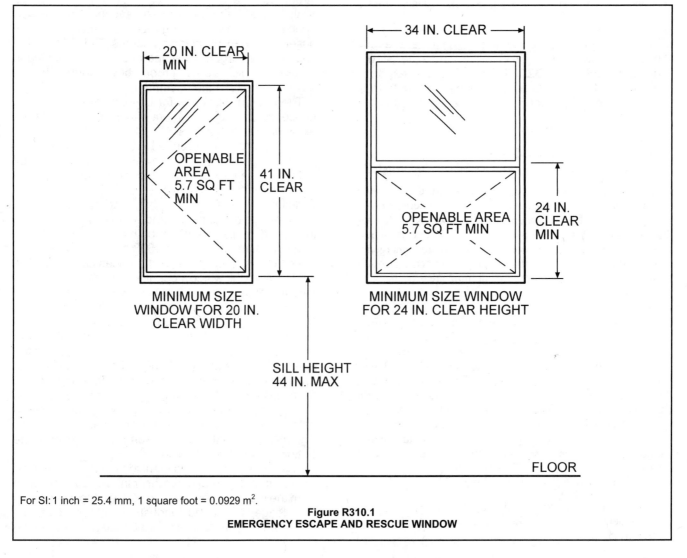

For SI: 1 inch = 25.4 mm, 1 square foot = 0.0929 m^2.

Figure R310.1
EMERGENCY ESCAPE AND RESCUE WINDOW

based on the minimum dimension of a fire fighter with full rescue equipment.

R310.1.4 Operational constraints. Emergency escape and rescue openings shall be operational from the inside of the room without the use of keys, tools or special knowledge.

❖ Openings for emergency escape and rescue must be operational from the inside. Keys, tools or special tools knowledge must not be needed to operate these openings. If keys or tools were necessary, they might not be readily available in an emergency or panic situation, and an individual might not be able to use them, so the opening would be unusable. Section R310.1 also requires the opening size to be obtained by the normal operation of the window. See Section R310.1 commentary.

R310.2 Window wells. The minimum horizontal area of the window well shall be 9 square feet (0.9 m²), with a minimum horizontal projection and width of 36 inches (914 mm). The area of the window well shall allow the emergency escape and rescue opening to be fully opened.

Exception: The ladder or steps required by Section R310.2.1 shall be permitted to encroach a maximum of 6 inches (152 mm) into the required dimensions of the window well.

❖ Window wells in front of emergency escape and rescue openings also have minimum size requirements. These provisions address those emergency escape windows that occur below grade. Just applying the standard emergency escape window criteria to these windows will result in an opening that occupants can get through, but the window well may actually trap the occupants against the building without providing for their escape from the window well or providing for a fire fighter to enter the residence.

The minimum size requirements in cross-section are similar to the emergency escape and opening criteria; that is, they are sufficient to provide a nominal size to allow for the escape of occupants or the entry of fire fighters (see Commentary Figure R310.2). The ladder or steps requirement is the main difference.

R310.2.1 Ladder and steps. Window wells with a vertical depth greater than 44 inches (1118 mm) shall be equipped with a permanently affixed ladder or steps usable with the window in the fully open position. Ladders or steps required by this section shall not be required to comply with Sections R311.5 and R311.6. Ladders or rungs shall have an inside width of at least 12 inches (305 mm), shall project at least 3 inches (76 mm) from the wall and shall be spaced not more than 18 inches (457 mm) on center vertically for the full height of the window well.

❖ When the depth of a window well exceeds 44 inches (1118 mm), a ladder or steps from the window is required. The details for construction of steps are not identified in the provisions; however, the design of the ladder is specifically addressed. Because ladders and steps in window wells are provided for emergency use only, they are not required to comply with the provisions for stairways found in Sections R311.5 and R311.6.

R310.3 Bulkhead enclosures. Bulkhead enclosures shall provide direct access to the basement. The bulkhead enclosure with the door panels in the fully open position shall provide the minimum net clear opening required by Section R310.1.1. Bulkhead enclosures shall also comply with Section R311.5.8.2.

❖ Bulkhead enclosures, when provided for access to below-grade openings used for emergency escape and rescue, must meet the net openable area provisions of Section R310.1.1. Also, bulkhead enclosures must comply with Section R311.5.8.2.

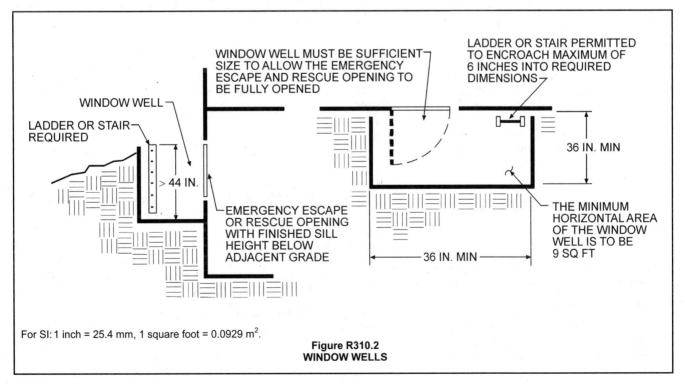

For SI: 1 inch = 25.4 mm, 1 square foot = 0.0929 m².

Figure R310.2
WINDOW WELLS

R310.4 Bars, grilles, covers and screens. Bars, grilles, covers, screens or similar devices are permitted to be placed over emergency escape and rescue openings, bulkhead enclosures, or window wells that serve such openings, provided the minimum net clear opening size complies with Sections R310.1.1 to R310.1.3, and such devices shall be releasable or removable from the inside without the use of a key, tool, special knowledge or force greater than that which is required for normal operation of the escape and rescue opening.

❖ The ever-increasing concern for security, particularly in residential buildings, has created a fairly large demand for security devices such as grilles, bars and steel shutters. Unless properly designed and constructed, the security devices over bedroom windows can completely defeat the purpose of the emergency escape and rescue opening. Therefore, the IRC makes provisions for security devices if the release mechanism has been approved and is operable from the inside without the use of a key, tool or force greater than that required for normal operation of the escape and rescue opening.

The essence of the requirement for emergency escape openings is that a person must be able to effect escape or be rescued in a short period of time because the fire might have spread to a point where all other exit routes are blocked. Thus, time cannot be wasted in figuring out means of opening rescue windows or obtaining egress through them. Any impediment to escape or rescue caused by security devices, inadequate window size or difficult operating mechanisms is not permitted by the code.

R310.5 Emergency escape windows under decks and porches. Emergency escape windows are allowed to be installed under decks and porches provided the location of the deck allows the emergency escape window to be fully opened and provides a path not less than 36 inches (914 mm) in height to a yard or court.

❖ The design of some homes makes the underside of decks the only location where an emergency escape and rescue window can be located. The 36 inch (914 mm) minimum height requirement allows a usable means of egress pathway. The 36 inch (914 mm) minimum area was based on the minimum window well size of 3 by 3 feet (914 mm by 914 mm).

SECTION R311
MEANS OF EGRESS

R311.1 General. Stairways, ramps, exterior egress balconies, hallways and doors shall comply with this section.

❖ Sections R311.2 through R311.6 contains the requirements for the exit and the means of egress components.

R311.2 Construction.

❖ Sections R311.2.1 and R311.2.2 contain the provisions for attaching means of egress components to the main structure and for fire protection of under-stair enclosures.

R311.2.1 Attachment. Required exterior egress balconies, exterior exit stairways and similar means of egress components shall be positively anchored to the primary structure to resist both vertical and lateral forces. Such attachment shall not be accomplished by use of toenails or nails subject to withdrawal.

❖ Exterior exit balconies, stairs and similar exit facilities must be properly attached to the primary structure so that their reaction to vertical and lateral forces will not cause separation from the structure. This is reiterating the requirement for a complete load path in Section R301.1. The reason for doing so is the need to maintain key elements of the egress system that are needed for emergency evacuations and the increased possibility of overlooking such connection. "Exterior exit balconies" and "exit facilities" are not defined or otherwise described in this code. The requirement for positive anchor to the primary structure applies to all exterior means of egress components used as part of an egress system whether part of the required exit or not.

R311.2.2 Under stair protection. Enclosed accessible space under stairs shall have walls, under stair surface and any soffits protected on the enclosed side with $^1/_2$-inch (13 mm) gypsum board.

❖ Often times the space under a stairway is used for storage because this space is often of little use for other purposes. The code permits the use of an open space beneath a stair without the need for any additional protection. Additionally, if the space is walled off and there is no access to the area, the code is also not concerned. If, however, the area beneath the stairway is enclosed and any type of access is provided into the space, the walls, soffits and ceilings of the enclosed space must be protected on the enclosed side with at least $^1/_2$ inch (12.7 mm) gypsum board.

R311.3 Hallways. The minimum width of a hallway shall be not less than 3 feet (914 mm).

❖ Hallways must be a minimum of 3 feet (914 mm) wide to accommodate moving furniture into rooms off the hallway and for safe egress from the structure.

R311.4 Doors.

❖ The requirements for doors are contained in Sections R311.4.1 through R311.4.4.

R311.4.1 Exit door required. Not less than one exit door conforming to this section shall be provided for each dwelling unit. The required exit door shall provide for direct access from the habitable portions of the dwelling to the exterior without requiring travel through a garage. Access to habitable levels not having an exit in accordance with this section shall be by a ramp in accordance with Section R311.6 or a stairway in accordance with Section R311.5.

❖ All dwelling units must to have at least one exit door that complies with the provisions of Sections R311.4.2, R311.4.3 and R311.4.4. This door must also access the exterior without the dwelling's occupants traveling through a garage where hazards could prevent a suitable means of egress. A ramp or stairway is required for access to habitable areas not having an exit.

R311.4.2 Door type and size. The required exit door shall be a side-hinged door not less than 3 feet (914 mm) in width and 6 feet 8 inches (2032 mm) in height. Other doors shall not be required to comply with these minimum dimensions.

❖ This provision applies only to a single exit door required by Section R311.4.1. Other exterior doors need not comply with provisions of this section.

R311.4.3 Landings at doors. There shall be a floor or landing on each side of each exterior door. The floor or landing at the exterior door shall be not more than 1.5 inches (38 mm) lower than the top of the threshold. The landing shall be permitted to have a slope not to exceed 0.25 unit vertical in 12 units horizontal (2-percent).

Exceptions:

1. Where a stairway of two or fewer risers is located on the exterior side of a door, other than the required exit door, a landing is not required for the exterior side of the door provided the door, other than an exterior storm or screen door does not swing over the stairway.

2. The exterior landing at an exterior doorway shall not be more than $7^3/_4$ inches (196 mm) below the top of the threshold, provided the door, other than an exterior storm or screen door does not swing over the landing.

3. The height of floors at exterior doors other than the exit door required by Section R311.4.1 shall not be more than $7^3/_4$ inches (186 mm) lower than the top of the threshold.

The width of each landing shall not be less than the door served. Every landing shall have a minimum dimension of 36 inches (914 mm) measured in the direction of travel.

❖ Landings are required for exterior doors and must be constructed on both the exterior and interior side of the door. Thresholds should not be higher than $1^1/_2$ inches (38 mm) above floor level (see Commentary Figure R311.4.3). The threshold height represents an important element in building construction. It has to be high enough to keep out snow accumulation and driving rain yet low enough not to represent a tripping hazard or become a barrier to entry. The landing must be reasonably level (a slope not exceeding 25:12) while still allowing enough of a slope for proper drainage.

The first exception would not require a landing when two or fewer stair risers are on the exterior side of a door, other than the required exit door. However, the door is not allowed to swing over the steps to take advantage of this exception.

The second exception permits the exterior landing of an exterior door, other than the exit door required in Section R311.4.1, to be a maximum of $7^3/_4$ inches (196 mm) below the top of the threshold (see Commentary Figure R311.4.3).

The third exception addresses the height of floors at exterior doors other than the required exit door. Landings must be the same width as the door they serve and must be at least 36 inches (914 mm) in length. The length of a landing is measured in the direction of travel.

R311.4.4 Type of lock or latch. All egress doors shall be readily openable from the side from which egress is to be made without the use of a key or special knowledge or effort.

❖ This section mandates that the required exit door be under the control of and operable by the person seeking egress; thus, the statement that the egress door be operable from the side from which egress is sought, without the need of a key or any special knowledge or effort. This section applies to all egress doors.

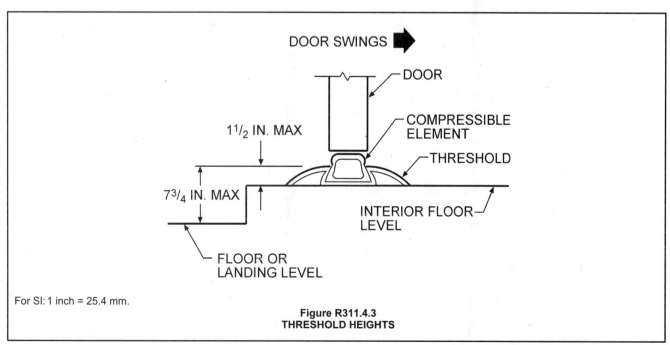

For SI: 1 inch = 25.4 mm.

Figure R311.4.3
THRESHOLD HEIGHTS

R311.5 Stairways.

❖ The requirements for stairways are contained in Sections R311.5.1 through R311.5.8.

R311.5.1 Width. Stairways shall not be less than 36 inches (914 mm) in clear width at all points above the permitted handrail height and below the required headroom height. Handrails shall not project more than 4.5 inches (114 mm) on either side of the stairway and the minimum clear width of the stairway at and below the handrail height, including treads and landings, shall not be less than 31.5 inches (787 mm) where a handrail is installed on one side and 27 inches (698 mm) where handrails are provided on both sides.

> **Exception:** The width of spiral stairways shall be in accordance with Section R311.5.8.

❖ Section R311.5 contains the code requirements for regulating stairways in an effort not only to make them usable, but also to help in reducing the number of injuries caused by stairway accidents.

Section R311.5.1 requires a minimum stairway width of 36 inches (914 mm). Generally, when the code specifies a required width of a component in the egress system, the width will be the clear, net, usable, unobstructed width. In this case, however, the width is specified as applying only to the area "above the permitted handrail height and below the required headroom height."

At and below the handrail height, the required width for the stairway, including treads and landings, is 27 inches (686 mm) if handrails are provided on each

side, and 31¹/₂ inches (800 mm) if there is a handrail installed on only one side. In essence, the code is not concerned about elements such as trim, stringers or other items that may be found below the level of the handrail, as long as they do not exceed the handrail's projection. This reduced width below the handrail is based on a body's movements as a person walks on a stair or other surface (see Commentary Figure R311.5.1). The exception and the provisions of Section R311.5.8 will permit a minimum width of 26 inches (660 mm) for spiral stairways. It is important to note that three of the key elements in the means of egress, halls, stairways and the exit door, have a required width of not less than 36 inches (914 mm).

R311.5.2 Headroom. The minimum headroom in all parts of the stairway shall not be less than 6 feet 8 inches (2036 mm) measured vertically from the sloped plane adjoining the tread nosing or from the floor surface of the landing or platform.

❖ A minimum headroom clearance of 6 feet 8 inches (2032 mm) is required in connection with every stairway. This includes not only the above-the-tread portion but also above any landings serving the stairway. The clearance is to be measured vertically above a plane that connects the stair nosings and also vertically above any landing or floor surface that is a part of the stairway (see Commentary Figure R311.5.2). This specific height requirement overrides the general ceiling height limitations of Section R305 and is modified for spiral stairways by Section R311.5.8.1.

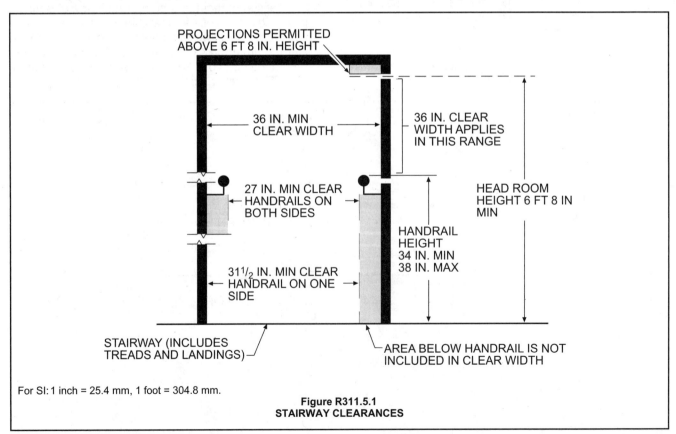

PROJECTIONS PERMITTED
ABOVE 6 FT 8 IN. HEIGHT

36 IN. MIN
CLEAR WIDTH

36 IN. CLEAR
WIDTH APPLIES
IN THIS RANGE

27 IN. MIN CLEAR
HANDRAILS ON
BOTH SIDES

HEAD ROOM
HEIGHT 6 FT 8 IN
MIN

HANDRAIL
HEIGHT
34 IN. MIN
38 IN. MAX

31¹/₂ IN. MIN CLEAR
HANDRAIL ON ONE
SIDE

STAIRWAY (INCLUDES
TREADS AND LANDINGS)

AREA BELOW HANDRAIL IS NOT
INCLUDED IN CLEAR WIDTH

For SI: 1 inch = 25.4 mm, 1 foot = 304.8 mm.

Figure R311.5.1
STAIRWAY CLEARANCES

R311.5.3 Stair treads and risers.

❖ The riser height, tread depth and profile requirements for stairways are specified in Sections R311.5.3.1 through R311.5.3.3.

R311.5.3.1 Riser height. The maximum riser height shall be 7³/₄ inches (196 mm). The riser shall be measured vertically between leading edges of the adjacent treads. The greatest riser height within any flight of stairs shall not exceed the smallest by more than ³/₈ inch (9.5 mm).

❖ The code establishes that the maximum riser height is 7³/₄ inches (197 mm). The IRC does not state a minimum riser height as does the IBC, where a 4-inch (102 mm) limit is specified. The provisions specify how the riser height is to be measured [see Commentary Figure R311.5.3.1(1)]. A significant safety factor for stairways is the uniformity of risers and treads in any flight of stairs. The section of a stairway leading from one landing to the next is defined as a flight of stairs. It is very important that any variation that would interfere

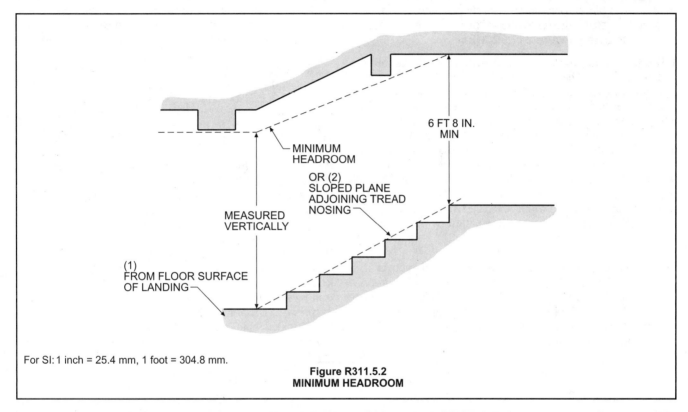

For SI: 1 inch = 25.4 mm, 1 foot = 304.8 mm.

Figure R311.5.2
MINIMUM HEADROOM

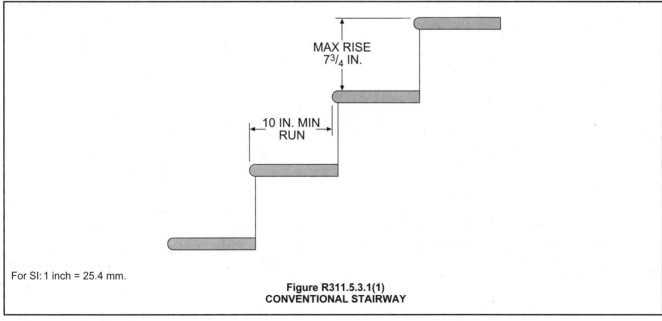

For SI: 1 inch = 25.4 mm.

Figure R311.5.3.1(1)
CONVENTIONAL STAIRWAY

with the rhythm of the stair user be avoided. Although it is true that adequate attention to the use of the stair can compensate for substantial variations in risers and treads, too frequently the stair user does not give the necessary attention.

To obtain the best uniformity possible in a flight of stairs, the maximum variation between the highest and lowest risers is limited to $^3/_8$ inch (9.5 mm). This tolerance is not to be used as a design variation, but its inclusion is in recognition that construction practices make it difficult to get exactly identical riser heights and tread dimensions in constructing a stairway in the field. Therefore, the code allows the variation indicated in Commentary Figure 311.5.3.1(2).

R311.5.3.2 Tread depth. The minimum tread depth shall be 10 inches (254 mm). The tread depth shall be measured horizontally between the vertical planes of the foremost projection of adjacent treads and at a right angle to the tread's leading edge. The greatest tread depth within any flight of stairs shall not exceed the smallest by more than $^3/_8$ inch (9.5 mm). Winder treads shall have a minimum tread depth of 10 inches (254 mm) measured as above at a point 12 inches (305 mm) from the side where the treads are narrower. Winder treads shall have a minimum tread depth of 6 inches (152 mm) at any point. Within any flight of stairs, the largest winder tread depth at the 12 inch (305 mm) walk line shall not exceed the smallest by more than $^3/_8$ inch (9.5 mm).

❖ The code establishes that the minimum tread depth is 10 inches (254 mm). The provisions specify how the tread depth is to be measured [See Commentary Figure R311.5.3.1(1)]. To obtain the best uniformity possible in a flight of stairs, the maximum variation between the greatest and smallest tread depth is limited to $^3/_8$ inch (9.5

mm). See commentary Section R311.5.3.1 for the discussion on uniformity. The same criterion applies to winder treads. However, the depth is to be measured at the "walk line". The "walk line" is at a point 12 inches (305 mm) from the side where the treads are narrower. Winder treads must have a minimum depth of 6 inches (152 mm) at any point. A stairway may consist of straight treads, or it may be constructed using winders. If winders are used, they can either be used for an entire flight of a stairway, as a portion of a stairway such as at a change of direction or to form a circular stairway. Winders consist of tapered treads that are narrow on one end and widen out, pie-shaped, toward the opposite side of the stairs. Because they are primarily used to change the direction of the stair, and they create a change in the rhythm for the stair user, it is important that winders comply with the specified dimensional criteria. See Commentary Figure R311.5.3.2(1) for examples of winders used as a portion of a stairway at a change of direction. See Commentary Figure R311.5.3.2(2) for an example of winders used to form a circular stairway.

R311.5.3.3 Profile. The radius of curvature at the leading edge of the tread shall be no greater than $^9/_{16}$ inch (14 mm). A nosing not less than $^3/_4$ inch (19 mm) but not more than $1^1/_4$ inch (32 mm) shall be provided on stairways with solid risers. The greatest nosing projection shall not exceed the smallest nosing projection by more than $^3/_8$ inch (9.5 mm) between two stories, including the nosing at the level of floors and landings. Beveling of nosing shall not exceed $^1/_2$ inch (12.7 mm). Risers shall be vertical or sloped from the underside of the leading edge of the tread above at an angle not more than 30 degrees (0.51 rad) from the vertical. Open risers are permitted, provided that the opening between treads does not permit the passage of a 4-inch diameter (102 mm) sphere.

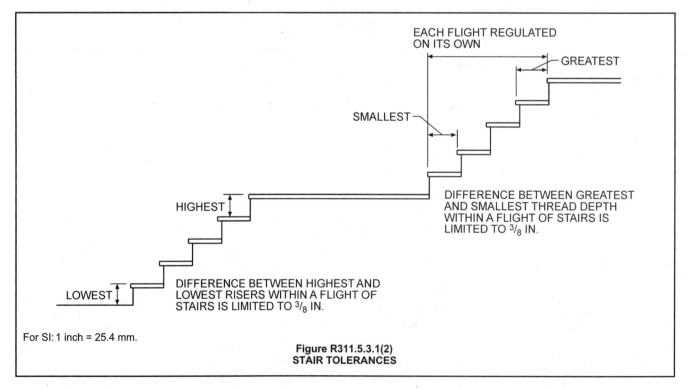

For SI: 1 inch = 25.4 mm.

Figure R311.5.3.1(2)
STAIR TOLERANCES

Exceptions:

1. A nosing is not required where the tread depth is a minimum of 11 inches (279 mm).

2. The opening between adjacent treads is not limited on stairs with a total rise of 30 inches (762 mm) or less.

❖ The leading edge or nosing of the treads is limited to a maximum radius of $^9/_{16}$ inch (14.3 mm). This varies from the $^1/_2$-inch (12.7 mm) maximum used in the IBC. These limitations apply to not only the nosings on the stair treads, but also to nosings at the level of floors or landings that are a part of the stairway. The IRC places a requirement for a minimum nosing of $^3/_4$ inch (19.1 mm) when the stair has a solid riser. This also varies from the IBC, which does not specify a minimum nosing requirement. However, the first exception will eliminate the nosing requirement if the stair tread is at least 11 inches (279 mm) in depth. Tread nosings are limited to a maximum projection of $1^1/_4$ inches (32 mm). The nosing projection is to be consistent from one story level to the next, instead of simply within each flight as is required for the treads and

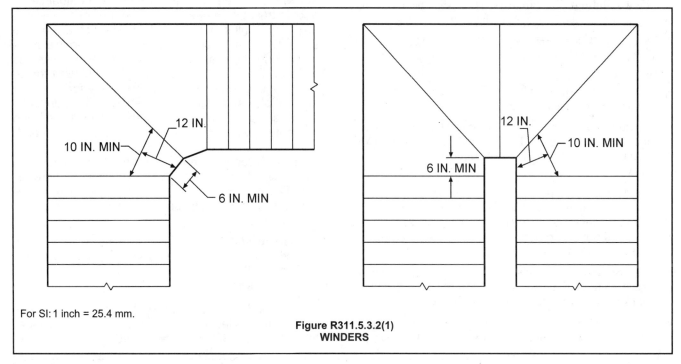

For SI: 1 inch = 25.4 mm.

Figure R311.5.3.2(1)
WINDERS

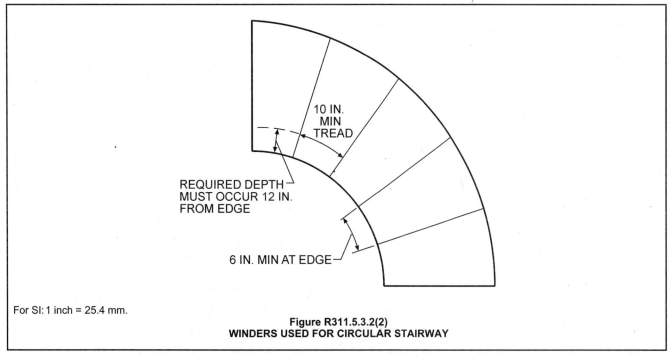

For SI: 1 inch = 25.4 mm.

Figure R311.5.3.2(2)
WINDERS USED FOR CIRCULAR STAIRWAY

risers. The greatest variation permitted for the nosing is $^3/_8$ inch (9.5 mm), which is a construction tolerance and not a design variation. See Commentary Figure R311.5.3.3 for examples of the profile provisions. The code does not require solid risers, but where the height of the stairway exceeds 30 inches (762 mm), either solid risers or another method to limit the opening between adjacent treads is needed. As with the guard provisions of Section R312, a 4-inch (102 mm) sphere is used to determine compliance for any type of open riser system.

R311.5.4 Landings for stairways. There shall be a floor or landing at the top and bottom of each stairway.

> **Exception:** A floor or landing is not required at the top of an interior flight of stairs, including stairs in an enclosed garage, provided a door does not swing over the stairs.

A flight of stairs shall not have a vertical rise larger than 12 feet (3658 mm) between floor levels or landings.

The width of each landing shall not be less than the width of the stairway served. Every landing shall have a minimum dimension of 36 inches (914 mm) measured in the direction of travel.

❖ A landing is required at the top and bottom of each stairway; however, a landing is not required at the top of interior stairways, including an enclosed garage, if a door does not swing over the stairway (see Commentary Figure R311.5.4).

R311.5.5 Stairway walking surface. The walking surface of treads and landings of stairways shall be sloped no steeper than one unit vertical in 48 inches horizontal (2-percent slope).

❖ The slope of the walking surfaces must provide drainage to stairs and landings that may be subjected to accumulation of liquids such as water, rain or melting snow. Also, it provides for a safe limit of the slope to maintain a safe walking surface. This requirement applies to all stairs and landings, both exterior and interior.

R311.5.6 Handrails. Handrails shall be provided on at least one side of each continuous run of treads or flight with four or more risers.

❖ One of the best means of creating safer stairs and assisting stairway users is to provide a graspable handrail which can be used for support and guidance as people travel from one level to the next or to assist in arresting a fall should they slip while they are on the stairs. Other than controlling the people who use stairs by making them pay attention and not carry things while on the stairs, the handrail probably will provide the greatest benefit in increased safety for the least amount of cost. It will never be known how many missteps, accidents, injuries or even fatalities have been prevented by having a properly installed, sturdy handrail.

The IRC requires that a handrail be installed on at least one side of any stairway that has four or more risers. The code does not have any exemption for the elimination of handrails once the stairway has four or more risers.

R311.5.6.1 Height. Handrail height, measured vertically from the sloped plane adjoining the tread nosing, or finish surface of ramp slope, shall be not less than 34 inches (864 mm) and not more than 38 inches (965 mm).

❖ Where handrails are required, they must be installed at a height of at least 34 inches (864 mm) and not more than 38 inches (965 mm), measured vertically from the

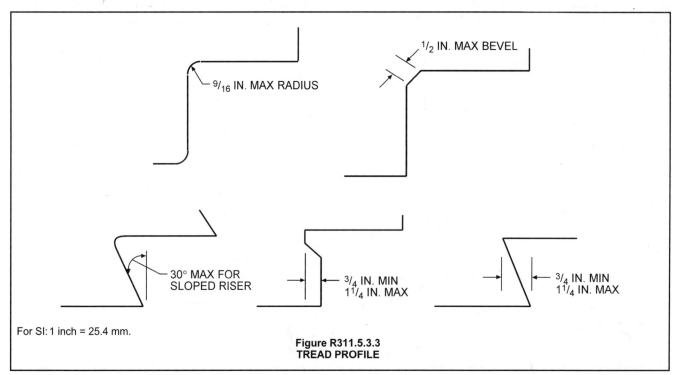

For SI: 1 inch = 25.4 mm.

Figure R311.5.3.3
TREAD PROFILE

nosing of the stair treads. This height should be measured to the top of the handrail at the point that it is directly above the nosing.

R311.5.6.2 Continuity. Handrails for stairways shall be continuous for the full length of the flight, from a point directly above the top riser of the flight to a point directly above the lowest riser of the flight. Handrail ends shall be returned or shall terminate in newel posts or safety terminals. Handrails adjacent to a wall shall have a space of not less than $1^1/_2$ inch (38 mm) between the wall and the handrails.

Exceptions:

1. Handrails shall be permitted to be interrupted by a newel post at the turn.

2. The use of a volute, turnout, starting easing or starting newel shall be allowed over the lowest tread.

❖ This required handrail is to be "continuous" for the length of the stairs (see Commentary Figure R311.5.6.2). The term "continuous" means not only that a single handrail must run from the top riser to the bottom riser, but it also indicates that users should be able to grasp the handrail and maintain their grasp without having to release the rail where it is supported. There is no requirement within the code for installation of a second handrail, but depending on the design and the placement of the required handrail, Section R316 and the requirement for a guard should be reviewed. The two exceptions to this section create situations where the graspable portion of the handrail may not end up being completely continuous from the top riser to the bottom riser. These traditional situations have routinely been accepted in the past and therefore are permitted by the code.

In view of the desire to make the handrail graspable, and considering the requirement that the handrail be continuous, a clear space of at least $1^1/_2$ inches (38 mm) is necessary between the handrail and any abutting wall. This distance will, under normal use, permit the fingers to slide past any adjacent rough surface that may cause injury, and it will provide an adequate distance so that the handrail may be quickly grabbed to help arrest a fall.

One important aspect of this section is the requirement for the ends of handrails to be returned to the wall or floor or to terminate in some type of end that will not catch clothing or limbs. It is still common to find many locations where handrails into unfinished basements or from raised deck areas do not have the ends properly protected.

R311.5.6.3 Handrail grip size. All required handrails shall be of one of the following types or provide equivalent graspability.

1. Type I. Handrails with a circular cross section shall have an outside diameter of at least $1^1/_4$ inches (32 mm) and not greater than 2 inches (51 mm). If the handrail is not circular it shall have a perimeter dimension of at least 4 inches (102 mm) and not greater than $6^1/_4$ inches (160 mm) with a maximum cross section of dimension of $2^1/_4$ inches (57 mm).

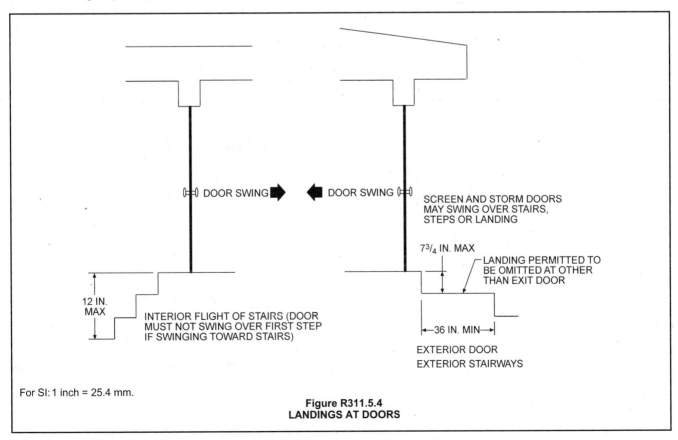

For SI: 1 inch = 25.4 mm.

**Figure R311.5.4
LANDINGS AT DOORS**

2. Type II. Handrails with a perimeter greater than $6^1/_4$ inches (160 mm) shall provide a graspable finger recess area on both sides of the profile. The finger recess shall begin within a distance of $^3/_4$ inch (19 mm) measured vertically from the tallest portion of the profile and achieve a depth of at least $^5/_{16}$ inch (8 mm) within $^7/_8$ inch (22 mm) below the widest portion of the profile. This required depth shall continue for at least $^3/_8$ inch (10 mm) to a level that is not less than $1^3/_4$ inches (45 mm) below the tallest portion of the profile. The minimum width of the handrail above the recess shall be $1^1/_4$ inches (32 mm) to a maximum of $2^3/_4$ inches (70 mm). Edges shall have a minimum radius of 0.01 inch (0.25 mm).

❖ To be effective, a handrail must be of a size that can easily be grasped by the vast majority of users. If it is too large, it is difficult for a user to get a strong enough grip to provide the needed support. Because of the variations in people's sizes and physical conditions, it is often very difficult to determine which types of commonly used handrails fall into this acceptable range.

The code specifies that the handrail be either a Type I or Type II. A type I can be either circular or noncircular in shape. See Commentary Figure R311.5.6.3(1) for examples of Type I handrails.

A Type II handrail has a perimeter larger than $6^1/_4$ inches (160 mm) with graspable finger recess area on both sides of the profile. See Commentary Figure R311.5.6.3(2) for the limitations of a Type II handrail.

R311.5.7 Illumination. All stairs shall be provided with illumination in accordance with Section R303.6.

❖ This section contains a reference to the illumination provisions of Section R303.6. The proper illumination of stairways is an important part of stairway safety. This lighting can assist users by making sure the level

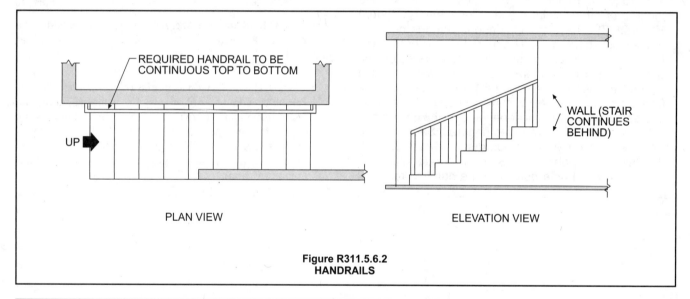

PLAN VIEW

ELEVATION VIEW

Figure R311.5.6.2
HANDRAILS

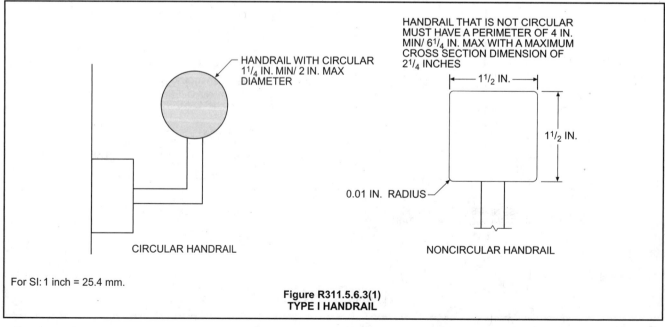

CIRCULAR HANDRAIL

NONCIRCULAR HANDRAIL

For SI: 1 inch = 25.4 mm.

Figure R311.5.6.3(1)
TYPE I HANDRAIL

changes do not occur in areas with shadows or in contrasting light, which would therefore make them difficult to see. See the discussion at Section R303.6 for additional information.

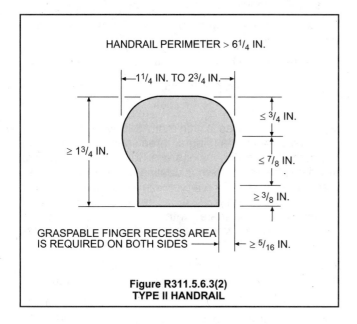

HANDRAIL PERIMETER > 6¹/₄ IN.

1¹/₄ IN. TO 2³/₄ IN.

≥ 1³/₄ IN.

≤ ³/₄ IN.

≤ ⁷/₈ IN.

≥ ³/₈ IN.

GRASPABLE FINGER RECESS AREA IS REQUIRED ON BOTH SIDES ⟶ ⟵ ≥ ⁵/₁₆ IN.

Figure R311.5.6.3(2)
TYPE II HANDRAIL

R311.5.8 Special stairways. Spiral stairways and bulkhead enclosure stairways shall comply with all requirements of Section R311.5 except as specified below.

❖ Sections R311.5.8.1 and R311.5.8.2 are exceptions to the general requirements for stairways as prescribed in Section R311.5.

R311.5.8.1 Spiral stairways. Spiral stairways are permitted, provided the minimum width shall be 26 inches (660 mm) with each tread having a 7¹/₂-inches (190 mm) minimum tread depth at 12 inches from the narrower edge. All treads shall be identical, and the rise shall be no more than 9¹/₂ inches (241 mm). A minimum headroom of 6 feet 6 inches (1982 mm) shall be provided.

❖ A spiral stairway is one of several types of special stairs that the code permits. Although a spiral stair may be difficult to use to move furniture on from one level to another, the code places no limitations on its use within the egress system if it meets the size requirements of this section. A spiral stairway that meets these requirements may provide the only means of egress from a level regardless of the occupant load or size of area served.

A spiral stairway is one in which the treads radiate from a central pole. Such a stair must provide a clear width of at least 26 inches (660 mm). Each tread must be identical and have a minimum dimension of 7¹/₂ inches (191 mm) at a point 12 inches (305 mm) from its narrow end. The stair must have at least 6 feet 6 inches (1981 mm) of headroom measured vertically from the leading edge of the tread. The rise between treads can be as much as, but not more than, 9¹/₂ inches (241 mm). Commentary Figure R311.5.8.1 shows the required dimensions of a spiral stairway.

R311.5.8.2 Bulkhead enclosure stairways. Stairways serving bulkhead enclosures, not part of the required building egress, providing access from the outside grade level to the basement shall be exempt from the requirements of Sections R311.4.3 and R311.5 where the maximum height from the basement finished floor level to grade adjacent to the stairway does not exceed 8 feet (2438 mm), and the grade level opening to the stairway is covered by a bulkhead enclosure with hinged doors or other approved means.

❖ This section exempts exterior "bulkhead enclosure stairways" from the landing stairway and handrail re-

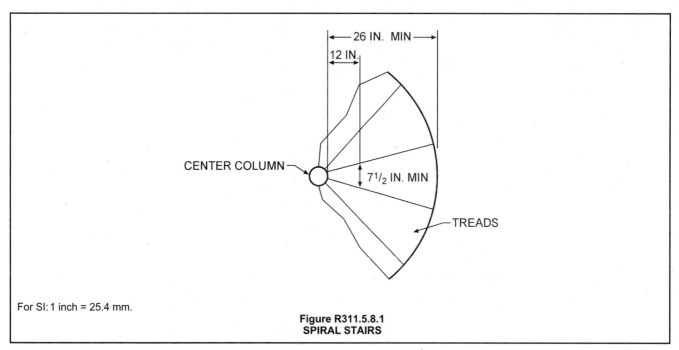

26 IN. MIN

12 IN.

CENTER COLUMN

7¹/₂ IN. MIN

TREADS

For SI: 1 inch = 25.4 mm.

Figure R311.5.8.1
SPIRAL STAIRS

quirements found in Chapter 3, and it therefore permits a situation that has been fairly common in some areas. See Commentary Figure R311.5.8.2 for an illustration of the requirements. Because these stairways are not a part of the building's egress system and serve only as a convenient way to access the basement from the exterior, the code exemption will not greatly affect the occupants' safety. Through this exemption, the size of the enclosure that is needed to provide weather protection for the stairway is greatly reduced.

R311.6 Ramps.

❖ Section R311.6 states the code requirements for ramps, which are defined by the IRC as being a walking surface that has a running slope steeper than one unit vertical in 20 units horizontal (5-percent slope).

R311.6.1 Maximum slope. Ramps shall have a maximum slope of one unit vertical in twelve units horizontal (8.3-percent slope).

> **Exception:** Where it is technically infeasible to comply because of site constraints, ramps may have a maximum slope of one unit vertical in eight horizontal (12.5 percent slope).

❖ Section R311.6.1 places a maximum slope of one unit vertical in 12 units horizontal (8.3-percent slope) on ramps. This requirement applies to all ramps, including those on circulation routes and those leading to an exit. This maximum slope matches what is permitted by the IBC for ramps that are not a part of the means of egress. Egress ramps under the IBC also have a maximum slope limit of 1:12 to comply with the current accessibility requirements.

R311.6.2 Landings required. A minimum 3-foot-by-3-foot (914 mm by 914 mm) landing shall be provided:

1. At the top and bottom of ramps.

2. Where doors open onto ramps.

3. Where ramps change direction.

❖ The code requires a minimum 3-foot-by-3-foot (914 mm by 914 mm) landing at three specific locations on ramps. These dimensions are not tied to the actual width of the ramp. Item 2, dealing with doors that open onto ramps, calls for a larger size landing if it is also required by Section R311.4.3. The specified landing dimensions coordinate with the requirements for nonaccessible dwelling units, which are found in exceptions in the IBC.

R311.6.3 Handrails required. Handrails shall be provided on at least one side of all ramps exceeding a slope of one unit vertical in 12 units horizontal (8.33-percent slope).

❖ Where a ramp exceeds a slope of one unit vertical in 12 units horizontal (8.3-percent slope) the code requires that a handrail be installed on at least one side

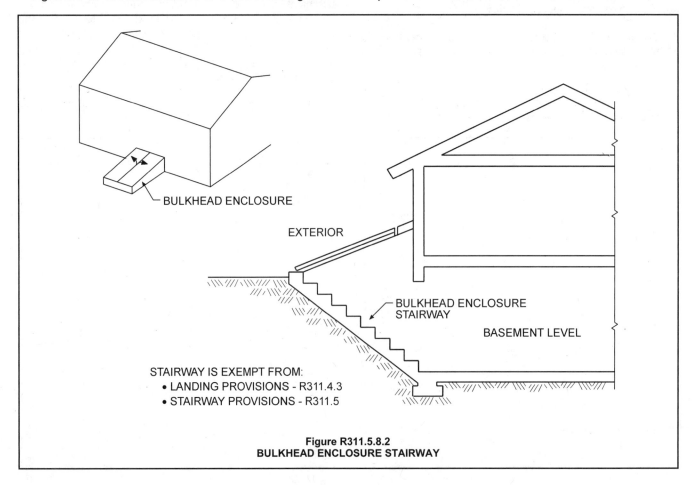

BULKHEAD ENCLOSURE

EXTERIOR

BULKHEAD ENCLOSURE STAIRWAY

BASEMENT LEVEL

STAIRWAY IS EXEMPT FROM:
- LANDING PROVISIONS - R311.4.3
- STAIRWAY PROVISIONS - R311.5

Figure R311.5.8.2
BULKHEAD ENCLOSURE STAIRWAY

to assist ramp users. This provision differs from that of the IBC, where a slope of one unit vertical in 20 units horizontal (5-percent slope) and a ramp rise of 6 inches (152 mm) establishes the limits.

R311.6.3.1 Height. Handrail height, measured above the finished surface of the ramp slope, shall be not less than 34 inches (864 mm) and not more than 38 inches (965 mm).

❖ Where handrails are required, they must be installed at a height of at least 34 inches (864 mm) and not more than 38 inches (965 mm), measured vertically from the finished surface of the ramp slope. This height should be measured to the top of the handrail.

R311.6.3.2 Handrail grip size. Handrails on ramps shall comply with Section R311.5.6.3.

❖ See the commentary for Section R311.5.6.3.

R311.6.3.3 Continuity. Handrails where required on ramps shall be continuous for the full length of the ramp. Handrail ends shall be returned or shall terminate in newel posts or safety terminals. Handrails adjacent to a wall shall have a space of not less than 1.5 inches (38 mm) between the wall and the handrails.

❖ The continuity requirement for the ramp handrail is similar to the continuity requirement for the stair handrail. See the commentary for Section R311.5.6.2

SECTION R312
GUARDS

R312.1 Guards. Porches, balconies, ramps or raised floor surfaces located more than 30 inches (762 mm) above the floor or grade below shall have guards not less than 36 inches (914 mm) in height. Open sides of stairs with a total rise of more than 30 inches (762 mm) above the floor or grade below shall have

guards not less than 34 inches (864 mm) in height measured vertically from the nosing of the treads.

Porches and decks which are enclosed with insect screening shall be equipped with guards where the walking surface is located more than 30 inches (762 mm) above the floor or grade below.

❖ The guard provisions of the IRC address the issue of protecting occupants from falling from any type of elevated walking surface. The provisions in Section R312 provide the scoping requirements as well as the general construction requirements for the guards. Besides this section, code users should be aware that Section R301.5 contains the design load criteria for guards.

Section R312.1 of the code establishes the requirement for and the minimum height requirements for guards. The code provides for guard protection at open sides along raised floor or walking surfaces such as those at balconies, mezzanines, stairways, ramps, porches and landings that are more than 30 inches (762 mm) above the grade or floor surface below.

The requirements for guards on stairs are different from other guard requirements in two ways. The first is the scoping requirements that establish the need for the guard, and the second is the required height of the guard. The scoping requirement for guards along open sides of stairs not only applies to the portion of a stairway that is more than 30 inches (762 mm) above the adjacent floor, but it will also apply to the entire open side of the stair, including the parts that are less than 30 inches (762 mm) above the floor. This requirement applies to the entire "open side" of the stairway, if any point of the open side is more than 30 inches (762 mm) high. See Commentary Figures R312.1(1) and (2) for examples of how this provision is applied.

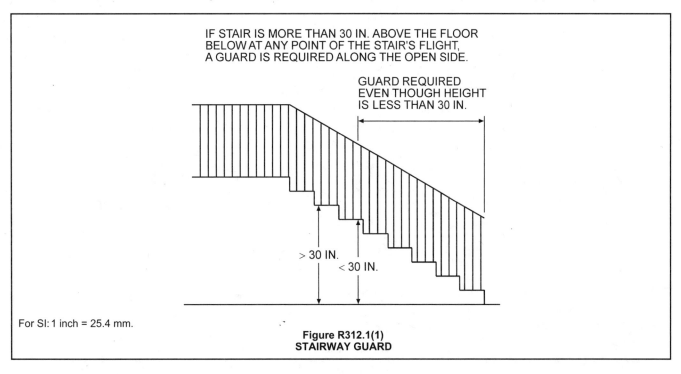

IF STAIR IS MORE THAN 30 IN. ABOVE THE FLOOR BELOW AT ANY POINT OF THE STAIR'S FLIGHT, A GUARD IS REQUIRED ALONG THE OPEN SIDE.

GUARD REQUIRED EVEN THOUGH HEIGHT IS LESS THAN 30 IN.

> 30 IN.

< 30 IN.

For SI: 1 inch = 25.4 mm.

Figure R312.1(1)
STAIRWAY GUARD

Where required, the guard must be of an adequate height to prevent someone from falling over the edge of the protected areas. It must also be designed to prevent someone, including small children, from falling through or under the top rail. Therefore, the code establishes 36 inches (914 mm) as the minimum acceptable height for guard protection. Although this height matches what has commonly been accepted within residential occupancies, it is less than the 42-inch (1067 mm) height that is generally required by the IBC and that places the top of the rail above the center of gravity for the vast majority of the general population. The height of a guard located on the open side of a stair can be a minimum of 34 inches (864 mm) when measured above the nosing of the tread [see Commentary Figure R312.1(3)]. As mentioned in the discussion of Section R311.5.6.1, this height should be measured to the top of the guard. This 34-inch height (864 mm) is acceptable on the stairs because the handrail height on a stairway is required to be between 34 and 38 inches (864 and 965 mm). Because reduced-height rails have been used on the open sides of stairs for years, the code has continued this practice, so a single element can serve as both a handrail and as a guard.

Insect screening lacks sufficient strength to prevent someone from falling under a top rail. For this reason a guard is required for porches and decks enclosed with insect screening where the walking surface is located more than 30 inches (762 mm) above a floor or grade below.

R312.2 Guard opening limitations. Required guards on open sides of stairways, raised floor areas, balconies and porches shall have intermediate rails or ornamental closures which do not allow passage of a sphere 4 inches (102mm) or more in diameter.

Exceptions:

1. The triangular openings formed by the riser, tread and bottom rail of a guard at the open side of a stairway are permitted to be of such a size that a sphere 6 inches (152 mm) cannot pass through.

2. Openings for required guards on the sides of stair treads shall not allow a sphere $4\,^3/_8$ inches (107 mm) to pass through.

❖ Guards must be constructed so that they not only prevent people from falling over them but also prevent smaller occupants such as children from falling through them. To prevent people from slipping through a guard, any required guard would need to have supports, spindles, intermediate rails or some type of ornamental pattern so that a 4-inch (102 mm) sphere cannot pass through it. This spacing was chosen after many years of discussion and because information was submitted to show that based on the size of a child's head, very few children would be able to crawl or walk through a 4-inch opening. The code does allow two exceptions for this spacing requirement, permitting the use of a 6-inch (152 mm) sphere for the triangular area formed by the riser, tread and bottom rail of a guard along the open side of a stair and $4^3/_8$ inch (111 mm) sphere for the guard on the open side of stair treads. See Commentary Figure R312.2 for an illustration of the guard requirements.

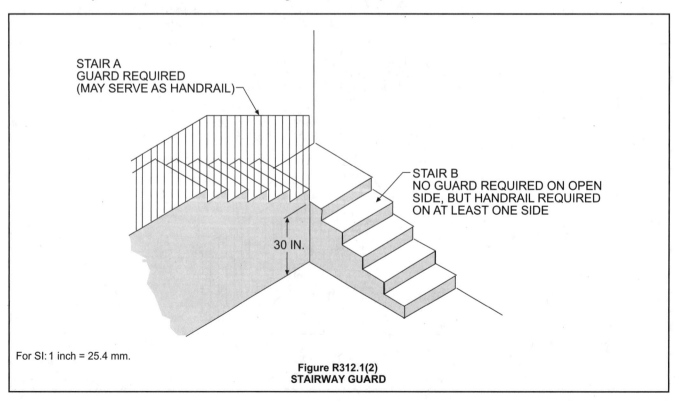

For SI: 1 inch = 25.4 mm.

Figure R312.1(2)
STAIRWAY GUARD

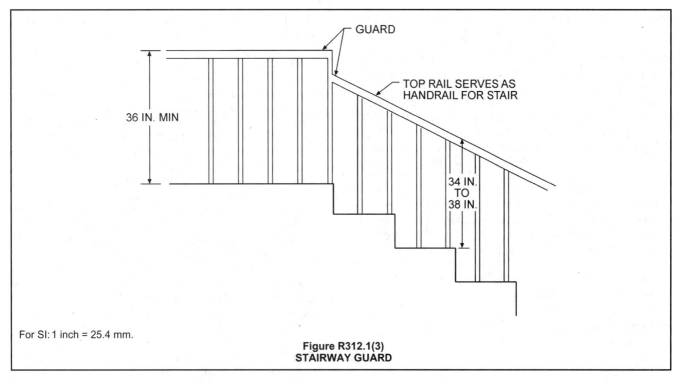

For SI: 1 inch = 25.4 mm.

**Figure R312.1(3)
STAIRWAY GUARD**

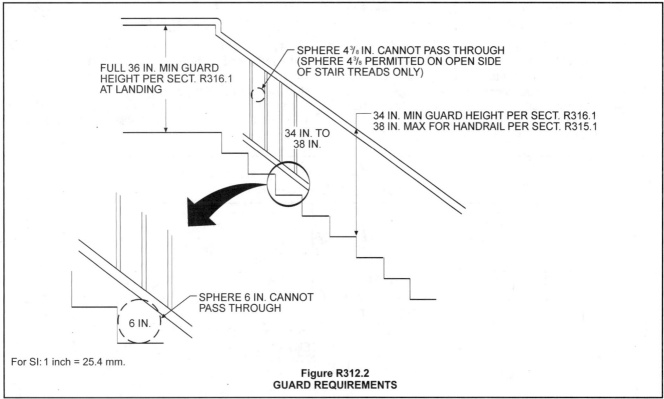

For SI: 1 inch = 25.4 mm.

**Figure R312.2
GUARD REQUIREMENTS**

SECTION R313
SMOKE ALARMS

R313.1 Smoke detection and notification. All smoke alarms shall be listed in accordance with UL 217 and installed in accordance with the provisions of this code and the household fire warning equipment provisions of NFPA 72.

Household fire alarm systems installed in accordance with NFPA 72 that include smoke alarms, or a combination of smoke detector and audible notification device installed as required by this section for smoke alarms, shall be permitted. The household fire alarm system shall provide the same level of smoke detection and alarm as required by this section for smoke alarms in the event the fire alarm panel is removed or the system is not connected to a central station.

❖ Section R313 provides the details of smoke detection and notification to alert occupants of potential problems. When asleep, the occupants of residential buildings will usually be unaware of a fire, and the fire will have an opportunity to spread before being detected. A majority of fire deaths occurring in residential buildings have occurred because of this delay in detection. It is for this reason that the IRC requires smoke alarms. This detection and notification system provides early warning to occupants of the building in the event of a fire, thereby providing a greater opportunity for everyone in the building to evacuate or relocate to a safe area. Of all of the provisions for safety features that have been placed within the code over the past few decades, the provisions for these detection and alarm devices have probably offered the greatest benefit in increasing safety and reducing the loss of life when compared to their minor expense.

R313.2 Location. Smoke alarms shall be installed in the following locations:

1. In each sleeping room.

2. Outside each separate sleeping area in the immediate vicinity of the bedrooms.

3. On each additional story of the dwelling, including basements but not including crawl spaces and uninhabitable attics. In dwellings or dwelling units with split levels and without an intervening door between the adjacent levels, a smoke alarm installed on the upper level shall suffice for the adjacent lower level provided that the lower level is less than one full story below the upper level.

When more than one smoke alarm is required to be installed within an individual dwelling unit the alarm devices shall be interconnected in such a manner that the actuation of one alarm will activate all of the alarms in the individual unit.

❖ So that all areas have at least some level of protection, and so that sleeping areas are adequately protected, Section R313.2 specifies where the devices are to be installed. The code requires that alarms be located within each sleeping room and outside each separate sleeping area but in the immediate vicinity of the bedrooms. The device within the bedroom will provide protection should the fire begin within that sleeping room, while the device outside of the room will provide early notification and protection should a problem develop in the area that generally will serve as the egress path for the bedroom. In addition, Item 3 will require installation of at least one smoke alarm on each story of the dwelling, including basements. The code does not require the installation of alarms within crawl spaces or within attics that are not habitable. This provides detection and notification within the areas of general occupancy but ignores spaces that are not occupied. See Commentary Figure R313.2(1) for an illustration of the required alarm locations.

Where split levels occur in a dwelling and the adjacent levels openly communicate with each other, the alarm may be placed on the upper portion of the split level if it is not more than one full story different in elevation. Commentary Figure R313.2(2) is an example of this provision. This requirement is based on the fact that any fire initiating on the lower portion of the level will send products of combustion up to the upper portion and that a detector there will provide a quick response and early warning.

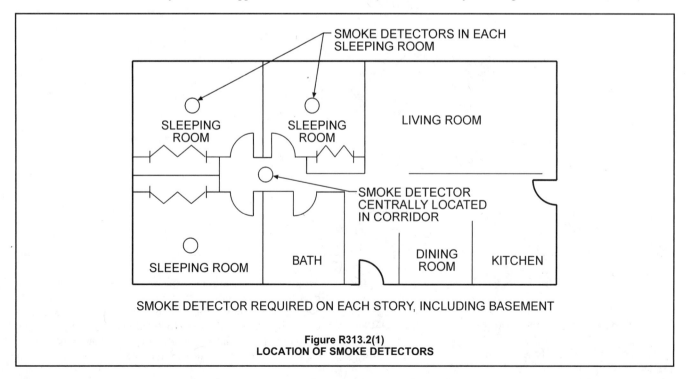

Figure R313.2(1)
LOCATION OF SMOKE DETECTORS

To assure that audible alarm notification is loud enough to alert the occupants of any problem within the unit, the code requires that two or more smoke alarms be interconnected so that if one device is activated, all alarms within the dwelling unit will be activated. This interconnection is required so that no matter where the smoke first develops or is detected, occupants throughout the unit will be made aware of the situation. One of the main concerns of the code is occupants who may be asleep and unaware of any developing fire. The IRC requires that the alarm signal be "clearly audible" in the bedroom area. If the smoke alarms are being installed in an existing building, see the commentary for Section R313.2.1, which contains an exception to the requirement for interconnection.

R313.2.1 Alterations, repairs and additions. When alterations, repairs or additions requiring a permit occur, or when one or more sleeping rooms are added or created in existing dwellings, the individual dwelling unit shall be equipped with smoke alarms located as required for new dwellings; the smoke alarms shall be interconnected and hard wired.

Exceptions:

1. Interconnection and hard-wiring of smoke alarms in existing areas shall not be required where the alterations or repairs do not result in the removal of interior wall or ceiling finishes exposing the structure, unless there is an attic, crawl space or basement available which could provide access for hard wiring and interconnection without the removal of interior finishes.

2. Work involving the exterior surfaces of dwellings, such as the replacement of roofing or siding, or the addition or replacement of windows or doors, or the addition of a porch or deck, are exempt from the requirements of this section.

❖ This section contains a unique provision in the code, applying the smoke alarm provisions to existing buildings when an addition, alteration or repair is made that will require a permit, or if any sleeping rooms are added or created. See the discussion at Section R105.2 regarding what types of repairs or alterations require a permit. The smoke alarms in these existing buildings are to be installed in the same manner as required for new dwellings. This would not only require their installation in the same locations within the dwelling but also that they be interconnected and receive their power from the building wiring. The commentary for Section R313.3 contains more discussion of the power source.

Two exceptions provide relief from the normal smoke alarm requirements in existing buildings that undergo some types of alteration, repair or addition. The first exception negates the requirement for hard-wired and interconnected alarms in areas where interior wall or ceiling finishes are not removed, making the placement of wiring difficult. In these cases use of a single-station battery operated alarm to provide the protection is acceptable. This exception will not apply if there is an attic, crawl space or unfinished basement available that would permit the wiring to be run on that side of the existing finish materials.

The second exception exempts "exterior surface" repairs from initiating the requirement for smoke alarms being placed in an existing dwelling. This exception exempts work that is done on the exterior only. The final determination of what type of work is included

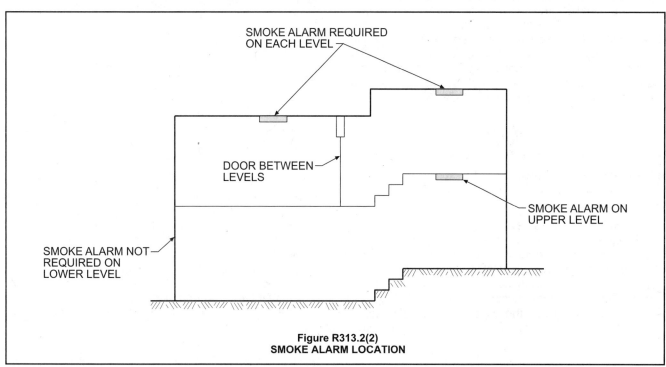

Figure R313.2(2)
SMOKE ALARM LOCATION

is left to the building official, but this would generally be viewed as covering reroofing, siding repairs or siding replacement and could possibly include some window replacements.

R313.3 Power source. In new construction, the required smoke alarms shall receive their primary power from the building wiring when such wiring is served from a commercial source, and when primary power is interrupted, shall receive power from a battery. Wiring shall be permanent and without a disconnecting switch other than those required for overcurrent protection. Smoke alarms shall be permitted to be battery operated when installed in buildings without commercial power or in buildings that undergo alterations, repairs or additions regulated by Section R313.2.1.

❖ Smoke alarms must use ac power as their primary source and battery power as a secondary source to enhance their reliability. For example, during a power outage, the probability of fire is increased because of the use of candles or lanterns for temporary light. Required backup battery power provides for continued performance of the smoke alarms. Smoke alarms are commonly designed to emit a recurring signal when batteries are low and need to be replaced. The last part of the section acknowledges that the code does not require that smoke alarms in all existing buildings be served from a commercial power source. Battery-operated smoke alarms may be the only power source when a commercial power source is not available or when extensive alterations or repairs are not being made. Where permanent building wiring can be installed without the removal of interior finishes (see Section R313.2.1, Exception 1), this section recognizes the increased reliability that the "hard wired" commercial power source with battery back-up can provide. Therefore, where feasible, permanent wiring is to be installed. It is also for the reliability issue that the code does not permit the alarms to be on any type of circuit that could be disconnected or turned off, such as a lighting circuit with a switch. The only way to disconnect power to the smoke alarms should be through the electrical panel box by either flipping a circuit breaker or removing the circuit's fuse.

SECTION R314
FOAM PLASTIC

R314.1 General. The provisions of this section shall govern the materials, design, application, construction and installation of foam plastic materials.

❖ Section R314 covers several topics related to the use and installation of various types of foam plastic materials used for insulation, trim and finishes. These requirements cover the acceptable uses of this combustible product and the associated protection needed to use it in building construction.

Section R314.1 lists the two basic issues that serve as the basis for the foam plastic insulation require-

ments: the flame spread rating of the material and the separation of the foam plastic insulation from the interior of the building.

Commentary Figure R314.1 shows a simple flowchart to help the code user more easily comply with the foam plastics sections of the IRC.

R314.2 Labeling and identification. Packages and containers of foam plastic insulation and foam plastic insulation components delivered to the job site shall bear the label of an approved agency showing the manufacturer's name, the product listing, product identification and information sufficient to determine that the end use will comply with the requirements.

❖ Foam plastics or packages of foam plastics delivered to the construction site must be labeled. Also, labels are required on containers (usually two components in 55-gallon (208 L) drums) of ingredients delivered for the production of foam plastic at the construction site. The label should include the name of the manufacturer or distributor, the type of foam plastic, the performance characteristics required to show code compliance and the name of the approved testing agency. The label may reference documents such as ICC-ES reports, approval agency certificates and other information that can be used to determine code-required performance characteristics. [Note, although not required by the IRC, the Federal Trade Commission (FTC) also places specific labeling requirements regarding the insulation power or *R*-value on all insulations, including foam plastic insulation, used in residential applications.]

R314.3 Surface burning characteristics. Unless otherwise allowed in Section R314.5 or R314.6, all foam plastic or foam plastic cores used as a component in manufactured assemblies used in building construction shall have a flame spread index of not more than 75 and shall have a smoke-developed index of not more than 450 when tested in the maximum thickness of 4 inches (102 mm), provided the end use is approved in accordance with Section R314.6 using the thickness and density intended for use.

Exception: Foam plastic insulation more than 4 inches thick shall have a maximum flame spread index of 75 and a smoke-developed index of 450 where tested at a minimum thickness of 4 inches, provided the end use is approved in accordance with Section R314.6 using the thickness and density intended for use.

❖ Unless otherwise allowed in Section R314.5 or R314.6, foam plastic or foam plastic cores used as a component in manufactured assemblies used in building construction must have a flame spread index of not more than 75 and a smoke-developed index of not more than 450 when tested in the maximum thickness intended for use in accordance with ASTM E 84. Loose-fill type foam plastic insulation must be tested as board stock for the flame spread index and smoke-developed index.

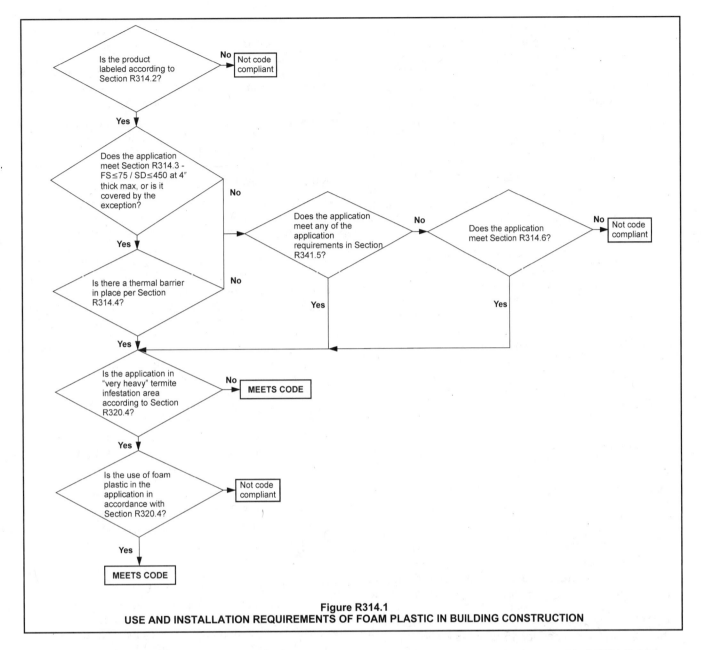

Figure R314.1
USE AND INSTALLATION REQUIREMENTS OF FOAM PLASTIC IN BUILDING CONSTRUCTION

Foam plastic insulation or foam plastic cores used as a component in a manufactured assembly are combustible and must be assessed for flame spread index (FS) and smoke developed index (SD). Foam plastic materials must be tested in accordance with test method ASTM E 84, unless they are:

1. Specifically exempted per Section R314.3;

2. One of several applications listed in Section R314.5 (i.e., roofing, foam-filled exterior doors, foam filled garage doors, interior trim, interior finish); or

3. Tested in an application that has been approved through Section R314.6 (testing under actual end-use configurations).

When testing in accordance with ASTM E 84 is required, the materials must be tested in the maximum thickness to be used [up to 4 inch (102 mm) thickness] with results of < 75 FS and < 450 SD unless otherwise specified. If a thicker foam is going to be used, it must still be tested to ASTM E 84 at 4 inch (102 mm) thickness with < 75 FS and < 450 SD results, as well as testing per Section R314.6 that is done at the actual foam thickness and density. If the material is loose-fill type foam plastic, it must be tested as board stock for flame spread and smoke-development performance.

The maximum flame spread value of 75 was chosen on the basis that it was lower than untreated wood (which usually was 100 to 165). The maximum smoke-developed rating of 450 was selected because, at the time, the code permitted interior finish materials that gave off "smoke no more dense than that given off by untreated wood." In selecting the maximum flame spread and smoke values, it was believed that a conservative approach was being taken by requiring an insulation material to meet the same requirements as interior finish, even though the insulation was intended to be covered with an interior finish material. The requirements for surface-burning characteristics of foam plastic apply to foam plastics used as cores of manufactured assemblies. The intent is that, even though the finished assemblies might or might not require testing for surface-burning characteristics, the foam-plastic core is not exempt from the general requirement; therefore, foam plastic is regulated in factory-manufactured assemblies the same as it is in field-fabricated applications.

R314.4 Thermal barrier. Unless otherwise allowed in Section R314.5 or Section R314.6, foam plastic shall be separated from the interior of a building by an approved thermal barrier of minimum 0.5 inch (12.7 mm) gypsum wallboard or an approved finish material equivalent to a thermal barrier material that will limit the average temperature rise of the unexposed surface to no more than 250°F (139°C) after 15 minutes of fire exposure complying with the ASTM E 119 standard time temperature curve. The thermal barrier shall be installed in such a manner that it will remain in place for 15 minutes based on NFPA 286 with the acceptance criteria of Section R315.4, FM 4880, UL 1040 or UL 1715.

❖ The use of an approved thermal barrier to separate foam plastics from the interior of a building is a basic requirement for the use of foam plastic as shown in this section of the code. The job of a thermal barrier is to isolate the foam plastic. An approved thermal barrier is defined as minimum $^1/_2$-inch (12.7 mm) gypsum wallboard or the equivalent. This section sets forth the test methods and performance criteria by which alternative thermal barriers are to be qualified (limiting the average temperature rise of the unexposed face to 250°F (121°C) for 15 minutes of fire exposure, while complying with the time-temperature conditions of ASTM E 119). The thermal barrier must be installed so that it remains in place for 15 minutes when exposed to fire.

Section R314.5 and Section R314.6 describe circumstances where the requirement for a thermal barrier is modified or eliminated.

Before 1975, experience had shown that foam plastics covered with plaster or $^1/_2$-inch (12.7 mm) gypsum wallboard had performed satisfactorily in building fires. For this reason, $^1/_2$-inch (12.7 mm) gypsum wallboard was included in the code as a minimum requirement. It was recognized that specifying a single material would not be desirable in a performance code; therefore, the words "or equivalent" were added. The thermal barrier test was selected as the appropriate method to determine equivalent performance. Although regular $^1/_2$-inch (12.7 mm) gypsum wallboard happened to have a 15-minute rating in the prescribed thermal barrier test, there was no intent that the 15-minute requirement for the thermal barrier would ensure 15 min-

utes of escape time in an actual building fire. This is related to the fact that the time-temperature conditions of ASTM E 119 may not reflect actual fire conditions. Gypsum wallboard is still the most commonly used thermal barrier, but equivalent materials are permitted and have been used when shown to be equivalent.

R314.5 Specific requirements. The following requirements shall apply to these uses of foam plastic unless specifically approved in accordance with Section R314.6 or by other sections of the code or the requirements of Sections R314.2 through R314.4 have been met.

❖ This "prescriptive" section can be used as another path to code compliance for foam plastics. As the flow chart in the commentary of Section R314.1 points out, if an application, including any listed below, meets the requirements in Sections R314.2, R314.3, R314.4 and R314.7, or Sections R314.2, R314.6 and R314.7, that application is code compliant and the requirements spelled out below do not apply. It is only when the requirements of Section R314.3 or Section R314.4 are not met that the applications spelled out below can be used to show code compliance. Many of the applications below modify or remove the flame spread and smoke-developed requirement of Section R314.3 or modify or remove the need for the thermal barrier specified in Section R314.4.

Two applications that have caused confusion in the past are foam backer board and foam insulation used in residential applications. These two applications, by definition, are examples of foam plastic insulation used on the exterior of a wall assembly. If the foam plastic being used meets the requirements of Section R314.3 (< 75 FS and < 450 SD) and Section R314.4 [thermal barrier of $^1/_2$ inch (12.7 mm) gypsum board or equivalent on the interior of the wall], the foam plastic insulation can be used up to the allowed thickness of 4 inches (102 mm) and Sections R314.5.7 and R314.5.8 do not apply.

R314.5.1 Masonry or concrete construction. The thermal barrier specified in Section R314.4 is not required in a masonry or concrete wall, floor or roof when the foam plastic insulation is separated from the interior of the building by a minimum 1-inch (25 mm) thickness of masonry or concrete.

❖ See Commentary Figure R314.5.1. No thermal barrier is required when 1 inch (25 mm) or more of masonry or concrete is placed between the foam plastic and the interior of the building. The intent is to accept 1 inch (25 mm) of masonry or concrete as adequate protection against ignition, even though the concrete does not necessarily meet the performance criteria for thermal barriers. This condition can arise when foam plastics are installed either within a wall or on one side of a wall. Some common examples are when foam plastics are installed:

- In the cavity of a hollow masonry wall,
- As the core of a concrete-faced panel,
- On the exterior face of a masonry wall and covered with an exterior finish,

- Within the cores of hollow masonry units or
- Encapsulated within a minimum of 1 inch (25 mm) concrete or masonry wall, floor or roof system, as in insulated tilt-up or pour-in-place concrete panels.

R314.5.2 Roofing. The thermal barrier specified in Section R314.4 is not required when the foam plastic in a roof assembly or under a roof covering is installed in accordance with the code and the manufacturer's installation instructions and is separated from the interior of the building by tongue-and-groove wood planks or wood structural panel sheathing in accordance with Section R803, not less than $^{15}/_{32}$ inch (11.9 mm) thick bonded with exterior glue and identified as Exposure 1, with edges supported by blocking or tongue-and-groove joints or an equivalent material. The smoke-developed index for roof applications shall not be limited.

❖ No thermal barrier is required when a foam plastic is incorporated into a roof assembly on the exterior side, over tongue and groove wood planks or wood structural panel sheathing, if the wood product meets all of the following:

- Used in accordance with Section R803,
- Identified as Exposure 1,
- Manufactured with exterior grade glue,
- Minimum $^{15}/_{32}$ inch thick (12 mm),
- Installed according to manufacturers instructions and
- Installed to provide adequate edge support (blocking when edges do not occur over framing members, tongue and groove joints or equivalent).

Also, the flame spread rating of the foam plastic used must comply with the requirements of Section R314.3, but the smoke-developed rating of the foam plastic is not limited.

R314.5.3 Attics. The thermal barrier specified in Section 314.4 is not required where attic access is required by Section R807.1 and where the space is entered only for service of utilities and when the foam plastic insulation is protected against ignition using one of the following ignition barrier materials:

1. 1.5-inch-thick (38 mm) mineral fiber insulation;
2. 0.25-inch-thick (6.4 mm) wood structural panels;
3. 0.375-inch (9.5 mm) particleboard;
4. 0.25-inch (6.4 mm) hardboard;
5. 0.375-inch (9.5 mm) gypsum board; or
6. Corrosion-resistant steel having a base metal thickness of 0.016 inch (0.406 mm).

The above ignition barrier is not required where the foam plastic insulation has been tested in accordance with Section R314.6.

❖ In an attic where access is required by Section R807.1 (where attic areas exceed 30 square feet (2.8 m²) and have a vertical height of 30 inches (762 mm) or more), and entry is only for service of utilities, and when foam plastics are used, an ignition barrier may be used in place of a thermal barrier to cover the foam plastic. Multiple materials are listed which can be used as the ignition barrier (see Commentary Figure R314.5.4). The foam plastic material, covered with the ignition barrier can be on the floor, wall (often called a knee wall or gable end) or the ceiling of the attic. The phrase "where entry is only for service of utilities" applies to attics that contain only mechanical equipment, electrical wiring, fans, plumbing, gas or electric hot water heaters, gas or electric furnaces, etc. The attic space cannot be used for storage. The reduced provision (from a thermal barrier to an ignition barrier) provides a barrier whose only purpose is to prevent the direct impingement of flame on the foam plastic insulation.

If the foam plastic insulation has passed testing, in the thickness and density intended for use, in accordance with Section R314.6, no thermal barrier or igni-

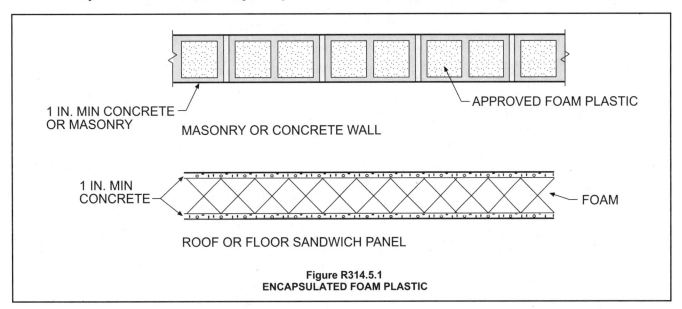

1 IN. MIN CONCRETE OR MASONRY

MASONRY OR CONCRETE WALL

APPROVED FOAM PLASTIC

1 IN. MIN CONCRETE

FOAM

ROOF OR FLOOR SANDWICH PANEL

Figure R314.5.1
ENCAPSULATED FOAM PLASTIC

tion barrier is required over the foam plastic insulation in an attic and this section of the code does not apply. It is important to note that the actual configuration must be tested. For example, a foam plastic insulation applied to the ceiling of the attic must be tested with the foam applied to the ceiling in a room corner test or in an assembly that reflects end use. The same restrictions would apply to those insulations applied to the walls, floors or combinations of surfaces.

R314.5.4 Crawl spaces. The thermal barrier specified in Section R314.4 is not required where crawlspace access is required by Section R408.3 and where entry is made only for service of utilities and the foam plastic insulation is protected against ignition using one of the following ignition barrier materials:

1. 1.5-inch-thick (38 mm) mineral fiber insulation;

2. 0.25-inch-thick (6.4 mm) wood structural panels;

3. 0.375-inch (9.5 mm) particleboard;

4. 0.25-inch (6.4 mm) hardboard;

5. 0.375-inch (9.5 mm) gypsum board; or

6. Corrosion-resistant steel having a base metal thickness of 0.016 inch (0.41 mm).

The above ignition barrier is not required where the foam plastic insulation has been tested in accordance with Section R314.6.

❖ In a crawl space where access is required by Section R408.4 (access shall be provided to all under-floor spaces) and entry is only for service of utilities, and when foam plastics are used, an ignition barrier may be used in place of a thermal barrier to cover the foam plastic. Multiple materials are listed which can be used as the ignition barrier (see Commentary Figure

R314.5.4). The foam plastic material, covered with the ignition barrier can be on the floor, wall or ceiling of the crawl space. The phrase "where entry is only for service of utilities" applies to crawl spaces that contain ONLY mechanical equipment, electrical wiring, fans, plumbing, gas or electric hot water heaters, gas or electric furnaces, etc. The crawl space cannot be used for storage. The reduced requirement (from a thermal barrier to an ignition barrier) provides a barrier whose only purpose is to prevent the direct impingement of flame on the foam plastic.

If the foam plastic insulation has passed testing, in the thickness and density intended for use, in accordance with Section R314.6, no thermal barrier or ignition barrier is required over the foam plastic insulation in the crawl space and this section of the code does not apply. It is important to note that the actual configuration must be tested. For example, a foam plastic insulation applied to the ceiling of the crawl space must be tested with the foam applied to the ceiling in a room corner test or in an assembly that reflects end use. The same restrictions would apply to those insulations applied to the walls, floors or combinations of surfaces.

R314.5.5 Foam-filled exterior doors. Foam-filled exterior doors are exempt from the requirements of Sections R314.3 and R314.4.

❖ No thermal barrier (Section R314.4) or surface burning characteristics testing (Section R314.3) is required for foam-filled exterior doors.

R314.5.6 Foam-filled garage doors. Foam-filled garage doors in attached or detached garages are exempt from the requirements of Sections R314.3 and R314.4.

❖ No thermal barrier (Section R314.4) or surface burning characteristics testing (Section R314.3) is required

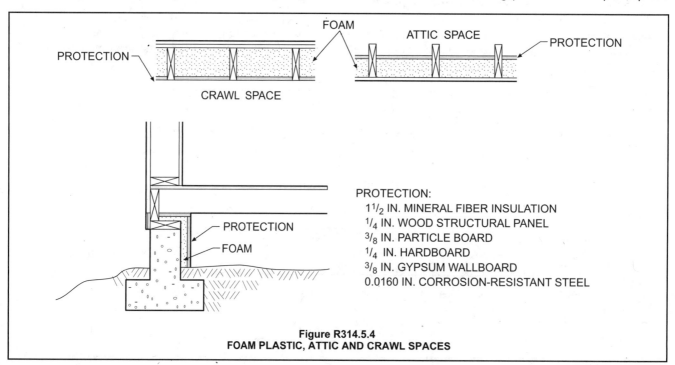

PROTECTION:
1 1/2 IN. MINERAL FIBER INSULATION
1/4 IN. WOOD STRUCTURAL PANEL
3/8 IN. PARTICLE BOARD
1/4 IN. HARDBOARD
3/8 IN. GYPSUM WALLBOARD
0.0160 IN. CORROSION-RESISTANT STEEL

Figure R314.5.4
FOAM PLASTIC, ATTIC AND CRAWL SPACES

for foam-filled garage doors in either an attached or detached garage.

R314.5.7 Foam backer board. The thermal barrier specified in Section R314.4 is not required where siding backer board foam plastic insulation has a maximum thickness of 0.5 inch (12.7 mm) and a potential heat of not more than 2000 Btu per square foot (22 720 kJ/m²) when tested in accordance with NFPA 259 provided that:

1. The foam plastic insulation is separated from the interior of the building by not less than 2 inches (51 mm) of mineral fiber insulation or

2. The foam plastic insulation is installed over existing exterior wall finish in conjunction with re-siding or

3. The foam plastic insulation has been tested in accordance with Section R314.6.

❖ The 2006 IRC has a new definition added for foam backer board: "Foam backer board." See Section R202. If these siding products are used on the exterior of a wall and the requirements of Sections R314.3 and R314.4 are met, this section of the code does not apply. If a thermal barrier is not used on the interior of the building, limitations are placed on the product and its use. In addition to the flame spread limitations of Section R314.3, other properties of the foam plastic portion of the product include maximum thickness of $^1/_2$ inch (12.7mm) and potential heat of < 2000 Btu per square foot (22 720 kJ/m²) when tested using NFPA 259. Limitation in siding/foam combination product use includes separation from the interior of the building by > 2 inches (51 mm) of mineral fiber insulation OR installation over an existing wall finish as part of re-siding OR the foam plastic insulation is tested in accordance with Section R314.6. The removal of the thermal barrier requirement in this section is reasonable considering the separation provided by the existing construction and the limitation of the potential heat of the foam plastic imposed by the code.

R314.5.8 Re-siding. The thermal barrier specified in Section R314.4 is not required where the foam plastic insulation is installed over existing exterior wall finish in conjunction with re-siding provided the foam plastic has a maximum thickness of 0.5 inch (12.7 mm) and a potential heat of not more than 2000 Btu per square foot (22 720 kJ/m²) when tested in accordance with NFPA 259.

❖ Foam plastic is frequently used in re-siding applications to provide a leveling surface for new siding, while also bringing additional insulation value to the wall assembly. If these products are used in a wall assembly and the requirements of Sections R314.3 and R314.4 are met, this section of the code does not apply. If a thermal barrier is not used between the foam plastic and the interior of the building, the foam insulation must meet the flame spread requirements of Section R314.3, is limited to a maximum thickness of $^1/_2$ inch (12.7 mm) and potential heat of < 2000 Btu per square foot (22 720 kJ/m²) when tested using NFPA 259. The

removal of the thermal barrier requirement in this section is reasonable considering the separation provided by the existing construction and the limitation of the potential heat of the foam plastic imposed by the code.

R314.5.9 Interior trim. The thermal barrier specified in Section R314.4 is not required for exposed foam plastic interior trim, provided all of the following are met:

1. The minimum density is 20 pounds per cubic foot (320 kg/m³).

2. The maximum thickness of the trim is 0.5 inch (12.7 mm) and the maximum width is 8 inches (204 mm).

3. The interior trim shall not constitute more than 10 percent of the aggregate wall and ceiling area of any room or space.

4. The flame spread index does not exceed 75 when tested per ASTM E 84. The smoke-developed index is not limited.

❖ Foam plastic interior trim is defined as exposed foam plastic used as picture molds, chair rails, crown moldings, baseboards, handrails, ceiling beams, door trim and window trim and similar decorative or protective materials.

For a foam plastic to qualify as interior trim, each of the four criteria listed in this section must be met. Because foam plastic in this application is left exposed, these criteria limit its use.

1. The density of materials must be at least 20 pounds per cubic foot (pcf) (320 kg/m³). The intent was to separate those materials used for trim from those intended for use as insulation. As a comparison, most foam plastic insulation is in the range of 1 to 2$^1/_2$ pounds per cubic foot (16.02 to 40 kg/m³) with very few materials over 5 pounds per cubic foot (81 kg/m³).

2. Even though other nonfoam plastic trim materials are not limited in dimension, the maximum thickness and width of foam plastic trim is limited to $^1/_2$ inch (12.7 mm) and 4 inches (102 mm), respectively.

3. Foam plastic trim cannot constitute more than 10 percent of the aggregate area of the walls and ceiling of a room.

4. The flame spread must not be higher than 75 when tested in accordance with ASTM E 84. The value of 75 was selected to be consistent with the requirement for foam plastic insulation, even though other materials used as trim are permitted to have flame spread indexes of up to 200 in many locations. The smoke-developed index is not regulated.

R314.5.10 Interior finish. Foam plastics shall be permitted as interior finish where approved in accordance with R314.6. Foam plastics that are used as interior finish shall also meet the flame spread and smoke-developed requirements of Section R315.

❖ Foam plastic used as interior finish must be approved through Section R314.6. This means that the foam plastic material has been tested to eliminate the thermal barrier in accordance with NFPA 286 and the acceptance criteria of Section R315.4, FM4880, UL 1040 or UL 1715, or fire tests related to actual end-use configurations (including the foam plastic thickness). The foam plastic must also meet the flame spread index requirements of Section R315 (FS < 200).

R314.5.11 Sill plates and headers. Foam plastic shall be permitted to be spray applied to a sill plate and header without the thermal barrier specified in Section R314.4 subject to all of the following:

1. The maximum thickness of the foam plastic shall be $3^1/_4$ inches (83 mm).

2. The density of the foam plastic shall be in the range of 1.5 to 2.0 pounds per cubic foot (24 to 32 kg/m³).

3. The foam plastic shall have a flame spread index of 25 or less and an accompanying smoke developed index of 450 or less when tested in accordance with ASTM E 84.

❖ No thermal barrier is required when a foam plastic is spray applied to the sill plate and joist header when all of the conditions listed in Section R314.5.11 are met. Because foam plastic insulation in this application is left exposed, the three conditions listed [thickness < $3^1/_4$ inches (82.6 mm), density 1.5 to 2.0 pcf (24 to 32 kg/m³), FS < 25 and SD < 450] control the spray applied foam plastic used in this application (see Commentary Figure R314.5.11).

R314.5.12 Sheathing. Foam plastic insulation used as sheathing shall comply with Section R314.3 and Section R314.4. Where the foam plastic sheathing is exposed to the attic space at a gable or kneewall, the provisions of Section R314.5.3 shall apply.

❖ Foam plastic used as sheathing is a very common application, adding insulation to the framing of the building. This section makes it clear that foam plastic insulation is code approved for use as a sheathing material. When used as a sheathing material, the foam plastic must meet Sections R314.3 and R314.4. Often foam plastic sheathing is used on the outside of an exterior wall, continuously covering the wall. In this example the living area of the building will have a thermal barrier such as $1/_2$ inch (12.7mm) gypsum board in place. That thermal barrier, combined with a foam plastic which has met the surface burning requirements of < 75 FS and < 450 SD, gives a code compliant application. The attic, in this example, will have foam plastic on the exterior side of the wall. When the foam plastic is exposed to the attic space at a gable or knee wall, the foam plastic insulation must meet the requirements of Section R314.5.3. For applications where foam plastic is used as sheathing, code approval follows the same paths outlined in Section R314.1.

If the foam plastic insulation has passed testing, in the thickness and density intended for use, in accordance with Section R314.6, no thermal barrier or ignition barrier is required over the foam plastic insulation and this section of the code does not apply. It is important to note that the actual configuration must be tested, including typical seams, joints and other details that will occur in the finished installation.

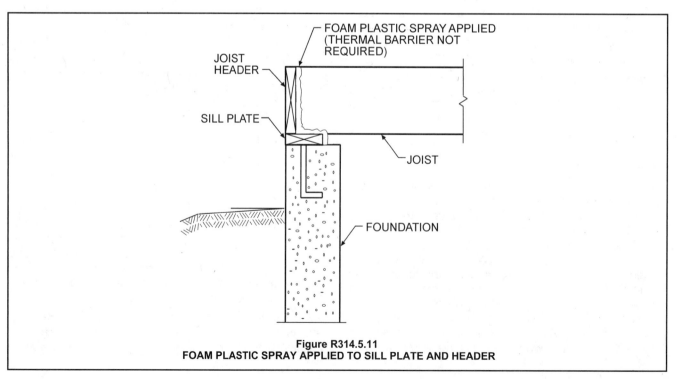

Figure R314.5.11
FOAM PLASTIC SPRAY APPLIED TO SILL PLATE AND HEADER

R314.6 Specific approval. Foam plastic not meeting the requirements of Sections R314.3 through R314.5 shall be specifically approved on the basis of one of the following approved tests: NFPA 286 with the acceptance criteria of Section R315.4, FM4880, UL 1040 or UL 1715, or fire tests related to actual end-use configurations. The specific approval shall be based on the actual end use configuration and shall be performed on the finished foam plastic assembly in the maximum thickness intended for use. Assemblies tested shall include seams, joints and other typical details used in the installation of the assembly and shall be tested in the manner intended for use.

❖ Foam plastic does not have to comply with the installation and use requirements of Sections R314.3 through R314.5 when specific approval is obtained in accordance with this section. This section lists examples of specific large scale tests, such as: FM 4880, UL 1040, NFPA 286 or UL 1715. Also, other large scale fire tests related to actual end-use configuration can be used. The intent is to require testing based on the proposed end-use configuration of the foam-plastic assembly with a fire exposure that is appropriate in size and location for the proposed application. These tests must be performed on full-scale assemblies. The tested assemblies must include typical seams, joints and other details that will occur in the finished installation. The foam plastic must be tested in the maximum thickness and density intended for use. Thorough testing provides an accurate depiction of the in-place fire performance of assemblies and systems using foam plastics.

There are two ways to show code compliance under Section R314.6. One method is to provide the actual test report that contains a description of the assembly and test results showing that the foam plastic, in the end use application, has passed the test. The second method is to obtain, from the ICC-ES, an Evaluation Report that covers the end use application.

R314.7 Termite damage. The use of foam plastics in areas of "very heavy" termite infestation probability shall be in accordance with Section R320.4.

❖ This section of the code refers to another section of the code, Section R320.4, that addresses the use of foam plastics in areas of "very heavy" termite infestation probability. When the structure is built in a area defined as "very heavy" termite infestation, Section R320.4 prohibits the use of foam plastics installed on the exterior face of below grade foundations walls or slab foundations, under exterior or interior foundation walls or slab foundations below grade or where located within 6 inches (152 mm) of exposed earth. Section R320.4 states three exceptions where foam plastics are permitted:

1. Foam plastics are permitted on the interior side of basement walls,

2. where the structural members of the building are either noncombustible or pressure-preservative treated wood, or

3. where, in addition to the requirements of Section R320.1 the foam plastic is adequately protected from subterranean termite damage.

SECTION R315
FLAME SPREAD AND SMOKE DENSITY

R315.1 Wall and ceiling. Wall and ceiling finishes shall have a flame-spread classification of not greater than 200.

Exception: Flame-spread requirements for finishes shall not apply to trim defined as picture molds, chair rails, baseboards and handrails; to doors and windows or their frames; or to materials that are less than $1/_{28}$ inch (0.91 mm) in thickness cemented to the surface of walls or ceilings if these materials have a flame-spread characteristic no greater than paper of this thickness cemented to a noncombustible backing.

❖ The control of interior finishes is an important aspect of fire protection. Section R315 contains the requirements for controlling fire growth within buildings by restricting interior finish materials. The dangers of unregulated interior finish include both the rapid spread of fire and the contribution of additional fuel to the fire. The rapid spread of fire presents a threat to the occupants of a building by limiting or denying their use of exitways within and outside the building. This can be caused by the rapid spread of the fire itself or by the production of large quantities of dense, black smoke, which obscures the exit path or makes movement difficult. Unregulated finish materials also have the potential for adding fuel to the fire, thus increasing its intensity and shortening the time available to the occupants to exit safely. However, based on the test standard that is used, the IRC does not regulate the fuel contribution of interior finish materials.

The IRC regulates interior finish materials on both walls and ceilings. These provisions do not address the floor or any coverings applied to the floor. The level of performance the code establishes for finish materials is a flame spread classification of 200 or less. The flame spread classification for a material is based on reviewing its performance under the test standard specified in Section R315.3. The limitation of 200 for the flame spread classification matches a "Class C" material in the IBC. Therefore, any material that has a flame spread rating of less than 200 may be used as a finish material on both walls and ceilings.

The exception will permit the installation of materials that will not significantly contribute to a fire. This includes various types of trim such as chair rails, door and window frames and baseboards, which because of their quantities and locations do not cause great concern. The actual size and quantity of such trim is not specified in the code because it has not been a concern. However, if some type of foam plastic is being used instead of other types of combustible trim, Section R314.2.6 does limit those situations. The exception also does not regulate thin materials such as wallpaper, which are less than $^1/_{28}$ inch (0.907 mm) thick when they are properly installed. These thin materials, when cemented to the surface of the wall or ceiling, behave essentially as the backing to which they are applied, and as a result are not regulated.

R315.2 Smoke-developed index. Wall and ceiling finishes shall have a smoke-developed index of not greater than 450.

❖ The development of smoke affects the occupants' safety. This section places a limitation of 450 on the density of smoke that is allowed from the wall and ceiling finishes. A product's smoke-developed index is established for the finish materials when they are tested under the standard specified in Section R315.3. The test measures only the obscurity caused by the smoke and does not consider the toxic content within the smoke.

R315.3 Testing. Tests shall be made in accordance with ASTM E 84.

❖ This section establishes that the standard test for flame spread and smoke-development characteristics is ASTM E 84, commonly known as the Steiner Tunnel Test. ASTM E 84 determines the relative burning behavior of materials on exposed surfaces, such as ceilings and walls, by visually observing the flame spread along the test specimen. Flame spread and smoke density are reported. The test method renders measurements of surface flame spread and smoke density in comparison with test results obtained by using select red oak and asbestos-cement board as control materials. Red oak is used for the furnace calibration because it is a fairly uniform grade of lumber that is readily available nationally, is uniform in thickness and moisture content, and generally gives consistent and reproducible results. Asbestos-cement board has a flame spread of zero, while the red oak is assigned a flame-spread rating of 100. All other materials are then given a classification based on a comparison with these two materials. Therefore, the flame spread rating of 200 that is permitted by Section R315.1 essentially means that the maximum flame spread for any finish material is twice that of the sample specimen of red oak.

R315.4 Alternate test method. As an alternate to having a flame-spread classification of not greater than 200 and a smoke developed index of not greater than 450 when tested in accordance with ASTM E 84, wall and ceiling finishes, other than textiles, shall be permitted to be tested in accordance with NFPA 286. Materials tested in accordance with NFPA 286 shall meet the following criteria:

During the 40 kW exposure, the interior finish shall comply with Item 1. During the 160 kW exposure, the interior finish shall comply with Item 2. During the entire test, the interior finish shall comply with Item 3.

1. During the 40 kW exposure, flames shall not spread to the ceiling.

2. During the 160 kW exposure, the interior finish shall comply with the following:

   2.1. Flame shall not spread to the outer extremity of the sample on any wall or ceiling.

   2.2. Flashover, as defined in NFPA 286, shall not occur.

3. The total smoke released throughout the NFPA 286 test shall not exceed 1,000 m².

❖ This section allows the use of NFPA 286 instead of ASTM E 84 for testing of wall and ceiling finishes other than textiles. NFPA 286 is known as a "room corner" fire test. In this test, a fire source consisting of a wood crib is placed in the corner of a compartment. The materials tested are then placed on the walls of the compartment (see Commentary Figure R315.4). This generally provides a more realistic understanding of the hazards involved with the materials.

Two levels of exposures are used during an NFPA 286 fire test to better represent a growing fire. The first is 40 kW fire size for 5 minutes and then 160 kW for 10 minutes. The 40 kW exposure represents the beginning of a fire where the initial spread is critical. Therefore the stated criterion is that the fire cannot spread to the ceiling. The 160 kW exposure is obviously a more intense fire and the criterion relates to preventing flashover (as defined by NFPA 286) and the extent of flame spread throughout the entire test assembly. There is also a total smoke production limit of 1000 m² for both levels of exposure.

It should be noted that the flashover criteria for NFPA 286 are as follows:

- Heat release exceeds 1 MW,
- Heat flux at the floor exceeds 20 kW/m²,
- Average upper layer temperature exceeds 600°C (1112°F),
- Flames exit the doorway and
- Autoignition of paper target on the floor occurs.

SECTION R316
INSULATION

R316.1 Insulation. Insulation materials, including facings, such as vapor retarders or vapor permeable membranes installed within floor-ceiling assemblies, roof-ceiling assemblies, wall assemblies, crawl spaces and attics shall have a flame-spread index not to exceed 25 with an accompanying smoke-developed index not to exceed 450 when tested in accordance with ASTM E 84.

Exceptions:

1. When such materials are installed in concealed spaces, the flame-spread and smoke-developed limitations do not apply to the facings, provided that the facing is installed in substantial contact with the unexposed surface of the ceiling, floor or wall finish.

2. Cellulose loose-fill insulation, which is not spray applied, complying with the requirements of Section

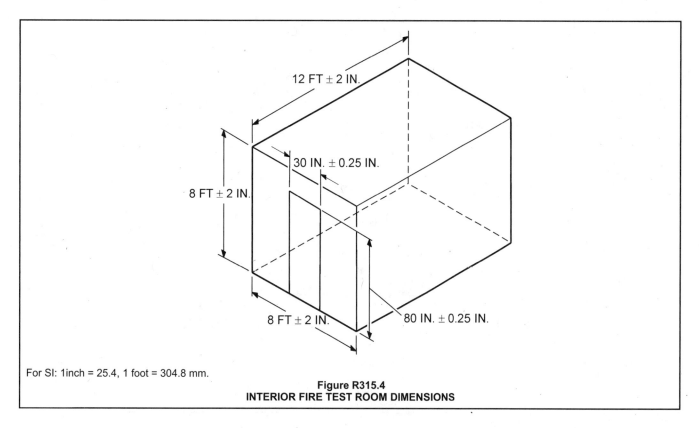

12 FT ± 2 IN.

30 IN. ± 0.25 IN.

8 FT ± 2 IN.

8 FT ± 2 IN.

80 IN. ± 0.25 IN.

For SI: 1inch = 25.4, 1 foot = 304.8 mm.

Figure R315.4
INTERIOR FIRE TEST ROOM DIMENSIONS

R316.3, shall only be required to meet the smoke-developed index of not more than 450.

❖ Section R316 addresses insulating materials installed in building spaces. Insulating materials can affect fire development and fire spread and therefore are regulated.

Section R316.1 addresses the various insulating materials that may be installed in building spaces, including insulating batts, blankets, fills (including vapor barriers and vapor permeable membranes) and other coverings. Exposed insulating materials represent the same fire exposure hazard as any other exposed material, such as an interior finish. The provisions of Sections R316.2, R316.3, R316.4, as well as the foam plastic provisions of Section R314 should also be reviewed based on the actual type of insulation and how it is installed. As a general requirement, insulation, including facings used as vapor retarders or as breather papers, must have a flame spread index not in excess of 25 and a smoke-developed index not in excess of 450. These values limit the contribution of the insulation to a fire. The flame spread requirement of 25 for the insulation will be more limiting than the 200, which is accepted for interior finishes by Section R315. The test method used to establish these limits is the ASTM E 84 standard. See the discussion of Section R315.3 for additional information.

The two exceptions address situations where, because of the way the material is installed or because of other imposed regulations, the material does not make any significant contribution to a fire. The first exception eliminates the flame spread and smoke-developed rating for the facing portion of the insulation if it is installed

"in substantial contact" with the unexposed surface of the ceiling, floor or wall finish. For example, when paper-backed insulation is placed directly on top of a ceiling, the paper facing is not required to meet the smoke and flame spread limits. If the same material is applied to the underside of a roof deck and the paper facing is exposed to the attic space, the paper facing must then meet the general criteria. The potential for flame spread is greatly diminished when the facings are installed in direct contact with the finish material because of the lack of airspace to support a fire if the facing were to be exposed to a source of ignition. See Commentary Figure R316.1 for an example of the various facing provisions.

The second exception addresses the fact that cellulose loose-fill insulation is federally regulated by the Consumer Product Safety Commission (CPSC). Parts 1209 and 1404 of CPSC 16 CFR contain various requirements that regulate the product to avoid excessive flammability or significant fire hazards. The smoke-developed index for cellulose loose-fill insulation must be determined by the ASTM E 84 test and must be 450 or less. This section requires that the smoke-developed index be measured using the ASTM E 84 test rather than the test procedures specified in Section R316.2.

R316.2 Loose-fill insulation. Loose-fill insulation materials that cannot be mounted in the ASTM E 84 apparatus without a screen or artificial supports shall comply with the flame spread and smoke-developed limits of Sections R316.1 and R316.4 when tested in accordance with CAN/ULC S102.2.

Exception: Cellulose loose-fill insulation shall not be required to comply with the flame spread index requirement

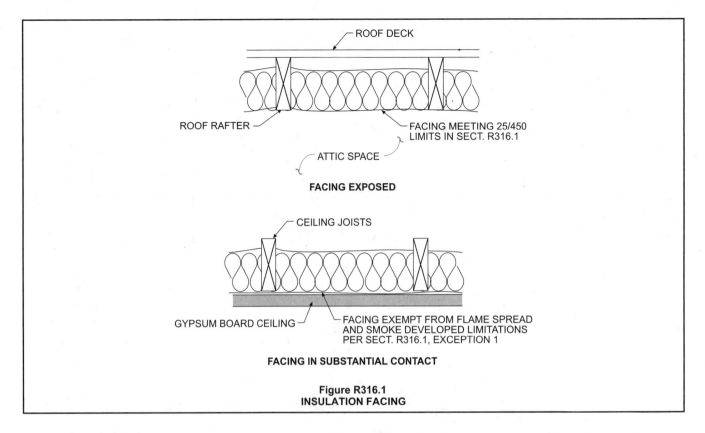

FACING EXPOSED

FACING IN SUBSTANTIAL CONTACT

Figure R316.1
INSULATION FACING

of CAN/ULC S102.2, provided such insulation complies with the requirements of Section R316.3.

❖ The main provision establishes that CAN/ULC-S102.2 is used as the test standard to determine the flame spread and smoke-developed ratings for loose-fill insulation materials that cannot be tested using the normal ASTM E 84 test method. The exception makes a distinction between cellulose insulation that is spray applied using a water-mist spray applicator, and cellulose loose-fill insulation that is poured or blown into place. Spray-applied cellulose insulation can be exposed on vertical and horizontal ceiling-type surfaces, so it is tested like any other insulating material. Cellulose loose-fill insulation that is poured or blown in is regulated by the Consumer Product Safety Commission requirements (see discussion at Section R316.1, Exception 2), and is exempt from the test procedure described in this section.

R316.3 Cellulose loose-fill insulation. Cellulose loose-fill insulation shall comply with CPSC 16 CFR, Parts 1209 and 1404. Each package of such insulating material shall be clearly labeled in accordance with CPSC 16 CFR, Parts 1209 and 1404.

❖ Because cellulose loose-fill insulation is federally regulated, this section provides the reference to the various federal regulations that are to be used. Because the federal regulations have precedence, the IRC cannot impose other requirements on this product. Therefore, this section and the reference to it from Section R316.1, Exception 2, establish the requirements for this product and provide the exemption from the normal ASTM E 84 test standard.

R316.4 Exposed attic insulation. All exposed insulation materials installed on attic floors shall have a critical radiant flux not less than 0.12 watt per square centimeter.

❖ This section provides the performance requirements for the test exposure that insulation must meet when it is exposed on the floor of an attic. It is tied to the testing provisions found in Section R316.5, which specifies that the ASTM E 970 test is to be used for determining the critical radiant flux. See the commentary for Section R316.5 regarding the application of this requirement to cellulose loose-fill insulation.

R316.5 Testing. Tests for critical radiant flux shall be made in accordance with ASTM E 970.

❖ ASTM E 970 is a test method developed by the insulation industry to evaluate the fire hazard of exposed attic insulation and is referenced in the material standards for insulation. Cellulose loose-fill insulation must comply with CPSC 16 CFR, Part 1209 (see Section R316.3), which requires testing by this standard. Spray-applied cellulose insulation that is not subject to the CPSC standard (see Section R316.1 and Exception 2 in that section) is also subject to the ASTM E 970 testing through Section R316.4.

SECTION R317
DWELLING UNIT SEPARATION

R317.1 Two-family dwellings. Dwelling units in two-family dwellings shall be separated from each other by wall and/or floor assemblies having not less than a 1-hour fire-resistance rating when tested in accordance with ASTM E 119. Fire-resistance-rated floor-ceiling and wall assemblies shall extend to and be tight against the exterior wall, and wall assemblies shall extend to the underside of the roof sheathing.

Exceptions:

1. A fire-resistance rating of $^1/_2$ hour shall be permitted in buildings equipped throughout with an automatic sprinkler system installed in accordance with NFPA 13.

2. Wall assemblies need not extend through attic spaces when the ceiling is protected by not less than $^5/_8$-inch (15.9 mm) Type X gypsum board and an attic draft stop constructed as specified in Section R502.12.1 is provided above and along the wall assembly separating the dwellings. The structural framing supporting the ceiling shall also be protected by not less than $^1/_2$-inch (12.7 mm) gypsum board or equivalent.

❖ Most of the nation's fires occur in residential buildings, particularly one- and two-family dwellings. These fires account for more than 80 percent of all deaths from fire in residential uses (including hotels, apartments, dormitories, etc.) and about two-thirds of all fire fatalities in any type of building. One- and two-family dwellings also account for more than 80 percent of residential property losses and more than one-half of all property losses from fire. Despite this poor fire record, there is wide-spread resistance to mandating much in the way of fire protection systems or methods because of our society's belief that people's homes are their castles. This viewpoint has limited the types of protection that

are imposed on these private homes to the installation of smoke alarms and the more recent requirement of dwelling unit separation. Section R317 provides a separation for protection of the occupants of one dwelling unit from the actions of their neighbor. The requirements of this section pertain to any structure regulated by this code other than a single-family dwelling unit. To accomplish this protection, the code addresses separation between the units, structural support and any openings or penetrations of the separation.

Depending on the layout of the various dwelling units, Section R317.1 requires that the walls and/or floor assemblies that divide one dwelling unit from the adjacent unit be of at least 1-hour fire-resistant construction. See Commentary Figure R317.1 for examples of the separation. The separation rating is to be determined by ASTM E 119 which is the normal test used for determining fire-resistance. Many tested assemblies are available for use in these locations. The provisions of the section also address the continuity of the separation, so that one dwelling unit is completely divided from the other. The horizontal aspect of the separation, which requires that the assemblies extend to and be tight against the exterior wall, is not difficult to comply with. It is most likely the vertical aspect (continuing a wall assembly to the underside of the roof sheathing) that will require some detailed planning, careful construction and careful inspection for the units to be separated.

Exception 1 grants a reduction in the required separation for those cases in which the building is equipped

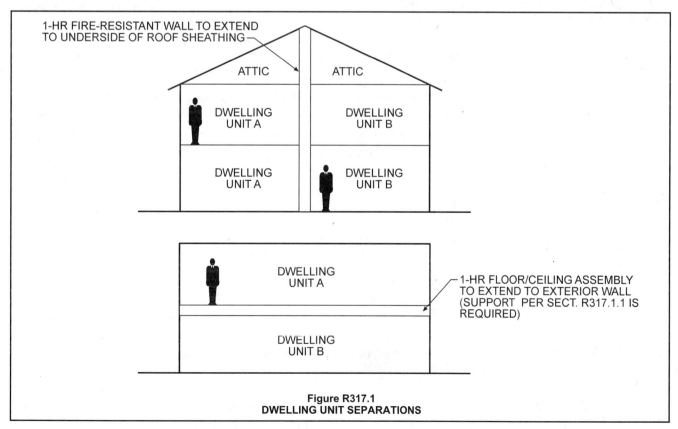

Figure R317.1
DWELLING UNIT SEPARATIONS

with an automatic sprinkler system. In these cases, a rating of $^1/_2$ hour is permitted versus a 1-hour fire-resistance rating. The sprinkler system must be "installed in accordance with NFPA 13," and is to be installed "throughout" the building. The type of sprinkler system used must meet the NFPA 13 standard and may not be installed to either the NFPA 13R or 13D standards, even though those two standards do address certain types of residential uses. The word "throughout" requires that the sprinkler system be installed in all portions of both dwelling units and any common spaces. The provisions of the NFPA 13 standard that permit omitting sprinklers in certain areas, such as small concealed spaces, are applicable. Therefore, the provision requires a complying sprinkler system "throughout" the building (that is, in all areas of the building that must be protected according to the standard), and it does not accept any partial system such as one installed in only one dwelling unit or only in the basement level of both units.

Exception 2 addresses separation in the area of the attic of two-family dwellings or duplexes. As long as the attic draft stop is present and meets the requirements in Section R502.12.1, there is a provision for the 1-hour fire separation to stop at a ceiling constructed of $^5/_8$ inch (15.9 mm) Type X gypsum board. Many times the type of truss or attic rafter and rafter tie/collar tie configuration will prohibit continuing construction of the 1-hour separation wall all the way up to the roof sheathing.

R317.1.1 Supporting construction. When floor assemblies are required to be fire-resistance-rated by Section R317.1, the supporting construction of such assemblies shall have an equal or greater fire-resistive rating.

❖ This provision applies to only the floor assemblies that form the separation between different dwelling units. When either all or portions of a dwelling unit separation are provided by a floor assembly, the code requires that the structural supports for the separation have a

rating equal to or higher than the floor. This is conceptually similar to the garage separation of Section R309.2. Without the supporting construction being protected, a fire on the lower level could lead to an early failure of the dwelling unit separation (see Commentary Figure R317.1.1).

R317.2 Townhouses. Each townhouse shall be considered a separate building and shall be separated by fire-resistance-rated wall assemblies meeting the requirements of Section R302 for exterior walls.

Exception: A common 2-hour fire-resistance-rated wall is permitted for townhouses if such walls do not contain plumbing or mechanical equipment, ducts or vents in the cavity of the common wall. Electrical installations shall be installed in accordance with Chapters 33 through 42. Penetrations of electrical outlet boxes shall be in accordance with Section R317.3.

❖ The application of this section has its basis in the exterior wall requirements found in Section R302 that deal with the building's location on the lot. The definition of a townhouse in Section R202 should be reviewed as well as the requirement for structural independence in Section R317.2.4. In general, because the "exterior wall" of the townhouse is essentially being constructed with no fire separation distance where one townhouse adjoins another, the code requires by Section R302 that the wall have not less than a 1-hour fire-resistance rating. The adjacent townhouse would have the same requirement. Therefore, the general requirement at this location (based on Sections R302.1 and R317.1) would be that each townhouse have its own "exterior wall." This would result in the construction of two separate 1-hour walls located side by side where one townhouse adjoins another.

Because of the difficulties involved in construction and the potential for unnecessary duplication, the exception offers an alternative to the two separate 1-hour walls by permitting the construction of a shared or

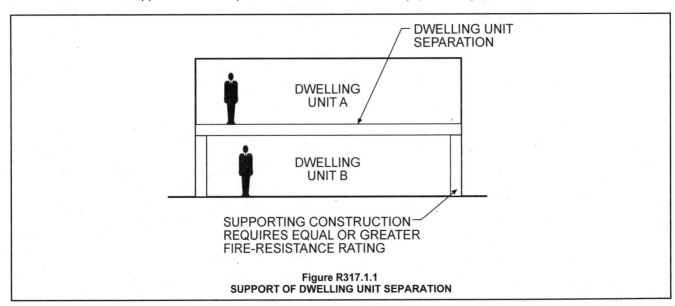

Figure R317.1.1
SUPPORT OF DWELLING UNIT SEPARATION

"common" 2-hour rated wall between the townhouses. See Commentary Figure R317.2 for an illustration of the two separate 1-hour walls and the common 2-hour wall. This exception has its basis in the actions of many building officials who permit this type of common wall as an alternative method of construction using provisions similar to those found in Section R104.11. Because the common wall has the potential to create an interconnection between the adjacent dwelling units and reduce the clear separation that would exist if two separate walls were constructed, the code places limits on services being located within the wall. This exception does not permit the inclusion of any type of plumbing, mechanical equipment, ducts or vents within the cavity of the common wall. This prohibition is applicable even if the penetrations or openings are protected by the penetration provisions of Section R317.3 or if a damper is installed in the duct or vent. The prohibition on plumbing includes all types of plumbing materials and systems as well as water supply and drainage piping of either combustible or noncombustible materials. However, the exception permits the cavity of the wall to be used for electrical installations if they comply with the electrical provisions of the code and the penetrations are properly protected.

R317.2.1 Continuity. The fire-resistance-rated wall or assembly separating townhouses shall be continuous from the foundation to the underside of the roof sheathing, deck or slab. The fire-resistance rating shall extend the full length of the wall or assembly, including wall extensions through and separating attached enclosed accessory structures.

❖ This section addresses the continuity of the fire resistance-rated wall or assembly separating townhouses

using the exception in Section R317.2. The requirements are conceptually similar to the continuity issues that exist in Section R317.1. These provisions, by regulating the extensions and terminations of the wall, can make possible the separation of dwelling units from each other. The provisions do not address or apply to the condition where two separate 1-hour fire-resistant "exterior" walls complying with Section R302 exist between the dwelling units.

R317.2.2 Parapets. Parapets constructed in accordance with Section R317.2.3 shall be constructed for townhouses as an extension of exterior walls or common walls in accordance with the following:

1. Where roof surfaces adjacent to the wall or walls are at the same elevation, the parapet shall extend not less than 30 inches (762 mm) above the roof surfaces.

2. Where roof surfaces adjacent to the wall or walls are at different elevations and the higher roof is not more than 30 inches (762 mm) above the lower roof, the parapet shall extend not less than 30 inches (762 mm) above the lower roof surface.

   Exception: A parapet is not required in the two cases above when the roof is covered with a minimum class C roof covering, and the roof decking or sheathing is of noncombustible materials or approved fire-retardant-treated wood for a distance of 4 feet (1219 mm) on each side of the wall or walls, or one layer of $^5/_8$-inch (15.9 mm) Type X gypsum board is installed directly beneath the roof decking or sheathing, supported by a minimum of nominal 2-inch (51 mm) ledgers attached to the sides of the roof framing members, for a minimum distance of 4 feet (1220 mm) on each side of the wall or walls.

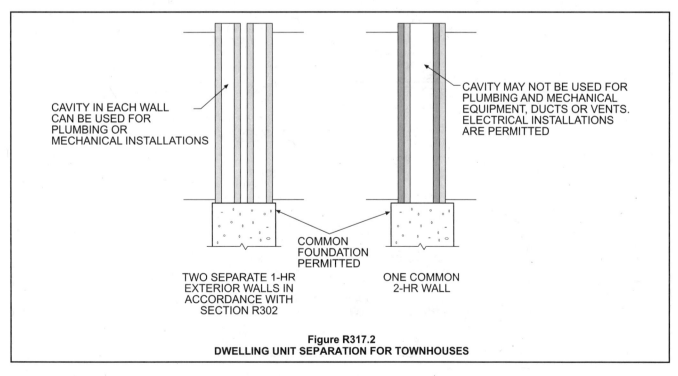

CAVITY IN EACH WALL CAN BE USED FOR PLUMBING OR MECHANICAL INSTALLATIONS

CAVITY MAY NOT BE USED FOR PLUMBING AND MECHANICAL EQUIPMENT, DUCTS OR VENTS. ELECTRICAL INSTALLATIONS ARE PERMITTED

COMMON FOUNDATION PERMITTED

TWO SEPARATE 1-HR EXTERIOR WALLS IN ACCORDANCE WITH SECTION R302

ONE COMMON 2-HR WALL

**Figure R317.2
DWELLING UNIT SEPARATION FOR TOWNHOUSES**

3. A parapet is not required where roof surfaces adjacent to the wall or walls are at different elevations and the higher roof is more than 30 inches (762 mm) above the lower roof. The common wall construction from the lower roof to the underside of the higher roof deck shall have not less than a 1-hour fire-resistance rating. The wall shall be rated for exposure from both sides.

❖ This section provides for the continuation of the dwelling-unit separation by requiring that a parapet be constructed above the common wall or the two "exterior walls" that occur between adjacent townhouses. This parapet requirement is applicable only to the wall or walls between townhouses and does not apply to the separation in a two-family dwelling, nor does it apply to other exterior walls on either townhouses or dwellings.

The code states three requirements that address the details of the parapet depending on the height of the adjacent roofs. The code also has an exception that can be used to eliminate the need for the parapet in the two conditions that require them. In general, parapets must extend at least 30 inches (762 mm) above the roof surfaces of the adjacent townhouses. This standard requirement is found in Item 1 and is applicable where the roof surfaces of the adjacent dwelling units are at the same level. The second item addresses the requirement for a parapet where the roofs are at different levels, but the difference is less that 30 inches (762 mm). Under this condition, the parapet height must still be 30 inches (762 mm), but it is measured only from the roof surface of the lower roof. The third item requires a parapet, but it may not be apparent to an observer because there is no requirement for anything to extend above the roof surfaces. In this case, the height difference between the roofs must be more than 30 inches (762 mm), and the code requires the common wall between the units to be rated to the height of the upper roof deck. The rating of this portion of the wall is a 1-hour fire-resistance rating and not the typical 2-hour (or two separate 1-hour walls) that separate the actual dwelling units. R317.2.2(1) for an illustration of the three requirements.

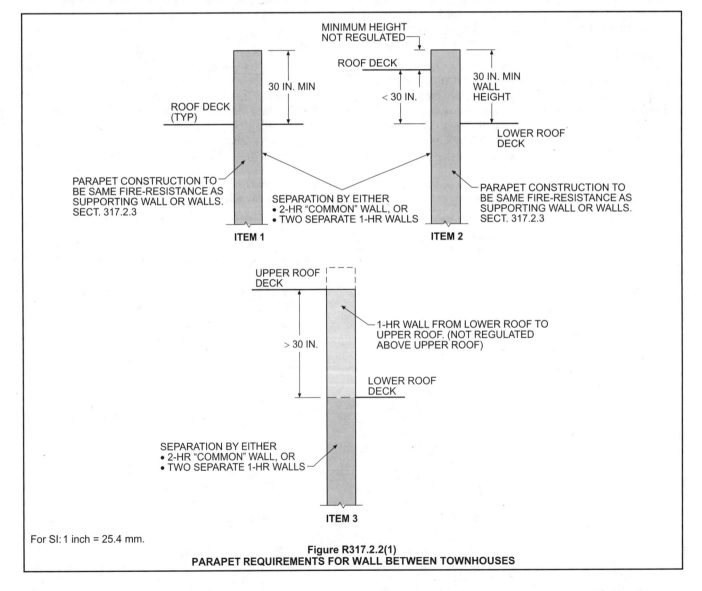

For SI: 1 inch = 25.4 mm.

Figure R317.2.2(1)
PARAPET REQUIREMENTS FOR WALL BETWEEN TOWNHOUSES

The exception found in this section applies only to the first two items of the section. See Commentary Figure R317.2.2(2) for an illustration of the exception. It does not apply where the height difference between the roofs is more than 30 inches (762 mm) and the provisions of Item 3 have been applied.

R317.2.3 Parapet construction. Parapets shall have the same fire-resistance rating as that required for the supporting wall or walls. On any side adjacent to a roof surface, the parapet shall have noncombustible faces for the uppermost 18 inches (457 mm), to include counterflashing and coping materials. Where the roof slopes toward a parapet at slopes greater than two units vertical in 12 units horizontal (16.7-percent slope), the parapet shall extend to the same height as any portion of the roof within a distance of 3 feet (914 mm), but in no case shall the height be less than 30 inches (762 mm).

❖ In addition to having the same degree of fire resistance as required for the wall or walls below, the surface of the parapet that faces the roof must be of noncombustible materials for the upper 18 inches (457 mm). Thus, a fire that might travel along the roof and reach the parapet would not be able to continue upward along the face of the parapet and over the top, exposing the adjacent building. The requirement applies to the upper 18 inches (457 mm) of the parapet to allow for extending the roof covering up the base of the parapet so that it can be effectively flashed. The 18-inch (457 mm) figure is based on a parapet height of at least 30 inches (762 mm) and would not be applicable to the face of the parapet that was toward the higher roof deck as constructed using Section R317.2.2, Item 2, nor would it be applicable where the parapet is not required based on the exception in Section R317.2.2 or on Item 3 of that section.

The 30-inch (762 mm) requirement is measured "above the roof surface" as stated in Section R317.2.2.

When a cricket or other element is installed adjacent to the parapet, the 30-inch (762 mm) dimension must be taken from the top of the roof surface on the cricket so the parapet truly extends 30 inches (762 mm) above the roof.

In those cases where the roof slopes upward away from the parapet and the slope exceeds 2 units vertical in 12 units horizontal (16.7 percent slope), the parapet must be extended to the same height as any portion of the roof that is within a distance of 3 feet. However, in no case can the height of the parapet be less than 30 inches (762 mm). See Commentary Figure R317.2.3 for an illustration of this requirement.

R317.2.4 Structural independence. Each individual townhouse shall be structurally independent.

Exceptions:

1. Foundations supporting exterior walls or common walls.

2. Structural roof and wall sheathing from each unit may fasten to the common wall framing.

3. Nonstructural wall coverings.

4. Flashing at termination of roof covering over common wall.

5. Townhouses separated by a common 2-hour fire-resistance-rated wall as provided in Section R317.2.

❖ Each townhouse must be structurally independent and capable of being removed without affecting the adjacent dwelling unit. This provision is applicable only to townhouses, not two-family dwellings. This independence is useful not only in the event of a fire within one unit, but also during any remodeling or alteration. The objective of this structural independence is that a com-

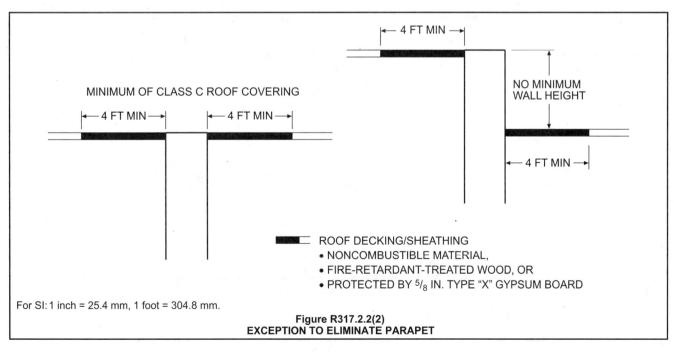

MINIMUM OF CLASS C ROOF COVERING

4 FT MIN

4 FT MIN

4 FT MIN

NO MINIMUM WALL HEIGHT

4 FT MIN

ROOF DECKING/SHEATHING
• NONCOMBUSTIBLE MATERIAL,
• FIRE-RETARDANT-TREATED WOOD, OR
• PROTECTED BY $^5/_8$ IN. TYPE "X" GYPSUM BOARD

For SI: 1 inch = 25.4 mm, 1 foot = 304.8 mm.

Figure R317.2.2(2)
EXCEPTION TO ELIMINATE PARAPET

plete burnout could occur on one side of the wall without causing the collapse of the adjacent townhouse. This condition occurs rarely. The provision also helps if there ever is a fire or other problem by creating a clear separation between the units. With separate ownership and each owner having a different insurance company, the ability to gain access or get repairs made can be difficult and time consuming. By having clearly separated units, it is much easier to determine who is responsible and to make any needed repairs.

The code lists five exceptions that waive the structural independence. A quick review of the exceptions shows that they generally deal with items that will not structurally affect the townhouses should a problem develop in the adjacent dwelling unit. Exception 1 is based on the norm within the industry for foundation construction. In the IRC, Section R402 lists only wood and concrete within the foundation materials section, although Section R404 accepts masonry foundation walls. In general, concrete and masonry are the most common types of foundations; wood foundations are viewed as unique. Given the performance of both masonry and concrete and the fact that these foundation systems must sustain loads from both the structure and the adjacent soils, it is reasonable to assume that the foundation will not be the item that fails in most situations. Permitting a common foundation also helps solve other problems that would arise if the structural independence issue were taken as an absolute. An example where requiring separate foundations would probably create more problems or difficulty is in the dampproofing or waterproofing of below-grade foundation walls.

If a wood foundation is used between adjacent units, what is the level of fire protection that may be needed? Because concrete and masonry foundations are the norm, it would be easy to forget or overlook protecting the foundation when it is constructed of wood. In these cases, it would seem appropriate to deal with the foundation as any other wall, and protect it on any exposed side. The level of fire resistance should be equal to that of the wall or walls that the foundation supports.

R317.3 Rated penetrations. Penetrations of wall or floor/ceiling assemblies required to be fire-resistance rated in accordance with Section R317.1 or R317.2 shall be protected in accordance with this section.

❖ This section addresses the specific requirements for maintaining the integrity of fire-resistance-rated assemblies at penetrations. If the penetration of a rated assembly is not properly constructed, the assembly itself is jeopardized and may not perform as intended. The provisions of this section apply to penetrations of fire-resistance-rated walls and floor/ceiling assemblies that are a part of the dwelling unit separation in either two-family dwellings or in townhouses. Penetrations of the rated assemblies range from combustible pipe and tubing to noncombustible wiring with combustible covering to noncombustible items such as pipe, tube, conduit and ductwork.

Each type of penetration requires a specific method of protection, which is based on the type of fire-resistance-rated assembly penetrated and the size and type of the penetrating item. The first step in determining the type of penetration protection required is to

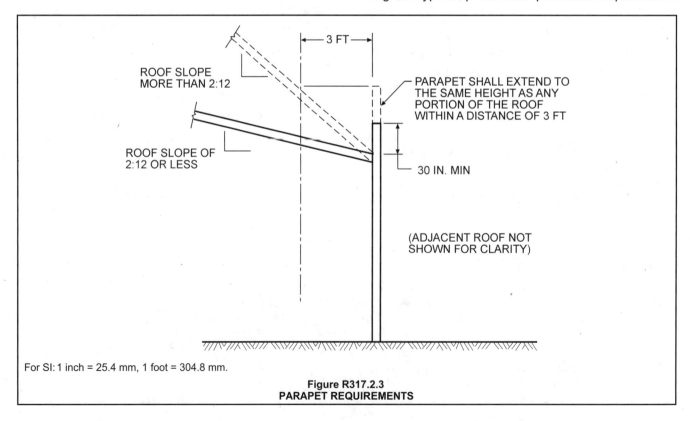

For SI: 1 inch = 25.4 mm, 1 foot = 304.8 mm.

Figure R317.2.3
PARAPET REQUIREMENTS

identify whether a wall or floor/ceiling assembly is being penetrated. The next step is to determine the type of penetrating item and whether it is a membrane or through-penetration. Once these factors are known, then the applicable section must be applied and the applicable method of protection must be decided upon.

R317.3.1 Through penetrations. Through penetrations of fire-resistance-rated wall or floor assemblies shall comply with Section R317.3.1.1 or R317.3.1.2.

Exception: Where the penetrating items are steel, ferrous or copper pipes, tubes or conduits, the annular space shall be protected as follows:

1. In concrete or masonry wall or floor assemblies where the penetrating item is a maximum 6 inches (152 mm) nominal diameter and the area of the opening through the wall does not exceed 144 square inches (92 900 mm²), concrete, grout or mortar is permitted where installed to the full thickness of the wall or floor assembly or the thickness required to maintain the fire-resistance rating.

2. The material used to fill the annular space shall prevent the passage of flame and hot gases sufficient to ignite cotton waste where subjected to ASTM E 119 time temperature fire conditions under a minimum positive pressure differential of 0.01 inch of water (3 Pa) at the location of the penetration for the time period equivalent to the fire resistance rating of the construction penetrated.

❖ This section contains the general requirements for through penetrations, which are openings that pass through an entire assembly. A through penetration is in contrast to a membrane penetration, which creates an opening through only one side of an assembly. Membrane penetrations are addressed later in Section R317.3.2. See Commentary Figure R317.3.1 for an illustration of these two types of penetrations.

Through penetrations must be protected to maintain the fire-resistance of the penetrated assembly. The code states two methods, found in Sections R317.3.1.1 and R317.3.1.2, that can be used to assure the adequacy of the penetration protection. The difference between these two is the test methodology used, but they both provide essentially the same results. The commentary for those sections is additional discussion of the differences.

Based on the history of these provisions and on the wealth of fire test data that exists concerning items such as conduit, water piping and other similar penetrations, the code provides two exceptions that permit protection by methods other than those generally required. The first permits the use of concrete, grout or mortar to protect certain penetrations of concrete and masonry wall or floor assemblies. The concrete, grout or mortar must be applied for the full thickness of the assembly unless evidence can be produced demonstrating that the required fire-resistance rating can be achieved with a lesser depth. Concrete, grout and mortar have traditionally been used as protection for the annular space in penetrations of concrete and masonry assemblies. Experience has shown this form of protection to be viable. However, caution must be used any time something such as a water pipe or conduit is placed in concrete or masonry. Sections

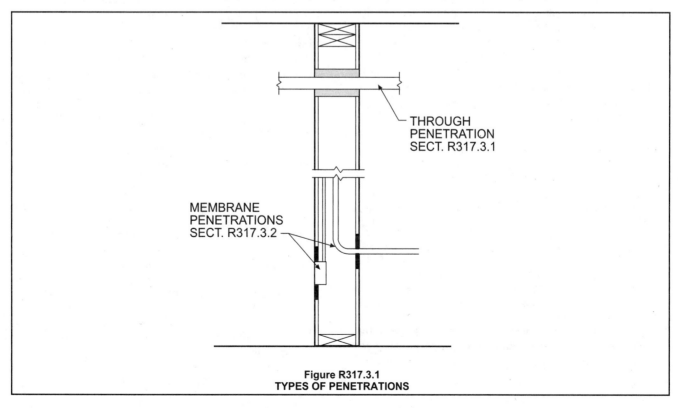

Figure R317.3.1
TYPES OF PENETRATIONS

P2603.3 and P2603.5 contain examples of protection of plumbing systems.

Exception 2 addresses the space between the penetrating item and the original assembly construction. This gap is called the annular space, and this exception provides a method to simply evaluate the performance of the material used to fill that space. It is often mistakenly believed that this exception permits a variety of untested items, but as can be seen from the provision itself, the materials need to meet a specific performance level. This exception requires that the ability of the material to prevent the passage of flame and hot gases sufficient to ignite cotton when subjected to the time-temperature criteria of the ASTM E 119 test standard be prequalified. This requirement is similar to provisions found within both the ASTM E 119 and ASTM E 814 standards, the standards used to evaluate fire-resistant assemblies and penetration protection. Because it is very likely that the penetration in the actual fire will be exposed to a positive pressure, this section specifies that the test-fire exposure include a positive pressure of 0.01 inch (0.25 mm) of water column as a further means to verify the performance of this protection method. Thus the protection will not be blown out or moved from its place during a fire.

R317.3.1.1 Fire-resistance-rated assembly. Penetrations shall be installed as tested in the approved fire-resistance-rated assembly.

❖ This section addresses situations in which the penetration is tested as a part of the regular full-scale test for the wall or floor/ceiling assembly. The penetration and proposed type of protection are evaluated as a part of the regular ASTM E 119 test, which evaluates the wall or floor/ceiling rating. This section and the option it provides are not used frequently because of the cost of conducting such full-scale tests and the limitations placed on application of the tested assembly. Because of these issues, penetrations are most often protected in accordance with one of the exceptions in Section R317.3.1 or on the provisions of Section R317.3.1.2.

R317.3.1.2 Penetration firestop system. Penetrations shall be protected by an approved penetration firestop system installed as tested in accordance with ASTM E 814 or UL 1479, with a minimum positive pressure differential of 0.01 inch of water (3 Pa) and shall have an F rating of not less than the required fire-resistance rating of the wall or floor/ceiling assembly penetrated.

❖ Through-penetration firestop systems consist of specific materials or an assembly of materials that are designed to restrict the passage of fire and hot gases for a prescribed period of time through openings made in fire-resistance-rated assemblies. To determine the effectiveness of a through-penetration firestop system in restricting the passage of fire, and to determine that the penetration has not jeopardized the original fire-resistant assembly, firestop systems must be subjected to fire testing using the ASTM E 814 or UL 1479 test standard. This is a small-scale test method developed specifically for the evaluation of a firestop system's ability to resist the passage of flame and hot gases, withstand thermal stresses and restrict the transfer of heat through the penetrated assembly. There are hundreds if not thousands of tested through-penetration firestop systems available today. The actual type of system used will depend on the type and construction of the assembly being penetrated, the material makeup and size of the penetrating item and the size of the annular space that exists between the penetrating item and the original assembly. Because there are a multitude of products available, and there is no "one size fits all" system available, it is helpful if the methods of protection are included on the construction documents as covered by Section R106.1.1.

The actual rating of the through-penetration firestop system (TPFS) is generated from the results of the testing and are reported as an "F" (flame) rating and a "T" (temperature) rating. The IRC requires only an "F" rating. The "F" rating indicates the period of time, in hours, that the TPFS tested remained in place without allowing the passage of fire during the fire exposure test, or the passage of water during the hose stream portion of the test. The required "F" rating of a TPFS must be equal to the fire-resistance rating of the wall or floor/ceiling assembly that is being penetrated. This means either a 1- or 2-hour rating, depending on the dwelling unit separation.

Two of the most common materials used in through-penetration firestop systems are intumescent and endothermic materials. Intumescent materials expand approximately 8 to 10 times their original volume when exposed to temperatures exceeding 250°F (121°C). The expansion of the material fills the voids or openings within the penetration to resist the passage of flame, while the outer layer of the expanded intumescent material forms an insulating charred layer that assists in limiting the transfer of heat. The expansion properties of intumescent materials allow them to seal openings left by combustible penetrating items that burn away during a fire, but they do not retard heat as well as endothermic materials. Intumescent materials are typically used with combustible penetrating items or where a higher "T" rating is not required.

Endothermic materials provide protection through chemically-bound water released in the form of steam when exposed to temperatures exceeding 600°F (316°C). This released water cools the penetration and retards heat transfer through the penetration. Endothermic materials tend to be superior in heat-transfer resistance and have higher "T" ratings, but they do not expand to fill voids left by combustible penetrating items that burn away during a fire. Therefore, endothermic materials are typically used with noncombustible penetrating items and where a higher "T" rating is required.

R317.3.2 Membrane penetrations. Membrane penetrations shall comply with Section R317.3.1. Where walls are required to have a fire-resistance rating, recessed fixtures shall be so installed such that the required fire resistance will not be reduced.

Exceptions:

1. Membrane penetrations of maximum 2-hour fire-resistance-rated walls and partitions by steel electrical boxes that do not exceed 16 square inches (0.0103 m²) in area provided the aggregate area of the openings through the membrane does not exceed 100 square inches (0.0645 m²) in any 100 square feet (9.29 m²) of wall area. The annular space between the wall membrane and the box shall not exceed ¹/₈ inch (3.1 mm). Such boxes on opposite sides of the wall shall be separated as follows:

   1.1. By a horizontal distance of not less than 24 inches (610 mm) except at walls or partitions constructed using parallel rows of studs or staggered studs;

   1.2. By a horizontal distance of not less than the depth of the wall cavity when the wall cavity is filled with cellulose loose-fill, rockwool or slag mineral wool insulation;

   1.3. By solid fire blocking in accordance with Section R602.8.1;

   1.4. By protecting both boxes with listed putty pads; or

   1.5. By other listed materials and methods.

2. Membrane penetrations by listed electrical boxes of any materials provided the boxes have been tested for use in fire-resistance-rated assemblies and are installed in accordance with the instructions included in the listing. The annular space between the wall membrane and the box shall not exceed ¹/₈ inch (3.1 mm) unless listed otherwise. Such boxes on opposite sides of the wall shall be separated as follows:

   2.1. By a horizontal distance of not less than 24 inches (610 mm) except at walls or partitions constructed using parallel rows of studs or staggered studs;

   2.2. By solid fire blocking in accordance with Section R602.8;

   2.3. By protecting both boxes with listed putty pads; or

   2.4. By other listed materials and methods.

3. The annular space created by the penetration of a fire sprinkler provided it is covered by a metal escutcheon plate.

❖ This section deals with instances where only a single side of the fire-resistance-rated assembly is penetrated. This would be the situation for items such as electrical outlet boxes or plumbing fixtures located on one side of the wall only. Commentary Figure R317.3.1 shows this type of penetration. For the most

part, a membrane penetration is to be protected by one of the previously described methods established for through penetrations. However, there are some penetrations that are allowed without a specific firestopping material in the annular space around them. These are addressed by the exceptions. This section also deals with the installation of recessed luminaires in fire-resistance-rated assemblies and states that their installation may not reduce the assembly's protection. Although these fixtures are common, they do represent a penetration of the assembly's protection and must be installed so that the assembly is not compromised.

Exception 1 allows penetrations of steel electrical outlet boxes under certain conditions. The criteria of this section limit the size of the box to 16 square inches (0.0103 m²) or less in area and to an aggregate area not to exceed 100 square inches (64 500 mm²) in 100 square feet (9.3 m²). Commentary Figure R317.3.2(1) shows some of the requirements of this section. The area limitations are consistent with the criteria from fire tests, which have shown that within these limitations, these penetrations will not adversely affect the fire-resistance rating of the assembly. However, the boxes are assumed to be installed as they were during the fire tests. In general the test requirements match the limitations shown by the code regarding their size and the need to be offset. An additional requirement, one that does not appear in the code, regulates the size of the annular space created around the outlet boxes. Both the Underwriter's Laboratory's Fire Resistance Directory and the Gypsum Association's Fire-Resistance Design Manual specify a maximum over-cut of ¹/₈ inch (3 mm) for the annular space around the outlet boxes. Additionally, Article 373 of the National Electrical Code (NEC) includes the size limitation of the over-cut. Therefore, the exception applies only when the boxes are installed as they were during the original fire tests, including the limited annular space. Because outlet boxes on both sides of a wall create penetrations of both layers of a wall assembly's protection, the code provides five methods to address this problem. This gives code users several options and does not limit them to the usual 24-inch (610 mm) offset.

Exception 2 permits using outlet boxes of nonmetallic materials if they have been specifically tested. Because many different types of nonmetallic boxes are available, it is important to determine that the boxes being used in the rated dwelling unit separation have been tested. Although the exception applies to nonmetallic electrical outlet boxes, the same concept would apply to steel boxes that exceed the sizes specified in Exception 1.

Exception 3 provides an alternative to the annular space protection provisions for a fire sprinkler that penetrates a single membrane. This exception is available if the annular space around the sprinkler is completely covered by an escutcheon plate of noncombustible material. The nature of the hazard posed by single-membrane penetrations of the sprinkler is limited by the size of the opening, the potential number of openings present and the presence of a sprinkler system. The installation of a noncombustible escutcheon provides protection against free passage of fire through the annular space and allows for the movement of the sprinkler piping without breaking during a seismic event [See Commentary Figure R317.3.2(2)].

SECTION R318
MOISTURE VAPOR RETARDERS

R318.1 Moisture control. In all framed walls, floors and roof/ceilings comprising elements of the building thermal envelope, a vapor retarder shall be installed on the warm-in-winter side of the insulation.

Exceptions:

1. In construction where moisture or freezing will not damage the materials.

2. Where the framed cavity or space is ventilated to allow moisture to escape.

3. In counties identified as in climate zones 1 through 4 in Table N1101.2.

❖ This section addresses the use of a vapor retarder for moisture control at the exterior thermal envelope; it is applicable to walls, floors and roof/ceiling assemblies of frame construction.

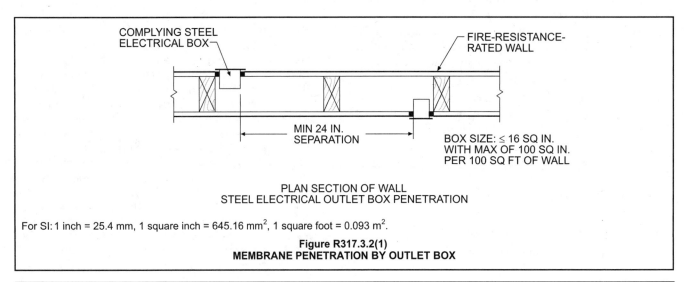

For SI: 1 inch = 25.4 mm, 1 square inch = 645.16 mm^2, 1 square foot = 0.093 m^2.

Figure R317.3.2(1)
MEMBRANE PENETRATION BY OUTLET BOX

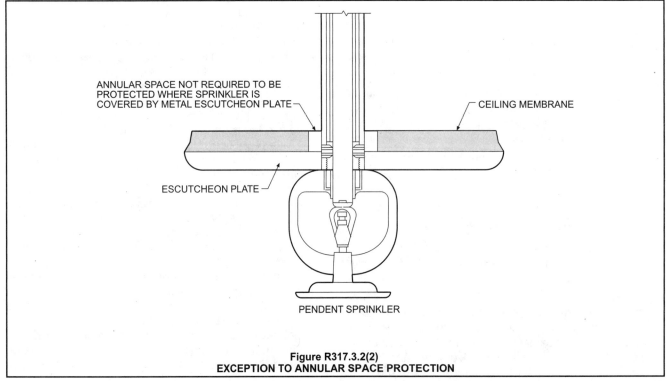

Figure R317.3.2(2)
EXCEPTION TO ANNULAR SPACE PROTECTION

Water condensation in an exterior framed element can diminish the insulation value and lead to early structural failure. The code takes steps to eliminate these detrimental effects by requiring a vapor retarder, also referred to as a vapor barrier. Because attic and under-floor spaces generally have adequate means of ventilation, the exterior wall cavities are the locations where a vapor retarder is necessary. Water vapor is part of the atmosphere and as such travels with the air flow. However, as does independent gas, it migrates by diffusion through air and materials according to its own pressure differentials. Once inside the wall cavity, the vapor can condense on surfaces that are below its dew point.

The location of the vapor barrier is important, but to understand why, we must examine how moisture enters the wall cavity. Moisture can enter in at least two ways, through a hole with direct air migration or by diffusion. Much more moisture can enter and be deposited in a wall cavity through air currents than by diffusion alone. The first step to effective moisture control is construction of a tight envelope.

Assuming a tight envelope is present, moisture flows through a wall by the process of diffusion. Diffusion is driven by the water-vapor pressure differences between indoor air and outdoor air. For most of the year, in a large part of the United States, the vapor pressure is greater indoors than outdoors. Under these conditions, the moisture diffuses through the wall from inside to outside. Therefore, the code provides for the vapor barrier on the warm-in-winter side of the wall, as shown in Commentary Figure R318.1.

Moisture migration is more complicated in air-conditioned buildings. A portion of the southern United States (along the Gulf of Mexico and the lower Atlantic coast) is a humid climate. In these locations, identified as climate zones 1 through 4 in Table N1101.2, no vapor retarder is required, as the exterior weather membrane provides the necessary protection.

SECTION R319
PROTECTION AGAINST DECAY

R319.1 Location required. Protection from decay shall be provided in the following locations by the use of naturally durable wood or wood that is preservative treated in accordance with AWPA U1 for the species, product, preservative and end use. Preservatives shall be listed in Section 4 of AWPA U1.

1. Wood joists or the bottom of a wood structural floor when closer than 18 inches (457 mm) or wood girders when closer than 12 inches (305 mm) to the exposed ground in crawl spaces or unexcavated area located within the periphery of the building foundation.

2. All wood framing members that rest on concrete or masonry exterior foundation walls and are less than 8 inches (203 mm) from the exposed ground.

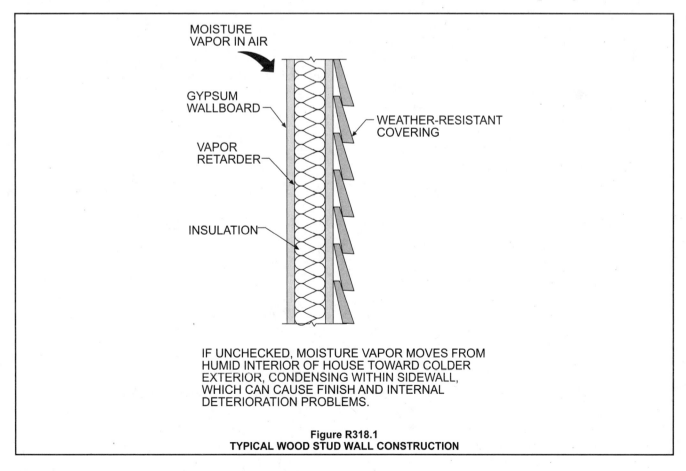

MOISTURE
VAPOR IN AIR

GYPSUM
WALLBOARD

VAPOR
RETARDER

INSULATION

WEATHER-RESISTANT
COVERING

IF UNCHECKED, MOISTURE VAPOR MOVES FROM
HUMID INTERIOR OF HOUSE TOWARD COLDER
EXTERIOR, CONDENSING WITHIN SIDEWALL,
WHICH CAN CAUSE FINISH AND INTERNAL
DETERIORATION PROBLEMS.

Figure R318.1
TYPICAL WOOD STUD WALL CONSTRUCTION

3. Sills and sleepers on a concrete or masonry slab that is in direct contact with the ground unless separated from such slab by an impervious moisture barrier.

4. The ends of wood girders entering exterior masonry or concrete walls having clearances of less than 0.5 inch (12.7 mm) on tops, sides and ends.

5. Wood siding, sheathing and wall framing on the exterior of a building having a clearance of less than 6 inches (152 mm) from the ground.

6. Wood structural members supporting moisture-permeable floors or roofs that are exposed to the weather, such as concrete or masonry slabs, unless separated from such floors or roofs by an impervious moisture barrier.

7. Wood furring strips or other wood framing members attached directly to the interior of exterior masonry walls or concrete walls below grade except where an approved vapor retarder is applied between the wall and the furring strips or framing members.

❖ This section addresses the need for minimum protection against decay damage for wood members located in certain locations.

For those portions of a wood-framed structure that are subject to damage by decay, the code mandates that the lumber be pressure-preservative treated or be naturally durable wood, or be of a species of wood having a natural resistance to decay. Naturally durable wood by definition is the heartwood of decay-resistant redwood, cedars, black locust and black walnut.

Crawl spaces and unexcavated areas under a building usually contain moisture-laden air. These spaces must be ventilated in accordance with Section R408 to remove as much moisture as possible before it causes decay. Wood placed a minimum specified distance above grade in unexcavated under-floor areas or crawl spaces as shown in Commentary Figure R319.1(1) need not be either preservative-treated wood or wood that is naturally decay resistant durable wood. These clearances below floor joists and beams are deemed to be the minimum necessary to allow adequate circulation and removal of moisture from the air and from the wood framing members. Such clearances apply within the exterior wall line of the building foundation.

Foundation walls will absorb moisture from the ground and by capillary action move it to framing members that are in contact with the foundation. Unless a minimum clearance of 8 inches (203 mm) is maintained from the finished grade to wood sills resting on concrete or masonry exterior foundation walls, decay-resistant or preservative-treated wood as shown in Commentary Figure R319.1(2) must be used. The 8-inch (203 mm) clearance specified in this section has been determined to be large enough to prevent wetting of wood framing members under most circumstances.

Concrete and masonry slabs that are in direct contact with the earth are very susceptible to moisture because of absorption of ground water. This can occur on interior slabs as well as at the perimeter. In the case of wood sills or sleepers placed on concrete or masonry slabs, decay-resistant wood or pressure-treated wood is required where the slabs are in direct contact with the ground, as illustrated in Commentary Figure R319.1(3). Concrete that is fully separated from the ground by a vapor barrier is not in direct contact with earth.

A minimum $^1/_2$-inch (12.7 mm) clearance along the top, sides and ends of wood members projecting into

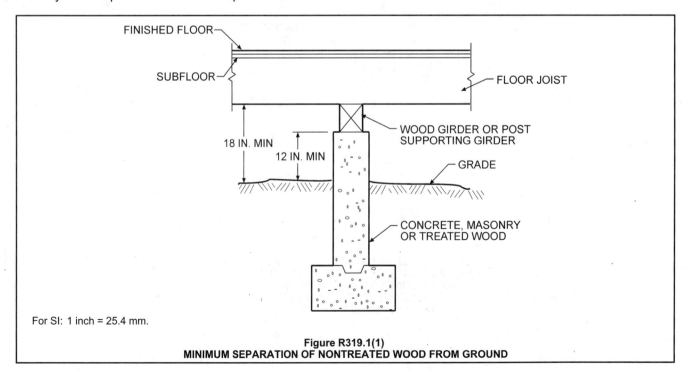

For SI: 1 inch = 25.4 mm.

Figure R319.1(1)
MINIMUM SEPARATION OF NONTREATED WOOD FROM GROUND

exterior masonry or concrete walls must be maintained as illustrated in Commentary Figure R319.1(4), unless the wood is treated or is of a species that is naturally decay resistant.

Experience has shown that wood siding may extend below the sill plate to within 6 inches (152 mm) of the earth without decaying. Commentary Figure R319.1(5) shows the required minimum 6-inch (152 mm) clearance from the ground for wood siding, sheathing and wall framing on the exterior of a building. It should not be in direct contact with the foundation wall.

R319.1.1 Field treatment. Field-cut ends, notches and drilled holes of preservative-treated wood shall be treated in the field in accordance with AWPA M4.

❖ The requirement for field treatment of cuts and holes is duplicated from Section R320.3.1, which addresses the same type of situation in treated wood applications for termite protection.

R319.1.2 Ground contact. All wood in contact with the ground, embedded in concrete in direct contact with the ground or embedded in concrete exposed to the weather that supports permanent structures intended for human occupancy shall be approved pressure-preservative-treated wood suitable for ground contact use, except untreated wood may be used where entirely below groundwater level or continuously submerged in fresh water.

❖ Wood members that are designed to be in contact with the ground and wood that is embedded in concrete in direct contact with the ground, or embedded in concrete exposed to the weather must be suitable for ground contact use. This provision applies to all wood members that support permanent structures designed for human occupancy. Untreated wood is permitted only where the wood members will be located below the groundwater level or where the members are continuously submerged in fresh water because fungus that decays wood and termites cannot survive in a water-only environment with no oxygen (see Commentary Figure R319.1.2).

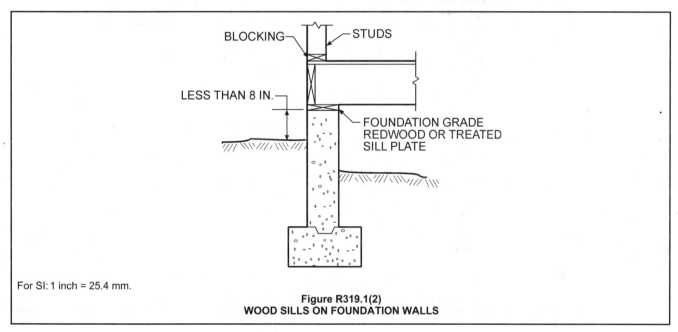

For SI: 1 inch = 25.4 mm.

Figure R319.1(2)
WOOD SILLS ON FOUNDATION WALLS

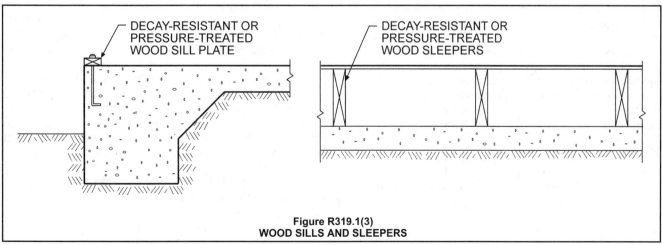

Figure R319.1(3)
WOOD SILLS AND SLEEPERS

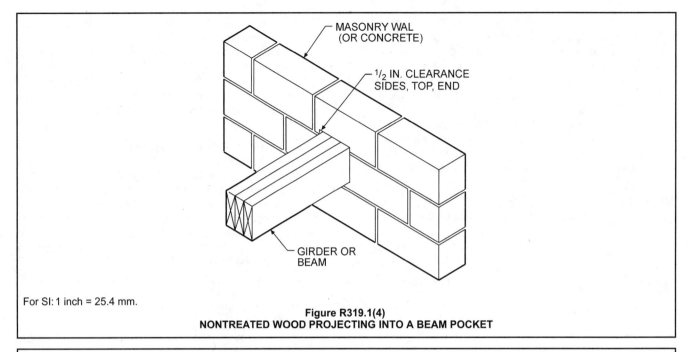

For SI: 1 inch = 25.4 mm.

Figure R319.1(4)
NONTREATED WOOD PROJECTING INTO A BEAM POCKET

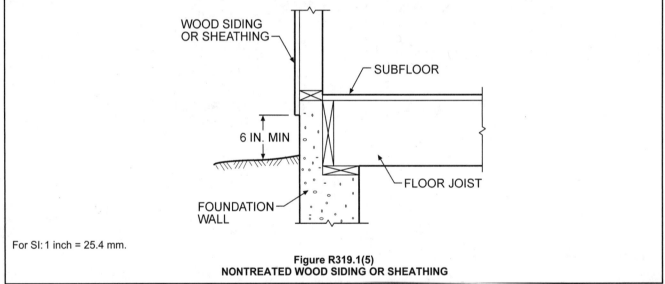

For SI: 1 inch = 25.4 mm.

Figure R319.1(5)
NONTREATED WOOD SIDING OR SHEATHING

R319.1.3 Geographical areas. In geographical areas where experience has demonstrated a specific need, approved naturally durable or pressure-preservative-treated wood shall be used for those portions of wood members that form the structural supports of buildings, balconies, porches or similar permanent building appurtenances when those members are exposed to the weather without adequate protection from a roof, eave, overhang or other covering that would prevent moisture or water accumulation on the surface or at joints between members. Depending on local experience, such members may include:

1. Horizontal members such as girders, joists and decking.

2. Vertical members such as posts, poles and columns.

3. Both horizontal and vertical members.

❖ This section gives the local jurisdiction the authority to require additional protection when the need can be documented by experience. Such experience may include high water tables, extended high-humidity conditions, etc.

Guidance is provided in this section for the protection of wood members other than those specifically identified elsewhere in this section that are exposed to the weather without benefit of protective elements. Included in the list of possible wood members that must be pressure-preservative treated or naturally durable wood are girders, joists, decking, columns and similar structural members of porches, balconies and other exterior building elements.

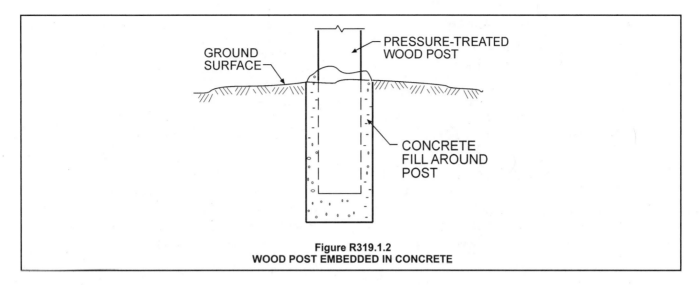

Figure R319.1.2
WOOD POST EMBEDDED IN CONCRETE

R319.1.4 Wood columns. Wood columns shall be approved wood of natural decay resistance or approved pressure-preservative-treated wood.

Exceptions:

1. Columns exposed to the weather or in basements when supported by concrete piers or metal pedestals projecting 1 inch (25.4 mm) above a concrete floor or 6 inches (152 mm) above exposed earth and the earth is covered by an approved impervious moisture barrier.

2. Columns in enclosed crawl spaces or unexcavated areas located within the periphery of the building when supported by a concrete pier or metal pedestal at a height more than 8 inches (203mm) from exposed earth and the earth is covered by an impervious moisture barrier.

❖ Commentary Figure R319.1.4 shows where protection against decay is required for wood structural members on permeable surfaces in direct contact with the ground. Concrete or masonry that is in direct contact with the ground is susceptible to moisture because of absorption of ground water. The exceptions permit the use of common framing lumber when the column is isolated from the earth or concrete. Generally, structural elements in basements or cellars are easier to inspect, which explains the lower clearances. Columns in crawl spaces must have a larger separation distance to safeguard against termites between inspections. In both exceptions, an impervious moisture barrier must be installed to ensure the column does not absorb water from the earth or concrete pier.

R319.1.5 Exposed glued-laminated timbers. The portions of glued-laminated timbers that form the structural supports of a building or other structure and are exposed to weather and not properly protected by a roof, eave or similar covering shall be pressure treated with preservative, or be manufactured from naturally durable or preservative-treated wood.

❖ It is common practice to design large glued-laminated arches that are connected to foundations near ground

level. Exterior walls are built inside the span of these arches, leaving the initial few feet of the wood arches exposed to the weather. Experience has shown that covering the tops of these arches with metal or other water seals is not sufficient to prevent decay of the timber. Therefore, arches and other exposed wood members not protected by roofs or similar covers must be laminated of naturally durable or preservative treated wood.

R319.2 Quality mark. Lumber and plywood required to be pressure-preservative-treated in accordance with Section R319.1 shall bear the quality mark of an approved inspection agency that maintains continuing supervision, testing and inspection over the quality of the product and that has been approved by an accreditation body that complies with the requirements of the American Lumber Standard Committee treated wood program.

❖ The quality of lumber or plywood that has been pressure-preservative treated in accordance with the code is identifiable through a quality mark, such as shown in Commentary Figure R319.2.1. Without the proper identifying mark, it would be impossible to determine in the field that preservative-treated wood conforms to the applicable standards. The identifying mark must be from an approved inspection agency that has continuous follow-up services. Additionally, the inspection agency must be certified as being competent by an approved organization. The required identifying mark is in addition to the requirements for grade marks to identify the species and grade as specified elsewhere in the code.

R319.2.1 Required information. The required quality mark on each piece of pressure-preservative-treated lumber or plywood shall contain the following information:

1. Identification of the treating plant.

2. Type of preservative.

3. The minimum preservative retention.

4. End use for which the product was treated.

5. Standard to which the product was treated.

6. Identity of the approved inspection agency.

7. The designation "Dry," if applicable.

> **Exception:** Quality marks on lumber less than 1 inch (25.4 mm) nominal thickness, or lumber less than nominal 1 inch by 5 inches (25.4 mm by 127 mm) or 2 inches by 4 inches (51 mm by 102 mm) or lumber 36 inches (914 mm) or less in length shall be applied by stamping the faces of exterior pieces or by end labeling not less than 25 percent of the pieces of a bundled unit.

❖ This section provides the information that must be shown on a quality mark for pressure-preservative treated lumber or plywood. Commentary Figure R319.2.1 illustrates how this information may be shown on a quality label.

R319.3 Fasteners. Fasteners for pressure-preservative and fire-retardant-treated wood shall be of hot-dipped zinc-coated galvanized steel, stainless steel, silicon bronze or copper. The coating weights for zinc-coated fasteners shall be in accordance with ASTM A 153.

Exceptions:

1. One-half-inch (12.7 mm) diameter or larger steel bolts.

2. Fasteners other than nails and timber rivets shall be permitted to be of mechanically deposited zinc-coated steel with coating weights in accordance with ASTM B 695, Class 55, minimum.

❖ Except for large steel bolts, all fasteners used in conjunction with pressure-preservative-treated wood must be corrosion resistant. Some chemicals used to preservative-treat or fire-retardant-treat wood may

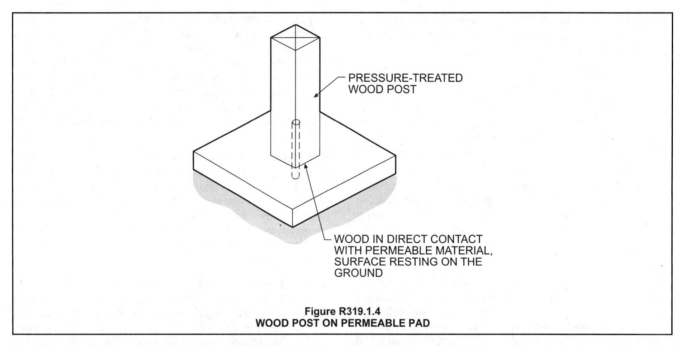

Figure R319.1.4
WOOD POST ON PERMEABLE PAD

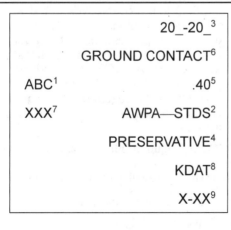

20_-20_ [3]

GROUND CONTACT[6]

ABC[1]   .40[5]

XXX[7]   AWPA—STDS[2]

PRESERVATIVE[4]

KDAT[8]

X-XX[9]

1- THE IDENTIFYING SYMBOL, LOGO OR NAME OF THE ACCREDITED AGENCY.

2- THE APPLICABLE AMERICAN WOOD PRESERVER'S ASSOCIATION (AWPA) COMMODITY STANDARD.

3- THE YEAR OF TREATMENT IF REQUIRED BY AWPA STANDARD.

4- THE PRESERVATIVE USED, WHICH MAY BE ABBREVIATED.

5- THE PRESERVATIVE RETENTION.

6- THE EXPOSURE CATEGORY (E.G., ABOVE GROUND ,GROUND CONTACT, ETC.).

7- THE PLANT NAME AND LOCATION; PLANT NAME AND NUMBER, OR PLANT NUMBER.

8- IF APPLICABLE, MOISTURE CONTENT AFTER TREATMENT.

9- IF APPLICABLE, LENGTH, AND/OR CLASS.

Figure R319.2.1
TYPICAL LABEL FOR PRESSURE-TREATED WOOD

have a corrosive effect on fasteners installed in locations that are high in moisture. Acceptable materials include stainless steel, silicon bronze, copper or hot-dipped zinc-coated galvanized steel. Such fasteners are also mandated for use with fire-retardant-treated wood members. Use of durable fasteners is required to decrease the likelihood that the load-carrying capacity of fasteners will be reduced by corrosion.

The second exception allows the use of a mechanical galvanization process in which the coating weight is mandatory as outlined in ASTM B 695.

SECTION R320
PROTECTION AGAINST
SUBTERRANEAN TERMITES

R320.1 Subterranean termite control methods. In areas subject to damage from termites as indicated by Table R301.2(1), methods of protection shall be one of the following methods or a combination of these methods:

1. Chemical termiticide treatment, as provided in Section R320.2.

2. Termite baiting system installed and maintained according to the label.

3. Pressure-preservative-treated wood in accordance with the AWPA standards listed in Section R319.1.

4. Naturally termite-resistant wood as provided in Section R320.3.

5. Physical barriers as provided in Section R320.4.

❖ This section establishes the rules for protection of structures from damage caused by termites. The methods of protection address not only wood members but also foam plastic materials.

Figure R301.2(6) illustrates those geographical areas where termite damage is probable. In those areas, the structure must be protected from termite damage in an appropriate manner. There are a number of methods permitted by the code to provide the necessary protection against termite damage. The most common method of termite control is soil poisoning. Alternatives include the use of pressure-preservative treated, naturally termite-resistant wood and barriers over perimeter walls. Often a combination of these methods is necessary to establish the required level of protection. The five acceptable methods of protection are broken out to make them easier to read.

R320.1.1 Quality mark. Lumber and plywood required to be pressure-preservative-treated in accordance with Section R320.1 shall bear the quality mark of an approved inspection agency which maintains continuing supervision, testing and inspection over the quality of the product and which has been approved by an accreditation body which complies with the requirements of the American Lumber Standard Committee treated wood program.

❖ See commentary for Section R319.2 and Commentary Figure R319.2.1.

R320.1.2 Field treatment. Field-cut ends, notches, and drilled holes of pressure-preservative-treated wood shall be retreated in the field in accordance with AWPA M4.

❖ When pressure-preservative-treated wood is used for protection against termites, any cuts, notches or bored holes done in the field must be retreated. The re-treatment methods and materials are to be in accordance with AWPA M4, which regulates the care of preservative-treated wood products.

R320.2 Chemical termiticide treatment. Chemical termiticide treatment shall include soil treatment and/or field applied wood treatment. The concentration, rate of application and method of treatment of the chemical termiticide shall be in strict accordance with the termiticide label.

❖ This language accurately reflects the current industry terminology in reference to termiticides and it emphasizes the necessity of using chemicals as required by their labels.

Where using a chemical poisoning of the soil for termite control, the chemical's instructions must be followed in detail. Treatment must be consistent with the recommendations of the chemical manufacturer.

R320.3 Naturally resistant wood. Heartwood of redwood and eastern red cedar shall be considered termite resistant.

❖ Only the heartwood of redwood and eastern red cedar are naturally resistant to termite damage.

Both treated wood and naturally resistant wood provide some added measure of resistance to attack by termites. Where such naturally termite-resistant wood is used as a physical barrier, it must be located where it can be inspected occasionally for signs of termite activity.

R320.4 Barriers. Approved physical barriers, such as metal or plastic sheeting or collars specifically designed for termite prevention, shall be installed in a manner to prevent termites from entering the structure. Shields placed on top of an exterior foundation wall are permitted to be used only if in combination with another method of protection.

❖ This section specifically lists several specific types of termite barriers. In addition, the proper use of a termite shield is addressed. Historically, termite shields were assumed to be effective in preventing termite penetration. However, the term "shield" is actually somewhat of a misnomer because they do not stop termites and were originally designed to make inspection and observation easier.

R320.5 Foam plastic protection. In areas where the probability of termite infestation is "very heavy" as indicated in Figure R301.2(6), extruded and expanded polystyrene, polyisocyanurate and other foam plastics shall not be installed on the exterior face or under interior or exterior foundation walls or slab foundations located below grade. The clearance between foam plastics installed above grade and exposed earth shall be at least 6 inches (152 mm).

Exceptions:

1. Buildings where the structural members of walls, floors, ceilings and roofs are entirely of noncombustible materials or pressure-preservative-treated wood.

2. When in addition to the requirements of Section R320.1, an approved method of protecting the foam plastic and structure from subterranean termite damage is used.

3. On the interior side of basement walls.

❖ In those geographical areas where there is a very high risk of termite infestation, the code prohibits the use of foam plastic materials below or adjacent to ground level. This prohibition includes foam plastics installed on the exterior face of foundation walls or slab foundations, under exterior or interior foundation walls or slab foundations below grade or where located within 6 inches (152 mm) of exposed earth. Foam plastics are permitted where they are adequately protected or where the structural members are either noncombustible or of pressure-preservative-treated wood.

SECTION R321
SITE ADDRESS

R321.1 Premises identification. Approved numbers or addresses shall be provided for all new buildings in such a position as to be plainly visible and legible from the street or road fronting the property.

❖ The code requires buildings to have plainly visible and legible address numbers posted on the building or in such a place on the property that the building may be identified by emergency services such as fire, medical and police. The primary concern is that emergency forces should be able to locate the building without going through a lengthy search procedure. In furthering the concept, the code states that the approved street numbers be placed in a location readily visible from the street or roadway fronting the property if a sign on the building would not be visible from the street.

SECTION R322
ACCESSIBILITY

R322.1 Scope. Where there are four or more dwelling units or sleeping units in a single structure, the provisions of Chapter 11 of the *International Building Code* for Group R-3 shall apply.

❖ Section 1103.2.4, of the IBC, specifies that detached one- and two-family dwellings, their accessory structures and their associated sites and facilities need not be accessible.

Multiple single-family dwellings (townhouses) containing four or more dwelling units in a single structure must meet the accessibility requirements of Chapter 11 of the IBC. In the IBC, townhouses are occupancy classification Group R 3.

Section 1107.5.4 of the IBC requires that, for Group R 3 with four or more dwelling units in a single structure, every dwelling unit must be a Type B dwelling unit. To determine the number of units in a single structure, the Fair Housing Act (FHA) considers all the units built under the same roof. The FHA does not recognize separating units with fire walls to create separate buildings. All dwelling units built in a single structure are counted together. If four or more units are built as a single structure, all units must be Type B dwelling units. Therefore, three townhouses built under the same roof are not required to meet the accessibility requirements.

SECTION R323
ELEVATORS AND PLATFORM LIFTS

R323.1 Elevators. Where provided, passenger elevators, limited-use/limited-application elevators or private residence elevators shall comply with ASME A17.1.

❖ ASME A17.1 is the safety code for elevators. Part V applies to elevators installed in or at a private residence. This part also applies to similar elevators installed in buildings as a means of access to private residences within such buildings if the elevators are installed so that they are not accessible to the general public or to other occupants of the building.

R323.2 Platform lifts. Where provided, platform lifts shall comply with ASME A18.1.

❖ The previous technical standard (i.e., ASME A17.1 - 1996) required key operation to platform lifts, which necessarily inhibits independent access b persons with physical disabilities. The requirements for platform lifts have been removed from the platform standard and now have their own standard, ASME 18.1, *Safety Standard for Platform Lifts and Stairway Chairlifts*, which is referenced in this section.

A platform lift is an electrically operated mechanical device designed to transport a person who cannot use stairs over a short vertical distance. Platform lifts must be sized to accommodate a wheelchair user. Platform lifts can be used by wheelchair users and persons with limited mobility and are sometimes also equipped with folding seats. A fold-down seat that moves up the stairway is not a platform lift. A stairway chair lift cannot serve as part of a required accessible route. A platform lift is most suitable for changes of elevation of one story or less where the installation of a ramp is not feasible.

There are two kinds of platform lifts: vertical lifts and inclined lifts. Vertical lifts are similar to elevators in that they travel only up and down in a fixed vertical space. Inclined platform lifts are usually installed in conjunction with a stairway and travel along the slope of the stairway. Inclined lifts are a design consideration for long flights of stairs where a vertical platform lift is not practical, where headroom is limited or where ceilings are low.

R323.3 Accessibility. Elevators or platform lifts that are part of an accessible route required by Chapter 11 of the *International Building Code*, shall comply with ICC A117.1.

❖ Section 1107 of the IBC includes scoping requirements for Accessible Type A and Type B dwelling units. Accessible routes may also be required to public or common areas on a site. Also see Section R322.1. This section requires that all elevators or platform lifts on an accessible route be useable by a person in a wheelchair without assistance. The referenced standard ICC/A117.1-2003 provides technical data on elevator and platform lift size, operation and controls. Limited access/limited use elevators (LULA) elevators and private residence elevators are considered a type of passenger elevator.

SECTION R324
FLOOD-RESISTANT CONSTRUCTION

R324.1 General. Buildings and structures constructed in whole or in part in flood hazard areas (including A or V Zones) as established in Table R301.2(1) shall be designed and constructed in accordance with the provisions contained in this section.

Exception: Buildings and structures located in whole or in part in identified floodways as established in Table R301.2(1) shall be designed and constructed as stipulated in the *International Building Code*.

❖ This section addresses additional requirements for all buildings and structures proposed to be located in whole or in part in areas designated as flood hazard areas. These areas are commonly referred to as "floodplains" and are shown on a community's Flood Insurance Rate Map (FIRM) prepared by the Federal Emergency Management Agency (FEMA) or other adopted flood hazard map. Through the adoption of the IRC, communities meet some of the requirements necessary to participate in the National Flood Insurance Program (NFIP).

To learn more about how communities may meet the requirements by adoption of the IRC and the IBC (with Appendix G or a comparable ordinance), see *Reducing Flood Losses Through the International Code Series: Meeting the Requirements of the National Flood Insurance Program.*

The NFIP was established to reduce flood losses, to better indemnify individuals from flood losses and to reduce federal expenditures for disaster assistance. A community that has flood hazard areas participates in the NFIP to protect health, safety and property, and so that its citizens can purchase federally-backed flood insurance. FEMA administers the NFIP and monitors community compliance with the flood plain management requirements of the NFIP.

Many states require that communities regulate flood plain development to a higher standard than the minimum standards established by the NFIP. Communities considering using the IRC and other *International Codes*® to meet the flood plain management requirements of the NFIP are advised to consult with their state NFIP coordinator or the appropriate FEMA regional office.

R324.1.1 Structural systems. All structural systems of all buildings and structures shall be designed, connected and anchored to resist flotation, collapse or permanent lateral movement due to structural loads and stresses from flooding equal to the design flood elevation.

❖ New buildings, new structures and substantial improvements to existing buildings and structures in flood hazard areas must be designed and constructed to resist flood forces to minimize damage. Buildings must be connected to structural foundation systems that are capable of resisting flood forces, including flotation, lateral (hydrostatic) pressures, moving water (hydrodynamic) pressures, wave impact and debris impact.

R324.1.2 Flood-resistant construction. All buildings and structures erected in areas prone to flooding shall be constructed by methods and practices that minimize flood damage.

❖ In addition to the structural requirement of Section R324.1.1, this section establishes a broad requirement that techniques used for construction in flood hazard areas must contribute to the flood resistance by minimizing flood damage.

R324.1.3 Establishing the design flood elevation. The design flood elevation shall be used to define areas prone to flooding, and shall describe, at a minimum, the base flood elevation at the depth of peak elevation of flooding (including wave height) which has a 1 percent (100-year flood) or greater chance of being equaled or exceeded in any given year.

❖ This section defines the "flood hazard area," which is the area predicted to be inundated during a flood that has a 1-in-100 (or 1-percent) chance of occurring in any given year. At a minimum, the flood plain is the Special Flood Hazard Area shown on the community's Flood Insurance Rate Map (FIRM). This section also defines the term "design flood elevation," which is used throughout the code. The design flood elevation is the greater of the base flood elevation shown on the FIRM or other flood hazard area map identified in Table R301.2(1).

R324.1.3.1 Determination of design flood elevations. If design flood elevations are not specified, the building official is authorized to require the applicant to:

1. Obtain and reasonably use data available from a federal, state or other source; or

2. Determine the design flood elevation in accordance with accepted hydrologic and hydraulic engineering practices used to define special flood hazard areas. Determina-

tions shall be undertaken by a registered design professional who shall document that the technical methods used reflect currently accepted engineering practice. Studies, analyses and computations shall be submitted in sufficient detail to allow thorough review and approval.

❖ Many FIRMs show flood hazard areas without specifying the base flood elevations (BFE). These areas, often referred to as "unnumbered A or V Zones," are subject to the floodplain management requirements of the code and the appendix, although the minimum height to which buildings and structures must be elevated has not been determined by FEMA. An important step in regulating these flood hazard areas is determination of the design flood elevation (DFE). As defined in Section R324.1.3, at a minimum the DFE is the BFE shown on a community's FIRM identified in Table R301.2(1). In some instances flood elevation information may have been developed by sources other than FEMA, including other federal or state agencies. The building official is to obtain the information if it is available or the building official can require the permit applicant to do so. Some communities develop flood hazard information and provide it to applicants so that all development in an area is based on the same level of risk. If flood elevation information is not available, the building official may require the applicant to develop the DFE in accordance with accepted engineering practices. Local officials unfamiliar with establishing DFEs in unnumbered A and V Zones are encouraged to contact the state National Flood Insurance Program (NFIP) coordinator or the appropriate FEMA regional office. For additional guidance, refer to FEMA 265, *Managing Floodplain Development in Approximate Zone Areas: A Guide for Obtaining and Developing Base (100-Year) Flood Elevations*.

R324.1.3.2 Determination of impacts. In riverine flood hazard areas where design flood elevations are specified but floodways have not been designated, the applicant shall demonstrate that the effect of the proposed buildings and structures on design flood elevations, including fill, when combined with all other existing and anticipated flood hazard area encroachments, will not increase the design flood elevation more than 1 foot (305 mm) at any point within the jurisdiction.

❖ Although FEMA has provided floodways along many rivers and streams shown on a community's FIRM, many other riverine flood hazard areas have BFEs but do not have designated floodways. In these areas the potential effects that floodplain activities may have on flood elevations have not been properly evaluated. If FEMA has not designated a regulatory floodway on a community's FIRM, the community is responsible for regulating development so as not to increase flood elevations by more than 1 foot (305 mm) at any point in the community. In effect, this means a community must either prepare a hydraulic analysis for proposed activities or require permit applicants to do so. Several states have more restrictive requirements, which can be determined by contacting the state NFIP coordinator.

R324.1.4 Lowest floor. The lowest floor shall be the floor of the lowest enclosed area, including basement, but excluding any unfinished flood-resistant enclosure that is useable solely for vehicle parking, building access or limited storage provided that such enclosure is not built so as to render the building or structure in violation of this section.

❖ The lowest floor is the most important reference point when designing and constructing a building or structure in a flood hazard area. The term is specifically defined to include basements, which are any areas that are below grade on all sides. This definition differs from the one in Chapter 2, which includes areas that are partly below grade.

It is important to understand the distinction between the lowest floor of a building or structure and how it relates to enclosures. The National Flood Insurance Program recognizes that elevated buildings may reasonably have certain enclosures that will be subject to flooding. Enclosures that meet certain provisions are deemed not to be the lowest floor, and federal flood insurance premium rates are based on the elevated portion of the building. For flood hazard areas known as "A zones," enclosure provisions are set forth in R324.2.2. For coastal high hazard areas including "V zones," provisions for enclosures are set forth in R324.3.3 and R324.3.4.

R324.1.5 Protection of mechanical and electrical systems. Electrical systems, equipment and components, and heating, ventilating, air conditioning and plumbing appliances, plumbing fixtures, duct systems, and other service equipment shall be located at or above the design flood elevation. If replaced as part of a substantial improvement, electrical systems, equipment and components, and heating, ventilating, air conditioning, and plumbing appliances, plumbing fixtures, duct systems, and other service equipment shall meet the requirements of this section. Systems, fixtures, and equipment and components shall not be mounted on or penetrate through walls intended to break away under flood loads.

Exception: Electrical systems, equipment and components, and heating, ventilating, air conditioning and plumbing appliances, plumbing fixtures, duct systems, and other service equipment are permitted to be located below the design flood elevation provided that they are designed and installed to prevent water from entering or accumulating within the components and to resist hydrostatic and hydrodynamic loads and stresses, including the effects of buoyancy, during the occurrence of flooding to the design flood elevation in compliance with the flood-resistant construction requirements of the *International Building Code*. Electrical wiring systems are permitted to be located below the design flood elevation provided they conform to the provisions of the electrical part of this code for wet locations.

❖ This section sets a broad requirement that electrical and mechanical system elements, including electrical, heating, ventilating, air conditioning, plumbing and other service equipment are to be elevated above the design flood elevation. The same requirement applies to replacement of electrical and mechanical system elements that are included as part of substantial improvements to existing buildings. However, if new or replacement system elements (including duct work) are permitted below the design flood elevations, they must be protected against flood damage. If a building's floor is at the design flood elevation (the minimum requirement), anything installed below the floor is automatically below the minimum elevation and therefore not in compliance with this section. For additional guidance, refer to FEMA 348, *Protecting Building Utilities from Flood Damage*.

The exception to this section provides criteria for placing specific equipment below the design flood elevation. It is important to be clear that to do so the equipment must be designed to prevent the entry of water. For example, a typical single family home has an exterior air conditioning unit located on the ground adjacent to the structure. This unit is typically not designed to withstand the entry of water and would not meet the requirements of this exception.

R324.1.6 Protection of water supply and sanitary sewage systems. New and replacement water supply systems shall be designed to minimize or eliminate infiltration of flood waters into the systems in accordance with the plumbing provisions of this code. New and replacement sanitary sewage systems shall be designed to minimize or eliminate infiltration of floodwaters into systems and discharges from systems into floodwaters in accordance with the plumbing provisions of this code and Chapter 3 of the *International Private Sewage Disposal Code*.

❖ Health concerns arise when water supply systems are exposed to flood waters, and contamination from flooded sewage systems can pose health and environmental risks. To reduce these risks, water supply systems and sanitary sewage systems must be designed to minimize or eliminate infiltration and discharges under flood conditions. For on-site sewage disposal systems, the effects of flooding may be reduced by locating the systems on the highest available ground. State health regulations may apply to on-site sewage disposal systems proposed in flood hazard areas.

R324.1.7 Flood-resistant materials. Building materials used below the design flood elevation shall comply with the following:

1. All wood, including floor sheathing, shall be pressure-preservative-treated in accordance with AWPA U1 for the species, product, preservative and end use or be the decay-resistant heartwood of redwood, black locust or cedars. Preservatives shall be listed in Section 4 of AWPA U1.

2. Materials and installation methods used for flooring and interior and exterior walls and wall coverings shall conform to the provisions of FEMA/FIA-TB

❖ To minimize flood damage, wood materials used below the design flood elevation must resist damage caused by flood waters. Even though certain enclosures may be allowed under otherwise properly elevated buildings, the materials used in those enclo-

sures must be resistant to water damage to minimize flood losses. For further guidance on pressure-preservative-treated lumber refer to AWPA U1 or for flood resistance requirements see FEMA TB-2 and FEMA FIA–TB #8.

R324.1.8 Manufactured housing. New or replacement manufactured housing shall be elevated in accordance with Section R324.2 and the anchor and tie-down requirements of Sections AE604 and AE605 of Appendix E shall apply. The foundation and anchorage of manufactured housing to be located in identified flood ways as established in Table R301.2(1) shall be designed and constructed in accordance with the applicable provisions in the *International Building Code*.

❖ The placement of new manufactured-housing units is subject to the same elevation requirements as other new construction. Replacement of existing manufactured homes is considered "new construction" and must comply with the provisions of this section. Manufactured units are to be elevated so that their lowest floors are at or above the design flood elevation. In part, this provision exceeds the minimum requirements of the NFIP, which allows locating replacement manufactured homes below the DFE in some limited situations in existing manufactured home parks and subdivisions. Because of anticipated flood forces, units must be anchored and tied down in accordance with Appendix E (AE 604 and AE 605). Within floodways, where water depths and velocities are anticipated to be greatest, foundation and anchorage must be designed in accordance with the IBC.

R324.1.9 As-built elevation documentation. A registered design professional shall prepare and seal documentation of the elevations specified in Section R324.2 or R324.3.

❖ Documentation of the "as-built" lowest floor elevation is to be submitted to verify compliance. Most communities require or request submission of FEMA's Elevation Certificate (FEMA Form 81-31, available on-line at www.fema.gov/nfip/forms.htm), which is designed to obtain the information necessary to show compliance. Building owners need elevation information to obtain flood insurance, and insurance agents use the Elevation Certificates to compute the proper flood insurance premium rates.

As called for in Section R109.1.3, the elevation information is to be provided when the floor elevation is set and before further significant vertical construction occurs. If an error is discovered, it will be significantly easier and less expensive to correct at this stage. The building official must obtain the lowest floor elevation when the building is completed and before the Certificate of Occupancy is issued to determine that the building is in compliance with the flood hazard provisions of the code. The building official must maintain copies of the elevation information on file and make elevation and related building information available for insurance rating and other purposes.

R324.2 Flood hazard areas (including A Zones). Areas that have been determined to be prone to flooding but not subject to high velocity wave action shall be designated as flood hazard areas. All buildings and structures constructed in whole or in part in flood hazard areas shall be designed and constructed in accordance with Sections R324.2.1 and R324.2.3.

❖ Flood hazard areas, often referred to as "A Zones," are found along nontidal waterways and lakes, and include the inland portions of coastal floodplains that are not subject to high velocity wave action [i.e., where waves are less than 3.0 feet (914 mm) in height]. The zone designation does not indicate anticipated velocities of river and stream flooding. Especially in areas where the slopes of streambeds are relatively steep, designers should check the Flood Insurance Study to see whether velocity information is available for use in considering the effects of hydrodynamic flood loads.

R324.2.1 Elevation requirements.

1. Buildings and structures shall have the lowest floors elevated to or above the design flood elevation.

2. In areas of shallow flooding (AO Zones), buildings and structures shall have the lowest floor (including basement) elevated at least as high above the highest adjacent grade as the depth number specified in feet (mm) on the FIRM, or at least 2 feet (610 mm) if a depth number is not specified.

3. Basement floors that are below grade on all sides shall be elevated to or above the design flood elevation.

Exception: Enclosed areas below the design flood elevation, including basements whose floors are not below grade on all sides, shall meet the requirements of Section R324.2.2.

❖ This section presents the required minimum floor elevations for all buildings and structures proposed for construction in whole or in part within flood hazard areas known as A zones, AE zones, AH zones and AO zones. The minimum requirement is that lowest floors must be elevated to or above the design flood elevation. AO zones are areas of shallow flooding where FEMA has specified a "depth" number rather than a flood elevation that is referenced to a datum. Buildings in AO zones must be elevated so the lowest floor, including the basement, is at least as high above grade as the depth number, or at least 2 feet (610 mm) if a depth number is not specified. The point where this depth is measured is at the highest natural grade adjacent to the proposed building. FEMA refers to some areas subject to shallow flooding as AH zones. AH zones have specified base flood elevations and they are treated the same as other A zones with base flood elevations.

This section emphasizes that basements that are below grade on all sides must be elevated to or above the design flood elevation. The NFIP defines a basement not by its common use, but by whether it is below grade on all sides. Even if below grade on all sides by only 1 inch (25 mm), such an area is, technically, a basement.

In many flood hazard areas it is common to place fill to elevate a building site above the design flood elevation. Fills that are exposed to flooding may become saturated and unstable. These fill slopes may be exposed to erosive velocities and waves during a flood. For additional guidance for development activities involved in the placement of fill in a flood hazard area, refer to FEMA FIA-TB #10. Building officials may require applicants to submit documentation regarding use of fill and whether structures on fill are reasonably safe from flooding.

The specification in Item 3 means that basements excavated into fill within a flood plain cannot be excavated below the design flood elevation. The exception to this section allows enclosures that, if designed and constructed to meet the use limitations and the flood opening specifications of Section R324.2.2, are not the lowest floor. Such enclosures include garages, building access, limited storage areas and crawl spaces.

R324.2.2 Enclosed area below design flood elevation. Enclosed areas, including crawl spaces, that are below the design flood elevation shall:

1. Be used solely for parking of vehicles, building access or storage.

2. Be provided with flood openings that meet the following criteria:

    2.1. There shall be a minimum of two openings on different sides of each enclosed area; if a building has more than one enclosed area below the design flood elevation, each area shall have openings on exterior walls.

    2.2. The total net area of all openings shall be at least 1 square inch (645 mm²) for each square foot (0.093 m²) of enclosed area, or the openings shall be designed and the construction documents shall include a statement that the design and installation will provide for equalization of hydrostatic flood forces on exterior walls by allowing for the automatic entry and exit of floodwaters.

    2.3. The bottom of each opening shall be 1 foot (305 mm) or less above the adjacent ground level.

    2.4. Openings shall be at least 3 inches (76 mm) in diameter.

    2.5. Any louvers, screens or other opening covers shall allow the automatic flow of floodwaters into and out of the enclosed area.

    2.6. Openings installed in doors and windows, that meet requirements 2.1 through 2.5, are acceptable; however, doors and windows without installed openings do not meet the requirements of this section.

❖ This section contains the use limitations and flood opening requirements that apply to enclosures below otherwise properly elevated buildings. If enclosures meet these requirements they are not considered the "lowest floor" for insurance purposes. The use limitation states that enclosures, including crawl spaces, are to be used only for parking of vehicles, building access or storage. Although the NFIP regulations do not define storage, most states interpret it to mean "limited" storage. Some states specify that hazardous materials must not be stored in enclosures. Experience has suggested that two design factors influence the use of enclosures and may discourage conversions by owners: the size of the enclosed area and utility service. Plumbing service is not appropriate for the allowable uses, and minimal electrical service can be provided (a light switch for a stairwell, for example, or electrical outlets dropped from the ceiling).

Openings intended to relieve differential hydrostatic pressure during the rise and fall of floodwaters must meet several specific requirements for location and size. Information about the number, size and location of flood openings is collected when FEMA's Elevation Certificate is completed (FEMA Form 81-31); the information is used by insurance agents to determine the lowest floor and to compute the proper federal flood insurance rates.

The requirements related to location of flood openings are clear. Openings are to be provided on different sides of the enclosed area and they are to be no more than 12 inches (305 mm) above the adjacent exterior ground level. The size of total net open area of the flood openings is determined prescriptively by computing the total area enclosed by the walls that extend below the DFE and providing 1 square inch of net open area for every square foot (6944 mm^2 per m^2) of enclosed area. Alternatively, flood openings may be designed to automatically equalize hydrostatic flood forces, and the construction documents must include a statement from a registered design professional that they meet this requirement.

It is common for devices intended for underfloor ventilation openings to be used as flood openings. Care should be taken to determine the net open area provided by such devices, taking into account obstructions such as face plates, grilles, and other coverings. Airflow vent devices are to be disabled in the "open" position to ensure automatic functioning in the event of a flood. For additional guidance on meeting the flood opening requirements of this section, see FEMA FIA-TB #1.

R324.2.3 Foundation design and construction. Foundation walls for all buildings and structures erected in flood hazard areas shall meet the requirements of Chapter 4.

Exception: Unless designed in accordance with Section R404:

1. The unsupported height of 6-inch (152 mm) plain masonry walls shall be no more than 3 feet (914 mm).

2. The unsupported height of 8-inch (203 mm) plain masonry walls shall be no more than 4 feet (1219 mm).

3. The unsupported height of 8-inch (203 mm) reinforced masonry walls shall be no more than 8 feet (2438 mm).

For the purpose of this exception, unsupported height is the distance from the finished grade of the under-floor space and the top of the wall.

❖ Flood conditions impose additional loads on foundation walls, including hydrostatic and hydrodynamic loads. Under a Cooperative Agreement with FEMA, the American Society of Civil Engineers (ASCE) analyzed flood loads and flood-load combinations to examine the structural adequacy of the foundation walls prescribed in Chapter 4. A range of flood depths and velocities was used to determine whether there were any height limitations for certain types of walls. The results suggested that unless the walls are designed in accordance with Section R404, it is appropriate to limit the unsupported height of certain masonry walls based on wall thickness and whether they are plain or reinforced.

R324.3 Coastal high-hazard areas (including V Zones). Areas that have been determined to be subject to wave heights in excess of 3 feet (914 mm) or subject to high-velocity wave action or wave-induced erosion shall be designated as coastal high-hazard areas. Buildings and structures constructed in whole or in part in coastal high-hazard areas shall be designated and constructed in accordance with Sections R324.3.1 through R324.3.6.

❖ Buildings and structures located in flood hazard areas known as "V zones," or coastal high hazard areas, are exposed to flooding and wind-driven waves in excess of 3 feet (914 mm). Waves impose additional loads on foundations and can cause erosion. Designers are referred to FEMA 55 and builders are referred to FEMA 499. Training on the design and construction of residential structures in coastal areas may be offered by the NFIP state coordinator or the appropriate FEMA regional office.

R324.3.1 Location and site preparation.

1. Buildings and structures shall be located landward of the reach of mean high tide.

2. For any alteration of sand dunes and mangrove stands the building official shall require submission of an engineering analysis which demonstrates that the proposed alteration will not increase the potential for flood damage.

❖ Buildings and structures must be sited such that they are not regularly affected by normal (mean) high tides. Many coastal locations benefit from natural storm protection because of the presence of sand dunes and mangrove stands. Those natural features must remain intact. However, alterations may be considered if specific analyses indicate that the alterations will not increase the potential for flood damage, including increasing the exposure of the site to greater flood depths and higher wave heights.

R324.3.2 Elevation requirements.

1. All buildings and structures erected within coastal high hazard areas shall be elevated so that the lowest portion of all structural members supporting the lowest floor, with the exception of mat or raft foundations, piling, pile caps, columns, grade beams and bracing, is located at or above the design flood elevation.

2. Basement floors that are below grade on all sides are prohibited.

3. The use of fill for structural support is prohibited.

4. The placement of fill beneath buildings and structures is prohibited.

Exception: Walls and partitions enclosing areas below the design flood elevation shall meet the requirements of Sections R324.3.4 and R324.3.5.

❖ Buildings and structures to be located in coastal high hazard areas (V zones) must be elevated so that the bottom of the lowest horizontal structural member will be at or above the design flood elevation. Elevation is one of the most important requirements to provide resistance to flood damage. It is also the main factor used in determining federal flood insurance premium rates for buildings and structures. In coastal high-hazard areas, the use of fill to provide structural support is prohibited because areas subject to wave action are more likely to experience flood-related erosion, which may lead to failure of foundations on fill. This section and the NFIP allow certain enclosures under buildings that otherwise meet the elevation requirements, and the exception notes that the requirements for such enclosures are found in Sections R324.3.3 and R324.3.5. Additional guidance about design and construction in coastal high hazard areas can be found in FEMA FIA–TB 5, FEMA FIA-TB #9, FEMA 499, and FEMA 55.

R324.3.3 Foundations. Buildings and structures erected in coastal high-hazard areas shall be supported on pilings or columns and shall be adequately anchored to those pilings or columns. Pilings shall have adequate soil penetrations to resist the combined wave and wind loads (lateral and uplift). Water loading values used shall be those associated with the design flood. Wind loading values shall be those required by this code. Pile embedment shall include consideration of decreased resistance capacity caused by scour of soil strata surrounding the piling. Pile systems design and installation shall be certified in accordance with Section R324.3.6. Mat, raft or other foundations that support columns shall not be permitted where soil investigations that are required in accordance with Section R401.4 indicate that soil material under the mat, raft or other foundation is subject to scour or erosion from wave-velocity flow conditions. Slabs, pools, pool decks and walkways shall be located and constructed to be structurally independent of buildings and structures and their foundations to prevent transfer of flood loads to the buildings and structures during conditions of flooding, scour or erosion from wave-velocity flow conditions, unless the buildings and structures and their foundation are designed to resist the additional flood load.

❖ Several provisions applicable to foundations for buildings and structures in coastal high hazard areas (V zones) are in this section. During coastal storms, both wind and water loads are significant, and buildings must be designed and constructed to withstand both loads. Buildings must be anchored to their foundations, and the foundations must withstand the anticipated forces. A limitation is outlined on the use of mat, raft and other foundations that support columns in areas where soil investigations indicate that scour or erosion is likely. It is common to have structures such as pools, pool decks and slabs built immediately adjacent to buildings in coastal communities. Such structures should be located and built so they do not obstruct waves and the flow of water or adversely impact buildings during a flood, including wave-induced scour and erosion.

R324.3.4 Walls below design flood elevation. Walls and partitions are permitted below the elevated floor, provided that such walls and partitions are not part of the structural support of the building or structure and:

1. Electrical, mechanical, and plumbing system components are not to be mounted on or penetrate through walls that are designed to break away under flood loads; and

2. Are constructed with insect screening or open lattice; or

3. Are designed to break away or collapse without causing collapse, displacement or other structural damage to the elevated portion of the building or supporting foundation system. Such walls, framing and connections shall have a design safe loading resistance of not less than 10 (479 Pa) and no more than 20 pounds per square foot (958 Pa); or

4. Where wind loading values of this code exceed 20 pounds per square foot (958 Pa), the construction documents shall include documentation prepared and sealed by a registered design professional that:

    4.1. The walls and partitions below the design flood elevation have been designed to collapse from a water load less than that which would occur during the design flood.

    4.2. The elevated portion of the building and supporting foundation system have been designed to withstand the effects of wind and flood loads acting simultaneously on all building components (structural and nonstructural). Water loading values used shall be those associated with the design flood. Wind loading values shall be those required by this code.

❖ Buildings and structures in coastal high-hazard areas must be elevated to or above the design flood elevation. However, certain enclosures are allowed below otherwise elevated buildings if they meet certain specifications. These specifications allow the walls surrounding the enclosures to fail or break away under certain loads without causing damage to the foundation. Post-flood investigations indicate that breakaway walls do not perform as intended if wires, pipes and other utility components are mounted on or penetrate through the walls. Walls other than lattice or insect screening must meet specific loading resistance; they are to be designed to fail under loads of not less than 10 (480 Pa) and no more than 20 pounds per square foot (960 Pa). If anticipated loads exceed 20 pounds per square foot (960 Pa), the wall design is to be certified as meeting certain specifications. For additional guidance, refer to NFIP Technical Bulletin #9, FEMA 499 and FEMA 55.

R324.3.5 Enclosed areas below design flood elevation. Enclosed areas below the design flood elevation shall be used solely for parking of vehicles, building access or storage.

❖ The limitations that are outlined in Section R324.2.2 on how enclosures below elevated buildings are used are repeated here. They apply to enclosures below elevated buildings in coastal high hazard areas. Enclosures are to be used only for parking, building access and storage. Although the NFIP regulations do not define storage, most states interpret it to mean "limited" storage. Some states specify that hazardous materials must not be stored in enclosures. Experience has suggested that two design factors influence the use of enclosures and may discourage conversions by owners: the size of the enclosed area and utility service.

To minimize the likelihood of conversion to other uses, some communities limit the size of enclosures. Another factor that influences enclosure use is the availability of utility services. Plumbing service is not associated with the allowable uses and in accordance with Section R324.1.5 must not be located below the design flood elevation. Minimal electrical service can be installed (a light switch for a stairwell, for example, or electrical outlets that are dropped from the ceiling).

R324.3.6 Construction documents. The construction documents shall include documentation that is prepared and sealed by a registered design professional that the design and methods of construction to be used meet the applicable criteria of this section.

❖ Documentation must be submitted to demonstrate that the design and proposed construction methods address the provisions of the code. This requirement recognizes that coastal high hazard areas subject to waves and erosion are dynamic environments that impose significant loads on buildings. Documentation of the elevation of the bottom of the lowest horizontal structural member is addressed in Section R324.1.9. The documentation provided by the registered design professional is applicable to the design criteria for Section R324.3 for coastal high-hazard areas in addition to the as-built elevations as required by Section R324.1.9. See commentary for Section R324.1.9.

Bibliography

The following resource materials are referenced in this chapter or are relevant to the subject matter addressed in this chapter.

16 CFR Part1201-(1977), *Safety Standard for Architectural Glazing.* Bethesda, MD: Consumer Product Safety Commission, 1977.

16 CFR Part 1209-(1979), *Interim Safety Standard for Cellulose Insulation.* Bethesda, MD: Consumer Product Safety Commission, 1979.

16 CFR Part 1404-(1979), *Cellulose Insulation.* Bethesda, MD: Consumer Product Safety Commission, 1979.

AAMA 101/I.S2/ A440-05, *Specifications for Windows, Doors and Unit Skylights.* Schaumburg, IL: American Architectural Manufactures Association, 2005.

AFPA WFCM-2001, *Wood Frame Construction Manual for One- and Two-family Dwellings.* Washington, DC: American Forest and Paper Association, 2001.

AISI PM-2001, *Standard for Cold-formed Steel Framing-Prescriptive Method for One- and Two-family Dwellings.* Washington, DC: American Iron and Steel Institute, 2001.

An Architect's Compendium to the UBC. Whittier, CA: International Conference of Building Officials, 2000.

ANSI Z97.1-84(R1994), *Safety Glazing Materials Used in Buildings—Safety Performance Specifications and Methods of Test (Reaffirmed 1994).* New York: American National Standards Institute, 1984.

APA PAS #473, *Subdivision Design in Flood Hazard Areas.* Washington, DC: American Planning Association, 1997.

ASCE 7-05, *Minimum Design Loads for Buildings and Other Structures.* Reston, VA: American Society of Civil Engineers, 2005.

ASCE 24-05, *Flood Resistant Design and Construction.* Reston, VA: American Society of Civil Engineers, 2005.

ASTM C 34-03, *Specification for Structural Clay Load-Bearing Wall Tile.* West Conshohocken, PA: ASTM International, 2003.

ASTM C 55-03, *Specification for Concrete Brick.* West Conshohocken, PA: ASTM International, 2003.

ASTM C 67-03, *Test Methods of Sampling and Testing Brick and Structural Clay Tile.* West Conshohocken, PA: ASTM International, 2003.

ASTM C 73-99a, *Specification for Calcium Silicate Face Brick (Sand Lime Brick).* West Conshohocken, PA: ASTM International, 1999.

ASTM C 90-03, *Specification for Load-bearing Concrete Masonry Units.* West Conshohocken, PA: ASTM International, 2003.

ASTM C 129-03, *Specification for Nonload-bearing Concrete Masonry Units and Related Units.* West Conshohocken, PA: ASTM International, 2003.

ASTM C 216-04a, *Specification for Facing Brick (Solid Masonry Units Made from Clay or Shale).* West Conshohocken, PA: ASTM International, 2004.

ASTM C 652-04a, *Specification for Hollow Brick (Hollow Masonry Units Made from Clay or Shale).* West Conshohocken, PA: ASTM International, 2004.

ASTM E 84-04, *Test Method for Surface Burning Characteristics of Building Materials.* West Conshohocken, PA: ASTM International, 2004.

ASTM E 119-00a, T*est Method for Fire Tests of Building Construction and Materials.* West Conshohocken, PA: ASTM International, 2000.

ASTM E 152-95, *Method of Fire Tests for Door Assemblies.* West Conshohocken, PA: ASTM International, 1995.

ASTM E 814-02, *Test Method for Fire Tests of Through-Penetration Fire Stops.* West Conshohocken, PA: ASTM International, 2002.

ASTM E 970-00, *Test Method for Critical Radiant Flux of Exposed Attic Floor Insulation Using a Radiant Heat Energy Source.* West Conshohocken, PA: ASTM International, 2000.

ASTM E 1886-04, *Test Method for Performance of Exterior Windows, Curtain Walls, Doors and Storm Shutters Impacted by Missile(s) and Exposed to Cyclic Pressure Differentials.* West Conshohocken, PA: ASTM International, 2004.

ASTM E 1996-04, *Specification for Performance of Exterior Windows, Curtain Walls, Doors and Storm Shutters Impacted by Windborne Debris in Hurricanes.* West Conshohocken, PA: ASTM International, 2004.

AWPA C1-00, *All Timber Products—Preservative Treatment by Pressure Processes.* Granbury, TX: American Wood-Preservers' Association, 2000.

AWPA C2-01, *Lumber, Timbers, Bridge Ties and Mine Ties—Preservative Treatment by Pressure Processes.* Granbury, TX: American Wood-Preservers' Association, 2001.

AWPA C3-99, *Piles—Preservative Treatment by Pressure Processes.* Granbury, TX: American Wood–Preservers' Association, 1999.

AWPA C4-99, *Poles—Preservative Treatment by Pressure Processes.* Granbury, TX: American Wood–Preservers' Association, 1999.

AWPA C9-00, *Plywood—Preservative Treatment by Pressure Processes.* Granbury, TX: American Wood-Preservers' Association, 2000.

AWPA C15-00, *Wood for Commercial–Residential Construction Preservative Treatment by Pressure Processes.* Granbury, TX: American Wood–Preservers' Association, 2000.

AWPA C18-99, *Standard for Pressure Treated Material in Marine Construction.* Granbury, TX: American Wood-Preservers' Association, 1999.

AWPA C22-96, *Lumber and Plywood for Permanent Wood Foundations—Preservative Treatment by Pressure Processes.* Granbury, TX: American Wood-Preservers' Association, 1996.

AWPA C23-00, *Round Poles and Posts Used in Building Construction—Preservative Treatment by Pressure Processes.* Granbury, TX: American Wood-Preservers' Association, 2000.

AWPA C24-96, *Sawn Timber Used to Support Residential and Commercial Structures.* Granbury, TX: American Wood-Preservers' Association, 1996.

AWPA C28-99, *Standard for Preservative Treatment by Pressure Process of Structural Glued Laminated Members and Laminations Before Gluing.* Granbury, TX: American Wood-Preservers' Association, 1999.

AWPA M4-02, *Standard for the Care of Preservative-treated Wood Products.* Granbury, TX: American Wood-Preservers' Association, 2002.

Fire Resistance Directory. Northbrook, IL: Underwriters Laboratories, Inc., 2003.

FM 4880-(2001), *American National Standard for Evaluating Insulated Wall or Wall and Roof/Ceiling Assemblies, Plastic Interior Finish Materials, Plastic Exterior Building Panels, Wall/Ceiling Coating Systems, Interior or Exterior Finish Systems.* Norwood, MA: Factory Mutual Global Research, 2001.

GA 600-00 *Fire-resistance Design Manual 16th Edition,* April, 2000. Washington, DC: Gypsum Association, 2000.

ULC S 102-M88, *Surface Burning Characteristics of Building Materials and Assemblies.* Scarborough, Ontario, Canada: Underwriters' Laboratories of Canada, 1988.

ICC SBCCI SSTD 10-99, *Standard for Hurricane Resistant Construction.* Falls Church, VA: International Code Council, 1999.

IBC-2006*, International Building Code.* Falls Church, VA: International Code Council, 2006.

IPC-2006, *International Plumbing Code.* Falls Church, VA: International Code Council, 2006.

Handbook to the International Building Code. Whittier, CA: International Conference of Building Officials, 2000.

NFPA 13-02, *Installation of Sprinkler Systems.* Quincy, MA: National Fire Protection Association, 2002.

NFPA 70-05, *National Electrical Code.* Quincy, MA: National Fire Protection Association, 2005.

NFPA 72-02, *National Fire Alarm Code.* Quincy, MA: National Fire Protection Association, 2002.

NFPA 259-04, *Standard Test Method for Potential Heat of Building Materials.* Quincy, MA: National Fire Protection Association, 2004.

NFPA 286-00, *Standard Methods of Fire Tests for Evaluating Contribution of Wall and Ceiling Interior Finish to Room Fire Growth.* Quincy, MA: National Fire Protection Association, 2000.

UL 325-02, *Door, Drapery, Louver and Window Operations and Systems—with Revisions through March 2003.* Northbrook, IL: Underwriters Laboratories, 2002.

UL 1040-96, *Fire Test of Insulated Wall Construction—with Revisions through April 2001.* Northbrook, IL: Underwriters Laboratories, 1996.

UL 1479-03, *Fire Tests of Through-Penetration Firestops.* Northbrook, IL: Underwriters Laboratories, 2003.

UL 1715-97, *Fire Test of Interior Finish Material.* Northbrook, IL: Underwriters Laboratories, 1997.

The following publications and forms may be obtained at no cost from the Federal Emergency Management Agency (FEMA). These and other materials may also be available online at http://www.fema.gov/library/lib06.htm, or ordered from:

Federal Emergency Management Agency
P.O. Box 2012
Jessup, Maryland 20794-2012
800-480-2520

FEMA 55CD (third edition): *Coastal Construction Manual.*

FEMA 213: *Answers to Questions about Substantially Damaged Buildings.*

FEMA 232: *Home Builder's Guide to Seismic Resistant Construction.*

FEMA 259: *Engineering Principles and Practices for Retrofitting Flood Prone Residential Buildings.*

FEMA 265: *Managing Flood plain Development in Approximate Zone A Areas: A Guide for Obtaining and Developing Base (100-Year) Flood Elevations.*

FEMA 311: *Guidance on Estimating Substantial Damage Using the NFIP Residential Substantial Damage Estimator.*

FEMA 303, *National Earthquake Hazards Reduction Program Recommended Provisions for Seismic Regulations for New Buildings and Other Structures. 1997 Edition. Part 2:* Commentary. Washington, DC: Building Seismic Safety Council, 1997.

FEMA 348: *Protecting Building Utilities From Flood Damage: Principles and Practices for the Design and Construction of Flood Resistant Building Utility Systems.*

FEMA 499: *Home Builder's Guide to Coastal Construction: Technical Fact Sheet Series.*

FEMA Form 81–31: Elevation Certificate. Available at http://www.fema.gov/library/elvcert.pdf

NFIP Technical Bulletin Series. Available at http://www.fema.gov/mit/techbul.htm

FEMA FIA–TB #1: *Openings in Foundation Walls for Buildings Located in Special Flood Hazard Areas.*

FEMA/FIA–TB-2-93: *Flood–resistant Materials Requirements.*

FEMA FIA–TB #3: *Non-Residential Floodproofing Requirements and Certification for Buildings Located in Special Flood Hazard Areas.* (Note: applies to non-residential only.)

FEMA FIA–TB #4: *Elevator Installation for Buildings Located in Special Flood Hazard Areas.*

FEMA FIA–TB #5: *Free-of-Obstruction Requirements for Buildings Located in Coastal High Hazard Areas.*

FEMA FIA–TB #6: *Below-Grade Parking Requirements for Buildings Located in Special Flood Hazard Areas.*

FEMA FIA–TB #7: *Wet Flood proofing Requirements for Structures Located in Special Flood Hazard Areas.* (Note: applies to non-residential only.)

FEMA FIA–TB #8: *Corrosion Protection for Metal Connectors In Coastal Areas for Structures Located in Special Flood Hazard Areas.*

FEMA FIA–TB #9: *Design and Construction Guidance for Breakaway Walls Below Elevated Coastal Buildings.*

FEMA FIA-TB #10: *Ensuring that Structures Built on Fill In or Near Special Flood Hazard Areas are Reasonably Safe from Flooding.*

FEMA FIA-TB #11: *Crawlspace Construction for Buildings Located in Special Flood Hazard Areas.*

FEMA 44 CFR Parts 59–73: *National Flood Insurance Program (NFIP).*

Reducing Flood Losses Through the International Code Series: Meeting the Requirements of National Flood Insurance Program.

Chapter 4:
Foundations

General Comments

Section R401 establishes the scope and applicability of the chapter. The criteria provide the details necessary for designing, building and inspecting most common foundation systems for homes. The selection of a particular type of foundation is based on many variables and it should result in the system that best suits the needs of a particular project. Also, the lot itself must be evaluated to determine the type of soil and the drainage pattern.

Section R402 specifies material requirements for wood and concrete foundations. Section R403 provides minimum requirements for concrete and masonry footings as well as wood foundations. Section R404 regulates foundation walls of concrete, masonry and wood as well as insulating concrete form foundation walls. Section R405 has important criteria for maintaining the drainage of water away from foundations. The requirements for waterproofing and dampproofing in Section R406 protect below-grade habitable spaces from moisture. Section R407 provides requirements for underfloor areas and for protecting columns. Section R408 includes provisions for proper ventilation, adequate access, and the removal of debris.

Purpose

Chapter 4 regulates foundations designed and constructed under the *International Residential Code*® (IRC®). A foundation system consists of two interdependent components: the foundation structure itself and the supporting soil. Proper design and construction of a foundation system is vital to the satisfactory performance of the entire structure that it supports.

The prescriptive provisions of this chapter provide requirements for constructing foundation walls and footings of concrete, masonry or wood. In addition to a foundation's ability to support the required design loads, this chapter addresses several other factors that can affect foundation performance. These include controlling surface water and subsurface drainage, requiring soil tests where conditions warrant, and evaluating proximity to slopes and minimum depth requirements. The chapter also provides requirements to minimize adverse effects of moisture, decay and pests in basements and crawl spaces.

SECTION R401
GENERAL

R401.1 Application. The provisions of this chapter shall control the design and construction of the foundation and foundation spaces for all buildings. In addition to the provisions of this chapter, the design and construction of foundations in areas prone to flooding as established by Table R301.2(1) shall meet the provisions of Section R324. Wood foundations shall be designed and installed in accordance with AF&PA Report No. 7.

Exception: The provisions of this chapter shall be permitted to be used for wood foundations only in the following situations:

1. In buildings that have no more than two floors and a roof.

2. When interior basement and foundation walls are constructed at intervals not exceeding 50 feet (15 240 mm).

Wood foundations in Seismic Design Category D₀, D₁ or D₂ shall be designed in accordance with accepted engineering practice.

❖ This chapter contains all requirements relating to the design and construction of foundations and underfloor spaces. Also, an explicit link to the flood-resistant construction provisions of Section R324 is provided. Thus, foundations for buildings to be located in areas prone

to flooding are to meet the provisions of Section R324 as well.

This section also specifically allows a wood foundation system designed and installed according to the American Forest & Paper Association manual Basic Requirements for Permanent Wood Foundation Systems, Technical Report No. 7. When this system is installed under the prescriptive code requirements of this chapter, the code limits its use to structures where the loads exerted on the foundation are relatively light and to areas of low or moderate seismic hazard. This type of foundation construction is essentially a below-grade, loadbearing, wood-frame system capable of providing support for light frame structures. It generally would consist of the following components.

1. Walls consisting of plywood fastened to wood studs (Section R404.2).

2. A composite footing made up of a continuous wood plate set on a bed of granular materials, such as sand, gravel, or crushed stone, which in turn supports the foundation walls and transmits their loads to the bearing soil below (Section R403.2).

3. Polyethylene film that serves as a vapor barrier and covers the exterior side of the plywood foun-

dation walls from grade level down to the footing plate (Section R406.3).

4. Caulking compounds used for sealing the joints in the plywood walls, bonding agents for attaching the polyethylene film to the plywood, and for the film joints (Section R406.3).

5. Metal fasteners made of silicon bronze, copper, or stainless steel or hot-dipped zinc-coated steel nails or staples (Section R402.1).

6. Pressure-treated plywood and lumber to protect the foundation material against decay, termites, and other insects (Section R402.1).

R401.2 Requirements. Foundation construction shall be capable of accommodating all loads according to Section R301 and of transmitting the resulting loads to the supporting soil. Fill soils that support footings and foundations shall be designed, installed and tested in accordance with accepted engineering practice. Gravel fill used as footings for wood and precast concrete foundations shall comply with Section R403.

❖ In order to fulfill its role in the complete load path, the foundation must support the required design loads and transmit these to the soil. The phrase "accepted engineering practice" in this section means common practice that is acceptable to the code official of the jurisdiction.

R401.3 Drainage. Surface drainage shall be diverted to a storm sewer conveyance or other approved point of collection so as to not create a hazard. Lots shall be graded to drain surface water away from foundation walls. The grade shall fall a minimum of 6 inches (152 mm) within the first 10 feet (3048 mm).

Exception: Where lot lines, walls, slopes or other physical barriers prohibit 6 inches (152 mm) of fall within 10 feet (3048 mm), the final grade shall slope away from the foundation at a minimum slope of 5 percent and the water shall be directed to drains or swales to ensure drainage away from

the structure. Swales shall be sloped a minimum of 2 percent when located within 10 feet (3048 mm) of the building foundation. Impervious surfaces within 10 feet (3048 mm) of the building foundation shall be sloped a minimum of 2 percent away from the building.

❖ Along with the proper support for a structure through the foundation system, adequate preparation of the building site is necessary to keep water drainage away from the supporting foundations. Proper site drainage is an important element in preventing wet basements, damp crawl spaces, eroded banks, and possible failure of a foundation system.

One of the most important considerations is the arrangement of structures on a building site in a manner that retains natural drainage patterns and minimizes the alteration or disturbance to existing grades. If the designer keeps such factors in mind, the result will be a reduction of ground surface stabilization problems and opportunities for differential settlement through the reduction in the use of fills. A detailed treatment of drainage design is beyond the scope of this document; therefore, only rough guidelines can be provided for areas where a more comprehensive set of grading regulations does not exist.

As illustrated in Commentary Figures R401.3(1) and R401.3(2), drainage patterns should result in adequate slopes to approved drainage devices that are capable of carrying concentrated runoff. In some cases, control of concentrated roof runoff by gutters and downspouts may be needed, and if gutters and downspouts are used, provisions should be made to discharge runoff in order to prevent soil erosion. Refer also to Section R801.3.

Cross-lot drainage and drainage over graded slopes should generally be avoided. Slopes should be designed with as moderate a grade as possible to minimize slope instability and erosion. The minimum slope gradients which should be used are a function of the

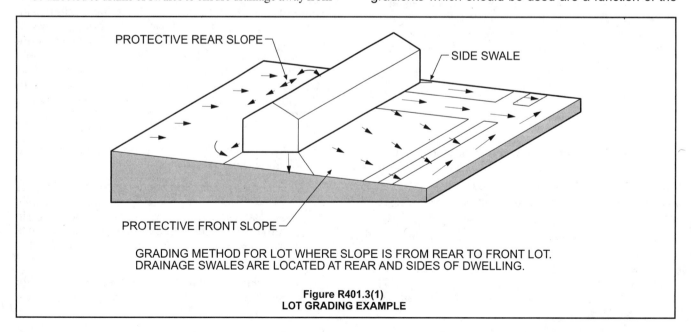

GRADING METHOD FOR LOT WHERE SLOPE IS FROM REAR TO FRONT LOT. DRAINAGE SWALES ARE LOCATED AT REAR AND SIDES OF DWELLING.

Figure R401.3(1)
LOT GRADING EXAMPLE

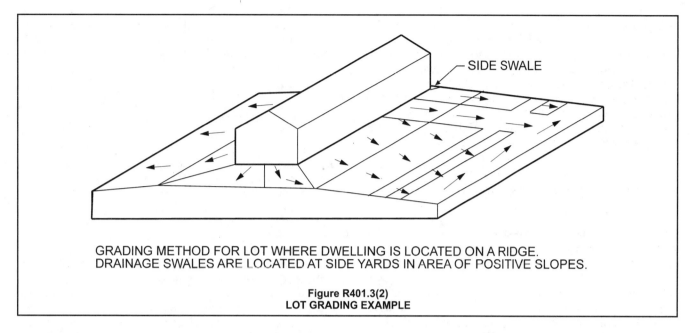

GRADING METHOD FOR LOT WHERE DWELLING IS LOCATED ON A RIDGE.
DRAINAGE SWALES ARE LOCATED AT SIDE YARDS IN AREA OF POSITIVE SLOPES.

Figure R401.3(2)
LOT GRADING EXAMPLE

combined ground frost and moisture conditions, soil type, geological features, and geographic condition. The slope away from the building is required to be 6 inches (152 mm) within the first 10 feet (3048 mm). The exception in this section allows for use of drains and swales if lot lines or physical barriers, such as a steeply sloping lot, do not allow for a 6-inch (152 mm) slope. The minimum slope is 2 percent.

R401.4 Soil tests. In areas likely to have expansive, compressible, shifting or other unknown soil characteristics, the building official shall determine whether to require a soil test to determine the soil's characteristics at a particular location. This test shall be made by an approved agency using an approved method.

❖ Loading conditions beyond the normal live, dead, and wind or earthquake loading generally include the types of loading attributable to special regional conditions. Such conditions might include soil instability, forces generated on foundations by expansive soils, and increased lateral pressures due to a high water table or surcharge loads from adjacent structures. Accounting for such conditions in the structure is generally considered to be beyond the scope of the conventional methods of foundation construction specified within this chapter. In some cases, a soil test may be required to evaluate the existing conditions.

Where the bearing capacity of the soil has not been determined by geotechnical evaluation such as borings, field load tests, laboratory tests, and engineering analysis, it is a common practice to use presumptive bearing values for the design of the foundation system. Section R401.4.1 requires that the presumptive bearing values in Table R401.4.1 be used if a geotechnical evaluation has not been performed.

R401.4.1 Geotechnical evaluation. In lieu of a complete geotechnical evaluation, the load-bearing values in Table R401.4.1 shall be assumed.

❖ See the commentary for Section R401.4.

TABLE R401.4.1
PRESUMPTIVE LOAD-BEARING VALUES OF FOUNDATION MATERIALS[a]

CLASS OF MATERIAL	LOAD-BEARING PRESSURE (pounds per square foot)
Crystalline bedrock	12,000
Sedimentary and foliated rock	4,000
Sandy gravel and/or gravel (GW and GP)	3,000
Sand, silty sand, clayey sand, silty gravel and clayey gravel (SW, SP, SM, SC, GM and GC)	2,000
Clay, sandy clay, silty clay, clayey silt, silt and sandy silt (CL, ML, MH and CH)	1,500[b]

For SI: 1 pound per square foot = 0.0479 kPa.

a. When soil tests are required by Section R401.4, the allowable bearing capacities of the soil shall be part of the recommendations.

b. Where the building official determines that in-place soils with an allowable bearing capacity of less than 1,500 psf are likely to be present at the site, the allowable bearing capacity shall be determined by a soils investigation.

❖ As explained in the commentary for Section R401.4, where a soils investigation is not required by the code, load-bearing values of Table R401.4.1 are to be used. The values vary by soil classifications, which are based on the Unified Soil Classification System. Table R405.1 also gives these soil descriptions in more detail. The allowable soil bearing values listed in the table are based on lengthy experience with the behavior of these materials in supporting loads from all types of structures.

R401.4.2 Compressible or shifting soil. Instead of a complete geotechnical evaluation, when top or subsoils are compressible or shifting, they shall be removed to a depth and width sufficient to assure stable moisture content in each active zone and shall not be used as fill or stabilized within each active zone by chemical, dewatering or presaturation.

❖ A geotechnical engineer should be consulted by the owner or the owner's builder if there is a concern that a compressible or shifting soil condition exists. Where these conditions are present, one of two methods must be used. The first of these deals with removal and replacement of the soil, thus reducing the forces associated with soil movement. The second approach is soil stabilization by chemical means, dewatering or presaturation.

SECTION R402
MATERIALS

R402.1 Wood foundations. Wood foundation systems shall be designed and installed in accordance with the provisions of this code.

❖ The wood foundation system requirements are from the American Forest & Paper Association manual Basic Requirements for the Permanent Wood Foundation System, Technical Report No. 7. Figures R403.1(2) and R403.1(3) of the code illustrate some typical details of this system. Also refer to the commentary for Section R401.1.

R402.1.1 Fasteners. Fasteners used below grade to attach plywood to the exterior side of exterior basement or crawlspace wall studs, or fasteners used in knee wall construction, shall be of Type 304 or 316 stainless steel. Fasteners used above grade to attach plywood and all lumber-to-lumber fasteners except those used in knee wall construction shall be of Type 304 or 316 stainless steel, silicon bronze, copper, hot-dipped galvanized (zinc coated) steel nails, or hot-tumbled galvanized (zinc coated) steel nails. Electro- galvanized steel nails and galvanized (zinc coated) steel staples shall not be permitted.

❖ Appropriate fasteners must be used in wood foundation construction because the presence of any moisture in combination with the preservative treatment can corrode incompatible fasteners.

R402.1.2 Wood treatment. All lumber and plywood shall be pressure-preservative treated and dried after treatment in accordance with AWPA U1 (Commodity Specification A, Use Category 4B and Section 5.2), and shall bear the label of an accredited agency. Where lumber and/or plywood is cut or drilled after treatment, the treated surface shall be field treated with copper naphthenate, the concentration of which shall contain a minimum of 2 percent copper metal, by repeated brushing, dipping or soaking until the wood absorbs no more preservative.

❖ Performance of the wood foundation system is dependent on the use of properly treated materials; thus, the code provision emphasizes the use of properly treated lumber and plywood. Verification of the proper materials is provided by identification showing the approval of an accredited inspection agency. An example of such identification is shown in Commentary Figure R402.1.2. Where treated lumber or plywood is field cut (exposing wood that is untreated), the code specifies the method of treating the cut surface.

R402.2 Concrete. Concrete shall have a minimum specified compressive strength of f'_c, as shown in Table R402.2. Concrete subject to moderate or severe weathering as indicated in Table R301.2(1) shall be air entrained as specified in Table R402.2. The maximum weight of fly ash, other pozzolans, silica fume, slag or blended cements that is included in concrete mixtures for garage floor slabs and for exterior porches, carport slabs and steps that will be exposed to deicing chemicals shall not exceed the percentages of the total weight of cementitious materials specified in Section 4.2.3 of ACI 318. Materials used

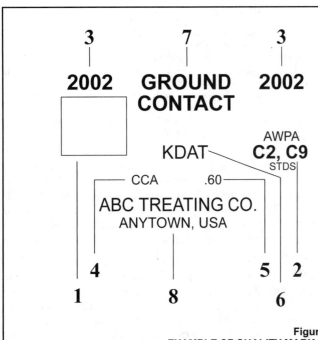

1	TRADEMARK OF INSPECTION AGENCY ACCREDITED BY AMERICAN LUMBER STANDARD COMMITTEE (ALSC)		
2	APPLICABLE AMERICAN WOOD PRESERVER'S ASSOCIATION (AWPA) STANDARD		
3	YEAR OF TREATMENT		
4	PRESERVATIVE USED FOR TREATMENT		
5	RETENTION LEVEL		
6	DRY OR KDAT, IF APPLICABLE		
7	PROPER EXPOSURE CONDITIONS		
8	TREATING COMPANY & LOCATION		

Figure R402.1.2
EXAMPLE OF QUALITY MARK FOR PRESSURE-TREATED LUMBER

to produce concrete and testing thereof shall comply with the applicable standards listed in Chapter 3 of ACI 318.

❖ The code specifies minimum concrete compressive strengths ranging from 2,500 to 3,000 psi (17.2 to 24.1 Mpa). Table R402.2 specifies the required compressive strength of concrete based on a locale's weathering potential and a building element's exposure. Section R301 uses the Weathering Probability Map for Concrete to classify an area's weathering potential as negligible, moderate or severe. For concrete that will be subject to freezing and thawing in a moist condition (i.e., weathering) or subject to direct or indirect application of deicing chemicals, Table R402.2 requires the use of higher-strength, air-entrained concrete. Freezing and thawing cycles can be the most destructive weathering factors for concrete when it is wet. Water freezing in the cement matrix, in the aggregate, or both causes deterioration of concrete that is not air-entrained. Studies

have documented that concrete with proper air entrainment is highly resistant to this deterioration.

All materials and testing must be in accordance with the American Concrete Institute's ACI 318, *Building Code Requirements for Structural Concrete.* Cementitious materials include portland cement (ASTM C 150), blended hydraulic cements (ASTM C 595 and ASTM C 1157), and expansive cement (ASTM C 845). They can also include pozzolanic materials such as fly ash and other raw or calcined natural pozzolan (ASTM C 618), ground granulated blast-furnace slag (ASTM C 989), and/or silica fume (ASTM C 1240) when used in combination with the cements. Most concrete contains admixtures such as fly ash or other pozzolans that are added as a replacement for a portion of the cement. Although such substitutions can be for economic reasons, in some cases they can also be used to improve certain concrete properties.

TABLE R402.2
MINIMUM SPECIFIED COMPRESSIVE STRENGTH OF CONCRETE

TYPE OR LOCATION OF CONCRETE CONSTRUCTION	MINIMUM SPECIFIED COMPRESSIVE STRENGTH[a] (f'_c)		
	Weathering Potential[b]		
	Negligible	Moderate	Severe
Basement walls, foundations and other concrete not exposed to the weather	2,500	2,500	2,500[c]
Basement slabs and interior slabs on grade, except garage floor slabs	2,500	2,500	2,500[c]
Basement walls, foundation walls, exterior walls and other vertical concrete work exposed to the weather	2,500	3,000[d]	3,000[d]
Porches, carport slabs and steps exposed to the weather, and garage floor slabs	2,500	3,000[d,e,f]	3,500[d,e,f]

For SI: 1 pound per square inch = 6.895 kPa.

a. Strength at 28 days psi.

b. See Table R301.2(1) for weathering potential.

c. Concrete in these locations that may be subject to freezing and thawing during construction shall be air-entrained concrete in accordance with Footnote d.

d. Concrete shall be air-entrained. Total air content (percent by volume of concrete) shall be not less than 5 percent or more than 7 percent.

e. See Section R402.2 for maximum cementitious materials content.

f. For garage floors with a steel troweled finish, reduction of the total air content (percent by volume of concrete) to not less than 3 percent is permitted if the specified compressive strength of the concrete is increased to not less than 4,000 psi.

❖ The code requires the use of higher-strength, air-entrained concrete for enclosed garage floor slabs in moderate and severe weathering regions. Even though these slabs may be enclosed and not subject to freezing and thawing conditions, they are likely to be subject to the deteriorating effects of deicing chemicals that drip from vehicles. The replacement of portland cement with other cementitious materials can be detrimental to the durability of the concrete if the concrete will be exposed to deicing chemicals. Thus, the code limits the total amount of these replacement materials to 50 percent of the total weight of all cementitious materials.

Note c to Table R402.2 requires air-entrained concrete for concrete elements not exposed to freezing and thawing conditions when the building is completed if it is likely that these elements will be subject to freezing and thawing during construction (also see commentary, Section R404.2).

Note f permits the air entrainment to be reduced to not less than 3 percent where the garage floor will have steel trowel finish. Field experience has shown that durable concrete may be obtained with a lesser amount of air-entrainment if the specified compressive strength of the concrete is increased to 4,000 psi (27.6 MPa). The higher concrete strength is accompanied by a denser cement-paste matrix, which results in the concrete being less permeable. This option has been successfully used on garage slabs with a steel trowel finish in areas of moderate and severe exposures.

R402.3 Precast concrete. Approved precast concrete foundations shall be designed and installed in accordance with the provisions of this code and the manufacturer's installation instructions.

❖ This section permits the design and installation of precast concrete foundations. All requirements in this code for concrete foundations apply to precast concrete as well. Also, any manufacturer's instructions must be complied with.

SECTION R403
FOOTINGS

R403.1 General. All exterior walls shall be supported on continuous solid or fully grouted masonry or concrete footings, wood foundations, or other approved structural systems which shall be of sufficient design to accommodate all loads according to Section R301 and to transmit the resulting loads to the soil within the limitations as determined from the character of the soil. Footings shall be supported on undisturbed natural soils or engineered fill.

❖ The provisions of this section include the general statement, also contained in Section R401.2, that a footing must be capable of supporting the required design loads. The section expands on this by referencing the character of the soil and the minimum extension below the frost line. The code mandates that footings be supported on undisturbed natural soil or engineered fill.

TABLE R403.1
MINIMUM WIDTH OF CONCRETE OR
MASONRY FOOTINGS (inches)[a]

	LOAD-BEARING VALUE OF SOIL (psf)			
	1,500	2,000	3,000	≥4,000
Conventional light–frame construction				
1-story	12	12	12	12
2-story	15	12	12	12
3-story	23	17	12	12
4-inch brick veneer over light frame or 8-inch hollow concrete masonry				
1-story	12	12	12	12
2-story	21	16	12	12
3-story	32	24	16	12
8-inch solid or fully grouted masonry				
1-story	16	12	12	12
2-story	29	21	14	12
3-story	42	32	21	16

For SI: 1 inch = 25.4 mm, 1 pound per square foot = 0.0479 kPa.

a. Where minimum footing width is 12 inches, use of a single wythe of solid or fully grouted 12-inch nominal concrete masonry units is permitted.

❖ Table R403.1 specifies minimum footing widths where normal soil conditions are encountered. The minimum footing widths are based on the soil load-bearing values in Table R401.4.1. The minimum values were calculated based on 20 feet (6096 mm) of tributary roof area, 16 feet (4877 mm) of tributary floor areas, 10-foot-high (3048 mm) first story walls, and 8-foot-high (2438 mm) second and third story walls. A 50-pound-per-square foot (2.4 kPa) snow load was used.

A minimum footing width of 12 inches (305 mm) is required so that footings can span over weak locations in the soil and to allow for minor misalignment of foundations. The footnote permits solid or fully grouted 12-inch-nominal concrete masonry block to be used for this minimum width of footing. Section R403.1.1 specifies the minimum thickness of footings. These provisions provide for the transfer of loads to the supporting soil without exceeding the capacity of the materials used to construct the footings.

If any of the following conditions exist in the area of the foundation, conventional spread footings should not be used, and a designed foundation may be necessary:

- Filled ground, except when properly compacted.
- Foundation soils subject to subsidence.
- Expansive soils such as those having a plasticity index greater than (see Section R403.1.8.1).
- Highly compressive clays.
- Unconfined sands and silts.

In cold weather, care must be taken that foundations are not placed on frozen soil, nor should such foundations be placed in freezing weather unless a method of ensuring that the underlying soil is free of frost is employed and the foundations are properly protected against the weather.

R403.1.1 Minimum size. Minimum sizes for concrete and masonry footings shall be as set forth in Table R403.1 and Figure R403.1(1). The footing width, W, shall be based on the load-bearing value of the soil in accordance with Table R401.4.1. Spread footings shall be at least 6 inches (152 mm) thick. Footing projections, P, shall be at least 2 inches (51 mm) and shall not exceed the thickness of the footing. The size of footings supporting piers and columns shall be based on the tributary load and allowable soil pressure in accordance with Table R401.4.1. Footings for wood foundations shall be in accordance with the details set forth in Section R403.2, and Figures R403.1(2) and R403.1(3).

❖ Table R403.1 specifies minimum footing widths based on soil pressure and the number of stories supported. Figure R403.1(1) illustrates masonry and concrete footings. To avoid construction that might overstress the footing, this section also specifies the minimum footing thickness and places a limitation on the maximum projection of the footing beyond the face of the foundation wall as shown in Figure R403.1(1). The size of isolated footings for piers and columns should be determined from the tributary (design) load supported using the presumptive soil values of Table R401.4.1.

The limitation on the maximum projection of the footing beyond the face of the foundation wall is particularly important for 6-inch-thick (152 mm) footings of plain concrete subjected to soil-bearing pressures on the order of 2,000 pounds per square foot (95.8 kPa). Projections in excess of the maximum allowable could result in the footing being cracked in a plane coinciding with the face of the foundation wall. A crack could occur if the allowable flexural tension stress of the concrete is exceeded.

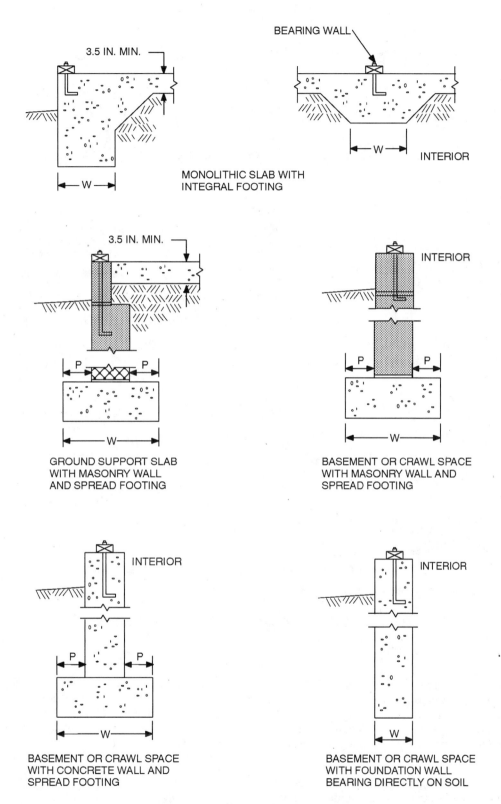

For SI: 1 inch = 25.4 mm.

FIGURE R403.1(1)
CONCRETE AND MASONRY FOUNDATIONS DETAILS

❖ See the commentary for Section R403.1.

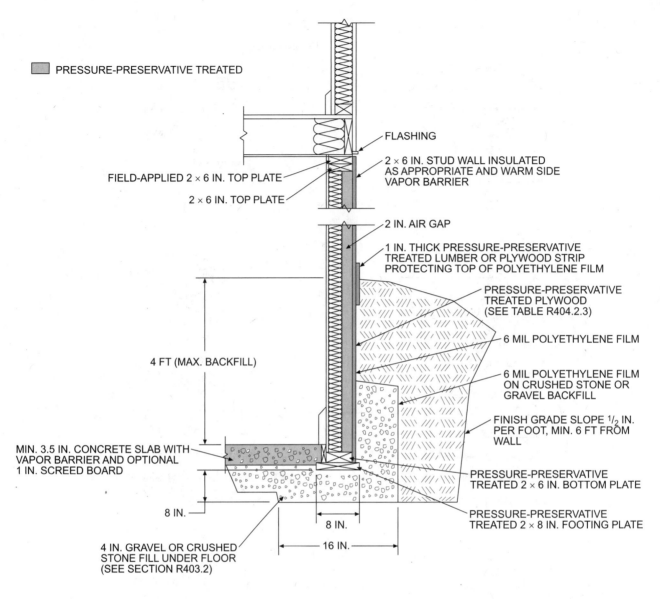

PRESSURE-PRESERVATIVE TREATED

FLASHING

2 × 6 IN. STUD WALL INSULATED AS APPROPRIATE AND WARM SIDE VAPOR BARRIER

FIELD-APPLIED 2 × 6 IN. TOP PLATE

2 × 6 IN. TOP PLATE

2 IN. AIR GAP

1 IN. THICK PRESSURE-PRESERVATIVE TREATED LUMBER OR PLYWOOD STRIP PROTECTING TOP OF POLYETHYLENE FILM

PRESSURE-PRESERVATIVE TREATED PLYWOOD (SEE TABLE R404.2.3)

6 MIL POLYETHYLENE FILM

4 FT (MAX. BACKFILL)

6 MIL POLYETHYLENE FILM ON CRUSHED STONE OR GRAVEL BACKFILL

FINISH GRADE SLOPE 1/2 IN. PER FOOT, MIN. 6 FT FROM WALL

MIN. 3.5 IN. CONCRETE SLAB WITH VAPOR BARRIER AND OPTIONAL 1 IN. SCREED BOARD

PRESSURE-PRESERVATIVE TREATED 2 × 6 IN. BOTTOM PLATE

8 IN.

8 IN.

PRESSURE-PRESERVATIVE TREATED 2 × 8 IN. FOOTING PLATE

4 IN. GRAVEL OR CRUSHED STONE FILL UNDER FLOOR (SEE SECTION R403.2)

16 IN.

For SI: 1 inch = 25.4 mm, 1 foot 304.8 mm, 1 mil = 0.0254 mm.

FIGURE R403.1(2)
PERMANENT WOOD FOUNDATION BASEMENT WALL SECTION

❖ See the commentary for Section R403.1.

R403.1.2 Continuous footing in Seismic Design Categories D₀, D₁ and D₂. The braced wall panels at exterior walls of buildings located in Seismic Design Categories D_0, D_1 and D_2 shall be supported by continuous footings. All required interior braced wall panels in buildings with plan dimensions greater than 50 feet (15 240 mm) shall also be supported by continuous footings.

❖ The code requires continuous footings for support of braced wall panels in exterior walls, a critical component of a structure's lateral-force-resisting system (see Chapter 6). Braced wall panels must be supported by a foundation that is continuous along the entire length of the braced wall line. For a structure classified as Seismic Design Category D_0, D_1 or D_2 the code requires continuous footings at all exterior walls as well as at interior braced wall lines in buildings with plan dimensions greater than 50 feet (15 240 mm). For buildings in Seismic Design Category D_2, additional continuous foundation requirements for interior braced wall lines must comply with Section R602.10.9 which requires continuous foundation for all interior braced wall lines for two-story buildings regardless of plan dimension (see commentary, Section R602.10.9).

R403.1.3 Seismic reinforcing. Concrete footings located in Seismic Design Categories D_0, D_1 and D_2, as established in Table R301.2(1), shall have minimum reinforcement. Bottom reinforcement shall be located a minimum of 3 inches (76 mm) clear from the bottom of the footing.

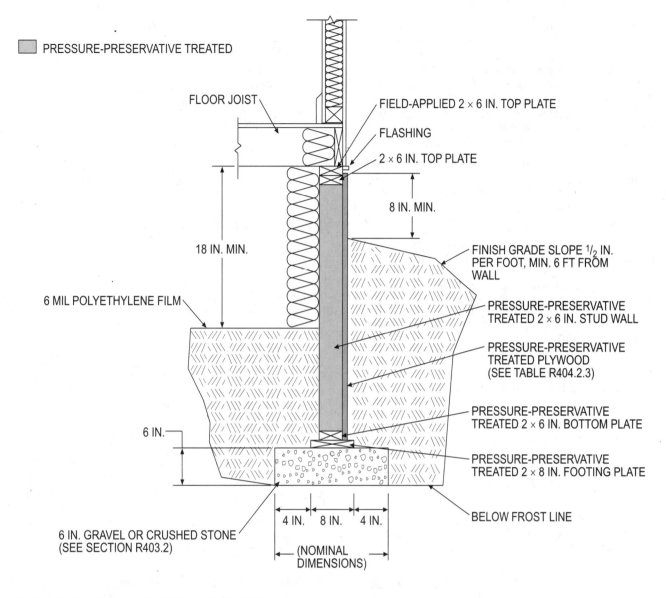

PRESSURE-PRESERVATIVE TREATED

FLOOR JOIST

FIELD-APPLIED 2 × 6 IN. TOP PLATE

FLASHING

2 × 6 IN. TOP PLATE

8 IN. MIN.

18 IN. MIN.

FINISH GRADE SLOPE 1/2 IN. PER FOOT, MIN. 6 FT FROM WALL

6 MIL POLYETHYLENE FILM

PRESSURE-PRESERVATIVE TREATED 2 × 6 IN. STUD WALL

PRESSURE-PRESERVATIVE TREATED PLYWOOD (SEE TABLE R404.2.3)

PRESSURE-PRESERVATIVE TREATED 2 × 6 IN. BOTTOM PLATE

6 IN.

PRESSURE-PRESERVATIVE TREATED 2 × 8 IN. FOOTING PLATE

BELOW FROST LINE

6 IN. GRAVEL OR CRUSHED STONE (SEE SECTION R403.2)

4 IN.   8 IN.   4 IN.

(NOMINAL DIMENSIONS)

For SI: 1 inch = 25.4 mm, 1 foot 304.8 mm, 1 mil = 0.0254 mm.

FIGURE R403.1(3)
PERMANENT WOOD FOUNDATION CRAWL SPACE SECTION

❖ See the commentary for Section R403.1.

In Seismic Design Categories D_0, D_1 and D_2 where a construction joint is created between a concrete footing and a stem wall, a minimum of one No. 4 bar shall be installed at not more than 4 feet (1219 mm) on center. The vertical bar shall extend to 3 inches (76 mm) clear of the bottom of the footing, have a standard hook and extend a minimum of 14 inches (357 mm) into the stem wall.

In Seismic Design Categories D_0, D_1 and D_2 where a grouted masonry stem wall is supported on a concrete footing and stem wall, a minimum of one No. 4 bar shall be installed at not more than 4 feet on center. The vertical bar shall extend to 3 inches (76 mm) clear of the bottom of the footing and have a standard hook.

In Seismic Design Categories D_0, D_1 and D_2 masonry stem walls without solid grout and vertical reinforcing are not permitted.

Exception: In detached one- and two-family dwellings which are three stories or less in height and constructed with stud bearing walls, plain concrete footings without longitudinal reinforcement supporting walls and isolated plain concrete footings supporting columns or pedestals are permitted.

❖ This section specifies minimal reinforcement for buildings classified as Seismic Design Category D_0, D_1 or D_2. Interconnection of stem walls and supporting foundations is necessary to resist the tendency of the wall to slip in an earthquake. In grouted masonry stem walls supported on concrete footings and at construction joints between concrete stem walls and their supporting footing, a minimum of one vertical No. 4 bar at

48 inches (1219 mm) spacing must be provided. A monolithically cast foundation and stem wall do not require this vertical reinforcement. Masonry stem walls must be grouted and provided with vertical reinforcement. However, plain concrete footings are permitted for detached one- and two-family dwellings. If an alternate braced wall panel is used in accordance with Section R602.10.6, that section contains more specific foundation requirements that also must be met.

R403.1.3.1 Foundations with stemwalls. Foundations with stem walls shall have installed a minimum of one No. 4 bar within 12 inches (305 mm) of the top of the wall and one No. 4 bar located 3 inches (76 mm) to 4 inches (102 mm) from the bottom of the footing.

❖ Where a stem wall exists, longitudinal reinforcement consisting of one No. 4 bar at the top of the wall and one No. 4 bar located 3 to 4 inches (76 to 102 mm) from the bottom of the footing must be provided for footings in Seismic Design Category D_0, D_1 or D_2.

R403.1.3.2 Slabs-on-ground with turned-down footings. Slabs-on-ground with turned-down footings shall have a minimum of one No. 4 bar at the top and bottom of the footing.

Exception: For slabs-on-ground cast monolithically with a footing, one No. 5 bar or two No. 4 bars shall be located in the middle third of the footing depth.

❖ For footings in Seismic Design Category D_0, D_1 or D_2, a slab-on-ground with a turned down footing requires a No. 4 bar at the top and bottom of the footing. If cast monolithically with the slab, either one No. 5 or two No. 4 bars should be placed in the middle third of the footing depth.

R403.1.4 Minimum depth. All exterior footings shall be placed at least 12 inches (305 mm) below the undisturbed ground surface. Where applicable, the depth of footings shall also conform to Sections R403.1.4.1 through R403.1.4.2.

❖ Footings are required to extend below the ground surface a minimum of 12 inches (305 mm). This is considered a minimum depth to protect the footing from movement of the soil caused by freezing and thawing in mild climate areas. See Section R403.1.4.1 for general frost protection requirements.

Footing depths may also be influenced by adjacent footings at different elevations; the footing at a higher elevation imposes surcharge pressures on lower adjacent footings. One method of alleviating the surcharge pressures on lower adjacent footings is the use of the guidelines illustrated in Commentary Figure R403.1.4. Other methods based on a soil and foundation analysis are also acceptable.

R403.1.4.1 Frost protection. Except where otherwise protected from frost, foundation walls, piers and other permanent supports of buildings and structures shall be protected from frost by one or more of the following methods:

1. Extended below the frost line specified in Table R301.2.(1);

2. Constructing in accordance with Section R403.3;

3. Constructing in accordance with ASCE 32; or

4. Erected on solid rock.

Exceptions:

1. Protection of freestanding accessory structures with an area of 600 square feet (56 m²) or less, of light-framed construction, with an eave height of 10 feet (3048 mm) or less shall not be required.

2. Protection of freestanding accessory structures with an area of 400 square feet (37 m²) or less, of other than light-framed construction, with an eave height of 10 feet (3048 mm) or less shall not be required.

3. Decks not supported by a dwelling need not be provided with footings that extend below the frost line.

Footings shall not bear on frozen soil unless the frozen condition is permanent.

❖ Buildings must not be founded on or within frozen ground. This provision prevents damage to the exterior walls and other walls bearing on the frozen soil. The volume changes (frost heave) that take place during freezing and thawing produce excessive stresses in the foundations and, as a result, may cause extensive damage to the walls that are supported.

Section R403.1.4.1, Frost Protection, contains the three acceptable ways of protecting the footings and foundations from frost: 1) extend below the frost line; 2) insulate the foundation according to the IRC prescriptive frost protected shallow foundation requirements or the ASCE 32, *Design and Construction of Frost Protected Shallow Foundations,* performance standard; and 3) erect the foundation on solid rock.

The frost-line depth is established by the authority having jurisdiction based on field experience and must be listed along with other design criteria in Table R301.2(1). Exterior foundations must extend below the frost line and a minimum of 12 inches (305 mm) below undisturbed soil. Foundations supported on rock or that are frost-protected in accordance with Section R403.3 are exempt from the extension below the frost line, but the 12-inch (305 mm) minimum still applies.

ASCE 32 contains requirements for:

1. A simplified method for heated buildings;

2. A more complicated performance method that permits trading off insulation for frost-susceptible fill or use of varying foundation depths;

3. Requirements for slabs-on-ground, crawl spaces (without outside vents), walk out basements and semi-heated buildings;

4. Requirements for unheated buildings that conserve heat from the ground with thick layers of insulation under the entire foundation and beyond it several feet; continuous walls and columns; and

5. Special design conditions.

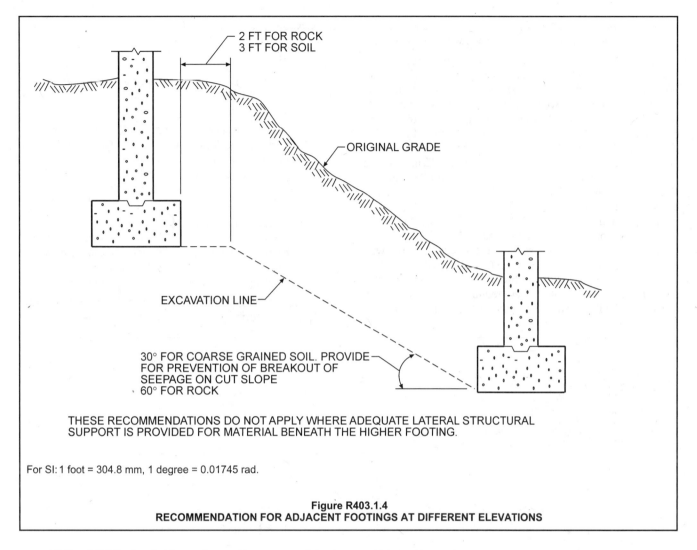

For SI: 1 foot = 304.8 mm, 1 degree = 0.01745 rad.

Figure R403.1.4
RECOMMENDATION FOR ADJACENT FOOTINGS AT DIFFERENT ELEVATIONS

All the ASCE standard requirements are consistent with theory, widespread practice in Scandinavia, and the successful construction of thousands of buildings in the U.S.

Foundations are not to be placed on frozen soil because when the ground thaws, uneven settlement of the structure is apt to occur, thereby causing structural damage. This section does, however, permit footings to be constructed on permanently frozen soil. In permafrost areas, special precautions are necessary to prevent heat from the structure from thawing the soil beneath the foundation.

R403.1.4.2 Seismic conditions. In Seismic Design Categories D_0, D_1 and D_2, interior footings supporting bearing or bracing walls and cast monolithically with a slab on grade shall extend to a depth of not less than 12 inches (305 mm) below the top of the slab.

❖ In buildings classified as Seismic Design Category D_0, D_1 or D_2, the code establishes the minimum depth below the top of slab for interior footings that support bearing or bracing walls and are poured monolithically with the slab. This allows for the installation of anchor bolts, tie-downs, etc.

R403.1.5 Slope. The top surface of footings shall be level. The bottom surface of footings shall not have a slope exceeding one unit vertical in 10 units horizontal (10-percent slope). Footings shall be stepped where it is necessary to change the elevation of the top surface of the footings or where the slope of the bottom surface of the footings will exceed one unit vertical in ten units horizontal (10-percent slope).

❖ The code requires that the top surface of footings for buildings be essentially level. However, a slope of 1 unit vertical in 10 units horizontal (10-percent slope) is permitted for the bottom bearing surface. If the slope is steeper, the foundation must be stepped. Commentary Figure R403.1.5 schematically shows a stepped foundation. Although the code places no restriction on a stepped foundation, the figure shows a recommended overlap of the top of the foundation wall beyond the step in the foundation. It is larger than the vertical step in the foundation wall at that point. This recommendation is based on possible crack propagation at an angle of 45 degrees (0.79 rad).

R403.1.6 Foundation anchorage. When braced wall panels are supported directly on continuous foundations, the wall wood sill plate or cold-formed steel bottom track shall be anchored to the foundation in accordance with this section.

The wood sole plate at exterior walls on monolithic slabs and wood sill plate shall be anchored to the foundation with anchor bolts spaced a maximum of 6 feet (1829 mm) on center. There shall be a minimum of two bolts per plate section with one bolt located not more than 12 inches (305 mm) or less than seven bolt diameters from each end of the plate section. In Seismic Design Categories D_0, D_1 and D_2, anchor bolts shall be spaced at 6 feet (1829 mm) on center and located within 12 inches (305 mm) of the ends of each plate section at interior braced wall lines when required by Section R602.10.9 to be supported on a continuous foundation. Bolts shall be at least $^1/_2$ inch (13 mm) in diameter and shall extend a minimum of 7 inches (178 mm) into masonry or concrete. Interior bearing wall sole plates on monolithic slab foundation shall be positively anchored with approved fasteners. A nut and washer shall be tightened on each bolt of the plate. Sills and sole plates shall be protected against decay and termites where required by Sections R319 and R320. Cold-formed steel framing systems shall be fastened to the wood sill plates or anchored directly to the foundation as required in Section R505.3.1 or R603.1.1.

Exceptions:

1. Foundation anchorage, spaced as required to provide equivalent anchorage to $^1/_2$-inch-diameter (13 mm) anchor bolts.

2. Walls 24 inches (610 mm) total length or shorter connecting offset braced wall panels shall be anchored to the foundation with a minimum of one anchor bolt located in the center third of the plate section and shall be attached to adjacent braced wall panels per Figure R602.10.5 at corners.

3. Walls 12 inches (305 mm) total length or shorter connecting offset braced wall panels shall be permitted to be connected to the foundation without anchor bolts. The wall shall be attached to adjacent braced wall panels per Figure R602.10.5 at corners.

❖ To prevent walls and floors from shifting under lateral loads, the code requires anchorage to the supporting

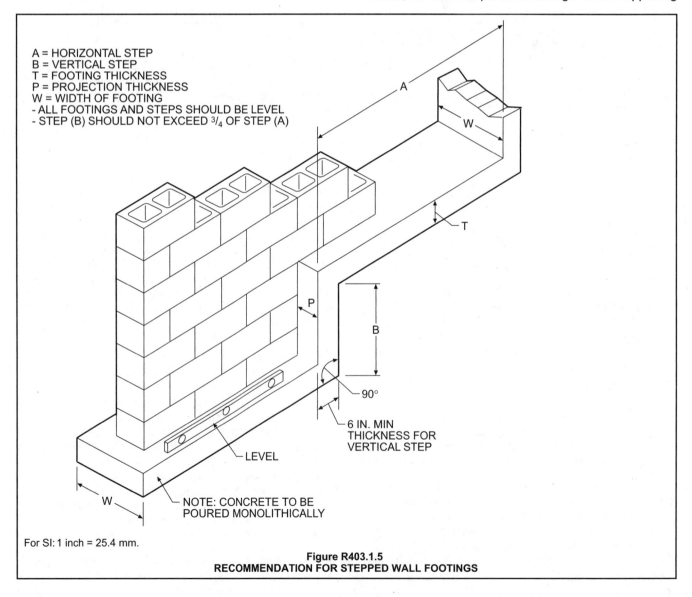

A = HORIZONTAL STEP
B = VERTICAL STEP
T = FOOTING THICKNESS
P = PROJECTION THICKNESS
W = WIDTH OF FOOTING
- ALL FOOTINGS AND STEPS SHOULD BE LEVEL
- STEP (B) SHOULD NOT EXCEED $^3/_4$ OF STEP (A)

90°

6 IN. MIN THICKNESS FOR VERTICAL STEP

LEVEL

NOTE: CONCRETE TO BE POURED MONOLITHICALLY

For SI: 1 inch = 25.4 mm.

Figure R403.1.5
RECOMMENDATION FOR STEPPED WALL FOOTINGS

foundation. Anchor bolts installed as specified in this section supply the minimum required capacity. Commentary Figure R403.1.6 illustrates anchorage of wood sill plates. This anchorage applies at exterior walls in any seismic design category. An exception explicitly allows other foundation anchorages if they are spaced to provide anchorage capacity equivalent to that of the $^1/_2$-inch (12.7 mm) diameter anchor bolts specified. In this case it would be necessary to use manufacturers' data, such as evaluation reports, to document the anchorage shear and tension capacities.

R403.1.6.1 Foundation anchorage in Seismic Design Categories C, D_0, D_1 and D_2. In addition to the requirements of Section R403.1.6, the following requirements shall apply to wood light-frame structures in Seismic Design Categories D_0, D_1 and D_2 and wood light-frame townhouses in Seismic Design Category C.

1. Plate washers conforming to Section R602.11.1 shall be provided for all anchor bolts over the full length of required braced wall lines. Properly sized cut washers shall be permitted for anchor bolts in wall lines not containing braced wall panels.

2. Interior braced wall plates shall have anchor bolts spaced at not more than 6 feet (1829 mm) on center and located within 12 inches (305 mm) of the ends of each plate section when supported on a continuous foundation.

3. Interior bearing wall sole plates shall have anchor bolts spaced at not more than 6 feet (1829 mm) on center and located within 12 inches (305 mm) of the ends of each plate section when supported on a continuous foundation.

4. The maximum anchor bolt spacing shall be 4 feet (1219 mm) for buildings over two stories in height.

5. Stepped cripple walls shall conform to Section R602.11.3.

6. Where continuous wood foundations in accordance with Section R404.2 are used, the force transfer shall have a capacity equal to or greater than the connections required by Section R602.11.1 or the braced wall panel shall be connected to the wood foundations in accordance with the braced wall panel-to-floor fastening requirements of Table R602.3(1).

❖ The additional anchorage requirements of this section apply to light-frame wood structures classified as Seismic Design Category D_0, D_1 or D_2 and wood light-frame townhouses in Seismic Design Category C. These provisions were added following observations of past earthquakes, especially the 1994 Northridge earthquake in California. Considerable longitudinal splitting of sill plates occurred in that earthquake, resulting in substantial damage to the plates and subsequent loss of lateral load capacity. Plate washers are required on each bolt along the full length of the braced wall line (see commentary, Section R602.11.1).

R403.1.7 Footings on or adjacent to slopes. The placement of buildings and structures on or adjacent to slopes steeper than 1 unit vertical in 3 units horizontal (33.3-percent slope) shall conform to Sections R403.1.7.1 through R403.1.7.4.

❖ The provisions of the referenced sections apply to buildings placed on or adjacent to slopes steeper than 1 unit vertical in 3 units horizontal (33.3-percent slope) only.

R403.1.7.1 Building clearances from ascending slopes. In general, buildings below slopes shall be set a sufficient distance from the slope to provide protection from slope drainage, erosion and shallow failures. Except as provided in Section R403.1.7.4 and Figure R403.1.7.1, the following criteria will be assumed to provide this protection. Where the existing slope is steeper than one unit vertical in one unit horizontal (100-percent slope), the toe of the slope shall be assumed to be at the intersection of a horizontal plane drawn from the top of the foundation and a plane drawn tangent to the slope at an angle of 45 degrees (0.79 rad) to the horizontal. Where a retaining wall is con-

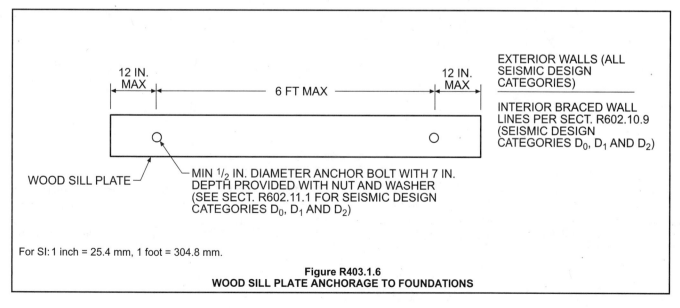

12 IN. MAX

6 FT MAX

12 IN. MAX

EXTERIOR WALLS (ALL SEISMIC DESIGN CATEGORIES)

INTERIOR BRACED WALL LINES PER SECT. R602.10.9 (SEISMIC DESIGN CATEGORIES D_0, D_1 AND D_2)

WOOD SILL PLATE

MIN $^1/_2$ IN. DIAMETER ANCHOR BOLT WITH 7 IN. DEPTH PROVIDED WITH NUT AND WASHER (SEE SECT. R602.11.1 FOR SEISMIC DESIGN CATEGORIES D_0, D_1 AND D_2)

For SI: 1 inch = 25.4 mm, 1 foot = 304.8 mm.

Figure R403.1.6
WOOD SILL PLATE ANCHORAGE TO FOUNDATIONS

structed at the toe of the slope, the height of the slope shall be measured from the top of the wall to the top of the slope.

❖ Figure R403.1.7.1 provides the criteria for the location of foundations adjacent to the toe of ascending slopes not exceeding one unit vertical in one unit horizontal. Commentary Figure R403.1.7.2(1) illustrates the criteria for the determination of the location of the toe of the slope where the slope exceeds one unit vertical in one unit horizontal.

R403.1.7.2 Footing setback from descending slope surfaces. Footings on or adjacent to slope surfaces shall be founded in material with an embedment and setback from the slope surface sufficient to provide vertical and lateral support for the footing without detrimental settlement. Except as provided for in Section R403.1.7.4 and Figure R403.1.7.1, the following setback is deemed adequate to meet the criteria. Where the slope is steeper than one unit vertical in one unit horizontal (100-percent slope), the required setback shall be measured from an imaginary plane 45 degrees (0.79 rad) to the horizontal, projected upward from the toe of the slope.

❖ In this section, the code restricts the placement of footings adjacent to or on descending slopes so that both vertical and lateral support are provided. The criteria for this condition are as shown in Figure R403.1.7.1. It is possible to locate buildings closer to the slope than the indicated setback in Figure R403.1.7.1, and in fact it is possible to locate the footing of the structure on the slope itself. In these two cases, it will be necessary to provide an adequate depth of embedment of the footing so that the face of the footing at the bearing plane is set

back from the edge of the slope at least the distance required by the code [H/3, but need not exceed 40 feet (12 192 mm)].

Commentary Figure R403.1.7.2(2) depicts the condition where the descending slope is steeper than one unit vertical in one unit horizontal and shows the proper location of the top of the slope as required by the code. The setback at the top of descending slopes will primarily provide lateral support for the foundations. The area so allocated also provides space for lot drainage away from the slope without creating too steep a drainage profile, which could create erosion problems. Furthermore, this space also provides for access around the building.

R403.1.7.3 Foundation elevation. On graded sites, the top of any exterior foundation shall extend above the elevation of the street gutter at point of discharge or the inlet of an approved drainage device a minimum of 12 inches (305 mm) plus 2 percent. Alternate elevations are permitted subject to the approval of the building official, provided it can be demonstrated that required drainage to the point of discharge and away from the structure is provided at all locations on the site.

❖ Commentary Figure R403.1.7.3 depicts the requirements of the code in this item for the elevation for exterior foundations with respect to the street, gutter or point of inlet of a drainage device. The elevation of the street or gutter shown is that point at which drainage from the site reaches the street or gutter.

This requirement protects the building from water encroachment in case of unusually heavy rains and may be modified on the approval of the building official

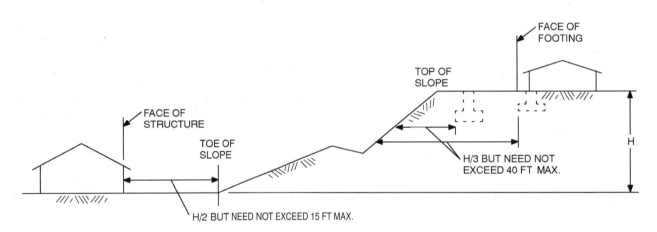

For SI: 1 foot = 304.8 mm.

FIGURE R403.1.7.1
FOUNDATION CLEARANCE FROM SLOPES

❖ The setback required by the code provides protection to the structure from shallow failures (sometimes referred to as sloughing) and protection from erosion and slope drainage. Furthermore, the space provided by the setback provides access around the building and helps to create a light and open-air environment. The dimension from the toe of the slope to the face of the structure need not exceed 15 feet (4572 mm). Also refer to the commentary for Section R403.1.7.1.

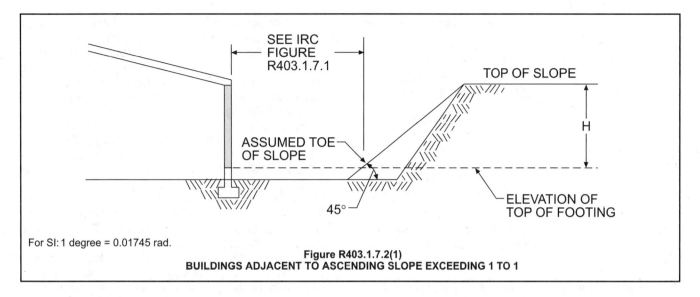

For SI: 1 degree = 0.01745 rad.

Figure R403.1.7.2(1)
BUILDINGS ADJACENT TO ASCENDING SLOPE EXCEEDING 1 TO 1

if the building official finds that positive drainage slopes are provided to drain water away from the building and that the drainage pattern is not subject to temporary flooding due to landscaping or other impediments to drainage.

R403.1.7.4 Alternate setback and clearances. Alternate setbacks and clearances are permitted, subject to the approval of the building official. The building official is permitted to require an investigation and recommendation of a qualified engineer to demonstrate that the intent of this section has been satisfied. Such an investigation shall include consideration of material, height of slope, slope gradient, load intensity and erosion characteristics of slope material.

❖ This item provides that the building official may approve alternate setbacks and clearances from slopes provided the building official is satisfied that the intent of this section has been met. The code gives the building official authority to require a foundation investigation by a qualified geotechnical engineer. This section also specifies which parameters must be considered by the geotechnical engineer in such an investigation.

R403.1.8 Foundations on expansive soils. Foundation and floor slabs for buildings located on expansive soils shall be designed in accordance with Section 1805.8 of the *International Building Code*.

Exception: Slab-on-ground and other foundation systems which have performed adequately in soil conditions similar to those encountered at the building site are permitted subject to the approval of the building official.

❖ Expansive soils are those that shrink and swell due to changes in moisture content. The resulting movements can be damaging to residential structures. The amount and depth of potential swelling that can occur in a clay material are to some extent functions of the cyclical moisture content in the soil. In dryer climates where the moisture content in the soil near the ground surface is low because of evaporation, there is a greater potential for extensive swelling than in the same soil in wetter cli-

mates where the variations of moisture content are not as severe.

When foundations or floor slabs are supported on soils determined to be expansive, the code defers to the *International Building Code* provisions that address the design of foundations and slabs on expansive soil. The exception permits the use of systems that have demonstrated adequate performance with the approval of the building official.

R403.1.8.1 Expansive soils classifications. Soils meeting all four of the following provisions shall be considered expansive, except that tests to show compliance with Items 1, 2 and 3 shall not be required if the test prescribed in Item 4 is conducted:

1. Plasticity Index (PI) of 15 or greater, determined in accordance with ASTM D 4318.

2. More than 10 percent of the soil particles pass a No. 200 sieve (75 mm), determined in accordance with ASTM D 422.

3. More than 10 percent of the soil particles are less than 5 micrometers in size, determined in accordance with ASTM D 422.

4. Expansion Index greater than 20, determined in accordance with ASTM D 4829.

❖ This section defines "expansive soil" as any plastic material with a PI of 15 or greater with more than 10 percent of the soil particles passing a No. 200 sieve and less than 5 micrometers in size. As an alternative, tests in accordance with ASTM D 4829 can be used to determine whether a soil is expansive. The expansion index is a measure of the swelling potential of the soil.

R403.2 Footings for wood foundations. Footings for wood foundations shall be in accordance with Figures R403.1(2) and R403.1(3). Gravel shall be washed and well graded. The maximum size stone shall not exceed $^3/_4$ inch (19.1 mm). Gravel shall be free from organic, clayey or silty soils. Sand shall be coarse, not smaller than $^1/_{16}$-inch (1.6 mm) grains and shall be free from organic, clayey or silty soils. Crushed stone shall have a maximum size of $^1/_2$ inch (12.7 mm).

❖ See Figures R403.1(2) and R403.1(3) and refer to the commentary for Section R401.1.

R403.3 Frost protected shallow foundations. For buildings where the monthly mean temperature of the building is maintained at a minimum of 64°F (18°C), footings are not required to extend below the frost line when protected from frost by insulation in accordance with Figure R403.3(1) and Table R403.3. Foundations protected from frost in accordance with Figure R403.3(1) and Table R403.3 shall not be used for unheated spaces such as porches, utility rooms, garages and carports, and shall not be attached to basements or crawl spaces that are not maintained at a minimum monthly mean temperature of 64°F (18°C).

Materials used below grade for the purpose of insulating footings against frost shall be labeled as complying with ASTM C 578.

❖ This section provides an alternative method of protecting foundations against frost heave, thus allowing foundations to be constructed above the frost line. Frost-protected foundations use insulation to reduce the heat loss at the slab edge. By holding heat from the dwelling in the ground under the foundation, the insulation, in effect, raises the frost line around the foundation.

The provisions require that the building be heated and that insulation be installed in accordance with the criteria specified in Figure R403.3(1) and Table R403.3 based on the air-freezing index established by Figure R403.3(2). As Figure R403.3(1) illustrates, these provisions apply only to slab-on-ground floors. Frost-protected shallow foundation (FPSF) designs for heated buildings are not to be used for unheated spaces. The provisions also prohibit the attachment of additions with FPSF to basements or crawl-spaces that are not heated because, although unlikely, in very cold climates frost could penetrate under FPSFs from unheated basements or ventilated crawlspaces. The insulation values specified in this section are only for the purposes of exercising this option, and the Chapter 11 energy conservation provisions could result in slab edge insulation with a greater degree of thermal resistance. Conversely, if slab-edge insulation is used to comply with Chapter 11, a builder may take advantage of that insulation to reduce the foundation depth in accordance with this section.

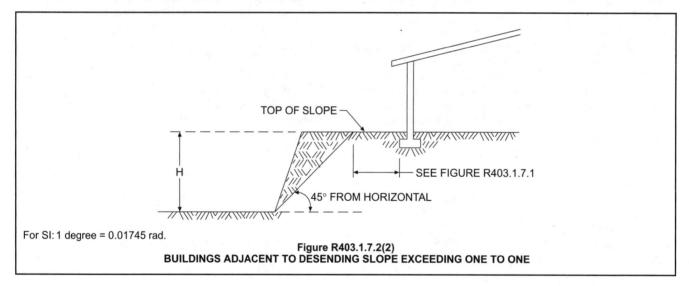

For SI: 1 degree = 0.01745 rad.

Figure R403.1.7.2(2)
BUILDINGS ADJACENT TO DESENDING SLOPE EXCEEDING ONE TO ONE

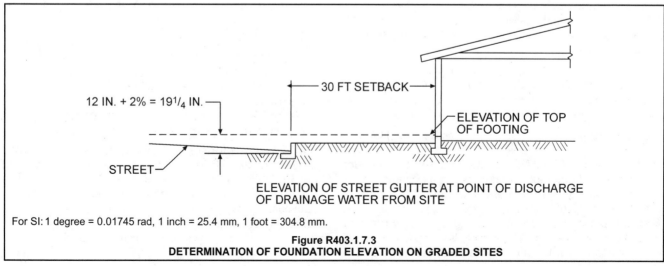

For SI: 1 degree = 0.01745 rad, 1 inch = 25.4 mm, 1 foot = 304.8 mm.

Figure R403.1.7.3
DETERMINATION OF FOUNDATION ELEVATION ON GRADED SITES

R403.3.1 Foundations adjoining frost protected shallow foundations. Foundations that adjoin frost protected shallow foundations shall be protected from frost in accordance with Section R403.1.4.

❖ This section requires that foundations that adjoin FPSFs be protected from frost according to the current requirements of Section R403.1.4, which requires that foundations extend below the frost line, be built on solid rock, or use FPSF designs. Each method prevents frost from heaving foundations. This language prevents the attachment of FPSFs to reinforced floating garage foundations, which may be permitted by some local codes to be placed at shallow depths without insulating for frost protection (i.e., the foundation moves up and down with frost).

R403.3.1.1 Attachment to unheated slab-on-ground structure. Vertical wall insulation and horizontal insulation of frost protected shallow foundations that adjoin a slab-on-ground foundation that does not have a monthly mean temperature maintained at a minimum of 64°F (18°C), shall be in accordance with Figure R403.3(3) and Table R403.3. Vertical wall insulation shall extend between the frost protected shallow foundation and the adjoining slab foundation. Required horizontal insulation shall be continuous under the adjoining slab foundation and through any foundation walls adjoining the frost protected shallow foundation. Where insulation passes through a foundation wall, it shall either be of a type complying with this section and having bearing capacity equal to or greater than the structural loads imposed by the building, or the building shall be designed and constructed using beams, lintels, cantilevers or other means of transferring building loads

such that the structural loads of the building do not bear on the insulation.

❖ This section requires that a building with a FPSF be insulated to current FPSF requirements even where an unheated garage or other slab-on-ground is attached to it. The vertical and horizontal insulation must be in accordance with Figure 403.3(3) and Table R403.3 and be continuous through the foundation of the unheated structure. The vertical wall insulation must extend between the FPSF and the adjoining slab foundation, and any required horizontal insulation must be continuous through any foundation walls adjoining the FPSF. No frost penetration or differential movement occurred with this construction in a HUD demonstration in Fargo, North Dakota. Deep foundations of unheated garages have been attached to hundreds of other buildings built on FPSFs in the U.S. In most cases the FPSF insulation has been placed continuously between the garage floor and the house FPSF; however, the insulation has not been continuous through the foundation walls of the unheated structures. In these buildings, no differential movement between the deep and shallow foundations has been reported.

Insulation passing through a foundation wall must comply with the requirements of Section R403.3, i.e., it must meet ASTM C578 and the requirements of Table R403.3, and the insulation must be of a type that has sufficient compressive strength to carry the load of the building. Alternatively, the building must be designed to carry those loads over the insulation and not bear on it.

TABLE R403.3
MINIMUM INSULATION REQUIREMENTS FOR FROST-PROTECTED FOOTINGS IN HEATED BUILDINGS[a]

AIR FREEZING INDEX (°F-days)[b]	VERTICAL INSULATION R-VALUE[c,d]	HORIZONTAL INSULATION R-VALUE[c,e]		HORIZONTAL INSULATION DIMENSIONS PER FIGURE R403.3(1) (inches)		
		Along walls	At corners	A	B	C
1,500 or less	4.5	Not required	Not required	Not required	Not required	Not required
2,000	5.6	Not required	Not required	Not required	Not required	Not required
2,500	6.7	1.7	4.9	12	24	40
3,000	7.8	6.5	8.6	12	24	40
3,500	9.0	8.0	11.2	24	30	60
4,000	10.1	10.5	13.1	24	36	60

a. Insulation requirements are for protection against frost damage in heated buildings. Greater values may be required to meet energy conservation standards. Interpolation between values is permissible.

b. See Figure R403.3(2) for Air Freezing Index values.

c. Insulation materials shall provide the stated minimum R–values under long-term exposure to moist, below-ground conditions in freezing climates. The following R–values shall be used to determine insulation thicknesses required for this application: Type II expanded polystyrene—2.4R per inch; Type IV extruded polystyrene—4.5R per inch; Type VI extruded polystyrene—4.5R per inch; Type IX expanded polystyrene—3.2R per inch; Type X extruded polystyrene—4.5R per inch.

d. Vertical insulation shall be expanded polystyrene insulation or extruded polystyrene insulation.

e. Horizontal insulation shall be extruded polystyrene insulation.

❖ This table provides the required R-value of horizontal and vertical insulation based on the air-freezing index. In addition, it gives the plan dimensions required for the horizontal insulation in Figure R403.3(1).

INSULATION DETAIL

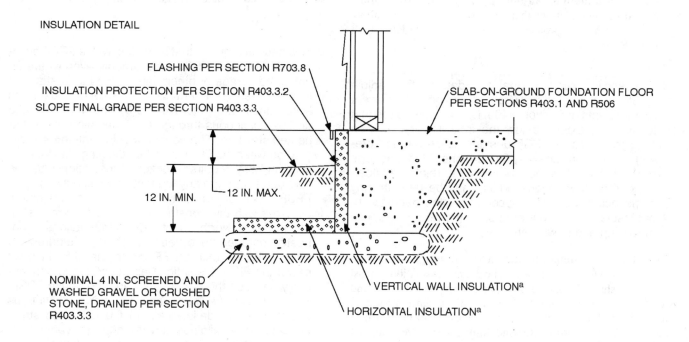

HORIZONTAL INSULATION PLAN

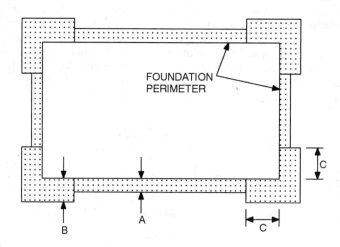

For SI: 1 inch = 25.4 mm.

a. See Table R403.3 for required dimensions and R-values for vertical and horizontal insulation.

FIGURE R403.3(1)
INSULATION PLACEMENT FOR FROST-PROTECTED FOOTINGS IN HEATED BUILDINGS

❖ See the commentary for Section R403.3.

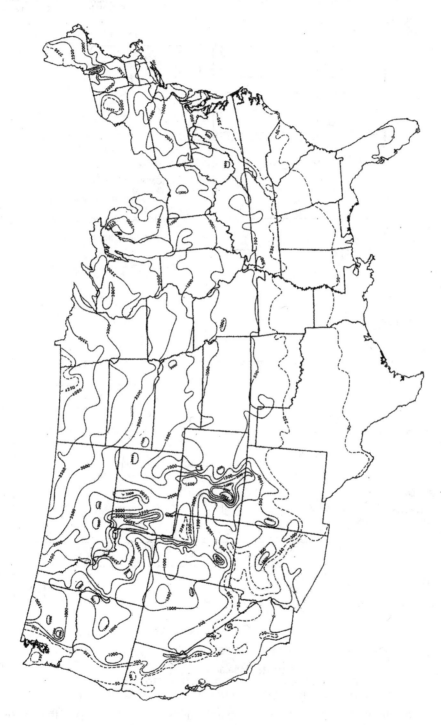

For SI: °C = [(°F)-32]/1.8.

Note: The air-freezing index is defined as cumulative degree days below 32°F. It is used as a measure of the combined magnitude and duration of air temperature below freezing. The index was computed over a 12-month period (July-June) for each of the 3,044 stations used in the above analysis. Data from the 1951-80 period were fitted to a Weibull probability distribution to produce an estimate of the 100-year return period.

FIGURE R403.3(2)
AIR-FREEZING INDEX
AN ESTIMATE OF THE 100-YEAR RETURN PERIOD

❖ This figure provides the criteria necessary to apply the minimum insulation requirements of Table R404.3. The note explains the term air-freezing index.

INSULATION DETAIL

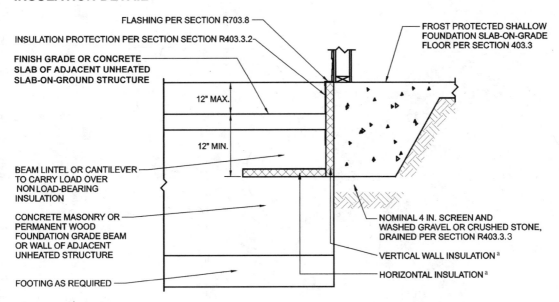

HORIZONTAL INSULATION PLAN

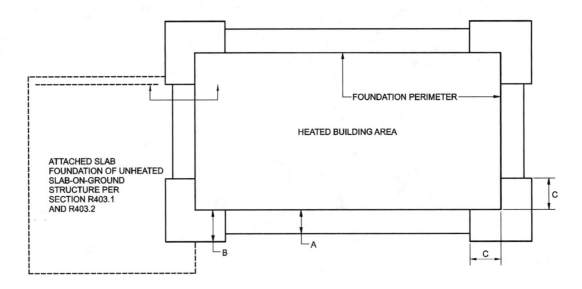

For SI: 1 inch = 25.4 mm.
a. See Table R403.3 for required dimensions and R-value for vertical and horizontal insulation.

FIGURE R403.3(3)
INSULATION PLACEMENT FOR FROST-PROTECTED FOOTING
ADJACENT TO UNCHEATED SLAB-ON-GROUND STRUCTURE

❖ This figure of insulation detail shows 1) the maximum vertical distance between to top of the FPSF and the top of slab of adjacent unheated slab-on-ground structure and 2) the minimum vertical distance from the top of the slab and the bottom of the insulation. The horizontal insulation plan shows placement of a FPSF where it meets an unheated structure.

R403.3.1.2 Attachment to heated structure. Where a frost protected shallow foundation abuts a structure that has a monthly mean temperature maintained at a minimum of 64°F (18°C), horizontal insulation and vertical wall insulation shall not be required between the frost protected shallow foundation and the adjoining structure. Where the frost protected shallow foundation abuts the heated structure, the horizontal insulation and vertical wall insulation shall extend along the adjoining foundation in accordance with Figure R403.3(4) a distance of not less than Dimension A in Table R403.3.

> **Exception:** Where the frost protected shallow foundation abuts the heated structure to form an inside corner, vertical insulation extending along the adjoining foundation is not required.

❖ This section stipulates that additions built on FPSFs need not be insulated where they join a heated structure because heat from both structures keeps the ground from freezing. However, where the FPSF abuts the heated building, the horizontal and vertical insulation must extend along the adjoining foundation to prevent cold from intruding under the slab through the foundation wall. The dimension of the insulation is related to the width of the horizontal insulation. This technique was shown to be effective by the U.S. Army Corps of Engineers Cold Regions Research and Engineering Laboratory (CRREL) on a FPSF addition to an airport control tower with a deep foundation in a severe Alaskan climate with a 13-foot frost line.

An abutment at an inside corner is an exception. In this condition, both the building and the addition heat the ground, and extra insulation is not needed.

R403.3.2 Protection of horizontal insulation below ground. Horizontal insulation placed less than 12 inches (305 mm) below the ground surface or that portion of horizontal insulation extending outward more than 24 inches (610 mm) from the foundation edge shall be protected against damage by use of a concrete slab or asphalt paving on the ground surface directly above the insulation or by cementitious board, plywood rated for below-ground use, or other approved materials placed below ground, directly above the top surface of the insulation.

❖ This protection for the insulation prevents damage due to excavating (e.g., for landscaping purposes).

R403.3.3 Drainage. Final grade shall be sloped in accordance with Section R401.3. In other than Group I Soils, as detailed in Table R405.1, gravel or crushed stone beneath horizontal insulation below ground shall drain to daylight or into an approved sewer system.

❖ See Figure R403.3(1) and the commentary for Section R401.3.

R403.3.4 Termite damage. The use of foam plastic in areas of "very heavy" termite infestation probability shall be in accordance with Section R320.5.

❖ Studies have shown that rigid board insulation installed below grade, particularly in areas where the hazard associated with termite infestation is very heavy, creates a pathway for termites that cannot be blocked with currently available termiticide treatments. Because it provides a place for termites to burrow in areas of very heavy termite infestation probability, the foam plastic insulation must be protected as stated in Section R320.5.

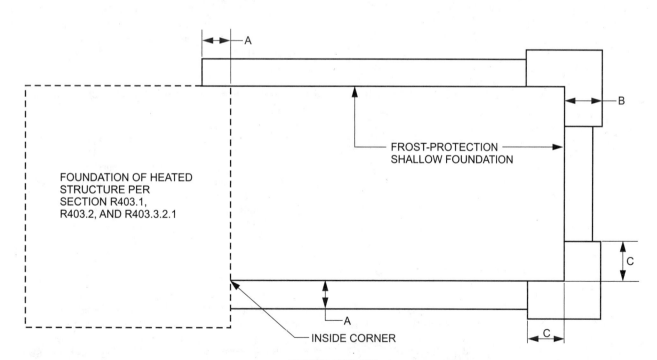

FIGURE R403.3(4)
INSULATION PLACEMENT FOR FROST-PROTECTED FOOTINGS
ADJACENT TO HEATED STRUCTURE

❖ This figure shows where extra insulation is required when a FPSF abuts a heated structure.

SECTION R404
FOUNDATION AND RETAINING WALLS

R404.1 Concrete and masonry foundation walls. Concrete and masonry foundation walls shall be selected and constructed in accordance with the provisions of Section R404 or in accordance with ACI 318, ACI 332, NCMA TR68–A or ACI 530/ASCE 5/TMS 402 or other approved structural standards. When ACI 318, ACI 332 or ACI 530/ASCE 5/TMS 402 or the provisions of Section R404 are used to design concrete or masonry foundation walls, project drawings, typical details and specifications are not required to bear the seal of the architect or engineer responsible for design, unless otherwise required by the state law of the jurisdiction having authority.

Foundation walls that meet all of the following shall be considered laterally supported:

1. Full basement floor shall be 3.5 inches (89 mm) thick concrete slab poured tight against the bottom of the foundation wall.

2. Floor joists and blocking shall be connected to the sill plate at the top of wall by the prescriptive method called out in Table R404.1(1), or; shall be connected with an approved connector with listed capacity meeting Table R404.1(1).

3. Bolt spacing for the sill plate shall be no greater than per Table R404.1(2).

4. Floor shall be blocked perpendicular to the floor joists. Blocking shall be full depth within two joist spaces of the foundation wall, and be flat-blocked with minimum 2-inch by 4-inch (51 mm by 102 mm) blocking elsewhere.

5. Where foundation walls support unbalanced load on opposite sides of the building, such as a daylight basement, the building aspect ratio, L/W, shall not exceed the value specified in Table R404.1(3). For such foundation walls, the rim board shall be attached to the sill with a 20 gage metal angle clip at 24 inches (610 mm) on center, with five 8d nails per leg, or an approved connector supplying 230 pounds per linear foot (3.36 kN/m) capacity.

❖ These provisions are primarily for masonry or concrete basement walls. Foundation walls should be in accordance with this section or designed under one of the following design documents:

- ACI 318: *Building Code Requirements for Structural Concrete*
- ACI 332: *Requirements for Residential Concrete Construction*
- NCMA TR68-A: *Concrete Masonry Foundation Walls*
- ACI 530/ASCE 5/TMS 402: *Building Code Requirements for Masonry Structures.*

Foundation walls are usually designed and constructed to carry the vertical loads from the structure above, resist wind and any lateral forces transmitted to the foundations, and sustain earth pressures exerted against the walls, including any forces that may be imposed by frost action.

Most states do not require the seal of an architect or engineer for small residential buildings. Check with the state authorities in the location of the project for local regulations.

Lateral support at the top and bottom of the foundation wall must be provided to resist the lateral loads imposed on the foundation wall.

The prescriptive bottom support is provided by a basement floor.

The top support is provided by connections and anchor bolts as prescribed in Tables R404.1(1) and (2).

The approach for providing lateral support follows the lead of SSTD-10 which contains similar lateral support details, and uses the same wall thickness and reinforcement tables as the IRC. (SSTD-10 is a standard recognized by the IRC for wind design.) This standard shows required floor blocking as described and metal clips bolted to sill and joist for lateral support.

Lateral forces at the top are provided in Table R404.1(1), such that approved connectors can be used in place of the prescriptive metal clips. The capacity of the 3- 8d nail connection was taken from the table in the Wood Framed Construction Manual Commentary, adjusted for nonwind condition. Capacity of the generic 20 gage clip was taken to be the same as a Simpson A35 clip. Capacities for the $1/_4$ inch (6 mm) steel angle clips were taken from NDS.

Foundation walls that support unbalanced loads on opposite sides of the building, such as a daylight basements, create another problem. Rather than the forces directly transferring across the floor through the blocking, the unbalanced portion of the forces must be transferred to the perpendicular walls by shear. The aspect ratio limitation, prescribed in Table R404.1(3), is due to the shear capacity of plywood nailed at 6 inches (152 mm) on center. Normal connection of the rim board to plate is not as strong as the allowable plywood shear. To get the capacity to a similar magnitude, additional clips are required.

R404.1.1 Masonry foundation walls. Concrete masonry and clay masonry foundation walls shall be constructed as set forth in Table R404.1.1(1), R404.1.1(2), R404.1.1(3) or R404.1.1(4) and shall also comply with the provisions of Section R404 and the applicable provisions of Sections R606, R607 and R608. In Seismic Design Categories D_0, D_1 and D_2, concrete masonry and clay masonry foundation walls shall also comply with Section R404.1.4. Rubble stone masonry foundation walls shall be constructed in accordance with Sections R404.1.8 and R607.2.2. Rubble stone masonry walls shall not be used in Seismic Design Categories D_0, D_1 and D_2.

❖ Minimum thickness of concrete masonry and clay masonry foundation walls listed in Table R404.1.1(1) are predicated on the type of soil (see Table R405.1), the maximum heights of wall and unbalanced backfill (see Commentary Figure R404.1.1), and the type of wall system that is to be supported. The tabulated values are based on the assumption that the lateral loads will be carried by the walls through bending in the vertical direction. The ASTM Standards for Masonry Units regulate this minimum thickness of the units. The minimum thickness may be as much as $^1/_2$ inch (12.7 mm) less than the required nominal thickness specified in the table.

Example:

From Table R404.1.1(1): an 8-foot (2438 mm) high foundation wall constructed of hollow-unit masonry in a gravel-sand-clay mixture (SM) with a backfill height above the basement floor of 5 feet (1524 mm) would have to be minimum nominal thickness of 10 inches (254 mm). The actual wall thickness could be a minimum of $9^1/_2$ inches (241 mm).

The reinforcement specified in Tables R404.1.1(2) through R404.1.1(4) of the code are based on the nominal wall thickness, the type of soil, the height of the wall and the height of unbalanced fill. The height of unbalanced fill is measured as shown in Commentary Figure R404.1.1. The tabulated values are based on the assumption that lateral loads will be carried by the walls through bending, primarily in the vertical direction.

Additional (general) requirements for masonry construction found in Sections R606 through R608 must be followed also. See Section R404.1.4 for buildings in Seismic Design Category D_0, D_1 or D_2, and for rubble stone masonry, see the commentary for Section R404.1.8.

R404.1.2 Concrete foundation walls. Concrete foundation walls shall be constructed as set forth in Table R404.1.1(5) and shall also comply with the provisions of Section R404 and the applicable provisions of Section R402.2. In Seismic Design Categories D_0, D_1 and D_2, concrete foundation walls shall also comply with Section R404.1.4.

❖ Minimum thicknesses of concrete walls listed in Table R404.1.1(5) are predicated on the type of soil (see Table R405.1), the maximum heights of wall and unbalanced backfill (see Commentary Figure R404.1.1), and the type of wall system that is to be supported. The tabulated values are based on the assumption that the lateral loads will be carried by the walls through bending in the vertical direction. The design of concrete foundation wall is based on the actual minimum thickness shown in Table R404.1.1(5). Unlike masonry units, which must be manufactured in accordance with an ASTM standard, there is no standard to regulate the minimum thickness of a nominal 6-, 8-, 10- or 12-inch concrete wall.

Example:

From Table R404.1.1(5): an 8-foot (2438 mm) high concrete foundation wall constructed in a gravel-sand-clay mixture (SM) with a backfill height above the basement floor of 6 feet (1828 mm) would have to have a minimum wall thickness of 5.5 inches (140 mm) (6 inches nominal) with vertical reinforcement of #5 bars at 37 inches (940 mm) on center. The minimum wall thickness of plain concrete would be 7.5 inches (191 mm) (8 inches nominal).

The height of unbalanced fill is measured as shown in Commentary Figure R404.1.1. The tabulated values are based on the assumption that lateral loads will be carried by the walls through bending, primarily in the vertical direction.

Additional (general) requirements found in Sections R402.2 and R404 must be followed also. See Section

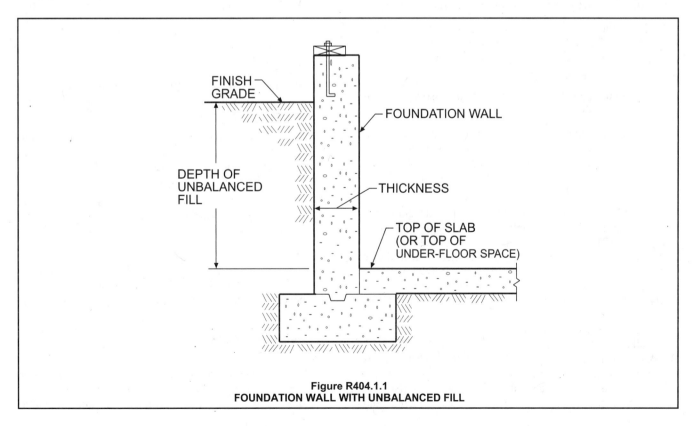

Figure R404.1.1
FOUNDATION WALL WITH UNBALANCED FILL

R404.1.4 for buildings in Seismic Design Category D_0, D_1 or D_2.

R404.1.3 Design required. Concrete or masonry foundation walls shall be designed in accordance with accepted engineering practice when either of the following conditions exists:

1. Walls are subject to hydrostatic pressure from groundwater.

2. Walls supporting more than 48 inches (1219 mm) of unbalanced backfill that do not have permanent lateral support at the top or bottom.

❖ Walls subjected to hydrostatic pressure from groundwater and walls without permanent lateral support at the top or bottom are not specifically included in the code. This section specifies conditions under which foundation walls require design. Section R404.1 cites the standards that are commonly used for "accepted engineering practice."

R404.1.4 Seismic Design Categories D_0, D_1 and D_2. In addition to the requirements of Tables R404.1.1(1) and R404.1.1(5), plain concrete and plain masonry foundation walls located in Seismic Design Categories D_0, D_1 and D_2, as established in Table R301.2(1), shall comply with the following.

1. Wall height shall not exceed 8 feet (2438 mm).

2. Unbalanced backfill height shall not exceed 4 feet (1219 mm).

3. Minimum reinforcement for plain concrete foundation walls shall consist of one No. 4 (No. 13) horizontal bar located in the upper 12 inches (305 mm) of the wall.

4. Minimum thickness for plain concrete foundation walls shall be 7.5 inches (191 mm) except that 6 inches (152 mm) is permitted when the maximum height is 4 feet, 6 inches (1372 mm).

5. Minimum nominal thickness for plain masonry foundation walls shall be 8 inches (203 mm).

6. Masonry stem walls shall have a minimum vertical reinforcement of one No. 3 (No. 10) bar located a maximum of 4 feet (1220 mm) on center in grouted cells. Vertical reinforcement shall be tied to the horizontal reinforcement in the footings.

Foundation walls located in Seismic Design Categories D_0, D_1 and D_2, as established in Table R301.2(1), supporting more than 4 feet (1219 mm) of unbalanced backfill or exceeding 8 feet (2438 mm) in height shall be constructed in accordance with Table R404.1.1(2), R404.1.1(3) or R404.1.1(4) for masonry, or Table R404.1.1(5) for concrete. Where Table R404.1.1(5) permits plain concrete walls, not less than No. 4 (No. 13) vertical bars at a spacing not exceeding 48 inches (1219 mm) shall be provided. Insulating concrete form foundation walls shall be reinforced as required in Table R404.4(1), R404.4(2), R404.4(3), R404.4(4) or R404.4(5). Where no vertical reinforcement is required by Table R404.4(2), R404.4(3) or R404.4(4) there shall be a minimum of one No. 4 (No. 13) bar at 48 inches (1220 mm) on center. All concrete and

masonry foundation walls shall have two No. 4 (No. 13) horizontal bars located in the upper 12 inches (305 mm) of the wall.

❖ This section places six restrictions on the requirements in Tables R404.1.1(1) and R404.1.1(5) for plain concrete and plain masonry foundation walls in Seismic Design Category D_0, D_1 or D_2.

The code requires minimum vertical reinforcement of masonry stem walls in order for them to resist lateral earthquake forces. Any wall over 8 feet (2338 mm) or supporting more than 4 feet (1219 mm) of unbalanced backfill must be constructed with minimum vertical reinforcement in accordance with Tables R404.1.1(2) through R404.1.1(5) and have the specified longitudinal reinforcement.

R404.1.5 Foundation wall thickness based on walls supported. The thickness of concrete and masonry foundation walls shall not be less than the thickness of the wall supported, except that foundation walls of at least 8-inch (203 mm) nominal thickness shall be permitted under brick- veneered frame walls and under 10-inch-wide (254 mm) cavity walls where the total height of the wall supported, including gables, is not more than 20 feet (6096 mm), provided the requirements of Sections R404.1.1 and R404.1.2 are met.

❖ Other than the two exceptions noted, concrete and masonry foundation walls should be equal to or greater than the thickness of the wall supported.

R404.1.5.1 Pier and curtain wall foundations. Use of Pier and curtain wall foundations shall be permitted to support light-frame construction not more than two stories in height, provided the following requirements are met:

1. All load-bearing walls shall be placed on continuous concrete footings placed integrally with the exterior wall footings.

2. The minimum actual thickness of a load-bearing masonry wall shall be not less than 4 inches (102 mm) nominal or $3^3/_8$ inches (92 mm) actual thickness, and shall be bonded integrally with piers spaced in accordance with Section R606.9.

3. Piers shall be constructed in accordance with Section R606.6 and Section R606.6.1, and shall be bonded into the load-bearing masonry wall in accordance with Section R608.1.1 or Section R608.1.1.2.

4. The maximum height of a 4-inch (102 mm) load-bearing masonry foundation wall supporting wood-frame walls and floors shall not be more than 4 feet (1219 mm).

5. Anchorage shall be in accordance with Section R403.1.6, Figure R404.1.5(1), or as specified by engineered design accepted by the building official.

6. The unbalanced fill for 4-inch (102 mm) foundation walls shall not exceed 24 inches (610 mm) for solid masonry or 12 inches (305 mm) for hollow masonry.

7. In Seismic Design Categories D_0, D_1 and D_2, prescriptive reinforcement shall be provided in the horizontal and vertical direction. Provide minimum horizontal joint rein-

forcement of two No.9 gage wires spaced not less than 6 inches (152 mm) or one $^1/_4$ inch (6.4 mm) diameter wire at 10 inches (254 mm) on center vertically. Provide minimum vertical reinforcement of one No. 4 bar at 48 inches (1220 mm) on center horizontally grouted in place.

❖ Pier and curtain wall foundations may be used to support light-frame construction for buildings in Seismic Design Category A, B, C, D$_0$, D$_1$ or D$_2$, provided several conditions are met. A curtain wall typically refers to a nonload-bearing wall, but within the limits established in this section, it may be used to support up to two stories of light-frame construction. Four-inch (102 mm) nominal masonry walls are limited to 4 feet (1219 mm) in height so that a thin wall does not fail laterally due to the vertical loads on the wall. In addition, no more than 2 feet (610 mm) of unbalanced fill is permitted for solid masonry and 1 foot (305 mm) of unbalanced fill for hollow masonry to prevent failure due to the pressures created by unbalanced fill. Additional requirements are shown in Figure R404.1.5(1). This type of foundation wall must be reinforced when located in Seismic Design Category D$_0$, D$_1$ or D$_2$.

TABLE R404.1(1)
TOP REACTIONS AND PRESCRIPTIVE SUPPORT FOR FOUNDATION WALLS[a]

MAXIMUM WALL HEIGHT (feet)	MAXIMUM UNBALANCED BACKFILL HEIGHT (feet)	HORIZONTAL REACTION TO TOP (plf)		
		Soil Classes (Letter indicates connection types[b])		
		GW, GP, SW and SP soils	GM, GC, SM-SC and ML soils	SC, MH, ML-CL and inorganic CL soils
7	4	45.7 A	68.6 A	91.4 A
	5	89.3 A	133.9 B	178.6 B
	6	154.3 B	231.4 C	308.6 C
	7	245.0 C	367.5 C	490.0 D
8	4	40.0 A	60.0 A	80.0 A
	5	78.1 A	117.2 B	156.3 B
	6	135.0 B	202.5 B	270.0 C
	7	214.0 B	321.6 C	428.8 C
	8	320.0 C	480.0 C	640.0 D
9	4	35.6 A	53.3 A	71.1 A
	5	69.4 A	104.2 B	138.9 B
	6	120.0 B	180.0 B	240.0 C
	7	190.6 B	285.8 C	381.1 C
	8	284.4 C	426.7 C	568.9 D
	9	405.0 C	607.5 D	810.0 D

For SI: 1 foot = 304.8 mm, 1 pound = 0.454 kg, 1 plf = pounds per linear foot = 1.488 kg/m.

a. Loads are pounds per linear foot of wall. Prescriptive options are limited to maximum joist and blocking spacing of 24 inches on center.

b. Prescriptive Support Requirements:

Type	Joist/blocking Attachment Requirement
A	3 8d per joist per Table R602.3(1).
B	1-20 gage angle clip each joist with 5-8d per leg.
C	1-$^1/_4$-inch thick steel angle. Horizontal leg attached to sill bolt adjacent to joist/blocking, vertical leg attached to joist/blocking with $^1/_2$-inch minimum diameter bolt.
D	2-$^1/_4$-inch thick steel, angles, one on each side of joist/blocking. Attach each angle to adjacent sill bolt through horizontal leg. Bolt to joist/blocking with $^1/_2$-inch minimum diameter bolt common to both angles.

❖ See the commentary for Section R404.1.

TABLE R404.1(2)
MAXIMUM PLATE ANCHOR-BOLT SPACING FOR SUPPORTED FOUNDATION WALL[a]

MAXIMUM WALL HEIGHT (feet)	MAXIMUM UNBALANCED BACKFILL HEIGHT (feet)	ANCHOR BOLT SPACING (inches)		
		Soil Classes		
		GW, GP, SW and SP soils	GM, GC, SM-SC and ML soils	SC, MH, ML-CL and inorganic CL soils
7	4	72	58	43
	5	44	30	22
	6	26	17	13
	7	16	11	8
8	4	72	66	50
	5	51	34	25
	6	29	20	15
	7	18	12	9
	8	12	8	6
9	4	72	72	56
	5	57	38	29
	6	33	22	17
	7	21	14	10
	8	14	9	7
	9	10	7	5

For SI: 1 inch = 25.4 mm, 1 foot = 304.8 mm.

a. Spacing is based on $^1/_2$-inch diameter anchor bolts. For $^5/_8$-inch diameter anchor bolts, spacing may be multiplied by 1.27, with a maximum spacing of 72 inches.

❖ See the commentary for Section R404.1.

TABLE R404.1(3)
MAXIMUM ASPECT RATIO, L/W FOR UNBALANCED FOUNDATIONS

MAXIMUM WALL HEIGHT (feet)	MAXIMUM UNBALANCED BACKFILL HEIGHT (feet)	SOIL CLASSES		
		GW, GP, SW and SP soils	GM, GC, SM-SC and ML soils	SC, MH, ML-CL and inorganic CL soils
7	4	4.0	4.0	4.0
	5	4.0	3.4	2.6
	6	3.0	2.0	1.5
	7	1.9	1.2	0.9
8	4	4.0	4.0	4.0
	5	4.0	3.9	2.9
	6	3.4	2.3	1.7
	7	2.1	1.4	1.1
	8	1.4	1.0	0.7
9	4	4.0	4.0	4.0
	5	4.0	4.0	3.3
	6	3.8	2.6	1.9
	7	2.4	1.6	1.2
	8	1.6	1.1	0.8
	9	1.1	0.8	0.6

For SI: 1 foot = 304.8 mm.

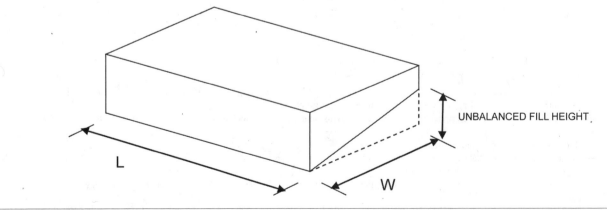

❖ See commentary for Section R404.1.

TABLE R404.1.1(1)
PLAIN MASONRY FOUNDATION WALLS

MAXIMUM WALL HEIGHT (feet)	MAXIMUM UNBALANCED BACKFILL HEIGHT[c] (feet)	PLAIN MASONRY[a] MINIMUM NOMINAL WALL THICKNESS (inches)		
		Soil classes[b]		
		GW, GP, SW and SP	GM, GC, SM, SM-SC and ML	SC, MH, ML-CL and inorganic CL
5	4	6 solid[d] or 8	6 solid[d] or 8	6 solid[d] or 8
	5	6 solid[d] or 8	8	10
6	4	6 solid[d] or 8	6 solid[d] or 8	6 solid[d] or 8
	5	6 solid[d] or 8	8	10
	6	8	10	12
7	4	6 solid[d] or 8	8	8
	5	6 solid[d] or 8	10	10
	6	10	12	10 solid[d]
	7	12	10 solid[d]	12 solid[d]
8	4	6 solid[d] or 8	6 solid[d] or 8	8
	5	6 solid[d] or 8	10	12
	6	10	12	12 solid[d]
	7	12	12 solid[d]	Footnote e
	8	10 solid[d]	12 solid[d]	Footnote e
9	4	6 solid[d] or 8	6 solid[d] or 8	8
	5	8	10	12
	6	10	12	12 solid[d]
	7	12	12 solid[d]	Footnote e
	8	12 solid[d]	Footnote e	Footnote e
	9	Footnote e	Footnote e	Footnote e

For SI: 1 inch = 25.4 mm, 1 foot = 304.8 mm, 1 pound per square inch = 6.895 Pa.

a. Mortar shall be Type M or S and masonry shall be laid in running bond. Ungrouted hollow masonry units are permitted except where otherwise indicated.

b. Soil classes are in accordance with the Unified Soil Classification System. Refer to Table R405.1.

c. Unbalanced backfill height is the difference in height between the exterior finish ground level and the lower of the top of the concrete footing that supports the foundation wall or the interior finish ground level. Where an interior concrete slab-on-grade is provided and is in contact with the interior surface of the foundation wall, measurement of the unbalanced backfill height from the exterior finish ground level to the top of the interior concrete slab is permitted.

d. Solid grouted hollow units or solid masonry units.

e. Wall construction shall be in accordance with Table R404.1.1(2) or a design shall be provided.

❖ See Section R404.1.1 and example.

R404.1.6 Height above finished grade. Concrete and masonry foundation walls shall extend above the finished grade adjacent to the foundation at all points a minimum of 4 inches (102 mm) where masonry veneer is used and a minimum of 6 inches (152 mm) elsewhere.

❖ The minimum distance above adjacent grade to which the foundation must be extended provides termite protection and minimizes the chance of decay resulting from moisture migrating to the wood framing. A reduced foundation extension is permitted when masonry veneer is used.

R404.1.7 Backfill placement. Backfill shall not be placed against the wall until the wall has sufficient strength and has been anchored to the floor above, or has been sufficiently braced to prevent damage by the backfill.

Exception: Bracing is not required for walls supporting less than 4 feet (1219 mm) of unbalanced backfill.

❖ Backfilling should not begin until a foundation wall is in suitable condition to resist lateral earth pressure. The wall must have sufficient strength, and the floor framing system should be in place and anchored to the foundation wall as shown in Commentary Figure

R404.1.7 to minimize the chance of damage to the foundation wall resulting from placement of backfill against the foundation wall.

R404.1.8 Rubble stone masonry. Rubble stone masonry foundation walls shall have a minimum thickness of 16 inches (406 mm), shall not support an unbalanced backfill exceeding 8 feet (2438 mm) in height, shall not support a soil pressure greater than 30 pounds per square foot per foot (4.71 kPa/m), and shall not be constructed in Seismic Design Categories D$_0$, D$_1$, D$_2$ or townhouses in Seismic Design Category C, as established in Figure R301.2(2).

❖ Rubble stone masonry is masonry walls constructed of roughly shaped stones. See the definitions in the *International Building Code*. Rubble stone masonry is almost never used for foundation walls. This type of wall construction is permitted only in Seismic Design Category A or B, and detached one- and two-family dwelling in Seismic Design Category C. Furthermore, it is restricted to a maximum unbalanced backfill height of 8 feet (2438 mm) and must not be subjected to lateral soil pressures more than 30 pounds per square foot (1436 Pa).

TABLE R404.1.1(2)
8-INCH MASONRY FOUNDATION WALLS WITH REINFORCING
WHERE d > 5 INCHES[a]

WALL HEIGHT	HEIGHT OF UNBALANCED BACKFILL[e]	MINIMUM VERTICAL REINFORCEMENT[b,c]		
		Soil classes and lateral soil load[d] (psf per foot below grade)		
		GW, GP, SW and SP soils 30	GM, GC, SM, SM-SC and ML soils 45	SC, ML-CL and inorganic CL soils 60
6 feet 8 inches	4 feet (or less)	#4 at 48″ o.c.	#4 at 48″ o.c.	#4 at 48″ o.c.
	5 feet	#4 at 48″ o.c.	#4 at 48″ o.c.	#4 at 48″ o.c.
	6 feet 8 inches	#4 at 48″ o.c.	#5 at 48″ o.c.	#6 at 48″ o.c.
7 feet 4 inches	4 feet (or less)	#4 at 48″ o.c.	#4 at 48″ o.c.	#4 at 48″ o.c.
	5 feet	#4 at 48″ o.c.	#4 at 48″ o.c.	#4 at 48″ o.c.
	6 feet	#4 at 48″ o.c.	#5 at 48″ o.c.	#5 at 48″ o.c.
	7 feet 4 inches	#5 at 48″ o.c.	#6 at 48″ o.c.	#6 at 40″ o.c.
8 feet	4 feet (or less)	#4 at 48″ o.c.	#4 at 48″ o.c.	#4 at 48″ o.c.
	5 feet	#4 at 48″ o.c.	#4 at 48″ o.c.	#4 at 48″ o.c.
	6 feet	#4 at 48″ o.c.	#5 at 48″ o.c.	#5 at 48″ o.c.
	7 feet	#5 at 48″ o.c.	#6 at 48″ o.c.	#6 at 40″ o.c.
	8 feet	#5 at 48″ o.c.	#6 at 48″ o.c.	#6 at 32″ o.c.
8 feet 8 inches	4 feet (or less)	#4 at 48″ o.c.	#4 at 48″ o.c.	#4 at 48″ o.c.
	5 feet	#4 at 48″ o.c.	#4 at 48″ o.c.	#5 at 48″ o.c.
	6 feet	#4 at 48″ o.c.	#5 at 48″ o.c.	#6 at 48″ o.c.
	7 feet	#5 at 48″ o.c.	#6 at 48″ o.c.	#6 at 40″ o.c.
	8 feet 8 inches	#6 at 48″ o.c.	#6 at 32″ o.c.	#6 at 24″ o.c.
9 feet 4 inches	4 feet (or less)	#4 at 48″ o.c.	#4 at 48″ o.c.	#4 at 48″ o.c.
	5 feet	#4 at 48″ o.c.	#4 at 48″ o.c.	#5 at 48″ o.c.
	6 feet	#4 at 48″ o.c.	#5 at 48″ o.c.	#6 at 48″ o.c.
	7 feet	#5 at 48″ o.c.	#6 at 48″ o.c.	#6 at 40″ o.c.
	8 feet	#6 at 48″ o.c.	#6 at 40″ o.c.	#6 at 24″ o.c.
	9 feet 4 inches	#6 at 40″ o.c.	#6 at 24″ o.c.	#6 at 16″ o.c.
10 feet	4 feet (or less)	#4 at 48″ o.c.	#4 at 48″ o.c.	#4 at 48″ o.c.
	5 feet	#4 at 48″ o.c.	#4 at 48″ o.c.	#5 at 48″ o.c.
	6 feet	#4 at 48″ o.c.	#5 at 48″ o.c.	#6 at 48″ o.c.
	7 feet	#5 at 48″ o.c.	#6 at 48″ o.c.	#6 at 32″ o.c.
	8 feet	#6 at 48″ o.c.	#6 at 32″ o.c.	#6 at 24″ o.c.
	9 feet	#6 at 40″ o.c.	#6 at 24″ o.c.	#6 at 16″ o.c.
	10 feet	#6 at 32″ o.c.	#6 at 16″ o.c.	#6 at 16″ o.c.

For SI: 1 inch = 25.4 mm, 1 foot = 304.8 mm, 1 pound per square foot per foot = 0.157 kPa/mm.

a. Mortar shall be Type M or S and masonry shall be laid in running bond.

b. Alternative reinforcing bar sizes and spacings having an equivalent cross-sectional area of reinforcement per lineal foot of wall shall be permitted provided the spacing of the reinforcement does not exceed 72 inches.

c. Vertical reinforcement shall be Grade 60 minimum. The distance from the face of the soil side of the wall to the center of vertical reinforcement shall be at least 5 inches.

d. Soil classes are in accordance with the Unified Soil Classification System and design lateral soil loads are for moist conditions without hydrostatic pressure. Refer to Table R405.1.

e. Unbalanced backfill height is the difference in height between the exterior finish ground level and the lower of the top of the concrete footing that supports the foundation wall or the interior finish ground level. Where an interior concrete slab-on-grade is provided and is in contact with the interior surface of the foundation wall, measurement of the unbalanced backfill height from the exterior finish ground level to the top of the interior concrete slab is permitted.

❖ See the commentary for Section R404.1.1.

TABLE R404.1.1(3)
10-INCH MASONRY FOUNDATION WALLS WITH REINFORCING
WHERE d > 6.75 INCHES[a]

WALL HEIGHT	HEIGHT OF UNBALANCED BACKFILL[e]	MINIMUM VERTICAL REINFORCEMENT[b, c]		
		Soil classes and later soil load[d] (psf per foot below grade)		
		GW, GP, SW and SP soils 30	GM, GC, SM, SM-SC and ML soils 45	SC, MH, ML-CL and inorganic CL soils 60
6 feet 8 inches	4 feet (or less)	#4 at 56″ o.c.	#4 at 56″ o.c.	#4 at 56″ o.c.
	5 feet	#4 at 56″ o.c.	#4 at 56″ o.c.	#4 at 56″ o.c.
	6 feet 8 inches	#4 at 56″ o.c.	#5 at 56″ o.c.	#5 at 56″ o.c.
7 feet 4 inches	4 feet (or less)	#4 at 56″ o.c.	#4 at 56″ o.c.	#4 at 56″ o.c.
	5 feet	#4 at 56″ o.c.	#4 at 56″ o.c.	#4 at 56″ o.c.
	6 feet	#4 at 56″ o.c.	#4 at 56″ o.c.	#5 at 56″ o.c.
	7 feet 4 inches	#4 at 56″ o.c.	#5 at 56″ o.c.	#6 at 56″ o.c.
8 feet	4 feet (or less)	#4 at 56″ o.c.	#4 at 56″ o.c.	#4 at 56″ o.c.
	5 feet	#4 at 56″ o.c.	#4 at 56″ o.c.	#4 at 56″ o.c.
	6 feet	#4 at 56″ o.c.	#4 at 56″ o.c.	#5 at 56″ o.c.
	7 feet	#4 at 56″ o.c.	#5 at 56″ o.c.	#6 at 56″ o.c.
	8 feet	#5 at 56″ o.c.	#6 at 56″ o.c.	#6 at 48″ o.c.
8 feet 8 inches	4 feet (or less)	#4 at 56″ o.c.	#4 at 56″ o.c.	#4 at 56″ o.c.
	5 feet	#4 at 56″ o.c.	#4 at 56″ o.c.	#4 at 56″ o.c.
	6 feet	#4 at 56″ o.c.	#4 at 56″ o.c.	#5 at 56″ o.c.
	7 feet	#4 at 56″ o.c.	#5 at 56″ o.c.	#6 at 56″ o.c.
	8 feet 8 inches	#5 at 56″ o.c.	#6 at 48″ o.c.	#6 at 32″ o.c.
9 feet 4 inches	4 feet (or less)	#4 at 56″ o.c.	#4 at 56″ o.c.	#4 at 56″ o.c.
	5 feet	#4 at 56″ o.c.	#4 at 56″ o.c.	#4 at 56″ o.c.
	6 feet	#4 at 56″ o.c.	#5 at 56″ o.c.	#5 at 56″ o.c.
	7 feet	#4 at 56″ o.c.	#5 at 56″ o.c.	#6 at 56″ o.c.
	8 feet	#5 at 56″ o.c.	#6 at 56″ o.c.	#6 at 40″ o.c.
	9 feet 4 inches	#6 at 56″ o.c.	#6 at 40″ o.c.	#6 at 24″ o.c.
10 feet	4 feet (or less)	#4 at 56″ o.c.	#4 at 56″ o.c.	#4 at 56″ o.c.
	5 feet	#4 at 56″ o.c.	#4 at 56″ o.c.	#4 at 56″ o.c.
	6 feet	#4 at 56″ o.c.	#5 at 56″ o.c.	#5 at 56″ o.c.
	7 feet	#5 at 56″ o.c.	#6 at 56″ o.c.	#6 at 48″ o.c.
	8 feet	#5 at 56″ o.c.	#6 at 48″ o.c.	#6 at 40″ o.c.
	9 feet	#6 at 56″ o.c.	#6 at 40″ o.c.	#6 at 24″ o.c.
	10 feet	#6 at 48″ o.c.	#6 at 32″ o.c.	#6 at 24″ o.c.

For SI: 1 inch = 25.4 mm, 1 foot = 304.8 mm, 1 pound per square foot per foot = 0.157 kPa/mm.

a. Mortar shall be Type M or S and masonry shall be laid in running bond.

b. Alternative reinforcing bar sizes and spacings having an equivalent cross-sectional area of reinforcement per lineal foot of wall shall be permitted provided the spacing of the reinforcement does not exceed 72 inches.

c. Vertical reinforcement shall be Grade 60 minimum. The distance from the face of the soil side of the wall to the center of vertical reinforcement shall be at least 6.75 inches.

d. Soil classes are in accordance with the Unified Soil Classification System and design lateral soil loads are for moist conditions without hydrostatic pressure. Refer to Table R405.1.

e. Unbalanced backfill height is the difference in height between the exterior finish ground level and the lower of the top of the concrete footing that supports the foundation wall or the interior finish ground level. Where an interior concrete slab-on-grade is provided and is in contact with the interior surface of the foundation wall, measurement of the unbalanced backfill height from the exterior finish ground level to the top of the interior concrete slab is permitted.

❖ See the commentary for Section R404.1.1.

TABLE R404.1.1(4)
12-INCH MASONRY FOUNDATION WALLS WITH REINFORCING
WHERE d > 8.75 INCHES[a]

WALL HEIGHT	HEIGHT OF UNBALANCED BACKFILL[e]	MINIMUM VERTICAL REINFORCEMENT[b, c]		
		Soil classes and lateral soil load[d] (psf per foot below grade)		
		GW, GP, SW and SP soils 30	GM, GC, SM, SM-SC and ML soils 45	SC, ML-CL and inorganic CL soils 60
6 feet 8 inches	4 feet (or less)	#4 at 72″ o.c.	#4 at 72″ o.c.	#4 at 72″ o.c.
	5 feet	#4 at 72″ o.c.	#4 at 72″ o.c.	#4 at 72″ o.c.
	6 feet 8 inches	#4 at 72″ o.c.	#4 at 72″ o.c.	#5 at 72″ o.c.
7 feet 4 inches	4 feet (or less)	#4 at 72″ o.c.	#4 at 72″ o.c.	#4 at 72″ o.c.
	5 feet	#4 at 72″ o.c.	#4 at 72″ o.c.	#4 at 72″ o.c.
	6 feet	#4 at 72″ o.c.	#4 at 72″ o.c.	#5 at 72″ o.c.
	7 feet 4 inches	#4 at 72″ o.c.	#5 at 72″ o.c.	#6 at 72″ o.c.
8 feet	4 feet (or less)	#4 at 72″ o.c.	#4 at 72″ o.c.	#4 at 72″ o.c.
	5 feet	#4 at 72″ o.c.	#4 at 72″ o.c.	#4 at 72″ o.c.
	6 feet	#4 at 72″ o.c.	#4 at 72″ o.c.	#5 at 72″ o.c.
	7 feet	#4 at 72″ o.c.	#5 at 72″ o.c.	#6 at 72″ o.c.
	8 feet	#5 at 72″ o.c.	#6 at 72″ o.c.	#6 at 64″ o.c.
8 feet 8 inches	4 feet (or less)	#4 at 72″ o.c.	#4 at 72″ o.c.	#4 at 72″ o.c.
	5 feet	#4 at 72″ o.c.	#4 at 72″ o.c.	#4 at 72″ o.c.
	6 feet	#4 at 72″ o.c.	#4 at 72″ o.c.	#5 at 72″ o.c.
	7 feet	#4 at 72″ o.c.	#5 at 72″ o.c.	#6 at 72″ o.c.
	8 feet 8 inches	#5 at 72″ o.c.	#7 at 72″ o.c.	#6 at 48″ o.c.
9 feet 4 inches	4 feet (or less)	#4 at 72″ o.c.	#4 at 72″ o.c.	#4 at 72″ o.c.
	5 feet	#4 at 72″ o.c.	#4 at 72″ o.c.	#4 at 72″ o.c.
	6 feet	#4 at 72″ o.c.	#5 at 72″ o.c.	#5 at 72″ o.c.
	7 feet	#4 at 72″ o.c.	#5 at 72″ o.c.	#6 at 72″ o.c.
	8 feet	#5 at 72″ o.c.	#6 at 72″ o.c.	#6 at 56″ o.c.
	9 feet 4 inches	#6 at 72″ o.c.	#6 at 48″ o.c.	#6 at 40″ o.c.
10 feet	4 feet (or less)	#4 at 72″ o.c.	#4 at 72″ o.c.	#4 at 72″ o.c.
	5 feet	#4 at 72″ o.c.	#4 at 72″ o.c.	#4 at 72″ o.c.
	6 feet	#4 at 72″ o.c.	#5 at 72″ o.c.	#5 at 72″ o.c.
	7 feet	#4 at 72″ o.c.	#6 at 72″ o.c.	#6 at 72″ o.c.
	8 feet	#5 at 72″ o.c.	#6 at 72″ o.c.	#6 at 48″ o.c.
	9 feet	#6 at 72″ o.c.	#6 at 56″ o.c.	#6 at 40″ o.c.
	10 feet	#6 at 64″ o.c.	#6 at 40″ o.c.	#6 at 32″ o.c.

For SI: 1 inch = 25.4 mm, 1 foot = 304.8 mm, 1 pound per square foot per foot = 0.157 kPa/mm.

a. Mortar shall be Type M or S and masonry shall be laid in running bond.

b. Alternative reinforcing bar sizes and spacings having an equivalent cross-sectional area of reinforcement per lineal foot of wall shall be permitted provided the spacing of the reinforcement does not exceed 72 inches.

c. Vertical reinforcement shall be Grade 60 minimum. The distance from the face of the soil side of the wall to the center of vertical reinforcement shall be at least 8.75 inches.

d. Soil classes are in accordance with the Unified Soil Classification System and design lateral soil loads are for moist conditions without hydrostatic pressure. Refer to Table R405.1.

e. Unbalanced backfill height is the difference in height between the exterior finish ground level and the lower of the top of the concrete footing that supports the foundation wall or the interior finish ground levels. Where an interior concrete slab-on-grade is provided and in contact with the interior surface of the foundation wall, measurement of the unbalanced backfill height is permitted to be measured from the exterior finish ground level to the top of the interior concrete slab is permitted.

❖ See the commentary for Section R404.1.1.

TABLE R404.1.1(5)
CONCRETE FOUNDATION WALLS[h, i, j, k]

MAXIMUM WALL HEIGHT (feet)	MAXIMUM UNBALANCED BACKFILL HEIGHT[b] (feet)	MINIMUM VERTICAL REINFORCEMENT SIZE AND SPACING[c, d, e, f, l]											
		Soil classes[a] and design lateral soil (psf per foot of depth)											
		GW, GP, SW and SP 30				GM, GC, SM, SM-SC and ML 45				SC, ML-CL and inorganic CL 60			
		Minimum wall thickness (inches)											
		5.5	7.5	9.5	11.5	5.5	7.5	9.5	11.5	5.5	7.5	9.5	11.5
5	4	PC	PC	PC	PC	PC	PC	PC	PC	PC	PC	PC	PC
	5	PC	PC	PC	PC	PC	PC	PC	PC	PC	PC	PC	PC
6	4	PC	PC	PC	PC	PC	PC	PC	PC	PC	PC	PC	PC
	5	PC	PC	PC	PC	PC	PCg	PC	PC	#4@35"	PCg	PC	PC
	6	PC	PC	PC	PC	#5@48"	PC	PC	PC	#5@36"	PC	PC	PC
7	4	PC	PC	PC	PC	PC	PC	PC	PC	PC	PC	PC	PC
	5	PC	PC	PC	PC	PC	PC	PC	PC	#5@47"	PC	PC	PC
	6	PC	PC	PC	PC	#5@42"	PC	PC	PC	#6@43"	#5@48"	PCg	PC
	7	#5@46"	PC	PC	PC	#6@42"	#5@46"	PCg	PC	#6@34"	#6@48"	PC	PC
8	4	PC	PC	PC	PC	PC	PC	PC	PC	PC	PC	PC	PC
	5	PC	PC	PC	PC	#4@38"	PCg	PC	PC	#5@43"	PC	PC	PC
	6	#4@37"	PCg	PC	PC	#5@37"	PC	PC	PC	#6@37"	#5@43"	PCg	PC
	7	#5@40"	PC	PC	PC	#6@37"	#5@41"	PC	PC	#6@34"	#6@43"	PC	PC
	8	#6@43"	#5@47"	PCg	PC	#6@34"	#6@43"	PC	PC	#6@27"	#6@32"	#6@44"	PC
9	4	PC	PC	PC	PC	PC	PC	PC	PC	PC	PC	PC	PC
	5	PC	PC	PC	PC	#4@35"	PCg	PC	PC	#5@40"	PC	PCe	PC
	6	#4@34"	PCg	PC	PC	#6@48"	PC	PC	PC	#6@36"	#5@39"	PCg	PC
	7	#5@36"	PC	PC	PC	#6@34"	#5@37"	PC	PC	#6@33"	#6@38"	#5@37"	PCg
	8	#6@38"	#5@41"	PCg	PC	#6@33"	#6@38"	#5@37"	PCg	#6@24"	#7@39"	#6@39"	#4@48"h
	9	#6@34"	#6@46"	PC	PC	#6@26"	#7@41"	#6@41"	PC	#6@19"	#7@31"	#7@41"	#6@39"
10	4	PC	PC	PC	PC	PC	PC	PC	PC	PC	PC	PC	PC
	5	PC	PC	PC	PC	#4@33"	PCg	PC	PC	#5@38"	PC	PC	PC
	6	#5@48"	PCg	PC	PC	#6@45"	PC	PC	PC	#6@34"	#5@37"	PC	PC
	7	#6@47"	PC	PC	PC	#6@34"	#6@48"	PC	PC	#6@30"	#6@35"	#6@48"	PCg
	8	#6@34"	#5@38"	PC	PC	#6@30"	#7@47"	#6@47"	PCg	#6@22"	#7@35"	#7@48"	#6@45"h
	9	#6@34"	#6@41"	#4@48"	PCg	#6@23"	#7@37"	#7@48"	#4@48"h	DR	#6@22"	#7@37"	#7@47"
	10	#6@28"	#7@45"	#6@45"	PC	DR	#7@31"	#7@40"	#6@38"	DR	#6@22"	#7@30"	#7@38"

For SI: 1 inch = 25.4 mm, 1 foot = 304.8 mm, 1 pound per square foot = 0.0479 kPa; 1 pound per square foot per foot = 0.157 kPa/mm.

a. Soil classes are in accordance with the United Soil Classification System. Refer to Table R405.1

b. Unbalanced backfill height is the difference in height of the exterior and interior finish ground levels. Where there is an interior concrete slab, the unbalanced backfill height shall be measured from the exterior finish ground level to the top of the interior concrete slab.

c. The size and spacing of vertical reinforcement shown in the table is based on the use of reinforcement with a minimum yield strength of 60,000 psi. Vertical reinforcement with a minimum yield strength of 40,000 psi or 50,000 psi is permitted, provided the same size bar is used and the spacing shown in the table is reduced by multiplying the spacing by 0.67 or 0.83, respectively.

d. Vertical reinforcement, when required, shall be placed nearest the inside face of the wall a distance d from the outside face (soil side) of the wall. The distance d is equal to the wall thickness, t, minus 1.25 inches plus one-half the bar diameter, d_b (d = t - (1.25 + d_b/2). The reinforcement shall be placed within a tolerance of ± $^3/_8$ inch where d is less than or equal to 8 inches, or ± $^1/_2$ inch where d is greater than 8 inches.

e. In lieu of the reinforcement shown, smaller reinforcing bar sizes and closer spacings resulting in an equivalent cross-sectional area of reinforcement per linear foot of wall are permitted.

f. Concrete cover for reinforcement measured from the inside face of the wall shall not be less than $^3/_4$ inch. Concrete cover for reinforcement measured from the outside face of the wall shall not be less than 1$^1/_2$ inches for No. 5 bars and smaller, and not less than 2 inches for larger bars.

g. The minimum thickness is permitted to be reduced 2 inches, provided the minimum specified compressive strength of concrete f_c', is 4,000 psi.

(continued)

TABLE R404.1.1(5)—continued
CONCRETE FOUNDATION WALLS[h, i, j, k]

h. A plain concrete wall with a minimum thickness of 11.5 inches is permitted, provided minimum specified compressive strength of concrete, f_c', is 3,500 psi.

i. Concrete shall have a specified compressive strength of not less than 2,500 psi at 28 days, unless a higher strength is required by note g or h.

j. "DR" means design is required in accordance with ACI 318 or ACI 332.

k. "PC" means plain concrete.

l. Where vertical reinforcement is required, horizontal reinforcement shall be provided in accordance with the requirements of Section R404.4.6.2 for ICF foundation walls.

❖ See the commentary for Section 404.1.2.

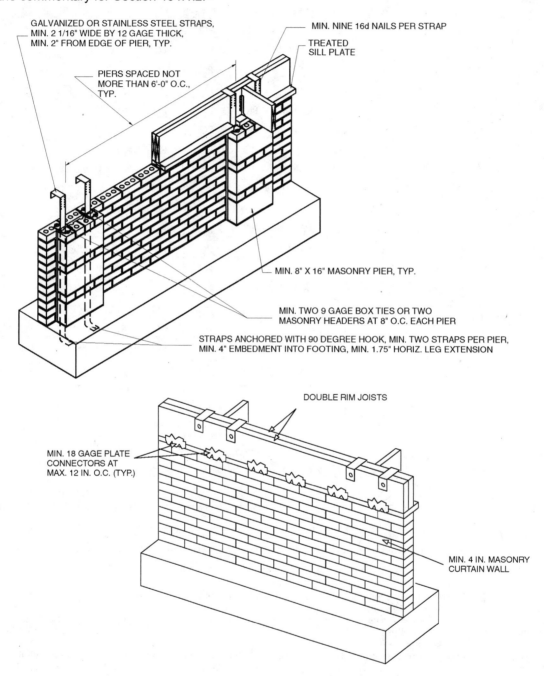

For SI: 1 inch = 25.4 mm, 1 foot = 304.8 mm, 1 degree = 0.79/45 rad.

FIGURE R404.1.5(1)
FOUNDATION WALL CLAY MASONRY CURTAIN WALL WITH CONCRETE MASONRY PIERS

❖ This figure illustrates several requirements for pier and curtain wall foundations described in Section R404.1.5.1. The pier spacing of not more than 6 feet, 0 inches (1829 mm) on center establishes minimum lateral (horizontal) wall support similar to that of Section R606.9 (also see commentary, Section R404.1.5.1).

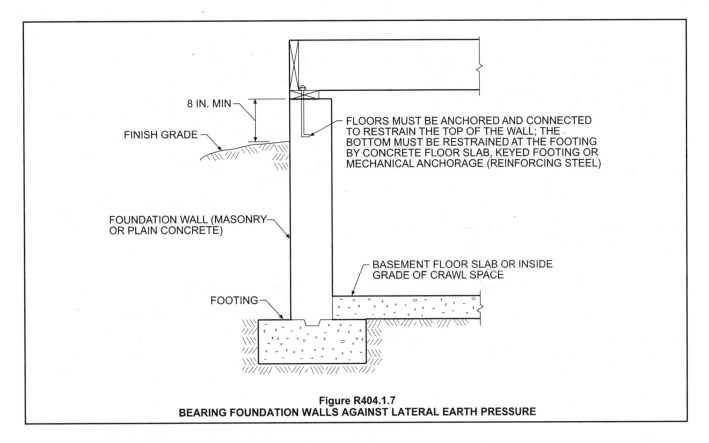

Figure R404.1.7
BEARING FOUNDATION WALLS AGAINST LATERAL EARTH PRESSURE

(Figure labels:)
8 IN. MIN
FINISH GRADE
FLOORS MUST BE ANCHORED AND CONNECTED TO RESTRAIN THE TOP OF THE WALL; THE BOTTOM MUST BE RESTRAINED AT THE FOOTING BY CONCRETE FLOOR SLAB, KEYED FOOTING OR MECHANICAL ANCHORAGE (REINFORCING STEEL)
FOUNDATION WALL (MASONRY OR PLAIN CONCRETE)
BASEMENT FLOOR SLAB OR INSIDE GRADE OF CRAWL SPACE
FOOTING

R404.2 Wood foundation walls. Wood foundation walls shall be constructed in accordance with the provisions of Sections R404.2.1 through R404.2.6 and with the details shown in Figures R403.1(2) and R403.1(3).

❖ This section covers the design and installation of a multi-component wood foundation system. The construction is essentially a below-grade load-bearing wood frame system that serves as the enclosure for basements and crawl spaces and also serves as the structural foundation for the support of light-frame structures. See Figures R403.1(2) and R403.1(3) for typical installation details for wood foundation walls (also see commentary, Section R401.1).

Certain cold-weather precautions should be taken in the installation of the wood foundation system. The composite footing consisting of a wood plate supported on a bed of stone or sand fill should not be placed on frozen ground. While the bottom of the wood plate footing would normally be placed below the frost line (i.e., basement construction) under certain drainage conditions, AF&PA TR7-87 permits the wood plate to be set above the frost level. The code official should verify that such an alternative design satisfies the intent of the code. Also important is the use of proper sealants during very cold weather. All manufacturers of sealants and bonding agents impose temperature restrictions on the use of their products. Only sealants and bonding agents specifically produced for cold weather conditions should be used.

R404.2.1 Identification. All load-bearing lumber shall be identified by the grade mark of a lumber grading or inspection

agency which has been approved by an accreditation body that complies with DOC PS 20. In lieu of a grade mark, a certificate of inspection issued by a lumber grading or inspection agency meeting the requirements of this section shall be accepted. Wood structural panels shall conform to DOC PS 1 or DOC PS 2 and shall be identified by a grade mark or certificate of inspection issued by an approved agency.

❖ See the commentary for Sections R502.1 and R503.2.1 as well as Figures R502.1 and R503.2.1.

R404.2.2 Stud size. The studs used in foundation walls shall be 2-inch by 6-inch (51 mm by 152 mm) members. When spaced 16 inches (406 mm) on center, a wood species with an F_b value of not less than 1,250 pounds per square inch (8612 kPa) as listed in AF&PA/NDS shall be used. When spaced 12 inches (305 mm) on center, an F_b of not less than 875 psi (6029 kPa) shall be required.

❖ This section specifies the minimum bending stress, F_b, for studs based on their spacing in a foundation wall. The appropriate wood species that meet this requirement can be found in the supplement to the AF&PA/NDS: *National Design Specification for Wood Construction.*

R404.2.3 Height of backfill. For wood foundations that are not designed and installed in accordance with AF&PA Report No.7, the height of backfill against a foundation wall shall not exceed 4 feet (1219 mm). When the height of fill is more than 12 inches (305 mm) above the interior grade of a crawl space or floor of a basement, the thickness of the plywood sheathing shall meet the requirements of Table R404.2.3.

❖ Table R404.2.3 specifies requirements for plywood based on height of backfill and plywood span (stud spacing) where the height of retained earth is more than 12 inches (305 mm). See Figure R403.1(2). The backfill limits of Table R404.2.3 do not apply to wood foundations designed and installed in accordance with AF&PA report No. 7.

R404.2.4 Backfilling. Wood foundation walls shall not be backfilled until the basement floor and first floor have been constructed or the walls have been braced. For crawl space construction, backfill or bracing shall be installed on the interior of the walls prior to placing backfill on the exterior.

❖ These provisions require that lateral support be in place prior to backfilling on the exterior.

R404.2.5 Drainage and dampproofing. Wood foundation basements shall be drained and dampproofed in accordance with Sections R405 and R406, respectively.

❖ See Sections R405.2 and R406.3 of this commentary.

R404.2.6 Fastening. Wood structural panel foundation wall sheathing shall be attached to framing in accordance with Table R602.3(1) and Section R402.1.1.

❖ The fastening schedule must be followed for wall sheathing attachment. Also see the commentary for Section R402.1.1.

R404.3 Wood sill plates. Wood sill plates shall be a minimum of 2-inch by 4-inch (51 mm by 102 mm) nominal lumber. Sill plate anchorage shall be in accordance with Sections R403.1.6 and R602.11.

TABLE R404.2.3
PLYWOOD GRADE AND THICKNESS FOR WOOD FOUNDATION CONSTRUCTION
(30 pcf equivalent-fluid weight soil pressure)

HEIGHT OF FILL (inches)	STUD SPACING (inches)	FACE GRAIN ACROSS STUDS			FACE GRAIN PARALLEL TO STUDS		
		Grade[a]	Minimum thickness (inches)	Span rating	Grade[a]	Minimum thickness (inches)[b,c]	Span rating
24	12	B	$15/32$	32/16	A	$15/32$	32/16
					B	$15/32$[c]	32/16
	16	B	$15/32$	32/16	A	$15/32$[c]	32/16
					B	$19/32$[c] (4, 5 ply)	40/20
36	12	B	$15/32$	32/16	A	$15/32$	32/16
					B	$15/32$[c] (4, 5 ply)	32/16
					B	$19/32$ (4, 5 ply)	40/20
	16	B	$15/32$[c]	32/16	A	$19/32$	40/20
					B	$23/32$	48/24
48	12	B	$15/32$	32/16	A	$15/32$[c]	32/16
					B	$19/32$[c] (4, 5 ply)	40/20
	16	B	$19/32$	40/20	A	$19/32$[c]	40/20
					A	$23/32$	48/24

For SI: 1 inch = 25.4 mm, 1 foot = 304.8 mm, 1 pound per cubic foot = 0.1572 kN/m³.

a. Plywood shall be of the following minimum grades in accordance with DOC PS 1 or DOC PS 2:

  1. DOC PS 1 Plywood grades marked:

    1.1. Structural I C-D (Exposure 1)

    1.2. C-D (Exposure 1)

  2. DOC PS 2 Plywood grades marked:

    2.1. Structural I Sheathing (Exposure 1)

    2.2. Sheathing (Exposure 1)

  3. Where a major portion of the wall is exposed above ground and a better appearance is desired, the following plywood grades marked exterior are suitable:

    3.1. Structural I A-C, Structural I B-C or Structural I C-C (Plugged) in accordance with DOC PS 1

    3.2. A-C Group 1, B-C Group 1, C-C (Plugged) Group 1 or MDO Group 1 in accordance with DOC PS 1

    3.3. Single Floor in accordance with DOC PS 1 or DOC PS 2

b. Minimum thickness $15/32$ inch, except crawl space sheathing may be $3/8$ inch for face grain across studs 16 inches on center and maximum 2-foot depth of unequal fill.

c. For this fill height, thickness and grade combination, panels that are continuous over less than three spans (across less than three stud spacings) require blocking 16 inches above the bottom plate. Offset adjacent blocks and fasten through studs with two 16d corrosion-resistant nails at each end.

❖ This table specifies the minimum grade and thickness of plywood based on the height of retained earth (fill) and the plywood span.

❖ Minimum 2-inch by 4-inch (51 mm by 102 mm) nominal sill plates are required, and they should be anchored to the foundation as required by Section R403.1.6.

R404.4 Insulating concrete form foundation walls. Insulating concrete form (ICF) foundation walls shall be designed and constructed in accordance with the provisions of this section or in accordance with the provisions of ACI 318. When ACI 318 or the provisions of this section are used to design insulating concrete form foundation walls, project drawings, typical details and specifications are not required to bear the seal of the architect or engineer responsible for design unless otherwise required by the state law of the jurisdiction having authority.

❖ This concrete forming system uses stay-in-place, rigid foam plastic forms for basement wall construction. The three systems covered in the code are flat wall, waffle-grid, and screen-grid. This section contains prescriptive requirements, and ACI 318 may also be used for design and construction of insulating concrete form foundation walls.

R404.4.1 Applicability limits. The provisions of this section shall apply to the construction of insulating concrete form foundation walls for buildings not more than 60 feet (18 288 mm) in plan dimensions, and floors not more than 32 feet (9754 mm) or roofs not more than 40 feet (12 192 mm) in clear span. Buildings shall not exceed two stories in height above grade with each story not more than 10 feet (3048 mm) high. Foundation walls constructed in accordance with the provisions of this section shall be limited to buildings subjected to a maximum ground snow load of 70 psf (3.35 kN/m²) and located in Seismic Design Category A, B or C. In Seismic Design Categories D_0, D_1 and D_2, foundation walls shall comply with Section R404.1.4. Insulating concrete form foundation walls supporting above-grade concrete walls shall be reinforced as required for the above-grade wall immediately above or the requirements in Tables R404.4(1), R404.4(2), R404.4(3), R404.4(4) or R404.4(5), whichever is greater.

❖ Figure R611.2 illustrates the applicability limits for buildings using insulating concrete formwall construction. In Chapter 6, the use of insulating concrete form walls is also limited to wind speeds of 150 mph (67 m/s) or less.

Section R611.2 limits the use of ICF walls to Seismic Design Categories A, B, C, D_0, D_1, and D_2. However, ICF foundation walls constructed in accordance with the provisions of Section R404.4, are limited to Seismic Design Categories A, B, and C. ICF foundation walls located in Seismic Design Categories D_0, D_1 and D_2 must comply with Section R404.1.4.

R404.4.2 Flat insulating concrete form wall systems. Flat ICF wall systems shall comply with Figure R611.3, shall have a minimum concrete thickness of 5.5 inches (140 mm), and shall have reinforcement in accordance with Table R404.4(1), R404.4(2) or R404.4(3). Alternatively, for 7.5-inch (191 mm) and 9.5-inch (241 mm) flat ICF wall systems, use of Table R404.1.1(5) shall be permitted, provided the vertical reinforcement is of the grade and located within the wall as required by that table.

❖ Figure R611.3 illustrates flat-insulating concrete form walls. They must be in accordance with the limits and

requirements of Tables R404.4(1) through R404.4(3). These tables are based on 2500 pounds per square inch (17.2 MPa) concrete and reinforcing steel with a minimum yield strength of 40,000 pounds per square inch (275.8 MPa). When reinforcing steel with a minimum yield strength or 60,000 pounds per square inch (413.7 MPa) is used, the spacing of the reinforcement can be increased as permitted by Note "a."

Table R404.1.1(5) is permitted to be used for 7.5 inch and 9.5 inch (191 and 241 mm) ICF foundation walls. The reinforcing steel must have a minimum yield strength of 60,000 pounds per square inch (413.7 MPa) and the vertical reinforcement must be located in accordance with Note "d." Tables R404.4(1), (2) and (3) are based on the assumption that the vertical reinforcement will be placed in the center of the walls. This is conservative. Table R404.1.1(5) requires that the vertical reinforcement be placed near the inside surface of the wall (away from the soil side) where flexural tensile stresses are the greatest. This alternate provides a more efficient use of the steel and concrete.

R404.4.3 Waffle-grid insulating concrete form wall systems. Waffle-grid wall systems shall have a minimum nominal concrete thickness of 6 inches (152 mm) for the horizontal and vertical concrete members (cores) and shall be reinforced in accordance with Table R404.4(4). The minimum core dimension shall comply with Table R611.2 and Figure R611.4.

❖ Figure R611.4 illustrates waffle-grid wall systems. They must be constructed in accordance with Table R404.4(4).

R404.4.4 Screen-grid insulating concrete form wall systems. Screen-grid ICF wall systems shall have a minimum nominal concrete thickness of 6 inches (152 mm) for the horizontal and vertical concrete members (cores). The minimum core dimensions shall comply with Table R611.2 and Figure R611.5. Walls shall have reinforcement in accordance with Table R404.4(5).

❖ Figure R611.5 illustrates screen-grid wall systems. They must be constructed in accordance with Table R404.4(5).

R404.4.5 Concrete material. Ready-mixed concrete for insulating concrete form walls shall be in accordance with Section R402.2. Maximum slump shall not be greater than 6 inches (152 mm) as determined in accordance with ASTM C 143. Maximum aggregate size shall not be larger than $^3/_4$ inch (19.1 mm).

Exception: Concrete mixes conforming to the ICF manufacturer's recommendations.

❖ The requirements for concrete in Section R402.2 must be followed for insulating concrete form walls. In addition, this section specifies maximum slump and aggregate size. Insulating concrete form manufacturer's recommendations for concrete mixes are also acceptable.

R404.4.6 Reinforcing steel.

R404.4.6.1 General. Reinforcing steel shall meet the requirements of ASTM A 615, A 706 or A 996. The minimum yield strength of reinforcing steel shall be 40,000 psi (Grade 40) (276 MPa). Vertical and horizontal wall reinforcements shall be placed no closer to the outside face of the wall than one-half the wall thickness. Steel reinforcement for foundation walls shall have concrete cover in accordance with ACI 318.

Exception: Where insulated concrete forms are used and the form remains in place as cover for the concrete, the minimum concrete cover for the reinforcing steel is permitted to be reduced to $^3/_4$ inch (19.1 mm).

❖ This section specifies the minimum grades of reinforcing steel that may be used. The cover must be in accordance with ACI 318, and the cover on the exterior wall face must be at least one-half of the wall thickness.

TABLE R404.4(1)
5.5-INCH THICK FLAT ICF FOUNDATION WALLS[a, b, c, d]

HEIGHT OF BASEMENT WALL (feet)	MAXIMUM UNBALANCED BACKFILL HEIGHT[e] (feet)	MINIMUM VERTICAL REINFORCEMENT SIZE AND SPACING		
		Soil classes[f] and design lateral soil load (psf per foot of depth)		
		GW, GP, SW and SP 30	GM, GC, SM, SM-SC and ML 45	SC, ML-CL and inorganic CL 60
8	4	#4@48″	#4@48″	#4@48″
	5	#4@48″	#3@12″; #4@22″; #5@32″	#3@8″; #4@14″; #5@20″; #6@26″
	6	#3@12″; #4@22″; #5@30″	#3@8″; #4@14″; #5@20″; #6@24″	#3@6″; #4@10″: #5@14″; #6@20″
	7	#3@8″; #4@14″; #5@22″; #6@26″	#3@5″; #4@10″; #5@14″; #6@18″	#3@4″; #4@6″; #5@10″; #6@14″
9	4	#4@48″	#4@48″	#4@48″
	5	#4@48″	#3@12″; #4@20″; #5@28″; #6@36″	#3@8″; #4@14″; #5@20″; #6@22″
	6	#3@10″; #4@20″; #5@28″; #6@34″	#3@6″; #4@12″; #5@18″; #6@20″	#4@8″; #5@14″; #6@16″
	7	#3@8″; #4@14″; #5@20″; #6@22″	#4@8″; #5@12″; #6@16″	#4@6″; #5@10″; #6@12″
	8	#3@6″; #4@10″; #5@14″; #6@16″	#4@6″; #5@10″; #6@12″	#4@4″; #5@6″; #6@8″
10	4	#4@48″	#4@48″	#4@48″
	5	#4@48″	#3@10″; #4@18″; #5@26″; #6@30″	#3@6″; #4@14″; #5@18″; #6@20″
	6	#3@10″; #4@18″; #5@24″; #6@30″	#3@6″; #4@12″; #5@16″; #6@18″	#3@4″; #4@8″; #5@12″; #6@14″
	7	#3@6″; #4@12″; #5@16″; #6@18″	#3@4″; #4@8″; #5@12″	#4@6″; #5@8″; #6@10″
	8	#4@8″; #5@12″; #6@14″	#4@6″; #5@8″; #6@12″	#4@4″; #5@6″; #6@8″
	9	#4@6″; #5@10″; #6@12″	#4@4″; #5@6″; #6@8″	#5@4″; #6@6″

For SI: 1 inch = 25.4 mm, 1 foot = 304.8 mm, 1 pound per square inch = 6.895 kPa, 1 pound per square foot = 0.0479 kPa.

a. This table is based on concrete with a minimum specified concrete strength of 2500 psi, reinforcing steel with a minimum yield strength of 40,000 psi. When reinforcing steel with a minimum yield strength of 60,000 psi is used, the spacing of the reinforcement shall be increased to 1.5 times the spacing value in the table but in no case greater than 48 inches on center.

b. This table is not intended to prohibit the use of an ICF manufacturer's tables based on engineering analysis in accordance with ACI 318.

c. Deflection criteria: $L/240$.

d. Interpolation between rebar sizes and spacing is not permitted.

e. Unbalanced backfill height is the difference in height of the exterior and interior finished ground. Where an interior concrete slab is provided, the unbalanced backfill height shall be measured from the exterior finished ground level to the top of the interior concrete slab.

f. Soil classes are in accordance with the Unified Soil Classification System. Refer to Table R405.1.

❖ Tables R404.4(1) through R404.4(3) specify vertical reinforcing requirements for flat insulating concrete form walls based on the basement wall height, unbalanced backfill height and the soil group as determined in accordance with Table R405.1.

R404.4.6.2 Horizontal reinforcement. When vertical reinforcement is required, ICF foundation walls shall have horizontal reinforcement in accordance with this section. ICF foundation walls up to 8 feet (2438 mm) in height shall have a minimum of one continuous No. 4 horizontal reinforcing bar placed at 48 inches (1219 mm) on center with one bar located within 12 inches (305 mm) of the top of the wall story. ICF Foundation walls greater than 8 feet (2438 mm) in height shall have a minimum of one continuous No. 4 horizontal reinforcing bar placed at 36 inches (914 mm) on center with one bar located within 12 inches (305 mm) of the top of the wall story.

❖ Where the insulating concrete form wall tables indicate a requirement for vertical reinforcement, horizontal reinforcement must also be provided in accordance with this section.

R404.4.6.3 Wall openings. Vertical wall reinforcement required by Section R404.4.2, R404.4.3 or R404.4.4 that is interrupted by wall openings shall have additional vertical reinforcement of the same size placed within 12 inches (305 mm) of each side of the opening.

❖ If a wall opening interrupts reinforcement, additional reinforcement of equivalent area must be provided. Although the code doesn't elaborate, a good practice is to distribute this equivalent reinforcing symmetrically on either side of the opening.

R404.4.7 Foam plastic insulation. Foam plastic insulation in insulating concrete foam construction shall comply with this section.

❖ Foam plastic insulation must be installed in accordance with Section R314.

R404.4.7.1 Material. Insulating concrete form material shall meet the surface burning characteristics of Section R314.3. A thermal barrier shall be provided on the building interior in accordance with Section R314.4.

❖ See the commentary for Section R314.3 and R314.4.

TABLE R404.4(2)
7.5-INCH-THICK FLAT ICF FOUNDATION WALLS[a, b, c, d, e]

HEIGHT OF BASEMENT WALL (feet)	MAXIMUM UNBALANCED BACKFILL HEIGHT[f] (feet)	MINIMUM VERTICAL REINFORCEMENT SIZE AND SPACING		
		Soil classes[g] and design lateral soil load (psf per foot of depth)		
		GW, GP, SW and SP 30	GM, GC, SM, SM-SC and ML 45	SC, ML-CL and inorganic CL 60
8	6	N/R	N/R	#3@6"; #4@12"; #5@18"; #6@24"
	7	N/R	#3@8"; #4@14"; #5@20"; #6@28"	#3@6"; #4@10"; #5@16"; #6@20"
9	6	N/R	N/R	#3@8"; #4@14"; #5@20"; #6@28"
	7	N/R	#3@6"; #4@12"; #5@18"; #6@26"	#3@4"; #4@8"; #5@14"; #6@18"
	8	#3@8"; #4@14"; #5@22"; #6@28"	#3@4"; #4@8"; #5@14"; #6@18"	#3@4"; #4@6"; #5@10"; #6@14"
10	6	N/R	N/R	#3@6"; #4@12"; #5@18"; #6@26"
	7	N/R	#3@6"; #4@12"; #5@18"; #6@24"	#3@4"; #4@8"; #5@12"; #6@18"
	8	#3@6"; #4@12"; #5@20"; #6@26"	#3@4"; #4@8"; #5@12"; #6@16"	#3@4"; #4@6"; #5@8"; #6@12"
	9	#3@6"; #4@10"; #5@14"; #6@20"	#3@4"; #4@6"; #5@10"; #6@12"	#4@4"; #5@6"; #6@10"

For SI: 1 inch = 25.4 mm, 1 foot = 304.8 mm, 1 pound per square inch = 6.895 kPa, 1 pound per square foot = 0.0479 kPa.

a. This table is based on concrete with a minimum specified concrete strength of 2500 psi, reinforcing steel with a minimum yield strength of 40,000 psi. When reinforcing steel with a minimum yield strength of 60,000 psi is used, the spacing of the reinforcement shall be increased to 1.5 times the spacing value in the table.

b. This table is not intended to prohibit the use of an ICF manufacturer's tables based on engineering analysis in accordance with ACI 318.

c. N/R denotes "not required."

d. Deflection criteria: $L/240$.

e. Interpolation between rebar sizes and spacing is not permitted.

f. Unbalanced backfill height is the difference in height of the exterior and interior finished ground. Where an interior concrete slab is provided, the unbalanced backfill height shall be measured from the exterior finished ground level to the top of the interior concrete slab.

g. Soil classes are in accordance with the Unified Soil Classification System. Refer to Table R405.1.

❖ See the commentary for Table R404.4(1).

TABLE R404.4(3)
9.5-INCH-THICK FLAT ICF FOUNDATION WALLS[a, b, c, d, e]

HEIGHT OF BASEMENT WALL (feet)	MAXIMUM UNBALANCED BACKFILL HEIGHT[f] (feet)	MINIMUM VERTICAL REINFORCEMENT SIZE AND SPACING		
		Soil classes[g] and design lateral soil load (psf per foot of depth)		
		GW, GP, SW and SP 30	GM, GC, SM, SM-SC and ML 45	SC, ML-CL and inorganic CL 60
8	7	N/R	N/R	N/R
9	6	N/R	N/R	N/R
	7	N/R	N/R	#3@6"; #4@12"; #5@18"; #6@26"
	8	N/R	#3@6"; #4@12"; #5@18"; #6@26"	#3@4"; #4@8"; #5@14"; #6@18"
10	5	N/R	N/R	N/R
	6	N/R	N/R	N/R
	7	N/R	N/R	#3@6"; #4@10"; #5@18"; #6@24"
	8	N/R	#3@6"; #4@12"; #5@16"; #6@24"	#3@4"; #4@8"; #5@12"; #6@16"
	9	#3@4"; #4@10"; #5@14"; #6@20"	#3@4"; #4@8"; #5@12"; #6@18"	#3@4"; #4@6"; #5@10"; #6@12"

For SI: 1 inch = 25.4 mm, 1 foot = 304.8 mm, 1 pound per square inch = 6.895 kPa, 1 pound per square foot = 0.0479 kPa.

a. This table is based on concrete with a minimum specified concrete strength of 2500 psi, reinforcing steel with a minimum yield strength of 40,000 psi. When reinforcing steel with a minimum yield strength of 60,000 psi is used, the spacing of the reinforcement shall be increased to 1.5 times the spacing value in the table.

b. This table is not intended to prohibit the use of an ICF manufacturer's tables based on engineering analysis in accordance with ACI 318.

c. N/R denotes "not required."

d. Deflection criteria: $L/240$.

e. Interpolation between rebar sizes and spacing is not permitted.

f. Unbalanced backfill height is the difference in height of the exterior and interior finished ground. Where an interior concrete slab is provided, the unbalanced backfill height shall be measured from the exterior finished ground level to the top of the interior concrete slab.

g. Soil classes are in accordance with the Unified Soil Classification System. Refer to Table R405.1.

❖ See the commentary for Table R404.4(1).

R404.4.7.2 Termite hazards. In areas where hazard of termite damage is very heavy in accordance with Figure R301.2(6), foam plastic insulation shall be permitted below grade on foundation walls in accordance with one of the following conditions:

1. When in addition to the requirements in Section R320.1, an approved method of protecting the foam plastic and structure from subterranean termite damage is provided.

2. The structural members of walls, floors, ceilings and roofs are entirely of noncombustible materials or pressure preservatively treated wood.

3. On the interior side of basement walls.

❖ Where the termite hazard is very heavy, foam plastic insulation may be installed below grade only if it installed on the interior or if all walls, floors and ceilings are noncombustible material or treated wood·or in accordance with Section R320.1. See the commentary for Section R404.3.4.

R404.4.8 Foundation wall thickness based on walls supported. The thickness of ICF foundation walls shall not be less than the thickness of the wall supported above.

❖ The thickness of insulating concrete form foundation walls must not be less then that of the wall supported.

R404.4.9 Height above finished ground. ICF foundation walls shall extend above the finished ground adjacent to the foundation at all points a minimum of 4 inches (102 mm) where masonry veneer is used and a minimum of 6 inches (152 mm) elsewhere.

❖ See the commentary for Section R404.1.6.

R404.4.10 Backfill placement. Backfill shall be placed in accordance with Section R404.1.7.

❖ See the commentary for Section R404.1.7.

R404.4.11 Drainage and dampproofing/waterproofing. ICF foundation basements shall be drained and dampproofed/waterproofed in accordance with Sections R405 and R406.

❖ See the requirements for concrete walls in Sections R405 and R406 of this commentary.

R404.5 Retaining walls. Retaining walls that are not laterally supported at the top and that retain in excess of 24 inches (610 mm) of unbalanced fill shall be designed to ensure stability against overturning, sliding, excessive foundation pressure and water uplift. Retaining walls shall be designed for a safety factor of 1.5 against lateral sliding and overturning.

❖ This section provides design considerations for a retaining wall that retains in excess of 24 inches (610 mm) of unbalanced fill. The lateral pressure of the soil against the retaining wall is greatly influenced by soil moisture. Backfill is usually kept from being saturated for an extended length of time by placing drains near the base of the retaining wall to remove the water in the soil behind it.

TABLE R404.4(4)
WAFFLE GRID ICF FOUNDATION WALLS[a, b, c, d, e]

MINIMUM NOMINAL WALL THICKNESS[f] (inches)	HEIGHT OF BASEMENT WALL (feet)	MAXIMUM UNBALANCED BACKFILL HEIGHT[g] (feet)	MINIMUM VERTICAL REINFORCEMENT SIZE AND SPACING Soil classes[h] and design lateral soil load (psf per foot of depth)		
			GW, GP, SW and SP 30	GM, GC, SM, SM-SC and ML 45	SC, ML-CL and inorganic CL 60
6	8	4	#4@48″	#3@12″; #4@24″	#3@12″
		5	#3@12″; #5@24″	#4@12″	#7@12″
		6	#4@12″	Design required	Design required
		7	#7@12″	Design required	Design required
	9	4	#4@48″	#3@12″; #5@24″	#3@12″
		5	#3@12″	#4@12″	Design required
		6	#5@12″	Design required	Design required
		7	Design required	Design required	Design required
	10	4	#4@48″	#4@12″	#5@12″
		5	#3@12″	Design required	Design required
		6	Design required	Design required	Design required
		7	Design required	Design required	Design required
8	8	4	N/R	N/R	N/R
		5	N/R	#3@12″; #4@24″; #5@36″	#3@12″; #5@24″
		6	#3@12″; #4@24″; #5@36″	#4@12″; #5@24″	#4@12″
		7	#3@12″; #6@24″	#4@12″	#5@12″
	9	4	N/R	N/R	N/R
		5	N/R	#3@12″; #5@24″	#3@12″; #5@24″
		6	#3@12″; #4@24″	#4@12″	#4@12″
		7	#4@12″; #5@24″	#5@12″	#5@12″
		8	#4@12″	#5@12″	#8@12″
	10	4	N/R	#3@12″; #4@24″; #6@36″	#3@12″; #5@24″
		5	N/R	#3@12″; #4@24″; #6@36″	#4@12″; #5@24″
		6	#3@12″; #5@24″	#4@12″	#5@12″
		7	#4@12″	#5@12″	#6@12″
		8	#4@12″	#6@12″	Design required
		9	#5@12″	Design required	Design required

For SI: 1 inch = 25.4 mm, 1 foot = 304.8 mm, 1 pound per square inch = 6.895 kPa, 1 pound per square foot = 0.0479 kPa.

a. This table is based on concrete with a minimum specified concrete strength of 2500 psi, reinforcing steel with a minimum yield strength of 40,000 psi. When reinforcing steel with a minimum yield strength of 60,000 psi is used, the spacing of the reinforcement shall be increased 12 inches but in no case greater than 48 inches on center.

b. This table is not intended to prohibit the use of an ICF manufacturer's tables based on engineering analysis in accordance with ACI 318.

c. N/R denotes "not required."

d. Deflection criteria: L/240.

e. Interpolation between rebar sizes and spacing is not permitted.

f. Refer to Table R611.4(2) for wall dimensions.

g. Unbalanced backfill height is the difference in height of the exterior and interior finished ground. Where an interior concrete slab is provided, the unbalanced backfill height shall be measured from the exterior finished ground level to the top of the interior concrete slab.

h. Soil classes are in accordance with the Unified Soil Classification System. Refer to Table R405.1.

❖ This table specifies vertical reinforcement requirements for waffle-grid basement walls based on the height and thickness of the wall, the height of unbalanced backfill and the soil group as determined in accordance with Table R405.1. Several table entries note that a design is required.

TABLE R404.4(5)
SCREEN-GRID ICF FOUNDATION WALLS[a, b c, d, e]

MINIMUM NOMINAL WALL THICKNESS[f] (inches)	HEIGHT OF BASEMENT WALL (feet)	MAXIMUM UNBALANCED BACKFILL HEIGHT[g] (feet)	MINIMUM VERTICAL REINFORCEMENT SIZE AND SPACING		
			Soil classes[h] and design lateral soil load (psf per foot of depth)		
			GW, GP, SW and SP 30	GM, GC, SM, SM-SC and ML 45	SC, ML-CL and inorganic CL 60
6	8	4	#4@48″	#3@12″; #4@24″; #5@36″	#3@12″; #5@24″
		5	#3@12″; #4@24″	#3@12″	#4@12″
		6	#4@12″	#5@12″	Design required
		7	#4@12″	Design required	Design required
	9	4	#4@48″	#3@12″; #4@24″	#3@12″; #6@24″
		5	#3@12″; #5@24″	#4@12″	#7@12″
		6	#4@12″	Design required	Design required
		7	Design required	Design required	Design required
		8	Design required	Design required	Design required
	10	4	#4@48″	#3@12″; #5@24″	#3@12″
		5	#3@12″	#4@12″	#7@12″
		6	#4@12″	Design required	Design required
		7	Design required	Design required	Design required
		8	Design required	Design required	Design required

For SI: 1 inch = 25.4 mm, 1 foot = 304.8 mm, 1 pound per square inch = 6.895 kPa, 1 pound per square foot = 0.0479 kPa.

a. This table is based on concrete with a minimum specified concrete strength of 2500 psi, reinforcing steel with a minimum yield strength of 40,000 psi. When reinforcing steel with a minimum yield strength of 60,000 psi is used, the spacing of the reinforcement in the shaded cells shall be increased 12 inches.

b. This table is not intended to prohibit the use of an ICF manufacturer's tables based on engineering analysis in accordance with ACI 318.

c. N/R denotes "not required."

d. Deflection criteria: *L*/240.

e. Interpolation between rebar sizes and spacing is not permitted.

f. Refer to Table R611.4(2) for wall dimensions.

g. Unbalanced backfill height is the difference in height of the exterior and interior finished ground. Where an interior concrete slab is provided, the unbalanced backfill height shall be measured from the exterior finished ground level to the top of the interior concrete slab.

h. Soil classes are in accordance with the Unified Soil Classification System. Refer to Table R405.1.

❖ This table provides insulating concrete form screen-grid wall requirements and is similar to Table R404.4(4).

SECTION R405
FOUNDATION DRAINAGE

R405.1 Concrete or masonry foundations. Drains shall be provided around all concrete or masonry foundations that retain earth and enclose habitable or usable spaces located below grade. Drainage tiles, gravel or crushed stone drains, perforated pipe or other approved systems or materials shall be installed at or below the area to be protected and shall discharge by gravity or mechanical means into an approved drainage system. Gravel or crushed stone drains shall extend at least 1 foot (305 mm) beyond the outside edge of the footing and 6 inches (152 mm) above the top of the footing and be covered with an approved filter membrane material. The top of open joints of drain tiles shall be protected with strips of building paper, and the drainage tiles or perforated pipe shall be placed on a minimum of 2 inches (51 mm) of washed gravel or crushed rock at least one sieve size larger than the tile joint opening or perforation and covered with not less than 6 inches (152 mm) of the same material.

Exception: A drainage system is not required when the foundation is installed on well-drained ground or sand-gravel mixture soils according to the Unified Soil Classification System, Group I Soils, as detailed in Table R405.1.

❖ To allow free groundwater that may be present adjacent to the foundation wall to be drained away, drains are usually placed around houses to remove water and prevent leakage into habitable or usable spaces below grade. The top of the gravel or crushed stone drain, with or without drainage tile or perforated pipe, must be covered with an approved filter membrane. Drainage tiles are extremely important in areas having moderate to heavy rainfall and in soils having a low percolation rate. The filter membrane allows water to pass through the perimeter drain tiles or pipes without allowing, or at least greatly reducing, the possibility of fine soil materials entering the drainage system. A few examples of such systems are illustrated in Commentary Figures R405.1(1) and R405.1(2). The exception allows omission of the drainage system in Group I soils, which by definition exhibit good drainage characteristics as shown in Table R405.1.

R405.2 Wood foundations. Wood foundations enclosing habitable or usable spaces located below grade shall be adequately drained in accordance with Sections R405.2.1 through R405.2.3.

❖ As does Section R405.1, this section gives specific drainage requirements for wood foundations. Also refer to the commentary for Section R401.1.

R405.2.1 Base. A porous layer of gravel, crushed stone or coarse sand shall be placed to a minimum thickness of 4 inches (102 mm) under the basement floor. Provision shall be made for automatic draining of this layer and the gravel or crushed stone wall footings.

❖ See Figure R403.1(2). This requirement for drainage under the basement floor is similar to those in Sections R504.2.1 (wood floor) and R406.2.2 (concrete floor).

R405.2.2 Moisture barrier. A 6-mil-thick (0.15 mm) polyethylene moisture barrier shall be applied over the porous layer with the basement floor constructed over the polyethylene.

❖ See Figure R403.1(2). Also see the commentary for Section R506.2.3.

R405.2.3 Drainage system. In other than Group I soils, a sump shall be provided to drain the porous layer and footings. The sump shall be at least 24 inches (610 mm) in diameter or 20 inches square (0.0129 m²), shall extend at least 24 inches (610 mm) below the bottom of the basement floor and shall be capable of positive gravity or mechanical drainage to remove any accumulated water. The drainage system shall discharge into an approved sewer system or to daylight.

❖ Except for Group I soils, which exhibit good drainage (see Table R405.1), a sump is necessary to drain the

TABLE R405.1
PROPERTIES OF SOILS CLASSIFIED ACCORDING TO THE UNIFIED SOIL CLASSIFICATION SYSTEM

SOIL GROUP	UNIFIED SOIL CLASSIFICATION SYSTEM SYMBOL	SOIL DESCRIPTION	DRAINAGE CHARACTERISTICS[a]	FROST HEAVE POTENTIAL	VOLUME CHANGE POTENTIAL EXPANSION[b]
Group I	GW	Well-graded gravels, gravel sand mixtures, little or no fines	Good	Low	Low
	GP	Poorly graded gravels or gravel sand mixtures, little or no fines	Good	Low	Low
	SW	Well-graded sands, gravelly sands, little or no fines	Good	Low	Low
	SP	Poorly graded sands or gravelly sands, little or no fines	Good	Low	Low
	GM	Silty gravels, gravel-sand-silt mixtures	Good	Medium	Low
	SM	Silty sand, sand-silt mixtures	Good	Medium	Low
Group II	GC	Clayey gravels, gravel-sand-clay mixtures	Medium	Medium	Low
	SC	Clayey sands, sand-clay mixture	Medium	Medium	Low
	ML	Inorganic silts and very fine sands, rock flour, silty or clayey fine sands or clayey silts with slight plasticity	Medium	High	Low
	CL	Inorganic clays of low to medium plasticity, gravelly clays, sandy clays, silty clays, lean clays	Medium	Medium	Medium to Low
Group III	CH	Inorganic clays of high plasticity, fat clays	Poor	Medium	High
	MH	Inorganic silts, micaceous or diatomaceous fine sandy or silty soils, elastic silts	Poor	High	High
Group IV	OL	Organic silts and organic silty clays of low plasticity	Poor	Medium	Medium
	OH	Organic clays of medium to high plasticity, organic silts	Unsatisfactory	Medium	High
	Pt	Peat and other highly organic soils	Unsatisfactory	Medium	High

For SI: 1 inch = 25.4 mm.

a. The percolation rate for good drainage is over 4 inches per hour, medium drainage is 2 inches to 4 inches per hour, and poor is less than 2 inches per hour.

b. Soils with a low potential expansion typically have a plasticity index (PI) of 0 to 15, soils with a medium potential expansion have a PI of 10 to 35 and soils with a high potential expansion have a PI greater than 20.

❖ This table provides soil properties based on the soil's classification in accordance with the United Soil Classification System. This method of identifying soil is determined in ASTM D 2487, *Standard Classification of Soils for Engineering Purposes.*

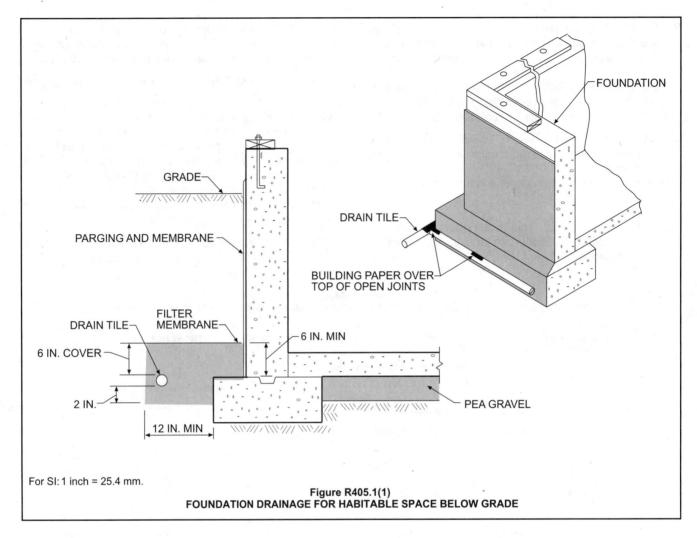

For SI: 1 inch = 25.4 mm.

Figure R405.1(1)
FOUNDATION DRAINAGE FOR HABITABLE SPACE BELOW GRADE

footing (Section R403.2) and porous layer (Section R405.2.1). Also see Figure R403.1(2).

SECTION R406
FOUNDATION WATERPROOFING AND DAMPPROOFING

R406.1 Concrete and masonry foundation dampproofing. Except where required by Section R406.2 to be waterproofed, foundation walls that retain earth and enclose interior spaces and floors below grade shall be dampproofed from the top of the footing to the finished grade. Masonry walls shall have not less than $^3/_8$ inch (9.5 mm) portland cement parging applied to the exterior of the wall. The parging shall be dampproofed in accordance with one of the following:

1. Bituminous coating.

2. 3 pounds per square yard (1.63 kg/m²) of acrylic modified cement.

3. 1/8-inch (3.2 mm) coat of surface-bonding cement complying with ASTM C 887.

4. Any material permitted for waterproofing in Section R406.2.

5. Other approved methods or materials.

> **Exception:** Parging of unit masonry walls is not required where a material is approved for direct application to the masonry.

Concrete walls shall be dampproofed by applying any one of the above listed dampproofing materials or any one of the waterproofing materials listed in Section R406.2 to the exterior of the wall.

❖ To minimize moisture in the form of water vapor from entering below-ground spaces from the outside, dampproofing of the exterior foundation walls is necessary, unless waterproofing is required by Section R406.2. Although the terms "waterproofing" and "dampproofing" both relate to moisture protection, they are sometimes misapplied in the sense that "waterproofing" is used when "dampproofing" is really meant. Dampproofing does not give the same degree of moisture protection as does waterproofing. Methods of dampproofing for masonry and concrete foundation walls are shown in Commentary Figures R406.1(1) and R406.1(2).

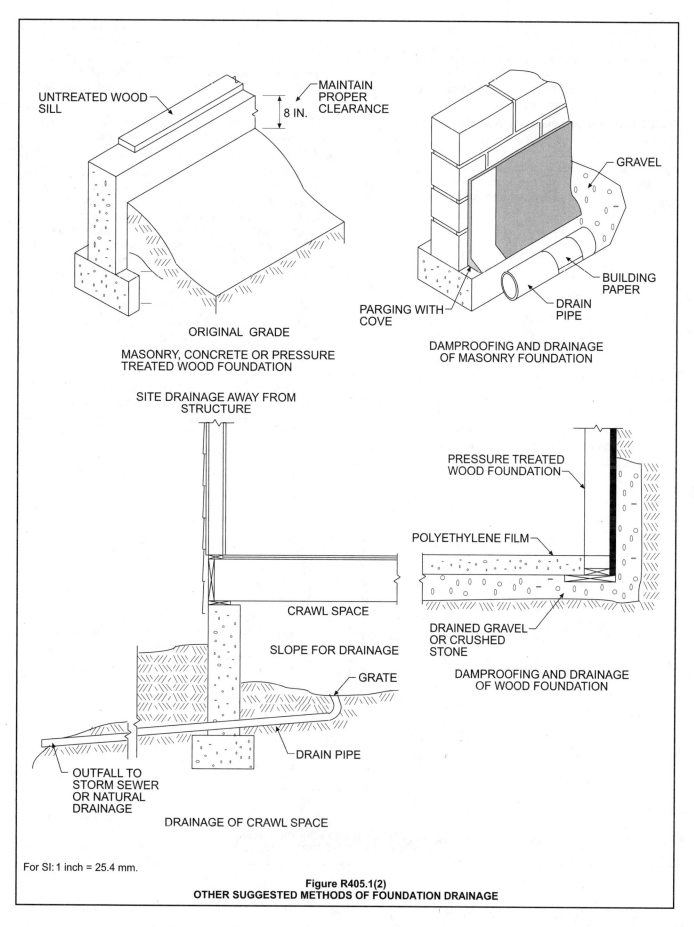

For SI: 1 inch = 25.4 mm.

Figure R405.1(2)
OTHER SUGGESTED METHODS OF FOUNDATION DRAINAGE

Dampproofing installations generally consist of the application of one or more coatings of impervious compounds that are intended to prevent the passage of water vapor through walls or other building elements. Dampproofing may also restrict the flow of water under slight pressure.

R406.2 Concrete and masonry foundation waterproofing. In areas where a high water table or other severe soil-water conditions are known to exist, exterior foundation walls that retain earth and enclose interior spaces and floors below grade shall be waterproofed from the top of the footing to the finished grade. Walls shall be waterproofed in accordance with one of the following:

1. 2-ply hot-mopped felts.

2. 55 pound (25 kg) roll roofing.

3. 6-mil (0.15 mm) polyvinyl chloride.

4. 6-mil (0.15 mm) polyethylene.

5. 40-mil (1 mm) polymer-modified asphalt.

6. 60-mil (1.5 mm) flexible polymer cement.

7. $^1/_8$ inch (3 mm) cement-based, fiber-reinforced, waterproof coating.

8. 60-mil (0.22 mm) solvent-free liquid-applied synthetic rubber.

Exception: Organic-solvent-based products such as hydrocarbons, chlorinated hydrocarbons, ketones and esters shall not be used for ICF walls with expanded polystyrene form material. Use of plastic roofing cements, acrylic coatings, latex coatings, mortars and pargings to seal ICF walls is permitted. Cold-setting asphalt or hot asphalt shall conform to type C of ASTM D 449. Hot asphalt shall be applied at a temperature of less than 200°F (93°C).

All joints in membrane waterproofing shall be lapped and sealed with an adhesive compatible with the membrane.

❖ Foundation walls that retain earth and enclose interior spaces and floors and extend below groundwater level (seasonal or otherwise), require a positive means of preventing moisture migration. Waterproofing installations consist of the application of a combination of sealing materials and impervious coatings used on walls or other building elements to prevent the passage of moisture in either a vapor or liquid form under conditions of significant hydrostatic pressure. Methods for waterproofing concrete and masonry foundation walls are shown in Commentary Figure R406.2.

R406.3 Dampproofing for wood foundations. Wood foundations enclosing habitable or usable spaces located below grade shall be dampproofed in accordance with Sections R406.3.1 through R406.3.4.

❖ The requirements for dampproofing wood foundation walls are shown in Figure R403.1(2). These requirements for sealing the plywood panel joints, installing a moisture barrier and installing porous fill can minimize

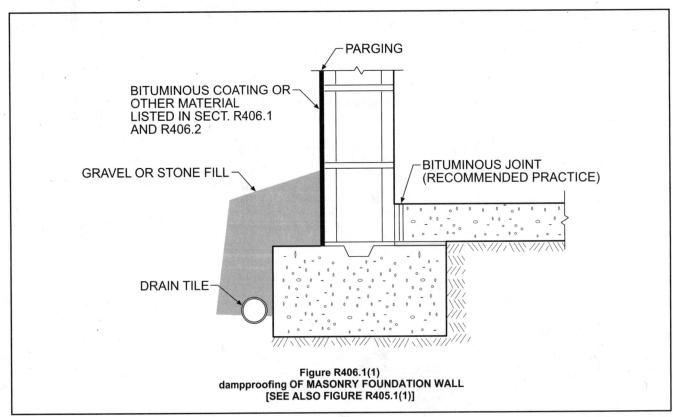

PARGING

BITUMINOUS COATING OR OTHER MATERIAL LISTED IN SECT. R406.1 AND R406.2

GRAVEL OR STONE FILL

BITUMINOUS JOINT (RECOMMENDED PRACTICE)

DRAIN TILE

Figure R406.1(1)
dampproofing OF MASONRY FOUNDATION WALL
[SEE ALSO FIGURE R405.1(1)]

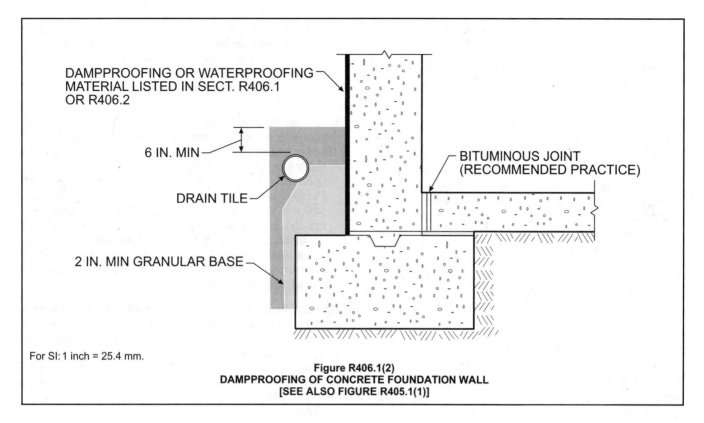

DAMPPROOFING OR WATERPROOFING
MATERIAL LISTED IN SECT. R406.1
OR R406.2

6 IN. MIN

DRAIN TILE

2 IN. MIN GRANULAR BASE

BITUMINOUS JOINT
(RECOMMENDED PRACTICE)

For SI: 1 inch = 25.4 mm.

Figure R406.1(2)
DAMPPROOFING OF CONCRETE FOUNDATION WALL
[SEE ALSO FIGURE R405.1(1)]

the intrusion of water into a basement or usable space. Also refer to the commentary for Section R401.1.

R406.3.1 Panel joint sealed. Plywood panel joints in the foundation walls shall be sealed full length with a caulking compound capable of producing a moisture-proof seal under the conditions of temperature and moisture content at which it will be applied and used.

❖ See the commentary for Section R406.3.

R406.3.2 Below-grade moisture barrier. A 6-mil-thick (0.15 mm) polyethylene film shall be applied over the below-grade portion of exterior foundation walls prior to backfilling. Joints in the polyethylene film shall be lapped 6 inches (152 mm) and sealed with adhesive. The top edge of the polyethylene film shall be bonded to the sheathing to form a seal. Film areas at grade level shall be protected from mechanical damage and exposure by a pressure preservatively treated lumber or plywood strip attached to the wall several inches above finish grade level and extending approximately 9 inches (229 mm) below grade. The joint between the strip and the wall shall be caulked full length prior to fastening the strip to the wall. Other coverings appropriate to the architectural treatment may also be used. The polyethylene film shall extend down to the bottom of the wood footing plate but shall not overlap or extend into the gravel or crushed stone footing.

❖ See the commentary for Section R406.3.

R406.3.3 Porous fill. The space between the excavation and the foundation wall shall be backfilled with the same material used for footings, up to a height of 1 foot (305 mm) above the footing for well-drained sites, or one-half the total back-fill

height for poorly drained sites. The porous fill shall be covered with strips of 30-pound (13.6 kg) asphalt paper or 6-mil (0.15 mm) polyethylene to permit water seepage while avoiding infiltration of fine soils.

❖ See the commentary for Section R406.3.

R406.3.4 Backfill. The remainder of the excavated area shall be backfilled with the same type of soil as was removed during the excavation.

❖ See the commentary for Sections R404.2.4 and R406.3.

SECTION R407
COLUMNS

R407.1 Wood column protection. Wood columns shall be protected against decay as set forth in Section R319.

❖ See the commentary for Section R319.

R407.2 Steel column protection. All surfaces (inside and outside) of steel columns shall be given a shop coat of rust-inhibitive paint, except for corrosion-resistant steel and steel treated with coatings to provide corrosion resistance.

❖ The requirements of this section protect steel columns from the adverse affects of corrosion.

R407.3 Structural requirements. The columns shall be restrained to prevent lateral displacement at the bottom end. Wood columns shall not be less in nominal size than 4 inches by 4 inches (102 mm by 102 mm) and steel columns shall not be less than 3-inch-diameter (76 mm) standard pipe or approved equivalent.

Exception: In Seismic Design Categories A, B and C columns no more than 48 inches (1219 mm) in height on a pier or footing are exempt from the bottom end lateral displacement requirement within underfloor areas enclosed by a continuous foundation.

❖ Minimum sizes for wood columns and steel columns are specified to reduce concerns about the structural capacity of slender columns. To minimize the chance of accidentally displacing columns supporting beams or girders, a means of mechanically anchoring a column is required. Accepted construction details that have performed adequately in the field are typically used in place of a structural analysis and design [see Commentary Figures R407.3(1), R407.3(2), and R407.3(3)]. If prefabricated metal-wood connectors are used, they should be installed in accordance with the manufacturer's installation instructions.

SECTION R408
UNDER-FLOOR SPACE

R408.1 Ventilation. The under-floor space between the bottom of the floor joists and the earth under any building (except space occupied by a basement) shall have ventilation openings through foundation walls or exterior walls. The minimum net area of ventilation openings shall not be less than 1 square foot (0.0929 m²) for each 150 square feet (14 m²) of under-floor space area. One such ventilating opening shall be within 3 feet (914 mm) of each corner of the building.

❖ Raised floor construction results in an under-floor space, commonly referred to as a crawl space. To control condensation within crawl space areas and thus reduce the chance of dry rot, natural ventilation of such spaces by reasonably distributed openings through foundation walls or exterior walls is required. Condensation is a function of the geographical location and the climatic conditions and, thus, the dependence on ventilating openings through the foundation wall or exterior wall may run counter to energy-conservation measures. Commentary Figure R408.1 illustrates the use of openings through the foundation walls.

R408.2 Openings for under-floor ventilation. The minimum net area of ventilation openings shall not be less than 1 square foot (0.0929 m²) for each 150 square feet (14 m²) of under-floor area. One ventilating opening shall be within 3 feet (914 mm) of each corner of the building. Ventilation openings shall be covered for their height and width with any of the following materials provided that the least dimension of the covering shall not exceed ¹/₄ inch (6.4 mm):

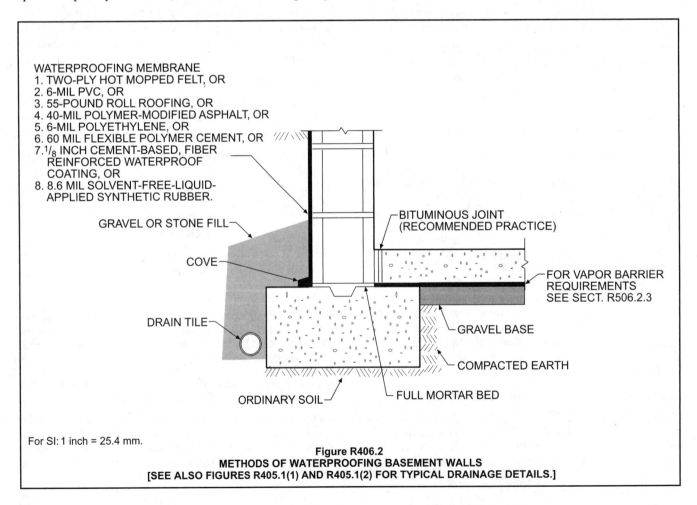

WATERPROOFING MEMBRANE
1. TWO-PLY HOT MOPPED FELT, OR
2. 6-MIL PVC, OR
3. 55-POUND ROLL ROOFING, OR
4. 40-MIL POLYMER-MODIFIED ASPHALT, OR
5. 6-MIL POLYETHYLENE, OR
6. 60 MIL FLEXIBLE POLYMER CEMENT, OR
7. ¹/₈ INCH CEMENT-BASED, FIBER REINFORCED WATERPROOF COATING, OR
8. 8.6 MIL SOLVENT-FREE-LIQUID-APPLIED SYNTHETIC RUBBER.

GRAVEL OR STONE FILL

COVE

DRAIN TILE

BITUMINOUS JOINT
(RECOMMENDED PRACTICE)

FOR VAPOR BARRIER REQUIREMENTS SEE SECT. R506.2.3

GRAVEL BASE

COMPACTED EARTH

ORDINARY SOIL

FULL MORTAR BED

For SI: 1 inch = 25.4 mm.

Figure R406.2
METHODS OF WATERPROOFING BASEMENT WALLS
[SEE ALSO FIGURES R405.1(1) AND R405.1(2) FOR TYPICAL DRAINAGE DETAILS.]

1. Perforated sheet metal plates not less than 0.070 inch (1.8 mm) thick.

2. Expanded sheet metal plates not less than 0.047 inch (1.2 mm) thick.

3. Cast-iron grill or grating.

4. Extruded load-bearing brick vents.

5. Hardware cloth of 0.035 inch (0.89 mm) wire or heavier.

6. Corrosion-resistant wire mesh, with the least dimension being $^1/_8$ inch (3.2 mm).

❖ Installing a covering material over the ventilation opening keeps animals such as rodents or vermin from entering the crawl space. Several options are provided. The area of the ventilation opening accounts for net free area of the covering material used to protect the ventilation opening. For some covering materials the net free area can be up to 50 percent less than the gross area of the ventilation opening.

R408.3 Unvented crawl space. Ventilation openings in under-floor spaces specified in Sections R408.1 and R408.2 shall not be required where:

1. Exposed earth is covered with a continuous vapor retarder. Joints of the vapor retarder shall overlap by 6 inches (152 mm) and shall be sealed or taped. The edges of the vapor retarder shall extend at least 6 inches (152 mm) up the stem wall and shall be attached and sealed to the stem wall; and

2. One of the following is provided for the under-floor space:

   2.1. Continuously operated mechanical exhaust ventilation at a rate equal to 1 cfm (0.47 L/s) for each 50 ft^2 (4.7 m^2) of crawlspace floor area, including an air pathway to the common area (such as a

duct or transfer grille), and perimeter walls insulated in accordance with Section N1102.2.8;

   2.2. Conditioned air supply sized to deliver at a rate equal to 1 cfm (0.47 L/s) for each 50 ft^2 (4.7 m^2) of under-floor area, including a return air pathway to the common area (such as a duct or transfer grille), and perimeter walls insulated in accordance with Section N1102.2.8;

   2.3. Plenum complying with Section M1601.4, if under-floor space is used as a plenum.

❖ This section lists several conditions under which the ventilation openings through the foundation walls are not required. The exposed earth of the under-floor area must be covered with a continuous vapor retarder. If the perimeter walls are insulated, either a continuously operating mechanical exhaust vent or conditioned air must be provided in order to eliminate the ventilation openings.

   Elimination of the ventilation openings is permitted when the under-floor space is used as a plenum.

R408.4 Access. Access shall be provided to all under-floor spaces. Access openings through the floor shall be a minimum of 18 inches by 24 inches (457 mm by 610 mm). Openings through a perimeter wall shall be not less than 16 inches by 24 inches (407 mm by 610 mm). When any portion of the through-wall access is below grade, an areaway not less than 16 inches by 24 inches (407 mm by 610 mm) shall be provided. The bottom of the areaway shall be below the threshold of the access opening. Through wall access openings shall not be located under a door to the residence. See Section M1305.1.4 for access requirements where mechanical equipment is located under floors.

❖ The provisions of this section require access to all under-floor spaces. Access is required for continuing

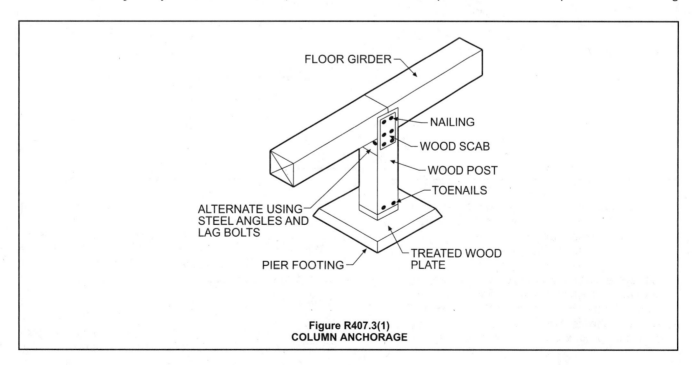

Figure R407.3(1)
COLUMN ANCHORAGE

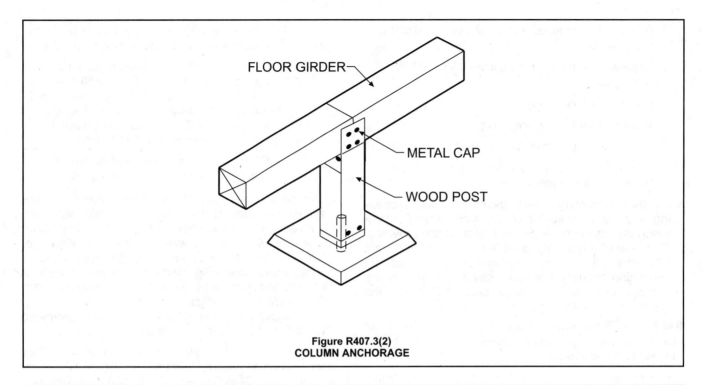

Figure R407.3(2)
COLUMN ANCHORAGE

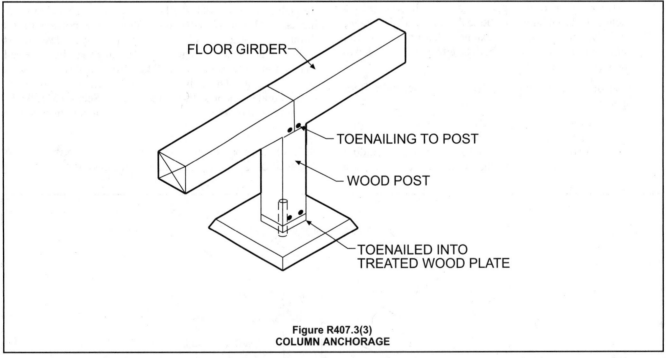

Figure R407.3(3)
COLUMN ANCHORAGE

maintenance of the building and for inspection and repair of such items as plumbing, mechanical systems or electrical system runs within the crawl space. This section addresses the different conditions encountered in accessing an under-floor space through the floor system or the foundation wall. Where the access opening is through the floor, an opening of not less than 18 inches by 24 inches (457 mm by 610 mm) is required. Where the access opening is through the foundation wall, an opening of not less than 16 inches by 24 inches (406 mm by 610 mm) is required. See Commentary Figure R408.1. The 16-and 24-inch dimensions are to work with standard CMU coursing. Where mechanical equipment is located under the floor, the minimum access opening must comply with Section M1305.1.4.

R408.5 Removal of debris. The under-floor grade shall be cleaned of all vegetation and organic material. All wood forms used for placing concrete shall be removed before a building is occupied or used for any purpose. All construction materials

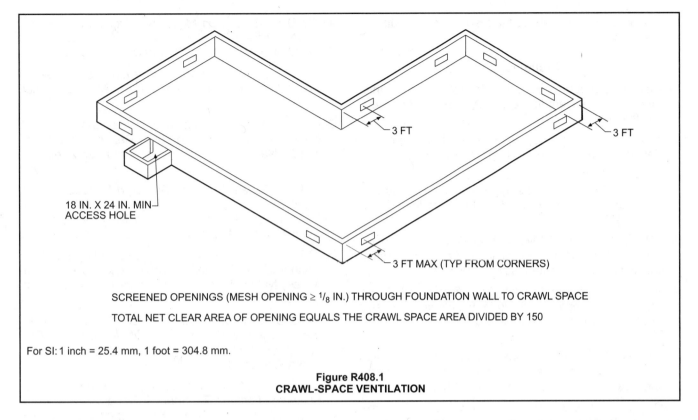

3 FT

3 FT

18 IN. X 24 IN. MIN
ACCESS HOLE

3 FT MAX (TYP FROM CORNERS)

SCREENED OPENINGS (MESH OPENING ≥ ¹/₈ IN.) THROUGH FOUNDATION WALL TO CRAWL SPACE

TOTAL NET CLEAR AREA OF OPENING EQUALS THE CRAWL SPACE AREA DIVIDED BY 150

For SI: 1 inch = 25.4 mm, 1 foot = 304.8 mm.

Figure R408.1
CRAWL-SPACE VENTILATION

shall be removed before a building is occupied or used for any purpose.

❖ Vegetation, stumps, roots and other matter left in an excavation around a building are major causes of termite infestation and moisture problems. As such material decays, the ground settles, negating the original drainage plan. Even before decay, the material provides pockets for water accumulation, which can have subsequent destructive impact on the structure. To eliminate a natural attraction to termites, insects or animals, all vegetation and organic material must be cleared.

R408.6 Finished grade. The finished grade of under-floor surface may be located at the bottom of the footings; however, where there is evidence that the groundwater table can rise to within 6 inches (152 mm) of the finished floor at the building perimeter or where there is evidence that the surface water does not readily drain from the building site, the grade in the under-floor space shall be as high as the outside finished grade, unless an approved drainage system is provided.

❖ To circumvent moisture accumulation in the crawl space area, the grade in the under-floor space must be at the same elevation as grade outside of the building if the groundwater table can rise to within 6 inches (152 mm) of the finished floor elevation or the surface water does not readily drain from the site. An alternative would be an approved drainage system.

R408.7 Flood resistance. For buildings located in areas prone to flooding as established in Table R301.2(1):

1. Walls enclosing the under-floor space shall be provided with flood openings in accordance with Section R324.2.2.

2. The finished ground level of the under-floor space shall be equal to or higher than the outside finished ground level.

> **Exception:** Under-floor spaces that meet the requirements of FEMA/FIA TB 11-1.

❖ To minimize hydrostatic loads by allowing the free inflow and outflow of floodwaters, buildings in flood hazard areas (A zones) that have walls enclosing underfloor spaces are to have flood openings in those walls. This provision alerts the designer that flood openings are to be provided in addition to standard ventilation openings. This requirement applies only to buildings in flood hazard areas not subject to high velocity wave action. Buildings in coastal high hazard areas subject to high velocity wave action may have enclosed areas below the design flood elevation, but only if those walls are intended to break away as specified in Section R324.3.4.

The provisions of this section require the elevation of the finished ground level of the under-floor space to be at or above the elevation of the outside finish ground level. The exception permits the use of FEMA/FIA-TB-11-01, which permits the elevation of the under-floor space to be below the elevation of the outside finish ground level. Use of FEMA/FIA-TB-11-01 will result in flood insurance rates higher than for buildings that have the finished ground level of the under-floor space at or above the elevation of the outside finish ground level.

Bibliography

The following resource materials are referenced in this chapter or are relevant to the subject matter addressed in this chapter.

ACI 318-05, *Building Code Requirements for Structural Concrete*. Farmington Hills, MI: American Concrete Institute, 2005.

ACI 332-05, *Requirements for Residential Concrete Construction*. Farmington Hills, MI: American Concrete Institute, 2005

ACI 530-05/ASCE 5-05/TMS 402-05, *Building Code Requirements for Masonry Structures*. Farmington Hills, MI: American Concrete Institute, 2005.

AFPA NDS-05, *National Design Specification (NDS) for Wood Construction—with 2005 Supplement*. Washington, DC: American Forest and Paper Association, 2005.

AFPA TR7-87, *Basic Requirements for Permanent Wood Foundation System*. Washington, D.C.: American Forest and Paper Association, 1987.

ASCE 32-01, *Design and Construction of Frost Protected Shallow Foundations*. Reston, VA: American Society of Civil Engineers, 2001.

ASTM A 615/A 615M-04a, *Specification for Deformed and Plain Billet Steel bars for Concrete Reinforcement*. West Conshohocken, PA: ASTM International, 2004.

ASTM A 706M-04a, *Specification for Low-Alloy Steel Deformed Plain Bars for Concrete Reinforcement*. West Conshohocken, PA: ASTM International, 2004.

ASTM C 143/C 0143M-03, *Test Method for Slump or Hydraulic Cement Concrete*. West Conshohocken, PA: ASTM International, 2003.

ASTM C578-04, *Specification for Rigid, Cellular Polystyrene Thermal Insulation*. West Conshohocken, PA: ASTM International, 2004.

ASTM C887-79(2001), *Specification for Packaged, Dry, Combined Materials for Surface Bonding Mortar*. West Conshohocken, PA: ASTM International, 2001.

ASTM C 1157-03, *Performance Specification for Hydraulic Cements*. West Conshohocken, PA: ASTM International, 2003.

ASTM D 422-63(2002), *Test Method for Particle-Size Analysis of Soils*. West Conshohocken, PA: ASTM International, 2002.

ASTM D 449-03, *Specification for Asphalt Used in Dampproofing and Waterproofing*. West Conshohocken, PA: ASTM International, 2003

ASTM D 4318-00, *Test Method for Liquid Limit, Plastic Limit and Plasticity Index of Soils*. West Conshohocken, PA: ASTM International, 2000.

ASTM D 4829-03, *Test Method for Expansion Index of Soils*. West Conshohocken, PA: ASTM International, 2003.

DOC PS 1-95, *Construction and Industrial Plywood*. Gaithersburg, MD: United States Department of Commerce, National Institute of Standards and Technology, 1995.

DOC PS 2-92, *Performance Standard for Wood-Based Structural-Use Panels*. Gaithersburg, MD: United States Department of Commerce, National Institute of Standards and Technology, 1992.

DOC PS 20-99, *American Softwood Lumber Standard*. Gaithersburg, MD: United States Department of Commerce, National Institute of Standards and Technology, 1999.

FEMA 303, *Recommended Provisions for Seismic Regulations for New Buildings and Other Structures*. Part 2: Commentary. Washington, DC: Building Seismic Safety Council, 1997.

FEMA/FIA-TB-11-01, *Crawlspace Construction for Buildings Located in Special Flood Hazard Area*. Washington, DC: Federal Emergency Management Agency, 2001.

NCMATR 68-A-75, *Design and Construction of Plain and Reinforced Concrete Masonry and Basement and Foundation Walls*. Herndon, VA: National Concrete Masonry Association, 1975.

Prescriptive Method for Insulating Concrete Forms in Residential Construction. Skokie, IL: Portland Cement Association, 1998.

Sculling, Michael. *Excavation and Grading Code Administration, Inspection and Enforcement*. Englewood Cliffs, NJ: Prentice Hall, 1990.

Chapter 5:
Floors

General Comments

Section R501 provides the scope of the chapter and states the general performance requirements for floor systems. The floor systems covered in the *International Residential Code®* (IRC®) consist of four different types: wood floor framing, wood floors on the ground, steel floor framing and concrete slabs on the ground. Section R502 addresses wood floor framing criteria and includes prescriptive tables from which the size of the joists can be determined based on the span and species of wood.

Section R503 specifies floor-sheathing requirements. Section R504 governs wood floors constructed on ground that would serve as part of a wood foundation system described in Chapter 4. This system relies on pressure-preservative-treated wood to resist the damage from exposure to moisture from contact with the ground. Section R505 contains steel floor framing provisions, which include prescriptive criteria that greatly simplify the use of steel framing in a floor system. Section R506 contains requirements for concrete slab-on-ground floors. Concrete-slab-on-grade construction requires a minimum 3.5-inch (89 mm) thick slab and a vapor retarder in most cases.

The primary consideration of this chapter is the structural integrity of the floor system. In a two-family dwelling, a floor that separates dwelling units must be a fire-resistance-rated assembly in accordance with Section R317.1. Such a floor must conform to the conditions of the floor assembly's listing in addition to the requirements of this chapter.

Purpose

Chapter 5 provisions govern the design and construction of floor systems that will be capable of supporting minimum required design loads. The allowable-span tables greatly simplify the determination of joist, girder and sheathing sizes for raised floor systems.

The floor system must also serve as a diaphragm; therefore, it plays a key role in resisting lateral loads from earthquakes and/or wind. Although this chapter does not base floor system requirements on these loads, floor systems constructed as required in this chapter will perform this function. Chapter 6 of the IRC accounts for these loads by varying the spacing of the supporting braced wall lines or shear walls based on seismic design category and basic wind speed or, in selected cases, specifying additional diaphragm requirements.

SECTION R501
GENERAL

R501.1 Application. The provisions of this chapter shall control the design and construction of the floors for all buildings including the floors of attic spaces used to house mechanical or plumbing fixtures and equipment.

❖ Floors (including attic floors) that house mechanical equipment or plumbing fixtures (see the definition in Chapter 2) must comply with this chapter.

R501.2 Requirements. Floor construction shall be capable of accommodating all loads according to Section R301 and of transmitting the resulting loads to the supporting structural elements.

❖ This is a general performance statement that requires the floor system to support the design loads and provide the necessary load path to supporting elements.

SECTION R502
WOOD FLOOR FRAMING

R502.1 Identification. Load-bearing dimension lumber for joists, beams and girders shall be identified by a grade mark of a lumber grading or inspection agency that has been approved by an accreditation body that complies with DOC PS 20. In lieu of a grade mark, a certificate of inspection issued by a lumber grading or inspection agency meeting the requirements of this section shall be accepted.

❖ Load-carrying wood members must be identified to verify their conformance to a minimum quality control standard. The required grade mark must reflect the following information:

   1. The species.

   2. The grade.

   3. The moisture content (MC) at time of grading: S-Dry or S-Gm, MC15, KD15, or KD19. S-Grn indicates that the moisture content was in excess of 19 percent at the time of grading, and KD indicates kiln dried.

   4. The grading agency.

   5. The mill name or mill number.

   6. In the case of independent agencies, the rules used for grading.

For examples of grade marks, see Commentary Figure R502.1.

Figure R502.1
GRADE MARK EXAMPLES

R502.1.1 Preservative-treated lumber. Preservative treated dimension lumber shall also be identified as required by Section R319.1.

❖ Grade marks as described in Section R502.1 must also include the quality mark of an agency accredited in the ALSC treated wood program. See Commentary Figure R502.1.1 for an example of a preservative-treated-lumber grade mark.

R502.1.2 Blocking and subflooring. Blocking shall be a minimum of utility grade lumber. Subflooring may be a minimum of utility grade lumber or No. 4 common grade boards.

❖ Allowing the use of lower lumber grades in blocking and subflooring in these instances results in possible economies.

R502.1.3 End-jointed lumber. Approved end-jointed lumber identified by a grade mark conforming to Section R502.1 may be used interchangeably with solid-sawn members of the same species and grade.

❖ End-jointed lumber is generally considered equivalent to solid sawn lumber, provided it is properly identified by a grade mark. The key is verifying proper end use. For instance, grade marks including the phrase "STUD USE ONLY" denote finger-jointed lumber that is limited to stud use and/or bending or tension stresses of short duration loads such as wind and seismic. Structural finger joints, on the other hand, allow horizontal as well as vertical applications.

R502.1.4 Prefabricated wood I-joists. Structural capacities and design provisions for prefabricated wood I-joists shall be established and monitored in accordance with ASTM D 5055.

❖ Prefabricated wood I-joists are some of the many engineered wood products currently available. These structural members consist of sawn lumber or structural composite lumber flanges bonded to wood structural panel webs, forming an I-shaped cross section. Because these products are widely used in residential floor construction ASTM D 5055, *Standard Specification for Establishing and Monitoring Structural Capacities of Prefabricated Wood I-joists,* is referenced as the applicable standard.

This standard does not specify the makeup of the component materials or how they must be assembled. Each manufacturer must develop product specifications and construction details for its products. These are subject to qualification tests, which are conducted or witnessed by qualified, independent agencies. The best results are used to establish the following design values:

• Allowable bending moment.
• Allowable shear.
• Member stiffness (EI).
• Sheer deflection factor (K).
• Allowable reactions.
• Bearing length requirements.
• Web stiffener requirements.
• A chart detailing the size, location, shape and spacing of any holes permitted in the web.

R502.1.5 Structural glued laminated timbers. Glued laminated timbers shall be manufactured and identified as required in AITC A190.1 and ASTM D 3737.

❖ This section requires that glulam timbers be manufactured following ANSI/AITC 90.1 and ASTM D 3737. Knowing these standards makes it easier to determine that the product found in the field will meet the design requirements.

R502.1.6 Structural log members. Stress grading of structural log members of nonrectangular shape, as typically used in log buildings, shall be in accordance with ASTM D 3957. Such

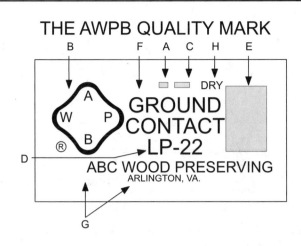

THE AWPB QUALITY MARK

GROUND CONTACT
LP-22
ABC WOOD PRESERVING
ARLINGTON, VA.

A. YEAR OF TREATMENT

B. AMERICAN WOOD PRESERVERS BUREAU TRADEMARK

C. THE PRESERVATIVE USED FOR TREATMENT

D. THE APPLICABLE AMERICAN WOOD PRESERVERS BUREAU QUALITY STANDARD

E. TRADEMARK OF THE AGENCY SUPERVISING THE TREATING PLANT

F. PROPER EXPOSURE CONDITIONS

G. TREATING COMPANY AND PLANT LOCATION

H. DRY OR KDAT IF APPLICABLE

Figure R502.1.1
TREATED LUMBER GRADE MARK

structural log members shall be identified by the grade mark of an approved lumber grading or inspection agency. In lieu of a grade mark on the material, a certificate of inspection as to species and grade issued by a lumber-grading or inspection agency meeting the requirements of this section shall be permitted to be accepted.

❖ This section addresses grading requirements for logs used as structural members. This subsection specifies the reference for acceptable methods for establishing structural capacities of logs and specifies the requirement for a grading stamp or alternate certification on structural logs. The grading of structural log members must be in accordance with ASTM D 3957, Standard Practices for Establishing Stress Grades for *Structural Members Used in Log Buildings*.

R502.2 Design and construction. Floors shall be designed and constructed in accordance with the provisions of this chapter, Figure R502.2 and Sections R319 and R320 or in accordance with AF&PA/NDS.

❖ Design and construction of wood floors must be in accordance with the prescriptive requirements of this section or where required in accordance with AF&PA/NDS, *National Design Specification for Wood Construction*. Figure R502.2 includes section references to provisions that apply to typical elements of floor systems.

R502.2.1 Framing at braced wall lines. A load path for lateral forces shall be provided between floor framing and braced wall panels located above or below a floor, as specified in Section R602.10.8.

❖ At braced wall panels, joists and blocking are an essential part of a complete lateral load path for either wind or seismic loads. These framing members are needed to accomplish the framing to framing connections specified in Table R602.3(1) and Section R602.10.8. Section R602.10.8 specifies the framing required for a complete load path for these conditions. Section R602.10.8 addresses the framing needed when braced wall panels occur above or below a floor and when floor framing is parallel to a braced wall panel.

R502.2.2 Decks. Where supported by attachment to an exterior wall, decks shall be positively anchored to the primary structure and designed for both vertical and lateral loads as applicable. Such attachment shall not be accomplished by the use of toenails or nails subject to withdrawal. Where positive connection to the primary building structure cannot be verified during inspection, decks shall be self-supporting. For decks with cantilevered framing members, connections to exterior walls or other framing members, shall be designed and constructed to resist uplift resulting from the full live load specified in Table R301.5 acting on the cantilevered portion of the deck.

❖ If an exterior wall is used to support a deck, the deck framing must be positively attached to the building structure. This connection design must include a consideration of both vertical and lateral loads, and the connection must be available for inspection. If it is not, this method of support is not permitted and the deck must be self-supporting.

If a deck has cantilevered framing, the framing must have a connection to its support that is designed to resist any uplift resulting from the full live load acting on the cantilevered span only. This load condition will produce maximum uplift at the support opposite the cantilevered end.

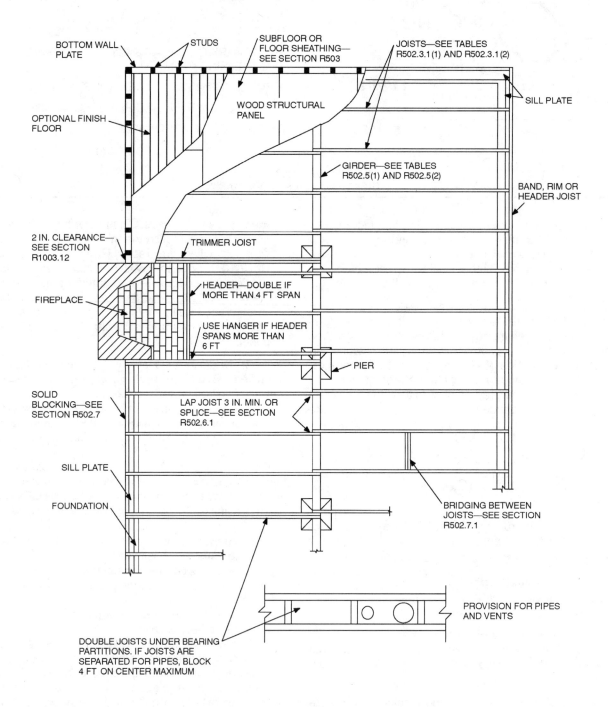

For SI: 1 inch = 25.4 mm, 1 foot = 304.8 mm.

**FIGURE R502.2
FLOOR CONSTRUCTION**

❖ This figure is a schematic plan of typical wood floor framing. It serves as a key, providing references to the applicable provisions of the code.

R502.3 Allowable joist spans. Spans for floor joists shall be in accordance with Tables R502.3.1(1) and R502.3.1(2). For other grades and species and for other loading conditions, refer to the AF&PA Span Tables for Joists and Rafters.

❖ Tables R502.3.1(1) and R502.3.1(2) contain allowable floor joist spans for common lumber species and grades based on design loads and joist spacing. The referenced standard may be used for grades and species of lumber not included in these tables.

Table R502.3.1(1) covers sleeping rooms where the design live load is 30 pounds per square foot (psf) (1.44 kPa) per Table R301.5. This includes attics with fixed stair access. It is important to identify the intended use because an attic space used as a bonus room, for example, could require use of a 40-pounds-per-square-foot (1.92 kPa) live load. Table 502.3.1(2) covers rooms other than sleeping rooms where the design load is 40 pounds per square foot (1.92 kPa) per Table R301.5.

Example:

A floor has a design live load of 40 psf (1.92 kPa) and a dead load of 10 psf (0.48 kPa). The joists span 17 feet, 6 inches (5334 mm). Find the required joist size using Douglas-Fir Larch #2 lumber and a joist spacing of 16 inches (406 mm).

Solution:

In Table R502.3.1(2), which is based on LL = 40 psf (1.92 kPa), find 16-inch (406 mm) joist spacing in the left hand column. Locate Douglas Fir-Larch #2 and under DL = 10 psf (0.48 kPa) look for an allowable span of 17 feet, 6 inches (5334 mm) or greater. Note that the allowable span for 2 inch by 12 inch (51 mm by 305 mm) is 17 feet, 10 inches (5436 mm). Therefore, use 2-inch by 12-inch (51 mm by 305 mm) joists at 16-inch (406 mm) spacing.

The span tables account for a uniform load condition. They will also permit isolated concentrated loads such as nonbearing partitions offset from a support by a distance less than or equal to the joist depth. They may not support large concentrated loads such as ones that result from an entire kitchen utility wall or bathtubs parallel to joists. In such instances, additional joists and other adequate supports must be installed.

R502.3.1 Sleeping areas and attic joists. Table R502.3.1(1) shall be used to determine the maximum allowable span of floor joists that support sleeping areas and attics that are accessed by means of a fixed stairway in accordance with Section R311.5 provided that the design live load does not exceed 30 psf (1.44 kPa) and the design dead load does not exceed 20 psf (0.96 kPa). The allowable span of ceiling joists that support attics used for limited storage or no storage shall be determined in accordance with Section R802.4.

❖ See the commentary for Section R502.3.

R502.3.2 Other floor joists. Table R502.3.1(2) shall be used to determine the maximum allowable span of floor joists that support all other areas of the building, other than sleeping rooms and attics, provided that the design live load does not exceed 40 psf (1.92 kPa) and the design dead load does not exceed 20 psf (0.96 kPa).

❖ See the commentary for Section R502.3.

R502.3.3 Floor cantilevers. Floor cantilever spans shall not exceed the nominal depth of the wood floor joist. Floor cantilevers constructed in accordance with Table R502.3.3(1) shall be permitted when supporting a light-frame bearing wall and roof only. Floor cantilevers supporting an exterior balcony are permitted to be constructed in accordance with Table R502.3.3(2).

❖ The cantilever span is permitted to be equal to the nominal depth of the joist without additional limitations. This provides for load transfer to the support by direct bearing so that shear and bending of the joist is not a concern.

Larger cantilevers are permitted in accordance with the limitations of the appropriate table for floor joists supporting an exterior balcony or supporting a light-frame bearing wall and roof.

R502.4 Joists under bearing partitions. Joists under parallel bearing partitions shall be of adequate size to support the load. Double joists, sized to adequately support the load, that are separated to permit the installation of piping or vents shall be full depth solid blocked with lumber not less than 2 inches (51 mm) in nominal thickness spaced not more than 4 feet (1219 mm) on center. Bearing partitions perpendicular to joists shall not be offset from supporting girders, walls or partitions more than the joist depth unless such joists are of sufficient size to carry the additional load.

❖ Where floor joists support bearing partitions, the joists must be of adequate size. Commentary Figure R502.4(1) shows an example of using adequate sized double joists. This provides added support for the additional load from the bearing partition above. Alternatively, a beam of adequate size may also be substituted for the double joist.

Where piping or vents must penetrate the floor system, the double joists must be separated to allow the pipe or vent to pass through. In this case, full depth solid blocking spaced a maximum of 4 feet (1219 mm) on center, as shown in Commentary Figure R502.4(2), is required so that the two joists will function as a combined member without twisting.

Bearing partitions oriented perpendicular to joists cannot be offset from supporting girders, walls or partitions more than the depth of the joists unless the joists are of sufficient size to carry the additional load.

TABLE R502.3.1(1)
FLOOR JOIST SPANS FOR COMMON LUMBER SPECIES
(Residential sleeping areas, live load = 30 psf, L/Δ = 360)[a]

JOIST SPACING (inches)	SPECIES AND GRADE		DEAD LOAD = 10 psf				DEAD LOAD = 20 psf			
			2×6	2×8	2×10	2×12	2×6	2×8	2×10	2×12
			Maximum floor joist spans							
			(ft - in.)	(ft - in.)	(ft - in.)	(ft - in.)	(ft - in.)	(ft - in.)	(ft - in.)	(ft - in.)
12	Douglas fir-larch	SS	12- 6	16- 6	21- 0	25- 7	12- 6	16- 6	21- 0	25- 7
	Douglas fir-larch	#1	12- 0	15-10	20- 3	24- 8	12- 0	15- 7	19- 0	22- 0
	Douglas fir-larch	#2	11-10	15- 7	19-10	23- 0	11- 6	14- 7	17- 9	20- 7
	Douglas fir-larch	#3	9- 8	12- 4	15- 0	17- 5	8- 8	11- 0	13- 5	15- 7
	Hem-fir	SS	11-10	15- 7	19-10	24- 2	11-10	15- 7	19-10	24- 2
	Hem-fir	#1	11- 7	15- 3	19- 5	23- 7	11- 7	15- 2	18- 6	21- 6
	Hem-fir	#2	11- 0	14- 6	18- 6	22- 6	11- 0	14- 4	17- 6	20- 4
	Hem-fir	#3	9- 8	12- 4	15- 0	17- 5	8- 8	11- 0	13- 5	15- 7
	Southern pine	SS	12- 3	16- 2	20- 8	25- 1	12- 3	16- 2	20- 8	25- 1
	Southern pine	#1	12- 0	15-10	20- 3	24- 8	12- 0	15-10	20- 3	24- 8
	Southern pine	#2	11-10	15- 7	19-10	24- 2	11-10	15- 7	18- 7	21- 9
	Southern pine	#3	10- 5	13- 3	15- 8	18- 8	9- 4	11-11	14- 0	16- 8
	Spruce-pine-fir	SS	11- 7	15- 3	19- 5	23- 7	11- 7	15- 3	19- 5	23- 7
	Spruce-pine-fir	#1	11- 3	14-11	19- 0	23- 0	11- 3	14- 7	17- 9	20- 7
	Spruce-pine-fir	#2	11- 3	14-11	19- 0	23- 0	11- 3	14- 7	17- 9	20- 7
	Spruce-pine-fir	#3	9- 8	12- 4	15- 0	17- 5	8- 8	11- 0	13- 5	15- 7
16	Douglas fir-larch	SS	11- 4	15- 0	19- 1	23- 3	11- 4	15- 0	19- 1	23- 0
	Douglas fir-larch	#1	10-11	14- 5	18- 5	21- 4	10- 8	13- 6	16- 5	19- 1
	Douglas fir-larch	#2	10- 9	14- 1	17- 2	19-11	9-11	12- 7	15- 5	17-10
	Douglas fir-larch	#3	8- 5	10- 8	13- 0	15- 1	7- 6	9- 6	11- 8	13- 6
	Hem-fir	SS	10- 9	14- 2	18- 0	21-11	10- 9	14- 2	18- 0	21-11
	Hem-fir	#1	10- 6	13-10	17- 8	20- 9	10- 4	13- 1	16- 0	18- 7
	Hem-fir	#2	10- 0	13- 2	16-10	19- 8	9-10	12- 5	15- 2	17- 7
	Hem-fir	#3	8- 5	10- 8	13- 0	15- 1	7- 6	9- 6	11- 8	13- 6
	Southern pine	SS	11- 2	14- 8	18- 9	22-10	11- 2	14- 8	18- 9	22-10
	Southern pine	#1	10-11	14- 5	18- 5	22- 5	10-11	14- 5	17-11	21- 4
	Southern pine	#2	10- 9	14- 2	18- 0	21- 1	10- 5	13- 6	16- 1	18-10
	Southern pine	#3	9- 0	11- 6	13- 7	16- 2	8- 1	10- 3	12- 2	14- 6
	Spruce-pine-fir	SS	10- 6	13-10	17- 8	21- 6	10- 6	13-10	17- 8	21- 4
	Spruce-pine-fir	#1	10- 3	13- 6	17- 2	19-11	9-11	12- 7	15- 5	17-10
	Spruce-pine-fir	#2	10- 3	13- 6	17- 2	19-11	9-11	12- 7	15- 5	17-10
	Spruce-pine-fir	#3	8- 5	10- 8	13- 0	15- 1	7- 6	9- 6	11- 8	13- 6
19.2	Douglas fir-larch	SS	10- 8	14- 1	18- 0	21-10	10- 8	14- 1	18- 0	21- 0
	Douglas fir-larch	#1	10- 4	13- 7	16- 9	19- 6	9- 8	12- 4	15- 0	17- 5
	Douglas fir-larch	#2	10- 1	12-10	15- 8	18- 3	9- 1	11- 6	14- 1	16- 3
	Douglas fir-larch	#3	7- 8	9- 9	11-10	13- 9	6-10	8- 8	10- 7	12- 4
	Hem-fir	SS	10- 1	13- 4	17- 0	20- 8	10- 1	13- 4	17- 0	20- 7
	Hem-fir	#1	9-10	13- 0	16- 4	19- 0	9- 6	12- 0	14- 8	17- 0
	Hem-fir	#2	9- 5	12- 5	15- 6	17- 1	8-11	11- 4	13-10	16- 1
	Hem-fir	#3	7- 8	9- 9	11-10	13- 9	6-10	8- 8	10- 7	12- 4
	Southern pine	SS	10- 6	13-10	17- 8	21- 6	10- 6	13-10	17- 8	21- 6
	Southern pine	#1	10- 4	13- 7	17- 4	21- 1	10- 4	13- 7	16- 4	19- 6
	Southern pine	#2	10- 1	13- 4	16- 5	19- 3	9- 6	12- 4	14- 8	17- 2
	Southern pine	#3	8- 3	10- 6	12- 5	14- 9	7- 4	9- 5	11- 1	13- 2
	Spruce-pine-fir	SS	9- 10	13- 0	16- 7	20- 2	9-10	13- 0	16- 7	19- 6
	Spruce-pine-fir	#1	9- 8	12- 9	15- 8	18- 3	9- 1	11- 6	14- 1	16- 3
	Spruce-pine-fir	#2	9- 8	12- 9	15- 8	18- 3	9- 1	11- 6	14- 1	16- 3
	Spruce-pine-fir	#3	7- 8	9- 9	11-10	13- 9	6-10	8- 8	10- 7	12- 4
24	Douglas fir-larch	SS	9-11	13- 1	16- 8	20- 3	9-11	13- 1	16- 2	18- 9
	Douglas fir-larch	#1	9- 7	12- 4	15- 0	17- 5	8- 8	11- 0	13- 5	15- 7
	Douglas fir-larch	#2	9- 1	11- 6	14- 1	16- 3	8- 1	10- 3	12- 7	14- 7
	Douglas fir-larch	#3	6-10	8- 8	10- 7	12- 4	6- 2	7- 9	9- 6	11- 0
	Hem-fir	SS	9- 4	12- 4	15- 9	19- 2	9- 4	12- 4	15- 9	18- 5
	Hem-fir	#1	9- 2	12- 0	14- 8	17- 0	8- 6	10- 9	13- 1	15- 2
	Hem-fir	#2	8- 9	11- 4	13-10	16- 1	8- 0	10- 2	12- 5	14- 4
	Hem-fir	#3	6-10	8- 8	10- 7	12- 4	6- 2	7- 9	9- 6	11- 0
	Southern pine	SS	9- 9	12-10	16- 5	19-11	9- 9	12-10	16- 5	19-11
	Southern pine	#1	9- 7	12- 7	16- 1	19- 6	9- 7	12- 4	14- 7	17- 5
	Southern pine	#2	9- 4	12- 4	14- 8	17- 2	8- 6	11- 0	13- 1	15- 5
	Southern pine	#3	7- 4	9- 5	11- 1	13- 2	6- 7	8- 5	9-11	11-10
	Spruce-pine-fir	SS	9- 2	12- 1	15- 5	18- 9	9- 2	12- 1	15- 0	17- 5
	Spruce-pine-fir	#1	8-11	11- 6	14- 1	16- 3	8- 1	10- 3	12- 7	14- 7
	Spruce-pine-fir	#2	8-11	11- 6	14- 1	16- 3	8- 1	10- 3	12- 7	14- 7
	Spruce-pine-fir	#3	6-10	8- 8	10- 7	12- 4	6- 2	7- 9	9- 6	11- 0

For SI: 1 inch = 25.4 mm, 1 foot = 304.8 mm, 1 pound per square foot = 0.0479 kPa.

NOTE: Check sources for availability of lumber in lengths greater than 20 feet.

a. Dead load limits for townhouses in Seismic Design Category C and all structures in Seismic Design Categories D_0, D_1 and D_2 shall be determined in accordance with Section R301.2.2.2.1.

❖ This table provides allowable joist spans for common lumber species, grades and joist sizes. The spans are applicable to residential sleeping areas. Use of this table is similar to the use described in the example in Section R502.3.

TABLE R502.3.1(2)
FLOOR JOIST SPANS FOR COMMON LUMBER SPECIES
(Residential living areas, live load = 40 psf, L/Δ = 360)[b]

JOIST SPACING (inches)	SPECIES AND GRADE		DEAD LOAD = 10 psf				DEAD LOAD = 20 psf			
			2×6	2×8	2×10	2×12	2×6	2×8	2×10	2×12
			— Maximum floor joist spans							
			(ft - in.)	(ft - in.)	(ft - in.)	(ft - in.)	(ft - in.)	(ft - in.)	(ft - in.)	(ft - in.)
12	Douglas fir-larch	SS	11- 4	15- 0	19- 1	23- 3	11- 4	15- 0	19- 1	23- 3
	Douglas fir-larch	#1	10-11	14- 5	18- 5	22- 0	10-11	14- 2	17- 4	20- 1
	Douglas fir-larch	#2	10- 9	14- 2	17- 9	20- 7	10- 6	13- 3	16- 3	18-10
	Douglas fir-larch	#3	8- 8	11- 0	13- 5	15- 7	7-11	10- 0	12- 3	14- 3
	Hem-fir	SS	10- 9	14- 2	18- 0	21-11	10- 9	14- 2	18- 0	21-11
	Hem-fir	#1	10- 6	13-10	17- 8	21- 6	10- 6	13-10	16-11	19- 7
	Hem-fir	#2	10- 0	13- 2	16-10	20- 4	10- 0	13- 1	16- 0	18- 6
	Hem-fir	#3	8- 8	11- 0	13- 5	15- 7	7-11	10- 0	12- 3	14- 3
	Southern pine	SS	11- 2	14- 8	18- 9	22-10	11- 2	14- 8	18- 9	22-10
	Southern pine	#1	10-11	14- 5	18- 5	22- 5	10-11	14- 5	18- 5	22- 5
	Southern pine	#2	10- 9	14- 2	18- 0	21- 9	10- 9	14- 2	16-11	19-10
	Southern pine	#3	9- 4	11-11	14- 0	16- 8	8- 6	10-10	12-10	15- 3
	Spruce-pine-fir	SS	10- 6	13-10	17- 8	21- 6	10- 6	13-10	17- 8	21- 6
	Spruce-pine-fir	#1	10- 3	13- 6	17- 3	20- 7	10- 3	13- 3	16- 3	18-10
	Spruce-pine-fir	#2	10- 3	13- 6	17- 3	20- 7	10- 3	13- 3	16- 3	18-10
	Spruce-pine-fir	#3	8- 8	11- 0	13- 5	15- 7	7-11	10- 0	12- 3	14- 3
16	Douglas fir-larch	SS	10- 4	13- 7	17- 4	21- 1	10- 4	13- 7	17- 4	21- 0
	Douglas fir-larch	#1	9-11	13- 1	16- 5	19- 1	9- 8	12- 4	15- 0	17- 5
	Douglas fir-larch	#2	9- 9	12- 7	15- 5	17-10	9- 1	11- 6	14- 1	16- 3
	Douglas fir-larch	#3	7- 6	9- 6	11- 8	13- 6	6-10	8- 8	10- 7	12- 4
	Hem-fir	SS	9- 9	12-10	16- 5	19-11	9- 9	12-10	16- 5	19-11
	Hem-fir	#1	9- 6	12- 7	16- 0	18- 7	9- 6	12- 0	14- 8	17- 0
	Hem-fir	#2	9- 1	12- 0	15- 2	17- 7	8-11	11- 4	13-10	16- 1
	Hem-fir	#3	7- 6	9- 6	11- 8	13- 6	6-10	8- 8	10- 7	12- 4
	Southern pine	SS	10- 2	13- 4	17- 0	20- 9	10- 2	13- 4	17- 0	20- 9
	Southern pine	#1	9-11	13- 1	16- 9	20- 4	9-11	13- 1	16- 4	19- 6
	Southern pine	#2	9- 9	12-10	16- 1	18-10	9- 6	12- 4	14- 8	17- 2
	Southern pine	#3	8- 1	10- 3	12- 2	14- 6	7- 4	9- 5	11- 1	13- 2
	Spruce-pine-fir	SS	9- 6	12- 7	16- 0	19- 6	9- 6	12- 7	16- 0	19- 6
	Spruce-pine-fir	#1	9- 4	12- 3	15- 5	17-10	9- 1	11- 6	14- 1	16- 3
	Spruce-pine-fir	#2	9- 4	12- 3	15- 5	17-10	9- 1	11- 6	14- 1	16- 3
	Spruce-pine-fir	#3	7- 6	9- 6	11- 8	13- 6	6-10	8- 8	10- 7	12- 4
19.2	Douglas fir-larch	SS	9- 8	12-10	16- 4	19-10	9- 8	12-10	16- 4	19- 2
	Douglas fir-larch	#1	9- 4	12- 4	15- 0	17- 5	8-10	11- 3	13- 8	15-11
	Douglas fir-larch	#2	9- 1	11- 6	14- 1	16- 3	8- 3	10- 6	12-10	14-10
	Douglas fir-larch	#3	6-10	8- 8	10- 7	12- 4	6- 3	7-11	9- 8	11- 3
	Hem-fir	SS	9- 2	12- 1	15- 5	18- 9	9- 2	12- 1	15- 5	18- 9
	Hem-fir	#1	9- 0	11-10	14- 8	17- 0	8- 8	10-11	13- 4	15- 6
	Hem-fir	#2	8- 7	11- 3	13-10	16- 1	8- 2	10- 4	12- 8	14- 8
	Hem-fir	#3	6-10	8- 8	10- 7	12- 4	6- 3	7-11	9- 8	11- 3
	Southern pine	SS	9- 6	12- 7	16- 0	19- 6	9- 6	12- 7	16- 0	19- 6
	Southern pine	#1	9- 4	12- 4	15- 9	19- 2	9- 4	12- 4	14-11	17- 9
	Southern pine	#2	9- 2	12- 1	14- 8	17- 2	8- 8	11- 3	13- 5	15- 8
	Southern pine	#3	7- 4	9- 5	11- 1	13- 2	6- 9	8- 7	10- 1	12- 1
	Spruce-pine-fir	SS	9- 0	11-10	15- 1	18- 4	9- 0	11-10	15- 1	17- 9
	Spruce-pine-fir	#	8- 9	11- 6	14- 1	16- 3	8- 3	10- 6	12-10	14-10
	Spruce-pine-fir	#2	8- 9	11- 6	14- 1	16- 3	8- 3	10- 6	12-10	14-10
	Spruce-pine-fir	#3	6-10	8- 8	10- 7	12- 4	6- 3	7-11	9- 8	11- 3
24	Douglas fir-larch	SS	9- 0	11-11	15- 2	18- 5	9- 0	11-11	14- 9	17- 1
	Douglas fir-larch	#1	8- 8	11- 0	13- 5	15- 7	7-11	10- 0	12- 3	14- 3
	Douglas fir-larch	#2	8- 1	10- 3	12- 7	14- 7	7- 5	9- 5	11- 6	13- 4
	Douglas fir-larch	#3	6- 2	7- 9	9- 6	11- 0	5- 7	7- 1	8- 8	10- 1
	Hem-fir	SS	8- 6	11- 3	14- 4	17- 5	8- 6	11- 3	14- 4	16-10[a]
	Hem-fir	#1	8- 4	10- 9	13- 1	15- 2	7- 9	9- 9	11-11	13-10
	Hem-fir	#2	7-11	10- 2	12- 5	14- 4	7- 4	9- 3	11- 4	13- 1
	Hem-fir	#3	6- 2	7- 9	9- 6	11- 0	5- 7	7- 1	8- 8	10- 1
	Southern pine	SS	8-10	11- 8	14-11	18- 1	8-10	11- 8	14-11	18- 1
	Southern pine	#1	8- 8	11- 5	14- 7	17- 5	8- 8	11- 3	13- 4	15-11
	Southern pine	#2	8- 6	11- 0	13- 1	15- 5	7- 9	10- 0	12- 0	14- 0
	Southern pine	#3	6- 7	8- 5	9-11	11-10	6- 0	7- 8	9- 1	10- 9
	Spruce-pine-fir	SS	8- 4	11- 0	14- 0	17- 0	8- 4	11- 0	13- 8	15-11
	Spruce-pine-fir	#1	8- 1	10- 3	12- 7	14- 7	7- 5	9- 5	11- 6	13- 4
	Spruce-pine-fir	#2	8- 1	10- 3	12- 7	14- 7	7- 5	9- 5	11- 6	13- 4
	Spruce-pine fir	#3	6- 2	7- 9	9- 6	11- 0	5- 7	7- 1	8- 8	10- 1

For SI: 1 inch = 25.4 mm, 1 foot = 304.8 mm, 1 pound per square foot = 0.0479 kPa.

NOTE: Check sources for availability of lumber in lengths greater than 20 feet.

a. End bearing length shall be increased to 2 inches.

b. Dead load limits for townhouses in Seismic Design Category C and all structures in Seismic Design Categories D_0, D_1, and D_2 shall be determined in accordance with Section R301.2.2.2.1.

❖ See the commentary for Section R502.3.

TABLE R502.3.3(1)
CANTILEVER SPANS FOR FLOOR JOISTS SUPPORTING LIGHT-FRAME EXTERIOR BEARING WALL AND ROOF ONLY[a, b, c, f, g, h]
(Floor Live Load ≤ 40 psf, Roof Live Load ≤ 20 psf)

Member & Spacing	Maximum Cantilever Span (Uplift Force at Backspan Support in Lbs.)[d, e]											
	Ground Snow Load											
	≤ 20 psf			30 psf			50 psf			70 psf		
	Roof Width			Roof Width			Roof Width			Roof Width		
	24 ft	32 ft	40 ft	24 ft	32 ft	40 ft	24 ft	32 ft	40 ft	24 ft	32 ft	40 ft
2 × 8 @ 12"	20" (177)	15" (227)	—	18" (209)	—	—	—	—	—	—	—	—
2 × 10 @ 16"	29" (228)	21" (297)	16" (364)	26" (271)	18" (354)	—	20" (375)	—	—	—	—	—
2 × 10 @ 12"	36" (166)	26" (219)	20" (270)	34" (198)	22" (263)	16" (324)	26" (277)	—	—	19" (356)	—	—
2 × 12 @ 16"	—	32" (287)	25" (356)	36" (263)	29" (345)	21" (428)	29" (367)	20" (484)	—	23" (471)	—	—
2 × 12 @ 12"	—	42" (209)	31" (263)	—	37" (253)	27" (317)	36" (271)	27" (358)	17" (447)	31" (348)	19" (462)	—
2 × 12 @ 8"	—	48" (136)	45" (169)	—	48" (164)	38" (206)	—	40" (233)	26" (294)	36" (230)	29" (304)	18" (379)

For SI: 1 inch = 25.4 mm, 1 pound per square foot = 0.0479 kPa.

a. Tabulated values are for clear-span roof supported solely by exterior bearing walls.

b. Spans are based on No. 2 Grade lumber of Douglas fir-larch, hem-fir, southern pine, and spruce-pine-fir for repetitive (3 or more) members.

c. Ratio of backspan to cantilever span shall be at least 3:1.

d. Connections capable of resisting the indicated uplift force shall be provided at the backspan support.

e. Uplift force is for a backspan to cantilever span ratio of 3:1. Tabulated uplift values are permitted to be reduced by multiplying by a factor equal to 3 divided by the actual backspan ratio provided (3/backspan ratio).

f. See Section R301.2.2.2.2, Item 1, for additional limitations on cantilevered floor joists for detached one- and two-family dwellings in Seismic Design Category D_0, D_1, or D_2 and townhouses in Seismic Design Category C, D_0, D_1, or D_2.

g. A full-depth rim joist shall be provided at the cantilevered end of the joists. Solid blocking shall be provided at the cantilever support.

h. Linear interpolation shall be permitted for building widths and ground snow loads other than shown.

❖ This table provides the prescriptive span and uplift requirements for cantilever wood floor joists supporting a light-frame bearing wall and roof only. The permitted cantilever span ranges from 15 to 48 inches (331 to 1219 mm) based on the limitations prescribed in the table and in the notes.

TABLE R502.3.3(2)
CANTILEVER SPANS FOR FLOOR JOISTS SUPPORTING EXTERIOR BALCONY[a, b, e, f]

Member Size	Spacing	Maximum Cantilever Span (Uplift Force at Backspan Support in lb)[c, d]		
		Ground Snow Load		
		≤ 30 psf	50 psf	70 psf
2 × 8	12"	42" (139)	39" (156)	34" (165)
2 × 8	16"	36" (151)	34" (171)	29" (180)
2 × 10	12"	61" (164)	57" (189)	49" (201)
2 × 10	16"	53" (180)	49" (208)	42" (220)
2 × 10	24"	43" (212)	40" (241)	34" (255)
2 × 12	16"	72" (228)	67" (260)	57" (268)
2 × 12	24"	58" (279)	54" (319)	47" (330)

For SI: 1 inch = 25.4 mm, 1 pound per square foot = 0.0479 kPa.

a. Spans are based on No. 2 Grade lumber of Douglas fir-larch, hem-fir, southern pine, and spruce-pine-fir for repetitive (3 or more) members.

b. Ratio of backspan to cantilever span shall be at least 2:1.

c. Connections capable of resisting the indicated uplift force shall be provided at the backspan support.

d. Uplift force is for a backspan to cantilever span ratio of 2:1. Tabulated uplift values are permitted to be reduced by multiplying by a factor equal to 2 divided by the actual backspan ratio provided (2/backspan ratio).

e. A full-depth rim joist shall be provided at the cantilevered end of the joists. Solid blocking shall be provided at the cantilevered support.

f. Linear interpolation shall be permitted for ground snow loads other than shown.

❖ This table provides the prescriptive span and uplift requirements for cantilever wood floor joists supporting an exterior balcony. The table was developed based on an engineered design consistent with the requirements of the IBC. A live load of 60 psf (2.87 kPa) is used for the cantilevered portion of the floor joists and is combined with anticipated dead loads and snow loads. Snow loads vary up to 70 psf (3.35 kPa) and include consideration of drifted snow from an adjacent roof. Deflections are limited to the values specified in IBC Table 1604.3. As indicated in Note b, the ratio of backspan to cantilever span is limited to a minimum of 2:1.

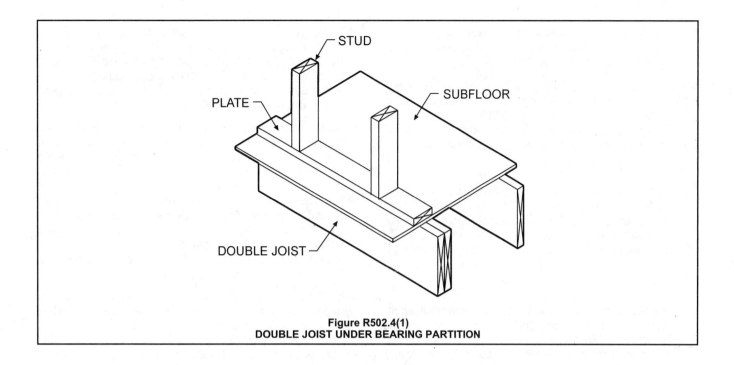

Figure R502.4(1)
DOUBLE JOIST UNDER BEARING PARTITION

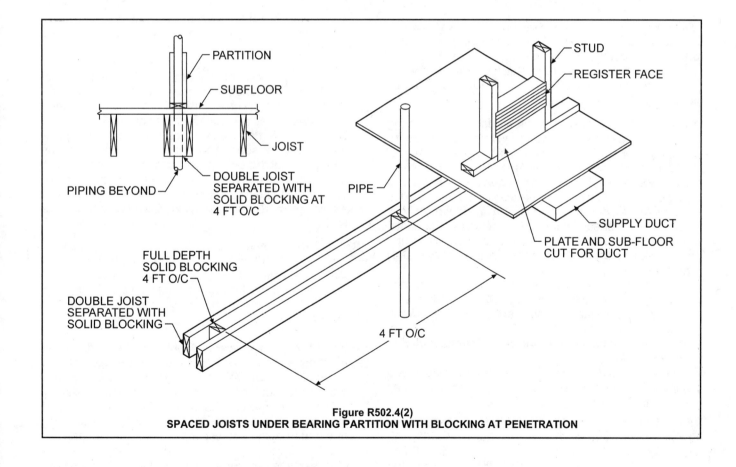

Figure R502.4(2)
SPACED JOISTS UNDER BEARING PARTITION WITH BLOCKING AT PENETRATION

R502.5 Allowable girder spans. The allowable spans of girders fabricated of dimension lumber shall not exceed the values set forth in Tables R502.5(1) and R502.5(2).

❖ Tables R502.5(1) and R502.5(2) list allowable spans for girders. These tables also apply to header spans and are referenced in Section R602.7 for that purpose. See Commentary Figure R502.5(1) and the following example for girder span-spacing relationships.

Example:

Assuming two equal floor joist spans of 10 feet (3048 mm) as indicated in Figure R502.5(1) (only one joist span and the two supporting girders are shown for clarity), determine the maximum span for a double 2-inch by 10-inch (51 mm by 254 mm) interior girder supporting one floor only.

Solution:

Table R502.5(2) contains the information for an interior girder supporting one floor only. For a double 2-inch by 10-inch (51 mm by 254 mm) girder and a building width of 20 feet (6096 mm), the maximum allowable girder span is read directly from Table R502.5(2) as 7 feet (2134 mm).

R502.6 Bearing. The ends of each joist, beam or girder shall have not less than 1.5 inches (38 mm) of bearing on wood or metal and not less than 3 inches (76 mm) on masonry or concrete except where supported on a 1-inch-by-4-inch (25.4 mm by 102 mm) ribbon strip and nailed to the adjacent stud or by the use of approved joist hangers.

❖ This section establishes minimum lengths of bearing for several alternative support systems to provide for the transfer of floor loads to supporting elements. These minimum requirements are based on anticipated loads and the allowable compressive stresses perpendicular to the grain for beam sizes and grades typical for wood-frame construction, in addition to consideration of shear failure of the masonry. For joists, beams or girders bearing on wood or metal, the minimum bearing shown in Commentary Figure R502.6(1) must be provided.

As an alternative, the members may be supported by a 1-inch by 4-inch (25 mm by 102 mm) ribbon strip when joists are nailed to adjacent studs as shown in Commentary Figure R502.6(3) or by the use of approved joist hangers as shown in Commentary Figure R502.6(4).

When joists, beams or girders bear on masonry, a minimum bearing of 3 inches (76 mm) is required, as shown in Commentary Figure R502.6(2).

R502.6.1 Floor systems. Joists framing from opposite sides over a bearing support shall lap a minimum of 3 inches (76 mm) and shall be nailed together with a minimum three 10d face nails. A wood or metal splice with strength equal to or greater than that provided by the nailed lap is permitted.

❖ To provide a concentric application of load from the joist to supporting beams or girders, joists framing from opposite sides of a beam or girder are required to lap at least 3 inches (76 mm) or the opposing joists are to be tied together with a wood or metal splice (see Commentary Figure R502.6.1).

R502.6.2 Joist framing. Joists framing into the side of a wood girder shall be supported by approved framing anchors or on ledger strips not less than nominal 2 inches by 2 inches (51 mm by 51 mm).

❖ For joists framed into the side of a wood beam or girder, the joists must be supported by approved framing anchors, as shown in Commentary Figure R502.6.2(1), or by ledger strips having a minimum nominal dimension of 2 inches (51 mm), as shown in Commentary Figure R502.6.2(2).

R502.7 Lateral restraint at supports. Joists shall be supported laterally at the ends by full-depth solid blocking not less than 2 inches (51 mm) nominal in thickness; or by attachment to a full-depth header, band or rim joist, or to an adjoining stud or shall be otherwise provided with lateral support to prevent rotation.

Exception: In Seismic Design Categories D_0, D_1 and D_2, lateral restraint shall also be provided at each intermediate support.

❖ Bridging, blocking or some other acceptable means of holding a joist in place is required so the floor joists do not twist out of the plane of the applied load. Lateral support at the ends of joists provide an additional function by transferring lateral loads to the supporting elements. Lateral support at ends may be provided by full-depth solid blocking not less than 2 inches (51 mm) in thickness, or the ends of joists may be nailed or bolted to a full-depth header, band or rim joist or to an adjoining stud as shown in Commentary Figure R502.7.

R502.7.1 Bridging. Joists exceeding a nominal 2 inches by 12 inches (51 mm by 305 mm) shall be supported laterally by solid blocking, diagonal bridging (wood or metal), or a continuous 1-inch-by-3-inch (25.4 mm by 76 mm) strip nailed across the bottom of joists perpendicular to joists at intervals not exceeding 8 feet (2438 mm).

❖ In addition to the lateral support at the ends, joists must have intermediate lateral support at intervals not exceeding 8 feet (2438 mm). Intermediate blocking is not required for joists 2 inches by 12 inches (51 mm by 305 mm) or smaller. The intermediate lateral support may be provided by solid blocking, diagonal bridging or wood bridging not less than 1 inch by 3 inches (25 mm by 76 mm) nominal, nailed to the bottom of the joist. Commentary Figure R502.7.1 illustrates the various alternatives for intermediate lateral support.

TABLE R502.5(1)
GIRDER SPANS[a] AND HEADER SPANS[a] FOR EXTERIOR BEARING WALLS
(Maximum spans for Douglas fir-larch, hem-fir, southern pine and spruce-pine-fir[b] and required number of jack studs)

GIRDERS AND HEADERS SUPPORTING	SIZE	GROUND SNOW LOAD (psf)[e]																	
		30						50						70					
		Building width[c] (feet)																	
		20		28		36		20		28		36		20		28		36	
		Span	NJ[d]	Span	NJ[d]	Span	NJ[d]	Span	NJ[d]	Span	NJ[d]	Span	NJ[d]	Span	NJ[d]	Span	NJ[d]	Span	NJ[d]
Roof and ceiling	2-2×4	3-6	1	3-2	1	2-10	1	3-2	1	2-9	1	2-6	1	2-10	1	2-6	1	2-3	1
	2-2×6	5-5	1	4-8	1	4-2	1	4-8	1	4-1	1	3-8	2	4-2	1	3-8	2	3-3	2
	2-2×8	6-10	1	5-11	2	5-4	2	5-11	2	5-2	2	4-7	2	5-4	2	4-7	2	4-1	2
	2-2×10	8-5	2	7-3	2	6-6	2	7-3	2	6-3	2	5-7	2	6-6	2	5-7	2	5-0	2
	2-2×12	9-9	2	8-5	2	7-6	2	8-5	2	7-3	2	6-6	2	7-6	2	6-6	2	5-10	3
	3-2×8	8-4	1	7-5	1	6-8	1	7-5	1	6-5	2	5-9	2	6-8	1	5-9	2	5-2	2
	3-2×10	10-6	1	9-1	2	8-2	2	9-1	2	7-10	2	7-0	2	8-2	2	7-0	2	6-4	2
	3-2×12	12-2	2	10-7	2	9-5	2	10-7	2	9-2	2	8-2	2	9-5	2	8-2	2	7-4	2
	4-2×8	9-2	1	8-4	1	7-8	1	8-4	1	7-5	1	6-8	1	7-8	1	6-8	1	5-11	2
	4-2×10	11-8	1	10-6	1	9-5	1	10-6	1	9-1	2	8-2	2	9-5	2	8-2	2	7-3	2
	4-2×12	14-1	1	12-2	2	10-11	2	12-2	2	10-7	2	9-5	2	10-11	2	9-5	2	8-5	2
Roof, ceiling and one center-bearing floor	2-2×4	3-1	1	2-9	1	2-5	1	2-9	1	2-5	1	2-2	1	2-7	1	2-3	1	2-0	1
	2-2×6	4-6	1	4-0	1	3-7	2	4-1	1	3-7	2	3-3	2	3-9	2	3-3	2	2-11	2
	2-2×8	5-9	2	5-0	2	4-6	2	5-2	2	4-6	2	4-1	2	4-9	2	4-2	2	3-9	2
	2-2×10	7-0	2	6-2	2	5-6	2	6-4	2	5-6	2	5-0	2	5-9	2	5-1	2	4-7	3
	2-2×12	8-1	2	7-1	2	6-5	2	7-4	2	6-5	2	5-9	3	6-8	2	5-10	3	5-3	3
	3-2×8	7-2	1	6-3	2	5-8	2	6-5	2	5-8	2	5-1	2	5-11	2	5-2	2	4-8	2
	3-2×10	8-9	2	7-8	2	6-11	2	7-11	2	6-11	2	6-3	2	7-3	2	6-4	2	5-8	2
	3-2×12	10-2	2	8-11	2	8-0	2	9-2	2	8-0	2	7-3	2	8-5	2	7-4	2	6-7	2
	4-2×8	8-1	1	7-3	1	6-7	1	7-5	1	6-6	1	5-11	2	6-10	1	6-0	2	5-5	2
	4-2×10	10-1	1	8-10	2	8-0	2	9-1	2	8-0	2	7-2	2	8-4	2	7-4	2	6-7	2
	4-2×12	11-9	2	10-3	2	9-3	2	10-7	2	9-3	2	8-4	2	9-8	2	8-6	2	7-7	2
Roof, ceiling and one clear span floor	2-2×4	2-8	1	2-4	1	2-1	1	2-7	1	2-3	1	2-0	1	2-5	1	2-1	1	1-10	1
	2-2×6	3-11	1	3-5	2	3-0	2	3-10	2	3-4	2	3-0	2	3-6	2	3-1	2	2-9	2
	2-2×8	5-0	2	4-4	2	3-10	2	4-10	2	4-2	2	3-9	2	4-6	2	3-11	2	3-6	2
	2-2×10	6-1	2	5-3	2	4-8	2	5-11	2	5-1	2	4-7	3	5-6	2	4-9	2	4-3	3
	2-2×12	7-1	2	6-1	3	5-5	3	6-10	2	5-11	3	5-4	3	6-4	2	5-6	3	5-0	3
	3-2×8	6-3	2	5-5	2	4-10	2	6-1	2	5-3	2	4-8	2	5-7	2	4-11	2	4-5	2
	3-2×10	7-7	2	6-7	2	5-11	2	7-5	2	6-5	2	5-9	2	6-10	2	6-0	2	5-4	2
	3-2×12	8-10	2	7-8	2	6-10	2	8-7	2	7-5	2	6-8	2	7-11	2	6-11	2	6-3	2
	4-2×8	7-2	1	6-3	2	5-7	2	7-0	1	6-1	2	5-5	2	6-6	1	5-8	2	5-1	2
	4-2×10	8-9	2	7-7	2	6-10	2	8-7	2	7-5	2	6-7	2	7-11	2	6-11	2	6-2	2
	4-2×12	10-2	2	8-10	2	7-11	2	9-11	2	8-7	2	7-8	2	9-2	2	8-0	2	7-2	2
Roof, ceiling and two center-bearing floors	2-2×4	2-7	1	2-3	1	2-0	1	2-6	1	2-2	1	1-11	1	2-4	1	2-0	1	1-9	1
	2-2×6	3-9	2	3-3	2	2-11	2	3-8	2	3-2	2	2-10	2	3-5	2	3-0	2	2-8	2
	2-2×8	4-9	2	4-2	2	3-9	2	4-7	2	4-0	2	3-8	2	4-4	2	3-9	2	3-5	2
	2-2×10	5-9	2	5-1	2	4-7	3	5-8	2	4-11	2	4-5	3	5-3	2	4-7	3	4-2	3
	2-2×12	6-8	2	5-10	3	5-3	3	6-6	2	5-9	2	5-2	3	6-1	3	5-4	3	4-10	3
	3-2×8	5-11	2	5-2	2	4-8	2	5-9	2	5-1	2	4-7	2	5-5	2	4-9	2	4-3	2
	3-2×10	7-3	2	6-4	2	5-8	2	7-1	2	6-2	2	5-7	2	6-7	2	5-9	2	5-3	2
	3-2×12	8-5	2	7-4	2	6-7	2	8-2	2	7-2	2	6-5	3	7-8	2	6-9	2	6-1	3
	4-2×8	6-10	1	6-0	2	5-5	2	6-8	1	5-10	2	5-3	2	6-3	2	5-6	2	4-11	2
	4-2×10	8-4	2	7-4	2	6-7	2	8-2	2	7-2	2	6-5	2	7-7	2	6-8	2	6-0	2
	4-2×12	9-8	2	8-6	2	7-8	2	9-5	2	8-3	2	7-5	2	8-10	2	7-9	2	7-0	2

(continued)

TABLE R502.5(1)—continued
GIRDER SPANS[a] AND HEADER SPANS[a] FOR EXTERIOR BEARING WALLS
(Maximum spans for Douglas fir-larch, hem-fir, southern pine and spruce-pine-fir[b] and required number of jack studs)

GIRDERS AND HEADERS SUPPORTING	SIZE	GROUND SNOW LOAD (psf)[e]																
		30						50						70				
		Building width[c] (feet)																
		20		28		36		20		28		36		20		28		36	
		Span	NJ[d]	Span	NJ[d]	Span	NJ[d]	Span	NJ[d]	Span	NJ[d]	Span	NJ[d]	Span	NJ[d]	Span	NJ[d]	Span	NJ[d]
Roof, ceiling, and two clear span floors	2-2×4	2-1	1	1-8	1	1-6	2	2-0	1	1-8	1	1-5	2	2-0	1	1-8	1	1-5	2
	2-2×6	3-1	2	2-8	2	2-4	2	3-0	2	2-7	2	2-3	2	2-11	2	2-7	2	2-3	2
	2-2×8	3-10	2	3-4	2	3-0	3	3-10	2	3-4	2	2-11	3	3-9	2	3-3	2	2-11	3
	2-2×10	4-9	2	4-1	3	3-8	3	4-8	2	4-0	3	3-7	3	4-7	3	4-0	3	3-6	3
	2-2×12	5-6	3	4-9	3	4-3	3	5-5	3	4-8	3	4-2	3	5-4	3	4-7	3	4-1	4
	3-2×8	4-10	2	4-2	2	3-9	2	4-9	2	4-1	2	3-8	2	4-8	2	4-1	2	3-8	2
	3-2×10	5-11	2	5-1	2	4-7	3	5-10	2	5-0	2	4-6	3	5-9	2	4-11	2	4-5	3
	3-2×12	6-10	2	5-11	3	5-4	3	6-9	2	5-10	3	5-3	3	6-8	2	5-9	3	5-2	3
	4-2×8	5-7	2	4-10	2	4-4	2	5-6	2	4-9	2	4-3	2	5-5	2	4-8	2	4-2	2
	4-2×10	6-10	2	5-11	2	5-3	2	6-9	2	5-10	2	5-2	2	6-7	2	5-9	2	5-1	2
	4-2×12	7-11	2	6-10	2	6-2	3	7-9	2	6-9	2	6-0	3	7-8	2	6-8	2	5-11	3

For SI: 1 inch = 25.4 mm, 1 pound per square foot = 0.0479 kPa.

a. Spans are given in feet and inches.

b. Tabulated values assume #2 grade lumber.

c. Building width is measured perpendicular to the ridge. For widths between those shown, spans are permitted to be interpolated.

d. NJ - Number of jack studs required to support each end. Where the number of required jack studs equals one, the header is permitted to be supported by an approved framing anchor attached to the full-height wall stud and to the header.

e. Use 30 psf ground snow load for cases in which ground snow load is less than 30 psf and the roof live load is equal to or less than 20 psf.

❖ In addition to providing the allowable spans for girders and headers, this table lists the number of jack studs required to support the girder or header. The header studs (jack studs) on which the headers rest should be continuous from the header to the sill plate of the wall. Cutting the header stud to support a sill is not allowed. Headers should be adequately nailed together and to the wall studs. The table is broken down into five loading conditions, illustrated in Commentary Figure R502.5(2).

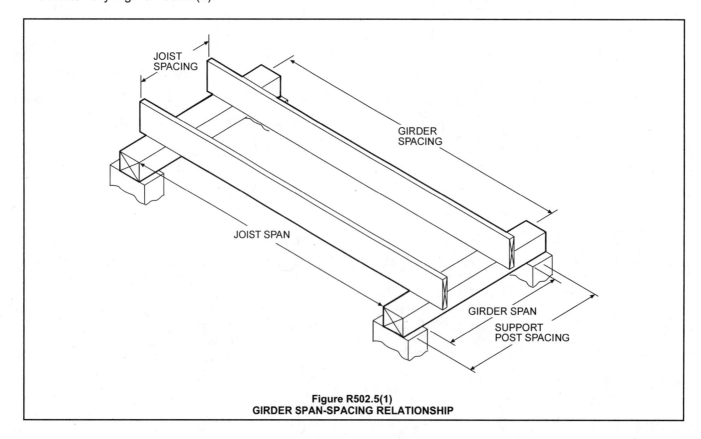

Figure R502.5(1)
GIRDER SPAN-SPACING RELATIONSHIP

TABLE R502.5(2)
GIRDER SPANS[a] AND HEADER SPANS[a] FOR INTERIOR BEARING WALLS
(Maximum spans for Douglas fir-larch, hem-fir, southern pine and spruce-pine-fir[b] and required number of jack studs)

HEADERS AND GIRDERS SUPPORTING	SIZE	BUILDING WIDTH[c] (feet)					
		20		28		36	
		Span	NJ[d]	Span	NJ[d]	Span	NJ[d]
One floor only	2-2×4	3-1	1	2-8	1	2-5	1
	2-2×6	4-6	1	3-11	1	3-6	1
	2-2×8	5-9	1	5-0	2	4-5	2
	2-2×10	7-0	2	6-1	2	5-5	2
	2-2×12	8-1	2	7-0	2	6-3	2
	3-2×8	7-2	1	6-3	1	5-7	2
	3-2×10	8-9	1	7-7	2	6-9	2
	3-2×12	10-2	2	8-10	2	7-10	2
	4-2×8	9-0	1	7-8	1	6-9	1
	4-2×10	10-1	1	8-9	1	7-10	2
	4-2×12	11-9	1	10-2	2	9-1	2
Two floors	2-2×4	2-2	1	1-10	1	1-7	1
	2-2×6	3-2	2	2-9	2	2-5	2
	2-2×8	4-1	2	3-6	2	3-2	2
	2-2×10	4-11	2	4-3	2	3-10	3
	2-2×12	5-9	2	5-0	3	4-5	3
	3-2×8	5-1	2	4-5	2	3-11	2
	3-2×10	6-2	2	5-4	2	4-10	2
	3-2×12	7-2	2	6-3	2	5-7	3
	4-2×8	6-1	1	5-3	2	4-8	2
	4-2×10	7-2	2	6-2	2	5-6	2
	4-2×12	8-4	2	7-2	2	6-5	2

For SI: 1 inch = 25.4 mm, 1 foot = 304.8 mm.

a. Spans are given in feet and inches.

b. Tabulated values assume #2 grade lumber.

c. Building width is measured perpendicular to the ridge. For widths between those shown, spans are permitted to be interpolated.

d. NJ - Number of jack studs required to support each end. Where the number of required jack studs equals one, the header is permitted to be supported by an approved framing anchor attached to the full-height wall stud and to the header.

❖ In addition to providing the allowable spans for girders and headers, this table lists the number of jack studs required based on the bearing necessary for the girder or header. The table is broken down into two loading conditions, illustrated in Commentary Figure R502.5(3).

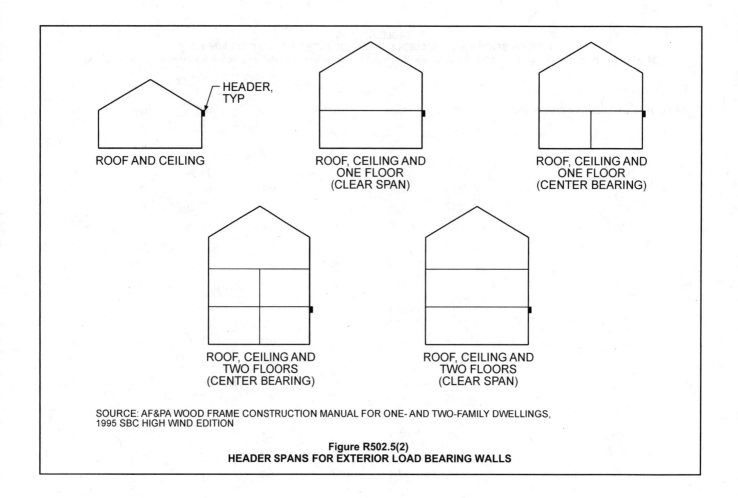

SOURCE: AF&PA WOOD FRAME CONSTRUCTION MANUAL FOR ONE- AND TWO-FAMILY DWELLINGS, 1995 SBC HIGH WIND EDITION

Figure R502.5(2)
HEADER SPANS FOR EXTERIOR LOAD BEARING WALLS

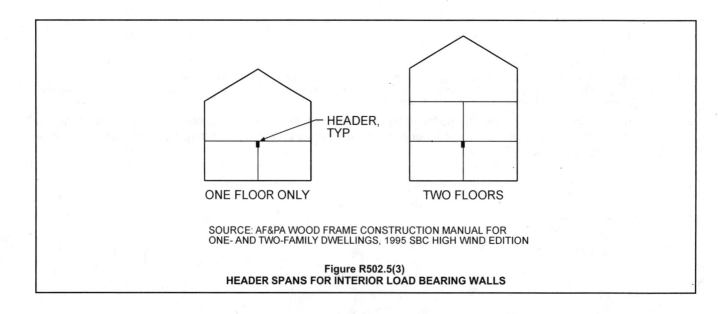

SOURCE: AF&PA WOOD FRAME CONSTRUCTION MANUAL FOR ONE- AND TWO-FAMILY DWELLINGS, 1995 SBC HIGH WIND EDITION

Figure R502.5(3)
HEADER SPANS FOR INTERIOR LOAD BEARING WALLS

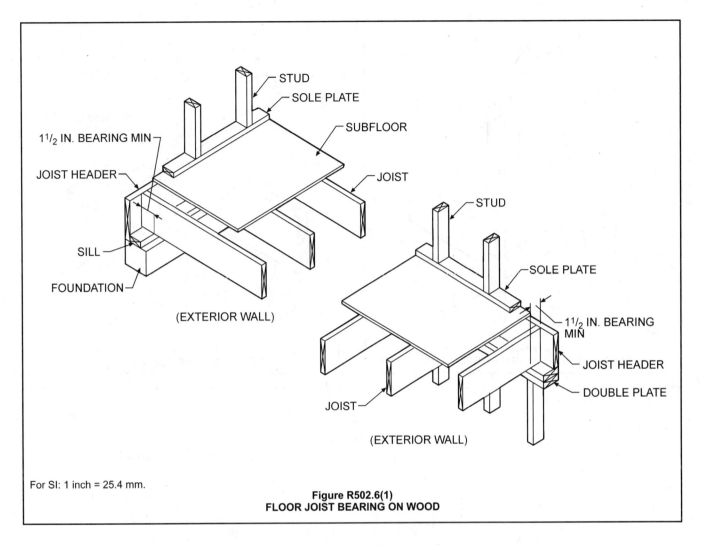

For SI: 1 inch = 25.4 mm.

Figure R502.6(1)
FLOOR JOIST BEARING ON WOOD

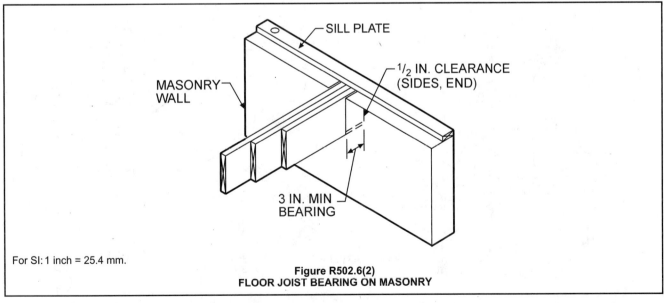

For SI: 1 inch = 25.4 mm.

Figure R502.6(2)
FLOOR JOIST BEARING ON MASONRY

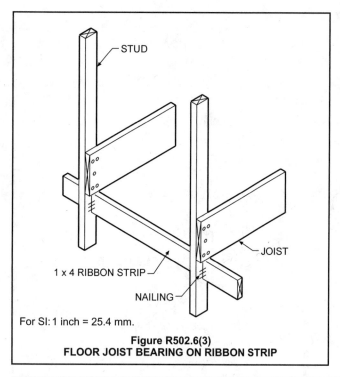

For SI: 1 inch = 25.4 mm.

Figure R502.6(3)
FLOOR JOIST BEARING ON RIBBON STRIP

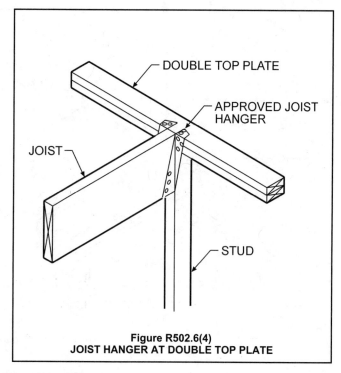

Figure R502.6(4)
JOIST HANGER AT DOUBLE TOP PLATE

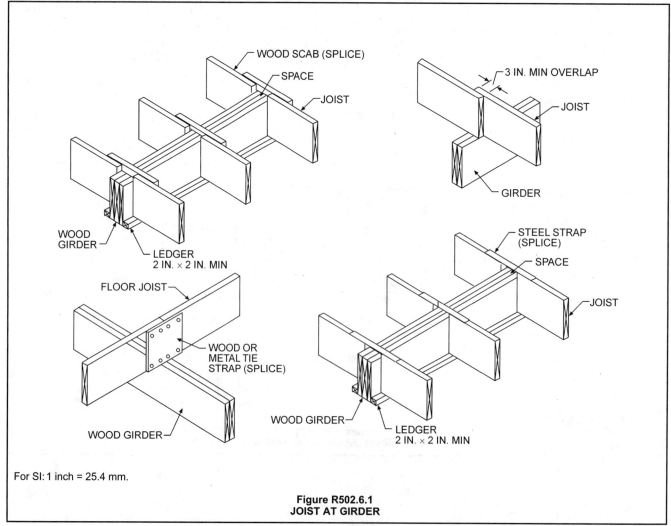

For SI: 1 inch = 25.4 mm.

Figure R502.6.1
JOIST AT GIRDER

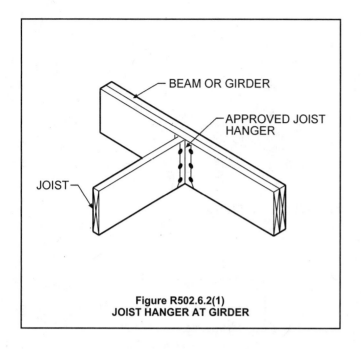

Figure R502.6.2(1)
JOIST HANGER AT GIRDER

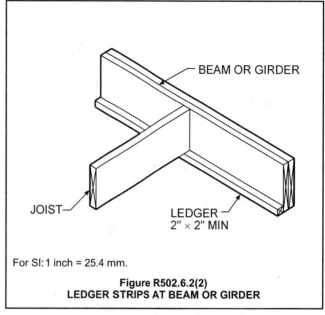

For SI: 1 inch = 25.4 mm.

Figure R502.6.2(2)
LEDGER STRIPS AT BEAM OR GIRDER

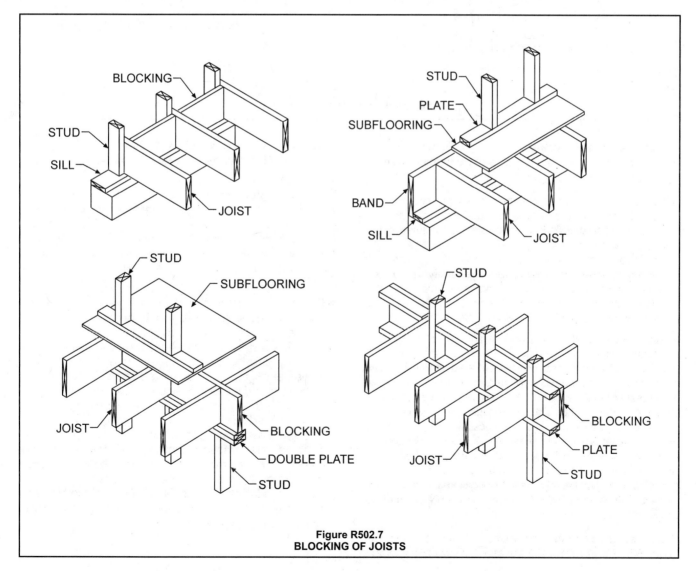

Figure R502.7
BLOCKING OF JOISTS

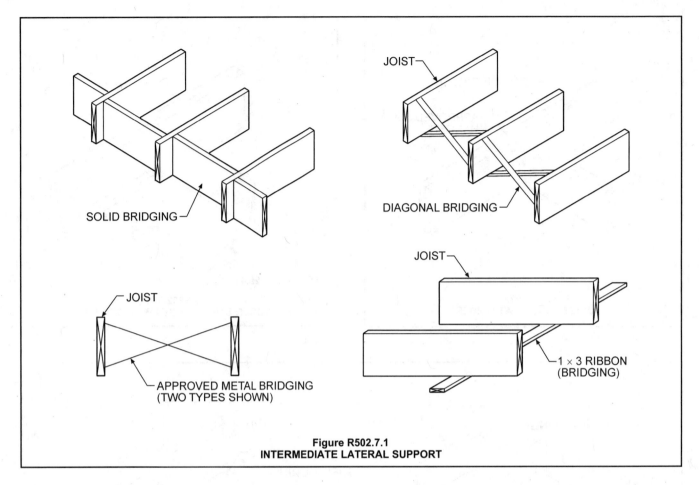

Figure R502.7.1
INTERMEDIATE LATERAL SUPPORT

R502.8 Drilling and notching. Structural floor members shall not be cut, bored or notched in excess of the limitations specified in this section. See Figure R502.8.

❖ Some designs and installation practices require that limited notching and cutting occur. Notching should be avoided when possible, and holes bored in beams and joists create the same problems as notches. When necessary, the holes should be located in areas with the least stress concentration, generally along the neutral axis of the joist. Limitations on the allowable cutting and notching of wood floor joists are meant to retain structural or functional integrity.

R502.8.1 Sawn lumber. Notches in solid lumber joists, rafters and beams shall not exceed one-sixth of the depth of the member, shall not be longer than one-third of the depth of the member and shall not be located in the middle one-third of the span. Notches at the ends of the member shall not exceed one-fourth the depth of the member. The tension side of members 4 inches (102 mm) or greater in nominal thickness shall not be notched except at the ends of the members. The diameter of holes bored or cut into members shall not exceed one-third the depth of the member. Holes shall not be closer than 2 inches (51 mm) to the top or bottom of the member, or to any other hole located in the member. Where the member is also notched, the hole shall not be closer than 2 inches (51 mm) to the notch.

❖ Cutting and notching limitations are illustrated in Figure R502.8. Additionally, the tension side of members

with a thickness of 4 inches (102 mm) or more can be notched at the ends only (see commentary, Section R502.8).

R502.8.2 Engineered wood products. Cuts, notches and holes bored in trusses, structural composite lumber, structural glue-laminated members or I-joists are prohibited except where permitted by the manufacturer's recommendations or where the effects of such alterations are specifically considered in the design of the member by a registered design professional.

❖ Cutting and notching limitations for sawn lumber do not apply to engineered wood products. Structural composite lumber is a generic term which encompasses a variety of engineered composite wood products including laminated veneer lumber (LVL). Also, included in the term are laminated strand lumber (LSL), parallel strand lumber (PSL), and oriented strand lumber (OSL). The prohibitions in this section apply to all of these products.

Engineered wood products must not be cut, notched or bored unless those alterations are considered in the design of the member. That consideration must come from either the manufacturer and be reflected in use recommendations (which is common in I-joists, permitting some limited alterations to webs) or from a registered design professional.

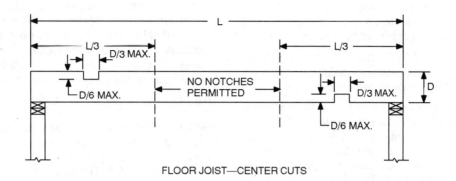

FLOOR JOIST—CENTER CUTS

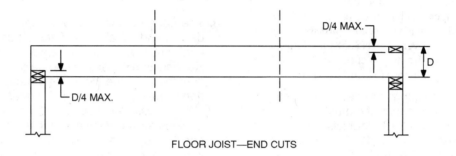

FLOOR JOIST—END CUTS

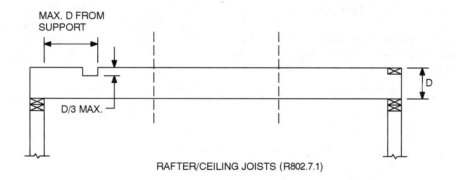

RAFTER/CEILING JOISTS (R802.7.1)

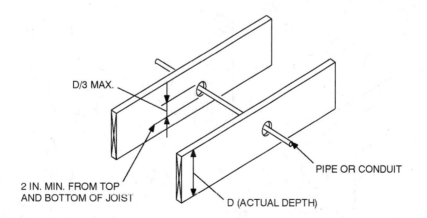

For SI: 1 inch = 25.4 mm.

FIGURE 502.8
CUTTING, NOTCHING AND DRILLING

R502.9 Fastening. Floor framing shall be nailed in accordance with Table R602.3(1). Where posts and beam or girder construction is used to support floor framing, positive connections shall be provided to ensure against uplift and lateral displacement.

❖ Commentary Figure R502.9 shows various methods of accomplishing the mandatory positive connection between post and beam or girder construction.

R502.10 Framing of openings. Openings in floor framing shall be framed with a header and trimmer joists. When the header joist span does not exceed 4 feet (1219 mm), the header joist may be a single member the same size as the floor joist. Single trimmer joists may be used to carry a single header joist that is located within 3 feet (914 mm) of the trimmer joist bearing. When the header joist span exceeds 4 feet (1219 mm), the trimmer joists and the header joist shall be doubled and of sufficient cross section to support the floor joists framing into the header. Approved hangers shall be used for the header joist to trimmer joist connections when the header joist span exceeds 6 feet (1829 mm). Tail joists over 12 feet (3658 mm) long shall be supported at the header by framing anchors or on ledger strips not less than 2 inches by 2 inches (51 mm by 51 mm).

❖ Where larger floor openings are necessary, adequate load-carrying transfer capability must be provided. Header joists the same size as floor joists may be used for spans not exceeding 4 feet (1219 mm), as shown in Commentary Figure R502.10(1). Trimmers must also be doubled to carry the additional load, unless the header joist is located within 3 feet (914 mm) of the trimmer joist bearing.

Where the header joist span exceeds 4 feet (1219 mm), a larger load-carrying capability is required. The header and trimmer joists must be doubled and be of

sufficient size to support floor joist framing as illustrated in Commentary Figure R502.10(2).

In some cases, nailing would be insufficient to transfer vertical loads. Therefore, positive connections as shown in Commentary Figure R502.10(3) must be used. These hangers are required when the header joist span exceeds 6 feet (1829 mm).

R502.11 Wood trusses.

R502.11.1 Design. Wood trusses shall be designed in accordance with approved engineering practice. The design and manufacture of metal plate connected wood trusses shall comply with ANSI/TPI 1. The truss design drawings shall be prepared by a registered professional where required by the statutes of the jurisdiction in which the project is to be constructed in accordance with Section R106.1.

❖ The code contains no prescriptive provisions for the design and installation of wood trusses. A design that conforms to accepted engineering practice is required. Recognizing the extensive use of trusses in residential construction, the code references ANSI/TPI 1, *National Design Standard for Metal-Plate-Connected Wood Truss Construction*. The standard contains regulations for the design and installation of metal-plate-connected wood trusses, including the procedures for full-scale tests and testing methods for evaluating metal plate connectors. In addition to adequate design, it is important that the trusses be handled and erected properly so that the performance capability of the trusses is not compromised. A truss member should never be cut without approval from the design engineer.

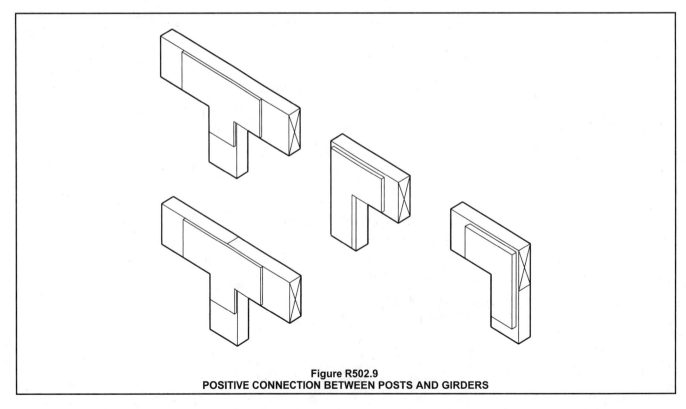

Figure R502.9
POSITIVE CONNECTION BETWEEN POSTS AND GIRDERS

Usually trusses are delivered in bundles, which reduces the potential for damage. Slings and spreader bars sized for the load should be used to reduce stresses caused by sway and bending.

Shop drawings showing the lumber schedule, design loads and panel point details (size, location and attachment of plates) should be filed with the building permit application and should be available at the time of inspection.

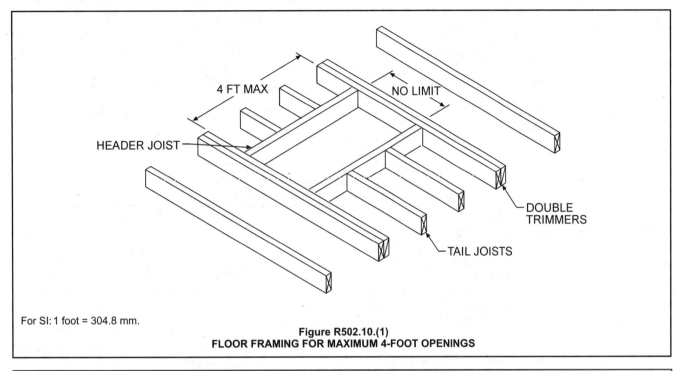

For SI: 1 foot = 304.8 mm.

Figure R502.10.(1)
FLOOR FRAMING FOR MAXIMUM 4-FOOT OPENINGS

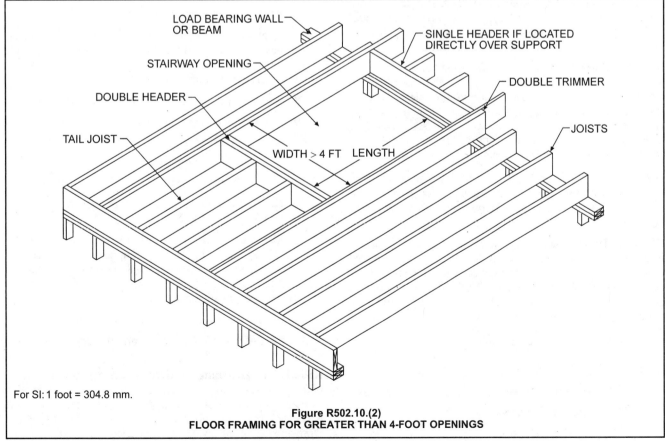

For SI: 1 foot = 304.8 mm.

Figure R502.10.(2)
FLOOR FRAMING FOR GREATER THAN 4-FOOT OPENINGS

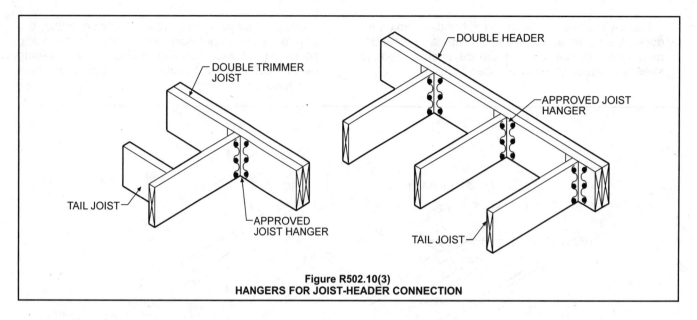

Figure R502.10(3)
HANGERS FOR JOIST-HEADER CONNECTION

R502.11.2 Bracing. Trusses shall be braced to prevent rotation and provide lateral stability in accordance with the requirements specified in the construction documents for the building and on the individual truss design drawings. In the absence of specific bracing requirements, trusses shall be braced in accordance with the Building Component Safety Information (BCSI 1-03) Guide to Good Practice for Handling, Installing & Bracing of Metal Plate Connected Wood Trusses.

❖ To prevent collapse during construction and until permanent bracing is installed, trusses should be adequately braced temporarily. When braced for use, trusses should be positioned as vertical as possible; tilted trusses will not perform as required.

R502.11.3 Alterations to trusses. Truss members and components shall not be cut, notched, spliced or otherwise altered in any way without the approval of a registered design professional. Alterations resulting in the addition of load (e.g., HVAC equipment, water heater, etc.), that exceed the design load for the truss, shall not be permitted without verification that the truss is capable of supporting the additional loading.

❖ Addition of loads in excess of the design load is allowed only if the additional capacity of the truss can be verified. Also see the commentary for Section R502.8.2.

R502.11.4 Truss design drawings. Truss design drawings, prepared in compliance with Section R502.11.1, shall be submitted to the building official and approved prior to installation. Truss design drawings shall be provided with the shipment of trusses delivered to the job site. Truss design drawings shall include, at a minimum, the information specified below:

1. Slope or depth, span and spacing.

2. Location of all joints.

3. Required bearing widths.

4. Design loads as applicable:

    4.1. Top chord live load;

    4.2. Top chord dead load;

    4.3. Bottom chord live load;

    4.4. Bottom chord dead load;

    4.5. Concentrated loads and their points of application; and

    4.6. Controlling wind and earthquake loads.

5. Adjustments to lumber and joint connector design values for conditions of use.

6. Each reaction force and direction.

7. Joint connector type and description, e.g., size, thickness or gauge, and the dimensioned location of each joint connector except where symmetrically located relative to the joint interface.

8. Lumber size, species and grade for each member.

9. Connection requirements for:

    9.1. Truss-to-girder-truss;

    9.2. Truss ply-to-ply; and

    9.3. Field splices.

10. Calculated deflection ratio and/or maximum description for live and total load.

11. Maximum axial compression forces in the truss members to enable the building designer to design the size, connections and anchorage of the permanent continuous lateral bracing. Forces shall be shown on the truss drawing or on supplemental documents.

12. Required permanent truss member bracing location.

❖ See the definition in Chapter 2 and the commentary for Section R502.11.1.

R502.12 Draftstopping required. When there is usable space both above and below the concealed space of a floor/ceiling assembly, draftstops shall be installed so that the area of the concealed space does not exceed 1,000 square feet (92.9 m²). Draftstopping shall divide the concealed space into approximately equal areas. Where the assembly is enclosed by a floor membrane above and a ceiling membrane below draftstopping shall be provided in floor/ceiling assemblies under the following circumstances:

1. Ceiling is suspended under the floor framing.

2. Floor framing is constructed of truss-type open-web or perforated members.

❖ Draftstopping is required to limit the spread of fire through combustible spaces in floor/ceiling assemblies when such spaces create a connected area beyond the normal joist cavity. For example, draftstopping is required when the ceiling is suspended under the floor framing, as illustrated in Commentary Figure R502.12(1), or when the floor framing is constructed of truss-type or open-web perforated members, as illustrated in Commentary Figure R502.12(2). When such space exceeds 1,000 square feet (92.9 m²) in area, the code requires the space between the ceiling membrane and the floor to be divided into two approximately equal areas with no area larger than 1,000 square feet (92.9 m²). A floor/ceiling assembly having a space exceeding 2,000 square feet (185.8 m²), but not exceeding 3,000 square feet (278.7 m²), would have to be divided into a minimum of three approximately equal areas.

R502.12.1 Materials. Draftstopping materials shall be less than $^1/_2$-inch (12.7 mm) gypsum board, $^3/_8$-inch (9.5 mm) wood structural panels, $^3/_8$-inch (9.5 mm) Type 2-M-W particleboard or other approved materials adequately supported. Draftstopping shall be installed parallel to the floor framing members unless otherwise approved by the building official. The integrity of all draftstops shall be maintained.

❖ Materials used to meet the draftstopping requirements are typically $^1/_2$-inch (12.7 mm) gypsum board, $^3/_8$-inch (9.5 mm) plywood or $^3/_8$-inch (9.5 mm) Type 2-M-W particle board adequately attached to the supporting members. Other materials may be used for draftstopping, if those materials produce an equivalent barrier.

The code places an additional restriction on the specific location of the draftstopping, mandating that the draftstopping be placed parallel to the main framing members. Commentary Figures R502.12(1) and (2) illustrate two examples of draftstopping orientation placed as required.

When a penetration through the draftstopping is necessary, the integrity of the draftstop must be maintained in a manner similar to firestopping penetrations.

R502.13 Fireblocking required. Fireblocking shall be provided in accordance with Section R602.8.

❖ See the commentary for Section R602.8.

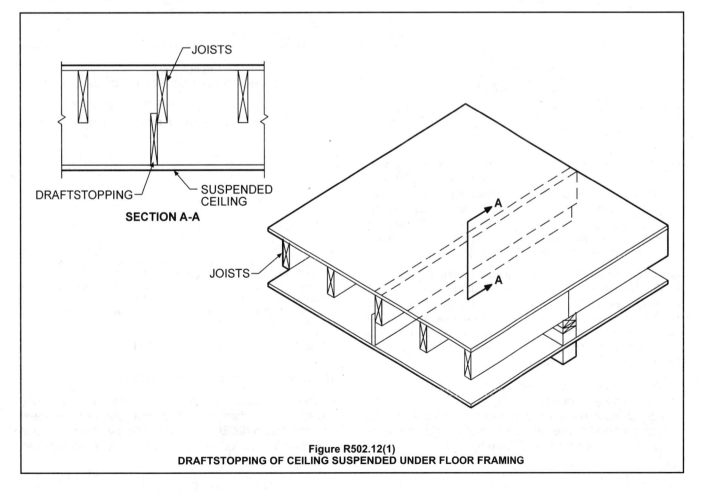

Figure R502.12(1)
DRAFTSTOPPING OF CEILING SUSPENDED UNDER FLOOR FRAMING

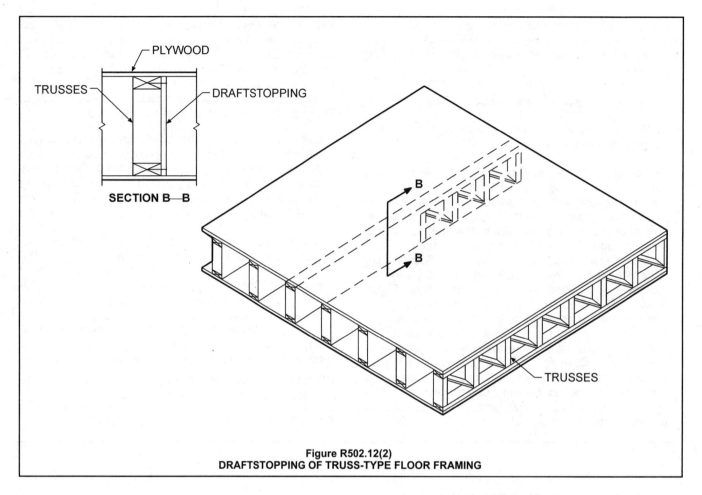

Figure R502.12(2)
DRAFTSTOPPING OF TRUSS-TYPE FLOOR FRAMING

SECTION R503
FLOOR SHEATHING

R503.1 Lumber sheathing. Maximum allowable spans for lumber used as floor sheathing shall conform to Tables R503.1, R503.2.1.1(1) and R503.2.1.1(2).

❖ Table R503.1 of the code sets forth the required thickness of lumber used as floor sheathing. The allowable spans are based on the floor sheathing thickness, joist spacing and the orientation of the sheathing with respect to the joist. Commentary Figures R503.1(1) and R503.1(2) show floor sheathing applications that are perpendicular and diagonal to the joist.

R503.1.1 End joints. End joints in lumber used as subflooring shall occur over supports unless end-matched lumber is used, in which case each piece shall bear on at least two joists. Subflooring may be omitted when joist spacing does not exceed 16 inches (406 mm) and a 1-inch (25.4 mm) nominal tongue-and-groove wood strip flooring is applied perpendicular to the joists.

❖ When lumber is to be used as subflooring, adequate load-carrying capability and continuity must be provided through either the proper placement of end joints over supports or the use of end-matched lumber as illustrated in Commentary Figure R503.1.1.

TABLE R503.1
MINIMUM THICKNESS OF LUMBER FLOOR SHEATHING

JOIST OR BEAM SPACING (inches)	MINIMUM NET THICKNESS	
	Perpendicular to joist	Diagonal to joist
24	$^{11}/_{16}$	$^3/_4$
16	$^5/_8$	$^5/_8$
48[a]		
54[b]	$1^1/_2$ T & G	N/A
60[c]		

For SI: 1 inch = 25.4 mm, 1 pound per square inch = 6.895 kPa.

a. For this support spacing, lumber sheathing shall have a minimum F_b of 675 and minimum E of 1,100,000 (see AF&PA/NDS).

b. For this support spacing, lumber sheathing shall have a minimum F_b of 765 and minimum E of 1,400,000 (see AF&PA/NDS).

c. For this support spacing, lumber sheathing shall have a minimum F_b of 855 and minimum E of 1,700,000 (see AF&PA/NDS).

❖ See the commentary for Section R503.1.

R503.2 Wood structural panel sheathing.

❖ Wood structural panels are manufactured with fully waterproof adhesive and include plywood, oriented strand board (OSB) and composite panels made up of a combination of wood veneers and reconstructed wood layers. Additional information on wood structural panels is available from the APA—The Engineered Wood Association (formerly the American Plywood Association), other trade groups and manufacturers.

Plywood is a wood structural panel manufactured by gluing three or more cross-laminated wood layers together. Plywood may be manufactured in accordance with PS 1 or PS 2. It is inspected, and if it is certified, the trademark of an approved testing and grading agency is stamped on it.

OSB panels are fabricated out of multiple layers of wood flakes or strands called "furnish." Like the layers in plywood, these layers of furnish are oriented 90 degrees (1.57 rad) to each other. This gives OSB panels properties that are very similar to those of plywood.

Composite panels are structural panels made up of a combination of wood veneers and wood–based materials. The wood veneer usually forms the two outer layers and may also be used in the core of the panel. These layers of veneer and furnish may be cross-laminated.

Both OSB and composite panels are manufactured in accordance with PS 2. They are inspected, and if they are certified, the trademark of an approved testing and grading agency is stamped on them. The grade stamps of plywood, OSB and composite panels are almost identical in appearance, and in most cases, wood structural panels can be specified without regard to specific panel type.

Wood structural panels can be manufactured with Exterior, Exposure 1, Exposure 2 and Interior expo-sure durability classifications. Exterior and Exposure 1 are by far the most common classifications and both are manufactured with the same class of fully waterproof adhesives. Exterior panels are intended for applications subject to permanent exposure to the weather or moisture. Exposure 1 panels are intended for applications where long construction delays may be expected prior to protection being installed. Exposure 2 panels may be used for protected applications that are not continuously exposed to high humidity conditions. Interior panels may be used only for permanently protected interior applications. Exposure 2 and Interior panels are not readily available.

The intended end use of wood structural panels is designated in their grade-marking stamp.

R503.2.1 Identification and grade. Wood structural panel sheathing used for structural purposes shall conform to DOC PS 1, DOC PS 2 or, when manufactured in Canada, CSA 0437 or CSA 0325. All panels shall be identified by a grade mark of certificate of inspection issued by an approved agency.

❖ Plywood that performs a load-carrying function in floor construction must conform to a known quality control standard; therefore, a method of identifying the limitations under which the plywood may be used is needed. This is accomplished by a grade mark or certificate of inspection issued by an approved agency. Examples of plywood grade marks are shown in Commentary Figure R503.2.1.

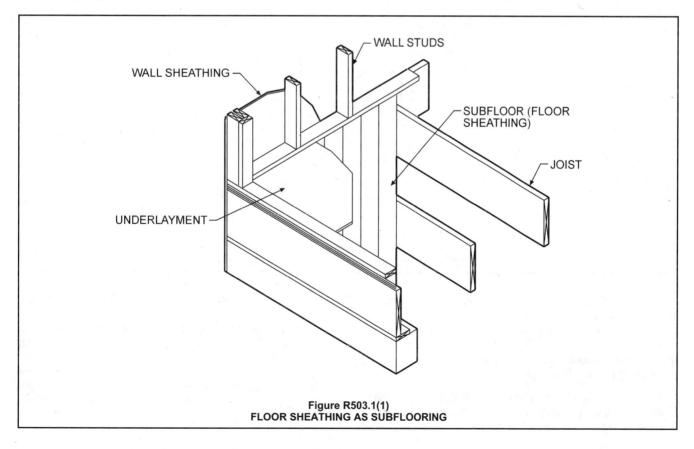

Figure R503.1(1)
FLOOR SHEATHING AS SUBFLOORING

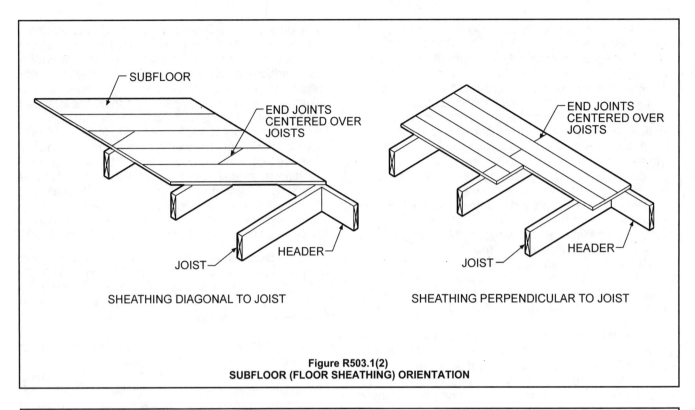

Figure R503.1(2)
SUBFLOOR (FLOOR SHEATHING) ORIENTATION

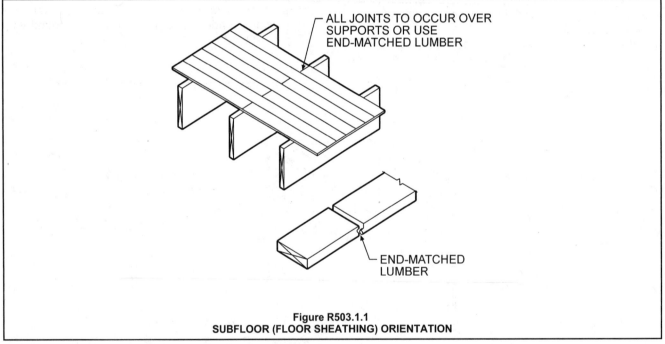

Figure R503.1.1
SUBFLOOR (FLOOR SHEATHING) ORIENTATION

R503.2.1.1 Subfloor and combined subfloor underlayment.
Where used as subflooring or combination subfloor
underlayment, wood structural panels shall be of one of the
grades specified in Table R503.2.1.1(1). When sanded plywood
is used as combination subfloor underlayment, the grade shall be
as specified in Table R503.2.1.1(2).

❖ This section specifies grades of wood structural panels when they are used as subflooring or combination
subflooring underlayment. It also addresses the use of

sanded plywood as a combination subfloor
underlayment.

A floor system could consist of three elements: (1) a
subfloor, (2) an underlayment or a combination
subfloor/underlayment system and (3) a finished floor
surface material. A finished surfacing material may be
wood-strip flooring; tongue-and-groove flooring or various types of resilient flooring coverings such as vinyl,
tile or carpeting.

Wood structural panels are manufactured for use as either structural subfloor or combination subfloor underlayment. The allowable spans for structural subflooring and combination subfloor underlayment are based on the wood structural panels' face grain strength axis parallel to its supporting member, or they are based on the panels being continuous over two or more spans, with a face grain placed perpendicular to the supports. These qualifications are critical in determining the permissible spans. Most wood structural panels are considerably stronger when their face grain is parallel to the supports and continuous over two or more spans. Panels with multiple spans have greater capacity than those that are simply supported between the two joists.

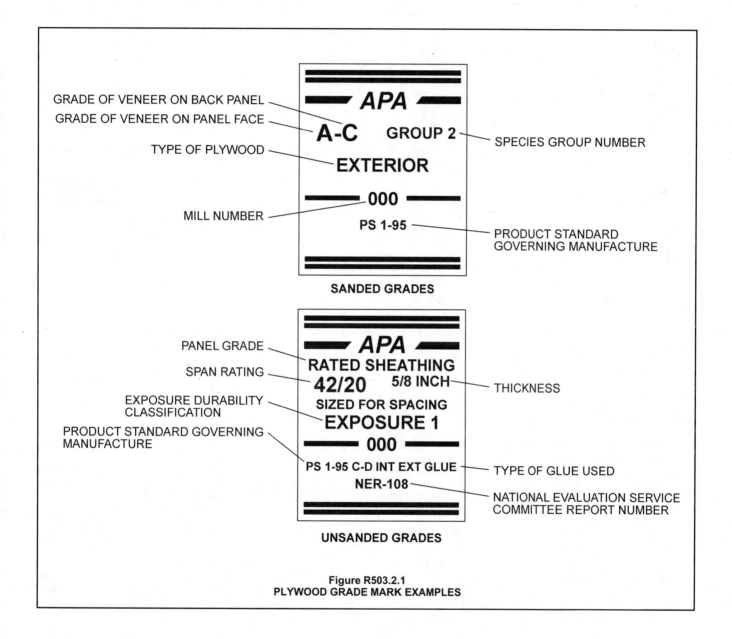

Figure R503.2.1
PLYWOOD GRADE MARK EXAMPLES

TABLE R503.2.1.1(1)
**ALLOWABLE SPANS AND LOADS FOR WOOD STRUCTURAL PANELS FOR ROOF
AND SUBFLOOR SHEATHING AND COMBINATION SUBFLOOR UNDERLAYMENT[a, b, c]**

SPAN RATING	MINIMUM NOMINAL PANEL THICKNESS (inch)	ALLOWABLE LIVE LOAD (psf)[h, l]		MAXIMUM SPAN (inches)		LOAD (pounds per square foot, at maximum span)		MAXIMUM SPAN (inches)
		SPAN @ 16" o.c.	SPAN @ 24" o.c.	With edge support[d]	Without edge support	Total load	Live load	
Sheathing[e]				Roof[f]				Subfloor[j]
12/0	$^5/_{16}$	—	—	12	12	40	30	0
16/0	$^5/_{16}$	30	—	16	16	40	30	0
20/0	$^5/_{16}$	50	—	20	20	40	30	0
24/0	$^3/_8$	100	30	24	20[g]	40	30	0
24/16	$^7/_{16}$	100	40	24	24	50	40	16
32/16	$^{15}/_{32}, ^1/_2$	180	70	32	28	40	30	16[h]
40/20	$^{19}/_{32}, ^5/_8$	305	130	40	32	40	30	20[h,i]
48/24	$^{23}/_{32}, ^3/_{48}$	—	175	48	36	45	35	24
60/32	$^7/_8$	—	305	60	48	45	35	32
Underlayment, C-C plugged, single floor[e]				Roof[f]				Combination subfloor underlayment[k]
16 o.c.	$^{19}/_{32}, ^5/_8$	100	40	24	24	50	40	16[i]
20 o.c.	$^{19}/_{32}, ^5/_8$	150	60	32	32	40	30	20[i,j]
24 o.c.	$^{23}/_{32}, ^3/_4$	240	100	48	36	35	25	24
32 o.c.	$^7/_8$	—	185	48	40	50	40	32
48 o.c.	$^{13}/_{32}, 1^1/_8$	—	290	60	48	50	40	48

For SI: 1 inch = 25.4 mm, 1 pound per square foot = 0.0479 kPa.

a. The allowable total loads were determined using a dead load of 10 psf. If the dead load exceeds 10 psf, then the live load shall be reduced accordingly.

b. Panels continuous over two or more spans with long dimension perpendicular to supports. Spans shall be limited to values shown because of possible effect of concentrated loads.

c. Applies to panels 24 inches or wider.

d. Lumber blocking, panel edge clips (one midway between each support, except two equally spaced between supports when span is 48 inches), tongue-and-groove panel edges, or other approved type of edge support.

e. Includes Structural 1 panels in these grades.

f. Uniform load deflection limitation: $^1/_{180}$ of span under live load plus dead load, $^1/_{240}$ of span under live load only.

g. Maximum span 24 inches for $^{15}/_{32}$-and $^1/_2$-inch panels.

h. Maximum span 24 inches where $^3/_4$-inch wood finish flooring is installed at right angles to joists.

i. Maximum span 24 inches where 1.5 inches of lightweight concrete or approved cellular concrete is placed over the subfloor.

j. Unsupported edges shall have tongue-and-groove joints or shall be supported with blocking unless minimum nominal $^1/_4$-inch thick underlayment with end and edge joints offset at least 2 inches or 1.5 inches of lightweight concrete or approved cellular concrete is placed over the subfloor, or $^3/_4$-inch wood finish flooring is installed at right angles to the supports. Allowable uniform live load at maximum span, based on deflection of $^1/_{360}$ of span, is 100 psf.

k. Unsupported edges shall have tongue-and-groove joints or shall be supported by blocking unless nominal $^1/_4$-inch-thick underlayment with end and edge joints offset at least 2 inches or $^3/_4$-inch wood finish flooring is installed at right angles to the supports. Allowable uniform live load at maximum span, based on deflection of $^1/_{360}$ of span, is 100 psf, except panels with a span rating of 48 on center are limited to 65 psf total uniform load at maximum span.

l. Allowable live load values at spans of 16" o.c. and 24" o.c taken from reference standard APA E30, APA Engineered Wood Construction Guide. Refer to reference standard for allowable spans not listed in the table.

❖ The maximum spans for wood structural panel floor and roof sheathing are limited by the stresses and deflection imposed by the design live loads. For convenience, the trademarks of the inspection agencies include a span rating, which appears as two numbers separated by a slash (32/16 or 48/24, for example). The first number represents the maximum recommended span for roof sheathing when the panels are applied with the long dimension across three or more supports and the edges are blocked or when other support required by the table is provided. The second number indicates the maximum recommended span when the panel is used for structural floor sheathing with the panels applied with the long dimension across three or more supports. Single panels intended for single floor applications will be marked with a single number representing this span.

The edges of the wood structural panels between floor supports are prevented from moving relative to each other by tongue and groove panel edges, by the addition of wood blocking or by the use of an approved underlayment or structural finished floor system.

TABLE R503.2.1.1(2)
ALLOWABLE SPANS FOR SANDED PLYWOOD
COMBINATION SUBFLOOR UNDERLAYMENT[a]

IDENTIFICATION	SPACING OF JOISTS (inches)		
	16	20	24
Species group[b]	—	—	—
1	$^1/_2$	$^5/_8$	$^3/_4$
2, 3	$^5/_8$	$^3/_4$	$^7/_8$
4	$^3/_4$	$^7/_8$	1

For SI: 1 inch = 25.4 mm, 1 pound per square foot = 0.0479 kPa.

a. Plywood continuous over two or more spans and face grain perpendicular to supports. Unsupported edges shall be tongue-and-groove or blocked except where nominal $^1/_4$-inch-thick underlayment or $^3/_4$-inch wood finish floor is used. Allowable uniform live load at maximum span based on deflection of $^1/_{360}$ of span is 100 psf.

b. Applicable to all grades of sanded exterior-type plywood.

❖ For exterior-type sanded plywood grades, the thickness required for a specific span is related to the species grouping of the panel used. For example, species Group 1 may be $^1/_2$ inch (12.7 mm) for 16-inch (406 mm) joist spacing, but Group 4 will require a thickness of $^3/_4$ inch (19.1 mm) for the same span.

When the panels are used as a combination subfloor/underlayment, extra precautions should be observed in attaching the panels to the floor framing. Joints in adjacent panels should not be continuous and should not occur at locations where the orientation of the joist supports is different.

R503.2.2 Allowable spans. The maximum allowable span for wood structural panels used as subfloor or combination subfloor underlayment shall be as set forth in Table R503.2.1.1(1), or APA E30. The maximum span for sanded plywood combination subfloor underlayment shall be as set forth in Table R503.2.1.1(2).

❖ Table R503.2.1.1(1) indicates spans for plywood and wood structural panels used as subflooring. This table also covers roof sheathing and is referenced in Section R803.2.2 for that purpose. The span limitations are predicated on the grade of plywood used. In the case of rated sheathing used as structural subflooring, the maximum span is easily identified through the panel span rating (identification index). Where the panel span rating is stamped on the sheet of plywood, the denominator represents the allowable span of the plywood floor sheathing [see Commentary Figure R503.2.2(1)]. An illustration of the limitations imposed by Notes j and k to Table R503.2.1.1(1) on combination subfloor underlayment applications is illustrated in Commentary Figure R503.2.2(2). Allowable increases in the maximum span for certain span ratings are permitted as described in Notes h and i to Table R503.2.1.1(1).

In lieu of using Table R503.2.1.1(1), the code permits the use of APA E30, *Engineered Wood Construction Guide*.

When plywood combination subfloor underlayment is to be used, the allowable span as listed in Table R503.2.1.1(2) is applicable to underlayment grade C-C plugged and sanded exterior-type plywood of specific species.

R503.2.3 Installation. Wood structural panels used as subfloor or combination subfloor underlayment shall be attached to wood framing in accordance with Table R602.3(1) and shall be attached to cold-formed steel framing in accordance with Table R505.3.1(2).

❖ This section refers to the wood framing fastener schedule in Chapter 6 and the cold-formed steel floor-fastening schedule in Table R505.3.1(2).

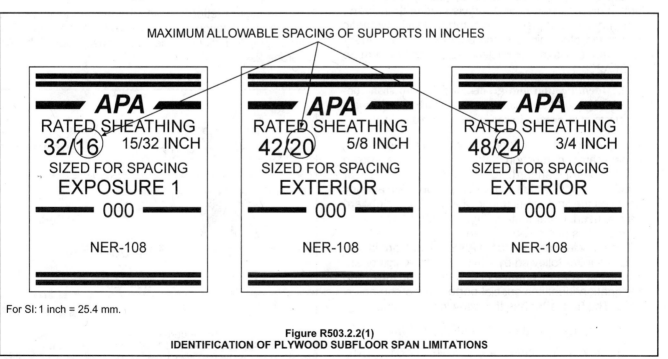

For SI: 1 inch = 25.4 mm.

Figure R503.2.2(1)
IDENTIFICATION OF PLYWOOD SUBFLOOR SPAN LIMITATIONS

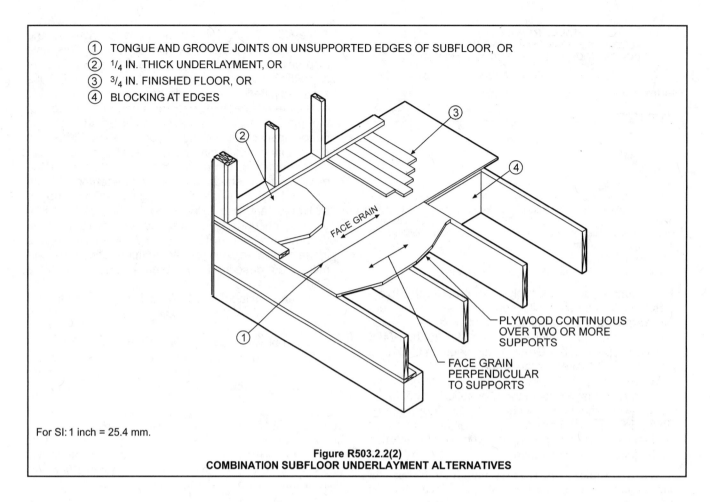

① TONGUE AND GROOVE JOINTS ON UNSUPPORTED EDGES OF SUBFLOOR, OR
② ¹/₄ IN. THICK UNDERLAYMENT, OR
③ ³/₄ IN. FINISHED FLOOR, OR
④ BLOCKING AT EDGES

FACE GRAIN

③ PLYWOOD CONTINUOUS OVER TWO OR MORE SUPPORTS

FACE GRAIN PERPENDICULAR TO SUPPORTS

For SI: 1 inch = 25.4 mm.

Figure R503.2.2(2)
COMBINATION SUBFLOOR UNDERLAYMENT ALTERNATIVES

R503.3 Particleboard.

❖ Particleboard is a generic term given to panels manufactured from cellulosic materials, usually wood in the form of discrete pieces and particles rather than fibers. The particles and pieces are combined with synthetic resins and other binders and bonded together under heat and pressure. Particleboard must conform to ANSI A208.1.

R503.3.1 Identification and grade. Particleboard shall conform to ANSI A208.1 and shall be so identified by a grade mark or certificate of inspection issued by an approved agency.

❖ Particleboard used in floor construction that performs a load-carrying function must conform to a known quality control standard. To this end, a grade mark or certificate of inspection issued by an approved agency identifies the limitations under which the particleboard may be used. Commentary Figure R503.3.1 shows an example of a particleboard grade mark.

The particleboard grades in ANSI A208.1 are identified by a letter followed by a hyphen and a digit or another letter. A letter, number or term following the grade designation identifies special performance characteristics. The first letter has the following meaning:

H: High density (above 50 lb/ft³)(801 kg/m³),

M: Medium density (40—50 lb/ft³) (641-801 kg/m³),

LD: Low density (below 40 lb/ft³) (641 kg/m³),

D: Manufactured home decking

PBU: Underlayment.

The digit or letter following the hyphen indicates the grade identification within a particular description. For example, 2 indicates medium density particleboard, Grade 2.

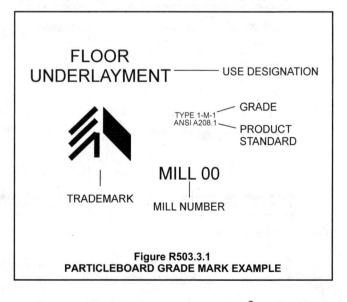

FLOOR UNDERLAYMENT —— USE DESIGNATION

TYPE 1-M-1 —— GRADE
ANSI A208.1 —— PRODUCT STANDARD

MILL 00

TRADEMARK   MILL NUMBER

Figure R503.3.1
PARTICLEBOARD GRADE MARK EXAMPLE

R503.3.2 Floor underlayment. Particleboard floor underlayment shall conform to Type PBU and shall not be less than $\frac{1}{4}$ inch (6.4 mm) in thickness.

❖ Particleboard is often used over a structural subfloor to provide a smooth, even surface under textile or resilient type finish floors. Particleboard must conform to Type PBU with a minimum thickness of $\frac{1}{4}$ inch (6.4 mm).

R503.3.3 Installation. Particleboard underlayment shall be installed in accordance with the recommendations of the manufacturer and attached to framing in accordance with Table R602.3(1).

❖ Particleboard used as underlayment only is not subject to span limitations because the plywood or lumber subflooring provides the load-carrying capability for the floor system; the particleboard has a nonstructural role. However, joints in particleboard should not occur over joints in the subfloor; the joints should be staggered.

SECTION R504
PRESSURE-PRESERVATIVE-TREATED WOOD FLOORS (ON GROUND)

R504.1 General. Pressure-preservative-treated wood basement floors and floors on ground shall be designed to withstand axial forces and bending moments resulting from lateral soil pressures at the base of the exterior walls and floor live and dead loads. Floor framing shall be designed to meet joist deflection requirements in accordance with Section R301.

❖ As indicated in the commentary for Section R404.1.1, lateral soil pressure loads are carried by the foundation walls through bending in the vertical direction. The resulting forces (reactions) are resisted by the basement floor and the first floor above. This section stipulates the requirements for basement wood-floor framing in order for it to resist the applied loads at the base of the foundation wall. It is implicit that this type of floor would be used in conjunction with wood foundations described in Sections R401.1.

R504.1.1 Unbalanced soil loads. Unless special provision is made to resist sliding caused by unbalanced lateral soil loads, wood basement floors shall be limited to applications where the differential depth of fill on opposite exterior foundation walls is 2 feet (610 mm) or less.

❖ The requirements of foundation walls permit up to 4 feet (1219 mm) of unbalanced fill (see Table R404.2.3). When exterior foundation walls on opposite sides of a building support an equal (or nearly equal) height of unbalanced fill, sliding is not a concern. When there is a difference in the amount of fill supported, however, there is a net horizontal force that must be resisted; otherwise, sliding may occur. To limit this unbalanced force, the height of differential fill is accordingly limited to 2 feet (610 mm) for this floor system.

R504.1.2 Construction. Joists in wood basement floors shall bear tightly against the narrow face of studs in the foundation wall or directly against a band joist that bears on the studs. Ply-wood subfloor shall be continuous over lapped joists or over butt joints between in-line joists. Sufficient blocking shall be provided between joists to transfer lateral forces at the base of the end walls into the floor system.

❖ These construction requirements provide a load path for the horizontal force described in the previous section.

R504.1.3 Uplift and buckling. Where required, resistance to uplift or restraint against buckling shall be provided by interior bearing walls or properly designed stub walls anchored in the supporting soil below.

❖ This is a general caution that the effects of uplift forces on the floor system, which could result from fluctuations in the water table, should be considered.

R504.2 Site preparation. The area within the foundation walls shall have all vegetation, topsoil and foreign material removed, and any fill material that is added shall be free of vegetation and foreign material. The fill shall be compacted to assure uniform support of the pressure preservatively treated-wood floor sleepers.

❖ This requirement is similar to Section R506.2. See the commentary for Section R506.2.2.

R504.2.1 Base. A minimum 4-inch-thick (102 mm) granular base of gravel having a maximum size of $\frac{3}{4}$ inch (19.1 mm) or crushed stone having a maximum size of $\frac{1}{2}$ inch (12.7 mm) shall be placed over the compacted earth.

❖ This requirement is similar to Section R506.2.2. See the commentary for Section R506.2.2.

R504.2.2 Moisture barrier. Polyethylene sheeting of minimum 6-mil (0.15 mm) thickness shall be placed over the granular base. Joints shall be lapped 6 inches (152 mm) and left unsealed. The polyethylene membrane shall be placed over the pressure preservatively treated-wood sleepers and shall not extend beneath the footing plates of the exterior walls.

❖ See the commentary for Section R506.2.2 concerning application of the moisture barrier.

R504.3 Materials. All framing materials, including sleepers, joists, blocking and plywood subflooring, shall be pressure-preservative treated and dried after treatment in accordance with AWPA U1 (Commodity Specification A, Use Category 4B and section 5.2), and shall bear the label of an accredited agency.

❖ To resist potential decay, wood floor material is to be treated in accordance with AWPA U1, use category system: *User Specification for Treated Wood.* The wood materials must be treated as Use Category 4B(UC4B), Commodity Specification A, in accordance with AWPA U1. Also the material must bear the label of an accredited agency to ensure that the treated wood is produced under an accredited quality auditing program.

SECTION R505
STEEL FLOOR FRAMING

R505.1 Cold-formed steel floor framing. Elements shall be straight and free of any defects that would significantly affect structural performance. Cold-formed steel floor framing members shall comply with the requirements of this section.

❖ The provisions of this section apply to the construction of floor systems using cold-formed steel framing, which is a type of construction made up in part or entirely of steel structural members cold formed to shape from sheet or strip steel. The general shape of the framing members is in the form of the letter "C" (C-shape) and is described in further detail in Section R505.2. The use of cold-formed steel-framed designs not consistent with these provisions is beyond the scope of this code, and an analysis by a design professional is recommended.

R505.1.1 Applicability limits. The provisions of this section shall control the construction of steel floor framing for buildings not greater than 60 feet (18,288 mm) in length perpendicular to the joist span, not greater than 40 feet (12 192 mm) in width parallel to the joist span, and not greater than two stories in height. Steel floor framing constructed in accordance with the provisions of this section shall be limited to sites subjected to a maximum design wind speed of 110 miles per hour (49

m/s), Exposure A, B, or C, and a maximum ground snow load of 70 psf (3.35 kPa).

❖ Commentary Figure R505.1.1 depicts building limitations. These limits are repeated in Section R603.1.1 for walls and in Section R804.1.1 for roofs. Steel floor framing constructed in accordance with the provisions of this section must be limited to sites with a maximum design wind speed of 110 miles per hour (49 m/s); Exposure A, B, or C and a maximum ground snow load of 70 pounds per square foot (3.35 kPa).

R505.1.2 In-line framing. When supported by steel-framed walls in accordance with Section R603, steel floor framing shall be constructed with floor joists located directly in-line with load-bearing studs located below the joists with a maximum tolerance of $^3/_4$ inch (19.1 mm) between the center lines of the joist and the stud.

❖ In-line framing is the preferred framing method because it provides a direct load path for the transfer of forces from joists to studs. In-line framing maximizes framing alignment in order to minimize secondary moments on the framing members, taking into account that the track cannot function as a load-transfer member. The $^3/_4$-inch (19 mm) alignment value is accepted industry practice and addresses those conditions where two framing members (e.g., roof rafter and ceil-

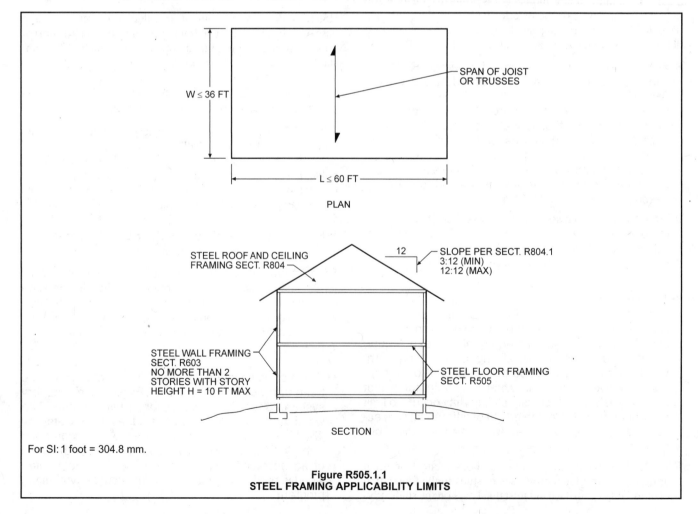

For SI: 1 foot = 304.8 mm.

Figure R505.1.1
STEEL FRAMING APPLICABILITY LIMITS

ing joist) are to be aligned over a wall stud. In the absence of in-line framing, a load distribution member, such as a structural track, may be required for this force transfer.

R505.1.3 Floor trusses. The design, quality assurance, installation and testing of cold-formed steel trusses shall be in accordance with the AISI Standard for Cold-formed Steel Framing-Truss Design (COFS/Truss). Truss members shall not be notched, cut or altered in any manner without an approved design.

❖ The code contains no prescriptive provisions for the design and installation of cold-formed steel trusses. A design is required that complies with accepted engineering practice. Recognizing the extensive use of trusses in residential construction, the code references AISI, *Standard for Cold-Formed Steel Framing —Truss Design.* The standard provides regulations for the design and installation of cold-formed steel trusses, including quality assurance and the procedures for full-scale tests. In addition to adequate design, it is important that the handling and erection of the trusses be performed properly so that the performance capability of the trusses is not compromised. A truss member must never be cut, notched or altered without approval from the design professional.

R505.2 Structural framing. Load-bearing floor framing members shall comply with Figure R505.2(1) and with the dimensional and minimum thickness requirements specified in Tables R505.2(1) and R505.2(2). Tracks shall comply with Figure R505.2(2) and shall have a minimum flange width of $1^{1}/_{4}$ inches (32 mm). The maximum inside bend radius for members shall be the larger of $^{3}/_{32}$ inch (2.4 mm) or twice the uncoated steel thickness. Holes in joist webs shall comply with all of the following conditions:

1. Holes shall conform to Figure R505.2(3);

2. Holes shall be permitted only along the centerline of the web of the framing member;

3. Holes shall have a center-to-center spacing of not less than 24 inches (610 mm);

4. Holes shall have a web hole width not greater than 0.5 times the member depth, or $2^{1}/_{2}$ inches (64.5 mm);

5. Holes shall have a web hole length not exceeding $4^{1}/_{2}$ inches (114 mm); and

6. Holes shall have a minimum distance between the edge of the bearing surface and the edge of the web hole of not less than 10 inches (254 mm).

Framing members with web holes not conforming to the above requirements shall be patched in accordance with Section R505.3.6 or designed in accordance with accepted engineering practices.

❖ Joists must comply with the dimensional and minimum thickness requirements of Tables R505.2(1) and R505.2(2), respectively. Note a to Table R505.2(1) explains the meaning of the alphanumeric member designation used for all steel framing members in the IRC. This system replaces the varied designation approaches that were used by each individual manufacturer. In addition, the designation is used to identify not only a specific steel framing member but also the section properties of that same member through the use of the product technical information document.

Steel thickness is expressed in mils, which pertains to the base metal thickness measured prior to painting or the application of metallic (corrosion resistant) coatings. The base metal thickness is typically stamped or embossed on the member by the manufacturer. Table R505.2(2) also references the equivalent thickness in inches and provides a further reference to the gage number. The use of gage numbers to define the thickness (a specific mean decimal thickness for which rolling tolerances were established) of steel sheet is a practice that has been discontinued by the steel industry.

To allow for routing utilities through steel-framed floors, holes (also referred to as penetrations, utility holes and punchouts) are permitted in the webs of floor joists. To avoid adversely affecting the strength of floor joists, the web holes are limited to the locations along the centerline of the web of the framing member as illustrated in Figure R505.2(3).

To avoid overstressing the joist web through high shear, these web holes must occur more than 10 inches (254 mm) from the edge of a load-bearing surface. Patching of web holes must extend at least 1 inch (25.4 mm) beyond all edges of the hole, and it must be fastened to the web with No. 8 screws having a mini-

TABLE R505.2(1)
COLD-FORMED STEEL JOIST SIZES

MEMBER DESIGNATION[a]	WEB DEPTH (inches)	MINIMUM FLANGE WIDTH (inches)	MAXIMUM FLANGE WIDTH (inches)	MINIMUM LIP SIZE (inches)
550S162-t	5.5	1.625	2	0.5
800S162-t	8	1.625	2	0.5
1000S162-t	10	1.625	2	0.5
1200S162-t	12	1.625	2	0.5

For SI: 1 inch = 25.4 mm, 1 mil = 0.0254 mm.

a. The member designation is defined by the first number representing the member depth in 0.01 inch, the letter "S" representing a stud or joist member, the second number representing the flange width in 0.01 inch, and the letter "t" shall be a number representing the minimum base metal thickness in mils [See Table R505.2(2)].

❖ See the commentary for Section R505.2.

mum edge distance of $^1/_2$ inch (12.7 mm). Framing members with web holes exceeding these limits must be patched or designed in accordance with accepted engineering practice (see commentary, Section R505.3.6).

Steel track sections are to serve in nonload-bearing applications only and are shown in Figure R505.2(2). They must have a minimum flange width of $1^1/_4$ inches (32 mm).

TABLE R505.2(2)
MINIMUM THICKNESS OF COLD-FORMED STEEL MEMBERS

DESIGNATION (mils)	MINIMUM UNCOATED THICKNESS (inches)	REFERENCE GAGE NUMBER
33	0.033	20
43	0.043	18
54	0.054	16
68	0.068	14

For SI: 1 inch = 25.4 mm, 1 mil = 0.0254 mm.

❖ See the commentary for Section R505.2.

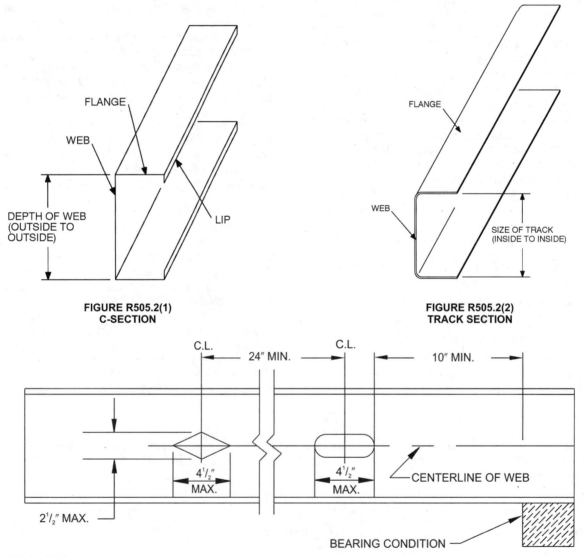

FIGURE R505.2(1)
C-SECTION

FIGURE R505.2(2)
TRACK SECTION

For SI: 1 inch = 25.4 mm.

FIGURE R505.2(3)
FLOOR JOIST WEB HOLES

❖ See the commentary for Section R505.2 for Figures R505.2(1) through (3).

R505.2.1 Material. Load-bearing members used in steel floor construction shall be cold-formed to shape from structural quality sheet steel complying with the requirements of one of the following:

1. ASTM A 653: Grades 33, 37, 40 and 50 (Class 1 and 3).

2. ASTM A 792: Grades 33, 37, 40 and 50A.

3. ASTM A 875: Grades 33, 37, 40 and 50 (Class 1 and 3).

4. ASTM A 1003: Grades 33, 37, 40 and 50.

❖ Load bearing steel framing members must have a legible label, stencil stamp, or embossment. This identification allows for verification that materials installed are consistent with the design and meet the provisions of the code.

R505.2.2 Identification. Load-bearing steel framing members shall have a legible label, stencil, stamp or embossment with the following information as a minimum:

1. Manufacturer's identification.

2. Minimum uncoated steel thickness in inches (mm).

3. Minimum coating designation.

4. Minimum yield strength, in kips per square inch (ksi) (kPa).

❖ Load-bearing steel framing members must have a legible label, stencil, stamp or embossment. This allows for verification that the materials installed are consistent with the design and meet the code requirements.

R505.2.3 Corrosion protection. Load-bearing steel framing shall have a metallic coating complying with one of the following:

1. A minimum of G 60 in accordance with ASTM A 653.

2. A minimum of AZ 50 in accordance with ASTM A 792.

3. A minimum of GF 60 in accordance with ASTM A 875.

❖ The specified metallic coatings correspond to requirements in the referenced material standards. The minimum coating designations assume normal exposure conditions that are best defined as having the framing members enclosed within a building envelope or wall assembly within a controlled environment. When more severe exposure conditions are probable, such as industrial or marine atmospheres, consideration should be given to specifying a heavier coating.

R505.2.4 Fastening requirements. Screws for steel-to-steel connections shall be installed with a minimum edge distance and center-to-center spacing of 0.5 inch (12.7 mm), shall be self-drilling tapping, and shall conform to SAE J78. Floor sheathing shall be attached to steel joists with minimum No. 8 self-drilling tapping screws that conform to SAE J78. Screws attaching floor-sheathing-to-steel joists shall have a minimum head diameter of 0.292 inch (7.4 mm) with countersunk heads and shall be installed with a minimum edge distance of 0.375 inch (9.5 mm). Gypsum board ceilings shall be attached to steel joists with minimum No. 6 screws conforming to ASTM C 954 and shall be installed in accordance with Section R702. For all connections, screws shall extend through the steel a minimum of three exposed threads. All self-drilling

tapping screws conforming to SAE J78 shall have a Type II coating in accordance with ASTMB 633.

Where No. 8 screws are specified in a steel to steel connection the required number of screws in the connection is permitted to be reduced in accordance with the reduction factors in Table R505.2.4 when larger screws are used or when one of the sheets of steel being connected is thicker than 33 mils (0.84 mm). When applying the reduction factor the resulting number of screws shall be rounded up.

❖ Fasteners meeting the referenced standards and installed in accordance with this section provide the necessary load capacity consistent with these prescriptive provisions. This section specifies self-drilling tapping screws conforming to SAE J78, *Steel Self-Drilling Tapping Screws* for all steel-to-steel connections as well as for fastening floor sheathing to steel joists. They must have a Type II coating in accordance with ASTM B 633F, *Standard Specification for Electrodeposited Coatings of Zinc on Iron and Steel.* Gypsum board must be attached to steel joists with No. 6 screws (minimum) conforming to ASTM C 954, *Standard Specification for Steel Drill Screws for the Application of Gypsum Panel Products or Metal Plaster Bases to Steel Studs from 0.033 in. (0.84 mm) to 0.112 in. (2.84 mm) in Thickness.* See Section R702 for installation. For all connections, screws must extend through the steel a minimum of three exposed threads.

TABLE R505.2.4
SCREW SUBSTITUTION FACTOR

SCREW SIZE	THINNEST CONNECTED STEEL SHEET (mils)	
	33	43
#8	1.0	0.67
#10	0.93	0.62
#12	0.86	0.56

For SI: 1 mil = 0.0254 mm.

❖ In providing prescriptive connection requirements, the code typically specifies the quantity of No. 8 screws necessary to resist the required design forces transferred through a given connection. If the quantity of screws required for a particular connection is large, it may be desirable to reduce the quantity of screws by substituting a larger screw size. Where No. 8 screws are specified for a steel-to-steel connection, the required quantity of screws may be reduced if a larger screw size is substituted in accordance with the reduction factors of Table R505.2.4. Likewise, if the thinnest of the connected sheets is at least 43 mils (1.1 mm), the required quantity of screws may be reduced.

R505.3 Floor construction. Cold-formed steel floors shall be constructed in accordance with this section and Figure R505.3.

❖ Load-bearing steel-framing members must be cold formed to shape from structural quality sheet steel complying with appropriate material standards. The steel grades specified have the ductility and strength to meet these provisions. For more detailed information refer to the AISI *Specification for Design of Cold-Formed Steel Structural Members Commentary.*

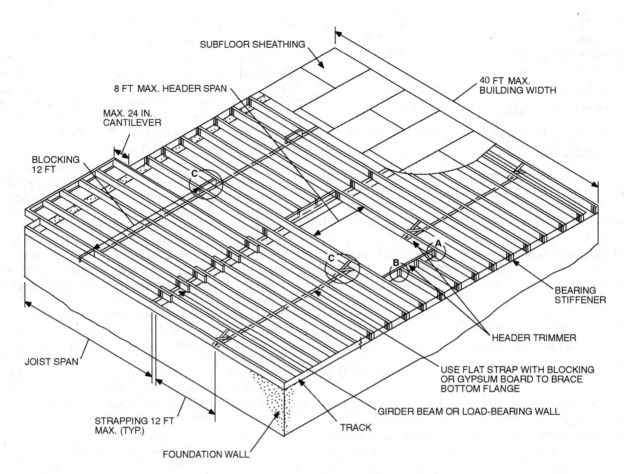

For SI: 1 inch = 25.4 mm, 1 foot = 304.8 mm.

FIGURE 505.3
STEEL CONSTRUCTION

(continued)

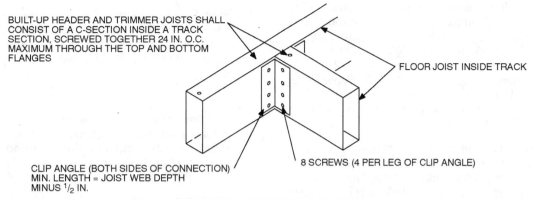

BUILT-UP HEADER AND TRIMMER JOISTS SHALL CONSIST OF A C-SECTION INSIDE A TRACK SECTION, SCREWED TOGETHER 24 IN. O.C. MAXIMUM THROUGH THE TOP AND BOTTOM FLANGES

FLOOR JOIST INSIDE TRACK

8 SCREWS (4 PER LEG OF CLIP ANGLE)

CLIP ANGLE (BOTH SIDES OF CONNECTION) MIN. LENGTH = JOIST WEB DEPTH MINUS 1/2 IN.

BUILT-UP HEADER OR TRIMMER JOIST

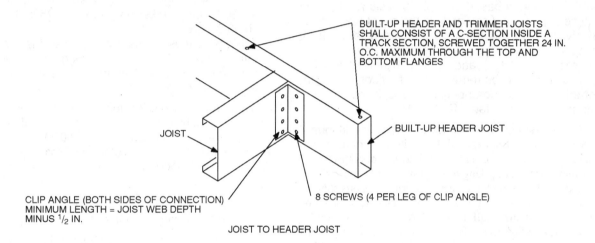

BUILT-UP HEADER AND TRIMMER JOISTS SHALL CONSIST OF A C-SECTION INSIDE A TRACK SECTION, SCREWED TOGETHER 24 IN. O.C. MAXIMUM THROUGH THE TOP AND BOTTOM FLANGES

JOIST

BUILT-UP HEADER JOIST

CLIP ANGLE (BOTH SIDES OF CONNECTION) MINIMUM LENGTH = JOIST WEB DEPTH MINUS 1/2 IN.

8 SCREWS (4 PER LEG OF CLIP ANGLE)

JOIST TO HEADER JOIST

1 SCREW THROUGH BRACE AT EACH FLANGE

0.75 IN. x 33 MIL

2 SCREWS THROUGH EACH LEG OF 33 MIL 2 IN. x 2 IN. x BLOCKING DEPTH, CLIP ANGLE

33 MIL TRACK OR C-SECTION MINIMUM HEIGHT OF JOIST DEPTH MINUS 2 IN.

OR

X-BRACING

BLOCKING OPTIONS

SOLID BLOCKING

For SI: 1 inch = 25.4 mm, 1 mil = 0.0254 mm.

FIGURE R505.3—continued
STEEL FLOOR CONSTRUCTION

❖ This figure is a schematic of a typical steel-framed floor. It illustrates several of the detailed provisions for steel floor framing.

R505.3.1 Floor to foundation or bearing wall connections.
Cold-formed steel floors shall be anchored to foundations, wood sills or load-bearing walls in accordance with Table R505.3.1(1) and Figure R505.3.1(1), R505.3.1(2), R505.3.1(3), R505.3.1(4), R505.3.1(5) or R505.3.1(6). Continuous steel joists supported by interior load-bearing walls shall be constructed in accordance with Figure R505.3.1(7). Lapped steel joists shall be constructed in accordance with Figure R505.3.1(8). Fastening of steel joists to other framing members shall be in accordance with Table R505.3.1(2).

❖ This section provides minimum fastening requirements for attachment of steel floor framing to foundations, wood sills and both exterior and interior load-bearing walls. Figures R505.3.1(1) through R505.3.1(6) illustrate exterior wall conditions; thus, the corresponding connection requirements in Table R505.3.1(1) vary based on a site's basic wind speed and seismic design category. In the event these criteria give rise to requirements that are different from those in the table, the more restrictive would govern. Figures R505.3.1(7) and R505.3.1(8) illustrate the minimum connection requirements for floor joists supported by interior load-bearing walls in conjunction with the first line of Table R505.3.1(2).

R505.3.2 Allowable joist spans. The clear span of cold-formed steel floor joists shall not exceed the limits set forth in Tables R505.3.2(1), R505.3.2(2), and R505.3.2(3). Floor joists shall have a minimum bearing length of 1.5 inches (38 mm). When continuous joists are used, the interior bearing supports shall be located within 2 feet (610 mm) of mid span of the steel joists, and the individual spans shall not exceed the span in Tables R505.3.2(2) and R505.3.2(3). Bearing stiffeners shall be installed at each bearing location in accordance with Section R505.3.4 and as shown in Figure R505.3.

Blocking is not required for continuous back-to-back floor joists at bearing supports. Blocking shall be installed between the joists for single continuous floor joists across bearing supports. Blocking shall be spaced at a maximum of 12 feet (3660 mm) on center. Blocking shall consist of C-shape or track section with a minimum thickness of 33 mils (0.84 mm). Blocking shall be fastened to each adjacent joist through a 33-mil (0.84 mm) clip angle, bent web of blocking or flanges of web stiffeners with two No. 8 screws on each side. The minimum depth of the blocking shall be equal to the depth of the joist minus 2 inches (51 mm). The minimum length of the angle shall be equal to the depth of the joist minus 2 inches (51 mm).

❖ Table R505.3.2(1) lists the allowable clear spans for single span steel floor joists. Tables R505.3.2(2) and R505.3.2(3) list the allowable individual spans when the joists are continuous over an interior bearing support. The tabulated spans are based on dead loads of 10 psf (0.48 kPa) and design live loads of 30 psf (1.44 kPa) or 40 psf (1.92 kPa). The following example illustrates use of the tables.

Example:

Steel floor joists spaced at 16 inches (406 mm) on center support a floor live load of 40 psf (1.92 kPa), a dead load of 10 psf (0.48 kPa) and span 17 feet, 6 inches (5334 mm). Determine the required joist size.

Solution:

In Table R505.3.2(1) under 40 psf (1.92 kPa), find the live load column headed 16 inches (406 mm). Read down this column to find the listed span of 17 feet, 6 inches (5334 mm) or greater. The required joist is 800S162-97.

When joists are continuous over interior bearing supports, the spans must be arranged as illustrated in Commentary Figure R505.3.2 and the spans must not exceed the limits in Tables R505.3.2(2) and R505.3.2(3).

Blocking must be installed between the joists, at the interior bearing support, for a single continuous floor joist. Blocking is not required at the bearing support for back-to-back floor joists.

R505.3.3 Joist bracing. The top flanges of steel joists shall be laterally braced by the application of floor sheathing fastened to the joists in accordance with Table R505.3.1(2). Floor joists with spans that exceed 12 feet (3658 mm) shall have the bottom flanges laterally braced in accordance with one of the following:

1. Gypsum board installed with minimum No. 6 screws in accordance with Section R702.

2. Continuous steel strapping installed in accordance with Figure R505.3. Steel straps shall be at least 1.5 inches (38 mm) in width and 33 mils (0.84 mm) in thickness. Straps shall be fastened to the bottom flange at each joist with at least one No. 8 screw and shall be fastened to blocking with at least two No. 8 screws. Blocking or bridging (X-bracing) shall be installed between joists in-line with straps at a maximum spacing of 12 feet (3658 mm) measured perpendicular to the joist run and at the termination of all straps.

❖ The application of floor sheathing to steel floor joists in accordance with Table R505.3.1(2) of the code allows the transfer of loads to the joists and provides the necessary lateral bracing for the top flanges of those joists. Steel floor joists with spans greater than 12 feet (3658 mm) are more prone to overturning and therefore must also have the bottom flanges laterally braced either with gypsum board or continuous steel strapping as illustrated in Figure R505.3 of the code. Strap bracing (see Figure R505.3) requires the installation of solid blocking or bridging (X-bracing) in line with the straps at the termination of all straps and at a maximum spacing of 12 feet (3658 mm) measured perpendicular to the joist run.

TABLE R505.3.1(1)
FLOOR TO FOUNDATION OR BEARING WALL CONNECTION REQUIREMENTS [a, b]

FRAMING CONDITION	WIND SPEED (mph) AND EXPOSURE	
	Up to 110 A/B or 85 C or Seismic Design Categories A, B, C	Up to 110 C
Floor joist to wall track of exterior steel load-bearing wall per Figure R505.3.1(1)	2-No. 8 screws	3-No. 8 screws
Floor joist track to wood sill per Figure R505.3.1(2)	Steel plate spaced at 3' o.c., with 4-No. 8 screws and 4-10d or 6-8d common nails	Steel plate, spaced at 2' o.c., with 4-No. 8 screws and 4-10d or 6-8d common nails
Floor joist track to foundation per Figure R505.3.1(3)	$^1/_2''$ minimum diameter anchor bolt and clip angle spaced at 6' o.c. with 8-No. 8 screws	$^1/_2''$ minimum diameter anchor bolt and clip angle spaced at 4' o.c. with 8-No. 8 screws
Joist cantilever to wall track per Figure R505.3.1(4)	2-No. 8 screws per stiffener or bent plate	3-No. 8 screws per stiffener or bent plate
Joist cantilever to wood sill per Figure R505.3.1(5)	Steel plate spaced at 3' o.c., with 4-No. 8 screws and 4-10d or 6-8d common nails	Steel plate spaced at 2' o.c., with 4-No. 8 screws and 4-10d or 6-8d common nails
Joist cantilever to foundation per Figure R505.3.1(6)	$^1/_2''$ minimum diameter anchor bolt and clip angle spaced at 6' o.c. with 8-No. 8 screws	$^1/_2''$ minimum diameter anchor bolt and clip angle spaced at 4' o.c. with 8-No. 8 screws

For SI: 1 inch = 25.4 mm, 1 foot = 304.8 mm, 1 mile per hour = 0.447 m/s.

a. Anchor bolts shall be located not more than 12 inches from corners or the termination of bottom tracks (e.g., at door openings). Bolts shall extend a minimum of 15 inches into masonry or 7 inches into concrete.

b. All screw sizes shown are minimum.

❖ See the commentary for Section R505.3.1 as well as Figures R505.3.1(1) through R505.3.1(6)

TABLE R505.3.1(2)
FLOOR FASTENING SCHEDULE [a]

DESCRIPTION OF BUILDING ELEMENTS	NUMBER AND SIZE OF FASTENERS	SPACING OF FASTENERS
Floor joist to track of an interior load-bearing wall per Figures R505.3.1(7) and R505.3.1(8)	2 No. 8 screws	Each joist
Floor joist to track at end of joist	2 No. 8 screws	One per flange or two per bearing stiffener
Subfloor to floor joists	No. 8 screws	6" o.c. on edges and 12" o.c. at intermediate supports

For SI: 1 inch = 25.4 mm.

a. All screw sizes shown are minimum.

❖ In addition to the other connections covered under Section R505.3.1, Table R505.3.1(2) provides the fastening requirements for tracks at the end of floor joists and for subfloor-to-floor joists.

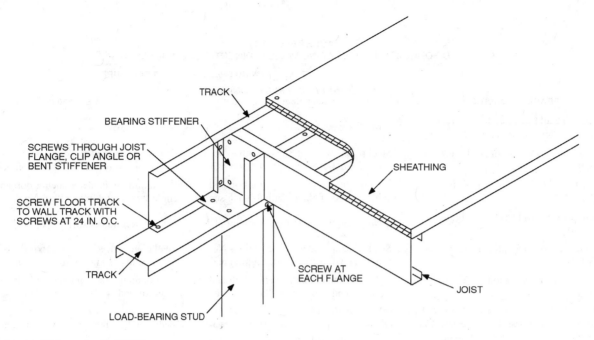

For SI: 1 inch = 25.4 mm, 1 mil = 0.0254 mm.

FIGURE R505.3.1(1)
FLOOR TO LOAD-BEARING WALL STUD CONNECTION

❖ Figure R505.3.1(1) depicts connection requirements for steel floor joists at exterior steel bearing walls. The minimum number of screws through either the joist flange, clip angle or bent stiffener should be selected from the first row of Table R505.3.1(1) based on the seismic design category or wind speed and exposure. Similarly, if these joists cantilever beyond the wall as illustrated in Figure R505.3.1(4), the required number of screws should be selected from the fourth row of Table R505.3.1(1).

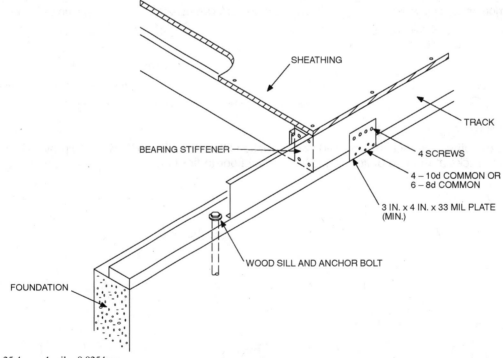

For SI: 1 inch = 25.4 mm, 1 mil = 0.0254 mm.

FIGURE R505.3.1(2)
FLOOR TO WOOD SILL CONNECTION

❖ Figures R505.3.1(2) and R505.3.1(5) depict the connection requirements of steel floor framing to a wood sill, which, in turn, is anchored to the foundation. Details such as the spacing of the clip angle and anchor bolt should be selected from the appropriate row of Table R505.3.1(1) [see commentary, Figure R505.3.1(1)].

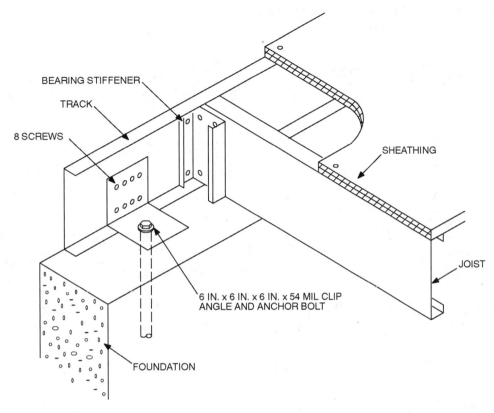

For SI: 1 inch = 25.4 mm, 1 mil = 0.0254 mm.

FIGURE R505.3.1(3)
FLOOR TO FOUNDATION CONNECTION

❖ Figures R505.3.1(3) and R505.3.1(6) depict requirements for anchoring steel floor framing directly to concrete or masonry foundations. Details such as the spacing of the clip angle and anchor bolt should be selected from the appropriate row of Table R505.3.1(1) [see commentary, Figure R505.3.1(1)].

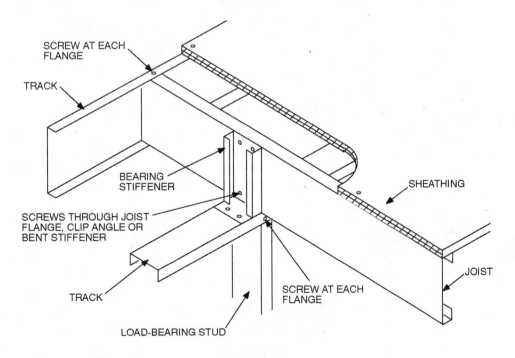

FIGURE R505.3.1(4)
FLOOR CANTILEVER TO LOAD-BEARING WALL CONNECTION

❖ See the commentary for Figure R505.3.1(1).

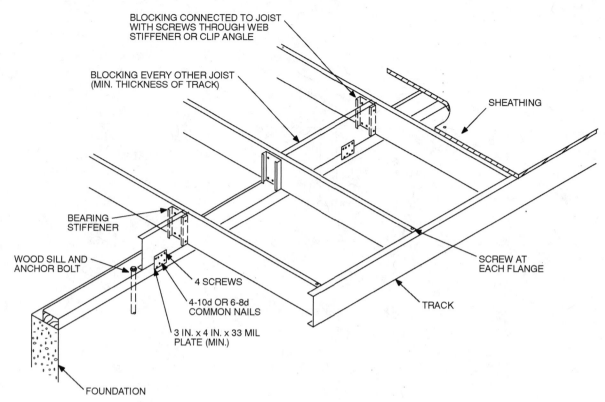

For SI   1 inch = 25.4 mm, 1 mil = 0.0254 mm.

FIGURE R505.3.1(5)
FLOOR CANTILEVER TO WOOD SILL CONNECTION

❖ See the commentary for Figure R505.3.1(1) and R505.3.1(2).

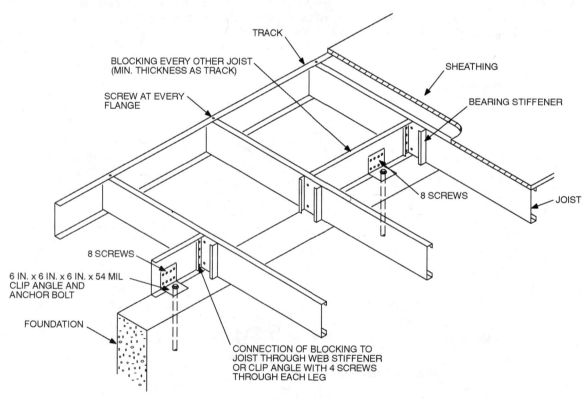

For SI   1 inch = 25.4 mm, 1 mil = 0.0254 mm.

FIGURE R505.3.1(6)
FLOOR CANTILEVER TO FOUNDATION CONNECTION

❖ See the commentary for Figure R505.3.1(1) and R505.3.1(3).

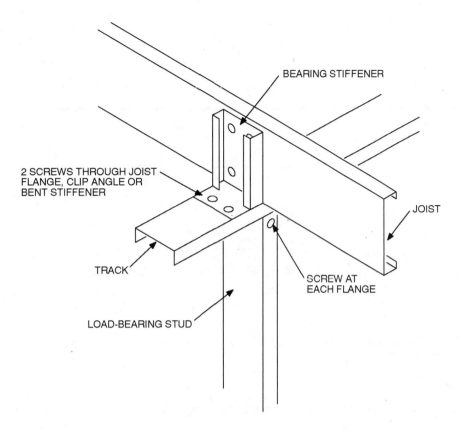

FIGURE R505.3.1(7)
CONTINUOUS JOIST SPAN SUPPORTED ON STUD

❖ See the commentary for Section R505.3.1.

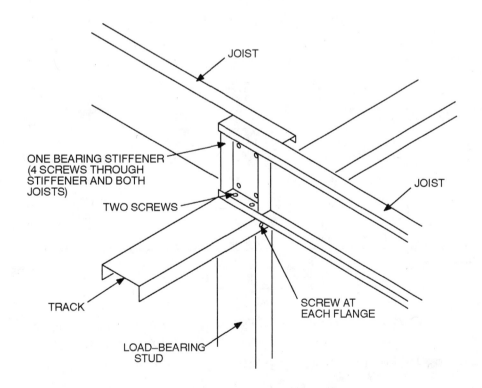

FIGURE R505.3.1(8)
LAPPED JOISTS SUPPORTED ON STUD

❖ See the commentary for Section R505.3.1.

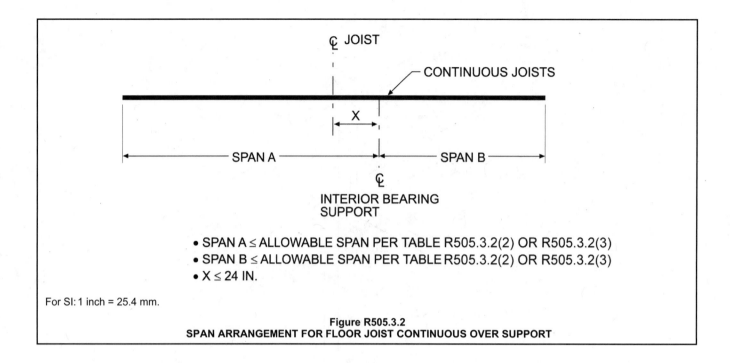

Figure R505.3.2
SPAN ARRANGEMENT FOR FLOOR JOIST CONTINUOUS OVER SUPPORT

TABLE R505.3.2(1)
ALLOWABLE SPANS FOR COLD-FORMED STEEL JOISTS—SINGLE SPANS[a, b] 33 ksi STEEL

| JOIST DESIGNATION | 30 PSF LIVE LOAD | | | | 40 PSF LIVE LOAD | | | |
| | Spacing (inches) | | | | Spacing (inches) | | | |
	12	16	19.2	24	12	16	19.2	24
550S162-33	11'-7"	10'-7"	9'-6"	8'-6"	10'-7"	9'-3"	8'-6"	7'-6"
550S162-43	12'-8"	11'-6"	10'-10"	10'-2"	11'-6"	10'-5"	9'-10"	9'-1"
550S162-54	13'-7"	12'-4"	11'-7"	10'-9"	12'-4"	11'-2"	10'-6"	9'-9"
550S162-68	14'-7"	13'-3"	12'-6"	11'-7"	13'-3"	12'-0"	11'-4"	10'-6"
800S162-97	16'-2"	14'-9"	13'-10"	12'-10"	14'-9"	13'-4"	12'-7"	11'-8"
800S162-33	15'-8"	13'-11"	12'-9"	11'-5"	14'-3"	12'-5"	11'-3"	9'-0"
800S162-43	17'-1"	15'-6"	14'-7"	13'-7"	15'-6"	14'-1"	13'-3"	12'-4"
800S162-54	18'-4"	16'-8"	15'-8"	14'-7"	16'-8"	15'-2"	14'-3"	13'-3"
800S162-68	19'-9"	17'-11"	16'-10"	15'-8"	17'-11"	16'-3"	15'-4"	14'-2"
800S162-97	22'-0"	20'-0"	16'-10"	17'-5"	20'-0"	18'-2"	17'-1"	15'-10"
1000S162-43	20'-6"	18'-8"	17'-6"	15'-8"	18'-8"	16'-11"	15'-6"	13'-11"
1000S162-54	22'-1"	20'-0"	18'-10"	17'-6"	20'-0"	18'-2"	17'-2"	15'-11"
1000S162-68	23'-9"	21'-7"	20'-3"	18'-10"	21'-7"	19'-7"	18'-5"	17'-1"
1000S162-97	26'-6"	24'-1"	22'-8"	21'-0"	24'-1"	21'-10"	20'-7"	19'-1"
1200S162-43	23'-9"	20'-10"	19'-0"	16'-8"	21'-5"	18'-6"	16'-6"	13'-2"
1200S162-54	25'-9"	23'-4"	22'-0"	20'-1"	23'-4"	21'-3"	20'-0"	17'-10"
1200S162-68	27'-8"	25'-1"	23'-8"	21'-11"	25'-1"	22'-10"	21'-6"	21'-1"
1200S162-97	30'-11"	28'-1"	26'-5"	24'-6"	28'-1"	25'-6"	24'-0"	22'-3"

For SI: 1 inch = 25.4 mm, 1 foot = 304.8 mm, 1 pound per square foot = 0.0479kPa.
a. Deflection criteria: $L/480$ for live loads, $L/240$ for total loads.
b. Floor dead load = 10 psf.

❖ See the commentary and example for Section R505.3.2.

TABLE R505.3.2(2)
ALLOWABLE SPANS FOR COLD-FORMED STEEL JOISTS—MULTIPLE SPANS[a, b] 33 ksi STEEL

JOIST DESIGNATION	30 PSF LIVE LOAD				40 PSF LIVE LOAD			
	Spacing (inches)				Spacing (inches)			
	12	16	19.2	24	12	16	19.2	24
550S162-33	12'-1"	10'-5"	9'-6"	8'-6"	10'-9"	9'-3"	8'-6"	7'-6"
550S162-43	14'-5"	12'-5"	11'-4"	10'-2"	12'-9"	11'-11"	10'-1"	9'-0"
550S162-54	16'-3"	14'-1"	12'-10"	11'-6"	14'-5"	12'-6"	11'-5"	10'-2"
550S162-68	19'-7"	17'-9"	16'-9"	15'-6"	17'-9"	16'-2"	15'-2"	14'-1"
800S162-97	21'-9"	19'-9"	18'-7"	17'-3"	19'-9"	17'-11"	16'-10"	15'-4"
800S162-33	14'-8"	11'-10"	10'-4"	8'-8"	12'-4"	9'-11"	8'-7"	7'-2"
800S162-43	20'-0"	17'-4"	15'-9"	14'-1"	17'-9"	15'-4"	14'-0"	12'-0"
800S162-54	23'-7"	20'-5"	18'-8"	16'-8"	21'-0"	18'-2"	16'-7"	14'-10"
800S162-68	26'-5"	23'-1"	21'-0"	18'-10"	23'-8"	20'-6"	18'-8"	16'-9"
800S162-97	29'-6"	26'-10"	25'-3"	22'-8"	26'-10"	24'-4"	22'-6"	20'-2"
1000S162-43	22'-2"	18'-3"	16'-0"	13'-7"	18'-11"	15'-5"	13'-6"	11'-5"
1000S162-54	26'-2"	22'-8"	20'-8"	18'-6"	23'-3"	20'-2"	18'-5"	16'-5"
1000S162-68	31'-5"	27'-2"	24'-10"	22'-2"	27'-11"	24'-2"	22'-1"	19'-9"
1000S162-97	35'-6"	32'-3"	29'-11"	26'-9"	32'-3"	29'-2"	26'-7"	23'-9"
1200S162-43	21'-8"	17'-6"	15'-3"	12'-10"	18'-3"	14'-8"	12'-8"	10'-6²
1200S162-54	28'-5"	24'-8"	22'-6"	19'-6"	25'-3"	21'-11"	19'-4"	16'-6"
1200S162-68	33'-7"	29'-1"	26'-6"	23'-9"	29'-10"	25'-10"	23'-7"	21'-1"
1200S162-97	41'-5"	37'-8"	34'-6"	30'-10"	37'-8"	33'-6"	30'-7"	27'-5"

For SI: 1 inch = 25.4 mm, 1 foot = 304.8 mm, 1 pound per square foot = 0.0479 kPa.

a. Deflection criteria: $L/480$ for live loads, $L/240$ for total loads.

b. Floor dead load = 10 psf.

❖ See the commentary for Section R505.3.2.

TABLE R505.3.2(3)
ALLOWABLE SPANS FOR COLD-FORMED STEEL JOISTS—MULTIPLE SPANS[a, b] 50 ksi STEEL

JOIST DESIGNATION	30 PSF LIVE LOAD				40 PSF LIVE LOAD			
	Spacing (inches)				Spacing (inches)			
	12	16	19.2	24	12	16	19.2	24
550S162-33	13'-11"	12'-0"	11'-0"	9'-3"	12'-3"	10'-8"	9'-7"	8'-4"
550S162-43	16'-3"	14'-1"	12'-10"	11'-6"	14'-6"	12'-6"	11'-5"	10'-3"
550S162-54	18'-2"	16'-6"	15'-4"	13'-8"	16'-6"	14'-11"	13'-7"	12'-2"
550S162-68	19'-6"	17'-9"	16'-8"	15'-6"	17'-9"	16'-1"	15'-2"	14'-0"
550S162-97	21'-9"	19'-9"	18'-6"	17'-2"	19'-8"	17'-10"	16'-8"	15'-8"
800S162-33	15'-6"	12'-6"	10'-10"	9'-1"	13'-0"	10'-5"	8'-11"	6'-9"
800S162-43	22'-0"	19'-1"	17'-5"	15'-0"	19'-7"	16'-11"	14'-10"	12'-8"
800S162-54	24'-6"	22'-4"	20'-6"	17'-11"	22'-5"	19'-9"	17'-11"	15'-10"
800S162-68	26'-6"	24'-1"	22'-8"	21'-0"	24'-1"	21'-10"	20'-7"	19'-2"
800S162-97	29'-9"	26'-8"	25'-2"	23'-5"	26'-8"	24'-3"	22'-11"	21'-4"
1000S162-43	23'-6"	19'-2"	16'-9"	14'-2"	19'-11"	16'-2"	14'-0"	11'-9"
1000S162-54	28'-2"	23'-10"	21'-7"	18'-11"	24'-8"	20'-11"	18'-9"	18'-4"
1000S162-68	31'-10"	28'-11"	27'-2"	25'-3"	28'-11"	26'-3"	24'-9"	22'-9"
1000S162-97	35'-4"	32'-1"	30'-3"	28'-1"	32'-1"	29'-2"	27'-6"	25'-6"
1200S162-43	22'-11"	18'-5"	16'-0"	13'-4"	19'-2"	15'-4"	13'-2"	10'-6"
1200S162-54	32'-8"	28'-1"	24'-9"	21'-2"	29'-0"	23'-10"	20'-11"	17'-9"
1200S162-68	37'-1"	32'-5"	29'-4"	25'-10"	33'-4"	28'-6"	25'-9"	22'-7"
1200S162-97	41'-2"	37'-6"	35'-3"	32'-9"	37'-6"	34'-1"	32'-1"	29'-9"

For SI: 1 inch = 25.4 mm, 1 foot = 304.8 mm, 1 pound per square foot = 0.0479kPa.

a. Deflection criteria: $L/480$ for live loads, $L/240$ for total loads.

b. Floor dead load = 10 psf.

❖ See the commentary for Section R505.3.2.

R505.3.4 Bearing stiffeners. Bearing stiffeners shall be installed at all bearing locations for steel floor joists. A bearing stiffener shall be fabricated from a minimum 33 mil (0.84 mm) C-section or 43 mil (1.09 mm) track section. Each stiffener shall be fastened to the web of the joist with a minimum of four No. 8 screws equally spaced as shown in Figure R505.3.4. Stiffeners shall extend across the full depth of the web and shall be installed on either side of the web.

❖ Stiffeners (also referred to as transverse stiffeners or web stiffeners) at points of bearing provide resistance to web crippling in the joist. These stiffeners are illustrated in Figure R505.3.4, but a stiffener can be installed on either side of the web of a joist as depicted in Figures R505.3.1(1) through R505.3.1(8).

R505.3.5 Cutting and notching. Flanges and lips of load-bearing steel floor framing members shall not be cut or notched.

❖ Flanges and lips of load-bearing steel floor framing members must not be cut or notched because this would affect their structural integrity. The only holes permitted in steel floor joists are those that conform to Section R505.2.

R505.3.6 Hole patching. Web holes not conforming to the requirements in Section R505.2 shall be designed in accordance with one of the following:

1. Framing members shall be replaced or designed in accordance with accepted engineering practices when web holes exceed the following size limits:

    1.1. The depth of the hole, measured across the web, exceeds 70 percent of the flat width of the web; or

    1.2. The length of the hole measured along the web, exceeds 10 inches (254 mm) or the depth of the web, whichever is greater.

2. Web holes not exceeding the dimensional requirements in Section R505.3.6, Item 1, shall be patched with a solid steel plate, stud section, or track section in accordance with Figure R505.3.6. The steel patch shall, as a minimum, be of the same thickness as the receiving member and shall extend at least 1 inch (25 mm) beyond all edges of the hole. The steel patch shall be fastened to the web of the receiving member with No.8 screws spaced no greater than 1 inch (25 mm) center-to-center along the edges of the patch with minimum edge distance of $^1/_2$ inch (13 mm).

❖ Framing members with web holes exceeding the limits, outlined in Item 1 of this section, must be replaced or designed in accordance with accepted engineering practice. Framing members with web holes within the limits of Item 1,but exceeding the limits of Section R505.2, must be patched in accordance with Item 2 of this section (see commentary, Section R505.2).

R505.3.7 Floor cantilevers. Floor cantilevers shall not exceed 24 inches (610 mm) as illustrated in Figure R505.3. The cantilever back-span shall extend a minimum of 6 feet (1830 mm) within the building, and shall be fastened to a bearing condition in accordance with Section R505.3.1. Floor cantilevers shall be permitted only on the second floor of a two-story building or the first floor of a one-story building. Floor framing that is cantilevered and supports the cantilevered floor only shall consist of single joist members in accordance with Section R505.3.2. Floor framing that is cantilevered and supports the cantilevered floor and the roof framing load above shall consist of double joist members of the same size and material thickness as that for single joist members in accordance with Section R505.3.2, and shall be fastened web-to-web with minimum No. 8 screws at 24 inches (610 mm) maximum on-center spacing top and bottom. Built-up floor framing consisting of a C-section inside a track section, fastened at the top and bottom flanges by minimum No. 8 screws at 24 inches (610 mm) maximum on center spacing, is permitted in lieu of the web-to-web double joist method.

❖ Figure R505.3 illustrates the maximum permitted cantilever. Floor cantilevers are permitted on the second floor of two-story buildings or the first floor of one-story buildings.

Floor framing that is cantilevered but supports only the cantilevered floor may be a single joist conforming to the allowable span requirements of Section R505.3.2. Cantilevered floor framing that also supports the roof framing above must consist of double joist members of the same size and material thickness as that for single joist members in accordance with the allowable span requirements of Section R505.3.2. Built-up floor framing consisting of a C-section inside a track section is permitted as an option to the web-to-web double joist member.

R505.3.8 Splicing. Joists and other structural members shall not be spliced. Splicing of tracks shall conform with Figure R505.3.8.

❖ Splicing of structural members is not permitted except when lapped joists occur at interior bearing points.

Splicing of tracks is allowed because they are not permitted to act as load-carrying members (see Figure R505.3.8). The exception to this would be under lateral loads from wind and earthquakes. In this case, the floor system must act as a diaphragm, with tracks serving as diaphragm chords. To provide continuity and carry the anticipated chord force, the splice must be as shown in Figure R505.3.8.

R505.3.9 Framing of openings. Openings in floor framing shall be framed with header and trimmer joists. Header joist spans shall not exceed 8 feet (2438 mm). Header and trimmer joists shall be fabricated from joist and track sections, which shall be of a minimum size and thickness as the adjacent floor joists and shall be installed in accordance with Figure R505.3. Each header joist shall be connected to trimmer joists with a minimum of four 2-inch-by-2-inch (51 mm by 51 mm) clip angles. Each clip angle shall be fastened to both the header and trimmer joists with four No. 8 screws, evenly spaced, through each leg of the clip angle. The clip angles shall have a steel thickness not less than that of the floor joist.

❖ Recognizing that floor openings are necessary for such things as stairways and utility chases, the code provides prescriptive criteria for framing these openings. These requirements for openings in floors are illustrated in Figure R505.3.

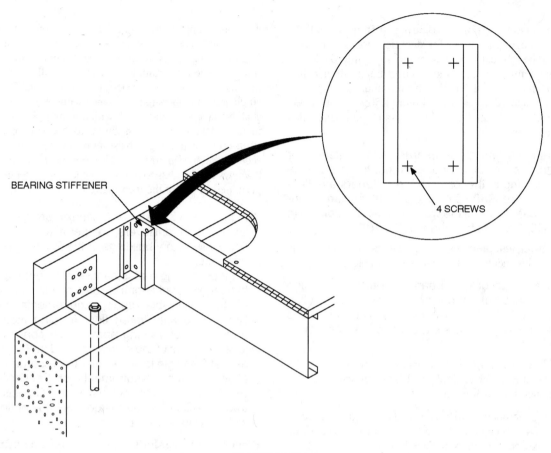

FIGURE R505.3.4
BEARING STIFFENER

❖ See the commentary for Section R505.3.4.

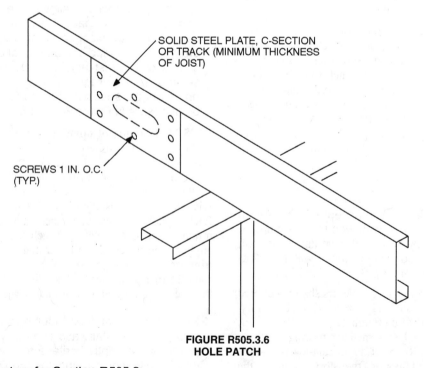

FIGURE R505.3.6
HOLE PATCH

❖ See the commentary for Section R505.2.

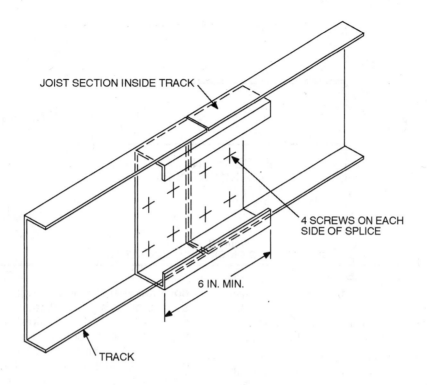

JOIST SECTION INSIDE TRACK

4 SCREWS ON EACH SIDE OF SPLICE

6 IN. MIN.

TRACK

For SI: 1 inch = 25.4 mm.

FIGURE R505.3.8
TRACK SPLICE

❖ See the commentary for Section R505.3.8.

SECTION R506
CONCRETE FLOORS (ON GROUND)

R506.1 General. Concrete slab-on-ground floors shall be a minimum 3.5 inches (89 mm) thick (for expansive soils, see Section R403.1.8). The specified compressive strength of concrete shall be as set forth in Section R402.2.

❖ For weathering requirements, see the commentary for Sections R301.2 and R402.2. Commentary Figure R506.1 shows the requirements of this section.

R506.2 Site preparation. The area within the foundation walls shall have all vegetation, top soil and foreign material removed.

❖ Removal of construction debris and foreign materials, such as lumber formwork, stakes, tree stumps and other vegetation, limits the attraction of termites, insects and vermin. Top soil and soil vegetation should also be removed because such top soil is generally loosely compacted or so full of vegetation that soil settlement will occur when the vegetation decays. For concrete slabs placed on uncompacted fill or on large quantities of foreign materials, differential settlement may take place as a result of subsequent compaction

of the soil, which can result in cracking of the floor slab and the interior wall/ceiling finishes.

R506.2.1 Fill. Fill material shall be free of vegetation and foreign material. The fill shall be compacted to assure uniform support of the slab, and except where approved, the fill depths shall not exceed 24 inches (610 mm) for clean sand or gravel and 8 inches (203 mm) for earth.

❖ To minimize differential settlement caused by consolidation of uncompacted fill and the problems associated with differential settlement, any fill beneath a concrete slab must be compacted. Properly compacted fill, besides minimizing settlement, increases the soil load-bearing characteristics and soil stability and reduces water penetration. The amount of compaction for fill is not specifically stipulated. Generally, fill soils should be compacted to 95-percent maximum density as determined by a Standard Proctor Test (ASTM D 698). Compaction requirements for expansive soils should be determined from an engineering analysis.

Soil may be compacted using equipment appropriate to the type of material being compacted, with lifts not exceeding 8 inches (203 mm). In general, thinner layers produce better compaction of fill, regardless of the type of soil being compacted.

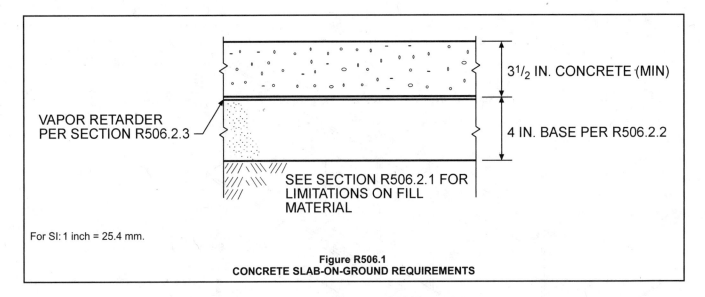

For SI: 1 inch = 25.4 mm.

Figure R506.1
CONCRETE SLAB-ON-GROUND REQUIREMENTS

R506.2.2 Base. A 4-inch-thick (102 mm) base course consisting of clean graded sand, gravel, crushed stone or crushed blast-furnace slag passing a 2-inch (51 mm) sieve shall be placed on the prepared subgrade when the slab is below grade.

> **Exception:** A base course is not required when the concrete slab is installed on well-drained or sand-gravel mixture soils classified as Group I according to the United Soil Classification System in accordance with Table R405.1.

❖ Slabs need to be protected from the penetration of water and water vapor from below to prevent damage to interior finish material. Additionally, the effectiveness of thermal insulation at the slab may be adversely affected because of moisture. A 4-inch (102 mm) minimum granular base course is placed over the fill or undisturbed soil for slabs below grade to provide a capillary stop for water rising through the soil and into the slab. A base course is especially necessary when the site soil is other than gravel or clean sand. Both surface and groundwater problems must be addressed. Proper site preparation, selection of adequate fill and base course materials, and the installation of a vapor barrier minimize potential moisture problems associated with slab-on-grade construction.

R506.2.3 Vapor retarder. A 6 mil (0.006 inch; 152 μm) polyethylene or approved vapor retarder with joints lapped not less than 6 inches (152 mm) shall be placed between the concrete floor slab and the base course or the prepared subgrade where no base course exists.

> **Exception:** The vapor retarder may be omitted:
>
> 1. From garages, utility buildings and other unheated accessory structures.
>
> 2. From driveways, walks, patios and other flatwork not likely to be enclosed and heated at a later date.
>
> 3. Where approved by the building official, based on local site conditions.

❖ Although good quality, uncracked concrete is practically impermeable to the passage of water (unless the water is under a considerable pressure), concrete is not impervious to the passage of water vapors. If the surface of the slab is not sealed, water vapor will pass through the slab. If a floor finish such as linoleum, vinyl tile, wood flooring or any type of covering is placed on top of the slab, the moisture will be trapped in the slab. If the floor finish is adhered to the concrete, it may eventually loosen and buckle or blister.

Many of the moisture problems associated with enclosed slabs-on-ground can be minimized by installing a vapor retarder. When required, vapor retarders of either single-layer membranes or multiple-layer membranes should be acceptable if the vapor retarder is properly installed with lapped joints and the barrier is not punctured during construction.

R506.2.4 Reinforcement support. Where provided in slabs on ground, reinforcement shall be supported to remain in place from the center to upper one third of the slab for the duration of the concrete placement.

❖ The code does not require reinforcement for the concrete slab on ground floors. When reinforcement is provided, common practice is to use welded wire fabric. The welded wire fabric frequently is left on the ground during the pour and not picked up. When lifted up by the construction personnel, it is often not in the proper location and not in a consistent location, with some on the ground and some near the top of the slab. This section does not require installation of reinforcement; however, if it is installed, it must be properly located in the slab, or it's benefit is lost.

Bibliography

The following resource materials are referenced in this chapter or are relevant to the subject matter addressed in this chapter.

AISI NAS-01, *North American Specification for Design of Cold-formed Steel Structural Members,* including 2004 Supplement. Washington, DC: American Iron and Steel Institute, 2004.

AISI Truss-04, *Standard for Cold-formed Steel Framing-Truss Design.* Washington, DC: American Iron and Steel Institute, 2004.

AFPA 93, *Span Tables for Joists and Rafters.* Washington, DC: American Forest and Paper Association, 1993.

AFPA NDS 05, *National Design Specification for Wood Construction* with 2005 Supplement. Washington, DC: American Forest and Paper Association, 2005.

ANSI A208.1-99, *Particleboard.* New York: American National Standards Institute, 1999.

APA E30-03, *Engineered Wood Construction Guide.* Tacoma, WA: APA-The Engineered Wood Association, 2003.

ASTM A 653/A 653M-04a, *Standard Specification for Steel Sheet, Zinc-Coated Galvanized or Zinc-Iron Alloy-Coated (Galvannealed) by the Hot-Dip Process.* West Conshohocken, PA: ASTM International, 2004.

ASTM A 792M-03, *Standard Specification for Steel Sheet, 55% Aluminum-Zinc Alloy-Coated by the Hot-Dip Process.* West Conshohocken, PA: ASTM International, 2003.

ASTM A 875/875M-02a, *Standard Specification for Steel Sheet, Zinc-5% Aluminum Alloy-Coated by the Hot-Dip Process.* West Conshohocken, PA: ASTM International, 2002.

ASTM B 633-98e[01], *Standard Specification for Electrodeposited Coatings of Zinc on Iron and Steel.* West Conshohocken, PA: ASTM International, 1998.

ASTM C 954-00, *Standard Specification for Steel Drill Screws for the Application of Gypsum Panel Products or Metal Plaster Bases to Steel Studs from 0.033 in. (0.84 mm) to 0.112 in. (2.84 mm) in Thickness.* West Conshohocken, PA: ASTM International, 2000.

ASTM D 3957-03, *Standard Practices for Establishing Stress Grades for Structural Members Used in Log Buildings.* West Conshohocken, PA: ASTM International, 2003.

ASTM D 5055-04, *Standard Specification for Establishing and Monitoring Structural Capacities of Prefabricated Wood I–Joists.* West Conshohocken, PA: ASTM International, 2004.

AWPA U1-04, *USE CATEGORY SYSTEM: User Specification for Treated Wood, except Section 6, Commodity Specification H.* Granbury, TX: American Wood-Preservers Association, 2004.

CSA 0325.0-92, *Construction Sheathing.* Rexdale, Ontario, Canada: Canadian Standards Association, 1992.

DOC PS 1-95, *Construction and Industrial Plywood.* Gaithersburg, MD: United States Department of Commerce, National Institute of Standards and Technology, 1995.

DOC PS 2-92, *Performance Standard for Wood-Based Structural-Use Panels.* Gaithersburg, MD: United States Department of Commerce, National Institute of Standards and Technology, 1992.

DOC PS 20-99, *American Softwood Lumber Standard.* Gaithersburg, MD: United States Department of Commerce, National Institute of Standards and Technology, 1999.

FEMA FIA-TB-2-93, *Flood Resistant Material Requirements.* Washington, DC: Federal Emergency Management Agency, 1993.

SAE J78-1998, *Steel Self-Drilling Tapping Screws. Warrendale,* PA: Society of Automotive Engineers, 1998.

TPI1-2002, *National Design Standard for Metal-Plate-Connected Wood Truss Construction.* Madison, WI: Truss Plate Institute, 2002.

The following and other materials are available online at http://www.fema.gov/mit/techbul.htm:

NFIP *Technical Bulletin Series.* Washington, DC: Federal Emergency Management Agency, National Flood Insurance Program.

Chapter 6:
Wall Construction

General Comments

Chapter 6 contains provisions that regulate the design and construction of walls. The primary concern of this chapter is the structural integrity of wall construction.

Section R601 contains the scope as well as the performance expectations for walls. Section R602 addresses wood wall framing. Section R603 addresses steel wall framing. Section R604 specifies requirements for wood-structural-panel wall sheathing. Section R605 provides requirements for particleboard used as wall sheathing. Section R606 states the general requirements pertaining to masonry wall construction. Section R607 addresses unit masonry walls. Section R608 covers walls of multiple-wythe masonry. Section R609 containing grouted masonry provisions. Section R610 specifies requirements for wall panels of glass unit masonry. Section R611 addresses prescriptive requirements for the Insulated Concrete Form system, while Section R612 contains provisions for concrete wall construction, including conventionally formed concrete. Section R613 contains the criteria for the performance of exterior windows and doors.

In certain instances a wall must have a fire-resistance rating. Where this is necessary, the wall system must be a tested assembly, and any conditions specific to the in-stallation of the assembly must apply as well. Examples of these provisions include:

- Exterior walls based on location on property (Section R302).
- Walls serving as dwelling unit separations in two-family dwellings (Section R317.1).
- Common wall between townhouses (Section R317.2, exception).

Purpose

Chapter 6 provisions for the design and construction of walls and partitions contain prescriptive requirements that provide for the support of the required dead, live and snow loads. In addition, these walls are a key component in resisting the lateral forces attributable to earthquakes and wind. Where the hazards are greater, the requirements for braced wall lines and/or shear walls are correspondingly more stringent.

So that windows and glass doors do not compromise the integrity of the building envelope, this chapter also includes performance expectations for these elements. The proper testing and labeling of these elements are key to verifying their performance.

SECTION R601
GENERAL

R601.1 Application. The provisions of this chapter shall control the design and construction of all walls and partitions for all buildings.

❖ This section establishes the scope of the chapter.

R601.2 Requirements. Wall construction shall be capable of accommodating all loads imposed according to Section R301 and of transmitting the resulting loads to the supporting structural elements.

❖ This is a general performance statement, requiring that walls support the required design loads and provide an adequate load path to supporting elements.

R601.2.1 Compressible floor-covering materials. Compressible floor-covering materials that compress more than $^1/_{32}$ inch (0.8 mm) when subjected to 50 pounds (23 kg) applied over 1 inch square (645 mm) of material and are greater than $^1/_8$ inch (3 mm) in thickness in the uncompressed state shall not extend beneath walls, partitions or columns, which are fastened to the floor.

❖ Although it is preferable to fasten walls directly to the supporting structure, this provision allows them to be installed over finish floor materials that meet the specified criteria.

SECTION R602
WOOD WALL FRAMING

R602.1 Identification. Load-bearing dimension lumber for studs, plates and headers shall be identified by a grade mark of a lumber grading or inspection agency that has been approved by an accreditation body that complies with DOC PS 20. In lieu of a grade mark, a certification of inspection issued by a lumber grading or inspection agency meeting the requirements of this section shall be accepted.

❖ Wood materials performing a load-carrying function must conform to a minimum quality control standard. To verify this, load-bearing lumber must be properly identified to indicate that the grades and species meet the minimum requirements specified in this chapter. Commentary Figure R602.1 shows the information contained in a grade mark. For more examples, see Commentary Figure R502.1. For examples of marks for wood structural panels, see Commentary Figure R503.2.1.

R602.1.1 End-jointed lumber. Approved end-jointed lumber identified by a grade mark conforming to Section R602.1 may be used interchangeably with solid-sawn members of the same species and grade.

❖ End-jointed lumber that has been fabricated in accordance with nationally recognized standards can be used interchangeably with solid-sawn members, provided that the species and grade of the wood are comparable. Commentary Figure 602.1.1 shows end-jointed lumber, which is also referred to as "finger-jointed." Also see the commentary for Section R502.1.3.

R602.1.2 Structural glued laminated timbers. Glued laminated timbers shall be manufactured and identified as required in AITC A190.1 and ASTM D 3737.

❖ Glued laminated (gluelam) timbers are engineered wood elements consisting of layers of plywood plies with varying grades of plywood through the thickness of the member. The strength of the members and the use of the members (beams, studs, etc.) are based on different combinations of plywood veneer grades, called layups. The design, testing and manufacture of different layups for different applications are standardized in the referenced standards. The standards also address the identification of these timbers, which is useful at the construction site during inspection.

R602.1.3 Structural log members. Stress grading of structural log members of nonrectangular shape, as typically used in log buildings, shall be in accordance with ASTM D 3957. Such structural log members shall be identified by the grade mark of an approved lumber grading or inspection agency. In lieu of a grade mark on the material, a certificate of inspection as to species and grade issued by a lumber-grading or inspection agency meeting the requirements of this section shall be permitted to be accepted.

❖ This section addresses grading requirements for logs used as structural members. This subsection specifies the reference for acceptable methods for establishing structural capacities of logs and specifies the requirement for a grading stamp or alternate certification n structural logs. Structural log members must be graded in accordance with ASTM D 3957, *Standard Practices for Establishing Stress Grades for Structural Members Used in Log Buildings*.

INTERPRETING GRADE STAMPS

MOST GRADE STAMPS, EXCEPT THOSE FOR ROUGH LUMBER OR HEAVY TIMBERS, CONTAIN FIVE BASIC ELEMENTS:

(B)
1 2 **STAND** (C)
A B C   S-DRY   /D\ (D)
(A)   (E)   /FIR\

A. THE TRADEMARK INDICATES AGENCY QUALITY SUPERVISION.

B. MILL IDENTIFICATION – FIRM NAME, BRAND OR ASSIGNED MILL NUMBER.

C. GRADE DESIGNATION – GRADE NAME, NUMBER OR ABBREVIATION.

D. SPECIAL IDENTIFICATION – INDICATES SPECIES INDIVIDUALLY OR IN COMBINATION.

E. CONDITION OF SEASONING AT TIME OF SURFACING:
S-DRY – 19% MAX MOISTURE CONTENT
MC 15 – 15% MAX MOISTURE CONTENT
S-GRN – OVER 19% MOISTURE CONTENT (UNSEASONED)

Figure R602.1
GRADE STAMP EXAMPLE FOR DIMENSIONAL LUMBER

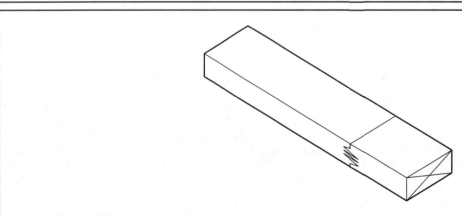

Figure R602.1.1
END-JOINTED LUMBER—FINGER JOINT

R602.2 Grade. Studs shall be a minimum No. 3, standard or stud grade lumber.

Exception: Bearing studs not supporting floors and nonbearing studs may be utility grade lumber, provided the studs are spaced in accordance with Table R602.3(5).

❖ A minimum grade of No. 3, standard, or stud grade lumber is specified for studs in conventional wood-frame walls to maintain minimum load-carrying capabilities. The maximum size of standard grade lumber is limited to 2 to 4 inches nominal (51 to 102 mm) in thickness and 4 inches (102 mm) in width. Thus, a 4-inch by 6-inch (102 mm by 152 mm) piece of dimensional lumber would not be available in the standard grade.

To allow more economical use of lumber for bearing studs not supporting floors and nonbearing studs, the exception permits the use of utility grade lumber, provided that the studs are spaced according to the limits established in Table R602.3(5). Note "a" of the table places a more restrictive limit on utility grade stud spacing where permitted in a load-bearing application.

R602.3 Design and construction. Exterior walls of wood-frame construction shall be designed and constructed in accordance with the provisions of this chapter and Figures R602.3(1) and R602.3(2) or in accordance with AF&PA's NDS. Components of exterior walls shall be fastened in accordance with Tables R602.3(1) through R602.3(4). Exterior walls covered with foam plastic sheathing shall be braced in accordance with Section R602.10. Structural sheathing shall be fastened directly to structural framing members.

❖ Figures R602.3(1) and R602.3(2) in conjunction with the minimum nailing requirements of Tables R602.3(1) and R602.3(2) provide typical wall-framing details for construction of exterior walls. Walls must be in accordance with this section or American Forest & Paper Association's National Design Specification (NDS).

Typically, foam-plastic material used for insulation is nailed to the outside face of walls as sheathing, but the materials are usually not structural sheathing and therefore provide no lateral bracing for the building. Thus, the code calls out requirements for structural sheathing to remind the user that foam-plastic sheathing should not be confused with elements that provide sufficient lateral bracing.

R602.3.1 Stud size, height and spacing. The size, height and spacing of studs shall be in accordance with Table R602.3.(5).

Exceptions:

1. Utility grade studs shall not be spaced more than 16 inches (406 mm) on center, shall not support more than a roof and ceiling, and shall not exceed 8 feet (2438 mm) in height for exterior walls and load-bearing walls or 10 feet (3048 mm) for interior nonload-bearing walls.

2. Studs more than 10 feet (3048 mm) in height which are in accordance with Table R602.3.1.

❖ Table R602.3(5) lists required stud spacing based on the stud size and loading condition.

Exception 1 addresses the fact that utility grade lumber is lower in strength and quality and therefore requires tighter spacing and shorter lengths in exterior walls. Note that the minimum grade lumber for application of Table R602.3(5) is, by implication, stud grade or standard grade.

As noted in Exception 2, if the bearing wall height exceeds 10 feet (3048 mm), stud size is determined using Table R602.3.1. A design is required if the limits given in the table's title are exceeded or if Note "a" applies.

R602.3.2 Top plate. Wood stud walls shall be capped with a double top plate installed to provide overlapping at corners and intersections with bearing partitions. End joints in top plates shall be offset at least 24 inches (610 mm). Joints in plates need not occur over studs. Plates shall be not less than 2-inches (51 mm) nominal thickness and have a width at least equal to the width of the studs.

Exception: A single top plate may be installed in stud walls, provided the plate is adequately tied at joints, corners and intersecting walls by a minimum 3-inch-by- 6-inch by a 0.036-inch-thick (76 mm by 152 mm by 0.914 mm) galvanized steel plate that is nailed to each wall or segment of wall by six 8d nails on each side, provided the rafters or joists are centered over the studs with a tolerance of no more than 1 inch (25 mm). The top plate may be omitted over lintels that are adequately tied to adjacent wall sections with steel plates or equivalent as previously described.

❖ Plates are the horizontal elements of walls capping the top of walls and framing the bottom of stud walls. They are called plates because they are laid flat, with the deeper dimension horizontal. In building construction applications, top plates must be a nominal 2-inch (51 mm) thickness.

Double top plates serve three major functions:

1. They overlap at corners and bearing wall intersections, providing a means of tying the building together. (Note that the overlap is not required at intersections with nonbearing interior walls.)

2. They serve as beams to support joists and rafters that are not located directly over the studs.

3. They serve as chords for floor and roof diaphragms.

Along with provisions for stud size and spacing limitations, this section requires the installation of double top plates to provide a continuous tie along the tops of the walls. With the advent of wider wall framing to accommodate increased thickness of insulation, a desire to save on material costs led to the allowance of a single top plate alternative. The exception permits the use of a single top plate in bearing and exterior walls as long as adequate top-plate ties are provided. In addition, joists or rafters framing into the wall must be placed more closely to the vertical stud below. This is necessary to limit the bending stress in the top plate. The single top plate exception is illustrated in Commentary Figure R602.3.2.

TABLE R602.3(1)
FASTENER SCHEDULE FOR STRUCTURAL MEMBERS

DESCRIPTION OF BUILDING ELEMENTS	NUMBER AND TYPE OF FASTENER[a,b,c]	SPACING OF FASTENERS
Joist to sill or girder, toe nail	3-8d $(2\text{-}1/_2'' \times 0.113'')$	—
$1'' \times 6''$ subfloor or less to each joist, face nail	2-8d $(2^1/_2'' \times 0.113'')$ 2 staples, $1^3/_4''$	— —
$2''$ subfloor to joist or girder, blind and face nail	2-16d $(3^1/_2'' \times 0.135'')$	—
Sole plate to joist or blocking, face nail	16d $(3^1/_2'' \times 0.135'')$	$16''$ o.c.
Top or sole plate to stud, end nail	2-16d $(3^1/_2'' \times 0.135'')$	—
Stud to sole plate, toe nail	3-8d $(2^1/_2'' \times 0.113'')$ or 2-16d $(3^1/_2'' \times 0.135'')$	—
Double studs, face nail	10d $(3'' \times 0.128'')$	$24''$ o.c.
Double top plates, face nail	10d $(3'' \times 0.128'')$	$24''$ o.c.
Sole plate to joist or blocking at braced wall panels	3-16d $(3^1/_2'' \times 0.135'')$	$16''$ o.c.
Double top plates, minimum 24-inch offset of end joints, face nail in lapped area	8-16d $(3^1/_2'' \times 0.135'')$	—
Blocking between joists or rafters to top plate, toe nail	3-8d $(2^1/_2'' \times 0.113'')$	—
Rim joist to top plate, toe nail	8d $(2^1/_2'' \times 0.113'')$	$6''$ o.c.
Top plates, laps at corners and intersections, face nail	2-10d $(3'' \times 0.128'')$	—
Built-up header, two pieces with $^1/_2''$ spacer	16d $(3^1/_2'' \times 0.135'')$	$16''$ o.c. along each edge
Continued header, two pieces	16d $(3^1/_2'' \times 0.135'')$	$16''$ o.c. along each edge
Ceiling joists to plate, toe nail	3-8d $(2^1/_2'' \times 0.113'')$	—
Continuous header to stud, toe nail	4-8d $(2^1/_2'' \times 0.113'')$	—
Ceiling joist, laps over partitions, face nail	3-10d $(3'' \times 0.128'')$	—
Ceiling joist to parallel rafters, face nail	3-10d $(3'' \times 0.128'')$	—
Rafter to plate, toe nail	2-16d $(3^1/_2'' \times 0.135'')$	—
$1''$ brace to each stud and plate, face nail	2-8d $(2^1/_2'' \times 0.113'')$ 2 staples, $1^3/_4''$	— —
$1'' \times 6''$ sheathing to each bearing, face nail	2-8d $(2^1/_2'' \times 0.113'')$ 2 staples, $1^3/_4''$	— —
$1'' \times 8''$ sheathing to each bearing, face nail	2-8d $(2^1/_2'' \times 0.113'')$ 3 staples, $1^3/_4''$	— —
Wider than $1'' \times 8''$ sheathing to each bearing, face nail	3-8d $(2^1/_2'' \times 0.113'')$ 4 staples, $1^3/_4''$	— —
Built-up corner studs	10d $(3'' \times 0.128'')$	$24''$ o.c.
Built-up girders and beams, 2-inch lumber layers	10d $(3'' \times 0.128'')$	Nail each layer as follows: $32''$ o.c. at top and bottom and staggered. Two nails at ends and at each splice.
$2''$ planks	2-16d $(3^1/_2'' \times 0.135'')$	At each bearing
Roof rafters to ridge, valley or hip rafters: toe nail face nail	 4-16d $(3^1/_2'' \times 0.135'')$ 3-16d $(3^1/_2'' \times 0.135'')$	 — —
Rafter ties to rafters, face nail	3-8d $(2^1/_2'' \times 0.113'')$	—
Collar tie to rafter, face nail, or $1^1/_4'' \times 20$ gage ridge strap	3-10d $(3'' \times 0.128'')$	—

(continued)

TABLE R602.3(1)—continued
FASTENER SCHEDULE FOR STRUCTURAL MEMBERS

DESCRIPTION OF BUILDING MATERIALS	DESCRIPTION OF FASTENER[b, c, e]	SPACING OF FASTENERS	
		Edges (inches)[i]	Intermediate supports[c,e] (inches)
Wood structural panels, subfloor, roof and wall sheathing to framing, and particleboard wall sheathing to framing			
$^5/_{16}$"-$^1/_2$"	6d common ($2'' \times 0.113''$) nail (subfloor, wall) 8d common ($2^1/_2'' \times 0.131''$) nail (roof)[f]	6	12[g]
$^{19}/_{32}$"-1"	8d common nail ($2^1/_2'' \times 0.131''$)	6	12[g]
$1^1/_8$"-$1^1/_4$"	10d common ($3'' \times 0.148''$) nail or 8d ($2^1/_2'' \times 0.131''$) deformed nail	6	12
Other wall sheathing[h]			
$^1/_2$" structural cellulosic fiberboard sheathing	$1^1/_2$" galvanized roofing nail 8d common ($2^1/_2'' \times 0.131''$) nail; staple 16 ga., $1^1/_2$" long	3	6
$^{25}/_{32}$" structural cellulosic fiberboard sheathing	$1^3/_4$" galvanized roofing nail 8d common ($2^1/_2'' \times 0.131''$) nail; staple 16 ga., $1^3/_4$" long	3	6
$^1/_2$" gypsum sheathing[d]	$1^1/_2$" galvanized roofing nail; 6d common ($2'' \times 0.131''$) nail; staple galvanized $1^1/_2$" long; $1^1/_4$" screws, Type W or S	4	8
$^5/_8$" gypsum sheathing[d]	$1^3/_4$" galvanized roofing nail; 8d common ($2^1/_2'' \times 0.131''$) nail; staple galvanized $1^5/_8$" long; $1^5/_8$" screws, Type W or S	4	8
Wood structural panels, combination subfloor underlayment to framing			
$^3/_4$" and less	6d deformed ($2'' \times 0.120''$) nail or 8d common ($2^1/_2'' \times 0.131''$) nail	6	12
$^7/_8$"-1"	8d common ($2^1/_2'' \times 0.131''$) nail or 8d deformed ($2^1/_2'' \times 0.120''$) nail	6	12
$1^1/_8$"-$1^1/_4$"	10d common ($3'' \times 0.148''$) nail or 8d deformed ($2^1/_2'' \times 0.120''$) nail	6	12

For SI: 1 inch = 25.4 mm, 1 foot = 304.8 mm, 1 mile per hour = 0.447 m/s; 1ksi = 6.895 MPa.

a. All nails are smooth-common, box or deformed shanks except where otherwise stated. Nails used for framing and sheathing connections shall have minimum average bending yield strengths as shown: 80 ksi for shank diameter of 0.192 inch (20d common nail), 90 ksi for shank diameters larger than 0.142 inch but not larger than 0.177 inch, and 100 ksi for shank diameters of 0.142 inch or less.

b. Staples are 16 gage wire and have a minimum $^7/_{16}$-inch on diameter crown width.

c. Nails shall be spaced at not more than 6 inches on center at all supports where spans are 48 inches or greater.

d. Four-foot-by-8-foot or 4-foot-by-9-foot panels shall be applied vertically.

e. Spacing of fasteners not included in this table shall be based on Table R602.3(2).

f. For regions having basic wind speed of 110 mph or greater, 8d deformed ($2^1/_2'' \times 0.120$) nails shall be used for attaching plywood and wood structural panel roof sheathing to framing within minimum 48-inch distance from gable end walls, if mean roof height is more than 25 feet, up to 35 feet maximum.

g. For regions having basic wind speed of 100 mph or less, nails for attaching wood structural panel roof sheathing to gable end wall framing shall be spaced 6 inches on center. When basic wind speed is greater than 100 mph, nails for attaching panel roof sheathing to intermediate supports shall be spaced 6 inches on center for minimum 48-inch distance from ridges, eaves and gable end walls; and 4 inches on center to gable end wall framing.

h. Gypsum sheathing shall conform to ASTM C 79 and shall be installed in accordance with GA 253. Fiberboard sheathing shall conform to ASTM C 208.

i. Spacing of fasteners on floor sheathing panel edges applies to panel edges supported by framing members and required blocking and at all floor perimeters only. Spacing of fasteners on roof sheathing panel edges applies to panel edges supported by framing members and required blocking. Blocking of roof or floor sheathing panel edges perpendicular to the framing members need not be provided except as required by other provisions of this code. Floor perimeter shall be supported by framing members or solid blocking.

❖ The fastener schedule provides minimum nailing requirements (i.e., size, spacing) for connecting building elements used in wood frame construction. For wood structural panels, both edge nailing and intermediate (field) nailing are specified. In addition to the nailing for wood structural panels, fasteners are specified for gypsum wall sheathing, cellulosic fiberboard wall sheathing and combination subfloor underlayment.

TABLE R602.3(2)
ALTERNATE ATTACHMENTS

NOMINAL MATERIAL THICKNESS (inches)	DESCRIPTION[a, b] OF FASTENER AND LENGTH (inches)	SPACING[c] OF FASTENERS	
		Edges (inches)	Intermediate supports (inches)
Wood structural panels subfloor, roof and wall sheathing to framing and particleboard wall sheathing to framing[f]			
up to $^1/_2$	Staple 15 ga. $1^3/_4$	4	8
	0.097 - 0.099 Nail $2^1/_4$	3	6
	Staple 16 ga. $1^3/_4$	3	6
$^{19}/_{32}$ and $^5/_8$	0.113 Nail 2	3	6
	Staple 15 and 16 ga. 2	4	8
	0.097 - 0.099 Nail $2^1/_4$	4	8
$^{23}/_{32}$ and $^3/_4$	Staple 14 ga. 2	4	8
	Staple 15 ga. $1^3/_4$	3	6
	0.097 - 0.099 Nail $2^1/_4$	4	8
	Staple 16 ga. 2	4	8
1	Staple 14 ga. $2^1/_4$	4	8
	0.113 Nail $2^1/_4$	3	6
	Staple 15 ga. $2^1/_4$	4	8
	0.097 - 0.099 Nail $2^1/_2$	4	8

NOMINAL MATERIAL THICKNESS (inches)	DESCRIPTION[a,b] OF FASTENER AND LENGTH (inches)	SPACING[c] OF FASTENERS	
		Edges (inches)	Body of panel[d] (inches)
Floor underlayment; plywood-hardboard-particleboard[f]			
Plywood			
$^1/_4$ and $^5/_{16}$	$1^1/_4$ ring or screw shank nail—minimum $12^1/_2$ ga. (0.099″) shank diameter	3	6
	Staple 18 ga., $^7/_8$, $^3/_{16}$ crown width	2	5
$^{11}/_{32}$, $^3/_8$, $^{15}/_{32}$, $^1/_2$ and $^{19}/_{32}$	$1^1/_4$ ring or screw shank nail—minimum $12^1/_2$ ga. (0.099″) shank diameter	6	8[e]
$^5/_8$, $^{23}/_{32}$ and $^3/_4$	$1^1/_2$ ring or screw shank nail—minimum $12^1/_2$ ga. (0.099″) shank diameter	6	8
	Staple 16 ga. $1^1/_2$	6	8
Hardboard[f]			
0.200	$1^1/_2$ long ring-grooved underlayment nail	6	6
	4d cement-coated sinker nail	6	6
	Staple 18 ga., $^7/_8$ long (plastic coated)	3	6
Particleboard			
$^1/_4$	4d ring-grooved underlayment nail	3	6
	Staple 18 ga., $^7/_8$ long, $^3/_{16}$ crown	3	6
$^3/_8$	6d ring-grooved underlayment nail	6	10
	Staple 16 ga., $1^1/_8$ long, $^3/_8$ crown	3	6
$^1/_2$, $^5/_8$	6d ring-grooved underlayment nail	6	10
	Staple 16 ga., $1^5/_8$ long, $^3/_8$ crown	3	6

For SI: 1 inch = 25.4 mm.

a. Nail is a general description and may be T-head, modified round head or round head.

b. Staples shall have a minimum crown width of $^7/_{16}$-inch on diameter except as noted.

c. Nails or staples shall be spaced at not more than 6 inches on center at all supports where spans are 48 inches or greater. Nails or staples shall be spaced at not more than 12 inches on center at intermediate supports for floors.

d. Fasteners shall be placed in a grid pattern throughout the body of the panel.

e. For 5-ply panels, intermediate nails shall be spaced not more than 12 inches on center each way.

f. Hardboard underlayment shall conform to ANSI/AHA A135.4.

❖ This table offers alternatives to the nailing specified for wood structural panels in Table R602.3(1).

TABLE R602.3(3)
WOOD STRUCTURAL PANEL WALL SHEATHING

PANEL SPAN RATING	PANEL NOMINAL THICKNESS (inch)	MAXIMUM STUD SPACING (inches)	
		Siding nailed to:[a]	
		Stud	Sheathing
12/0, 16/0, 20/0, or wall —16 o.c.	$^5/_{16}$, $^3/_8$	16	16[b]
24/0, 24/16, 32/16 or wall—24 o.c.	$^3/_8$, $^7/_{16}$, $^{15}/_{32}$, $^1/_2$	24	24[c]

For SI: 1 inch = 25.4 mm.
a. Blocking of horizontal joints shall not be required.
b. Plywood sheathing $^3/_8$-inch thick or less shall be applied with long dimension across studs.
c. Three-ply plywood panels shall be applied with long dimension across studs.

❖ Table R602.3(3) gives maximum permissible stud spacing based on the thicknesses of wood structural panels used as sheathing. Although the span ratings are listed, they are not applicable for determining the spacing of vertical framing members. Some wood structural panels are manufactured for use as wall sheathing only. The maximum stud spacing for these specialty panels is specified on the trademark. All wood structural panels may be installed with the face veneers either parallel or perpendicular to the framing, except where thinner panels are used as a nailing base for cladding, as indicated in Notes b and c Face veneers must be perpendicular to studs.

TABLE R602.3(4)
ALLOWABLE SPANS FOR PARTICLEBOARD WALL SHEATHING[a]

THICKNESS (inch)	GRADE	STUD SPACING (inches)	
		When siding is nailed to studs	When siding is nailed to sheathing
$^3/_8$	—1 Exterior glue	16	—
$^1/_2$	—2 Exterior glue	16	16

For SI: 1 inch = 25.4 mm.
a. Wall sheathing not exposed to the weather. If the panels are applied horizontally, the end joints of the panel shall be offset so that four panels corners will not meet. All panel edges must be supported. Leave a $^1/_{16}$-inch gap between panels and nail no closer than $^3/_8$ inch from panel edges.

❖ The allowable spans for particleboard wall sheathing listed in Table R602.3(4) are for wall sheathing that is not exposed to weather. The panels may be applied with the long dimension parallel or perpendicular to the studs.

TABLE R602.3(5)
SIZE, HEIGHT AND SPACING OF WOOD STUDS[a]

STUD SIZE (inches)	BEARING WALLS					NONBEARING WALLS	
	Laterally unsupported stud height[a] (feet)	Maximum spacing when supporting roof and ceiling only (inches)	Maximum spacing when supporting one floor, roof and ceiling (inches)	Maximum spacing when supporting two floors, roof and ceiling (inches)	Maximum spacing when supporting one floor only (inches)	Laterally unsupported stud height[a] (feet)	Maximum spacing (inches)
2 × 3[b]	—	—	—	—	—	10	16
2 × 4	10	24	16	—	24	14	24
3 × 4	10	24	24	16	24	14	24
2 × 5	10	24	24	—	24	16	24
2 × 6	10	24	24	16	24	20	24

For SI: 1 inch = 25.4 mm.
a. Listed heights are distances between points of lateral support placed perpendicular to the plane of the wall. Increases in unsupported height are permitted where justified by analysis.
b. Shall not be used in exterior walls.

❖ This table provides the allowed stud spacing in bearing walls up to 10 feet (3048 mm) high as stated in Section R602.3.1. As described in Note a increases in unsupported height are permitted where justified by an engineering analysis. The "unsupported height" of a stud is really the distance between points of lateral support, which generally occurs at each floor level.

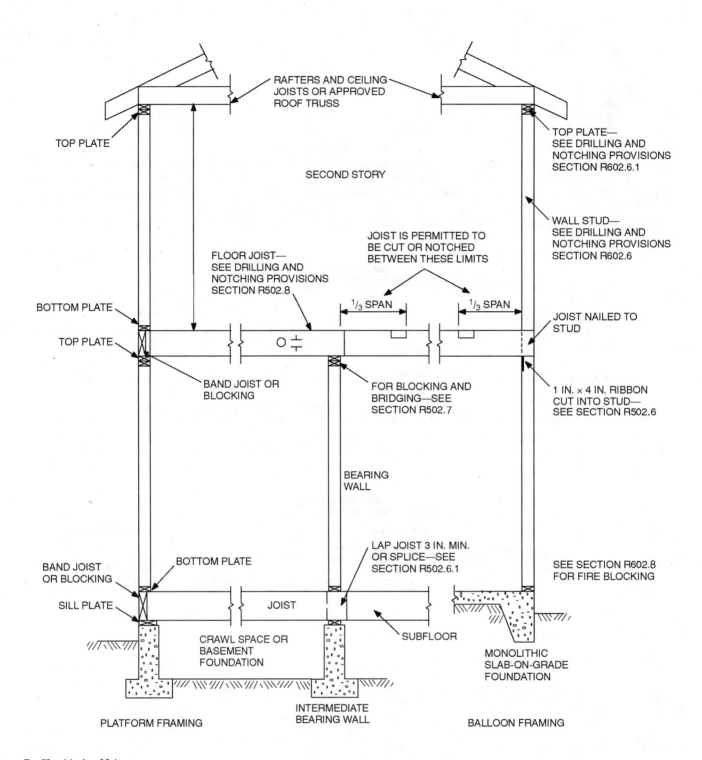

RAFTERS AND CEILING
JOISTS OR APPROVED
ROOF TRUSS

TOP PLATE

SECOND STORY

TOP PLATE—
SEE DRILLING AND
NOTCHING PROVISIONS
SECTION R602.6.1

WALL STUD—
SEE DRILLING AND
NOTCHING PROVISIONS
SECTION R602.6

FLOOR JOIST—
SEE DRILLING AND
NOTCHING PROVISIONS
SECTION R502.8

JOIST IS PERMITTED TO
BE CUT OR NOTCHED
BETWEEN THESE LIMITS

$^1/_3$ SPAN $^1/_3$ SPAN

BOTTOM PLATE

TOP PLATE

JOIST NAILED TO
STUD

BAND JOIST OR
BLOCKING

FOR BLOCKING AND
BRIDGING—SEE
SECTION R502.7

1 IN. × 4 IN. RIBBON
CUT INTO STUD—
SEE SECTION R502.6

BEARING
WALL

BAND JOIST
OR BLOCKING

BOTTOM PLATE

LAP JOIST 3 IN. MIN.
OR SPLICE—SEE
SECTION R502.6.1

SEE SECTION R602.8
FOR FIRE BLOCKING

SILL PLATE

JOIST

CRAWL SPACE OR
BASEMENT
FOUNDATION

SUBFLOOR

MONOLITHIC
SLAB-ON-GRADE
FOUNDATION

PLATFORM FRAMING

INTERMEDIATE
BEARING WALL

BALLOON FRAMING

For SI: 1 inch = 25.4 mm.

FIGURE R602.3(1)
TYPICAL WALL, FLOOR AND ROOF FRAMING

❖ This is a schematic wood-frame building section that shows code provisions for wood floor systems as well as wall construction.

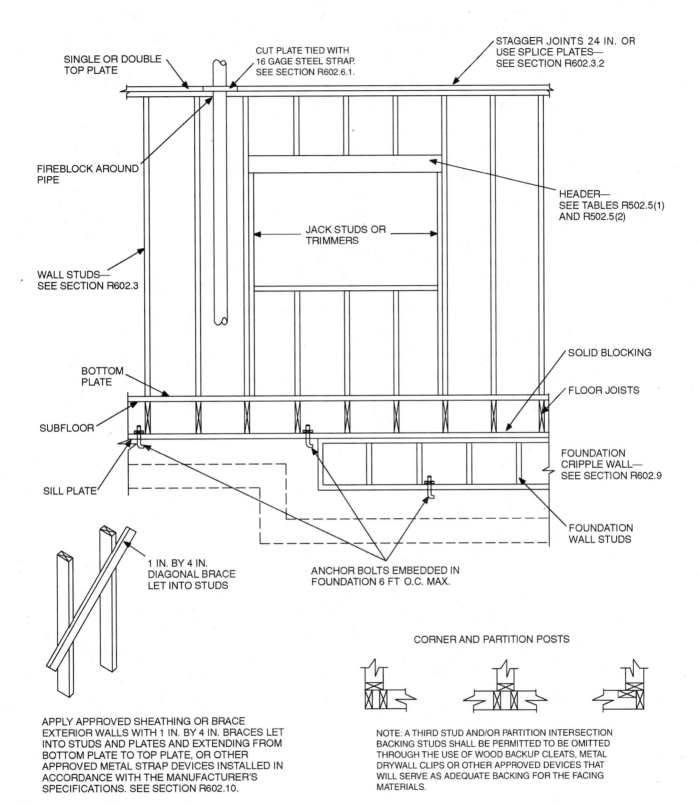

SINGLE OR DOUBLE
TOP PLATE

CUT PLATE TIED WITH
16 GAGE STEEL STRAP.
SEE SECTION R602.6.1.

STAGGER JOINTS 24 IN. OR
USE SPLICE PLATES—
SEE SECTION R602.3.2

FIREBLOCK AROUND
PIPE

HEADER—
SEE TABLES R502.5(1)
AND R502.5(2)

JACK STUDS OR
TRIMMERS

WALL STUDS—
SEE SECTION R602.3

SOLID BLOCKING

BOTTOM
PLATE

FLOOR JOISTS

SUBFLOOR

FOUNDATION
CRIPPLE WALL—
SEE SECTION R602.9

SILL PLATE

FOUNDATION
WALL STUDS

1 IN. BY 4 IN.
DIAGONAL BRACE
LET INTO STUDS

ANCHOR BOLTS EMBEDDED IN
FOUNDATION 6 FT O.C. MAX.

CORNER AND PARTITION POSTS

APPLY APPROVED SHEATHING OR BRACE
EXTERIOR WALLS WITH 1 IN. BY 4 IN. BRACES LET
INTO STUDS AND PLATES AND EXTENDING FROM
BOTTOM PLATE TO TOP PLATE, OR OTHER
APPROVED METAL STRAP DEVICES INSTALLED IN
ACCORDANCE WITH THE MANUFACTURER'S
SPECIFICATIONS. SEE SECTION R602.10.

NOTE: A THIRD STUD AND/OR PARTITION INTERSECTION
BACKING STUDS SHALL BE PERMITTED TO BE OMITTED
THROUGH THE USE OF WOOD BACKUP CLEATS, METAL
DRYWALL CLIPS OR OTHER APPROVED DEVICES THAT
WILL SERVE AS ADEQUATE BACKING FOR THE FACING
MATERIALS.

For SI: 1 inch = 25.4 mm, 1 foot = 304.8 mm.

**FIGURE R602.3(2)
FRAMING DETAILS**

❖ This figure is an elevation of typical wood-frame wall construction showing several code provisions as well as illus-
trating features such as cripple walls.

TABLE R602.3.1
MAXIMUM ALLOWABLE LENGTH OF WOOD WALL STUDS EXPOSED TO WIND SPEEDS OF 100 mph OR LESS
IN SEISMIC DESIGN CATEGORIES A, B, C and D_0, D_1, AND D_2[b, c]

HEIGHT (feet)	ON-CENTER SPACING (inches)			
	24	16	12	8
Supporting a roof only				
>10	2 × 4	2 × 4	2 × 4	2 × 4
12	2 × 6	2 × 4	2 × 4	2 × 4
14	2 × 6	2 × 6	2 × 6	2 × 4
16	2 × 6	2 × 6	2 × 6	2 × 4
18	NA[a]	2 × 6	2 × 6	2 × 6
20	NA[a]	NA[a]	2 × 6	2 × 6
24	NA[a]	NA[a]	NA[a]	2 × 6
Supporting one floor and a roof				
>10	2 × 6	2 × 4	2 × 4	2 × 4
12	2 × 6	2 × 6	2 × 6	2 × 4
14	2 × 6	2 × 6	2 × 6	2 × 6
16	NA[a]	2 × 6	2 × 6	2 × 6
18	NA[a]	2 × 6	2 × 6	2 × 6
20	NA[a]	NA[a]	2 × 6	2 × 6
24	NA[a]	NA[a]	NA[a]	2 × 6
Supporting two floors and a roof				
>10	2 × 6	2 × 6	2 × 4	2 × 4
12	2 × 6	2 × 6	2 × 6	2 × 6
14	2 × 6	2 × 6	2 × 6	2 × 6
16	NA[a]	NA[a]	2 × 6	2 × 6
18	NA[a]	NA[a]	2 × 6	2 × 6
20	NA[a]	NA[a]	NA[a]	2 × 6
22	NA[a]	NA[a]	NA[a]	NA[a]
24	NA[a]	NA[a]	NA[a]	NA[a]

For SI: 1 inch = 25.4 mm, 1 foot = 304.8 mm, 1 pound per square foot = 0.0479kPa,
   1 pound per square inch = 6.895 kPa, 1 mile per hour = 0.447 m/s.

a. Design required.

b. Applicability of this table assumes the following: Snow load not exceeding 25 psf, f_b not less than 1310 psi determined by multiplying the AF&PA NDS tabular base design value by the repetitive use factor, and by the size factor for all species except southern pine, E not less than 1.6×10^6 psi, tributary dimensions for floors and roofs not exceeding 6 feet, maximum span for floors and roof not exceeding 12 feet, eaves not over 2 feet in dimension and exterior sheathing. Where the conditions are not within these parameters, design is required.

c. Utility, standard, stud and No. 3 grade lumber of any species are not permitted.

(continued)

TABLE R602.3.1—continued
MAXIMUM ALLOWABLE LENGTH OF WOOD WALL STUDS EXPOSED TO WIND SPEEDS OF 100 mph OR LESS IN SEISMIC DESIGN CATEGORIES A, B, C, D$_0$, D$_1$, and D$_2$

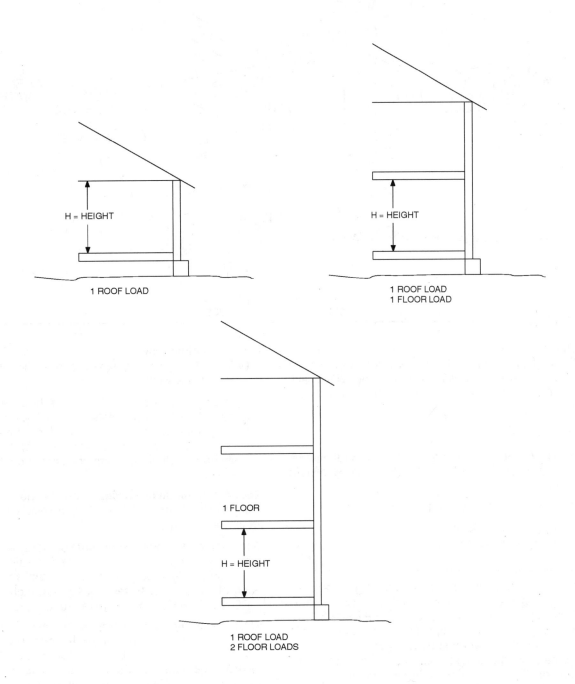

H = HEIGHT

1 ROOF LOAD

H = HEIGHT

1 ROOF LOAD
1 FLOOR LOAD

1 FLOOR

H = HEIGHT

1 ROOF LOAD
2 FLOOR LOADS

❖ This table provides the stud spacing limits in bearing walls over 10 feet (3048 mm) high as stated in Section R602.3.1. Notes b and c make it clear that the lumber grades specified in Section R602.2 are not applicable, but instead must be determined by using AF&PA NDS and applying the limitations of Note b. Conditions noted as "NA" in the table require a design per Note a.

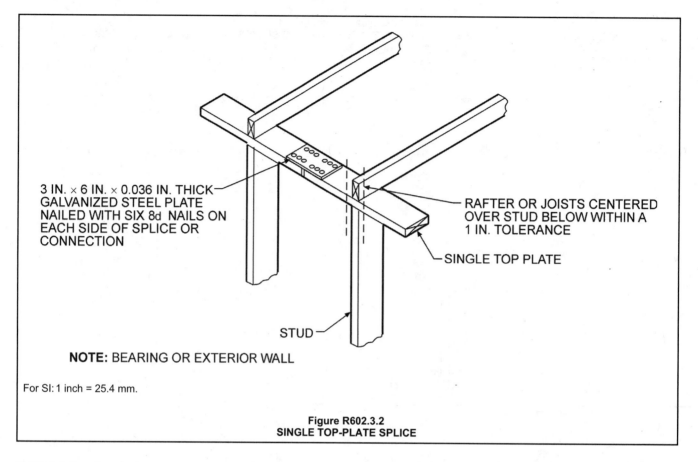

3 IN. × 6 IN. × 0.036 IN. THICK GALVANIZED STEEL PLATE NAILED WITH SIX 8d NAILS ON EACH SIDE OF SPLICE OR CONNECTION

RAFTER OR JOISTS CENTERED OVER STUD BELOW WITHIN A 1 IN. TOLERANCE

SINGLE TOP PLATE

STUD

NOTE: BEARING OR EXTERIOR WALL

For SI: 1 inch = 25.4 mm.

Figure R602.3.2
SINGLE TOP-PLATE SPLICE

R602.3.3 Bearing studs. Where joists, trusses or rafters are spaced more than 16 inches (406 mm) on center and the bearing studs below are spaced 24 inches (610 mm) on center, such members shall bear within 5 inches (127 mm) of the studs beneath.

Exceptions:

1. The top plates are two 2-inch by 6-inch (38 mm by 140 mm) or two 3-inch by 4-inch (64 mm by 89 mm) members.

2. A third top plate is installed.

3. Solid blocking equal in size to the studs is installed to reinforce the double top plate.

❖ The bottom plate serves to anchor the wall to the floor, and studs are attached to it by end nailing or toe-nailing. The basic requirements of this section are shown in Commentary Figures R602.3.3(1) and R602.3.3(3). The code permits three options (exceptions) to account for loading that must be resisted by the double top plate when the stud spacing is 24 inches (610 mm) on center. The first option is an increase in the minimum top plate size. The second option is to provide a third plate. The third option is solid blocking used as reinforcement for a double plate, installed so that the load from the framing members above is transferred to the supporting studs. Commentary Figures R602.3.3(2) and R602.3.3(4) illustrate the first and third options.

R602.3.4 Bottom (sole) plate. Studs shall have full bearing on a nominal 2-by (38 mm) or larger plate or sill having a width at least equal to the width of the studs.

❖ This provision requires that studs bear on an adequately sized bottom plate. This allows the bottom plate to serve as a nailing surface for the wall sheathing while controlling compression perpendicular to the grain on the sill. If too great, this stress could result in crushing.

R602.4 Interior load-bearing walls. Interior load-bearing walls shall be constructed, framed and fireblocked as specified for exterior walls.

❖ Interior load-bearing walls must be designed and constructed using the criteria for exterior walls. Thus, Table R602.3(5) should be used to establish stud spacing of walls up to 10 feet (3048 mm) high, and Table R602.3.1 applies to walls over 10 feet (3048 mm).

R602.5 Interior nonbearing walls. Interior nonbearing walls shall be permitted to be constructed with 2-inch-by-3-inch (51 mm by 76 mm) studs spaced 24 inches (610 mm) on center or, when not part of a braced wall line, 2-inch-by-4-inch (51 mm by 102 mm) flat studs spaced at 16 inches (406 mm) on center. Interior nonbearing walls shall be capped with at least a single top plate. Interior nonbearing walls shall be fireblocked in accordance with Section R602.8.

❖ The code permits reduced stud sizes, increased stud spacing and capping with a single top plate in interior nonbearing partitions. These allowances are based on minimal superimposed vertical loads.

R602.6 Drilling and notching–studs. Drilling and notching of studs shall be in accordance with the following:

1. Notching. Any stud in an exterior wall or bearing partition may be cut or notched to a depth not exceeding 25 percent of its width. Studs in nonbearing partitions may be notched to a depth not to exceed 40 percent of a single stud width.

2. Drilling. Any stud may be bored or drilled, provided that the diameter of the resulting hole is no more than 60 percent of the stud width, the edge of the hole is no more than $5/_8$ inch (16 mm) to the edge of the stud, and the hole

is not located in the same section as a cut or notch. Studs located in exterior walls or bearing partitions drilled over 40 percent and up to 60 percent shall also be doubled with no more than two successive doubled studs bored. See Figures R602.6(1) and R602.6(2).

Exception: Use of approved stud shoes is permitted when they are installed in accordance with the manufacturer's recommendations.

❖ This section addresses the allowable drilling and notching of studs used to frame partitions. See Figures R602.6(1) and R602.6(2) for examples of permitted drilling and notching of studs in exterior walls, bearing walls and nonbearing walls. These limitations retain the structural integrity of the studs. Where stud shoes

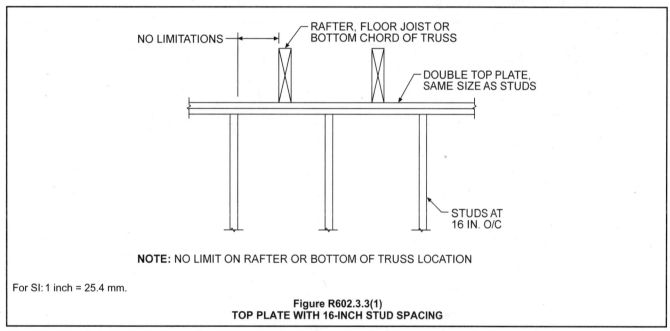

NO LIMITATIONS

RAFTER, FLOOR JOIST OR BOTTOM CHORD OF TRUSS

DOUBLE TOP PLATE, SAME SIZE AS STUDS

STUDS AT 16 IN. O/C

NOTE: NO LIMIT ON RAFTER OR BOTTOM OF TRUSS LOCATION

For SI: 1 inch = 25.4 mm.

Figure R602.3.3(1)
TOP PLATE WITH 16-INCH STUD SPACING

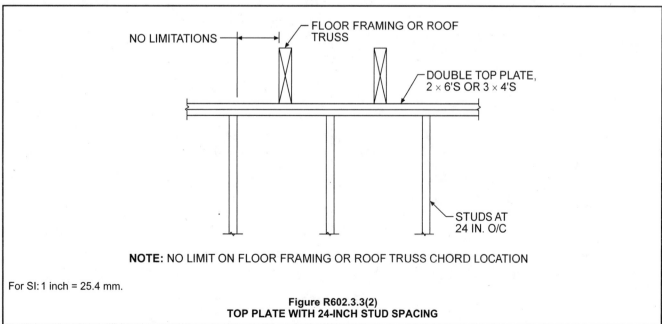

NO LIMITATIONS

FLOOR FRAMING OR ROOF TRUSS

DOUBLE TOP PLATE, 2 × 6'S OR 3 × 4'S

STUDS AT 24 IN. O/C

NOTE: NO LIMIT ON FLOOR FRAMING OR ROOF TRUSS CHORD LOCATION

For SI: 1 inch = 25.4 mm.

Figure R602.3.3(2)
TOP PLATE WITH 24-INCH STUD SPACING

are used, the exception allows drilling and notching to be in accordance with the approved manufacturer's instructions.

R602.6.1 Drilling and notching of top plate. When piping or ductwork is placed in or partly in an exterior wall or interior load-bearing wall, necessitating cutting, drilling or notching of the top plate by more than 50 percent of its width, a galvanized metal tie of not less than 0.054 inch thick (1.37 mm) (16 ga) and 1 ¹/₂ inches (38 mm) wide shall be fastened across and to the plate at each side of the opening with not less than eight 16d nails at each side or equivalent. See Figure R602.6.1.

Exception: When the entire side of the wall with the notch or cut is covered by wood structural panel sheathing.

❖ In many cases, drilling or notching of the top plate is necessary to allow plumbing, heating or other pipes to

be placed within the exterior walls and load-bearing interior walls. When drilling or notching the top plate by more than one-half its width is necessary, strapping across the plates as illustrated in Figure R602.6.1 is required to provide top plate continuity and to retain the structural integrity of the wall system as a whole. In the case of a double top plate, only the upper of the two plates needs to have a strap installed as described.

R602.7 Headers. For header spans see Tables R502.5(1) and R502.5(2).

❖ At a wall opening, headers transfer loads received from the wall and floors/roof to the foundation. Openings such as doors or windows within bearing walls must be framed with headers of sufficient size to span

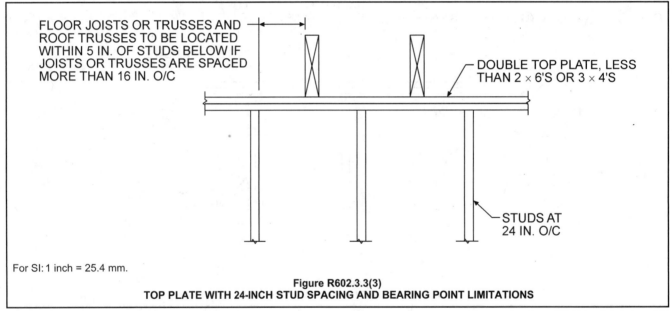

For SI: 1 inch = 25.4 mm.

Figure R602.3.3(3)
TOP PLATE WITH 24-INCH STUD SPACING AND BEARING POINT LIMITATIONS

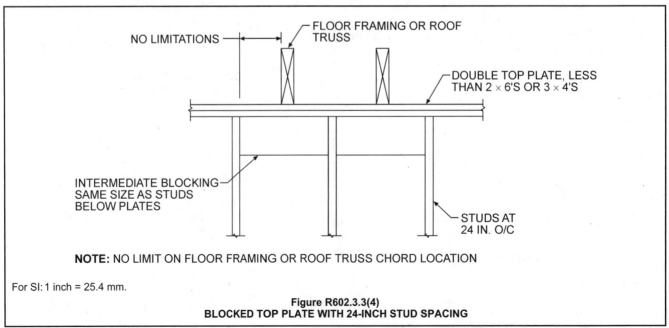

NOTE: NO LIMIT ON FLOOR FRAMING OR ROOF TRUSS CHORD LOCATION

For SI: 1 inch = 25.4 mm.

Figure R602.3.3(4)
BLOCKED TOP PLATE WITH 24-INCH STUD SPACING

the opening and transfer loads to jamb studs (header studs). Commentary Figure R602.7 illustrates a double header over a door opening. Tables R502.5(1) and R502.5(2) of the code are used to determine allowable spans for headers as well as girders. These tables are applicable to openings in bearing walls, but are not applicable to nonbearing walls. See Section R602.7.2 for headers in nonbearing walls. Table R502.5(1) is to be used for headers over openings in exterior bearing walls. Table R502.5(2) is to be used for headers over openings in interior bearing walls. Where wood girders are used in a basement of a one-story dwelling, Table R502.5(2) may be used to prescriptively size the basement girders (the space between supporting columns may be viewed as openings).

Example:

You wish to size a header in the exterior wall of a one-story house that is 30 feet (9144 mm) wide. The opening size is 8 feet (2338 mm). No. 2 grade southern pine is common for the area, and the ground-snow loading is 20 pounds per square foot (958 Pa).

Solution:

Table R502.5(1) lists, for a 30 psf (1436 Pa) snow load and a building width of 36 feet (10,973 mm), a triple 2 x 10 (51 mm x 254 mm) header; it will span 8 feet, 2 inches (2489 mm).

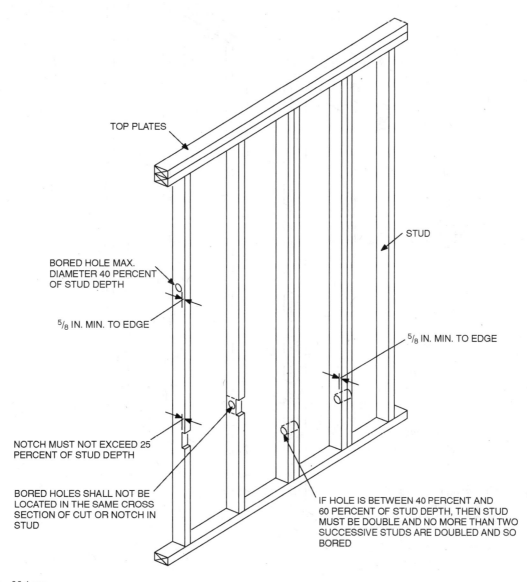

TOP PLATES

STUD

BORED HOLE MAX. DIAMETER 40 PERCENT OF STUD DEPTH

5/8 IN. MIN. TO EDGE

5/8 IN. MIN. TO EDGE

NOTCH MUST NOT EXCEED 25 PERCENT OF STUD DEPTH

BORED HOLES SHALL NOT BE LOCATED IN THE SAME CROSS SECTION OF CUT OR NOTCH IN STUD

IF HOLE IS BETWEEN 40 PERCENT AND 60 PERCENT OF STUD DEPTH, THEN STUD MUST BE DOUBLE AND NO MORE THAN TWO SUCCESSIVE STUDS ARE DOUBLED AND SO BORED

For SI: 1 inch = 25.4 mm.
Note: Condition for exterior and bearing walls.

FIGURE R602.6(1)
NOTCHING AND BORED HOLE LIMITATIONS FOR EXTERIOR WALLS AND BEARING WALLS

❖ See the commentary for Section R602.6.

R602.7.1 Wood structural panel box headers. Wood structural panel box headers shall be constructed in accordance with Figure R602.7.2 and Table R602.7.2.

❖ Wood structural panel box headers are used in some construction and are an efficient way of carrying load over a opening. The box header must be built in accordance with the criteria specified in Figures R602.7.2 and Table R602.7.2. Spans of box headers are established based on the construction of the header, the header depth, and the depth of the house. Note "a" to the table explains the depth criterion, which is based on a single story with a clear-span trussed roof or a two- story in which part of the load from the floors and roofs is supported by interior bearing walls. House depth is based on the span from the header to an interior bearing wall or the opposing exterior wall.

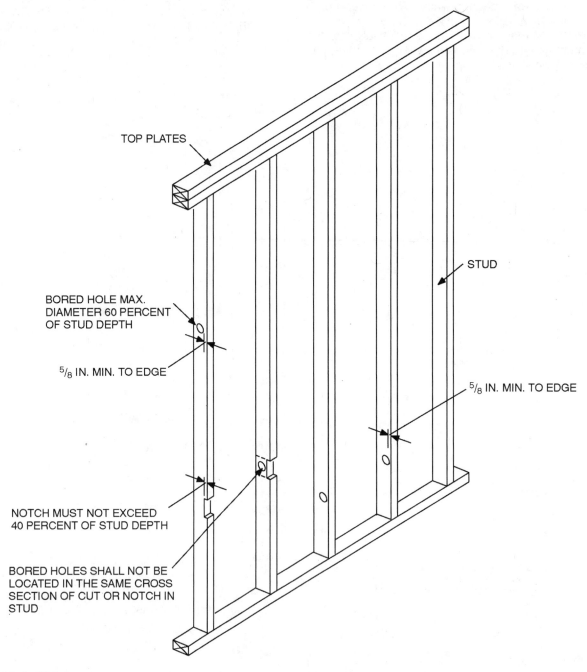

TOP PLATES

STUD

BORED HOLE MAX. DIAMETER 60 PERCENT OF STUD DEPTH

⁵/₈ IN. MIN. TO EDGE

⁵/₈ IN. MIN. TO EDGE

NOTCH MUST NOT EXCEED 40 PERCENT OF STUD DEPTH

BORED HOLES SHALL NOT BE LOCATED IN THE SAME CROSS SECTION OF CUT OR NOTCH IN STUD

For SI: 1 inch = 25.4 mm.

FIGURE R602.6(2)
NOTCHING AND BORED HOLE LIMITATIONS FOR INTERIOR NONBEARING WALLS

❖ See the commentary for Section R602.6.

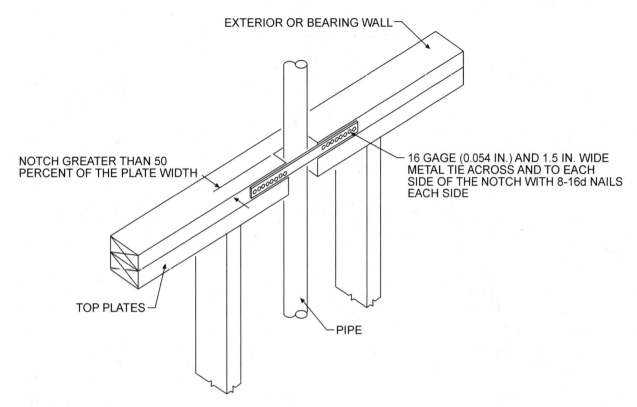

NOTCH GREATER THAN 50 PERCENT OF THE PLATE WIDTH

EXTERIOR OR BEARING WALL

16 GAGE (0.054 IN.) AND 1.5 IN. WIDE METAL TIE ACROSS AND TO EACH SIDE OF THE NOTCH WITH 8-16d NAILS EACH SIDE

TOP PLATES

PIPE

For SI: 1 inch = 25.4 mm.

FIGURE R602.6.1
TOP PLATE FRAMING TO ACCOMMODATE PIPING

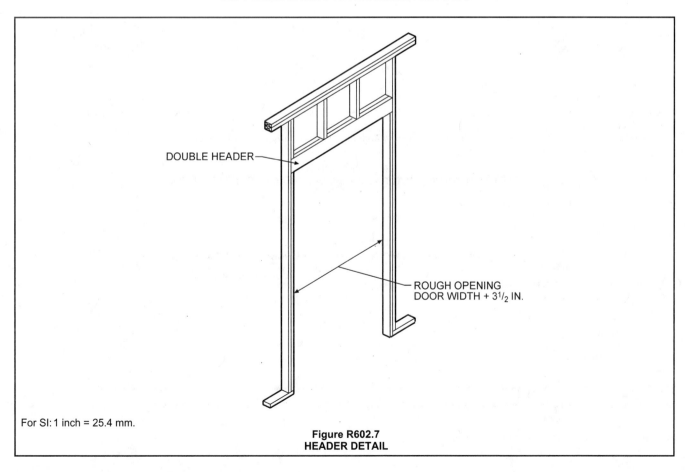

DOUBLE HEADER

ROUGH OPENING
DOOR WIDTH + $3^1/_2$ IN.

For SI: 1 inch = 25.4 mm.

Figure R602.7
HEADER DETAIL

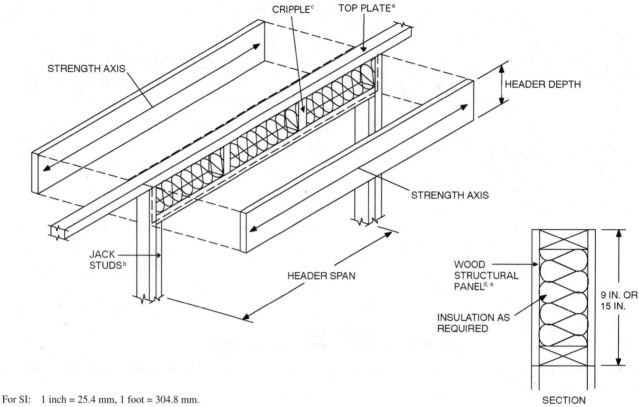

For SI: 1 inch = 25.4 mm, 1 foot = 304.8 mm.

NOTES:

a. The top plate shall be continuous over header.

b. Jack studs shall be used for spans over 4 feet.

c. Cripple spacing shall be the same as for studs.

d. Wood structural panel faces shall be single pieces of $^{15}/_{32}$-inch-thick Exposure 1 (exterior glue) or thicker, installed on the interior or exterior or both sides of the header.

e. Wood structural panel faces shall be nailed to framing and cripples with 8d common or galvanized box nails spaced 3 inches on center, staggering alternate nails $^{1}/_{2}$ inch. Galvanized nails shall be hot-dipped or tumbled.

FIGURE R602.7.2
TYPICAL WOOD STRUCTURAL PANEL BOX HEADER CONSTRUCTION

❖ See the commentary for Section R602.7.1.

TABLE R602.7.2
MAXIMUM SPANS FOR WOOD STRUCTURAL PANEL BOX HEADERS[a]

HEADER CONSTRUCTION[b]	HEADER DEPTH (inches)	HOUSE DEPTH (feet)				
		24	26	28	30	32
Wood structural panel—one side	9	4	4	3	3	—
	15	5	5	4	3	3
Wood structural panel—both sides	9	7	5	5	4	3
	15	8	8	7	7	6

For SI: 1 inch = 25.4 mm, 1 foot = 304.8 mm.

a. Spans are based on single story with clear-span trussed roof or two-story with floor and roof supported by interior-bearing walls.

b. See Figure R602.7.2 for construction details.

❖ See the commentary for Section R602.7.1.

R602.7.2 Nonbearing walls. Load-bearing headers are not required in interior or exterior nonbearing walls. A single flat 2-inch-by-4-inch (51 mm by 102 mm) member may be used as a header in interior or exterior nonbearing walls for openings up to 8 feet (2438 mm) in width if the vertical distance to the parallel nailing surface above is not more than 24 inches (610 mm). For such nonbearing headers, no cripples or blocking are required above the header.

❖ In walls that are not supporting significant loads, the code provides a simple (more economical) option for constructing headers.

R602.8 Fireblocking required. Fireblocking shall be provided to cut off all concealed draft openings (both vertical and horizontal) and to form an effective fire barrier between stories, and between a top story and the roof space. Fireblocking shall be provided in wood-frame construction in the following locations.

1. In concealed spaces of stud walls and partitions, including furred spaces and parallel rows of studs or staggered studs; as follows:

    1.1. Vertically at the ceiling and floor levels.

    1.2. Horizontally at intervals not exceeding 10 feet (3048 mm).

2. At all interconnections between concealed vertical and horizontal spaces such as occur at soffits, drop ceilings and cove ceilings.

3. In concealed spaces between stair stringers at the top and bottom of the run. Enclosed spaces under stairs shall comply with Section R311.2.2.

4. At openings around vents, pipes, ducts, cables and wires at ceiling and floor level, with an approved material to resist the free passage of flame and products of combustion.

5. For the fireblocking of chimneys and fireplaces, see Section R1003.19.

6. Fireblocking of cornices of a two-family dwelling is required at the line of dwelling unit separation.

❖ To restrict the movement of flame and gasses to other areas of a building through concealed passages in building components such as floors, walls and stairs, fireblocking of these concealed combustible spaces is required to form a barrier between stories and between a top story and the roof space. For example, the following locations must be firestopped in wood-frame construction:

- In concealed spaces of stud walls and partitions, including spaces at the ceiling and floor levels [see Commentary Figures R602.8(1) and R602.8(2)].

- At all interconnections between concealed vertical and horizontal spaces such as soffits [Commentary Figure R602.8(3)], dropped ceilings [Commentary Figure R602.8(4)], and cove ceilings [Commentary Figure R602.8(5)]. Interconnections shown in Commentary Figure

R602.8(6) for a bathtub installation must also be firestopped.

- In concealed spaces between stair stringers at the top and bottom of the run [see Commentary Figure R602.8(7)].

- At openings around vent pipes, ducts, chimneys, and fireplaces at ceiling and floor levels with noncombustible materials [see Commentary Figures R602.8(8) and R602.8(9)]. Commentary Figure R602.8(9) illustrates fireblocking at chimneys and fireplaces. The fireblocking at ductwork would be similar.

R602.8.1 Materials. Except as provided in Section R602.8, Item 4, fireblocking shall consist of 2-inch (51 mm) nominal lumber, or two thicknesses of 1-inch (25.4 mm) nominal lumber with broken lap joints, or one thickness of $^{23}/_{32}$-inch (19.8 mm) wood structural panels with joints backed by $^{23}/_{32}$-inch (19.8 mm) wood structural panels or one thickness of $^{3}/_{4}$-inch (19.1 mm) particleboard with joints backed by $^{3}/_{4}$-inch (19.1 mm) particleboard, $^{1}/_{2}$-inch (12.7 mm) gypsum board, or $^{1}/_{4}$-inch (6.4 mm) cement-based millboard. Batts or blankets of mineral wool or glass fiber or other approved materials installed in such a manner as to be securely retained in place shall be permitted as an acceptable fire block. Batts or blankets of mineral or glass fiber or other approved nonrigid materials shall be permitted for compliance with the 10 foot horizontal fireblocking in walls constructed using parallel rows of studs or staggered studs. Loose-fill insulation material shall not be used as a fire block unless specifically tested in the form and manner intended for use to demonstrate its ability to remain in place and to retard the spread of fire and hot gases.

❖ This section specifies the material required for fireblocking. Fireblocking around chimneys and fireplaces must be noncombustible according to Section R1001.6 as specified in Item 5 of Section R602.8. The fireblocking at openings around vents, pipes, ducts and wires at the ceiling and floor level is required only to be adequate to resist the free passage of flames and smoke according to Item 4 of Section R602.8. So that it is not accidentally displaced, all fireblocking material must be securely fastened in place.

R602.8.1.1 Unfaced fiberglass. Unfaced fiberglass batt insulation used as fireblocking shall fill the entire cross section of the wall cavity to a minimum height of 16 inches (406 mm) measured vertically. When piping, conduit or similar obstructions are encountered, the insulation shall be packed tightly around the obstruction.

❖ Fiberglass batts may serve as fireblocking if installed per this section.

R602.8.1.2 Fireblocking integrity. The integrity of all fireblocks shall be maintained.

❖ Piping, ducts or other similar items that pass through firestops must be installed so that the integrity of the firestop is maintained. This may be accomplished by packing an oversized hole with an acceptable fireblocking material.

R602.9 Cripple walls. Foundation cripple walls shall be framed of studs not smaller than the studding above. When exceeding 4 feet (1219 mm) in height, such walls shall be framed of studs having the size required for an additional story.

Cripple walls with a stud height less than 14 inches (356 mm) shall be sheathed on at least one side with a wood structural panel that is fastened to both the top and bottom plates in accordance with Table R602.3(1), or the cripple walls shall be constructed of solid blocking. Cripple walls shall be supported on continuous foundations.

❖ A cripple wall is a framed stud extending from the top of foundation to the underside of the floor framing above the foundation. Should these foundation studs exceed 4 feet (1219 mm) in height, the partial-height wall must, for purposes of stud sizing, be sized as if the partial-height wall were an additional story. Such a condition may occur in buildings of a split-level type with floor levels partially below grade or in buildings with under-floor crawl spaces larger than the minimum specified. See Figure R602.3(2). The rationale for the wall being considered an additional story is based on the load capacity of studs of a particular size changing as the unsupported length is increased.

The minimum length of 14 inches (356 mm) for crip-

ple wall studs provides sufficient clear space for required nailing of the framing. If the wood-framed foundation wall does not permit the installation of foundation studs of such a length, wood structural panel sheathing must be applied, or a solid-blocking method of construction may be used.

R602.10 Wall bracing. All exterior walls shall be braced in accordance with this section. In addition, interior braced wall lines shall be provided in accordance with Section R602.10.1.1. For buildings in Seismic Design Categories D_0, D_1 and D_2, walls shall be constructed in accordance with the additional requirements of Sections R602.10.9, R602.10.11, and R602.11.

❖ Wall construction must include bracing to resist imposed lateral loads resulting from wind or seismic loading and to provide stability to the structure. When subjected to wind loads, the upper portion of the structure moves horizontally while the lower portion is restrained at ground level. During an earthquake, the ground motion displaces the foundation while the top portion of the structure tries to remain stationary. In both of these cases the bracing resists the differential movement and thus prevents or limits damage to the building.

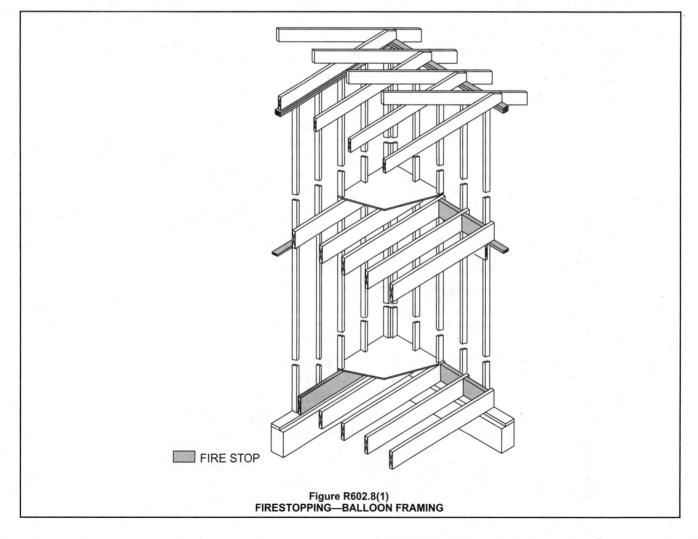

☐ FIRE STOP

Figure R602.8(1)
FIRESTOPPING—BALLOON FRAMING

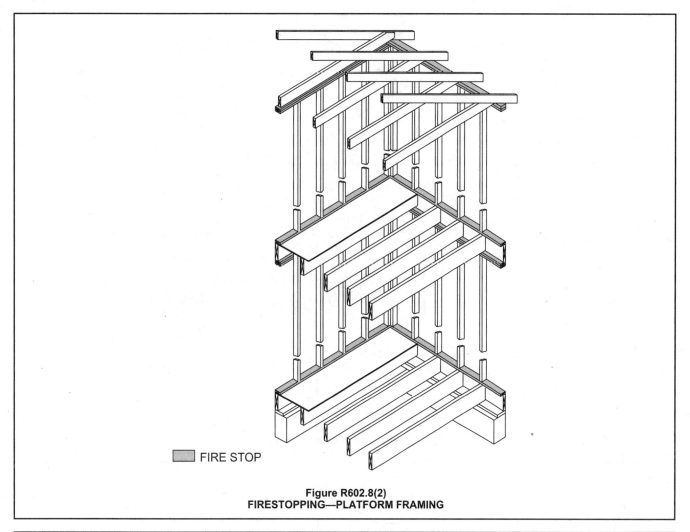

FIRE STOP

**Figure R602.8(2)
FIRESTOPPING—PLATFORM FRAMING**

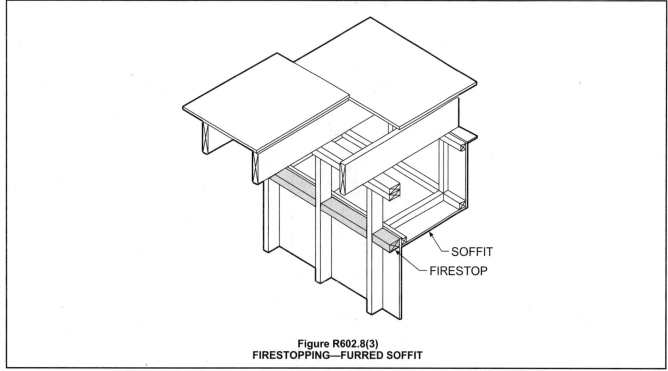

SOFFIT

FIRESTOP

**Figure R602.8(3)
FIRESTOPPING—FURRED SOFFIT**

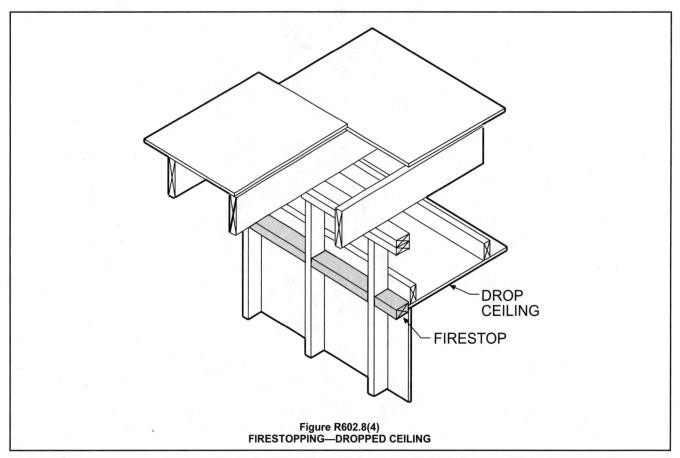

Figure R602.8(4)
FIRESTOPPING—DROPPED CEILING

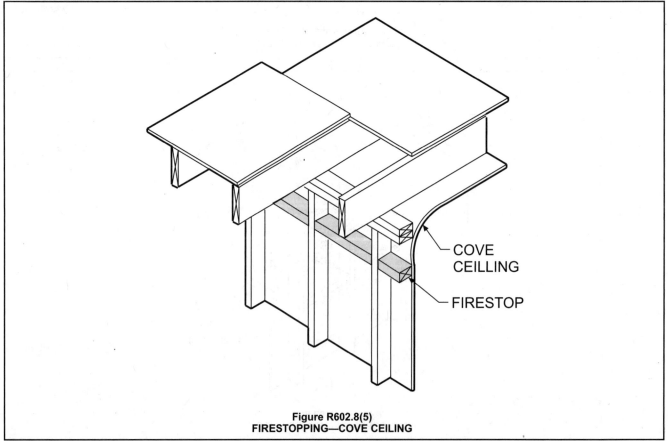

Figure R602.8(5)
FIRESTOPPING—COVE CEILLING

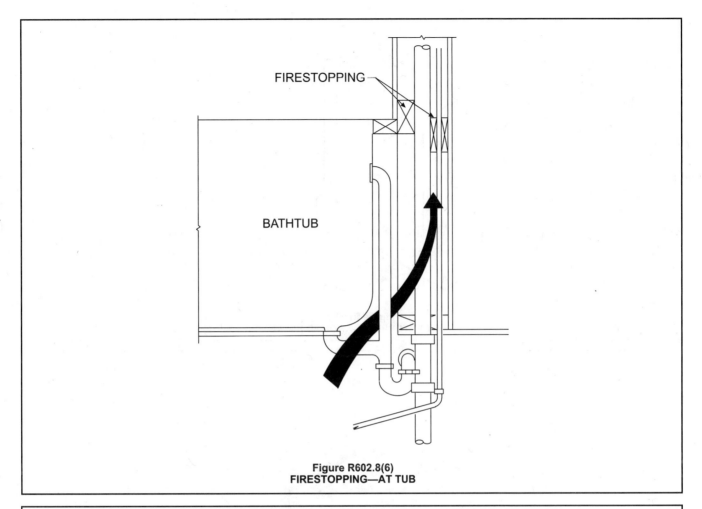

Figure R602.8(6)
FIRESTOPPING—AT TUB

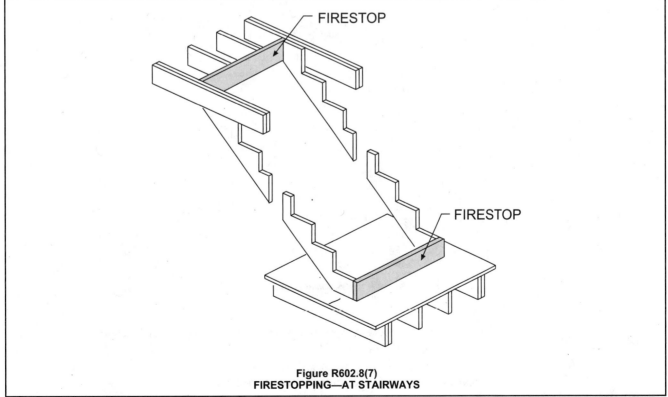

Figure R602.8(7)
FIRESTOPPING—AT STAIRWAYS

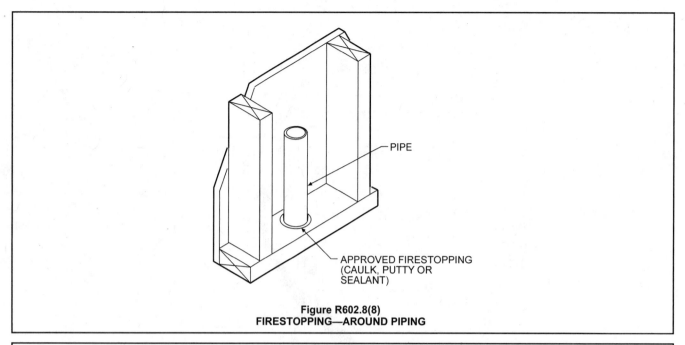

PIPE

APPROVED FIRESTOPPING
(CAULK, PUTTY OR
SEALANT)

**Figure R602.8(8)
FIRESTOPPING—AROUND PIPING**

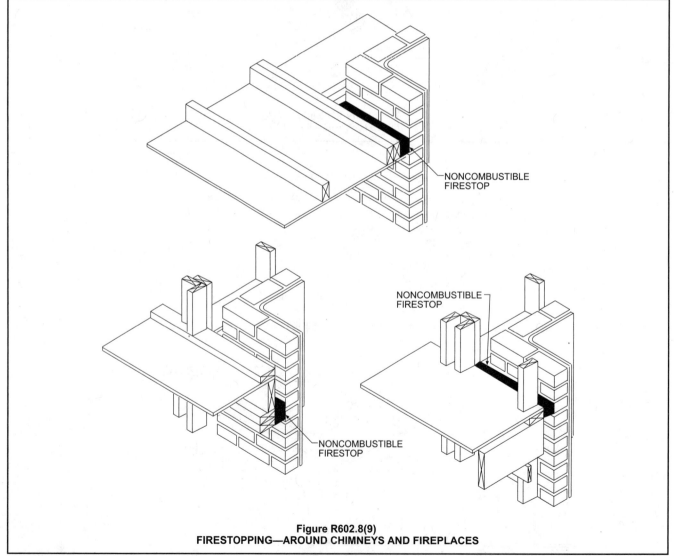

NONCOMBUSTIBLE
FIRESTOP

NONCOMBUSTIBLE
FIRESTOP

NONCOMBUSTIBLE
FIRESTOP

**Figure R602.8(9)
FIRESTOPPING—AROUND CHIMNEYS AND FIREPLACES**

R602.10.1 Braced wall lines. Braced wall lines shall consist of braced wall panel construction in accordance with Section R602.10.3. The amount and location of bracing shall be in accordance with Table R602.10.1 and the amount of bracing shall be the greater of that required by the seismic design category or the design wind speed. Braced wall panels shall begin no more than 12.5 feet (3810 mm) from each end of a braced wall line. Braced wall panels that are counted as part of a braced wall line shall be in line, except that offsets out-of-plane of up to 4 feet (1219 mm) shall be permitted provided that the total out-to-out offset dimension in any braced wall line is not more than 8 feet (2438 mm).

❖ Wall bracing in wood wall construction must consist of braced wall panels constructed by the methods specified in Section R602.10.3. The amount of wall bracing is determined from Table R602.10.1. The percentage and spacing of wall bracing will vary based on wind speed and seismic design category (use the larger value determined), location within the structure and the method of wall panel construction. These are illustrated in Commentary Figure R602.10.1. Let-in braces (Method 1) have very limited applicability.

Ideally, the braced wall panels forming a braced wall line should be in the same plane, but the code does allow out-of-plane offsets within the braced wall line. A wall with multiple offsets may be analyzed as a single braced wall line as long as any single offset does not exceed 4 feet (1219 mm), and the total out to out distance between the innermost and outermost offset walls (as measured perpendicular to the braced wall line) does not exceed 8 feet (2438 mm). Where an offset exceeds 4 feet (1219 mm) or the total out-to-out offset exceeds 8 feet (2438 mm), the walls on either side of such an offset must be analyzed as separate braced wall lines.

While Table R602.10.1 seems to require bracing to start at the end of each braced wall line, this section clarifies that the first panel may be located up to $12^1/_2$ feet (3810 mm) from the end of the braced wall line. This is exactly one-half of the typical center-to-center spacing of panels. Also see Section R602.10.11 for further modifications and limitations in Seismic Design Categories D_0, D_1 and D_2.

Table R602.10.1. See next page.

❖ See the commentary for Section R602.10.1 and Commentary Figure R602.10.1. The amount of wall bracing required for buildings is a function of basic wind speed or seismic design category. This must be based on the most restrictive of these criteria. For example, a building classified as Seismic Design Category A but having a basic wind speed of 105 mph (47 m/s) would need to comply with the bracing amounts listed under Seismic Design Category C or less than 110 mph (49 m/s).

Notes a and d allow decreases to the tabulated amounts of bracing based on seismic design categories, while Note e requires an increase if a roof/ceiling dead load exceeds 15 psf (0.72 kPa). In these cases, wind speed criteria remain unchanged.

Note that the prescriptive bracing requirements of Table R602.10.1 may be overly restrictive for relatively short braced wall lines which are over 4 feet (1219 mm) in length (if under 4 feet (1219 mm), they may be considered part of the adjacent braced wall line in accordance with Section R602.10). No openings, or only minimal openings, may be permitted in such walls where the code is strictly applied. However, engineering analysis will often prove that bracing in such short walls is not required because the roof, ceiling and/or floor diaphragms typically tied to such short walls will help them resist lateral forces because these diaphragms are also typically tied to other exterior and/or interior walls which are provided with lateral bracing. Therefore, the building official should exercise discretion before requesting engineering for relatively short walls over 4 feet (1219 mm) in length.

R602.10.1.1 Spacing. Spacing of braced wall lines shall not exceed 35 feet (10 668 mm) on center in both the longitudinal and transverse directions in each story.

Exception: Spacing of braced wall lines not exceeding 50 feet shall be permitted where:

1. The wall bracing installed equals or exceeds the amount of bracing required by Table R602.10.1 multiplied by a factor equal to the braced wall line spacing divided by 35 feet and

2. The length-to-width ratio for the floor or roof diaphragm does not exceed 3:1.

❖ Buildings that are not in Seismic Design Categories D_0, D_1 or D_2 must be braced in accordance with this section. The requirements for wall bracing vary, depending upon the level of seismic activity for the region in which the building is located and whether the story in question is supporting other stories. The bracing requirements increase for higher seismic design categories and for walls that support the weight of additional stories above. Various types of structural sheathing materials are permitted, although not every material is permitted in every instance.

The spacing of the braced wall lines is can be increased if both of the items in the exception are met. The concept is that the spacing can be increased where additional length of wall bracing is provided versus the minimum length required by the code. This increased spacing is allowed only for residential buildings that are not long in relation to their width. The overall length-to-width ratio of the floor and roof diaphragms, as measured between the outermost exterior walls, must not exceed 3 to 1.

Nailing requirements are critical to the performance of wall bracing. To ensure that the bracing material is effective in resisting racking of the wall, it must be installed in the manner prescribed (see Commentary Figure 602.10.1.1).

TABLE R602.10.1
WALL BRACING

SEISMIC DESIGN CATEGORY OR WIND SPEED	CONDITION	TYPE OF BRACE[b, c]	AMOUNT OF BRACING[a, d, e]
Category A and B ($S_s \leq 0.35g$ and $S_{ds} \leq 0.33g$) or 100 mph or less	One story Top of two or three story	Methods 1, 2, 3, 4, 5, 6, 7 or 8	Located in accordance with Section R602.10 and at least every 25 feet on center but not less than 16% of braced wall line for Methods 2 through 8.
	First story of two story Second story of three story	Methods 1, 2, 3, 4, 5, 6, 7 or 8	Located in accordance with Section R602.10 and at least every 25 feet on center but not less than 16% of braced wall line for Method 3 or 25% of braced wall line for Methods 2, 4, 5, 6, 7 or 8.
	First story of three story	Methods 2, 3, 4, 5, 6, 7 or 8	Located in accordance with Section R602.10 and at least every 25 feet on center but not less than 25% of braced wall line for Method 3 or 35% of braced wall line for Methods 2, 4, 5, 6, 7 or 8.
Category C ($S_s \leq 0.6g$ and $S_{ds} \leq 0.50g$) or less than 110 mph	One story Top of two or three story	Methods 1, 2, 3, 4, 5, 6, 7 or 8	Located in accordance with Section R602.10 and at least every 25 feet on center but not less than 16% of braced wall line for Method 3 or 25% of braced wall line for Methods 2, 4, 5, 6, 7 or 8.
	First story of two story Second story of three story	Methods 2, 3, 4, 5, 6, 7 or 8	Located in accordance with Section R602.10 and at least every 25 feet on center but not less than 30% of braced wall line for Method 3 or 45% of braced wall line for Methods 2, 4, 5, 6, 7 or 8.
	First story of three story	Methods 2, 3, 4, 5, 6, 7 or 8	Located in accordance with Section R602.10 and at least every 25 feet on center but not less than 45% of braced wall line for Method 3 or 60% of braced wall line for Methods 2, 4, 5, 6, 7 or 8.
Categories D_0 and D_1 ($S_s \leq 1.25g$ and $S_{ds} \leq 0.83g$) or less than 110 mph	One story Top of two or three story	Methods 2, 3, 4, 5, 6, 7 or 8	Located in accordance with Section R602.10 and at least every 25 feet on center but not less than 20% of braced wall line for Method 3 or 30% of braced wall line for Methods 2, 4, 5, 6, 7 or 8.
	First story of two story Second story of three story	Methods 2, 3, 4, 5, 6, 7 or 8	Located in accordance with Section R602.10 and at least every 25 feet on center but not less than 45% of braced wall line for Method 3 or 60% of braced wall line for Methods 2, 4, 5, 6, 7 or 8.
	First story of three story	Methods 2, 3, 4, 5, 6, 7 or 8	Located in accordance with Section R602.10 and at least every 25 feet on center but not less than 60% of braced wall line for Method 3 or 85% of braced wall line for Methods 2, 4, 5, 6, 7 or 8.
Category D_2 or less than 110 mph	One story Top of two story	Methods 2, 3, 4, 5, 6, 7 or 8	Located in accordance with Section R602.10 and at least every 25 feet on center but not less than 25% of braced wall line for Method 3 or 40% of braced wall line for Methods 2, 4, 5, 6, 7 or 8.
	First story of two story	Methods 2, 3, 4, 5, 6, 7 or 8	Located in accordance with Section R602.10 and at least every 25 feet on center but not less than 55% of braced wall line for Method 3 or 75% of braced wall line for Methods 2, 4, 5, 6, 7 or 8.
	Cripple walls	Method 3	Located in accordance with Section R602.10 and at least every 25 feet on center but not less than 75% of braced wall line.

For SI: 1 inch = 25.4 mm, 1 foot = 304.8 mm, 1 pound per square foot = 0.0479 kPa, 1 mile per hour = 0.477 m/s.

a. Wall bracing amounts are based on a soil site class "D." Interpolation of bracing amounts between the S_{ds} values associated with the seismic design categories shall be permitted when a site specific S_{ds} value is determined in accordance with Section 1613.5 of the *International Building Code.*

b. Foundation cripple wall panels shall be braced in accordance with Section R602.10.2.

c. Methods of bracing shall be as described in Section R602.10.3. The alternate braced wall panels described in Section R602.10.6.1 or R602.10.6.2 shall also be permitted.

d. The bracing amounts for Seismic Design Categories are based on a 15 psf wall dead load. For walls with a dead load of 8 psf or less, the bracing amounts shall be permitted to be multiplied by 0.85 provided that the adjusted bracing amount is not less than that required for the site's wind speed. The minimum length of braced panel shall not be less than required by Section R602.10.3.

e. When the dead load of the roof/ceiling exceeds 15 psf, the bracing amounts shall be increased in accordance with Section R301.2.2.2.1. Bracing required for a site's wind speed shall not be adjusted.

2006 INTERNATIONAL RESIDENTIAL CODE® COMMENTARY

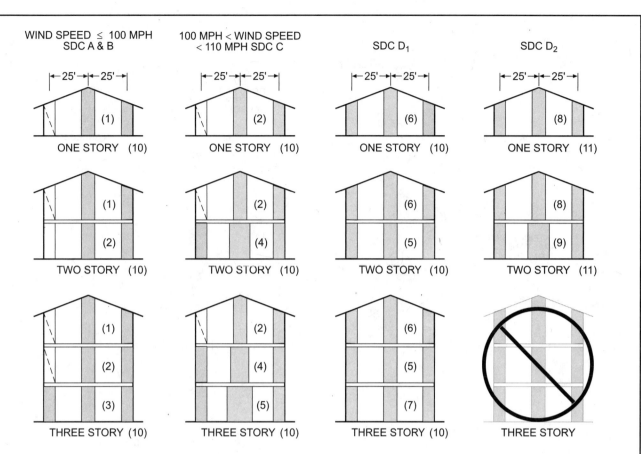

LEGEND:

\   LET-IN BRACE – (METHOD 1, SECT. R602.10.3)

▢   BRACED WALL PANEL CONSTRUCTION IN ACCORDANCE WITH SECTIONS R602.10.3 AND R602.10.4

NOTES	BRACED WALL PANELS PER SECT. R602.10.4		CONTINUOUS WOOD STRUCTURAL PANEL SHEATHING PER SECT. R602.10.5	
	BRACED WALL PANEL CONSTRUCTION		MAXIMUM OPENING HEIGHT	
	METHODS 2, 4 - 8	METHOD 3	85% OF WALL HT.	67% OF WALL HT.
(1)	16%	16%	14%	13%
(2)	25%	16%	14%	13%
(3)	35%	25%	23%	20%
(4)	45%	30%	27%	24%
(5)	60%	45%	41%	36%
(6)	30%	20%	18%	16%
(7)	85%	60%	54%	48%
(8)	40%	25%	23%	20%
(9)	75%	55%	50%	44%
	CRIPPLE WALLS			
(10)	PER SECT. R602.10.2			
(11)	NOT ALLOWED	75%		

Figure R602.10.1
REQUIRED WALL BRACING, PERCENT OF BRACED WALL LINE LENGTH

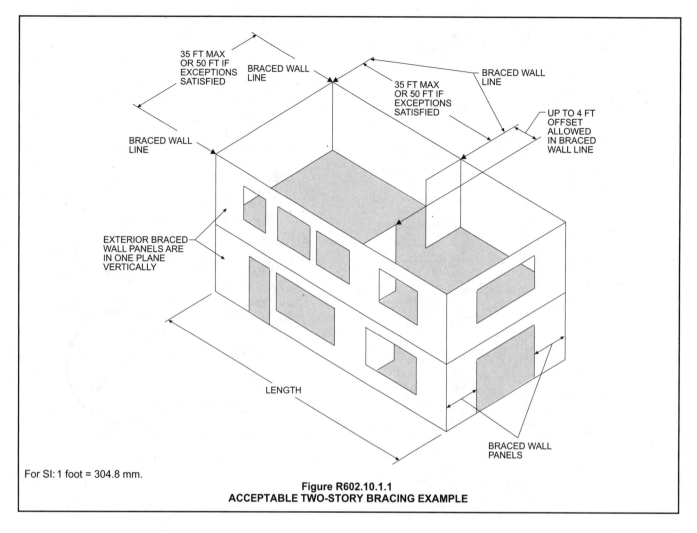

For SI: 1 foot = 304.8 mm.

Figure R602.10.1.1
ACCEPTABLE TWO-STORY BRACING EXAMPLE

R602.10.2 Cripple wall bracing.

❖ Cripple walls participate in the load path necessary to resist earthquake and wind loads. See the commentaries for Section R602.9 and Figure R602.3(2). This section outlines wall bracing requirements for buildings with cripple walls based on the seismic design category.

R602.10.2.1 Seismic design categories other than D_2. In Seismic Design Categories other than D_2, cripple walls shall be braced with an amount and type of bracing as required for the wall above in accordance with Table R602.10.1 with the following modifications for cripple wall bracing:

1. The percent bracing amount as determined from Table R602.10.1 shall be increased by 15 percent and

2. The wall panel spacing shall be decreased to 18 feet (5486 mm) instead of 25 feet (7620 mm).

❖ The bracing requirement for the wall (story) above is increased 15 percent for the supporting cripple wall, and the braced wall panel spacing is reduced from 25 feet to 18 feet (7620 mm to 5486 mm). A cripple wall, in effect, has the impact of being an additional story in seismic performance of a building. The intent of the provision is to provide additional lateral stability to a

building with a cripple wall to compensate for the possible overstress that could occur in this situation.

R602.10.2.2 Seismic Design Category D_2. In Seismic Design Category D_2, cripple walls shall be braced in accordance with Table R602.10.1.

❖ Table R602.10.1 contains specific requirements for bracing of cripple walls in Seismic Design Category D_2. This is done to ensure stability for the structure in a serious seismic event.

R602.10.2.3 Redesignation of cripple walls. In any seismic design category, cripple walls are permitted to be redesignated as the first story walls for purposes of determining wall bracing requirements. If the cripple walls are redesignated, the stories above the redesignated story shall be counted as the second and third stories, respectively.

❖ The provisions of Sections R602.10.2.1 and R602.10.2.2 are predicated on the fact that the designer is not treating the cripple wall as a story. If the cripple wall is designed as a first story in the considerations for bracing (with the first story being counted as the second story, the second story as the third story, etc.), the concerns regarding less stability related to cripple walls becomes moot.

R602.10.3 Braced wall panel construction methods. The construction of braced wall panels shall be in accordance with one of the following methods:

1. Nominal 1-inch-by-4-inch (25 mm by 102 mm) continuous diagonal braces let in to the top and bottom plates and the intervening studs or approved metal strap devices installed in accordance with the manufacturer's specifications. The let-in bracing shall be placed at an angle not more than 60 degrees (1.06 rad) or less than 45 degrees (0.79 rad) from the horizontal.

2. Wood boards of $^5/_8$ inch (16 mm) net minimum thickness applied diagonally on studs spaced a maximum of 24 inches (610 mm). Diagonal boards shall be attached to studs in accordance with Table R602.3(1).

3. Wood structural panel sheathing with a thickness not less than $^5/_{16}$ inch (8 mm) for 16-inch (406 mm) stud spacing and not less than $^3/_8$ inch (9 mm) for 24-inch (610 mm) stud spacing. Wood structural panels shall be installed in accordance with Table R602.3(3).

4. One-half-inch (13 mm) or $^{25}/_{32}$-inch (20 mm) thick structural fiberboard sheathing applied vertically or horizontally on studs spaced a maximum of 16 inches (406 mm) on center. Structural fiberboard sheathing shall be installed in accordance with Table R602.3(1).

5. Gypsum board with minimum $^1/_2$-inch (13 mm) thickness placed on studs spaced a maximum of 24 inches (610 mm) on center and fastened at 7 inches (178 mm) on center with the size nails specified in Table R602.3(1) for sheathing and Table R702.3.5 for interior gypsum board.

6. Particleboard wall sheathing panels installed in accordance with Table R602.3(4).

7. Portland cement plaster on studs spaced a maximum of 16 inches (406 mm) on center and installed in accordance with Section R703.6.

8. Hardboard panel siding when installed in accordance with Table R703.4.

Exception: Alternate braced wall panels constructed in accordance with Section R602.10.6.1 or R602.10.6.2 shall be permitted to replace any of the above methods of braced wall panels.

❖ Braced wall panels must be constructed according to one of these methods and in the locations and amounts required in Table R602.10.1. In accordance with the exception, an alternative braced panel may be substituted where allowed by Section R602.10.6.

R602.10.4 Length of braced panels. For Methods 2, 3, 4, 6, 7 and 8 above, each braced wall panel shall be at least 48 inches (1219 mm) in length, covering a minimum of three stud spaces where studs are spaced 16 inches (406 mm) on center and covering a minimum of two stud spaces where studs are spaced 24 inches (610 mm) on center. For Method 5 above, each braced wall panel shall be at least 96 inches (2438 mm) in length where

applied to one face of a braced wall panel and at least 48 inches (1219 mm) where applied to both faces.

Exceptions:

1. Lengths of braced wall panels for continuous wood structural panel sheathing shall be in accordance with Section R602.10.5.

2. Lengths of alternate braced wall panels shall be in accordance with Section R602.10.6.1 or Section R602.10.6.2.

❖ For virtually all of the bracing methods specified, braced wall panels of structural sheathing are required to be at least 48 inches (1219 mm) long, as measured horizontally, covering three 16-inch (406 mm) stud spaces or two 24-inch (610 mm) stud spaces. However, where gypsum board is intended to act as structural sheathing (in accordance with Method 5), and it is applied to only one side of the braced wall, each required braced wall panel must be at least 96 inches (2438 mm) long. Required braced wall panels must extend vertically from the bottom to the top plate of the wall, including walls at gable ends. Structural sheathing fastening should include edge nailing and field nailing (i.e., at intermediate supports) in accordance with Table R602.3(1) or R602.3(2). Where sheets of braced wall sheathing material are located immediately adjacent to other such sheets, they may be evaluated as a single braced wall panel, as long as the sheets are attached in accordance with Table R602.3(1) or R602.3(2).

The exceptions to Section R602.10.4 allow the specified braced panel lengths to be reduced in accordance with the provisions of Section R602.10.5 or R602.10.6. For other braced panel length reductions specific to panels located adjacent to overhead garage door openings, see Notes b and c to Table R602.10.5 or Section R602.10.6.2.

R602.10.5 Continuous wood structural panel sheathing. When continuous wood structural panel sheathing is provided in accordance with Method 3 of Section R602.10.3 on all sheathable areas of all exterior walls, and interior braced wall lines, where required, including areas above and below openings, bracing wall panel lengths shall be in accordance with Table R602.10.5. Wood structural panel sheathing shall be installed at corners in accordance with Figure R602.10.5. The bracing amounts in Table R602.10.1 for Method 3 shall be permitted to be multiplied by a factor of 0.9 for wall with a maximum opening height that does not exceed 85 percent of the wall height or a factor of 0.8 for walls with a maximum opening height that does not exceed 67 percent of the wall height.

❖ This section allows two reductions to the requirements for braced wall lines that use wood structural panel sheathing installed continuously in accordance with method 3 of Section R602.10.3. The first is an exception to the 48-inch (1219 mm) length required by Section R602.10.4. Figure R602.10.5 illustrates the reduced wall panel lengths of Table R602.10.5. Where the opening is 100 percent of the wall height, no reduction is permitted for the braced wall panels immediately adjacent to the opening. In addition, the total required length of braced wall panels from Table R602.10.1 may be reduced by 10 percent if none of the openings in the braced wall line exceed 67 percent of the bottom-to-top-plate wall height, or may be reduced by 20 percent if none of the openings in the braced wall line exceed 85 percent of the wall height. Required braced wall locations and amounts, including allowable reduced amounts after applying the appropriate multipliers, are shown in Commentary Figure R602.10.1. Note that, where Section R602.10.5 is applied, the structural sheathing at corners must be fastened in accordance with Figure R602.10.5 and such bracing is truly required at the corners [not up to 12 feet (3658 mm) away as permitted by Section R602.10.1].

R602.10.6 Alternate braced wall panel construction methods. Alternate braced wall panels shall be constructed in accordance with Sections R602.10.6.1 and R602.10.6.2.

❖ This section refers to its subsections for the design of alternate braced wall panels with reduced panel lengths. These alternate methods are best suited for the limited width typically available adjacent to garage doors or where a significant number of openings are desired in a braced wall line.

R602.10.6.1 Alternate braced wall panels. Alternate braced wall lines constructed in accordance with one of the following provisions shall be permitted to replace each 4 feet (1219 mm) of braced wall panel as required by Section R602.10.4. The maximum height and minimum width of each panel shall be in accordance with Table R602.10.6:

1. In one-story buildings, each panel shall be sheathed on one face with $3/8$-inch-minimum-thickness (10 mm) wood structural panel sheathing nailed with 8d common or galvanized box nails in accordance with Table R602.3(1) and blocked at all wood structural panel sheathing edges. Two anchor bolts installed in accordance with Figure R403.1(1) shall be provided in each panel. Anchor bolts shall be placed at panel quarter points. Each panel end stud shall have a tie-down device fastened to the foundation, capable of providing an uplift capacity in accordance with Table R602.10.6. The tie down device shall be installed in accordance with the manufacturer's recommendations. The panels shall be supported directly on a foundation or on floor framing supported directly on a foundation which is continuous across the entire length of the braced wall line. This foundation shall be reinforced with not less than one No. 4 bar top and bottom. When the continuous foundation is required to have a depth greater than 12 inches (305 mm), a minimum 12-inch-by-12-inch (305 mm by 305 mm) continuous footing or turned down slab edge is permitted at door openings in the braced wall line. This continuous footing or turned down slab edge shall be reinforced with not less than one No. 4 bar top and bottom. This reinforcement shall be lapped 15 inches (381 mm) with the reinforcement required in the continuous foundation located directly under the braced wall line.

2. In the first story of two-story buildings, each braced wall panel shall be in accordance with Item 1 above, except

TABLE R602.10.5
LENGTH REQUIREMENTS FOR BRACED WALL PANELS IN A CONTINUOUSLY SHEATHED WALL[a, b, c]

MINIMUM LENGTH OF BRACED WALL PANEL (inches)			MAXIMUM OPENING HEIGHT NEXT TO THE BRACED WALL PANEL (% of wall height)
8-foot wall	9-foot wall	10-foot wall	
48	54	60	100
32	36	40	85
24	27	30	65

For SI: 1 inch = 25.4 mm, 1 foot = 305 mm, 1 pound per square foot = 0.0479 kPa.

a. Linear interpolation shall be permitted.

b. Full-height sheathed wall segments to either side of garage openings that support light frame roofs only, with roof covering dead loads of 3 psf or less shall be permitted to have a 4:1 aspect ratio.

c. Walls on either or both sides of openings in garages attached to fully sheathed dwellings shall be permitted to be built in accordance with Section R602.10.6.2 and Figure R602.10.6.2 except that a single bottom plate shall be permitted and two anchor bolts shall be placed at 1/3 points. In addition, tie-down devices shall not be required and the vertical wall segment shall have a maximum 6:1 height-to-width ratio (with height being measured from top of header to the bottom of the sill plate). This option shall be permitted for the first story of two-story applications in Seismic Design Categories A through C.

❖ See the commentary for Section R610.10.5. Note b provides an exception to the braced wall panel lengths located on either side of a garage door opening. It allows an aspect ratio of 4:1. While "aspect ratio" is not defined in the *International Residential Code®* (IRC®), for a shear wall it represents the height-width ratio; see Section 2305.3.4 of the *International Building Code®* (IRC®). The height is the clear height from the bottom wall plate to the bottom of the diaphragm framing above. Typically the latter would also represent the top plate of the wall. The shear wall (segment) width is the horizontal dimension of full-height sheathing adjacent to unrestrained openings. Note c allows narrow panels, with an aspect ratio of 6:1, when using the provisions of Section R602.10.6.2.

that the wood structural panel sheathing shall be installed on both faces, sheathing edge nailing spacing shall not exceed 4 inches (102 mm) on center, at least three anchor bolts shall be placed at one-fifth points.

❖ This section allows reduced braced panel lengths (as measured horizontally) in accordance with Table R602.10.6. Minimum $^3/_8$-inch (9.5 mm) wood structural panels and hold-down fasteners at each end of the panel are required. This method of construction provides for a very rigid and strong shear wall that is positively anchored to the foundation to resist the forces caused by wind or earthquakes.

R602.10.6.2 Alternate braced wall panel adjacent to a door or window opening. Alternate braced wall panels constructed in accordance with one of the following provisions are also permitted to replace each 4 feet (1219 mm) of braced wall panel as required by Section R602.10.4 for use adjacent to a window or door opening with a full-length header:

1. In one-story buildings, each panel shall have a length of not less than 16 inches (406 mm) and a height of not more than 10 feet (3048 mm). Each panel shall be sheathed on one face with a single layer of $^3/_8$-inch-minimum-thickness (10 mm) wood structural panel sheathing nailed with 8d common or galvanized box nails in accordance

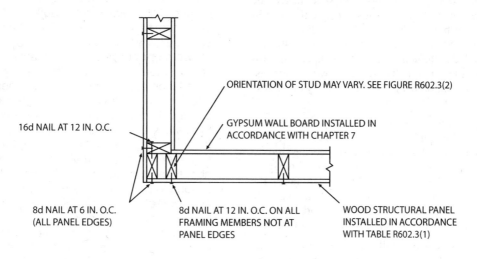

16d NAIL AT 12 IN. O.C.

ORIENTATION OF STUD MAY VARY. SEE FIGURE R602.3(2)

GYPSUM WALL BOARD INSTALLED IN ACCORDANCE WITH CHAPTER 7

8d NAIL AT 6 IN. O.C. (ALL PANEL EDGES)

8d NAIL AT 12 IN. O.C. ON ALL FRAMING MEMBERS NOT AT PANEL EDGES

WOOD STRUCTURAL PANEL INSTALLED IN ACCORDANCE WITH TABLE R602.3(1)

(a) OUTSIDE CORNER DETAIL

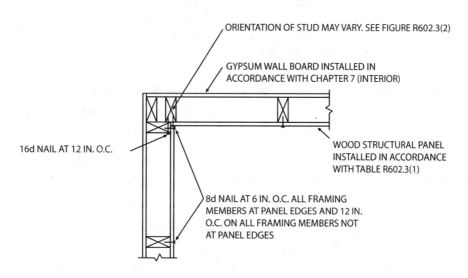

ORIENTATION OF STUD MAY VARY. SEE FIGURE R602.3(2)

GYPSUM WALL BOARD INSTALLED IN ACCORDANCE WITH CHAPTER 7 (INTERIOR)

16d NAIL AT 12 IN. O.C.

WOOD STRUCTURAL PANEL INSTALLED IN ACCORDANCE WITH TABLE R602.3(1)

8d NAIL AT 6 IN. O.C. ALL FRAMING MEMBERS AT PANEL EDGES AND 12 IN. O.C. ON ALL FRAMING MEMBERS NOT AT PANEL EDGES

(b) INSIDE CORNER DETAIL

For SI: 1 inch = 25.4 mm.

FIGURE R602.10.5
EXTERIOR CORNER FRAMING

❖ This figure illustrates corner framing details specifically required by Sections R602.10.5 and R602.10.11 (exception).

with Figure R602.10.6.2. The wood structural panel sheathing shall extend up over the solid sawn or glued-laminated header and shall be nailed in accordance with Figure R602.10.6.2. Use of a built-up header consisting of at least two 2 x 12s and fastened in accordance with Table R602.3(1) shall be permitted. A spacer, if used, shall be placed on the side of the built-up beam opposite the wood structural panel sheathing. The header shall extend between the inside faces of the first full-length outer studs of each panel. The clear span of the header between the inner studs of each panel shall be not less than 6 feet (1829 mm) and not more than 18 feet (5486 mm) in length. A strap with an uplift capacity of not less than 1000 pounds (4448 N) shall fasten the header to the side of the inner studs opposite the sheathing. One anchor bolt not less than $^5/_8$-inch-diameter (16 mm) and installed in accordance with Section R403.1.6 shall be installed in the center of each sill plate. The studs at each end of the panel shall have a tie-down device fastened to the foundation with an uplift capacity of not less than 4,200 pounds (18 683 N).

Where a panel is located on one side of the opening, the header shall extend between the inside face of the first full-length stud of the panel and the bearing studs at the other end of the opening. A strap with an uplift capacity of not less than 1000 pounds (4448 N) shall fasten the header to the bearing studs. The bearing studs shall also have a tie-down device fastened to the foundation with an uplift capacity of not less than 1000 pounds (4448 N).

The tie-down devices shall be an embedded- strap type, installed in accordance with the manufacturer's recommendations. The panels shall be supported directly on a foundation which is continuous across the entire length of the braced wall line. The foundation shall be reinforced with not less than one No. 4 bar top and bottom.

Where the continuous foundation is required to have a depth greater than 12 inches (305 mm), a minimum 12-inch-by-12-inch (305 mm by 305 mm) continuous footing or turned down slab edge is permitted at door openings in the braced wall line. This continuous footing or turned down slab edge shall be reinforced with not less than one No. 4 bar top and bottom. This reinforcement shall be lapped not less than 15 inches (381 mm) with the reinforcement required in the continuous foundation located directly under the braced wall line.

2. In the first story of two-story buildings, each wall panel shall be braced in accordance with Item 1 above, except that each panel shall have a length of not less than 24 inches (610 mm).

❖ This additional bracing method allows the lengths of braced wall panels located adjacent to door or window openings, to be reduced to 16 inches (406 mm) in one story buildings, or 24 inches (610 mm) in the first story of two-story buildings, The alternate braced wall assembly must bear on and be directly connected to the foundation as prescribed and be sheathed on at least one face with a layer of $^3/_8$-inch-thick (9.5 mm) wood structural sheathing. The required overlap of the structural sheathing, studs and header, along with specific fastening requirements, including foundation connection requirements, provides for a more rigid frame about the opening as compared to other methods prescribed in the code. The header may span between 6 and 18 feet (1829 and 5486 mm) and must be composed of at least two 2- by 12-inch (51 by 305 mm) members or be in accordance with Table R502.5(1), whichever is more restrictive. Headers that exceed the spans permitted in Table R502.5(1) must be engineered. This method of construction provides for a very rigid and strong braced frame that is positively anchored to the foundation to resist the forces caused by wind or earthquakes. Because this bracing method requires fastening to the foundation, it cannot be used in the upper floors of two- or three-story structures.

TABLE R602.10.6
MINIMUM WIDTHS AND TIE-DOWN FORCES OF ALTERNATE BRACED WALL PANELS

SEISMIC DESIGN CATEGORY AND WINDSPEED	TIE-DOWN FORCE (lb)	HEIGHT OF BRACED WALL PANEL				
		Sheathed Width				
		8 ft. 2' - 4"	9 ft. 2' - 8"	10 ft. 2' - 8"	11 ft. 3' - 2"	12 ft. 3' - 6"
SDC A, B, and C Windspeed < 110 mph	R602.10.6.1, Item 1	1800	1800	1800	2000	2200
	R602.10.6.1, Item 2	3000	3000	3000	3300	3600
SDC D$_0$, D$_1$ and D$_2$ Windspeed < 110 mph		Sheathed Width				
		2' - 8"	2' - 8"	2' - 8"	Note a	Note a
	R602.10.6.1, Item 1	1800	1800	1800	—	—
	R602.10.6.1, Item 2	3000	3000	3000	—	—

For SI: 1 inch = 25.4 mm, 1 foot = 304.8 mm.

a. Not permitted because maximum height is 10 feet.

❖ See the commentary for Section R602.10.6.1. This table is to be used only in association with Section R602.10.6.1.

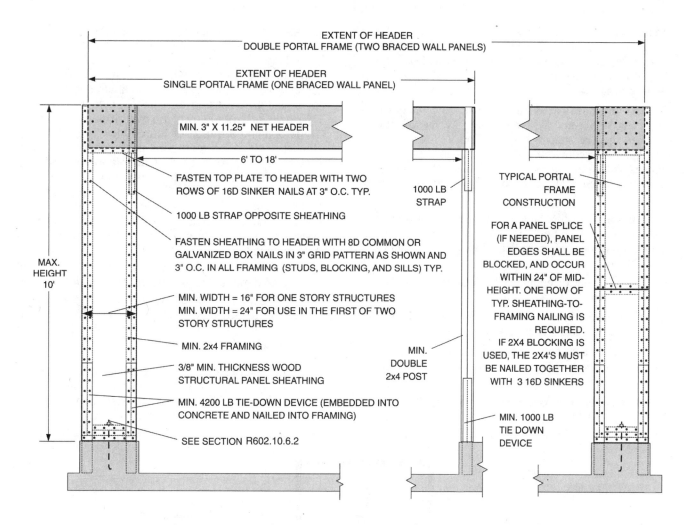

FIGURE R602.10.6.2
ALTERNATE BRACED WALL PANEL ADJACENT TO A DOOR OR WINDOW OPENING

❖ See the commentary for Section R602.10.6.2.

R602.10.7 Panel joints. All vertical joints of panel sheathing shall occur over, and be fastened to, common studs. Horizontal joints in braced wall panels shall occur over, and be fastened to, common blocking of a minimum 1¹/₂ inch (38 mm) thickness.

Exception: Blocking is not required behind horizontal joints in Seismic Design Categories A and B and detached dwellings in Seismic Design Category C when constructed in accordance with Section R602.10.3, braced-wall-panel construction method 3 and Table R602.10.1, method 3, or where permitted by the manufacturer's installation requirements for the specific sheathing material.

❖ To provide a surface for panel nailing, vertical joints of panel sheathing should occur at common studs (i.e., adjacent sheathing panels should be attached to the same stud at the vertical joint), and horizontal joints should occur over blocking. The exception recognizes that in lesser seismic areas the blocking behind horizontal joints in Method 3 is unnecessary to ensure stability of the structure. This exception does not apply to townhouses in Seismic Design Category C.

R602.10.8 Connections. Braced wall line sole plates shall be fastened to the floor framing and top plates shall be connected to the framing above in accordance with Table R602.3(1). Sills shall be fastened to the foundation or slab in accordance with Sections R403.1.6 and R602.11. Where joists are perpendicular to the braced wall lines above, blocking shall be provided under and in line with the braced wall panels. Where joists are perpendicular to braced wall lines below, blocking shall be provided over and in line with the braced wall panels. Where joists are parallel to braced wall lines above or below, a rim joist or other parallel framing member shall be provided at the wall to permit fastening per Table R602.3(1).

❖ The cross references provided in this section refer the reader to the various sections specifying connections that provide a continuous load path for lateral loads as described in Section R301. For the wall/floor interface, the connection is simply required to be in accordance the fastening schedule provided [Table R602.3(1)]. Wood framing to foundation or floor slab connections must be in accordance with Sections R403.1.6 and R602.11. There is no intent to require connections or fastening beyond the prescriptive provisions of these referenced sections. Blocking or in-line framing is required at framed floor and ceiling systems both above and below the braced wall panels, but not at the entire braced wall line.

R602.10.9 Interior braced wall support. In one-story buildings located in Seismic Design Category D_2, interior braced wall lines shall be supported on continuous foundations at intervals not exceeding 50 feet (15 240 mm). In two- story buildings located in Seismic Design Category D_2, all interior braced wall panels shall be supported on continuous foundations.

Exception: Two-story buildings shall be permitted to have interior braced wall lines supported on continuous foundations at intervals not exceeding 50 feet (15 240 mm) provided that:

1. The height of cripple walls does not exceed 4 feet (1219 mm).

2. First-floor braced wall panels are supported on doubled floor joists, continuous blocking or floor beams.

3. The distance between bracing lines does not exceed twice the building width measured parallel to the braced wall line.

❖ When an interior braced wall line (as required by Section R602.10.11) is not supported by a continuous foundation, the supporting floor (diaphragm) must distribute the lateral force to a resisting element (foundation). If a floor diaphragm is overloaded, it could lead to a failure, which in turn results in an incomplete load path. This section limits this condition for buildings classified as Seismic Design Category D_2. However, where any of the overall plan dimensions of the building are greater than 50 feet (15 240 mm), Section R403.1.2 supersedes this section (it is more restrictive) and requires that all interior braced wall lines be supported on continuous footings in any building located in Seismic Design Category D_0, D_1 or D_2. Braced wall lines (as required by Section R602.10.11.1) in structures located in Seismic Design Category D_2 must be supported on continuous footings except:

• One-story buildings where none of the overall plan dimensions exceed 50 feet (15 240 mm), or

• Two-story buildings where none of the overall plan dimensions exceed 50 feet (15 240 mm) and all three items of the Exception to Section R602.10.9 are satisfied.

Commentary Figure R602.10.9(1) shows the limitations for one- and two-story buildings. Commentary Figure R602.10.9(2) shows the exception to the requirement for two-story buildings.

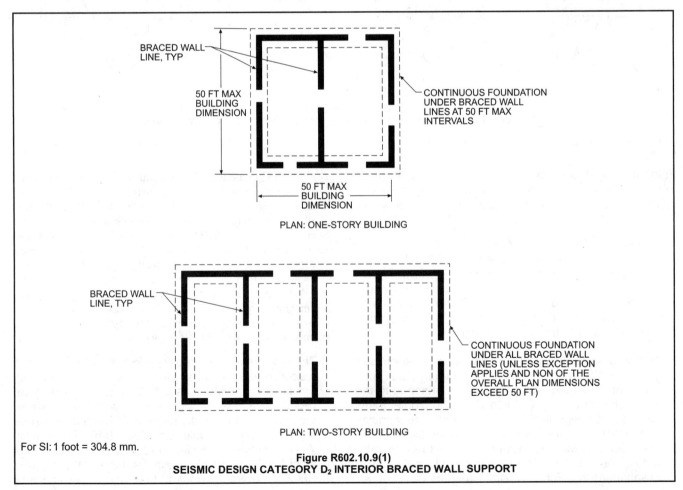

For SI: 1 foot = 304.8 mm.

Figure R602.10.9(1)
SEISMIC DESIGN CATEGORY D_2 INTERIOR BRACED WALL SUPPORT

R602.10.10 Design of structural elements. Where a building, or portion thereof, does not comply with one or more of the bracing requirements in this section, those portions shall be designed and constructed in accordance with accepted engineering practice.

❖ This section essentially reiterates Section R301.1.2 in requiring design of elements not conforming to conventional light-frame construction requirements.

R602.10.11 Bracing in Seismic Design Categories D_0, D_1 and D_2. Structures located in Seismic Design Categories D_0, D_1 and D_2 shall have exterior and interior braced wall lines.

❖ The subsections of this section provide braced wall line spacing and braced wall panel location requirements specific to Seismic Design Categories D_0, D_1 and D_2.

R602.10.11.1 Braced wall line spacing. Spacing between braced wall lines in each story shall not exceed 25 feet (7620 mm) on center in both the longitudinal and transverse directions.

Exception: In one- and two-story buildings, spacing between two adjacent braced wall lines shall not exceed 35 feet (10 363 mm) on center in order to accommodate one single room not exceeding 900 square feet (84 m²) in each

dwelling unit. Spacing between all other braced wall lines shall not exceed 25 feet (7620 mm).

❖ The conventional light-frame provisions assume that every exterior wall serves as a braced wall line; this is considered sufficient for Seismic Design Categories A, B and C, except where additional interior braced wall lines are required by Section R602.10.1.1. By establishing a maximum spacing for braced wall lines in Seismic Design Categories D_0, D_1 and D_2, the code is essentially limiting the span of floor or roof diaphragms as well as the loading on the braced wall panels. The exception permits this limit to be increased to 35 feet (10 668 mm) to provide one large room [900 square feet (83.6 m²)] per dwelling.

R602.10.11.2 Braced wall panel location. Exterior braced wall lines shall have a braced wall panel at each end of the braced wall line.

Exception: For braced wall panel construction Method 3 of Section R602.10.3, the braced wall panel shall be permitted to begin no more than 8 feet (2438 mm) from each end of the braced wall line provided the following is satisfied:

1. A minimum 24-inch-wide (610 mm) panel is applied to each side of the building corner and the two 24-inch

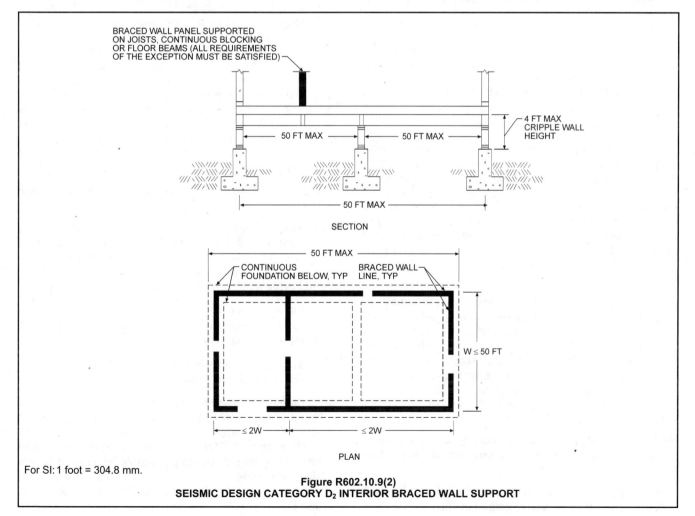

For SI: 1 foot = 304.8 mm.

Figure R602.10.9(2)
SEISMIC DESIGN CATEGORY D₂ INTERIOR BRACED WALL SUPPORT

(610 mm) panels at the corner shall be attached to framing in accordance with Figure R602.10.5; or

2. The end of each braced wall panel closest to the corner shall have a tie-down device fastened to the stud at the edge of the braced wall panel closest to the corner and to the foundation or framing below. The tie-down device shall be capable of providing an uplift allowable design value of at least 1,800 pounds (8 kN). The tie-down device shall be installed in accordance with the manufacturer's recommendations.

❖ For Seismic Design Categories D_0, D_1 and D_2, this section modifies the basic requirements of Section R602.10.1 by requiring installation of a braced wall panel at each end of an exterior braced wall line, not 12 feet (3658 mm) from the corner or end as permitted by Section R602.10.1, unless a designed collector is provided in accordance with Section R602.10.11.3. The exception permits braced wall panels of wood structural panel sheathing (see Section R602.10, Method 3) to begin up to 8 feet (2438 mm) from the end of a braced wall line as long as either of the two conditions is met. Additionally, a designed collector in accordance with Section R602.10.11.3 must be provided if this braced wall panel begins more than 8 feet (2438 mm) from the end of the exterior braced wall line.

R602.10.11.3 Collectors. A designed collector shall be provided if a braced wall panel is not located at each end of a braced wall line as indicated in Section R602.10.11.2, or, when using the Section R602.10.11.2 exception, if a braced wall panel is more than 8 feet (2438 mm) from each end of a braced wall line.

❖ See the commentary to Section R602.11.2.

R602.10.11.4 Cripple wall bracing. In addition to the requirements of Section R602.10.2, where interior braced wall lines occur without a continuous foundation below, the length of parallel exterior cripple wall bracing shall be one and one-half times the length required by Table R602.10.1. Where cripple walls braced using Method 3 of Section R602.10.3 cannot provide this additional length, the capacity of the sheathing shall be increased by reducing the spacing of fasteners along the perimeter of each piece of sheathing to 4 inches (102 mm) on center.

❖ Section R602.10.2 contains the basic requirements for bracing cripple walls for all seismic design categories. This section requires additional cripple wall bracing in Seismic Design Categories D_0, D_1 and D_2 where installation of an interior braced wall line is permitted without being supported on a continuous foundation. Commentary Figure R602.10.9(1) shows an example of this condition that would require the added cripple wall bracing.

R602.10.11.5 Sheathing attachment. Adhesive attachment of wall sheathing shall not be permitted in Seismic Design Categories C, D_0, D_1 and D_2.

❖ Generally, sheathing must be attached to the wall framing using fasteners specified in Table R602.3(1).

Attachment with adhesives, although not covered under the prescriptive fastening requirements, is certainly a possibility as an alternative method under Section R104.11. However, attachment of sheathing using adhesives results in a considerable increase in the strength and stiffness of the wall assembly. Even though there may be advantages to this approach, the added stiffness may generate higher seismic loads. See the discussion of inertial force caused by earthquakes in Section R301.2.2. The higher seismic forces could, in turn, overload other elements in the lateral-force-resisting system, resulting in premature failure of those elements. Therefore, in buildings classified as Seismic Design Category C, D_0, D_1 or D_2, this method of wall sheathing attachment is not permitted.

R602.11 Framing and connections for Seismic Design Categories D_0, D_1 and D_2. The framing and connections details of buildings located in Seismic Design Categories D_0, D_1 and D_2 shall be in accordance with Sections R602.11.1 through R602.11.3.

❖ This section provides the connection details necessary to transfer lateral forces in a building classified as Seismic Design Category D_0, D_1 or D_2.

R602.11.1 Wall anchorage. Braced wall line sills shall be anchored to concrete or masonry foundations in accordance with Sections R403.1.6 and R602.11. For all buildings in Seismic Design Categories D_0, D_1 and D_2 and townhouses in Seismic Design Category C, plate washers, a minimum of 0.229 inch by 3 inches by 3 inches (5.8 mm by 76 mm by 76 mm) in size, shall be installed between the foundation sill plate and the nut. The hole in the plate washer is permitted to be diagonally slotted with a width of up to $^3/_{16}$ inch (5 mm) larger than the bolt diameter and a slot length not to exceed $1^3/_4$ inches (44 mm), provided a standard cut washer is placed between the plate washer and the nut.

❖ This section reiterates the requirements of Section R403.1.6.1.

R602.11.2 Interior braced wall panel connections. Interior braced wall lines shall be fastened to floor and roof framing in accordance with Table R602.3(1), to required foundations in accordance with Section R602.11.1, and in accordance with the following requirements:

1. Floor joists parallel to the top plate shall be toe-nailed to the top plate with at least 8d nails spaced a maximum of 6 inches (152 mm) on center.

2. Top plate laps shall be face-nailed with at least eight 16d nails on each side of the splice.

❖ To provide the required lateral load path, fastening of braced wall panels to floor and roof framing must be according to the fastener schedule in Table R602.3(1). Anchorage to the foundation is also specified. These are typical requirements applicable to all seismic design categories. The provisions of Items 1 and 2, however, are specific to Seismic Design Categories D_0, D_1 and D_2.

R602.11.3 Stepped foundations. Where stepped foundations occur, the following requirements apply:

1. Where the height of a required braced wall panel that extends from foundation to floor above varies more than 4 feet (1220 mm), the braced wall panel shall be constructed in accordance with Figure R602.11.3.

2. Where the lowest floor framing rests directly on a sill bolted to a foundation not less than 8 feet (2440 mm) in length along a line of bracing, the line shall be considered as braced. The double plate of the cripple stud wall beyond the segment of footing that extends to the lowest framed floor shall be spliced by extending the upper top plate a minimum of 4 feet (1219 mm) along the foundation. Anchor bolts shall be located a maximum of 1 foot and 3 feet (305 and 914 mm) from the step in the foundation.

3. Where cripple walls occur between the top of the foundation and the lowest floor framing, the bracing requirements for a story shall apply.

4. Where only the bottom of the foundation is stepped and the lowest floor framing rests directly on a sill bolted to the foundations, the requirements of Section R602.11.1 shall apply.

❖ Where foundations are stepped (see Section R403.1.5), this section provides additional criteria for Seismic Design Category Categories D_0, D_1 or D_2. See Figure R602.11.3.

This section is applicable only where the elevation of the top of the foundation varies by more than 4 feet (1219 mm).

Also, if the floor framing is supported directly on a sill plate that is anchored to a foundation 8 feet (2438 mm) or more in length, the braced wall line is adequately braced. If not, Item 3 applies, and the cripple wall must be considered a story.

SECTION R603
STEEL WALL FRAMING

R603.1 General. Elements shall be straight and free of any defects that would significantly affect structural performance. Cold-formed steel wall framing members shall comply with the requirements of this section.

❖ The provisions of this section apply to the construction of load-bearing wall systems using cold-formed steel framing, which is that type of construction made up in part or entirely of steel structural members cold-formed to shape from sheet or strip steel. The framing members are generally in the form of the letter "C" (C-shape) which is described in further detail in Section R603.2. The use of cold-formed steel-framed designs not consistent with these provisions is beyond the scope of this code. An analysis by a design professional is recommended.

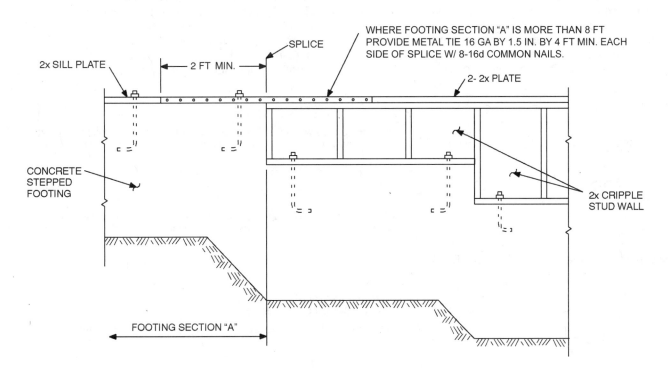

For SI: 1 inch = 25.4 mm, 1 foot = 304.8 mm.

Note: Where footing Section "A" is less than 8 feet long in a 25-foot-long wall, install bracing at cripple stud wall.

FIGURE R602.11.3
STEPPED FOUNDATION CONSTRUCTION

❖ See the commentary for Section R602.11.3.

R603.1.1 Applicability limits. The provisions of this section shall control the construction of exterior steel wall framing and interior load-bearing steel wall framing for buildings not more than 60 feet (18 288 mm) long perpendicular to the joist or truss span, not more than 40 feet (12 192 mm) wide parallel to the joist or truss span, and not more than two stories in height. All exterior walls installed in accordance with the provisions of this section shall be considered as load-bearing walls. Steel walls constructed in accordance with the provisions of this section shall be limited to sites subjected to a maximum design wind speed of 110 miles per hour (49 m/s) Exposure A, B or C and a maximum ground snow load of 70 psf (3.35 kPa).

❖ See the commentary for Section R505.1.1.

R603.1.2 In-line framing. Load-bearing steel studs constructed in accordance with Section R603 shall be located directly in-line with joists, trusses and rafters with a maximum tolerance of $^3/_4$ inch (19.1 mm) between their center lines. Interior load-bearing steel stud walls shall be supported on foundations or shall be located directly above load-bearing walls with a maximum tolerance of $^3/_4$ inch (19 mm) between the centerline of the studs.

❖ In-line framing is the preferred framing method. The advantage of in-line framing is that it provides a direct load path for the transfer of forces from joists to studs. In-line framing maximizes framing alignment to minimize secondary moments on the framing members, taking into account that the track cannot function as a load transfer member. The $^3/_4$-inch (19 mm) alignment value is accepted industry practice, and addresses those conditions where two framing members (roof rafter and ceiling joist, for example) are to be aligned over a wall stud. In the absence of in-line framing, a load distribution member, such as a structural track, may be required for this force transfer.

R603.2 Structural framing. Load-bearing steel wall framing members shall comply with Figure R603.2(1) and with the dimensional and minimum thickness requirements specified in Tables R603.2(1) and R603.2(2). Tracks shall comply with Figure R603.2(2) and shall have a minimum flange width of $1^1/_4$ inches (32 mm). The maximum inside bend radius for members shall be the greater of $^3/_{32}$ inch (2.4 mm) or twice the uncoated steel thickness. Holes in wall studs and other structural members shall comply with all of the following conditions:

1. Holes shall conform to Figure R603.2(3);

2. Holes shall be permitted only along the centerline of the web of the framing member;

3. Holes shall have a center-to-center spacing of not less than 24 inches (610 mm);

4. Holes shall have a width not greater than 0.5 times the member depth, or $1^1/_2$ inches (38.1 mm);

5. Holes shall have a length not exceeding $4^1/_2$ inches (114 mm); and

6. Holes shall have a minimum distance between the edge of the bearing surface and the edge of the hole of not less than 10 inches (254 mm).

Framing members with web holes violating the above requirements shall be patched in accordance with Section R603.3.5 or designed in accordance with accepted engineering practices.

❖ Stud framing must comply with the dimensional and minimum thickness requirements specified in Tables R603.2(1) and R603.2(2), respectively. Note a to Table R603.2(1) explains the meaning of the alphanumeric member designation used for all steel framing members in the code. This system replaces the varied designation approaches that were produced by each indi-

TABLE R603.2(1)
LOAD-BEARING COLD-FORMED STEEL STUD SIZES

MEMBER DESIGNATION[a]	WEB DEPTH (inches)	MINIMUM FLANGE WIDTH (inches)	MAXIMUM FLANGE WIDTH (inches)	MINIMUM LIP SIZE (inches)
350S162-t	3.5	1.625	2	0.5
550S162-t	5.5	1.625	2	0.5

For SI: 1 inch = 25.4 mm; 1 mil = 0.0254 mm.

a. The member designation is defined by the first number representing the member depth in hundredths of an inch "S" representing a stud or joist member, the second number representing the flange width in hundredths of an inch, and the letter "t" shall be a number representing the minimum base metal thickness in mils [See Table R603.2(2)].

❖ See the commentary for Section R603.2.

TABLE R603.2(2)
MINIMUM THICKNESS OF COLD-FORMED STEEL STUDS

DESIGNATION (mils)	MINIMUM UNCOATED THICKNESS (inches)	REFERENCE GAGE NUMBER
33	0.033	20
43	0.043	18
54	0.054	16
68	0.068	14

For SI: 1 inch = 25.4 mm, 1 mil = 0.0254 mm.

❖ See the commentary for Section R603.2.

vidual manufacturer. In addition, the designation is used to identify not only a specific steel framing member but also the section properties of that same member through the use of the product technical information document.

Steel thickness is expressed in mils, which pertains to the base metal thickness measured prior to painting or the application of metallic (corrosion resistant) coatings. The base metal thickness is typically stamped or embossed on the member by the manufacturer. Table R603.2(2) also references the equivalent thickness in inches and provides a further reference to gage number. The use of gage numbers to define the thickness (a specific mean decimal thickness for which rolling tolerances were established) of steel sheet is a practice that has been discontinued by the steel industry.

To allow for routing utilities, web holes (also referred to as "penetrations," "utility holes" and "punchouts") are permitted in the webs of studs. To avoid adversely affecting the strength of wall framing, web holes must meet all the requirements of Items 1 through 6. Web holes are limited to the locations along the centerline of the web of the framing member as illustrated in Figure R603.2(3). Where these holes occur within 10 inches (254 mm) of the edge of a member, they must be patched as illustrated in Figure R603.3.5. Unless engineered, patching of web holes must extend at least 1 inch (25.4 mm) beyond all edges of the hole, and it must be fastened to the web with No. 8 screws having a minimum edge distance of $^1/_2$ inch (12.7 mm).

Steel track sections are to serve in nonload–bearing applications only and are shown in Figure R603.2(2). They must have a minimum flange width of $1^1/_4$ inches (32 mm).

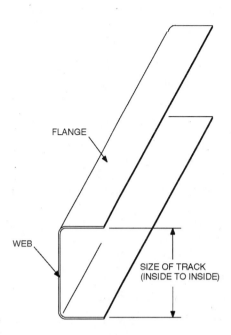

FIGURE R603.2(2)
TRACK SECTION

❖ See the commentary for Section R603.2.

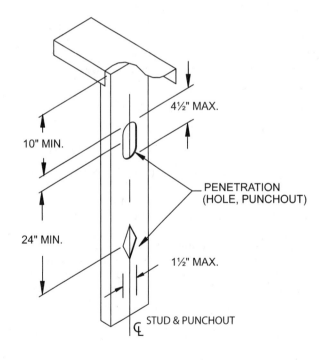

For SI: 1 inch = 25.4 mm.

FIGURE R603.2(3)
WEB HOLES

❖ Holes are permitted in the web of steel wall studs as shown to allow for the installation of plumbing wiring, etc. in a wall. The limits on size, location and spacing of these holes accommodate these utilities without adversely affecting the stud's load-carrying capacity.

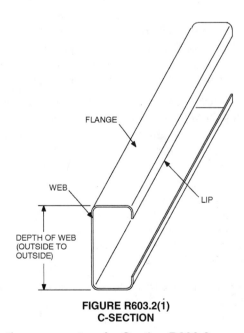

FIGURE R603.2(1)
C-SECTION

❖ See the commentary for Section R603.2.

R603.2.1 Material. Load-bearing steel framing members shall be cold-formed to shape from structural quality sheet steel complying with the requirements of one of the following:

1. ASTM A 653: Grades 33, 37, 40 and 50 (Class 1 and 3).

2. ASTM A 792: Grades 33, 37, 40 and 50A.

3. ASTM A 875: Grades 33, 37, 40 and 50 (Class 1 and 3).

4. ASTM A 1003: Grades 33, 37, 40 and 50.

❖ Load-bearing steel framing members must be cold-formed to shape from structural-quality sheet applicable material standards. The steel grades specified have the ductility and strength to meet the intent of these provisions.

R603.2.2 Identification. Load-bearing steel framing members shall have a legible label, stencil, stamp or embossment with the following information as a minimum:

1. Manufacturer's identification.

2. Minimum uncoated steel thickness in inches (mm).

3. Minimum coating designation.

4. Minimum yield strength, in kips per square inch (ksi) (kN).

❖ Load-bearing steel framing members must have a legible label, stencil, stamp or embossment. This identification allows for verification that materials installed are consistent with the design and meet the intent of the code.

R603.2.3 Corrosion protection. Load-bearing steel framing shall have a metallic coating complying with one of the following:

1. A minimum of G 60 in accordance with ASTM A 653.

2. A minimum of AZ 50 in accordance with ASTM A 792.

3. A minimum of GF 60 in accordance with ASTM A 875.

❖ The metallic coatings specified correspond to requirements in the referenced material standards. The minimum coating designations assume normal exposure conditions that are best defined as having the framing members enclosed within a building envelope or wall assembly within a controlled environment. When more severe exposure conditions are probable, such as industrial or marine atmospheres, consideration should be given to specifying a heavier coating.

R603.2.4 Fastening requirements. Screws for steel-to-steel connections shall be installed with a minimum edge distance and center-to-center spacing of $^1/_2$ inch (12.7 mm), shall be self-drilling tapping and shall conform to SAE J 78. Structural sheathing shall be attached to steel studs with minimum No. 8 self-drilling tapping screws that conform to SAE J 78. Screws for attaching structural sheathing to steel wall framing shall have a minimum head diameter of 0.292 inch (7.4 mm) with countersunk heads and shall be installed with a minimum edge distance of $^3/_8$ inch (9.5 mm). Gypsum board shall be attached to steel wall framing with minimum No. 6 screws conforming to ASTM C 954 and shall be

installed in accordance with Section R702. For all connections, screws shall extend through the steel a minimum of three exposed threads. All self-drilling tapping screws conforming to SAE J 78 shall have a Type II coating in accordance with ASTM B 633.

Where No. 8 screws are specified in a steel-to-steel connection the required number of screws in the connection is permitted to be reduced in accordance with the reduction factors in Table R603.2.4, when larger screws are used or when one of the sheets of steel being connected is thicker than 33 mils (0.84 mm). When applying the reduction factor the resulting number of screws shall be rounded up.

❖ Fasteners meeting the referenced standards and installed in accordance with this section will provide a load capacity consistent with these prescriptive provisions.

TABLE R603.2.4
SCREW SUBSTITUTION FACTOR

SCREW SIZE	THINNEST CONNECTED STEEL SHEET (mils)	
	33	43
#8	1.0	0.67
#10	0.93	0.62
#12	0.86	0.56

For SI: 1 mil = 0.0254 mm.

❖ In providing prescriptive connection requirements, the code typically specifies the quantity of No. 8 screws necessary to resist the required design forces transferred through a given connection. If the quantity of screws required for a particular connection is large, it may be desirable to reduce the quantity of screws by substituting a larger screw size. Where No. 8 screws are specified for a steel-to-steel connection, the required quantity of screws may be reduced if a larger screw size is substituted in accordance with the reduction factors of Table R603.2.4. Likewise, if the thinnest of the connected sheets is at least 43 mils (1.1mm), the required quantity of screws may be reduced.

R603.3 Wall construction. All exterior steel framed walls and interior load-bearing steel framed walls shall be constructed in accordance with the provisions of this section and Figure R603.3.

❖ The balance of this section consists of the prescriptive design provisions for steel-framed walls.

R603.3.1 Wall to foundation or floor connections. Steel framed walls shall be anchored to foundations or floors in accordance with Table R603.3.1 and Figure R603.3.1(1) or R603.3.1(2).

❖ Steel-framed walls anchored directly to foundations must be in accordance with Figure R603.3.1(1) with $^1/_2$-inch (12.7 mm) diameter anchor bolts spaced as required by Table R603.3.1 based on either the seismic design category or the wind speed/exposure. Wall tracks connected to wood sills must be in accordance with Figure R603.3.1(2) and Table R603.3.1. This connection is the same as required for floors in Figure R505.3.1(2).

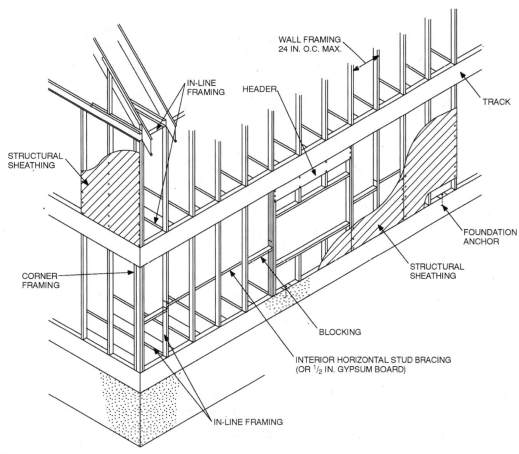

For SI: 1 inch = 25.4 mm.

**FIGURE R603.3
STEEL WALL CONSTRUCTION**

❖ This figure illustrates typical steel-framed wall construction and many of the steel-framed wall elements regulated by this section.

**TABLE R603.3.1
WALL TO FOUNDATION OR FLOOR CONNECTION REQUIREMENTS[a,b,c]**

FRAMING CONDITION	BASIC WIND SPEED (mph) AND EXPOSURE		
	85 A/B or Seismic Design Categories A, B and C	85 C or less than 110 A/B	Less than 110 C
Wall bottom track to floor joist or track	1-No. 8 screw at 12″ o.c.	1-No. 8 screw at 12″o.c.	2-No. 8 screw at 12″ o.c.
Wall bottom track to wood sill per Figure R603.3.1(2)	Steel plate spaced at 4′ o.c., with 4-No. 8 screws and 4-10d or 6-8d common nails	Steel plate spaced at 3′ o.c., with 4-No. 8 screws and 4-10d or 6-8d common nails	Steel plate spaced at 2′ o.c., with 4-No. 8 screws and 4-10d or 6-8d common nails
Wall bottom track to foundation per Figure R603.3.1(1)	$^1/_2$″ minimum diameter anchor bolt at 6′ o.c.	$^1/_2$″ minimum diameter anchor bolt at 6′ o.c.	$^1/_2$″ minimum diameter anchor bolt at 4′ o.c.
Wind uplift connector capacity for 16-inch stud spacing[c]	N/R	N/R	65 lb
Wind uplift connector capacity for 24-inch stud spacing[c]	N/R	N/R	100 lb

For SI: 1 inch = 25.4 mm, 1 foot = 304.8 mm, 1 mile per hour = 0.447 m/s, 1 pound = 4.4 N.

a. Anchor bolts shall be located not more than 12 inches from corners or the termination of bottom tracks (e.g., at door openings or corners). Bolts shall extend a minimum of 7 inches into concrete or masonry.

b. All screw sizes shown are minimum.

c. N/R = uplift connector not required. Uplift connectors are in addition to other connection requirements and shall be applied in accordance with Section R603.8.

❖ This table specifies connections for wall tracks, which will provide appropriate load capacity based on the seismic design category and the basic wind speed.

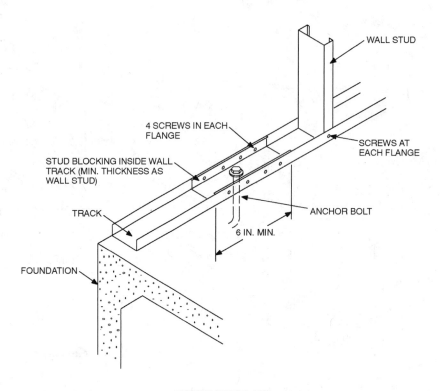

For SI: 1 inch = 25.4 mm.

FIGURE R603.3.1(1)
WALL TO FOUNDATION CONNECTION

❖ This connection detail depicts the item on the third line of Table R603.3.1.

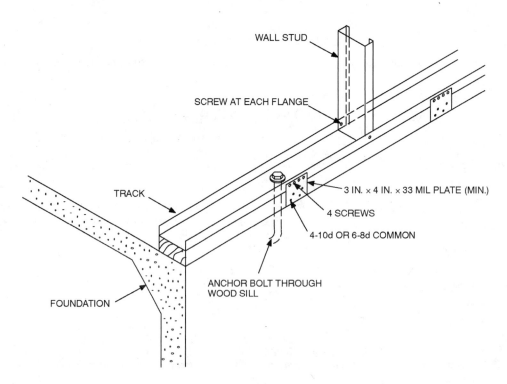

For SI: 1 inch = 25.4 mm.

FIGURE R603.3.1(2)
WALL TO WOOD SILL CONNECTION

❖ This connection detail illustrates the item on the second line of Table R603.3.1.

R603.3.2 Load-bearing walls. Steel studs shall comply with Tables R603.3.2(2) through R603.3.2(21). The tabulated stud thickness for structural walls shall be used when the attic load is 10 psf (0.48 kPa) or less. When an attic storage load is greater than 10 psf (0.48 kPa) but less than or equal to 20 psf (0.96 kPa), the next higher snow load column value from Tables R603.3.2(2) through R603.3.2(21) shall be used to select the stud size. The tabulated stud thickness for structural walls supporting one floor, roof and ceiling shall be used when the second floor live load is 30 psf (1.44 kPa). When the second floor live load is greater than 30 psf (1.44 kPa) but less than or equal to 40 psf (1.92 kPa) the design value in the next higher snow load column from Tables R603.2(12) through R603.3.2(21) shall be used to select the stud size.

Fastening requirements shall be in accordance with Section R603.2.4 and Table R603.3.2(1). Tracks shall have the same minimum thickness as the wall studs. Exterior walls with a minimum of $^{1}/_{2}$-inch (13 mm) gypsum board installed in accordance with Section R702 on the interior surface and wood structural panels of minimum $^{7}/_{16}$-inch thick (11mm) oriented-strand board or $^{15}/_{32}$-inch thick (12 mm) plywood installed in accordance with Table R603.3.2(1) on the outside surface shall be permitted to use the next thinner stud from Tables R603.3.2(2) through R603.3.2(13) but not less than 33 mils (0.84 mm). Interior load-bearing walls with a minimum $^{1}/_{2}$-inch (13 mm) gypsum board installed in accordance with Section R702 on both sides of the wall shall be permitted to use the next thinner stud from Tables R603.3.2(2) through R603.3.2(13) but not less than 33 mils (0.84 mm).

❖ Steel studs must comply with Tables R603.3.2(2) through R603.3.2(21). These tables specify the minimum steel stud thickness based on floor/ceilings supported, wall height, wind speed/exposure, ground snow load, building width and stud depth and spacing. Although the tabular values are calculated based upon attic live loads of 10 psf (0.479 kPa) or less, values for attic live loads of up to 20 psf (0.958 kPa) may be determined by simply moving to the next higher snow load column. Similarly, although the tabular values are based upon floor live loads of 30 psf (1.44 kPa) of less, values for floor live loads of up to 40 psf (1.29 kPa) may be determined by moving to the next higher snow load column.

In recognition of the stiffening effects of sheathing, the code allows the required stud thickness to be reduced to the next thinner size for exterior walls with minimum $^{1}/_{2}$-inch (12.7 mm) gypsum board installed on the interior surface and wood structural panels [minimum $^{7}/_{16}$-inch (11.1 mm) thick oriented strand board or $^{15}/_{32}$-inch (11.9 mm) thick plywood] installed on the exterior. Stud thickness must not be less than 33 mils (0.84 mm). Similarly, interior load-bearing walls with minimum $^{1}/_{2}$-inch (12.7 mm) gypsum board installed on both sides of the wall may use the next thinner stud, but not less than 33 mils (0.84 mm).

Fastening of studs to tracks and structural sheathing to studs must be as specified in Section R603.2.4 and Table R603.3.2(1). Track thickness must be no less than that of the wall studs.

R603.3.3 Stud bracing. The flanges of steel studs shall be laterally braced in accordance with one of the following:

1. Gypsum board installed with minimum No. 6 screws in accordance with Section R702 or structural sheathing installed in accordance with Table R603.3.2(1).

2. Horizontal steel strapping installed in accordance with Figure R603.3 at mid-height for 8-foot (2438 mm) walls, and one-third points for 9-foot and 10-foot (2743 mm and 3048 mm) walls. Steel straps shall be at least 1.5 inches in width and 33 mils in thickness (38 mm by 0.84 mm). Straps shall be attached to the flanges of studs with at least one No. 8 screw. In-line blocking shall be installed between studs at the termination of all straps. Straps shall be fastened to the blocking with at least two No. 8 screws.

3. Sheathing on one side and strapping on the other side. Sheathing shall be installed in accordance with Method #1 above. Steel straps shall be installed in accordance with Method #2 above.

❖ To provide resistance against buckling, lateral bracing of steel-stud flanges must be provided by gypsum board installed according to Section R702, structural sheathing installed in accordance with Table R603.3.2(1), horizontal steel strapping as indicated in Figure R603.3 or sheathing in accordance with Item 1 on one side and steel straps in accordance with Item 2 on the other side. Steel strap bracing must be installed at mid-height for 8-foot (2438 mm) walls as shown in Figure R603.3 and at one-third points for 9-foot (2743 mm) and 10-foot (3048 mm) walls (two rows of bracing). These straps must be attached to stud flanges with at least one No. 8 screw. In-line blocking must be installed between studs at the termination of all straps. Straps must be fastened to this blocking with at least two No. 8 screws.

TABLE R603.3.2(1)
WALL FASTENING SCHEDULE[a]

DESCRIPTION OF BUILDING ELEMENT	NUMBER AND SIZE OF FASTENERS[a]	SPACING OF FASTENERS
Floor joist to track of load-bearing wall	2-No. 8 screws	Each joist
Wall stud to top or bottom track	2-No. 8 screws	Each end of stud, one per flange
Structural sheathing to wall studs	No. 8 screws	6″ o.c. on edges and 12″ o.c. at intermediate supports
Roof framing to wall	Approved design or tie down in accordance with Section R802.11	

For SI: 1 inch = 25.4 mm.

a. All screw sizes shown are minimum.

❖ This table provides fastener size and spacing for various portions of steel stud walls.

TABLE R603.3.2(2)
24-FOOT-WIDE BUILDING SUPPORTING ROOF AND CEILING ONLY[a, b, c]
33 ksi STEEL

WIND SPEED		MEMBER SIZE	STUD SPACING (inches)	MINIMUM STUD THICKNESS (mils)											
				8-Foot Studs				9-Foot Studs				10-Foot Studs			
				Ground Snow Load (psf)											
Exp. A/B	Exp. C			20	30	50	70	20	30	50	70	20	30	50	70
85 mph	—	350S162	16	33	33	33	33	33	33	33	33	33	33	33	33
			24	33	33	33	33	33	33	33	33	33	33	33	43
		550S162	16	33	33	33	33	33	33	33	33	33	33	33	33
			24	33	33	33	33	33	33	33	33	33	33	33	33
90 mph	—	350S162	16	33	33	33	33	33	33	33	33	33	33	33	33
			24	33	33	33	33	33	33	33	33	33	33	33	43
		550S162	16	33	33	33	33	33	33	33	33	33	33	33	33
			24	33	33	33	33	33	33	33	33	33	33	33	33
100 mph	85 mph	350S162	16	33	33	33	33	33	33	33	33	33	33	33	33
			24	33	33	33	33	33	33	33	33	33	33	33	43
		550S162	16	33	33	33	33	33	33	33	33	33	33	33	33
			24	33	33	33	33	33	33	33	33	33	33	33	33
110 mph	90 mph	350S162	16	33	33	33	33	33	33	33	33	33	33	33	33
			24	33	33	33	33	33	33	33	43	33	33	33	43
		550S162	16	33	33	33	33	33	33	33	33	33	33	33	33
			24	33	33	33	33	33	33	33	33	33	33	33	33
—	100 mph	350S162	16	33	33	33	33	33	33	33	33	33	33	33	33
			24	33	33	33	43	33	33	33	43	43	43	43	43
		550S162	16	33	33	33	33	33	33	33	33	33	33	33	33
			24	33	33	33	33	33	33	33	33	33	33	33	33
—	110 mph	350S162	16	33	33	33	33	33	33	33	33	33	33	33	33
			24	33	33	33	43	43	43	43	43	54	54	54	54
		550S162	16	33	33	33	33	33	33	33	33	33	33	33	33
			24	33	33	33	33	33	33	33	33	33	33	33	33

For SI: 1 inch = 25.4 mm, 1 foot = 304.8 mm, 1 mil = 0.0254 mm, 1 mile per hour = 0.447 m/s, 1 pound per square foot = 0.0479 kPa,
1 ksi = 1000 psi = 6.895 MPa.

a. Deflection criterion: $L/240$.
b. Design load assumptions:
   Roof/ceiling dead load is 12 psf.
   Attic live load is 10 psf.
c. Building width is in the direction of horizontal framing members supported by the wall studs.

❖ See the commentary for Section R603.3.2.

TABLE R603.3.2(3)
24-FOOT-WIDE BUILDING SUPPORTING ROOF AND CEILING ONLY[a, b, c]
50 ksi STEEL

WIND SPEED		MEMBER SIZE	STUD SPACING (inches)	MINIMUM STUD THICKNESS (mils)											
				8-Foot Studs				9-Foot Studs				10-Foot Studs			
Exp. A/B	Exp. C			Ground Snow Load (psf)											
				20	30	50	70	20	30	50	70	20	30	50	70
85 mph	—	350S162	16	33	33	33	33	33	33	33	33	33	33	33	33
			24	33	33	33	33	33	33	33	33	33	33	33	33
		550S162	16	33	33	33	33	33	33	33	33	33	33	33	33
			24	33	33	33	33	33	33	33	33	33	33	33	33
90 mph	—	350S162	16	33	33	33	33	33	33	33	33	33	33	33	33
			24	33	33	33	33	33	33	33	33	33	33	33	33
		550S162	16	33	33	33	33	33	33	33	33	33	33	33	33
			24	33	33	33	33	33	33	33	33	33	33	33	33
100 mph	85 mph	350S162	16	33	33	33	33	33	33	33	33	33	33	33	33
			24	33	33	33	33	33	33	33	33	33	33	33	33
		550S162	16	33	33	33	33	33	33	33	33	33	33	33	33
			24	33	33	33	33	33	33	33	33	33	33	33	33
110 mph	90 mph	350S162	16	33	33	33	33	33	33	33	33	33	33	33	33
			24	33	33	33	33	33	33	33	33	33	33	33	33
		550S162	16	33	33	33	33	33	33	33	33	33	33	33	33
			24	33	33	33	33	33	33	33	33	33	33	33	33
—	100 mph	350S162	16	33	33	33	33	33	33	33	33	33	33	33	33
			24	33	33	33	33	33	33	33	33	43	43	43	43
		550S162	16	33	33	33	33	33	33	33	33	33	33	33	33
			24	33	33	33	33	33	33	33	33	33	33	33	33
—	110 mph	350S162	16	33	33	33	33	33	33	33	33	33	33	33	33
			24	33	33	33	33	33	33	33	33	54	54	54	54
		550S162	16	33	33	33	33	33	33	33	33	33	33	33	33
			24	33	33	33	33	33	33	33	33	33	33	33	33

For SI: 1 inch = 25.4 mm, 1 foot = 304.8 mm, 1 mil = 0.0254 mm, 1 mile per hour = 0.447 m/s, 1 pound per square foot = 0.0479 kPa,
1 ksi = 1000 psi = 6.895 MPa.

a. Deflection criterion: $L/240$.
b. Design load assumptions:
   Roof/ceiling dead load is 12 psf.
   Attic live load is 10 psf.
c. Building width is in the direction of horizontal framing members supported by the wall studs.

❖ See the commentary for Section R603.3.2.

TABLE R603.3.2(4)
28-FOOT-WIDE BUILDING SUPPORTING ROOF AND CEILING ONLY[a, b, c]
33 ksi STEEL

WIND SPEED		MEMBER SIZE	STUD SPACING (inches)	MINIMUM STUD THICKNESS (mils)											
				8-Foot Studs				9-Foot Studs				10-Foot Studs			
Exp. A/B	Exp. C			Ground Snow Load (psf)											
				20	30	50	70	20	30	50	70	20	30	50	70
85 mph	—	350S162	16	33	33	33	33	33	33	33	33	33	33	33	33
			24	33	33	33	43	33	33	33	43	33	33	33	43
		550S162	16	33	33	33	33	33	33	33	33	33	33	33	33
			24	33	33	33	33	33	33	33	33	33	33	33	33
90 mph	—	350S162	16	33	33	33	33	33	33	33	33	33	33	33	33
			24	33	33	33	43	33	33	33	43	33	33	33	43
		550S162	16	33	33	33	33	33	33	33	33	33	33	33	33
			24	33	33	33	33	33	33	33	33	33	33	33	33
100 mph	85 mph	350S162	16	33	33	33	33	33	33	33	33	33	33	33	33
			24	33	33	33	43	33	33	33	43	33	33	43	43
		550S162	16	33	33	33	33	33	33	33	33	33	33	33	33
			24	33	33	33	33	33	33	33	33	33	33	33	33
110 mph	90 mph	350S162	16	33	33	33	33	33	33	33	33	33	33	33	33
			24	33	33	33	43	33	33	33	43	33	33	43	43
		550S162	16	33	33	33	33	33	33	33	33	33	33	33	33
			24	33	33	33	33	33	33	33	33	33	33	33	33
—	100 mph	350S162	16	33	33	33	33	33	33	33	33	33	33	33	33
			24	33	33	33	43	33	33	43	43	43	43	43	43
		550S162	16	33	33	33	33	33	33	33	33	33	33	33	33
			24	33	33	33	33	33	33	33	33	33	33	33	33
—	110 mph	350S162	16	33	33	33	33	33	33	33	33	33	33	33	33
			24	33	33	33	43	43	43	43	43	54	54	54	54
		550S162	16	33	33	33	33	33	33	33	33	33	33	33	33
			24	33	33	33	33	33	33	33	33	33	33	33	33

For SI: 1 inch = 25.4 mm, 1 foot = 304.8 mm, 1 mil = 0.0254 mm, 1 mile per hour = 0.447 m/s, 1 pound per square foot = 0.0479 kPa,
1 ksi = 1000 psi = 6.895 MPa.

a. Deflection criterion: $L/240$.
b. Design load assumptions:
    Roof/ceiling dead load is 12 psf.
    Attic live load is 10 psf.
c. Building width is in the direction of horizontal framing members supported by the wall studs.

❖ See the commentary for Section R603.3.2.

TABLE R603.3.2(5)
28-FOOT-WIDE BUILDING SUPPORTING ROOF AND CEILING ONLY[a, b, c]
50 ksi STEEL

WIND SPEED		MEMBER SIZE	STUD SPACING (inches)	MINIMUM STUD THICKNESS (mils)											
				8-Foot Studs				9-Foot Studs				10-Foot Studs			
Exp. A/B	Exp. C			Ground Snow Load (psf)											
				20	30	50	70	20	30	50	70	20	30	50	70
85 mph	—	350S162	16	33	33	33	33	33	33	33	33	33	33	33	33
			24	33	33	33	33	33	33	33	33	33	33	33	43
		550S162	16	33	33	33	33	33	33	33	33	33	33	33	33
			24	33	33	33	33	33	33	33	33	33	33	33	33
90 mph	—	350S162	16	33	33	33	33	33	33	33	33	33	33	33	33
			24	33	33	33	33	33	33	33	33	33	33	33	43
		550S162	16	33	33	33	33	33	33	33	33	33	33	33	33
			24	33	33	33	33	33	33	33	33	33	33	33	33
100 mph	85 mph	350S162	16	33	33	33	33	33	33	33	33	33	33	33	33
			24	33	33	33	33	33	33	33	33	33	33	33	43
		550S162	16	33	33	33	33	33	33	33	33	33	33	33	33
			24	33	33	33	33	33	33	33	33	33	33	33	33
110 mph	90 mph	350S162	16	33	33	33	33	33	33	33	33	33	33	33	33
			24	33	33	33	33	33	33	33	33	33	33	43	43
		550S162	16	33	33	33	33	33	33	33	33	33	33	33	33
			24	33	33	33	33	33	33	33	33	33	33	33	33
—	100 mph	350S162	16	33	33	33	33	33	33	33	33	33	33	33	33
			24	33	33	33	33	33	33	33	33	43	43	43	43
		550S162	16	33	33	33	33	33	33	33	33	33	33	33	33
			24	33	33	33	33	33	33	33	33	33	33	33	33
—	110 mph	350S162	16	33	33	33	33	33	33	33	33	33	33	33	33
			24	33	33	33	43	33	33	33	43	54	54	54	54
		550S162	16	33	33	33	33	33	33	33	33	33	33	33	33
			24	33	33	33	33	33	33	33	33	33	33	33	33

For SI: 1 inch = 25.4 mm, 1 foot = 304.8 mm, 1 mil = 0.0254 mm, 1 mile per hour = 0.447 m/s, 1 pound per square foot = 0.0479 kPa, 1 ksi = 1000 psi = 6.895 MPa.

a. Deflection criterion: $L/240$.
b. Design load assumptions:
   Roof/ceiling dead load is 12 psf.
   Attic live load is 10 psf.
c. Building width is in the direction of horizontal framing members supported by the wall studs.

❖ See the commentary for Section R603.3.2.

TABLE R603.3.2(6)
32-FOOT-WIDE BUILDING SUPPORTING ROOF AND CEILING ONLY[a, b, c]
33 ksi STEEL

WIND SPEED		MEMBER SIZE	STUD SPACING (inches)	MINIMUM STUD THICKNESS (mils)											
				8-Foot Studs				9-Foot Studs				10-Foot Studs			
Exp. A/B	Exp. C			Ground Snow Load (psf)											
				20	30	50	70	20	30	50	70	20	30	50	70
85 mph	—	350S162	16	33	33	33	33	33	33	33	33	33	33	33	33
			24	33	33	33	43	33	33	33	43	33	33	43	43
		550S162	16	33	33	33	33	33	33	33	33	33	33	33	33
			24	33	33	33	33	33	33	33	33	33	33	33	33
90 mph	—	350S162	16	33	33	33	33	33	33	33	33	33	33	33	33
			24	33	33	33	43	33	33	33	43	33	33	43	43
		550S162	16	33	33	33	33	33	33	33	33	33	33	33	33
			24	33	33	33	33	33	33	33	33	33	33	33	33
100 mph	85 mph	350S162	16	33	33	33	33	33	33	33	33	33	33	33	33
			24	33	33	33	43	33	33	33	43	33	33	43	43
		550S162	16	33	33	33	33	33	33	33	33	33	33	33	33
			24	33	33	33	33	33	33	33	33	33	33	33	33
110 mph	90 mph	350S162	16	33	33	33	33	33	33	33	33	33	33	33	33
			24	33	33	33	43	33	33	33	43	33	33	43	43
		550S162	16	33	33	33	33	33	33	33	33	33	33	33	33
			24	33	33	33	33	33	33	33	33	33	33	33	33
—	100 mph	350S162	16	33	33	33	33	33	33	33	33	33	33	33	43
			24	33	33	43	43	33	43	43	43	43	43	43	54
		550S162	16	33	33	33	33	33	33	33	33	33	33	33	33
			24	33	33	33	33	33	33	33	33	33	33	33	43
—	110 mph	350S162	16	33	33	33	33	33	33	33	33	33	33	33	43
			24	33	33	43	43	33	43	43	43	54	54	54	54
		550S162	16	33	33	33	33	33	33	33	33	33	33	33	33
			24	33	33	33	33	33	33	33	33	33	33	33	43

For SI: 1 inch = 25.4 mm, 1 foot = 304.8 mm, 1 mil = 0.0254 mm, 1 mile per hour = 0.447 m/s, 1 pound per square foot = 0.0479 kPa, 1 ksi = 1000 psi = 6.895 MPa.

a. Deflection criterion: $L/240$.
b. Design load assumptions:
   Roof/ceiling dead load is 12 psf.
   Attic live load is 10 psf.
c. Building width is in the direction of horizontal framing members supported by the wall studs.

❖ See the commentary for Section R603.3.2.

TABLE R603.3.2(7)
32-FOOT-WIDE BUILDING SUPPORTING ROOF AND CEILING ONLY[a, b, c]
50 ksi STEEL

WIND SPEED		MEMBER SIZE	STUD SPACING (inches)	MINIMUM STUD THICKNESS (mils)											
				8-Foot Studs				9-Foot Studs				10-Foot Studs			
Exp. A/B	Exp. C			Ground Snow Load (psf)											
				20	30	50	70	20	30	50	70	20	30	50	70
85 mph	—	350S162	16	33	33	33	33	33	33	33	33	33	33	33	33
			24	33	33	33	33	33	33	33	33	33	33	33	43
		550S162	16	33	33	33	33	33	33	33	33	33	33	33	33
			24	33	33	33	33	33	33	33	33	33	33	33	33
90 mph	—	350S162	16	33	33	33	33	33	33	33	33	33	33	33	33
			24	33	33	33	33	33	33	33	33	33	33	33	43
		550S162	16	33	33	33	33	33	33	33	33	33	33	33	33
			24	33	33	33	33	33	33	33	33	33	33	33	33
100 mph	85 mph	350S162	16	33	33	33	33	33	33	33	33	33	33	33	33
			24	33	33	33	43	33	33	33	33	33	33	33	43
		550S162	16	33	33	33	33	33	33	33	33	33	33	33	33
			24	33	33	33	33	33	33	33	33	33	33	33	33
110 mph	90 mph	350S162	16	33	33	33	33	33	33	33	33	33	33	33	33
			24	33	33	33	43	33	33	33	33	33	33	33	43
		550S162	16	33	33	33	33	33	33	33	33	33	33	33	33
			24	33	33	33	33	33	33	33	33	33	33	33	33
—	100 mph	350S162	16	33	33	33	33	33	33	33	33	33	33	33	33
			24	33	33	33	43	33	33	33	43	43	43	43	43
		550S162	16	33	33	33	33	33	33	33	33	33	33	33	33
			24	33	33	33	33	33	33	33	33	33	33	33	33
—	110 mph	350S162	16	33	33	33	33	33	33	33	33	33	33	33	33
			24	33	33	33	43	33	33	33	43	54	54	54	54
		550S162	16	33	33	33	33	33	33	33	33	33	33	33	33
			24	33	33	33	33	33	33	33	33	33	33	33	33

For SI: 1 inch = 25.4 mm, 1 foot = 304.8 mm, 1 mil = 0.0254 mm, 1 mile per hour = 0.447 m/s, 1 pound per square foot = 0.0479 kPa,
   1 ksi = 1000 psi = 6.895 MPa.
a. Deflection criterion: L/240.
b. Design load assumptions:
   Roof/ceiling dead load is 12 psf.
   Attic live load is 10 psf.
c. Building width is in the direction of horizontal framing members supported by the wall studs.

❖ See the commentary for Section R603.3.2.

TABLE R603.3.2(8)
36-FOOT-WIDE BUILDING SUPPORTING ROOF AND CEILING ONLY[a, b, c]
33 ksi STEEL

WIND SPEED		MEMBER SIZE	STUD SPACING (inches)	MINIMUM STUD THICKNESS (mils)											
				8-Foot Studs				9-Foot Studs				10-Foot Studs			
				Ground Snow Load (psf)											
Exp. A/B	Exp. C			20	30	50	70	20	30	50	70	20	30	50	70
85 mph	—	350S162	16	33	33	33	33	33	33	33	33	33	33	33	33
			24	33	33	43	43	33	33	33	43	33	33	43	43
		550S162	16	33	33	33	33	33	33	33	33	33	33	33	33
			24	33	33	33	33	33	33	33	33	33	33	33	33
90 mph	—	350S162	16	33	33	33	33	33	33	33	33	33	33	33	33
			24	33	33	43	43	33	33	33	43	33	33	43	43
		550S162	16	33	33	33	33	33	33	33	33	33	33	33	33
			24	33	33	33	33	33	33	33	33	33	33	33	33
100 mph	85 mph	350S162	16	33	33	33	33	33	33	33	33	33	33	33	33
			24	33	33	33	43	33	33	43	43	33	33	43	54
		550S162	16	33	33	33	33	33	33	33	33	33	33	33	33
			24	33	33	33	33	33	33	33	33	33	33	33	43
110 mph	90 mph	350S162	16	33	33	33	33	33	33	33	33	33	33	33	33
			24	33	33	43	43	33	33	43	43	33	43	43	54
		550S162	16	33	33	33	33	33	33	33	33	33	33	33	33
			24	33	33	33	33	33	33	33	33	33	33	33	43
—	100 mph	350S162	16	33	33	33	33	33	33	33	33	33	33	33	43
			24	33	33	43	43	43	43	43	43	43	43	43	54
		550S162	16	33	33	33	33	33	33	33	33	33	33	33	33
			24	33	33	33	43	33	33	33	43	33	33	33	43
—	110 mph	350S162	16	33	33	33	33	33	33	33	33	33	33	33	43
			24	33	33	43	54	43	43	43	54	54	54	54	54
		550S162	16	33	33	33	33	33	33	33	33	33	33	33	33
			24	33	33	33	43	33	33	33	43	33	33	33	33

For SI: 1 inch = 25.4 mm, 1 foot = 304.8 mm, 1 mil = 0.0254 mm, 1 mile per hour = 0.447 m/s, 1 pound per square foot = 0.0479 kPa, 1 ksi = 1000 psi = 6.895 MPa.

a. Deflection criterion: $L/240$.
b. Design load assumptions:
   Roof/ceiling dead load is 12 psf.
   Attic live load is 10 psf.
c. Building width is in the direction of horizontal framing members supported by the wall studs.

❖ See the commentary for Section R603.3.2.

TABLE R603.3.2(9)
36-FOOT-WIDE BUILDING SUPPORTING ROOF AND CEILING ONLY[a, b, c]
50 ksi STEEL

WIND SPEED		MEMBER SIZE	STUD SPACING (inches)	MINIMUM STUD THICKNESS (mils)											
				8-Foot Studs				9-Foot Studs				10-Foot Studs			
Exp. A/B	Exp. C			Ground Snow Load (psf)											
				20	30	50	70	20	30	50	70	20	30	50	70
85 mph	—	350S162	16	33	33	33	33	33	33	33	33	33	33	33	33
			24	33	33	33	43	33	33	33	43	33	33	33	43
		550S162	16	33	33	33	33	33	33	33	33	33	33	33	33
			24	33	33	33	33	33	33	33	33	33	33	33	33
90 mph	—	350S162	16	33	33	33	33	33	33	33	33	33	33	33	33
			24	33	33	33	43	33	33	33	43	33	33	33	43
		550S162	16	33	33	33	33	33	33	33	33	33	33	33	33
			24	33	33	33	33	33	33	33	33	33	33	33	33
100 mph	85 mph	350S162	16	33	33	33	33	33	33	33	33	33	33	33	33
			24	33	33	33	43	33	33	33	43	33	33	33	43
		550S162	16	33	33	33	33	33	33	33	33	33	33	33	33
			24	33	33	33	33	33	33	33	33	33	33	33	33
110 mph	90 mph	350S162	16	33	33	33	33	33	33	33	33	33	33	33	33
			24	33	33	33	43	33	33	33	43	33	33	43	43
		550S162	16	33	33	33	33	33	33	33	33	33	33	33	33
			24	33	33	33	33	33	33	33	33	33	33	33	33
—	100 mph	350S162	16	33	33	33	33	33	33	33	33	33	33	33	33
			24	33	33	33	43	33	33	33	43	43	43	43	43
		550S162	16	33	33	33	33	33	33	33	33	33	33	33	33
			24	33	33	33	33	33	33	33	33	33	33	33	33
—	110 mph	350S162	16	33	33	33	33	33	33	33	33	33	33	33	33
			24	33	33	33	43	33	33	33	43	54	54	54	54
		550S162	16	33	33	33	33	33	33	33	33	33	33	33	33
			24	33	33	33	33	33	33	33	33	33	33	33	33

For SI: 1 inch = 25.4 mm, 1 foot = 304.8 mm, 1 mil = 0.0254 mm, 1 mile per hour = 0.447 m/s, 1 pound per square foot = 0.0479 kPa, 1 ksi = 1000 psi = 6.895 MPa.

a. Deflection criterion: $L/240$.
b. Design load assumptions:
   Roof/ceiling dead load is 12 psf.
   Attic live load is 10 psf.
c. Building width is in the direction of horizontal framing members supported by the wall studs.

❖ See the commentary for Section R603.3.2.

TABLE R603.3.2(10)
40-FOOT-WIDE BUILDING SUPPORTING ROOF AND CEILING ONLY[a, b, c]
33 ksi STEEL

WIND SPEED		MEMBER SIZE	STUD SPACING (inches)	MINIMUM STUD THICKNESS (mils)											
				8-Foot Studs				9-Foot Studs				10-Foot Studs			
Exp. A/B	Exp. C			Ground Snow Load (psf)											
				20	30	50	70	20	30	50	70	20	30	50	70
85 mph	—	350S162	16	33	33	33	33	33	33	33	33	33	33	33	43
			24	33	33	43	43	33	33	43	43	33	33	43	54
		550S162	16	33	33	33	33	33	33	33	33	33	33	33	33
			24	33	33	33	43	33	33	33	33	33	33	33	43
90 mph	—	350S162	16	33	33	33	33	33	33	33	33	33	33	33	43
			24	33	33	43	43	33	33	43	43	33	33	43	54
		550S162	16	33	33	33	33	33	33	33	33	33	33	33	33
			24	33	33	33	43	33	33	33	33	33	33	33	43
100 mph	85 mph	350S162	16	33	33	33	33	33	33	33	33	33	33	33	43
			24	33	33	43	43	33	33	43	43	33	33	43	54
		550S162	16	33	33	33	33	33	33	33	33	33	33	33	33
			24	33	33	33	43	33	33	33	43	33	33	33	43
110 mph	90 mph	350S162	16	33	33	33	33	33	33	33	33	33	33	33	43
			24	33	33	43	54	33	33	43	43	43	43	43	54
		550S162	16	33	33	33	33	33	33	33	33	33	33	33	33
			24	33	33	33	43	33	33	33	43	33	33	33	43
—	100 mph	350S162	16	33	33	33	43	33	33	33	43	33	33	33	43
			24	33	33	43	54	43	43	43	54	43	43	43	54
		550S162	16	33	33	33	33	33	33	33	33	33	33	33	33
			24	33	33	33	43	33	33	33	43	33	33	33	43
—	110 mph	350S162	16	33	33	33	43	33	33	33	43	33	33	43	43
			24	33	43	43	54	33	43	43	54	54	54	54	54
		550S162	16	33	33	33	33	33	33	33	33	33	33	33	33
			24	33	33	33	43	33	33	33	43	33	33	33	43

For SI: 1 inch = 25.4 mm, 1 foot = 304.8 mm, 1 mil = 0.0254 mm, 1 mile per hour = 0.447 m/s, 1 pound per square foot = 0.0479 kPa, 1 ksi = 1000 psi = 6.895 MPa.

a. Deflection criterion: $L/240$.
b. Design load assumptions:
Roof/ceiling dead load is 12 psf.
Attic live load is 10 psf.
c. Building width is in the direction of horizontal framing members supported by the wall studs.

❖ See the commentary for Section R603.3.2.

TABLE R603.3.2(11)
40-FOOT-WIDE BUILDING SUPPORTING ROOF AND CEILING ONLY[a, b, c]
50 ksi STEEL

WIND SPEED		MEMBER SIZE	STUD SPACING (inches)	MINIMUM STUD THICKNESS (mils)											
				8-Foot Studs				9-Foot Studs				10-Foot Studs			
Exp. A/B	Exp. C			Ground Snow Load (psf)											
				20	30	50	70	20	30	50	70	20	30	50	70
85 mph	—	350S162	16	33	33	33	33	33	33	33	33	33	33	33	33
			24	33	33	33	43	33	33	33	43	33	33	43	43
		550S162	16	33	33	33	33	33	33	33	33	33	33	33	33
			24	33	33	33	33	33	33	33	33	33	33	33	33
90 mph	—	350S162	16	33	33	33	33	33	33	33	33	33	33	33	33
			24	33	33	33	43	33	33	33	43	33	33	43	43
		550S162	16	33	33	33	33	33	33	33	33	33	33	33	33
			24	33	33	33	33	33	33	33	33	33	33	33	33
100 mph	85 mph	350S162	16	33	33	33	33	33	33	33	33	33	33	33	33
			24	33	33	33	43	33	33	33	43	33	33	43	43
		550S162	16	33	33	33	33	33	33	33	33	33	33	33	33
			24	33	33	33	33	33	33	33	33	33	33	33	33
110 mph	90 mph	350S162	16	33	33	33	33	33	33	33	33	33	33	33	33
			24	33	33	33	43	33	33	33	43	33	33	43	43
		550S162	16	33	33	33	33	33	33	33	33	33	33	33	33
			24	33	33	33	33	33	33	33	33	33	33	33	33
—	100 mph	350S162	16	33	33	33	33	33	33	33	33	33	33	33	33
			24	33	33	43	43	33	33	33	43	43	43	43	54
		550S162	16	33	33	33	33	33	33	33	33	33	33	33	33
			24	33	33	33	33	33	33	33	33	33	33	33	33
—	110 mph	350S162	16	33	33	33	33	33	33	33	33	33	33	33	43
			24	33	33	43	43	33	33	43	43	54	54	54	54
		550S162	16	33	33	33	33	33	33	33	33	33	33	33	33
			24	33	33	33	33	33	33	33	33	33	33	33	33

For SI: 1 inch = 25.4 mm, 1 foot = 304.8 mm, 1 mil = 0.0254 mm, 1 mile per hour = 0.447 m/s, 1 pound per square foot = 0.0479 kPa,
1 ksi = 1000 psi = 6.895 MPa.
a. Deflection criterion: $L/240$.
b. Design load assumptions:
   Roof/ceiling dead load is 12 psf.
   Attic live load is 10 psf.
c. Building width is in the direction of horizontal framing members supported by the wall studs.

❖ See the commentary for Section R603.3.2.

TABLE R603.3.2(12)
24-FOOT-WIDE BUILDING SUPPORTING ONE FLOOR, ROOF AND CEILING[a, b, c]
33 ksi STEEL

WIND SPEED		MEMBER SIZE	STUD SPACING (inches)	MINIMUM STUD THICKNESS (mils)											
				8-Foot Studs				9-Foot Studs				10-Foot Studs			
Exp. A/B	Exp. C			Ground Snow Load (psf)											
				20	30	50	70	20	30	50	70	20	30	50	70
85 mph	—	350S162	16	33	33	33	33	33	33	33	33	33	33	33	33
			24	33	33	33	43	33	33	33	43	43	43	43	43
		550S162	16	33	33	33	33	33	33	33	33	33	33	33	33
			24	33	33	33	33	33	33	33	33	33	33	33	33
90 mph	—	350S162	16	33	33	33	33	33	33	33	33	33	33	33	33
			24	33	33	33	43	33	33	33	43	43	43	43	43
		550S162	16	33	33	33	33	33	33	33	33	33	33	33	33
			24	33	33	33	33	33	33	33	33	33	33	33	33
100 mph	85 mph	350S162	16	33	33	33	33	33	33	33	33	33	33	33	33
			24	33	33	33	43	33	33	43	43	43	43	43	43
		550S162	16	33	33	33	33	33	33	33	33	33	33	33	33
			24	33	33	33	33	33	33	33	33	33	33	33	33
110 mph	90 mph	350S162	16	33	33	33	33	33	33	33	33	33	33	33	33
			24	33	33	43	43	43	43	43	43	43	43	43	43
		550S162	16	33	33	33	33	33	33	33	33	33	33	33	33
			24	33	33	33	33	33	33	33	33	33	33	33	33
—	100 mph	350S162	16	33	33	33	33	33	33	33	33	33	33	43	43
			24	43	43	43	43	43	43	43	43	43	43	54	54
		550S162	16	33	33	33	33	33	33	33	33	33	33	33	33
			24	33	33	33	33	33	33	33	33	33	33	33	33
—	110 mph	350S162	16	33	33	33	33	33	33	33	33	43	43	43	43
			24	43	43	43	43	43	43	43	43	54	54	54	54
		550S162	16	33	33	33	33	33	33	33	33	33	33	33	33
			24	33	33	33	33	33	33	33	33	33	33	33	43

For SI: 1 inch = 25.4 mm, 1 foot = 304.8 mm, 1 mil = 0.0254 mm, 1 mile per hour = 0.447 m/s, 1 pound per square foot = 0.0479 kPa, 1 ksi = 1000 psi = 6.895 MPa.

a. Deflection criterion: L/240.
b. Design load assumptions:
   Second floor dead load is 10 psf.
   Second floor live load is 30 psf.
   Roof/ceiling dead load is 12 psf.
   Attic live load is 10 psf.
c. Building width is in the direction of horizontal framing members supported by the wall studs.

❖ See the commentary for Section R603.3.2.

TABLE R603.3.2(13)
24-FOOT-WIDE BUILDING SUPPORTING ONE FLOOR, ROOF AND CEILING[a, b, c]
50 ksi STEEL

WIND SPEED		MEMBER SIZE	STUD SPACING (inches)	MINIMUM STUD THICKNESS (mils)											
				8-Foot Studs				9-Foot Studs				10-Foot Studs			
Exp. A/B	Exp. C			Ground Snow Load (psf)											
				20	30	50	70	20	30	50	70	20	30	50	70
85 mph	—	350S162	16	33	33	33	33	33	33	33	33	33	33	33	33
			24	33	33	33	33	33	33	33	33	33	33	33	43
		550S162	16	33	33	33	33	33	33	33	33	33	33	33	33
			24	33	33	33	33	33	33	33	33	33	33	33	33
90 mph	—	350S162	16	33	33	33	33	33	33	33	33	33	33	33	33
			24	33	33	33	33	33	33	33	33	33	33	33	43
		550S162	16	33	33	33	33	33	33	33	33	33	33	33	33
			24	33	33	33	33	33	33	33	33	33	33	33	33
100 mph	85 mph	350S162	16	33	33	33	33	33	33	33	33	33	33	33	33
			24	33	33	33	33	33	33	33	33	33	33	33	43
		550S162	16	33	33	33	33	33	33	33	33	33	33	33	33
			24	33	33	33	33	33	33	33	33	33	33	33	33
110 mph	90 mph	350S162	16	33	33	33	33	33	33	33	33	33	33	33	33
			24	33	33	33	33	33	33	33	33	33	33	43	43
		550S162	16	33	33	33	33	33	33	33	33	33	33	33	33
			24	33	33	33	33	33	33	33	33	33	33	33	33
—	100 mph	350S162	16	33	33	33	33	33	33	33	33	33	33	33	33
			24	33	33	33	43	33	33	33	43	43	43	43	43
		550S162	16	33	33	33	33	33	33	33	33	33	33	33	33
			24	33	33	33	33	33	33	33	33	33	33	33	33
—	110 mph	350S162	16	33	33	33	33	33	33	33	33	33	33	33	33
			24	33	33	33	43	33	43	43	43	54	54	54	54
		550S162	16	33	33	33	33	33	33	33	33	33	33	33	33
			24	33	33	33	33	33	33	33	33	33	33	33	33

For SI: 1 inch = 25.4 mm, 1 foot = 304.8 mm, 1 mil = 0.0254 mm, 1 mile per hour = 0.447 m/s, 1 pound per square foot = 0.0479 kPa, 1 ksi = 1000 psi = 6.895 MPa.

a. Deflection criterion: $L/240$.
b. Design load assumptions:
   Second floor dead load is 10 psf.
   Second floor live load is 30 psf.
   Roof/ceiling dead load is 12 psf.
   Attic live load is 10 psf.
c. Building width is in the direction of horizontal framing members supported by the wall studs.

❖ See the commentary for Section R603.3.2.

TABLE R603.3.2(14)
28-FOOT-WIDE BUILDING SUPPORTING ONE FLOOR, ROOF AND CEILING[a, b, c]
33 ksi STEEL

WIND SPEED		MEMBER SIZE	STUD SPACING (inches)	MINIMUM STUD THICKNESS (mils)											
				8-Foot Studs				9-Foot Studs				10-Foot Studs			
				Ground Snow Load (psf)											
Exp. A/B	Exp. C			20	30	50	70	20	30	50	70	20	30	50	70
85 mph	—	350S162	16	33	33	33	33	33	33	33	33	33	33	33	33
			24	33	43	43	43	33	33	43	43	43	43	43	43
		550S162	16	33	33	33	33	33	33	33	33	33	33	33	33
			24	33	33	33	33	33	33	33	33	33	33	33	33
90 mph	—	350S162	16	33	33	33	33	33	33	33	33	33	33	33	33
			24	33	43	43	43	33	33	43	43	43	43	43	43
		550S162	16	33	33	33	33	33	33	33	33	33	33	33	33
			24	33	33	33	33	33	33	33	33	33	33	33	33
100 mph	85 mph	350S162	16	33	33	33	33	33	33	33	33	33	33	33	33
			24	33	43	43	43	43	43	43	43	43	43	43	43
		550S162	16	33	33	33	33	33	33	33	33	33	33	33	33
			24	33	33	33	33	33	33	33	33	33	33	33	33
110 mph	90 mph	350S162	16	33	33	33	33	33	33	33	33	33	33	33	33
			24	43	43	43	43	43	43	43	43	43	43	43	54
		550S162	16	33	33	33	33	33	33	33	33	33	33	33	33
			24	33	33	33	33	33	33	33	33	33	33	33	33
—	100 mph	350S162	16	33	33	33	33	33	33	33	33	43	43	43	43
			24	43	43	43	43	43	43	43	43	54	54	54	54
		550S162	16	33	33	33	33	33	33	33	33	33	33	33	33
			24	33	33	33	33	33	33	33	33	33	33	33	43
—	110 mph	350S162	16	33	33	33	33	33	33	33	43	43	43	43	43
			24	43	43	43	43	43	43	54	54	54	54	54	54
		550S162	16	33	33	33	33	33	33	33	33	33	33	33	33
			24	33	33	33	33	33	33	33	33	33	43	43	43

For SI: 1 inch = 25.4 mm, 1 foot = 304.8 mm, 1 mil = 0.0254 mm, 1 mile per hour = 0.447 m/s, 1 pound per square foot = 0.0479 kPa, 1 ksi = 1000 psi = 6.895 MPa.

a. Deflection criterion: L/240.
b. Design load assumptions:
   Second floor dead load is 10 psf.
   Second floor live load is 30 psf.
   Roof/ceiling dead load is 12 psf.
   Attic live load is 10 psf.
c. Building width is in the direction of horizontal framing members supported by the wall studs.

❖ See the commentary for Section R603.3.2.

TABLE R603.3.2(15)
28-FOOT-WIDE BUILDING SUPPORTING ONE FLOOR, ROOF AND CEILING[a, b, c]
50 ksi STEEL

WIND SPEED Exp. A/B	Exp. C	MEMBER SIZE	STUD SPACING (inches)	8-Foot Studs Ground Snow Load (psf) 20	30	50	70	9-Foot Studs Ground Snow Load (psf) 20	30	50	70	10-Foot Studs Ground Snow Load (psf) 20	30	50	70
85 mph	—	350S162	16	33	33	33	33	33	33	33	33	33	33	33	33
			24	33	33	33	43	33	33	33	33	33	33	33	43
		550S162	16	33	33	33	33	33	33	33	33	33	33	33	33
			24	33	33	33	33	33	33	33	33	33	33	33	33
90 mph	—	350S162	16	33	33	33	33	33	33	33	33	33	33	33	33
			24	33	33	33	43	33	33	33	33	33	33	33	43
		550S162	16	33	33	33	33	33	33	33	33	33	33	33	33
			24	33	33	33	33	33	33	33	33	33	33	33	33
100 mph	85 mph	350S162	16	33	33	33	33	33	33	33	33	33	33	33	33
			24	33	33	33	43	33	33	33	33	33	33	43	43
		550S162	16	33	33	33	33	33	33	33	33	33	33	33	33
			24	33	33	33	33	33	33	33	33	33	33	33	33
110 mph	90 mph	350S162	16	33	33	33	33	33	33	33	33	33	33	33	33
			24	33	33	33	43	33	33	33	43	43	43	43	43
		550S162	16	33	33	33	33	33	33	33	33	33	33	33	33
			24	33	33	33	33	33	33	33	33	33	33	33	33
—	100 mph	350S162	16	33	33	33	33	33	33	33	33	33	33	33	33
			24	33	33	43	43	33	33	43	43	43	43	43	43
		550S162	16	33	33	33	33	33	33	33	33	33	33	33	33
			24	33	33	33	33	33	33	33	33	33	33	33	33
—	110 mph	350S162	16	33	33	33	33	33	33	33	33	33	33	33	43
			24	33	43	43	43	43	43	43	43	54	54	54	54
		550S162	16	33	33	33	33	33	33	33	33	33	33	33	33
			24	33	33	33	33	33	33	33	33	33	33	33	33

For SI: 1 inch = 25.4 mm, 1 foot = 304.8 mm, 1 mil = 0.0254 mm, 1 mile per hour = 0.447 m/s, 1 pound per square foot = 0.0479 kPa,
1 ksi = 1000 psi = 6.895 MPa.

a. Deflection criterion: $L/240$.
b. Design load assumptions:
Second floor dead load is 10 psf.
Second floor live load is 30 psf.
Roof/ceiling dead load is 12 psf.
Attic live load is 10 psf.
c. Building width is in the direction of horizontal framing members supported by the wall studs.

❖ See the commentary for Section R603.3.2.

TABLE R603.3.2(16)
32-FOOT-WIDE BUILDING SUPPORTING ONE FLOOR, ROOF AND CEILING[a, b, c]
33 ksi STEEL

WIND SPEED		MEMBER SIZE	STUD SPACING (inches)	MINIMUM STUD THICKNESS (mils)											
				8-Foot Studs				9-Foot Studs				10-Foot Studs			
Exp. A/B	Exp. C			Ground Snow Load (psf)											
				20	30	50	70	20	30	50	70	20	30	50	70
85 mph	—	350S162	16	33	33	33	33	33	33	33	33	33	33	33	33
			24	43	43	43	43	43	43	43	43	43	43	43	54
		550S162	16	33	33	33	33	33	33	33	33	33	33	33	33
			24	33	33	33	43	33	33	33	33	33	33	33	43
90 mph	—	350S162	16	33	33	33	33	33	33	33	33	33	33	33	33
			24	43	43	43	43	43	43	43	43	43	43	43	54
		550S162	16	33	33	33	33	33	33	33	33	33	33	33	33
			24	33	33	33	43	33	33	33	33	33	33	33	43
100 mph	85 mph	350S162	16	33	33	33	33	33	33	33	33	33	33	33	43
			24	43	43	43	43	43	43	43	43	43	43	43	54
		550S162	16	33	33	33	33	33	33	33	33	33	33	33	33
			24	33	33	33	43	33	33	33	33	33	33	33	43
110 mph	90 mph	350S162	16	33	33	33	33	33	33	33	33	33	33	33	43
			24	43	43	43	43	43	43	43	43	43	43	43	54
		550S162	16	33	33	33	33	33	33	33	33	33	33	33	33
			24	33	33	33	43	33	33	33	33	33	33	33	43
—	100 mph	350S162	16	33	33	33	33	33	33	33	33	43	43	43	43
			24	43	43	43	54	43	43	43	54	54	54	54	54
		550S162	16	33	33	33	33	33	33	33	33	33	33	33	33
			24	33	33	33	43	33	33	33	43	33	33	43	43
—	110 mph	350S162	16	33	33	33	43	33	33	43	43	43	43	43	43
			24	43	43	43	54	43	54	54	54	54	54	54	68
		550S162	16	33	33	33	33	33	33	33	33	33	33	33	33
			24	33	33	33	43	33	33	33	43	43	43	43	43

For SI: 1 inch = 25.4 mm, 1 foot = 304.8 mm, 1 mil = 0.0254 mm, 1 mile per hour = 0.447 m/s, 1 pound per square foot = 0.0479 kPa,
1 ksi = 1000 psi = 6.895 MPa.

a. Deflection criterion: $L/240$.

b. Design load assumptions:
Second floor dead load is 10 psf.
Second floor live load is 30 psf.
Roof/ceiling dead load is 12 psf.
Attic live load is 10 psf.

c. Building width is in the direction of horizontal framing members supported by the wall studs.

❖ See the commentary for Section R603.3.2.

TABLE R603.3.2(17)
32-FOOT-WIDE BUILDING SUPPORTING ONE FLOOR, ROOF AND CEILING[a, b, c]
50 ksi STEEL

WIND SPEED		MEMBER SIZE	STUD SPACING (inches)	MINIMUM STUD THICKNESS (mils)											
				8-Foot Studs				9-Foot Studs				10-Foot Studs			
Exp. A/B	Exp. C			Ground Snow Load (psf)											
				20	30	50	70	20	30	50	70	20	30	50	70
85 mph	—	350S162	16	33	33	33	33	33	33	33	33	33	33	33	33
			24	33	43	43	43	33	33	33	43	33	33	43	43
		550S162	16	33	33	33	33	33	33	33	33	33	33	33	33
			24	33	33	33	33	33	33	33	33	33	33	33	33
90 mph	—	350S162	16	33	33	33	33	33	33	33	33	33	33	33	33
			24	33	43	43	43	33	33	33	43	33	33	43	43
		550S162	16	33	33	33	33	33	33	33	33	33	33	33	33
			24	33	33	33	33	33	33	33	33	33	33	33	33
100 mph	85 mph	350S162	16	33	33	33	33	33	33	33	33	33	33	33	33
			24	33	43	43	43	33	33	33	43	43	43	43	43
		550S162	16	33	33	33	33	33	33	33	33	33	33	33	33
			24	33	33	33	33	33	33	33	33	33	33	33	33
110 mph	90 mph	350S162	16	33	33	33	33	33	33	33	33	33	33	33	33
			24	33	43	43	43	33	33	33	43	43	43	43	54
		550S162	16	33	33	33	33	33	33	33	33	33	33	33	33
			24	33	33	33	33	33	33	33	33	33	33	33	33
—	100 mph	350S162	16	33	33	33	33	33	33	33	33	33	33	33	33
			24	33	43	43	43	43	43	43	43	43	43	43	54
		550S162	16	33	33	33	33	33	33	33	33	33	33	33	33
			24	33	33	33	33	33	33	33	33	33	33	33	33
—	110 mph	350S162	16	33	33	33	33	33	33	33	33	33	33	33	43
			24	43	43	43	43	43	43	43	43	54	54	54	54
		550S162	16	33	33	33	33	33	33	33	33	33	33	33	33
			24	33	33	33	33	33	33	33	33	33	33	33	33

For SI: 1 inch = 25.4 mm, 1 foot = 304.8 mm, 1 mil = 0.0254 mm, 1 mile per hour = 0.447 m/s, 1 pound per square foot = 0.0479 kPa, 1 ksi = 1000 psi = 6.895 MPa.

a. Deflection criterion: $L/240$.
b. Design load assumptions:
   Second floor dead load is 10 psf.
   Second floor live load is 30 psf.
   Roof/ceiling dead load is 12 psf.
   Attic live load is 10 psf.
c. Building width is in the direction of horizontal framing members supported by the wall studs.

❖ See the commentary for Section R603.3.2.

TABLE R603.3.2(18)
36-FOOT-WIDE BUILDING SUPPORTING ONE FLOOR, ROOF AND CEILING[a, b, c]
33 ksi STEEL

WIND SPEED Exp. A/B	Exp. C	MEMBER SIZE	STUD SPACING (inches)	8-Foot Studs				9-Foot Studs				10-Foot Studs			
				20	30	50	70	20	30	50	70	20	30	50	70
85 mph	—	350S162	16	33	33	33	33	33	33	33	33	33	33	33	43
			24	43	43	43	54	43	43	43	43	43	43	43	54
		550S162	16	33	33	33	33	33	33	33	33	33	33	33	33
			24	43	43	43	43	33	33	33	43	33	33	33	43
90 mph	—	350S162	16	33	33	33	33	33	33	33	33	33	33	33	43
			24	43	43	43	54	43	43	43	43	43	43	43	54
		550S162	16	33	33	33	33	33	33	33	33	33	33	33	33
			24	43	43	43	43	33	33	33	43	33	33	43	43
100 mph	85 mph	350S162	16	33	33	33	33	33	33	33	33	33	33	33	43
			24	43	43	43	54	43	43	43	54	43	43	54	54
		550S162	16	33	33	33	33	33	33	33	33	33	33	33	33
			24	43	43	43	43	33	33	33	43	33	33	43	43
110 mph	90 mph	350S162	16	33	33	33	43	33	33	33	33	33	33	43	43
			24	43	43	43	54	43	43	43	54	43	43	54	54
		550S162	16	33	33	33	33	33	33	33	33	33	33	33	33
			24	43	43	43	43	33	33	33	43	33	33	43	43
—	100 mph	350S162	16	33	33	33	43	33	33	43	43	43	43	43	43
			24	43	43	43	54	43	43	54	54	54	54	54	54
		550S162	16	33	33	33	33	33	33	33	33	33	33	33	33
			24	43	43	43	43	33	33	33	43	43	43	43	43
—	110 mph	350S162	16	33	33	33	43	33	33	43	43	43	43	43	43
			24	43	43	43	54	43	54	54	54	54	54	54	68
		550S162	16	33	33	33	33	33	33	33	33	33	33	33	33
			24	33	33	33	43	33	33	33	43	43	43	43	43

For SI: 1 inch = 25.4 mm, 1 foot = 304.8 mm, 1 mil = 0.0254 mm, 1 mile per hour = 0.447 m/s, 1 pound per square foot = 0.0479 kPa, 1 ksi = 1000 psi = 6.895 MPa.

a. Deflection criterion: $L/240$.
b. Design load assumptions:
   Second floor dead load is 10 psf.
   Second floor live load is 30 psf.
   Roof/ceiling dead load is 12 psf.
   Attic live load is 10 psf.
c. Building width is in the direction of horizontal framing members supported by the wall studs.

❖ See the commentary for Section R603.3.2.

TABLE R603.3.2(19)
36-FOOT-WIDE BUILDING SUPPORTING ONE FLOOR, ROOF AND CEILING[a, b, c]
50 ksi STEEL

WIND SPEED		MEMBER SIZE	STUD SPACING (inches)	MINIMUM STUD THICKNESS (mils)											
				8-Foot Studs				9-Foot Studs				10-Foot Studs			
Exp. A/B	Exp. C			Ground Snow Load (psf)											
				20	30	50	70	20	30	50	70	20	30	50	70
85 mph	—	350S162	16	33	33	33	33	33	33	33	33	33	33	33	33
			24	43	43	43	43	33	33	43	43	43	43	43	43
		550S162	16	33	33	33	33	33	33	33	33	33	33	33	33
			24	33	33	33	43	33	33	33	33	33	33	33	33
90 mph	—	350S162	16	33	33	33	33	33	33	33	33	33	33	33	33
			24	43	43	43	43	33	33	43	43	43	43	43	43
		550S162	16	33	33	33	33	33	33	33	33	33	33	33	33
			24	33	43	33	43	33	33	33	33	33	33	33	33
100 mph	85 mph	350S162	16	33	33	33	33	33	33	33	33	33	33	33	33
			24	43	43	43	43	33	33	43	43	43	43	43	43
		550S162	16	33	33	33	33	33	33	33	33	33	33	33	33
			24	33	33	33	43	33	33	33	33	33	33	33	33
110 mph	90 mph	350S162	16	33	33	33	33	33	33	33	33	33	33	33	33
			24	43	43	43	43	33	33	43	43	43	43	43	54
		550S162	16	33	33	33	33	33	33	33	33	33	33	33	33
			24	43	33	33	43	33	33	33	33	33	33	33	33
—	100 mph	350S162	16	33	33	33	33	33	33	33	33	33	33	33	43
			24	43	43	43	43	43	43	43	43	43	43	54	54
		550S162	16	33	33	33	33	33	33	33	33	33	33	33	33
			24	33	33	33	43	33	33	33	33	33	33	33	43
—	110 mph	350S162	16	33	33	33	33	33	33	33	33	33	33	43	43
			24	43	43	43	54	43	43	43	43	54	54	54	54
		550S162	16	33	33	33	33	33	33	33	33	33	33	33	33
			24	33	33	33	33	33	33	33	33	33	33	33	43

For SI: 1 inch = 25.4 mm, 1 foot = 304.8 mm, 1 mil = 0.0254 mm, 1 mile per hour = 0.447 m/s 1 pound per square foot = 0.0479 kPa,
        1 ksi = 1000 psi = 6.895 MPa.
a. Deflection criterion: $L/240$.
b. Design load assumptions:
      Second floor dead load is 10 psf.
      Second floor live load is 30 psf.
      Roof/ceiling dead load is 12 psf.
      Attic live load is 10 psf.
c. Building width is in the direction of horizontal framing members supported by the wall studs.

❖ See the commentary for Section R603.3.2.

TABLE R603.3.2(20)
40-FOOT-WIDE BUILDING SUPPORTING ONE FLOOR, ROOF AND CEILING[a, b, c]
33 ksi STEEL

WIND SPEED		MEMBER SIZE	STUD SPACING (inches)	MINIMUM STUD THICKNESS (mils)											
				8-Foot Studs				9-Foot Studs				10-Foot Studs			
				Ground Snow Load (psf)											
Exp. A/B	Exp. C			20	30	50	70	20	30	50	70	20	30	50	70
85 mph	—	350S162	16	33	33	33	43	33	33	33	43	33	33	33	43
			24	43	54	54	54	43	43	43	54	43	43	54	54
		550S162	16	33	33	33	33	33	33	33	33	33	33	33	33
			24	43	43	43	43	43	43	43	43	33	43	43	43
90 mph	—	350S162	16	33	33	33	43	33	33	33	43	33	33	33	43
			24	43	54	54	54	43	43	43	54	43	43	54	54
		550S162	16	33	33	33	33	33	33	33	33	33	33	33	33
			24	43	43	43	43	43	43	43	43	33	43	43	43
100 mph	85 mph	350S162	16	33	33	33	43	33	33	33	43	33	33	43	43
			24	43	54	54	54	43	43	43	54	43	43	54	54
		550S162	16	33	33	33	33	33	33	33	33	33	33	33	33
			24	43	43	43	43	43	43	43	43	33	43	43	43
110 mph	90 mph	350S162	16	33	33	33	43	33	33	33	43	33	43	43	43
			24	43	54	54	54	43	43	43	54	43	54	54	54
		550S162	16	33	33	33	33	33	33	33	33	33	33	33	33
			24	43	43	43	43	43	43	43	43	33	43	43	43
—	100 mph	350S162	16	33	33	33	43	33	33	43	43	33	43	43	43
			24	43	54	54	54	54	54	54	54	43	54	54	68
		550S162	16	33	33	33	33	33	33	33	33	33	33	33	33
			24	43	43	43	43	43	43	43	43	33	43	43	43
—	110 mph	350S162	16	33	33	43	43	43	43	43	43	33	43	43	43
			24	54	54	54	54	54	54	54	54	54	68	68	68
		550S162	16	33	33	33	33	33	33	33	33	33	33	33	33
			24	43	43	43	43	43	43	43	43	33	43	43	43

For SI: 1 inch = 25.4 mm, 1 foot = 304.8 mm, 1 mil = 0.0254 mm, 1 mile per hour = 0.447 m/s, 1 pound per square foot = 0.0479 kPa, 1 ksi = 1000 psi = 6.895 MPa.

a. Deflection criterion: $L/240$.
b. Design load assumptions:
   Second floor dead load is 10 psf.
   Second floor live load is 30 psf.
   Roof/ceiling dead load is 12 psf.
   Attic live load is 10 psf.
c. Building width is in the direction of horizontal framing members supported by the wall studs.

❖ See the commentary for Section R603.3.2.

TABLE R603.3.2(21)
40-FOOT-WIDE BUILDING SUPPORTING ONE FLOOR, ROOF AND CEILING[a, b, c]
50 ksi STEEL

WIND SPEED		MEMBER SIZE	STUD SPACING (inches)	MINIMUM STUD THICKNESS (mils)											
				8-Foot Studs				9-Foot Studs				10-Foot Studs			
Exp. A/B	Exp. C			Ground Snow Load (psf)											
				20	30	50	70	20	30	50	70	20	30	50	70
85 mph	—	350S162	16	33	33	33	33	33	33	33	33	33	33	33	33
			24	43	43	43	54	43	43	43	43	43	43	43	54
		550S162	16	33	33	33	33	33	33	33	33	33	33	33	33
			24	33	33	43	43	33	33	33	33	33	33	33	43
90 mph	—	350S162	16	33	33	33	33	33	33	33	33	33	33	33	33
			24	43	43	43	54	43	43	43	43	43	43	43	54
		550S162	16	33	33	33	33	33	33	33	33	33	33	33	33
			24	33	33	43	43	33	33	33	33	33	33	33	43
100 mph	85 mph	350S162	16	33	33	33	33	33	33	33	33	33	33	33	33
			24	43	43	43	54	43	43	43	43	43	43	43	54
		550S162	16	33	33	33	33	33	33	33	33	33	33	33	33
			24	33	33	43	43	33	33	33	33	33	33	33	43
110 mph	90 mph	350S162	16	33	33	33	33	33	33	33	33	33	33	33	43
			24	43	43	43	54	43	43	43	43	43	43	43	54
		550S162	16	33	33	33	33	33	33	33	33	33	33	33	33
			24	33	33	43	43	33	33	33	33	33	33	33	43
—	100 mph	350S162	16	33	33	33	33	33	33	33	33	33	33	33	43
			24	43	43	43	54	43	43	43	43	43	54	54	54
		550S162	16	33	33	33	33	33	33	33	33	33	33	33	33
			24	33	33	43	43	33	33	33	33	33	33	33	43
—	110 mph	350S162	16	33	33	33	43	33	33	33	43	33	43	43	43
			24	43	43	43	54	43	43	43	54	54	54	54	54
		550S162	16	33	33	33	33	33	33	33	33	33	33	33	33
			24	33	33	43	43	33	33	43	43	33	33	43	43

For SI: 1 inch = 25.4 mm, 1 foot = 304.8 mm, 1 mil = 0.0254 mm, 1 mile per hour = 0.447 m/s, 1 pound per square foot = 0.0479 kPa, 1 ksi = 1000 psi = 6.895 MPa.

a. Deflection criterion: $L/240$.
b. Design load assumptions:
   Second floor dead load is 10 psf.
   Second floor live load is 30 psf.
   Roof/ceiling dead load is 12 psf.
   Attic live load is 10 psf.
c. Building width is in the direction of horizontal framing members supported by the wall studs.

❖ See the commentary for Section R603.3.2.

R603.3.4 Cutting and notching. Flanges and lips of steel studs and headers shall not be cut or notched.

❖ Flanges and lips of load-bearing wall framing, to retain standard integrity, must not be cut or notched. The only holes permitted in steel wall framing are those conforming to Section R603.2.

R603.3.5 Hole patching. Web holes violating the requirements in Section R603.2 shall be designed in accordance with one of the following:

1. Framing members shall be replaced or designed in accordance with accepted engineering practices when web holes exceed the following size limits:

   1.1. The depth of the hole, measured across the web, exceeds 70 percent of the flat width of the web; or

   1.2. The length of the hole measured along the web exceeds 10 inches (254 mm) or the depth of the web, whichever is greater.

2. Web holes not exceeding the dimensional requirements in R603.3.5(1) shall be patched with a solid steel plate, stud section, or track section in accordance with Figure R603.3.5. The steel patch shall be as a minimum the same thickness as the receiving member and shall extend at least 1 inch (25 mm) beyond all edges of the hole. The

steel patch shall be fastened to the web of the receiving member with No. 8 screws spaced no more than 1 inch (25 mm) center-to-center along the edges of the patch with a minimum edge distance of $^1/_2$ inch (13 mm).

❖ See the commentary for Section R603.2.

R603.3.6 Splicing. Steel studs and other structural members shall not be spliced. Tracks shall be spliced in accordance with Figure R603.3.6.

❖ Splicing of structural members, such as studs, is not permitted. Splicing of tracks is allowed since they do not serve as gravity load-carrying members. The splice shown in Figure R603.3.6 provides nominal load-carrying capability; however, it is necessary to resist the lateral loads produced by moderate wind and earthquake loads. In these cases, the floor system must act as a diaphragm with the tracks serving as the diaphragm chords. See Section R603.8.1.4 for additional track splicing requirements in areas of high seismicity or high-wind regions.

R603.4 Corner framing. Corner studs and the top tracks shall be installed in accordance with Figure R603.4.

❖ Figure R603.4 illustrates the required configuration of corner studs and overlapping of tracks at corners.

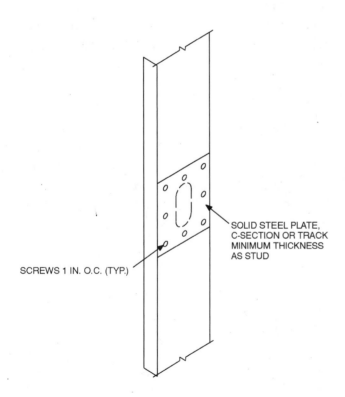

SOLID STEEL PLATE, C-SECTION OR TRACK MINIMUM THICKNESS AS STUD

SCREWS 1 IN. O.C. (TYP.)

For SI: 1 inch = 25.4 mm.

FIGURE R603.3.5
HOLE PATCH

❖ See the commentary for Section R603.3.5.

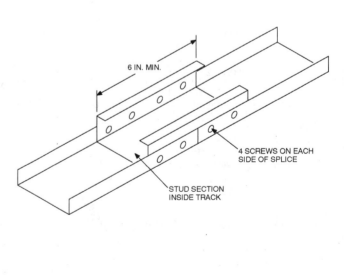

6 IN. MIN.

4 SCREWS ON EACH SIDE OF SPLICE

STUD SECTION INSIDE TRACK

For SI: 1 inch = 25.4 mm.

FIGURE R603.3.6
TRACK SPLICE

❖ See the commentary for Section R603.3.6.

R603.5 Exterior wall covering. The method of attachment of exterior wall covering materials to cold-formed steel stud wall framing shall conform to the manufacturer's installation instructions.

❖ Attachment of exterior wall coverings must be according to the manufacturer's instructions.

R603.6 Headers. Headers shall be installed above wall openings in all exterior walls and interior load-bearing walls in accordance with Figure R603.6 and Tables R603.6(1) through R603.6(8), or shall be designed in accordance with the AISI Standard for Cold-formed Steel Framing–Header Design (COFS/Header Design).

❖ Headers must be located immediately below the ceiling or roof framing (at the top track, for example) above wall openings in all exterior walls and interior load-bearing walls in accordance with Figure R603.6. This figure illustrates the two types of design op-

tions—either back-to-back or as a box beam. Tables R603.6(1) through R603.6(8) specify allowable header spans for various combinations of loading and header sizes. These tables give a very limited range of design options and are very conservative in many design scenarios. Therefore, the code provides the option to use the referenced AISI standard, which will result in a more economical design in many circumstances.

R603.6.1 Jack and king studs, and head track. The number of jack and king studs shall comply with Table R603.6(9). King and jack studs shall be of the same dimension and thickness as the adjacent wall studs. Headers constructed of C-shape framing members shall be connected to king studs in accordance with Table R603.6.(10). One-half the total number of screws shall be applied to the header and one-half to the king stud by use of a minimum 2-inch by 2-inch (51 mm by 51 mm) clip angle or 4-inch-wide (102 mm) steel plate. The clip angle or

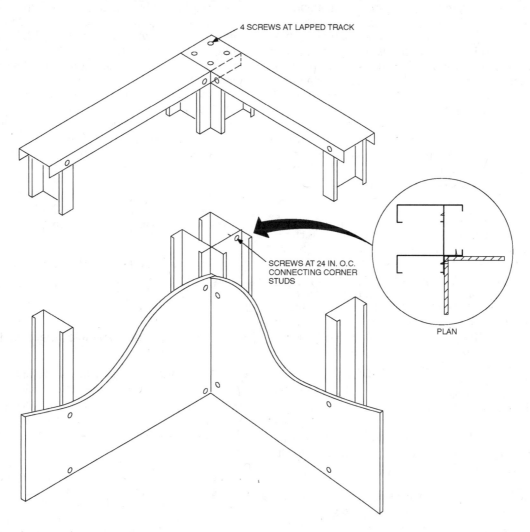

For SI: 1 inch = 25.4 mm.

FIGURE R603.4
CORNER FRAMING

❖ See the commentary to Section R603.4.

plate shall extend the depth of the header minus $^1/_2$ inch (13 mm) and shall have a minimum thickness of the header members or the wall studs, whichever is thicker.

Head track spans shall comply with Table R603.6(11) and shall be in accordance with Figures R603.3 and R603.6. Increasing the head track tabular value shall not be prohibited when in accordance with one of the following:

1. For openings less than 4 feet (1219 mm) in height that have a top and bottom head track, multiply the tabular value by 1.75; or

2. For openings less than 6 feet (1829 mm) in height that have a top and bottom head track, multiply the tabular value by 1.50.

❖ Table R603.6(9) specifies the number of jack and king studs based on the header span. King and jack studs must be the same dimension as adjacent wall studs, with a steel thickness not less than that of the adjacent wall studs. The size and number of screws required for the header-to-king-stud connection are listed in Table R603.6(10). One-half of the screws are to be applied to the header and one-half to the king stud with a clip angle as shown in Figure R603.6. The clip angle must be a minimum 2-inch by 2-inch (51 mm by 51 mm) angle with a minimum steel thickness not less than the header members and the wall studs.

Head track spans must comply with Table R603.6.11 and Figure R603.6. Where both a top and a bottom track are provided, the tabular head track spans may be increased by the multipliers indicated in Items 1 and 2.

R603.7 Structural sheathing. In areas where the basic wind speed is less than 110 miles per hour (49 m/s), wood structural panel sheathing shall be installed on all exterior walls of buildings in accordance with this section. Wood structural panel sheathing shall consist of minimum $^7/_{16}$-inch-thick (11 mm) oriented-strand board or $^{15}/_{32}$-inch-thick (12 mm) plywood and shall be installed on all exterior wall surfaces in accordance with Section R603.7.1 and Figure R603.3. The minimum length of full height sheathing on exterior walls shall be determined in accordance with Table R603.7, but shall not be less than 20 percent of the braced wall length in any case. The minimum percentage of full height sheathing in Table R603.7 shall include only those sheathed wall sections, uninterrupted by openings, which are a minimum of 48 inches (1120 mm) wide. The minimum percentage of full-height structural sheathing shall be multiplied by 1.10 for 9-foot-high (2743 mm) walls and multiplied by 1.20 for 10-foot-high (3048 mm) walls. In addition, structural sheathing shall:

1. Be installed with the long dimension parallel to the stud framing and shall cover the full vertical height of studs, from the bottom of the bottom track to the top of the top track of each story.

2. Be applied to each end (corners) of each of the exterior walls with a minimum 48-inch-wide (1219 mm) panel.

❖ Wood structural panel sheathing must be installed on all exterior wall surfaces as shown in Figure R603.3. Where the basic wind speed is less than 110 miles per hour (49 m/s), and the building is classified as Seismic Design Category A, B or C, wood structural panel sheathing must be installed in accordance with this section.

Wood structural-panel sheathing must be a minimum $^7/_{16}$-inch (11.1 mm) thick oriented strand board or $^{15}/_{32}$-inch (11.9 mm) thick plywood fastened to studs and tracks in accordance with Table R603.3.2.

The minimum length of full-height sheathing required on exterior walls is determined from Table R603.7. The minimum tabulated lengths are expressed as a percentage of the wall length. Because the table is based on an 8-foot (2438 mm) wall height, the lengths must be increased by the appropriate multiplier for 9-foot (2743 mm) and 10-foot (3048 mm) high walls as well as the building aspect ratio found in Note c of the table. Additional adjustments are permitted for buildings with hipped roofs and for hold-downs in accordance with Section R603.7.2. After all adjustments are applied to these tabulated values, the resulting minimum percentage of full-height sheathing must not be less than 20 percent of the wall length.

Only sheathed wall sections that are a minimum of 48 inches (1219 mm) wide and uninterrupted by openings may be considered in meeting the minimum length requirement. This structural sheathing must be installed with the long dimension parallel to the stud framing, and it must cover the full vertical height of studs, from the bottom of the bottom track to the top of the top track of each story. Additionally, structural sheathing must be applied to each end (corners) of all exterior walls with a minimum 48-inch (1219 mm) wide panel.

R603.7.1 Structural sheathing fastening. All edges and interior areas of wood structural panel sheathing shall be fastened to a framing member and tracks in accordance with Table R603.3.2(1).

❖ The wall-fastening schedule specifies the required size and spacing of screws.

R603.7.2 Hold-down requirements. Multiplying the percentage of structural sheathing required in Table R603.7 by 0.6 is permitted where a hold-down anchor with a capacity of 4,300 pounds (19 kN) is provided at each end of exterior walls. Installations of a single hold-down anchor at wall corners is permitted.

❖ This section affords the option of using shorter braced-wall lengths and providing hold-down anchorage at the ends of the braced walls to resist uplift. The length of wall may be reduced by up to 40 percent where the specified hold-down capacities are installed. In accordance with Note "e" of Table R603.7, a reduction to the required full height sheathing is permitted if hold-downs are installed.

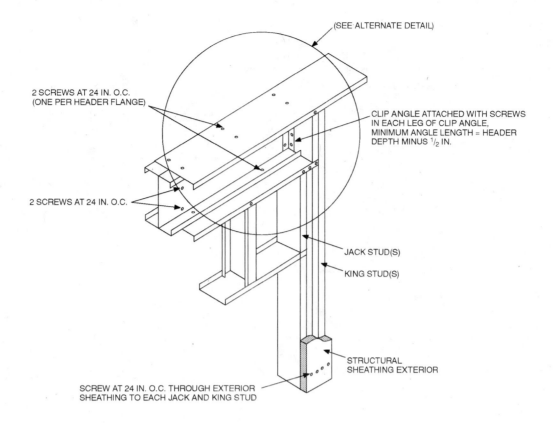

2 SCREWS AT 24 IN. O.C.
(ONE PER HEADER FLANGE)

(SEE ALTERNATE DETAIL)

CLIP ANGLE ATTACHED WITH SCREWS
IN EACH LEG OF CLIP ANGLE,
MINIMUM ANGLE LENGTH = HEADER
DEPTH MINUS $^1/_2$ IN.

2 SCREWS AT 24 IN. O.C.

JACK STUD(S)

KING STUD(S)

STRUCTURAL
SHEATHING EXTERIOR

SCREW AT 24 IN. O.C. THROUGH EXTERIOR
SHEATHING TO EACH JACK AND KING STUD

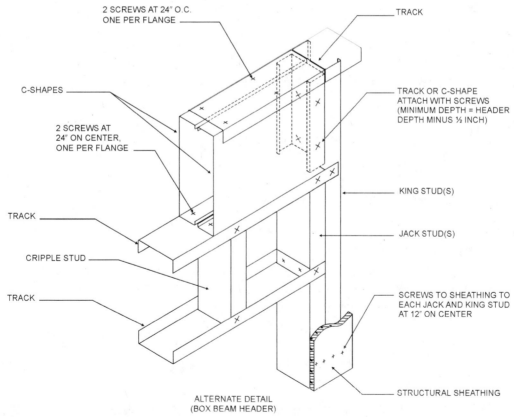

2 SCREWS AT 24" O.C.
ONE PER FLANGE

TRACK

C-SHAPES

TRACK OR C-SHAPE
ATTACH WITH SCREWS
(MINIMUM DEPTH = HEADER
DEPTH MINUS ½ INCH)

2 SCREWS AT
24" ON CENTER,
ONE PER FLANGE

KING STUD(S)

TRACK

JACK STUD(S)

CRIPPLE STUD

TRACK

SCREWS TO SHEATHING TO
EACH JACK AND KING STUD
AT 12" ON CENTER

ALTERNATE DETAIL
(BOX BEAM HEADER)

STRUCTURAL SHEATHING

For SI: 1 inch = 25.4 mm.

FIGURE R603.6
HEADER DETAIL

❖ See the commentary to Section 603.6.

TABLE R603.6(1)
BOX-BEAM HEADER SPANS
Headers supporting roof and ceiling only (33 ksi steel)[a, b, c]

MEMBER DESIGNATION	GROUND SNOW LOAD (20 psf)					GROUND SNOW LOAD (30 psf)				
	Building width[c]					Building width[c]				
	24'	28'	32'	36'	40'	24'	28'	32'	36'	40'
2-350S162-33	3'-10"	3'-5"	3'-0"	2'-6"	2'-2"	3'-3"	2'-9"	2'-4"	—	—
2-350S162-43	5'-1"	4'-8"	4'-4"	4'-0"	3'-7"	4'-6"	4'-2"	3'-8"	3'-4"	2'-11"
2-350S162-54	5'-9"	5'-4"	5'-0"	4'-9"	4'-5"	5'-3"	4'-10"	4'-6"	4'-2"	3'-10"
2-350S162-68	6'-7"	6'-1"	5'-9"	5'-5"	5'-1"	6'-0"	5'-6"	5'-2"	4'-10"	4'-7"
2-350S162-97	8'-0"	7'-5"	7'-0"	6'-6"	6'-3"	7'-3"	6'-9"	6'-4"	6'-0"	5'-7"
2-550S162-33	5'-8"	5'-0"	4'-5"	3'-11"	3'-4"	4'-9"	4'-1"	3'-6"	2'-11"	—
2-550S162-43	7'-2"	6'-8"	6'-3"	5'-8"	5'-2"	6'-6"	5'-11"	5'-3"	4'-9"	4'-3"
2-550S162-54	8'-2"	7'-7"	7'-2"	6'-9"	6'-5"	7'-5"	6'-11"	6'-6"	6'-0"	5'-6"
2-550S162-68	9'-3"	8'-7"	8'-0"	7'-8"	7'-3"	8'-5"	7'-10"	7'-4"	7'-0"	6'-7"
2-550S162-97	11'-2"	10'-5"	9'-10"	9'-3"	8'-11"	10'-2"	9'-6"	9'-1"	8'-5"	8'-0"
2-800S162-33	6'-9"	5'-11"	5'-2"	4'-6"	3'-10"	5'-6"	4'-6"	4'-0"	—	—
2-800S162-43	9'-0"	8'-5"	7'-8"	7'-0"	6'-4"	8'-1"	7'-3"	6'-6"	5'-9"	5'-2"
2-800S162-54	10'-9"	10'-0"	9'-5"	8'-11"	8'-4"	9'-9"	9'-1"	8'-6"	7'-9"	7'-1"
2-800S162-68	12'-2"	11'-4"	10'-8"	10'-2"	9'-7"	11'-1"	10'-4"	9'-9"	9'-3"	8'-9"
2-800S162-97	14'-9"	13'-9"	13'-0"	12'-3"	11'-7"	13'-5"	12'-6"	11'-10"	11'-2"	10'-7"
2-1000S162-43	10'-0"	9'-2"	8'-4"	7'-6"	6'-9"	8'-9"	7'-10"	7'-0"	6'-2"	5'-5"
2-1000S162-54	12'-0"	11'-2"	10'-6"	9'-11"	9'-2"	10'-11"	10'-2"	9'-3"	8'-6"	7'-9"
2-1000S162-68	14'-5"	13'-6"	12'-8"	12'-0"	11'-5"	13'-2"	12'-3"	11'-6"	11'-0"	10'-4"
2-1000S162-97	17'-5"	16'-4"	15'-4"	14'-6"	13'-11"	16'-0"	14'-11"	14'-0"	13'-3"	12'-7"
2-1200S162-43	10'-10"	9'-9"	8'-9"	7'-11"	7'-1"	9'-3"	8'-2"	7'-2"	6'-4"	5'-6"
2-1200S162-54	13'-0"	12'-2"	11'-6"	10'-7"	9'-9"	11'-11"	11'-0"	10'-0"	9'-0"	8'-2"
2-1200S162-68	15'-5"	14'-5"	13'-6"	12'-11"	12'-3"	14'-0"	13'-2"	12'-4"	11'-9"	10'-11"
2-1200S162-97	20'-1"	18'-9"	17'-9"	16'-9"	16'-0"	18'-4"	17'-2"	16'-2"	15'-3"	14'-7"

For SI: 1 inch = 25.4 mm, 1 foot = 304.8 mm, 1 pound per square foot = 0.0479 kPa, 1 pound per square inch = 6.895 kPa.

a. Deflection criteria: $L/360$ for live loads, $L/240$ for total loads.

b. Design load assumptions:
   Roof/Ceiling dead load is 12 psf.
   Attic dead load is 10 psf.

c. Building width is in the direction of horizontal framing members supported by the header.

❖ See the commentary to Section R603.6.

TABLE R603.6(2)
BOX-BEAM HEADER SPANS
Headers supporting roof and ceiling only (33 ksi steel)[a, b, c]

MEMBER DESIGNATION	GROUND SNOW LOAD (50 psf)					GROUND SNOW LOAD (70 psf)				
	Building width[c]					Building width[c]				
	24'	28'	32'	36'	40'	24'	28'	32'	36'	40'
2-350S162-33	—	—	—	—	—	—	—	—	—	—
2-350S162-43	3'-2"	2'-7"	2'-2"	—	—	2'-0"	—	—	—	—
2-350S162-54	4'-1"	3'-6"	3'-1"	2'-8"	2'-3"	3'-0"	2'-6"	—	—	—
2-350S162-68	4'-9"	4'-5"	4'-0"	3'-7"	3'-3"	4'-0"	3'-4"	3'-0"	2'-6"	2'-1"
2-350S162-97	5'-10"	5'-5"	5'-1"	4'-9"	4'-6"	5'-0"	4'-7"	4'-4"	4'-0"	3'-9"
2-550S162-33	2'-9"	—	—	—	—	—	—	—	—	—
2-550S162-43	4'-7"	3'-11"	3'-3"	—	—	3'-2"	—	—	—	—
2-550S162-54	5'-10"	5'-2"	4'-6"	4'-0"	3'-6"	4'-5"	3'-9"	3'-1"	—	—
2-550S162-68	6'-10"	6'-4"	5'-9"	5'-3"	4'-9"	5'-7"	5'-0"	4'-4"	3'-9"	3'-3"
2-550S162-97	8'-4"	7'-9"	7'-3"	6'-10"	6'-6"	7'-2"	6'-8"	6'-3"	5'-11"	5'-7"
2-800S162-33	—	—	—	—	—	—	—	—	—	—
2-800S162-43	5'-7"	4'-9"	3'-11"	—	—	—	—	—	—	—
2-800S162-54	7'-7"	6'-8"	5'-11"	5'-2"	4'-6"	5'-9"	4'-10"	—	—	—
2-800S162-68	9'-1"	8'-4"	7'-6"	6'-10"	6'-3"	7'-4"	6'-6"	5'-9"	5'-0"	4'-4"
2-800S162-97	11'-0"	10'-4"	9'-8"	9'-2"	8'-9"	9'-6"	8'-11"	8'-4"	7'-11"	7'-6"
2-1000S162-43	6'-0"	4'-11"	—	—	—	—	—	—	—	—
2-1000S162-54	8'-4"	7'-4"	6'-4"	5'-7"	4'-9"	6'-3"	5'-2"	—	—	—
2-1000S162-68	10'-9"	9'-9"	8'-10"	8'-0"	7'-3"	8'-7"	7'-7"	6'-7"	5'-9"	5'-0"
2-1000S162-97	13'-1"	12'-3"	11'-6"	10'-11"	10'-4"	11'-4"	10'-7"	10'-0"	9'-5"	8'-11"
2-1200S162-43	6'-1"	—	—	—	—	—	—	—	—	—
2-1200S162-54	8'-9"	7'-8"	6'-7"	5'-9"	—	6'-6"	—	—	—	—
2-1200S162-68	11'-6"	10'-4"	9'-4"	8'-4"	7'-7"	9'-1"	8'-0"	6'-11"	6'-0"	—
2-1200S162-97	15'-1"	14'-1"	13'-3"	12'-7"	12'-0"	13'-1"	12'-3"	11'-6"	11'-0"	10'-2"

For SI: 1 inch = 25.4 mm, 1 foot = 304.8 mm, 1 pound per square foot = 0.0479kPa, 1 pound per square inch = 6.895 kPa.

a. Deflection criteria: $L/360$ for live loads, $L/240$ for total loads.

b. Design load assumptions:
   Roof/Ceiling dead load is 12 psf.
   Attic dead load is 10 psf.

c. Building width is in the direction of horizontal framing members supported by the header

❖ See the commentary to Section R603.6.

TABLE R603.6(3)
BOX-BEAM HEADER SPANS
Headers supporting one floor, roof and ceiling (33 ksi steel)[a, b, c]

MEMBER DESIGNATION	GROUND SNOW LOAD (20 psf)					GROUND SNOW LOAD (30 psf)				
	Building width[c]					Building width[c]				
	24'	28'	32'	36'	40'	24'	28'	32'	36'	40'
2-350S162-33	—	—	—	—	—	—	—	—	—	—
2-350S162-43	2'-6"	—	—	—	—	2'-5"	—	—	—	—
2-350S162-54	3'-6"	3'-0"	2'-6"	—	—	3'-4"	2'-10"	2'-4"	—	—
2-350S162-68	4'-4"	3'-11"	3'-5"	3'-0"	2'-7"	4'-3"	3'-9"	3'-3"	2'-10"	2'-6"
2-350S162-97	5'-4"	5'-0"	4'-7"	4'-4"	4'-1"	5'-4"	4'-11"	4'-6"	4'-3"	4'-0"
2-550S162-33	—	—	—	—	—	—	—	—	—	—
2-550S162-43	3'-9"	3'-0"	—	—	—	3'-7"	2'-11"	—	—	—
2-550S162-54	5'-0"	4'-4"	3'-9"	3'-2"	—	4'-10"	4'-2"	3'-6"	3'-0"	—
2-550S162-68	6'-3"	5'-6"	5'-0"	4'-5"	4'-0"	6'-1"	5'-5"	4'-9"	4'-3"	3'-9"
2-550S162-97	7'-8"	7'-2"	6'-8"	6'-4"	6'-0"	7'-6"	7'-0"	6'-6"	6'-2"	5'-10"
2-800S162-33	—	—	—	—	—	—	—	—	—	—
2-800S162-43	4'-6"	—	—	—	—	4'-4"	—	—	—	—
2-800S162-54	6'-6"	5'-7"	4'-10"	4'-1"	—	6'-4"	5'-5"	4'-7"	—	—
2-800S162-68	8'-2"	7'-3"	6'-6"	5'-10"	5'-2"	8'-0"	7'-0"	6'-4"	5'-6"	5'-0"
2-800S162-97	10'-1"	9'-6"	8'-11"	8'-6"	8'-0"	10'-0"	9'-4"	8'-9"	8'-3"	7'-11"
2-1000S162-43	4'-9"	—	—	—	—	—	—	—	—	—
2-1000S162-54	7'-1"	6'-0"	5'-2"	—	—	6'-10"	5'-10"	4'-11"	—	—
2-1000S162-68	9'-7"	8'-6"	7'-7"	6'-9"	6'-0"	9'-4"	8'-4"	7'-4"	6'-6"	5'-9"
2-1000S162-97	12'-0"	11'-3"	10'-7"	10'-0"	9'-6"	11'-11"	11'-1"	10'-5"	9'-11"	9'-5"
2-1200S162-43	—	—	—	—	—	—	—	—	—	—
2-1200S162-54	7'-6"	6'-4"	—	—	—	7'-2"	6'-0"	—	—	—
2-1200S162-68	10'-1"	9'-0"	8'-0"	7'-0"	6'-2"	9'-11"	8'-9"	7'-9"	6'-9"	6'-0"
2-1200S162-97	14'-0"	13'-0"	12'-3"	11'-7"	11'-0"	13'-9"	12'-10"	12'-0"	11'-6"	10'-11"

For SI: 1 inch = 25.4 mm, 1 foot = 304.8 mm, 1 pound per square foot = 0.0479kPa, 1 pound per square inch = 6.895 kPa.
a. Deflection criteria: $L/360$ for live loads, $L/240$ for total loads.
b. Design load assumptions:
   Roof/Ceiling dead load is 12 psf.
   Attic dead load is 10 psf.
c. Building width is in the direction of horizontal framing members supported by the header.

❖ See the commentary to Section R603.6.

TABLE R603.6(4)
BOX-BEAM HEADER SPANS
Headers supporting one floor, roof and ceiling (33 ksi steel)[a, b, c]

MEMBER DESIGNATION	GROUND SNOW LOAD (50 psf)					GROUND SNOW LOAD (70 psf)				
	Building width[c]					Building width[c]				
	24'	28'	32'	36'	40'	24'	28'	32'	36'	40'
2-350S162-33	—	—	—	—	—	—	—	—	—	—
2-350S162-43	—	—	—	—	—	—	—	—	—	—
2-350S162-54	2'-6"	2'-1"	—	—	—	—	—	—	—	—
2-350S162-68	3'-6"	3'-0"	2'-6"	2'-2"	—	2'-9"	2'-2"	—	—	—
2-350S162-97	4'-9"	4'-5"	4'-1"	3'-10"	3'-7"	4'-2"	3'-11"	3'-7"	3'-4"	2'-11"
2-550S162-33	—	—	—	—	—	—	—	—	—	—
2-550S162-43	—	—	—	—	—	—	—	—	—	—
2-550S162-54	3'-11"	3'-4"	—	—	—	2'-10"	—	—	—	—
2-550S162-68	5'-2"	4'-6"	3'-11"	3'-4"	2'-10"	4'-1"	3'-5"	2'-9"	—	—
2-550S162-97	6'-10"	6'-4"	6'-0"	5'-7"	5'-4"	6'-1"	5'-7"	5'-4"	4'-9"	4'-4"
2-800S162-33	—	—	—	—	—	—	—	—	—	—
2-800S162-43	—	—	—	—	—	—	—	—	—	—
2-800S162-54	5'-1"	4'-2"	—	—	—	—	—	—	—	—
2-800S162-68	6'-9"	6'-1"	5'-2"	4'-5"	—	5'-5"	4'-6"	—	—	—
2-800S162-97	9'-1"	8'-6"	8'-0"	7'-6"	7'-1"	8'-2"	7'-7"	7'-0"	6'-5"	5'-10"
2-1000S162-43	—	—	—	—	—	—	—	—	—	—
2-1000S162-54	5'-6"	—	—	—	—	—	—	—	—	—
2-1000S162-68	7'-10"	6'-11"	6'-0"	5'-2"	—	6'-4"	5'-4"	—	—	—
2-1000S162-97	10'-10"	10'-1"	9'-6"	9'-0"	8'-4"	9'-9"	9'-2"	8'-4"	7'-7"	7'-0"
2-1200S162-43	—	—	—	—	—	—	—	—	—	—
2-1200S162-54	5'-7"	—	—	—	—	—	—	—	—	—
2-1200S162-68	8'-4"	7'-2"	6'-2"	—	—	6'-6"	—	—	—	—
2-1200S162-97	12'-6"	11'-8"	11'-0"	10'-4"	9'-6"	11'-3"	10'-6"	9'-6"	8'-8"	8'-0"

For SI: 1 inch = 25.4 mm, 1 foot = 304.8 mm, 1 pound per square foot = 0.0479 kPa, 1 pound per square inch = 6.895 kPa.

a. Deflection criteria: $L/360$ for live loads, $L/240$ for total loads.

b. Design load assumptions:
  Roof/Ceiling dead load is 12 psf.
  Attic dead load is 10 psf.

c. Building width is in the direction of horizontal framing members supported by the header.

❖ See the commentary to Section R603.6.

TABLE R603.6(5)
BACK-TO-BACK HEADER SPANS
Headers supporting roof and ceiling only (33 ksi steel)[a, b, c]

MEMBER DESIGNATION	GROUND SNOW LOAD (20 psf) Building width[c]					GROUND SNOW LOAD (30 psf) Building width[c]				
	24'	28'	32'	36'	40'	24'	28'	32'	36'	40'
2-350S162-33	3'-7"	3'-1"	2'-8"	2'-4"	—	2'-11"	2'-6"	—	—	—
2-350S162-43	5'-0"	4'-8"	4'-4"	3'-10"	3'-7"	4'-6"	4'-0"	3'-8"	3'-4"	2'-11"
2-350S162-54	5'-9"	5'-5"	5'-0"	4'-9"	4'-6"	5'-3"	4'-10"	4'-6"	4'-4"	3'-11"
2-350S162-68	6'-7"	6'-2"	5'-9"	5'-5"	5'-2"	5'-11"	5'-7"	5'-2"	4'-10"	4'-7"
2-350S162-97	7'-11"	7'-6"	6'-11"	6'-7"	6'-6"	7'-4"	6'-9"	6'-4"	5'-11"	5'-8"
2-550S162-33	5'-5"	4'-9"	4'-4"	3'-9"	3'-5"	4'-7"	3'-11"	3'-5"	2'-11"	—
2-550S162-43	7'-3"	6'-8"	6'-2"	5'-8"	5'-4"	6'-6"	5'-10"	5'-5"	4'-10"	4'-6"
2-550S162-54	8'-2"	7'-8"	7'-2"	6'-9"	6'-5"	7'-5"	6'-10"	6'-6"	6'-1"	5'-9"
2-550S162-68	9'-4"	8'-8"	8'-7"	7'-8"	7'-4"	8'-6"	7'-10"	7'-5"	6'-11"	6'-7"
2-550S162-97	11'-3"	10'-6"	9'-11"	9'-4"	8'-10"	10'-3"	9'-6"	8'-11"	8'-6"	8'-0"
2-800S162-33	6'-9"	5'-11"	5'-5"	4'-9"	4'-4"	6'-9"	5'-0"	4'-5"	3'-9"	—
2-800S162-43	9'-1"	8'-6"	7'-9"	7'-3"	6'-8"	8'-3"	7'-6"	6'-9"	6'-3"	5'-8"
2-800S162-54	10'-9"	10'-1"	9'-6"	8'-11"	8'-6"	9'-9"	9'-2"	8'-7"	8'-2"	7'-8"
2-800S162-68	12'-3"	11'-5"	10'-9"	10'-2"	9'-8"	11'-2"	10'-5"	9'-9"	9'-4"	8'-9"
2-800S162-97	14'-9"	13'-9"	13'-0"	12'-4"	11'-8"	13'-6"	12'-7"	11'-10"	11'-2"	10'-8"
2-1000S162-43	10'-1"	9'-5"	8'-8"	8'-0"	7'-6"	9'-1"	8'-4"	7'-7"	6'-11"	6'-5"
2-1000S162-54	12'-0"	11'-5"	10'-7"	10'-0"	9'-6"	11'-0"	10'-3"	9'-7"	9'-1"	8'-6"
2-1000S162-68	14'-6"	13'-6"	12'-8"	12'-0"	11'-6"	13'-2"	12'-4"	11'-7"	10'-11"	10'-6"
2-1000S162-97	17'-6"	16'-5"	15'-5"	14'-7"	13'-10"	16'-0"	14'-10"	14'-0"	13'-4"	12'-8"
2-1200S162-43	11'-0"	10'-4"	9'-6"	8'-9"	8'-2"	10'-0"	9'-1"	8'-4"	7'-7"	7'-0"
2-1200S162-54	13'-1"	12'-3"	11'-6"	10'-10"	10'-5"	11'-10"	11'-1"	10'-6"	9'-10"	9'-4"
2-1200S162-68	15'-6"	14'-6"	13'-7"	12'-10"	12'-3"	14'-1"	13'-2"	12'-5"	11'-9"	11'-2"
2-1200S162-97	20'-2"	18'-9"	17'-9"	16'-9"	16'-0"	18'-4"	17'-2"	16'-2"	15'-5"	14'-7"

For SI: 1 inch = 25.4 mm, 1 foot = 304.8 mm, 1 pound per square foot = 0.0479kPa, 1 pound per square inch = 6.895 kPa.
a. Deflection criteria: $L/360$ for live loads, $L/240$ for total loads.
b. Design load assumptions:
    Roof/Ceiling dead load is 12 psf.
    Attic dead load is 10 psf.
c. Building width is in the direction of horizontal framing members supported by the header.

❖ See the commentary to Section R603.6.

TABLE R603.6(6)
BACK-TO-BACK HEADER SPANS
Headers supporting roof and ceiling only (33 ksi steel)[a, b, c]

| MEMBER DESIGNATION | GROUND SNOW LOAD (50 psf) | | | | | GROUND SNOW LOAD (70 psf) | | | | |
| | Building width[c] | | | | | Building width[c] | | | | |
	24'	28'	32'	36'	40'	24'	28'	32'	36'	40'
2-350S162-33	—	—	—	—	—	—	—	—	—	—
2-350S162-43	3'-2"	2'-8"	2'-4"	—	—	2'-3"	—	—	—	—
2-350S162-54	4'-3"	3'-8"	3'-5"	2'-11"	2'-8"	3'-4"	2'-9"	2'-5"	2'-0"	1'-7"
2-350S162-68	4'-9"	4'-6"	4'-2"	3'-10"	3'-7"	4'-1"	3'-9"	3'-5"	3'-1"	2'-8"
2-350S162-97	5'-10"	5'-6"	5'-2"	4'-10"	4'-7"	5'-0"	4'-8"	4'-5"	4'-2"	3'-10"
2-550S162-33	2'-9"	—	—	—	—	—	—	—	—	—
2-550S162-43	4'-9"	4'-2"	3'-8"	3'-3"	2'-8"	3'-6"	2'-10"	—	—	—
2-550S162-54	6'-0"	5'-6"	4'-11"	4'-6"	4'-1"	4'-10"	4'-4"	3'-10"	3'-6"	2'-11"
2-550S162-68	6'-10"	6'-5"	5'-11"	5'-8"	5'-4"	5'-11"	5'-6"	4'-11"	4'-7"	4'-3"
2-550S162-97	8'-5"	7'-7"	4'-4"	6'-10"	6'-7"	7'-3"	6'-8"	6'-4"	5'-11"	5'-8"
2-800S162-33	3'-8"	—	—	—	—	—	—	—	—	—
2-800S162-43	6'-1"	5'-5"	4'-9"	4'-2"	3'-8"	4'-7"	3'-9"	—	—	—
2-800S162-54	8'-0"	7'-4"	6'-8"	6'-1"	5'-7"	6'-6"	5'-8"	5'-3"	4'-8"	4'-3"
2-800S162-68	9'-1"	8'-6"	7'-11"	7'-7"	7'-1"	7'-10"	7'-5"	6'-9"	6'-3"	5'-9"
2-800S162-97	11'-1"	10'-4"	9'-8"	9'-2"	8'-9"	9'-8"	8'-11"	8'-5"	7'-11"	7'-7"
2-1000S162-43	6'-9"	6'-0"	5'-5"	4'-9"	4'-2"	5'-2"	4'-5"	—	—	—
2-1000S162-54	8'-11"	8'-2"	7'-6"	6'-10"	6'-6"	7'-4"	6'-7"	5'-10"	5'-4"	4'-9"
2-1000S162-68	10'-9"	10'-1"	9'-6"	8'-11"	8'-6"	9'-5"	8'-9"	8'-1"	7'-6"	6'-10"
2-1000S162-97	13'-1"	12'-4"	11'-6"	10'-10"	10'-5"	11'-5"	10'-7"	9'-11"	9'-6"	8'-11"
2-1200S162-43	7'-6"	6'-7"	5'-10"	5'-2"	4'-7"	5'-8"	4'-10"	—	—	—
2-1200S162-54	9'-9"	8'-10"	8'-1"	7'-6"	6'-10"	7'-11"	7'-2"	6'-6"	5'-9"	5'-3"
2-1200S162-68	11'-7"	10'-9"	10'-2"	9'-7"	9'-1"	10'-2"	9'-6"	8'-7"	7'-11"	7'-5"
2-1200S162-97	15'-1"	14'-1"	13'-4"	12'-7"	12'-0"	13'-2"	12'-4"	11'-7"	10'-11"	10'-6"

For SI: 1 inch = 25.4 mm, 1 foot = 304.8 mm, 1 pound per square foot = 0.0479 kPa, 1 pound per square inch = 6.895 kPa.

a. Deflection criteria: $L/360$ for live loads, $L/240$ for total loads.

b. Design load assumptions:
   Roof/Ceiling dead load is 12 psf.
   Attic dead load is 10 psf.

c. Building width is in the direction of horizontal framing members supported by the header.

❖ See the commentary to Section R603.6.

TABLE R603.6(7)
BACK-TO-BACK HEADER SPANS
Headers supporting one floor, roof and ceiling (33 ksi steel)[a, b, c]

MEMBER DESIGNATION	GROUND SNOW LOAD (20 psf)					GROUND SNOW LOAD (30 psf)				
	Building width[c]					Building width[c]				
	24'	28'	32'	36'	40'	24'	28'	32'	36'	40'
2-350S162-33	—	—	—	—	—	—	—	—	—	—
2-350S162-43	2'-7"	2'-1"	—	—	—	2'-6"	—	—	—	—
2-350S162-54	3'-8"	3'-3"	2'-9"	2'-6"	2'-1"	3'-7"	3'-1"	2'-8"	2'-5"	1'-11"
2-350S162-68	4'-5"	4'-0"	3'-9"	3'-6"	3'-1"	4'-4"	3'-11"	3'-8"	3'-5"	3'-0"
2-350S162-97	5'-5"	4'-11"	4'-8"	4'-5"	4'-1"	5'-4"	4'-10"	4'-7"	4'-4"	4'-0"
2-550S162-33	—	—	—	—	—	—	—	—	—	—
2-550S162-43	4'-1"	3'-6"	2'-10"	2'-5"	—	3'-11"	3'-5"	2'-9"	—	—
2-550S162-54	5'-5"	4'-9"	4'-4"	3'-10"	3'-6"	5'-3"	4'-8"	4'-3"	3'-9"	3'-5"
2-550S162-68	6'-6"	5'-10"	5'-6"	5'-1"	4'-8"	6'-2"	5'-9"	5'-5"	4'-11"	4'-7"
2-550S162-97	7'-8"	7'-2"	6'-8"	6'-5"	5'-11"	7'-7"	7'-0"	6'-7"	6'-3"	5'-10"
2-800S162-33	—	—	—	—	—	—	—	—	—	—
2-800S162-43	5'-4"	4'-7"	3'-10"	3'-4"	—	5'-1"	4'-5"	3'-9"	—	—
2-800S162-54	7'-3"	6'-6"	5'-10"	5'-4"	4'-9"	6'-11"	6'-4"	5'-8"	5'-3"	4'-8"
2-800S162-68	8'-5"	7'-9"	7'-5"	6'-9"	6'-5"	8'-4"	7'-8"	7'-3"	6'-8"	6'-3"
2-800S162-97	10'-2"	9'-6"	8'-11"	8'-6"	8'-1"	10'-0"	9'-5"	8'-9"	8'-5"	7'-10"
2-1000S162-43	5'-9"	5'-1"	4'-5"	3'-9"	—"	5'-8"	4'-11"	4'-4"	—	—
2-1000S162-54	7'-11"	7'-3"	6'-7"	5'-11"	5'-5"	7'-9"	7'-1"	6'-5"	5'-9"	5'-4"
2-1000S162-68	9'-11"	9'-4"	8'-9"	8'-1"	7'-7"	9'-10"	9'-2"	8'-7"	7'-11"	7'-5"
2-1000S162-97	12'-1"	11'-4"	10'-8"	10'-0"	9'-7"	11'-11"	11'-1"	10'-6"	9'-10"	9'-6"
2-1200S162-43	6'-6"	5'-8"	4'-10"	4'-2"	—	6'-4"	5'-6"	4'-8"	—	—
2-1200S162-54	8'-8"	7'-10"	7'-2"	6'-6"	5'-11"	8'-6"	7'-8"	6'-11"	6'-5"	5'-9"
2-1200S162-68	10'-8"	9'-11"	9'-5"	8'-8"	8'-1"	10'-6"	9'-9"	9'-3"	8'-6"	7'-11"
2-1200S162-97	13'-11"	13'-0"	12'-4"	11'-7"	11'-1"	13'-9"	12'-10"	12'-1"	11'-6"	10'-10"

For SI: 1 inch = 25.4 mm, 1 foot = 304.8 mm, 1 pound per square foot = 0.0479kPa, 1 pound per square inch = 6.895 kPa.

a. Deflection criteria: $L/360$ for live loads, $L/240$ for total loads.

b. Design load assumptions:
   Second floor dead load is 10 psf.
   Roof/Ceiling dead load is 12 psf.
   Second floor live load is 30 psf.
   Roof/ceiling load is 12 psf.
   Attic dead load is 10 psf.

c. Building width is in the direction of horizontal framing members supported by the header.

❖ See the commentary to Section R603.6.

TABLE R603.6(8)
BACK-TO-BACK HEADER SPANS
Headers supporting one floor, roof and ceiling (33 ksi steel)[a, b, c]

MEMBER DESIGNATION	GROUND SNOW LOAD (50 psf)					GROUND SNOW LOAD (70 psf)				
	Building width[c]					Building width[c]				
	24'	28'	32'	36'	40'	24'	28'	32'	36'	40'
2-350S162-33	—	—	—	—	—	—	—	—	—	—
2-350S162-43	—	—	—	—	—	—	—	—	—	—
2-350S162-54	2'-11"	2'-6"	2'-2"	—	—	2'-4"	—	—	—	—
2-350S162-68	3'-10"	3'-6"	3'-2"	2'-9"	2'-6"	3'-4"	2'-10"	2'-6"	2'-3"	1'-10"
2-350S162-97	4'-9"	4'-5"	4'-2"	3'-10"	3'-8"	4'-3"	3'-10"	3'-8"	3'-5"	3'-2"
2-550S162-33	—	—	—	—	—	—	—	—	—	—
2-550S162-43	3'-1"	2'-5"	—	—	—	—	—	—	—	—
2-550S162-54	4'-6"	3'-10"	3'-6"	3'-0"	2'-7"	3'-8"	3'-1"	2'-7"	2'-1"	—
2-550S162-68	5'-7"	5'-1"	4'-8"	4'-4"	3'-11"	4'-10"	4'-5"	3'-10"	3'-6"	3'-3"
2-550S162-97	6'-10"	6'-5"	5'-11"	5'-8"	5'-5"	6'-2"	5'-8"	5'-5"	3'-1"	4'-9"
2-800S162-33	—	—	—	—	—	—	—	—	—	—
2-800S162-43	4'-1"	3'-5"	—	—	—	—	—	—	—	—
2-800S162-54	6'-0"	5'-5"	4'-9"	4'-4"	3'-9"	4'-11"	4'-5"	3'-9"	3'-2"	—
2-800S162-68	7'-6"	6'-10"	6'-5"	5'-10"	5'-5"	6'-7"	5'-10"	5'-5"	4'-10"	4'-6"
2-800S162-97	9'-1"	8'-6"	8'-0"	7'-7"	7'-4"	8'-3"	7'-8"	7'-3"	6'-9"	6'-6"
2-1000S162-43	4'-8"	3'-10"	—	—	—	—	—	—	—	—
2-1000S162-54	6'-9"	6'-0"	5'-5"	4'-10"	4'-4"	5'-8"	4'-10"	4'-4"	3'-8"	—
2-1000S162-68	8'-10"	8'-3"	7'-7"	6'-11"	6'-6"	7'-9"	7'-1"	6'-6"	5'-10"	5'-5"
2-1000S162-97	10'-10"	10'-3"	9'-7"	9'-1"	8'-8"	9'-9"	9'-3"	8'-7"	8'-3"	7'-9"
2-1200S162-43	5'-1"	4'-4[2]	—	—	—	—	—	—	—	—
2-1200S162-54	7'-5"	6'-6"	5'-10"	5'-4"	4'-9"	6'-3"	5'-5"	4'-8"	4'-1"	—
2-1200S162-68	9'-7"	8'-9"	8'-1"	7'-6"	6'-11"	8'-8"	7'-7"	6'-11"	6'-4"	5'-9"
2-1200S162-97	12'-6"	11'-8"	11'-1"	10'-6"	9'-11"	11'-4"	10'-7"	10'-0"	9'-6"	9'-0"

For SI: 1 inch = 25.4 mm, 1 foot = 304.8 mm, 1 pound per square foot = 0.0479kPa, 1 pound per square inch = 6.895 kPa.

a. Deflection criteria: $L/360$ for live loads, $L/240$ for total loads.

b. Design load assumptions:
> Second floor dead load is 10 psf.
> Roof/Ceiling dead load is 12 psf.
> Second floor live load is 30 psf.
> Roof/ceiling dead load is 12 psf.
> Attic dead load is 10 psf.

c. Building width is in the direction of horizontal framing members supported by the header.

❖ See the commentary to Section R603.6.

TABLE R603.6(9)
TOTAL NUMBER OF JACK AND KING STUDS REQUIRED AT EACH END OF AN OPENING

SIZE OF OPENING (feet-inches)	24″ O.C. STUD SPACING		16″ O.C. STUD SPACING	
	No. of jack studs	No. of king studs	No. of jack studs	No. of king studs
Up to 3′-6″	1	1	1	1
> 3′-6″ to 5′-0″	1	2	1	2
> 5′-0″ to 5′-6″	1	2	2	2
> 5′-6″ to 8′-0″	1	2	2	2
> 8′-0″ to 10′-6″	2	2	2	3
> 10′-6″ to 12′-0″	2	2	3	3
> 12′-0″ to 13′-0″	2	3	3	3
> 13′-0″ to 14′-0″	2	3	3	4
> 14′-0″ to 16′-0″	2	3	3	4
> 16′-0″ to 18′-0″	3	3	4	4

For SI: 1 inch = 25.4 mm, 1 foot = 304.8 mm.

❖ See the commentary to Section R603.6.1.

TABLE R603.6(10)
HEADER TO KING STUD CONNECTION REQUIREMENTS[a, b, c, d]

HEADER SPAN (feet)	BASIC WIND SPEED (mph), EXPOSURE		
	85 A/B or Seismic Design Categories A, B, C, D_0, D_1 and D_2	85 C or less than 110 A/B	Less than 110 C
≤ 4′	4-No. 8 screws	4-No. 8 screws	6-No. 8 screws
> 4′ to 8′	4-No. 8 screws	4-No. 8 screws	8-No. 8 screws
> 8′ to 12′	4-No. 8 screws	6-No. 8 screws	10-No. 8 screws
> 12′to 16′	4-No. 8 screws	8-No. 8 screws	12-No. 8 screws

For SI: 1 inch = 25.4 mm, 1 foot = 304.8 mm, 1 mile per hour = 0.447 m/s, 1 pound = 4.448 N.

a. All screw sizes shown are minimum.

b. For headers located on the first floor of a two-story building, the total number of screws may be reduced by two screws, but the total number of screws shall be no less than four.

c. For roof slopes of 6:12 or greater, the required number of screws may be reduced by half, but the total number of screws shall be no less than four.

d. Screws can be replaced by an uplift connector which has a capacity of the number of screws multiplied by 164 pounds (e.g., 12-No. 8 screws can be replaced by an uplift connector whose capacity exceeds 12 × 164 pounds = 1,968 pounds).

❖ See the commentary to Section R603.6.1.

TABLE R603.6(11)
HEAD TRACK SPAN (33 ksi Steel)

BASIC WIND SPEED (mph)		ALLOWABLE HEAD TRACK SPAN[a, b] (ft-in)					
Exposure		Track Designation					
A/B	C	350T125-33	350T125-43	350T125-54	550T125-33	550T125-43	550T125-54
85		5'-0"	5'-7"	6'-2"	5'-10"	6'-8"	7'-0"
90		4'-10"	5'-5"	6'-0"	5'-8"	6'-3"	6'-10"
100	85	4'-6"	5'-1"	5'-8"	5'-4"	5'-11"	6'-5"
110	90	4'-2"	4'-9"	5'-4"	5'-1"	5'-7"	6'-1"
	100	3'-11"	4'-6"	5'-0"	4'-10"	5'-4"	5'-10"
	110	3'-8"	4'-2"	4'-9"	4'-1"	5'-1"	5'-7"

For SI: 1 inch = 25.4 mm, 1 foot = 304.8 mm, 1 mile per hour = 0.447 m/s.

a. Deflection Limit: $L/240$

b. Head track spans are based on components and cladding wind speeds and a 49-inch tributary span.

❖ See the commentary to Section R603.6.1.

TABLE R603.7
MINIMUM PERCENTAGE OF FULL HEIGHT STRUCTURAL SHEATHING ON EXTERIOR WALLS[a, b, c, d, e]

WALL SUPPORTING	ROOF SLOPE	WIND SPEED (mph) AND EXPOSURE				
		85 A/B	100 A/B	110 A/B or 85 C	100 C	110 C
Roof and ceiling only	3:12	8	9	12	16	20
	6:12	12	15	20	26	35
	9:12	21	25	30	50	58
	12:12	30	35	40	66	75
One story, roof and ceiling	3:12	24	30	35	50	66
	6:12	25	30	40	58	74
	9:12	35	40	55	74	91
	12:12	40	50	65	100	115

For SI: 1 mile per hour = 0.447 m/s.

a. Linear interpolation shall be permitted.

b. Bracing amount shall not be less than 20 percent of the wall length after all applicable adjustments are made.

c. Minimum percentages are based on a building aspect ratio of 1:1. Minimum percentages for the shorter walls of a building shall be multiplied by a factor of 1.5 and 2.0 for building aspect ratios of 1.5:1 and 2:1 respectively.

d. For hip roofed homes with continuous structural sheathing, the amount of bracing shall be permitted to be multiplied by a factor of 0.95 for roof slopes not exceeding 7:12 and a factor of 0.9 for roof slopes greater than 7:12.

e. Sheathing percentages are permitted to be reduced in accordance with Section R603.7.2.

❖ See the commentary to Section R603.7.

SECTION R604
WOOD STRUCTURAL PANELS

R604.1 Identification and grade. Wood structural panels shall conform to DOC PS 1 or DOC PS 2. All panels shall be identified by a grade mark or certificate of inspection issued by an approved agency.

❖ To verify the acceptability of wood structural panel products, this section requires labeling in accordance with the referenced standards. The purpose is the same as the purpose of Section R602.1 for dimension lumber. This label provides the inspector with the necessary information to determine the acceptability of the wood structural panel (see commentary, Sections R503.2 and R503.2.1).

R604.2 Allowable spans. The maximum allowable spans for wood structural panel wall sheathing shall not exceed the values set forth in Table R602.3(3).

❖ The allowable span for wood structural panel sheathing is a function of its thickness and grade. Table R602.3(3) of the code provides allowable spans based on the thickness and the panel index.

R604.3 Installation. Wood structural panel wall sheathing shall be attached to framing in accordance with Table R602.3(1). Wood structural panels marked Exposure 1 or Exterior are considered water-repellent sheathing under the code.

❖ Table R602.3(1) of the code contains the fastening requirements for proper wood structural panel installation.

SECTION R605
PARTICLEBOARD

R605.1 Identification and grade. Particleboard shall conform to ANSI A208.1 and shall be so identified by a grade mark or certificate of inspection issued by an approved agency. Particleboard shall comply with the grades specified in Table R602.3(4).

❖ As with wood structural panels, particleboard must be labeled in accordance with the referenced standard. Table R602.3.(4) provides prescriptive requirements relative to allowable spans. As with wood structural panel spans, these spans are a function of the thickness and grade (see commentary, Sections R503.3 and R603.3.1).

SECTION R606
GENERAL MASONRY CONSTRUCTION

R606.1 General. Masonry construction shall be designed and constructed in accordance with the provisions of this section or in accordance with the provisions of ACI 530/ASCE 5/TMS 402.

❖ Masonry wall construction under the code must comply with this section or ACI 530/ASCE 5/TMS 402, *Building Code Requirements for Masonry Structures.*

R606.1.1 Professional registration not required. When the empirical design provisions of ACI 530/ASCE 5/TMS 402 Chapter 5 or the provisions of this section are used to design masonry, project drawings, typical details and specifications are not required to bear the seal of the architect or engineer responsible for design, unless otherwise required by the state law of the jurisdiction having authority.

❖ These empirical provisions are meant for use without the services of a professional engineer or architect. State law, however, may dictate otherwise and would take precedence.

R606.2 Thickness of masonry. The nominal thickness of masonry walls shall conform to the requirements of Section R606.2.1 through R606.2.4.

❖ This section establishes minimum thicknesses for various masonry wall constructions.

R606.2.1 Minimum thickness. The minimum thickness of masonry bearing walls more than one story high shall be 8 inches (203 mm). Solid masonry walls of one-story dwellings and garages shall not be less than 6 inches (152 mm) in thickness when not greater than 9 feet (2743 mm) in height, provided that when gable construction is used, an additional 6 feet (1829 mm) is permitted to the peak of the gable. Masonry walls shall be laterally supported in either the horizontal or vertical direction at intervals as required by Section R606.9.

❖ The minimum thickness of bearing walls and exterior nonbearing walls of masonry construction more than one story high is limited to a nominal 8 inches (203 mm), which means a net thickness of not less than $7^1/_2$ inches (191 mm). A lesser thickness of 6-inch (152 mm) nominal or $5^1/_2$-inch (140 mm) net of solid masonry is permitted if the wall does not exceed 9 feet (2743 mm) in height other than at the peak of a gable, where such height may extend to 15 feet (4572 mm) as shown in Commentary Figure R606.2.1.

R606.2.2 Rubble stone masonry wall. The minimum thickness of rough, random or coursed rubble stone masonry walls shall be 16 inches (406 mm).

❖ Rubble stone walls must be at least 16 inches (406 mm) thick (see commentary, Section R404.1.8).

R606.2.3 Change in thickness. Where walls of masonry of hollow units or masonry-bonded hollow walls are decreased in thickness, a course of solid masonry shall be constructed between the wall below and the thinner wall above, or special units or construction shall be used to transmit the loads from face shells or wythes above to those below.

❖ Where the thickness of hollow masonry wall construction is decreased, a solid masonry course provides bearing for the face shells of the thinner hollow unit above.

R606.2.4 Parapet walls. Unreinforced solid masonry parapet walls shall not be less than 8 inches (203 mm) thick and their height shall not exceed four times their thickness. Unreinforced hollow unit masonry parapet walls shall be not less than 8 inches (203 mm) thick, and their height shall not exceed three times their thickness. Masonry parapet walls in areas subject to wind loads of 30 pounds per square foot (1.44 kPa) located in Seismic Design Category D_0, D_1 or D_2, or on townhouses in Seismic Design Category C shall be reinforced in accordance with Section R606.12.

❖ A parapet is an extension of a wall that extends above the roof line. Generally, parapet walls are required to comply only with fire-resistance requirements of the code. Parapet walls are required to have a certain minimum thickness with height limitations based on the thickness and reinforcement in the wall.

R606.3 Corbeled masonry. Solid masonry units shall be used for corbeling. The maximum corbeled projection beyond the face of the wall shall not be more than one-half of the wall thickness or one-half the wythe thickness for hollow walls; the maximum projection of one unit shall not exceed one-half the height of the unit or one-third the thickness at right angles to the wall. When corbeled masonry is used to support floor or roof-framing members, the top course of the corbel shall be a header course or the top course bed joint shall have ties to the vertical wall. The hollow space behind the corbeled masonry shall be filled with mortar or grout.

❖ Corbeling of masonry walls is permitted within certain limitations. The maximum projection for each unit is limited to one-third the unit bed depth (D) or one-half the unit height (H), whichever is less, as shown in Commentary Figure R606.3(1). The total horizontal projection of the corbeled courses is limited to one-half the thickness of a solid wall or one-half the thickness of a wythe of a cavity wall as shown in Commentary Figure R606.3(2).

R606.4 Support conditions. Bearing and support conditions shall be in accordance with Sections R606.4.1 and R606.4.2.

❖ Masonry bearing and support conditions must be in accordance with the subsections of Section R606.4.

R606.4.1 Bearing on support. Each masonry wythe shall be supported by at least two-thirds of the wythe thickness.

❖ At least two thirds the thickness of the bottom course of all masonry wythes must bear directly on the supporting construction (i.e., bear on the supporting foundation wall, lintel or header, etc.).

R606.4.2 Support at foundation. Cavity wall or masonry veneer construction may be supported on an 8-inch (203 mm) foundation wall, provided the 8-inch (203 mm) wall is corbeled with solid masonry to the width of the wall system above. The total horizontal projection of the corbel shall not exceed 2 inches (51 mm) with individual corbels projecting not more than one-third the thickness of the unit or one-half the height of the unit.

❖ Foundations supporting masonry cavity walls must be at least 8 inches thick and, where the wall supported is thicker, the foundation wall must be corbelled to attain a thickness at least equal to that of the wall supported. Corbeling limitations of this section are shown in commentary Figure R606.4.2.

R606.5 Allowable stresses. Allowable compressive stresses in masonry shall not exceed the values prescribed in Table R606.5. In determining the stresses in masonry, the effects of all loads and conditions of loading and the influence of all forces affecting the design and strength of the several parts shall be taken into account.

❖ Masonry compressive stresses are determined from the type and compressive strength of the unit and the type of mortar used in construction. Table R606.5 lists the maximum permitted stresses. Stresses must be computed based on the actual dimensions of the masonry.

Example:

How does a Type S versus Type N mortar affect the allowable compressive stress in unit masonry for a

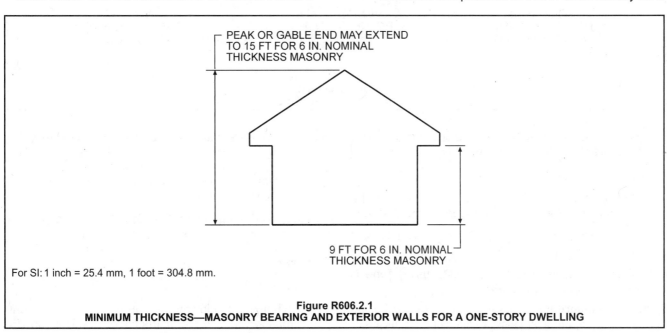

PEAK OR GABLE END MAY EXTEND TO 15 FT FOR 6 IN. NOMINAL THICKNESS MASONRY

9 FT FOR 6 IN. NOMINAL THICKNESS MASONRY

For SI: 1 inch = 25.4 mm, 1 foot = 304.8 mm.

Figure R606.2.1
MINIMUM THICKNESS—MASONRY BEARING AND EXTERIOR WALLS FOR A ONE-STORY DWELLING

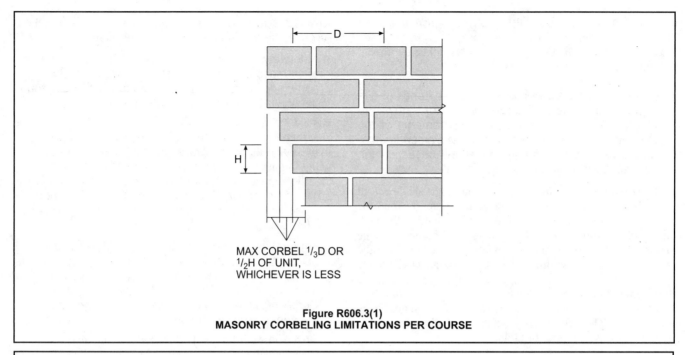

Figure R606.3(1)
MASONRY CORBELING LIMITATIONS PER COURSE

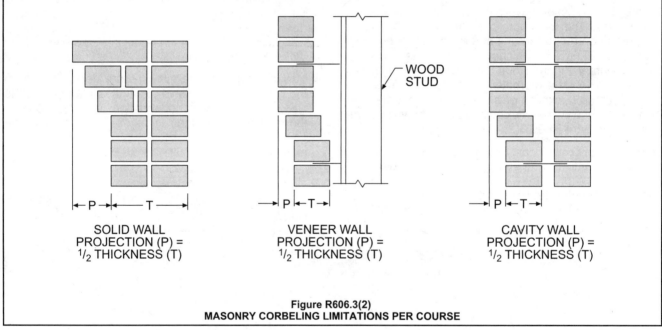

Figure R606.3(2)
MASONRY CORBELING LIMITATIONS PER COURSE

wall composed of solid concrete brick having a unit compressive stress of 2,500 psi (17 235 kPa)?

Solution:

From Table R606.5, the allowable compressive strength is a function of the mortar and would be:

2,500 psi solid units (17 235 kPa): using Type S mortar = 160 psi (1103 kPa).

2,500 psi solid units (17 235 kPa):using Type N mortar = 140 psi (965 kPa).

R606.5.1 Combined units. In walls or other structural members composed of different kinds or grades of units, materials or mortars, the maximum stress shall not exceed the allowable stress for the weakest of the combination of units, materials and mortars of which the member is composed. The net thickness of any facing unit that is used to resist stress shall not be less than 1.5 inches (38 mm).

❖ Walls containing different grades of units, materials or mortars must be limited to the maximum compressive stresses for the weakest combination of units and mortar.

Example:

What is the allowable compressive stress in unit masonry for a wall constructed of 8,000 psi (55 152 kPa) solid brick units, 2,000 psi (13 788 kPa) hollow masonry units, and Type S mortar?

Solution:

From Table R606.5, the allowable compressiveness would be limited to the smaller of:

8,000 psi (55 152 kPa) solid brick units using Type S mortar = 350 psi (2413 kPa);

2,000 psi (13 788 kPa) hollow masonry units using Type S mortar = 140 psi (965 kPa).

The allowable unit masonry stress used in design would be the weaker of the two materials constructed with Type S mortar. Therefore, the allowable stress would be 140 psi (965 kPa).

R606.6 Piers. The unsupported height of masonry piers shall not exceed ten times their least dimension. When structural clay tile or hollow concrete masonry units are used for isolated piers to support beams and girders, the cellular spaces shall be filled solidly with concrete or Type M or S mortar, except that unfilled hollow piers may be used if their unsupported height is not more than four times their least dimension. Where hollow masonry units are solidly filled with concrete or Type M, S or N mortar, the allowable compressive stress shall be permitted to be increased as provided in Table R606.5.

❖ Isolated masonry piers conforming to this section may be used to support beams and girders. The minimum pier area should be based on the design load supported and the allowable stress of the construction per Table R606.5.

R606.6.1 Pier cap. Hollow piers shall be capped with 4 inches (102 mm) of solid masonry or concrete or shall have cavities of the top course filled with concrete or grout or other approved methods.

❖ See the commentary for Section R606.6.

R606.7 Chases. Chases and recesses in masonry walls shall not be deeper than one-third the wall thickness, and the maximum length of a horizontal chase or horizontal projection shall not exceed 4 feet (1219 mm), and shall have at least 8 inches (203 mm) of masonry in back of the chases and recesses and between adjacent chases or recesses and the jambs of openings. Chases and recesses in masonry walls shall be designed and constructed so as not to reduce the required strength or required fire resistance of the wall and in no case shall a chase or recess be permitted within the required area of a pier. Masonry directly above chases or recesses wider than 12 inches (305 mm) shall be supported on noncombustible lintels.

❖ A chase is a continuous recess in a masonry wall that receives a pipe, conduit, etc. Generally, chases are vertical. It is best to construct chases as the masonry wall is built so the strength of the wall is not reduced.

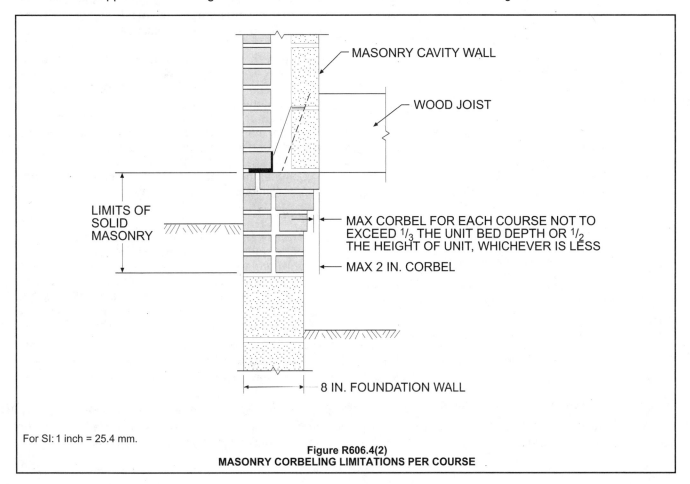

For SI: 1 inch = 25.4 mm.

Figure R606.4(2)
MASONRY CORBELING LIMITATIONS PER COURSE

TABLE R606.5
ALLOWABLE COMPRESSIVE STRESSES FOR
EMPIRICAL DESIGN OF MASONRY

CONSTRUCTION; COMPRESSIVE STRENGTH OF UNIT, GROSS AREA	ALLOWABLE COMPRESSIVE STRESSES[a] GROSS CROSS-SECTIONAL AREA[b]	
	Type M or S mortar	Type N mortar
Solid masonry of brick and other solid units of clay or shale; sand-lime or concrete brick:		
8,000 + psi	350	300
4,500 psi	225	200
2,500 psi	160	140
1,500 psi	115	100
Grouted[c] masonry, of clay or shale; sand-lime or concrete:		
4,500+ psi	225	200
2,500 psi	160	140
1,500 psi	115	100
Solid masonry of solid concrete masonry units:		
3,000+ psi	225	200
2,000 psi	160	140
1,200 psi	115	100
Masonry of hollow load-bearing units:		
2,000+ psi	140	120
1,500 psi	115	100
1,000 psi	75	70
700 psi	60	55
Hollow walls (cavity or masonry bonded[d]) solid units:		
2,500+ psi	160	140
1,500 psi	115	100
Hollow units	75	70
Stone ashlar masonry:		
Granite	720	640
Limestone or marble	450	400
Sandstone or cast stone	360	320
Rubble stone masonry:		
Coarse, rough or random	120	100

For SI: 1 pound per square inch = 6.895 kPa.

a. Linear interpolation shall be used for determining allowable stresses for masonry units having compressive strengths that are intermediate between those given in the table.

b. Gross cross-sectional area shall be calculated on the actual rather than nominal dimensions.

c. See Section R608.

d. Where floor and roof loads are carried upon one wythe, the gross cross-sectional area is that of the wythe under load; if both wythes are loaded, the gross cross-sectional area is that of the wall minus the area of the cavity between the wythes. Walls bonded with metal ties shall be considered as cavity walls unless the collar joints are filled with mortar or grout.

❖ See the commentary to Section R606.5

R606.8 Stack bond. In unreinforced masonry where masonry units are laid in stack bond, longitudinal reinforcement consisting of not less than two continuous wires each with a minimum aggregate cross-sectional area of 0.017 square inch (11 mm²) shall be provided in horizontal bed joints spaced not more than 16 inches (406 mm) on center vertically.

❖ To control cracking where masonry units are placed in a stack bond and where the wall is of unreinforced masonry construction, longitudinal reinforcement must be used in the horizontal bed joints as shown in Commentary Figure R606.8(1). Premanufactured ladder or truss-type reinforcement as shown in Commentary Figure R606.8(2) is generally used to meet the requirements for longitudinal steel.

R606.9 Lateral support. Masonry walls shall be laterally supported in either the horizontal or the vertical direction. The maximum spacing between lateral supports shall not exceed the distances in Table R606.9. Lateral support shall be provided by cross walls, pilasters, buttresses or structural frame members when the limiting distance is taken horizontally, or by floors or roofs when the limiting distance is taken vertically.

❖ The limitations on the maximum unsupported height or length of masonry walls specified in Table R606.9 provides reasonable performance. For purposes of applying the unsupported height limitations, Figures R606.11(1), R606.11(2) and R606.11(3) provide details recognized as appropriate methods of anchorage. At the base of the wall, footings are a lateral support point. Thus, the unsupported height from the footing to the anchorage point at the floor or roof is the unsupported height, which must be limited to the values in Table R606.9.

Instead of unsupported height limitations being measured vertically from footing to supporting floor or roof, the span limitations in Table R606.9 of the code may be met with the use of pilasters, columns, piers, cross walls or similar elements whose relative stiffness is greater than that of the wall. These elements are anchored to the roof or floor structural elements in a manner that transmits imposed lateral forces.

Commentary Figure R606.9 illustrates the lateral support limitations specified in the table.

Example:

The height or length limitation between lateral supports (see Commentary Figure R606.9) for a solid-grouted masonry bearing wall of 8-inch (203 mm) nominal units would be 8 x 20 = 160 inches (4064 mm). For a nonbearing interior wall using two wythes of 4-inch (102 mm) brick, the allowable limitation would be (4 + 4) x 36 = 288 inches (7315 mm).

R606.9.1 Horizontal lateral support. Lateral support in the horizontal direction provided by intersecting masonry walls shall be provided by one of the methods in Section R606.9.1.1 or Section R606.9.1.2.

❖ When masonry walls span horizontally between intersecting walls, anchorage is achieved using the method in either of the following sections.

TABLE R606.9
SPACING OF LATERAL SUPPORT FOR MASONRY WALLS

CONSTRUCTION	MAXIMUM WALL LENGTH TO THICKNESS OR WALL HEIGHT TO THICKNESS[a,b]
Bearing walls:	
Solid or solid grouted	20
All other	18
Nonbearing walls:	
Exterior	18
Interior	36

For SI: 1 foot = 304.8 mm.

a. Except for cavity walls and cantilevered walls, the thickness of a wall shall be its nominal thickness measured perpendicular to the face of the wall. For cavity walls, the thickness shall be determined as the sum of the nominal thicknesses of the individual wythes. For cantilever walls, except for parapets, the ratio of height to nominal thickness shall not exceed 6 for solid masonry, or 4 for hollow masonry. For parapets, see Section R606.2.4.

b. An additional unsupported height of 6 feet is permitted for gable end walls.

❖ Limitations on ratios of distance between lateral supports to wall thickness listed in this table are primarily traditional ratios based on successful performance. The thickness in this table is based on the nominal thickness of the wall for other than cavity wall construction and the sum of the nominal thickness of the wythes not including the cavity for cavity walls; see Note a. To allow for additional height at gable-end walls, Note b permits an additional 6 feet (1829 mm). This is similar to the allowance discussed in the commentary to Figure R606.2.1.

R606.9.1.1 Bonding pattern. Fifty percent of the units at the intersection shall be laid in an overlapping masonry bonding pattern, with alternate units having a bearing of not less than 3 inches (76 mm) on the unit below.

❖ Using this method, anchorage is accomplished by off-setting (overlapping) alternating courses of masonry.

R606.9.1.2 Metal reinforcement. Interior nonload-bearing walls shall be anchored at their intersections, at vertical intervals of not more than 16 inches (406 mm) with joint reinforcement of at least 9 gage [0.148 in. (4mm)], or $^1/_4$ inch (6 mm) galvanized mesh hardware cloth. Intersecting masonry walls, other than interior nonloadbearing walls, shall be anchored at vertical intervals of not more than 8 inches (203 mm) with joint reinforcement of at least 9 gage and shall extend at least 30 inches (762 mm) in each direction at the intersection. Other metal ties, joint reinforcement or anchors, if used, shall be spaced to provide equivalent area of anchorage to that required by this section.

❖ See Commentary Figure R606.9.1.2.

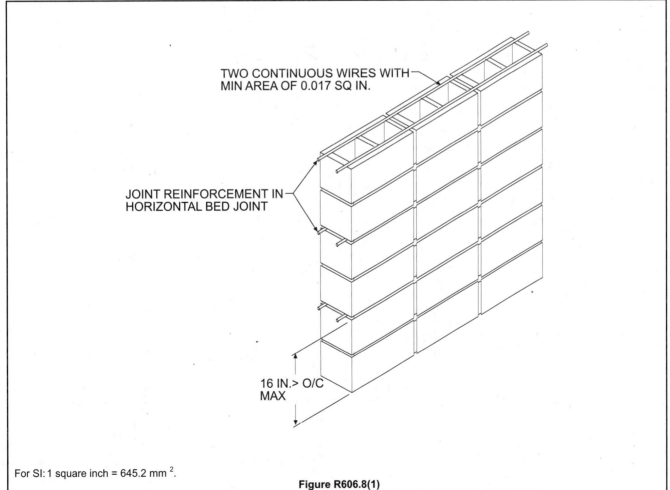

TWO CONTINUOUS WIRES WITH MIN AREA OF 0.017 SQ IN.

JOINT REINFORCEMENT IN HORIZONTAL BED JOINT

16 IN.> O/C MAX

For SI: 1 square inch = 645.2 mm ^{2}.

Figure R606.8(1)
MINIMUM HORIZONTAL JOINT REINFORCEMENT FOR STACK-BOND MASONRY

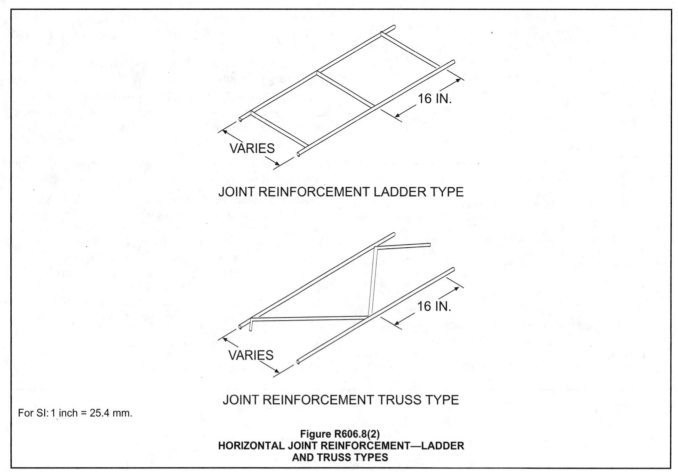

JOINT REINFORCEMENT LADDER TYPE

JOINT REINFORCEMENT TRUSS TYPE

For SI: 1 inch = 25.4 mm.

Figure R606.8(2)
HORIZONTAL JOINT REINFORCEMENT—LADDER
AND TRUSS TYPES

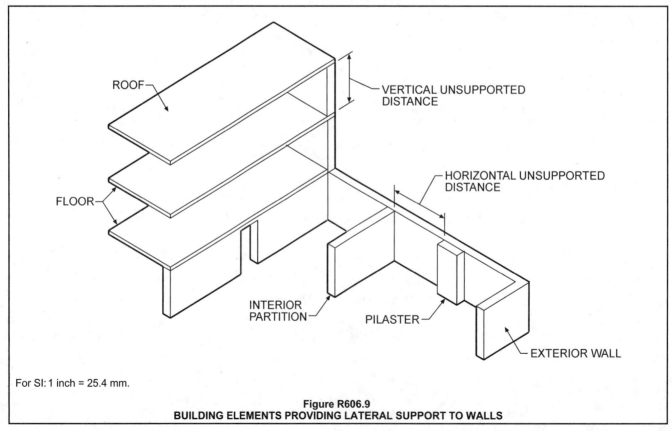

For SI: 1 inch = 25.4 mm.

Figure R606.9
BUILDING ELEMENTS PROVIDING LATERAL SUPPORT TO WALLS

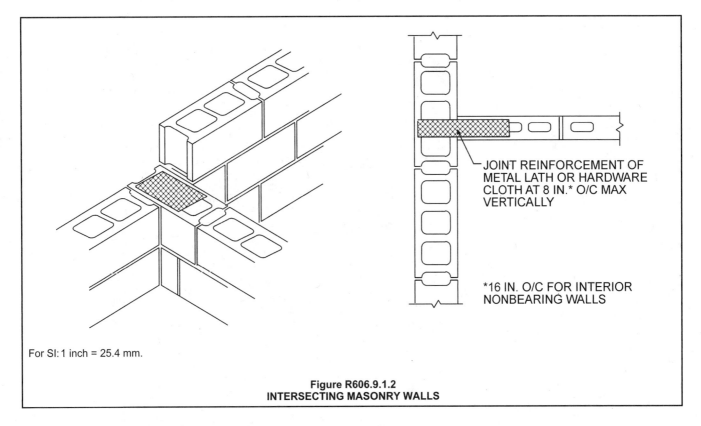

For SI: 1 inch = 25.4 mm.

JOINT REINFORCEMENT OF
METAL LATH OR HARDWARE
CLOTH AT 8 IN.* O/C MAX
VERTICALLY

*16 IN. O/C FOR INTERIOR
NONBEARING WALLS

Figure R606.9.1.2
INTERSECTING MASONRY WALLS

R606.9.2 Vertical lateral support. Vertical lateral support of masonry walls in Seismic Design Category A, B or C shall be provided in accordance with one of the methods in Section R606.9.2.1 or Section R606.9.2.2.

❖ Where walls span vertically in buildings classified as Seismic Design Category A, B or C, masonry walls must be anchored to the floor and roof diaphragms to transmit the anticipated lateral forces. The unsupported height from the footing to the floor or roof must be within the limits of Table R606.9.

R606.9.2.1 Roof structures. Masonry walls shall be anchored to roof structures with metal strap anchors spaced in accordance with the manufacturer's instructions, $1/2$-inch (13 mm) bolts spaced not more than 6 feet (1829 mm) on center, or other approved anchors. Anchors shall be embedded at least 16 inches (406 mm) into the masonry, or be hooked or welded to bond beam reinforcement placed not less than 6 inches (152 mm) from the top of the wall.

❖ If $1/2$-inch (12.7 mm) diameter anchor bolts are used, they must be spaced no farther apart than 6 feet (1829 mm) on center. Where approved metal strap anchors are used, the spacing must be in accordance with the manufacturer's recommendations. Anchors are embedded in the masonry wall and connected to the wooden ledger as illustrated in Commentary Figures R606.9.2.1(1) and R606.9.2.1(2).

R606.9.2.2 Floor diaphragms. Masonry walls shall be anchored to floor diaphragm framing by metal strap anchors spaced in accordance with the manufacturer's instructions, $1/2$-inch-diameter (13 mm) bolts spaced at intervals not to exceed 6 feet (1829 mm) and

installed as shown in Figure R606.11(1), or by other approved methods.

❖ Masonry walls must be anchored to the floor diaphragm with bolts spaced no farther apart than 6 feet (1829 mm) on center. Fasteners may be either metal strap anchors installed per the manufacturer's recommendations or bolts that are embedded in the masonry walls and connected to the wooden ledger. Ledge fasteners transfer shear forces when loading is in the plane of the wall; when forces are out of plane, ledge fasteners prevent the wall and flooring system from separating.

R606.10 Lintels. Masonry over openings shall be supported by steel lintels, reinforced concrete or masonry lintels or masonry arches, designed to support load imposed.

❖ Masonry wall openings require a structural member designed to support the masonry above. Note that wood is not permitted for a lintel supporting a masonry wall.

R606.11 Anchorage. Masonry walls shall be anchored to floor and roof systems in accordance with the details shown in Figure R606.11(1), R606.11(2) or R606.11(3). Footings may be considered as points of lateral support.

❖ Masonry walls depend on floors and roofs for out-of-plane lateral support. Inadequate anchorage of masonry walls in areas of high, and even moderate, seismicity can be problematic. The referenced figures show anchorage requirements that vary based on seismic design category. They illustrate details that provide adequate load transfer under lateral loads.

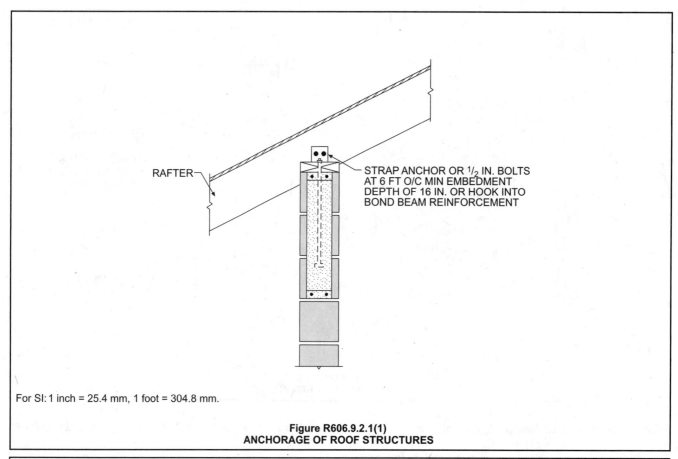

RAFTER

STRAP ANCHOR OR ¹/₂ IN. BOLTS
AT 6 FT O/C MIN EMBEDMENT
DEPTH OF 16 IN. OR HOOK INTO
BOND BEAM REINFORCEMENT

For SI: 1 inch = 25.4 mm, 1 foot = 304.8 mm.

Figure R606.9.2.1(1)
ANCHORAGE OF ROOF STRUCTURES

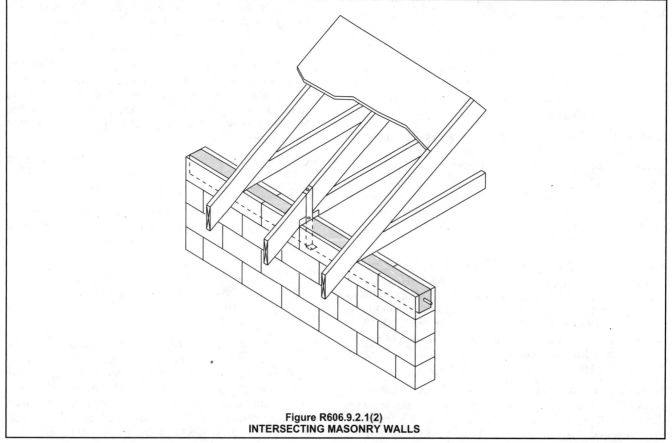

Figure R606.9.2.1(2)
INTERSECTING MASONRY WALLS

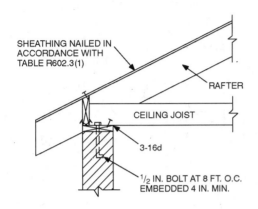

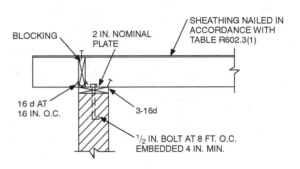

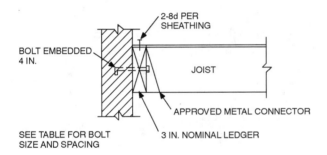

**LEDGER BOLT
SIZE AND SPACING**

JOIST SPAN	BOLT SIZE AND SPACING	
	ROOF	FLOOR
10 FT.	$^1/_2$ AT 2 FT. 6 IN. $^7/_8$ AT 3 FT. 6 IN.	$^1/_2$ AT 2 FT. 0 IN. $^7/_8$ AT 2 FT. 9 IN.
10–15 FT.	$^1/_2$ AT 1 FT. 9 IN. $^7/_8$ AT 2 FT. 6 IN.	$^1/_2$ AT 1 FT. 4 IN. $^7/_8$ AT 2 FT. 0 IN.
15-20 FT.	$^1/_2$ AT 1 FT. 3 IN. $^7/_8$ AT 2 FT. 0 IN.	$^1/_2$ AT 1 FT. 0 IN. $^7/_8$ AT 1 FT. 6 IN.

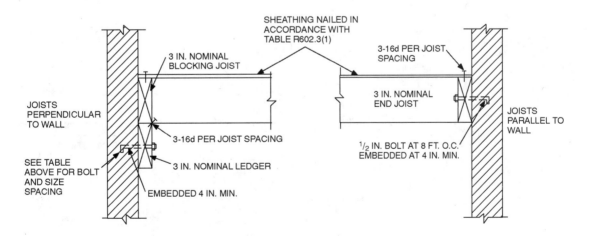

NOTE: Where bolts are located in hollow masonry, the cells in the courses receiving the bolt shall be grouted solid.

For SI: 1 inch = 25.4 mm, 1foot = 304.8 mm, 1 pound per square foot = 0.0479 kPa

**FIGURE R606.11(1)
ANCHORAGE REQUIREMENTS FOR MASONRY WALLS LOCATED IN SEISMIC DESIGN CATEGORY
A, B OR C AND WHERE WIND LOADS ARE LESS THAN 30 PSF**

❖ See the commentary for Sections R606.11 and R606.12.1.

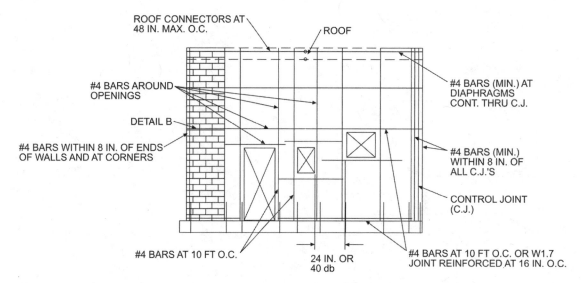

MINIMUM REINFORCEMENT FOR MASONRY WALLS

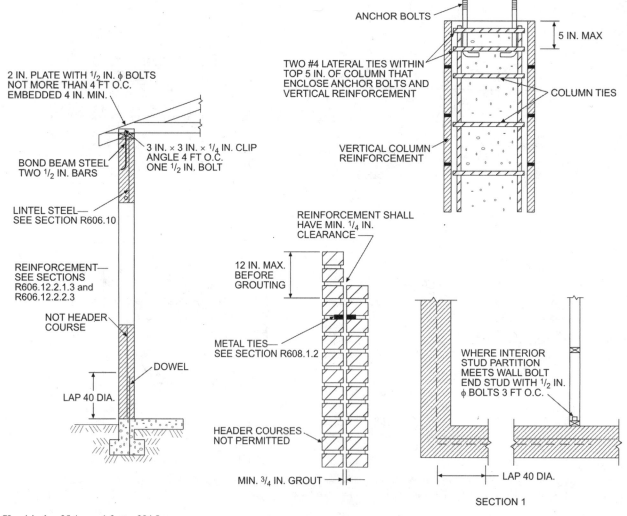

For SI: 1 inch = 25.4 mm, 1 foot = 304.8 mm.

FIGURE R606.11(2)
REQUIREMENTS FOR REINFORCED GROUTED MASONRY CONSTRUCTION IN SEISMIC DESIGN CATEGORY C

❖ See the commentary for Sections R606.11 and R606.12.1.

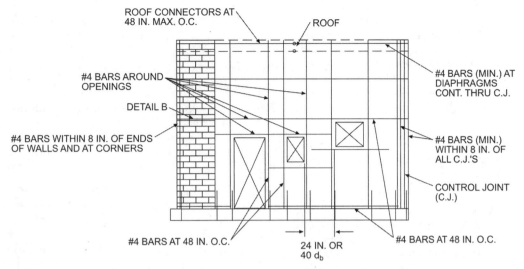

MINIMUM REINFORCEMENT FOR MASONRY WALLS

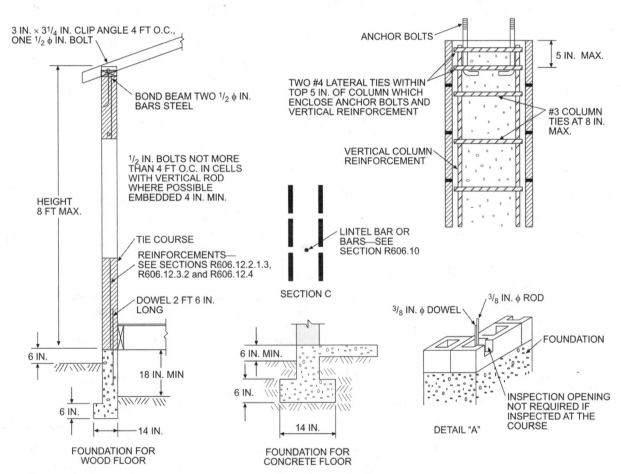

Note: A full bed joint must be provided. All cells containing vertical bars are to be filled to the top of wall and provide inspection opening as shown on detail "A." Horizontal bars are to be laid as shown on detail "B." Lintel bars are to be laid as shown on Section C.

NOTE: A full bed joint must be provided. All cells containing vertical bars are to be filled to the top of wall and provide inspection opening as shown on detail "A." Horizontal bars are to be laid as shown on detail "B." Lintel bars are to be laid as shown on Section C.

For SI: 1 inch = 25.4 mm, 1 foot = 304.8 mm.

FIGURE R606.11(3)
REQUIREMENTS FOR REINFORCED MASONRY CONSTRUCTION IN SEISMIC DESIGN CATEGORY D_0, D_1 OR D_2

❖ See the commentary for Sections R606.11 and R606.12.1.

R606.12 Seismic requirements. The seismic requirements of this section shall apply to the design of masonry and the construction of masonry building elements located in Seismic Design Category D_0, D_1 or D_2. Townhouses in Seismic Design Category C shall comply with the requirements of Section R606.12.2. These requirements shall not apply to glass unit masonry conforming to Section R610 or masonry veneer conforming to Section R703.7.

❖ Although the code generally permits unreinforced walls, this section requires reinforcement in walls of buildings classified as Seismic Design Category C, D_0, D_1 or D_2.

R606.12.1 General. Masonry structures and masonry elements shall comply with the requirements of Sections R606.12.2 through R606.12.4 based on the seismic design category established in Table R301.2(1). Masonry structures and masonry elements shall comply with the requirements of Section R606.12 and Figures R606.11(1), R606.11(2) and R606.11(3) or shall be designed in accordance with ACI 530/ASCE 5/TMS 402.

❖ Masonry walls must comply with this section unless a design is provided. The three cited figures show the reinforcing requirements as a function of the seismic design category and wind load.

R606.12.1.1 Floor and roof diaphragm construction. Floor and roof diaphragms shall be constructed of wood structural panels attached to wood framing in accordance with Table R602.3(1) or to cold-formed steel floor framing in accordance with Table R505.3.1(2) or to cold-formed steel roof framing in accordance with Table R804.3. Additionally, sheathing panel edges perpendicular to framing members shall be backed by blocking, and sheathing shall be connected to the blocking with fasteners at the edge spacing. For Seismic Design Categories C, D_0, D_1 and D_2, where the width-to-thickness dimension of the diaphragm exceeds 2-to-1, edge spacing of fasteners shall be 4 inches (102 mm) on center.

❖ This section reiterates the requirements of Chapters 5 and 8 for fastening structural wood sheathing to floor framing and roof framing. Additional blocking is specified for the structural wood sheathing edges perpendicular to the floor or roof framing, thus providing a blocked diaphragm. Edge fasteners as specified for panel edges must be used at this blocking. Furthermore, if a diaphragm's depth-to-span ratio is greater than 2 in a building classified as Seismic Design Category C, D_0, D_1 or D_2, the edge nail spacing must be reduced to 4 inches (102 mm) on center. These requirements result in added strength and stiffness for diaphragms that support masonry walls.

R606.12.2 Seismic Design Category C. Townhouses located in Seismic Design Category C shall comply with the requirements of this section.

❖ Townhouses classified as Seismic Design Category C must conform to the requirements of this section as well as Figures R606.11(1) and R606.11(2). The requirements are based on whether or not an element is part of the lateral force-resisting system. They are aimed at providing ductility by specifying minimum reinforcement. This section is similar to Section 2106.4 of the *International Building Code*.

R606.12.2.1 Design of elements not part of the lateral force-resisting system.

R606.12.2.1.1 Load-bearing frames or columns. Elements not part of the lateral-force-resisting system shall be analyzed to determine their effect on the response of the system. The frames or columns shall be adequate for vertical load carrying capacity and induced moment caused by the design story drift.

❖ Load-bearing elements not part of a lateral-force-resisting system must be analyzed for their effect on the response of the system and must be capable of supporting all loads in combination with drift-induced bending stresses. These empirical masonry provisions do not cover the design of elements referred to in this section. They would require an engineering design using ACI 530/ASCE 5/TMS 402. See the commentary for Section R606.12.2.

R606.12.2.1.2 Masonry partition walls. Masonry partition walls, masonry screen walls and other masonry elements that are not designed to resist vertical or lateral loads, other than those induced by their own weight, shall be isolated from the structure so that vertical and lateral forces are not imparted to these elements. Isolation joints and connectors between these elements and the structure shall be designed to accommodate the design story drift.

❖ Nonload-bearing elements that are not designed to resist vertical or lateral loads from earthquake effects on the building lateral force system must be isolated from the structure with isolation joints and connectors that can accommodate the design drift. The "design story drift" should be taken as one percent of the story height as allowed in Section 12.14.8.5 of ASCE 7-05 unless an analysis is performed.

R606.12.2.1.3 Reinforcement requirements for masonry elements. Masonry elements listed in Section R606.12.2.1.2 shall be reinforced in either the horizontal or vertical direction as shown in Figure R606.11(2) and in accordance with the following:

1. Horizontal reinforcement. Horizontal joint reinforcement shall consist of at least two longitudinal W1.7 wires spaced not more than 16 inches (406 mm) for walls greater than 4 inches (102 mm) in width and at least one longitudinal W1.7 wire spaced not more than 16 inches (406 mm) for walls not exceeding 4 inches (102 mm) in width; or at least one No. 4 bar spaced not more than 48 inches (1219 mm). Where two longitudinal wires of joint reinforcement are used, the space between these wires shall be the widest that the mortar joint will accommodate. Horizontal reinforcement shall be provided within 16 inches (406 mm) of the top and bottom of these masonry elements.

2. Vertical reinforcement. Vertical reinforcement shall consist of at least one No. 4 bar spaced not more than 48 inches (1219 mm). Vertical reinforcement shall be

located within 16 inches (406 mm) of the ends of masonry walls.

❖ This section specifies reinforcement for masonry walls that are not part of the building lateral force system, as described in the previous section (see commentary, Section R606.12.2).

R606.12.2.2 Design of elements part of the lateral-force-resisting system.

R606.12.2.2.1 Connections to masonry shear walls. Connectors shall be provided to transfer forces between masonry walls and horizontal elements in accordance with the requirements of Section 2.1.8 of ACI 530/ASCE 5/TMS 402. Connectors shall be designed to transfer horizontal design forces acting either perpendicular or parallel to the wall, but not less than 200 pounds per linear foot (2919 N/m) of wall. The maximum spacing between connectors shall be 4 feet (1219 mm). Such anchorage mechanisms shall not induce tension stresses perpendicular to grain in ledgers or nailers.

❖ The connection of horizontal elements such as diaphragms to masonry walls or columns must be designed in accordance with ACI 530/ASCE 5/TMS 402. These connections should also be in accordance with Figure R606.11(2) and meet the minimum requirements specified in this section. Because these connections have been a source of earthquake-related failures, the code does not allow diaphragm anchorage mechanisms that cause cross-grain bending. This addresses the problem of premature failure of wood ledgers (see commentary, Section R606.12.2).

R606.12.2.2.2 Connections to masonry columns. Connectors shall be provided to transfer forces between masonry columns and horizontal elements in accordance with the requirements of Section 2.1.8 of ACI 530/ASCE 5/TMS 402. Where anchor bolts are used to connect horizontal elements to the tops of columns, the bolts shall be placed within lateral ties. Lateral ties shall enclose both the vertical bars in the column and the anchor bolts. There shall be a minimum of two No. 4 lateral ties provided in the top 5 inches (127 mm) of the column.

❖ These empirical masonry provisions do not cover the design of masonry columns. A design per ACI 530/ASCE 5/TMS 402 must be provided. In that event, this section provides a reference to the appropriate section of the standard for the column-to-diaphragm connection design (see commentary, Section R606.12.2).

R606.12.2.2.3 Minimum reinforcement requirements for masonry shear walls. Vertical reinforcement of at least one No. 4 bar shall be provided at corners, within 16 inches (406 mm) of each side of openings, within 8 inches (203 mm) of each side of movement joints, within 8 inches (203 mm) of the ends of walls, and at a maximum spacing of 10 feet (3048 mm).

Horizontal joint reinforcement shall consist of at least two wires of W1.7 spaced not more than 16 inches (406 mm); or bond beam reinforcement of at least one No. 4 bar spaced not more than 10 feet (3048 mm) shall be provided. Horizontal reinforcement shall also be provided at the bottom and top of wall openings and shall extend not less than 24 inches (610 mm) nor less than 40 bar diameters past the opening; continuously at structurally connected roof and floor levels; and within 16 inches (406 mm) of the top of walls.

❖ Masonry shear walls resist the lateral loads from wind or earthquakes. This section specifies reinforcing requirements, which are illustrated in Figure R606.11(2) (see commentary, Section R606.12.2).

R606.12.3 Seismic Design Category D_0 or D_1. Structures in Seismic Design Category D_0 or D_1 shall comply with the requirements of Seismic Design Category C and the additional requirements of this section.

❖ Buildings classified as Seismic Design Category D_0 or D_1 must conform to the requirements of this section as well as the requirements for Seismic Design Category C [see Section R606.12.2 and Figure R606.11(2)]. This section contains provisions that are similar to those in Section 2106.5 of the *International Building Code*.

R606.12.3.1 Design requirements. Masonry elements other than those covered by Section R606.12.2.1.2 shall be designed in accordance with the requirements of Chapter 1 and Sections 2.1 and 2.3 of ACI 530/ASCE 5/TMS 402 and shall meet the minimum reinforcement requirements contained in Sections R606.12.3.2 and R606.12.3.2.1.

Exception: Masonry walls limited to one story in height and 9 feet (2743 mm) between lateral supports need not be designed provided they comply with the minimum reinforcement requirements of Sections R606.12.3.2 and R606.12.3.2.1.

❖ This section specifies the design criteria for masonry elements in Seismic Design Category D_0 or D_1.

Except for nonload-bearing elements that are not part of the lateral-force-resisting system, masonry elements must be designed using ACI 530/ASCE 5/TMS 402. The exception allows certain one-story masonry walls to satisfy the specified prescriptive code requirements rather than being designed to other engineering standards.

R606.12.3.2 Minimum reinforcement requirements for masonry walls. Masonry walls other than those covered by Section R606.12.2.1.3 shall be reinforced in both the vertical and horizontal direction. The sum of the cross-sectional area of horizontal and vertical reinforcement shall be at least 0.002 times the gross cross-sectional area of the wall, and the minimum cross-sectional area in each direction shall be not less than 0.0007 times the gross cross-sectional area of the wall. Reinforcement shall be uniformly distributed. Table R606.12.3.2 shows the minimum reinforcing bar sizes required for varying thicknesses of masonry walls. The maximum spacing of reinforcement shall be 48 inches (1219 mm) provided that the walls are solid grouted and constructed of hollow open-end units, hollow units laid with full head joints or two wythes of solid units. The maximum spacing of reinforcement shall be 24 inches (610 mm) for all other masonry.

❖ The sum of horizontal and vertical reinforcement must be at least 0.2 percent of the cross-sectional area of the wall, and the reinforcing should be distributed as shown in Table R606.12.3.2 [see commentary, Figure R606.11(3) and Section R606.12.3].

R606.12.3.2.1 Shear wall reinforcement requirements. The maximum spacing of vertical and horizontal reinforcement shall be the smaller of one-third the length of the shear wall, one-third the height of the shear wall, or 48 inches (1219 mm). The minimum cross-sectional area of vertical reinforcement shall be one-third of the required shear reinforcement. Shear reinforcement shall be anchored around vertical reinforcing bars with a standard hook.

❖ See the commentary for Section R606.12.3.

R606.12.3.3 Minimum reinforcement for masonry columns. Lateral ties in masonry columns shall be spaced not more than 8 inches (203 mm) on center and shall be at least $^3/_8$ inch (9.5 mm) diameter. Lateral ties shall be embedded in grout.

❖ See Figure R606.11(3). Also see the commentary for Sections R606.12.2.2.2 and R606.12.3.

R606.12.3.4 Material restrictions. Type N mortar or masonry cement shall not be used as part of the lateral-force-resisting system.

❖ This restates the requirement of Section R607.1.3.

R606.12.3.5 Lateral tie anchorage. Standard hooks for lateral tie anchorage shall be either a 135-degree (2.4 rad) standard hook or a 180-degree (3.2 rad) standard hook.

❖ Lateral ties (required by Section R606.12.3.3) must be anchored as specified. Also see the commentary for Section R606.12.3.

R606.12.4 Seismic Design Category D_2. All structures in Seismic Design Category D_2 shall comply with the requirements of Seismic Design Category D_1 and to the additional requirements of this section.

❖ Buildings classified as Seismic Design Category D_2 must conform to the requirements of this section as well as all the requirements for Seismic Design Categories D_0 and D_1 and Figure R606.11(3).

R606.12.4.1 Design of elements not part of the lateral-force-resisting system. Stack bond masonry that is not part of the lateral-force-resisting system shall have a horizontal cross-sectional area of reinforcement of at least 0.0015 times the gross cross-sectional area of masonry. Table R606.12.4.1 shows minimum reinforcing bar sizes for masonry walls. The maximum spacing of horizontal reinforcement shall be 24 inches (610 mm). These elements shall be solidly grouted and shall be constructed of hollow open-end units or two wythes of solid units.

❖ For elements that are not part of the lateral force-resisting system, horizontal reinforcement must be at least 0.0015 times the cross-sectional area of the wall as shown in Table R606.12.4.1.

TABLE R606.12.4.1
MINIMUM REINFORCING FOR STACKED BONDED MASONRY WALLS IN SEISMIC DESIGN CATEGORY D_2

NOMINAL WALL THICKNESS (inches)	MINIMUM BAR SIZE SPACED AT 24 INCHES
6	#4
8	#5
10	#5
12	#6

For SI: 1 inch = 25.4 mm.

❖ See the commentary for Section R606.12.4.1.

R606.12.4.2 Design of elements part of the lateral-force-resisting system. Stack bond masonry that is part of the lateral-force-resisting system shall have a horizontal cross-sectional area of reinforcement of at least 0.0025 times the gross cross-sectional area of masonry. Table R606.12.4.2 shows minimum reinforcing bar sizes for masonry walls. The maximum spacing of horizontal reinforcement shall be 16 inches (406 mm). These elements shall be solidly grouted and shall be constructed of hollow open-end units or two wythes of solid units.

❖ Elements that are part of the lateral-force-resisting system must have horizontal reinforcement of at least 0.0025 times the cross-sectional area of the wall as shown in Table R606.12.4.2. These elements may be constructed of solid grouted hollow open-end units or of two wythes of solid units solidly grouted.

TABLE R606.12.3.2
MINIMUM DISTRIBUTED WALL REINFORCEMENT FOR BUILDING ASSIGNED TO SEISMIC DESIGN CATEGORY D_0 or D_1

NOMINAL WALL THICKNESS (inches)	MINIMUM SUM OF THE VERTICAL AND HORIZONTAL REINFORCEMENT AREAS[a] (square inches per foot)	MINIMUM REINFORCEMENT AS DISTRIBUTED IN BOTH HORIZONTAL AND VERTICAL DIRECTIONS[b] (square inches per foot)	MINIMUM BAR SIZE FOR REINFORCEMENT SPACED AT 48 INCHES
6	0.135	0.047	#4
8	0.183	0.064	#5
10	0.231	0.081	#6
12	0.279	0.098	#6

For SI: 1 inch = 25.4 mm, 1 foot = 304.8 mm, 1 square inch per foot = 2064 mm²/m.

a. Based on the minimum reinforcing ratio of 0.002 times the gross cross-sectional area of the wall.

b. Based on the minimum reinforcing ratio each direction of 0.0007 times the gross cross-sectional area of the wall.

❖ This table lists the minimum reinforcement areas specified in Section R606.12.3.2. The minimum bar size shown in the far right column satisfies the requirement for 0.0007 times the gross cross-sectional area of the wall. This must be provided as a minimum in both the vertical and horizontal directions. Additional reinforcement is necessary in either direction to meet the requirement for a total of 0.002 times the gross cross-sectional area.

TABLE R606.12.4.2
MINIMUM REINFORCING FOR STACKED BONDED MASONRY WALLS IN SEISMIC DESIGN CATEGORY D$_2$

NOMINAL WALL THICKNESS (inches)	MINIMUM BAR SIZE SPACED AT 16 INCHES
6	#4
8	#5
10	#5
12	#6

For SI: 1 inch = 25.4 mm.

❖ See the commentary for Section R606.12.4.2.

R606.13 Protection for reinforcement. Bars shall be completely embedded in mortar or grout. Joint reinforcement embedded in horizontal mortar joints shall not have less than $^5/_8$-inch (15.9 mm) mortar coverage from the exposed face. All other reinforcement shall have a minimum coverage of one bar diameter over all bars, but not less than $^3/_4$ inch (19 mm), except where exposed to weather or soil, in which case the minimum coverage shall be 2 inches (51 mm).

❖ Commentary Figure R606.13 shows placement restrictions for reinforcement relative to exposed faces and coverage.

R606.14 Beam supports. Beams, girders or other concentrated loads supported by a wall or column shall have a bearing of at least 3 inches (76 mm) in length measured parallel to the beam upon solid masonry not less than 4 inches (102 mm) in thickness, or upon a metal bearing plate of adequate design and dimensions to distribute the load safely, or upon a continuous reinforced masonry member projecting not less than 4 inches (102 mm) from the face of the wall.

❖ To provide for the transfer of vertical loads from beams, girders or other elements to the masonry wall or column, a minimum bearing as shown in Commentary Figure R606.14 is required. An alternative to the minimum bearing is the use of metal bearing plates that are based on structural design.

R606.14.1 Joist bearing. Joists shall have a bearing of not less than $1^1/_2$ inches (38 mm), except as provided in Section R606.14, and shall be supported in accordance with Figure R606.11(1).

❖ Where joists bear on masonry wall or column elements, the minimum bearings as shown in Figure R606.11(1) and Commentary Figure R606.14.1 are considered to be adequate. The code does not provide guidance distinguishing whether a member is a joist or a beam; however, joists are generally considered to be members that are a nominal 2 inches (51 mm) maximum in thickness and placed not more than 24 inches (610 mm) on center.

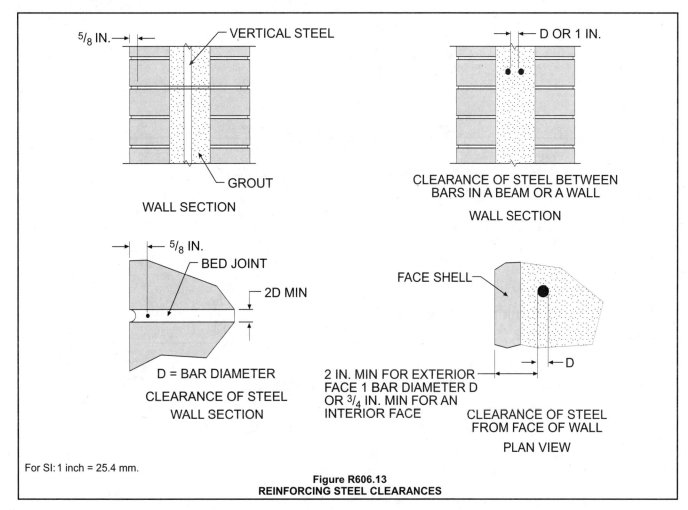

For SI: 1 inch = 25.4 mm.

Figure R606.13
REINFORCING STEEL CLEARANCES

R606.15 Metal accessories. Joint reinforcement, anchors, ties and wire fabric shall conform to the following: ASTM A 82 for wire anchors and ties; ASTM A 36 for plate, headed and bent-bar anchors; ASTM A 510 for corrugated sheet metal anchors and ties; ASTM A 951 for joint reinforcement; ASTM B 227 for copper-clad steel wire ties; or ASTM A 167 for stainless steel hardware.

❖ This section provides reference standards for metal accessories used in masonry wall construction.

R606.15.1 Corrosion protection. Minimum corrosion protection of joint reinforcement, anchor ties and wire fabric for use in masonry wall construction shall conform to Table R606.15.1.

❖ Table R606.15.1 contains reference standards for protecting metal accessories used in masonry wall construction.

The corrosion protection requirements are dependent on type of steel and exposure. Because of its high

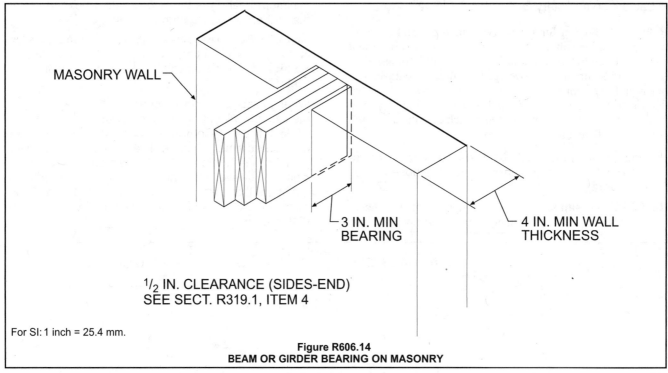

3 IN. MIN BEARING

4 IN. MIN WALL THICKNESS

MASONRY WALL

1/2 IN. CLEARANCE (SIDES-END)
SEE SECT. R319.1, ITEM 4

For SI: 1 inch = 25.4 mm.

Figure R606.14
BEAM OR GIRDER BEARING ON MASONRY

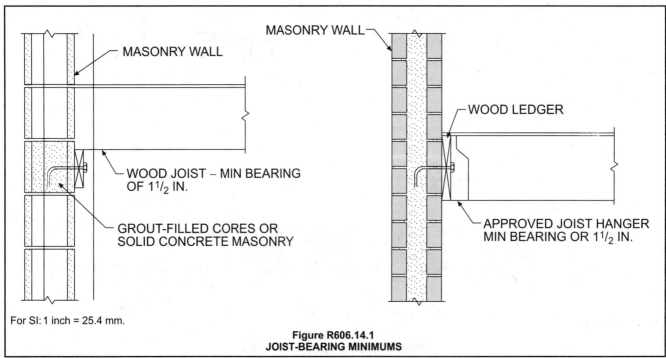

MASONRY WALL

MASONRY WALL

WOOD JOIST – MIN BEARING OF 1 1/2 IN.

GROUT-FILLED CORES OR SOLID CONCRETE MASONRY

WOOD LEDGER

APPROVED JOIST HANGER MIN BEARING OR 1 1/2 IN.

For SI: 1 inch = 25.4 mm.

Figure R606.14.1
JOIST-BEARING MINIMUMS

resistance to corrosion, stainless steel need not be coated. Joint reinforcement, anchors, ties and accessories of other than stainless steel must be protected by zinc coatings (galvanizing). The protective value of the zinc coating increases with increasing coating thickness; therefore, the amount of galvanizing required increases with the potential severity of exposure.

TABLE R606.15.1
MINIMUM CORROSION PROTECTION

MASONRY METAL ACCESSORY	STANDARD
Joint reinforcement, interior walls	ASTM A 641, Class 1
Wire ties or anchors in exterior walls completely embedded in mortar or grout	ASTM A 641, Class 3
Wire ties or anchors in exterior walls not completely embedded in mortar or grout	ASTM A 153, Class B-2
Joint reinforcement in exterior walls or interior walls exposed to moist environment	ASTM A 153, Class B-2
Sheet metal ties or anchors exposed to weather	ASTM A 153, Class B-2
Sheet metal ties or anchors completely embedded in mortar or grout	ASTM A 653, Coating Designation G60
Stainless steel hardware for any exposure	ASTM A 167, Type 304

❖ See the commentary for Section R606.15.1.

SECTION R607
UNIT MASONRY

R607.1 Mortar. Mortar for use in masonry construction shall comply with ASTM C 270. The type of mortar shall be in accordance with Sections R607.1.1, R607.1.2 and R607.1.3 and shall meet the proportion specifications of Table R607.1 or the property specifications of ASTM C 270.

❖ This section specifies the requirements for mortar used in masonry construction. Mortar is the bonding agent that separates masonry units while bonding them together. It is an integral part of any masonry wall and must be strong, durable and capable of keeping the wall intact. In addition, it should help to create a moisture-resistant barrier.

Mortar makes up approximately 25 percent of a standard modular brick wall and about 10 percent of a block wall. It is therefore a significant part of the structure. ASTM C 270 discusses the materials and methods for mixing mortar. For purposes of economy, a good rule of thumb is to specify the lowest strength mortar that will satisfy the structural requirements of the project. Mortar consists of cementitious materials and well-graded sand with sufficient fines. Mortar is used for the following purposes:

- It is a bedding or seating materials for the masonry unit.

- It allows the unit to be leveled and properly placed.
- It bonds the units together.
- It provides compressive strength.
- It provides shear strength, particularly parallel to the wall.
- It allows some movement and elasticity between units.
- It seals irregularities of the masonry unit and provides a weather-tight wall, preventing penetration of wind and water into and through the wall. It can provide color to the wall when a mineral color additive is used.
- It can provide an architectural appearance by using various types of joints.

R607.1.1 Foundation walls. Masonry foundation walls constructed as set forth in Tables R404.1.1(1) through R404.1.1(4) and mortar shall be Type M or S.

❖ This section specifies the applicable tables in Chapter 4 that apply to masonry foundation walls.

R607.1.2 Masonry in Seismic Design Categories A, B and C. Mortar for masonry serving as the lateral-force-resisting system in Seismic Design Categories A, B and C shall be Type M, S or N mortar.

❖ Masonry construction in buildings classified as Seismic Design Category A, B or C must use Type M, S or N mortar.

R607.1.3 Masonry in Seismic Design Categories D_0, D_1 and D_2. Mortar for masonry serving as the lateral-force-resisting system in Seismic Design Categories D_0, D_1 and D_2 shall be Type M or S portland cement-lime or mortar cement mortar.

❖ In buildings classified as Seismic Design Category D_0, D_1 or D_2, in walls which are intended to resist lateral forces, and in foundation walls in any building, only Type M or S mortar is acceptable.

R607.2 Placing mortar and masonry units.

R607.2.1 Bed and head joints. Unless otherwise required or indicated on the project drawings, head and bed joints shall be $^3/_8$ inch (10 mm) thick, except that the thickness of the bed joint of the starting course placed over foundations shall not be less than $^1/_4$ inch (7 mm) and not more than $^3/_4$ inch (19 mm).

❖ The provisions of Section R607 set forth specific requirements that are unique to construction using hollow-unit masonry. Hollow masonry units are to be laid and set in mortar. The properties of masonry walls, including strength and appearance, vary significantly depending on the thickness of the mortar joints. The initial bed joint, which is the bed joint between the first course of masonry and the foundation wall, is limited to a thickness of not less than $^1/_4$ inch (6.4 mm) and not greater than $^3/_4$ inch (19 mm). This larger variation is to allow for the inherent unevenness of concrete or concrete masonry foundation walls. The remainders of bed joints, which can be controlled by the mason, are limited to $^3/_8$ inch (9.5 mm), unless specified otherwise on the project drawings.

R607.2.1.1 Mortar joint thickness tolerance. Mortar joint thickness shall be within the following tolerances from the specified dimensions:

1. Bed joint: + $^1/_8$ inch (3 mm).

2. Head joint: $^1/_4$ inch (7 mm), + $^3/_8$ inch (10 mm).

3. Collar joints: $^1/_4$ inch (7 mm), + $^3/_8$ inch (10 mm).

Exception: Nonload-bearing masonry elements and masonry veneers designed and constructed in accordance with Section R703.7 are not required to meet these tolerances.

❖ Joint-thickness tolerances limit the eccentricity of applied loads and reflect typical industry practice. They also provide a more uniform appearance for exposed masonry. The bed joint is a horizontal joint; the head joint is a vertical joint.

R607.2.2 Masonry unit placement. The mortar shall be sufficiently plastic and units shall be placed with sufficient pressure to extrude mortar from the joint and produce a tight joint. Deep furrowing of bed joints that produces voids shall not be permitted. Any units disturbed to the extent that initial bond is broken after initial placement shall be removed and relaid in fresh mortar. Surfaces to be in contact with mortar shall be clean and free of deleterious materials.

❖ This section gives placement requirements meant to provide adequate bonding between mortar and masonry units. Mortar is the bonding agent that integrates masonry units into a wall; it is used to bind masonry units into a single element by developing a complete, strong, and durable bond. Mortar is usually placed between absorbent masonry units and loses water upon contact with the units. Mortars have a high water-cement ratio when mixed, but the ratio decreases when the mortar comes into contact with the absorbent unit.

Once the mortar has begun to set or harden, tapping or attempting to otherwise move masonry units can be detrimental to the bond. Movement at this time will break the bond between the masonry unit and mortar. The partially dried mortar will not have sufficient plasticity to re-adhere sufficiently to the masonry units. Af-

TABLE R607.1
MORTAR PROPORTIONS[a, b]

| MORTAR | TYPE | Portland cement or blended cement | Mortar cement | | | Masonry cement | | | Hydrated lime[c] or lime putty | Aggregate ratio (measured in damp, loose conditions) |
			M	S	N	M	S	N		
Cement-lime	M	1	—	—	—	—	—	—	$^1/_4$	
	S	1	—	—	—	—	—	—	over $^1/_4$ to $^1/_2$	
	N	1	—	—	—	—	—	—	over $^1/_2$ to $1^1/_4$	
	O	1	—	—	—	—	—	—	over $1^1/_4$ to $2^1/_2$	
Mortar cement	M	1	—	—	1	—	—	—		Not less than $2^1/_4$ and not more than 3 times the sum of separate volumes of lime, if used, and cement
	M	—	1	—	—	—	—	—	—	
	S	$^1/_2$	—	—	1	—	—	—		
	S	—	—	1	—	—	—	—		
	N	—	—	—	1	—	—	—		
	O	—	—	—	1	—	—	—		
Masonry cement	M	1				—	—	1		
	M	—				1	—	—	—	
	S	$^1/_2$				—	—	1		
	S	—				—	1	—		
	N	—				—	—	1		
	O	—				—	—	1		

For SI: 1 cubic foot = 0.0283 m³, 1 pound = 0.454 kg.

a. For the purpose of these specifications, the weight of 1 cubic foot of the respective materials shall be considered to be as follows:

Portland Cement	94 pounds	Masonry Cement	Weight printed on bag
Mortar Cement	Weight printed on bag	Hydrated Lime	40 pounds
Lime Putty (Quicklime)	80 pounds	Sand, damp and loose	80 pounds of dry sand

b. Two air-entraining materials shall not be combined in mortar.

c. Hydrated lime conforming to the requirements of ASTM C 207.

❖ The mortar required by Sections R607.1.1 through R607.1.3 must be proportioned as shown in this table or ASTM C 270. Type M mortar is suited for structures below or against grade such as retaining walls. It is also suited for masonry construction subject to high compressive loads, severe frost action or high lateral loads from earth pressures, hurricane winds or earthquakes. Type S mortar is appropriate for use in structures requiring high flexural bond strength that are subject to compressive and lateral loads. Type N mortar is meant for general use in above-grade masonry, residential basement construction, interior walls and partitions, masonry veneer and nonstructural masonry partitions.

ter the initial bond has been broken, the masonry units should be removed and replaced in fresh mortar.

R607.2.2.1 Solid masonry. Solid masonry units shall be laid with full head and bed joints and all interior vertical joints that are designed to receive mortar shall be filled.

❖ This section applies to solid masonry construction (see the definition in Section 202).

R607.2.2.2 Hollow masonry. For hollow masonry units, head and bed joints shall be filled solidly with mortar for a distance in from the face of the unit not less than the thickness of the face shell.

❖ See the definition for "Masonry Unit, Hollow" in Section R202. Normally, cross webs are not mortared. When individual cells are to be grouted, mortar should be placed on the webs on both sides of the cell to be grouted to prevent leakage of the grout.

R607.3 Installation of wall ties. The installation of wall ties shall be as follows:

1. The ends of wall ties shall be embedded in mortar joints. Wall tie ends shall engage outer face shells of hollow units by at least $^1/_2$ inch (13 mm). Wire wall ties shall be embedded at least $1^1/_2$ inches (38 mm) into the mortar bed of solid masonry units or solid grouted hollow units.

2. Wall ties shall not be bent after being embedded in grout or mortar.

❖ Proper installation of masonry wall ties is necessary to achieve bonding of multiple wythes (also see Section R608.1.2.1). In addition, wall ties should not be bent after placement in the mortar joints. This can break the initial bond between the masonry unit and mortar and possibly lead to an increased amount of moisture infiltration into the wall assembly. Bending the wall ties after initial placement can also be detrimental to the strength of the wall system because the ties lose their ability to transfer lateral loads to the backing wall assembly.

SECTION R608
MULTIPLE WYTHE MASONRY

R608.1 General. The facing and backing of multiple wythe masonry walls shall be bonded in accordance with Section R608.1.1, R608.1.2 or R608.1.3. In cavity walls, neither the facing nor the backing shall be less than 3 inches (76 mm) nominal in thickness and the cavity shall not be more than 4 inches (102 mm) nominal in width. The backing shall be at least as thick as the facing.

Exception: Cavities shall be permitted to exceed the 4-inch (102 mm) nominal dimension provided tie size and tie spacing have been established by calculation.

❖ The provisions of Section R608 apply to masonry walls more than one masonry unit in thickness. In cases where a wall thickness is made up of multiple wythes of masonry units, bonding is required so that the wythes are tied together to act as a unit. Section R608 contains the specifics of what constitutes an acceptable bonding mechanism. Two basic types of bonding are permissible: the first involves various overlapping arrangements; the second is achieved by installation of corrosion-resistant metal ties as stipulated for cavity-wall masonry construction.

R608.1.1 Bonding with masonry headers. Bonding with solid or hollow masonry headers shall comply with Sections R608.1.1.1 and R608.1.1.2.

❖ There are two methods of bonding multiwythe walls through the use of headers. They can be solid or hollow unit, depending on the wall system in question.

R608.1.1.1 Solid units. Where the facing and backing (adjacent wythes) of solid masonry construction are bonded by means of masonry headers, no less than 4 percent of the wall surface of each face shall be composed of headers extending not less than 3 inches (76 mm) into the backing. The distance between adjacent full-length headers shall not exceed 24 inches (610 mm) either vertically or horizontally. In walls in which a single header does not extend through the wall, headers from the opposite sides shall overlap at least 3 inches (76 mm), or headers from opposite sides shall be covered with another header course overlapping the header below at least 3 inches (76 mm).

❖ See the commentary for Section R608.1.1 and Commentary Figure R608.1.1.1.

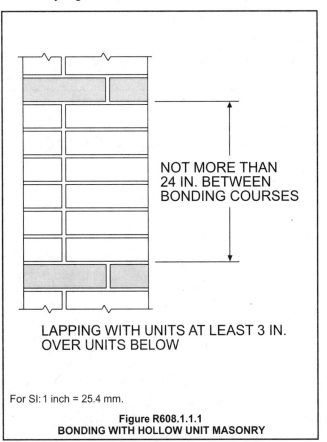

NOT MORE THAN 24 IN. BETWEEN BONDING COURSES

LAPPING WITH UNITS AT LEAST 3 IN. OVER UNITS BELOW

For SI: 1 inch = 25.4 mm.

Figure R608.1.1.1
BONDING WITH HOLLOW UNIT MASONRY

R608.1.1.2 Hollow units. Where two or more hollow units are used to make up the thickness of a wall, the stretcher courses shall be bonded at vertical intervals not exceeding 34 inches (864 mm) by lapping at least 3 inches (76 mm) over the unit below, or by lapping at vertical intervals not exceeding 17 inches (432 mm) with units that are at least 50 percent thicker than the units below.

❖ See the commentary for Section R608.1.1 and Commentary Figure R608.1.1.2.

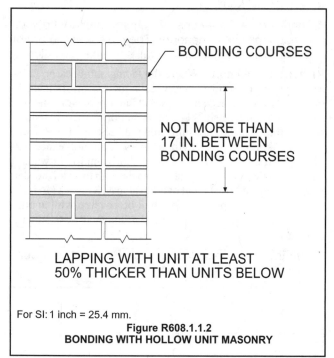

For SI: 1 inch = 25.4 mm.

Figure R608.1.1.2
BONDING WITH HOLLOW UNIT MASONRY

R608.1.2 Bonding with wall ties or joint reinforcement. Bonding with wall ties or joint reinforcement shall comply with Sections R608.1.2.1 through R608.1.2.3.

❖ See the commentary for Section R608.1.

R608.1.2.1 Bonding with wall ties. Bonding with wall ties, except as required by Section R610, where the facing and backing (adjacent wythes) of masonry walls are bonded with $^3/_{16}$-inch-diameter (5 mm) wall ties embedded in the horizontal mortar joints, there shall be at least one metal tie for each 4.5 square feet (0.418 m²) of wall area. Ties in alternate courses shall be staggered. The maximum vertical distance between ties shall not exceed 24 inches (610 mm), and the maximum horizontal distance shall not exceed 36 inches (914 mm). Rods or ties bent to rectangular shape shall be used with hollow masonry units laid with the cells vertical. In other walls, the ends of ties shall be bent to 90-degree (0.79 rad) angles to provide hooks no less than 2 inches (51 mm) long. Additional bonding ties shall be provided at all openings, spaced not more than 3 feet (914 mm) apart around the perimeter and within 12 inches (305 mm) of the opening.

❖ Multiwythe walls can be bonded with wire-type ties. Generally, this type of tie consists of metal Z ties or rectangular ties. The code requires certain maximum spacing for wire tires similar to that for masonry headers. This section provides minimum wire diameters and requires spacing based on a square footage area of the wall in question. Additional bonding ties are required around wall openings, such as windows or doors. These wire ties are used to anchor wythes of masonry together to resist loads or to transfer loads across air spaces to the backing wall material [see Commentary Figures R608.1.2.1(1) and R608.1.2.1(2)].

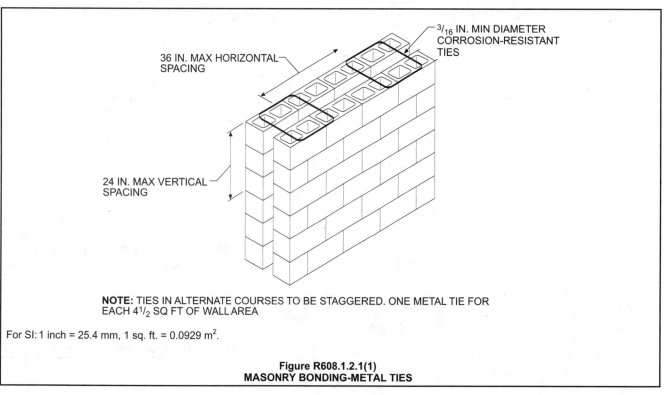

NOTE: TIES IN ALTERNATE COURSES TO BE STAGGERED. ONE METAL TIE FOR EACH 4$^1/_2$ SQ FT OF WALL AREA

For SI: 1 inch = 25.4 mm, 1 sq. ft. = 0.0929 m².

Figure R608.1.2.1(1)
MASONRY BONDING-METAL TIES

R608.1.2.2 Bonding with adjustable wall ties. Where the facing and backing (adjacent wythes) of masonry are bonded with adjustable wall ties, there shall be at least one tie for each 2.67 square feet (0.248 m²) of wall area. Neither the vertical nor the horizontal spacing of the adjustable wall ties shall exceed 24 inches (610 mm). The maximum vertical offset of bed joints from one wythe to the other shall be 1.25 inches (32 mm). The maximum clearance between connecting parts of the ties shall be ¹/₁₆ inch (2 mm). When pintle legs are used, ties shall have at least two ³/₁₆-inch-diameter (5 mm) legs.

❖ This section pertains to the use of adjustable ties for anchoring the facing to the backing. Reduced tie spacing is required when these types of ties are specified. Maximum vertical offsets are prescribed so that the adjustable tie will not disengage through use. Many forms of adjustable wall ties are available for masonry construction. See Commentary Figure R608.1.2.2.

R608.1.2.3 Bonding with prefabricated joint reinforcement. Where the facing and backing (adjacent wythes) of masonry are bonded with prefabricated joint reinforcement, there shall be at least one cross wire serving as a tie for each 2.67 square feet (0.248 m²) of wall area. The vertical spacing of the joint reinforcement shall not exceed 16 inches (406 mm). Cross wires on prefabricated joint reinforcement shall not be smaller than No. 9 gage. The longitudinal wires shall be embedded in the mortar.

❖ Adjacent wythes of masonry could also be bonded with prefabricated joint reinforcements. They can be used in these wall types when air spaces exist between two wythes of masonry, or they can be used in double-wythe grouted-wall construction. Cross wires must be of minimum size to provide proper load transfer. This reinforcement is placed in the bed joints. Commentary Figure R608.1.2.3 illustrates two of the more common types available.

R608.1.3 Bonding with natural or cast stone. Bonding with natural and cast stone shall conform to Sections R608.1.3.1 and R608.1.3.2.

❖ Ashlar masonry and rubble stone masonry require bonding through masonry headers. Although quite different from the usual masonry headers for clay or concrete masonry walls, bonder stones are required based on the square footage of wall area. Bonder stones must be embedded in mortar and have the proper extension into the wall surfaces on both sides. Each masonry type has separate bonding stone spacing requirements based on the construction. The bonding requirements in this section are primarily drawn from ACI 530/ASCE 5/TMS402.

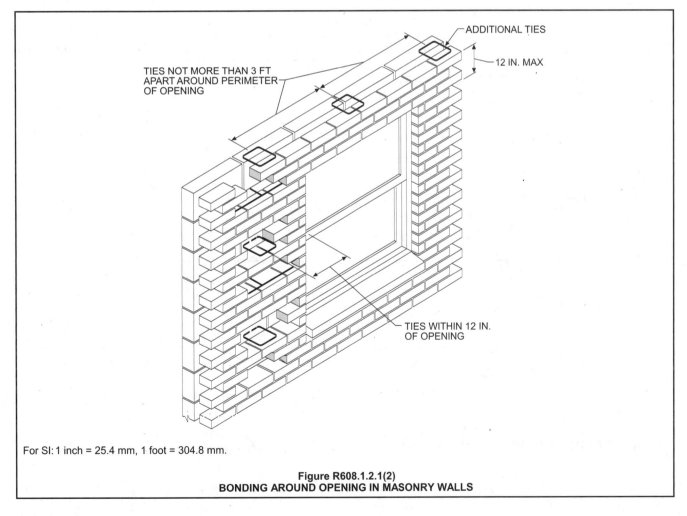

For SI: 1 inch = 25.4 mm, 1 foot = 304.8 mm.

Figure R608.1.2.1(2)
BONDING AROUND OPENING IN MASONRY WALLS

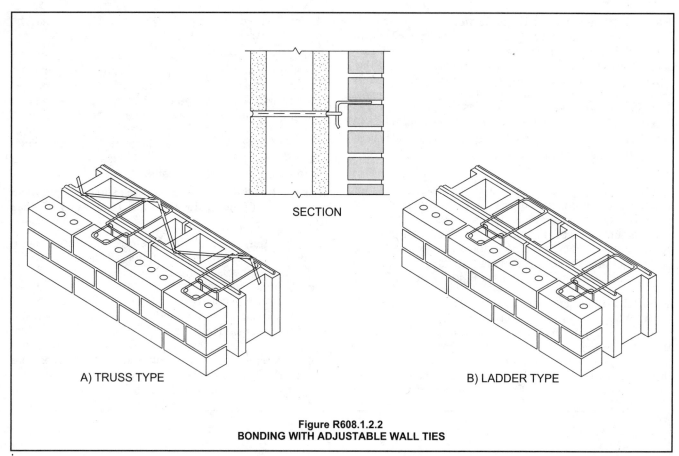

SECTION

A) TRUSS TYPE

B) LADDER TYPE

Figure R608.1.2.2
BONDING WITH ADJUSTABLE WALL TIES

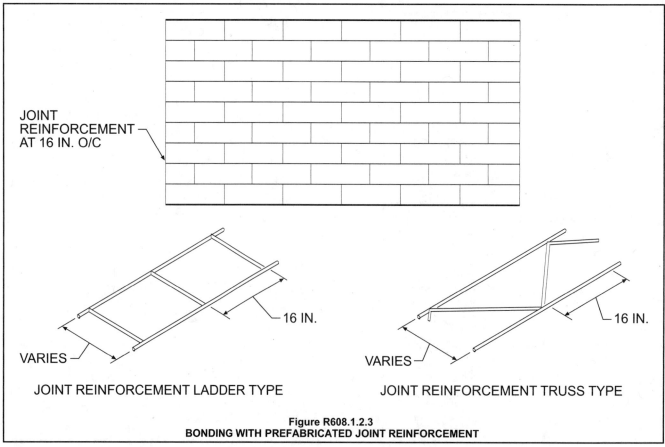

JOINT
REINFORCEMENT
AT 16 IN. O/C

16 IN.

16 IN.

VARIES

VARIES

JOINT REINFORCEMENT LADDER TYPE

JOINT REINFORCEMENT TRUSS TYPE

Figure R608.1.2.3
BONDING WITH PREFABRICATED JOINT REINFORCEMENT

R608.1.3.1 Ashlar masonry. In ashlar masonry, bonder units, uniformly distributed, shall be provided to the extent of not less than 10 percent of the wall area. Such bonder units shall extend not less than 4 inches (102 mm) into the backing wall.

❖ When one or both of the wythes in a multi-wythe masonry wall system are laid in an ashlar pattern, the wythes must be tied together with masonry bonder units spaced as prescribed in this section.

R608.1.3.2 Rubble stone masonry. Rubble stone masonry 24 inches (610 mm) or less in thickness shall have bonder units with a maximum spacing of 3 feet (914 mm) vertically and 3 feet (914 mm) horizontally, and if the masonry is of greater thickness than 24 inches (610 mm), shall have one bonder unit for each 6 square feet (0.557 m²) of wall surface on both sides.

❖ See the commentary for Section R608.1.3.

R608.2 Masonry bonding pattern. Masonry laid in running and stack bond shall conform to Section R608.2.1 and R608.2.2.

❖ Masonry must be laid in running bond (see the definition in Chapter 2) be reinforced horizontally as specified in this section.

R608.2.1 Masonry laid in running bond. In each wythe of masonry laid in running bond, head joints in successive courses shall be offset by not less than one-fourth the unit length, or the masonry walls shall be reinforced longitudinally as required in Section R608.2.2.

❖ When masonry block is placed with an offset, it is called running bond [see Commentary Figure R608.1.2.1(1)]. Running bond is the most commonly used bonding pattern for masonry wall construction.

R608.2.2 Masonry laid in stack bond. Where unit masonry is laid with less head joint offset than in Section R607.2.1, the minimum area of horizontal reinforcement placed in mortar bed joints or in bond beams spaced not more than 48 inches (1219 mm) apart, shall be 0.0007 times the vertical cross-sectional area of the wall.

❖ Blocks stacked vertically are in the stack bond pattern. Refer to the definition in Section R202 of the code and Commentary Figure R606.8(1). The amount of steel in this section is an arbitrary amount to provide continuity across the head joints.

SECTION R609
GROUTED MASONRY

R609.1 General. Grouted multiple-wythe masonry is a form of construction in which the space between the wythes is solidly filled with grout. It is not necessary for the cores of masonry units to be filled with grout. Grouted hollow unit masonry is a form of construction in which certain cells of hollow units are continuously filled with grout.

❖ The provisions of Section R609 pertain to grouted masonry. The two types of grouted masonry, grouted multiple-wythe and grouted hollow unit masonry, are defined in the wording of this section and illustrated in Commentary Figure R609.1. To provide for an ade-

quate bond between the masonry units and the grout, the units must be free of excessive dust or dirt.

R609.1.1 Grout. Grout shall consist of cementitious material and aggregate in accordance with ASTM C 476 and the proportion specifications of Table R609.1.1. Type M or Type S mortar to which sufficient water has been added to produce pouring consistency can be used as grout.

❖ ASTM C 476 contains the requirements for grout. The code identifies two types of grout for masonry construction: fine grout and coarse grout. They differ primarily in the maximum allowable size of aggregates. The proportions in Table R609.1.1 provide a minimum compressive strength of 2,000 psi (13 780 kPa). Type M or Type S mortar (see Section R607.1) may also be used as grout if sufficient water is added to produce a pouring consistency.

R609.1.2 Grouting requirements. Maximum pour heights and the minimum dimensions of spaces provided for grout placement shall conform to Table R609.1.2. If the work is stopped for one hour or longer, the horizontal construction joints shall be formed by stopping all tiers at the same elevation and with the grout 1 inch (25 mm) below the top.

❖ The selection of fine or coarse grout is made based on the size of grout space and the height of grout pour. Table R609.1.2 covers requirements for selection of grout type. Fine grout is required where grout space is small, narrow or too congested with reinforcing steel. Fine grout also can be used where coarse grout is permitted. It is preferable, however, to use coarse grout, because it reduces shrinkage and is less expensive. See Commentary Figure R609.1.2 for an illustration of grout placement.

R609.1.3 Grout space (cleaning). Provision shall be made for cleaning grout space. Mortar projections that project more than 0.5 inch (13 mm) into grout space and any other foreign matter shall be removed from grout space prior to inspection and grouting.

❖ See the commentary for Section R609.1.2.

R609.1.4 Grout placement. Grout shall be a plastic mix suitable for pumping without segregation of the constituents and shall be mixed thoroughly. Grout shall be placed by pumping or by an approved alternate method and shall be placed before any initial set occurs and in no case more than 1¹/₂ hours after water has been added. Grouting shall be done in a continuous pour, in lifts not exceeding 5 feet (1524 mm). It shall be consolidated by puddling or mechanical vibrating during placing and reconsolidated after excess moisture has been absorbed but before plasticity is lost.

❖ See Commentary Figure R609.1.2.

R609.1.4.1 Grout pumped through aluminum pipes. Grout shall not be pumped through aluminum pipes.

❖ Aluminum can adversely affect grout and unprotected reinforcing steel through galvanic action. This restriction protects against possible contamination during grout placement.

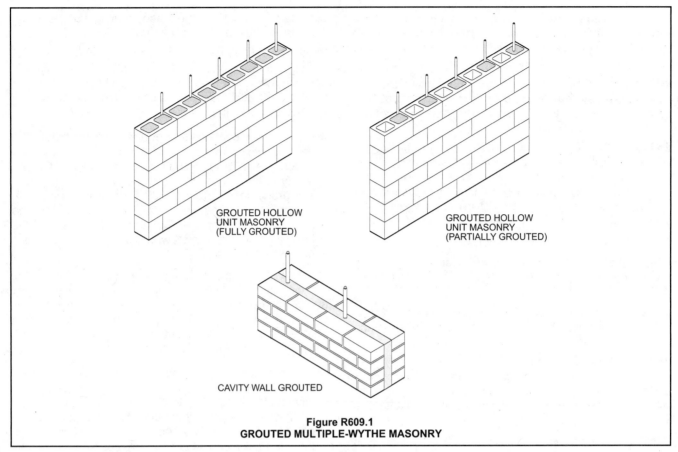

GROUTED HOLLOW
UNIT MASONRY
(FULLY GROUTED)

GROUTED HOLLOW
UNIT MASONRY
(PARTIALLY GROUTED)

CAVITY WALL GROUTED

Figure R609.1
GROUTED MULTIPLE-WYTHE MASONRY

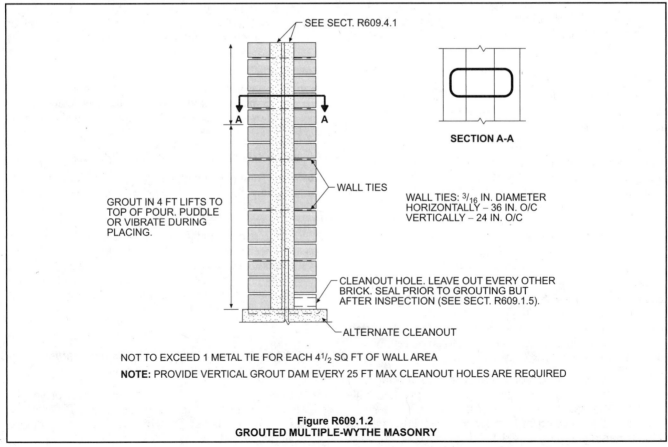

SEE SECT. R609.4.1

SECTION A-A

GROUT IN 4 FT LIFTS TO
TOP OF POUR. PUDDLE
OR VIBRATE DURING
PLACING.

WALL TIES

WALL TIES: $^3/_{16}$ IN. DIAMETER
HORIZONTALLY – 36 IN. O/C
VERTICALLY – 24 IN. O/C

CLEANOUT HOLE. LEAVE OUT EVERY OTHER
BRICK. SEAL PRIOR TO GROUTING BUT
AFTER INSPECTION (SEE SECT. R609.1.5).

ALTERNATE CLEANOUT

NOT TO EXCEED 1 METAL TIE FOR EACH 4$^1/_2$ SQ FT OF WALL AREA

NOTE: PROVIDE VERTICAL GROUT DAM EVERY 25 FT MAX CLEANOUT HOLES ARE REQUIRED

Figure R609.1.2
GROUTED MULTIPLE-WYTHE MASONRY

R609.1.5 Cleanouts. Where required by the building official, cleanouts shall be provided as specified in this section. The cleanouts shall be sealed before grouting and after inspection.

❖ Cleanouts should be furnished as described in this section. These aid the grout-spacing cleaning required by Section R609.1.2.

R609.1.5.1 Grouted multiple-wythe masonry. Cleanouts shall be provided at the bottom course of the exterior wythe at each pour of grout where such pour exceeds 5 feet (1524 mm) in height.

❖ See the commentary for Section R609.1.5.

R609.1.5.2 Grouted hollow unit masonry. Cleanouts shall be provided at the bottom course of each cell to be grouted at each pour of grout, where such pour exceeds 4 feet (1219 mm) in height.

❖ See the commentary for Section R609.1.5.

R609.2 Grouted multiple-wythe masonry. Grouted multiple-wythe masonry shall conform to all the requirements specified in Section R609.1 and the requirements of this section.

❖ Grouted multiple-wythe masonry walls must comply with this section.

R609.2.1 Bonding of backup wythe. Where all interior vertical spaces are filled with grout in multiple-wythe construction, masonry headers shall not be permitted. Metal wall ties shall be used in accordance with Section R608.1.2 to prevent spreading of the wythes and to maintain the vertical alignment of the wall. Wall ties shall be installed in accordance with Section R608.1.2 when the backup wythe in multiple-wythe construction is fully grouted.

❖ Masonry headers are not permitted in double-wythe grouted masonry construction because they span across the wythes and would interfere with the placement of the grout. Metal ties in accordance with Section R608.1.2 of the code prevent the spreading of wythes and maintain alignment during the grouting process.

R609.2.2 Grout spaces. Fine grout shall be used when interior vertical space to receive grout does not exceed 2 inches (51 mm) in thickness. Interior vertical spaces exceeding 2 inches (51 mm) in thickness shall use coarse or fine grout.

❖ Fine grout must be used for grout spaces not more than 2 inches wide (51 mm). This clearance should account for any horizontal wall reinforcing. ASTM C 476 differentiates between fine and coarse grout.

TABLE R609.1.1
GROUT PROPORTIONS BY VOLUME FOR MASONRY CONSTRUCTION

| TYPE | PORTLAND CEMENT OR BLENDED CEMENT SLAG CEMENT | HYDRATED LIME OR LIME PUTTY | AGGREGATE MEASURED IN A DAMP, LOOSE CONDITION | |
			Fine	Coarse
Fine	1	0 to 1/10	$2^1/_4$ to 3 times the sum of the volume of the cementitious materials	—
Coarse	1	0 to 1/10	$2^1/_4$ to 3 times the sum of the volume of the cementitious materials	1 to 2 times the sum of the volumes of the cementitious materials

❖ See the commentary for Section R609.1.1.

TABLE R609.1.2
GROUT SPACE DIMENSIONS AND POUR HEIGHTS

GROUT TYPE	GROUT POUR MAXIMUM HEIGHT (feet)	MINIMUM WIDTH OF GROUT SPACES[a,b] (inches)	MINIMUM GROUT[b,c] SPACE DIMENSIONS FOR GROUTING CELLS OF HOLLOW UNITS (inches x inches)
Fine	1	0.75	1.5 × 2
	5	2	2 × 3
	12	2.5	2.5 × 3
	24	3	3 × 3
Coarse	1	1.5	1.5 × 3
	5	2	2.5 × 3
	12	2.5	3 × 3
	24	3	3 × 4

For SI: 1 inch = 25.4 mm, 1 foot = 304.8 mm.

a. For grouting between masonry wythes.

b. Grout space dimension is the clear dimension between any masonry protrusion and shall be increased by the horizontal projection of the diameters of the horizontal bars within the cross section of the grout space.

c. Area of vertical reinforcement shall not exceed 6 percent of the area of the grout space.

❖ See the commentary for Section R609.1.2.

R609.2.3 Grout barriers. Vertical grout barriers or dams shall be built of solid masonry across the grout space the entire height of the wall to control the flow of the grout horizontally. Grout barriers shall not be more than 25 feet (7620 mm) apart. The grouting of any section of a wall between control barriers shall be completed in one day with no interruptions greater than one hour.

❖ Grout barriers allow large walls to be properly grouted in sections.

R609.3 Reinforced grouted multiple-wythe masonry. Reinforced grouted multiple-wythe masonry shall conform to all the requirements specified in Sections R609.1 and R609.2 and the requirements of this section.

❖ Reinforced grouted multiple-wythe masonry walls must comply with this section in addition to Section R609.2.

R609.3.1 Construction. The thickness of grout or mortar between masonry units and reinforcement shall not be less than $^{1}/_{4}$ inch (7 mm), except that $^{1}/_{4}$-inch (7 mm) bars may be laid in horizontal mortar joints at least $^{1}/_{2}$ inch (13 mm) thick, and steel wire reinforcement may be laid in horizontal mortar joints at least twice the thickness of the wire diameter.

❖ When reinforcing is used in grouted masonry, the dimensional constraints given in this section provide clearances necessary for the adequate placement of the grout within the cavity as well as the proper bond between the reinforcement and the grout. These minimums are illustrated in Commentary Figure R609.3.1.

R609.4 Reinforced hollow unit masonry. Reinforced hollow unit masonry shall conform to all the requirements of Section R609.1 and the requirements of this section.

❖ Reinforced hollow-unit masonry uses hollow masonry units with reinforcement embedded in mortar or grout within certain cells throughout the wall length.

R609.4.1 Construction. Requirements for construction shall be as follows:

1. Reinforced hollow-unit masonry shall be built to preserve the unobstructed vertical continuity of the cells to be filled. Walls and cross webs forming cells to be filled shall be full-bedded in mortar to prevent leakage of grout. Head and end joints shall be solidly filled with mortar for a distance in from the face of the wall or unit not less than the thickness of the longitudinal face shells. Bond shall be provided by lapping units in successive vertical courses.

2. Cells to be filled shall have vertical alignment sufficient to maintain a clear, unobstructed continuous vertical cell of dimensions prescribed in Table R609.1.2.

3. Vertical reinforcement shall be held in position at top and bottom and at intervals not exceeding 200 diameters of the reinforcement.

4. Cells containing reinforcement shall be filled solidly with grout. Grout shall be poured in lifts of 8-foot (2438 mm) maximum height. When a total grout pour exceeds 8 feet (2438 mm) in height, the grout shall be placed in lifts not exceeding 5 feet (1524 mm) and special inspection during grouting shall be required.

5. Horizontal steel shall be fully embedded by grout in an uninterrupted pour.

❖ The general purpose of the five requirements contained in the section is to provide for the proper placement of reinforcing steel and the grouting of cells so that the construction will act structurally as a unit.

Item 1 provides for proper laying of hollow masonry units with appropriate mortar beds, which contain the grout (prevents leakage). The item also contains requirements for the grout space in which reinforcement may be placed [see Commentary Figure R609.4.1(1)].

Item 2 specifies minimum vertical cell-dimension limitations to provide for sufficient clear space in which to place the grout as illustrated in Commentary Figure R609.4.1(2).

Following the provision in Item 3 is necessary in order to maintain the vertical alignment of reinforcing steel during the grout pour. Also, the reinforcement will be maintained in a proper position relative to the block thickness as required by the design if the vertical reinforcement is secured at the top, bottom, and intervening points so that the distance between such anchorage does not exceed 200 bar diameters. Commentary Fig-

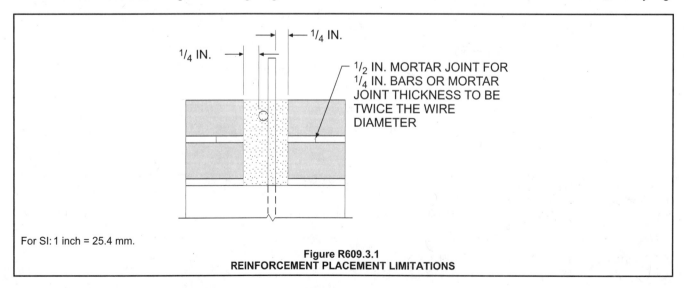

For SI: 1 inch = 25.4 mm.

Figure R609.3.1
REINFORCEMENT PLACEMENT LIMITATIONS

ure R609.4.1(3) illustrates methods of securing the re-inforcement.

Items 4 and 5 provide limitations on the method of grouting the cells containing reinforcement so that the cells to be grouted will be filled solidly and the grout will be reasonably consolidated around the reinforcing steel. Refer to Commentary Figures R609.4.1(4) and R609.4.1(5) for illustrations of the provisions.

SECTION R610
GLASS UNIT MASONRY

R610.1 General. Panels of glass unit masonry located in load-bearing and nonload-bearing exterior and interior walls shall be constructed in accordance with this section.

❖ Code provisions for glass unit masonry are empirical. By definition (see the definition for "Masonry Unit, Glass" in Section R202) these units are nonload-bearing. Therefore, when they are used in a wall opening, a properly designed header or lintel must be installed. Furthermore, panels of glass unit masonry must be isolated from the structure so that in-plane loads are not transferred to the glass.

R610.2 Materials. Hollow glass units shall be partially evacuated and have a minimum average glass face thickness of $^3/_{16}$ inch (5 mm). The surface of units in contact with mortar shall be treated with a polyvinyl butyral coating or latex-based paint. The use of reclaimed units is prohibited.

❖ Glass masonry units may be hollow or solid. Hollow units must have the required minimum face thickness.

The materials specified for treating edges in contact with mortar improve the bonding between glass blocks and mortar.

R610.3 Units. Hollow or solid glass block units shall be standard or thin units.

❖ Hollow or solid-glass units are classified as either standard units or thin units.

R610.3.1 Standard units. The specified thickness of standard units shall be at least $3^7/_8$ inches (98 mm).

❖ See the commentary for Section R610.3.

R610.3.2 Thin units. The specified thickness of thin units shall be at least $3^1/_8$ inches (79 mm) for hollow units and at least 3 inches (76 mm) for solid units.

❖ See the commentary for Section R610.3.

R610.4 Isolated panels. Isolated panels of glass unit masonry shall conform to the requirements of this section.

❖ The code limits hollow glass unit masonry panel sizes based on structural and performance considerations. Height limits are more restrictive than length limits, primarily as a result to historical requirements rather than actual field experience or engineering principles. Exterior panels constructed of standard units may be up to 144 square feet (13.4 m²). Exterior panels constructed of thin units are limited to 85 square feet (7.9 m²). Interior panels are limited to 250 square feet (23.2 m²) for standard units and 150 square feet (13.9 m²) for thin units. These limitations are derived from ACI 530/ASCE5/ TMS402.

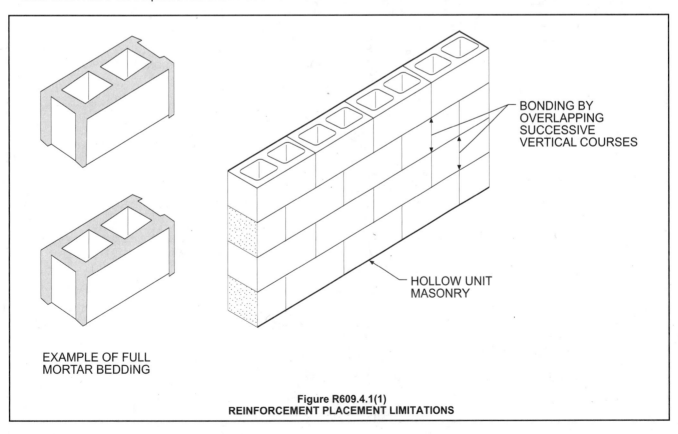

EXAMPLE OF FULL
MORTAR BEDDING

BONDING BY
OVERLAPPING
SUCCESSIVE
VERTICAL COURSES

HOLLOW UNIT
MASONRY

Figure R609.4.1(1)
REINFORCEMENT PLACEMENT LIMITATIONS

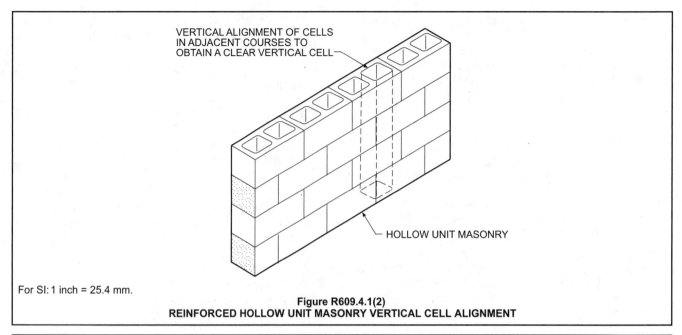

For SI: 1 inch = 25.4 mm.

Figure R609.4.1(2)
REINFORCED HOLLOW UNIT MASONRY VERTICAL CELL ALIGNMENT

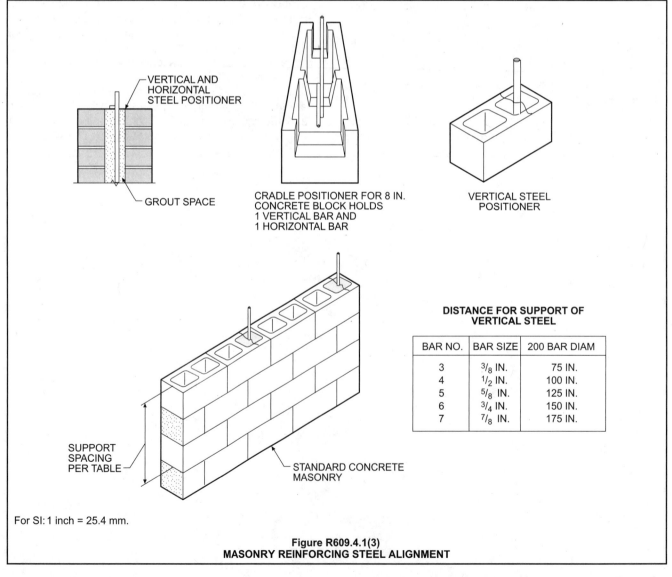

VERTICAL ALIGNMENT OF CELLS IN ADJACENT COURSES TO OBTAIN A CLEAR VERTICAL CELL

HOLLOW UNIT MASONRY

VERTICAL AND HORIZONTAL STEEL POSITIONER

GROUT SPACE

CRADLE POSITIONER FOR 8 IN. CONCRETE BLOCK HOLDS 1 VERTICAL BAR AND 1 HORIZONTAL BAR

VERTICAL STEEL POSITIONER

SUPPORT SPACING PER TABLE

STANDARD CONCRETE MASONRY

DISTANCE FOR SUPPORT OF VERTICAL STEEL

BAR NO.	BAR SIZE	200 BAR DIAM
3	$3/8$ IN.	75 IN.
4	$1/2$ IN.	100 IN.
5	$5/8$ IN.	125 IN.
6	$3/4$ IN.	150 IN.
7	$7/8$ IN.	175 IN.

For SI: 1 inch = 25.4 mm.

Figure R609.4.1(3)
MASONRY REINFORCING STEEL ALIGNMENT

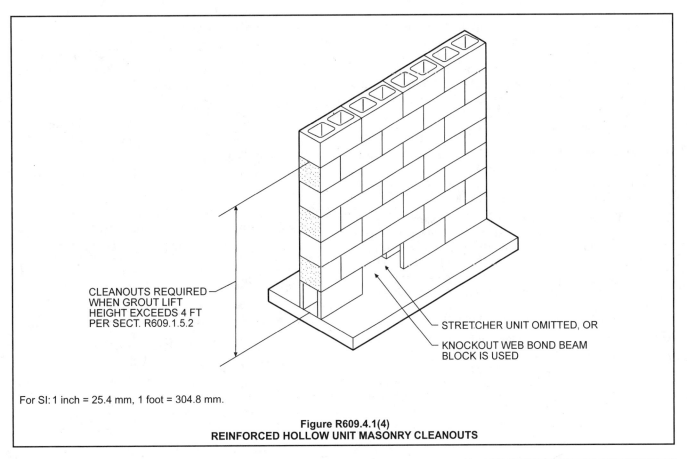

CLEANOUTS REQUIRED
WHEN GROUT LIFT
HEIGHT EXCEEDS 4 FT
PER SECT. R609.1.5.2

STRETCHER UNIT OMITTED, OR

KNOCKOUT WEB BOND BEAM
BLOCK IS USED

For SI: 1 inch = 25.4 mm, 1 foot = 304.8 mm.

Figure R609.4.1(4)
REINFORCED HOLLOW UNIT MASONRY CLEANOUTS

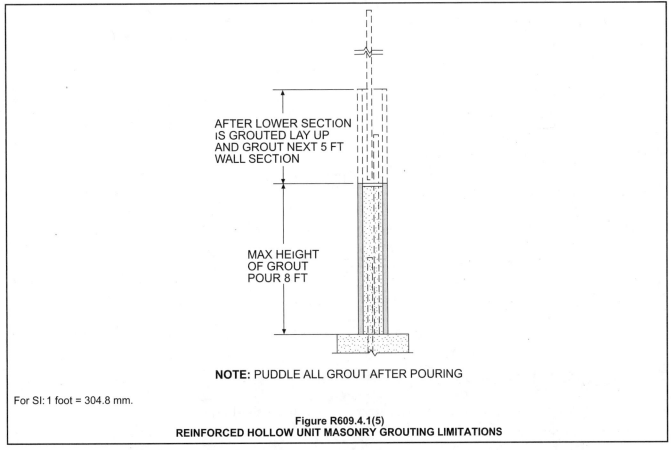

AFTER LOWER SECTION
IS GROUTED LAY UP
AND GROUT NEXT 5 FT
WALL SECTION

MAX HEIGHT
OF GROUT
POUR 8 FT

NOTE: PUDDLE ALL GROUT AFTER POURING

For SI: 1 foot = 304.8 mm.

Figure R609.4.1(5)
REINFORCED HOLLOW UNIT MASONRY GROUTING LIMITATIONS

R610.4.1 Exterior standard-unit panels. The maximum area of each individual standard-unit panel shall be 144 square feet (13.4 m²) when the design wind pressure is 20 psf (958 Pa). The maximum area of such panels subjected to design wind pressures other than 20 psf (958 Pa) shall be in accordance with Figure R610.4.1. The maximum panel dimension between structural supports shall be 25 feet (7620 mm) in width or 20 feet (6096 mm) in height.

❖ The area limit specified in this section corresponds to a wind pressure of 20 psf (958 Pa). The area should be adjusted using Figure R610.4.1 for other values of wind pressure. See the commentary for Section R610.4.

R610.4.2 Exterior thin-unit panels. The maximum area of each individual thin-unit panel shall be 85 square feet (7.9 m²). The maximum dimension between structural supports shall be 15 feet (4572 mm) in width or 10 feet (3048 mm) in height. Thin units shall not be used in applications where the design wind pressure as stated in Table R301.2(1) exceeds 20 psf (958 Pa).

❖ See the commentary for Section R610.4.

R610.4.3 Interior panels. The maximum area of each individual standard-unit panel shall be 250 square feet (23.2 m²). The maximum area of each thin-unit panel shall be 150 square feet (13.9 m²). The maximum dimension between structural supports shall be 25 feet (7620 mm) in width or 20 feet (6096 mm) in height.

❖ See the commentary for Section R610.4.

R610.4.4 Curved panels. The width of curved panels shall conform to the requirements of Sections R610.4.1, R610.4.2 and R610.4.3, except additional structural supports shall be provided at locations where a curved section joins a straight section, and at inflection points in multicurved walls.

❖ Support is required for standard and thin units, exterior and interior panels, at locations where a curved panel joins a straight section, and at inflection points. See the commentary for Section R610.4.

R610.5 Panel support. Glass unit masonry panels shall conform to the support requirements of this section.

❖ Structural members supporting glass-unit masonry panels must meet the deflection limit of this section to minimize the potential for cracking in the glass panels. Lateral support is required for the top and sides of panels and may be accomplished by anchors or channel-type restraints.

R610.5.1 Deflection. The maximum total deflection of structural members that support glass unit masonry shall not exceed $1/_{600}$.

❖ See the commentary for Section R610.5.

R610.5.2 Lateral support. Glass unit masonry panels shall be laterally supported along the top and sides of the panel. Lateral supports for glass unit masonry panels shall be designed to resist a minimum of 200 pounds per lineal feet (2918 N/m) of panel, or the actual applied loads, whichever is greater. Except

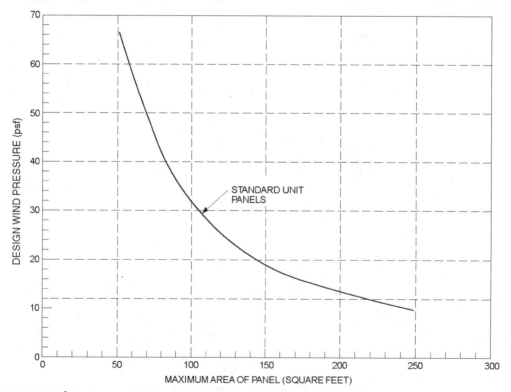

For SI: 1 square foot = 0.0929 m², 1 pound per square foot = 0.0479 kPa.

FIGURE R610.4.1
GLASS UNIT MASONRY DESIGN WIND LOAD RESISTANCE

❖ See the commentary for Section R610.4.1.

for single unit panels, lateral support shall be provided by panel anchors along the top and sides spaced a maximum of 16 inches (406 mm) on center or by channel-type restraints. Single unit panels shall be supported by channel-type restraints.

Exceptions:

1. Lateral support is not required at the top of panels that are one unit wide.

2. Lateral support is not required at the sides of panels that are one unit high.

❖ See the commentary for Section R610.5.

R610.5.2.1 Panel anchor restraints. Panel anchors shall be spaced a maximum of 16 inches (406 mm) on center in both jambs and across the head. Panel anchors shall be embedded a minimum of 12 inches (305 mm) and shall be provided with two fasteners so as to resist the loads specified in Section R610.5.2.

❖ See Commentary Figure R610.5.2.1.

R610.5.2.2 Channel-type restraints. Glass unit masonry panels shall be recessed at least 1 inch (25 mm) within channels and chases. Channel-type restraints shall be oversized to accommodate expansion material in the opening, packing and sealant between the framing restraints, and the glass unit masonry perimeter units.

❖ This type of lateral support is an option to the anchors described in the previous section. No specific material is specified for the channel, which means that any channel-type restraint providing the lateral support specified by Section R610.5.2 is acceptable.

R610.6 Sills. Before bedding of glass units, the sill area shall be covered with a water base asphaltic emulsion coating. The coating shall be a minimum of $^1/_8$ inch (3 mm) thick.

❖ Application of asphaltic emulsion is recommended by glass block manufacturers (see Commentary Figure R610.5.2.1).

R610.7 Expansion joints. Glass unit masonry panels shall be provided with expansion joints along the top and sides at all structural supports. Expansion joints shall be a minimum of $^3/_8$ inch (10 mm) in thickness and shall have sufficient thickness to accommodate displacements of the supporting structure. Expansion joints shall be entirely free of mortar and other debris and shall be filled with resilient material.

❖ See Commentary Figure R610.5.2.1.

R610.8 Mortar. Glass unit masonry shall be laid with Type S or N mortar. Mortar shall not be retempered after initial set. Mortar unused within $1^1/_2$ hours after initial mixing shall be discarded.

❖ See the commentary for Section R607.1.

R610.9 Reinforcement. Glass unit masonry panels shall have horizontal joint reinforcement spaced a maximum of 16 inches (406 mm) on center located in the mortar bed joint. Horizontal joint reinforcement shall extend the entire length of the panel but shall not extend across expansion joints. Longitudinal wires shall be lapped a minimum of 6 inches (152 mm) at splices. Joint reinforcement shall be placed in the bed joint

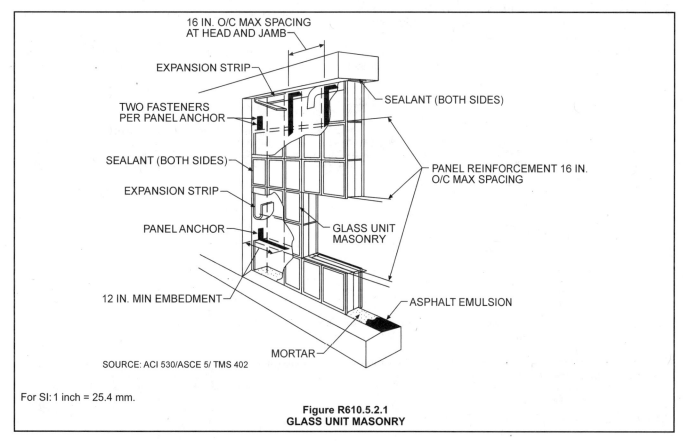

16 IN. O/C MAX SPACING AT HEAD AND JAMB

EXPANSION STRIP

TWO FASTENERS PER PANEL ANCHOR

SEALANT (BOTH SIDES)

EXPANSION STRIP

PANEL ANCHOR

12 IN. MIN EMBEDMENT

MORTAR

SEALANT (BOTH SIDES)

PANEL REINFORCEMENT 16 IN. O/C MAX SPACING

GLASS UNIT MASONRY

ASPHALT EMULSION

SOURCE: ACI 530/ASCE 5/ TMS 402

For SI: 1 inch = 25.4 mm.

**Figure R610.5.2.1
GLASS UNIT MASONRY**

immediately below and above openings in the panel. The reinforcement shall have not less than two parallel longitudinal wires of size W1.7 or greater, and have welded cross wires of size W1.7 or greater.

❖ See Commentary Figure R610.5.2.1.

R610.10 Placement. Glass units shall be placed so head and bed joints are filled solidly. Mortar shall not be furrowed. Head and bed joints of glass unit masonry shall be $^1/_4$ inch (6.4 mm) thick, except that vertical joint thickness of radial panels shall not be less than $^1/_8$ inch (3 mm) or greater than $^5/_8$ inch (16 mm). The bed joint thickness tolerance shall be minus $^1/_{16}$ inch (1.6 mm) and plus $^1/_8$ inch (3 mm). The head joint thickness tolerance shall be plus or minus $^1/_8$ inch (3 mm).

❖ This section specifies the joint thicknesses and tolerances consistent with ACI 530/ASCE 5/TMS 402.

SECTION R611
INSULATING CONCRETE FORM
WALL CONSTRUCTION

R611.1 General. Insulating Concrete Form (IFC) walls shall be designed and constructed in accordance with the provisions of this section or in accordance with the provisions of ACI 318. When ACI 318 or the provisions of this section are used to design insulating concrete form walls, project drawings, typical details and specifications are not required to bear the seal of the architect or engineer responsible for design, unless otherwise required by the state law of the jurisdiction having authority.

❖ Insulating Concrete Forms (IFC) refer to hollow blocks, planks, or panels of rigid foam plastic insulation that are used as stay-in-place forms for concrete walls. These provisions cover three types of ICF walls; flat, waffle-grid and screen-grid. Wall type descriptions are based on the shape of the concrete inside the form, since it is the concrete that provides the load-carrying capability. This section provides prescriptive requirements for design and construction of ICF walls. Designs per ACI 318 are acceptable alternatives.

R611.2 Applicability limits. The provisions of this section shall apply to the construction of insulating concrete form walls for buildings not greater than 60 feet (18 288 mm) in plan dimensions, and floors not greater than 32 feet (9754 mm) or roofs not greater than 40 feet (12 192 mm) in clear span. Buildings shall not exceed two stories in height above-grade. ICF walls shall comply with the requirements in Table R611.2. Walls constructed in accordance with the provisions of this section shall be limited to buildings subjected to a maximum design wind speed of 150 miles per hour (67 m/s), and Seismic Design Categories A, B, C, D_0, D_1 and D_2. The provisions of this section shall not apply to the construction of ICF walls for buildings or portions of buildings considered irregular as defined in Section R301.2.2.2.2.

For townhouses in Seismic Design Category C and all buildings in Seismic Design Category D_0, D_1 or D_2, the provisions of this section shall apply only to buildings meeting the following requirements.

1. Rectangular buildings with a maximum building aspect ratio of 2:1. The building aspect ratio shall be determined by dividing the longest dimension of the building by the shortest dimension of the building.

2. Walls are aligned vertically with the walls below.

3. Cantilever and setback construction shall not be permitted.

4. The weight of interior and exterior finishes applied to ICF walls shall not exceed 8 psf (380 Pa).

TABLE R611.2
REQUIREMENTS FOR ICF WALLS[b]

WALL TYPE AND NOMINAL SIZE	MAXIMUM WALL WEIGHT (psf)[c]	MINIMUM WIDTH OF VERTICAL CORE (inches)[a]	MINIMUM THICKNESS OF VERTICAL CORE (inches)[a]	MAXIMUM SPACING OF VERTICAL CORES (inches)	MAXIMUM SPACING OF HORIZONTAL CORES (inches)	MINIMUM WEB THICKNESS (inches)
3.5″ Flat[d]	44[d]	N/A	N/A	N/A	N/A	N/A
5.5″ Flat	69	N/A	N/A	N/A	N/A	N/A
7.5″ Flat	94	N/A	N/A	N/A	N/A	N/A
9.5″ Flat	119	N/A	N/A	N/A	N/A	N/A
6″ Waffle-Grid	56	6.25	5	12	16	2
8″ Waffle-Grid	76	7	7	12	16	2
6″ Screen-Grid	53	5.5	5.5	12	12	N/A

For SI: 1 inch = 25.4 mm; 1 pound per cubic foot = 16.018 kg/m^3; 1 pound per square foot = 0.0479 kPa.

a. For width "W", thickness "T", spacing, and web thickness, refer to Figures R611.4 and R611.5.

b. N/A indicates not applicable.

c. Wall weight is based on a unit weight of concrete of 150 pcf. The tabulated values do not include any allowance for interior and exterior finishes.

d. For all buildings in Seismic Design Category A or B, and detached one- and two-family dwellings in Seismic Design Category C the actual wall thickness is permitted to be up to 1 inch thicker than shown and the maximum wall weight to be 56 psf. Construction requirements and other limitations within Section R611 for 3.5-inch flat ICF walls shall apply. Interpolation between provisions for 3.5-inch and 5.5-inch flat ICF walls is not permitted.

❖ See the commentary for Section R611.2.

5. The gable portion of ICF walls shall be constructed of light-frame construction.

❖ To keep the prescriptive criteria from becoming overly complex, this section establishes practical limits, which, in general, are related to the complexity of the structure in dealing with seismic design or the overall size of the structure (see Commentary Figure R611.2). An engineered design in accordance with ACI 318 is required for buildings that fall outside these limits. The prescriptive details given in Section R611 do not include irregular buildings because such buildings require a higher level of design complexity in seismic design than can be adequately dealt with in the prescriptive requirements of this section.

The code contains even more restrictions for townhouses in Seismic Design Category C and all buildings in Seismic Design Categories D_0, D_1 and D_2. These structures require a higher level of design complexity in higher seismic categories. Therefore, these limitations are required to limit the number of permutations possible in the design as practical limits for the number of prescriptive tables and details that can be reasonably installed in this code. Again, buildings outside of these limitations would be possible, but an engineered design in accordance with ACI 318 would be required.

R611.3 Flat insulating concrete form wall systems. Flat ICF wall systems shall comply with Figure R611.3 and shall have reinforcement in accordance with Tables R611.3(1) and R611.3(2) and Section R611.7.

❖ Figure R611.3 illustrates flat ICF wall construction. Vertical wall reinforcing should be provided where required by Table 611.3(2). Table 611.3(1) provides conversion of wind speed to design wind pressure for different wind exposure categories. These pressures are used in the design wind pressure column of Table 611.3(2). Reinforcing requirements vary based on basic wind speed, wall height, elements being supported and wall thickness.

R611.4 Waffle-grid insulating concrete form wall systems. Waffle-grid wall systems shall comply with Figure R611.4 and shall have reinforcement in accordance with Tables R611.3(1) and R611.4(1) and Section R611.7. The minimum core dimensions shall comply with Table R611.2.

❖ Figure R611.4 illustrates waffle-grid ICF construction, which must conform to the dimensions in Table R611.2. Vertical wall reinforcing should be provided where required by Table R611.4(1). Table 611.3(1) provides conversion of wind speed to design wind pressure for different wind exposure categories. These pressures are used in the design wind pressure column of Table R611.4(1). Reinforcing requirements vary based on basic wind speed, wall height, elements being supported and wall thickness.

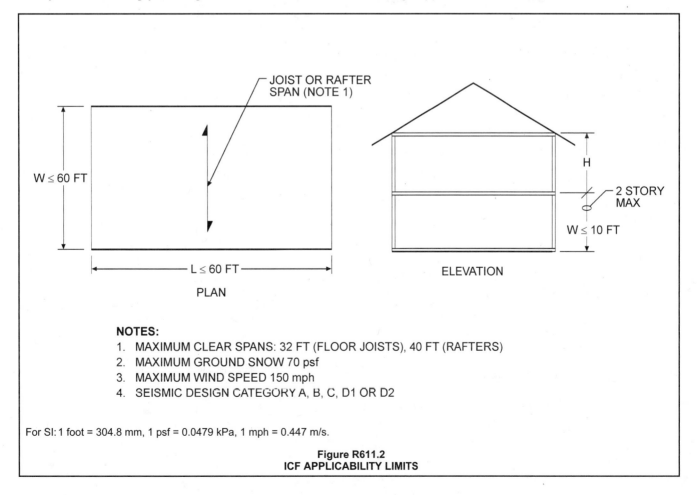

NOTES:
1. MAXIMUM CLEAR SPANS: 32 FT (FLOOR JOISTS), 40 FT (RAFTERS)
2. MAXIMUM GROUND SNOW 70 psf
3. MAXIMUM WIND SPEED 150 mph
4. SEISMIC DESIGN CATEGORY A, B, C, D1 OR D2

For SI: 1 foot = 304.8 mm, 1 psf = 0.0479 kPa, 1 mph = 0.447 m/s.

Figure R611.2
ICF APPLICABILITY LIMITS

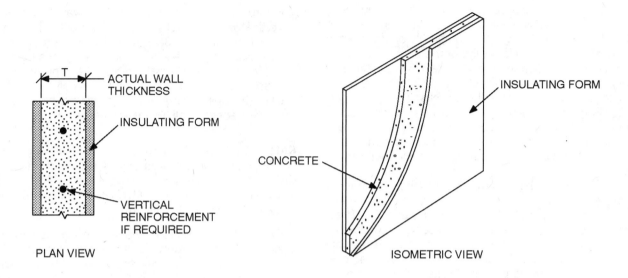

FIGURE R611.3
FLAT ICF WALL SYSTEM

❖ See the commentary for Section R611.3.

TABLE R611.3(1)
DESIGN WIND PRESSURE FOR USE WITH TABLES R611.3(2), R611.4(1), AND R611.5 FOR ABOVE GRADE WALLS[a]

WIND SPEED (mph)[e]	DESIGN WIND PRESSURE (psf)					
	Enclosed[b]			Partially Enclosed[b]		
	Exposure[c]			Exposure[c]		
	B	C	D	B	C	D
85	18	24	29	23	31	37
90	20	27	32	25	35	41
100	24	34	39	31	43	51
110	29	41	48	38	52	61
120	35	48	57	45	62	73
130	41	56	66	53	73	85[d]
140	47	65	77	61	84[d]	99[d]
150	54	75	88[d]	70	96[d]	114[d]

For SI: 1 pound per square foot = 0.0479 kPa; 1 mile per hour = 0.447 m/s; 1 foot = 304.8 mm; 1 square foot = 0.0929 m².

a. This table is based on ASCE 7-98 components and cladding wind pressures using a mean roof height of 35 ft and a tributary area of 10 ft².

b. Buildings in wind-borne debris regions as defined in Section R202 shall be considered as "Partially Enclosed" unless glazed openings are protected in accordance with Section R301.2.1.2, in which case the building shall be considered as "Enclosed." All other buildings shall be classified as "Enclosed."

c. Exposure Categories shall be determined in accordance with Section R301.2.1.4.

d. For wind pressures greater than 80 psf, design is required in accordance with ACI 318 and approved manufacturer guidelines.

e. Interpolation is permitted between wind speeds.

❖ See the commentary for Sections R611.3 through R611.5.

TABLE R611.3(2)
MINIMUM VERTICAL WALL REINFORCEMENT FOR FLAT ICF ABOVE-GRADE WALLS[a, b, c, d]

Design Wind Pressure [Table R611.3(1)] (psf)	Maximum Unsupported Wall Height (feet)	Minimum Vertical Reinforcement[d, e, f]					
		Nonload-Bearing Wall or Supporting Roof		Supporting Light-Framed Second Story and Roof		Supporting ICF Second Story and Roof	
		Minimum Wall Thickness (inches)					
		3.5[g]	5.5	3.5[g]	5.5	3.5[g]	5.5
20	8	#4@48	#4@48	#4@48	#4@48	#4@48	#4@48
	9	#4@48	#4@48	#4@48	#4@48	#4@48	#4@48
	10	#4@38	#4@48	#4@40	#4@48	#4@42	#4@48
30	8	#4@42	#4@48	#4@46	#4@48	#4@48	#4@48
	9	#4@32; #5@48	#4@48	#4@34; #5@48	#4@48	#4@34; #5@48	#4@48
	10	Design Required	#4@48	Design Required	#4@48	Design Required	#4@48
40	8	#4@30; #5@48	#4@48	#4@30; #5@48	#4@48	#4@32; #5@48	#4@48
	9	Design Required	#4@42	Design Required	#4@46	Design Required	#4@48
	10	Design Required	#4@32; #5@48	Design Required	#4@34; #5@48	Design Required	#4@38
50	8	#4@20; #5@30	#4@42	#4@22; #5@34	#4@46	#4@24; #5@36	#4@48
	9	Design Required	#4@34; #5@48	Design Required	#4@34; #5@48	Design Required	#4@38
	10	Design Required	#4@26; #5@38	Design Required	#4@26; #5@38	Design Required	#4@28; #5@46
60	8	Design Required	#4@34; #5@48	Design Required	#4@36	Design Required	#4@40
	9	Design Required	#4@26; #5@38	Design Required	#4@28; #5@46	Design Required	#4@34; #5@48
	10	Design Required	#4@22; #5@34	Design Required	#4@22; #5@34	Design Required	#4@26; #5@38
70	8	Design Required	#4@28; #5@46	Design Required	#4@30; #5@48	Design Required	#4@34; #5@48
	9	Design Required	#4@22; #5@34	Design Required	#4@22; #5@34	Design Required	#4@24; #5@36
	10	Design Required	#4@16; #5@26	Design Required	#4@18; #5@28	Design Required	#4@20; #5@30
80	8	Design Required	#4@26; #5@38	Design Required	#4@26; #5@38	Design Required	#4@28; #5@46
	9	Design Required	#4@20; #5@30	Design Required	#4@20; #5@30	Design Required	#4@21; #5@34
	10	Design Required	#4@14; #5@24	Design Required	#4@14; #5@24	Design Required	#4@16; #5@26

For SI: 1 inch = 25.4 mm; 1 foot = 304.8 mm; 1 mile per hour = 0.447 m/s; 1 pound per square inch = 6.895 kPa.

a. This table is based on reinforcing bars with a minimum yield strength of 40,000 psi and concrete with a minimum specified compressive strength of 2,500 psi. For Seismic Design Categories D_0, D_1 and D_2, reinforcing bars shall have a minimum yield strength of 60,000 psi. See Section R611.6.2.

b. Deflection criterion is $L/240$, where L is the height of the wall story in inches.

c. Interpolation shall not be permitted.

d. Reinforcement spacing for 3.5 inch walls shall be permitted to be multiplied by 1.6 when reinforcing steel with a minimum yield strength of 60,000 psi is used. Reinforcement shall not be less than one #4 bar at 48 inches (1.2 m) on center.

e. Reinforcement spacing for 5.5 inch (139.7 mm) walls shall be permitted to be multiplied by 1.5 when reinforcing steel with a minimum yield strength of 60,000 psi is used. Reinforcement shall not be less than one #4 bar at 48 inches on center.

f. See Section R611.7.1.2 for limitations on maximum spacing of vertical reinforcement in Seismic Design Categories C, D_0, D_1 and D_2.

g. A 3.5-inch wall shall not be permitted if wood ledgers are used to support floor or roof loads. See Section R611.8.

❖ See the commentary for Section R611.3.

TABLE R611.4(1)
MINIMUM VERTICAL WALL REINFORCEMENT FOR WAFFLE-GRID ICF ABOVE-GRADE WALLS[a, b, c]

Design Wind Pressure [Table R611.3(1)] (psf)	Maximum Unsupported Wall Height (feet)	MINIMUM VERTICAL REINFORCEMENT[d, e]					
		Nonload-Bearing Wall or Supporting Roof		Supporting Light-Framed Second Story and Roof		Supporting ICF Second Story and Roof	
		Minimum Wall Thickness (inches)					
		6	8	6	8	6	8
20	8	#4@48	#4@48	#4@48	#4@48	#4@48	#4@48
	9	#4@48	#4@48	#4@48	#4@48	#4@48	#4@48
	10	#4@48	#4@48	#4@48	#4@48	#4@48	#4@48
30	8	#4@48	#4@48	#4@48	#4@48	#4@48	#4@48
	9	#4@48	#4@48	#4@48	#4@48	#4@48	#4@48
	10	#4@36; #5@48	#4@48	#4@36; #5@48	#4@48	#4@36; #5@48	#4@48
40	8	#4@36; #5@48	#4@48	#4@48	#4@48	#4@48	#4@48
	9	#4@36; #5@48	#4@48	#4@36; #5@48	#4@48	#4@36; #5@48	#4@48
	10	#4@24; #5@36	#4@36; #5@48	#4@24; #5@36	#4@48	#4@24; #5@36	#4@48
50	8	#4@36; #5@48	#4@48	#4@36; #5@48	#4@48	#4@36; #5@48	#4@48
	9	#4@24; #5@36	#4@36; #5@48	#4@24; #5@36	#4@48	#4@24; #5@48	#4@48
	10	Design Required	#4@36; #5@48	Design Required	#4@36; #5@48	Design Required	#4@36; #5@48
60	8	#4@24; #5@36	#4@48	#4@24; #5@36	#4@48	#4@24; #5@48	#4@48
	9	Design Required	#4@36; #5@48	Design Required	#4@36; #5@48	Design Required	#4@36; #5@48
	10	Design Required	#4@24; #5@36	Design Required	#4@24; #5@36	Design Required	#4@24; #5@48
70	8	#4@24; #5@36	#4@36; #5@48	#4@24; #5@36	#4@36; #5@48	#4@24; #5@36	#4@48
	9	Design Required	#4@24; #5@36	Design Required	#4@24; #5@48	Design Required	#4@24; #5@48
	10	Design Required	#4@12; #5@36	Design Required	#4@24; #5@36	Design Required	#4@24; #5@36
80	8	#4@12; #5@24	#4@24; #5@48	#4@12; #5@24	#4@24; #5@48	#4@12; #5@24	#4@36; #5@48
	9	Design Required	#4@24; #5@36	Design Required	#4@24; #5@36	Design Required	#4@24; #5@36
	10	Design Required	#4@12; #5@24	Design Required	#4@12; #5@24	Design Required	#4@12; #5@24

For SI: 1 foot = 304.8 mm; 1 inch = 25.4 mm; 1 mile per hour = 0.447 m/s; 1 pound per square inch = 6.895 MPa.

a. This table is based on reinforcing bars with a minimum yield strength of 40,000 psi and concrete with a minimum specified compressive strength of 2,500 psi. For Seismic Design Categories D_0, D_1 and D_2, reinforcing bars shall have a minimum yield strength of 60,000 psi. See Section R611.6.2.

b. Deflection criterion is $L/240$, where L is the height of the wall story in inches.

c. Interpolation shall not be permitted.

d. Increasing reinforcement spacing by 12 inches shall be permitted when reinforcing steel with a minimum yield strength of 60,000 psi is used or substitution of No. 4 reinforcing bars for #5 bars shall be permitted when reinforcing steel with a minimum yield strength of 60,000 psi is used at the same spacing required for #5 bars. Reinforcement shall not be less than one #4 bar at 48 inches on center.

e. See Section R611.7.1.2 for limitations on maximum spacing of vertical reinforcement in Seismic Design Categories C, D_0, D_1 and D_2.

❖ See the commentary for Section R611.4.

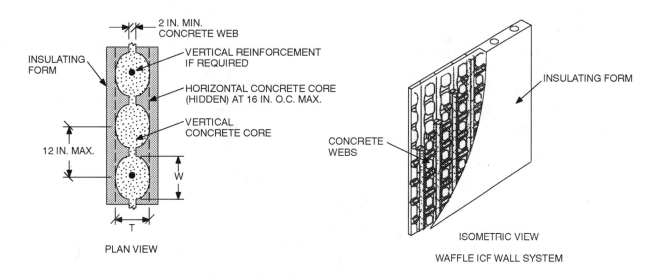

FIGURE R611.4
WAFFLE-GRID ICF WALL SYSTEM

❖ See the commentary for Section R611.4.

R611.5 Screen-grid insulating concrete form wall systems.
Screen-grid ICF wall systems shall comply with Figure R611.5 and shall have reinforcement in accordance with Tables R611.3(1) and R611.5 and Section R611.7. The minimum core dimensions shall comply with Table R611.2.

❖ Figure R611.5 illustrates screen-grid ICF construction, which must conform to the dimensions in Table R611.2. Vertical wall reinforcing should be installed where required by Table R611.5. Table 611.3(1) contains conversion of wind speed to design wind pressure for different wind exposure categories. These pressures are used in the design wind pressure column of Table R611.5. Reinforcing requirements vary based on basic wind speed, wall height, elements being supported and wall thickness.

R611.6 Material. Insulating concrete form wall materials shall comply with this section.

❖ ICFs are composed of three basic materials: concrete, reinforcing steel for concrete and foam plastic insulation. This section contains the scoping text for the three subsections that follow.

R611.6.1 Concrete material. Ready-mixed concrete for insulating concrete form walls shall be in accordance with Section R402.2. Maximum slump shall not be greater than 6 inches (152 mm) as determined in accordance with ASTM C 143. Maximum aggregate size shall not be larger than $^3/_4$ inch (19 mm).

Exception: Concrete mixes conforming to the ICF manufacturer's recommendations.

In Seismic Design Categories D_0, D_1 and D_2, the minimum concrete compressive strength shall be 3,000 psi (20.5 MPa).

❖ Concrete strengths are determined using Section R402.2 and the requirements of this section. The limits on concrete slump and aggregate size are consistent with current practice in ICF construction. These limits consider ease of concrete placement and the ability to completely fill forms while limiting the fluid pressure on these forms from the freshly poured concrete. Minimum concrete strengths for Seismic Design Categories D_0, D_1 and D_2 are increased because of the greater loads imposed on these structures in these seismic design categories.

R611.6.2 Reinforcing steel. Reinforcing steel shall meet the requirements of ASTM A 615, A 706, or A 996. Except in Seismic Design Categories D_0, D_1 and D_2, the minimum yield strength of reinforcing steel shall be 40,000 psi (Grade 40) (276 MPa). In Seismic Design Categories D_0, D_1 and D_2, reinforcing steel shall meet the requirements of ASTM A 706 for low-alloy steel with a minimum yield strength of 60,000 psi (Grade 60) (414 Mpa).

❖ The prescriptive reinforcement requirements for ICF walls of buildings in Seismic Design Categories A, B and C were developed using Grade 40 reinforcing steel. Because this grade is not readily available, it is likely that Grade 60 steel will be furnished. In this case, the reinforcement required by the tables may be conservative. Because of the higher loads imposed on buildings in Seismic Design Categories D_0, D_1 and D_2, Grade 60 reinforcing steel is required. In addition, the material specification must be ASTM A 706, for low-alloy steel. This is consistent with ACI 318 requirements for design in high seismic areas. The A 706 material is more consistent in the actual yield stress of the material produced compared to the required one. It is important to avoid using a material with a tensile strength that is too high, which could lead to undesirable brittle failure in concrete.

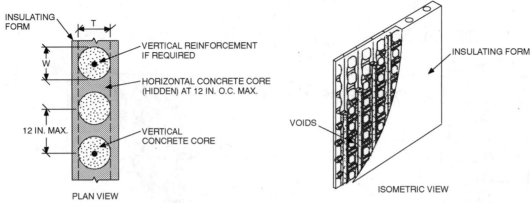

For SI: 1 inch = 25.4 mm.

FIGURE R611.5
SCREEN-GRID IFC WALL SYSTEM

TABLE R611.5
MINIMUM VERTICAL WALL REINFORCEMENT FOR SCREEN-GRID ICF ABOVE-GRADE WALLS[a, b, c]

DESIGN WIND PRESSURE [TABLE R611.3(1)] (psf)	MAXIMUM UNSUPPORTED WALL HEIGHT (feet)	MINIMUM VERTICAL REINFORCEMENT[d,e]		
		Nonload-Bearing Wall or Supporting Roof	Supporting Light-Framed Second Story and Roof	Supporting ICF Second Story and Roof
20	8	#4@48	#4@48	#4@48
	9	#4@48	#4@48	#4@48
	10	#4@48	#4@48	#4@48
30	8	#4@48	#4@48	#4@48
	9	#4@48	#4@48	#4@48
	10	#4@36; #5@48	#4@48	#4@48
40	8	#4@48	#4@48	#4@48
	9	#4@36; #5@48	#4@36; #5@48	#4@48
	10	#4@24; #5@48	#4@24; #5@48	#4@24; #5@48
50	8	#4@36; #5@48	#4@36; #5@48	#4@48
	9	#4@24; #5@48	#4@24; #5@48	#4@24; #5@48
	10	Design Required	Design Required	Design Required
60	8	#4@24; #5@48	#4@24; #5@48	#4@36; #5@48
	9	#4@24; #5@36	#4@24; #5@36	#4@24; #5@36
	10	Design Required	Design Required	Design Required
70	8	#4@24; #5@36	#4@24; #5@36	#4@24; #5@36
	9	Design Required	Design Required	Design Required
	10	Design Required	Design Required	Design Required
080	8	#4@12; #5@36	#4@24; #5@36	#4@24; #5@36
	9	Design Required	Design Required	Design Required
	10	Design Required	Design Required	Design Required

For SI: 1 inch = 25.4 mm, 1 foot = 304.8 mm, 1 mile per hour = 0.447 m/s; 1 pound per square inch = 6.895 kPa.

a. This table is based on reinforcing bars with a minimum yield strength of 40,000 psi and concrete with a minimum specified compressive strength of 2,500 psi. For Seismic Design Categories D_0, D_1 and D_2, reinforcing bars shall have a minimum yield strength of 60,000 psi. See Section R611.6.2.

b. Deflection criterion is $L/240$, where L is the height of the wall story in inches.

c. Interpolation shall not be permitted.

d. Increasing reinforcement spacing by 12 inches shall be permitted when reinforcing steel with a minimum yield strength of 60,000 psi is used. Reinforcement shall not be less than one #4 bar at 48 inches on center.

e. See Section R611.7.1.2 for limitations on maximum spacing of vertical reinforcement in Seismic Design Categories C, D_0, D_1 and D_2.

❖ See the commentary for Section R611.5.

R611.6.3 Insulation materials. Insulating concrete forms material shall meet the surface burning characteristics of Section R314.3. A thermal barrier shall be provided on the building interior in accordance with Section R314.4 or Section R702.3.4.

❖ Insulating form material must meet the surface-burning characteristics specified in Section R314.3, and a thermal barrier is required in accordance with Section R314.4.

R611.7 Wall construction. Insulating concrete form walls shall be constructed in accordance with the provisions of this section and Figure R611.7(1).

❖ Figure R611.7(1) illustrates correct ICF wall construction and the story limits of Section R611.2. The section provides references to sections that specify details for anchoring roof and floor assemblies.

R611.7.1 Reinforcement.

R611.7.1.1 Location. Vertical and horizontal wall reinforcement shall be placed within the middle third of the wall. Steel reinforcement shall have a minimum concrete cover in accordance with ACI 318.

Exception: Where insulated concrete forms are used and the form remains in place as cover for the concrete, the minimum concrete cover for the reinforcing steel is permitted to be reduced to $^3/_4$ inch (19 mm).

❖ The limits placed on the location of wall reinforcement are consistent with current practice. Reinforcing must have the minimum cover required in ACI 318. The cover may be reduced to $^3/_4$ inch (19.1 mm) where insulating concrete forms are used and are left in place.

R611.7.1.2 Vertical steel. Above-grade concrete walls shall have reinforcement in accordance with Sections R611.3, R611.4, or R611.5 and R611.7.2. Where the design wind pressure exceeds 40 psf (1.92 kPa) in accordance with Table R611.3(1) or for townhouses in Seismic Design Category C and all buildings in Seismic Design Categories D_0, D_1 and D_2, vertical wall reinforcement in the top-most ICF story shall terminate with a 90-degree (1.57 rad) standard hook in accordance with Section R611.7.1.5. The free end of the hook shall be within 4 inches (102 mm) of the top of the ICF wall and shall be oriented parallel to the horizontal steel in the top of the wall.

For townhouses in Seismic Design Category C, the minimum vertical reinforcement shall be one No. 5 bar at 24 inches (610 mm) on center or one No. 4 at 16 inches (407 mm) on center. For all buildings in Seismic Design Categories D_0, D_1 and D_2, the minimum vertical reinforcement shall be one No. 5 bar

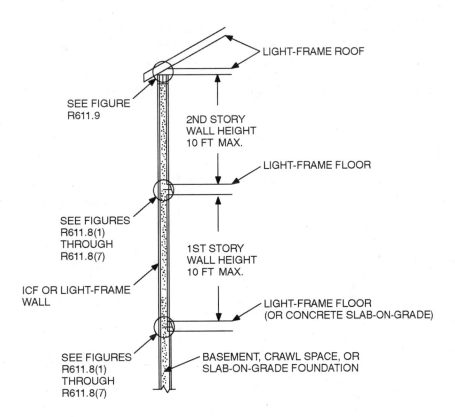

For SI: 1 foot = 304.8 mm.
NOTE: Section cut through flat wall or vertical core of waffle- or screen-grid walls.

FIGURE R611.7(1)
ICF WALL CONSTRUCTION

❖ See the commentary for Section R611.7.4.

at 18 inches (457 mm) on center or one No. 4 at 12 inches (305 mm) on center.

Above-grade ICF walls shall be supported on concrete foundations reinforced as required for the above-grade wall immediately above, or in accordance with Tables R404.4(1) through R404.4(5), whichever requires the greater amount of reinforcement.

Vertical reinforcement shall be continuous from the bottom of the foundation wall to the roof. Lap splices, if required, shall comply with Section R611.7.1.4. Where vertical reinforcement in the above-grade wall is not continuous with the foundation wall reinforcement, dowel bars with a size and spacing to match the vertical ICF wall reinforcement shall be embedded 40 d_b into the foundation wall and shall be lap spliced with the above-grade wall reinforcement. Alternatively, for No. 6 and larger bars, the portion of the bar embedded in the foundation wall shall be embedded 24 inches in the foundation wall and shall have a standard hook.

❖ Vertical reinforcing must be installed according to the tables and the type of ICF wall construction. In addition, wall openings must be reinforced as shown in Figure R611.7(2).

For higher seismic design categories and higher seismic performance demands, the amount of vertical reinforcement needs to be greater to deal with the larger demands on the performance of the wall.

R611.7.1.3 Horizontal reinforcement. Concrete walls with a minimum thickness of 4 inches (102 mm) shall have a minimum of one continuous No. 4 horizontal reinforcing bar placed at 32 inches (812 mm) on center with one bar within 12 inches (305 mm) of the top of the wall story. Concrete walls 5.5 inches (140 mm) thick or more shall have a minimum of one continuous No. 4 horizontal reinforcing bar placed at 48 inches (1219

mm) on center with one bar located within 12 inches (305 mm) of the top of the wall story.

For townhouses in Seismic Design Category C, the minimum horizontal reinforcement shall be one No. 5 bar at 24 inches (610 mm) on center or one No. 4 at 16 inches (407 mm) on center. For all buildings in Seismic Design Categories D_0, D_1 and D_2, the minimum horizontal reinforcement shall be one No. 5 bar at 18 inches (457 mm) on center or one No. 4 at 12 inches (305 mm) on center.

Horizontal reinforcement shall be continuous around building corners using corner bars or by bending the bars. In either case, the minimum lap splice shall be 24 inches (610 mm). For townhouses in Seismic Design Category C and for all buildings in Seismic Design Categories D_0, D_1 and D_2, each end of all horizontal reinforcement shall terminate with a standard hook or lap splice.

❖ These minimums are provided for temperature and shrinkage. In more demanding seismic design categories the additional horizontal steel is needed to improve the performance of walls in resisting horizontal movement.

R611.7.1.4 Lap splices. Where lap splicing of vertical or horizontal reinforcing steel is necessary, the lap splice shall be in accordance with Figure R611.7.1.4 and a minimum of 40 d_b, where d_b is the diameter of the smaller bar. The maximum distance between noncontact parallel bars at a lap splice shall not exceed 8 d_b.

❖ Where continuous bars are not provided, lap splices are necessary to allow forces to be transferred between sections of rebar. This section specifies the minimum lap splice length as 40 times the bar diameter. Noncontact lap splices are permitted provided the spacing of lapped bars does not exceed eight bar diameters. These requirements are based on ACI 318. Figure R611.7.1.4 illustrates the code text requirements.

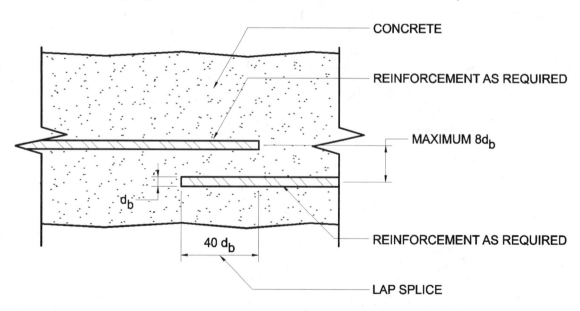

CONCRETE

REINFORCEMENT AS REQUIRED

MAXIMUM 8d_b

d_b

40 d_b

REINFORCEMENT AS REQUIRED

LAP SPLICE

FIGURE R611.7.1.4
LAP SPLICES

❖ This figure illustrates the requirements of Section R611.7.1.4.

R611.7.1.5 Standard hook. Where the free end of a reinforcing bar is required to have a standard hook, the hook shall be a 180-degree bend plus 4 d_b extension but not less than $2^1/_2$ inches, or a 90-degree bend plus 12 d_b extension.

❖ Requirements for hooks are consistent with the requirements of ACI 318.

R611.7.2 Wall openings. Wall openings shall have a minimum of 8 inches (203 mm) of depth of concrete for flat and waffle-grid ICF walls and 12 inches (305 mm) for screen-grid walls over the length of the opening. When the depth of concrete above the opening is less than 12 inches for flat or waffle-grid walls, lintels in accordance with Section R611.7.3 shall be provided. Reinforcement around openings shall be provided in accordance with Table R611.7(1) and Figure R611.7(2). Reinforcement placed horizontally above or below an opening shall extend a minimum of 24 inches (610 mm) beyond the limits of the opening. Wall opening reinforcement shall be provided in addition to the reinforcement required by Sections R611.3, R611.4, R611.5 and R611.7.1. The perimeter of all wall openings shall be framed with a minimum 2-inch by 4-inch plate, anchored to the wall with $^1/_2$-inch (13 mm) diameter anchor bolts spaced a maximum of 24 inches (610 mm) on center. The bolts shall be embedded into the concrete a minimum of 4 inches (102 mm) and have a minimum of $1^1/_2$ inches (38 mm) of concrete cover to the face of the wall.

> **Exception:** The 2-inch by 4-inch plate is not required where the wall is formed to provide solid concrete around the perimeter of the opening with a minimum depth of 4 inches (102 mm) for the full thickness of the wall.

❖ See Table R611.7(1) for a listing of the minimum reinforcement required at wall openings. The requirements vary based on the length of the opening.

R611.7.3 Lintels.

R611.7.3.1 General requirements. Lintels shall be provided over all openings greater than or equal to 2 feet (610 mm) in width. Lintels for flat ICF walls shall be constructed in accordance with Figure R611.7(3) and Table R611.7(2) or R611.7(3). Lintels for waffle-grid ICF walls shall be constructed in accordance with Figure R611.7(4) or Figure R611.7(5) and Table R611.7(4) or R611.7(5). Lintels for screen-grid ICF walls shall be constructed in accordance with Figure R611.7(6) or Figure R611.7(7). Lintel construction in accordance with Figure R611.7(3) shall be permitted with waffle-grid and screen-grid ICF wall construction. Lintel depths are permitted to be increased by the height of the ICF wall located directly above the opening, provided that the lintel depth spans the entire length of the opening.

❖ To support the loads from above, any wall opening of 2 feet (1219 mm) or more requires a lintel in accordance with Section R611.7.3. Figures R611.7(2) through R611.7(7) illustrate these requirements. By implication, therefore, openings smaller than 2 feet can be installed without the use of a lintel to transfer loads from above as a beam. The referenced figures indicate the need for horizontal reinforcement and stirrups for transverse reinforcement depending on the type of ICF wall being constricted and the size of the opening. Commentary Table

R611.7.3 provides a summary of the referenced figures and tables in Sections R611.7.3.1 through R611.7.3.5.

R611.7.3.2 Stirrups. Where required, No. 3 stirrups shall be installed in flat, waffle-grid and screen-grid wall lintels in accordance with the following:

1. For flat walls the stirrups shall be spaced at a maximum spacing of $d/2$ where d equals the depth of the lintel (D) minus the bottom cover of concrete as shown in Figure R611.7(3). Stirrups shall not be required in the middle portion of the span (A) per Figure R611.7(2), for flat walls for a length not to exceed the values shown in parenthesis in Tables R611.7(2) and R611.7(3) or for spans in accordance with Table R611.7(8).

2. For waffle-grid walls a minimum of two No. 3 stirrups shall be placed in each vertical core of waffle-grid lintels. Stirrups shall not be required in the middle portion of the span (A) per Figure R611.7(2), for waffle-grid walls for a length not to exceed the values shown in parenthesis in Tables R611.7(4) and R611.7(5) or for spans in accordance with Table R611.7(8).

3. For screen-grid walls one No. 3 stirrup shall be placed in each vertical core of screen-grid lintels.

   > **Exception:** Stirrups are not required in screen-grid lintels meeting the following requirements:
   >
   > 1. Lintel Depth (D) = 12 inches (305 mm) - spans less than or equal 3 feet 7 inches.
   >
   > 2. Lintel Depth (D) = 24 inches (610 mm) - spans less than or equal 4 feet 4 inches.

❖ Stirrups are required for lintels in load-bearing walls for transverse reinforcement and shear reinforcement consistent with the requirements of ACI 318. Each type of ICF form has a different cross-sectional dimension that produces a different configuration of concrete member. Thus, the text and figures referenced provide varying sizes and locations of stirrups. See the commentary for Section R611.7.3.1.

R611.7.3.3 Horizontal reinforcement. One No. 4 horizontal bar shall be provided in the top of the lintel. Horizontal reinforcement placed within 12 inches (305 mm) of the top of the wall in accordance with Section R611.7.1.3 shall be permitted to serve as the top or bottom reinforcement in the lintel provided the reinforcement meets the location requirements in Figure R611.7(2), R611.7(3), R611.7(4), R611.7(5), R611.7(6), or R611.7(7), and the size requirements in Tables R611.7(2), R611.7(3), R611.7(4), R611.7(5), R611.7(6), R611.7(7), or R611.7(8).

❖ Horizontal reinforcement for lintels is necessary for the design of the lintel to transfer loads from above across a door or window opening. The text of Section R611.7.3.3 along with the tables and figures give prescriptive requirements based on the size of the building and the amount of design snow load. In other words, the size of the tributary load being supported by the exterior wall impacts the needed load carrying capacity of the lintels that carry this load over an opening. See the commentary for Section R611.7.3.1.

R611.7.3.4 Load-bearing walls. Lintels in flat ICF load-bearing walls shall comply with Table R611.7(2), Table RR611.7(3) or Table R611.7(8). Lintels in waffle-grid ICF load-bearing walls shall comply with Table R611.7(4), Table R611.7(5) or Table R611.7(8). Lintels in screen-grid ICF load-bearing walls shall comply with Table R611.7(6) or Table R611.7(7).

Where spans larger than those permitted in Table R611.7(2), Table R611.7(3), Table R611.7(4), Table R611.7(5), R611.7(6), R611.7(7) or R611.7(8) are required, the lintels shall comply with Table R611.7 (9).

❖ This section references several allowable span tables for lintels based on the type of ICF wall construction, lintel depth and thickness, reinforcing, number of stories supported, and snow load. Table R611.7(6) accommodates larger openings such as would be required for garage doors.

R611.7.3.5 Nonload-bearing walls. Lintels in nonload-bearing flat, waffle-grid and screen-grid ICF walls shall comply with Table R611.7 (10). Stirrups are not required.

❖ For nonload-bearing walls, a single table provides allowable clear spans (opening length) based on lintel depth and support conditions.

R611.7.4 Minimum length of wall without openings. The wind velocity pressures of Table R611.7.4 shall be used to determine the minimum amount of solid wall length in accordance with Tables R611.7(9A) through R611.7(10B) and Figure R611.7.4. Table R611.7(11) shall be used to determine the minimum amount of solid wall length for townhouses in Seismic Design Category C, and all buildings in Seismic Design Categories D_0, D_1 and D_2 for all types of ICF walls. The greater amount of solid wall length required by wind loading or seismic loading shall apply. The minimum percentage of solid wall length shall include only those solid wall segments that are a minimum of 24 inches (610 mm) in length. The maximum distance between wall segments included in determining solid wall length shall not exceed 18 feet (5486 mm). A minimum length of 24 inches (610 mm) of solid wall segment, extending the full height of each wall story, shall occur at all interior and exterior corners of exterior walls.

❖ To provide lateral force resistance, the exterior walls must have the minimum lengths of solid wall specified in Tables R611.7(9A) through R611.7(10B) based on wind speed. Attached dwellings (townhouses) classified as Seismic Design Category C and all buildings in Seismic Design Categories D_0, D_1 and D_2 must also comply with the wall length required by Table R611.7(11). Figure 611.7.4 illustrates the requirements. The lengths in each of these tables are specified as a percentage of the total length of wall under consideration.

Table R611.7.3
APPLICABLE REQUIREMENTS FOR ICF LINTELS

Type of ICF Wall	Applicable Figures	Applicable Tables
Flat	R611.7(3)	R611.7(2) and R611.7(3) No stirrups: R611.7(8) Large spans: R611.7(9) Non-loadbearing: R611.7(10)
Waffle-grid	R611.7(4) and 611.7(5)	R611.7(4) and R611.7(5) No stirrups: R611.7(8) Large spans: R611.7(9) Non-loadbearing: R611.7(10)
Screen-grid	R611.7(6) and R611.7(7)	R611.7(6) and R611.7(7) Large spans: R611.7(9) Non-loadbearing: R611.7(10)

TABLE R611.7(1)
MINIMUM WALL OPENING REINFORCEMENT REQUIREMENTS IN ICF WALLS[a]

WALL TYPE AND OPENING WIDTH (L) (feet)	MINIMUM HORIZONTAL OPENING REINFORCEMENT	MINIMUM VERTICAL OPENING REINFORCEMENT
Flat, Waffle-, and Screen-Grid: $L < 2$	None required	None required
Flat, Waffle-, and Screen-Grid: $L \geq 2$	Provide lintels in accordance with Section R611.7.3. Provide one No. 4 bar within 12 inches from the bottom of the opening. Top and bottom lintel reinforcement shall extend a minimum of 24 inches beyond the limits of the opening.	In locations with wind speeds less than or equal to 110 mph or in Seismic Design Categories A and B, provide one No. 4 bar for the full height of the wall story within 12 inches of each side of the opening. In locations with wind speeds greater than 110 mph, townhouses in Seismic Design Category C, or all buildings in Seismic Design Categories D_0, D_1 and D_2, provide two No. 4 bars or one No. 5 bar for the full height of the wall story within 12 inches of each side of the opening.

For SI: 1 inch = 25.4 mm, 1 foot = 304.8 mm, 1 mile per hour = 0.447 m/s; 1 pound per square inch = 6.895 kPa.

a. This table is based on concrete with a minimum specified compressive strength of 2,500 psi, reinforcing steel with a minimum yield strength of 40,000 psi and an assumed equivalent rectangular cross section. This table is not intended to prohibit the use of ICF manufacturer's tables based on engineering analysis in accordance with ACI 318.

❖ See the commentary for Section R611.7.2.

TABLE R611.7(2)
MAXIMUM ALLOWABLE CLEAR SPANS FOR ICF LINTELS FOR FLAT LOAD-BEARING WALLS[a, b, c, d, f]
NO. 4 BOTTOM BAR SIZE

MINIMUM LINTEL THICKNESS, T (inches)	LINTEL DEPTH, D (inches)	MAXIMUM CLEAR SPAN, (feet-inches) (Number is Middle of Span, A)[e]					
		Supporting Roof Only		Supporting Light-Framed 2nd Story and Roof		Supporting ICF Second Story and Roof	
		Ground Snow Load					
		30 psf	70 psf	30 psf	70 psf	30 psf	70 psf
3.5	8	4-9 (1-2)	4-2 (0-9)	3-10 (0-8)	3-4 (0-6)	3-5 (0-6)	3-1 (0-5)
	12	6-8 (1-11)	5-5 (1-3)	5-0 (1-1)	4-5 (0-10)	4-6 (0-10)	4-0 (0-8)
	16	7-11 (2-9)	6-5 (1-9)	6-0 (1-6)	5-3 (1-2)	5-4 (1-2)	4-10 (1-0)
	20	8-11 (3-5)	7-4 (2-3)	6-9 (1-11)	6-0 (1-6)	6-1 (1-7)	5-6 (1-3)
	24	9-10 (4-1)	8-1 (2-9)	7-6 (2-4)	6-7 (1-10)	6-9 (1-11)	6-1 (1-6)
5.5	8	5-2 (1-10)	4-2 (1-2)	3-10 (1-0)	3-5 (0-9)	3-5 (0-10)	3-1 (0-8)
	12	6-8 (3-0)	5-5 (2-0)	5-0 (1-9)	4-5 (1-4)	4-6 (1-4)	4-1 (1-1)
	16	7-10 (4-1)	6-5 (2-9)	6-0 (2-5)	5-3 (1-10)	5-4 (1-11)	4-10 (1-7)
	20	8-10 (5-3)	7-3 (3-6)	6-9 (3-1)	6-0 (2-4)	6-1 (2-5)	5-6 (2-0)
	24	9-8 (6-3)	8-0 (4-3)	7-5 (3-8)	6-7 (2-11)	6-8 (3-0)	6-0 (2-5)
7.5	8	5-2 (2-6)	4-2 (1-8)	3-11 (1-5)	3-5 (1-1)	3-6 (1-1)	3-2 (0-11)
	12	6-7 (4-0)	5-5 (2-8)	5-0 (2-4)	4-5 (1-10)	4-6 (1-10)	4-1 (1-6)
	16	7-9 (5-5)	6-5 (3-8)	5-11 (3-3)	5-3 (2-6)	5-4 (2-7)	4-10 (2-2)
	20	8-8 (6-10)	7-2 (4-8)	6-8 (4-2)	5-11 (3-3)	6-0 (3-4)	5-5 (2-9)
	24	9-6 (8-2)	7-11 (5-8)	7-4 (5-1)	6-6 (3-11)	6-7 (4-1)	6-0 (3-4)
9.5	8	5-2 (3-1)	4-2 (2-1)	3-11 (1-9)	3-5 (1-5)	3-6 (1-5)	3-2 (1-2)
	12	6-7 (5-0)	5-5 (3-4)	5-0 (3-0)	4-5 (2-4)	4-6 (2-5)	4-1 (1-11)
	16	7-8 (6-9)	6-4 (4-7)	5-11 (4-2)	5-3 (3-3)	5-4 (3-4)	4-10 (2-8)

For SI: 1 inch = 25.4 mm, 1 foot = 304.8 mm, 1 pound per square inch = 6.895 kPa, 1 pound per square foot = 0.0479 kPa.

a. This table is based on concrete with a minimum specified compressive strength of 2,500 psi, reinforcing steel with a minimum yield strength of 40,000 psi and an assumed equivalent rectangular cross section. When reinforcement with a minimum yield strength of 60,000 psi is used, the span lengths in the shaded cells shall be increased by 1.2 times the table values.

b. This table is not intended to prohibit the use of ICF manufacturer's tables based on engineering analysis in accordance with ACI 318.

c. Deflection criterion: $L/240$.

d. Design load assumptions:
   Floor dead load is 10 psf
   Floor live load is 30 psf
   Building width is 32 feet
   Light-framed wall dead load is 10 psf
   Attic live load is 20 psf
   Roof dead load is 15 psf
   ICF wall dead load is 69 psf

e. No. 3 stirrups are required at $d/2$ spacing except no stirrups are required for the distance, (A), shown in the middle portion of the span in accordance with Figure R611.7(2) and Section R611.7.3.2.

f. Interpolation is permitted between ground snow loads and between lintel depths.

❖ See the commentary for Section R611.7.3.1.

TABLE R611.7(3)
MAXIMUM ALLOWABLE CLEAR SPANS FOR ICF LINTELS FOR FLAT LOAD-BEARING WALLS[a, b, c, d, f]
NO. 5 BOTTOM BAR SIZE

MINIMUM LINTEL THICKNESS, T (inches)	LINTEL DEPTH, D (inches)	MAXIMUM CLEAR SPAN, (feet-inches) (Number is Middle of Span, A)[e]					
		Supporting Roof		Supporting Light-Framed 2nd Story and Roof		Supporting ICF Second Story and Roof	
		Ground Snow Load					
		30 psf	70 psf	30 psf	70 psf	30 psf	70 psf
3.5	8	4-9 (1-2)	4-2 (0-9)	3-11 (0-8)	3-7 (0-6)	3-7 (0-6)	3-5 (0-5)
	12	7-2 (1-11)	6-3 (1-3)	5-11 (1-1)	5-5 (0-10)	5-5 (0-10)	5-0 (0-8)
	16	9-6 (2-9)	8-0 (1-9)	7-4 (1-6)	6-6 (1-2)	6-7 (1-2)	5-11 (1-0)
	20	11-1 (3-5)	9-1 (2-3)	8-4 (1-11)	7-5 (1-6)	7-6 (1-7)	6-9 (1-3)
	24	12-2 (4-1)	10-0 (2-9)	9-3 (2-4)	8-2 (1-10)	8-4 (1-11)	7-6 (1-6)
5.5	8	5-6 (1-10)	4-10 (1-2)	4-7 (1-0)	4-2 (0-9)	4-2 (0-10)	3-10 (0-8)
	12	8-3 (3-0)	6-9 (2-0)	6-3 (1-9)	5-6 (1-4)	5-7 (1-4)	5-0 (1-1)
	16	9-9 (4-1)	8-0 (2-9)	7-5 (2-5)	6-6 (1-10)	6-7 (1-11)	6-0 (1-7)
	20	10-11 (5-3)	9-0 (3-6)	8-4 (3-1)	7-5 (2-4)	7-6 (2-5)	6-9 (2-0)
	24	12-0 (6-3)	9-11 (4-3)	9-3 (3-8)	8-2 (2-11)	8-3 (3-0)	7-6 (2-5)
7.5	8	6-1 (2-6)	5-2 (1-8)	4-9 (1-5)	4-3 (1-1)	4-3 (1-1)	3-10 (0-11)
	12	8-2 (4-0)	6-9 (2-8)	6-3 (2-4)	5-6 (1-10)	5-7 (1-10)	5-0 (1-6)
	16	9-7 (5-5)	7-11 (3-8)	7-4 (3-3)	6-6 (2-6)	6-7 (2-7)	6-0 (2-2)
	20	10-10 (6-10)	8-11 (4-8)	8-4 (4-2)	7-4 (3-3)	7-6 (3-4)	6-9 (2-9)
	24	11-10 (8-2)	9-10 (5-8)	9-2 (5-1)	8-1 (3-11)	8-3 (4-1)	7-5 (3-4)
9.5	8	6-4 (3-1)	5-2 (2-1)	4-10 (1-9)	4-3 (1-5)	4-4 (1-5)	3-11 (1-2)
	12	8-2 (5-0)	6-8 (3-4)	6-2 (3-0)	5-6 (2-4)	5-7 (2-5)	5-0 (1-11)
	16	9-6 (6-9)	7-11 (4-7)	7-4 (4-2)	6-6 (3-3)	6-7 (3-4)	5-11 (2-8)
	20	10-8 (8-4)	8-10 (5-10)	8-3 (5-4)	7-4 (4-2)	7-5 (4-3)	6-9 (3-6)
	24	11-7 (10-0)	9-9 (6-11)	9-0 (6-5)	8-1 (5-0)	8-2 (5-2)	7-5 (4-3)

For SI: 1 inch = 25.4 mm, 1 foot = 304.8 mm, 1 pound per square inch = 6.895 kPa, 1 pound per square foot = 0.0479 kPa.

a. This table is based on concrete with a minimum specified compressive strength of 2,500 psi, reinforcing steel with a minimum yield strength of 40,000 psi and an assumed equivalent rectangular cross section. When reinforcement with a minimum yield strength of 60,000 psi is used the span lengths in the shaded cells shall be increased by 1.2 times the table values.

b. This table is not intended to prohibit the use of ICF manufacturer's tables based on engineering analysis in accordance with ACI 318.

c. Deflection criterion: $L/240$.

d. Design load assumptions:
   Floor dead load is 10 psf   Attic live load is 20 psf
   Floor live load is 30 psf   Roof dead load is 15 psf
   Building width is 32 feet   ICF wall dead load is 69 psf
   Light-framed wall dead load is 10 psf

e. No. 3 stirrups are required at $d/2$ spacing except no stirrups are required for the distance, (A), shown in the middle portion of the span in accordance with Figure R611.7(2) and Section R611.7.3.2.

f. Interpolation is permitted between ground snow loads and between lintel depths.

❖ See the commentary for Section R611.7.3.1.

TABLE R611.7(4)
MAXIMUM ALLOWABLE CLEAR SPANS FOR WAFFLE-GRID ICF WALL LINTELS[a, b, c, d, f]
NO. 4 BOTTOM BAR SIZE

NOMINAL LINTEL THICKNESS T[g,h] (inches)	LINTEL DEPTH D (inches)	MAXIMUM CLEAR SPAN (feet-inches) (Number is Middle of Span, A)[e]					
		Supporting Roof		Supporting Light-Framed 2nd Story and Roof		Supporting ICF Second Story and Roof	
		Ground Snow Load					
		30 psf	70 psf	30 psf	70 psf	30 psf	70 psf
6	8	5-2 (0-10)	4-2 (0-7)	3-10 (0-6)	3-5 (0-4)	3-6 (0-5)	3-2 (0-4)
	12	6-8 (1-5)	5-5 (0-11)	5-0 (0-9)	4-5 (0-7)	4-7 (0-8)	4-2 (0-6)
	16	7-11 (1-11)	6-6 (1-4)	6-0 (1-1)	5-3 (0-10)	5-6 (0-11)	4-11 (0-9)
	20	8-11 (2-6)	7-4 (1-8)	6-9 (1-5)	6-0 (1-1)	6-3 (1-2)	5-7 (0-11)
	24	9-10 (3-0)	8-1 (2-0)	7-6 (1-9)	6-7 (1-4)	6-10 (1-5)	6-2 (1-2)
8	8	5-2 (0-10)	4-3 (0-7)	3-11 (0-6)	3-5 (0-4)	3-7 (0-5)	3-2 (0-4)
	12	6-8 (1-5)	5-5 (0-11)	5-1 (0-9)	4-5 (0-7)	4-8 (0-8)	4-2 (0-6)
	16	7-10 (1-11)	6-5 (1-4)	6-0 (1-1)	5-3 (0-10)	5-6 (0-11)	4-11 (0-9)
	20	8-10 (2-6)	7-3 (1-8)	6-9 (1-5)	6-0 (1-1)	6-2 (1-2)	5-7 (0-11)
	24	9-8 (3-0)	8-0 (2-0)	7-5 (1-9)	6-7 (1-4)	6-10 (1-5)	6-2 (1-2)

For SI: 1 inch = 25.4 mm, 1 foot = 304.8 mm, 1 psi = 6.895 kPa, 1 psf = 0.0479 kPa.

a. This table is based on concrete with a minimum specified compressive strength of 2,500 psi, reinforcing steel with a minimum yield strength of 40,000 psi and an assumed equivalent rectangular cross section. When reinforcement with a minimum yield strength of 60,000 psi is used the span lengths in the shaded cells shall be increased by 1.2 times the table values.

b. This table is not intended to prohibit the use of ICF manufacturer's tables based on engineering analysis in accordance with ACI 318.

c. Deflection criterion: $L/240$.

d. Design load assumptions:

   Floor dead load is 10 psf   Attic live load is 20 psf
   Floor live load is 30 psf   Roof dead load is 15 psf
   Building width is 32 feet   ICF wall dead load is 55 psf
   Light-framed wall dead load is 10 psf

e. No. 3 stirrups are required at $d/2$ spacing except no stirrups are required for the distance, (A), shown in the middle portion of the span in accordance with Figure R611.7(2) and Section R611.7.3.2.

f. Interpolation is permitted between ground snow loads and between lintel depths.

g. For actual wall lintel width, refer to Table R611.2.

h. Lintel width corresponds to the nominal waffle-grid ICF wall thickness with a minimum thickness of 2 inches.

❖ See the commentary for Section R611.7.3.1.

TABLE R611.7(5)
MAXIMUM ALLOWABLE CLEAR SPANS FOR WAFFLE-GRID ICF WALL LINTELS[a, b, c, d, f]
NO. 5 BOTTOM BAR SIZE

NOMINAL LINTEL THICKNESS, T[g, h] (inches)	LINTEL DEPTH D (inches)	MAXIMUM CLEAR SPAN (feet-inches) (Number is Middle of Span, A)[e]					
		Supporting Roof		Supporting Light-Framed 2nd Story and Roof		Supporting ICF Second Story and Roof	
		Ground Snow Load					
		30 psf	70 psf	30 psf	70 psf	30 psf	70 psf
6	8	5-4 (0-10)	4-8 (0-7)	4-5 (0-6)	4-1 (0-4)	4-5 (0-5)	3-10 (0-4)
	12	8-0 (1-5)	6-9 (0-11)	6-3 (0-9)	5-6 (0-7)	6-3 (0-8)	5-1 (0-6)
	16	9-9 (1-11)	8-0 (1-4)	7-5 (1-1)	6-6 (0-10)	7-5 (0-11)	6-1 (0-9)
	20	11-0 (2-6)	9-1 (1-8)	8-5 (1-5)	7-5 (1-1)	8-5 (1-2)	6-11 (0-11)
	24	12-2 (3-0)	10-0 (2-0)	9-3 (1-9)	8-2 (1-4)	9-3 (1-5)	7-8 (1-2)
8	8	6-0 (0-10)	5-2 (0-7)	4-9 (0-6)	4-3 (0-4)	4-9 (0-5)	3-11 (0-4)
	12	8-3 (1-5)	6-9 (0-11)	6-3 (0-9)	5-6 (0-7)	6-3 (0-8)	5-2 (0-6)
	16	9-9 (1-11)	8-0 (1-4)	7-5 (1-1)	6-6 (0-10)	7-5 (0-11)	6-1 (0-9)
	20	10-11 (2-6)	9-0 (1-8)	8-4 (1-5)	7-5 (1-1)	8-4 (1-2)	6-11 (0-11)
	24	12-0 (3-0)	9-11 (2-0)	9-2 (1-9)	8-2 (1-4)	9-2 (1-5)	7-8 (1-2)

For SI: 1 inch = 25.4 mm, 1 foot = 304.8 mm, 1 psi = 6.895 kPa, 1 psf = 0.0479 kPa.

a. This table is based on concrete with a minimum specified compressive strength of 2,500 psi, reinforcing steel with a minimum yield strength of 40,000 psi and an assumed equivalent rectangular cross section. When reinforcement with a minimum yield strength of 60,000 psi is used the span lengths in the shaded cells shall be increased by 1.2 times the table values.

b. This table is not intended to prohibit the use of ICF manufacturer's tables based on engineering analysis in accordance with ACI 318.

c. Deflection criterion: $L/240$.

d. Design load assumptions:

Floor dead load is 10 psf   Attic live load is 20 psf
Floor live load is 30 psf   Roof dead load is 15 psf
Building width is 32 feet   ICF wall dead load is 53 psf
Light-framed wall dead load is 10 psf

e. No. 3 stirrups are required at $d/2$ spacing except no stirrups are required for the distance, (A), shown in the middle portion of the span in accordance with Figure R611.7(2) and Section R611.7.3.2.

f. Interpolation is permitted between ground snow loads and between lintel depths.

g. For actual wall lintel width, refer to Table R611.2.

h. Lintel width corresponds to the nominal waffle-grid ICF wall thickness with a minimum thickness of 2 inches.

❖ See the commentary for Section R611.7.3.1.

TABLE R611.7(6)
MAXIMUM ALLOWABLE CLEAR SPANS FOR SCREEN-GRID ICF LINTELS IN LOAD-BEARING WALLS[a, b, c, d,e, f, g]
NO. 4 BOTTOM BAR SIZE

MINIMUM LINTEL THICKNESS, T (inches)[h,i]	MINIMUM LINTEL DEPTH, D (inches)	MAXIMUM CLEAR SPAN (feet-inches)					
		Supporting Roof		Supporting Light-Framed Second Story and Roof		Supporting ICF Second Story and Roof	
		Maximum Ground Snow Load (psf)					
		30	70	30	70	30	70
6	12	3-7	2-10	2-5	2-0	2-0	NA
	24	9-10	8-1	7-6	6-7	6-11	6-2

For SI: 1 inch = 25.4 mm, 1 foot = 304.8 mm, 1 psi = 6.895 kPa, 1 psf = 0.0479 kPa.

a. This table is based on concrete with a minimum specified compressive strength of 2,500 psi, reinforcing steel with a minimum yield strength of 40,000 psi and an assumed equivalent rectangular cross section. When reinforcement with a minimum yield strength of 60,000 psi is used the span lengths in the shaded cells shall be increased by 1.2 times the table values.

b. This table is not intended to prohibit the use of ICF manufacturer's tables based on engineering analysis in accordance with ACI 318.

c. Deflection criterion: $L/240$.

d Design load assumptions:
  Floor dead load is 10 psf Attic live load is 20 psf
  Floor live load is 30 psf Roof dead load is 15 psf
  Maximum floor clear span is 32 ft ICF wall dead load is 53 psf
  Light-frame wall dead load is 10 psf

e. Stirrup requirements:
  Stirrups are not required for lintels 12 inches deep.
  One No. 3 stirrup is required in each vertical core for lintels 24 inches deep.

f. Interpolation is permitted between ground snow loads.

g. Flat ICF lintels may be used in lieu of screen-grid lintels.

h. For actual wall lintel width, refer to Table R611.2.

i. Lintel width corresponds to the nominal screen-grid ICF wall thickness.

❖ See the commentary for Section R611.7.3.1 and R611.7.3.4.

TABLE R611.7(7)
MAXIMUM ALLOWABLE CLEAR SPANS FOR SCREEN-GRID ICF LINTELS IN LOAD-BEARING WALLS[a, b, c, d, e, f, g]
NO. 5 BOTTOM BAR SIZE

MINIMUM LINTEL THICKNESS, T (inches)[h,i]	MINIMUM LINTEL DEPTH, D (inches)	MAXIMUM CLEAR SPAN (feet-inches)					
		Supporting Roof		Supporting Light-Framed Second Story and Roof		Supporting ICF Second Story and Roof	
		Maximum Ground Snow Load (psf)					
		30	70	30	70	30	70
6	12	3-7	2-10	2-5	2-0	2-0	NA
	24	12-3	10-0	9-3	8-3	8-7	7-8

For SI: 1 inch = 25.4 mm, 1 foot = 304.8 mm, 1 pound per square inch = 6.895 kPa, 1 pound per square foot = 0.0479 kPa.

a. This table is based on concrete with a minimum specified compressive strength of 2,500 psi, reinforcing steel with a minimum yield strength of 40,000 psi and an assumed equivalent rectangular cross section. When reinforcement with a minimum yield strength of 60,000 psi is used the span lengths in the shaded cells shall be increased by 1.2 times the table values.

b. This table is not intended to prohibit the use of ICF manufacturer's tables based on engineering analysis in accordance with ACI 318.

c. Deflection criterion: $L/240$.

d. Design load assumptions:
  Floor dead load is 10 psf Attic live load is 20 psf
  Floor live load is 30 psf Roof dead load is 15 psf
  Maximum floor clear span is 32 ft ICF wall dead load is 53 psf
  Light-frame wall dead load is 10 psf

e. Stirrup requirements:
  Stirrups are not required for lintels 12 inches deep.
  One No. 3 stirrup is required in each vertical core for lintels 24 inches deep.

f. Interpolation is permitted between ground snow loads.

g. Flat ICF lintels may be used in lieu of screen-grid lintels.

h. For actual wall lintel width, refer to Table R611.2.

i. Lintel width corresponds to the nominal screen-grid ICF wall thickness.

❖ See the commentary for Sections R611.7.3.1 and R611.7.3.4.

TABLE R611.7(8)
MAXIMUM ALLOWABLE CLEAR SPANS FOR ICF LINTELS WITHOUT STIRRUPS IN LOAD-BEARING WALLS[a, b, c, d, e, f, g, h]
(NO. 4 OR NO. 5) BOTTOM BAR SIZE

MINIMUM LINTEL THICKNESS, T (inches)	MINIMUM LINTEL DEPTH, D (inches)	MAXIMUM CLEAR SPAN (feet-inches)					
		Supporting Roof Only		Supporting Light-Framed Second Story and Roof		Supporting ICF Second Story and Roof	
		MAXIMUM GROUND SNOW LOAD (psf)					
		30	70	30	70	30	70
Flat ICF Lintel							
3.5	8	2-6	2-6	2-6	2-4	2-5	2-2
	12	4-2	4-2	4-1	3-10	3-10	3-7
	16	4-11	4-8	4-6	4-2	4-2	3-11
	20	6-3	5-3	4-11	4-6	4-6	4-3
	24	7-7	6-4	6-0	5-6	5-6	5-2
5.5	8	2-10	2-6	2-6	2-5	2-6	2-2
	12	4-8	4-4	4-3	3-11	3-10	3-7
	16	6-5	5-1	4-8	4-2	4-3	3-11
	20	8-2	6-6	6-0	5-4	5-5	5-0
	24	9-8	7-11	7-4	6-6	6-7	6-1
7.5	8	3-6	2-8	2-7	2-5	2-5	2-2
	12	5-9	4-5	4-4	4-0	3-10	3-7
	16	7-9	6-1	5-7	4-10	4-11	4-5
	20	8-8	7-2	6-8	5-11	6-0	5-5
	24	9-6	7-11	7-4	6-6	6-7	6-0
9.5	8	4-2	3-1	2-9	2-5	2-5	2-2
	12	6-7	5-1	4-7	3-11	4-0	3-7
	16	7-10	6-4	5-11	5-3	5-4	4-10
	20	8-7	7-2	6-8	5-11	6-0	5-5
	24	9-4	7-10	7-3	6-6	6-7	6-0
Waffle-Grid ICF Lintel							
6 or 8	8	2-6	2-6	2-6	2-4	2-4	2-2
	12	4-2	4-2	4-1	3-8	3-9	3-7
	16	5-9	5-8	5-7	5-1	5-2	4-8
	20	7-6	7-4	6-9	6-0	6-3	5-7
	24	9-2	8-1	7-6	6-7	6-10	6-2

For SI: 1 inch = 25.4 mm; 1 foot = 304.8 mm; 1 pound per square foot = 0.0479 kPa; 1 pound per square inch = 6.895 kPa.

a. Table values are based on tensile reinforcement with a minimum yield strength of 40,000 psi (276 MPa), concrete with a minimum specified compressive strength of 2,500 psi, and a building width (clear span) of 32 feet.

b. Spans located in shaded cells shall be permitted to be multiplied by 1.05 when concrete with a minimum compressive strength of 3,000 psi is used or by 1.1 when concrete with a minimum compressive strength of 4,000 psi is used.

c. Deflection criterion is $L/240$, where L is the clear span of the lintel in inches.

d. Linear interpolation shall be permitted between ground snow loads and between lintel depths.

e. Lintel depth, D, shall be permitted to include the available height of ICF wall located directly above the lintel, provided that the increased lintel depth spans the entire length of the opening.

f. Spans shall be permitted to be multiplied by 1.05 for a building width (clear span) of 28 feet.

g. Spans shall be permitted to be multiplied by 1.1 for a building width (clear span) of 24 feet or less.

h. ICF wall dead load is 69 psf.

❖ See the commentary for Sections R611.7.3.1 and R611.7.3.4.

TABLE R611.7(9)
MINIMUM BOTTOM BAR ICF LINTEL REINFORCEMENT FOR LARGE CLEAR SPANS IN LOAD-BEARING WALLS[a, b, c, d, e, f, h]

MINIMUM LINTEL THICKNESS, $T^{e,g}$ (inches)	MINIMUM LINTEL DEPTH, D (inches)	MINIMUM BOTTOM LINTEL REINFORCEMENT					
		Supporting Light-Frame Roof Only		Supporting Light-Framed Second Story and Roof		Supporting ICF Second Story and Light-Frame Roof	
		Maximum Ground Snow Load (psf)					
		30	70	30	70	30	70
Flat ICF Lintel, 12 feet- 3 inches Maximum Clear Span							
3.5	24	1 #5	1 #7	D/R	D/R	D/R	D/R
5.5	20	1 #6	1 #7	D/R	D/R	D/R	D/R
5.5	24	1 #5	1 #7	1 #7	1 #8	1 #8	D/R
7.5	16	1 #7; 2 #5	D/R	D/R	D/R	D/R	D/R
7.5	20	1 #6; 2 #4	1#7; 2 #5	1 #8; 2 #6	D/R	D/R	D/R
7.5	24	1 #6; 2 #4	1 #7; 2 #5	1 #7; 2 #5	1 #8; 2 #6	1 #8; 2 #6	1 #8; 2 #6
9.5	16	1 #7; 2 #5	D/R	D/R	D/R	D/R	D/R
9.5	20	1 #6; 2 #4	1 #7; 2 #5	1 #8; 2 #6	1 #8; 2 #6	1 #8; 2 #6	1 #9; 2 #6
9.5	24	1 #6; 2 #4	1 #7; 2 #5	1 #7; 2 #5	1 #7; 2 #6	1 #8; 2 #6	1 #9; 2 #6
Flat ICF Lintel, 16 feet-3 inches Maximum Clear Span							
5.5	24	1 #7	D/R	D/R	D/R	D/R	D/R
7.5	24	1 #7; 2 #5	D/R	D/R	D/R	D/R	D/R
9.5	24	1 #7; 2 #5	1 #9; 2 #6	1 #9; 2 #6	D/R	D/R	D/R
Waffle-Grid ICF Lintel, 12 feet-3 inches Maximum Clear Span							
6	20	1 #6	D/R	D/R	D/R	D/R	D/R
6	24	1 #5	1 #7; 2 #5	1 #7; 2 #5	1 #8; 2 #6	1 #8; 2 #6	D/R
8	16	1 #7; 2 #5	D/R	D/R	D/R	D/R	D/R
8	20	1 #6; 2 #4	1 #7; 2 #5	1 #8; 2 #6	D/R	D/R	D/R
8	24	1 #5	1 #7; 2 #5	1 #7; 2 #5	1 #8; 2 #6	1 #8; 2 #6	1 #8; 2 #6
Screen-Grid ICF Lintel, 12 feet-3 inches Maximum Clear Span							
6	24	1 #5	1 #7	D/R	D/R	D/R	D/R

For SI: 1 inch = 25.4 mm, 1 foot = 304.8 mm, 1 psi = 6.895 kPa, 1 psf = 0.0479 kPa.

a. This table is based on concrete with a minimum specified compressive strength of 2,500 psi, reinforcing steel with a minimum yield strength of 40,000 psi and an assumed equivalent rectangular cross section. When reinforcement with a minimum yield strength of 60,000 psi is used the span lengths in the shaded cells shall be increased by 1.2 times the table values.

b. This table is not intended to prohibit the use of ICF manufacturers tables based on engineering analysis in accordance with ACI 318.

c. D/R indicates design is required.

d. Deflection criterion: $L/240$.

e. Interpolation is permitted between ground snow loads and between lintel depths.

f. No. 3 stirrups are required a maximum d/2 spacing for spans greater than 4 feet.

g. Actual thickness is shown for flat lintels; nominal thickness is given for waffle-grid and screen-grid lintels. Lintel thickness corresponds to the nominal waffle-grid and screen-grid ICF wall thickness. Refer to Table R611.2 for actual wall thickness.

h. ICF wall dead load varies based on wall thickness using 150 pcf concrete density.

❖ See the commentary for Sections R611.7.3.1 and R611.7.3.4.

TABLE R611.7(9A)
MINIMUM SOLID END WALL LENGTH REQUIREMENTS FOR FLAT ICF WALLS (WIND PERPENDICULAR TO RIDGE)[a, b, c]

WALL CATEGORY	BUILDING SIDE WALL LENGTH, L (feet)	Roof Slope	WIND VELOCITY PRESSURE FROM TABLE R611.7.4 (psf)							
			20	25	30	35	40	45	50	60
			Minimum Solid Wall Length on Building End Wall (feet)							
One-Story or Top Story of Two-Story	16	≤ 1:12	4.00	4.00	4.00	4.00	4.00	4.00	4.00	4.00
		5:12	4.00	4.00	4.00	4.00	4.00	4.00	4.25	4.50
		7:12[d]	4.00	4.25	4.25	4.50	4.75	4.75	5.00	5.50
		12:12[d]	4.25	4.50	4.75	5.00	5.25	5.50	5.75	6.25
	24	≤ 1:12	4.00	4.00	4.00	4.00	4.00	4.00	4.25	4.50
		5:12	4.00	4.00	4.00	4.25	4.25	4.50	4.50	4.75
		7:12[d]	4.25	4.50	4.75	5.00	5.25	5.50	5.75	6.25
		12:12[d]	4.75	5.00	5.25	5.75	6.00	6.50	6.75	7.50
	32	≤ 1:12	4.00	4.00	4.00	4.00	4.25	4.25	4.50	4.75
		5:12	4.00	4.00	4.25	4.50	4.50	4.75	5.00	5.25
		7:12[d]	4.50	5.00	5.25	5.50	6.00	6.25	6.50	7.25
		12:12[d]	5.00	5.50	6.00	6.50	7.00	7.25	7.75	8.75
	40	≤ 1:12	4.00	4.00	4.25	4.25	4.50	4.50	4.75	5.00
		5:12	4.00	4.25	4.50	4.75	4.75	5.00	5.25	5.50
		7:12[d]	4.75	5.25	5.75	6.00	6.50	7.00	7.25	8.00
		12:12[d]	5.50	6.00	6.50	7.25	7.75	8.25	8.75	10.0
	50	≤ 1:12	4.00	4.25	4.25	4.50	4.75	4.75	5.00	5.50
		5:12	4.25	4.50	4.75	5.00	5.25	5.50	5.75	6.00
		7:12[d]	5.25	5.75	6.25	6.75	7.25	7.75	8.25	9.25
		12:12[d]	6.00	6.75	7.50	8.00	8.75	9.50	10.25	11.5
	60	≤ 1:12	4.00	4.25	4.50	4.75	5.00	5.25	5.25	5.75
		5:12	4.50	4.75	5.00	5.25	5.50	5.75	6.00	6.75
		7:12[d]	5.50	6.25	6.75	7.50	8.00	8.50	9.25	10.25
		12:12[d]	6.50	7.25	8.25	9.00	9.75	10.5	11.5	13.0
First Story of Two-Story	16	≤ 1:12	4.00	4.25	4.50	4.75	5.00	5.25	5.25	5.75
		5:12	4.50	4.75	5.00	5.25	5.50	5.75	6.00	6.75
		7:12[d]	4.50	5.00	5.25	5.75	6.00	6.25	6.75	7.25
		12:12[d]	5.00	5.25	5.75	6.25	6.50	7.00	7.25	8.25
	24	≤ 1:12	4.50	4.75	5.00	5.25	5.50	5.75	6.00	6.75
		5:12	4.75	5.25	5.50	6.00	6.25	6.75	7.00	7.75
		7:12[d]	5.25	5.75	6.25	6.75	7.00	7.50	8.00	9.00
		12:12[d]	5.50	6.25	6.75	7.25	8.00	8.50	9.00	10.25
	32	≤ 1:12	4.75	5.00	5.50	5.75	6.25	6.50	6.75	7.50
		5:12	5.25	5.75	6.25	6.75	7.25	7.50	8.00	9.00
		7:12[d]	5.75	6.50	7.00	7.75	8.25	9.00	9.50	10.75
		12:12[d]	6.25	7.00	7.75	8.50	9.25	10.0	10.75	12.25
	40	≤ 1:12	5.00	5.50	5.75	6.25	6.75	7.25	7.50	8.50
		5:12	5.50	6.25	6.75	7.25	8.00	8.50	9.00	10.25
		7:12[d]	6.25	7.00	7.75	8.75	9.50	10.25	11.0	12.5
		12:12[d]	7.00	8.00	8.75	9.75	10.75	11.5	12.5	14.25
	50	≤ 1:12	5.50	6.00	6.50	7.00	7.50	8.00	8.50	9.50
		5:12	6.00	6.75	7.50	8.25	9.00	9.75	10.5	11.75
		7:12[d]	7.00	8.00	9.00	10.0	10.75	11.75	12.75	14.5
		12:12[d]	7.75	9.00	10.0	11.25	12.25	13.50	14.75	17.0
	60	≤ 1:12	5.75	6.50	7.00	7.50	8.25	8.75	9.50	10.75
		5:12	6.75	7.50	8.25	9.25	10.0	10.75	11.75	13.25
		7:12[d]	7.75	9.00	10.0	11.0	12.25	13.25	14.5	16.75
		12:12[d]	8.75	10.0	11.5	12.75	14.0	15.5	16.75	19.5

(continued)

Footnotes to Table R611.7 (9A)

For SI: 1 foot = 304.8 mm; 1 inch = 25.4 mm; 1 pound per square foot = 0.0479 kPa.

a. Table values are based on a 3.5 in thick flat wall. For a 5.5 in thick flat wall, multiply the table values by 0.9. The adjusted values shall not result in solid wall lengths less than 4ft.

b. Table values are based on a maximum unsupported wall height of 10 ft.

c. Linear interpolation shall be permitted.

d. The minimum solid wall lengths shown in the table are based on a building with an end wall length "W" of 60 feet and a roof slope of less than 7:12. For roof slopes of 7:12 or greater and end wall length "W" greater than 30 feet, the minimum solid wall length determined from the table shall be multiplied by:
   $1 + 0.4[(W-30)/30]$.

❖ See the commentary for Sections R611.7.3.1 and R611.7.4.

TABLE R611.7(9B)
MINIMUM SOLID SIDEWALL LENGTH REQUIREMENTS FOR FLAT ICF WALLS (WIND PARALLEL TO RIDGE) [a, b, c, d]

WALL CATEGORY	BUILDING END WALL WIDTH, W (feet)	WIND VELOCITY PRESSURE FROM TABLE R611.7.4 (psf)							
		20	25	30	35	40	45	50	60
		Minimum Solid Wall Length on Building Side Wall (feet)							
One-Story or Top Story of Two-Story	16	4.00	4.00	4.00	4.00	4.25	4.25	4.50	4.75
	24	4.00	4.25	4.50	4.75	4.75	5.00	5.25	5.50
	32	4.50	4.75	5.00	5.25	5.50	6.00	6.25	6.75
	40	5.00	5.50	5.75	6.25	6.75	7.00	7.50	8.25
	50	5.75	6.25	7.00	7.50	8.25	8.75	9.50	10.75
	60	6.50	7.50	8.25	9.25	10.0	10.75	11.75	13.25
First Story of Two-Story	16	4.25	4.50	4.75	5.00	5.25	5.50	5.75	6.50
	24	4.75	5.25	5.50	6.00	6.25	6.75	7.00	8.00
	32	5.50	6.00	6.50	7.00	7.50	8.00	8.75	9.75
	40	6.25	7.00	7.50	8.25	9.00	9.75	10.5	12.0
	50	7.25	8.25	9.25	10.25	11.25	12.25	13.25	15.25
	60	8.50	9.75	11.0	12.25	13.5	15.0	16.25	18.75

For SI: 1 foot = 304.8 mm; 1 inch = 25.4 mm; 1 pound per square foot = 0.0479 kPa.

a. Table values are based on a 3.5 in thick flat wall. For a 5.5 in thick flat wall, multiply the table values by 0.9. The adjusted values shall not result in solid wall lengths less than 4ft.

b. Table values are based on a maximum unsupported wall height of 10 ft.

c. Table values are based on a maximum 12:12 roof pitch.

d. Linear interpolation shall be permitted.

❖ See the commentary for Sections R611.7.3.1 and R611.7.4.

TABLE R611.7(10)
MAXIMUM ALLOWABLE CLEAR SPANS FOR ICF LINTELS IN NONLOAD-BEARING WALLS WITHOUT STIRRUPS[a,b,c,d]
NO. 4 BOTTOM BAR

MINIMUM LINTEL THICKNESS, T (inches)	MINIMUM LINTEL DEPTH, D (inches)	MAXIMUM CLEAR SPAN	
		Supporting Light-Framed Nonbearing Wall (feet-inches)	Supporting ICF Second Story and Nonbearing Wall (feet-inches)
Flat ICF Lintel			
3.5	8	11-1	3-1
	12	15-11	5-1
	16	16-3	6-11
	20	16-3	8-8
	24	16-3	10-5
5.5	8	16-3	4-4
	12	16-3	7-0
	16	16-3	9-7
	20	16-3	12-0
	24	16-3	14-3
7.5	8	16-3	5-6
	12	16-3	8-11
	16	16-3	12-2
	20	16-3	15-3
	24	16-3	16-3
9.5	8	16-3	6-9
	12	16-3	10-11
	16	16-3	14-10
	20	16-3	16-3
	24	16-3	16-3
Waffle-Grid ICF Lintel			
6 or 8	8	9-1	2-11
	12	13-4	4-10
	16	16-3	6-7
	20	16-3	8-4
	24	16-3	9-11
Screen-Grid Lintel			
6	12	5-8	4-1
	24	16-3	9-1

For SI: 1 foot = 304.8 mm; 1 inch = 25.4 mm; 1 pounds per square foot = 0.0479 kPa.

a. This table is based on concrete with a minimum specified compressive strength of 2,500 psi, reinforcing steel with a minimum yield strength of 40,000 psi and an assumed equivalent rectangular cross section.

b. This table is not intended to prohibit the use of ICF manufacturers tables based on engineering analysis in accordance with ACI 318.

c. Deflection criterion is $L/240$, where L is the clear span of the lintel in inches.

d. Linear interpolation is permitted between lintel depths.

❖ See the commentary for Sections R611.7.3.1 and R611.7.3.5.

**TABLE R611.7(10A)
MINIMUM SOLID END WALL LENGTH REQUIREMENTS FOR WAFFLE AND
SCREEN-GRID ICF WALLS (WIND PERPENDICULAR TO RIDGE)[a, b, c]**

WALL CATEGORY	BUILDING SIDE WALL LENGTH, L (feet)	ROOF SLOPE	WIND VELOCITY PRESSURE FROM TABLE R611.7.4							
			20	25	30	35	40	45	50	60
			Minimum Solid Wall Length on Building End Wall (feet)							
One-Story or Top Story of Two-Story	16	≤1:12	4.00	4.00	4.00	4.00	4.00	4.00	4.00	4.25
		5:12	4.00	4.00	4.00	4.00	4.00	4.25	4.25	4.50
		7:12[d]	4.00	4.25	4.50	4.75	5.00	5.25	5.50	6.00
		12:12[d]	4.25	4.75	5.00	5.50	5.75	6.00	6.50	7.00
	24	≤1:12	4.00	4.00	4.00	4.00	4.00	4.25	4.25	4.50
		5:12	4.00	4.00	4.00	4.25	4.50	4.50	4.75	5.00
		7:12[d]	4.50	4.75	5.00	5.50	5.75	6.25	6.50	7.25
		12:12[d]	5.00	5.50	6.00	6.50	7.00	7.25	7.75	8.75
	32	≤1:12	4.00	4.00	4.00	4.25	4.25	4.50	4.75	5.00
		5:12	4.00	4.00	4.25	4.50	4.75	5.00	5.25	5.75
		7:12[d]	4.75	5.25	5.75	6.25	6.50	7.00	7.50	8.50
		12:12[d]	5.50	6.25	6.75	7.50	8.00	8.75	9.25	10.5
	40	≤1:12	4.00	4.00	4.25	4.50	4.50	4.75	5.00	5.50
		5:12	4.00	4.25	4.50	5.00	5.25	5.50	5.75	6.25
		7:12[d]	5.25	5.75	6.25	7.00	7.50	8.00	8.50	9.75
		12:12[d]	6.00	6.75	7.75	8.50	9.25	10.0	10.75	12.25
	50	≤1:12	4.00	4.25	4.50	4.75	5.00	5.25	5.50	6.00
		5:12	4.25	4.75	5.00	5.25	5.50	6.00	6.25	7.00
		7:12[d]	5.75	6.50	7.00	7.75	8.50	9.25	9.75	11.25
		12:12[d]	6.75	7.75	8.75	9.75	10.75	11.5	12.5	14.5
	60	≤1:12	4.25	4.50	4.75	5.00	5.25	5.50	5.75	6.50
		5:12	4.50	5.00	5.25	5.75	6.00	6.50	6.75	7.75
		7:12[d]	6.25	7.00	8.00	8.75	9.50	10.25	11.25	12.75
		12:12[d]	7.50	8.75	9.75	11.0	12.0	13.25	14.25	16.5
First Story of Two-Story	16	≤1:12	4.25	4.50	4.75	5.00	5.25	5.50	5.75	6.50
		5:12	4.50	5.00	5.25	5.75	6.00	6.50	6.75	7.75
		7:12[d]	4.75	5.25	5.75	6.25	6.75	7.25	7.75	8.75
		12:12[d]	5.25	5.75	6.50	7.00	7.50	8.00	8.75	9.75
	24	≤1:12	4.50	5.00	5.25	5.75	6.25	6.50	7.00	7.75
		5:12	5.00	5.75	6.25	6.75	7.25	7.75	8.25	9.25
		7:12[d]	5.75	6.25	7.00	7.75	8.25	9.00	9.75	11.0
		12:12[d]	6.25	7.00	7.75	8.50	9.50	10.25	11.0	12.75
	32	≤1:12	5.00	5.50	6.00	6.50	7.00	7.50	8.00	9.00
		5:12	5.75	6.25	7.00	7.75	8.25	9.00	9.75	11.0
		7:12[d]	6.50	7.25	8.25	9.00	10.0	10.75	11.75	13.5
		12:12[d]	7.25	8.25	9.25	10.25	11.25	12.5	13.5	15.5
	40	≤1:12	5.50	6.00	6.50	7.25	7.75	8.50	9.00	10.25
		5:12	6.25	7.00	7.75	8.75	9.50	10.25	11.0	12.75
		7:12[d]	7.25	8.25	9.25	10.5	11.5	12.5	13.75	15.75
		12:12[d]	8.00	9.50	10.75	12.0	13.25	14.5	15.75	18.25

(continued)

TABLE R611.7(10A)—continued
MINIMUM SOLID END WALL LENGTH REQUIREMENTS FOR WAFFLE AND
SCREEN-GRID ICF WALLS (WIND PERPENDICULAR TO RIDGE)[a, b, c]

WALL CATEGORY	BUILDING SIDE WALL LENGTH, L (feet)	ROOF SLOPE	WIND VELOCITY PRESSURE FROM TABLE R611.7.4							
			20	25	30	35	40	45	50	60
			Minimum Solid Wall Length on Building End Wall (feet)							
First Story of Two-Story	50	≤ 1:12	6.00	6.75	7.50	8.00	8.75	9.50	10.25	11.75
		5:12	7.00	8.00	9.00	10.0	11.0	12.0	13.0	14.75
		7:12[d]	8.25	9.50	10.75	12.25	13.5	14.75	16.0	18.75
		12:12[d]	9.25	11.0	12.5	14.0	15.5	17.25	18.75	22.0
	60	≤ 1:12	6.50	7.25	8.25	9.00	10.0	10.75	11.75	13.25
		5:12	7.75	8.75	10.0	11.25	12.25	13.5	14.75	17.0
		7:12[d]	9.25	10.75	12.25	14.0	15.5	17.0	18.5	21.75
		12:12[d]	10.5	12.25	14.25	16.25	18.0	20.0	21.75	25.5

For SI: 1 foot = 304.8 mm; 1 inch = 25.4 mm; 1 pound per square foot = 0.0479kPa.

a. Table values are based on a 6 in (152.4 mm) thick nominal waffle-grid wall. For a 8 in thick nominal waffle-grid wall, multiply the table values by 0.90.

b. Table values are based on a maximum unsupported wall height of 10 ft.

c. Linear interpolation is permitted.

d. The minimum solid wall lengths shown in the table are based on a building with an end wall length "W" of 60 feet and a roof slope of less than 7:12. For roof slopes of 7:12 or greater and end wall length "W" greater than 30 feet, the minimum solid wall length determined from the table shall be multiplied by: 1 + 0.4 [(W-30)/30].

❖ See the commentary for Section R611.7.4.

TABLE R611.7(10B)
MINIMUM SOLID SIDE WALL LENGTH REQUIREMENTS FOR 6-INCH WAFFLE AND
SCREEN-GRID ICF WALLS (WIND PARALLEL TO RIDGE)[a, b, c, d]

WALL CATEGORY	BUILDING END WALL WIDTH, W (feet)	WIND VELOCITY PRESSURE FROM TABLE R611.7.4 (psf)							
		20	25	30	35	40	45	50	60
		Minimum Solid Wall Length on Building Side Wall (feet)							
One-Story or Top Story of Two-Story	16	4.00	4.00	4.00	4.25	4.25	4.50	4.75	5.00
	24	4.00	4.25	4.50	5.00	5.25	5.50	5.75	6.25
	32	4.50	5.00	5.50	5.75	6.25	6.75	7.00	8.00
	40	5.25	6.00	6.50	7.00	7.75	8.25	8.75	10.0
	50	6.50	7.25	8.00	9.00	9.75	10.75	11.5	13.25
	60	7.75	8.75	10.0	11.25	12.25	13.5	14.5	17.0
First Story of Two-Story	16	4.50	4.75	5.25	5.50	5.75	6.25	6.50	7.25
	24	5.00	5.75	6.25	6.75	7.25	7.75	8.25	9.50
	32	6.00	6.75	7.50	8.25	9.00	9.75	10.5	12.0
	40	7.00	8.00	9.00	10.0	11.0	12.0	13.0	15.0
	50	8.50	9.75	11.25	12.5	14.0	15.25	16.75	19.5
	60	10.25	12.0	13.75	15.5	17.25	19.0	21.0	24.5

For SI: 1 foot = 304.8 mm; 1 inch = 25.4 mm; 1 pound per square foot = 0.0479kPa.

a. Table values are based on a 6 in thick nominal waffle-grid wall. For a 8 in thick nominal waffle-grid wall, multiply the table values by 0.90.

b. Table values are based on a maximum unsupported wall height of 10 ft.

c. Table values are based on a maximum 12:12 roof pitch.

d. Linear interpolation shall be permitted.

❖ See the commentary for Section R611.7.4.

TABLE R611.7(11)
MINIMUM PERCENTAGE OF SOLID WALL LENGTH ALONG EXTERIOR WALL LINES FOR TOWNHOUSES IN SEISMIC DESIGN CATEGORY C AND ALL BUILDINGS IN SEISMIC DESIGN CATEGORIES D₀, D₁, AND D₂[a, b]

SEISMIC DESIGN CATEGORY (SDC)	MINIMUM SOLID WALL LENGTH (percent)		
	One-Story or Top Story of Two-Story	Wall Supporting Light-Framed Second Story and Roof	Wall Supporting ICF Second Story and Roof
Townhouses in SDC C[c]	20 percent	25 percent	35 percent
D₁[d]	25 percent	30 percent	40 percent
D₂[d]	30 percent	35 percent	45 percent

For SI: 1 inch = 25.4 mm; 1 mile per hour = 0.447 m/s.

a. Base percentages are applicable for maximum unsupported wall height of 10-feet, light-frame gable construction, and all ICF wall types. These percentages assume that the maximum weight of the interior and exterior wall finishes applied to ICF walls do not exceed 8 psf.

b. For all walls, the minimum required length of solid walls shall be based on the table percent value multiplied by the minimum dimension of a rectangle inscribing the overall building plan.

c. Walls shall be reinforced with a minimum No. 5 bar (Grade 40 or 60) spaced a maximum of 24 inches on center each way or a No. 4 bar spaced a maximum of 16 inches on center each way. (Grade 40 or 60) spaced at a maximum of 16 inches on center each way.

d. Walls shall be constructed with a minimum concrete compressive strength of 3,000 psi and reinforced with minimum #5 rebar (Grade 60 ASTM A 706) spaced a maximum of 18 inches on center each way or No. 4 rebar (Grade 60 ASTM A706) spaced at a maximum of 12 inches (304.8 mm) on center each way. The minimum thickness of flat ICF walls shall be 5.5 inches.

❖ See the commentary for Section R611.7.4.

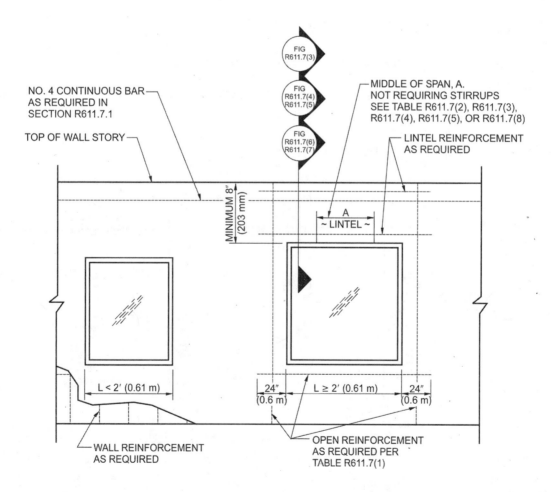

FIGURE R611.7(2)
REINFORCEMENT OF OPENING

❖ This figure illustrates the reinforcement of openings specified in Sections R611.7.2 and R611.7.3.

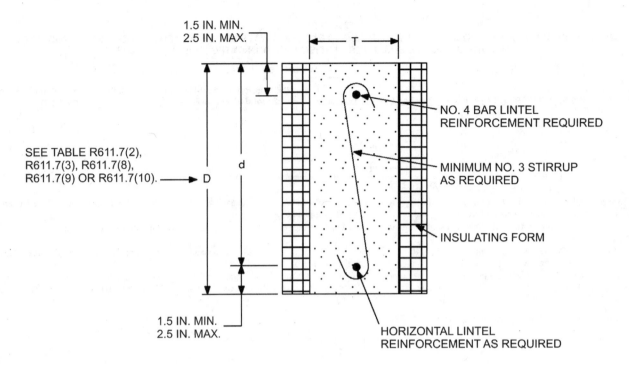

For SI: 1 inch = 25.4 mm.
NOTE: Section cut through flat wall.

FIGURE R611.7(3)
ICF LINTELS FOR FLAT AND SCREEN-GRID WALLS

❖ See the commentary for Section R611.7.3.1.

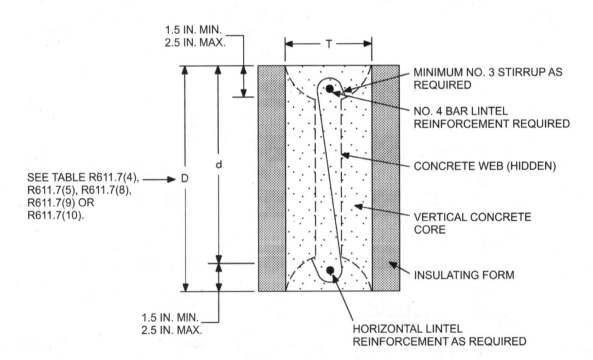

For SI: 1 inch = 25.4 mm.
NOTE: Section cut through vertical core of a waffle-grid lintel.

FIGURE R611.7(4)
SINGLE FORM HEIGHT WAFFLE-GRID LINTEL

❖ See the commentary for Section R611.7.3.1.

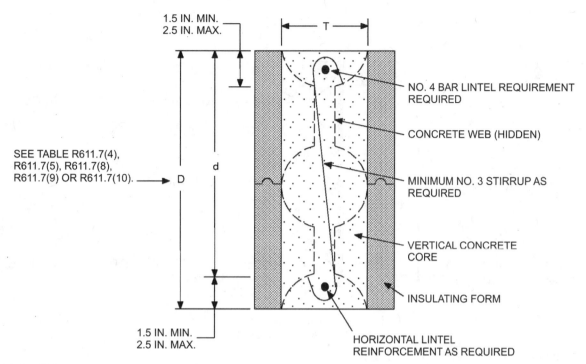

For SI: 1 inch = 25.4 mm.
NOTE: Section cut through vertical core of a waffle-grid lintel.

FIGURE R611.7(5)
DOUBLE FORM HEIGHT WAFFLE-GRID LINTEL

❖ See the commentary for section R611.7.3.1.

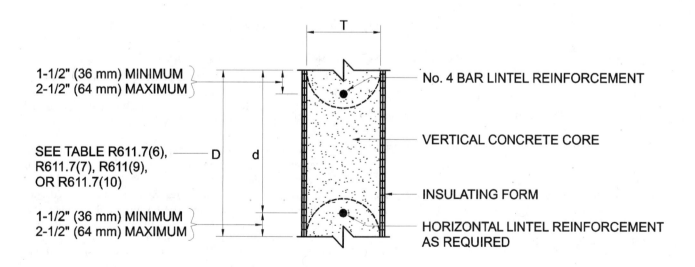

FIGURE R611.7(6)
SINGLE FORM HEIGHT SCREEN-GRID LINTEL

❖ See the commentary for Section R611.7.3.1

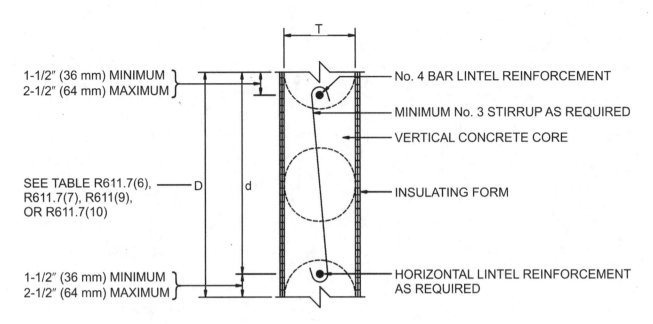

1-1/2″ (36 mm) MINIMUM
2-1/2″ (64 mm) MAXIMUM

No. 4 BAR LINTEL REINFORCEMENT

MINIMUM No. 3 STIRRUP AS REQUIRED

VERTICAL CONCRETE CORE

SEE TABLE R611.7(6),
R611.7(7), R611(9),
OR R611.7(10)

INSULATING FORM

1-1/2″ (36 mm) MINIMUM
2-1/2″ (64 mm) MAXIMUM

HORIZONTAL LINTEL REINFORCEMENT
AS REQUIRED

FIGURE R611.7(7)
DOUBLE FORM HEIGHT SCREEN-GRID LINTEL

❖ See the commentary for Section R611.7.3.1

TABLE R611.7.4
WIND VELOCITY PRESSURE FOR DETERMINATION OF
MINIMUM SOLID WALL LENGTH[a]

WIND SPEED (mph)[d]	VELOCITY PRESSURE (psf)		
	Exposure[b]		
	B	C	D
85	14	19	23
90	16	21	25
100	19	26	31
110	23	32	37
120	27	38	44
130	32	44	52
140	37	51	60
150	43	59	69[c]

For SI: 1 pound per square foot = 0.0479 kPa; 1 mile per hour = 0.447 m/s.

a. Table values are based on ASCE 7-98 Figure 6-4 using a mean roof height of 35 ft.
b. Exposure Categories shall be determined in accordance with Section R301.2.1.4.
c. Design is required in accordance with ACI 318 and approved manufacturer guidelines.
d. Interpolation is permitted between wind speeds.

❖ See the commentary for Section R611.7.4.

R611.8 ICF wall-to-floor connections.

R611.8.1 Top bearing. Floors bearing on the top of ICF foundation walls in accordance with Figure R611.8(1) shall have the wood sill plate anchored to the ICF wall with minimum $^1/_2$-inch (13 mm) diameter bolts embedded a minimum of 7 inches (178 mm) and placed at a maximum spacing of 6 feet (1829 mm) on center and not more than 12 inches (305 mm) from corners.

Anchor bolts for waffle-grid and screen-grid walls shall be located in the cores. In conditions where wind speeds are in excess of 90 miles per hour (40 m/s), the $^1/_2$-inch (13 mm) diameter anchor bolts shall be placed at a maximum spacing of 4 feet (1219 mm) on center. Bolts shall extend a minimum of 7 inches (178 mm) into concrete. Sill plates shall be protected against decay where required by Section R319. Cold-formed steel framing systems shall be anchored to the concrete in accordance with Section R505.3.1 or Section R603.3.1.

❖ This section specifies requirements for bearings and connection of floors on top of ICF walls and illustrates details for a variety of ledger conditions.

R611.8.1.1 Top bearing requirements for Seismic Design Categories C, D_0, D_1 and D_2. For townhouses in Seismic Design Category C, wood sill plates attached to ICF walls shall be anchored with Grade A 307, $^3/_8$-inch-diameter (10 mm) headed anchor bolts embedded a minimum of 7 inches (178 mm) and placed at a maximum spacing of 36 inches (914 mm) on center. For all buildings in Seismic Design Category D_0 or D_1, wood sill plates attached to ICF walls shall be anchored with ASTM A 307, Grade A, $^3/_8$-inch-diameter (10 mm) headed anchor bolts embedded a minimum of 7 inches (178 mm) and placed at a maximum spacing of 24 inches (610 mm) on center. For all buildings in Seismic Design Category D_2, wood sill plates attached to ICF walls shall be anchored with ASTM A 307, Grade A, $^3/_8$-inch-diameter (10 mm) headed anchor bolts embedded a minimum of 7 inches (178 mm) and placed at a maximum spacing of 16 inches (406 mm) on center. Larger diameter bolts than specified herein shall not be used.

For townhouses in Seismic Design Category C, each floor joist perpendicular to an ICF wall shall be attached to the sill plate with an 18-gage [(0.0478 in.) (1.2 mm)] angle bracket using 3 - 8d common nails per leg in accordance with Figure

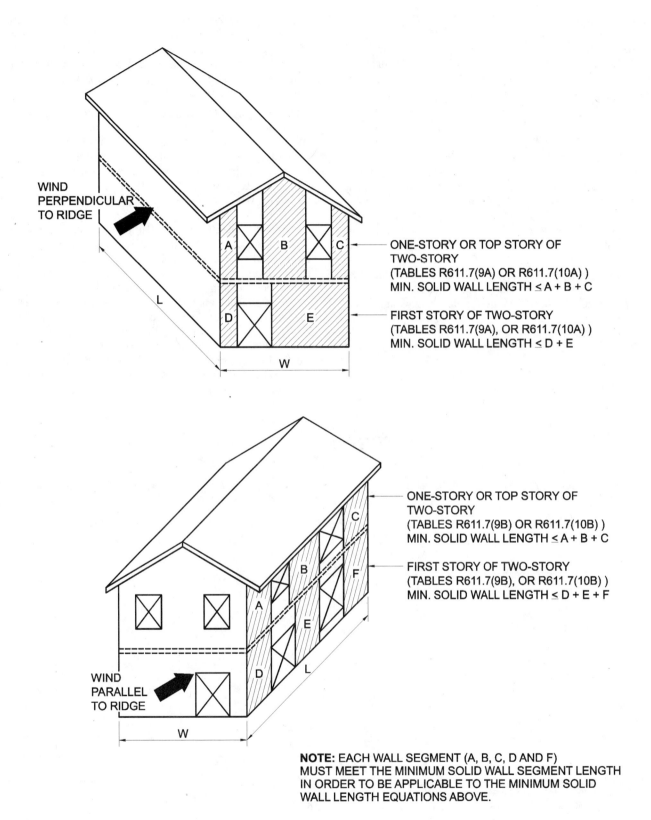

WIND PERPENDICULAR TO RIDGE

ONE-STORY OR TOP STORY OF TWO-STORY (TABLES R611.7(9A) OR R611.7(10A)) MIN. SOLID WALL LENGTH ≤ A + B + C

FIRST STORY OF TWO-STORY (TABLES R611.7(9A), OR R611.7(10A)) MIN. SOLID WALL LENGTH ≤ D + E

WIND PARALLEL TO RIDGE

ONE-STORY OR TOP STORY OF TWO-STORY (TABLES R611.7(9B) OR R611.7(10B)) MIN. SOLID WALL LENGTH ≤ A + B + C

FIRST STORY OF TWO-STORY (TABLES R611.7(9B), OR R611.7(10B)) MIN. SOLID WALL LENGTH ≤ D + E + F

NOTE: EACH WALL SEGMENT (A, B, C, D AND F) MUST MEET THE MINIMUM SOLID WALL SEGMENT LENGTH IN ORDER TO BE APPLICABLE TO THE MINIMUM SOLID WALL LENGTH EQUATIONS ABOVE.

FIGURE R611.7.4
MINIMUM SOLID WALL LENGTH

❖ See the commentary for Section R611.7.4.

R611.8(1). For all buildings in Seismic Design Category D_0 or D_1, each floor joist perpendicular to an ICF wall shall be attached to the sill plate with an 18-gage [(0.0478 in.) (1.2 mm)] angle bracket using 4 - 8d common nails per leg in accordance with Figure R611.8(1). For all buildings in Seismic Design Category D_2, each floor joist perpendicular to an ICF wall shall be attached to the sill plate with an 18-gage [(0.0478 in.) (1.2 mm)] angle bracket using 6 - 8d common nails per leg in accordance with Figure R611.8(1).

For ICF walls parallel to floor framing in townhouses in Seismic Design Category C, full depth blocking shall be placed at 24 inches (610 mm) on center and shall be attached to the sill plate with an 18-gage [(0.0478 in.) (1.2 mm)] angle bracket using 5 - 8d common nails per leg in accordance with Figure R611.8(6). For ICF walls parallel to floor framing for all buildings in Seismic Design Category D_0 or D_1, full depth blocking shall be placed at 24 inches (610 mm) on center and shall be attached to the sill plate with an 18-gage [(0.0478 in.) (1.2 mm)] angle bracket using 6 - 8d common nails per leg in accordance with Figure R611.8(6). For ICF walls parallel to floor framing for all buildings in Seismic Design Category D_2, full depth blocking shall be placed at 24 inches (610 mm) on center and shall be attached to the sill plate with an 18-gage [(0.0478 in.) (1.2 mm)] angle bracket using 9 - 8d common nails per leg in accordance with Figure R611.8(6).

❖ This section states additional requirements for bearings and connections of floors on top of an ICF wall in a building used for a townhouse in Seismic Design Category C and in all buildings in Seismic Design Categories D_0, D_1 and D_2. As they do with all other structural provisions of this code, these buildings require a more rigorous set of detail requirements to perform acceptably in the higher seismic conditions.

R611.8.2 Ledger bearing. Wood ledger boards supporting bearing ends of joists or trusses shall be anchored to flat ICF walls with minimum thickness of 5.5 inches (140 mm) and to waffle- or screen-grid ICF walls with minimum nominal thickness of 6 inches (152 mm) in accordance with Figure R611.8(2), R611.8(3), R611.8(4) or R611.8(5) and Table R611.8(1). Wood ledger boards supporting bearing ends of joists or trusses shall be anchored to flat ICF walls with minimum thickness of 3.5 inches (140 mm) in accordance with Figure R611.8(5) and Table R611.8(1). The ledger shall be a minimum 2 by 8, No. 2 Southern Yellow Pine or No. 2 Douglas Fir. Ledgers anchored to nonload-bearing walls to support floor or roof sheathing shall be attached with $^1/_2$ inch (12.7 mm) diameter or headed anchor bolts spaced a maximum of 6 feet (1829 mm) on center. Anchor bolts shall be embedded a minimum of 4 inches (102 mm) into the concrete measured from the inside face of the insulating form. For insulating forms with a face shell thickness of 1.5 inches (38 mm) or less, the hole in the form shall be a minimum of 4 inches (102 mm) in diameter. For insulating forms with a face shell thicker than 1.5 inches

TABLE R611.8(1)
FLOOR LEDGER-ICF WALL CONNECTION (SIDE-BEARING CONNECTION) REQUIREMENTS[a, b, c]

MAXIMUM FLOOR CLEAR SPAN[d] (feet)	MAXIMUM ANCHOR BOLT SPACING[e] (inches)			
	Staggered $^1/_2$-inch-diameter anchor bolts	Staggered $^5/_8$-inch-diameter anchor bolts	Two $^1/_2$-inch-diameter anchor bolts[f]	Two $^5/_8$-inch-diameter anchor bolts[f]
8	18	20	36	40
10	16	18	32	36
12	14	18	28	36
14	12	16	24	32
16	10	14	20	28
18	9	13	18	26
20	8	11	16	22
22	7	10	14	20
24	7	9	14	18
26	6	9	12	18
28	6	8	12	16
30	5	8	10	16
32	5	7	10	14

For SI: 1 inch = 25.4 mm, 1 foot = 304.8 mm.

a. Minimum ledger board nominal depth shall be 8 inches. The thickness of the ledger board shall be a minimum of 2 inches. Thickness of ledger board is in nominal lumber dimensions. Ledger board shall be minimum No. 2 Grade.

b. Minimum edge distance shall be 2 inches for $^1/_2$-inch-diameter anchor bolts and 2.5 inches for $^5/_8$-inch-diameter anchor bolts.

c. Interpolation is permitted between floor spans.

d. Floor span corresponds to the clear span of the floor structure (i.e., joists or trusses) spanning between load-bearing walls or beams.

e. Anchor bolts shall extend through the ledger to the center of the flat ICF wall thickness or the center of the horizontal or vertical core thickness of the waffle-grid or screen-grid ICF wall system.

f. Minimum vertical distance between bolts shall be 1.5 inches for $^1/_2$-inch-diameter anchor bolts and 2 inches for $^5/_8$-inch-diameter anchor bolts.

❖ See the commentary for Section R611.8.1.

(38 mm), the diameter of the hole in the form shall be increased by 1 inch (25 mm) for each $1/2$ inch (13 mm) of additional insulating form face shell thickness. The ledger board shall be in direct contact with the concrete at each bolt location.

❖ This section specifies requirements for connections and bearing of wood ledger boards supporting joists or trusses on top of ICF walls and illustrates details for a variety of ledger conditions.

R611.8.2.1 Ledger bearing requirements for Seismic Design Categories C, D_0, D_1 and D_2. Additional anchorage mechanisms connecting the wall to the floor system shall be installed at a maximum spacing of 6 feet (1829 mm) on center for townhouses in Seismic Design Category C and 4 feet (1220 mm) on center for all buildings in Seismic Design Categories D_0, D_1 and D_2. The additional anchorage mechanisms shall be attached to the ICF wall reinforcement and joist rafters or blocking in accordance with Figures R611.8(1) through R611.8(7). The additional anchorage shall be installed through an oversized hole in the ledger board that is $1/2$ inch (13 mm) larger than the anchorage mechanism diameter to prevent combined tension and shear in the mechanism. The blocking shall be attached to floor or roof sheathing in accordance with edge fastener spacing. Such additional anchorage shall not be accomplished by the use of toe nails or nails subject to withdrawal nor shall such anchorage mechanisms induce tension stresses perpendicular to grain in ledgers or nailers. The capacity of such anchors shall result in connections capable of resisting the design values listed in Table R611.8(2). The diaphragm sheathing fasteners applied directly to a ledger shall not be considered effective in providing the additional anchorage required by this section.

Where the additional anchorage mechanisms consist of threaded rods with hex nuts or headed bolts complying with ASTM A 307, Grade A or ASTM F 1554, Grade 36, the design tensile strengths shown in Table R611.9 shall be equal to or greater than the product of the design values listed in Table R611.8(2) and the spacing of the bolts in feet (mm). Anchor bolts shall be embedded as indicated in Table R611.9. Bolts with hooks shall not be used.

❖ This section contains additional requirements for bearings and connections of wood ledger boards supporting joists or trusses on top of ICF walls in a building used for a townhouse in Seismic Design Category C and in all buildings in Seismic Design Categories D_0, D_1 and D_2. As they do with all other structural provisions of this code, these buildings require a more rigorous set of detail requirements to perform acceptably in the higher seismic conditions.

TABLE R611.8(2)
DESIGN VALUES (PLF) FOR FLOOR JOIST-TO-WALL ANCHORS REQUIRED FOR TOWNHOUSES IN SEISMIC DESIGN CATEGORY C AND ALL BUILDINGS IN SEISMIC DESIGN CATEGORIES D_0, D_1 AND D_2[a, b]

WALL TYPE	SEISMIC DESIGN CATEGORY		
	C	D_0 or D_1	D_2
Flat 3.5	193	NP	NP
Flat 5.5	303	502	708
Flat 7.5	413	685	965
Flat 9.5	523	867	1,223
Waffle 6	246	409	577
Waffle 8	334	555	782
Screen 6	233	387	546

For SI: 1 pound per linear foot = 1.488 kg/m.

NP = Not Permitted

a. Table values are based on IBC Equation 16–64 using a tributary wall height of 11 feet. Table values shall be permitted to be reduced for tributary wall heights less than 11 feet by multiplying the table values by X/11, where X is the tributary wall height.

b. Values may be reduced by 30 percent when used for ASD.

❖ See the commentary for Section R611.8.1.

TABLE R611.9
DESIGN TENSILE STRENGTH OF HEADED BOLTS CAST IN CONCRETE[a]

DIAMETER OF BOLT (inches)	MINIMUM EMBEDMENT DEPTH (inches)	DESIGN TENSILE STRENGTH[b] (pounds)
$1/4$	2	1040
$3/8$ with washer[c]	$2^3/_4$[d]	2540
$1/2$ with washer[c]	4[d]	4630

For SI: 1 pound per square inch = 6.895 kPa.

a. Applicable to concrete of all strengths. See Notes (c) and (d).

b. Values are based on ASTM F 1554, Grade 36 bolts. Where ASTM A 307, Grade A headed bolts are used, the strength shall be increased by 1.034.

c. A hardened washer shall be installed at the nut embedded in the concrete or head of the bolt to increase the bearing area. The washer is not required where the concrete strength is 4000 psi or more.

d. Embedment depth shall be permitted to be reduced $1/4$-inch where 4000 psi concrete is used.

❖ See the commentary to Section R611.9.1.

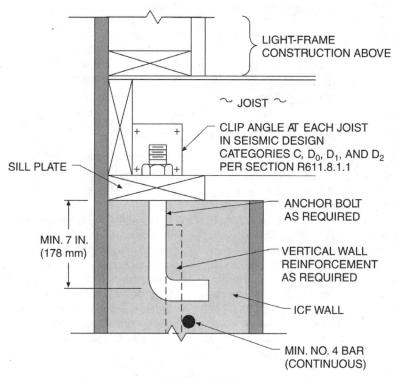

FIGURE R611.8(1)
SECTION CUT THROUGH FLAT WALL OR VERTICAL CORE
OF WAFFLE- OR SCREEN-GRID WALL

❖ See the commentary for Section R611.8.1.

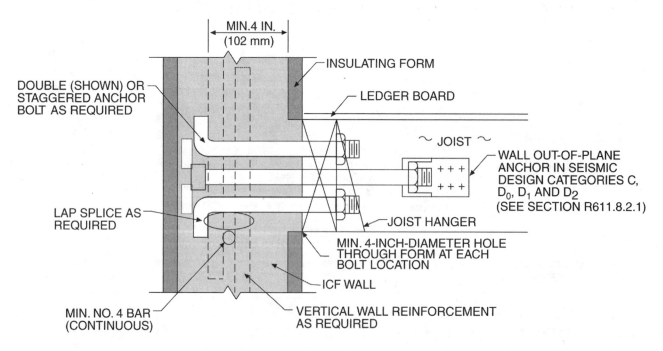

For SI: 1 inch = 25.4 mm.
NOTE: Section cut through flat wall or vertical core of a waffle- or screen-grid wall.

FIGURE R611.8(2)
FLOOR LEDGER—ICF WALL CONNECTION (SIDE-BEARING CONNECTION)

❖ See the commentary for Section R611.8.1.

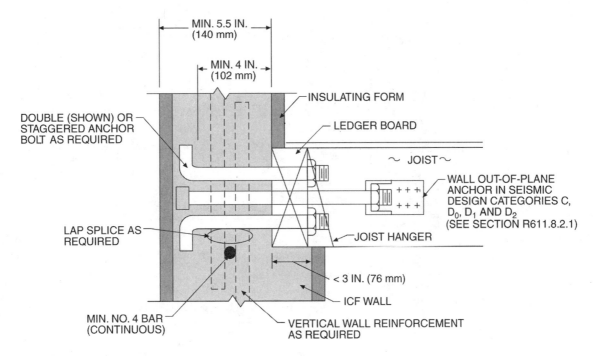

For SI: 1 inch = 25.4 mm.
NOTE: Section cut through flat wall or vertical core of a waffle- or screen-grid wall.

FIGURE R611.8(3)
FLOOR LEDGER—ICF WALL CONNECTION (LEDGE-BEARING CONNECTION)

❖ See the commentary for Section R611.8.2.

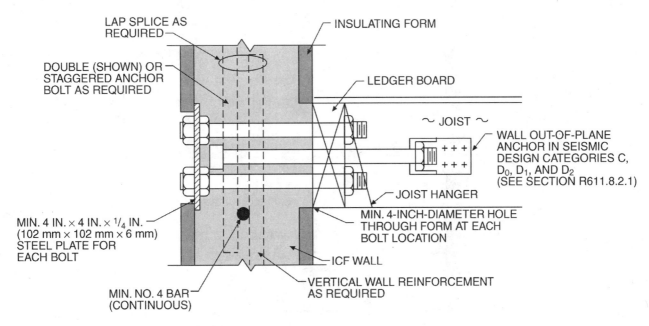

For SI: 1 inch = 25.4 mm.
NOTE: Section cut through flat wall.

FIGURE R611.8(4)
WOOD FLOOR LEDGER—ICF WALL SYSTEM CONNECTION
(THROUGH-BOLT SIDE-BEARING CONNECTION)

❖ See the commentary for Section R611.8.2.

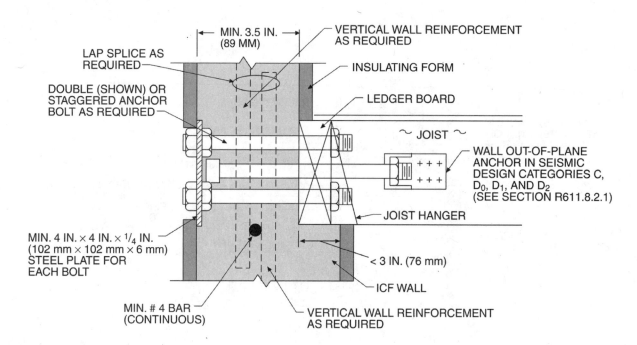

For SI: 1 inch = 25.4 mm.
NOTE: Section cut through flat wall.

FIGURE R611.8(5)
FLOOR LEDGER—ICF WALL CONNECTION

❖ See the commentary for Section R611.8.2.

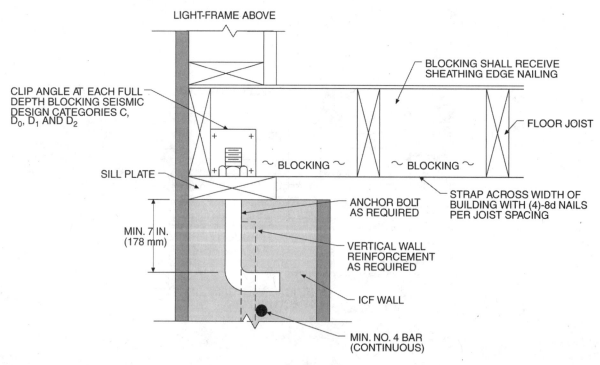

FIGURE R611.8(6)
ANCHORAGE REQUIREMENTS FOR TOP BEARING WALLS FOR TOWNHOUSES IN SEISMIC DESIGN CATEGORY C
AND ALL BUILDINGS IN SEISMIC DESIGN CATEGORIES D_0, D_1, AND D_2 FOR FLOOR FRAMING PARALLEL TO WALL

❖ See the commentary for Section R611.8.1.1.

2006 INTERNATIONAL RESIDENTIAL CODE® COMMENTARY

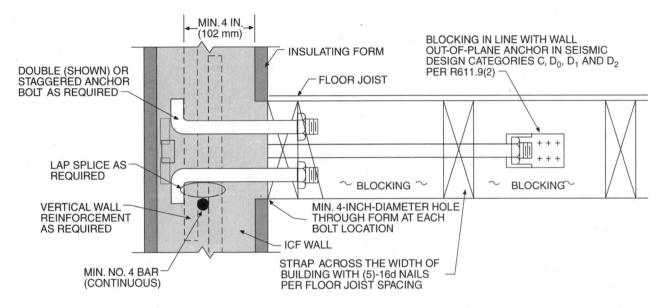

BLOCKING SHALL EXTEND THE WIDTH
OF THE BUILDING AND SHALL RECEIVE
SHEATHING EDGE NAILING

MIN. 4 IN.
(102 mm)

INSULATING FORM

BLOCKING IN LINE WITH WALL
OUT-OF-PLANE ANCHOR IN SEISMIC
DESIGN CATEGORIES C, D₀, D₁ AND D₂
PER R611.9(2)

DOUBLE (SHOWN) OR
STAGGERED ANCHOR
BOLT AS REQUIRED

FLOOR JOIST

LAP SPLICE AS
REQUIRED

~ BLOCKING ~

~ BLOCKING~

VERTICAL WALL
REINFORCEMENT
AS REQUIRED

MIN. 4-INCH-DIAMETER HOLE
THROUGH FORM AT EACH
BOLT LOCATION

ICF WALL

MIN. NO. 4 BAR
(CONTINUOUS)

STRAP ACROSS THE WIDTH OF
BUILDING WITH (5)-16d NAILS
PER FLOOR JOIST SPACING

FIGURE R611.8(7)
ANCHORAGE REQUIREMENTS FOR LEDGER BEARING WALLS FOR TOWNHOUSES IN SEISMIC DESIGN CATEGORY C AND ALL
BUILDINGS IN SEISMIC DESIGN CATEGORIES D₀, D₁ AND D₂ FOR FLOOR FRAMING PARALLEL TO WALL

❖ See the commentary for Section R611.8.2.1.

R611.8.3 Floor and roof diaphragm construction. Floor and roof diaphragms shall be constructed of wood structural panel sheathing attached to wood framing in accordance with Table R602.3(1) or Table R602.3(2) or to cold-formed steel floor framing in accordance with Table R505.3.1(2) or to cold-formed steel roof framing in accordance with Table R804.3.

❖ The requirements of this section are similar to those for masonry wall construction in Section R606.12.1.1. The addition of blocking for panel edges perpendicular to the framing provides added diaphragm capacity.

R611.8.3.1 Floor and roof diaphragm construction requirements in Seismic Design Categories D₀, D₁ and D₂. The requirements of this section shall apply in addition to those required by Section R611.8.3. Edge spacing of fasteners in floor and roof sheathing shall be 4 inches (102 mm) on center for Seismic Design Category D₀ or D₁ and 3 inches (76 mm) on center for Seismic Design Category D₂. In Seismic Design Categories D₀, D₁ and D₂, all sheathing edges shall be attached to framing or blocking. Minimum sheathing fastener size shall be 0.113 inch (3 mm) diameter with a minimum penetration of $1^3/_8$-inches (35 mm) into framing members supporting the sheathing. Minimum wood structural panel thickness shall be $^7/_{16}$ inch (11 mm) for roof sheathing and $^{23}/_{32}$ inch (18 mm) for floor sheathing. Vertical offsets in floor framing shall not be permitted.

❖ This section states additional requirements for design and detailing of floor and roof diaphragm construction in a building used for a townhouse in Seismic Design Category C and in all buildings in Seismic Design Categories D₀, D₁ and D₂. As they do with all other struc-

tural provisions of this code, these buildings require more rigorous requirements to perform acceptably in the higher seismic conditions.

R611.9 ICF wall to top sill plate (roof) connections. Wood sill plates attaching roof framing to ICF walls shall be anchored with minimum $^1/_2$ inch (13 mm) diameter anchor bolt embedded a minimum of 7 inches (178 mm) and placed at 6 feet (1829 mm) on center in accordance with Figure R611.9. Anchor bolts shall be located in the cores of waffle-grid and screen-grid ICF walls. Roof assemblies subject to wind uplift pressure of 20 pounds per square foot (1.44 kPa) or greater as established in Table R301.2(2) shall have rafter or truss ties provided in accordance with Table R802.11.

❖ These provisions address the uplift forces on roofs caused by wind. See the commentary for Section R802.11.

R611.9.1 ICF wall to top sill plate (roof) connections for Seismic Design Categories C, D₀, D₁ and D₂. The requirements of this section shall apply in addition to those required by Section R611.9. The top of an ICF wall at a gable shall be attached to an attic floor in accordance with Section R611.8.1.1. For townhouses in Seismic Design Category C, attic floor diaphragms shall be constructed of structural wood sheathing panels attached to wood framing in accordance with Table R602.3(1) or Table R602.3(2). Edge spacing of fasteners in attic floor sheathing shall be 4 inches (102 mm) on center for Seismic Design Category D₀ or D₁ and 3 inches (76 mm) on center for Seismic Design Category D₂. In Seismic Design Categories D₀, D₁ and

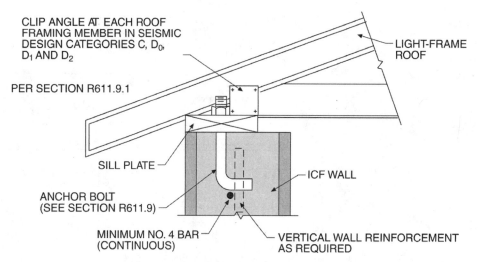

NOTE: Section cut through flat wall or vertical core of a waffle- or screen-grid wall.

FIGURE R611.9
ROOF SILL PLATE—ICF WALL CONNECTION

❖ See the commentary to Sections R611.9 and R802.11.

D_2, all sheathing edges shall be attached to framing or blocking. Minimum sheathing fastener size shall be 0.113 inch (2.8 mm) diameter with a minimum penetration of $1^3/_8$ inches (35 mm) into framing members supporting the sheathing. Minimum wood structural panel thickness shall be $^7/_{16}$ inch (11 mm) for the attic floor sheathing. Where hipped roof construction is used, the use of a structural attic floor is not required.

For townhouses in Seismic Design Category C, wood sill plates attached to ICF walls shall be anchored with ASTM A 307, Grade A, $^3/_8$-inch (10 mm) diameter anchor bolts embedded a minimum of 7 inches (178 mm) and placed at a maximum spacing of 36 inches (914 mm) on center. For all buildings in Seismic Design Category D_0 or D_1, wood sill plates attached to ICF walls shall be anchored with ASTM A 307, Grade A, $^3/_8$-inch (10 mm) diameter anchor bolts embedded a minimum of 7 inches (178 mm) and placed at a maximum spacing of 16 inches (406 mm) on center. For all buildings in Seismic Design Category D_2, wood sill plates attached to ICF walls shall be anchored with ASTM A 307, Grade A, $^3/_8$-inch (10 mm) diameter anchor bolts embedded a minimum of 7 inches (178 mm) and placed at a maximum spacing of 16 inches (406 mm) on center.

For townhouses in Seismic Design Category C, each floor joist shall be attached to the sill plate with an 18-gage [(0.0478 in.) (1.2 mm)] angle bracket using 3 - 8d common nails per leg in accordance with Figure R611.8(1). For all buildings in Seismic Design Category D_0 or D_1, each floor joist shall be attached to the sill plate with an 18-gage [(0.0478 in.) (1.2 mm)] angle bracket using 4 - 8d common nails per leg in accordance with Figure R611.8(1). For all buildings in Seismic Design Category D_2, each floor joist shall be attached to the sill plate with an 18-gage [(0.0478 in.) (1.2 mm)] angle bracket using 6-8d common nails per leg in accordance with Figure R611.8(1).

Where hipped roof construction is used without an attic floor, the following shall apply. For townhouses in Seismic Design Category C, each rafter shall be attached to the sill plate with an 18-gage [(0.0478 in.) (1.2 mm)] angle bracket using 3 - 8d common nails per leg in accordance with Figure R611.9. For all buildings in Seismic Design Category D_0 or D_1, each rafter shall be attached to the sill plate with an 18-gage [(0.0478 in.) (1.2 mm)] angle bracket using 4 - 8d common nails per leg in accordance with Figure R611.9. For all buildings in Seismic Design Category D_2, each rafter shall be attached to the sill plate with an 18-gage [(0.0478 in.) (1.2 mm)] angle bracket using 6-8d common nails per leg in accordance with Figure R611.9.

❖ This section describes additional requirements for bearings and connections of roofs on top of an ICF wall in a building used for a townhouse in Seismic Design Category C and in all buildings in Seismic Design Categories D_0, D_1 and D_2. As they do with all other structural provisions of this code, these buildings require a more rigorous set of detail requirements to perform acceptably in the higher seismic conditions.

SECTION R612
CONVENTIONALLY FORMED CONCRETE WALL CONSTRUCTION

R612.1 General. Conventionally formed concrete walls with flat surfaces shall be designed and constructed in accordance with the provisions of Section R611 for Flat ICF walls or in accordance with the provisions of ACI 318.

❖ Conventionally formed concrete walls may be constructed under the provisions for Flat ICF wall systems in Section R611 or in accordance with ACI 318.

SECTION R613
EXTERIOR WINDOWS AND GLASS DOORS

R613.1 General. This section prescribes performance and construction requirements for exterior window systems installed in wall systems. Windows shall be installed and flashed in accordance with the manufacturer's written installation instructions. Written installation instructions shall be provided by the manufacturer for each window.

❖ Glass doors and windows are components of the exterior wall. Accordingly, this section specifies performance criteria for exterior windows and glass doors as well as their supporting elements to protect against glass breakage caused by high wind pressure.

R613.2 Window sills. In dwelling units, where the opening of an operable window is located more than 72 inches (1829 mm) above the finished grade or surface below, the lowest part of the clear opening of the window shall be a minimum of 24 inches (610 mm) above the finished floor of the room in which the window is located. Glazing between the floor and 24 inches (610 mm) shall be fixed or have openings through which a 4-inch-diameter (102 mm) sphere cannot pass.

Exceptions:

1. Windows whose openings will not allow a 4-inch-diameter (102 mm) sphere to pass through the opening when the opening is in its largest opened position.

2. Openings that are provided with window guards that comply with ASTM F 2006 or F 2090.

❖ This section is not applicable to fixed or stationary windows. If any part of the clear opening area of an operable window is located more than 72 inches (1829 mm) above the finished grade, this section requires that the lowest part of the clear opening be at least 24 inches (610 mm) above the floor surface of the room in which it is located. Windows may be located less than 24 inches (610 mm)above the interior floor surface only if they meet any one of the following criteria: 1) are fixed, 2) are located 72 inches (1829 mm) or less above grade, 3) have openings which will not allow passage of a 4 inch (102 mm) diameter sphere, or 4) are equipped with guards in accordance with ASTM F 2006 or F 2090. ASTM F 2090 addresses window fall prevention devices specifically associated with emergency escape and rescue openings, while ASTM F 2006 addresses devices at non-emergency escape and rescue openings.

R613.3 Performance. Exterior windows and doors shall be designed to resist the design wind loads specified in Table R301.2(2) adjusted for height and exposure per Table R301.2(3).

❖ Exterior glazing is subject to wind loading as specified in Section R301.2.1. The code requires use of the component and cladding pressures of Table R301.2(2), illustrated in Figure R301.2(7).

R613.4 Testing and labeling. Exterior windows and sliding doors shall be tested by an approved independent laboratory, and bear a label identifying manufacturer, performance characteristics and approved inspection agency to indicate compliance with AAMA/WDMA/CSA 101/I.S.2/A440. Exterior side-hinged doors shall be tested and labeled as conforming to AAMA/WDMA/CSA 101/I.S.2/A440 or comply with Section R613.6.

Exception: Decorative glazed openings.

❖ AAMA/WDMA 101/I.S.2/A440 is the standard referenced by the code for evaluating exterior glazing components. Section R613.6 is an exception to the original scope of section R613.4, giving requirements for all exterior window, skylight and door assemblies regardless of whether they are glass or nonglass. All of these doors, skylights and windows must be tested in accordance with ASTM E 330, as referenced in Section R613.6, because they are outside the scope of the referenced standard AAMA/WDMA 101/I.S.2/A440. ASTM E 330 will require testing to 1.5 times the design pressure.

R613.4.1 Comparative analysis. Structural wind load design pressures for window and door units smaller than the size tested in accordance with Section R613.4 shall be permitted to be higher than the design value of the tested unit provided such higher pressures are determined by accepted engineering analysis. All components of the small unit shall be the same as those of the tested unit. Where such calculated design pressures are used, they shall be validated by an additional test of the window or door unit having the highest allowable design pressure.

❖ Where a window, skylight or door is tested in accordance with the requirements of Section R613.4, smaller units of identical construction need not be tested. Instead, their properties may be determined by comparison to the tested unit and the application of accepted engineering analysis. However, where the analysis determines that the smaller units have a higher allowable design pressure, the unit with the highest allowable calculated design pressure must be tested to validate the findings, or the design pressure of the largest tested unit must be used.

R613.5 Vehicular access doors. Vehicular access doors shall be tested in accordance with either ASTM E 330 or ANSI/DASMA 108, and shall meet the acceptance criteria of ANSI/DASMA 108.

❖ Overhead garage doors for vehicle access must be tested in accordance with either one of the standards indicated. In addition, they must also meet the acceptance criteria of ANSI/DASMA 108. See Section R613.6 for an exception to these requirements.

R613.6 Other exterior window and door assemblies. Exterior windows and door assemblies not included within the scope of Section R613.4 or Section R613.5 shall be tested in accordance with ASTM E 330. Glass in assemblies covered by this exception shall comply with Section R308.5.

❖ This section is, in essence, an exception to Sections R613.4 and R613.5 which allows ASTM E 330 as an alternate test standard. ASTM E 330 will require testing to 1.5 times the design pressure.

R613.7 Wind-borne debris protection. Protection of exterior windows and glass doors in buildings located in wind-borne debris regions shall be in accordance with Section R301.2.1.2.

❖ See the commentary for Sections R301.2.1.2 and R613.7.1.

R613.7.1 Fenestration testing and labeling. Fenestration shall be tested by an approved independent laboratory, listed by an approved entity, and bear a label identifying manufacturer, performance characteristics, and approved inspection agency to indicate compliance with the requirements of the following specification:

1. ASTM E 1886 and ASTM E 1996; or

2. AAMA 506.

❖ In buildings located in windborne debris regions as established in accordance with Section R301.2.1, exterior windows, skylights and doors must be protected in accordance with Section R301.2.1.2, which mirrors some of the requirements of Section R613.7.1. Fenestration regulated by Section R613.7.1 includes storm shutters not in accordance with the prescriptive requirements of the Exception to Section R301.2.1.2, and exterior windows, skylights and doors not equipped with storm shutters. Such fenestrations must bear a label indicating that they have been tested in accordance with AAMA 506, or both ASTM E 1886 and ASTM E 1996. These tests determine the ability of the fenestration to resist cyclic pressures and impact by windborne debris (also see commentary, Section R301.2.1.2).

R613.8 Anchorage methods. The methods cited in this section apply only to anchorage of window and glass door assemblies to the main force-resisting system.

❖ These requirements provide a mechanism for transferring the wind forces from the door or window to a lateral force-resisting element. Anchorage should be according to the manufacturer's recommendations. Variations should provide equal or greater anchorage performance.

R613.8.1 Anchoring requirements. Window and glass door assemblies shall be anchored in accordance with the published manufacturer's recommendations to achieve the design pressure specified. Substitute anchoring systems used for substrates not specified by the fenestration manufacturer shall provide equal or greater anchoring performance as demonstrated by accepted engineering practice.

❖ See the commentary for Section R613.8.

R613.8.2 Anchorage details. Products shall be anchored in accordance with the minimum requirements illustrated in Figures R613.8(1), R613.8(2), R613.8(3), R613.8(4), R613.8(5), R613.8(6), R613.8(7) and R613.8(8).

❖ The code incorporates prescriptive anchorage methods that provide for transfer of wind loads to the supporting construction. See the commentary for Section R613.8.

R613.8.2.1 Masonry, concrete or other structural substrate. Where the wood shim or buck thickness is less than $1^{1}/_{2}$ inches (38 mm), window and glass door assemblies shall be anchored through the jamb, or by jamb clip and anchors shall be embedded directly into the masonry, concrete or other substantial substrate material. Anchors shall adequately transfer load from the window or door frame into the rough opening substrate [see Figures R613.8(1) and R613.8(2).]

Where the wood shim or buck thickness is $1^{1}/_{2}$ inches (38 mm) or more, the buck is securely fastened to the masonry, concrete or other substantial substrate, and the buck extends beyond the interior face of the window or door frame, window and glass door assemblies shall be anchored through the jamb, or by jamb clip, or through the flange to the secured wood buck. Anchors shall be embedded into the secured wood buck to adequately transfer load from the window or door frame assembly [Figures R613.8(3), R613.8(4) and R613.8(5)].

❖ When a wood member at least 1.5 inches (38 mm) thick is used as a shim, it must be anchored to the substrate. Otherwise window and door assemblies must be anchored directly to the substrate. See Figures R613.8(1) through R613.8(5).

R613.8.2.2 Wood or other approved framing material. Where the framing material is wood or other approved framing material, window and glass door assemblies shall be anchored through the frame, or by frame clip, or through the flange. Anchors shall be embedded into the frame construction to adequately transfer load [Figures R613.8(6), R613.8(7) and R613.8(8)].

❖ Door and window assemblies must be anchored directly to wood framing. See Figures R613.8(6) through R613.8(8).

R613.9 Mullions occurring between individual window and glass door assemblies.

R613.9.1 Mullions. Mullions shall be tested by an approved testing laboratory in accordance with AAMA 450, or be engineered in accordance with accepted engineering practice. Mullions tested as stand-alone units or qualified by engineering shall use performance criteria cited in Sections R613.9.2, R613.9.3 and R613.9.4. Mullions qualified by an actual test of an entire assembly shall comply with Sections R613.9.2 and R613.9.4.

❖ Mullions provide support for adjacent window and/or door assemblies. Generally, they must be tested and perform as required for windows and glass doors. Deflection limits and a factor of safety maintain serviceability and provide minimum strength.

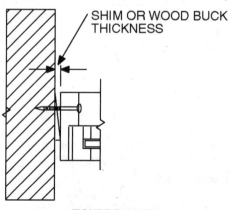

FIGURE R613.8(1)
THROUGH THE FRAME

❖ See the commentary for Sections R613.8 and R613.8.2.1.

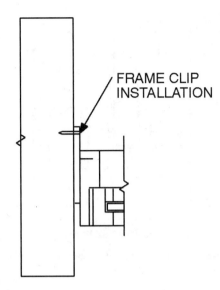

APPLY FRAME CLIP TO WINDOW OR DOOR IN ACCORDANCE
WITH PUBLISHED MANUFACTURER'S RECOMMENDATIONS.

FIGURE R613.8(2)
FRAME CLIP

❖ See the commentary for Sections R613.8 and
R613.8.2.1.

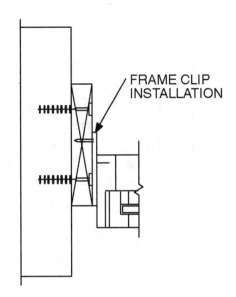

APPLY FRAME CLIP TO WINDOW OR DOOR FRAME IN ACCOR-
DANCE WITH PUBLISHED MANUFACTURER'S RECOMMENDA-
TIONS. ANCHORS SHALL BE PROVIDED TO TRANSFER LOAD
FROM THE FRAME CLIP INTO THE ROUGH OPENING SUB-
STRATE.

FIGURE R613.8(4)
FRAME CLIP

❖ See the commentary for Sections R613.8 and
R613.8.2.1.

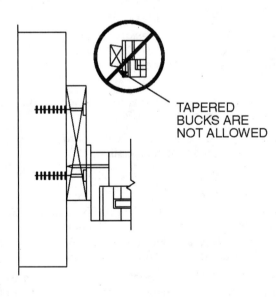

THROUGH THE FRAME ANCHORING METHOD. ANCHORS SHALL
BE PROVIDED TO TRANSFER LOAD FROM THE WINDOW OR
DOOR FRAME INTO THE ROUGH OPENING SUBSTRATE.

FIGURE R613.8(3)
THROUGH THE FRAME

❖ See the commentary for Sections R613.8 and
R613.8.2.1.

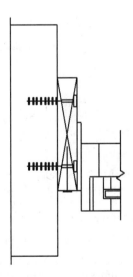

APPLY ANCHORS THROUGH FLANGE IN ACCORDANCE WITH
PUBLISHED MANUFACTURER'S RECOMMENDATIONS.

FIGURE R613.8(5)
THROUGH THE FLANGE

❖ See the commentary for Sections R613.8 and
R613.8.2.1.

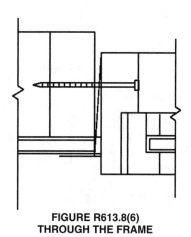

FIGURE R613.8(6)
THROUGH THE FRAME

❖ See the commentary for Sections R613.8 and R613.8.2.1.

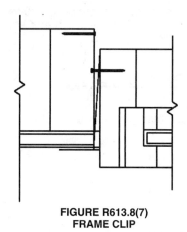

FIGURE R613.8(7)
FRAME CLIP

❖ See the commentary for Sections R613.8 and R613.8.2.1.

R613.9.2 Load transfer. Mullions shall be designed to transfer the design pressure loads applied by the window and door assemblies to the rough opening substrate.

❖ See the commentary for Section R613.9.1.

R613.9.3 Deflection. Mullions shall be capable of resisting the design pressure loads applied by the window and door assemblies to be supported without deflecting more than $L/175$, where L is the span of the mullion in inches.

❖ See the commentary for Section R613.9.1.

R613.9.4 Structural safety factor. Mullions shall be capable of resisting a load of 1.5 times the design pressure loads applied by the window and door assemblies to be supported without exceeding the appropriate material stress levels. If tested by an approved laboratory, the 1.5 times the design pressure load shall be sustained for 10 seconds, and the permanent deformation shall not exceed 0.4 percent of the mullion span after the 1.5 times design pressure load is removed.

❖ See the commentary for Section R613.9.1.

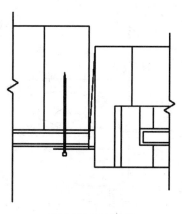

FIGURE R613.8(8)
THROUGH THE FLANGE

❖ See the commentary for Sections R613.8 and R613.8.2.1.

Bibliography

The following resource materials are referenced in this chapter or are relevant to the subject matter addressed in this chapter.

ACI 318-05, *Building Code Requirements for Structural Concrete.* Farmington Hills, MI: American Concrete Institute, 2005.

ACI 530-05, *Building Code Requirements for Masonry Structures.* Farmington Hills, MI: American Concrete Institute, 2005.

AFPA NDS-05, *National Design Specification (NDS) for Wood Construction—with 2001 Supplement.* Washington DC: American Forest and Paper Association, 2005.

AHA A 135.4-04, *Basic Hardware.* Palatine, IL: American Hardboard Association, 2004.

AHA A 194.1-85, *Cellulosic Fiber Board.* Palatine, IL: American Hardboard Association, 1985.

ANSI A208.1-99, *Particleboard.* New York: American National Standards Institute, 1999.

ANSI/AAMA/ADMA 101/I.S.2/A440-05, *Voluntary Specifications for Aluminum Vinyl PVC and Wood Windows and Glass Doors.* Schaumburg, IL: American Architectural Manufacturers Association, 2005.

ASTM A 36/A 36M-04, *Specification for Carbon Structural Steel.* West Conshohocken, PA: ASTM International, 2004.

ASTM A 153-03, *Specification for Zinc Coating (HotDip) on Iron and Steel Hardware.* West Conshohocken, PA: ASTM International, 2003.

ASTM A 167-99, *Specification for Stainless and Heat-Resisting Chromium-nickel Steel Plate, Sheet, and Strip.* West Conshohocken, PA: ASTM International, 1999.

ASTM A 510M-03, *Specification for General Requirements for Wire Rods and Coarse Round Wire, Carbon Steel Metric.* West Conshohocken, PA: ASTM International, 2003.

ASTM A 615/A 0615M-04a, *Specification for Deformed and Plain Billet-steel Bars for Concrete Reinforcement.* West Conshohocken, PA: ASTM International, 2004.

ASTM A 641/A 0641M-03, *Specification for Zinc-Coated (Galvanized) Carbon Steel Wire.* West Conshohocken, PA: ASTM International, 2003.

ASTM A 653/A 0653M-4a, *Specification for Steel Sheet, Zinc-Coated Galvanized) Zinc-Iron Alloy-Coated (Galvannealed) by the Hot-Dip Process.* West Conshohocken, PA: ASTM International, 2004.

ASTM A 706M-04a, *Specification for Low-alloy Steel Deformed and Plain Bars for Concrete Reinforcement.* West Conshohocken, PA: ASTM International, 2004.

ASTM A 792M-03, *Specification for Steel Sheet, 55% Aluminum-zinc Alloy-coated by the Hot-dip Process.* West Conshohocken, PA: ASTM International, 2003.

ASTM A 875/A 0875M-02a, *Specification for Steel Sheet, Zinc-5% Aluminum Alloy-Coated by the Hot-Dip Process.* West Conshohocken, PA: ASTM International, 2002.

ASTM A 951-02, *Specification for Masonry Joint Reinforcement.* West Conshohocken, PA: ASTM International, 2002.

ASTM B 227-04, *Specification for Hard-drawn Copperclad Steel Wire.* West Conshohocken, PA: ASTM International, 2004.

ASTM B 633-98e01, *Specification for Electrodeposited Coatings of Zinc on Iron and Steel.* West Conshohocken, PA: ASTM International, 2001.

ASTM C 79-04a, *Specification for Treated Core and Nontreated Core Gypsum Sheathing Board.* West Conshohocken, PA: ASTM International, 2004.

ASTM C 143/C 0143M-03, *Test Method for Slump or Hydraulic Cement Concrete.* West Conshohocken, PA: ASTM International, 2003.

ASTM C 208-95(2001), *Specification for Cellulosic Fiber Insulating Board.* West Conshohocken, PA: ASTM International, 2001.

ASTM C 270-04, *Specification for Mortar for Unit Masonry.* West Conshohocken, PA: ASTM International, 2004.

ASTM C 476-02, *Specification for Grout for Masonry.* West Conshohocken, PA: ASTM International, 2002.

ASTM C 954-00, *Specification for Steel Drill Screws for the Application of Gypsum Panel Products or Metal Plaster Bases to Steel Studs From 0.033 in. (0.84 mm) to 0.112 in. (2.84 mm) in Thickness.* West Conshohocken, PA: ASTM International, 2000.

ASTM E 330-02, *Test Method fo* _mance of Exterior Windows, Cur_ _by Uniform Static Air Pressure_ _ Conshohocken, PA: ASTM International, _

DOC PS 1-95, *Construction and Industrial Plywood.* Gaithersburg, MD: United States Department of Commerce, National Institute of Standards and Technology, 1995.

DOC PS 2-92, *Performance Standard for Wood-based Structural-use Panels.* Gaithersburg, MD: United States Department of Commerce, National Institute of Standards and Technology, 1992.

DOC PS 20-99, *American Softwood Lumber Standard.* Gaithersburg, MD: United States Department of Commerce, National Institute of Standards and Technology, 1999.

GA 253-99, *Recommended Standard Specification for the Application of Gypsum Sheathing.* Washington, DC: Gypsum Association, 1999.

SAE J 78-(1998), *Steel Self-drilling Tapping Screws.* Warrendale, PA: Society of Automotive Engineers, 1998.

Chapter 7:
Wall Covering

General Comments

Interior wall coverings are used for a variety of purposes. Often designed simply as an aesthetic element to finish the interior of the building, these wall coverings may also protect the structural elements from impact or moisture damage. Exterior wall coverings, in great part, provide the weather-resistant exterior envelope that protects the building's interior from the elements. As evidenced by the common use of exterior insulation finish systems in various regions of the country, new methods and materials are constantly introduced to provide for different appearances, improved insulating quality, sound transmission control and fire resistance. The code has developed prescriptive and performance regulations to control these aspects and the types and thickness of exterior wall coverings.

This chapter contains provisions for the design and construction of interior and exterior wall coverings. In addition to identifying the various types of wall covering regulated by the code, Chapter 7 references a number of material standards related to the specific types of wall covering materials. Of primary consideration are the application methods prescribed by the code for those wall-covering materials conforming to their applicable reference standard.

Section R701 establishes the scope of the chapter and states a general requirement dealing with weather pro-

tection for wall covering materials sensitive to adverse weather conditions. Section R702 establishes the various types of materials, materials standards and methods of application permitted for use as interior coverings, including interior plaster, gypsum board, ceramic tile, wood veneer paneling, hardboard paneling, wood shakes and wood shingles. Section R703 addresses exterior wall coverings as well as the water-resistive barrier required beneath the exterior materials. Exterior wall coverings regulated by this section include wood, hardboard and wood structural panel siding, wood shakes and shingles, exterior plaster, stone and masonry veneer and exterior insulation finish systems.

Purpose

This chapter provides the minimum requirements applicable to wall covering materials used both in exterior and interior applications. It specifies the types of wall coverings addressed by the code, including such common materials as gypsum board and ceramic tile used on a building's interior, as well as horizontal siding and wood structural panel siding installed as the exterior membrane. This chapter also recognizes a relatively new exterior wall covering, the exterior insulation finish system (EIFS).

SECTION R701
GENERAL

R701.1 Application. The provisions of this chapter shall control the design and construction of the interior and exterior wall covering for all buildings.

❖ Chapter 7 deals with both interior wall coverings and exterior wall coverings. A variety of interior coverings are regulated, including interior plaster, gypsum board, ceramic tile, wood veneer paneling, hardboard paneling, wood shakes and wood shingles. Wood siding, hardboard siding, wood structural panel siding, wood shakes, wood shingles, exterior plaster, stone veneer and masonry veneer are addressed where used as exterior wall coverings.

R701.2 Installation. Products sensitive to adverse weather shall not be installed until adequate weather protection for the installation is provided. Exterior sheathing shall be dry before applying exterior cover.

❖ Because many materials are subject to deterioration from the effects of moisture or other adverse weather conditions, the code limits the installation of such materials until proper weather protection is in place. Gyp-

sum plaster systems, gypsum board and interior grade plywood are particularly sensitive to moisture. It is also critical that exterior sheathing be dry prior to installation of the surface cover materials. Entrapment of moisture within the exterior wall assembly can result in a reduction in the effectiveness of the wall over time.

To assist in the application of the provisions of this chapter, it is important to understand the terminology that is used. Commentary Figures R701.2(1) and R701.2(2) provide pictorial descriptions of several of these terms.

SECTION R702
INTERIOR COVERING

R702.1 General. Interior coverings or wall finishes shall be installed in accordance with this chapter and Table R702.1(1), Table R702.1(2), Table R702.1(3) and Table R702.3.5. Interior masonry veneer shall comply with the requirements of Section R703.7.1 for support and Section R703.7.4 for anchorage, except an air space is not required. Interior finishes and materials shall conform to the flame spread and smoke-density requirements of Section R315.

❖ This section contains the installation requirements for interior wall coverings. Interior wall coverings of plaster and gypsum board must be installed in accordance with the applicable tables. The installation requirements include material orientation, backing support, spacing and size as well as method of attachment to supports. These requirements can result in wall coverings that will perform as intended from durability, appearance and structural points of view. A reference is made to Section R315, which addresses the flame spread and smoke-density provisions applicable to interior wall finishes.

TABLE R702.1(1)
THICKNESS OF PLASTER

PLASTER BASE	FINISHED THICKNESS OF PLASTER FROM FACE OF LATH, MASONRY, CONCRETE (inches)	
	Gypsum plaster	Portland cement mortar
Expanded metal lath	$5/8$, minimum[a]	$5/8$, minimum[a]
Wire lath	$5/8$, minimum[a]	$3/4$, minimum (interior)[b] $7/8$, minimum (exterior)[b]
Gypsum lath[g]	$1/2$, minimum	$3/4$, minimum (interior)[b]
Masonry walls[c]	$1/2$, minimum	$1/2$, minimum
Monolithic concrete walls[c,d]	$5/8$, maximum	$7/8$, maximum
Monolithic concrete ceilings[c,d]	$3/8$, maximum[e]	$1/2$, maximum
Gypsum veneer base[f,g]	$1/16$, minimum	$3/4$, minimum (interior)[b]
Gypsum sheathing[g]	—	$3/4$, minimum (interior)[b] $7/8$, minimum (exterior)[b]

For SI: 1 inch = 25.4 mm.
a. When measured from back plane of expanded metal lath, exclusive of ribs, or self-furring lath, plaster thickness shall be $3/4$ inch minimum.
b. When measured from face of support or backing.
c. Because masonry and concrete surfaces may vary in plane, thickness of plaster need not be uniform.
d. When applied over a liquid bonding agent, finish coat may be applied directly to concrete surface.
e. Approved acoustical plaster may be applied directly to concrete or over base coat plaster, beyond the maximum plaster thickness shown.
f. Attachment shall be in accordance with Table R702.3.5.
g. Where gypsum board is used as a base for portland cement plaster, weather-resistant sheathing paper complying with Section R703.2 shall be provided.

❖ Based on several variables, this table sets forth the required plaster thickness measured from the face of the lath, masonry or concrete to the plaster surface. Both gypsum plaster and portland cement mortar are addressed, with the option of the plaster application occurring over a variety of plaster bases. In most cases, the minimum plaster thickness is mandated; however, the maximum thickness is regulated where plaster is applied over a monolithic concrete wall or ceiling.

TABLE R702.1(2)
GYPSUM PLASTER PROPORTIONS[a]

NUMBER	COAT	PLASTER BASE OR LATH	MAXIMUM VOLUME AGGREGATE PER 100 POUNDS NEAT PLASTER[b] (cubic feet)	
			Damp loose sand[a]	Perlite or vermiculite[c]
Two-coat work	Base coat	Gypsum lath	2.5	2
	Base coat	Masonry	3	3
Three-coat work	First coat	Lath	2[d]	2
	Second coat	Lath	3[d]	2[e]
	First and second coats	Masonry	3	3

For SI: 1 inch = 25.4 mm, 1 cubic foot = 0.0283 m³, 1 pound = 0.454 kg.
a. Wood-fibered gypsum plaster may be mixed in the proportions of 100 pounds of gypsum to not more than 1 cubic foot of sand where applied on masonry or concrete.
b. When determining the amount of aggregate in set plaster, a tolerance of 10 percent shall be allowed.
c. Combinations of sand and lightweight aggregate may be used, provided the volume and weight relationship of the combined aggregate to gypsum plaster is maintained.
d. If used for both first and second coats, the volume of aggregate may be 2.5 cubic feet.
e. Where plaster is 1 inch or more in total thickness, the proportions for the second coat may be increased to 3 cubic feet.

❖ This table covers the maximum permitted amount of aggregate for gypsum plaster, measured in cubic feet per 100 pounds. Applicable for both damp loose sand, perlite and vermiculite, the limits vary based on the type of plaster base or lath. The maximum proportions for each coat of both two-coat and three-coat plaster systems are identified.

TABLE R702.1(3)
PORTLAND CEMENT PLASTER

Coat	MAXIMUM VOLUME AGGREGATE PER VOLUME CEMENTITIOUS MATERIAL[a]				MINIMUM PERIOD MOIST COATS	MINIMUM INTERVAL BETWEEN
	Portland cement plaster[b] maximum volume aggregate per volume cement	Portland cement-lime plaster[c]				
		Maximum volume lime per volume cement	Maximum volume sand per volume cement and lime	Approximate minimum thickness[d] curing (inches)		
First	4	$^3/_4$	4	$^3/_8$[e]	48 Hours[f]	48 Hours[g]
Second	5	$^3/_4$	5	First and second coats	48 Hours	7 Days[h]
Finish	3[i]	—	3[i]	$^1/_8$	—	Note h

For SI: 1 inch = 25.4 mm, 1 pound = 0.454 kg.

a. When determining the amount of aggregate in set plaster, a tolerance of 10 percent may be allowed.
b. From 10 to 20 pounds of dry hydrated lime (or an equivalent amount of lime putty) may be added as a plasticizing agent to each sack of Type I and Type II standard portland cement in base coat plaster.
c. No plasticizing agents shall be added.
d. See Table R702.1(1).
e. Measured from face of support or backing to crest of scored plaster.
f. Twenty-four-hour minimum period for moist curing of interior portland cement plaster.
g. Twenty-four hour minimum interval between coats of interior portland cement plaster.
h. Finish coat plaster may be applied to interior portland cement base coats after a 48-hour period.
i. For finish coat, plaster up to an equal part of dry hydrated lime by weight (or an equivalent volume of lime putty) may be added to Type I, Type II and Type III standard portland cement.

❖ Where portland cement plaster is used rather than gypsum plaster, this table is applicable for determining the maximum permitted volume of aggregate. In addition, the minimum period for moist curing is established for interior portland cement plaster as is the minimum required interval between coats.

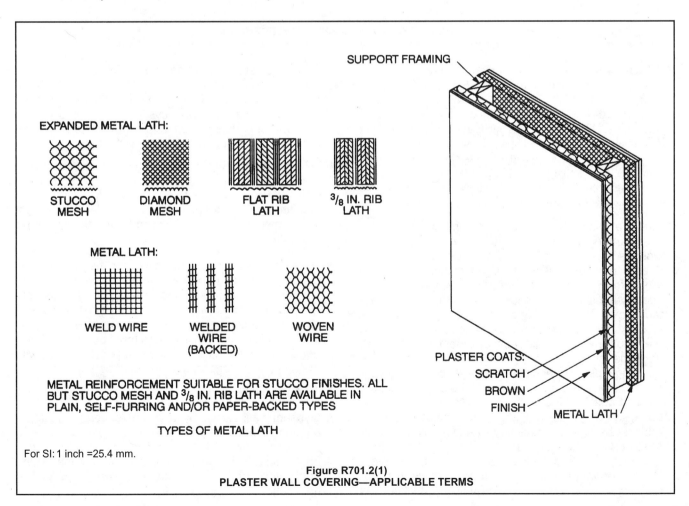

For SI: 1 inch =25.4 mm.

Figure R701.2(1)
PLASTER WALL COVERING—APPLICABLE TERMS

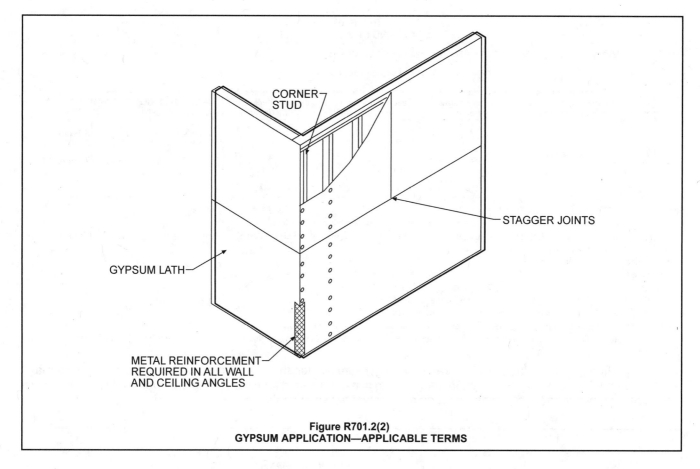

Figure R701.2(2)
GYPSUM APPLICATION—APPLICABLE TERMS

R702.2 Interior plaster. Gypsum plaster or portland cement plastering materials shall conform to ASTM C 5, C 28, C 35, C 37, C 59, C 61, C 587, C 588, C 631, C 847, C 897, C 933, C 1032 and C 1047, and shall be installed or applied in conformance with ASTM C 843, C 844 and C 1063. Plaster shall not be less than three coats when applied over metal lath and not less than two coats when applied over other bases permitted by this section, except that veneer plaster may be applied in one coat not to exceed $^3/_{16}$ inch (5 mm) thickness, provided the total thickness is as set forth in Table R702.1(1).

❖ Multicoat plastering has been commonplace for the past century. The industry generally views multicoat work as necessary for control of plaster thickness and density, particularly when applied by hand. Because of the uniformity created where plaster is applied in thin, successive layers, the code requires three-coat plastering over metal lath and two-coat work over other plaster bases approved for use by the code. Reducing the requirement for plaster bases other than metal lath depends on the rigidity of the plaster base itself. More-rigid plaster bases are not as susceptible to variations in thickness and flatness of the surface. In fact, the first coat applied in three-coat work on a flexible base, such as metal lath, is used to stiffen that base to provide the rigidity necessary to attain uniform thickness and surface flatness.

Both a gypsum plaster and a portland cement plaster interior wall covering system are shown in Commentary Figure R702.2(1). The finished thicknesses

for both plastering methods determined by the type of base materials are set forth in Table R702.1(1). Commentary Figure R702.2(2) illustrates the veneer plaster interior wall covering system. This section also cites the materials standards to which gypsum plaster and portland cement plaster must conform.

R702.2.1 Support. Support spacing for gypsum or metal lath on walls or ceilings shall not exceed 16 inches (406 mm) for $^3/_8$ inch thick (10 mm) or 24 inches (610 mm) for $^1/_2$-inch-thick (13 mm) plain gypsum lath. Gypsum lath shall be installed at right angles to support framing with end joints in adjacent courses staggered by at least one framing space.

❖ To provide a relatively firm base for the first layer of plaster, the code mandates a maximum span between supporting elements of the lath. Applicable to both gypsum and metal lath for either wall or ceiling installations, the maximum spacing of supports is based on the lath thickness. Gypsum lath must be installed perpendicular to the direction of the support framing, with the end joints staggered in adjacent courses.

R702.3 Gypsum board.

❖ Gypsum wallboard is by far the most common interior covering in use today. The variety of types and sizes of the material allow its use in almost every construction application. This section sets forth the criteria for the application of gypsum board in wall and ceiling applications.

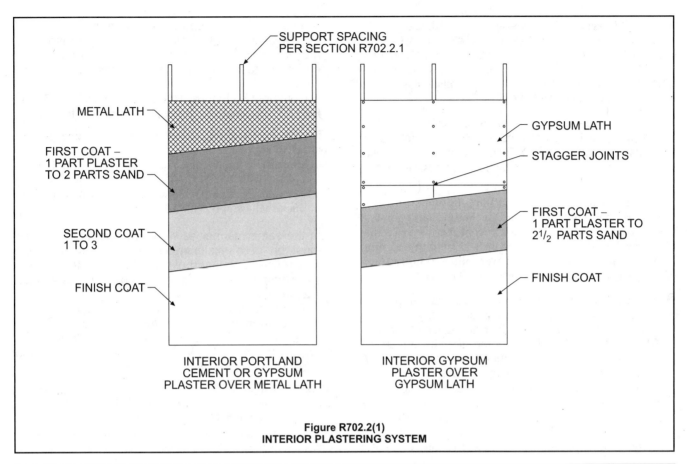

Figure R702.2(1)
INTERIOR PLASTERING SYSTEM

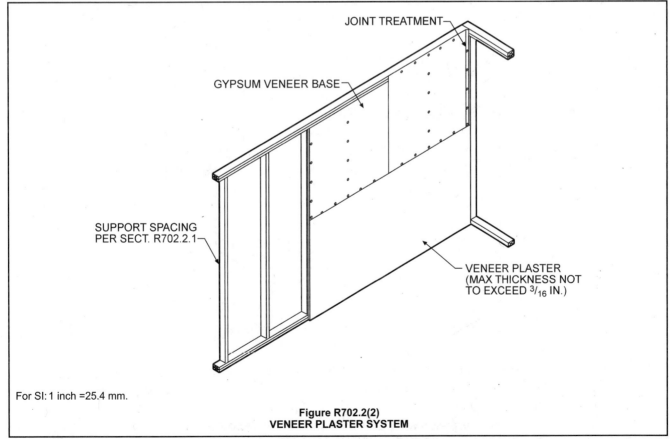

For SI: 1 inch = 25.4 mm.

Figure R702.2(2)
VENEER PLASTER SYSTEM

R702.3.1 Materials. All gypsum board materials and accessories shall conform to ASTM C 36, C 79, C 475, C 514, C 630, C 931, C 960, C 1002, C 1047, C 1177, C 1178, C 1278, C 1395 or C 1396 and shall be installed in accordance with the provisions of this section. Adhesives for the installation of gypsum board shall conform to ASTM C 557.

❖ This section sets forth the standards applicable to gypsum wallboard materials and accessories. Included in the list of standards is ASTM C 557, which deals with the use of adhesives to attach gypsum wallboard to wood framing members.

R702.3.2 Wood framing. Wood framing supporting gypsum board shall not be less than 2 inches (51 mm) nominal thickness in the least dimension except that wood furring strips not less than 1-inch-by-2 inch (25 mm by 51 mm) nominal dimension may be used over solid backing or framing spaced not more than 24 inches (610 mm) on center.

❖ For adequate structural backing, vertical support for gypsum wallboard used as an interior wall covering is required to have a least nominal dimension of 2 inches (51 mm). Dimensional lumber meeting this 2-inch (51 mm) nominal dimension would have an actual thickness of approximately $1^{1}/_{2}$ inches (38 mm).

Wood stripping used as furring over vertical supports on which the interior wall covering is to be applied is also required to meet the 2-inch (51 mm) nominal dimension in any direction unless the framing members are spaced at 24 inches (610 mm) on center or less. Where the spacing of the framing does not exceed 24 inches (610 mm) or where the furring strips are placed on solid backing, the minimum size requirements are reduced to 1 inch by 2 inches (25.4 mm by 51 mm).

Commentary Figures R702.3.2(1), R702.3.2(2) and R702.3.2(3) illustrate the dimensional constraints.

R702.3.3 Steel framing. Steel framing supporting gypsum board shall not be less than 1.25 inches (32 mm) wide in the least dimension. Light-gage nonload-bearing steel framing shall comply with ASTM C 645. Load-bearing steel framing and steel framing from 0.033 inch to 0.112 inch (1 mm to 3 mm) thick shall comply with ASTM C 955.

❖ In the same manner as wood framing, steel framing must be of a minimum cross-sectional dimension where used to support gypsum board. In the case of steel members, the least dimension must be at least 1.25 inches (32 mm).

This section also references two standards relating to steel framing members. ASTM C 645 addresses nonstructural steel framing members, while ASTM C 955 deals with the load-bearing steel studs, tracks and bracing involving gypsum panels.

R702.3.4 Insulating concrete form walls. Foam plastics for insulating concrete form walls constructed in accordance with Sections R404.4 and R611 on the interior of habitable spaces shall be covered in accordance with Section R314.4. Use of adhesives in conjunction with mechanical fasteners is permit-

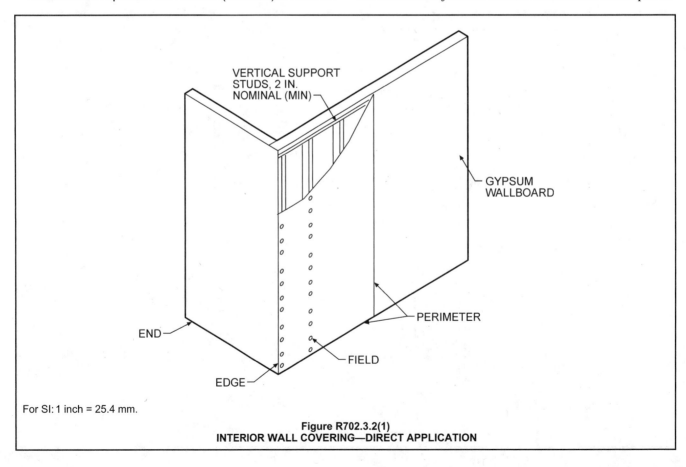

For SI: 1 inch = 25.4 mm.

Figure R702.3.2(1)
INTERIOR WALL COVERING—DIRECT APPLICATION

ted. Adhesives used for interior and exterior finishes shall be compatible with the insulating form materials.

❖ Reinforcing the provisions of Section R314.4 regarding the thermal barrier requirement for foam plastic insulation, this section references the requirements in

Chapters 4 and 6 regulating insulating concrete form foundation walls and insulating concrete form wall systems. Foam plastic insulation used in these wall applications must be separated from the interior of the building by a minimum $\frac{1}{2}$-inch (12.7 mm) gypsum board or equivalent material.

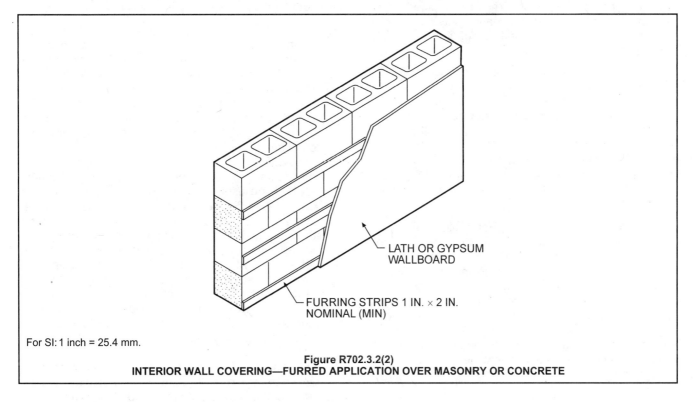

For SI: 1 inch = 25.4 mm.

Figure R702.3.2(2)
INTERIOR WALL COVERING—FURRED APPLICATION OVER MASONRY OR CONCRETE

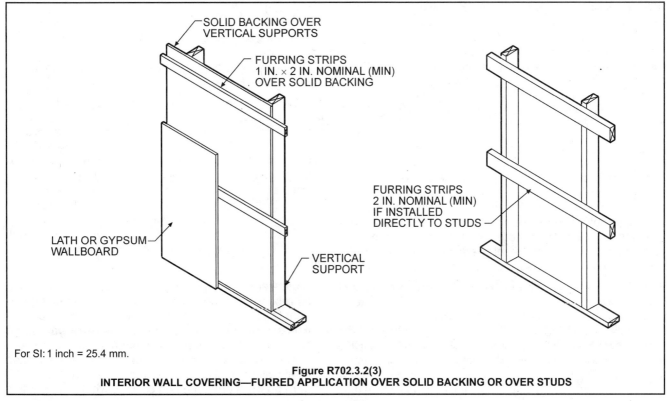

For SI: 1 inch = 25.4 mm.

Figure R702.3.2(3)
INTERIOR WALL COVERING—FURRED APPLICATION OVER SOLID BACKING OR OVER STUDS

Adhesives may be used to fasten the protective membrane (typically gypsum board) over the insulating concrete form wall. In all cases, however, mechanical fasteners must supplement the adhesive in accordance with Section R314.4. A total reliance on the adhesive to ensure the continued protection afforded by the gypsum board during a fire is misguided. Where used in conjunction with mechanical fasteners, adhesives must also be of a type that is compatible with the insulating form materials.

R702.3.5 Application. Maximum spacing of supports and the size and spacing of fasteners used to attach gypsum board shall comply with Table R702.3.5. Gypsum sheathing shall be attached to exterior walls in accordance with Table R602.3(1). Gypsum board shall be applied at right angles or parallel to framing members. All edges and ends of gypsum board shall occur on the framing members, except those edges and ends that are perpendicular to the framing members. Interior gypsum board shall not be installed where it is directly exposed to the weather or to water.

❖ To provide adequate backing for nailing the gypsum wallboard to the support framing, all edges and ends except those that are perpendicular to framing support members must occur on framing members. Gypsum wallboard may be installed in either a perpendicular or parallel manner. This is illustrated in Commentary Figure R702.3.5(1). Already addressed in a general sense in Section R701.2, the installation of interior gypsum wallboard must not be done while the interior of the building is exposed to weather. Direct exposure of interior gypsum board to weather or water is not permitted.

R702.3.6 Fastening. Screws for attaching gypsum board to wood framing shall be Type W or Type S in accordance with ASTM C 1002 and shall penetrate the wood not less than $\frac{5}{8}$ inch (16 mm). Screws for attaching gypsum board to light-gage steel framing shall be Type S in accordance with ASTM C 1002 and shall penetrate the steel not less than $\frac{3}{8}$ inch (10 mm). Screws for attaching gypsum board to steel framing 0.033 inch to 0.112 inch (1 mm to 3 mm) thick shall comply with ASTM C 954.

❖ Screws may be used in place of nails in the application of gypsum wallboard. This section restates the provisions of Table R702.3.5 regarding the use of screws. Gypsum wallboard screws, either Type S or Type W, are to be spaced in accordance with the table and must penetrate wood framing members a minimum of $\frac{5}{8}$ inch (15.9 mm). Where gypsum wallboard is attached to light-gage steel framing members, the screws must penetrate a minimum of $\frac{3}{8}$ inch (9.5 mm).

R702.3.7 Horizontal gypsum board diaphragm ceilings. Use of gypsum board shall be permitted on wood joists to create a horizontal diaphragm in accordance with Table R702.3.7. Gypsum board shall be installed perpendicular to ceiling framing members. End joints of adjacent courses of board shall not occur on the same joist. The maximum allowable diaphragm proportions shall be $1\frac{1}{2}$:1 between shear resisting elements. Rotation or cantilever conditions shall not be permitted. Gyp-

sum board shall not be used in diaphragm ceilings to resist lateral forces imposed by masonry or concrete construction. All perimeter edges shall be blocked using wood members not less than 2-inch (51 mm) by 6-inch (152 mm) nominal dimension. Blocking material shall be installed flat over the top plate of the wall to provide a nailing surface not less than 2 inches (51 mm) in width for the attachment of the gypsum board.

❖ Generally a gypsum board ceiling does not serve as a load-carrying structural element for other than its own weight. This section provides installation requirements [refer to Commentary Figure R702.3.7(1)] that allow gypsum board to be used as a membrane in horizontal diaphragm ceilings. When installed according to this section, it provides an economical and aesthetically pleasing finished surface that can also resist horizontal shear and wind forces.

Table R702.3.7 provides the shear capacity, maximum fastener spacing and minimum edge distance that is to be used in designing a horizontal gypsum board ceiling diaphragm. Because the material thickness and fastener size do not vary, the capacity is strictly a function of the framing member spacing. Further qualifications on the tabulated shear capacities that are stated in the footnotes include: ceiling diaphragm capacities are not cumulative with other diaphragm capacities; the capacities given are only for "short-term" loads due to wind and earthquakes; in structures classified as seismic design category D_0, D_1, D_2, or E the capacities must be reduced by 50 percent.

Limiting the diaphragm proportions and prohibiting cantilevers and rotation controls the deflection and distortion of the ceiling diaphragm [see Commentary Figure R702.3.7(1)].

The use of a gypsum board ceiling diaphragm to resist lateral loads from masonry or concrete is prohibited.

All perimeter edges must be blocked [see commentary Figure R702.3.7(2)].

R702.3.8 Water-resistant gypsum backing board. Gypsum board used as the base or backer for adhesive application of ceramic tile or other required nonabsorbent finish material shall conform to ASTM C 630 or C 1178. Use of water-resistant gypsum backing board shall be permitted on ceilings where framing spacing does not exceed 12 inches (305 mm) on center for $\frac{1}{2}$-inch-thick (13 mm) or 16 inches (406 mm) for $\frac{5}{8}$-inch-thick (16 mm) gypsum board. Water-resistant gypsum board shall not be installed over a vapor retarder in a shower or tub compartment. Cut or exposed edges, including those at wall intersections, shall be sealed as recommended by the manufacturer.

❖ Water-resistant gypsum backing board must be used as a base for ceramic tile or other required nonabsorbent finish materials where attached with an adhesive. It must be used in conjunction with the tile or wall panels because of its moisture-resistant qualities.

The use of water-resistant backing board is limited in a ceiling application, because $\frac{1}{2}$-inch-thick (12.7 mm) board is prohibited on ceilings where the framing members are spaced in excess of 12 inches (305 mm) on center. In addition, $5/8$-inch-thick water-resistant gypsum backing board is permitted only where ceiling members are spaced at 16 inches (406 mm) on center.

Gypsum board installed on the walls and ceilings at shower and bathtub areas must be finished to prevent moisture from penetrating the wall or ceiling finish and contacting the gypsum board. The finish applied to the exposed face of the gypsum board must create a wa-ter-resistant barrier that not only stops water from getting to the gypsum board but also prevents the release of moisture from within the wall or the gypsum board itself. For this reason, gypsum board must not be installed over the outboard side of any vapor barrier or retarder. This will create a waterproof membrane on both faces of the gypsum board, causing moisture to be trapped in the gypsum board that will ultimately cause it to decompose and fail.

TABLE R702.3.5
MINIMUM THICKNESS AND APPLICATION OF GYPSUM BOARD

THICKNESS OF GYPSUM BOARD (inches)	APPLICATION	ORIENTATION OF GYPSUM BOARD TO FRAMING	MAXIMUM SPACING OF FRAMING MEMBERS (inches o.c.)	MAXIMUM SPACING OF FASTENERS (inches)		SIZE OF NAILS FOR APPLICATION TO WOOD FRAMING[c]
				Nails[a]	Screws[b]	
Application without adhesive						
$3/8$	Ceiling[d]	Perpendicular	16	7	12	13 gage, $1^1/4''$ long, $19/64''$ head; 0.098'' diameter, $1^1/4''$ long, annular-ringed; or 4d cooler nail, 0.080'' diameter, $1^3/8''$ long, $7/32''$ head.
	Wall	Either direction	16	8	16	
$1/2$	Ceiling	Either direction	16	7	12	13 gage, $1^3/8''$ long, $19/64''$ head; 0.098'' diameter, $1^1/4''$ long, annular-ringed; 5d cooler nail, 0.086'' diameter, $1^5/8''$ long, $15/64''$ head; or gypsum board nail, 0.086'' diameter, $1^5/8''$ long, $9/32''$ head.
	Ceiling[d]	Perpendicular	24	7	12	
	Wall	Either direction	24	8	12	
	Wall	Either direction	16	8	16	
$5/8$	Ceiling	Either direction	16	7	12	13 gage, $1^5/8''$ long, $19/64''$ head; 0.098'' diameter, $1^3/8''$ long, annular-ringed; 6d cooler nail, 0.092'' diameter, $1^7/8''$ long, $1/4''$ head; or gypsum board nail, 0.0915'' diameter, $1^7/8''$ long, $19/64''$ head.
	Ceiling[e]	Perpendicular	24	7	12	
	Wall	Either direction	24	8	12	
	Wall	Either direction	16	8	16	
Application with adhesive						
$3/8$	Ceiling[d]	Perpendicular	16	16	16	Same as above for $3/8''$ gypsum board
	Wall	Either direction	16	16	24	
$1/2$ or $5/8$	Ceiling	Either direction	16	16	16	Same as above for $1/2''$ and $5/8''$ gypsum board, respectively
	Ceiling[d]	Perpendicular	24	12	16	
	Wall	Either direction	24	16	24	
Two $3/8$ layers	Ceiling	Perpendicular	16	16	16	Base ply nailed as above for $1/2''$ gypsum board; face ply installed with adhesive
	Wall	Either direction	24	24	24	

For SI: 1 inch = 25.4 mm.

a. For application without adhesive, a pair of nails spaced not less than 2 inches apart or more than $2^1/2$ inches apart may be used with the pair of nails spaced 12 inches on center.

b. Screws shall be Type S or W per ASTM C 1002 and shall be sufficiently long to penetrate wood framing not less than $5/8$ inch and metal framing not less than $3/8$ inch.

c. Where metal framing is used with a clinching design to receive nails by two edges of metal, the nails shall be not less than $5/8$ inch longer than the gypsum board thickness and shall have ringed shanks. Where the metal framing has a nailing groove formed to receive the nails, the nails shall have barbed shanks or be 5d, $13^1/2$ gage, $1^5/8$ inches long, $15/64$-inch head for $1/2$-inch gypsum board; and 6d, 13 gage, $1^7/8$ inches long, $15/64$-inch head for $5/8$-inch gypsum board.

d. Three-eighths-inch-thick single-ply gypsum board shall not be used on a ceiling where a water-based textured finish is to be applied, or where it will be required to support insulation above a ceiling. On ceiling applications to receive a water-based texture material, either hand or spray applied, the gypsum board shall be applied perpendicular to framing. When applying a water-based texture material, the minimum gypsum board thickness shall be increased from $3/8$ inch to $1/2$ inch for 16-inch on center framing, and from $1/2$ inch to $5/8$ inch for 24-inch on center framing or $1/2$-inch sag-resistant gypsum ceiling board shall be used.

e. Type X gypsum board for garage ceilings beneath habitable rooms shall be installed perpendicular to the ceiling framing and shall be fastened at maximum 6 inches o.c. by minimum $1^7/8$ inches 6d coated nails or equivalent drywall screws.

❖ Table R702.3.5 is a comprehensive table identifying the minimum thickness and fastening requirements for gypsum wallboard. The wallboard thickness [$3/8$-inch, $1/2$-inch or $5/8$-inch (10 mm, 12.7 mm or 16 mm)], location of the wallboard (wall or ceiling), orientation of the wallboard to the framing members (parallel or perpendicular), spacing of framing members [16 inches or 24 inches (406 mm or 610 mm)] on center and type of fasteners (nails or screws) are all set forth in the table. Where gypsum sheathing is to be attached to exterior walls, the provisions of Table R602.3(1) apply. Commentary Figure R702.3.5(2) shows a single-nailing method of attachment; however, Note a of Table R702.3.5 permits a double-nailing method as illustrated in Commentary Figure 702.3.5(3).

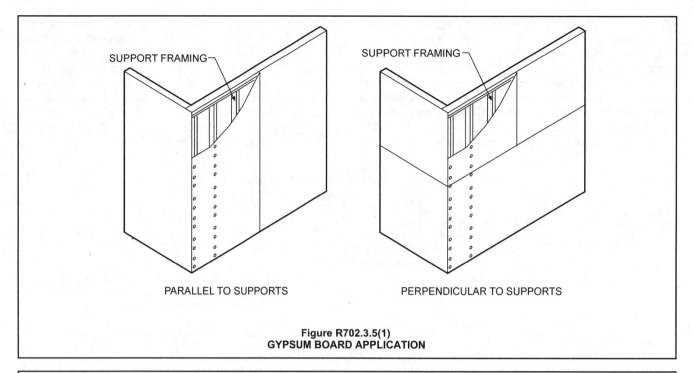

Figure R702.3.5(1)
GYPSUM BOARD APPLICATION

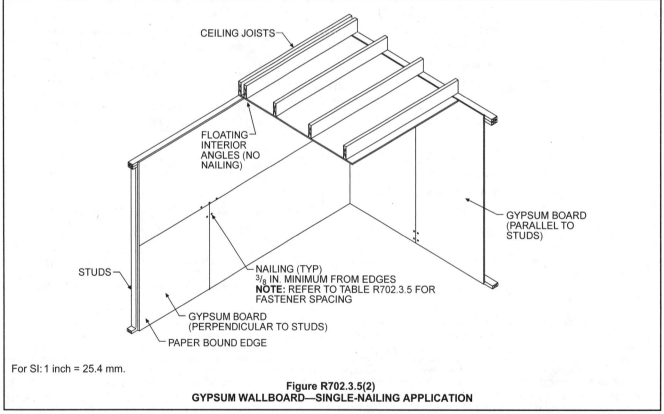

For SI: 1 inch = 25.4 mm.

Figure R702.3.5(2)
GYPSUM WALLBOARD—SINGLE-NAILING APPLICATION

R702.3.8.1 Limitations. Water resistant gypsum backing board shall not be used where there will be direct exposure to water, or in areas subject to continuous high humidity.

❖ Although many gypsum board sheet products are manufactured and approved for use in wet areas or areas exposed to moisture or humidity, there are still some extreme conditions where even water-resistant gypsum board will not provide the level of moisture protection necessary.

Water-resistant gypsum board is not to be used in areas that will be subject to a continuous exposure to moisture or humidity at locations such as saunas, steam rooms, gang showers or indoor pools. Gypsum board products, including the water-resistant type, are not intended for these extreme conditions and will not perform satisfactorily. Nongypsum wall and ceiling materials such as concrete masonry, ceramic tile on cement backer board, cement plaster (stucco) or other materials designed and recommended for high-moisture exposure must be used in these locations.

R702.4 Ceramic tile.

❖ Based on the code's requirement for nonabsorbent surfaces in bathtub and shower areas, ceramic tile is one of several materials used in such areas subject to water splash. Ceramic tile is regulated wherever installed, particularly when gypsum board is used as a base or backer board. This section regulates the installation of ceramic tile in residential construction.

R702.4.1 General. Ceramic tile surfaces shall be installed in accordance with ANSI A108.1, A108.4, A108.5, A108.6, A108.11, A118.1, A118.3, A136.1 and A137.1.

❖ This section references a number of ANSI standards that apply to the installation of ceramic tile. Commentary Figure R702.4.1 illustrates two possible methods of ceramic tile application.

R702.4.2 Cement, fiber-cement and glass mat gypsum backers. Cement, fiber-cement or glass mat gypsum backers in compliance with ASTM C 1288, C 1325 or C 1178 and installed in accordance with manufacturers' recommendations shall be used as backers for wall tile in tub and shower areas and wall panels in shower areas.

❖ Paper-faced gypsum board (i.e. green board) must not be used in areas that are in direct contact with water.

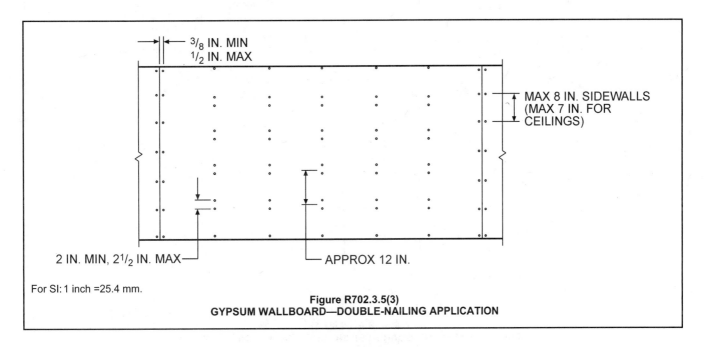

3/8 IN. MIN
1/2 IN. MAX

MAX 8 IN. SIDEWALLS
(MAX 7 IN. FOR
CEILINGS)

2 IN. MIN, 2 1/2 IN. MAX — APPROX 12 IN.

For SI: 1 inch = 25.4 mm.

Figure R702.3.5(3)
GYPSUM WALLBOARD—DOUBLE-NAILING APPLICATION

TABLE R702.3.7
SHEAR CAPACITY FOR HORIZONTAL WOOD-FRAMED
GYPSUM BOARD DIAPHRAGM CEILING ASSEMBLIES

MATERIAL	THICKNESS OF MATERIAL (min.) (in.)	SPACING OF FRAMING MEMBERS (max.) (in.)	SHEAR VALUE[a, b] (plf of ceiling)	MINIMUM FASTENER SIZE[c, d]
Gypsum Board	1/2	16 o.c.	90	5d cooler or wallboard nail; 1 5/8-inch long; 0.086-inch shank; 15/64-inch head
Gypsum Board	1/2	24 o.c.	70	5d cooler or wallboard nail; 1 5/8-inch long; 0.086-inch shank; 15/64-inch head

For SI: 1 inch = 25.4 mm, 1 pound per linear foot = 1.488 kg/m.

a. Values are not cumulative with other horizontal diaphragm values and are for short-term loading caused by wind or seismic loading. Values shall be reduced 25 percent for normal loading.

b. Values shall be reduced 50 percent in Seismic Design Categories D_0, D_1, D_2 and E.

c. 1 1/4", #6 Type S or W screws may be substituted for the listed nails.

d. Fasteners shall be spaced not more than 7 inches on center at all supports, including perimeter blocking, and not less than 3/8 inch from the edges and ends of the gypsum board.

❖ See the commentary for Section R702.3.7

When this material gets wet through cracks in the grout joints, deteriorated caulking between the tile and tub assembly or improper flashing between the tub and tile interface, the water is absorbed into the material and can cause the paper facing to delaminate. The delamination can cause damage to the tile surface. In addition, water that has been trapped in paper-faced gypsum board has been known to create mold problems. The use of cement, fiber-cement and glass mat backers ensures that tile backers installed in extreme wet conditions will be moisture resistant, even if water penetrates through the grout joints or failed caulk joints in tub or shower wall panels.

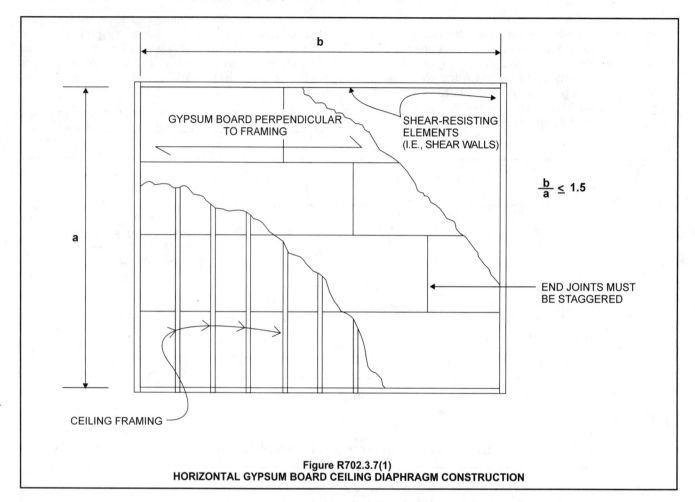

Figure R702.3.7(1)
HORIZONTAL GYPSUM BOARD CEILING DIAPHRAGM CONSTRUCTION

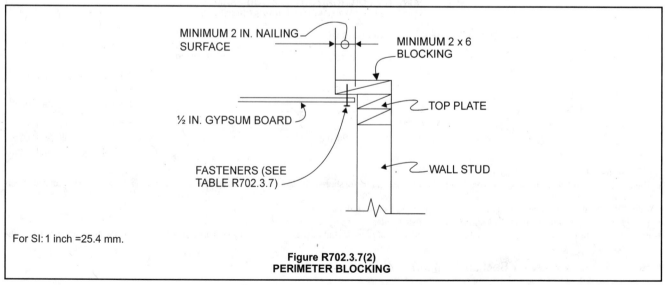

For SI: 1 inch =25.4 mm.

Figure R702.3.7(2)
PERIMETER BLOCKING

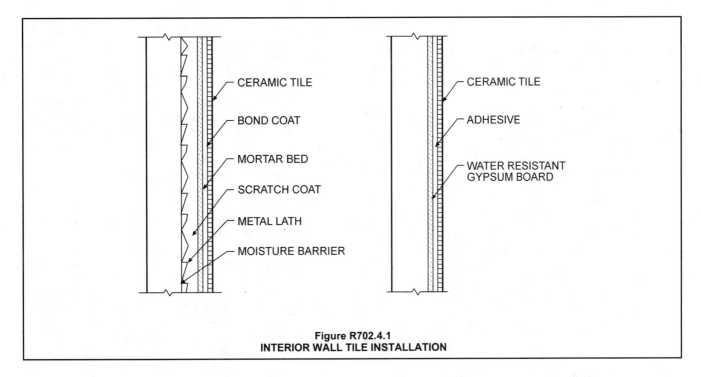

Figure R702.4.1
INTERIOR WALL TILE INSTALLATION

R702.5 Other finishes. Wood veneer paneling and hardboard paneling shall be placed on wood or cold-formed steel framing spaced not more than 16 inches (406 mm) on center. Wood veneer and hard board paneling less than $^1/_4$ inch (6 mm) nominal thickness shall not have less than a $^3/_8$-inch (10 mm) gypsum board backer. Wood veneer paneling not less than $^1/_4$-inch (6 mm) nominal thickness shall conform to ANSI/HPVA HP-1. Hardboard paneling shall conform to ANSI/AHA A135.5.

❖ This section contains requirements for wood veneer and hardboard paneling that are not addressed in previous sections. Modular home construction will generally use hardboard paneling applied directly to the studs. For this reason, there are specific provisions for wood veneer and hardboard siding. Wall finishes with adequate thickness will resist excessive bending and possible failure.

R702.6 Wood shakes and shingles. Wood shakes and shingles shall conform to CSSB *Grading Rules for Wood Shakes and Shingles* and shall be permitted to be installed directly to the studs with maximum 24 inches (610 mm) on-center spacing.

❖ Complying wood shakes and wood shingles may be attached directly to the studs in an interior application provided the studs are spaced at no more than 24 inches (610 mm) on center. The shakes and shingles must conform to the appropriate grading rules set forth by the Cedar Shake and Shingle Bureau.

R702.6.1 Attachment. Nails, staples or glue are permitted for attaching shakes or shingles to the wall, and attachment of the shakes or shingles directly to the surface shall be permitted provided the fasteners are appropriate for the type of wall surface material. When nails or staples are used, two fasteners shall be provided and shall be placed so that they are covered by the course above.

❖ Because there is no concern for weather protection, the shakes or shingles may be attached directly to the surface with nails, staples or glue. At least two fasteners are required per shake or shingle where nails or staples are used, located so that the course above will cover the fasteners.

R702.6.2 Furring strips. Where furring strips are used, they shall be 1 inch by 2 inches or 1 inch by 3 inches (25 mm by 51 mm or 25 mm by 76 mm), spaced a distance on center equal to the desired exposure, and shall be attached to the wall by nailing through other wall material into the studs.

❖ In addition to direct attachment of wood shakes and shingles to studs or a wall surface, the use of furring strips is permitted under specific conditions. Furring strips should be either 1 inch by 2 inches (25.4 mm by 51 mm) or 1 inch by 3 inches (25.4 mm by 76 mm). The spacing of the strips is determined by the desired exposure of the shake or shingle, attached by nailing through other wall surface materials into the studs.

SECTION R703
EXTERIOR COVERING

R703.1 General. Exterior walls shall provide the building with a weather-resistant exterior wall envelope. The exterior wall envelope shall include flashing as described in Section R703.8. The exterior wall envelope shall be designed and constructed in a manner that prevents the accumulation of water within the wall assembly by providing a water-resistant barrier behind the exterior veneer as required by Section R703.2. and a means of draining water that enters the assembly to the exterior. Protection against condensation in the exterior wall assembly shall be provided in accordance with Chapter 11 of this code.

Exceptions:

1. A weather-resistant exterior wall envelope shall not be required over concrete or masonry walls designed in accordance with Chapter 6 and flashed according to Section R703.7 or R703.8.

2. Compliance with the requirements for a means of drainage, and the requirements of Section R703.2 and Section R703.8, shall not be required for an exterior wall envelope that has been demonstrated to resist wind-driven rain through testing of the exterior wall envelope, including joints, penetrations and intersections with dissimilar materials, in accordance with ASTM E 331 under the following conditions:

    2.1. Exterior wall envelope test assemblies shall include at least one opening, one control joint, one wall/eave interface and one wall sill. All tested openings and penetrations shall be representative of the intended end-use configuration.

    2.2. Exterior wall envelope test assemblies shall be at least 4 feet (1219 mm) by 8 feet (2438 mm) in size.

    2.3. Exterior wall assemblies shall be tested at a minimum differential pressure of 6.24 pounds per square foot (299 Pa).

    2.4. Exterior wall envelope assemblies shall be subjected to a minimum test exposure duration of 2 hours.

    The exterior wall envelope design shall be considered to resist wind-driven rain where the results of testing indicate that water did not penetrate: control joints in the exterior wall envelope; joints at the perimeter of openings penetration; or intersections of terminations with dissimilar materials.

❖ Exterior walls of buildings must be protected against damage caused by precipitation, wind and other weather conditions. The main text of this section prescribes three basic components of a weather-resistive exterior wall assembly: a water-resistive barrier installed over the building substrate; flashings at penetrations and terminations of the exterior wall finish and a means of draining moisture that may penetrate behind the finish back to the exterior. Section R703.2 is referenced for the requirements of the water-resistive barrier and Section R703.8 is referenced for requirements for the flashings (see commentary, Sections R703.2 and R703.8). This section does not, however, contain a prescriptive requirement for the means of drainage to be provided. The method to provide the means of drainage is a performance criterion and must be evaluated based on the ability to allow moisture that may penetrate behind the exterior wall covering to drain back to the exterior. This may be as complicated as a rain-screen pressure-equalized type of exterior assembly or as simple as providing discontinuities or gaps between the surface of the substrate and the back side of the finish, such as through the use of noncorrodible furring.

For common types of construction, such as vinyl siding or brick veneer, the typical practice of installing building paper and weeps will comply with the intent of this section. Discontinuities between the exterior covering and substrate must be such that they encourage the flow of moisture via gravity or capillary action to a location where the water may exit, such as at flashings and weeps. The absence of a means of drainage may result in the accumulation of moisture that becomes trapped between the finish and the substrate. Over time, extended exposure to moisture may contribute to the degradation of the finish, building substrate or even the structural elements of the exterior wall.

Exception 1 states that where the exterior wall envelope is designed and constructed of concrete or masonry materials in accordance with the requirements of Chapter 6 and Sections R703.7 or R703.8, respectively, the water-resistive barrier and means of drainage may be omitted. This is because the penetration of moisture behind the exterior wall finish is not detrimental to concrete and masonry substrates.

Exception 2 permits the use of exterior wall finishes that do not meet the prescriptive requirements of Sections R703.2 and R703.8, provided that the system, with penetration details, is tested for wind-driven rain resistance. The test specimen(s) must incorporate those penetration and termination details intended for use, and the system will be limited to use with those details that successfully pass the test. The minimum panel size specified represents that which is commonly used in testing to ASTM E 331; however, this does not preclude the testing of larger panels if desired. The modifications to the test pressure differential and test duration are intended to represent more closely conditions that will be encountered in service. The pass/fail criterion is based on the visual observation of moisture on the rear side of the wall assembly. In cavity-type assemblies, such as stud walls, this requires the observation of locations such as the rear face of the exterior wall sheathing and wall framing members for the presence of moisture. The test method is intended to assess the performance of the method(s) intended for use in sealing the interface between the termination of the exterior wall finish and the penetration items or abutting construction. The test is not necessarily intended to test the performance of the penetrating item.

Walls designed and constructed in accordance with this chapter must also comply with the requirements of Chapter 11 of this code. This requires that wall cavities be protected from moisture infiltration from the building interior through the use of a vapor retarder (see Commentary Figure R703.1) or by the ventilation of the wall cavity.

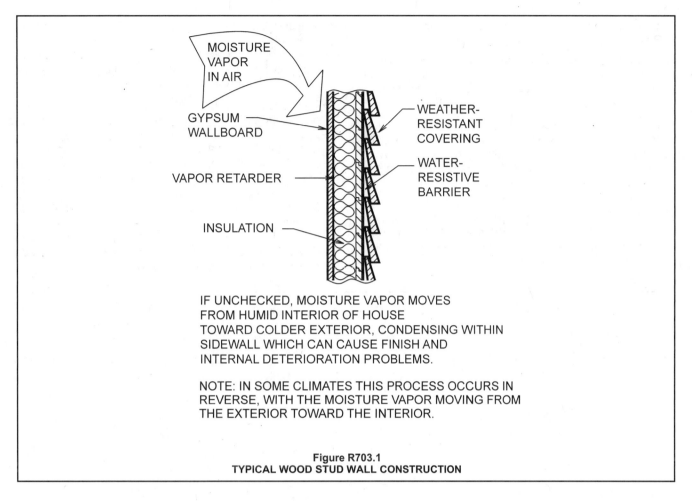

IF UNCHECKED, MOISTURE VAPOR MOVES
FROM HUMID INTERIOR OF HOUSE
TOWARD COLDER EXTERIOR, CONDENSING WITHIN
SIDEWALL WHICH CAN CAUSE FINISH AND
INTERNAL DETERIORATION PROBLEMS.

NOTE: IN SOME CLIMATES THIS PROCESS OCCURS IN
REVERSE, WITH THE MOISTURE VAPOR MOVING FROM
THE EXTERIOR TOWARD THE INTERIOR.

Figure R703.1
TYPICAL WOOD STUD WALL CONSTRUCTION

R703.2 Water-resistive barrier. One layer of No. 15 asphalt felt, free from holes and breaks, complying with ASTM D 226 for Type 1 felt or other approved water-resistive barrier shall be applied over studs or sheathing of all exterior walls. Such felt or material shall be applied horizontally, with the upper layer lapped over the lower layer not less than 2 inches (51 mm). Where joints occur, felt shall be lapped not less than 6 inches (152 mm). The felt or other approved material shall be continuous to the top of walls and terminated at penetrations and building appendages in a manner to meet the requirements of the exterior wall envelope as described in Section R703.1.

Exception: Omission of the water-resistive barrier is permitted in the following situations:

1. In detached accessory buildings.

2. Under exterior wall finish materials as permitted in Table R703.4.

3. Under paperbacked stucco lath when the paper backing is an approved weather-resistive sheathing paper.

❖ Asphalt-saturated felt or any other approved water-resistive material is required behind all types of siding material because of the possibility of moisture penetrating behind the siding. This felt or other material protects the wall construction from potential rotting.

The water-resistive membrane may be omitted where there is a low possibility of moisture penetration or the potential for moisture penetration is not a great concern. The water-resistive membrane may be eliminated when approved paperbacked stucco lath is used; the paper backing functions as the membrane if the backing paper is not punctured. The paper backing must be an approved weather-resistive sheathing paper.

R703.3 Wood, hardboard and wood structural panel siding.

❖ This section addresses the use of wood siding products as exterior wall coverings. Both horizontal siding and panel siding are regulated.

R703.3.1 Panel siding. Joints in wood, hardboard or wood structural panel siding shall be made as follows unless otherwise approved. Vertical joints in panel siding shall occur over framing members, unless wood or wood structural panel sheathing is used, and shall be shiplapped or covered with a batten. Horizontal joints in panel siding shall be lapped a minimum of 1 inch (25 mm) or shall be shiplapped or shall be flashed with Z-flashing and occur over solid blocking, wood or wood structural panel sheathing.

❖ Plywood is acceptable as an exterior wall covering if the plywood is approved for use in an exterior location and the joints are made waterproof. Typical joint treatments include: 1) lapping joints horizontally, and 2) having the joints occur over framing and protecting those joints with a continuous wood batten using approved caulking, flashing or vertical or horizontal shiplaps. See Commentary Figures R703.3.1(1) and R703.3.1(2) for illustrations of joint treatment.

R703.3.2 Horizontal siding. Horizontal lap siding shall be lapped a minimum of 1 inch (25 mm), or 0.5 inch (13 mm) if rabbeted, and shall have the ends caulked, covered with a batten, or sealed and installed over a strip of flashing.

❖ To maintain the necessary weather protection, horizontal siding must have a minimum lap of at least 1 inch (25.4 mm), unless the siding is rabbeted, in which case the joint need be only $^1/_2$ inch (12.7 mm). Various methods can be used to protect the ends of horizontal siding, including the use of flashing, caulking or battens.

R703.4 Attachments. Unless specified otherwise, all wall coverings shall be securely fastened in accordance with Table R703.4 or with other approved aluminum, stainless steel, zinc-coated or other approved corrosion-resistive fasteners.

Where the basic wind speed per Figure R301.2(4) is 110 miles per hour (49 m/s) or higher, the attachment of wall coverings shall be designed to resist the component and cladding loads specified in Table R301.2(2), adjusted for height and exposure in accordance with Table R301.2(3).

❖ The presence of frequent moisture in the atmosphere necessitates that wall covering be fastened with approved corrosion-resistant fasteners such as aluminum, stainless steel, zinc or zinc-coated fasteners. Table R703.4 establishes the size of fasteners and other attachment requirements. The attachment provisions in Table R703.4 are not intended for construction where wind speeds meet or exceed 110 mph (49 m/s) unless verified through design.

For wind speeds 110 mph (49 m/s) or greater Tables R301.2(2) and R301.2(3) must be used for wind loads for the attachment of wall coverings.

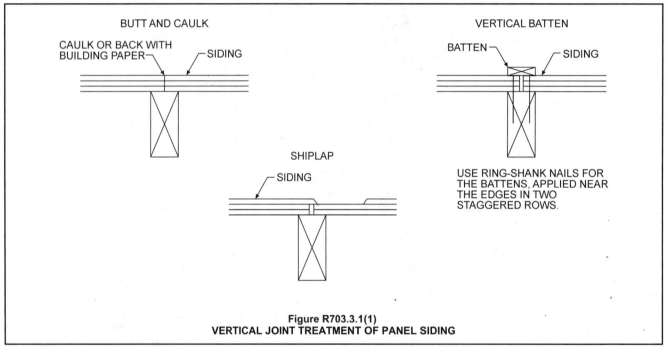

Figure R703.3.1(1)
VERTICAL JOINT TREATMENT OF PANEL SIDING

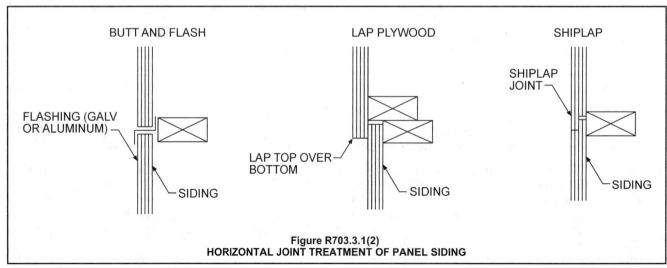

Figure R703.3.1(2)
HORIZONTAL JOINT TREATMENT OF PANEL SIDING

TABLE R703.4
WEATHER–RESISTANT SIDING ATTACHMENT AND MINIMUM THICKNESS

SIDING MATERIAL		NOMINAL THICKNES[a] (inches)	JOINT TREATMENT	WATER-RESISTIVE BARRIER REQUIRED	TYPE OF SUPPORTS FOR THE SIDING MATERIAL AND FASTENERS[b,c,d]					Number or spacing of fasteners
					Wood or wood structural panel sheathing	Fiberboard sheathing into stud	Gypsum sheathing into stud	Foam plastic sheathing into stud	Direct to studs	
Horizontal aluminum[e]	Without insulation	0.019[f]	Lap	Yes	0.120 nail 1$^1/_2$" long	0.120 nail 2" long	0.120 nail 2" long	0.120 nail[y]	Not allowed	Same as stud spacing
		0.024	Lap	Yes	0.120 nail 1$^1/_2$" long	0.120 nail 2" long	0.120 nail 2" long	0.120 nail[y]	Not allowed	
	With insulation	0.019	Lap	Yes	0.120 nail 1$^1/_2$" long	0.120 nail 2$^1/_2$" long	0.120 nail 2$^1/_2$" long	0.120 nail[y]	0.120 nail 1$^1/_2$" long	
Brick veneer[z] Concrete masonry veneer[z]		2 2	Section R703	Yes (Note l)	See Section R703 and Figure R703.7[g]					
Hardboard[k] Panel siding-vertical		$^7/_{16}$	—	Yes	Note n	Note n	Note n	Note n	Note n	6" panel edges 12" inter. sup.[o]
Hardboard[k] Lap-siding-horizontal		$^7/_{16}$	Note q	Yes	Note p	Note p	Note p	Note p	Note p	Same as stud spacing 2 per bearing
Steel[h]		29 ga.	Lap	Yes	0.113 nail 1$^3/_4$" Staple–1$^3/_4$"	0.113 nail 2$^3/_4$" Staple–2$^1/_2$"	0.113 nail 2$^1/_2$" Staple–2$^1/_4$"	0.113 nail[y] Staple[y]	Not allowed	Same as stud spacing
Stone veneer		2	Section R703	Yes (Note l)	See Section R703 and Figure R703.7[g]					
Particleboard panels		$^3/_8$ – $^1/_2$	—	Yes	6d box nail (2" × 0.099")	6d box nail (2" × 0.099")	6d box nail (2" × 0.099")	box nail[y]	6d box nail (2" × 0.099"), $^3/_8$ not allowed	6" panel edge, 12" inter. sup.
		$^5/_8$	—	Yes	6d box nail (2" × 0.099")	8d box nail (2$^1/_2$" × 0.113")	8d box nail (2$^1/_2$" × 0.113")	box nail[y]	6d box nail (2" × 0.099")	
Plywood panel[i] (exterior grade)		$^3/_8$	—	Yes	0.099 nail–2"	0.113 nail–2$^1/_2$"	0.099 nail–2"	0.113 nail[y]	0.099 nail–2"	6" on edges, 12" inter. sup.
Vinyl siding[m]		0.035	Lap	Yes	0.120 nail 1$^1/_2$" Staple–1$^3/_4$"	0.120 nail 2" Staple–2$^1/_2$"	0.120 nail 2" Staple–2$^1/_2$"	0.120 nail[y] Staple[y]	Not allowed	Same as stud spacing
Wood[j] rustic, drop		$^3/_8$ Min	Lap	Yes	Fastener penetration into stud–1"				0.113 nail– 2$^1/_2$" Staple–2"	Face nailing up to 6" widths, 1 nail per bearing; 8" widths and over, 2 nails per bearing
Shiplap		$^{19}/_{32}$ Average	Lap	Yes						
Bevel		$^7/_{16}$								
Butt tip		$^3/_{16}$	Lap	Yes						
Fiber cement panel siding[r]		$^5/_{16}$	Note s	Yes Note x	6d corrosion-resistant nail[t]	6d corrosion-resistant nail[t]	6d corrosion-resistant nail[t]	6d corrosion-resistant nail[t, y]	4d corrosion-resistant nail[u]	6" o.c. on edges, 12" o.c. on intermed. studs
Fiber cement lap siding[r]		$^5/_{16}$	Note v	Yes Note x	6d corrosion-resistant nail[t]	6d corrosion-resistant nail[t]	6d corrosion-resistant nail[t]	6d corrosion-resistant nail[t, y]	6d corrosion-resistant nail[w]	Note w

For SI: 1 inch = 25.4 mm.

a. Based on stud spacing of 16 inches on center where studs are spaced 24 inches, siding shall be applied to sheathing approved for that spacing.

b. Nail is a general description and shall be T-head, modified round head, or round head with smooth or deformed shanks.

c. Staples shall have a minimum crown width of $^7/_{16}$-inch outside diameter and be manufactured of minimum 16 gage wire.

d. Nails or staples shall be aluminum, galvanized, or rust-preventative coated and shall be driven into the studs for fiberboard or gypsum backing.

e. Aluminum nails shall be used to attach aluminum siding.

f. Aluminum (0.019 inch) shall be unbacked only when the maximum panel width is 10 inches and the maximum flat area is 8 inches. The tolerance for aluminum siding shall be +0.002 inch of the nominal dimension.

g. All attachments shall be coated with a corrosion-resistant coating.

h. Shall be of approved type.

(continued)

Footnotes to Table R703.4—continued

i. Three-eighths-inch plywood shall not be applied directly to studs spaced more than 16 inches on center when long dimension is parallel to studs. Plywood $^1/_2$-inch or thinner shall not be applied directly to studs spaced more than 24 inches on center. The stud spacing shall not exceed the panel span rating provided by the manufacturer unless the panels are installed with the face grain perpendicular to the studs or over sheathing approved for that stud spacing.

j. Wood board sidings applied vertically shall be nailed to horizontal nailing strips or blocking set 24 inches on center. Nails shall penetrate $1^1/_2$ inches into studs, studs and wood sheathing combined, or blocking. A weather-resistive membrane shall be installed weatherboard fashion under the vertical siding unless the siding boards are lapped or battens are used.

k. Hardboard siding shall comply with AHA A135.6.

l. For masonry veneer, a weather-resistive sheathing paper is not required over a sheathing that performs as a weather-resistive barrier when a 1-inch air space is provided between the veneer and the sheathing. When the 1-inch space is filled with mortar, a weather-resistive sheathing paper is required over studs or sheathing.

m. Vinyl siding shall comply with ASTM D 3679.

n. Minimum shank diameter of 0.092 inch, minimum head diameter of 0.225 inch, and nail length must accommodate sheathing and penetrate framing $1^1/_2$ inches.

o. When used to resist shear forces, the spacing must be 4 inches at panel edges and 8 inches on interior supports.

p. Minimum shank diameter of 0.099 inch, minimum head diameter of 0.240 inch, and nail length must accommodate sheathing and penetrate framing $1^1/_2$ inches.

q. Vertical end joints shall occur at studs and shall be covered with a joint cover or shall be caulked.

r. Fiber cement siding shall comply with the requirements of ASTM C 1186.

s. See Section R703.10.1.

t. Minimum 0.102″ smooth shank, 0.255″ round head.

u. Minimum 0.099″ smooth shank, 0.250″ round head.

v. See Section R703.10.2.

w. Face nailing: 2 nails at each stud. Concealed nailing: one 11 gage $1^1/_2$ galv. roofing nail (0.371″ head diameter, 0.120″ shank) or 6d galv. box nail at each stud.

x. See Section R703.2 exceptions.

y. Minimum nail length must accommodate sheathing and penetrate framing $1^1/_2$ inches.

z. Adhered masonry veneer shall comply with the requirements in Sections 6.1 and 6.3 of ACI 530/ASCE 5/TMS-402.

❖ This table should be used in addition to all other applicable requirements of the code for the specific material under consideration. Testing and experience have determined that the minimum thicknesses tabulated will be durable and protect the building against the elements for relatively long periods when the siding is attached and maintained as indicated.

R703.5 Wood shakes and shingles. Wood shakes and shingles shall conform to CSSB *Grading Rules for Wood Shakes and Shingles.*

❖ Grading rules established by the Cedar Shake and Shingle Bureau are applicable to wood shakes and shingles used as exterior wall covering materials. The provisions of this section deal with the specific application and attachment of wood shakes and shingles to the exterior of a building.

R703.5.1 Application. Wood shakes or shingles shall be applied either single-course or double-course over nominal $^1/_2$-inch (13 mm) wood-based sheathing or to furring strips over $^1/_2$-inch (13 mm) nominal nonwood sheathing . A permeable water-resistive barrier shall be provided over all sheathing, with horizontal overlaps in the membrane of not less than 2 inches (51mm) and vertical overlaps of not less than 6 inches (152 mm). Where furring strips are used, they shall be 1 inch by 3 inches or 1 inch by 4 inches (25 mm by 76 mm or 25 mm by 102 mm) and shall be fastened horizontally to the studs with 7d or 8d box nails and shall be spaced a distance on center equal to the actual weather exposure of the shakes or shingles, not to exceed the maximum exposure specified in Table R703.5.2. The spacing between adjacent shingles to allow for expansion shall not exceed $^1/_4$ inch (6 mm), and between adjacent shakes, it shall not exceed $^1/_2$ inch (13 mm). The offset spacing between joints in adjacent courses shall be a minimum of $1^1/_2$ inches (38 mm).

❖ Where wood shakes and shingles are applied as an exterior wall covering, the sheathing must be covered with a water-resistive permeable membrane prior to shingle or shake installation. The application of wood shakes and shingles may be over either solid sheathing or furring strips applied directly to the studs. The minimum sheathing thickness and furring strip size and fastening is prescribed by the code. To allow for expansion, a small space should be established between each shake or shingle. The spacing is limited to $^1/_4$ inch (6.4 mm) between adjacent shingles and $^1/_2$ inch (12.7 mm) between adjacent shakes. To avoid channels between the shakes or shingles where water penetration might occur, joints must be offset at least $1^1/_2$ inches (38 mm) in adjacent courses.

R703.5.2 Weather exposure. The maximum weather exposure for shakes and shingles shall not exceed that specified in Table R703.5.2.

❖ Table R703.5.2 specifies the maximum weather exposure permitted for wood shakes and wood shingles installed on exterior walls. Where a single course of wood shingles is used, the maximum exposure varies from $7^1/_2$ inches to $11^1/_2$ inches (190.5 mm to 292 mm), depending on the length of the shingles. Maximum wood shake exposures are either $8^1/_2$ inches or $11^1/_2$ inches (216 mm to 292 mm) for a single course application. In double course applications, the maximum exposures are increased for both shingles and shakes.

TABLE R703.5.2
MAXIMUM WEATHER EXPOSURE FOR WOOD SHAKES AND SHINGLES ON EXTERIOR WALLS[a,b,c]
(Dimensions are in inches)

LENGTH	EXPOSURE FOR SINGLE COURSE	EXPOSURE FOR DOUBLE COURSE
Shingles[a]		
16	$7^1/_2$	12[b]
18	$8^1/_2$	14[c]
24	$11^1/_2$	16
Shakes[a]		
18	$8^1/_2$	14
24	$11^1/_2$	18

For SI: 1 inch = 25.4 mm.

a. Dimensions given are for No. 1 grade.

b. A maximum 10-inch exposure is permitted for No. 2 grade.

c. A maximum 11-inch exposure is permitted for No. 2 grade.

❖ Depending on the length of the shingle and the number of courses, the table specifies the maximum length of exposure for weathering purposes. Shingles are available in three different lengths, while shakes are regulated for two lengths. Both a single-course application and double-course application are addressed.

R703.5.3 Attachment. Each shake or shingle shall be held in place by two hot-dipped zinc-coated, stainless steel, or aluminum nails or staples. The fasteners shall be long enough to penetrate the sheathing or furring strips by a minimum of $^1/_2$ inch (13 mm) and shall not be overdriven.

❖ Hot-dipped zinc-coated, stainless steel, or aluminum fasteners are to be used to attach wood shakes and shingles. Either nails or staples are permitted, with two fasteners used for each shake or shingle. The length of the nails or staples must be adequate to penetrate sheathing or furring strips a minimum of $^1/_2$ inch (12.7 mm). As the nails or staples are driven, it is important that they do not excessively penetrate the shake or shingle, which could result in reduced holding force.

R703.5.3.1 Staple attachment. Staples shall not be less than 16 gage and shall have a crown width of not less than $^7/_{16}$ inch (11 mm), and the crown of the staples shall be parallel with the butt of the shake or shingle. In single-course application, the fasteners shall be concealed by the course above and shall be driven approximately 1 inch (25 mm) above the butt line of the succeeding course and $^3/_4$ inch (19 mm) from the edge. In double-course applications, the exposed shake or shingle shall be face-nailed with two casing nails, driven approximately 2 inches (51 mm) above the butt line and $^3/_4$ inch (19 mm) from each edge. In all applications, staples shall be concealed by the course above. With shingles wider than 8 inches (203 mm) two additional nails shall be required and shall be nailed approximately 1 inch (25 mm) apart near the center of the shingle.

❖ This section describes in detail the proper attachment of either staples or nails where used as fasteners for wood shakes and shingles applied to exterior walls. Both single-course and double-course application methods are described. The general requirement of Section R703.5.3 for two fasteners per shingle is modified by this section, as two additional fasteners are needed for shingles having a width of more than 8 inches (203 mm).

R703.5.4 Bottom courses. The bottom courses shall be doubled.

❖ In all cases, the bottom course of shakes and shingles in an exterior wall application is to be doubled.

R703.6 Exterior plaster. Installation of these materials shall be in compliance with ASTM C 926 and ASTM C 1063 and the provisions of this code.

❖ Portland cement plaster is the only material approved by the code for exterior plaster. Gypsum plaster deteriorates under conditions of weather and moisture, which are prevalent on the exterior surfaces of buildings. Commentary Figures R703.6(1) and R703.6(2) illustrate exterior plastering systems. ASTM C 926 is the standard specification for the installation of portland-cement-based plaster. ASTM C 926 provides the minimum requirements for the application of portland cement-based plaster for exterior (stucco) along with the tables necessary for proportioning various plaster mixes and thicknesses. ASTM C 1063 is the standard specification for exterior lathing and furring for portland-cement-based plastering, as specified in ASTM C 926. Installation of exterior plaster must also comply with the provisions of this code.

R703.6.1 Lath. All lath and lath attachments shall be of corrosion-resistant materials. Expanded metal or woven wire lath shall be attached with $1^1/_2$-inch-long (38 mm), 11 gage nails having a $^7/_{16}$-inch (11.1 mm) head, or $^7/_8$-inch-long (22.2 mm), 16 gage staples, spaced at no more than 6 inches (152 mm), or as otherwise approved.

❖ Lath and its attachments must be of corrosion-resistant materials. Unless another fastening method is approved, lath must be attached to framing members with the nails or staples specified in this section and spaced at a maximum of 6 inches (152 mm) on center along the framing member.

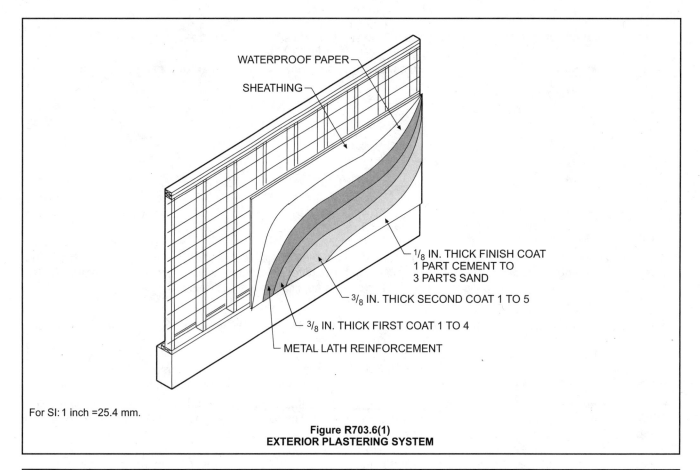

For SI: 1 inch = 25.4 mm.

Figure R703.6(1)
EXTERIOR PLASTERING SYSTEM

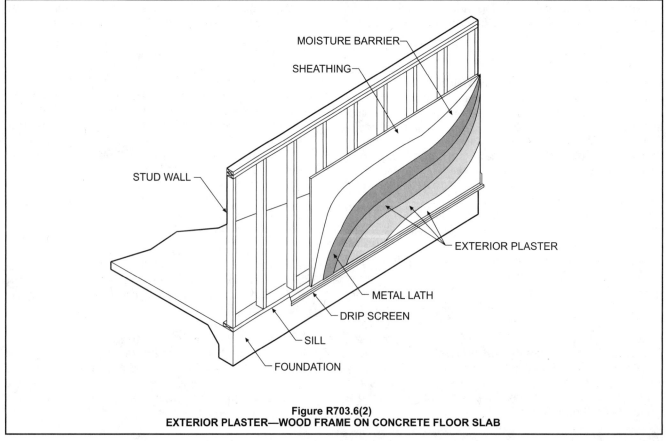

Figure R703.6(2)
EXTERIOR PLASTER—WOOD FRAME ON CONCRETE FLOOR SLAB

R703.6.2 Plaster. Plastering with portland cement plaster shall be not less than three coats when applied over metal lath or wire lath and shall be not less than two coats when applied over masonry, concrete, pressure-preservative treated wood or decay-resistant wood as specified in Section R319.1 or gypsum backing. If the plaster surface is completely covered by veneer or other facing material or is completely concealed, plaster application need be only two coats, provided the total thickness is as set forth in Table R702.1(1).

On wood-frame construction with an on-grade floor slab system, exterior plaster shall be applied to cover, but not extend below, lath, paper and screed.

The proportion of aggregate to cementitious materials shall be as set forth in Table R702.1(3).

❖ The code requires that exterior portland cement plaster be applied in not less than three coats when applied over metal or wire-fabric lath, for the same reasons as discussed for interior plaster. Where the portland cement plaster is applied over other approved plaster bases, the code requires only two-coat work. The code permits plaster work that is completely concealed to be of only two coats provided the total thickness is that required by Table R702.1(1), because the finish coat of plaster is to provide a surface for exterior finishes and to provide an aesthetic appearance. Thus, where the plaster surface is to be completely concealed, it is not necessary to apply a finish coat.

The code requires that the exterior plaster be installed to completely cover, but not extend below, the lath and paper-on-wood exterior wall construction supported by an on-grade concrete slab. This requirement, combined with the presence of a weep screed, prevents the entrapment of free moisture and the subsequent channeling of the moisture to the interior of the building.

R703.6.2.1 Weep screeds. A minimum 0.019-inch (0.5 mm) (No. 26 galvanized sheet gage), corrosion-resistant weep screed or plastic weep screed, with a minimum vertical attachment flange of $3^1/_2$ inches (89 mm) shall be provided at or below the foundation plate line on exterior stud walls in accordance with ASTM C 926. The weep screed shall be placed a minimum of 4 inches (102 mm) above the earth or 2 inches (51 mm) above paved areas and shall be of a type that will allow trapped water to drain to the exterior of the building. The weather-resistant barrier shall lap the attachment flange. The exterior lath shall cover and terminate on the attachment flange of the weep screed.

❖ Water and moisture can penetrate an exterior plaster wall for a variety of reasons and in a number of ways. Some moisture will penetrate the plaster in an exterior wall; therefore, the design of the wall should include a weep screed, which will provide a way to release the moisture [see Commentary Figure R703.6(3)]. Once water or moisture penetrates the plaster, it will migrate down the exterior face of the weather-resistive barrier until it reaches the sill plate or mud sill. At this point, the water will seek a way out of the wall. If the exterior plaster system is not detailed and constructed with

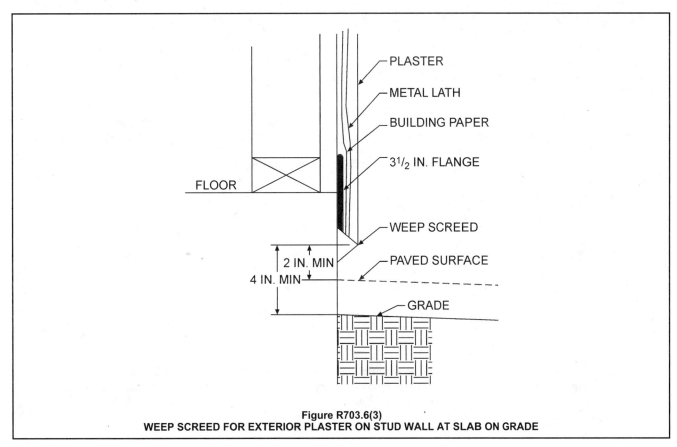

Figure R703.6(3)
WEEP SCREED FOR EXTERIOR PLASTER ON STUD WALL AT SLAB ON GRADE

provisions to allow the moisture to escape to the exterior, it will find its own way out. This exit will almost certainly be through the interior of the wall and cause leaking, and therefore damage, to the interior of the building. For this reason, the code requires a continuous weep screed at the bottom of exterior walls to permit the moisture to escape to the exterior of the building.

R703.6.3 Water-resistive barriers. Water-resistive barriers shall be installed as required in Section R703.2 and, where applied over wood-based sheathing, shall include a water-resistive vapor-permeable barrier with a performance at least equivalent to two layers of Grade D paper.

Exception: Where the water-resistive barrier that is applied over wood-based sheathing has a water resistance equal to or greater than that of 60 minute Grade D paper and is separated from the stucco by an intervening, substantially nonwater-absorbing layer or designed drainage space.

❖ The code requires that a water-resistive barrier be installed behind exterior plaster for the reasons provided in Section R703.2. The code also requires that when the barrier is applied over wood-based sheathing such as plywood, the barrier is to be two layers of Grade D building paper. This requirement is based on the observed problems where one layer of typical No. 15, Type I felt is applied over wood sheathing. The wood sheathing eventually exhibits dry rot as a result of the penetration of moisture. Cracking is then created in the plaster by movement of the sheathing caused by alternate expansion and contraction. Field experience has shown that where two layers of building paper are used, the penetration of moisture is considerably decreased, as is the cracking of the plaster resulting from movement of the sheathing caused by the wet and dry cycles. Grade D building paper is specified because it has the proper water vapor permeability to prevent entrapment of moisture between the paper and the sheathing.

The exception permits the application of alternative standard practices by recognizing stucco systems in which one of the two layers of water-resistive barrier is replaced by a layer that although not a "Grade D Paper" provides separation from the wet stucco application and provides a barrier to moisture from the stucco to the water-resistive barrier.

R703.7 Stone and masonry veneer, general. Stone and masonry veneer shall be installed in accordance with this chapter, Table R703.4 and Figure R703.7. These veneers installed over a backing of wood or cold-formed steel shall be limited to the first story above-grade and shall not exceed 5 inches (127 mm) in thickness.

Exceptions:

1. For all buildings in Seismic Design Categories A, B and C, exterior stone or masonry veneer, as specified in Table R703.7(1), with a backing of wood or steel framing shall be permitted to the height specified in Table R703.7(1) above a noncombustible foundation.

Wall bracing at exterior and interior braced wall lines shall be in accordance with Section R602.10 or R603.7, and the additional requirements of Table R703.7(1).

2. For detached one- or two-family dwellings in Seismic Design Categories D_0, D_1 and D_2, exterior stone or masonry veneer, as specified in Table R703.7(2), with a backing of wood framing shall be permitted to the height specified in Table R703.7(2) above a noncombustible foundation. Wall bracing and hold downs at exterior and interior braced wall lines shall be in accordance with Sections R602.10 and R602.11 and the additional requirements of Table R703.7(2). In Seismic Design Categories D_0, D_1 and D_2, cripple walls shall not be permitted, and required interior braced wall lines shall be supported on continuous foundations.

❖ For the purposes of this section, stone and masonry veneer are nonstructural facing materials providing ornamentation, protection or insulation. To be considered a veneer, the material cannot act structurally with the backing insofar as the structural strength of the assembly is concerned. Stone and masonry veneer can be used as either an interior or exterior wall finish.

In addition to the provisions of this section, Table R703.4 and Figure R703.7 provide details for the installation of stone and or masonry veneer. Because of their weight, stone or masonry veneer can impose lateral loads from seismic events that are beyond the loading considered acceptable for wood or cold-formed steel construction. For this reason, stone or masonry veneer is limited to the first story above grade and a maximum thickness of 5 inches (127 mm) where installed over a backing of wood or cold-formed steel.

The use of stone or masonry veneer above the first story above grade is permitted in accordance with the exceptions. Exception 1 allows stone or masonry veneer, with wood or cold-formed steel backing, either two or three stories above grade on structures classified as Seismic Design Category A, B, or C under certain conditions that include increasing the amount of wall bracing. See Table R703.7.1.

Exception 1 applies to stone or masonry veener on wood or cold-formed steel frame detached one- or two-family dwellings and townhouses.

Exception 2 applies only to stone or masonry veener on wood frame detached one- and two-family dwellings.

Exception 2 allows stone or masonry veneer, with wood backing, three stories above grade on structures classified as Seismic Design Category D0 or D1 and two stories above grade for Seismic Design Category D2 under certain conditions that include increasing the amount of wall bracing and providing hold down connectors. See Table R703.7(2) and Figure R703.7(1).

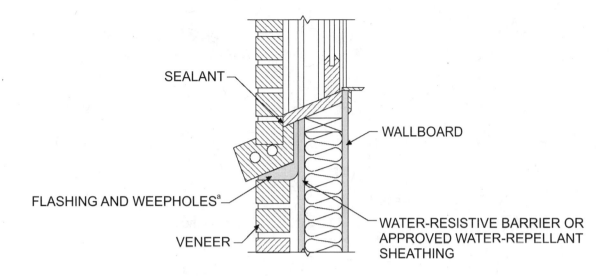

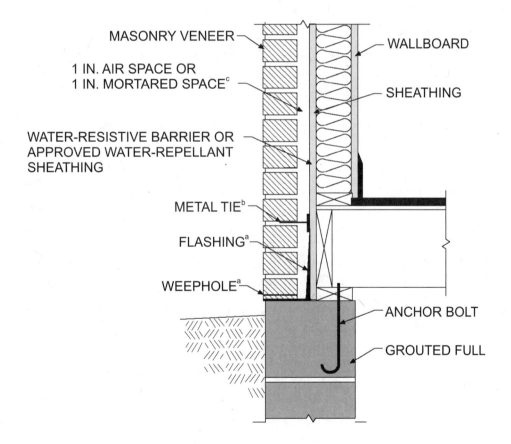

For SI: 1 inch = 25.4 mm.

FIGURE R703.7
MASONRY VENEER WALL DETAILS

(continued)

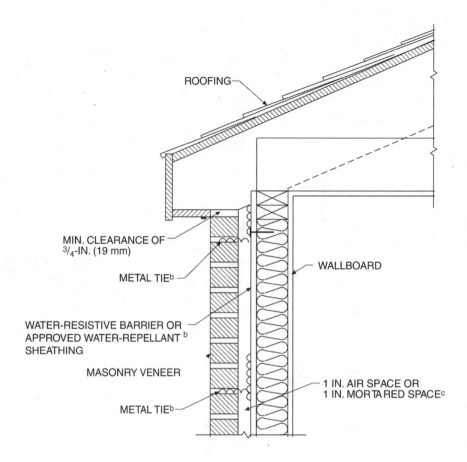

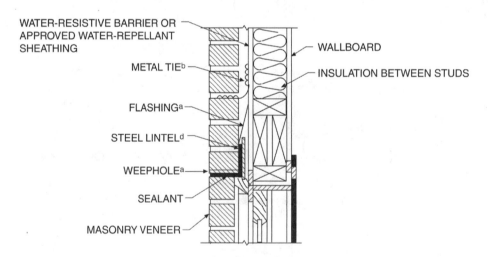

For SI: 1 inch =25.4 mm.
a. See Sections R703.7.5, R703.7.6 and R703.8.
b. See Sections R703.2 and 703.7.4.
c. See Sections R703.7.4.2 and R703.7.4.3.
d. See Section R703.7.3.

FIGURE R703.7—continued
MASONRY VENEER WALL DETAILS

❖ This figure outlines a number of methods for attaching masonry veneer to framed walls. The fundamental application and attachment methods are listed, including the use of a water-resistive barrier or water-repellant sheathing, the provision for weepholes and the need for an airspace unless mortared. References to the appropriate code requirements are listed in the notes.

TABLE R703.7(1)
STONE OR MASONRY VENEER LIMITATIONS AND REQUIREMENTS, WOOD
OR STEEL FRAMING, SEISMIC DESIGN CATEGORIES A, B AND C

SEISMIC DESIGN CATEGORY	NUMBER OF WOOD OR STEEL FRAMED STORIES	MAXIMUM HEIGHT OF VENEER ABOVE NONCOMBUSTIBLE FOUNDATION[a] (feet)	MAXIMUM NOMINAL THICKNESS OF VENEER (inches)	MAXIMUM WEIGHT OF VENEER (psf)[b]	WOOD OR STEEL FRAMED STORY	MINIMUM SHEATHING AMOUNT (percent of braced wall line length)[c]
A or B	Steel: 1 or 2 Wood: 1, 2 or 3	30	5	50	all	Table R602.10.1 or Table R603.7
C	1	30	5	50	1 only	Table R602.10.1 or Table R603.7
	2	30	5	50	top	Table R602.10.1 or Table R603.7
					bottom	1.5 times length required by Table R602.10.1 or 1.5 times length required by Table R603.7
	Wood only: 3	30	5	50	top	Table R602.10.1
					middle	1.5 times length required by Table R602.10.1
					bottom	1.5 times length required by Table R602.10.1

For SI: 1 inch = 25.4 mm, 1 foot = 304.8 mm, 1 pound per square foot = 0.479 kPa.

a. An Additional 8 feet is permitted for gable end walls. See also story height limitations of Section R301.3.

b. Maximum weight is installed weight and includes weight of mortar, grout, lath and other materials used for installation. Where veneer is placed on both faces of a wall, the combined weight shall not exceed that specified in this table.

c. Applies to exterior and interior braced wall lines.

❖ See the commentary for Section R703.7.

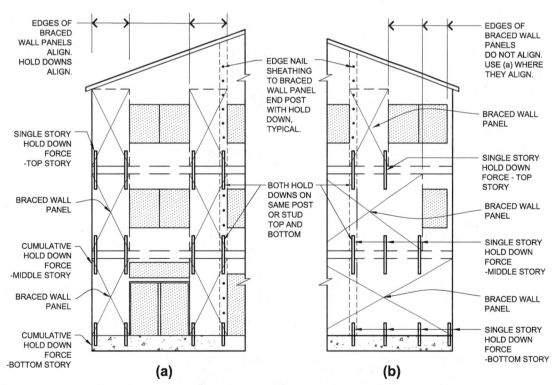

(a) Braced wall panels stacked (aligned story to story). Use cumulative hold down force.
(b) Braced wall panels not stacked. Use single story hold down force.

FIGURE R703.7(1)
HOLD DOWNS AT EXTERIOR AND INTERIOR BRACED WALL PANELS WHEN USING STONE OR MASONRY VENEER

❖ See the commentary for Section R703.7.

R703.7.1 Interior veneer support. Veneers used as interior wall finishes shall be permitted to be supported on wood or cold-formed steel floors that are designed to support the loads imposed.

❖ Where designed to support the weight of the veneer, a wood-framed or cold-formed steel floor system may be used to support stone or masonry veneer used in an interior application.

R703.7.2 Exterior veneer support. Except in Seismic Design Categories D_0, D_1 and D_2, exterior masonry veneers having an installed weight of 40 pounds per square foot (195 kg/m²) or less shall be permitted to be supported on wood or cold-formed steel construction. When masonry veneer supported by wood or cold-formed steel construction adjoins masonry veneer supported by the foundation, there shall be a movement joint between the veneer supported by the wood or cold-formed steel construction and the veneer supported by the foundation. The wood or cold-formed steel construction supporting the masonry veneer shall be designed to limit the deflection to $1/600$ of the span for the supporting members. The design of the wood or cold-formed steel construction shall consider the weight of the veneer and any other loads.

❖ Two criteria must be met to support exterior masonry veneer directly on wood or cold-formed steel construction rather than on a concrete foundation wall. First, the construction must be in Seismic Design Category A, B or C. Second, the total installed weight of the veneer must not exceed 40 pounds per square foot (195 kg/m²). The prescriptive provisions of the code are not designed to address conditions where the potential for extensive lateral loading from earthquakes is relatively high or where the weight of the masonry veneer is excessive. Even in those seismic design categories permitted by this section, lateral movement must be addressed. Therefore, a movement joint must be

TABLE R703.7(2)
STONE OR MASONRY VENEER LIMITATIONS AND REQUIREMENTS, ONE- AND TWO-FAMILY DETACHED DWELLINGS, WOOD FRAMING, SEISMIC DESIGN CATEGORIES D_0, D_1 AND D_2

SEISMIC DESIGN CATEGORY	NUMBER OF WOOD FRAMED STORIES[a]	MAXIMUM HEIGHT OF VENEER ABOVE NONCOMBUSTIBLE FOUNDATION OR FOUNDATION WALL (feet)	MAXIMUM NOMINAL THICKNESS OF VENEER (inches)	MAXIMUM WEIGHT OF VENEER (psf)[b]	WOOD FRAMED STORY	MINIMUM SHEATHING AMOUNT (percent of braced wall line length)[c]	MINIMUM SHEATHING THICKNESS AND FASTENING	SINGLE STORY HOLD DOWN FORCE (lb)[d]	CUMULATIVE HOLD DOWN FORCE (lb)[e]
D_0	1	20[f]	4	40	1 only	35	$7/16$-inch wood structural panel sheathing with 8d common nails spaced at 4 inches on center at panel edges, 12 inches on center at intermediate supports. 8d common nails at 4 inches on center at braced wall panel end posts with hold down attached.	N/A	—
	2	20[f]	4	40	top	35		1900	—
					bottom	45		3200	5100
	3	30[g]	4	40	top	40		1900	—
					middle	45		3500	5400
					bottom	60		3500	8900
D_1	1	20[f]	4	40	1 only	45		2100	—
	2	20[f]	4	40	top	45		2100	—
					bottom	45		3700	5800
	3	20[f]	4	40	top	45		2100	—
					middle	45		3700	5800
					bottom	60		3700	9500
D_2	1	20[f]	3	30	1 only	55		2300	—
	2	20[f]	3	30	top	55		2300	—
					bottom	55		3900	6200

For SI: 1 inch = 25.4 mm, 1 foot = 304.8 mm, 1 pound per square foot = 0.479 kPa, 1 pound-force = 4.448 N.

a. Cripple walls are not permitted in Seismic Design Categories D_0, D_1 and D_2.

b. Maximum weight is installed weight and includes weight of mortar, grout and lath, and other materials used for installation.

c. Applies to exterior and interior braced wall lines.

d. Hold down force is minimum allowable stress design load for connector providing uplift tie from wall framing at end of braced wall panel at the noted story to wall framing at end of braced wall panel at the story below, or to foundation or foundation wall. Use single story hold down force where edges of braced wall panels do not align; a continuous load path to the foundation shall be maintained. [See Figure R703.7(1)(b)].

e. Where hold down connectors from stories above align with stories below, use cumulative hold down force to size middle and bottom story hold down connectors. [See Figure R703.7(1)(a)].

f. The veneer shall not exceed 20 feet in height above a noncombustible foundation, with an additional 8 feet permitted for gable end walls, or 30 feet in height with an additional 8 feet for gable end walls where the lower 10 feet has a backing of concrete or masonry wall. See also story height limitations of Section R301.3.

g. The veneer shall not exceed 30 feet in height above a noncombustible foundation, with an additional 8 feet permitted for gable end walls. See also story height limitations of Section R301.3.

❖ See the commentary for Section R703.7.

provided between the exterior veneer supported by wood or steel and the adjacent exterior veneer supported by the foundation. The deflection of the wood or cold-formed steel members supporting the masonry veneer is also limited to a maximum of $^1/_{600}$ of the span of the supporting members.

R703.7.2.1 Support by steel angle. A minimum 6 inches by 4 inches by $^5/_{16}$ inch (152 mm by 102 mm by 8 mm) steel angle, with the long leg placed vertically, shall be anchored to double 2 inches by 4 inches (51 mm by 102 mm) wood studs at a maximum on-center spacing of 16 inches (406 mm). Anchorage of the steel angle at every double stud spacing shall be a minimum of two $^7/_{16}$ inch (11 mm) diameter by 4 inch (102 mm) lag screws. The steel angle shall have a minimum clearance to underlying construction of $^1/_{16}$ inch (2 mm). A minimum of two-thirds the width of the masonry veneer thickness shall bear on the steel angle. Flashing and weep holes shall be located in the masonry veneer wythe in accordance with Figure R703.7.2.1. The maximum height of masonry veneer above the steel angle support shall be 12 feet, 8 inches (3861 mm). The air space separating the masonry veneer from the wood backing shall be in accordance

with Sections R703.7.4 and R703.7.4.2. The method of support for the masonry veneer on wood construction shall be constructed in accordance with Figure R703.7.2.1.

The maximum slope of the roof construction without stops shall be 7:12. Roof construction with slopes greater than 7:12 but not more than 12:12 shall have stops of a minimum 3 inch × 3 inch × $^1/_4$ inch (76 mm × 76 mm × 6 mm) steel plate welded to the angle at 24 inches (610 mm) on center along the angle or as approved by the building official.

❖ As illustrated in Figure R703.7.2.1, a steel angle may be used to support exterior masonry veneer in wood-frame construction. The minimum size of the angle is specified, as are the details of anchorage and the minimum bearing length. Two limits are placed on this method of veneer support: the maximum veneer height bearing on the steel angle must be 12 feet, 8 inches (3861 mm), and the roof slope is limited to 7:12. A roof slope greater than 7:12 is permitted if stops are welded to the steel angle to prevent the veneer from sliding down the angle.

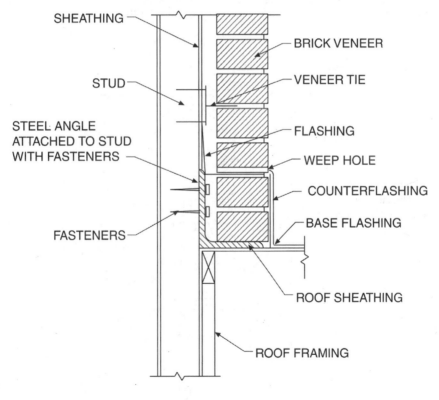

SHEATHING

STUD

STEEL ANGLE ATTACHED TO STUD WITH FASTENERS

FASTENERS

BRICK VENEER

VENEER TIE

FLASHING

WEEP HOLE

COUNTERFLASHING

BASE FLASHING

ROOF SHEATHING

ROOF FRAMING

SUPPORT BY STEEL ANGLE

FIGURE R703.7.2.1
EXTERIOR MASONRY VENEER SUPPORT BY STEEL ANGLES

❖ Under most conditions, exterior masonry veneer will be supported by the foundation of the building. Where support of the veneer is by wood construction, the code describes special methods of veneer support. Figures R703.7.2.1 and R703.7.2.2 illustrate support by either steel angle or roof members. The figures show the fundamental elements of attachment, while Sections R703.7.2.1 and R703.7.2.2 contain the details of the installation materials and methods.

R703.7.2.2 Support by roof construction. A steel angle shall be placed directly on top of the roof construction. The roof supporting construction for the steel angle shall consist of a minimum of three 2-inch by 6-inch (51 mm by 152 mm) wood members. The wood member abutting the vertical wall stud construction shall be anchored with a minimum of three $^5/_8$-inch (16 mm) diameter by 5-inch (127 mm) lag screws to every wood stud spacing. Each additional roof member shall be anchored by the use of two 10d nails at every wood stud spacing. A minimum of two-thirds the width of the masonry veneer thickness shall bear on the steel angle. Flashing and weep holes shall be located in the masonry veneer wythe in accordance with Figure R703.7.2.2. The maximum height of the masonry veneer above the steel angle support shall be 12 feet, 8 inches (3861 mm). The air space separating the masonry veneer from the wood backing shall be in accordance with Sections R703.7.4 and R703.7.4.2. The support for the masonry veneer on wood construction shall be constructed in accordance with Figure R703.7.2.2.

The maximum slope of the roof construction without stops shall be 7:12. Roof construction with slopes greater than 7:12 but not more than 12:12 shall have stops of a minimum 3 inch × 3 inch × $^1/_4$ inch (76 mm × 76 mm × 6 mm) steel plate welded to the angle at 24 inches (610 mm) on center along the angle or as approved by the building official.

❖ In addition to the option of having a steel angle support exterior masonry veneer, the code also permits the wood roof construction itself to support the exterior veneer. Detailed in Figure R703.7.2.2, the steel angle supporting the masonry veneer must be fully supported by at least three 2-inch by 6-inch (51 mm by 152 mm) wood members. This section also addresses the attachment details and bearing conditions. As described for masonry veneer supported by steel angles, the maximum veneer height and maximum roof slope are also regulated. The roof slope is limited to 7:12; however, a roof slope greater than 7:12 is permitted if stops are welded to the steel angle to prevent the veneer from sliding down the angle.

R703.7.3 Lintels. Masonry veneer shall not support any vertical load other than the dead load of the veneer above. Veneer above openings shall be supported on lintels of noncombustible materials and the allowable span shall not exceed the value set forth in Table R703.7.3. The lintels shall have a length of bearing not less than 4 inches (102 mm).

❖ Masonry veneer, like all veneers, is a nonload-bearing wall covering. It is not intended to support any loads, except for the dead load of the masonry veneer above.

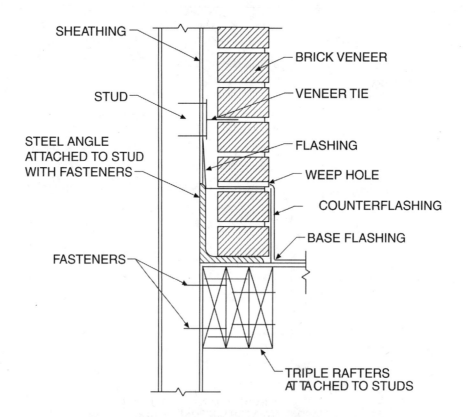

SHEATHING

BRICK VENEER

STUD

VENEER TIE

STEEL ANGLE ATTACHED TO STUD WITH FASTENERS

FLASHING

WEEP HOLE

COUNTERFLASHING

BASE FLASHING

FASTENERS

TRIPLE RAFTERS ATTACHED TO STUDS

SUPPORT BY ROOF MEMBERS

FIGURE R703.7.2.2
EXTERIOR MASONRY VENEER SUPPORT BY ROOF MEMBERS

❖ See the commentary for Section R703.7.2.1.

Where masonry veneer occurs above an opening created for a doorway, window or similar feature in a building, the veneer must be supported on a noncombustible lintel with a minimum length-of-bearing of 4 inches (102 mm) on each end. Table R703.7.3 identifies the maximum spans permitted for four different sizes of steel lintels supporting masonry veneer. The allowable spans vary based on the number of stories of veneer supported by the lintel.

R703.7.4 Anchorage. Masonry veneer shall be anchored to the supporting wall with corrosion-resistant metal ties. Where veneer is anchored to wood backings by corrugated sheet metal ties, the distance separating the veneer from the sheathing material shall be a maximum of a nominal 1 inch (25 mm). Where the veneer is anchored to wood backings using metal strand wire ties, the distance separating the veneer from the sheathing material shall be a maximum of $4^1/_2$ inches (114 mm). Where the veneer is anchored to cold-formed steel backings, adjustable metal strand wire ties shall be used. Where veneer is anchored to cold-formed steel backings, the distance separating the veneer from the sheathing material shall be a maximum of $4^1/_2$ inches (114 mm).

❖ To maintain the attachment of the masonry veneer to the wall construction, proper anchorage must be provided. In all cases, corrosion-resistant metal ties must be used. Two types of ties are addressed for attachment to wood construction: corrugated sheet metal ties and metal strand wire ties. The distance between the veneer and the sheathing is limited and varies with the type of tie used. Only adjustable metal strand wire ties may be used where the backing is of cold-formed steel. The distance between the steel backing and the masonry veneer is also limited.

R703.7.4.1 Size and spacing. Veneer ties, if strand wire, shall not be less in thickness than No. 9 U.S. gage [(0.148 in.) (4

mm)] wire and shall have a hook embedded in the mortar joint, or if sheet metal, shall be not less than No. 22 U.S. gage by [(0.0299 in.)(0.76 mm)] $^7/_8$ inch (22 mm) corrugated. Each tie shall be spaced not more than 24 inches (610 mm) on center horizontally and vertically and shall support not more than 2.67 square feet (0.25 m²) of wall area.

Exception: In Seismic Design Category D_0, D_1 or D_2 or townhouses in Seismic Design Category C or in wind areas of more than 30 pounds per square foot pressure (1.44 kPa), each tie shall support not more than 2 square feet (0.2 m²) of wall area.

❖ This section sets forth the minimum size of metal veneer ties for use with masonry veneer applications. The spacing of the ties is also limited, with a maximum horizontal distance between ties of 24 inches (610 mm). In addition, no single tie can support more than 2.67 square feet (0.248 m²) of wall area. An increased number of ties are required in those areas having a high seismic design category or a high wind load.

R703.7.4.1.1 Veneer ties around wall openings. Veneer ties around wall openings. Additional metal ties shall be provided around all wall openings greater than 16 inches (406 mm) in either dimension. Metal ties around the perimeter of openings shall be spaced not more than 3 feet (9144 mm) on center and placed within 12 inches (305 mm) of the wall opening.

❖ Because of the discontinuity of the veneer adjacent to wall openings such as windows and doors, it is often necessary to increase the number of metal ties around wall openings. The location of ties around openings is specified, including the maximum spacing and the maximum distance from the opening. Openings less than 16 inches in either dimension are exempt from these provisions.

TABLE R703.7.3
ALLOWABLE SPANS FOR LINTELS SUPPORTING MASONRY VENEER[a, b, c]

SIZE OF STEEL ANGLE[a, c] (inches)	NO STORY ABOVE	ONE STORY ABOVE	TWO STORIES ABOVE	NO. OF $^1/_2$" OR EQUIVALENT REINFORCING BARS[c]
$3 \times 3 \times ^1/_4$	6'-0"	4'-6"	3'-0"	1
$4 \times 3 \times ^1/_4$	8'-0"	6'-0"	4'-6"	1
$5 \times 3^1/_2 \times ^5/_{16}$	10'-0"	8'-0"	6'-0"	2
$6 \times 3^1/_2 \times ^5/_{16}$	14'-0"	9'-6"	7'-0"	2
$2\text{-}6 \times 3^1/_2 \times ^5/_{16}$	20'-0"	12'-0"	9'-6"	4

For SI: 1 inch = 25.4 mm, 1 foot = 304.8 mm.

a. Long leg of the angle shall be placed in a vertical position.

b. Depth of reinforced lintels shall not be less than 8 inches and all cells of hollow masonry lintels shall be grouted solid. Reinforcing bars shall extend not less than 8 inches into the support.

c. Steel members indicated are adequate typical examples; other steel members meeting structural design requirements may be used.

❖ Where masonry veneer extends above an opening in a wall, it must be supported by a noncombustible lintel. The maximum span for the lintel is based on the number of stories located above the opening. This measurement provides a reasonable estimate of the amount of veneer to be carried. The allowable span is increased based on the size of the steel angle used as the lintel. The table addresses conditions for one-, two- and three-story masonry veneer applications and describes the required reinforcing necessary based on the size of the steel angle.

R703.7.4.2 Air space. The veneer shall be separated from the sheathing by an air space of a minimum of a nominal 1 inch (25 mm) but not more than 4$^1/_2$ inches (114 mm).

❖ There must be an air space of at least a nominal 1 inch (25.4 mm) between the veneer and the sheathing unless the space is filled with mortar or grout per Section R703.7.4.3. The maximum air-space width of 4$^1/_2$ inches (114 mm) is based on the maximum permitted separation between the sheathing material and metal strand wire ties. Where the air space is provided, it is not necessary to provide a water-resistive barrier over any sheathing materials that perform as a water-resistive barrier.

R703.7.4.3 Mortar or grout fill. As an alternate to the air space required by Section R703.7.4.2, mortar or grout shall be permitted to fill the air space .When the air space is filled with mortar, a water-resistive barrier is required over studs or sheathing. When filling the air space, replacing the sheathing and water-resistive barrier with a wire mesh and approved water-resistive barrier or an approved water-resistive barrier-backed reinforcement attached directly to the studs is permitted.

❖ Where an air space complying with Section R703.7.4.2 is not provided between the masonry veneer and the supporting sheathing, mortar or grout may be used to fill the void. Under these conditions, it is important to cover the studs or sheathing with a water-resistive barrier, a system of wire mesh and approved paper, or an approved paper-backed reinforcement attached directly to the studs.

R703.7.5 Flashing. Flashing shall be located beneath the first course of masonry above finished ground level above the foundation wall or slab and at other points of support, including structural floors, shelf angles and lintels when masonry veneers are designed in accordance with Section R703.7. See Section R703.8 for additional requirements.

❖ Flashing is necessary to close off the points of water entry at the first course of masonry above the finished ground level, as well as at other points of support such as at shelf angles and lintels. As always, flashing must be of an approved corrosion-resistant material.

R703.7.6 Weepholes. Weepholes shall be provided in the outside wythe of masonry walls at a maximum spacing of 33 inches (838 mm) on center. Weepholes shall not be less than $^3/_{16}$ inch (5 mm) in diameter. Weepholes shall be located immediately above the flashing.

❖ At a point directly above the flashing mandated by Section R703.7.5, weepholes must be provided to allow for the escape of any moisture that may have penetrated the masonry veneer. Because moisture will adversely affect the integrity of the wall if not removed from the wall assembly, weepholes must be installed within the maximum spacing specified in this section. The minimum diameter of the weepholes is also regulated.

R703.8 Flashing. Approved corrosion-resistant flashing shall be applied shingle-fashion in such a manner to prevent entry of water into the wall cavity or penetration of water to the building structural framing components. The flashing shall extend to the surface of the exterior wall finish. Approved corrosion-resistant flashings shall be installed at all of the following locations:

1. Exterior window and door openings. Flashing at exterior window and door openings shall extend to the surface of the exterior wall finish or to the water-resistive barrier for subsequent drainage.

2. At the intersection of chimneys or other masonry construction with frame or stucco walls, with projecting lips on both sides under stucco copings.

3. Under and at the ends of masonry, wood or metal copings and sills.

4. Continuously above all projecting wood trim.

5. Where exterior porches, decks or stairs attach to a wall or floor assembly of wood-frame construction.

6. At wall and roof intersections.

7. At built-in gutters.

❖ The code requires that all points subject to the entry of moisture be appropriately flashed. Roof and wall intersections and parapets create significant challenges, as do exterior wall openings exposed to the weather. Where wind-driven rain is expected, the concerns are even greater. Although the code identifies a number of locations where flashing is specifically required, the entire exterior envelope must be weather-tight to protect the interior from weather. Therefore, any location on the exterior envelope that provides a route for the admission of water or moisture into the building must be properly protected. Commentary Figure R703.8 illustrates examples of flashing.

R703.9 Exterior insulation finish systems, general. All Exterior Insulation Finish Systems (EIFS) shall be installed in accordance with the manufacturer's installation instructions and the requirements of this section. Decorative trim shall not be face nailed through the EIFS. The EIFS shall terminate not less than 6 inches (152 mm) above the finished ground level.

❖ The installation requirements for exterior insulation finish systems, commonly referred to as EIFS, are for the most part provided by the manufacturer of the specific system under consideration. Manufacturer's installation instructions must be followed for a durable, weather-tight exterior envelope. The code sets forth additional general requirements that must be followed.

As a basic requirement, EIFS must not be located within 6 inches of the finished ground level. In addition, any decorative trim applied over the EIFS must not be nailed through the EIFS. Both provisions prevent the penetration of moisture into the exterior wall system.

R703.9.1 Water-resistive barrier. All EIFS shall have a water-resistive barrier applied between the underlying water-sensitive building components and the exterior insulation, and a means of draining water to the exterior of the veneer. A water-resistive barrier shall be compliant with ASTM D 226 Type I asphalt saturated felt or equivalent, shall be applied horizontally with the upper layer lapped over the lower layer not less than 2 inches (51 mm), and shall have all vertical joints lapped not less than 6 inches (152 mm).

❖ The use of a water-resistive barrier with EIFS is necessary between the structural components and the wall covering, as it is with most other wall-covering systems. In addition, there must be a means of removal for any moisture that penetrates into the wall system. The type and installation of the required water-resistive barrier for EIFS is consistent with that required for other types of exterior wall systems. The code requires drainable EIFS.

R703.9.2 Flashing, general. Flashing of EIFS shall be provided in accordance with the requirements of Section R703.8.

❖ The general provisions for flashing at the exterior wall envelope are also applicable to exterior insulation finish systems.

R703.10 Fiber cement siding.

❖ Fiber cement siding is a recent addition to the list of material used for weather-resistant siding. It is manufactured from fiber-reinforced cement, and this code permits its use as either panel siding or horizontal lap siding.

R703.10.1 Panel siding. Panels shall be installed with the long dimension parallel to framing. Vertical joints shall occur over framing members and shall be sealed with caulking or covered with battens. Horizontal joints shall be flashed with Z-flashing and blocked with solid wood framing.

❖ Panel siding of fiber cement is acceptable exterior wall covering if the joints are made waterproof. Typical methods for accomplishing the joint treatment include (1) having the joints occur over framing and protecting the joints with caulking or a continuous batten and (2) flashing and blocking horizontal joints. See the commentary for Figures R703.3.1(1) and R703.3.1(2) for examples of these joint treatments.

R703.10.2 Horizontal lap siding. Lap siding shall be lapped a minimum of $1^1/_4$ inches (32 mm) and shall have the ends sealed with caulking, covered with an H-section joint cover, or located over a strip of flashing. Lap siding courses may be installed with the fastener heads exposed or concealed, according to approved manufacturers' installation instructions.

❖ To maintain the necessary weather protection, horizontal siding must have a minimum lap of at least $1^1/_4$ inches (31.8 mm). The end joints must be treated with one of the following methods: (1) sealed with caulk, (2) covered with an H-section joint cover or (3) located over a strip of flashing. Fastener heads may be exposed or concealed in accordance with approved manufacturers' installation instructions.

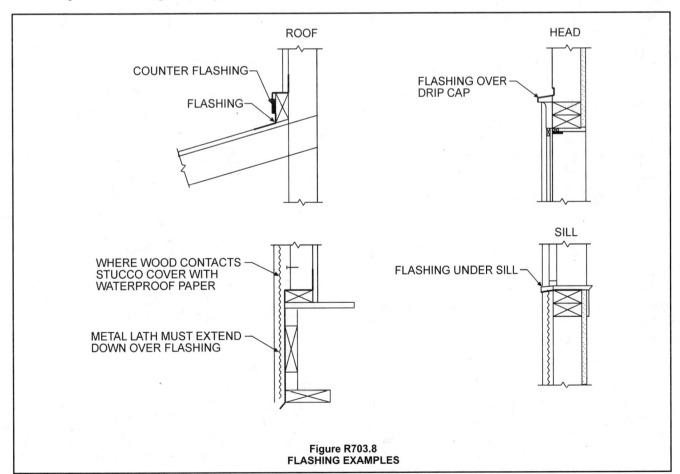

Figure R703.8
FLASHING EXAMPLES

R703.11 Vinyl siding. Vinyl siding shall be certified and labeled as conforming to the requirements of ASTM D 3679 by an approved quality control agency.

❖ Polyvinyl chloride (PVC) siding is specifically addressed here as an exterior wall covering. PVC siding must conform to the provisions of ASTM D 3679. The vinyl siding must bear a label, which means that the manufacturer must have regular inspections by a third-party quality control agency.

R703.11.1 Installation. Vinyl siding, soffit and accessories shall be installed in accordance with the manufacturer's installation instructions.

❖ Vinyl siding must be applied to conform to the water-resistive barrier requirements of Sections R703.1 and R703.2 and the attachment requirements of Section R703.4.(See the commentary for Sections R703.1, R703.2 and R703.4)

The installation must also comply with the manufacturer's instructions.

ASTM D 3679 requires installation of the siding in accordance with Practice D 4756 (ASTM D 4756) and the manufacturer's instructions. These instructions are necessary for compliance with the performance requirements of this chapter.

Bibliography

The following resource materials are referenced in this chapter or are relevant to the subject matter addressed in this chapter.

ANSI/AHA A135.5-04, *Prefinished Hardboard Paneling*. Palatine, IL: American Hardboard Association, 2004.

ANSI/AHA A135.6-98, *Hardboard Siding*. Palatine, IL: American Hardboard Association, 1998.

ANSI A108.1A-99, *Installation of Ceramic Tile in the Wet-set Method, with Portland Cement Mortar*. New York: American National Standards Institute, 1999.

ANSI A108.1B-99, *Installation of Ceramic Tile, Quarry Tile on a Cured Portland Cement Mortar Setting Bed with Dry-set or Latex-Portland Mortar*. New York: American National Standards Institute, 1999.

ANSI A108.4-99, *Installation of Ceramic Tile with Organic Adhesives or Water Cleanable Tile-setting Epoxy Adhesive*. New York: American National Standards Institute, 1999.

ANSI A108.5-99, *Installation of Ceramic Tile with Dry-set Portland Cement Mortar or Latex-Portland Cement Mortar*. New York: American National Standards Institute, 1999.

ANSI A108.6-99, *Installation of Ceramic Tile with Chemical Resistant, Water Cleanable Tile-setting and -grouting Epoxy*. New York: American National Standards Institute, 1999.

ANSI A108.11-99, *Interior Installation of Cementitious Backer Units*. New York: American National Standards Institute, 1999.

ANSI A118.1-99, *American National Standard Specifications for Dry-set Portland Cement Mortar*. New York: American National Standards Institute, 1999.

ANSI A118.3-99, *American National Standard Specification for Chemical Resistant, Water Cleanable Tile-setting and Grouting Epoxy and Water Cleanable Tile-setting Epoxy Adhesive*. New York: American National Standards Institute, 1999.

ANSI A136.1-99, *American National Standard Specifications for Organic Adhesives for Installation of Ceramic Tile*. New York: American National Standards Institute, 1999.

ANSI A137.1-88, *American National Standard Specifications for Ceramic Tile*. New York: American National Standards Institute, 1988.

ANSI/HPVA HP-1-2000, *American National Standard for Hardwood and Decorative Plywood*. Reston, VA: Hardwood Plywood & Veneer Association, 2000.

ASTM C 5-03, *Standard Specification for Quicklime for Structural Purposes*. West Conshohocken, PA: ASTM International, 2003.

.ASTM C 28/C28M-00e01, *Specification for Gypsum Plasters*. West Conshohocken, PA: ASTM International, 2001.

ASTM C 35-95(2001), *Specification for Inorganic Aggregates for Use in Gypsum Plaster*. West Conshohocken, PA: ASTM International, 2001.

ASTM C 36/C0036M-03, *Specification for Gypsum Wallboard*. West Conshohocken, PA: ASTM International, 2003.

ASTM C 37/C0037M-01, *Specification for Gypsum Lath*. West Conshohocken, PA: ASTM International, 2001.

ASTM C 59/C0059M-00, *Specification for Gypsum Casting Plaster and Gypsum Molding Plaster*. West Conshohocken, PA: ASTM International, 2000.

ASTM C 61/C0061M-00, *Specification for Gypsum Keene's Cement*. West Conshohocken, PA: ASTM International, 2000.

ASTM C 79-04a, *Specification for Treated Core and Nontreated Core Gypsum Sheathing Board*. West Conshohocken, PA: ASTM International, 2004.

ASTM C 475/C 475M-02, *Specification for Joint Compound and Joint Tape for Finishing Gypsum Board*. West Conshohocken, PA: ASTM International, 2002.

ASTM C 514-01, *Specification for Nails for the Application of Gypsum Wallboard*. West Conshohocken, PA: ASTM International, 2001.

ASTM C 557-03, *Specification for Adhesives for Fastening Gypsum Wallboard to Wood Framing*. West Conshohocken, PA: ASTM International, 2003.

ASTM C 587-02, *Specification for Gypsum Veneer Plaster*. West Conshohocken, PA: ASTM International, 2002.

ASTM C 588/C588M-01, *Specification for Gypsum Base for Veneer Plasters*. West Conshohocken, PA: ASTM International, 2001.

ASTM C 630/0630M-03, *Specification for Water-resistant Gypsum Backing Board*. West Conshohocken, PA: ASTM International, 2003.

ASTM C 631-(2000), *Specification for Bonding Compounds for Interior Gypsum Plastering*. West Conshohocken, PA: ASTM International, 2000.

ASTM C 645-04, *Specification for Nonstructural Steel Framing Members*. West Conshohocken, PA: ASTM International, 2004.

ASTM C 843-99e01, *Specification for Application of Gypsum Veneer Plaster*. West Conshohocken, PA: ASTM International, 1999

ASTM C 844-99, *Specification for Application of Gypsum Base to Receive Gypsum Veneer Plaster*. West Conshohocken, PA: ASTM International, 1999.

ASTM C 847-(2000), *Specification for Metal Lath*. West Conshohocken, PA: ASTM International, 2000.

ASTM C 897-00, *Specification for Aggregate for Job-Mixed Portland Cement-Based Plasters*. West Conshohocken, PA: ASTM International, 2000.

ASTM C 926-98a, *Specification for Application of Portland Cement Based-Plaster*. West Conshohocken, PA: ASTM International, 1998.

ASTM C 933-04, *Specification for Welded Wire Lath*. West Conshohocken, PA: ASTM International, 2004.

ASTM C 954-00, *Specification for Steel Drill Screws for the Application of Gypsum Panel Products or Metal Plaster Bases to Steel Studs From 0.033 in. (0.84 mm) to 0.112 in. (2.84 mm) in Thickness*. West Conshohocken, PA: ASTM International, 2000.

ASTM C 955-03, *Specification for Load-bearing Transverse and Axial Steel Studs, Runner Tracks, and Bracing or Bridging for Screw Application of Gypsum Panel Products and Metal Plaster Bases*. West Conshohocken, PA: ASTM International, 2003.

ASTM C 960/C 960M-04, *Specification for Predecorated Gypsum Board*. West Conshohocken, PA: ASTM International, 2004.

ASTM C 1002-01, *Specification for Steel Drill Screws for the Application of Gypsum Panel Products or Metal Plaster Bases*. West Conshohocken, PA: ASTM International, 2001.

ASTM C 1032-04, *Specification for Woven Wire Plaster Base*. West Conshohocken, PA: ASTM International, 2004.

ASTM C 1047-99, *Specification for Accessories for Gypsum Wallboard and Gypsum Veneer Base*. West Conshohocken, PA: ASTM International, 1999.

ASTM C 1063-03, *Specification for Installation of Lathing and Furring to Receive Interior and Exterior Portland Cement-Based Plaster*. West Conshohocken, PA: ASTM International, 2003.

ASTM C 1177/C 1177M-04, *Specification for Glass Mat Gypsum Substrate for Use as Sheathing*. West Conshohocken, PA: ASTM International, 2004.

ASTM C 1178/C 1178M-04, *Specification for Glass Mat Water–resistant Gypsum Backing Panel*. West Conshohocken, PA: ASTM International, 2004.

ASTM C 1278/C 1278M-03, *Specification for Fiber–reinforced Gypsum Panels*. West Conshohocken, PA: ASTM International, 2003.

ASTM C 1395/C 1395M-04, *Specification for Gypsum Ceiling Board*. West Conshohocken, PA: ASTM International, 2004.

ASTM D 226–97a, *Specification for Asphalt–saturated Organic Felt Used in Roofing and Waterproofing*. West Conshohocken, PA: ASTM International, 1997.

ASTM D 3679-04, *Specification for Rigid Poly (Vinyl Chloride) (PVC) Siding*. West Conshohocken, PA: ASTM International, 2004.

ASTM D 4756-03, *Practice for the Installation of Rigid Poly (Vinyl-Chloride) (PVC) Siding and Soffits*. West Conshohocken, PA: ASTM International, 2003.

ASTM E 331-00, *Test Method for Water Penetration of Exterior Windows, Skylights, Doors and Curtain Walls by Uniform Static Air Pressure Difference*. West Conshohocken, PA: ASTM International, 2000.

CSSB-97, *Grading and Packing Rules for Western Red Cedar Shakes and Western Red Shingles of the Cedar Shake and Shingle Bureau*. Bellevue, WA: Cedar Shake and Shingle Bureau, 1997.

Chapter 8:
Roof-ceiling Construction

General Comments

Section R801 establishes the scope of the chapter as well as performance requirements for roof and ceiling construction. There are two roof-ceiling framing systems: wood framing and steel framing. Section R802 addresses wood roof and ceiling framing, while Section R804 deals with steel roof and ceiling framing. Section R803 specifies requirements for roof sheathing to be used with either of these framing systems. Other topics covered in this chapter are the application of ceiling finishes in Section R805, the proper ventilation of concealed spaces in roofs (e.g., enclosed attics and rafter spaces) in Section R806, the "access into attic" criteria in Section R807 and the proper clearance of combustible insulation from heat-producing devices in Section R808.

Purpose

Chapter 8 regulates the design and construction of roof-ceiling systems. Proper roof-member design provides for the support of the required design live and snow loads as well as providing for the transfer of these loads to supporting walls. Attics must be identified as "no storage" or "limited storage" zones so that ceiling joists can be appropriately sized. The allowable span tables are formatted to simplify the selection of rafter and ceiling joist size.

Roof systems must resist wind uplift, and tiedowns must be installed when necessary. Also, the roof system comprised of sheathing fastened to either wood or steel framing serves as a diaphragm, which is a key lateral load path element in resisting the forces of wind and earthquakes.

This chapter also includes roof ventilation requirements, aimed at minimizing the adverse effects of moisture. The requirements for insulation clearance protect against combustible insulation coming into contact with a heat source.

SECTION R801
GENERAL

R801.1 Application. The provisions of this chapter shall control the design and construction of the roof-ceiling system for all buildings.

❖ The provisions of Chapter 8 address conventionally framed roof-ceiling construction.

R801.2 Requirements. Roof and ceiling construction shall be capable of accommodating all loads imposed according to Section R301 and of transmitting the resulting loads to the supporting structural elements.

❖ This section states the performance expectations for roof and ceiling construction.

R801.3 Roof drainage. In areas where expansive or collapsible soils are known to exist, all dwellings shall have a controlled method of water disposal from roofs that will collect and discharge roof drainage to the ground surface at least 5 feet (1524 mm) from foundation walls or to an approved drainage system.

❖ Saturated expansive or collapsible soils can lead to foundation failures because their additional loads are imposed on the foundation wall. To minimize the potential for the soil adjacent to the foundation wall to become saturated by roof drainage, the code requires that the roofs drain 5 feet (1524 mm) from the foundation. This requirement applies to foundation walls only and does not apply to slab-on-grade foundations.

SECTION R802
WOOD ROOF FRAMING

R802.1 Identification. Load-bearing dimension lumber for rafters, trusses and ceiling joists shall be identified by a grade mark of a lumber grading or inspection agency that has been approved by an accreditation body that complies with DOC PS 20. In lieu of a grade mark, a certificate of inspection issued by a lumber grading or inspection agency meeting the requirements of this section shall be accepted.

❖ See the commentary for Section R502.1 and Figure R502.1.

R802.1.1 Blocking. Blocking shall be a minimum of utility grade lumber.

❖ Because blocking is a less demanding task than load carrying, the code permits the use of utility grade lumber.

R802.1.2 End-jointed lumber. Approved end-jointed lumber identified by a grade mark conforming to Section R802.1 may be used interchangeably with solid-sawn members of the same species and grade.

❖ See the commentary for Section R502.1.3.

R802.1.3 Fire-retardant-treated wood. Fire-retardant-treated wood (FRTW) is any wood product which, when impregnated with chemicals by a pressure process or other means during manufacture, shall have, when tested in accordance with ASTM E 84, a listed flame spread index of 25 or less and shows no evidence of significant progressive combustion when the test is continued for an additional 20-minute period. In addition, the flame front shall not progress more than

10.5 feet (3200 mm) beyond the center line of the burners at any time during the test.

❖ Fire-retardant-treated wood (FRTW) is plywood and lumber that has been pressure impregnated with chemicals to improve its flame spread characteristics beyond those of untreated wood. The effectiveness of the pressure-impregnated fire retardant treatment is determined by subjecting the material to tests conducted in accordance with ASTM E 84, with the modification that the test is extended to 30 minutes rather than 15 minutes. Using this procedure, a flame spread index is established during the standard 10-minute test period. The test is continued for an additional 20 minutes. During this added time period, there must not be any significant flame spread. At no time must the flame spread more than $10^1/_2$ feet (3200 mm) past the centerline of the burners.

The result of impregnating wood with fire retardant chemicals is a chemical reaction at certain temperature ranges. This reaction reduces the release of certain intermediate products that contribute to the flaming of wood, and the reaction also results in the formation of a greater percentage of charcoal and water. Some chemicals are effective in reducing the oxidation rate for charcoal residue. Fire retardant chemicals also reduce the heat release rate of FRTW when it is burning over a wide range of temperatures. This section gives provisions for the treatment and use of FRTW.

R802.1.3.1 Labeling. Fire-retardant-treated lumber and wood structural panels shall be labeled. The label shall contain:

1. The identification mark of an approved agency in accordance with Section 1703.5 of the *International Building Code.*

2. Identification of the treating manufacturer.

3. The name of the fire-retardant treatment.

4. The species of wood treated.

5. Flame spread and smoke-developed rating.

6. Method of drying after treatment.

7. Conformance to appropriate standards in accordance with Sections R802.1.3.2 through R802.1.3.5.

8. For FRTW exposed to weather, or a damp or wet location, the words "No increase in the listed classification when subjected to the Standard Rain Test" (ASTM D 2898).

❖ For continued quality, each piece of fire-retardant-treated wood must be identified by an approved agency having a reinspection service. The identification must show the performance rating of the material, including the 30-minute ASTM E 84 test results determined in Section R802.1.3 and the design adjustment values determined in Section R802.1.3.2. The third-party agency that provides the FRTW label is also required to state on the label that the FRTW complies with the requirements of Section R802.1.3 and that design adjustment values have been determined for the FRTW in compliance with the provisions of Section R802.1.3.2.

The FRTW label must be distinct from the grading label to avoid confusion between the two. The grading label provides information about the properties of wood before it is fire-retardant treated; the FRTW label shows properties of the wood after FRTW treatment. It is imperative that the FRTW label be presented in a manner that complements the grading label and does not create confusion over which label takes precedence.

R802.1.3.2 Strength adjustments. Design values for untreated lumber and wood structural panels as specified in Section R802.1 shall be adjusted for fire-retardant-treated wood. Adjustments to design values shall be based upon an approved method of investigation which takes into consideration the effects of the anticipated temperature and humidity to which the fire-retardant-treated wood will be subjected, the type of treatment and redrying procedures.

❖ Experience has shown that certain factors can affect the physical properties of FRTW. Among these factors are the pressure treatment and redrying processes used and the extremes of temperature and humidity that the FRTW will be subjected to once installed. The design values for all FRTW must be adjusted for the effects of the treatment and environmental conditions, such as high temperature and humidity in attic installations. This section requires the determination of these design adjustment values, based on an investigation procedure that includes subjecting the FRTW to similar temperatures and humidities and that has been approved by the code official. The FRTW tested must be identical to that which is produced. Items to be considered by the code official reviewing the test procedure include species and grade of the untreated wood and conditioning of the wood, such as drying before the fire retardant treatment process. A fire-retardant wood treater may choose to have its treatment process evaluated by model code evaluation services.

The FRTW is required by Section R802.1.3.1 to be labeled with the design adjustment values. These can take the form of factors that are multiplied by the original design values of the untreated wood to determine its allowable stresses or new allowable stresses that have already been factored down in consideration of the FRTW treatment.

R802.1.3.2.1 Wood structural panels. The effect of treatment and the method of redrying after treatment, and exposure to high temperatures and high humidities on the flexure properties of fire-retardant-treated softwood plywood shall be determined in accordance with ASTM D 5516. The test data developed by ASTM D 5516 shall be used to develop adjustment factors, maximum loads and spans, or both for untreated plywood design values in accordance with ASTM D 6305. Each manufacturer shall publish the allowable maximum loads and spans for service as floor and roof sheathing for their treatment.

❖ This section references the test standard developed to evaluate the flexural properties of fire-retardant-treated plywood that is exposed to high temperatures. Note that while the section title refers to wood structural panels, the standard is specifically for softwood plywood. Therefore, judgment is required in determining the effects of elevated temperature and humidity on other types of wood structural panels.

R802.1.3.2.2 Lumber. For each species of wood treated, the effect of the treatment and the method of redrying after treatment and exposure to high temperatures and high humidities on the allowable design properties of fire-retardant-treated lumber shall be determined in accordance with ASTM D 5664. The test data developed by ASTM D 5664 shall be used to develop modification factors for use at or near room temperature and at elevated temperatures and humidity in accordance with ASTM D 6841. Each manufacturer shall publish the modification factors for service at temperatures of not less than 80°F (27°C) and for roof framing. The roof framing modification factors shall take into consideration the climatological location.

❖ This section references the test standards developed to determine the necessary adjustments to design values for lumber that has been fire-retardant treated and includes the effects of elevated temperatures and humidity.

R802.1.3.3 Exposure to weather. Where fire-retardant-treated wood is exposed to weather or damp or wet locations, it shall be identified as "Exterior" to indicate there is no increase in the listed flame spread index as defined in Section R802.1.3 when subjected to ASTM D 2898.

❖ Some fire-retardant treatments are soluble when exposed to the weather or used under high-humidity conditions. The humidity threshold established for interior applications in Section R802.1.3.4 is 92 percent. Therefore FRTW used in an interior location that will exceed this threshold must comply with this section. When a FRTW product is to be exposed to any of the conditions noted in this section, it must be further tested in accordance with ASTM D 2898. Testing requires the material to meet the performance criteria listed in Section R802.1.3. The material is then subjected to the ASTM weathering test and retested after drying. There must not be any significant differences in the performance recorded before and after the weathering test.

R802.1.3.4 Interior applications. Interior fire-retardant-treated wood shall have a moisture content of not over 28 percent when tested in accordance with ASTM D 3201 procedures at 92 percent relative humidity. Interior fire-retardant-treated wood shall be tested in accordance with Section R802.1.3.2.1 or R802.1.3.2.2. Interior fire-retardant-treated wood designated as Type A shall be tested in accordance with the provisions of this section.

❖ The environment in which the FRTW is used can affect its performance. To make sure that performance will be adequate in a humid interior condition, the testing in accordance with ASTM D 3201 as well as testing specified in Sections R802.1.3.2.1 or R802.1.3.2.2 is required for all interior wood. Requiring all interior wood to be tested for the effects of high temperature and humidity reduces the chance of a premature failure resulting from improper use of FRTW.

R802.1.3.5 Moisture content. Fire-retardant-treated wood shall be dried to a moisture content of 19 percent or less for lumber and 15 percent or less for wood structural panels before use. For wood kiln dried after treatment (KDAT) the kiln temperatures shall not exceed those used in kiln drying the lumber and plywood submitted for the tests described in Section R802.1.3.2.1 for plywood and R802.1.3.2.2 for lumber.

❖ These moisture content thresholds are necessary to prevent leaching of the fire retardant from the wood. Section R802.1.3.2 requires that the strength adjustments consider the redrying procedure, and this section clarifies that the drying temperatures for wood kiln-dried after treatment must be consistent with those on which the adjustment factors were based.

R802.1.4 Structural glued laminated timbers. Glued laminated timbers shall be manufactured and identified as required in AITC A190.1 and ASTM D 3737.

❖ This section states that glulam timbers must be manufactured following ANSI/AITC 90.1 and ASTM D 3737. Knowing the standard these products must meet makes it easier to determine that the product found in the field will meet the design requirements.

R802.1.5 Structural log members. Stress grading of structural log members of nonrectangular shape, as typically used in log buildings, shall be in accordance with ASTM D 3957. Such structural log members shall be identified by the grade mark of an approved lumber grading or inspection agency. In lieu of a grade mark on the material, a certificate of inspection as to species and grade issued by a lumber-grading or inspection agency meeting the requirements of this section shall be permitted to be accepted.

❖ This section addresses grading requirements for logs used as structural members. This subsection specifies the reference for acceptable methods for establishing structural capacities of logs and specifies the requirement for a grading stamp or alternate certification on structural logs. Structural log members must be graded in accordance with ASTM D 3957.

R802.2 Design and construction. The framing details required in Section R802 apply to roofs having a minimum slope of three units vertical in 12 units horizontal (25-percent slope) or greater. Roof-ceilings shall be designed and constructed in accordance with the provisions of this chapter and Figures R606.11(1), R606.11(2) and R606.11(3) or in accordance with AFPA/NDS. Components of roof-ceilings shall be fastened in accordance with Table R602.3(1).

❖ The prescriptive framing requirements of Section R802 apply only to roofs having a minimum slope of 3 units vertical in 12 units horizontal.

Wood roof-ceiling construction must comply with Section R802 or AF&PA/NDS, *National Design Specification for Wood Construction*. The references to Chapter 6 figures are meant for buildings using masonry wall construction.

R802.3 Framing details. Rafters shall be framed to ridge board or to each other with a gusset plate as a tie. Ridge board shall be at least 1-inch (25 mm) nominal thickness and not less in depth than the cut end of the rafter. At all valleys and hips there shall be a valley or hip rafter not less than 2-inch (51 mm) nominal thickness and not less in depth than the cut end of the rafter. Hip and valley rafters shall be supported at the ridge by a brace to a bearing partition or be designed to carry and distribute the specific load at that point. Where the roof pitch is less than three units vertical in 12 units horizontal (25-percent slope), structural members that support rafters and ceiling joists, such as ridge beams, hips and valleys, shall be designed as beams.

❖ Traditional practice is to provide a ridgeboard between opposite rafters as a nailing base and to provide a full bearing for the rafter. Rafters must be placed directly opposite each other, and the ridgeboard must have a depth equal to the end of the rafter as illustrated in Commentary Figure R802.3(1). Commentary Figure R802.3(2) shows an option for framing opposing rafters using a gusset plate.

R802.3.1 Ceiling joist and rafter connections. Ceiling joists and rafters shall be nailed to each other in accordance with Table R802.5.1(9), and the rafter shall be nailed to the top wall plate in accordance with Table R602.3(1). Ceiling joists shall be continuous or securely joined in accordance with Table R802.5.1(9) where they meet over interior partitions and are nailed to adjacent rafters to provide a continuous tie across the building when such joists are parallel to the rafters.

Where ceiling joists are not connected to the rafters at the top wall plate, joists connected higher in the attic shall be installed as rafter ties, or rafter ties shall be installed to provide a continuous tie. Where ceiling joists are not parallel to rafters, rafter ties shall be installed. Rafter ties shall be a minimum of 2-inch

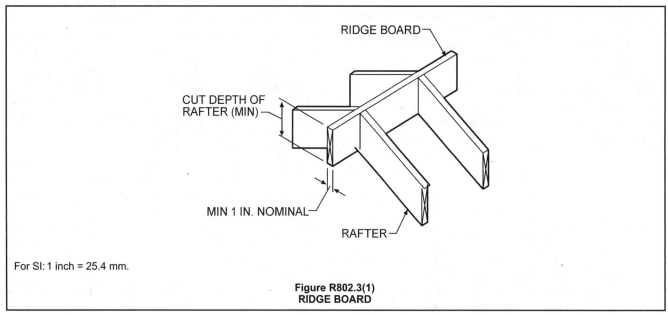

For SI: 1 inch = 25.4 mm.

Figure R802.3(1)
RIDGE BOARD

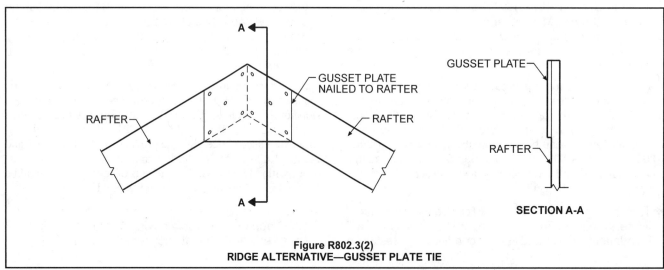

Figure R802.3(2)
RIDGE ALTERNATIVE—GUSSET PLATE TIE

by 4-inch (51 mm by 102 mm) (nominal), installed in accordance with the connection requirements in Table R802.5.1(9), or connections of equivalent capacities shall be provided. Where ceiling joists or rafter ties are not provided, the ridge formed by these rafters shall be supported by a wall or girder designed in accordance with accepted engineering practice.

Collar ties or ridge straps to resist wind uplift shall be connected in the upper third of the attic space in accordance with Table R602.3(1).

Collar ties shall be a minimum of 1-inch by 4-inch (25 mm by 102 mm) (nominal), spaced not more than 4 feet (1219 mm) on center.

❖ The requirements of this Section describe connections to resist horizontal thrust from gravity loads and ridge uplift from wind loads. Ceiling joists or rafter ties are the framing members used to resist the horizontal thrust from gravity loads. A ceiling joist or rafter tie is required at every rafter. Collar ties or ridge straps are used to resist the ridge uplift caused by the wind load. Collar ties or ridge straps must be spaced not further than every 4 feet (1219 mm) on center.

Ceiling joists located at the top plate and parallel to the rafters must be connected as shown in Commentary Figure R802.3.1(1). The fastener schedule of Table R602.3(1) applies only if the roof live load is 20 pounds per square (958 Pa) foot or less. For a roof live load greater than 20 pounds per square foot (958 Pa), the connection must be in accordance with Table R802.5.1(9).

Ceiling joists located at the top plate and parallel to the rafters must be continuous or lap joined as shown in Commentary Figures R802.3.1(3) and R802.3.2(1). The fastener schedule of Table R602.3(1) applies only if the roof live load is 20 pounds per square foot or less (958 Pa). For a roof live load greater than 20 pounds per square foot (958 Pa), the connection must be in accordance with Table R802.5.1(9).

Where ceiling joists or rafter ties are located above the top plate or the ceiling joists are not parallel to the rafters, as shown in Commentary Figure R802.3.1(2), the connections must be in accordance with Table R802.5.1(9).

Cathedral ceilings (where ceiling joists or rafter ties are not installed) must be designed so the wall or ridge beam carries the full load of the roof. This would require the walls to be supported by a continuous foundation and/or beams and girders. The ridge beam must also be capable of supporting the full load exerted by the tributary area of the rafters. The ridge beam must be stiff enough to minimize the deflection from the rafter loads so that rafter thrust does not displace the walls.

Collar ties or ridge straps must be located in the upper $^1/_3$ of the attic space, be a minimum of 1 inch by 4 inches (25 mm by 102 mm), be spaced not more than 4 feet (1219 mm) on center and connected in accordance with Table R602.3(1). The minimum prescriptive connections

in Table R602.3(1) for collar ties and ridge straps are based on minimum connection requirements using the 2001 Wood Frame Construction Manual for slopes greater than 3:12, at a windspeed of 100 mph (45 m/s) or less, for a roof span of 36 feet (10 973 mm) or less [see Commentary Figures R802.3.1(2) and R802.3.1(3)].

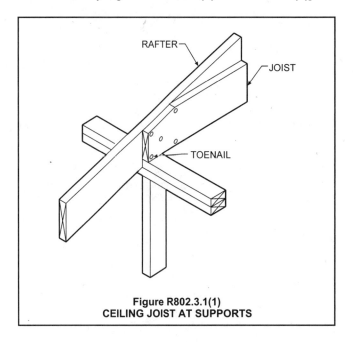

Figure R802.3.1(1)
CEILING JOIST AT SUPPORTS

R802.3.2 Ceiling joists lapped. Ends of ceiling joists shall be lapped a minimum of 3 inches (76 mm) or butted over bearing partitions or beams and toenailed to the bearing member. When ceiling joists are used to provide resistance to rafter thrust, lapped joists shall be nailed together in accordance with Table R602.3(1) and butted joists shall be tied together in a manner to resist such thrust.

❖ Where ceiling joists function as a structural element, providing resistance to rafter thrust, proper connection through a lapped joint or butted joint connection to form a continuous tie is required [see Commentary Figures R802.3.2(1) and R802.3.2(2)].

The lap splice nailing of Table R602.3(1) applies only if the roof live load is 20 pounds per square foot (958 Pa) or less. For a roof live load greater than 20 pounds per square foot (958 Pa), the connection must be in accordance with Note e of Table R802.5.1(9). See the Commentary for Section R802.3.1.

R802.4 Allowable ceiling joist spans. Spans for ceiling joists shall be in accordance with Tables R802.4(1) and R802.4(2). For other grades and species and for other loading conditions, refer to the AF&PA Span Tables for Joists and Rafters.

❖ Tables R802.4(1) and R802.4(2) list allowable ceiling joist spans for common lumber sizes, species and grades based on spacing and design loads. These tables are similar to the rafter tables explained in the commentary for Section R802.5.

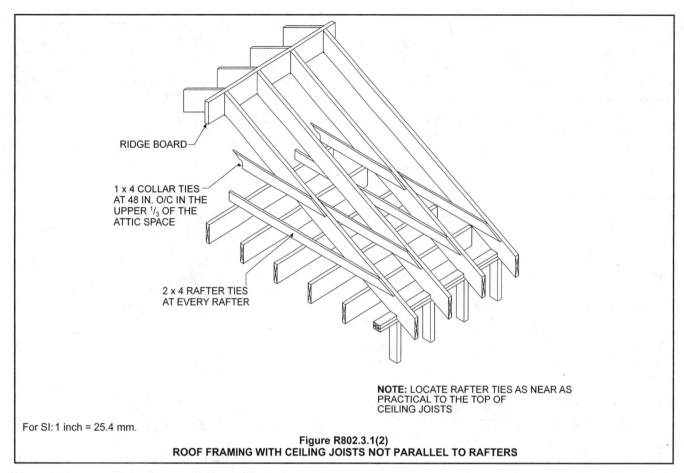

RIDGE BOARD

1 x 4 COLLAR TIES
AT 48 IN. O/C IN THE
UPPER $^1/_3$ OF THE
ATTIC SPACE

2 x 4 RAFTER TIES
AT EVERY RAFTER

NOTE: LOCATE RAFTER TIES AS NEAR AS
PRACTICAL TO THE TOP OF
CEILING JOISTS

For SI: 1 inch = 25.4 mm.

Figure R802.3.1(2)
ROOF FRAMING WITH CEILING JOISTS NOT PARALLEL TO RAFTERS

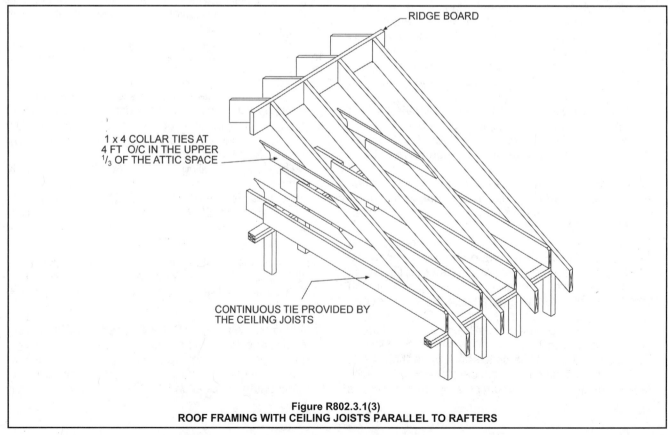

RIDGE BOARD

1 x 4 COLLAR TIES AT
4 FT O/C IN THE UPPER
$^1/_3$ OF THE ATTIC SPACE

CONTINUOUS TIE PROVIDED BY
THE CEILING JOISTS

Figure R802.3.1(3)
ROOF FRAMING WITH CEILING JOISTS PARALLEL TO RAFTERS

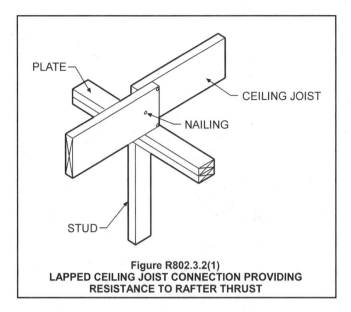

Figure R802.3.2(1)
LAPPED CEILING JOIST CONNECTION PROVIDING
RESISTANCE TO RAFTER THRUST

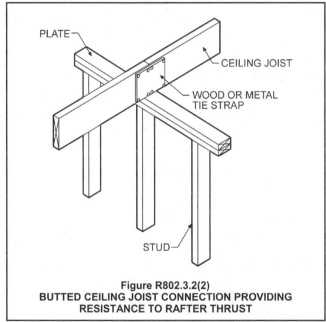

Figure R802.3.2(2)
BUTTED CEILING JOIST CONNECTION PROVIDING
RESISTANCE TO RAFTER THRUST

R802.5 Allowable rafter spans. Spans for rafters shall be in accordance with Tables R802.5.1(1) through R802.5.1(8). For other grades and species and for other loading conditions, refer to the AF&PA Span Tables for Joists and Rafters. The span of each rafter shall be measured along the horizontal projection of the rafter.

❖ Tables R802.5.1(1) through R802.5.1(8) list allowable rafter spans for common lumber sizes, species and grades based on spacing and design loads. Tables R802.5.1(2) through R802.5.1(8) are based on ground snow loads of 30 pounds per square foot (1.44 kPa), 50 pounds per square foot (2.39 kPa) and 70 pounds per square foot (3.35 kPa). The snow loads are based on ASCE 7-95 criteria as follows:

for balanced condition $p_s = Cs \cdot p_f = 0.7 \cdot Ce \cdot Ct \cdot Cs \cdot p_g = 0.77 \cdot p_g$ where $Ce = 1.9$, $Ct = 1.1$, and $Cs = 1.0$.

for unbalanced condition $p_s = 1.3\ p_f/Ce = 0.7 \cdot 1.3 \cdot Ct \cdot Cs \cdot p_g = 1.00 \cdot p_g$ where $Ct = 1.1$ and $Cs = 1.0$.

The unbalanced condition controls. By inspection, the unbalanced condition is equal to the ground snow load.

Example 1 in the commentary for Table R802.5.1(1) explains the allowable spans for clear-span roof rafters using the code tables. As illustrated in Figure R802.5.1, the horizontal projection of the rafter must be used for the span length rather than using the true length of the rafter measured along the slope of the roof.

R802.5.1 Purlins. Installation of purlins to reduce the span of rafters is permitted as shown in Figure R802.5.1. Purlins shall be sized no less than the required size of the rafters that they support. Purlins shall be continuous and shall be supported by 2-inch by 4-inch (51 mm by 102 mm) braces installed to bearing walls at a slope not less than 45 degrees from the horizontal. The braces shall be spaced not more than 4 feet (1219 mm) on center and the unbraced length of braces shall not exceed 8 feet (2438 mm).

❖ This sections contains specific instructions for the installation of purlins that will provide intermediate support for the rafters and thus reduce the rafter span. The purlins must be at least the same size as the rafters which they support. The rafters must bear on the purlin to provide proper support for the rafters. See Commentary Figure R802.5.1(1) for an example of typical installation of purlins. Example 2 in the commentary for Table R802.5.1(1) shows the effect on rafter design and roof purlin requirements when braces to interior bearing walls support the roof purlins.

R802.6 Bearing. The ends of each rafter or ceiling joist shall have not less than $1^1/_2$ inches (38 mm) of bearing on wood or metal and not less than 3 inches (76 mm) on masonry or concrete.

❖ The requirements for roof rafters and ceiling joists bearing on wood, metal and masonry are similar to the requirements for floor joists. See the commentary for Section R502.6 relative to the bearing requirements for wood, metal and masonry.

R802.6.1 Finished ceiling material. If the finished ceiling material is installed on the ceiling prior to the attachment of the ceiling to the walls, such as in construction at a factory, a compression strip of the same thickness as the finish ceiling material shall be installed directly above the top plate of bearing walls if the compressive strength of the finish ceiling material is less than the loads it will be required to withstand. The compression strip shall cover the entire length of such top plate and shall be at least one-half the width of the top plate. It shall be of material capable of transmitting the loads transferred through it.

❖ In prefabricated (panelized) construction where the finished ceiling material is attached to the roof assembly before it is attached to the walls, care must be taken to provide a compression strip to line up with the top plate of the supporting walls in order to have an adequate bearing surface.

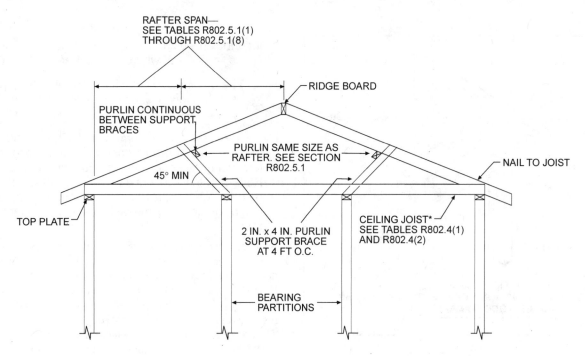

For SI: 1 inch = 25.4 mm, 1 foot = 304.8 mm, 1 degree = 0.018 rad.
Note: Where ceiling joints run perpendicular to the rafters, rafter ties shall be nailed to each rafter near the top of the ceiling joist.

FIGURE R802.5.1
BRACED RAFTER CONSTRUCTION

❖ This figure illustrates the use of continuous purlins to reduce rafter spans as described in Section R802.5.1.

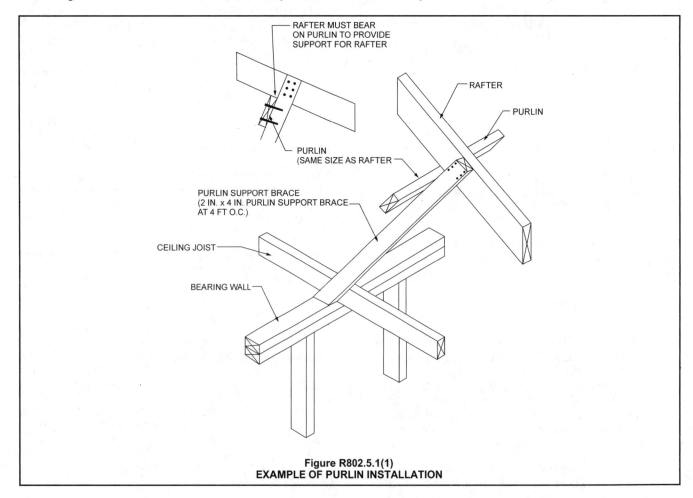

Figure R802.5.1(1)
EXAMPLE OF PURLIN INSTALLATION

TABLE R802.4(1)
CEILING JOIST SPANS FOR COMMON LUMBER SPECIES
(Uninhabitable attics without storage, live load = 10 psf, L/Δ = 240)

CEILING JOIST SPACING (inches)	SPECIES AND GRADE		DEAD LOAD = 5 psf			
			2 × 4	2 × 6	2 × 8	2 × 10
			Maximum ceiling joist spans			
			(feet - inches)	(feet - inches)	(feet - inches)	(feet - inches)
12	Douglas fir-larch	SS	13-2	20-8	Note a	Note a
	Douglas fir-larch	#1	12-8	19-11	Note a	Note a
	Douglas fir-larch	#2	12-5	19-6	25-8	Note a
	Douglas fir-larch	#3	10-10	15-10	20-1	24-6
	Hem-fir	SS	12-5	19-6	25-8	Note a
	Hem-fir	#1	12-2	19-1	25-2	Note a
	Hem-fir	#2	11-7	18-2	24-0	Note a
	Hem-fir	#3	10-10	15-10	20-1	24-6
	Southern pine	SS	12-11	20-3	Note a	Note a
	Southern pine	#1	12-8	19-11	Note a	Note a
	Southern pine	#2	12-5	19-6	25-8	Note a
	Southern pine	#3	11-6	17-0	21-8	25-7
	Spruce-pine-fir	SS	12-2	19-1	25-2	Note a
	Spruce-pine-fir	#1	11-10	18-8	24-7	Note a
	Spruce-pine-fir	#2	11-10	18-8	24-7	Note a
	Spruce-pine-fir	#3	10-10	15-10	20-1	24-6
16	Douglas fir-larch	SS	11-11	18-9	24-8	Note a
	Douglas fir-larch	#1	11-6	18-1	23-10	Note a
	Douglas fir-larch	#2	11-3	17-8	23-0	Note a
	Douglas fir-larch	#3	9-5	13-9	17-5	21-3
	Hem-fir	SS	11-3	17-8	23-4	Note a
	Hem-fir	#1	11-0	17-4	22-10	Note a
	Hem-fir	#2	10-6	16-6	21-9	Note a
	Hem-fir	#3	9-5	13-9	17-5	21-3
	Southern pine	SS	11-9	18-5	24-3	Note a
	Southern pine	#1	11-6	18-1	23-1	Note a
	Southern pine	#2	11-3	17-8	23-4	Note a
	Southern pine	#3	10-0	14-9	18-9	22-2
	Spruce-pine-fir	SS	11-0	17-4	22-10	Note a
	Spruce-pine-fir	#1	10-9	16-11	22-4	Note a
	Spruce-pine-fir	#2	10-9	16-11	22-4	Note a
	Spruce-pine-fir	#3	9-5	13-9	17-5	21-3
19.2	Douglas fir-larch	SS	11-3	17-8	23-3	Note a
	Douglas fir-larch	#1	10-10	17-0	22-5	Note a
	Douglas fir-larch	#2	10-7	16-7	21-0	25-8
	Douglas fir-larch	#3	8-7	12-6	15-10	19-5
	Hem-fir	SS	10-7	16-8	21-11	Note a
	Hem-fir	#1	10-4	16-4	21-6	Note a
	Hem-fir	#2	9-11	15-7	20-6	25-3
	Hem-fir	#3	8-7	12-6	15-10	19-5
	Southern -pine	SS	11-0	17-4	22-10	Note a
	Southern pine	#1	10-10	17-0	22-5	Note a
	Southern pine	#2	10-7	16-8	21-11	Note a
	Southern pine	#3	9-1	13-6	17-2	20-3
	Spruce-pine-fir	SS	10-4	16-4	21-6	Note a
	Spruce-pine-fir	#1	10-2	15-11	21-0	25-8
	Spruce-pine-fir	#2	10-2	15-11	21-0	25-8
	Spruce-pine-fir	#3	8-7	12-6	15-10	19-5

(continued)

TABLE R802.4(1)—continued
CEILING JOIST SPANS FOR COMMON LUMBER SPECIES
(Uninhabitable attics without storage, live load = 10 psf, L/Δ = 240)

CEILING JOIST SPACING (inches)	SPECIES AND GRADE		DEAD LOAD = 5 psf			
			2 × 4	2 × 6	2 × 8	2 × 10
			Maximum ceiling joist spans			
			(feet - inches)	(feet - inches)	(feet - inches)	(feet - inches)
24	Douglas fir-larch	SS	10-5	16-4	21-7	Note a
	Douglas fir-larch	#1	10-0	15-9	20-1	24-6
	Douglas fir-larch	#2	9-10	14-10	18-9	22-11
	Douglas fir-larch	#3	7-8	11-2	14-2	17-4
	Hem-fir	SS	9-10	15-6	20-5	Note a
	Hem-fir	#1	9-8	15-2	19-7	23-11
	Hem-fir	#2	9-2	14-5	18-6	22-7
	Hem-fir	#3	7-8	11-2	14-2	17-4
	Southern pine	SS	10-3	16-1	21-2	Note a
	Southern pine	#1	10-0	15-9	20-10	Note a
	Southern pine	#2	9-10	15-6	20-1	23-11
	Southern pine	#3	8-2	12-0	15-4	18-1
	Spruce-pine-fir	SS	9-8	15-2	19-11	25-5
	Spruce-pine-fir	#1	9-5	14-9	18-9	22-11
	Spruce-pine-fir	#2	9-5	14-9	18-9	22-11
	Spruce-pine-fir	#3	7-8	11-2	14-2	17-4

Check sources for availability of lumber in lengths greater than 20 feet.

For SI: 1 inch = 25.4 mm, 1 foot = 304.8 mm, 1 pound per square foot = 0.0479kPa.

a. Span exceeds 26 feet in length.

❖ See the commentary for Section R802.4.

TABLE R802.4(2)
CEILING JOIST SPANS FOR COMMON LUMBER SPECIES
(Uninhabitable attics with limited storage, live load = 20 psf, L/Δ = 240)

CEILING JOIST SPACING (inches)	SPECIES AND GRADE		DEAD LOAD = 10 psf			
			2 × 4	2 × 6	2 × 8	2 × 10
			Maximum ceiling joist spans			
			(feet - inches)	(feet - inches)	(feet - inches)	(feet - inches)
12	Douglas fir-larch	SS	10-5	16-4	21-7	Note a
	Douglas fir-larch	#1	10-0	15-9	20-1	24-6
	Douglas fir-larch	#2	9-10	14-10	18-9	22-11
	Douglas fir-larch	#3	7-8	11-2	14-2	17-4
	Hem-fir	SS	9-10	15-6	20-5	Note a
	Hem-fir	#1	9-8	15-2	19-7	23-11
	Hem-fir	#2	9-2	14-5	18-6	22-7
	Hem-fir	#3	7-8	11-2	14-2	17-4
	Southern pine	SS	10-3	16-1	21-2	Note a
	Southern pine	#1	10-0	15-9	20-10	Note a
	Southern pine	#2	9-10	15-6	20-1	23-11
	Southern pine	#3	8-2	12-0	15-4	18-1
	Spruce-pine-fir	SS	9-8	15-2	19-11	25-5
	Spruce-pine-fir	#1	9-5	14-9	18-9	22-11
	Spruce-pine-fir	#2	9-5	14-9	18-9	22-11
	Spruce-pine-fir	#3	7-8	11-2	14-2	17-4
16	Douglas fir-larch	SS	9-6	14-11	19-7	25-0
	Douglas fir-larch	#1	9-1	13-9	17-5	21-3
	Douglas fir-larch	#2	8-9	12-10	16-3	19-10
	Douglas fir-larch	#3	6-8	9-8	12-4	15-0
	Hem-fir	SS	8-11	14-1	18-6	23-8
	Hem-fir	#1	8-9	13-5	16-10	20-8
	Hem-fir	#2	8-4	12-8	16-0	19-7
	Hem-fir	#3	6-8	9-8	12-4	15-0
	Southern pine	SS	9-4	14-7	19-3	24-7
	Southern pine	#1	9-1	14-4	18-11	23-1
	Southern pine	#2	8-11	13-6	17-5	20-9
	Southern pine	#3	7-1	10-5	13-3	15-8
	Spruce-pine-fir	SS	8-9	13-9	18-1	23-1
	Spruce-pine-fir	#1	8-7	12-10	16-3	19-10
	Spruce-pine-fir	#2	8-7	12-10	16-3	19-10
	Spruce-pine-fir	#3	6-8	9-8	12-4	15-0
19.2	Douglas fir-larch	SS	8-11	14-0	18-5	23-4
	Douglas fir-larch	#1	8-7	12-6	15-10	19-5
	Douglas fir-larch	#2	8-0	11-9	14-10	18-2
	Douglas fir-larch	#3	6-1	8-10	11-3	13-8
	Hem-fir	SS	8-5	13-3	17-5	22-3
	Hem-fir	#1	8-3	12-3	15-6	18-11
	Hem-fir	#2	7-10	11-7	14-8	17-10
	Hem-fir	#3	6-1	8-10	11-3	13-8
	Southern pine	SS	8-9	13-9	18-1	23-1
	Southern pine	#1	8-7	13-6	17-9	21-1
	Southern pine	#2	8-5	12-3	15-10	18-11
	Southern pine	#3	6-5	9-6	12-1	14-4
	Spruce-pine-fir	SS	8-3	12-11	17-1	21-8
	Spruce-pine-fir	#1	8-0	11-9	14-10	18-2
	Spruce-pine-fir	#2	8-0	11-9	14-10	18-2
	Spruce-pine-fir	#3	6-1	8-10	11-3	13-8

(continued)

TABLE R802.4(2)—continued
CEILING JOIST SPANS FOR COMMON LUMBER SPECIES
(Uninhabitable attics with limited storage, live load = 20 psf, L/Δ = 240)

CEILING JOIST SPACING (inches)	SPECIES AND GRADE		DEAD LOAD = 10 psf			
			2 × 4	2 × 6	2 × 8	2 × 10
			Maximum Ceiling Joist Spans			
			(feet - inches)	(feet - inches)	(feet - inches)	(feet - inches)
24	Douglas fir-larch	SS	8-3	13-0	17-1	20-11
	Douglas fir-larch	#1	7-8	11-2	14-2	17-4
	Douglas fir-larch	#2	7-2	10-6	13-3	16-3
	Douglas fir-larch	#3	5-5	7-11	10-0	12-3
	Hem-fir	SS	7-10	12-3	16-2	20-6
	Hem-fir	#1	7-6	10-11	13-10	16-11
	Hem-fir	#2	7-1	10-4	13-1	16-0
	Hem-fir	#3	5-5	7-11	10-0	12-3
	Southern pine	SS	8-1	12-9	16-10	21-6
	Southern pine	#1	8-0	12-6	15-10	18-10
	Southern pine	#2	7-8	11-0	14-2	16-11
	Southern pine	#3	5-9	8-6	10-10	12-10
	Spruce-pine-fir	SS	7-8	12-0	15-10	19-5
	Spruce-pine-fir	#1	7-2	10-6	13-3	16-3
	Spruce-pine-fir	#2	7-2	10-6	13-3	16-3
	Spruce-pine-fir	#3	5-5	7-11	10-0	12-3

Check sources for availability of lumber in lengths greater than 20 feet.

For SI: 1 inch = 25.4 mm, 1 foot = 304.8 mm, 1 pound per square foot = 0.0479 kPa.

a. Span exceeds 26 feet in length.

❖ See the commentary for Section R802.4.

TABLE R802.5.1(1)
RAFTER SPANS FOR COMMON LUMBER SPECIES
(Roof live load=20 psf, ceiling not attached to rafters, L/Δ = 180)

RAFTER SPACING (inches)	SPECIES AND GRADE		DEAD LOAD = 10 psf					DEAD LOAD = 20 psf				
			2 × 4	2 × 6	2 × 8	2 × 10	2 × 12	2 × 4	2 × 6	2 × 8	2 × 10	2 × 12
						Maximum rafter spans[a]						
			(feet - inches)	(feet - inches)	(feet - inches)	(feet - inches)	(feet - inches)	(feet - inches)	(feet - inches)	(feet - inches)	(feet - inches)	(feet - inches)
12	Douglas fir-larch	SS	11-6	18-0	23-9	Note b	Note b	11-6	18-0	23-5	Note b	Note b
	Douglas fir-larch	#1	1-1	17-4	22-5	Note b	Note b	10-6	15-4	19-5	23-9	Note b
	Douglas fir-larch	#2	1-10	16-7	21-0	25-8	Note b	9-10	14-4	18-2	22-3	25-9
	Douglas fir-larch	#3	8-7	12-6	15-10	19-5	22-6	7-5	10-10	13-9	16-9	19-6
	Hem-fir	SS	10-10	17-0	22-5	Note b	Note b	10-10	17-0	22-5	Note b	Note b
	Hem-fir	#1	10 -7	16-8	21-10	Note b	Note b	10-3	14-11	18-11	23-2	Note b
	Hem-fir	#2	10-1	15-11	20-8	25-3	Note b	9-8	14-2	17-11	21-11	25-5
	Hem-fir	#3	8-7	12-6	15-10	19-5	22-6	7-5	10-10	13-9	16-9	19-6
	Southern pine	SS	11-3	17 8	23-4	Note b	Note b	11-3	17-8	23-4	Note b	Note b
	Southern pine	#1	11-1	17-4	22-11	Note b	Note b	11-1	17-3	21-9	25-10	Note b
	Southern pine	#2	10-10	17-0	22-5	Note b	Note b	10-6	15-1	19-5	23-2	Note b
	Southern pine	#3	9-1	13-6	17-2	20-3	24-1	7-11	11-8	14-10	17-6	20-11
	Spruce-pine-fir	SS	10-7	16-8	21-11	Note b	Note b	10-7	16-8	21-9	Note b	Note b
	Spruce-pine-fir	#1	10-4	16-3	21-0	25-8	Note b	9-10	14-4	18-2	22-3	25-9
	Spruce-pine-fir	#2	10-4	16-3	21-0	25-8	Note b	9-10	14-4	18-2	22-3	25-9
	Spruce-pine-fir	#3	8-7	12-6	15-10	19-5	22-6	7-5	10-10	13-9	16-9	19-6
16	Douglas fir-larch	SS	10-5	16-4	21-7	Note b	Note b	10-5	16-0	20-3	24-9	Note b
	Douglas fir-larch	#1	10-0	15-4	19-5	23-9	Note b	9-1	13-3	16-10	20-7	23-10
	Douglas fir-larch	#2	9-10	14-4	18-2	22-3	25-9	8-6	12-5	15-9	19-3	22-4
	Douglas fir-larch	#3	7-5	10-10	13-9	16-9	19-6	6-5	9-5	11-11	14-6	16-10
	Hem-fir	SS	9-10	15-6	20-5	Note b	Note b	9-10	15-6	19-11	24-4	Note b
	Hem-fir	#1	9-8	14-11	18-11	23-2	Note b	8-10	12-11	16-5	20-0	23-3
	Hem-fir	#2	9-2	14-2	17-11	21-11	25-5	8-5	12-3	15-6	18-11	22-0
	Hem-fir	#3	7-5	10-10	13-9	16-9	19-6	6-5	9-5	11-11	14-6	16-10
	Southern pine	SS	10-3	16-1	21-2	Note b	Note b	10-3	16-1	21-2	Note b	Note b
	Southern pine	#1	10-0	15-9	20-10	25-10	Note b	10-0	15-0	18-10	22-4	Note b
	Southern pine	#2	9-10	15-1	19-5	23-2	Note b	9-1	13-0	16-10	20-1	23-7
	Southern pine	#3	7-11	11-8	14-10	17-6	20-11	6-10	10-1	12-10	15-2	18-1
	Spruce-pine-fir	SS	9-8	15-2	19-11	25-5	Note b	9-8	14-10	18-10	23-0	Note b
	Spruce-pine-fir	#1	9-5	14-4	18-2	22-3	25-9	8-6	12-5	15-9	19-3	22-4
	Spruce-pine-fir	#2	9-5	14-4	18-2	22-3	25-9	8-6	12-5	15-9	19-3	22-4
	Spruce-pine-fir	#3	7-5	10-10	13-9	16-9	19-6	6-5	9-5	11-11	14-6	16-10
19.2	Douglas fir-larch	SS	9-10	15-5	20-4	25-11	Note b	9-10	14-7	18-6	22-7	Note b
	Douglas fir-larch	#1	9-5	14-0	17-9	21-8	25-2	8-4	12-2	15-4	18-9	21-9
	Douglas fir-larch	#2	8-11	13-1	16-7	20-3	23-6	7-9	11-4	14-4	17-7	20-4
	Douglas fir-larch	#3	6-9	9-11	12-7	15-4	17-9	5-10	8-7	10-10	13-3	15-5
	Hem-fir	SS	9-3	14-7	19-2	24-6	Note b	9-3	14-4	18-2	22-3	25-9
	Hem-fir	#1	9-1	13-8	17-4	21-1	24-6	8-1	11-10	15-0	18-4	21-3
	Hem-fir	#2	8-8	12-11	16-4	20-0	23-2	7-8	11-2	14-2	17-4	20-1
	Hem-fir	#3	6-9	9-11	12-7	15-4	17-9	5-10	8-7	10-10	13-3	15-5
	Southern pine	SS	9-8	15-2	19-11	25-5	Note b	9-8	15-2	19-11	25-5	Note b
	Southern pine	#1	9-5	14-10	19-7	23-7	Note b	9-3	13-8	17-2	20-5	24-4
	Southern pine	#2	9-3	13-9	17-9	21-2	24-10	8-4	11-11	15-4	18-4	21-6
	Southern pine	#3	7-3	10-8	13-7	16-0	19-1	6-3	9-3	11-9	13-10	16-6
	Spruce-pine-fir	SS	9-1	14-3	18-9	23-11	Note b	9-1	13-7	17-2	21-0	24-4
	Spruce-pine-fir	#1	8-10	13-1	16-7	20-3	23-6	7-9	11-4	14-4	17-7	20-4
	Spruce-pine-fir	#2	8-10	13-1	16-7	20-3	23-6	7-9	11-4	14-4	17-7	20-4
	Spruce-pine-fir	#3	6-9	9-11	12-7	15-4	17-9	5-10	8-7	10-10	13-3	15-5

(continued)

TABLE R802.5.1(1)—continued
RAFTER SPANS FOR COMMON LUMBER SPECIES
(Roof live load=20 psf, ceiling not attached to rafters, L/Δ = 180)

RAFTER SPACING (inches)	SPECIES AND GRADE		DEAD LOAD = 10 psf					DEAD LOAD = 20 psf				
			2 × 4	2 × 6	2 × 8	2 × 10	2 × 12	2 × 4	2 × 6	2 × 8	2 × 10	2 × 12
			Maximum rafter spans[a]									
			(feet - inches)	(feet - inches)	(feet - inches)	(feet - inches)	(feet - inches)	(feet - inches)	(feet - inches)	(feet - inches)	(feet - inches)	(feet - inches)
24	Douglas fir-larch	SS	9-1	14-4	18-10	23-4	Note b	8-11	13-1	16-7	20-3	23-5
	Douglas fir-larch	#1	8-7	12-6	15-10	19-5	22-6	7-5	10-10	13-9	16-9	19-6
	Douglas fir-larch	#2	8-0	11-9	14-10	18-2	21-0	6-11	10-2	12-10	15-8	18-3
	Douglas fir-larch	#3	6-1	8-10	11-3	13-8	15-11	5-3	7-8	9-9	11-10	13-9
	Hem-fir	SS	8-7	13-6	17-10	22-9	Note b	8-7	12-10	16-3	19-10	23-0
	Hem-fir	#1	8-4	12-3	15-6	18-11	21-11	7-3	10-7	13-5	16-4	19-0
	Hem-fir	#2	7-11	11-7	14-8	17-10	20-9	6-10	10-0	12-8	15-6	17-11
	Hem-fir	#3	6-1	8-10	11-3	13-8	15-11	5-3	7-8	9-9	11-10	13-9
	Southern pine	SS	8-11	14-1	18-6	23-8	Note b	8-11	14-1	18-6	22-11	Note b
	Southern pine	#1	8-9	13-9	17-9	21-1	25-2	8-3	12-3	15-4	18-3	21-9
	Southern pine	#2	8-7	12-3	15-10	18-11	22-2	7-5	10-8	13-9	16-5	19-3
	Southern pine	#3	6-5	9-6	12-1	14-4	17-1	5-7	8-3	10-6	12-5	14-9
	Spruce-pine-fir	SS	8-5	13-3	17-5	21-8	25-2	8-4	12-2	15-4	18-9	21-9
	Spruce-pine-fir	#1	8-0	11-9	14-10	18-2	21-0	6-11	10-2	12-10	15-8	18-3
	Spruce-pine-fir	#2	8-0	11-9	14-10	18-2	21-0	6-11	10-2	12-10	15-8	18-3
	Spruce-pine-fir	#3	6-1	8-10	11-3	13-8	15-11	5-3	7-8	9-9	11-10	13-9

Check sources for availability of lumber in lengths greater than 20 feet.

For SI: 1 inch = 25.4 mm, 1 foot = 304.8 mm, 1 pound per square foot = 0.0479 kPa.

a. The tabulated rafter spans assume that ceiling joists are located at the bottom of the attic space or that some other method of resisting the outward push of the rafters on the bearing walls, such as rafter ties, is provided at that location. When ceiling joists or rafter ties are located higher in the attic space, the rafter spans shall be multiplied by the factors given below:

H_C/H_R	Rafter Span Adjustment Factor
1/3	0.67
1/4	0.76
1/5	0.83
1/6	0.90
1/7.5 or less	1.00

where:

H_C = Height of ceiling joists or rafter ties measured vertically above the top of the rafter support walls.

H_R = Height of roof ridge measured vertically above the top of the rafter support walls.

b. Span exceeds 26 feet in length.

❖ Adjustment factors , in Note a, for rafter span tables R802.5(1-8) are limited to cases where the ceiling joists or rafter ties are in the lower third of the attic space. When the ceiling joist or rafter ties are located higher in the attic space, lateral deflection of the rafter below the rafter ties can become excessive and require additional engineering analysis. The following examples assume that the ceiling joists are located at the top plate.

Example 1:

You wish to design a rafter system using Douglas-Fir-Larch #2 lumber for a 20 psf (958 Pa) live load, with rafters spaced 24 inches (610 mm) on center and having a horizontal projection of the rafter span of 14 feet, 6 inches (4420 mm). The rafters do not support a ceiling, and the roof slope is $2^1/_2$ units vertical in 12 units horizontal. Assume the roof dead load is 10 psf (479 Pa).

Solution:

The object is to locate a tabulated span value of 14 feet, 6 inches (4420 mm) or more. Going to Table R802.5.1(1) of the code under the portion headed "DEAD LOAD = 10 psf," note that 2-inch by 8-inch (51 mm by 208 mm) rafters at 24 inches (610 mm) on center will span 14 feet, 10 inches (4521 mm).

Example 2:

To find a substitute for the 2-inch by 8-inch (51 mm by 208 mm) rafters of Example 1, determine the required rafter size for a condition that uses a purlin and 2-inch by 4-inch (51 mm by 102 mm) braces (located at mid span of the rafters) at 4 feet (1219 mm) on center as depicted in Figure R802.5.1. Assume the same loads, rafter spacing, and 14-feet, 6-inch (4420 mm) horizontal rafter projection. Between the exterior wall and ridge, use Hem Fir #3 lumber.

Solution:

The rafter span, identified in Figure R802.5.1, is reduced to 7 feet, 3 inches (2210 mm) by adding the purlin at midspan. In Table R802.5.1(1) of the code, 2-inch by 6-inch (51 mm by 152 mm) rafters at 24 inches (610 mm) on center will span 8 feet, 10 inches (2692 mm).

TABLE R802.5.1(2)
RAFTER SPANS FOR COMMON LUMBER SPECIES
(Roof live load=20 psf, ceiling attached to rafters, L/Δ = 240)

RAFTER SPACING (inches)	SPECIES AND GRADE		DEAD LOAD = 10 psf					DEAD LOAD = 20 psf				
			2 × 4	2 × 6	2 × 8	2 × 10	2 × 12	2 × 4	2 × 6	2 × 8	2 × 10	2 × 12
			Maximum rafter spans[a]									
			(feet-inches)	(feet-inches)	(feet-inches)	(feet-inches)	(feet-inches)	(feet-inches)	(feet-inches)	(feet-inches)	(feet-inches)	(feet-inches)
12	Douglas fir-larch	SS	10-5	16-4	21-7	Note b	Note b	10-5	16-4	21-7	Note b	Note b
	Douglas fir-larch	#1	10-0	15-9	20-10	Note b	Note b	10-0	15-4	19-5	23-9	Note b
	Douglas fir-larch	#2	9-10	15-6	20-5	25-8	Note b	9-10	14-4	18-2	22-3	25-9
	Douglas fir-larch	#3	8-7	12-6	15-10	19-5	22-6	7-5	10-10	13-9	16-9	19-6
	Hem-fir	SS	9-10	15-6	20-5	Note b	Note b	9-10	15-6	20-5	Note b	Note b
	Hem-fir	#1	9-8	15-2	19-11	25-5	Note b	9-8	14-11	18-11	23-2	Note b
	Hem-fir	#2	9-2	14-5	19-0	24-3	Note b	9-2	14-2	17-11	21-11	25-5
	Hem-fir	#3	8-7	12-6	15-10	19-5	22-6	7-5	10-10	13-9	16-9	19-6
	Southern pine	SS	10-3	16-1	21-2	Note b	Note b	10-3	16-1	21-2	Note b	Note b
	Southern pine	#1	10-0	15-9	20-10	Note b	Note b	10-0	15-9	20-10	25-10	Note b
	Southern pine	#2	9-10	15-6	20-5	Note b	Note b	9-10	15-1	19-5	23-2	Note b
	Southern pine	#3	9-1	13-6	17-2	20-3	24-1	7-11	11-8	14-10	17-6	20-11
	Spruce-pine-fir	SS	9-8	15-2	19-11	25-5	Note b	9-8	15-2	19-11	25-5	Note b
	Spruce-pine-fir	#1	9-5	14-9	19-6	24-10	Note b	9-5	14-4	18-2	22-3	25-9
	Spruce-pine-fir	#2	9-5	14-9	19-6	24-10	Note b	9-5	14-4	18-2	22-3	25-9
	Spruce-pine-fir	#3	8-7	12-6	15-10	19-5	22-6	7-5	10-10	13-9	16-9	19-6
16	Douglas fir-larch	SS	9-6	14-11	19-7	25-0	Note b	9-6	14-11	19-7	24-9	Note b
	Douglas fir-larch	#1	9-1	14-4	18-11	23-9	Note b	9-1	13-3	16-10	20-7	23-10
	Douglas fir-larch	#2	8-11	14-1	18-2	22-3	25-9	8-6	12-5	15-9	19-3	22-4
	Douglas fir-larch	#3	7-5	10-10	13-9	16-9	19-6	6-5	9-5	11-11	14-6	16-10
	Hem-fir	SS	8-11	14-1	18-6	23-8	Note b	8-11	14-1	18-6	23-8	Note b
	Hem-fir	#1	8-9	13-9	18-1	23-1	Note b	8-9	12-11	16-5	20-0	23-3
	Hem-fir	#2	8-4	13-1	17-3	21-11	25-5	8-4	12-3	15-6	18-11	22-0
	Hem-fir	#3	7-5	10-10	13-9	16-9	19-6	6-5	9-5	11-11	14-6	16-10
	Southern pine	SS	9-4	14-7	19-3	24-7	Note b	9-4	14-7	19-3	24-7	Note b
	Southern pine	#1	9-1	14-4	18-11	24-1	Note b	9-1	14-4	18-10	22-4	Note b
	Southern pine	#2	8-11	14-1	18-6	23-2	Note b	8-11	13-0	16-10	20-1	23-7
	Southern pine	#3	7-11	11-8	14-10	17-6	20-11	6-10	10-1	12-10	15-2	18-1
	Spruce-pine-fir	SS	8-9	13-9	18-1	23-1	Note b	8-9	13-9	18-1	23-0	Note b
	Spruce-pine-fir	#1	8-7	13-5	17-9	22-3	25-9	8-6	12-5	15-9	19-3	22-4
	Spruce-pine-fir	#2	8-7	13-5	17-9	22-3	25-9	8-6	12-5	15-9	19-3	22-4
	Spruce-pine-fir	#3	7-5	10-10	13-9	16-9	19-6	6-5	9-5	11-11	14-6	16-10
19.2	Douglas fir-larch	SS	8-11	14-0	18-5	23-7	Note b	8-11	14-0	18-5	22-7	Note b
	Douglas fir-larch	#1	8-7	13-6	17-9	21-8	25-2	8-4	12-2	15-4	18-9	21-9
	Douglas fir-larch	#2	8-5	13-1	16-7	20-3	23-6	7-9	11-4	14-4	17-7	20-4
	Douglas fir-larch	#3	6-9	9-11	12-7	15-4	17-9	5-10	8-7	10-10	13-3	15-5
	Hem-fir	SS	8-5	13-3	17-5	22-3	Note b	8-5	13-3	17-5	22-3	25-9
	Hem-fir	#1	8-3	12-11	17-1	21-1	24-6	8-1	11-10	15-0	18-4	21-3
	Hem-fir	#2	7-10	12-4	16-3	20-0	23-2	7-8	11-2	14-2	17-4	20-1
	Hem-fir	#3	6-9	9-11	12-7	15-4	17-9	5-10	8-7	10-10	13-3	15-5
	Southern pine	SS	8-9	13-9	18-1	23-1	Note b	8-9	13-9	18-1	23-1	Note b
	Southern pine	#1	8-7	13-6	17-9	22-8	Note b	8-7	13-6	17-2	20-5	24-4
	Southern pine	#2	8-5	13-3	17-5	21-2	24-10	8-4	11-11	15-4	18-4	21-6
	Southern pine	#3	7-3	10-8	13-7	16-0	19-1	6-3	9-3	11-9	13-10	16-6
	Spruce-pine-fir	SS	8-3	12-11	17-1	21-9	Note b	8-3	12-11	17-1	21-0	24-4
	Spruce-pine-fir	#1	8-1	12-8	16-7	20-3	23-6	7-9	11-4	14-4	17-7	20-4
	Spruce-pine-fir	#2	8-1	12-8	16-7	20-3	23-6	7-9	11-4	14-4	17-7	20-4
	Spruce-pine-fir	#3	6-9	9-11	12-7	15-4	17-9	5-10	8-7	10-10	13-3	15-5

(continued)

TABLE R802.5.1(2)—continued
RAFTER SPANS FOR COMMON LUMBER SPECIES
(Roof live load=20 psf, ceiling attached to rafters, L/Δ = 240)

RAFTER SPACING (inches)	SPECIES AND GRADE		DEAD LOAD = 10 psf					DEAD LOAD = 20 psf				
			2 × 4	2 × 6	2 × 8	2 × 10	2 × 12	2 × 4	2 × 6	2 × 8	2 × 10	2 × 12
			Maximum rafter spans[a]									
			(feet - inches)	(feet - inches)	(feet - inches)	(feet - inches)	(feet - inches)	(feet - inches)	(feet - inches)	(feet - inches)	(feet - inches)	(feet - inches)
24	Douglas fir-larch	SS	8-3	13-0	17-2	21-10	Note b	8-3	13-0	16-7	20-3	23-5
	Douglas fir-larch	#1	8-0	12-6	15-10	19-5	22-6	7-5	10-10	13-9	16-9	19-6
	Douglas fir-larch	#2	7-10	11-9	14-10	18-2	21-0	6-11	10-2	12-10	15-8	18-3
	Douglas fir-larch	#3	6-1	8-10	11-3	13-8	15-11	5-3	7-8	9-9	11-10	13-9
	Hem-fir	SS	7-10	12-3	16-2	20-8	25-1	7-10	12-3	16-2	19-10	23-0
	Hem-fir	#1	7-8	12-0	15-6	18-11	21-11	7-3	10-7	13-5	16-4	19-0
	Hem-fir	#2	7-3	11-5	14-8	17-10	20-9	6-10	10-0	12-8	15-6	17-11
	Hem-fir	#3	6-1	8-10	11-3	13-8	15-11	5-3	7-8	9-9	11-10	13-9
	Southern pine	SS	8-1	12-9	16-10	21-6	Note b	8-1	12-9	16-10	21-6	Note b
	Southern pine	#1	8-0	12-6	16-6	21-1	25-2	8-0	12-3	15-4	18-3	21-9
	Southern pine	#2	7-10	12-3	15-10	18-11	22-2	7-5	10-8	13-9	16-5	19-3
	Southern pine	#3	6-5	9-6	12-1	14-4	17-1	5-7	8-3	10-6	12-5	14-9
	Spruce-pine-fir	SS	7-8	12-0	15-10	20-2	24-7	7-8	12-0	15-4	18-9	21-9
	Spruce-pine-fir	#1	7-6	11-9	14-10	18-2	21-0	6-11	10-2	12-10	15-8	18-3
	Spruce-pine-fir	#2	7-6	11-9	14-10	18-2	21-0	6-11	10-2	12-10	15-8	18-3
	Spruce-pine-fir	#3	6-1	8-10	11-3	13-8	15-11	5-3	7-8	9-9	11-10	13-9

Check sources for availability of lumber in lengths greater than 20 feet.

For SI: 1 inch = 25.4 mm, 1 foot = 304.8 mm, 1 pound per square foot = 0.0479 kPa.

a. The tabulated rafter spans assume that ceiling joists are located at the bottom of the attic space or that some other method of resisting the outward push of the rafters on the bearing walls, such as rafter ties, is provided at that location. When ceiling joists or rafter ties are located higher in the attic space, the rafter spans shall be multiplied by the factors given below:

H_C/H_R	Rafter Span Adjustment Factor
1/3	0.67
1/4	0.76
1/5	0.83
1/6	0.90
1/7.5 or less	1.00

where:

H_C = Height of ceiling joists or rafter ties measured vertically above the top of the rafter support walls.

H_R = Height of roof ridge measured vertically above the top of the rafter support walls.

b. Span exceeds 26 feet in length.

❖ See the commentary for Sections R802.5 and R802.5.1 and Table R802.5.1(1).

TABLE R802.5.1(3)
RAFTER SPANS FOR COMMON LUMBER SPECIES
(Ground snow load=30 psf, ceiling not attached to rafters, L/Δ = 180)

RAFTER SPACING (inches)	SPECIES AND GRADE		DEAD LOAD = 10 psf					DEAD LOAD = 20 psf				
			2 × 4	2 × 6	2 × 8	2 × 10	2 × 12	2 × 4	2 × 6	2 × 8	2 × 10	2 × 12
			Maximum rafter spans[a]									
			(feet - inches)	(feet - inches)	(feet - inches)	(feet - inches)	(feet - inches)	(feet - inches)	(feet - inches)	(feet - inches)	(feet - inches)	(feet - inches)
12	Douglas fir-larch	SS	10-0	15-9	20-9	Note b	Note b	10-0	15-9	20-1	24-6	Note b
	Douglas fir-larch	#1	9-8	14-9	18-8	22-9	Note b	9-0	13-2	16-8	20-4	23-7
	Douglas fir-larch	#2	9-5	13-9	17-5	21-4	24-8	8-5	12-4	15-7	19-1	22-1
	Douglas fir-larch	#3	7-1	10-5	13-2	16-1	18-8	6-4	9-4	11-9	14-5	16-8
	Hem-fir	SS	9-6	14-10	19-7	25-0	Note b	9-6	14-10	19-7	24-1	Note b
	Hem-fir	#1	9-3	14-4	18-2	22-2	25-9	8-9	12-10	16-3	19-10	23-0
	Hem-fir	#2	8-10	13-7	17-2	21-0	24-4	8-4	12-2	15-4	18-9	21-9
	Hem-fir	#3	7-1	10-5	13-2	16-1	18-8	6-4	9-4	11-9	14-5	16-8
	Southern pine	SS	9-10	15-6	20-5	Note b	Note b	9-10	15-6	20-5	Note b	Note b
	Southern pine	#1	9-8	15-2	20-0	24-9	Note b	9-8	14-10	18-8	22-2	Note b
	Southern pine	#2	9-6	14-5	18-8	22-3	Note b	9-0	12-11	16-8	19-11	23-4
	Southern pine	#3	7-7	11-2	14-3	16-10	20-0	6-9	10-0	12-9	15-1	17-11
	Spruce-pine-fir	SS	9-3	14-7	19-2	24-6	Note b	9-3	14-7	18-8	22-9	Note b
	Spruce-pine-fir	#1	9-1	13-9	17-5	21-4	24-8	8-5	12-4	15-7	19-1	22-1
	Spruce-pine-fir	#2	9-1	13-9	17-5	21-4	24-8	8-5	12-4	15-7	19-1	22-1
	Spruce-pine-fir	#3	7-1	10-5	13-2	16-1	18-8	6-4	9-4	11-9	14-5	16-8
16	Douglas fir-larch	SS	9-1	14-4	18-10	23-9	Note b	9-1	13-9	17-5	21-3	24-8
	Douglas fir-larch	#1	8-9	12-9	16-2	19-9	22-10	7-10	11-5	14-5	17-8	20-5
	Douglas fir-larch	#2	8-2	11-11	15-1	18-5	21-5	7-3	10-8	13-6	16-6	19-2
	Douglas fir-larch	#3	6-2	9-0	11-5	13-11	16-2	5-6	8-1	10-3	12-6	14-6
	Hem-fir	SS	8-7	13-6	17-10	22-9	Note b	8-7	13-6	17-1	20-10	24-2
	Hem-fir	#1	8-5	12-5	15-9	19-3	22-3	7-7	11-1	14-1	17-2	19-11
	Hem-fir	#2	8-0	11-9	14-11	18-2	21-1	7-2	10-6	13-4	16-3	18-10
	Hem-fir	#3	6-2	9-0	11-5	13-11	16-2	5-6	8-1	10-3	12-6	14-6
	Southern pine	SS	8-11	14-1	18-6	23-8	Note b	8-11	14-1	18-6	23-8	Note b
	Southern pine	#1	8-9	13-9	18-1	21-5	25-7	8-8	12-10	16-2	19-2	22-10
	Southern pine	#2	8-7	12-6	16-2	19-3	22-7	7-10	11-2	14-5	17-3	20-2
	Southern pine	#3	6-7	9-8	12-4	14-7	17-4	5-10	8-8	11-0	13-0	15-6
	Spruce-pine-fir	SS	8-5	13-3	17-5	22-1	25-7	8-5	12-9	16-2	19-9	22-10
	Spruce-pine-fir	#1	8-2	11-11	15-1	18-5	21-5	7-3	10-8	13-6	16-6	19-2
	Spruce-pine-fir	#2	8-2	11-11	15-1	18-5	21-5	7-3	10-8	13-6	16-6	19-2
	Spruce-pine-fir	#3	6-2	9-0	11-5	13-11	16-2	5-6	8-1	10-3	12-6	14-6
19.2	Douglas fir-larch	SS	8-7	13-6	17-9	21-8	25-2	8-7	12-6	15-10	19-5	22-6
	Douglas fir-larch	#1	7-11	11-8	14-9	18-0	20-11	7-1	10-5	13-2	16-1	18-8
	Douglas fir-larch	#2	7-5	10-11	13-9	16-10	19-6	6-8	9-9	12-4	15-1	17-6
	Douglas fir-larch	#3	5-7	8-3	10-5	12-9	14-9	5-0	7-4	9-4	11-5	13-2
	Hem-fir	SS	8-1	12-9	16-9	21-4	24-8	8-1	12-4	15-7	19-1	22-1
	Hem-fir	#1	7-9	11-4	14-4	17-7	20-4	6-11	10-2	12-10	15-8	18-2
	Hem-fir	#2	7-4	10-9	13-7	16-7	19-3	6-7	9-7	12-2	14-10	17-3
	Hem-fir	#3	5-7	8-3	10-5	12-9	14-9	5-0	7-4	9-4	11-5	13-2
	Southern pine	SS	8-5	13-3	17-5	22-3	Note b	8-5	13-3	17-5	22-0	25-9
	Southern pine	#1	8-3	13-0	16-6	19-7	23-4	7-11	11-9	14-9	17-6	20-11
	Southern pine	#2	7-11	11-5	14-9	17-7	20-7	7-1	10-2	13-2	15-9	18-5
	Southern pine	#3	6-0	8-10	11-3	13-4	15-10	5-4	7-11	10-1	11-11	14-2
	Spruce-pine-fir	SS	7-11	12-5	16-5	20-2	23-4	7-11	11-8	14-9	18-0	20-11
	Spruce-pine-fir	#1	7-5	10-11	13-9	16-10	19-6	6-8	9-9	12-4	15-1	17-6
	Spruce-pine-fir	#2	7-5	10-11	13-9	16-10	19-6	6-8	9-9	12-4	15-1	17-6
	Spruce-pine-fir	#3	5-7	8-3	10-5	12-9	14-9	5-0	7-4	9-4	11-5	13-2

(continued)

TABLE R802.5.1(3)—continued
RAFTER SPANS FOR COMMON LUMBER SPECIES
(Ground snow load=30 psf, ceiling not attached to rafters, L/Δ = 180)

RAFTER SPACING (inches)	SPECIES AND GRADE		DEAD LOAD = 10 psf					DEAD LOAD = 20 psf				
			2 × 4	2 × 6	2 × 8	2 × 10	2 × 12	2 × 4	2 × 6	2 × 8	2 × 10	2 × 12
			Maximum rafter spans[a]									
			(feet - inches)	(feet - inches)	(feet - inches)	(feet - inches)	(feet - inches)	(feet - inches)	(feet - inches)	(feet - inches)	(feet - inches)	(feet - inches)
24	Douglas fir-larch	SS	7-11	12-6	15-10	19-5	22-6	7-8	11-3	14-2	17-4	20-1
	Douglas fir-larch	#1	7-1	10-5	13-2	16-1	18-8	6-4	9-4	11-9	14-5	16-8
	Douglas fir-larch	#2	6-8	9-9	12-4	15-1	17-6	5-11	8-8	11-0	13-6	15-7
	Douglas fir-larch	#3	5-0	7-4	9-4	11-5	13-2	4-6	6-7	8-4	10-2	11-10
	Hem-fir	SS	7-6	11-10	15-7	19-1	22-1	7-6	11-0	13-11	17-0	19-9
	Hem-fir	#1	6-11	10-2	12-10	15-8	18-2	6-2	9-1	11-6	14-0	16-3
	Hem-fir	#2	6-7	9-7	12-2	14-10	17-3	5-10	8-7	10-10	13-3	15-5
	Hem-fir	#3	5-0	7-4	9-4	11-5	13-2	4-6	6-7	8-4	10-2	11-10
	Southern pine	SS	7-10	12-3	16-2	20-8	25-1	7-10	12-3	16-2	19-8	23-0
	Southern pine	#1	7-8	11-9	14-9	17-6	20-11	7-1	10-6	13-2	15-8	18-8
	Southern pine	#2	7-1	10-2	13-2	15-9	18-5	6-4	9-2	11-9	14-1	16-6
	Southern pine	#3	5-4	7-11	10-1	11-11	14-2	4-9	7-1	9-0	10-8	12-8
	Spruce-pine-fir	SS	7-4	11-7	14-9	18-0	20-11	7-1	10-5	13-2	16-1	18-8
	Spruce-pine-fir	#1	6-8	9-9	12-4	15-1	17-6	5-11	8-8	11-0	13-6	15-7
	Spruce-pine-fir	#2	6-8	9-9	12-4	15-1	17-6	5-11	8-8	11-0	13-6	15-7
	Spruce-pine-fir	#3	5-0	7-4	9-4	11-5	13-2	4-6	6-7	8-4	10-2	11-10

Check sources for availability of lumber in lengths greater than 20 feet.

For SI: 1 inch = 25.4 mm, 1 foot = 304.8 mm, 1 pound per square foot = 0.0479 kPa.

a. The tabulated rafter spans assume that ceiling joists are located at the bottom of the attic space or that some other method of resisting the outward push of the rafters on the bearing walls, such as rafter ties, is provided at that location. When ceiling joists or rafter ties are located higher in the attic space, the rafter spans shall be multiplied by the factors given below:

H_C/H_R	Rafter Span Adjustment Factor
1/3	0.67
1/4	0.76
1/5	0.83
1/6	0.90
1/7.5 or less	1.00

where:

H_C = Height of ceiling joists or rafter ties measured vertically above the top of the rafter support walls.

H_R = Height of roof ridge measured vertically above the top of the rafter support walls.

b. Span exceeds 26 feet in length.

❖ See the commentary for Sections R802.5 and R802.5.1 and Table 802.5.1(1).

TABLE R802.5.1(4)
RAFTER SPANS FOR COMMON LUMBER SPECIES
(Ground snow load=50 psf, ceiling not attached to rafters, L/Δ = 180)

RAFTER SPACING (inches)	SPECIES AND GRADE		DEAD LOAD = 10 psf					DEAD LOAD = 20 psf				
			2 × 4	2 × 6	2 × 8	2 × 10	2 × 12	2 × 4	2 × 6	2 × 8	2 × 10	2 × 12
			Maximum rafter spans[a]									
			(feet - inches)	(feet - inches)	(feet - inches)	(feet - inches)	(feet - inches)	(feet - inches)	(feet - inches)	(feet - inches)	(feet - inches)	(feet - inches)
12	Douglas fir-larch	SS	8-5	13-3	17-6	22-4	26-0	8-5	13-3	17-0	20-9	24-0
	Douglas fir-larch	#1	8-2	12-0	15-3	18-7	21-7	7-7	11-2	14-1	17-3	20-0
	Douglas fir-larch	#2	7-8	11-3	14-3	17-5	20-2	7-1	10-5	13-2	16-1	18-8
	Douglas fir-larch	#3	5-10	8-6	10-9	13-2	15-3	5-5	7-10	10-0	12-2	14-1
	Hem-fir	SS	8-0	12-6	16-6	21-1	25-6	8-0	12-6	16-6	20-4	23-7
	Hem-fir	#1	7-10	11-9	14-10	18-1	21-0	7-5	10-10	13-9	16-9	19-5
	Hem-fir	#2	7-5	11-1	14-0	17-2	19-11	7-0	10-3	13-0	15-10	18-5
	Hem-fir	#3	5-10	8-6	10-9	13-2	15-3	5-5	7-10	10-0	12-2	14-1
	Southern pine	SS	8-4	13-0	17-2	21-11	Note b	8-4	13-0	17-2	21-11	Note b
	Southern pine	#1	8-2	12-10	16-10	20-3	24-1	8-2	12-6	15-9	18-9	22-4
	Southern pine	#2	8-0	11-9	15-3	18-2	21-3	7-7	10-11	14-1	16-10	19-9
	Southern pine	#3	6-2	9-2	11-8	13-9	16-4	5-9	8-5	10-9	12-9	15-2
	Spruce-pine-fir	SS	7-10	12-3	16-2	20-8	24-1	7-10	12-3	15-9	19-3	22-4
	Spruce-pine-fir	#1	7-8	11-3	14-3	17-5	20-2	7-1	10-5	13-2	16-1	18-8
	Spruce-pine-fir	#2	7-8	11-3	14-3	17-5	15-2	7-1	10-5	13-2	16-1	18-8
	Spruce-pine-fir	#3	5-10	8-6	10-9	13-2	20-3	5-5	7-10	10-0	12-2	14-1
16	Douglas fir-larch	SS	7-8	12-1	15-10	19-5	22-6	7-8	11-7	14-8	17-11	20-10
	Douglas fir-larch	#1	7-1	10-5	13-2	16-1	18-8	6-7	9-8	12-2	14-11	17-3
	Douglas fir-larch	#2	6-8	9-9	12-4	15-1	17-6	6-2	9-0	11-5	13-11	16-2
	Douglas fir-larch	#3	5-0	7-4	9-4	11-5	13-2	4-8	6-10	8-8	10-6	12-3
	Hem-fir	SS	7-3	11-5	15-0	19-1	22-1	7-3	11-5	14-5	17-8	20-5
	Hem-fir	#1	6-11	10-2	12-10	15-8	18-2	6-5	9-5	11-11	14-6	16-10
	Hem-fir	#2	6-7	9-7	12-2	14-10	17-3	6-1	8-11	11-3	13-9	15-11
	Hem-fir	#3	5-0	7-4	9-4	11-5	13-2	4-8	6-10	8-8	10-6	12-3
	Southern pine	SS	7-6	11-10	15-7	19-11	24-3	7-6	11-10	15-7	19-11	23-10
	Southern pine	#1	7-5	11-7	14-9	17-6	20-11	7-4	10-10	13-8	16-2	19-4
	Southern pine	#2	7-1	10-2	13-2	15-9	18-5	6-7	9-5	12-2	14-7	17-1
	Southern pine	#3	5-4	7-11	10-1	11-11	14-2	4-11	7-4	9-4	11-0	13-1
	Spruce-pine-fir	SS	7-1	11-2	14-8	18-0	20-11	7-1	10-9	13-8	15-11	19-4
	Spruce-pine-fir	#1	6-8	9-9	12-4	15-1	17-6	6-2	9-0	11-5	13-11	16-2
	Spruce-pine-fir	#2	6-8	9-9	12-4	15-1	17-6	6-2	9-0	11-5	13-11	16-2
	Spruce-pine-fir	#3	5-0	7-4	9-4	11-5	13-2	4-8	6-10	8-8	10-6	12-3
19.2	Douglas fir-larch	SS	7-3	11-4	14-6	17-8	20-6	7-3	10-7	13-5	16-5	19-0
	Douglas fir-larch	#1	6-6	9-6	12-0	14-8	17-1	6-0	8-10	11-2	13-7	15-9
	Douglas fir-larch	#2	6-1	8-11	11-3	13-9	15-11	5-7	8-3	10-5	12-9	14-9
	Douglas fir-larch	#3	4-7	6-9	8-6	10-5	12-1	4-3	6-3	7-11	9-7	11-2
	Hem-fir	SS	6-10	10-9	14-2	17-5	20-2	6-10	10-5	13-2	16-1	18-8
	Hem-fir	#1	6-4	9-3	11-9	14-4	16-7	5-10	8-7	10-10	13-3	15-5
	Hem-fir	#2	6-0	8-9	11-1	13-7	15-9	5-7	8-1	10-3	12-7	14-7
	Hem-fir	#3	4-7	6-9	8-6	10-5	12-1	4-3	6-3	7-11	9-7	11-2
	Southern pine	SS	7-1	11-2	14-8	18-9	22-10	7-1	11-2	14-8	18-7	21-9
	Southern pine	#1	7-0	10-8	13-5	16-0	19-1	6-8	9-11	12-5	14-10	17-8
	Southern pine	#2	6-6	9-4	12-0	14-4	16-10	6-0	8-8	11-2	13-4	15-7
	Southern pine	#3	4-11	7-3	9-2	10-10	12-11	4-6	6-8	8-6	10-1	12-0
	Spruce-pine-fir	SS	6-8	10-6	13-5	16-5	19-1	6-8	9-10	12-5	15-3	17-8
	Spruce-pine-fir	#1	6-1	8-11	11-3	13-9	15-11	5-7	8-3	10-5	12-9	14-9
	Spruce-pine-fir	#2	6-1	8-11	11-3	13-9	15-11	5-7	8-3	10-5	12-9	14-9
	Spruce-pine-fir	#3	4-7	6-9	8-6	10-5	12-1	4-3	6-3	7-11	9-7	11-2

(continued)

TABLE R802.5.1(4)—continued
RAFTER SPANS FOR COMMON LUMBER SPECIES
(Ground snow load=50 psf, ceiling not attached to rafters, L/Δ = 180)

RAFTER SPACING (inches)	SPECIES AND GRADE		DEAD LOAD = 10 psf					DEAD LOAD = 20 psf				
			2 × 4	2 × 6	2 × 8	2 × 10	2 × 12	2 × 4	2 × 6	2 × 8	2 × 10	2 × 12
			Maximum rafter spans[a]									
			(feet - inches)	(feet - inches)	(feet - inches)	(feet - inches)	(feet - inches)	(feet - inches)	(feet - inches)	(feet - inches)	(feet - inches)	(feet - inches)
24	Douglas fir-larch	SS	6-8	10-	13-0	15-10	18-4	6-6	9-6	12-0	14-8	17-0
	Douglas fir-larch	#1	5-10	8-6	10-9	13-2	15-3	5-5	7-10	10-0	12-2	14-1
	Douglas fir-larch	#2	5-5	7-11	10-1	12-4	14-3	5-0	7-4	9-4	11-5	13-2
	Douglas fir-larch	#3	4-1	6-0	7-7	9-4	10-9	3-10	5-7	7-1	8-7	10-0
	Hem-fir	SS	6-4	9-11	12-9	15-7	18-0	6-4	9-4	11-9	14-5	16-8
	Hem-fir	#1	5-8	8-3	10-6	12-10	14-10	5-3	7-8	9-9	11-10	13-9
	Hem-fir	#2	5-4	7-10	9-11	12-1	14-1	4-11	7-3	9-2	11-3	13-0
	Hem-fir	#3	4-1	6-0	7-7	9-4	10-9	3-10	5-7	7-1	8-7	10-0
	Southern pine	SS	6-7	10-4	13-8	17-5	21-0	6-7	10-4	13-8	16-7	19-5
	Southern pine	#1	6-5	9-7	12-0	14-4	17-1	6-0	8-10	11-2	13-3	15-9
	Southern pine	#2	5-10	8-4	10-9	12-10	15-1	5-5	7-9	10-0	11-11	13-11
	Southern pine	#3	4-4	6-5	8-3	9-9	11-7	4-1	6-0	7-7	9-0	10-8
	Spruce-pine-fir	SS	6-2	9-6	12-0	14-8	17-1	6-0	8-10	11-2	13-7	15-9
	Spruce-pine-fir	#1	5-5	7-11	10-1	12-4	14-3	5-0	7-4	9-4	11-5	13-2
	Spruce-pine-fir	#2	5-5	7-11	10-1	12-4	14-3	5-0	7-4	9-4	11-5	13-2
	Spruce-pine-fir	#3	4-1	6-0	7-7	9-4	10-9	3-10	5-7	7-1	8-7	10-0

Check sources for availability of lumber in lengths greater than 20 feet.

For SI: 1 inch = 25.4 mm, 1 foot = 304.8 mm, 1 pound per square foot = 0.0479kPa.

a. The tabulated rafter spans assume that ceiling joists are located at the bottom of the attic space or that some other method of resisting the outward push of the rafters on the bearing walls, such as rafter ties, is provided at that location. When ceiling joists or rafter ties are located higher in the attic space, the rafter spans shall be multiplied by the factors given below:

H_C/H_R	Rafter Span Adjustment Factor
1/3	0.67
1/4	0.76
1/5	0.83
1/6	0.90
1/7.5 or less	1.00

where:

H_C = Height of ceiling joists or rafter ties measured vertically above the top of the rafter support walls.

H_R = Height of roof ridge measured vertically above the top of the rafter support walls.

b. Span exceeds 26 feet in length.

❖ See the commentary for Section R802.5 and R802.5.1 and Table 802.5.1(1).

TABLE R802.5.1(5)
RAFTER SPANS FOR COMMON LUMBER SPECIES
(Ground snow load=30 psf, ceiling attached to rafters, L/Δ = 240)

RAFTER SPACING (inches)	SPECIES AND GRADE		DEAD LOAD = 10 psf					DEAD LOAD = 20 psf				
			2 × 4	2 × 6	2 × 8	2 × 10	2 × 12	2 × 4	2 × 6	2 × 8	2 × 10	2 × 12
			Maximum rafter spans[a]									
			(feet - inches)	(feet - inches)	(feet - inches)	(feet - inches)	(feet - inches)	(feet - inches)	(feet - inches)	(feet - inches)	(feet - inches)	(feet - inches)
12	Douglas fir-larch	SS	9-1	14-4	18-10	24-1	Note b	9-1	14-4	18-10	24-1	Note b
	Douglas fir-larch	#1	8-9	13-9	18-2	22-9	Note b	8-9	13-2	16-8	20-4	23-7
	Douglas fir-larch	#2	8-7	13-6	17-5	21-4	24-8	8-5	12-4	15-7	19-1	22-1
	Douglas fir-larch	#3	7-1	10-5	13-2	16-1	18-8	6-4	9-4	11-9	14-5	16-8
	Hem-fir	SS	8-7	13-6	17-10	22-9	Note b	8-7	13-6	17-10	22-9	Note b
	Hem-fir	#1	8-5	13-3	17-5	22-2	25-9	8-5	12-10	16-3	19-10	23-0
	Hem-fir	#2	8-0	12-7	16-7	21-0	24-4	8-0	12-2	15-4	18-9	21-9
	Hem-fir	#3	7-1	10-5	13-2	16-1	18-8	6-4	9-4	11-9	14-5	16-8
	Southern pine	SS	8-11	14-1	18-6	23-8	Note b	8-11	14-1	18-6	23-8	Note b
	Southern pine	#1	8-9	13-9	18-2	23-2	Note b	8-9	13-9	18-2	22-2	Note b
	Southern pine	#2	8-7	13-6	17-10	22-3	Note b	8-7	12-11	16-8	19-11	23-4
	Southern pine	#3	7-7	11-2	14-3	16-10	20-0	6-9	10-0	12-9	15-1	17-11
	Spruce-pine-fir	SS	8-5	13-3	17-5	22-3	Note b	8-5	13-3	17-5	22-3	Note b
	Spruce-pine-fir	#1	8-3	12-11	17-0	21-4	24-8	8-3	12-4	15-7	19-1	22-1
	Spruce-pine-fir	#2	8-3	12-11	17-0	21-4	24-8	8-3	12-4	15-7	19-1	22-1
	Spruce-pine-fir	#3	7-1	10-5	13-2	16-1	18-8	6-4	9-4	11-9	14-5	16-8
16	Douglas fir-larch	SS	8-3	13-0	17-2	21-10	Note b	8-3	13-0	17-2	21-3	24-8
	Douglas fir-larch	#1	8-0	12-6	16-2	19-9	22-10	7-10	11-5	14-5	17-8	20-5
	Douglas fir-larch	#2	7-10	11-11	15-1	18-5	21-5	7-3	10-8	13-6	16-6	19-2
	Douglas fir-larch	#3	6-2	9-0	11-5	13-11	16-2	5-6	8-1	10-3	12-6	14-6
	Hem-fir	SS	7-10	12-3	16-2	20-8	25-1	7-10	12-3	16-2	20-8	24-2
	Hem-fir	#1	7-8	12-0	15-9	19-3	22-3	7-7	11-1	14-1	17-2	19-11
	Hem-fir	#2	7-3	11-5	14-11	18-2	21-1	7-2	10-6	13-4	16-3	18-10
	Hem-fir	#3	6-2	9-0	11-5	13-11	16-2	5-6	8-1	10-3	12-6	14-6
	Southern pine	SS	8-1	12-9	16-10	21-6	Note b	8-1	12-9	16-10	21-6	Note b
	Southern pine	#1	8-0	12-6	16-6	21-1	25-7	8-0	12-6	16-2	19-2	22-10
	Southern pine	#2	7-10	12-3	16-2	19-3	22-7	7-10	11-2	14-5	17-3	20-2
	Southern pine	#3	6-7	9-8	12-4	14-7	17-4	5-10	8-8	11-0	13-0	15-6
	Spruce-pine-fir	SS	7-8	12-0	15-10	20-2	24-7	7-8	12-0	15-10	19-9	22-10
	Spruce-pine-fir	#1	7-6	11-9	15-1	18-5	21-5	7-3	10-8	13-6	16-6	19-2
	Spruce-pine-fir	#2	7-6	11-9	15-1	18-5	21-5	7-3	10-8	13-6	16-6	19-2
	Spruce-pine-fir	#3	6-2	9-0	11-5	13-11	16-2	5-6	8-1	10-3	12-6	14-6
19.2	Douglas fir-larch	SS	7-9	12-3	16-1	20-7	25-0	7-9	12-3	15-10	19-5	22-6
	Douglas fir-larch	#1	7-6	11-8	14-9	18-0	20-11	7-1	10-5	13-2	16-1	18-8
	Douglas fir-larch	#2	7-4	10-11	13-9	16-10	19-6	6-8	9-9	12-4	15-1	17-6
	Douglas fir-larch	#3	5-7	8-3	10-5	12-9	14-9	5-0	7-4	9-4	11-5	13-2
	Hem-fir	SS	7-4	11-7	15-3	19-5	23-7	7-4	11-7	15-3	19-1	22-1
	Hem-fir	#1	7-2	11-4	14-4	17-7	20-4	6-11	10-2	12-10	15-8	18-2
	Hem-fir	#2	6-10	10-9	13-7	16-7	19-3	6-7	9-7	12-2	14-10	17-3
	Hem-fir	#3	5-7	8-3	10-5	12-9	14-9	5-0	7-4	9-4	11-5	13-2
	Southern pine	SS	7-8	12-0	15-10	20-2	24-7	7-8	12-0	15-10	20-2	24-7
	Southern pine	#1	7-6	11-9	15-6	19-7	23-4	7-6	11-9	14-9	17-6	20-11
	Southern pine	#2	7-4	11-5	14-9	17-7	20-7	7-1	10-2	13-2	15-9	18-5
	Southern pine	#3	6-0	8-10	11-3	13-4	15-10	5-4	7-11	10-1	11-11	14-2
	Spruce-pine-fir	SS	7-2	11-4	14-11	19-0	23-1	7-2	11-4	14-9	18-0	20-11
	Spruce-pine-fir	#1	7-0	10-11	13-9	16-10	19-6	6-8	9-9	12-4	15-1	17-6
	Spruce-pine-fir	#2	7-0	10-11	13-9	16-10	19-6	6-8	9-9	12-4	15-1	17-6
	Spruce-pine-fir	#3	5-7	8-3	10-5	12-9	14-9	5-0	7-4	9-4	11-5	13-2

(continued)

TABLE R802.5.1(5)—continued
RAFTER SPANS FOR COMMON LUMBER SPECIES
(Ground snow load=30 psf, ceiling attached to rafters, L/Δ = 240)

RAFTER SPACING (inches)	SPECIES AND GRADE		DEAD LOAD = 10 psf					DEAD LOAD = 20 psf				
			2 × 4	2 × 6	2 × 8	2 × 10	2 × 12	2 × 4	2 × 6	2 × 8	2 × 10	2 × 12
			Maximum rafter spans[a]									
			(feet-inches)	(feet-inches)	(feet-inches)	(feet-inches)	(feet-inches)	(feet-inches)	(feet-inches)	(feet-inches)	(feet-inches)	(feet-inches)
24	Douglas fir-larch	SS	7-3	11-4	15-0	19-1	22-6	7-3	11-3	14-2	17-4	20-1
	Douglas fir-larch	#1	7-0	10-5	13-2	16-1	18-8	6-4	9-4	11-9	14-5	16-8
	Douglas fir-larch	#2	6-8	9-9	12-4	15-1	17-6	5-11	8-8	11-0	13-6	15-7
	Douglas fir-larch	#3	5-0	7-4	9-4	11-5	13-2	4-6	6-7	8-4	10-2	11-10
	Hem-fir	SS	6-10	10-9	14-2	18-0	21-11	6-10	10-9	13-11	17-0	19-9
	Hem-fir	#1	6-8	10-2	12-10	15-8	18-2	6-2	9-1	11-6	14-0	16-3
	Hem-fir	#2	6-4	9-7	12-2	14-10	17-3	5-10	8-7	10-10	13-3	15-5
	Hem-fir	#3	5-0	7-4	9-4	11-5	13-2	4-6	6-7	8-4	10-2	11-10
	Southern pine	SS	7-1	11-2	14-8	18-9	22-10	7-1	11-2	14-8	18-9	22-10
	Southern pine	#1	7-0	10-11	14-5	17-6	20-11	7-0	10-6	13-2	15-8	18-8
	Southern pine	#2	6-10	10-2	13-2	15-9	18-5	6-4	9-2	11-9	14-1	16-6
	Southern pine	#3	5-4	7-11	10-1	11-11	14-2	4-9	7-1	9-0	10-8	12-8
	Spruce-pine-fir	SS	6-8	10-6	13-10	17-8	20-11	6-8	10-5	13-2	16-1	18-8
	Spruce-pine-fir	#1	6-6	9-9	12-4	15-1	17-6	5-11	8-8	11-0	13-6	15-7
	Spruce-pine-fir	#2	6-6	9-9	12-4	15-1	17-6	5-11	8-8	11-0	13-6	15-7
	Spruce-pine-fir	#3	5-0	7-4	9-4	11-5	13-2	4-6	6-7	8-4	10-2	11-10

Check sources for availability of lumber in lengths greater than 20 feet.

For SI: 1 inch = 25.4 mm, 1 foot = 304.8 mm, 1 pound per square foot = 0.0479 kPa.

a. The tabulated rafter spans assume that ceiling joists are located at the bottom of the attic space or that some other method of resisting the outward push of the rafters on the bearing walls, such as rafter ties, is provided at that location. When ceiling joists or rafter ties are located higher in the attic space, the rafter spans shall be multiplied by the factors given below:

H_C/H_R	Rafter Span Adjustment Factor
1/3	0.67
1/4	0.76
1/5	0.83
1/6	0.90
1/7.5 or less	1.00

where:

H_C = Height of ceiling joists or rafter ties measured vertically above the top of the rafter support walls.

H_R = Height of roof ridge measured vertically above the top of the rafter support walls.

b. Span exceeds 26 feet in length.

❖ See the commentary for Sections R802.5 and R802.5.1 and Table R802.5.1(1).

TABLE R802.5.1(6)
RAFTER SPANS FOR COMMON LUMBER SPECIES
(Ground snow load=50 psf, ceiling attached to rafters, L/Δ = 240)

RAFTER SPACING (inches)	SPECIES AND GRADE		DEAD LOAD = 10 psf					DEAD LOAD = 20 psf				
			2 × 4	2 × 6	2 × 8	2 × 10	2 × 12	2 × 4	2 × 6	2 × 8	2 × 10	2 × 12
			Maximum rafter spans[a]									
			(feet-inches)	(feet-inches)	(feet-inches)	(feet-inches)	(feet-inches)	(feet-inches)	(feet-inches)	(feet-inches)	(feet-inches)	(feet-inches)
12	Douglas fir-larch	SS	7-8	12-1	15-11	20-3	24-8	7-8	12-1	15-11	20-3	24-0
	Douglas fir-larch	#1	7-5	11-7	15-3	18-7	21-7	7-5	11-2	14-1	17-3	20-0
	Douglas fir-larch	#2	7-3	11-3	14-3	17-5	20-2	7-1	10-5	13-2	16-1	18-8
	Douglas fir-larch	#3	5-10	8-6	10-9	13-2	15-3	5-5	7-10	10-0	12-2	14-1
	Hem-fir	SS	7-3	11-5	15-0	19-2	23-4	7-3	11-5	15-0	19-2	23-4
	Hem-fir	#1	7-1	11-2	14-8	18-1	21-0	7-1	10-10	13-9	16-9	19-5
	Hem-fir	#2	6-9	10-8	14-0	17-2	19-11	6-9	10-3	13-0	15-10	18-5
	Hem-fir	#3	5-10	8-6	10-9	13-2	15-3	5-5	7-10	10-0	12-2	14-1
	Southern pine	SS	7-6	11-10	15-7	19-11	24-3	7-6	11-10	15-7	19-11	24-3
	Southern pine	#1	7-5	11-7	15-4	19-7	23-9	7-5	11-7	15-4	18-9	22-4
	Southern pine	#2	7-3	11-5	15-0	18-2	21-3	7-3	10-11	14-1	16-10	19-9
	Southern pine	#3	6-2	9-2	11-8	13-9	16-4	5-9	8-5	10-9	12-9	15-2
	Spruce-pine-fir	SS	7-1	11-2	14-8	18-9	22-10	7-1	11-2	14-8	18-9	22-4
	Spruce-pine-fir	#1	6-11	10-11	14-3	17-5	20-2	6-11	10-5	13-2	16-1	18-8
	Spruce-pine-fir	#2	6-11	10-11	14-3	17-5	20-2	6-11	10-5	13-2	16-1	18-8
	Spruce-pine-fir	#3	5-10	8-6	10-9	13-2	15-3	5-5	7-10	10-0	12-2	14-1
16	Douglas fir-larch	SS	7-0	11-0	14-5	18-5	22-5	7-0	11-0	14-5	17-11	20-10
	Douglas fir-larch	#1	6-9	10-5	13-2	16-1	18-8	6-7	9-8	12-2	14-11	17-3
	Douglas fir-larch	#2	6-7	9-9	12-4	15-1	17-6	6-2	9-0	11-5	13-11	16-2
	Douglas fir-larch	#3	5-0	7-4	9-4	11-5	13-2	4-8	6-10	8-8	10-6	12-3
	Hem-fir	SS	6-7	10-4	13-8	17-5	21-2	6-7	10-4	13-8	17-5	20-5
	Hem-fir	#1	6-5	10-2	12-10	15-8	18-2	6-5	9-5	11-11	14-6	16-10
	Hem-fir	#2	6-2	9-7	12-2	14-10	17-3	6-1	8-11	11-3	13-9	15-11
	Hem-fir	#3	5-0	7-4	9-4	11-5	13-2	4-8	6-10	8-8	10-6	12-3
	Southern pine	SS	6-10	10-9	14-2	18-1	22-0	6-10	10-9	14-2	18-1	22-0
	Southern pine	#1	6-9	10-7	13-11	17-6	20-11	6-9	10-7	13-8	16-2	19-4
	Southern pine	#2	6-7	10-2	13-2	15-9	18-5	6-7	9-5	12-2	14-7	17-1
	Southern pine	#3	5-4	7-11	10-1	11-11	14-2	4-11	7-4	9-4	11-0	13-1
	Spruce-pine-fir	SS	6-5	10-2	13-4	17-0	20-9	6-5	10-2	13-4	16-8	19-4
	Spruce-pine-fir	#1	6-4	9-9	12-4	15-1	17-6	6-2	9-0	11-5	13-11	16-2
	Spruce-pine-fir	#2	6-4	9-9	12-4	15-1	17-6	6-2	9-0	11-5	13-11	16-2
	Spruce-pine-fir	#3	5-0	7-4	9-4	11-5	13-2	4-8	6-10	8-8	10-6	12-3
19.2	Douglas fir-larch	SS	6-7	10-4	13-7	17-4	20-6	6-7	10-4	13-5	16-5	19-0
	Douglas fir-larch	#1	6-4	9-6	12-0	14-8	17-1	6-0	8-10	11-2	13-7	15-9
	Douglas fir-larch	#2	6-1	8-11	11-3	13-9	15-11	5-7	8-3	10-5	12-9	14-9
	Douglas fir-larch	#3	4-7	6-9	8-6	10-5	12-1	4-3	6-3	7-11	9-7	11-2
	Hem-fir	SS	6-2	9-9	12-10	16-5	19-11	6-2	9-9	12-10	16-1	18-8
	Hem-fir	#1	6-1	9-3	11-9	14-4	16-7	5-10	8-7	10-10	13-3	15-5
	Hem-fir	#2	5-9	8-9	11-1	13-7	15-9	5-7	8-1	10-3	12-7	14-7
	Hem-fir	#3	4-7	6-9	8-6	10-5	12-1	4-3	6-3	7-11	9-7	11-2
	Southern pine	SS	6-5	10-2	13-4	17-0	20-9	6-5	10-2	13-4	17-0	20-9
	Southern pine	#1	6-4	9-11	13-1	16-0	19-1	6-4	9-11	12-5	14-10	17-8
	Southern pine	#2	6-2	9-4	12-0	14-4	16-10	6-0	8-8	11-2	13-4	15-7
	Southern pine	#3	4-11	7-3	9-2	10-10	12-11	4-6	6-8	8-6	10-1	12-0
	Spruce-pine-fir	SS	6-1	9-6	12-7	16-0	19-1	6-1	9-6	12-5	15-3	17-8
	Spruce-pine-fir	#1	5-11	8-11	11-3	13-9	15-11	5-7	8-3	10-5	12-9	14-9
	Spruce-pine-fir	#2	5-11	8-11	11-3	13-9	15-11	5-7	8-3	10-5	12-9	14-9
	Spruce-pine-fir	#3	4-7	6-9	8-6	10-5	12-1	4-3	6-3	7-11	9-7	11-2

(continued)

TABLE R802.5.1(6)—continued
RAFTER SPANS FOR COMMON LUMBER SPECIES
(Ground snow load=50 psf, ceiling attached to rafters, L/∆ = 240)

RAFTER SPACING (inches)	SPECIES AND GRADE		DEAD LOAD = 10 psf					DEAD LOAD = 20 psf				
			2 × 4	2 × 6	2 × 8	2 × 10	2 × 12	2 × 4	2 × 6	2 × 8	2 × 10	2 × 12
			Maximum rafter spans[a]									
			(feet-inches)	(feet-inches)	(feet-inches)	(feet-inches)	(feet-inches)	(feet-inches)	(feet-inches)	(feet-inches)	(feet-inches)	(feet-inches)
24	Douglas fir-larch	SS	6-1	9-7	12-7	15-10	18-4	6-1	9-6	12-0	14-8	17-0
	Douglas fir-larch	#1	5-10	8-6	10-9	13-2	15-3	5-5	7-10	10-0	12-2	14-1
	Douglas fir-larch	#2	5-5	7-11	10-1	12-4	14-3	5-0	7-4	9-4	11-5	13-2
	Douglas fir-larch	#3	4-1	6-0	7-7	9-4	10-9	3-10	5-7	7-1	8-7	10-0
	Hem-fir	SS	5-9	9-1	11-11	15-2	18-0	5-9	9-1	11-9	14-5	15-11
	Hem-fir	#1	5-8	8-3	10-6	12-10	14-10	5-3	7-8	9-9	11-10	13-9
	Hem-fir	#2	5-4	7-10	9-11	12-1	14-1	4-11	7-3	9-2	11-3	13-0
	Hem-fir	#3	4-1	6-0	7-7	9-4	10-9	3-10	5-7	7-1	8-7	10-0
	Southern pine	SS	6-0	9-5	12-5	15-10	19-3	6-0	9-5	12-5	15-10	19-3
	Southern pine	#1	5-10	9-3	12-0	14-4	17-1	5-10	8-10	11-2	13-3	15-9
	Southern pine	#2	5-9	8-4	10-9	12-10	15-1	5-5	7-9	10-0	11-11	13-11
	Southern pine	#3	4-4	6-5	8-3	9-9	11-7	4-1	6-0	7-7	9-0	10-8
	Spruce-pine-fir	SS	5-8	8-10	11-8	14-8	17-1	5-8	8-10	11-2	13-7	15-9
	Spruce-pine-fir	#1	5-5	7-11	10-1	12-4	14-3	5-0	7-4	9-4	11-5	13-2
	Spruce-pine-fir	#2	5-5	7-11	10-1	12-4	14-3	5-0	7-4	9-4	11-5	13-2
	Spruce-pine-fir	#3	4-1	6-0	7-7	9-4	10-9	3-10	5-7	7-1	8-7	10-0

Check sources for availability of lumber in lengths greater than 20 feet.

For SI: 1 inch = 25.4 mm, 1 foot = 304.8 mm, 1 pound per square foot = 0.0479 kPa.

a. The tabulated rafter spans assume that ceiling joists are located at the bottom of the attic space or that some other method of resisting the outward push of the rafters on the bearing walls, such as rafter ties, is provided at that location. When ceiling joists or rafter ties are located higher in the attic space, the rafter spans shall be multiplied by the factors given below:

H_C/H_R	Rafter Span Adjustment Factor
1/3	0.67
1/4	0.76
1/5	0.83
1/6	0.90
1/7.5 or less	1.00

where:

H_C = Height of ceiling joists or rafter ties measured vertically above the top of the rafter support walls.

H_R = Height of roof ridge measured vertically above the top of the rafter support walls.

❖ See the commentary for Sections R802.5 and R802.5.1 and Table R802.5.1(1).

TABLE R802.5.1(7)
RAFTER SPANS FOR 70 PSF GROUND SNOW LOAD
(Ceiling not attached to rafters, L/Δ = 180)

RAFTER SPACING (inches)	SPECIES AND GRADE		DEAD LOAD = 10 psf					DEAD LOAD = 20 psf				
			2 × 4	2 × 6	2 × 8	2 × 10	2 × 12	2 × 4	2 × 6	2 × 8	2 × 10	2 × 12
			Maximum Rafter Spans[a]									
			(feet-inches)	(feet-inches)	(feet-inches)	(feet-inches)	(feet-inches)	(feet-inches)	(feet-inches)	(feet-inches)	(feet-inches)	(feet-inches)
12	Douglas fir-larch	SS	7-7	11-10	15-8	19-5	22-6	7-7	11-10	15-0	18-3	21-2
	Douglas fir-larch	#1	7-1	10-5	13-2	16-1	18-8	6-8	9-10	12-5	15-2	17-7
	Douglas fir-larch	#2	6-8	9-9	12-4	15-1	17-6	6-3	9-2	11-8	14-2	16-6
	Douglas fir-larch	#3	5-0	7-4	9-4	11-5	13-2	4-9	6-11	8-9	10-9	12-5
	Hem-fir	SS	7-2	11-3	14-9	18-10	22-1	7-2	11-3	14-8	18-0	20-10
	Hem-fir	#1	6-11	10-2	12-10	15-8	18-2	6-6	9-7	12-1	14-10	17-2
	Hem-fir	#2	6-7	9-7	12-2	14-10	17-3	6-2	9-1	11-5	14-0	16-3
	Hem-fir	#3	5-0	7-4	9-4	11-5	13-2	4-9	6-11	8-9	10-9	12-5
	Southern pine	SS	7-5	11-8	15-4	19-7	23-10	7-5	11-8	15-4	19-7	23-10
	Southern pine	#1	7-3	11-5	14-9	17-6	20-11	7-3	11-1	13-11	16-6	19-8
	Southern pine	#2	7-1	10-2	13-2	15-9	18-5	6-8	9-7	12-5	14-10	17-5
	Southern pine	#3	5-4	7-11	10-1	11-11	14-2	5-1	7-5	9-6	11-3	13-4
	Spruce-pine-fir	SS	7-0	11-0	14-6	18-0	20-11	7-0	11-0	13-11	17-0	19-8
	Spruce-pine-fir	#1	6-8	9-9	12-4	15-1	17-6	6-3	9-2	11-8	14-2	16-6
	Spruce-pine-fir	#2	6-8	9-9	12-4	15-1	17-6	6-3	9-2	11-8	14-2	16-6
	Spruce-pine-fir	#3	5-0	7-4	9-4	11-5	13-2	4-9	6-11	8-9	10-9	12-5
16	Douglas fir-larch	SS	6-10	10-9	13-9	16-10	19-6	6-10	10-3	13-0	15-10	18-4
	Douglas fir-larch	#1	6-2	9-0	11-5	13-11	16-2	5-10	8-6	10-9	13-2	15-3
	Douglas fir-larch	#2	5-9	8-5	10-8	13-1	15-2	5-5	7-11	10-1	12-4	14-3
	Douglas fir-larch	#3	4-4	6-4	8-1	9-10	11-5	4-1	6-0	7-7	9-4	10-9
	Hem-fir	SS	6-6	10-2	13-5	16-6	19-2	6-6	10-1	12-9	15-7	18-0
	Hem-fir	#1	6-0	8-9	11-2	13-7	15-9	5-8	8-3	10-6	12-10	14-10
	Hem-fir	#2	5-8	8-4	10-6	12-10	14-11	5-4	7-10	9-11	12-1	14-1
	Hem-fir	#3	4-4	6-4	8-1	9-10	11-5	4-1	6-0	7-7	9-4	10-9
	Southern pine	SS	6-9	10-7	14-0	17-10	21-8	6-9	10-7	14-0	17-10	21-0
	Southern pine	#1	6-7	10-2	12-9	15-2	18-1	6-5	9-7	12-0	14-4	17-1
	Southern pine	#2	6-2	8-10	11-5	13-7	16-0	5-10	8-4	10-9	12-10	15-1
	Southern pine	#3	4-8	6-10	8-9	10-4	12-3	4-4	6-5	8-3	9-9	11-7
	Spruce-pine-fir	SS	6-4	10-0	12-9	15-7	18-1	6-4	9-6	12-0	14-8	17-1
	Spruce-pine-fir	#1	5-9	8-5	10-8	13-1	15-2	5-5	7-11	10-1	12-4	14-3
	Spruce-pine-fir	#2	5-9	8-5	10-8	13-1	15-2	5-5	7-11	10-1	12-4	14-3
	Spruce-pine-fir	#3	4-4	6-4	8-1	9-10	11-5	4-1	6-0	7-7	9-4	10-9
19.2	Douglas fir-larch	SS	6-5	9-11	12-7	15-4	17-9	6-5	9-4	11-10	14-5	16-9
	Douglas fir-larch	#1	5-7	8-3	10-5	12-9	14-9	5-4	7-9	9-10	12-0	13-11
	Douglas fir-larch	#2	5-3	7-8	9-9	11-11	13-10	5-0	7-3	9-2	11-3	13-0
	Douglas fir-larch	#3	4-0	5-10	7-4	9-0	10-5	3-9	5-6	6-11	8-6	9-10
	Hem-fir	SS	6-1	9-7	12-4	15-1	17-4	6-1	9-2	11-8	14-2	15-5
	Hem-fir	#1	5-6	8-0	10-2	12-5	14-5	5-2	7-7	9-7	11-8	13-7
	Hem-fir	#2	5-2	7-7	9-7	11-9	13-7	4-11	7-2	9-1	11-1	12-10
	Hem-fir	#3	4-0	5-10	7-4	9-0	10-5	3-9	5-6	6-11	8-6	9-10
	Southern pine	SS	6-4	10-0	13-2	16-9	20-4	6-4	10-0	13-2	16-5	19-2
	Southern pine	#1	6-3	9-3	11-8	13-10	16-6	5-11	8-9	11-0	13-1	15-7
	Southern pine	#2	5-7	8-1	10-5	12-5	14-7	5-4	7-7	9-10	11-9	13-9
	Southern pine	#3	4-3	6-3	8-0	9-5	11-2	4-0	5-11	7-6	8-10	10-7
	Spruce-pine-fir	SS	6-0	9-2	11-8	14-3	16-6	5-11	8-8	11-0	13-5	15-7
	Spruce-pine-fir	#1	5-3	7-8	9-9	11-11	13-10	5-0	7-3	9-2	11-3	13-0
	Spruce-pine-fir	#2	5-3	7-8	9-9	11-11	13-10	5-0	7-3	9-2	11-3	13-0
	Spruce-pine-fir	#3	4-0	5-10	7-4	9-0	10-5	3-9	5-6	6-11	8-6	9-10

(continued)

TABLE R802.5.1(7)—continued
RAFTER SPANS FOR 70 PSF GROUND SNOW LOAD
(Ceiling not attached to rafters, L/Δ = 180)

RAFTER SPACING (inches)	SPECIES AND GRADE		DEAD LOAD = 10 psf					DEAD LOAD = 20 psf				
			2 × 4	2 × 6	2 × 8	2 × 10	2 × 12	2 × 4	2 × 6	2 × 8	2 × 10	2 × 12
			Maximum rafter spans[a]									
			(feet-inches)	(feet-inches)	(feet-inches)	(feet-inches)	(feet-inches)	(feet-inches)	(feet-inches)	(feet-inches)	(feet-inches)	(feet-inches)
24	Douglas fir-larch	SS	6-0	8-10	11-3	13-9	15-11	5-9	8-4	10-7	12-11	15-0
	Douglas fir-larch	#1	5-0	7-4	9-4	11-5	13-2	4-9	6-11	8-9	10-9	12-5
	Douglas fir-larch	#2	4-8	6-11	8-9	10-8	12-4	4-5	6-6	8-3	10-0	11-8
	Douglas fir-larch	#3	3-7	5-2	6-7	8-1	9-4	3-4	4-11	6-3	7-7	8-10
	Hem-fir	SS	5-8	8-8	11-0	13-6	13-11	5-7	8-3	10-5	12-4	12-4
	Hem-fir	#1	4-11	7-2	9-1	11-1	12-10	4-7	6-9	8-7	10-6	12-2
	Hem-fir	#2	4-8	6-9	8-7	10-6	12-2	4-4	6-5	8-1	9-11	11-6
	Hem-fir	#3	3-7	5-2	6-7	8-1	9-4	3-4	4-11	6-3	7-7	8-10
	Southern pine	SS	5-11	9-3	12-2	15-7	18-2	5-11	9-3	12-2	14-8	17-2
	Southern pine	#1	5-7	8-3	10-5	12-5	14-9	5-3	7-10	9-10	11-8	13-11
	Southern pine	#2	5-0	7-3	9-4	11-1	13-0	4-9	6-10	8-9	10-6	12-4
	Southern pine	#3	3-9	5-7	7-1	8-5	10-0	3-7	5-3	6-9	7-11	9-5
	Spruce-pine-fir	SS	5-6	8-3	10-5	12-9	14-9	5-4	7-9	9-10	12-0	12-11
	Spruce-pine-fir	#1	4-8	6-11	8-9	10-8	12-4	4-5	6-6	8-3	10-0	11-8
	Spruce-pine-fir	#2	4-8	6-11	8-9	10-8	12-4	4-5	6-6	8-3	10-0	11-8
	Spruce-pine-fir	#3	3-7	5-2	6-7	8-1	9-4	3-4	4-11	6-3	7-7	8-10

Check sources for availability of lumber in lengths greater than 20 feet.

For SI: 1 inch = 25.4 mm, 1 foot = 304.8 mm, 1 pound per square foot = 0.0479kPa.

a. The tabulated rafter spans assume that ceiling joists are located at the bottom of the attic space or that some other method of resisting the outward push of the rafters on the bearing walls, such as rafter ties, is provided at that location. When ceiling joists or rafter ties are located higher in the attic space, the rafter spans shall be multiplied by the factors given below:

H_C/H_R	Rafter Span Adjustment Factor
1/3	0.67
1/4	0.76
1/5	0.83
1/6	0.90
1/7.5 or less	1.00

where:

H_C = Height of ceiling joists or rafter ties measured vertically above the top of the rafter support walls.

H_R = Height of roof ridge measured vertically above the top of the rafter support walls.

❖ See the commentary for Sections R802.5 and R802.5.1 and Table R802.5.1(1).

TABLE R802.5.1(8)
RAFTER SPANS FOR 70 PSF GROUND SNOW LOAD
(Ceiling attached to rafters, L/Δ = 240)

RAFTER SPACING (inches)	SPECIES AND GRADE		DEAD LOAD = 10 psf					DEAD LOAD = 20 psf				
			2 × 4	2 × 6	2 × 8	2 × 10	2 × 12	2 × 4	2 × 6	2 × 8	2 × 10	2 × 12
			Maximum rafter spans[a]									
			(feet - inches)	(feet - inches)	(feet - inches)	(feet - inches)	(feet - inches)	(feet - inches)	(feet - inches)	(feet - inches)	(feet - inches)	(feet - inches)
12	Douglas fir-larch	SS	6-10	10-9	14-3	18-2	22-1	6-10	10-9	14-3	18-2	21-2
	Douglas fir-larch	#1	6-7	10-5	13-2	16-1	18-8	6-7	9-10	12-5	15-2	17-7
	Douglas fir-larch	#2	6-6	9-9	12-4	15-1	17-6	6-3	9-2	11-8	14-2	16-6
	Douglas fir-larch	#3	5-0	7-4	9-4	11-5	13-2	4-9	6-11	8-9	10-9	12-5
	Hem-fir	SS	6-6	10-2	13-5	17-2	20-10	6-6	10-2	13-5	17-2	20-10
	Hem-fir	#1	6-4	10-0	12-10	15-8	18-2	6-4	9-7	12-1	14-10	17-2
	Hem-fir	#2	6-1	9-6	12-2	14-10	17-3	6-1	9-1	11-5	14-0	16-3
	Hem-fir	#3	5-0	7-4	9-4	11-5	13-2	4-9	6-11	8-9	10-9	12-5
	Southern pine	SS	6-9	10-7	14-0	17-10	21-8	6-9	10-7	14-0	17-10	21-8
	Southern pine	#1	6-7	10-5	13-8	17-6	20-11	6-7	10-5	13-8	16-6	19-8
	Southern pine	#2	6-6	10-2	13-2	15-9	18-5	6-6	9-7	12-5	14-10	17-5
	Southern pine	#3	5-4	7-11	10-1	11-11	14-2	5-1	7-5	9-6	11-3	13-4
	Spruce-pine-fir	SS	6-4	10-0	13-2	16-9	20-5	6-4	10-0	13-2	16-9	19-8
	Spruce-pine-fir	#1	6-2	9-9	12-4	15-1	17-6	6-2	9-2	11-8	14-2	16-6
	Spruce-pine-fir	#2	6-2	9-9	12-4	15-1	17-6	6-2	9-2	11-8	14-2	16-6
	Spruce-pine-fir	#3	5-0	7-4	9-4	11-5	13-2	4-9	6-11	8-9	10-9	12-5
16	Douglas fir-larch	SS	6-3	9-10	12-11	16-6	19-6	6-3	9-10	12-11	15-10	18-4
	Douglas fir-larch	#1	6-0	9-0	11-5	13-11	16-2	5-10	8-6	10-9	13-2	15-3
	Douglas fir-larch	#2	5-9	8-5	10-8	13-1	15-2	5-5	7-11	10-1	12-4	14-3
	Douglas fir-larch	#3	4-4	6-4	8-1	9-10	11-5	4-1	6-0	7-7	9-4	10-9
	Hem-fir	SS	5-11	9-3	12-2	15-7	18-11	5-11	9-3	12-2	15-7	18-0
	Hem-fir	#1	5-9	8-9	11-2	13-7	15-9	5-8	8-3	10-6	12-10	14-10
	Hem-fir	#2	5-6	8-4	10-6	12-10	14-11	5-4	7-10	9-11	12-1	14-1
	Hem-fir	#3	4-4	6-4	8-1	9-10	11-5	4-1	6-0	7-7	9-4	10-9
	Southern pine	SS	6-1	9-7	12-8	16-2	19-8	6-1	9-7	12-8	16-2	19-8
	Southern pine	#1	6-0	9-5	12-5	15-2	18-1	6-0	9-5	12-0	14-4	17-1
	Southern pine	#2	5-11	8-10	11-5	13-7	16-0	5-10	8-4	10-9	12-10	15-1
	Southern pine	#3	4-8	6-10	8-9	10-4	12-3	4-4	6-5	8-3	9-9	11-7
	Spruce-pine-fir	SS	5-9	9-1	11-11	15-3	18-1	5-9	9-1	11-11	14-8	17-1
	Spruce-pine-fir	#1	5-8	8-5	10-8	13-1	15-2	5-5	7-11	10-1	12-4	14-3
	Spruce-pine-fir	#2	5-8	8-5	10-8	13-1	15-2	5-5	7-11	10-1	12-4	14-3
	Spruce-pine-fir	#3	4-4	6-4	8-1	9-10	11-5	4-1	6-0	7-7	9-4	10-9
19.2	Douglas fir-larch	SS	5-10	9-3	12-2	15-4	17-9	5-10	9-3	11-10	14-5	16-9
	Douglas fir-larch	#1	5-7	8-3	10-5	12-9	14-9	5-4	7-9	9-10	12-0	13-11
	Douglas fir-larch	#2	5-3	7-8	9-9	11-11	13-10	5-0	7-3	9-2	11-3	13-0
	Douglas fir-larch	#3	4-0	5-10	7-4	9-0	10-5	3-9	5-6	6-11	8-6	9-10
	Hem-fir	SS	5-6	8-8	11-6	14-8	17-4	5-6	8-8	11-6	14-2	15-5
	Hem-fir	#1	5-5	8-0	10-2	12-5	14-5	5-2	7-7	9-7	11-8	13-7
	Hem-fir	#2	5-2	7-7	9-7	11-9	13-7	4-11	7-2	9-1	11-1	12-10
	Hem-fir	#3	4-0	5-10	7-4	9-0	10-5	3-9	5-6	6-11	8-6	9-10
	Southern pine	SS	5-9	9-1	11-11	15-3	18-6	5-9	9-1	11-11	15-3	18-6
	Southern pine	#1	5-8	8-11	11-8	13-10	16-6	5-8	8-9	11-0	13-1	15-7
	Southern pine	#2	5-6	8-1	10-5	12-5	14-7	5-4	7-7	9-10	11-9	13-9
	Southern pine	#3	4-3	6-3	8-0	9-5	11-2	4-0	5-11	7-6	8-10	10-7
	Spruce-pine-fir	SS	5-5	8-6	11-3	14-3	16-6	5-5	8-6	11-0	13-5	15-7
	Spruce-pine-fir	#1	5-3	7-8	9-9	11-11	13-10	5-0	7-3	9-2	11-3	13-0
	Spruce-pine-fir	#2	5-3	7-8	9-9	11-11	13-10	5-0	7-3	9-2	11-3	13-0
	Spruce-pine-fir	#3	4-0	5-10	7-4	9-0	10-5	3-9	5-6	6-11	8-6	9-10

(continued)

TABLE R802.5.1(8)—continued
RAFTER SPANS FOR 70 PSF GROUND SNOW LOAD[a]
(Ceiling attached to rafters, L/Δ = 240)

RAFTER SPACING (inches)	SPECIES AND GRADE		DEAD LOAD = 10 psf					DEAD LOAD = 20 psf				
			2 × 4	2 × 6	2 × 8	2 × 10	2 × 12	2 × 4	2 × 6	2 × 8	2 × 10	2 × 12
			Maximum rafter spans[a]									
			(feet - inches)	(feet - inches)	(feet - inches)	(feet - inches)	(feet - inches)	(feet - inches)	(feet - inches)	(feet - inches)	(feet - inches)	(feet - inches)
24	Douglas fir-larch	SS	5-5	8-7	11-3	13-9	15-11	5-5	8-4	10-7	12-11	15-0
	Douglas fir-larch	#1	5-0	7-4	9-4	11-5	13-2	4-9	6-11	8-9	10-9	12-5
	Douglas fir-larch	#2	4-8	6-11	8-9	10-8	12-4	4-5	6-6	8-3	10-0	11-8
	Douglas fir-larch	#3	3-7	5-2	6-7	8-1	9-4	3-4	4-11	6-3	7-7	8-10
	Hem-fir	SS	5-2	8-1	10-8	13-6	13-11	5-2	8-1	10-5	12-4	12-4
	Hem-fir	#1	4-11	7-2	9-1	11-1	12-10	4-7	6-9	8-7	10-6	12-2
	Hem-fir	#2	4-8	6-9	8-7	10-6	12-2	4-4	6-5	8-1	9-11	11-6
	Hem-fir	#3	3-7	5-2	6-7	8-1	9-4	3-4	4-11	6-3	7-7	8-10
	Southern pine	SS	5-4	8-5	11-1	14-2	17-2	5-4	8-5	11-1	14-2	17-2
	Southern pine	#1	5-3	8-3	10-5	12-5	14-9	5-3	7-10	9-10	11-8	13-11
	Southern pine	#2	5-0	7-3	9-4	11-1	13-0	4-9	6-10	8-9	10-6	12-4
	Southern pine	#3	3-9	5-7	7-1	8-5	10-0	3-7	5-3	6-9	7-11	9-5
	Spruce-pine-fir	SS	5-0	7-11	10-5	12-9	14-9	5-0	7-9	9-10	12-0	12-11
	Spruce-pine-fir	#1	4-8	6-11	8-9	10-8	12-4	4-5	6-6	8-3	10-0	11-8
	Spruce-pine-fir	#2	4-8	6-11	8-9	10-8	12-4	4-5	6-6	8-3	10-0	11-8
	Spruce-pine-fir	#3	3-7	5-2	6-7	8-1	9-4	3-4	4-11	6-3	7-7	8-10

Check sources for availability of lumber in lengths greater than 20 feet.

For SI: 1 inch = 25.4 mm, 1 foot = 304.8 mm, 1 pound per square foot = 0.0479kPa.

a. The tabulated rafter spans assume that ceiling joists are located at the bottom of the attic space or that some other method of resisting the outward push of the rafters on the bearing walls, such as rafter ties, is provided at that location. When ceiling joists or rafter ties are located higher in the attic space, the rafter spans shall be multiplied by the factors given below:

H_C/H_R	Rafter Span Adjustment Factor
1/3	0.67
1/4	0.76
1/5	0.83
1/6	0.90
1/7.5 or less	1.00

where:

H_C = Height of ceiling joists or rafter ties measured vertically above the top of the rafter support walls.

H_R = Height of roof ridge measured vertically above the top of the rafter support walls.

❖ See the commentary for Sections R802.5 and R802.5.1 and Table R802.5.1(1).

TABLE R802.5.1(9)
RAFTER/CEILING JOIST HEEL JOINT CONNECTIONS[a, b, c, d, e, f, g]

RAFTER SLOPE	RAFTER SPACING (inches)	GROUND SNOW LOAD (psf)											
		30				50				70			
		Roof span (feet)											
		12	20	28	36	12	20	28	36	12	20	28	36
		Required number of 16d common nails[a,b] per heel joint splices[c,d,e,f]											
3:12	12	4	6	8	11	5	8	12	15	6	11	15	20
	16	5	8	11	14	6	11	15	20	8	14	20	26
	24	7	11	16	21	9	16	23	30	12	21	30	39
4:12	12	3	5	6	8	4	6	9	11	5	8	12	15
	16	4	6	8	11	5	8	12	15	6	11	15	20
	24	5	9	12	16	7	12	17	22	9	16	23	29
5:12	12	3	4	5	7	3	5	7	9	4	7	9	12
	16	3	5	7	9	4	7	9	12	5	9	12	16
	24	4	7	10	13	6	10	14	18	7	13	18	23
7:12	12	3	3	4	5	3	4	5	7	3	5	7	9
	16	3	4	5	6	3	5	7	9	4	6	9	11
	24	3	5	7	9	4	7	10	13	5	9	13	17
9:12	12	3	3	3	4	3	3	4	5	3	4	5	7
	16	3	3	4	5	3	4	5	7	3	5	7	9
	24	3	4	6	7	3	6	8	10	4	7	10	13
12:12	12	3	3	3	3	3	3	3	4	3	3	4	5
	16	3	3	3	4	3	3	4	5	3	4	5	7
	24	3	3	4	6	3	4	6	8	3	6	8	10

For SI: 1 inch = 25.4 mm, 1 foot = 304.8 mm, 1 pound per square foot = 0.0479kPa.

a. 40d box nails shall be permitted to be substituted for 16d common nails.

b. Nailing requirements shall be permitted to be reduced 25 percent if nails are clinched.

c. Heel joint connections are not required when the ridge is supported by a load-bearing wall, header or ridge beam.

d. When intermediate support of the rafter is provided by vertical struts or purlins to a loadbearing wall, the tabulated heel joint connection requirements shall be permitted to be reduced proportionally to the reduction in span.

e. Equivalent nailing patterns are required for ceiling joist to ceiling joist lap splices.

f. When rafter ties are substituted for ceiling joists, the heel joint connection requirement shall be taken as the tabulated heel joint connection requirement for two-thirds of the actual rafter-slope.

g. Tabulated heel joint connection requirements assume that ceiling joists or rafter ties are located at the bottom of the attic space. When ceiling joists or rafter ties are located higher in the attic, heel joint connection requirements shall be increased by the following factors:

H_C/H_R	Heel Joint Connection Adjustment Factor
1/3	1.5
1/4	1.33
1/5	1.25
1/6	1.2
1/10 or less	1.11

where:

H_C = Height of ceiling joists or rafter ties measured vertically above the top of the rafter support walls.

H_R = Height of roof ridge measured vertically above the top of the rafter support walls.

❖ This table contains requirements for the connection between roof rafters and ceiling joists to be used in areas with ground snow loads of 30 psf (1.44 kPa) up to 70 psf (3.35 kPa). The connection is essential to resist the thrust in the roof rafter where the ridge board does not provide vertical support of the roof rafter. Section R802.3 requires a designed ridge beam if the roof slope is less than 3:12.

Adjustment factors for the heel joint connection, in Note g are limited to cases where the ceiling joists or rafter ties are in the lower third of the attic space. When the ceiling joists or rafter ties are located higher in the attic space, lateral deflection of the rafter below the rafter ties can become excessive and require additional engineering analysis (see commentary, Section R802.3.1).

R802.7 Cutting and notching. Structural roof members shall not be cut, bored or notched in excess of the limitations specified in this section.

❖ The limitations on cutting and notching roof rafters and ceiling joists are similar to those for cutting and notching of floor joists (see commentary, Section R502.8).

R802.7.1 Sawn lumber. Notches in solid lumber joists, rafters and beams shall not exceed one-sixth of the depth of the member, shall not be longer than one-third of the depth of the member and shall not be located in the middle one-third of the span. Notches at the ends of the member shall not exceed one-fourth the depth of the member. The tension side of members 4 inches (102 mm) or greater in nominal thickness shall not be notched except at the ends of the members. The diameter of the holes bored or cut into members shall not exceed one-third the depth of the member. Holes shall not be closer than 2 inches (51 mm) to the top or bottom of the member, or to any other hole located in the member. Where the member is also notched, the hole shall not be closer than 2 inches (51 mm) to the notch.

Exception: Notches on cantilevered portions of rafters are permitted provided the dimension of the remaining portion of the rafter is not less than 4-inch nominal (102 mm) and the length of the cantilever does not exceed 24 inches (610 mm).

❖ Figure R502.8 illustrates cutting and notching limitations. The tension side of a member 4 inches (102 mm) or more in thickness can be notched only at the ends (also see commentary, Section R502.8).

R802.7.2 Engineered wood products. Cuts, notches and holes bored in trusses, structural composite lumber, structural glue-laminated members or I-joists are prohibited except where permitted by the manufacturer's recommendations or where the effects of such alterations are specifically considered in the design of the member by a registered design professional.

❖ The cutting and notching limitations for sawn lumber do not apply to engineered wood products. Structural composite lumber is a generic term which encompasses a variety of engineered composite wood products including laminated veneer lumber (LVL). Also, included in the term are laminated strand lumber (LSL), parallel strand lumber (PSL) and oriented strand lumber (OSL). The prohibitions in this section apply to all of these products.

Engineered wood products must not be cut, notched or bored unless those alterations are considered in the design of the member. That consideration is to be made either by the manufacturer and reflected in use recommendations (which is common in I-joists, permitting some limited alterations to webs) or by a registered design professional.

R802.8 Lateral support. Rafters and ceiling joists having a depth-to-thickness ratio exceeding 5 to 1 based on nominal dimensions shall be provided with lateral support at points of bearing to prevent rotation.

❖ Rafters and ceiling joists must be laterally supported to prevent twisting at the supports.

R802.8.1 Bridging. Rafters and ceiling joists having a depth-to-thickness ratio exceeding 6 to 1 based on nominal dimensions shall be supported laterally by solid blocking, diagonal bridging (wood or metal) or a continuous 1-inch by 3-inch (25 mm by 76 mm) wood strip nailed across the rafters or ceiling joists at intervals not exceeding 8 feet (2438 mm).

❖ Bridging must be installed for rafters and ceiling joists. This is similar to the requirements for floor joists in Section R502.7.1.

R802.9 Framing of openings. Openings in roof and ceiling framing shall be framed with header and trimmer joists. When the header joist span does not exceed 4 feet (1219 mm), the header joist may be a single member the same size as the ceiling joist or rafter. Single trimmer joists may be used to carry a single header joist that is located within 3 feet (914 mm) of the trimmer joist bearing. When the header joist span exceeds 4 feet (1219 mm), the trimmer joists and the header joist shall be doubled and of sufficient cross section to support the ceiling joists or rafter framing into the header. Approved hangers shall be used for the header joist to trimmer joist connections when the header joist span exceeds 6 feet (1829 mm). Tail joists over 12 feet (3658 mm) long shall be supported at the header by framing anchors or on ledger strips not less than 2 inches by 2 inches (51 mm by 51 mm).

❖ Requirements for framing at roof openings are similar to those of Section R502.10 for floor openings.

R802.10 Wood trusses.

R802.10.1 Truss design drawings. Truss design drawings, prepared in conformance to Section R802.10.1, shall be provided to the building official and approved prior to installation. Truss design drawings shall include, at a minimum, the information specified below. Truss design drawing shall be provided with the shipment of trusses delivered to the job site.

1. Slope or depth, span and spacing.

2. Location of all joints.

3. Required bearing widths.

4. Design loads as applicable.

    4.1. Top chord live load (as determined from Section R301.6).

    4.2. Top chord dead load.

    4.3. Bottom chord live load.

    4.4. Bottom chord dead load.

    4.5. Concentrated loads and their points of application.

    4.6. Controlling wind and earthquake loads.

5. Adjustments to lumber and joint connector design values for conditions of use.

6. Each reaction force and direction.

7. Joint connector type and description (e.g., size, thickness or gage) and the dimensioned location of each

joint connector except where symmetrically located relative to the joint interface.

8. Lumber size, species and grade for each member.

9. Connection requirements for:

    9.1. Truss to girder-truss.

    9.2. Truss ply to ply.

    9.3. Field splices.

10. Calculated deflection ratio and/or maximum description for live and total load.

11. Maximum axial compression forces in the truss members to enable the building designer to design the size, connections and anchorage of the permanent continuous lateral bracing. Forces shall be shown on the truss design drawing or on supplemental documents.

12. Required permanent truss member bracing location.

❖ See the definition in Chapter 2 of "Truss Design Drawing." Also see the commentary for Section R802.10.2.

R802.10.2 Design. Wood trusses shall be designed in accordance with accepted engineering practice. The design and manufacture of metal-plate-connected wood trusses shall comply with ANSI/TPI 1. The truss design drawings shall be prepared by a registered professional where required by the statutes of the jurisdiction in which the project is to be constructed in accordance with Section R106.1.

❖ The code contains no prescriptive provisions for the design and installation of wood trusses. A design is required in accordance with accepted engineering practice. For snow load, the truss design must be checked for both the balanced and unbalanced condition. Although the unbalanced condition controls for rafter design, that is not necessarily the case for truss design. The balanced condition could control. See the commentary for Section R802.5 for the snow load criteria. In recognition of the extensive use of trusses in residential construction, the code references ANSI/TPI 1, *National Design Standard for Metal-Plate-Connected Wood Truss Construction*. This standard provides regulations for the design and installation of metal-plate-connected wood trusses, including the procedures for full-scale tests and testing methods for evaluating metal-plate connectors. In addition to adequate design, the trusses must be handled and erected properly so the performance capabilities of the trusses are not compromised. A truss member should never be cut without approval from the design engineer.

Usually trusses are delivered in bundles, which reduces the potential for damage. Sufficient slings and spreader bars should be used to reduce stresses caused by sway and bending.

Shop drawings showing the lumber schedule, design loads and panel point details (size, location and attachment of plates) should be filed with the building permit application and should be available at the time of inspection.

R802.10.2.1 Applicability limits. The provisions of this section shall control the design of truss roof framing when snow controls for buildings not greater than 60 feet (18 288 mm) in length perpendicular to the joist, rafter or truss span, not greater than 36 feet (10 973 mm) in width parallel to the joist span or truss, not greater than two stories in height with each story not greater than 10 feet (3048 mm) high, and roof slopes not smaller than 3:12 (25-percent slope) or greater than 12:12 (100-percent slope). Truss roof framing constructed in accordance with the provisions of this section shall be limited to sites subjected to a maximum design wind speed of 110 miles per hour (49 m/s), Exposure A, B or C, and a maximum ground snow load of 70 psf (3352 Pa). Roof snow load is to be computed as: $0.7\ p_g$.

❖ This section defines the applicability limits of the snow load for wood trusses to be consistent with the snow load applicable to wood or steel rafters.

The roof snow load is based on a maximum ground snow load of 70 pounds per square foot (3.35 kPa). The roof snow load is taken as $0.7\ P_g$, where P_g is equal to the ground snow load.

Applied roof snow loads were calculated by multiplying the ground snow load by a 0.7 conversion factor in accordance with ASCE 7 (ASCE, 1998). No further reductions were made for special cases.

The sloped roof snow load, $P_s = Cs * P_f$, where P_f is the flat roof snow load. $P_f = 0.7 * C_e * C_t * I * P_g$.

Unbalanced snow loads, sliding snow loads, and snow drifts on lower roofs were not considered due to the lack of evidence for damage from unbalanced loads on homes and the lack of data to typify the statistical uncertainties associated with this load pattern on residential structures. Rain-on-snow surcharge load was also not considered in the calculations. Roof slopes in this document exceed the $^1\!/_2$-inch per foot requirement by ASCE 7 for the added load to be considered. Therefore, roof snow load was computed as: $1.0 * 0.7 * 1.0 * 1.0 * P_g = 0.7\ P_g$.

R802.10.3 Bracing. Trusses shall be braced to prevent rotation and provide lateral stability in accordance with the requirements specified in the construction documents for the building and on the individual truss design drawings. In the absence of specific bracing requirements, trusses shall be braced in accordance with the Building Component Safety Information (BCSI 1-03) Guide to Good Practice for Handling, Installing & Bracing of Metal Plate Connected Wood Trusses.

❖ To prevent their collapse during construction, and until permanent bracing is installed, trusses should be adequately braced temporarily. When braced for use, trusses should be positioned as close to vertical as possible; tilted trusses will not perform as required.

The construction documents and the individual truss design drawings should specify the bracing.

R802.10.4 Alterations to trusses. Truss members shall not be cut, notched, drilled, spliced or otherwise altered in any way without the approval of a registered design professional. Alterations resulting in the addition of load (e.g., HVAC equipment, water heater) that exceeds the design load for the truss shall not be permitted without verification that the truss is capable of supporting such additional loading.

❖ The addition of loads in excess of the design load is allowed only if the truss is shown to have the additional capacity. See the commentary for Section R502.8.2.

R802.10.5 Truss to wall connection. Trusses shall be connected to wall plates by the use of approved connectors having a resistance to uplift of not less than 175 pounds (779 N) and shall be installed in accordance with the manufacturer's specifications. For roof assemblies subject to wind uplift pressures of 20 pounds per square foot (960 Pa) or greater, as established in Table R301.2(2), adjusted for height and exposure per Table R301.2(3), see section R802.11.

❖ In conventional rafter-ceiling joist construction, the ceiling joist is toenailed to the top plate, and the rafter is toenailed to the top plate with the ceiling joist face nailed to the rafter. Toenailing of trusses to the top plate does not provide an equivalent connection. Toenailing may splinter the truss chord and compromise truss stability. This section requires that the proper framing anchor be used for the truss-to-wall connection to maintain the same structural integrity as expected in a conventionally framed roof.

R802.11 Roof tie-down.

❖ Because roof uplift caused by wind can be a significant factor in roof-system damage, the code requires roof-to-wall connections capable of resisting this force.

R802.11.1 Uplift resistance. Roof assemblies which are subject to wind uplift pressures of 20 pounds per square foot (960 Pa) or greater shall have roof rafters or trusses attached to their supporting wall assemblies by connections capable of providing the resistance required in Table R802.11. Wind uplift pressures shall be determined using an effective wind area of 100 square feet (9.3 m2) and Zone 1 in Table R301.2(2), as adjusted for height and exposure per Table R301.2(3).

A continuous load path shall be designed to transmit the uplift forces from the rafter or truss ties to the foundation.

❖ This section requires compliance with Table R602.3(1) for rafters and Section R802.10.5 for trusses in addition to Table R802.11, which specifies the minimum uplift resistance to be provided when the roof assemblies are subjected to wind uplift pressures of 20 pounds per square foot (960 Pa) or more. The toenail connections required by Table R602.3(1) could conceivably provide enough uplift resistance to satisfy Table R802.11 for short roof spans in the lower wind speed areas, but in almost all instances, additional uplift resistance is required in the form of approved connectors.

Basic wind speed for any site is determined from Figure R301.2(4) and recorded in Table R301.2(1). Using this wind speed, the designer selects the applicable uplift pressure from Table R301.2(2) for Zone 1 [see Figure R301.2(8)] using an effective wind area of 100 square feet (9.3 m²). This pressure must then be adjusted for height and exposure (see Section R301.2) using Table 301.2(3). If the resulting adjusted wind-uplift pressure is 20 psf (960 Pa) or more, uplift resistance must be provided in accordance with Table R802.11. Determination of the need for uplift resistance is illustrated in the following example. The connection of the rafter or truss to the wall must have the minimum uplift strength specified in Table R802.11.

Example:

A roof has an 18-degree (0.31 rad) slope and a mean height of 26 feet (7925 mm). The basic wind speed is 100 mph (45 m/s), and the site is Exposure B. Determine the adjusted wind-uplift pressure. Is uplift resistance required?

TABLE R802.11
REQUIRED STRENGTH OF TRUSS OR RAFTER CONNECTIONS TO RESIST WIND UPLIFT FORCES[a, b, c, e, f]
(Pounds per connection)

BASIC WIND SPEED (mph) (3-second gust)	ROOF SPAN (feet)							OVERHANGS[d] (pounds/foot)
	12	20	24	28	32	36	40	
85	-72	-120	-145	-169	-193	-217	-241	-38.55
90	-91	-151	-181	-212	-242	-272	-302	-43.22
100	-131	-218	-262	-305	-349	-393	-436	-53.36
110	-175	-292	-351	-409	-467	-526	-584	-64.56

For SI: 1 inch = 25.4 mm, 1 foot = 305 mm, 1 mph = 0.447 m/s, 1 pound/foot = 14.5939 N/m, 1 pound = 0.454 kg.

a. The uplift connection requirements are based on a 30 foot mean roof height located in Exposure B. For Exposures C and D and for other mean roof heights, multiply the above loads by the Adjustment Coefficients in Table R301.2(3).

b. The uplift connection requirements are based on the framing being spaced 24 inches on center. Multiply by 0.67 for framing spaced 16 inches on center and multiply by 0.5 for framing spaced 12 inches on center.

c. The uplift connection requirements include an allowance for 10 pounds of dead load.

d. The uplift connection requirements do not account for the effects of overhangs. The magnitude of the above loads shall be increased by adding the overhang loads found in the table. The overhang loads are also based on framing spaced 24 inches on center. The overhang loads given shall be multiplied by the overhang projection and added to the roof uplift value in the table.

e. The uplift connection requirements are based on wind loading on end zones as defined in Figure 6-2 of ASCE 7. Connection loads for connections located a distance of 20% of the least horizontal dimension of the building from the corner of the building are permitted to be reduced by multiplying the table connection value by 0.7 and multiplying the overhang load by 0.8.

f. For wall-to-wall and wall-to-foundation connections, the capacity of the uplift connector is permitted to be reduced by 100 pounds for each full wall above. (For example, if a 600-pound rated connector is used on the roof framing, a 500-pound rated connector is permitted at the next floor level down).

❖ See the commentary for Section R802.11.1.

Solution:

Using Table R301.2(2), find the section covering a roof slope greater than 10 through 30 degrees (0.175 through 0.524 rad). For Zone 1, with an effective wind area of 100 square feet (9.3 m²), the component pressures for 100 mph (45 m/s) are + 10 pounds per square foot and -14.9 pounds per square foot (+ 479 Pa and - 713 Pa). The negative values in this table represent uplift, and only these values should be used in complying with this section. The adjustment factor for Exposure B with a 26-foot (7925 mm) mean roof height is 1.00 from Table R301.2(3). Therefore, the adjusted uplift pressure = (14.9) (1.00) = 14.9 psf.

Because the adjusted uplift pressure is less than 20 pounds per square foot (960 Pa), no uplift resistance is required.

When uplift connectors are required by the table, one is required on every rafter or truss to the stud below, assuming the roof framing is spaced 24 inches (610 mm) on center, in accordance with Note b.

Additionally, the other notes to Table R802.11 may modify the tabular requirements. Also note that the text of this section requires that in addition to having the roof framing tied to the wall below, a load path must then be established to the foundation. Note f modifies the load requirements for the connections that establish the load path.

SECTION R803
ROOF SHEATHING

R803.1 Lumber sheathing. Allowable spans for lumber used as roof sheathing shall conform to Table R803.1. Spaced lumber sheathing for wood shingle and shake roofing shall conform to the requirements of Sections R905.7 and R905.8. Spaced lumber sheathing is not allowed in Seismic Design Category D₂.

❖ Table R803.1 specifies the minimum thickness of lumber roof sheathing based on rafter spacing. Spaced lumber sheathing used in conjunction with wood shingle or shake roofs is permitted when installed in accordance with Section R905. Because spaced lumber sheathing does not provide a roof diaphragm capable of transferring high in-plane loads to the supporting shear walls or braced wall lines, it is prohibited in structures classified as Seismic Design Category D₂.

TABLE R803.1
MINIMUM THICKNESS OF LUMBER ROOF SHEATHING

RAFTER OR BEAM SPACING (inches)	MINIMUM NET THICKNESS (inches)
24	$^5/_8$
48[a]	$1^1/_2$ T & G
60[b]	
72[c]	

For SI: 1 inch = 25.4 mm.
a. Minimum 270 F_b, 340,000 E.
b. Minimum 420 F_b, 660,000 E.
c. Minimum 600 F_b, 1,150,000 E.

❖ See the commentary for Section R803.1.

R803.2 Wood structural panel sheathing.

R803.2.1 Identification and grade. Wood structural panels shall conform to DOC PS 1, DOC PS 2 or, when manufactured in Canada, CSA 0437, and shall be identified by a grade mark or certificate of inspection issued by an approved agency. Wood structural panels shall comply with the grades specified in Table R503.2.1.1(1).

❖ Span rating and installation requirements for wood structural panels used as roof sheathing are similar to those of Section R503.2 for floor sheathing.

R803.2.1.1 Exposure durability. All wood structural panels, when designed to be permanently exposed in outdoor applications, shall be of an exterior exposure durability. Wood structural panel roof sheathing exposed to the underside may be of interior type bonded with exterior glue, identified as Exposure 1.

❖ Exterior-type wood structural panels are required when the wood structural panels are exposed to the environment. Exterior wood structural panels use a moisture-resistant glue that may penetrate the outer veneer lamination. Additionally, exterior wood structural panels typically use a higher grade of veneer for the inner plies, which are compatible with the exterior glue-line performance. Wood structural panel roof sheathing exposed only on the underside, such as eave overhangs, is permitted to be of the interior type with exterior glue because of the decrease in exposure hazard relative to moisture.

R803.2.1.2 Fire-retardant-treated plywood. The allowable unit stresses for fire-retardant-treated plywood, including fastener values, shall be developed from an approved method of investigation that considers the effects of anticipated temperature and humidity to which the fire-retardant-treated plywood will be subjected, the type of treatment and redrying process. The fire-retardant-treated plywood shall be graded by an approved agency.

❖ See the commentary for Section R802.1.3.

R803.2.2 Allowable spans. The maximum allowable spans for wood structural panel roof sheathing shall not exceed the values set forth in Table R503.2.1.1(1), or APA E30.

❖ Table 503.2.1.1(1) and APA E30 list allowable spans and loads for wood structural panel roof sheathing. Also see the commentary for Section R503.2.2.

R803.2.3 Installation. Wood structural panel used as roof sheathing shall be installed with joints staggered or not staggered in accordance with Table R602.3(1), or APA E30 for wood roof framing or with Table R804.3 for steel roof framing.

❖ This section refers to the fastening schedule in Chapter 6 for attachment to wood roof framing. Notes f and g to Table R602.3(1) establish zones requiring increased nailing based on criteria such as basic wind speed and/or mean roof height. APA E30, *APA Design and Construction Guide* is referenced as an alternative to the Chapter 6 fastening schedule.

This section also gives a cross reference to the Roof Framing Fastening Schedule, Table R804.3, for attachments to cold-formed steel roof framing.

SECTION R804
STEEL ROOF FRAMING

R804.1 General. Elements shall be straight and free of any defects that would significantly affect their structural performance. Cold-formed steel roof framing members shall comply with the requirements of this section.

❖ The provisions of this section govern cold-formed steel roof and ceiling framing.

R804.1.1 Applicability limits. The provisions of this section shall control the construction of steel roof framing for buildings not greater than 60 feet (18 288 mm) perpendicular to the joist, rafter or truss span, not greater than 40 feet (12 192 mm) in width parallel to the joist span or truss, not greater than two stories in height and roof slopes not smaller than 3:12 (25-percent slope) or greater than 12:12 (100 percent slope). Steel roof framing constructed in accordance with the provisions of this section shall be limited to sites subjected to a maximum design wind speed of 110 miles per hour (49 m/s), Exposure A, B, or C, and a maximum ground snow load of 70 pounds per square foot (3350 Pa).

❖ Other than providing roof slope limits, this section is identical to Section R505.1.1 and the limitations illustrated in Figure R505.1.1.

R804.1.2 In-line framing. Steel roof framing constructed in accordance with Section R804 shall be located directly in line with load-bearing studs below with a maximum tolerance of $^3/_4$ inch (19 mm) between the centerline of the stud and the roof joist/rafter.

❖ The code does not anticipate loads on the top plate of a wall that would cause bending of the top plate. Therefore, roof-framing members must be located directly in line with load-bearing studs of the supporting steel-framed walls as indicated in Figure R603.3. The permitted offset provides some tolerance for rafter location while not adversely affecting the top plate.

R804.1.3 Roof trusses. The design, quality assurance, installation and testing of cold-formed steel trusses shall be in accordance with the AISI Standard for Cold-formed Steel Framing-Truss Design (COFS/Truss). Truss members shall not be notched, cut or altered in any manner without an approved design.

❖ The code contains no prescriptive provisions for the design and installation of cold-formed steel trusses. A design is required in accordance with accepted engineering practice. Recognizing the extensive use of trusses in residential construction, the code references AISI, *Standard for Cold-Formed Steel Framing—Truss Design*. The standard provides regulations for the design and installation of cold-formed steel trusses, including quality assurance and the procedures for full-scale tests. In addition to adequate design, it is important that the trusses be handled and erected properly so that the performance capability of the trusses is not compromised. A truss member must never be cut, notched or altered without approval from the design professional.

R804.2 Structural framing. Load-bearing steel roof framing members shall comply with Figure R804.2(1) and with the dimensional and minimum thickness requirements specified in Tables R804.2(1) and R804.2(2). Tracks shall comply with Figure R804.2(2) and shall have a minimum flange width of $1^1/_4$ inches (32 mm). The maximum inside bend radius for load-bearing members shall be the greater of $^3/_{32}$ inch (2.4 mm) or twice the uncoated steel thickness. Holes in roof framing members shall comply with all of the following conditions:

1. Holes shall conform to Figure R804.2(3);

TABLE R804.2(1)
LOAD-BEARING COLD-FORMED STEEL MEMBER SIZES

NOMINAL MEMBER SIZE MEMBER DESIGNATION[a]	WEB DEPTH (inches)	MINIMUM FLANGE WIDTH (inches)	MAXIMUM FLANGE WIDTH (inches)	MINIMUM LIP SIZE (inches)
350S162-t	3.5	1.625	2	0.5
550S162-t	5.5	1.625	2	0.5
800S162-t	8	1.625	2	0.5
1000S162-t	10	1.625	2	0.5
1200S162-t	12	1.625	2	0.5

For SI: 1 inch = 25.4 mm.

a. The member designation is defined by the first number representing the member depth in hundredths of an inch, the letter "s" representing a stud or joist member, the second number representing the flange width in hundredths of an inch, and the letter "t" shall be a number representing the minimum base metal thickness in mils [see Table R804.2(2)].

❖ See the commentary for Section R804.2.

TABLE R804.2(2)
MINIMUM THICKNESS OF COLD-FORMED STEEL ROOF FRAMING MEMBERS

DESIGNATION (mils)	MINIMUM UNCOATED THICKNESS (inches)	REFERENCED GAGE NUMBER
33	0.033	20
43	0.043	18
54	0.054	16
68	0.068	14

For SI: 1 inch = 25.4 mm, 1 mil = 0.0254 mm.

❖ See the commentary for Section R804.2.

2. Holes shall be permitted only along the centerline of the web of the framing member;

3. Holes shall have a center-to-center spacing of not less than 24 inches (610 mm);

4. Holes shall have a width not greater than 0.5 times the member depth, or $2^1/_2$ inches (64 mm);

5. Holes shall have a length not exceeding $4^1/_2$ inches (114 mm); and

6. Holes shall have a minimum distance between the edge of the bearing surface and the edge of the hole of not less than 10 inches (254 mm).

Framing members with web holes not conforming to these requirements shall be patched in accordance with Section R804.3.6 or designed in accordance with accepted engineering practices.

❖ Figure R804.2(1) shows a typical load-bearing steel roof or ceiling member, which must comply with the dimensional and minimum thickness requirements specified in Tables R804.2(1) and R804.2(2). These tables are similar to those in Section R603 for walls and those in Section R505 for floors. Note "a" to Table R804.2(1) explains the member designation used for all steel framing members in the code. Steel thickness is expressed in mils, and Table R804.2(2) gives the equivalent thickness in inches as well as gage number. The referenced thickness pertains to the base metal thickness measured prior to painting or the application of galvanized coatings. The base metal thickness is typically stamped or embossed on the member by the manufacturer.

Holes are permitted only along the centerline of the web of roof-framing members. Their location and size

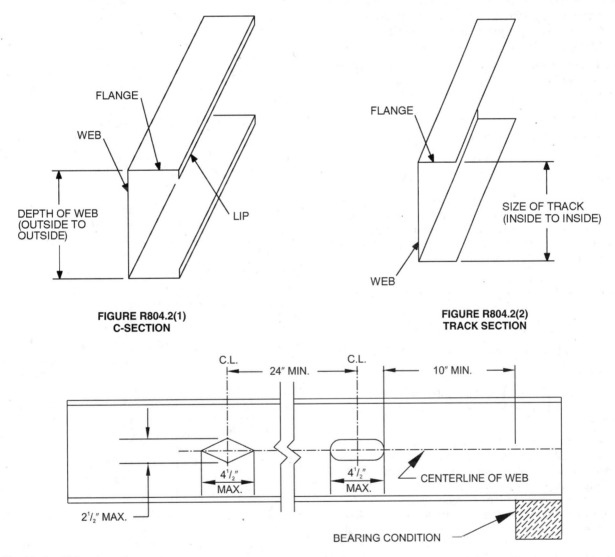

FIGURE R804.2(1)
C-SECTION

FIGURE R804.2(2)
TRACK SECTION

For SI: 1 inch = 25.4 mm.

FIGURE R804.2(3)
WEB HOLES

❖ See the commentary for Section R804.2 for Figures R804.2(1) through R804.2(3).

are limited as shown in Figure R804.2(3). Framing members with web holes exceeding these limits must be patched or designed in accordance with accepted engineering practices (see commentary for Section R804.3.6). These patching requirements are similar to those of Section R505.3.6. Also see the commentary for Section R505.2.

R804.2.1 Material. Load-bearing steel framing members shall be cold-formed to shape from structural quality sheet steel complying with the requirements of one of the following:

1. ASTM A 653: Grades 33, 37, 40 and 50 (Class 1 and 3).

2. ASTM A 792: Grades 33, 37, 40 and 50A.

3. ASTM A 875: Grades 33, 37, 40 and 50 (Class 1 and 3).

4. ASTM A 1003: Grades 33, 37, 40 and 50.

❖ Load-bearing steel framing members must be cold-formed to shape from structural quality sheet steel complying with the applicable material standards. The steel grades specified have the ductility and strength to meet the intent of these provisions.

R804.2.2 Identification. Load-bearing steel framing members shall have a legible label, stencil, stamp or embossment with the following information as a minimum:

1. Manufacturer's identification.

2. Minimum uncoated steel thickness in inches (mm).

3. Minimum coating designation.

4. Minimum yield strength, in kips per square inch (ksi).

❖ Load-bearing steel framing members must have a legible label, stencil, stamp or embossment. This identification allows for verification that materials that are installed are consistent with the design and meet the intent of the code.

R804.2.3 Corrosion protection. Load-bearing steel framing shall have a metallic coating complying with one of the following:

1. A minimum of G 60 in accordance with ASTM A 653.

2. A minimum of AZ 50 in accordance with ASTM A 792.

3. A minimum of GF 60 in accordance with ASTM A 875.

❖ The metallic coatings specified correspond to requirements in the referenced material standards in Section R804.2.1.

R804.2.4 Fastening requirements. Screws for steel-to-steel connections shall be installed with a minimum edge distance and center-to-center spacing of $^1/_2$ inch (13 mm), shall be self-drilling tapping, and shall conform to SAE J78. Structural sheathing shall be attached to roof rafters with minimum No. 8 self-drilling tapping screws that conform to SAE J78. Screws for attaching structural sheathing to steel roof framing shall have a minimum head diameter of 0.292 inch (7.4 mm) with countersunk heads and shall be installed with a minimum edge distance of $^3/_8$ inch (10 mm). Gypsum board ceilings shall be attached to steel joists with minimum No. 6 screws conforming to ASTM C 954 and shall be installed in accordance with Section R805. For all connections, screws shall extend through the

steel a minimum of three exposed threads. All self-drilling tapping screws conforming to SAE J78 shall have a minimum Type II coating in accordance with ASTM B 633.

Where No. 8 screws are specified in a steel-to-steel connection, reduction of the required number of screws in the connection is permitted in accordance with the reduction factors in Table R804.2.4 when larger screws are used or when one of the sheets of steel being connected is thicker that 33 mils (0.84 mm). When applying the reduction factor, the resulting number of screws shall be rounded up.

❖ Fasteners meeting the referenced standards and installed in accordance with this section will provide a load capacity consistent with these prescriptive provisions.

TABLE R804.2.4
SCREW SUBSTITUTION FACTOR

SCREW SIZE	THINNEST CONNECTED STEEL SHEET (mils)	
	33	43
#8	1.0	0.67
#10	0.93	0.62
#12	0.86	0.56

For SI: 1 mil = 0.0254 mm.

❖ See the commentary for Table R505.2.4.

R804.3 Roof construction. Steel roof systems constructed in accordance with the provisions of this section shall consist of both ceiling joists and rafters in accordance with Figure R804.3 and fastened in accordance with Table R804.3.

❖ Figure R804.3 illustrates steel roof system components. The roof-framing fastening schedule in Table R804.3 specifies the minimum connection requirements for components of the roof-framing system.

R804.3.1 Allowable ceiling joist spans. The clear span of cold-formed steel ceiling joists shall not exceed the limits set forth in Tables R804.3.1(1) through R804.3.1(8). Ceiling joists shall have a minimum bearing length of 1.5 inches (38 mm) and shall be connected to rafters (heel joint) in accordance with Figure R804.3.1(1) and Table R804.3.1. When continuous joists are framed across interior bearing supports, the interior bearing supports shall be located within 24 inches (610 mm) of midspan of the ceiling joist, and the individual spans shall not exceed the applicable spans in Tables R804.3.1(2), R804.3.1(4), R804.3.1(6), R804.3.1(8). Where required in Tables R804.3.1(1) through R804.3.1(8), bearing stiffeners shall be installed at each bearing location in accordance with Section R804.3.8 and Figure R804.3.8. When the attic is to be used as an occupied space, the ceiling joists shall be designed in accordance with Section R505.

❖ Allowable spans for ceiling joists are given in Tables R804.3.1(1), R804.3.1(2), R804.3.1(5) and R804.3.1(6) for live loads of 10 pounds per square foot (479 Pa) and Tables R804.3.1(3), R804.3.1(4), R804.3.1(7) and R804.3.1(8) for live loads of 20 psf (958 Pa). An attic with occupied space requires the use of higher design live loads, and the ceiling joists must be designed in accordance with Section R505.3.2. Ceiling joists must be connected to rafters in accordance with Figure R804.3.1(1) and Table R804.3.1(3). Use of the allowable span tables is illustrated below.

Example:

Steel ceiling joists at 16 inches (406 mm) on center support a ceiling with limited attic storage and are braced at mid-span. Their span is 17 feet, 6 inches (5334 mm) in a 35-foot wide (10 668 mm) building with a roof slope of 8:12 and 30 pounds per square foot (1436 Pa) snow load. Determine the required joist size and the required connection to rafters.

Solution:

The live load is 20 pounds per square foot (958 Pa) for limited attic storage. In Table R804.3.1(3), under the column headed mid-span bracing, look under 16-inch joist spacing. Note that 550S162-97 has an allowable span of 18 feet, 5 inches (5613 mm), which is greater than the required span.

The joist-rafter connection in Table R804.3.1 requires 4 No. 10 screws fastened per figure R804.3.1(1).

When continuous joists are framed across interior bearing supports, the interior bearing supports must be located within 24 inches (610 mm) of midspan of the ceiling joist, and the individual spans must not exceed the applicable spans in Tables R804.3.1(2), R804.3.1(4), R804.3.1(6) or R804.3.1(8).

Bearing stiffeners are required at each bearing point as well as at concentrated loads. See Note c to Tables R804.3.1(1) through R804.3.1(8). Bearing stiffeners must be installed in accordance with Section R804.3.8.

R804.3.2 Ceiling joist bracing. The bottom flanges of steel ceiling joists shall be laterally braced in accordance with Section R702. The top flanges of steel ceiling joists shall be laterally braced with a minimum of 33 mil (0.84 mm) C-section, 33 mil (0.84 mm) track section or $1^1/_2$ inch by 33 mil (38 mm by 0.84 mm) continuous steel strapping as required in Tables R804.3.1(1) through R804.3.1(8). Lateral bracing shall be installed in accordance with Figure R804.3. C-section, tracks or straps shall be fastened to the top flange at each joist with at least one No. 8 screw and shall be fastened to blocking with at least two No. 8 screws. Blocking or bridging (X-bracing) shall be installed between joists in line with strap bracing at a maximum spacing of 12 feet (3658 mm) measured perpendicular to the joists, and at the termination of all straps. The third-point bracing span values from Tables R804.3.1(1) through R804.3.1(8) shall be used for straps installed at closer spacings than third-point bracing, or when sheathing is applied to the top of the ceiling joists.

❖ The bottom flanges of steel ceiling joists must be laterally braced by a finished ceiling in accordance with Section R702. Tables R804.3.1(1) through R804.3.1(8) provide allowable spans for ceiling joists that are unbraced, braced at the midpoint (see Figure R804.3) or braced at the third-points. Top flanges of steel ceiling joists must be laterally braced according to the bracing

option selected from these tables and as shown in Figure R804.3. The allowable spans for third-point bracing should be used when top flange bracing is installed at closer intervals or when sheathing fastened to the top flange provides bracing. Strap bracing requires the installation of blocking or bridging similar to that of floor joists described in the commentary for Section R505.3.

R804.3.3 Allowable rafter spans. The horizontal projection of the rafter span, as shown in Figure R804.3, shall not exceed the limits set forth in Table R804.3.3(1). Wind speeds shall be converted to equivalent ground snow loads in accordance with Table R804.3.3(2). Rafter spans shall be selected based on the higher of the ground snow load or the equivalent snow load converted from the wind speed. When required, a rafter support brace shall be a minimum of 350S162-33 C-section with maximum length of 8 feet (2438 mm) and shall be connected to a ceiling joist and rafter with four No. 10 screws at each end.

❖ Table R804.3.3(1) lists allowable rafter spans based on ground snow loads. To account for wind loading on rafters, Table R804.3.3(2) must be used to convert wind speed to an equivalent ground snow load. Then the allowable rafter span is determined from Table R804.3.3(1) based on the larger of either the ground snow load or the equivalent snow load converted from the wind speed. As in Section R505.3.2 (for floor joists), a table must be used to determine allowable rafter spans.

When installed as illustrated in Figure R804.3, a rafter brace may be used to reduce the rafter span.

R804.3.3.1 Rafter framing. Rafters shall be connected to a parallel ceiling joist to form a continuous tie between exterior walls in accordance with Figures R804.3 and R804.3.1(1) and Table R804.3.1. Rafters shall be connected to a ridge member with a minimum 2-inch by 2-inch (51 mm by 51 mm) clip angle fastened with minimum No. 10 screws to the ridge member in accordance with Figure R804.3.3.1 and Table R804.3.3.1. The clip angle shall have a minimum steel thickness as the rafter member and shall extend the full depth of the rafter member. The ridge member shall be fabricated from a C-section and a track section, which shall be of a minimum size and steel thickness as the adjacent rafters and shall be installed in accordance with Figure R804.3.3.1.

❖ Ceiling joists connected to rafters as required in Section R804.3.1.1 provide the required continuous tie between exterior walls [see Figures R804.3 and R804.3.1(1)]. Rafters must be connected to a ridge member as illustrated in Figure R804.3.3.1.

R804.3.3.2 Roof cantilevers. Roof cantilevers shall not exceed 24 inches (610 mm) in accordance with Figure R804.3. Roof cantilevers shall be supported by a header in accordance with Section R603.6 or shall be supported by the floor framing in accordance with Section R505.3.7.

❖ Figure R804.3 shows the permitted roof cantilevers. In addition to limiting the gravity load, this provision places a cap on the amount of wind uplift on the cantilevered member, its hold-down and the supporting construction.

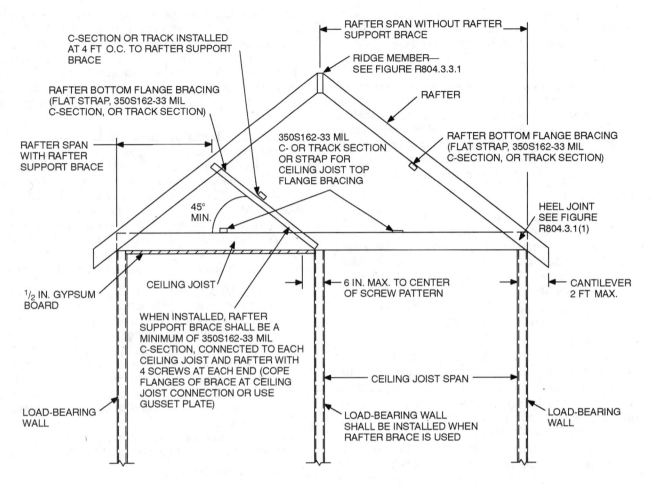

For SI: 1 inch = 25.4 mm, 1 foot = 304.8 mm, 1 mil = 0.0254 mm.

FIGURE R804.3
STEEL ROOF CONSTRUCTION

❖ This figure is a schematic of typical steel-framed room construction, providing section references to the code requirements for many of the roof components.

TABLE R804.3
ROOF FRAMING FASTENING SCHEDULE[a,b]

DESCRIPTION OF BUILDING ELEMENTS	NUMBER AND SIZE OF FASTENERS	SPACING OF FASTENERS
Ceiling joist to top track of load-bearing wall	2 No. 10 screws	Each joist
Roof sheathing (oriented strand board or plywood) to rafters	No. 8 screws	6″ o.c. on edges and 12″ o.c. at interior supports. 6″ o.c. at gable end truss
Truss to bearing wall[a]	2 No. 10 screws	Each truss
Gable end truss to endwall top track	No. 10 screws	12″ o.c.
Rafter to ceiling joist	Minimum No. 10 screws, per Table R804.3.1	Evenly spaced, not less than $^1/_2$″ from all edges.

For SI: 1 inch = 25.4 mm, 1 foot = 304.8 mm, 1 pound per square foot = 0.0479 kPa, 1 mil = 0.0254 mm.

a. Screws shall be applied through the flanges of the truss or ceiling joist or a 54 mil clip angle shall be used with two No. 10 screws in each leg. See Section R804.4 for additional requirements to resist uplift forces.

b. Spacing of fasteners on roof sheathing panel edges applies to panel edges supported by framing members and at all roof plane perimeters. Blocking of roof sheathing panel edges perpendicular to the framing members shall not be required except at the intersection of adjacent roof planes. Roof perimeter shall be supported by framing members or cold-formed blocking of the same depth and gage as the floor members.

❖ See the commentary for Section R804.3.

TABLE R804.3.1(1)
CEILING JOIST SPANS
SINGLE SPANS WITH BEARING STIFFENERS
10 lb per sq ft LIVE LOAD (NO ATTIC STORAGE)[a, b, c] 33 ksi STEEL

MEMBER DESIGNATION	ALLOWABLE SPAN (feet-inches)					
	Lateral Support of Top (Compression) Flange					
	Unbraced		Mid-Span Bracing		Third-Point Bracing	
	Ceiling Joist Spacing (inches)					
	16	24	16	24	16	24
350S162-33	9'-5"	8'-6"	12'-2"	10'-4"	12'-2"	10'-7"
350S162-43	10'-3"	9'-2"	12'-10"	11'-2"	12'-10"	11'-2"
350S162-54	11'-1"	9'-11"	13'-9"	12'-0"	13'-9"	12'-0"
350S162-68	12'-1"	10'-9"	14'-8"	12'-10"	14'-8"	12'-10"
350S162-97	14'-4"	12'-7"	16'-4"	14'-3"	16'-4"	14'-3"
550S162-33	10'-7"	9'-6"	14'-10"	12'-10"	15'-11"	13'-4"
550S162-43	11'-8"	10'-6"	16'-4"	14'-3"	17'-10"	15'-3"
550S162-54	12'-6"	11'-2"	17'-7"	15'-7"	19'-5"	16'-10"
550S162-68	13'-6"	12'-1"	19'-2"	17'-1"	21'-0"	18'-4"
550S162-97	15'-9"	13'-11"	21'-8"	19'-3"	23'-5"	20'-5"
800S162-33	12'-2"	10'-11"	17'-8"	15'-10"	19'-10"	17'-1"
800S162-43	13'-0"	11'-9"	18'-10"	17'-0"	21'-6"	19'-1"
800S162-54	13'-10"	12'-5"	20'-0"	18'-0"	22'-9"	20'-4"
800S162-68	14'-11"	13'-4"	21'-3"	19'-1"	24'-1"	21'-8"
800S162-97	17'-1"	15'-2"	23'-10"	21'-3"	26'-7"	23'-10"
1000S162-43	13'-11"	12'-6"	20'-2"	18'-3"	23'-1"	20'-9"
1000S162-54	14'-9"	13'-3"	21'-4"	19'-3"	24'-4"	22'-0"
1000S162-68	15'-10"	14'-2"	22'-8"	20'-5"	25'-9"	23'-2"
1000S162-97	18'-0"	16'-0"	25'-3"	22'-7"	28'-3"	25'-4"
1200S162-43	14'-8"	13'-3"	21'-4"	19'-3"	24'-5"	21'-8"
1200S162-54	15'-7"	14'-0"	22'-6"	20'-4"	25'-9"	23'-2"
1200S162-68	16'-8"	14'-11"	23'-11"	21'-6"	27'-2"	24'-6"
1000S162-97	18'-9"	16'-9"	26'-6"	23'-8"	29'-9"	26'-9"

For SI: 1 inch = 25.4 mm, 1 foot = 304.8 mm, 1 pound per square foot = 0.0479 kPa.

a. Deflection criterion: L/240 for total loads.

b. Ceiling dead load = 5 psf.

c. Bearing stiffeners are required at all bearing points and concentrated load locations.

❖ See the commentary for Section R804.3.1.

TABLE R804.3.1(2)
CEILING JOIST SPANS
TWO EQUAL SPANS WITH BEARING STIFFENERS
10 lb per sq ft LIVE LOAD (NO ATTIC STORAGE)[a, b, c] 33 ksi STEEL

MEMBER DESIGNATION	ALLOWABLE SPAN (feet-inches)					
	Lateral Support of Top (Compression) Flange					
	Unbraced		Mid-Span Bracing		Third-Point Bracing	
	Ceiling Joist Spacing (inches)					
	16	24	16	24	16	24
350S162-33	12'-11"	10'-11"	13'-5"	10'-11"	13'-5"	10'-11"
350S162-43	14'-2"	12'-8"	15'-10"	12'-11"	15'-10"	12'-11"
350S162-54	15'-6"	13'-10"	17'-1"	14'-6"	17'-9"	14'-6"
350S162-68	17'-3"	15'-3"	18'-6"	16'-1"	19'-8"	16'-1"
350S162-97	20'-10"	18'-4"	21'-5"	18'-10"	21'-11"	18'-10"
550S162-33	14'-4"	12'-11"	16'-7"	14'-1"	17'-3"	14'-1"
550S162-43	16'-0"	14'-1"	17'-11"	16'-1"	20'-7"	16'-10"
550S162-54	17'-4"	15'-6"	19'-5"	17'-6"	23'-2"	19'-0"
550S162-68	19'-1"	16'-11"	20'-10"	18'-8"	25'-2"	21'-5"
550S162-97	22'-8"	19'-9"	23'-6"	20'-11"	27'-11"	25'-1"
800S162-33	16'-5"	14'-10"	19'-2"	17'-3"	23'-1"	18'-3"
800S162-43	17'-9"	15'-11"	20'-6"	18'-5"	25'-0"	22'-6"
800S162-54	19'-1"	17'-1"	21'-8"	19'-6"	26'-4"	23'-9"
800S162-68	20'-9"	18'-6"	23'-1"	20'-9"	28'-0"	25'-2"
800S162-97	24'-5"	21'-6"	26'-0"	23'-2"	31'-1"	27'-9"
1000S162-43	18'-11"	17'-0"	21'-11"	19'-9"	26'-8"	24'-1"
1000S162-54	20'-3"	18'-2"	23'-2"	20'-10"	28'-2"	25'-5"
1000S162-68	21'-11"	19'-7"	24'-7"	22'-2"	29'-10"	26'-11"
1000S162-97	25'-7"	22'-7"	27'-6"	24'-6"	33'-0"	29'-7"
1200S162-43	19'-11"	17'-11"	23'-1"	20'-10"	28'-3"	25'-6"
1200S162-54	21'-3"	19'-1"	24'-5"	22'-0"	29'-9"	26'-10"
1200S162-68	23'-0"	20'-7"	25'-11"	23'-4"	31'-6"	28'-4"
1000S162-97	26'-7"	23'-6"	28'-9"	25'-10"	34'-8"	31'-1"

For SI: 1 inch = 25.4 mm, 1 foot = 304.8 mm, 1 pound per square foot = 0.0479 kPa.
a. Deflection criterion: L/240 for total loads.
b. Ceiling dead load = 5 psf.
c. Bearing stiffeners are required at all bearing points and concentrated load locations.

❖ See the commentary for Section R804.3.1.

TABLE R804.3.1(3)
CEILING JOIST SPANS
SINGLE SPANS WITH BEARING STIFFENERS
20 lb per sq ft LIVE LOAD (LIMITED ATTIC STORAGE)[a, b, c] 33 ksi STEEL

MEMBER DESIGNATION	ALLOWABLE SPAN (feet-inches)					
	Lateral Support of Top (Compression) Flange					
	Unbraced		Mid-Span Bracing		Third-Point Bracing	
	Ceiling Joist Spacing (inches)					
	16	24	16	24	16	24
350S162-33	8'-2"	7'-2"	9'-9"	8'-1"	9'-11"	8'-1"
350S162-43	8'-10"	7'-10"	11'-0"	9'-5"	11'-0"	9'-7"
350S162-54	9'-6"	8'-6"	11'-9"	10'-3"	11'-9"	10'-3"
350S162-68	10'-4"	9'-2"	12'-7"	11'-0"	12'-7"	11'-0"
350S162-97	12'-1"	10'-8"	14'-0"	12'-0"	14'-0"	12'-0"
550S162-33	9'-2"	8'-3"	12'-2"	10'-2"	12'-6"	10'-5"
550S162-43	10'-1"	9'-1"	13'-7"	11'-7"	14'-5"	12'-2"
550S162-54	10'-9"	9'-8"	14'-10"	12'-10"	15'-11"	13'-6"
550S162-68	11'-7"	10'-4"	16'-4"	14'-0"	17'-5"	14'-11"
550S162-97	13'-4"	11'-10"	18'-5"	16'-2"	20'-1"	17'-1"
800S162-33	10'-7"	9'-6"	15'-1"	13'-0"	16'-2"	13'-7"
800S162-43	11'-4"	10'-2"	16'-5"	14'-6"	18'-2"	15'-9"
800S162-54	12'-0"	10'-9"	17'-4"	15'-6"	19'-6"	17'-0"
800S162-68	12'-10"	11'-6"	18'-5"	16'-6"	20'-10"	18'-3"
800S162-97	14'-7"	12'-11"	20'-5"	18'-3"	22'-11"	20'-5"
1000S162-43	12'-1"	10'-11"	17'-7"	15'-10"	19'-11"	17'-3"
1000S162-54	12'-10"	11'-6"	18'-7"	16'-9"	21'-2"	18'-10"
1000S162-68	13'-8"	12'-3"	19'-8"	17'-8"	22'-4"	20'-1"
1000S162-97	15'-4"	13'-8"	21'-8"	19'-5"	24'-5"	21'-11"
1200S162-43	12'-9"	11'-6"	18'-7"	16'-6"	20'-9"	18'-2"
1200S162-54	13'-6"	12'-2"	19'-7"	17'-8"	22'-5"	20'-2"
1200S162-68	14'-4"	12'-11"	20'-9"	18'-8"	23'-7"	21'-3"
1000S162-97	16'-1"	14'-4"	22'-10"	20'-6"	25'-9"	23'-2"

For SI: 1 inch = 25.4 mm, 1 foot = 304.8 mm, 1 pound per square foot = 0.0479 kPa.

a. Deflection criterion: $L/240$ for total loads.

b. Ceiling dead load = 5 psf.

c. Bearing stiffeners are required at all bearing points and concentrated load locations.

❖ See the commentary for Section R804.3.1.

TABLE R804.3.1(4)
CEILING JOIST SPANS
TWO EQUAL SPANS WITH BEARING STIFFENERS
20 lb per sq ft LIVE LOAD (LIMITED ATTIC STORAGE)[a, b, c] 33 ksi STEEL

MEMBER DESIGNATION	ALLOWABLE SPAN (feet-inches)					
	Lateral Support of Top (Compression) Flange					
	Unbraced		Mid-Span Bracing		Third-Point Bracing	
	Ceiling Joist Spacing (inches)					
	16	24	16	24	16	24
350S162-33	10'-2"	8'-4"	10'-2"	8'-4"	10'-2"	8'-4"
350S162-43	12'-1"	9'-10"	12'-1"	9'-10"	12'-1"	9'-10"
350S162-54	13'-3"	11'-0"	13'-6"	11'-0"	13'-6"	11'-0"
350S162-68	14'-7"	12'-3"	15'-0"	12'-3"	15'-0"	12'-3"
350S162-97	17'-6"	14'-3"	17'-6"	14'-3"	17'-6"	14'-3"
550S162-33	12'-5"	10'-9"	13'-2"	10'-9"	13'-2"	10'-9"
550S162-43	13'-7"	12'-1"	15'-6"	12'-9"	15'-8"	12'-9"
550S162-54	14'-11"	13'-4"	16'-10"	14'-5"	17'-9"	14'-5"
550S162-68	16'-3"	14'-5"	18'-0"	16'-1"	20'-0"	16'-4"
550S162-97	19'-1"	16'-10"	20'-3"	18'-0"	23'-10"	19'-5"
800S162-33	14'-3"	12'-4"	16'-7"	12'-4"	16'-7"	12'-4"
800S162-43	15'-4"	13'-10"	17'-9"	16'-0"	21'-8"	17'-9"
800S162-54	16'-5"	14'-9"	18'-10"	16'-11"	22'-11"	20'-6"
800S162-68	17'-9"	15'-11"	20'-0"	18'-0"	24'-3"	21'-10"
800S162-97	20'-8"	18'-3"	22'-3"	19'-11"	26'-9"	24'-0"
1000S162-43	16'-5"	14'-9"	19'-0"	17'-2"	23'-3"	18'-11"
1000S162-54	17'-6"	15'-8"	20'-1"	18'-1"	24'-6"	22'-1"
1000S162-68	18'-10"	16'-10"	21'-4"	19'-2"	25'-11"	23'-4"
1000S162-97	21'-8"	19'-3"	23'-7"	21'-2"	28'-5"	25'-6"
1200S162-43	17'-3"	15'-7"	20'-1"	18'-2"	24'-6"	18'-3"
1200S162-54	18'-5"	16'-6"	21'-3"	19'-2"	25'-11"	23'-5"
1200S162-68	19'-9"	17'-8"	22'-6"	20'-3"	27'-4"	24'-8"
1000S162-97	22'-7"	20'-1"	24'-10"	22'-3"	29'-11"	26'-11"

For SI: 1 inch = 25.4 mm, 1 foot = 304.8 mm, 1 pound per square foot = 0.0479 kPa.

a. Deflection criterion: $L/240$ for total loads.

b. Ceiling dead load = 5 psf.

c. Bearing stiffeners are required at all bearing points and concentrated load locations.

❖ See the commentary for Section R804.1(4).

TABLE R804.3.1(5)
CEILING JOIST SPANS
SINGLE SPANS WITHOUT BEARING STIFFENERS
10 lb per sq ft LIVE LOAD (NO ATTIC STORAGE)[a, b] 33 ksi STEEL

MEMBER DESIGNATION	ALLOWABLE SPAN (feet-inches)					
	Lateral Support of Top (Compression) Flange					
	Unbraced		Mid-Span Bracing		Third-Point Bracing	
	Ceiling Joist Spacing (inches)					
	16	24	16	24	16	24
350S162-33	9'-5"	8'-6"	12'-2"	10'-4"	12'-2"	10'-7"
350S162-43	10'-3"	9'-12"	13'-2"	11'-6"	13'-2"	11'-6"
350S162-54	11'-1"	9'-11"	13'-9"	12'-0"	13'-9"	12'-0"
350S162-68	12'-1"	10'-9"	14'-8"	12'-10"	14'-8"	12'-10"
350S162-97	14'-4"	12'-7"	16'-10"	14'-3"	16'-4"	14'-3"
550S162-33	10'-7"	9'-6"	14'-10"	12'-10"	15'-11"	13'-4"
550S162-43	11'-8"	10'-6"	16'-4"	14'-3"	17'-10"	15'-3"
550S162-54	12'-6"	11'-2"	17'-7"	15'-7"	19'-5"	16'-10"
550S162-68	13'-6"	12'-1"	19'-2"	17'-0"	21'-0"	18'-4"
550S162-97	15'-9"	13'-11"	21'-8"	19'-3"	23'-5"	20'-5"
800S162-33	—	—	—	—	—	—
800S162-43	13'-0"	11'-9"	18'-10"	17'-0"	21'-6"	19'-0"
800S162-54	13'-10"	12'-5"	20'-0"	18'-0"	22'-9"	20'-4"
800S162-68	14'-11"	13'-4"	21'-3"	19'-1"	24'-1"	21'-8"
800S162-97	17'-1"	15'-2"	23'-10"	21'-3"	26'-7"	23'-10"
1000S162-43	—	—	—	—	—	—
1000S162-54	14'-9"	13'-3"	21'-4"	19'-3"	24'-4"	22'-0"
1000S162-68	15'-10"	14'-2"	22'-8"	20'-5"	25'-9"	23'-2"
1000S162-97	18'-0"	16'-0"	25'-3"	22'-7"	28'-3"	25'-4"
1200S162-43	—	—	—	—	—	—
1200S162-54	—	—	—	—	—	—
1200S162-68	16'-8"	14'-11"	23'-11"	21'-6"	27'-2"	24'-6"
1000S162-97	18'-9"	16'-9"	26'-6"	23'-8"	29'-9"	26'-9"

For SI: 1 inch = 25.4 mm, 1 foot = 304.8 mm, 1 pound per square foot = 0.0479 kPa.

a. Deflection criterion: $L/240$ for total loads.

b. Ceiling dead load = 5 psf.

❖ See the commentary for Section R804.3.1.

TABLE R804.3.1(6)
CEILING JOIST SPANS
TWO EQUAL SPANS WITHOUT BEARING STIFFENERS
10 lb per sq ft LIVE LOAD (NO ATTIC STORAGE)[a, b] 33 ksi STEEL

MEMBER DESIGNATION	ALLOWABLE SPAN (feet-inches)					
	Lateral Support of Top (Compression) Flange					
	Unbraced		Mid-Span Bracing		Third-Point Bracing	
	Ceiling Joist Spacing (inches)					
	16	24	16	24	16	24
350S162-33	11'-9"	8'-11"	11'-9"	8'-11"	11'-9"	8'-11"
350S162-43	14'-2"	11'-7"	14'-11"	11'-7"	14'-11"	11'-7"
350S162-54	15'-6"	13'-10"	17'-1"	13'-10"	17'-7"	13'-10"
350S162-68	17'-3"	15'-3"	18'-6"	16'-1"	19'-8"	16'-1"
350S162-97	20'-10"	18'-4"	21'-5"	18'-9"	21'-11"	18'-9"
550S162-33	13'-4"	9'-11"	13'-4"	9'-11"	13'-4"	9'-11"
550S162-43	16'-0"	13'-6"	17'-9"	13'-6"	17'-9"	13'-6"
550S162-54	17'-4"	15'-6"	19'-5"	16'-10"	21'-9"	16'-10"
550S162-68	19'-1"	16'-11"	20'-10"	18'-8"	24'-11"	20'-6"
550S162-97	22'-8"	20'-0"	23'-9"	21'-1"	28'-2"	25'-1"
800S162-33	—	—	—	—	—	—
800S162-43	17'-9"	15'-7"	20'-6"	15'-7"	21'-0"	15'-7"
800S162-54	19'-1"	17'-1"	21'-8"	19'-6"	26'-4"	23'-10"
800S162-68	20'-9"	18'-6"	23'-1"	20'-9"	28'-0"	25'-2"
800S162-97	24'-5"	21'-6"	26'-0"	23'-2"	31'-1"	27'-9"
1000S162-43	—	—	—	—	—	—
1000S162-54	20'-3"	18'-2"	23'-2"	20'-10"	28'-2"	21'-2"
1000S162-68	21'-11"	19'-7"	24'-7"	22'-2"	29'-10"	26'-11"
1000S162-97	25'-7"	22'-7"	27'-6"	24'-6"	33'-0"	29'-7"
1200S162-43	—	—	—	—	—	—
1200S162-54	—	—	—	—	—	—
1200S162-68	23'-0"	20'-7"	25'-11"	23'-4"	31'-6"	28'-4"
1000S162-97	26'-7"	23'-6"	28'-9"	25'-10"	34'-8"	31'-1"

For SI: 1 inch = 25.4 mm, 1 foot = 304.8 mm, 1 pound per square foot = 0.0479 kPa.
a. Deflection criterion: *L*/240 for total loads.
b. Ceiling dead load = 5 psf.

❖ See the commentary for Section R804.3.1.

TABLE R804.3.1(7)
CEILING JOIST SPANS
SINGLE SPANS WITHOUT BEARING STIFFENERS
20 lb per sq ft LIVE LOAD (LIMITED ATTIC STORAGE)[a, b] 33 ksi STEEL

MEMBER DESIGNATION	ALLOWABLE SPAN (feet-inches)					
	Lateral Support of Top (Compression) Flange					
	Unbraced		Mid-Span Bracing		Third-Point Bracing	
	Ceiling Joist Spacing (inches)					
	16	24	16	24	16	24
350S162-33	8'-2"	6'-10"	9'-9"	6'-10"	9'-11"	6'-10"
350S162-43	8'-10"	7'-10"	11'-0"	9'-5"	11'-0"	9'-7"
350S162-54	9'-6"	8'-6"	11'-9"	10'-3"	11'-9"	10'-3"
350S162-68	10'-4"	9'-2"	12'-7"	11'-0"	12'-7"	11'-0"
350S162-97	12'-10"	10'-8"	13'-9"	12'-0"	13'-9"	12'-0"
550S162-33	9'-2"	8'-3"	12'-2"	8'-5"	12'-6"	8'-5"
550S162-43	10'-1"	9'-1"	13'-7"	11'-8"	14'-5"	12'-2"
550S162-54	10'-9"	9'-8"	14'-10"	12'-10"	15'-11"	13'-6"
550S162-68	11'-7"	10'-4"	16'-4"	14'-0"	17'-5"	14'-11"
550S162-97	13'-4"	11'-10"	18'-5"	16'-2"	20'-1"	17'-4"
800S162-33	—	—	—	—	—	—
800S162-43	11'-4"	10'-1"	16'-5"	13'-6"	18'-1"	13'-6"
800S162-54	20'-0"	10'-9"	17'-4"	15'-6"	19'-6"	27'-0"
800S162-68	12'-10"	11'-6"	18'-5"	16'-6"	20'-10"	18'-3"
800S162-97	14'-7"	12'-11"	20'-5"	18'-3"	22'-11"	20'-5"
1000S162-43	—	—	—	—	—	—
1000S162-54	12'-10"	11'-6"	18'-7"	16'-9"	21'-2"	15'-5"
1000S162-68	13'-8"	12'-3"	19'-8"	17'-8"	22'-4"	20'-1"
1000S162-97	15'-4"	13'-8"	21'-8"	19'-5"	24'-5"	21'-11"
1200S162-43	—	—	—	—	—	—
1200S162-54	—	—	—	—	—	—
1200S162-68	14'-4"	12'-11"	20'-9"	18'-8"	23'-7"	21'-3"
1000S162-97	16'-1"	14'-4"	22'-10"	20'-6"	25'-9"	23'-2"

For SI: 1 inch = 25.4 mm, 1 foot = 304.8 mm, 1 pound per square foot = 0.0479 kPa.

a. Deflection criterion: *L*/240 for total loads.

b. Ceiling dead load = 5 psf.

❖ See the commentary for Section R804.3.1.

TABLE R804.3.1(8)
CEILING JOIST SPANS
TWO EQUAL SPANS WITHOUT BEARING STIFFENERS
20 lb per sq ft LIVE LOAD (LIMITED ATTIC STORAGE)[a, b] 33 ksi STEEL

MEMBER DESIGNATION	ALLOWABLE SPAN (feet-inches)					
	Lateral Support of Top (Compression) Flange					
	Unbraced		Mid-Span Bracing		Third-Point Bracing	
	Ceiling Joist Spacing (inches)					
	16	24	16	24	16	24
350S162-33	8'-1"	6'-1"	8'-1"	6'-1"	8'-1"	6'-1"
350S162-43	10'-7"	8'-1"	10'-7"	8'-1"	10'-7"	8'-1"
350S162-54	12'-8"	9'-10"	12'-8"	9'-10"	12'-8"	9'-10"
350S162-68	14'-7"	11'-10"	14'-11"	11'-10"	14'-11"	11'-10"
350S162-97	17'-6"	14'-3"	17'-6"	14'-3"	17'-6"	14'-3"
550S162-33	8'-11"	6'-8"	8'-11"	6'-8"	8'-11"	6'-8"
550S162-43	12'-3"	9'-2"	12'-3"	9'-2"	12'-3"	9'-2"
550S162-54	14'-11"	11'-8"	15'-4"	11'-8"	15'-4"	11'-8"
550S162-68	16'-3"	14'-5"	18'-0"	15'-8"	18'-10"	14'-7"
550S162-97	19'-1"	16'-10"	20'-3"	18'-0"	23'-9"	19'-5"
800S162-33	—	—	—	—	—	—
800S162-43	13'-11"	9'-10"	13'-11"	9'-10"	13'-11"	9'-10"
800S162-54	16'-5"	13'-9"	18'-8"	13'-9"	18'-8"	13'-9"
800S162-68	17'-9"	15'-11"	20'-0"	18'-0"	24'-1"	18'-3"
800S162-97	20'-8"	18'-3"	22'-3"	19'-11"	26'-9"	24'-0"
1000S162-43	—	—	—	—	—	—
1000S162-54	17'-6"	13'-11"	19'-1"	13'-11"	19'-1"	13'-11"
1000S162-68	18'-10"	16'-10"	21'-4"	19'-2"	25'-11"	19'-7"
1000S162-97	21'-8"	19'-3"	23'-7"	21'-2"	28'-5"	25'-6"
1200S162-43	—	—	—	—	—	—
1200S162-54	—	—	—	—	—	—
1200S162-68	19'-9"	17'-8"	22'-6"	19'-8"	26'-8"	19'-8"
1000S162-97	22'-7"	20'-1"	24'-10"	22'-3"	29'-11"	26'-11"

For SI: 1 inch = 25.4 mm, 1 foot = 304.8 mm, 1 pound per square foot = 0.0479 kPa.
a. Deflection criterion: *L*/240 for total loads.
b. Ceiling dead load = 5 psf.

❖ See the commentary for Section R804.3.1.

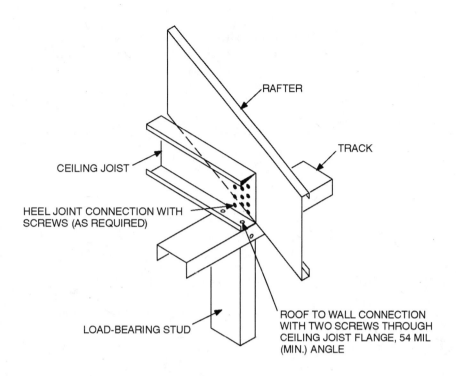

For SI: 1 mil = 0.0254 mm.

FIGURE R804.3.1(1)
JOIST TO RAFTER CONNECTION

❖ The fastening to the wall track is illustrated. See the commentary for Table R804.3.1(3) for a discussion of the joist-to-rafter connection.

TABLE R804.3.1
NUMBER OF SCREWS REQUIRED FOR CEILING JOIST TO RAFTER CONNECTION[a]

ROOF SLOPE	NUMBER OF SCREWS																			
	Building width (feet)																			
	24				28				32				36				40			
	Ground snow load (psf)																			
	20	30	50	70	20	30	50	70	20	30	50	70	20	30	50	70	20	30	50	70
3/12	5	6	9	11	5	7	10	13	6	8	11	15	7	8	13	17	8	9	14	19
4/12	4	5	7	9	4	5	8	10	5	6	9	12	5	7	10	13	6	7	11	14
5/12	3	4	6	7	4	4	6	8	4	5	7	10	5	5	8	11	5	6	9	12
6/12	3	3	5	6	3	4	6	7	4	4	6	8	4	5	7	9	4	5	8	10
7/12	3	3	4	6	3	3	5	7	3	4	6	7	4	4	6	8	4	5	7	9
8/12	2	3	4	5	3	3	5	6	3	4	5	7	3	4	6	8	4	4	6	8
9/12	2	3	4	5	3	3	4	6	3	3	5	6	3	4	5	7	3	4	6	8
10/12	2	2	4	5	2	3	4	5	3	3	5	6	3	3	5	7	3	4	6	7
11/12	2	2	3	4	2	3	4	5	3	3	4	6	3	3	5	6	3	4	5	7
12/12	2	2	3	4	2	3	4	5	2	3	4	5	3	3	5	6	3	4	5	7

For SI: 1 inch = 25.4 mm, 1 foot = 304.8 mm, 1 pound per square foot = 0.0479 kPa.
a. Screws shall be No. 10.

❖ This table lists the connections required between the roof rafter and the ceiling joists. The connection is essential to resist the thrust in the roof rafter where the ridge board does not provide vertical support to the roof rafter.

TABLE R804.3.3(1)
ALLOWABLE HORIZONTAL RAFTER SPANS[a, b, c] 33 ksi STEEL

MEMBER DESIGNATION	ALLOWABLE SPAN MEASURED HORIZONTALLY (feet-inches)							
	Ground Snow Load							
	20 psf		30 psf		50 psf		70 psf	
	Rafter spacing (in)							
	16	24	16	24	16	24	16	24
550S162-33	14'-0"	11'-5"	11'-10"	9'-8"	9'-5"	7'-8"	8'-1"	6'-7"
550S162-43	16'-6"	13'-10"	14'-4"	11'-9"	11'-5"	9'-4"	9'-10"	8'-0"
550S162-54	17'-9"	15'-6"	15'-6"	13'-2"	12'-11"	10'-6"	11'-1"	9'-0"
550S162-68	19'-0"	16'-7"	16'-8"	14'-7"	14'-1"	11'-10"	12'-5"	10'-2"
550S162-97	21'-2"	18'-6"	18'-7"	16'-2"	15'-8"	13'-8"	14'-0"	12'-2"
800S162-33	17'-0"	13'-11"	14'-5"	11'-9"	11'-6"	7'-9"	8'-6"	5'-8"
800S162-43	21'-1"	17'-3"	17'-10"	14'-7"	14'-3"	11'-7"	12'-2"	9'-11"
800S162-54	23'-11"	20'-4"	21'-0"	17'-3"	16'-10"	13'-9"	14'-5"	11'-9"
800S162-68	25'-9"	22'-6"	22'-7"	19'-5"	19'-0"	15'-6"	16'-3"	13'-3"
800S162-97	28'-9"	25'-1"	25'-2"	22'-0"	21'-3"	18'-7"	19'-0"	16'-0"
1000S162-43	23'-4"	19'-1"	19'-9"	16'-2"	15'-9"	12'-11"	13'-6"	10'-0"
1000S162-54	27'-8"	22'-7"	23'-5"	19'-1"	18'-8"	15'-3"	16'-0"	13'-1"
1000S162-68	30'-11"	27'-0"	27'-2"	22'-11"	22'-5"	18'-3"	19'-2"	15'-8"
1000S162-97	34'-7"	30'-2"	30'-4"	26'-6"	25'-7"	22'-1"	22'-10"	18'-11"
1200S162-43	25'-5"	20'-9"	21'-6"	17'-6"	17'-1"	11'-5"	12'-6"	8'-6"
1200S162-54	30'-0"	24'-6"	25'-5"	20'-9"	20'-3"	16'-7"	17'-5"	14'-2"
1200S162-68	35'-5"	28'-11"	30'-0"	24'-6"	23'-11"	19'-6"	20'-6"	16'-9"
1200S162-97	40'-4"	35'-3"	35'-5"	30'-11"	29'-10"	25'-5"	26'-8"	21'-9"

For SI: 1 inch = 25.4 mm, 1 foot = 304.8 mm, 1 pound per square foot = 0.0479 kPa.

a. Table provides maximum horizontal rafter spans in feet and inches for slopes between 3:12 and 12:12.

b. Deflection criterion: $L/240$ for live loads and $L/180$ for total loads.

c. Roof dead load = 12 psf.

❖ See the commentary for Section R804.3.3.

TABLE R804.3.3(2)
BASIC WIND SPEED TO EQUIVALENT SNOW LOAD CONVERSION

BASIC WIND SPEED AND EXPOSURE		EQUIVALENT GROUND SNOW LOAD (psf)									
		Roof slope									
Exp. A/B	Exp. C	3:12	4:12	5:12	6:12	7:12	8:12	9:12	10:12	11:12	12:12
85 mph	—	20	20	20	20	20	20	30	30	30	30
100 mph	85 mph	20	20	20	20	30	30	30	30	50	50
110 mph	100 mph	20	20	20	20	30	50	50	50	50	50
—	110 mph	30	30	30	50	50	50	70	70	70	—

For SI: 1 mile per hour = 0.447 m/s, 1 pound per square foot = 0.0479 kN/m².

❖ See the commentary for Section R804.3.3.

TABLE R804.3.3.1
NUMBER OF SCREWS REQUIRED AT EACH LEG OF CLIP
ANGLE FOR RAFTER TO RIDGE MEMBER CONNECTION[a]

BUILDING WIDTH (feet)	NUMBER OF SCREWS			
	Ground snow load (psf)			
	0 to 20	21 to 30	31 to 50	51 to 70
24	2	2	3	4
28	2	3	4	5
32	2	3	4	5
36	3	3	5	6
40	3	4	5	7

For SI: 1 inch = 25.4 mm, 1 foot = 304.8 mm, 1 pound per square foot = 0.0479 kPa.
a. Screws shall be No. 10 minimum.

❖ See the commentary for Section R804.3.3.1.

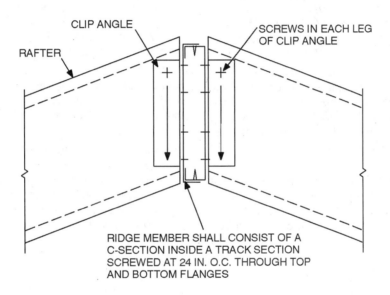

RAFTER
CLIP ANGLE
SCREWS IN EACH LEG OF CLIP ANGLE
RIDGE MEMBER SHALL CONSIST OF A C-SECTION INSIDE A TRACK SECTION SCREWED AT 24 IN. O.C. THROUGH TOP AND BOTTOM FLANGES

For SI: 1 inch = 25.4 mm.

FIGURE R804.3.3.1
RIDGE BOARD CONNECTION

❖ See the commentary for Section R804.3.3.1.

R804.3.4 Rafter bottom flange bracing. The bottom flanges of steel rafters shall be continuously braced with a minimum 33-mil (0.84 mm) C-section, 33-mil (0.84 mm) track section, or a 1^1/$_2$-inch by 33-mil (38 mm by 0.84 mm) steel strapping at a maximum spacing of 8 feet (2438 mm) as measured parallel to the rafters. Bracing shall be installed in accordance with Figure R804.3. The C-section, track section, or straps shall be fastened to blocking with at least two No. 8 screws. Blocking or bridging (X-bracing) shall be installed between rafters in-line with the continuous bracing at a maximum spacing of 12 feet (3658 mm) measured perpendicular to the rafters and at the termination of all straps. The ends of continuous bracing shall be fastened to blocking with at least two No. 8 screws.

❖ The bottom flanges of steel rafters require continuous bracing as illustrated in Figure R804.3. Installation of bracing is necessary to provide resistance to twisting and/or lateral displacement of the rafters.

R804.3.5 Cutting and notching. Flanges and lips of load-bearing steel roof framing members shall not be cut or notched. Holes in webs shall be in accordance with Section R804.2.

❖ Flanges and lips of load-bearing steel roof framing members must not be cut or notched because this would affect their structural integrity.

R804.3.6 Hole patching. Web holes not conforming to the requirements in Section R804.2 shall be designed in accordance with one of the following:

1. Framing members shall be replaced or designed in accordance with accepted engineering practices when web holes exceed the following size limits:

    1.1. The depth of the hole, measured across the web, exceeds 70 percent of the flat width of the web; or,

1.2. The length of the hole, measured along the web, exceeds 10 inches (254 mm) or the depth of the web, whichever is greater.

2. Web holes not exceeding the dimensional requirements in Section R804.3.6, Item 1 shall be patched with a solid steel plate, stud section, or track section in accordance with Figure R804.3.6. The steel patch shall be of a minimum thickness as the receiving member and shall extend at least 1 inch (25 mm) beyond all edges of the hole. The steel patch shall be fastened to the web of the receiving member with No. 8 screws spaced no greater than 1 inch (25 mm) center-to-center along the edges of the patch with minimum edge distance of $^1/_2$ inch (13 mm).

❖ Framing members with web holes exceeding the limits, outlined in Item 1 of this section, must be replaced or designed in accordance with accepted engineering practice. Framing members with web holes within the limits of Item 1, but exceeding the limits of Section R804.2, must be patched in accordance with Item 2 of this section (see commentary, Section R804.2).

R804.3.7 Splicing. Rafters and other structural members, except ceiling joists, shall not be spliced. Splices in ceiling joists shall only be permitted at interior bearing points and shall be constructed in accordance with Figure R804.3.7(1). Spliced ceiling joists shall be connected with the same number and size of screws on connection. Splicing of tracks shall conform to Figure R804.3.7(2).

❖ Splicing of load-carrying steel framing members is not permitted except when lapped ceiling joists occur at interior bearing points in accordance with Figure R804.3.7(1). Because Section R804.3.3.1 requires that ceiling joists provide a continuous tie between the exterior walls, the number of screws required for this splice is the same as required by Table R804.3.1(3) for the heel joint connection shown in Figure R804.3.1(1).

Splicing of tracks is allowed because, in accordance with Section R804.2, they are not permitted to act as load-carrying members [see Figure R804.3.7(2)].

R804.3.8 Bearing stiffener. A bearing stiffener shall be fabricated from a minimum 33-mil (0.84 mm) C-section or track section. Each stiffener shall be fastened to the web of the ceiling joist with a minimum of four No. 8 screws equally spaced as shown in Figure R804.3.8. Stiffeners shall extend across the full depth of the web and shall be installed on either side of the web.

❖ A bearing stiffener (also referred to as a "transverse stiffener" or "web stiffener") must be installed as shown in Figure R804.3.8. The stiffener at the joist/rafter connection is similar to the requirements in Section R505.3.4, except a 33 mil (0.84 mm) track may be used rather than the 43 mil (1.09 mm) required in Section R505.3.

R804.3.9 Headers. Roof-ceiling framing above wall openings shall be supported on headers. The allowable spans for headers in bearing walls shall not exceed the values set forth in Table R603.6(1).

❖ Headers must be located immediately below the ceiling or roof framing (for example, at the top of the track), above wall openings, in all exterior walls and in interior load-bearing walls in accordance with Figure R603.6. Table R603.6(1) gives the allowable spans for headers in bearing walls.

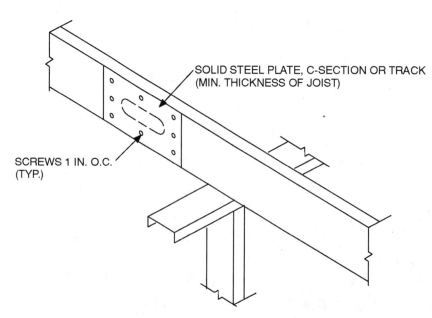

For SI: 1 inch = 25.4 mm.

FIGURE R804.3.6
HOLE PATCHING

❖ See the commentary for Section R804.2.

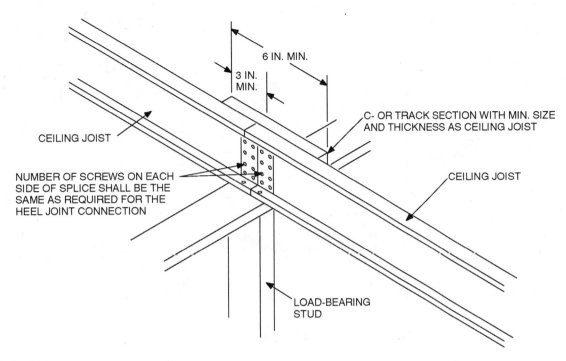

For SI: 1 inch = 25.4 mm.

**FIGURE R804.3.7(1)
SPLICED CEILING JOISTS**

❖ See the commentary for Section R804.3.7.

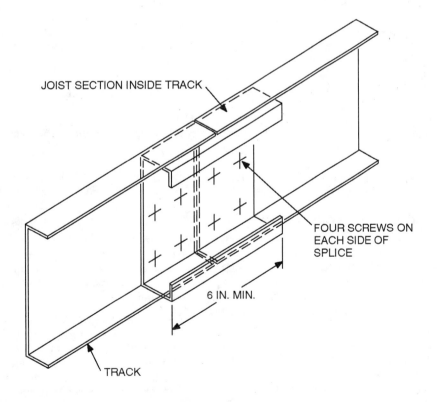

For SI: 1 inch = 25.4 mm.

**FIGURE R804.3.7(2)
TRACK SPLICE**

❖ See the commentary for Section R804.3.7.

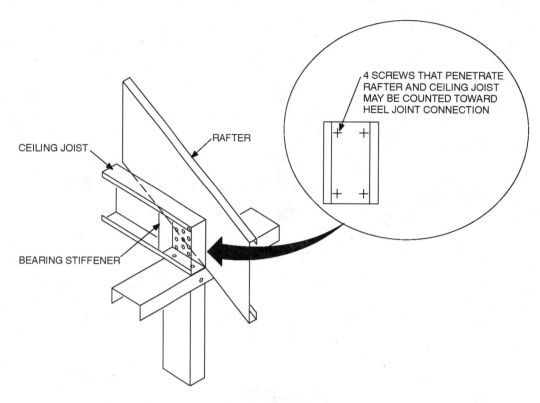

FIGURE R804.3.8
BEARING STIFFNER

❖ See the commentary for Section R804.3.8.

R804.3.10 Framing of opening. Openings in roof and ceiling framing shall be framed with headers and trimmers between ceiling joists or rafters. Header joist spans shall not exceed 4 feet (1219 mm). Header and trimmer joists shall be fabricated from joist and track sections, which shall be of a minimum size and thickness in accordance with Figures R804.3.10(1) and R804.3.10(2). Each header joist shall be connected to a trimmer joist with a minimum of four 2-inch by 2-inch (51 by 51 mm) clip angles. Each clip angle shall be fastened to both the header and trimmer joists with four No. 8 screws, evenly spaced, through each leg of the clip angle. The clip angles shall have a steel thickness not less than that of the floor joist.

❖ Roof openings must be framed as illustrated in Figures R804.3.10(1) and R804.3.10(2). Each header joist must be connected to the trimmer joists with a minimum of four 2-inch by 2-inch (51 mm by 51 mm) clip angles, which must be fastened to both the header and trimmer joists with four No. 8 screws, evenly spaced, through each leg. The clip angle's thickness must not be less than that of the floor joist.

R804.4 Roof tie-down. Roof assemblies subject to wind uplift pressures of 20 pounds per square foot (0.96 kN/m²) or greater, as established in Table R301.2(2), shall have rafter-to-bearing wall ties provided in accordance with Table R802.11.

❖ Steel-framed roof assemblies require rafter-to-bearing wall ties with a capacity determined in the manner described in Section R802.11.

SECTION R805
CEILING FINISHES

R805.1 Ceiling installation. Ceilings shall be installed in accordance with the requirements for interior wall finishes as provided in Section R702.

❖ This section provides a reference to the ceiling installation requirements found in Section R702.

SECTION R806
ROOF VENTILATION

R806.1 Ventilation required. Enclosed attics and enclosed rafter spaces formed where ceilings are applied directly to the underside of roof rafters shall have cross ventilation for each separate space by ventilating openings protected against the entrance of rain or snow. Ventilating openings shall be provided with corrosion-resistant wire mesh, with ¹/₈ inch (3.2 mm) minimum to ¹/₄ inch (6 mm) maximum openings.

❖ Large amounts of water vapor migrate by air movement or diffusion through the building envelope materials because of a vapor pressure difference. The sources of water vapor include cooking, laundering, bathing and human breathing and perspiration. These can account for an average daily production of 25 pounds (11.3 kg) of water vapor in a typical family-of-four dwelling. The average can be much higher where appliances such as humidifiers, washers and dryers are used.

As the vapor moves into the attic, it may reach its dew point, condensing on wood roof components. This wetting and drying action will cause rotting and decay. To avoid this, the attic must be ventilated to prevent the accumulation of water on building components. The installation of a vapor retarder acts to prevent the passage of moisture to the attic, and an effective vapor retarder allows a decrease in ventilation. Vapor retarders are ineffective when openings in the barrier allow moisture to be carried by air into the attic. This is also the reason exhaust fans must terminate outdoors and not in the attic. Care should be exercised to assure that attic vent openings remain unobstructed.

To minimize condensation problems within attic and enclosed rafter spaces, free-flow ventilation of such spaces is required. Ventilation openings must be screened to prevent the entry of animals.

R806.2 Minimum area. The total net free ventilating area shall not be less than $^1/_{150}$ of the area of the space ventilated except that reduction of the total area to $^1/_{300}$ is permitted, provided that at least 50 percent and not more than 80 percent of the required ventilating area is provided by ventilators located in the upper portion of the space to be ventilated at least 3 feet (914 mm) above the eave or cornice vents with the balance of the required ventilation provided by eave or cornice vents. As an alternative, the net free cross-ventilation area may be reduced to $^1/_{300}$ when a vapor barrier having a transmission rate not exceeding 1 perm (5.7×10^{-11} kg/s · m^2 · Pa) is installed on the warm-in-winter side of the ceiling.

❖ The attic vent size required by the code should not be overlooked. The net-free area can be as much as 50 percent less than the gross opening area. For example, one manufacturer's 24-inch square (610 mm) gable vent [gross area equals 576 square inches (0.37 m^2)] is listed in their catalog as having a net free area of 308 square inches (0.20 m^2), which is about 53 percent of the gross area. The manufacturer's literature should be consulted to obtain free-area information.

R806.3 Vent and insulation clearance. Where eave or cornice vents are installed, insulation shall not block the free flow of air. A minimum of a 1-inch (25 mm) space shall be provided between the insulation and the roof sheathing and at the location of the vent.

❖ Vent openings must be maintained clear, and they must not block the free flow of air; therefore, the code requires that insulation be held back from the vent opening a minimum of 1 inch (25.4 mm). The 1 inch (25 mm) clearance must be maintained not only at the vent but throughout the attic and rafter spaces.

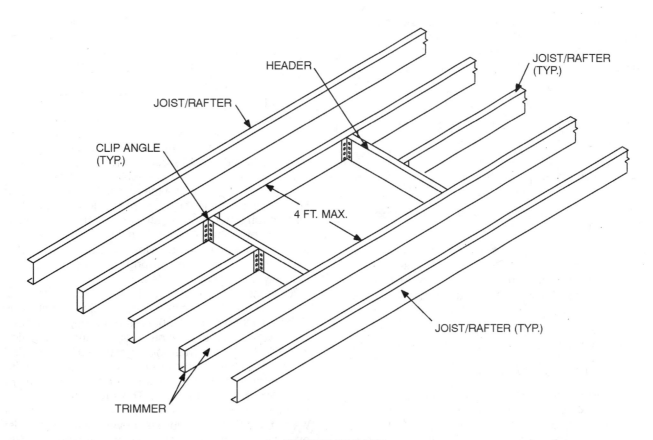

FIGURE R804.3.10(1)
ROOF OPENING

❖ See the commentary for Section R804.3.10.

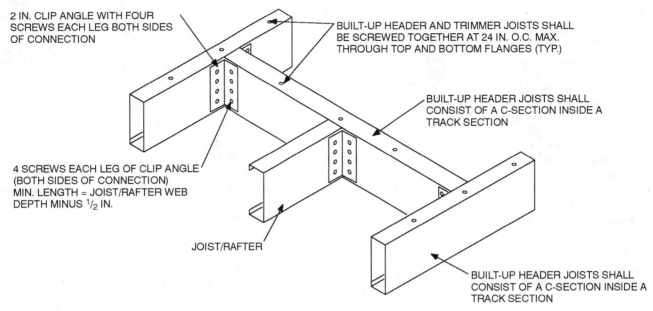

2 IN. CLIP ANGLE WITH FOUR
SCREWS EACH LEG BOTH SIDES
OF CONNECTION

BUILT-UP HEADER AND TRIMMER JOISTS SHALL
BE SCREWED TOGETHER AT 24 IN. O.C. MAX.
THROUGH TOP AND BOTTOM FLANGES (TYP.)

BUILT-UP HEADER JOISTS SHALL
CONSIST OF A C-SECTION INSIDE A
TRACK SECTION

4 SCREWS EACH LEG OF CLIP ANGLE
(BOTH SIDES OF CONNECTION)
MIN. LENGTH = JOIST/RAFTER WEB
DEPTH MINUS $^1/_2$ IN.

JOIST/RAFTER

BUILT-UP HEADER JOISTS SHALL
CONSIST OF A C-SECTION INSIDE A
TRACK SECTION

For SI: 1 inch

FIGURE R804.3.10(2)
ROOF OPENING

❖ See the commentary for Section R804.3.10.

R806.4 Conditioned attic assemblies. Unvented conditioned attic assemblies (spaces between the ceiling joists of the top story and the roof rafters) are permitted under the following conditions:

1. No interior vapor retarders are installed on the ceiling side (attic floor) of the unvented attic assembly.

2. An air-impermeable insulation is applied in direct contact to the underside/interior of the structural roof deck. "Air-impermeable" shall be defined by ASTM E 283.

    Exception: In Zones 2B and 3B, insulation is not required to be air impermeable.

3. In the warm humid locations as defined in Section N1101.2.1:

    3.1. For asphalt roofing shingles: A 1-perm (5.7 × 10^{-11} kg/s · m^2 · Pa) or less vapor retarder (determined using Procedure B of ASTM E 96) is placed to the exterior of the structural roof deck; that is, just above the roof structural sheathing.

    3.2. For wood shingles and shakes: a minimum continuous $^1/_4$-inch (6 mm) vented air space separates the shingles/shakes and the roofing felt placed over the structural sheathing.

4. In Zones 3 through 8 as defined in Section N1101.2, sufficient insulation is installed to maintain the monthly average temperature of the condensing surface above 45°F (7°C). The condensing surface is defined as either the structural roof deck or the interior surface of an air-impermeable insulation applied in direct contact with the underside/interior of the structural roof deck.

"Air-impermeable" is quantitatively defined by ASTM E 283. For calculation purposes, an interior temperature of 68°F (20°C) is assumed. The exterior temperature is assumed to be the monthly average outside temperature.

❖ This section describes attics where the insulation and air barrier are above instead of below the attic space. Moving the insulation and placing an air-impermeable barrier above the attic moderates attic conditions so they are similar to the conditions of the residential space below. The primary benefit of having the insulation and air barrier above the attic is that ducts and/or HVAC equipment in the attic are not delivering cooled air through a hot summer attic and heated air through a cold winter attic. Another benefit is to eliminate the attic vents that sometimes allow moisture to condense inside the attic, admit rain during extreme weather and possibly admit sparks in fires.

Because this space is inside the building's thermal envelope, the traditional attic ventilation required by Sections R806.1 and R806.2 is not required. It obviously would not make sense to require such ventilation in a living room, so therefore, it can also be excluded from a conditioned attic area. Unvented attics require water/moisture control. Water moves in (or out) of buildings three main ways. The greatest amount of moisture is moved as bulk water (rain or any kind of water flow). Less moisture is moved by moving moist air such as with infiltration. The least amount of moisture is moved by moisture migration through materials. As with any attic, the roof itself is the main barrier for keeping water from entering the attic.

The provisions of this section can be applied to any attic area which is in compliance with this section. The attic is a traditional attic space, with the exception that it need not be ventilated and it will not get as hot or as cold as an attic that is open to the exterior.

It is very important that all of the four listed conditions be reviewed and considered for each building that uses the provisions of this section.

Item 1, which applies to all climate zones, prohibits the installation of a vapor retarder where it is typically installed at the ceiling level (attic floor) of a traditional ventilated attic. This assures that no barrier is installed which would separate the conditioned attic area from the remaining portion of the home. This requirement gives the attic space a limited potential to dry into the space beneath the attic so that small amounts of excess moisture can be removed from the attic. A sheet of polyurethane film or any material with a foil film facing are examples of vapor barriers that are not permitted on the attic floor.

Item 2 is one of the main provisions of this section and is applicable to virtually every climate zone. The requirement for an "air-impermeable" insulation will assure that air and the moisture that it can contain will not pass through the insulation to reach a point where it could condense because of the temperature. This item specifies that air-impermeable insulation be in direct contact with the interior side of the roof deck. Air-impermeable insulation prevents the movement of moist air that comes in as infiltration through the roof into the interior of the attic. For this section, air-impermeable material could be taken as a material that has an air permanence of 0.02 L/s-m^2 (at 75 Pa pressure) or less. Expanding spray foams and insulated sheathing (hard-foam sheathing board) are common types of air-impermeable insulation. When using insulated sheathing, attention to the details of completing the air sealing is required as the sheathing is installed in the roof. Fiberglass and cellulose are common types of air-permeable insulation. Air-impermeable insulation is not required in dry southern climates (2B and 3B) because the air is dryer in these dry climates.

Item 3 contains special requirements which apply to two types of shingle roofs when the home is located in a warm humid location. See Table N1101.2.1 for a list of the counties and territories that are considered as being warm and humid. Asphalt shingles require a vapor retarder between the shingles and the roof decking. Wood shakes and shingles require a vented space under them to allow the wood to dry after it gets wet from rain.

Item 4, which is applicable in the more northern zones, requires sufficient insulation to keep moisture from condensing on the "condensing surface" inside the attic in "average conditions." The insulation works to prevent condensation by keeping the condensing surface above the temperature where condensation will occur. Small amounts of condensation may occasionally occur at more extreme conditions; however,

this is not a concern. The condensing surface is the interior side of the roof deck for air-permeable insulation (see item 2 exception for Climate Zones 2B and 3B) and the interior of the insulation for air-impermeable insulation. The condensing surfaces differ because attic air can circulate through air-permeable insulation to contact the roof deck but can get only to the interior of air-impermeable insulation.

Note that the insulation required by item 4 may be more or less than the insulation required for energy efficiency in Chapter 11. If the amount of insulation required by Item 4 varies from the amount specified in Chapter 11, the provisions of Section R102.1 would apply, and the higher insulation value would be required. The insulation provided to comply with item 4 may be considered as contributing to the insulation required in Chapter 11 (See Section N1102.1.1).

It is important to realize that this section cannot be viewed as modifying or eliminating requirements found elsewhere in the code. Examples of sections which still affect these attic areas include Sections R314, R316, R807, R808, N1102.1 and others. Because the insulation typically used with this provision is some type of foam plastic, the requirements of Section R314 must be applied. The provisions of Section R806.4 do not in any way modify or eliminate the requirements for a thermal barrier (Section R314.4) or protection from ignition (Section R314.5.3). See the commentary for Sections R314 and R316 for a complete discussion regarding these requirements and the options available. Ducts in this unvented attic construction would be considered as being inside the building thermal envelope and would not require insulation (Section N1103.2.1).

The provisions of this section consider the attic assembly as a "conditioned" space; there is no requirement for the space to be provided with conditioned air supply. The attic space is considered indirectly conditioned because of omission of the air barrier, insulation at the ceiling and leakage around the attic access opening. Although the IRC does contain a definition for conditioned space, the definition is not directly applicable to "Conditioned attic assemblies" prescribed in this section. The definition is applicable only "for energy purposes" and "for mechanical purposes." An attic assembly complying with Section R806.4 will generally fall within the temperature ranges specified in the definition.

The key concept of this section is to move the thermal envelope (insulation) above the attic, resulting in the attic being in a conditioned (or sometimes semi-conditioned) space. Direct air supply to the attic is not required if the attic floor is not insulated; the attic temperature would be similar to interior conditioned spaces. Ducts and/or HVAC equipment in the attic also help moderate the attic conditions.

SECTION R807
ATTIC ACCESS

R807.1 Attic access. Buildings with combustible ceiling or roof construction shall have an attic access opening to attic areas that exceed 30 square feet (2.8 m²) and have a vertical height of 30 inches (762 mm) or more.

The rough-framed opening shall not be less than 22 inches by 30 inches (559 mm by 762 mm) and shall be located in a hallway or other readily accessible location. A 30-inch (762 mm) minimum unobstructed headroom in the attic space shall be provided at some point above the access opening. See Section M1305.1.3 for access requirements where mechanical equipment is located in attics.

❖ The requirement for an attic access is predicated on the likelihood that during the life of the structure, access to an attic space for repair of piping, electrical and mechanical systems will be required.

SECTION R808
INSULATION CLEARANCE

R808.1 Combustible insulation. Combustible insulation shall be separated a minimum of 3 inches (76 mm) from recessed luminaires, fan motors and other heat-producing devices.

Exception: Where heat-producing devices are listed for lesser clearances, combustible insulation complying with the listing requirements shall be separated in accordance with the conditions stipulated in the listing.

Recessed luminaires installed in the building thermal envelope shall meet the requirements of Section N1102.4.3.

❖ The minimum clearance required between combustible insulation and light fixtures, fan motors, etc., minimizes the chance of accidental ignition of the insulation. If a heat-producing device is listed for less than the minimum clearance specified in this section and all conditions of the listing are satisfied, the reduced clearance would apply.

Bibliography

The following resource materials are referenced in this chapter or are relevant to the subject matter addressed in this chapter.

AFPA NDS-05, *National Design Specification (NDS) for Wood Construction—with 2005 Supplement*. Washington, DC: American Forest and Paper Association, 2005.

AFPA 93, *Span Tables for Joists and Rafters*. Washington, DC: American Forest and Paper Association, 1993.

AISI Truss-04, *Standard for Cold-formed Steel Framing Truss Design*. Washington, DC: American Iron and Steel Institute, 2004.

AITC A 190.1-02, *Structural Glued Laminated Timber*. Englewood, CO: American Institute of Timber Construction, 2002.

APA E30-03, *Engineered Wood Construction Guide*. Tacoma, WA: APA—The Engineered Wood Association, 2003.

ASCE 7-98, *Minimum Design Loads for Buildings and Other Structures*, Reston, VA: American Society of Civil Engineers, 1998.

ASTM A 653/0653M–04a, *Specification for Steel Sheet, Zinc-coated (Galvanized) or Zinc-iron Alloy-coated (Galvanized) by the Hot-dip Process*. West Conshohocken, PA: ASTM International, 2004.

ASTM A 792/A 792M-03, *Specification for Steel Sheet, 55% Aluminum-Zinc Alloy-Coated by the Hot-Dip Process*. West Conshohocken, PA: ASTM International, 2003.

ASTM A 875/0875M-02a , *Specification for Steel Sheet, Zinc-5% Aluminum Alloy-Coated by the Hot-Dip Process*. West Conshohocken, PA: ASTM International, 2002.

ASTM C 954-00, *Specification for Steel Drill Screws for the Application of Gypsum Panel Products or Metal Plaster Bases to Steel Studs from 0.033 in. (0.84 mm) to 0.112 in. (2.84 mm) in Thickness*. West Conshohocken, PA: ASTM International, 2000.

ASTM D 2898-94(1999), *Test Methods for Accelerated Weathering of Fire-retardant-treated Wood for Fire Testing*. West Conshohocken, PA: ASTM International, 1999.

ASTM D 3201-94(2003), *Test Method for Hygroscopic Properties of Fire-retardant Wood and Wood-base Products*. West Conshohocken, PA: ASTM International, 2003.

ASTM D 3737-03, *Practice for Establishing Allowable Properties for Structural Glued Laminated Timber (Glulam)*. West Conshohocken, PA: ASTM International, 2003.

ASTM D 3957-03, *Standard Practices for Establishing Stress Grades for Structural Members Used in Log Buildings*. West Conshohocken, PA: ASTM International, 2003.

ASTM E 84-04, *Test Method for Surface Burning Characteristics of Building Materials*. West Conshohocken, PA: ASTM International, 2004.

CSA 0437-Series 93, *Standards on OSB and Waferboard (Reaffirmed 2001)*. Toronto, Ontario, Canada: Canadian Standards Association, 1993.

DOC PS 1-95, *Construction and Industrial Plywood*. Gaithersburg, MD: United States Department of Commerce, National Institute of Standards and Technology, 1995.

DOC PS 2-92, *Performance Standard for Wood-Based Structural-Use Panels*. Gaithersburg, MD: United States Department of Commerce, National Institute of Standards and Technology, 1992.

DOC PS 20-99, *American Softwood Lumber Standard.* Gaithersburg, MD: United States Department of Commerce, National Institute of Standards and Technology, 1999.

SAE J78-(1998), *Steel Self-Drilling Tapping Screws.* Warrendale, PA: Society of Automotive Engineers, 1998.

TPI 1-2002, *National Design Standard for Metal-Plate-Connected Wood Truss Construction.* Madison, WI: Truss Plate Institute, 2002.

Chapter 9:
Roof Assemblies

General Comments

Although a small portion of this chapter regulates the fire classification of roof covering materials, the major emphasis is on the materials and installation methods that will result in a weathertight exterior envelope. The *International Residential Code*® (IRC®) limits the requirements for fire-retardant roof coverings to two scenarios: 1) where the local or state ordinances in effect mandate a classified roof covering and 2) where the edge of the roof is in close proximity to a property line. Otherwise, the provisions of the chapter focus on the weather-protection features of a roof assembly.

The chapter contains a number of specific code requirements for various types of roof covering materials. These requirements may place limitations on roof slope, identify the proper attachment methods, mandate the use of underlayment and address flashing at appropriate points. In addition, the code often references a specification or installation standard for a particular type of roof covering. In all cases, it is important that the manufacturer's installation instructions be followed.

This chapter contains provisions for the design and construction of roof assemblies, primarily focusing on the materials and installation of roof coverings. The general requirements for roof assemblies found in the IRC deal with the role of the roof as a portion of the exterior envelope that provides weather protection for the building. Fire classifications, roof drainage, roof insulation and reroofing operations are also addressed.

Section R901 states the scope of the chapter, including both the materials and the quality of roof construction. Section R902 deals with fire-classification requirements for roof covering materials where the roof extends very close to an adjoining property line. Section R903 establishes the general requirements for weather protection, including provisions for flashing and roof drainage.

Section R904 establishes general requirements for roofing materials. Section R905 discusses the many different types of roof covering materials regulated by the IRC, including asphalt shingles, clay and concrete tile, metal roof shingles, mineral-surfaced roll roofing, wood shakes and shingles, slate and slate-type shingles, built-up roofs, metal roof panels, modified bitumen roofing, thermoset and thermoplastic single-ply roofing, sprayed polyurethane foam roofing and liquid-applied coatings. Section R906 allows the installation of above-deck thermal installation where appropriately listed, and Section R907 establishes the methods and materials acceptable for recovering or replacing an existing roof.

Purpose

Roof assemblies fulfill a variety of functions; however, this chapter deals primarily with roof coverings for weather protection. The requirements address the design and construction of roof assemblies, with a focus on the materials and quality of installation. Although a number of requirements are specified directly from the code, the majority of the regulations applicable to roof coverings come from the material standards. In addition, the manufacturer's installation instructions should always be followed for the specific roof covering material under consideration. Flashing and roof drainage criteria are also important aspects of the weather-protection features. Many of the common roof coverings are addressed by detailed provisions regulating the roof deck, roof slope, underlayment, application and attachment. Where a new roof is required for an existing building, the code sets specific guidelines for either recovering or replacing the existing roof covering.

SECTION R901
GENERAL

R901.1 Scope. The provisions of this chapter shall govern the design, materials, construction and quality of roof assemblies.

❖ The focus of this chapter is on roof coverings. A variety of roof covering materials are included, including asphalt shingles, clay and concrete tile, metal roof shingles, wood shingles and shakes and metal roof panels. Roof classification, drainage, insulation and flashing are also regulated, as is the reroofing of an existing roof.

SECTION R902
ROOF CLASSIFICATION

R902.1 Roofing covering materials. Roofs shall be covered with materials as set forth in Sections R904 and R905. Class A, B or C roofing shall be installed in areas designated by law as requiring their use or when the edge of the roof is less than 3 feet (914 mm) from a property line. Classes A, B and C roofing required to be listed by this section shall be tested in accordance with UL 790 or ASTM E 108. Roof assemblies with coverings of brick, masonry, slate, clay or concrete roof tile, exposed concrete roof deck, ferrous or copper shingles or sheets, and metal sheets and shingles, shall be considered Class A roof coverings.

❖ The IRC identifies two conditions under which roof assembly classifications A, B, or C are required. First, where local, state or other governing ordinance requires the use of Class A, B or C roofing, it must be installed as designated by the law. Second, on roofs that project within 3 feet (914 mm) of the property line, a Class A, B or C roof must be installed. Testing for such roofing must be in accordance with either ASTM E 108 or UL 790.

Roof coverings classified as Class A, B or C have been shown through testing to provide protection of the roof against severe, moderate and light fire exposures, respectively. These exposures are external and are generally created by fires in adjoining structures, wild fires and fires from the subject building that extend up the exterior and onto the top surface of the roof. Wild fires and some structure fires create flying and flaming brands that can ignite nonclassified roof coverings.

By their nature, roof coverings composed of brick, masonry, slate, clay tile, concrete tile, ferrous or copper shingles or sheets and metal sheets or shingles are considered excellent materials for resisting fire spread and are classified as Class A. An exposed concrete roof deck also qualifies for this classification.

R902.2 Fire-retardant-treated shingles and shakes. Fire-retardant-treated wood shakes and shingles shall be treated by impregnation with chemicals by the full-cell vacuum-pressure process, in accordance with AWPA C1. Each bundle shall be marked to identify the manufactured unit and the manufacturer, and shall also be labeled to identify the classification of the material in accordance with the testing required in Section R902.1, the treating company and the quality control agency.

❖ This section recognizes the process of impregnating wood shakes and shingles with chemicals in order to make them fire retardant. AWPA C1, regulating the preservative treatment of timber products by pressure processes, is the referenced standard. During the impregnation process the cells of wood shakes and shingles are exposed to a vacuum that draws all of the air and moisture from the material. Fire-retardant chemicals are then impregnated into the shakes and shingles by pressurization. After proper curing, the shakes and shingles are then classified. The fire classification must be marked on each bundle of materials, along with other manufacturing information. Identification will also include the name of the company performing the pressure-treatment process, as well as the quality control agency reviewing for compliance with the test standard.

SECTION R903
WEATHER PROTECTION

R903.1 General. Roof decks shall be covered with approved roof coverings secured to the building or structure in accordance with the provisions of this chapter. Roof assemblies shall be designed and installed in accordance with this code and the approved manufacturer's installation instructions such that the roof assembly shall serve to protect the building or structure.

❖ In all cases, a roof must be designed to provide protection from the elements. For the roof to adequately perform this function, it must be designed in accordance with this chapter. This section requires flashing where the roof intersects vertical elements such as walls, chimneys, dormers, plumbing stacks, plumbing vents and other penetrations of the weather-protective barrier. Roof drainage to remove water from the roof to an approved location is also regulated.

Materials and installation methods for roof decks and supporting elements are regulated by Chapter 8 of the IRC. A complying roof deck must then be covered in an approved manner to provide the necessary weather protection mandated by the code. This chapter includes the necessary provisions to provide such a weatherproof barrier. In addition, and just as important, the approved manufacturer's installation instructions must be adhered to.

R903.2 Flashing. Flashings shall be installed in a manner that prevents moisture from entering the wall and roof through joints in copings, through moisture permeable materials and at intersections with parapet walls and other penetrations through the roof plane.

❖ As with flashing required for other locations at the building's exterior, it is critical that those locations where moisture may penetrate the exterior membrane at the roof be adequately protected. General locations identified include joints at the coping, intersections with parapet walls and other penetrations through the roof plane.

R903.2.1 Locations. Flashings shall be installed at wall and roof intersections, wherever there is a change in roof slope or direction and around roof openings. Where flashing is of metal, the metal shall be corrosion resistant with a thickness of not less than 0.019 inch (0.5 mm) (No. 26 galvanized sheet).

❖ This section identifies specific roof locations for the installation of flashings. In those cases where metal is used as the flashing material, it must be corrosion-resistant and a minimum No. 26 galvanized sheet.

R903.3 Coping. Parapet walls shall be properly coped with noncombustible, weatherproof materials of a width no less than the thickness of the parapet wall.

❖ The top of an exterior wall, typically a masonry parapet wall, must have a coping. It is typically sloped to prevent standing water, and it must be of noncombustible, weatherproofed materials. To maintain protection at the top of the wall, the coping must extend the full thickness of the parapet wall.

R903.4 Roof drainage. Unless roofs are sloped to drain over roof edges, roof drains shall be installed at each low point of the roof. Where required for roof drainage, scuppers shall be placed level with the roof surface in a wall or parapet. The scupper shall be located as determined by the roof slope and contributing roof area.

❖ In most residential construction, drainage water from the roof simply flows down the roof's slope to and over the roof edge, often to a gutter and downspout system that carries the water away from the building. In situations where the roof design does not allow for water flow over the roof edges, roof drains must be installed in the roof surface. The drains will be at the low points of the roof, sized and located to remove all roof water. In a case where the low points occur at parapets of the exterior walls, scuppers through the parapets must be provided, constructed with the flow points level with the roof surface at the low points and located based on the roof slope and contributing roof area.

R903.4.1 Overflow drains and scuppers. Where roof drains are required, overflow drains having the same size as the roof drains shall be installed with the inlet flow line located 2 inches (51 mm) above the low point of the roof, or overflow scuppers having three times the size of the roof drains and having a minimum opening height of 4 inches (102 mm) shall be installed in the adjacent parapet walls with the inlet flow located 2 inches (51 mm) above the low point of the roof served. The installation and sizing of overflow drains, leaders and conductors shall comply with the *International Plumbing Code*.

Overflow drains shall discharge to an approved location and shall not be connected to roof drain lines.

❖ It is quite possible that roof drains may become clogged or blocked by debris, resulting in insufficient drainage. This could cause structural failure of portions of the roof system resulting from the increased weight on the roof caused by the impounded water. This section addresses such concerns by requiring overflow drains or overflow scuppers for those roofs that do not drain over the roof edges.

Although the required roof drains must be located at the low points of the roof, the overflow drain inlets need to be located 2 inches (51 mm) vertically above the low points. The same criterion applies where overflow scuppers are provided. In addition, overflow scuppers must be sized with a minimum opening height of 4 inches (102 mm) and have an opening area at least three times that of the roof drains. The *International Plumbing Code* sets forth the specific details for the installation and sizing of overflow drains, leaders and conductors.

The overflow drains must not be connected to the roof drains. In the event a blockage occurs in the drainage line beyond the point of connection, both the roof drain and the overflow drain would be affected. The overflow drain should discharge to an approved location, typically a highly visible location that would indicate the overflow system is in use. This will provide an alert that there is some sort of blockage in the primary roof drain system.

R903.5 Hail exposure. Hail exposure, as specified in Sections R903.5.1 and R903.5.2, shall be determined using Figure R903.5.

❖ Most types of residential roof coverings are especially susceptible to damage from hail. A recent study

conducted by the Institute for Business and Home Safety (IBHS) has shown that approximately 44 percent of all "nonimpact-resistant" single family residential roofs investigated in the study needed repair or replacement after being struck by hailstones with diameters between 1 and 2 inches (25 and 51 mm). On a nationwide basis, this amounts to an average annual insured loss of about $1.6 billion for all roof types combined.

The design life of residential roofing systems can generally be assumed to be at least 20 years. Throughout large portions of the Great Plains and Southeastern US, hailstorms producing hail with diameters of 1.5 inches (38 mm) or larger are expected at mean recurrence intervals of 20 years or less. Thus, in such regions, damaging hail is expected within a period of time less than or equal to the lifetime of most residential roofing systems.

The map shown in Figure R903.5 was developed with statistical modeling software, using actual hail data from the National Climatic Data Center.

R903.5.1 Moderate hail exposure. One or more hail days with hail diameters larger than 1.5 inches (38 mm) in a 20-year period.

❖ See the commentary for Section R903.5.

R903.5.2 Severe hail exposure. One or more hail days with hail diameters larger than or equal to 2.0 inches (51 mm) in a 20-year period.

❖ See the commentary for Section R903.5.

SECTION R904
MATERIALS

R904.1 Scope. The requirements set forth in this section shall apply to the application of roof covering materials specified herein. Roof assemblies shall be applied in accordance with this chapter and the manufacturer's installation instructions. Installation of roof assemblies shall comply with the applicable provisions of Section R905.

❖ Roof assemblies and roof covering materials must conform to the provisions of this chapter, the applicable reference standards listed in Chapter 43, and the manufacturer's installation instructions. This section provides general conditions for the application of roof covering materials referenced elsewhere in Chapter 9. A cross reference to Section R905 is provided for requirements dealing with the installation of specific roof coverings.

R904.2 Compatibility of materials. Roof assemblies shall be of materials that are compatible with each other and with the building or structure to which the materials are applied.

❖ The materials that comprise a roof assembly must be compatible in avoid a decrease or failure in the performance of the roof. Not only must the roofing materials be compatible with each other, they must also be suitable for the building as a whole.

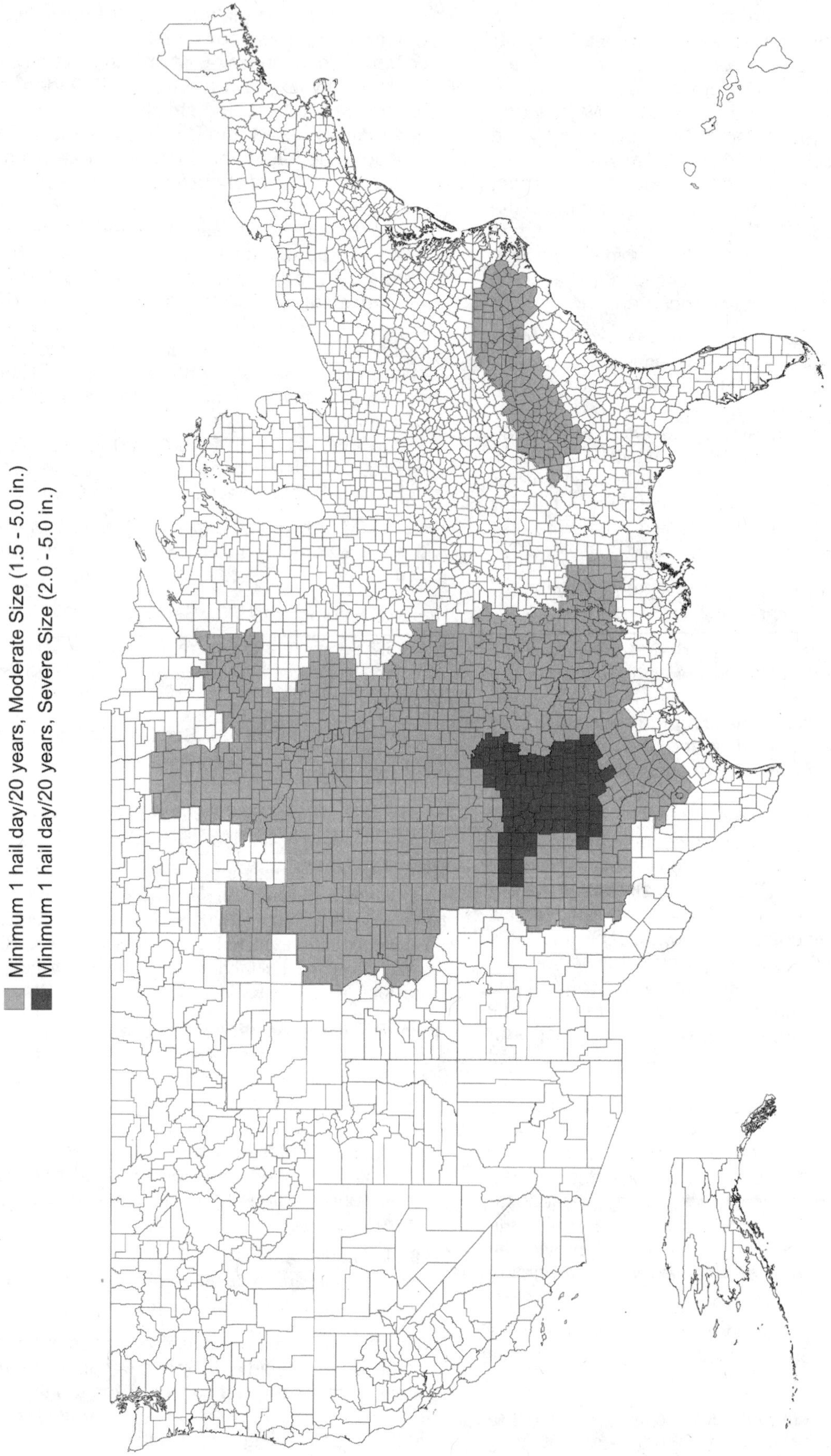

Minimum 1 hail day/20 years, Moderate Size (1.5 - 5.0 in.)

Minimum 1 hail day/20 years, Severe Size (2.0 - 5.0 in.)

**FIGURE R903.5
HAIL EXPOSURE MAP**

❖ See the commentary for Section R903.5.

R904.3 Material specifications and physical characteristics. Roof covering materials shall conform to the applicable standards listed in this chapter. In the absence of applicable standards or where materials are of questionable suitability, testing by an approved testing agency shall be required by the building official to determine the character, quality and limitations of application of the materials.

❖ In most cases, materials used as roof coverings are regulated by an applicable standard. Identified in this chapter, they are also listed in Chapter 43. If a roof covering material is not regulated by a specific standard, the building official must have substantiating data to show that the materials are consistent with the intent of the code regarding quality and application limitations. Typically, the roof covering material under consideration must be tested to provide the information necessary to show code compliance.

R904.4 Product identification. Roof covering materials shall be delivered in packages bearing the manufacturer's identifying marks and approved testing agency labels when required. Bulk shipments of materials shall be accompanied by the same information issued in the form of a certificate or on a bill of lading by the manufacturer.

❖ Identification of roof covering materials provides much of the information necessary for acceptance of the materials by the building official. Because it is often difficult, if not impossible, to identify complying roof covering materials by visual observation, other proper identification of the materials is mandatory. Packaging of roof covering materials must include the manufacturer's identification marks. If the code or applicable standard requires review by an approved testing agency, the identifying marks of the testing agency must also be shown. The code also addresses the identification requirements for bulk shipments of materials.

SECTION R905
REQUIREMENTS FOR ROOF COVERINGS

R905.1 Roof covering application. Roof coverings shall be applied in accordance with the applicable provisions of this section and the manufacturer's installation instructions. Unless otherwise specified in this section, roof coverings shall be installed to resist the component and cladding loads specified in Table R301.2(2), adjusted for height and exposure in accordance with Table R301.2(3).

❖ In addition to the minimum requirements specified for roof coverings in the code, manuals published by various associations provide detailed discussions of the proper methods of installing roof coverings. These methods have been established based on many years of experience with the materials and their performance. Although the provisions in these documents are not specific code requirements, this section mandates the use of the manufacturer's installation instructions. It is important that roof coverings remain in-

tact and in place when subjected to wind. Without an intact roof covering, the building would be subjected to either water damage, which could reduce its structural stability, or to higher wind pressures than the building is designed for. Unless Section R905 specifies otherwise, roof coverings must be installed to resist the wind pressures determined from Table R301.2(2). See the Commentary for Tables R301.2(2) and R301.2(3).

R905.2 Asphalt shingles. The installation of asphalt shingles shall comply with the provisions of this section.

❖ This section regulates asphalt shingles composed of organic felt or glass felt and coated with mineral granules. Provisions address requirements for sheathing, roof slope, underlayment, fasteners and attachment.

R905.2.1 Sheathing requirements. Asphalt shingles shall be fastened to solidly sheathed decks.

❖ The code requires a solid roof surface for the installation of asphalt shingles. Section R803 regulates solid sheathing.

R905.2.2 Slope. Asphalt shingles shall be used only on roof slopes of two units vertical in 12 units horizontal (2:12) or greater. For roof slopes from two units vertical in 12 units horizontal (2:12) up to four units vertical in 12 units horizontal (4:12), double underlayment application is required in accordance with Section R905.2.7.

❖ The performance of all roof coverings is based in part on the slope of the roof surface. As the slope of the roof decreases, water drainage is slowed, and the potential for water intrusion increases because of the greater potential of water back-up under the roofing. Asphalt shingles, because of their configuration and installation methods, are restricted to use on roofs having a minimum slope of 2:12. Where the slope is no steeper than 4:12, the underlayment must be doubled to provide a greater barrier to leakage. Section R905.2.7 specifies the method of such double underlayment. Commentary Figure R905.2.2(1) shows an example of asphalt roof shingles installed on a high-slope roof (4:12 minimum), while Commentary Figure R905.2.2(2) depicts a low-slope roof installation (between 2:12 and 4:12).

R905.2.3 Underlayment. Unless otherwise noted, required underlayment shall conform to ASTM D 226 Type I, ASTM D 4869 Type I, or ASTM D 6757.

Self-adhering polymer modified bitumen sheet shall comply with ASTM D 1970.

❖ Four types of underlayment are recognized for use with asphalt shingle roof coverings: asphalt-saturated organic felts as regulated by ASTM D 226, Type I; asphalt-saturated organic felt shingles, per ASTM D 4869, Type I; inorganic underlayment for use with steep slope roof, per ASTM D 6757; and self-adhering polymer modified bitumen sheet materials are addressed in ASTM D 1970. Section R905.2.7 contains the methods prescribed for the installation of underlayment for asphalt shingles.

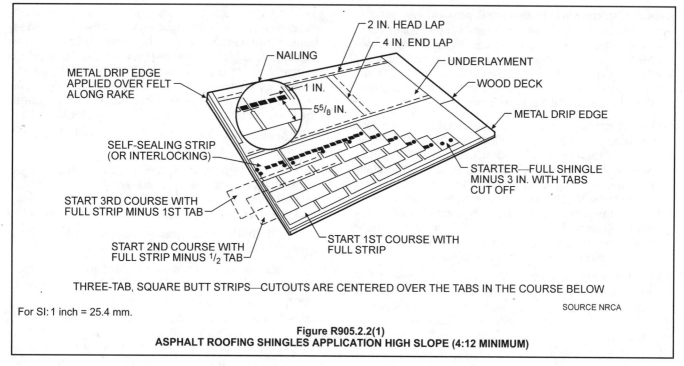

For SI: 1 inch = 25.4 mm.

Figure R905.2.2(1)
ASPHALT ROOFING SHINGLES APPLICATION HIGH SLOPE (4:12 MINIMUM)

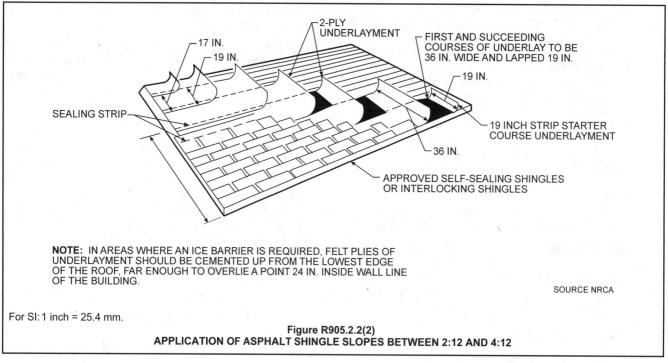

NOTE: IN AREAS WHERE AN ICE BARRIER IS REQUIRED, FELT PLIES OF UNDERLAYMENT SHOULD BE CEMENTED UP FROM THE LOWEST EDGE OF THE ROOF, FAR ENOUGH TO OVERLIE A POINT 24 IN. INSIDE WALL LINE OF THE BUILDING.

SOURCE NRCA

For SI: 1 inch = 25.4 mm.

Figure R905.2.2(2)
APPLICATION OF ASPHALT SHINGLE SLOPES BETWEEN 2:12 AND 4:12

R905.2.4 Asphalt shingles. Asphalt shingles shall have self-seal strips or be interlocking, and comply with ASTM D 225 or D 3462.

❖ The IRC permits two types of asphalt shingles: interlocking and those with self-seal strips. Although the method of installation varies for these two types, the result is the same: a weather-resistant barrier between the elements and the interior of the building. Two test standards regulate asphalt shingles. ASTM D 225 ad-

dresses asphalt shingles made from organic felt, while ASTM D 3462 deals with shingles made from glass felt. Both shingle types are surfaced with mineral granules.

R905.2.4.1 Wind resistance of asphalt shingles. Asphalt shingles shall be installed in accordance with Section R905.2.6. Shingles classified using ASTM D 3161 are acceptable for use in wind zones less than 110 mph (49 m/s). Shingles classified using ASTM D 3161, Class F, are acceptable for use in all cases where special fastening is required.

❖ Asphalt shingles tested in accordance with ASTM D 3161, Class A, D or F, are acceptable for use where the basic wind speed is less than 110 mph (49 m/s). ASTM D 3161, Class A and D, requires the shingles to pass at a test velocity of 60 mph (27 m/s) and 90 mph (40 m/s) respectively. ASTM D 3161, Class F, requires the shingles to pass at a test velocity of 110 mph (49 m/s). For all cases where special fastening is required, shingles tested in accordance with ASTM D 3161, Class F, are acceptable.

R905.2.5 Fasteners. Fasteners for asphalt shingles shall be galvanized steel, stainless steel, aluminum or copper roofing nails, minimum 12 gage [0.105 inch (3 mm)] shank with a minimum $^3/_8$-inch (10 mm) diameter head, ASTM F 1667, of a length to penetrate through the roofing materials and a minimum of $^3/_4$ inch (19 mm) into the roof sheathing. Where the roof sheathing is less than $^3/_4$ inch (19 mm) thick, the fasteners shall penetrate through the sheathing. Fasteners shall comply with ASTM F 1667.

❖ Roofing nails must be of a corrosion-resistant material, specified in this section as either galvanized steel, stainless steel, aluminum or copper. A roofing nail must have a minimum 12 gage shank with a head at least $^3/_8$ inch (9.5 mm) in diameter. To provide the necessary holding power, roofing nails must penetrate the roof sheathing a minimum of $^3/_4$ inch (19.1 mm). For roof sheathing having a thickness less than $^3/_4$ inch (19.1 mm), the nails must penetrate completely through the sheathing. The material specification standard for roofing nails is ASTM F 1667.

R905.2.6 Attachment. Asphalt shingles shall have the minimum number of fasteners required by the manufacturer. For normal application, asphalt shingles shall be secured to the roof with not less than four fasteners per strip shingle or two fasteners per individual shingle. Where the roof slope exceeds 20 units vertical in 12 units horizontal (167 percent slope), special methods of fastening are required. For roofs located where the basic wind speed per Figure R301.2(4) is 110 mph (49 m/s) or higher, special methods of fastening are required. Special fastening methods shall be tested in accordance with ASTM D 3161, Class F. Asphalt shingle wrappers shall bear a label indicating compliance with ASTM D 3161, Class F.

❖ In all cases, the minimum number of fasteners for asphalt shingles must be the number required by the shingle manufacturer. It is typical for asphalt strip shingles to be attached to the roof sheathing by at least four fasteners for each strip shingle. If individual shingles are installed, at least two fasteners are required per shingle. For very steep roof slopes, the manufacturer's installation instructions should be reviewed for any special fastening methods.

Where the basic wind speeds in accordance with Figure R301.2(4) are greater than or equal to 110 mph (49 m/s), special testing is required to determine how such roofing should be fastened.

Asphalt shingles that are to be used in an area where the basic wind speed is 110 mph (49 m/s) or greater must be labeled to demonstrate compliance with ASTM D 3161, Class F.

R905.2.7 Underlayment application. For roof slopes from two units vertical in 12 units horizontal (17-percent slope), up to four units vertical in 12 units horizontal (33-percent slope), underlayment shall be two layers applied in the following manner. Apply a 19-inch (483 mm) strip of underlayment felt parallel to and starting at the eaves, fastened sufficiently to hold in place. Starting at the eave, apply 36-inch-wide (914 mm) sheets of underlayment, overlapping successive sheets 19 inches (483 mm), and fastened sufficiently to hold in place. Distortions in the underlayment shall not interfere with the ability of the shingles to seal. For roof slopes of four units vertical in 12 units horizontal (33-percent slope) or greater, underlayment shall be one layer applied in the following manner. Underlayment shall be applied shingle fashion, parallel to and starting from the eave and lapped 2 inches (51 mm), fastened sufficiently to hold in place. Distortions in the underlayment shall not interfere with the ability of the shingles to seal. End laps shall be offset by 6 feet (1829 mm).

❖ This section specifies the underlayment requirements for asphalt shingles installed on both low-slope and high-slope roofs. Low-slope roofs present a potential problem because water drains slowly, and this creates the opportunity for water back-up. Wind-driven rain can also pose a problem. Therefore, a special underlayment application method is used so that the roof remains weathertight. All portions of the roof will be protected by a minimum of two layers of underlayment if the installer follows the instructions found in this section. Commentary Figure R905.2.7 illustrates the required application of underlayment for a low-slope application. Only one layer of underlayment, applied in shingle fashion in accordance with this section, is mandated for high-slope roofs of asphalt shingles. For both low-slope and high-slope roofs the underlayment must not have distortions that will interfere with the ability of the shingles to lie flat and seal. Installation of shingles over a distorted surface can result in reduced wind resistance and unacceptable aesthetics.

R905.2.7.1 Ice barrier. In areas where there has been a history of ice forming along the eaves causing a backup of water as designated in Table R301.2(1), an ice barrier that consists of a least two layers of underlayment cemented together or of a self-adhering polymer modified bitumen sheet, shall be used in lieu of normal underlayment and extend from the lowest edges of all roof surfaces to a point at least 24 inches (610 mm) inside the exterior wall line of the building.

Exception: Detached accessory structures that contain no conditioned floor area.

❖ Where ice dams may be formed along the eave because snow continually freezes and thaws or frozen slush backs up in gutters, the underlayment application in the area of the eaves must be modified to prevent ice dams from forcing water under the roofing, which could damage ceilings, walls and insulation [see Commentary Figure R905.2.7.1(1)]. Two layers of underlayment should be cemented together with asphalt cement from the lowest edge of the roof and continue up the roof to a point that is at least 24 inches (610 mm) inside the interior wall line of the building as shown in Commentary Figure R905.2.7.1(2). The environment within the envelope of the building provides adequate warmth to prevent ice dams from forming above the heated space; therefore, the two layers of cemented underlayment are permitted to terminate 24 inches (610 mm) inside the interior wall line of the building. The local jurisdiction is responsible for determining whether the ice barrier is required based on weather records, and it must so indicate in Table R301.2(1).

An exception to this section exempts accessory buildings from such restrictions because they are unheated structures where the need for protection against ice dams is unnecessary. The same exception is found in Sections R905.4.3.1, R905.5.3.1, R905.6.3.1, R905.7.3.1, and R905.8.3.1.

R905.2.7.2 Underlayment and high wind. Underlayment applied in areas subject to high winds [above 110 mph (49 m/s) per Figure R301.2(4)] shall be applied with corrosion-resistant fasteners in accordance with manufacturer's installation instructions. Fasteners are to be applied along the overlap not farther apart than 36 inches (914 mm) on center.

❖ In high-wind areas, corrosion-resistant fasteners must be located in accordance with the manufacturer's installation instructions, but in no case can they be more than 36 inches (914 mm) on center. The general requirement for sufficient fastening to hold the underlayment in place is not adequate where increased wind loads are anticipated.

R905.2.8 Flashing. Flashing for asphalt shingles shall comply with this section.

❖ This section sets forth the special conditions for flashing installed as a part of an asphalt shingle roof system. Specific flashing locations addressed include base and cap flashing, valleys, crickets and saddles and sidewall flashing.

R905.2.8.1 Base and cap flashing. Base and cap flashing shall be installed in accordance with manufacturer's installation instructions. Base flashing shall be of either corrosion-resistant metal of minimum nominal 0.019-inch (0.5 mm) thickness or mineral surface roll roofing weighing a minimum of 77 pounds per 100 square feet (4 kg/m²). Cap flashing shall be corrosion-resistant metal of minimum nominal 0.019-inch (0.5 mm) thickness.

❖ If metal is used as cap or base flashing, it must be corrosion resistant and have a minimum nominal thickness of 0.019-inch (0.483 mm). Mineral surface roll roofing may also be used as base flashing, if it has a minimum weight of 77 pounds per 100 square feet (4 kg/m²).

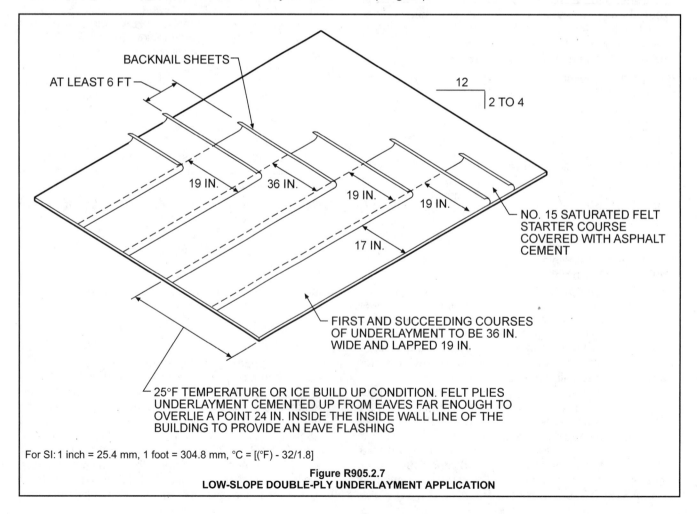

BACKNAIL SHEETS

AT LEAST 6 FT

12
2 TO 4

19 IN.   36 IN.   19 IN.   19 IN.

17 IN.

NO. 15 SATURATED FELT STARTER COURSE COVERED WITH ASPHALT CEMENT

FIRST AND SUCCEEDING COURSES OF UNDERLAYMENT TO BE 36 IN. WIDE AND LAPPED 19 IN.

25°F TEMPERATURE OR ICE BUILD UP CONDITION. FELT PLIES UNDERLAYMENT CEMENTED UP FROM EAVES FAR ENOUGH TO OVERLIE A POINT 24 IN. INSIDE THE INSIDE WALL LINE OF THE BUILDING TO PROVIDE AN EAVE FLASHING

For SI: 1 inch = 25.4 mm, 1 foot = 304.8 mm, °C = [(°F) - 32/1.8]

**Figure R905.2.7
LOW-SLOPE DOUBLE-PLY UNDERLAYMENT APPLICATION**

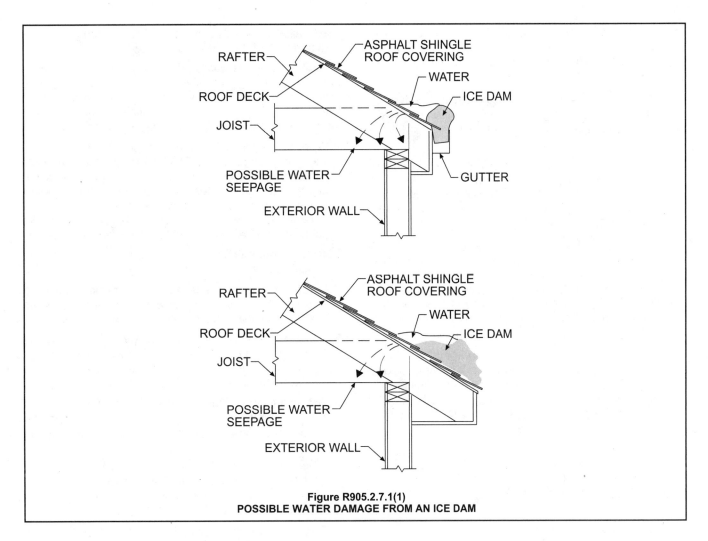

Figure R905.2.7.1(1)
POSSIBLE WATER DAMAGE FROM AN ICE DAM

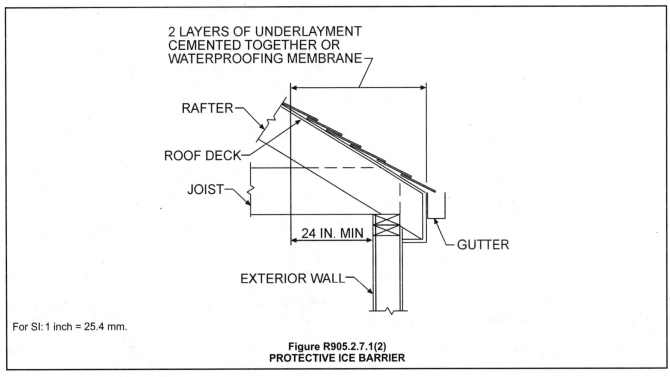

For SI: 1 inch = 25.4 mm.

Figure R905.2.7.1(2)
PROTECTIVE ICE BARRIER

R905.2.8.2 Valleys. Valley linings shall be installed in accordance with the manufacturer's installation instructions before applying shingles. Valley linings of the following types shall be permitted:

1. For open valley (valley lining exposed) lined with metal, the valley lining shall be at least 24 inches (610 mm) wide and of any of the corrosion-resistant metals in Table R905.2.8.2.

2. For open valleys, valley lining of two plies of mineral surfaced roll roofing, complying with ASTM D 3909 or ASTM D 6380 Class M, shall be permitted. The bottom layer shall be 18 inches (457mm) and the top layer a minimum of 36 inches (914 mm) wide.

3. For closed valleys (valley covered with shingles), valley lining of one ply of smooth roll roofing complying with ASTM D 6380 Class S Type III, Class M Type II, or ASTM D 3909 and at least 36 inches wide (914 mm) or valley lining as described in Items 1 and 2 above shall be permitted. Specialty underlayment complying with ASTM D 1970 may be used in lieu of the lining material.

❖ Open valley linings may be of either metal or mineral surface roll roofing as set forth in this section. Closed valleys are also permitted with a number of lining alternatives available. Commentary Figures R905.2.8.2(1) and R905.2.8.2(2) illustrate typical valley flashings for open and closed (woven) valleys, respectively.

TABLE R905.2.8.2
VALLEY LINING MATERIAL

MATERIAL	MINIMUM THICKNESS (inches)	GAGE	WEIGHT (pounds)
Cold–rolled copper	0.0216 nominal	—	ASTM B 370, 16 oz. per square foot
Lead–coated copper	0.0216 nominal	—	ASTM B 101, 16 oz. per square foot
High–yield copper	0.0162 nominal	—	ASTM B 370, 12 oz. per square foot
Lead–coated high–yield copper	0.0162 nominal	—	ASTM B 101, 12 oz. per square foot
Aluminum	0.024	—	—
Stainless steel	—	28	—
Galvanized steel	0.0179	26 (zinc coated G90)	—
Zinc alloy	0.027	—	—
Lead	—	—	$2^{1}/_{2}$
Painted terne	—	—	20

For SI: 1 inch = 25.4 mm, 1 pound = 0.454 kg.

❖ If exposed corrosion-resistant metal materials are used as valley linings, they are regulated by this table. The table identifies a variety of acceptable metal linings, including copper, aluminum, stainless steel, galvanized steel,

zinc alloy, lead and painted terne. The materials are regulated by their thickness, gage or weight.

R905.2.8.3 Crickets and saddles. A cricket or saddle shall be installed on the ridge side of any chimney or penetration more than 30 inches (762 mm) wide as measured perpendicular to the slope. Cricket or saddle coverings shall be sheet metal or of the same material as the roof covering.

❖ Where a chimney is more than 30 inches (762 mm) wide, as measured perpendicular to the slope, a cricket or saddle must be installed on the ridge side of the chimney to divert water away from the chimney penetration of the roof. A chimney of such width creates the potential for standing water that, despite appropriate flashing, can be detrimental to the roof's performance.

R905.2.8.4 Sidewall flashing. Flashing against a vertical sidewall shall be by the step-flashing method.

❖ Step flashing must be used to flash along the intersection of a roof and a vertical sidewall. Commentary Figure R905.2.8.4 shows an example of flashing at a side wall.

R905.2.8.5 Other flashing. Flashing against a vertical front wall, as well as soil stack, vent pipe and chimney flashing, shall be applied according to the asphalt shingle manufacturer's printed instructions.

❖ The installation of flashing at vertical roof penetrations such as chimneys, soil stacks and vent pipes must be in conformance with the instructions provided by the asphalt shingle manufacturer. Commentary Figure R905.2.8.5 shows an example of chimney flashing.

R905.3 Clay and concrete tile. The installation of clay and concrete shall comply with the provisions of this section. Clay roof tile shall comply with ASTM C 1167.

❖ This section addresses both concrete tile and clay tile. The specifications for clay roof tiles must be in conformance with ASTM C 1167. Commentary Figure R905.3 depicts several examples of clay and concrete roof tile.

R905.3.1 Deck requirements. Concrete and clay tile shall be installed only over solid sheathing or spaced structural sheathing boards.

❖ Both solid sheathing and spaced structural sheathing boards are permitted as a base for the installation of clay or concrete roof tile.

R905.3.2 Deck slope. Clay and concrete roof tile shall be installed on roof slopes of two and one-half units vertical in 12 units horizontal ($2^{1}/_{2}$:12) or greater. For roof slopes from two and one-half units vertical in 12 units horizontal ($2^{1}/_{2}$:12) to four units vertical in 12 units horizontal (4:12), double underlayment application is required in accordance with Section R905.3.3.

❖ To keep the roof covering weathertight and promote proper drainage, clay and concrete tile cannot be installed on a roof having a slope of less than $2^{1}/_{2}$:12. The application of two layers of underlayment is mandated on roofs of 4:12 and less. The provisions of Section R905.3.3 address the full requirements for underlayment on concrete or clay tile roofs.

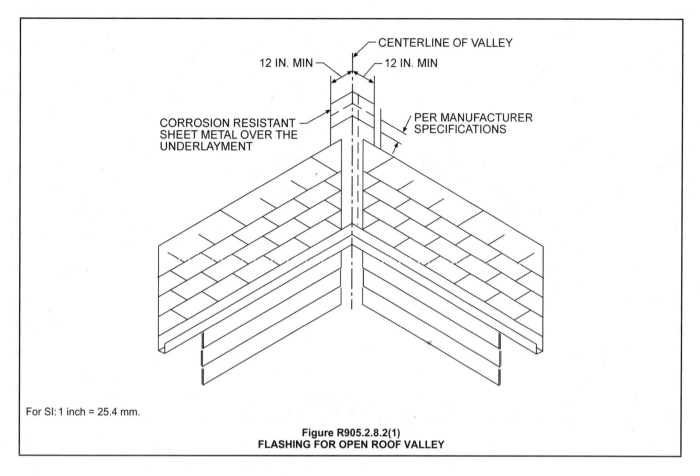

For SI: 1 inch = 25.4 mm.

Figure R905.2.8.2(1)
FLASHING FOR OPEN ROOF VALLEY

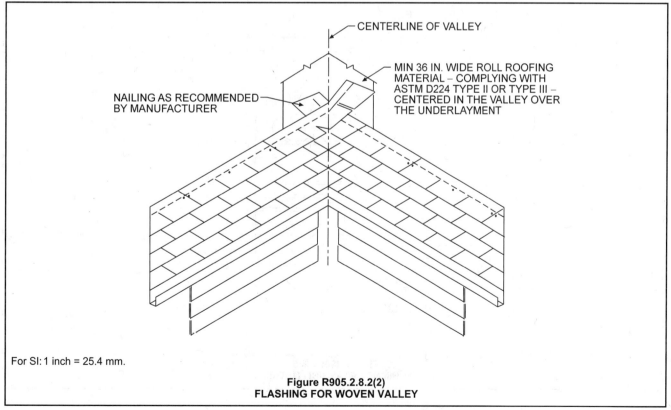

For SI: 1 inch = 25.4 mm.

Figure R905.2.8.2(2)
FLASHING FOR WOVEN VALLEY

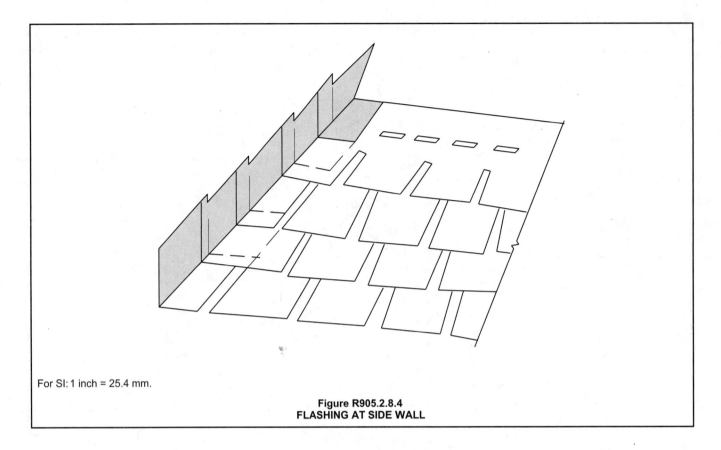

For SI: 1 inch = 25.4 mm.

Figure R905.2.8.4
FLASHING AT SIDE WALL

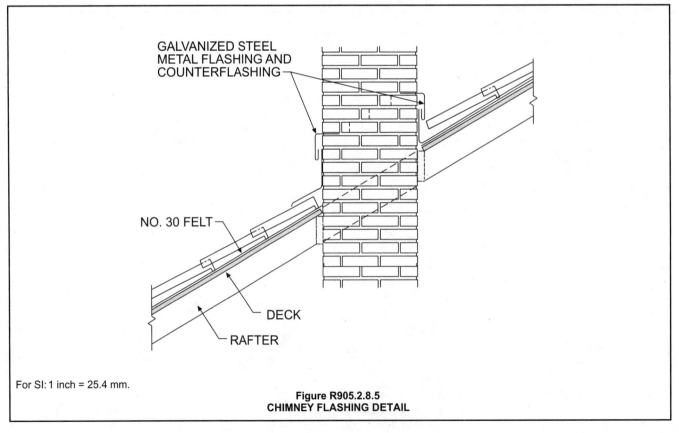

GALVANIZED STEEL
METAL FLASHING AND
COUNTERFLASHING

NO. 30 FELT

DECK

RAFTER

For SI: 1 inch = 25.4 mm.

Figure R905.2.8.5
CHIMNEY FLASHING DETAIL

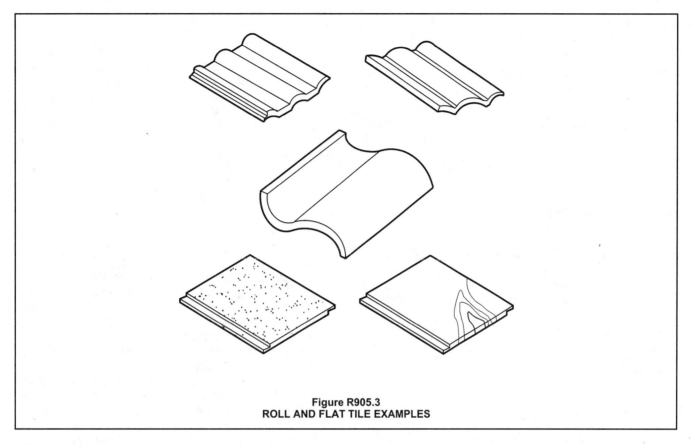

**Figure R905.3
ROLL AND FLAT TILE EXAMPLES**

R905.3.3 Underlayment. Unless otherwise noted, required underlayment shall conform to ASTM D 226 Type II; ASTM D 2626 Type I; or ASTM D 6380 Class M mineral surfaced roll roofing.

❖ The code recognizes three types of underlayment for use with clay or concrete tile roof covering: asphalt-saturated organic felts as regulated by ASTM D 226, Type II; asphalt-saturated and coated organic felt base sheets per ASTM D 2626, Type I, and asphalt organic felt roll roofing surfaced with mineral granules addressed in ASTM D 6380, Class M. This section also contains the methods prescribed for the installation of underlayment for clay and concrete tile.

R905.3.3.1 Low slope roofs. For roof slopes from two and one-half units vertical in 12 units horizontal ($2^1/_2$:12), up to four units vertical in 12 units horizontal (4:12), underlayment shall be a minimum of two layers underlayment applied as follows:

1. Starting at the eave, a 19-inch (483 mm) strip of underlayment shall be applied parallel with the eave and fastened sufficiently in place.

2. Starting at the eave, 36-inch-wide (914 mm) strips of underlayment felt shall be applied, overlapping successive sheets 19 inches (483 mm), and fastened sufficiently in place.

❖ Where the roof covering consists of clay tile or concrete tile, the roof is considered low-slope if the slope is at least $2^1/_2$:12 but no steeper than 4:12. Low-slope

roofs present a potential problem because water drains slowly; this creates an opportunity for water back-up. Wind-driven rain can also pose a problem. Therefore, a special underlayment application method is used so that the roof remains weathertight. All portions of the roof will be protected by a minimum of two layers of underlayment if the installer follows the instructions found in this section.

R905.3.3.2 High slope roofs. For roof slopes of four units vertical in 12 units horizontal (4:12) or greater, underlayment shall be a minimum of one layer of underlayment felt applied shingle fashion, parallel to and starting from the eaves and lapped 2 inches (51 mm), fastened sufficiently in place.

❖ Where the roof covering consists of clay tile or concrete tile, the roof is considered high-slope if the slope is at least 4:12. Only one layer of underlayment felt, applied in shingle fashion in accordance with this section, is mandated for high-slope roofs of concrete or clay tile.

R905.3.3.3 Underlayment and high wind. Underlayment applied in areas subject to high wind [over 110 miles per hour (49 m/s) per Figure R301.2(4)] shall be applied with corrosion-resistant fasteners in accordance with manufacturer's installation instructions. Fasteners are to be applied along the overlap not farther apart than 36 inches (914 mm) on center.

❖ In high-wind areas, corrosion-resistant fasteners must be located in accordance with the manufacturer's installation instructions, but in no case can they be more than 36 inches (914 mm) on center. The general requirement for sufficient fastening to hold the underlayment in place is not adequate where increased wind loads are anticipated.

R905.3.4 Tile. Clay roof tile shall comply with ASTM C 1167.

❖ ASTM C 1167 regulates clay roof tiles manufactured from clay, shale or similar natural earthy substances. The standard addresses a variety of performance requirements, including durability, strength and permeability.

R905.3.5 Concrete tile. Concrete roof tile shall comply with ASTM C 1492.

❖ ASTM C 1492 regulates concrete roof tiles manufactured from portland cement, water and mineral aggregates with or without the inclusion of other materials. Lightweight or normal weight aggregates or both are used in the manufacture of concrete tiles. The tiles are shaped during manufacturing by molding, pressing or extrusion. The tiles are usually planar or undulating rectangular shapes available in a variety of cross-sectional areas, profiles, shapes, sizes, surface textures and colors. The standard addresses a variety of performance tests including dimensional tolerances, freeze/thaw, transverse strength, permeability and water absorption.

R905.3.6 Fasteners. Nails shall be corrosion resistant and not less than 11 gage, $5/_{16}$-inch (11 mm) head, and of sufficient length to penetrate the deck a minimum of $3/_4$ inch (19 mm) or through the thickness of the deck, whichever is less. Attaching wire for clay or concrete tile shall not be smaller than 0.083 inch (2 mm). Perimeter fastening areas include three tile courses but not less than 36 inches (914 mm) from either side of hips or ridges and edges of eaves and gable rakes.

❖ Roofing nails for fastening clay or concrete tiles must be corrosion resistant. A roofing nail must have a minimum 11 gage shank with a head at least $5/_{16}$-inch (10.6 mm) in diameter. To provide the necessary holding power, roofing nails must penetrate the roof sheathing a minimum of $3/_4$ inch (19.1 mm). For roof sheathing having a thickness less than $3/_4$ inch (19.1 mm), the nails must penetrate completely through the sheathing. The portions of the roof considered as perimeter fastening areas include a minimum of three tile courses, but in no case are they less than 36 inches (914 mm) from the edges of eaves and gable rakes or from either side of hips and ridges.

R905.3.7 Application. Tile shall be applied in accordance with this chapter and the manufacturer's installation instructions, based on the following:

1. Climatic conditions.

2. Roof slope.

3. Underlayment system.

4. Type of tile being installed.

Clay and concrete roof tiles shall be fastened in accordance with this section and the manufacturer's installation instructions. Perimeter tiles shall be fastened with a minimum of one fastener per tile. Tiles with installed weight less than 9 pounds per square foot (0.4 kg/m^2) require a minimum of one fastener per tile regardless of roof slope. Clay and concrete roof tile attachment shall be in accordance with the manufacturer's installation instructions where applied in areas where the wind speed exceeds 100 miles per hour (45 m/s) and on buildings where the roof is located more than 40 feet (12 192 mm) above grade. In areas subject to snow, a minimum of two fasteners per tile is required. In all other areas, clay and concrete roof tiles shall be attached in accordance with Table R905.3.7.

❖ This section identifies four key issues that must be taken into consideration when using the code and the manufacturer's installation instructions in the proper application of clay and concrete roof tile. At least one fastener is required for each roof tile for those tiles at the perimeter of the roof, as well as any tiles weighing less than 9 pounds per square foot (0.4 kg/m^2). In areas where the roof is exposed to a roof snow load, at least two fasteners are required for each tile. Because of the special concerns in high-wind areas, or where the roof height is well above grade, the manufacturer's installation instructions must be referred to in the determination of the minimum number of fasteners.

If none of the above-mentioned conditions occur, Table R905.3.7 must be used to determine proper fastening. The type of sheathing (spaced or solid), the presence of battens and the slope of the roof all impact the minimum number of fasteners required. Commentary Figures R905.3.7(1) through (4) illustrate some of the common shapes of tile and installation techniques.

TABLE R905.3.7
CLAY AND CONCRETE TILE ATTACHMENT

SHEATHING	ROOF SLOPE	NUMBER OF FASTENERS
Solid without battens	All	One per tile
Spaced or solid with battens and slope < 5:12	Fasteners not required	—
Spaced sheathing without battens	5:12 ≤ slope < 12:12	One per tile/every other row.
	12:12 ≤ slope < 24:12	One per tile

❖ This table sets forth the number of fasteners required to attach clay or concrete tile to various types of sheathing. The table is not applicable where the wind speed exceeds 100 miles per hour (45 m/s), where the building's roof is located more than 40 feet (12 192 mm) above grade or in locales subject to snow. Under all other conditions, the table indicates the number of fasteners required based on the type of sheathing and, in some cases, the slope of the roof.

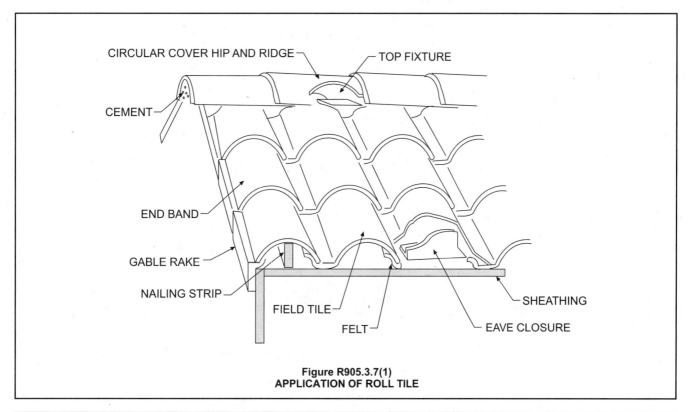

Figure R905.3.7(1)
APPLICATION OF ROLL TILE

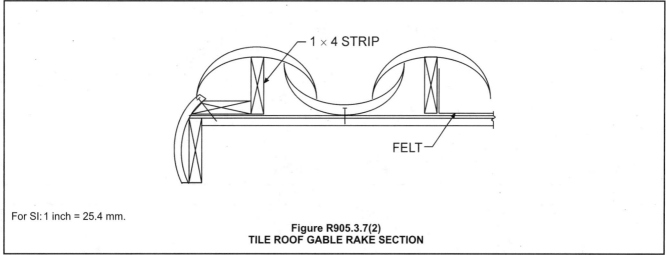

For SI: 1 inch = 25.4 mm.

Figure R905.3.7(2)
TILE ROOF GABLE RAKE SECTION

R905.3.8 Flashing. At the juncture of roof vertical surfaces, flashing and counterflashing shall be provided in accordance with this chapter and the manufacturer's installation instructions and, where of metal, shall not be less than 0.019 inch (0.5 mm) (No. 26 galvanized sheet gage) corrosion-resistant metal. The valley flashing shall extend at least 11 inches (279 mm) from the centerline each way and have a splash diverter rib not less than 1 inch (25 mm) high at the flow line formed as part of the flashing. Sections of flashing shall have an end lap of not less than 4 inches (102 mm). For roof slopes of three units vertical in 12 units horizontal (25-percent slope) and greater, valley flashing shall have a 36-inch-wide (914 mm) underlayment of one layer of Type I underlayment running the full length of the valley, in addition to other required underlayment. In areas where the average daily temperature in January is 25°F (-4°C) or less, metal valley flashing underlayment shall be solid-cemented to the roofing underlayment for slopes less than seven units vertical in 12 units horizontal (58-percent slope) or be of self-adhering polymer modified bitumen sheet.

❖ Flashing for clay and concrete roof tile is important because the shape of the tile may terminate 2 or 3 inches (51 or 76 mm) above the roof sheathing. Flashing details, therefore, must be constructed in accordance with the code and the manufacturer's installation instructions for the tile shape being used. Some common methods of providing flashing at different locations are illustrated in Commentary Figures R905.3.8(1) through (4).

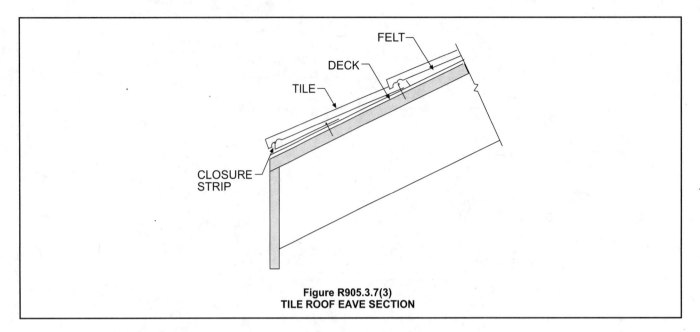

Figure R905.3.7(3)
TILE ROOF EAVE SECTION

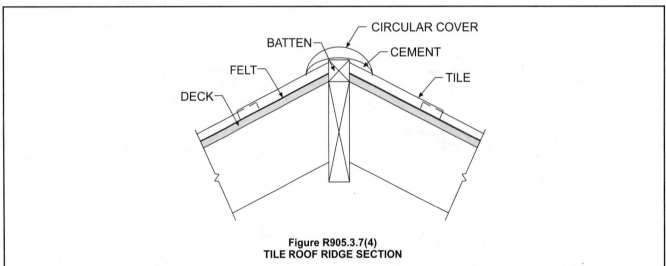

Figure R905.3.7(4)
TILE ROOF RIDGE SECTION

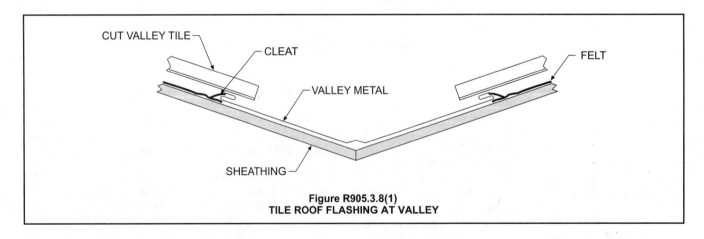

Figure R905.3.8(1)
TILE ROOF FLASHING AT VALLEY

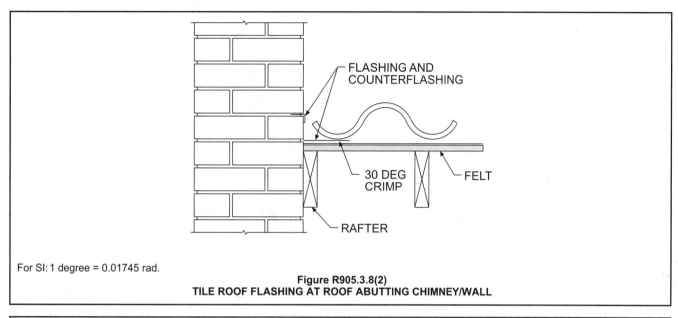

For SI: 1 degree = 0.01745 rad.

Figure R905.3.8(2)
TILE ROOF FLASHING AT ROOF ABUTTING CHIMNEY/WALL

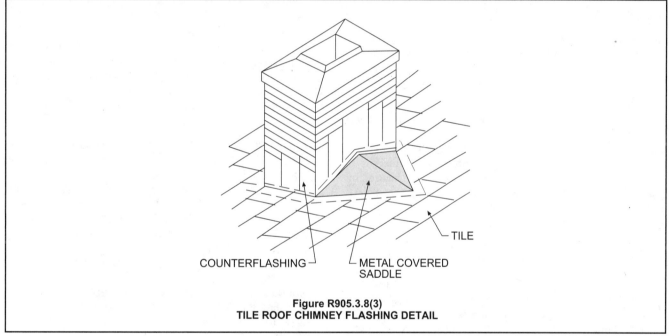

Figure R905.3.8(3)
TILE ROOF CHIMNEY FLASHING DETAIL

R905.4 Metal roof shingles. The installation of metal roof shingles shall comply with the provisions of this section.

❖ Metal roof shingles differ from the metal roof panels regulated by Section R905.10. By definition, a metal roof shingle must have an installed weather exposure of less than 3 square feet (0.28 m²) per shingle or sheet. While decking and slope limitations differ to some degree, the provisions for underlayment and flashing are very similar to those for other types of shingle applications.

R905.4.1 Deck requirements. Metal roof shingles shall be applied to a solid or closely fitted deck, except where the roof covering is specifically designed to be applied to spaced sheathing.

❖ Unless manufacturers of metal shingles indicate that their particular shingle may be placed over a roof deck having spaced sheathing, metal roof shingles must be installed on a solid or closely fitted deck.

R905.4.2 Deck slope. Metal roof shingles shall not be installed on roof slopes below three units vertical in 12 units horizontal (25-percent slope).

❖ Based on the characteristics of shape, size and method of attachment, metal roof shingles must not be applied to roofs having a slope of less than 3:12. A lesser slope increases the potential for leakage to an unacceptable level.

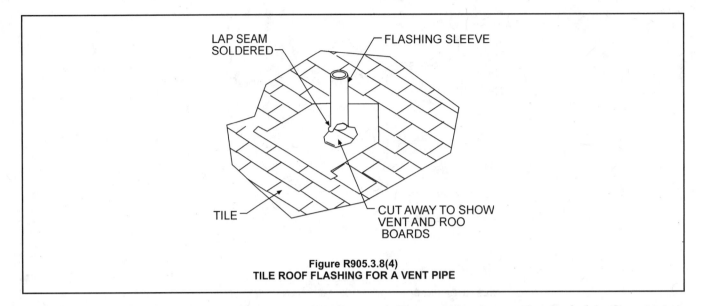

LAP SEAM SOLDERED

FLASHING SLEEVE

TILE

CUT AWAY TO SHOW VENT AND ROO BOARDS

Figure R905.3.8(4)
TILE ROOF FLASHING FOR A VENT PIPE

R905.4.3 Underlayment. Underlayment shall comply with ASTM D 226, Type I or ASTM D 4869, Type I or II.

❖ Two types of underlayment are recognized for use with metal shingle roof coverings; asphalt-saturated organic felt as regulated by ASTM D 226, Type I, or asphalt-saturated organic felt shingles per ASTM D 4869, Type I or II.

R905.4.3.1 Ice barrier. In areas where there has been a history of ice forming along the eaves causing a backup of water as designated in Table R301.2(1), an ice barrier that consists of at least two layers of underlayment cemented together or a self-adhering polymer modified bitumen sheet shall be used in place of normal underlayment and extend from the lowest edges of all roof surfaces to a point at least 24 inches (610 mm) inside the exterior wall line of the building.

Exception: Detached accessory structures that contain no conditioned floor area.

❖ Where ice dams may be formed along the eave because snow continually freezes and thaws or frozen slush backs up in gutters, the underlayment application in the area of the eaves must be modified to prevent ice dams from forcing water under the roofing, which could damage ceilings, walls and insulation. Two layers of underlayment should be cemented together with asphalt cement from the lowest edge of the roof and continue up the roof to a point that is at least 24 inches (610 mm) inside the interior wall line of the building as shown in Commentary Figure R905.2.7.1(2). The environment within the envelope of the building provides adequate warmth to prevent ice dams from forming above the heated space; therefore, the two layers of cemented underlayment are permitted to terminate 24 inches (610 mm) inside the interior wall line of the building. The local jurisdiction is responsible for determining whether the ice barrier is required based on weather records, and it must so indicate in Table R301.2(1).

The exception is used throughout Chapter 9 for unconditioned accessory buildings. See the commentary for Section R905.2.7.1.

R905.4.4 Material standards. Metal roof shingle roof coverings shall comply with Table R905.10.3(1). The materials used for metal roof shingle roof coverings shall be naturally corrosion resistant or be made corrosion resistant in accordance with the standards and minimum thicknesses listed in Table R905.10.3(2).

❖ Table R905.10.3(1) list the metal roof covering types and their material standards. Many of the materials listed are inherently corrosion resistant, such as aluminum, copper and lead. For these materials, listing a material standard for the base material is sufficient. However, for steel roofing products, coatings are added to the base material to provide the necessary corrosion resistance.

Table R905.10.3(2) lists the minimum coating thickness required to provide minimum corrosion resistance for the base steel. Note a to Table R905.10.3(2) specifies that the paint systems are intended to be applied to steel having one of the corrosion resistant coatings listed.

R905.4.5 Application. Metal roof shingles shall be secured to the roof in accordance with this chapter and the approved manufacturer's installation instructions.

❖ The proper application of metal roof shingles is typically based on the installation instructions of the shingle manufacturer. Where the IRC sets forth specific requirements for the attachment of metal shingles, those requirements must be met.

R905.4.6 Flashing. Roof valley flashing shall be of corrosion-resistant metal of the same material as the roof covering or shall comply with the standards in Table R905.10.3(1). The valley flashing shall extend at least 8 inches (203 mm) from the center line each way and shall have a splash diverter rib not less than ³/₄ inch (19 mm) high at the flow line formed as part of the flashing. Sections of flashing shall have an end lap of not less than 4 inches (102 mm). The metal valley flashing shall have a 36-inch-wide (914 mm) underlayment directly under it consisting of one layer of underlayment running the full length of the valley, in addition to underlayment required for metal roof shingles. In areas where the average daily temperature in January is 25°F (-4°C) or less , the metal valley flashing underlayment shall be solid cemented to the roofing underlayment for roof slopes under seven units vertical in 12 units horizontal (58-percent slope) or self-adhering polymer modified bitumen sheet.

❖ This section specifies two approaches to roof valley flashing for metal-shingle roofs. The first is the use of material consistent with that of the shingles for the flashing. As an option, the flashing must be accomplished in compliance with the applicable material standard listed in Table R905.10.3. This table identifies the various types of metal roof covering and the corresponding application standard. This section specifies installation details for the valley flashing as well as the requirement for underlayment running the full length of the valley. In cold-weather areas, the special underlayment techniques to eliminate ice dams apply to valley flashing underlayment.

R905.5 Mineral-surfaced roll roofing. The installation of mineral-surfaced roll roofing shall comply with this section.

❖ This section addresses the decking requirements and slope, underlayment, materials standards and application methods for mineral-surfaced roll roofing.

R905.5.1 Deck requirements. Mineral-surfaced roll roofing shall be fastened to solidly sheathed roofs.

❖ The deck must be solidly sheathed for mineral-surfaced roll roofing to be installed.

R905.5.2 Deck slope. Mineral-surfaced roll roofing shall not be applied on roof slopes below one unit vertical in 12 units horizontal (8-percent slope).

❖ Although mineral-surfaced roll roofing provides a virtually weathertight roof surface, it is still necessary that water flow be positive toward the drainage points. Therefore, a minimum slope of 1:12 is mandated for roof decks that are covered with mineral-surfaced roll roofing.

R905.5.3 Underlayment. Underlayment shall comply with ASTM D 226, Type I or ASTM D 4869, Type I or II.

❖ Two types of underlayment are recognized for use with mineral-surfaced roll roof coverings; asphalt-saturated organic felt as regulated by ASTM D 226, Type I, or asphalt-saturated organic felt shingles per ASTM D 4869, Type I or II.

R905.5.3.1 Ice barrier. In areas where there has been a history of ice forming along the eaves causing a backup of water as designated in Table R301.2(1), an ice barrier that consists of at least two layers of underlayment cemented together or a self-adhering polymer modified bitumen sheet shall be used in place of normal underlayment and extend from the lowest edges of all roof surfaces to a point at least 24 inches (610 mm) inside the exterior wall line of the building.

Exception: Detached accessory structures that contain no conditioned floor area.

❖ Where ice dams may be formed along the eave because snow continually freezes and thaws or frozen slush backs up in gutters, the underlayment application in the area of the eaves is modified to prevent ice dams from forcing water under the roofing, which could damage ceilings, walls and insulation. Two layers of underlayment should be cemented together with asphalt cement from the lowest edge of the roof and continue up the roof to a point that is at least 24 inches (610 mm) inside the interior wall line of the building as shown in Commentary Figure R905.2.7.1(2). The environment within the envelope of the building provides adequate warmth to prevent ice dams from forming above the heated space; therefore, the two layers of cemented underlayment are permitted to terminate 24 inches (610 mm) inside the interior wall line of the building. The local jurisdiction is responsible for determining whether the ice barrier is required based on weather records, and it must so indicate in Table R301.2(1).

The exception is used throughout Chapter 9 for unconditioned accessory buildings (see commentary, Section R905.2.7.1).

R905.5.4 Material standards. Mineral-surfaced roll roofing shall conform to ASTM D 3909 or ASTM D 6380, Class M.

❖ The code recognizes two materials standards for the regulation of mineral-surfaced roll roofing, ASTM D 3909 for asphalt roll roofing of glass felt surfaced with mineral granules and ASTM D 6380, Class M for asphalt roll roofing of organic felt surfaced with mineral granules.

R905.5.5 Application. Mineral-surfaced roll roofing shall be installed in accordance with this chapter and the manufacturer's installation instructions.

❖ The general provisions of this chapter apply to the installation of mineral-surfaced roll roofing, as do the installation instructions of the manufacturer of the specific roofing system.

R905.6 Slate and slate-type shingles. The installation of slate and slate-type shingles shall comply with the provisions of this section.

❖ This section establishes the materials standards and installation criteria for shingles of slate and slate-type materials. It includes specific provisions related to deck slope, underlayment, shingle application and flashing.

R905.6.1 Deck requirements. Slate shingles shall be fastened to solidly sheathed roofs.

❖ Only solidly-sheathed roofs can be used as the deck for the installation of slate and slate-type shingles.

R905.6.2 Deck slope. Slate shingles shall be used only on slopes of four units vertical in 12 units horizontal (33-percent slope) or greater.

❖ Shingles of slate or slate-type materials pose special problems because of the inconsistency of the shingles' shape and size. To keep the roof covering weather-tight and promote proper drainage, slate and slate-type shingles cannot be installed on a roof having a slope of less than 4:12. Lower-sloped roofs present a potential problem because water drains slowly and, thus, creates the opportunity for water back-up. Wind-driven rain can also pose a problem.

R905.6.3 Underlayment. Underlayment shall comply with ASTM D 226, Type I or ASTM D 4869, Type I or II.

❖ Two types of underlayment are recognized for use with slate and slate-type shingles' roof coverings; asphalt-saturated organic felt as regulated by ASTM D 226, Type I, or asphalt-saturated organic felt shingles per ASTM D 4869, Type I or II.

R905.6.3.1 Ice barrier. In areas where there has been a history of ice forming along the eaves causing a backup of water as designated in Table R301.2(1), an ice barrier that consists of at least two layers of underlayment cemented together or a self-adhering polymer modified bitumen sheet shall be used in lieu of normal underlayment and extend from the lowest edges of all roof surfaces to a point at least 24 inches (610 mm) inside the exterior wall line of the building.

> **Exception:** Detached accessory structures that contain no conditioned floor area.

❖ Where ice dams may be formed along the eave because snow continually freezes and thaws or frozen slush backs up in gutters, the underlayment application in the area of the eaves is modified to prevent ice dams from forcing water under the roofing, which could damage ceilings, walls and insulation. Two layers of underlayment should be cemented together with asphalt cement from the lowest edge of the roof and continue up the roof to a point that is at least 24 inches (610 mm) inside the interior wall line of the building as shown in Commentary Figure R905.2.7.1(2). The required underlayment must comply with ASTM D 226, Type II. This is the rare condition under which Type II underlayment is mandated. The referenced underlayment for most other types of roof covering materials is Type I. The environment within the envelope of the building provides adequate warmth to prevent ice dams from forming above the heated space; therefore, the two layers of cemented underlayment are permitted to terminate 24 inches (610 mm) inside the interior wall line of the building. The local jurisdiction is responsible for determining whether the ice barrier is required based on weather records, and it must so indicate in Table R301.2(1).

The exception is used throughout Chapter 9 for unconditioned accessory buildings (see commentary, Section R905.2.7.1).

R905.6.4 Material standards. Slate shingles shall comply with ASTM C 406.

❖ The standard applicable to roofing slate is ASTM C 406.

R905.6.5 Application. Minimum headlap for slate shingles shall be in accordance with Table R905.6.5. Slate shingles shall be secured to the roof with two fasteners per slate. Slate shingles shall be installed in accordance with this chapter and the manufacturer's installation instructions.

❖ The general provisions of this chapter apply to the installation of slate and slate-type shingles, as do the installation instructions of the manufacturer of the specific roofing materials. In addition, this section sets forth two requirements specific to slate roofing. First, two fasteners must be used to secure each slate shingle in place. Second, Table R905.6.5 mandates the minimum headlap at each shingle course. Taking into account the effects of both water traveling down the roof's surface and wind-driven rain, the minimum headlap is reduced as the slope of the roof increases.

TABLE R905.6.5
SLATE SHINGLE HEADLAP

SLOPE	HEADLAP (inches)
4:12 ≤ slope < 8:12	4
8:12 ≤ slope < 20:12	3
Slope ≤ 20:12	2

For SI: 1 inch = 25.4 mm.

❖ If slate shingles are installed, they must have a minimum lap at each course as required by this table. Because of the potential for water infiltration, as the slope of the roof decreases, the minimum headlap increases. The extended lap aids in resisting the effects of slow-traveling water and wind-driven rain.

R905.6.6 Flashing. Flashing and counterflashing shall be made with sheet metal. Valley flashing shall be a minimum of 15 inches (381 mm) wide. Valley and flashing metal shall be a minimum uncoated thickness of 0.0179-inch (0.5 mm) zinc coated G90. Chimneys, stucco or brick walls shall have a minimum of two plies of felt for a cap flashing consisting of a 4-inch-wide (102 mm) strip of felt set in plastic cement and extending 1 inch (25 mm) above the first felt and a top coating of plastic cement. The felt shall extend over the base flashing 2 inches (51 mm).

❖ Special requirements are necessary for the sheet metal flashing and counterflashing of roofs covered with slate or slate-type shingles. The provisions address flashing for roof valleys and for the roof intersections with chimneys and stucco or brick walls.

R905.7 Wood shingles. The installation of wood shingles shall comply with the provisions of this section.

❖ Wood shingles differ from wood shakes in that wood shingles are sawed materials and have a uniform butt thickness across the individual shingle length. Wood shakes are split from logs, shaped by the manufacturer, and have varying butt thicknesses. This section establishes the material standards and installation criteria for wood shingles. Also included are specific provisions related to deck requirements and slope, underlayment, shingle application and valley flashing.

R905.7.1 Deck requirements. Wood shingles shall be installed on solid or spaced sheathing. Where spaced sheathing is used, sheathing boards shall not be less than 1-inch by 4-inch (25.4 mm by 102 mm) nominal dimensions and shall be spaced on centers equal to the weather exposure to coincide with the placement of fasteners.

❖ Either solid or spaced sheathing is permitted as a deck for the installation of wood shingles. If spaced sheathing is used, the minimum 1-inch by 4-inch (25.4 mm by 102 mm) sheathing boards are required. The placement of the spaced sheathing is important, with the center-to-center spacing of the sheathing boards equivalent to the weather exposure of the wood shingles. This method allows a consistent fastening pattern for applying the wood-shingle roof-covering materials. Table R905.7.5 contains the maximum permitted weather exposure for wood shingles.

R905.7.1.1 Solid sheathing required. In areas where the average daily temperature in January is 25°F (-4°C) or less, solid sheathing is required on that portion of the roof requiring the application of an ice barrier.

❖ In cold-weather climates where an ice shield is required in accordance with Section R905.7.3, solid sheathing must be installed for decking. The solid sheathing need be located only where the ice shield is applied, typically a point from the eave's edge to at least 24 inches (610 mm) inside the exterior wall line of the building

R905.7.2 Deck slope. Wood shingles shall be installed on slopes of three units vertical in 12 units horizontal (25-percent slope) or greater.

❖ Wood shingle roofs pose special problems for weather protection because of the texture of the materials. To keep the roof covering weathertight and to promote proper drainage, wood shingles cannot be installed on a roof having a slope of less than 3:12. Lower-sloped roofs present a potential problem because water drains slowly, creating an opportunity for water back-up. Wind-driven rain can also pose a problem.

R905.7.3 Underlayment. Underlayment shall comply with ASTM D 226, Type I or ASTM D 4869, Type I or II.

❖ Two types of underlayment are recognized for use with wood shingles roof coverings; asphalt-saturated organic felt as regulated by ASTM D 226, Type I, or asphalt-saturated organic felt shingles per ASTM D 4869. Type I or II.

R905.7.3.1 Ice barrier. In areas where there has been a history of ice forming along the eaves causing a backup of water as designated in Table R301.2(1), an ice barrier that consists of at least two layers of underlayment cemented together or a self-adhering polymer modified bitumen sheet shall be used in lieu of normal underlayment and extend from the lowest edges of all roof surfaces to a point at least 24 inches (610 mm) inside the exterior wall line of the building.

Exception: Detached accessory structures that contain no conditioned floor area.

❖ Where ice dams may be formed along the eave because snow continually freezes and thaws or frozen slush backs up in gutters, the underlayment application in the area of the eaves is modified to prevent ice dams from forcing water under the roofing, which could damage ceilings, walls and insulation. Two layers of underlayment should be cemented together with asphalt cement from the lowest edge of the roof and continue up the roof to a point that is at least 24 inches (610 mm) inside the interior wall line of the building as shown in Commentary Figure R905.2.7.1(2). The environment within the envelope of the building provides adequate warmth to prevent ice dams from forming above the heated space; therefore, the two layers of cemented underlayment are permitted to terminate 24 inches (610 mm) inside the interior wall line of the building. The local jurisdiction is responsible for determining whether the ice barrier is required based on weather records, and it must so indicate in Table R301.2(1).

The exception is used throughout Chapter 9 for unconditioned accessory buildings (see commentary, Section R905.2.7.1).

R905.7.4 Material standards. Wood shingles shall be of naturally durable wood and comply with the requirements of Table R905.7.4.

❖ Wood shingles are manufactured in 24-, 18- and 16-inch (610-, 457- and 406-mm) lengths and are graded into three categories: 1, 2 and 3. The Cedar Shake and Shingle Bureau establishes the grading rules for wood shingles. Table R905.7.4 describes each grade, and Commentary Figure R905.7.4 shows an example label.

TABLE R905.7.4
WOOD SHINGLE MATERIAL REQUIREMENTS

MATERIAL	MINIMUM GRADES	APPLICABLE GRADING RULES
Wood shingles of naturally durable wood	1, 2 or 3	Cedar Shake and Shingle Bureau

❖ This table identifies the basic material requirements for wood roof shingles. It states that the shingles must be of naturally durable wood; graded as No. 1, No. 2 or No. 3 and subject to the applicable grading rules of the Cedar Shake and Shingle Bureau. Each grade is mentioned in Table R905.7.4. Commentary Figure R905.7.4 is an example of the required identification label of an approved grading or inspection bureau or agency.

R905.7.5 Application. Wood shingles shall be installed according to this chapter and the manufacturer's installation instructions. Wood shingles shall be laid with a side lap not less than $1\frac{1}{2}$ inches (38 mm) between joints in courses, and no two joints in any three adjacent courses shall be in direct alignment. Spacing between shingles shall not be less than $\frac{1}{4}$ inch to $\frac{3}{8}$ inch (6 mm to 10 mm). Weather exposure for wood shingles shall not exceed those set in Table R905.7.5. Fasteners for wood shingles shall be corrosion resistant with a minimum penetration of $\frac{1}{2}$ inch (13 mm) into the sheathing. For sheathing less than $\frac{1}{2}$ inch (13 mm) in thickness, the fasteners shall extend through the sheathing.

Wood shingles shall be attached to the roof with two fasteners per shingle, positioned no more than $3/4$ inch (19 mm) from each edge and no more than 1 inch (25 mm) above the exposure line.

❖ The general provisions of this chapter apply to the installation of wood shingles, as do the installation instructions of the manufacturer of the specific roofing materials. In addition, this section sets forth specific requirements regulating the installation method for wood shingles. Many of these requirements are shown in Commentary Figure R905.7.5(1).

Fasteners for applying wood shingles must be corrosion resistant. Only two nails or staples are to be used for each shingle, and they must be placed approximately $3/4$ inch (19 mm) from each side edge and 1 inch (25 mm) above the exposure line. The depth of fastener penetration must be a minimum of $1/2$ inch (12.7 mm) into the sheathing or completely through sheathing less than $1/2$ inch (12.7 mm) thick. Care in spacing and driving the fasteners is necessary to provide maximum service from the roof covering [see Commentary Figure R905.7.5(2)].

TABLE R905.7.5
WOOD SHINGLE WEATHER EXPOSURE AND ROOF SLOPE

ROOFING MATERIAL	LENGTH (inches)	GRADE	EXPOSURE (inches)	
			3:12 pitch to < 4:12	4:12 pitch or steeper
Shingles of naturally durable wood	16	No. 1	$3^3/_4$	5
		No. 2	$3^1/_2$	4
		No. 3	3	$3^1/_2$
	18	No. 1	$4^1/_4$	$5^1/_2$
		No. 2	4	$4^1/_2$
		No. 3	$3^1/_2$	4
	24	No. 1	$5^3/_4$	$7^1/_2$
		No. 2	$5^1/_2$	$6^1/_2$
		No. 3	5	$5^1/_2$

For SI: 1 inch = 25.4 mm.

❖ Table R905.7.5 specifies wood shingle exposure. Depending on the grade of the material, the total shingle length and the slope of the roof deck, the table specifies the maximum length of exposure for weathering purposes. The grade of the shingle is listed in the label required for wood shingles. Each of the three different grades is available in three different lengths. For lower-sloped roofs (3:12 to 4:12) where wind uplift is more significant, the maximum weather exposure is always less than for the steeper sloped roofs (4:12 and greater).

R905.7.6 Valley flashing. Roof flashing shall be not less than No. 26 gage [0.019 inches (0.5 mm)] corrosion-resistant sheet metal and shall extend 10 inches (254 mm) from the centerline each way for roofs having slopes less than 12 units vertical in 12 units horizontal (100-percent slope), and 7 inches (178 mm) from the centerline each way for slopes of 12 units vertical in

12 units horizontal and greater. Sections of flashing shall have an end lap of not less than 4 inches (102 mm).

❖ Valleys formed by the intersection of two sloping roofs are vulnerable to leakage because of the high concentration of water; therefore, valley flashing must be maintained to prevent blockage. This section requires the use of minimum No. 26 gage (0.019 inch/0.48 mm) corrosion-resistant sheet metal for this purpose. Commentary Figure R905.7.6 illustrates the code provisions for valley flashing.

R905.7.7 Label required. Each bundle of shingles shall be identified by a label of an approved grading or inspection bureau or agency.

❖ To verify compliance with the material standards set forth in Section R905.7.4, every bundle of wood shingles must be labeled. The label, a sample of which is shown in Commentary Figure R905.7.4, must indicate the name of the approved grading or inspection bureau or agency.

R905.8 Wood shakes. The installation of wood shakes shall comply with the provisions of this section.

❖ Wood shakes differ from wood shingles in that wood shakes are split from logs, shaped by the manufacturer and have varying butt thicknesses. Wood shingles are sawed materials and have a uniform butt thickness across the individual shingle length. This section establishes the materials standards and installation criteria for wood shakes. Also included are specific provisions related to deck requirements and slope, underlayment, interlayment, shingle application and valley flashing.

R905.8.1 Deck requirements. Wood shakes shall be used only on solid or spaced sheathing. Where spaced sheathing is used, sheathing boards shall not be less than 1-inch by 4-inch (25 mm by 102 mm) nominal dimensions and shall be spaced on centers equal to the weather exposure to coincide with the placement of fasteners. Where 1-inch by 4-inch (25 mm by 102 mm) spaced sheathing is installed at 10 inches (254 mm) on center, additional 1-inch by 4-inch (25 mm by 102 mm) boards shall be installed between the sheathing boards.

❖ Either solid or spaced sheathing is permitted as a deck for the installation of wood shakes. If spaced sheathing is used, the minimum 1-inch by 4-inch (25.4 mm by 102 mm) sheathing boards are required. The placement of the spaced sheathing is important, with the center-to-center spacing of the sheathing boards equivalent to the weather exposure of the wood shakes. This method allows for a consistent fastening pattern for applying wood shake roof covering materials. Table R905.8.6 contains the maximum permitted weather exposure for wood shakes. If the greatest allowable exposure of 10 inches is used per Table R905.8.6, an additional 1-inch by 4-inch (25.4 mm by 102 mm) board must be installed between the sheathing boards.

Figure R905.7.4
WOOD SHINGLE GRADE DESCRIPTION AND EXAMPLE LABEL

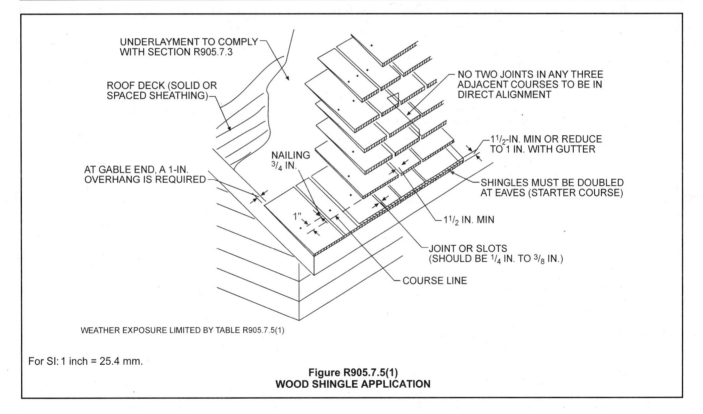

For SI: 1 inch = 25.4 mm.

Figure R905.7.5(1)
WOOD SHINGLE APPLICATION

R905.8.1.1 Solid sheathing required. In areas where the average daily temperature in January is 25°F (-4°C) or less, solid sheathing is required on that portion of the roof requiring an ice barrier.

❖ In cold-weather climates where an ice shield is required in accordance with Section R905.7.3, solid sheathing must be installed for decking. The solid sheathing need be located only where the ice shield is applied, typically a point from the eave's edge to at least 24 inches (610 mm) inside the exterior wall line of the building.

R905.8.2 Deck slope. Wood shakes shall only be used on slopes of three units vertical in 12 units horizontal (25-percent slope) or greater.

❖ Wood shake roofs pose special problems for weather protection because of the texture of the materials. To keep the roof covering weathertight and to promote proper drainage, wood shakes cannot be installed on a roof having a slope of less than 3:12. Lower-sloped roofs present a potential problem because water drains slowly, creating an opportunity for water back-up. Wind-driven rain can also pose a problem. The use of an interlayment as installed per Section R905.8.7 also plays an important part in maintaining the weathertightness of the roof.

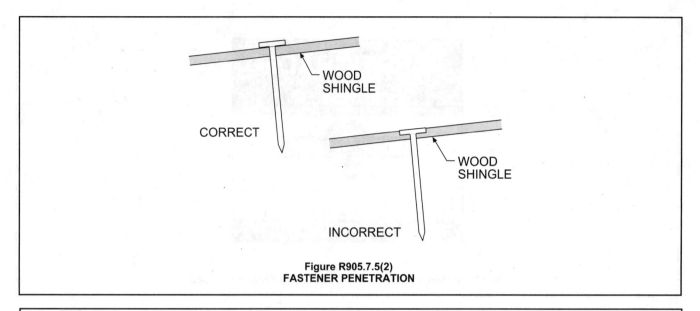

Figure R905.7.5(2)
FASTENER PENETRATION

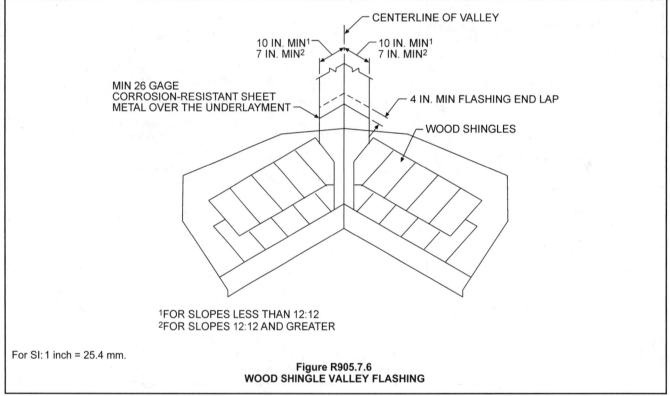

¹FOR SLOPES LESS THAN 12:12
²FOR SLOPES 12:12 AND GREATER

For SI: 1 inch = 25.4 mm.

Figure R905.7.6
WOOD SHINGLE VALLEY FLASHING

R905.8.3 Underlayment. Underlayment shall comply with ASTM D 226, Type I or ASTM D 4869, Type I or II.

❖ Two types of underlayment are recognized for use with wood-shingle roof coverings; asphalt-saturated organic felt as regulated by ASTM D 226, Type I, or asphalt-saturated organic felt shingles per ASTM D 4869, Type I or II.

R905.8.3.1 Ice barrier. In areas where there has been a history of ice forming along the eaves causing a backup of water as designated in Table R301.2(1), an ice barrier that consists of at least two layers of underlayment cemented together or a self-adhering polymer modified bitumen sheet shall be used in place of normal underlayment and extend from the lowest edges of all roof surfaces to a point at least 24 inches (610 mm) inside the exterior wall line of the building.

Exception: Detached accessory structures that contain no conditioned floor area.

❖ Where ice dams may be formed along the eave because snow continually freezes and thaws or frozen slush backs up in gutters, the underlayment application in the area of the eaves is modified to prevent ice dams from forcing water under the roofing, which could damage ceilings, walls and insulation. Two layers of underlayment should be cemented together with asphalt cement from the lowest edge of the roof and continue up the roof to a point that is at least 24 inches (610 mm) inside the interior wall line of the building as shown in Commentary Figure R905.2.7.1(2). The environment within the envelope of the building provides adequate warmth to prevent ice dams from forming above the heated space; therefore, the two layers of cemented underlayment are permitted to terminate 24 inches (610 mm) inside the interior wall line of the building. The local jurisdiction is responsible for determining whether the ice barrier is required based on weather records, and it must so indicate in Table R301.2(1).

The exception is used throughout Chapter 9 for unconditioned accessory buildings (see commentary, Section R905.2.7.1).

R905.8.4 Interlayment. Interlayment shall comply with ASTM D 226, Type I.

❖ Wood shakes must be installed with an interlayment of minimum No. 30 felt shingled between each course in the manner described by Section R905.8.7. The interlayment must be in compliance with ASTM D 226, Type I, for asphalt-saturated organic felt.

R905.8.5 Material standards. Wood shakes shall comply with the requirements of Table R905.8.5.

❖ Wood shakes are divided into five material classifications in the IRC, identified in Table R905.8.5. There are two grades of wood shakes, No. 1 and No. 2. The shakes are cut into 18-inch (457 mm) and 24-inch (610 mm) lengths.

TABLE R905.8.5
WOOD SHAKE MATERIAL REQUIREMENTS

MATERIAL	MINIMUM GRADES	APPLICABLE GRADING RULES
Wood shakes of naturally durable wood	1	Cedar Shake and Shingle Bureau
Taper sawn shakes of naturally durable wood	1 or 2	Cedar Shake and Shingle Bureau
Preservative-treated shakes and shingles of naturally durable wood	1	Cedar Shake and Shingle Bureau
Fire-retardant-treated shakes and shingles of naturally durable wood	1	Cedar Shake and Shingle Bureau
Preservative-treated taper sawn shakes of Southern pine treated in accordance with AWPA Standard U1 (Commodity Specification A, Use Category 3B and Section 5.6)	1 or 2	Forest Products Laboratory of the Texas Forest Services

❖ Wood shakes are regulated based on the type of shake material, the treatment applied to the shake and the cut of the shake. Most of the grading rules applicable to wood shakes come from the Cedar Shake and Shingle Bureau. If preservative-treated tapersawn southern yellow pine wood shakes are to be installed, they are governed under the grading rules of the Forest Products Laboratory of the Texas Forest Services. The minimum grade required by the table must be shown on the approved label required on each bundle of wood shakes.

R905.8.6 Application. Wood shakes shall be installed according to this chapter and the manufacturer's installation instructions. Wood shakes shall be laid with a side lap not less than $1^1/_2$ inches (38 mm) between joints in adjacent courses. Spacing between shakes in the same course shall be $^1/_8$ inch to $^5/_8$ inch (3 mm to 16 mm) for shakes and tapersawn shakes of naturally durable wood and shall be $^1/_4$ inch to $^3/_8$ inch (6 mm to 10 mm) for preservative treated taper sawn shakes. Weather exposure for wood shakes shall not exceed those set forth in Table R905.8.6. Fasteners for wood shakes shall be corrosion-resistant, with a minimum penetration of $^1/_2$ inch (12.7 mm) into the sheathing. For sheathing less than $^1/_2$ inch (12.7 mm) in thickness, the fasteners shall extend through the sheathing. Wood shakes shall be attached to the roof with two fasteners per shake, positioned no more than 1 inch (25 mm) from each edge and no more than 2 inches (51 mm) above the exposure line.

❖ The general provisions of this chapter apply to the installation of wood shakes, as do the installation instructions of the manufacturer of the specific roofing materials. In addition, this section sets forth specific requirements regulating the installation method for wood shakes. Commentary Figure R905.8.6 shows many of these requirements.

Fasteners for applying wood shakes must be corrosion-resistant. Only two nails or staples are to be used for each shake, and they must be placed approximately 1 inch (25.4 mm) from each side edge and no more than 2 inches (51 mm) above the exposure line. The depth of fastener penetration must be a minimum of $^1/_2$ inch (12.7 mm) into the sheathing, or completely through sheathing less than $^1/_2$ inch (12.7 mm) thick. Care in spacing and driving is necessary to obtain maximum service from the roof covering [see Commentary Figure R905.7.5(2)].

R905.8.7 Shake placement. The starter course at the eaves shall be doubled and the bottom layer shall be either 15-inch (381 mm), 18-inch (457 mm) or 24-inch (610 mm) wood shakes or wood shingles. Fifteen-inch (381 mm) or 18-inch (457 mm) wood shakes may be used for the final course at the ridge. Shakes shall be interlaid with 18-inch-wide (457 mm) strips of not less than No. 30 felt shingled between each course in such a manner that no felt is exposed to the weather by positioning the lower edge of each felt strip above the butt end of the shake it covers a distance equal to twice the weather exposure.

❖ To provide as weathertight a roof as possible where using wood shakes as the roof covering material, a special installation method is described in this section. At the eaves, the starter course must be two layers of shakes. Going up the roof toward the ridge, the shakes must be supplemented with 18-inch (457 mm) interlayment strips of minimum No. 30 felt. The manner of interlayment must prevent any felt being exposed to the weather. This is illustrated in Commentary Figure R905.8.7.

TABLE R905.8.6
WOOD SHAKE WEATHER EXPOSURE AND ROOF SLOPE

ROOFING MATERIAL	LENGTH (inches)	GRADE	EXPOSURE (inches) 4:12 pitch or steeper
Shakes of naturally durable wood	18	No. 1	$7^1/_2$
	24	No. 1	10^a
Preservative-treated taper sawn shakes of Southern Yellow Pine	18	No. 1	$7^1/_2$
	24	No. 1	10
	18	No. 2	$5^1/_2$
	24	No. 2	$7^1/_2$
Taper-sawn shakes of naturally durable wood	18	No. 1	$7^1/_2$
	24	No. 1	10
	18	No. 2	$5^1/_2$
	24	No. 2	$7^1/_2$

For SI: 1 inch = 25.4 mm.
a. For 24-inch by $^3/_8$-inch handsplit shakes, the maximum exposure is $7^1/_2$ inches.

❖ Table R905.8.6 specifies wood shake exposure. Depending on the grade of the material and the total shingle length, the table specifies the maximum length of exposure for weathering purposes. The grade of the shingle will be listed in the label required for wood shingles.

R905.8.8 Valley flashing. Roof valley flashing shall not be less than No. 26 gage [0.019 inch (0.5 mm)] corrosion-resistant sheet metal and shall extend at least 11 inches (279 mm) from the centerline each way. Sections of flashing shall have an end lap of not less than 4 inches (102 mm).

❖ Valleys formed by the intersection of two sloping roofs are vulnerable to leakage because of the high concentration of water; therefore, valley flashing must be maintained to prevent blockage. This section requires the use of minimum No. 26 gage (0.019 inch/0.48 mm) corrosion-resistant sheet metal for this purpose.

R905.8.9 Label required. Each bundle of shakes shall be identified by a label of an approved grading or inspection bureau or agency.

❖ To verify compliance with the material standards set forth in Section R905.7.4, every bundle of wood shakes must be labeled. The label, a sample of which is shown in Commentary Figure R905.8.9, must show the name of the approved grading or inspection bureau or agency.

R905.9 Built-up roofs. The installation of built-up roofs shall comply with the provisions of this section.

❖ A built-up roof is a roof covering system built at the job site. The provisions of this section deal with the fundamental requirements for such roofs. With all of the different products available, the possible combination of built-up roof systems is almost endless.

Asphalt and coal tar are the bitumens used for a built-up roof. They become fluid when heated and bond the felt layers together. To fuse the roof felts properly, the bitumens must be applied at the proper temperature. Too low a temperature will cause inadequate fusion, and overheating may change the properties of the bitumens or cause a fire hazard.

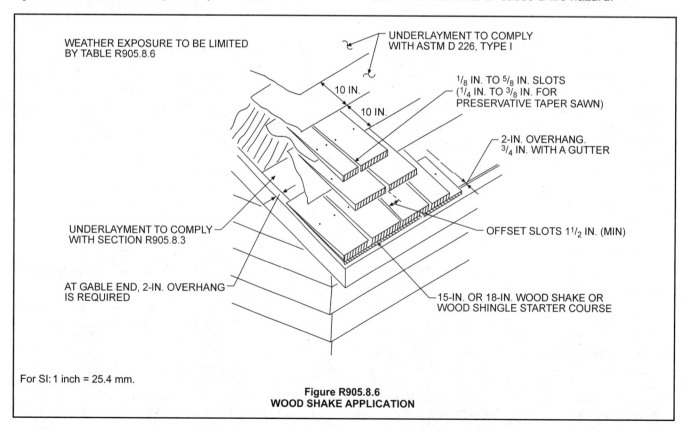

For SI: 1 inch = 25.4 mm.

Figure R905.8.6
WOOD SHAKE APPLICATION

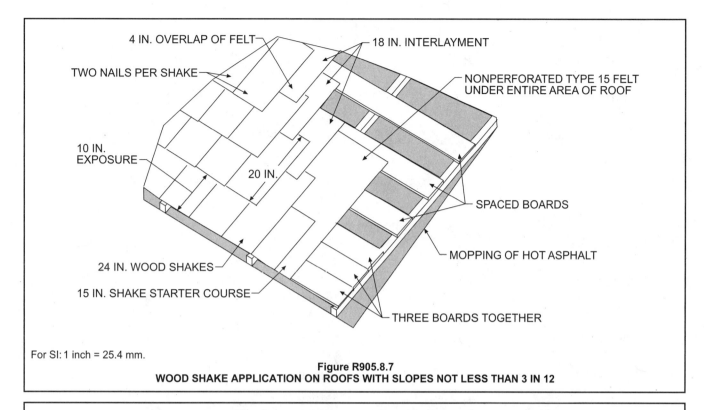

For SI: 1 inch = 25.4 mm.

Figure R905.8.7
WOOD SHAKE APPLICATION ON ROOFS WITH SLOPES NOT LESS THAN 3 IN 12

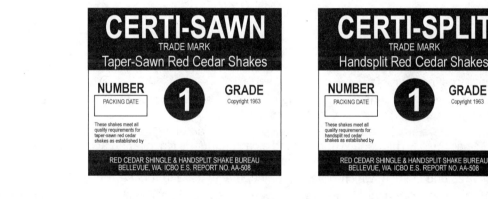

Figure R905.8.9
WOOD SHAKE DESCRIPTION AND EXAMPLE LABELS

R905.9.1 Slope. Built-up roofs shall have a design slope of a minimum of one-fourth unit vertical in 12 units horizontal (2-percent slope) for drainage, except for coal-tar built-up roofs, which shall have a design slope of a minimum one-eighth unit vertical in 12 units horizontal (1-percent slope).

❖ Built-up roofs are often chosen for very low-slope applications because they can be installed on roof decks having a minimum slope of $^1/_4$ inch per foot ($^1/_4$:12), just enough to allow positive drainage. Only where coal-tar built-up roofs are installed must the roof deck slope to a greater degree. A minimum roof slope of 1:12 is required for coal-tar built-up roofs.

R905.9.2 Material standards. Built-up roof covering materials shall comply with the standards in Table R905.9.2.

❖ Table R905.9.2 of the IRC sets forth the appropriate material standards for the various types of built-up roofs. Eighteen different types of roofing systems are identified, and their corresponding specification standard is noted.

R905.9.3 Application. Built-up roofs shall be installed according to this chapter and the manufacturer's installation instructions.

❖ The general provisions of this chapter apply to the installation of built-up roofs, as do the installation instructions of the manufacturer of the specific roofing system. Commentary Figures R905.9.3(1) through (3) illustrate examples of the application of the base ply to nailable and nonnailable decks.

TABLE R905.9.2
BUILT-UP ROOFING MATERIAL STANDARDS

MATERIAL STANDARD	STANDARD
Acrylic coatings used in roofing	ASTM D 6083
Aggregate surfacing	ASTM D 1863
Asphalt adhesive used in roofing	ASTM D 3747
Asphalt cements used in roofing	ASTM D 3019; D 2822; D 4586
Asphalt-coated glass fiber base sheet	ASTM D 4601
Asphalt coatings used in roofing	ASTM D 1227; D 2823; D 2824; D 4479
Asphalt glass felt	ASTM D 2178
Asphalt primer used in roofing	ASTM D 41
Asphalt-saturated and asphalt-coated organic felt base sheet	ASTM D 2626
Asphalt-saturated organic felt (perforated)	ASTM D 226
Asphalt used in roofing	ASTM D 312
Coal tar cements used in roofing	ASTM D 4022; D 5643
Coal-tar primer used in roofing, dampproofing and waterproofing	ASTM D 43
Coal-tar saturated organic felt	ASTM D 227
Coal-tar used in roofing	ASTM D 450, Types I or II
Glass mat, coal tar	ASTM D 4990
Glass mat, venting type	ASTM D 4897
Mineral-surfaced inorganic cap sheet	ASTM D 3909
Thermoplastic fabrics used in roofing	ASTM D 5665; D 5726

❖ The types of materials used in built-up roofing must comply with the following standards:

- ASTM D 6083, which covers a liquid-applied water-dispersed 100-percent acrylic elastomeric latex coating used as a protective coating for roofs.
- ASTM D 1863, which covers the quality and grading of crushed stone, crushed slag and water-worn gravel suitable for use as aggregate surfacing on built-up roofs.
- ASTM D 3747, which covers emulsified asphalt adhesive for use in adhering preformed roof insulation to steel roof decks with inclines up to 33 percent. When applied as a continuous film over an acceptable deck surface, the emulsion functions as both an adhesive and a vapor retarder.
- ASTM D 3019, D 2822 and D 4586, which cover asphalt cements used in roofing.
  - ASTM D 3019 covers lap cement consisting of asphalt dissolved in a volatile petroleum solvent with or without mineral or other stabilizers, or both, for use with roll roofing.
  - ASTM D 2822 covers asphalt roof cement used for trowel application to roofings and flashings.
  - ASTM D 4586 covers asbestos-free asphalt roof cement suitable for trowel application to roofings and flashings.
- ASTM D 4601, which covers asphalt-impregnated and coated glass fiber base sheet, with or without perforations, for use as the first ply of the built-up roofing. When not perforated, this sheet may be used as a vapor retarder under or between roof insulation with a solid top coating of asphaltic material.
- ASTM D 1227, D 2823, D 2824 and D 4479, which cover asphalt coatings used in roofing.
  - ASTM D1227 covers emulsified asphalt suitable for use as a protective coating for built-up roofs and other exposed surfaces with inclines of not less than 4 percent.
  - ASTM D 2823 covers asphalt roof coatings of brushing or spraying consistency.
  - ASTM D 2824 covers aluminum-pigmented asphalt roof coating, including nonfibered, asbestos-fibered and fibered without asbestos.
  - ASTM D 4479 covers asbestos-free asphalt roof coating of brushing or spraying consistency.
- ASTM D 2178, which covers glass felt impregnated to varying degrees with asphalt, which may be used with asphalts conforming to the requirements of ASTM D 312 in the construction of built-up roofs, and with asphalts conforming to the requirements of ASTM D 449 in the membrane system of waterproofing.

- ASTM D 41, which covers asphaltic primer suitable for use with asphalt in roofing, dampproofing and waterproofing below or above ground level, for application to concrete, masonry, metal and asphalt surfaces.
- ASTM D 2626, which covers base sheet with fine mineral surfacing on the top side, with or without perforations, for use as the first ply of a built-up roof. When not perforated, this sheet may be used as a vapor retarder under roof insulation.
- ASTM D 226, which covers asphalt-saturated organic felts, with perforations, that may be used with asphalts conforming to the requirements of ASTM D 312 in the construction of built-up roofs, and with asphalts conforming to the requirements of ASTM D 449 in the membrane system of waterproofing.
- ASTM D 312, which covers four types of asphalt intended for use in built-up roof construction. The specification is for general classification purposes only and does not imply restrictions on the slope at which an asphalt must be used. There are four classifications:
  - Type I includes asphalts that are relatively susceptible to flow at roof temperatures with good adhesive and self-sealing properties.
  - Type II includes asphalts that are moderately susceptible to flow at roof temperatures.
  - Type III includes asphalts that are relatively not susceptible to flow at roof temperatures for use in built-up roof construction on slope inclines from 8.3 percent to 25 percent.
  - Type IV includes asphalts that are generally not susceptible to flow at roof temperatures for use in built-up roof construction on slope inclines from approximately 16.7 percent to 50 percent.
- ASTM D 4022 and D 5643, which cover coal tar cements used in roofing.
  - ASTM D 4022 covers coal-tar roof cement suitable for trowel application in coal-tar roofing and flashing systems.
  - ASTM D 5643 covers asbestos-free coal-tar roof cement suitable for trowel application in coal-tar roofing and flashing systems.
- ASTM D 227, which covers coal-tar saturated organic felt that may be used with coal-tar pitches conforming to the requirements of ASTM D 450 in the construction of built-up roofs and in the membrane system of waterproofing.
- ASTM D 43 covers coal tar primer suitable for use with coal tar pitch in roofing, dampproofing and waterproofing below or above ground level for application to concrete, masonry and coal tar surfaces.
- ASTM D 450, Types I and II, which covers coal-tar pitch suitable for use in the construction of built-up roofing, dampproofing and membrane waterproofing systems.
  - Type I is suitable for use in built-up roofing systems with felts conforming to the requirements of ASTM D 227 or as specified by the manufacturer.
  - Type II is suitable for use in dampproofing and in membrane waterproofing systems.
- ASTM D 4990, which covers glass felt impregnated with coal tar intended to be used with coal-tar pitch conforming to the requirements of ASTM D 450 in construction of built-up roofs and in the construction waterproofing systems.
- ASTM D 4897, which covers asphalt-impregnated and coated glass fiber base sheet with mineral surfacing on the top side and coarse mineral granules on the bottom side for use as the first ply of a roofing membrane. These base sheets provide for the lateral release of pressure in roofing systems because they are not solidly attached, and the coarse granular surface provides an open, porous channel in the horizontal plane beneath the membrane. The base sheets can be with or without perforations or embossings.
- ASTM D 3909, which covers asphalt-impregnated and coated glass felt roll roofing surfaced on the weather side with mineral granules, for use as a cap sheet in the construction of built-up roofs.
- ASTM D 5665 and D 5726, which cover thermoplastic fabrics used in roofing.
  - ASTM D 5665 covers thermoplastic fabrics such as polyester, polyester/polyamide bicomponent or composites with fiberglass or polyester scrims that can be used during the construction of cold-applied roofing and waterproofing.
  - ASTM D 5726 covers thermoplastic fabrics such as polyester, polyester/polyamide bicomponent or composites with fiberglass or polyester scrims that can be used during the construction of hot-applied roofing and waterproofing.

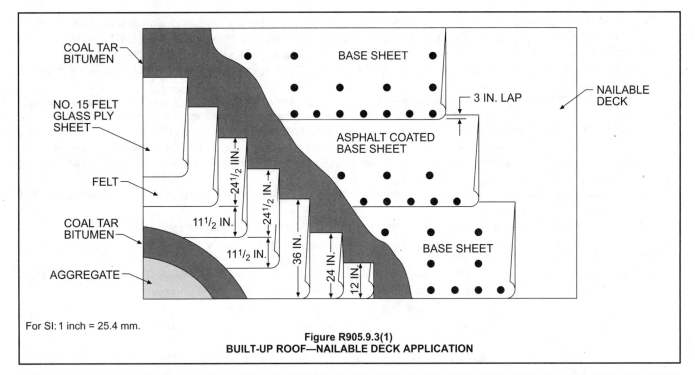

For SI: 1 inch = 25.4 mm.

Figure R905.9.3(1)
BUILT-UP ROOF—NAILABLE DECK APPLICATION

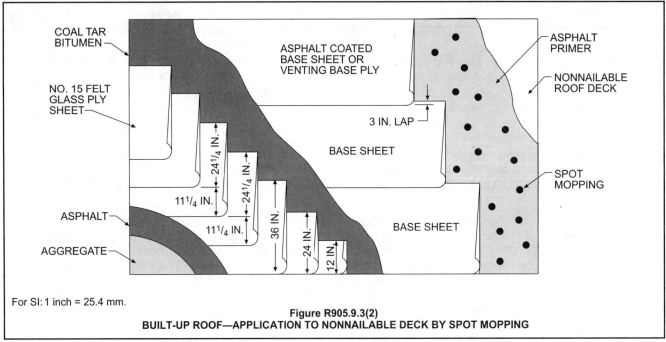

For SI: 1 inch = 25.4 mm.

Figure R905.9.3(2)
BUILT-UP ROOF—APPLICATION TO NONNAILABLE DECK BY SPOT MOPPING

R905.10 Metal roof panels. The installation of metal roof panels shall comply with the provisions of this section.

❖ Metal roof panels differ from metal roof shingles regulated by Section R905.4. By definition, a metal roof panel must have an installed weather exposure of at least 3 square feet (0.28 m²) per shingle or sheet. This section addresses the deck requirements, deck slope and attachment requirements for metal roof panels.

R905.10.1 Deck requirements. Metal roof panel roof coverings shall be applied to solid or spaced sheathing, except where the roof covering is specifically designed to be applied to spaced supports.

❖ Unless specifically limited, metal roof panels may be applied to either solidly sheathed decks or spaced sheathing. In those cases where the specific metal panels are designed to be installed only on spaced supports, the manufacturer's installation instructions govern.

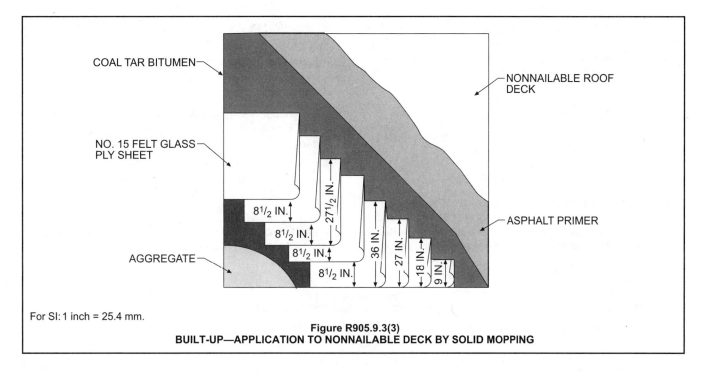

COAL TAR BITUMEN

NONNAILABLE ROOF DECK

NO. 15 FELT GLASS PLY SHEET

8¹/₂ IN.

27¹/₂ IN.

8¹/₂ IN.

36 IN.

27 IN.

18 IN.

9 IN.

8¹/₂ IN.

ASPHALT PRIMER

AGGREGATE

8¹/₂ IN.

For SI: 1 inch = 25.4 mm.

Figure R905.9.3(3)
BUILT-UP—APPLICATION TO NONNAILABLE DECK BY SOLID MOPPING

R905.10.2 Slope. Minimum slopes for metal roof panels shall comply with the following:

1. The minimum slope for lapped, nonsoldered-seam metal roofs without applied lap sealant shall be three units vertical in 12 units horizontal (25-percent slope).

2. The minimum slope for lapped, nonsoldered-seam metal roofs with applied lap sealant shall be one-half vertical unit in 12 units horizontal (4-percent slope). Lap sealants shall be applied in accordance with the approved manufacturer's installation instructions.

3. The minimum slope for standing-seam roof systems shall be one-quarter unit vertical in 12 units horizontal (2-percent slope).

❖ The minimum slope permitted for a roof deck supporting metal roof panels varies based on the type of panels being installed. Three different types of metal roof panel systems are regulated: lapped, nonsoldered seam metal roofs without applied lap sealant; lapped, nonsoldered seam metal roofs with applied lap sealant and standing seam metal roof systems. The slope limitations vary based on the ability of the metal panel roof system to shed water without water intrusion. This section allows lower minimum roof slopes for lapped nonsoldered seam metal roofs when lap sealants are used. To help ensure that lap sealants perform as part of the overall roofing system the lap sealant must be installed in accordance with the approved manufacturer's installation instructions.

R905.10.3 Material standards. Metal-sheet roof covering systems that incorporate supporting structural members shall be designed in accordance with the *International Building Code*. Metal-sheet roof coverings installed over structural decking shall comply with Table R905.10.3(1). The materials used for metal-sheet roof coverings shall be naturally corrosion resistant or provided with corrosion resistance in accordance with the standards and minimum thicknesses shown in Table R905.10.3(2).

❖ If the metal roof panel system uses the structural members of the roof construction for support, it is necessary to design the roof system is accordance with the *International Building Code*® (IBC®). Only where the metal roof coverings are installed and supported by structural decking may the provisions of the IRC be used. Under such conditions, IRC Tables R905.10.3(1) and R905.10.3(2) must be consulted.

Table R905.10.3(1) lists the metal roof covering types and their material standards. Many of the materials listed are inherently corrosion resistant, such as aluminum, copper and lead. For these materials, listing a material standard for the base material is sufficient. However, for steel roofing products, coatings are added to the base material to provide the necessary corrosion resistance.

Table R905.10.3(2) lists the minimum coating thickness required to provide minimum corrosion resistance for the base steel. Note a to Table R905.10.3(2) specifies that the paint systems are intended to be applied to steel having one of the corrosion-resistant coatings listed.

R905.10.4 Attachment. Metal roof panels shall be secured to the supports in accordance with this chapter and the manufacturer's installation instructions. In the absence of manufacturer's installation instructions, the following fasteners shall be used:

1. Galvanized fasteners shall be used for steel roofs.

2. Three hundred series stainless steel fasteners shall be used for copper roofs.

3. Stainless steel fasteners are acceptable for metal roofs.

❖ The general provisions of this chapter apply to the installation of metal panel roofing, as do the installation instructions of the manufacturer of the specific roofing system. The code requires three specific types of fasteners unless the manufacturer's installation instructions specify otherwise.

R905.11 Modified bitumen roofing. The installation of modified bitumen roofing shall comply with the provisions of this section.

❖ This section establishes the materials standards and installation criteria for modified bitumen roofing systems.

R905.11.1 Slope. Modified bitumen membrane roofs shall have a design slope of a minimum of one-fourth unit vertical in 12 units horizontal (2-percent slope) for drainage.

❖ Where a modified bitumen roofing system is to be installed, the roof design slope must be a minimum of 1:48, more often recognized as $^1/_4$ inch per foot.

TABLE R905.10.3(1)
METAL ROOF COVERINGS STANDARDS

ROOF COVERING TYPE	STANDARD APPLICATION RATE/THICKNESS
Galvanized Steel	ASTM A 653 G90 Zinc Coated
Stainless Steel	ASTM A 240, 300 Series Alloys
Steel	ASTM A 924
Lead-coated Copper	ASTM B 101
Cold Rolled Copper	ASTM B 370 minimum 16 oz/square ft and 12 oz/square ft high yield copper for metal-sheet roof-covering systems; 12 oz/square ft for preformed metal shingle systems.
Hard Lead	2 lb/ sq ft
Soft Lead	3 lb/ sq ft
Aluminum	ASTM B 209, 0.024 minimum thickness for rollformed panels and 0.019 inch minimum thickness for pressformed shingles.
Terne (tin) and terne-coated stainless	Terne coating of 40 lb per double base box, field painted where applicable in accordance with manufacturer's installation instructions.
Zinc	0.027 inch minimum thickness: 99.995% electrolytic high grade zinc with alloy additives of copper (0.08 - 0.20%), titanium (0.07% - 0.12%) and aluminum (0.015%).

For SI: 1 ounce per square foot = 0.305 kg/m², 1 pound per square foot = 4.214 kg/m², 1 inch = 25.4 mm, 1 pound = 0.454 kg.

❖ See the commentary for Sections R905.10.3 and R905.4.4. CDA 4115-1929, *Copper in Architecture—Design Handbook* is no longer listed as a reference for copper roof covering; however, the provisions of CDA 4115-1929 may be useful as a guideline.

TABLE R905.10.3(2)
MINIMUM CORROSION RESISTANCE

55% Aluminum-zinc alloy coated steel	ASTM A 792 AZ 50
5% aluminum alloy-coated steel	ASTM A 875 GF60
Aluminum-coated steel	ASTM A 463 T2 65
Galvanized steel	ASTM A 653 G-90
Prepainted steel	ASTM A 755[a]

a. Paint systems in accordance with ASTM A 755 shall be applied over steel products with corrosion-resistant coatings complying with ASTM A 792, ASTM A 875, ASTM A 463, or ASTM A 653.

❖ See the commentary for Sections R905.10.3 and R905.4.4.

R905.11.2 Material standards. Modified bitumen roof coverings shall comply with the standards in Table R905.11.2.

❖ Table R905.11.2 must be referenced for the installer to determine the appropriate specification standard for various materials that may be used in the application of modified bitumen roofing. The table identifies reference standards for the membrane, primer, cement, adhesive and coatings that may be a part of the roof installation.

TABLE R905.11.2
MODIFIED BITUMEN ROOFING MATERIAL STANDARDS

MATERIAL	STANDARD
Acrylic coating	ASTM D 6083
Asphalt adhesive	ASTM D 3747
Asphalt cement	ASTM D 3019
Asphalt coating	ASTM D 1227; D 2824
Asphalt primer	ASTM D 41
Modified bitumen roof membrane	ASTM D 6162; D 6163; D 6164; D 6222; D 6223; D 6298; CGSB 37–56M

❖ The materials used in modified bitumen roof coverings must comply with the following standards: ASTM D 6162, D 6163, D 6164, D 6222, D 6223, D6298 and CGSB 37-56M, which cover modified bitumen roof membranes.

  • ASTM D 6162 covers prefabricated modified bituminous sheet materials reinforced with a combination of polyester fabric and glass fiber, with or without granules, which use styrene-butadiene-styrene (SBS) thermoplastic elastomer as the primary modifier and are intended for use in the fabrication of multiple-ply roofing and waterproofing membranes.

  • ASTM D 6163 covers prefabricated modified bituminous sheet materials with glass fiber reinforcement, with or without granules, which use styrene-butadiene-styrene (SBS) thermoplastic elastomer as the primary modifier and are intended for use in the fabrication of

multiple-ply roofing and waterproofing membranes.

  • ASTM D 6164 covers prefabricated modified bituminous sheet materials reinforced with polyester fabric, with or without granules, which use styrene-butadiene-styrene (SBS) thermoplastic elastomer as the primary modifier and are intended for use in the fabrication of multiple-ply roofing and waterproofing membranes.

  • ASTM D 6222 covers atactic polypropelene (APP) modified bituminous sheet materials using polyester reinforcement.

  • ASTM D 6223 covers atactic polypropelene (APP) modified bituminous sheet materials using a combination of polyester and glass fiber reinforcement.

  • ASTM D 6298 covers fiberglass reinforced modified bituminous sheet materials that use styrene-butadiene-styrene (SBS) thermoplastic elastomer as the primary modifier and are surfaced with a factory-applied continuous metal foil.

  • CGSB 37-56M covers prefabricated modified bituminous membrane reinforced for roofing applications.

• ASTM D 41 covers asphaltic primer suitable for use with asphalt in roofing, dampproofing and waterproofing below or above ground level, for application to concrete, masonry, metal and asphalt surfaces.

• ASTM D 3019 covers lap cement consisting of asphalt dissolved in a volatile petroleum solvent with or without mineral or other stabilizers, or both, for use with roll roofing.

• ASTM D 3747 covers emulsified asphalt adhesive for use in adhering preformed roof insulation to steel roof decks with inclines up to 33 percent. When applied as a continuous film over an acceptable deck surface, the emulsion functions as both an adhesive and a vapor retarder.

• ASTM D 1227 and D 2824 cover asphalt roof coatings.

  • ASTM D 1227 covers emulsified asphalt suitable for use as a protective coating for built-up roofs and other exposed surfaces with inclines of not less than 4 percent.

  • ASTM D 2824 covers aluminum-pigmented asphalt roof coatings, including nonfibered, asbestos-fibered, and asbestos-free fibered materials.

• ASTM D 6083 covers a liquid-applied water-dispersed 100 percent acrylic elastomeric latex coating used as a protective coating for roofs.

R905.11.3 Application. Modified bitumen roofs shall be installed according to this chapter and the manufacturer's installation instructions.

❖ The general provisions of this chapter apply to the installation of modified bitumen roofing, as do the installation instructions of the manufacturer of the specific roofing materials or system.

R905.12 Thermoset single-ply roofing. The installation of thermoset single-ply roofing shall comply with the provisions of this section.

❖ This section establishes the materials standards and installation criteria for thermoset single-ply roofing systems

R905.12.1 Slope. Thermoset single-ply membrane roofs shall have a design slope of a minimum of one-fourth unit vertical in 12 units horizontal (2-percent slope) for drainage.

❖ If a thermoset single-ply roof system is to be installed, the roof design slope must be a minimum of 1:48, more often recognized as $1/4$-inch per foot.

R905.12.2 Material standards. Thermoset single-ply roof coverings shall comply with ASTM D 4637, ASTM D 5019 or CGSB 37-GP-52M.

❖ The code recognizes three materials standards for the regulation of thermoset single-ply membrane roofs. The standards include ASTM D 4637 for ethylene-propylene-diene-terpolymer (EPDM) sheets, ASTM D 5019 for reinforced nonvulcanized polymeric sheet made from chlorosulfonated polyethylene (CSPE) and poly-isobutylene (PIB) and CGSB 37-GP-52M for sheet-applied elastomeric roofing and waterproofing membrane.

R905.12.3 Application. Thermoset single-ply roofs shall be installed according to this chapter and the manufacturer's installation instructions.

❖ The general provisions of this chapter apply to the installation of thermoset single-ply roofing, as do the installation instructions of the manufacturer of the specific roofing system.

R905.13 Thermoplastic single-ply roofing. The installation of thermoplastic single-ply roofing shall comply with the provisions of this section.

❖ This section sets forth the materials standards and installation criteria for single-ply roof systems consisting of polyvinyl chloride material.

R905.13.1 Slope. Thermoplastic single-ply membrane roofs shall have a design slope of a minimum of one-fourth unit vertical in 12 units horizontal (2-percent slope).

❖ If a thermoplastic single-ply roof system is to be installed, the roof design slope must be a minimum of 1:48, more often recognized as $1/4$-inch per foot.

R905.13.2 Material standards. Thermoplastic single-ply roof coverings shall comply with ASTM D 4434, ASTM D 6754, ASTM D 6878, or CGSB CAN/CGSB 37.54.

❖ Four materials standards, ASTM D 4434, ASTM D 6754, ASTM D 6878 and CGSB CAN/CGSB 37.54, are recognized for the regulation of thermoplastic single-ply membrane roofs and are acceptable standards in determining the appropriateness of the roofing system.

R905.13.3 Application. Thermoplastic single-ply roofs shall be installed according to this chapter and the manufacturer's installation instructions.

❖ The general provisions of this chapter apply to the installation of thermoplastic single-ply roofing, as do the installation instructions of the manufacturer of the roofing system.

R905.14 Sprayed polyurethane foam roofing. The installation of sprayed polyurethane foam roofing shall comply with the provisions of this section.

❖ This section establishes the materials standards and installation criteria for sprayed polyurethane foam roofing systems.

R905.14.1 Slope. Sprayed polyurethane foam roofs shall have a design slope of a minimum of one-fourth unit vertical in 12 units horizontal (2-percent slope) for drainage.

❖ Where a sprayed polyurethane foam roof system is to be installed, the roof design slope must be a minimum of 1:48, more often recognized as $1/4$-inch per foot.

R905.14.2 Material standards. Spray-applied polyurethane foam insulation shall comply with ASTM C 1029.

❖ ASTM C 1029 is the standard for the regulation of spray-applied rigid cellular polyurethane thermal insulation.

R905.14.3 Application. Foamed-in-place roof insulation shall be installed in accordance with this chapter and the manufacturer's installation instructions. A liquid-applied protective coating that complies with Section R905.15 shall be applied no less than 2 hours nor more than 72 hours following the application of the foam.

❖ The general provisions of this chapter address the installation of spray-applied rigid cellular polyurethane thermal membrane roofing, as do the installation instructions of the manufacturer of the roofing system. In addition, a liquid-applied protective coating must be provided to protect the exterior surface of the foam. The protective coating, regulated in greater detail by Section R905.15, must be applied between 2 and 72 hours following the installation of the foam insulation.

R905.14.4 Foam plastics. Foam plastic materials and installation shall comply with Section R314.

❖ The foam-plastics provisions of Section R314, particularly those of Section R314.5.2 specific to roof-covering assemblies, are applicable to roof assemblies containing sprayed polyurethane foam. The use of wood structural-panel roof sheathing is permitted in lieu of a thermal barrier for separating the interior of the building from the sprayed polyurethane foam-plastic roofing material. Sections R314.5.2 and R803 contain specific requirements for wood structural-panel sheathing.

R905.15 Liquid-applied coatings. The installation of liquid-applied coatings shall comply with the provisions of this section.

❖ This section establishes the materials standards and installation criteria for liquid-applied roof coatings.

R905.15.1 Slope. Liquid-applied roofs shall have a design slope of a minimum of one-fourth unit vertical in 12 units horizontal (2-percent slope).

❖ If a liquid-applied roof coating is to be installed, the roof design slope must be a minimum of 1:48, more often recognized as $1/4$-inch per foot.

R905.15.2 Material standards. Liquid-applied roof coatings shall comply with ASTM C 836, C 957, D 1227, D 3468, D 6083 or D 6694.

❖ The code recognizes several materials standards for the regulation of liquid-applied roof coatings. They include ASTM C 836 for cold-liquid applied elastomeric waterproofing membranes that require the installation of separate wearing surfaces, ASTM C 957 for cold-liquid applied elastomeric waterproofing membranes that provide integral wearing surfaces, ASTM D 1227 for emulsified asphalt used as a protective coating for built-up roofing systems, ASTM D 3468 for liquid-applied neoprene and chlorosulfanated polyethylene materials used in roofing and waterproofing ASTM D 6083 for liquid-applied acrylic roof coatings and ASTM D 6694 for silicone-based elastomeric coating for spray polyethylene foam insulation.

R905.15.3 Application. Liquid-applied roof coatings shall be installed according to this chapter and the manufacturer's installation instructions.

❖ The general provisions of this chapter apply to the installation of liquid-applied roof coatings, as do the installation instructions of the manufacturer of the roofing system.

SECTION R906
ROOF INSULATION

R906.1 General. The use of above-deck thermal insulation shall be permitted provided such insulation is covered with an approved roof covering and passes FM 4450 or UL 1256.

❖ During these days of energy consciousness, roof insulation has become more and more prevalent. It has distinct benefits not only in energy conservation but also in building-occupant comfort. Insulation also provides a smooth, uniform substrate for application of the roofing materials. The code requires that above-deck thermal insulation be covered by an approved roof covering and be in compliance with FM 4450 or UL 1256. Although UL 1256 specifies the testing methods for determining the fire resistance of roof covering materials, FM 4450 deals with Class I insulated steel deck roofs only.

R906.2 Material standards. Above-deck thermal insulation board shall comply with the standards in Table R906.2.

❖ The referenced material standards provide additional guidance on the physical properties of roof insulation when used as above-deck components of roof assemblies.

TABLE R906.2
MATERIAL STANDARDS FOR ROOF INSULATION

Cellular glass board	ASTM C 552
Composite boards	ASTM C 1289, Type III, IV, V, or VI
Expanded polystyrene	ASTM C 578
Extruded polystyrene board	ASTM C 578
Perlite Board	ASTM C 728
Polyisocyanurate Board	ASTM C 1289, Type I or Type II
Wood fiberboard	ASTM C 208

❖ This table incorporates industry-recognized material standards into the code for materials commonly used in above roof deck insulation practices.

SECTION R907
REROOFING

R907.1 General. Materials and methods of application used for re-covering or replacing an existing roof covering shall comply with the requirements of Chapter 9.

Exception: Reroofing shall not be required to meet the minimum design slope requirement of one-quarter unit vertical in 12 units horizontal (2-percent slope) in Section R905 for roofs that provide positive roof drainage.

❖ This section addresses the concerns associated with unregulated reroofing operations. The provisions require that when an existing building is reroofed, the existing roof be structurally sound and in a proper condition to receive the new roofing. It is often necessary to remove the existing roof covering prior to installing the new roofing materials. This section identifies those situations where the reroofing cannot occur over an existing roof covering.

If new roof covering materials are to be installed as a replacement for an existing roof covering, or if the new roof covering is applied directly over an existing roof, the roof covering materials and the application process must conform to the provisions for a new roof covering installation.

There may be occasions where only a small portion of the existing roof covering is in need of replacement or a limited amount of the roof is in need of repair. In those situations, it is acceptable to use the provisions for existing buildings in Chapter 34 of the IBC, particularly Section 3402.3. This section of the IBC permits the alteration or repair of nonstructural elements using the same materials that are already in place. In other words, the repair or replacement of existing roof covering may be made using the same materials and methods found on the existing roof. It is assumed that the repairs will not cause the building to be reduced in code compliance from the previous level.

To avoid a situation in which significant portions of a roof covering are repaired or replaced in small increments without being required to comply with the current provisions of the IRC, the code limits the amount of exempted work that can be done in any one-year period. Where more than 25 percent of a building's roof covering is replaced within a 12-month period, the roof covering must be in compliance with the provisions for new roofing materials and application methods.

R907.2 Structural and construction loads. The structural roof components shall be capable of supporting the roof covering system and the material and equipment loads that will be encountered during installation of the roof covering system.

❖ A fundamental requirement throughout the code is the recognition of the impact of any new repair or replacement work on the structural system. If a new roof covering system is installed, the structural members of the roof and any additional structural members that carry roof loads must be reviewed for their ability to support the loads that will be imposed during the installation process. This includes the weight of the new roof covering materials as well as any installation equipment that is placed on the roof.

R907.3 Re-covering versus replacement. New roof coverings shall not be installed without first removing existing roof coverings where any of the following conditions occur:

1. Where the existing roof or roof covering is water-soaked or has deteriorated to the point that the existing roof or roof covering is not adequate as a base for additional roofing.

2. Where the existing roof covering is wood shake, slate, clay, cement or asbestos-cement tile.

3. Where the existing roof has two or more applications of any type of roof covering.

4. For asphalt shingles, when the building is located in an area subject to moderate or severe hail exposure according to Figure R903.5.

Exceptions:

1. Complete and separate roofing systems, such as standing-seam metal roof systems, that are designed to transmit the roof loads directly to the building's structural system and that do not rely on existing roofs and roof coverings for support, shall not require the removal of existing roof coverings.

2. Installation of metal panel, metal shingle, and concrete and clay tile roof coverings over existing wood shake roofs shall be permitted when the application is in accordance with Section R907.4.

3. The application of new protective coating over existing spray polyurethane foam roofing systems shall be permitted without tear-off of existing roof coverings.

❖ The base for application of new roofing materials must provide a sound and consistent surface on which to install the new materials. The code will not permit the installation of new roof coverings over existing roof coverings where there is a potential for future problems with the roof's effectiveness. Therefore, the code lists four specific conditions where the existing roof covering materials must be removed prior to the installation of new roofing.

If the existing roof or roof covering is water-soaked, the concealment of the existing roof in such a condition will lead to problems of deterioration and failure in the future. Any situation where the existing roof construction does not provide for an acceptable base is reason to remove all existing roof covering materials prior to reroofing. In addition, roofing materials such as wood shakes, clay tiles and similar materials are not generally acceptable as a base for a new roof covering because of the variation in their surfaces. Exception 2 permits the installation of a new roof covering over wood shakes or shingle roofs only if the surface below the new roofing is properly protected per Section R907.4.

A maximum of two roof coverings is permitted by the code, based primarily on the dead load weight that is accumulated on the roof for every layer of roofing materials. Where two or more applications of any roof covering are present, the existing materials must be removed prior to installation of the new roof covering. Asphalt shingles present a special concern in those areas likely to have hail. Where moderate or severe hail damage can be expected, based on the map shown in Figure R903.5, it is always necessary to remove an existing asphalt shingle roof covering prior to the application of any new roof covering. See the commentary for Section R903.5.

Exception 1 states that new roofing systems that are designed to transmit all roof loads directly to the structural supports of the building do not necessitate that the existing roofing system be removed.

Exception 3 permits recoating of an existing spray polyurethane foam roofing system without removal of the spray polyurethane foam roof covering. Recoating does not add significant weight to the roof assembly or compromise the long term performance of the roofing assembly. Industry practices for the recoat of an existing spray polyurethane foam roofing system are detailed in ASTM D 6705, *Standard Guide for the Repair and Recoat of Spray Polyurethane Foam Roofing Systems.*

R907.4 Roof recovering. Where the application of a new roof covering over wood shingle or shake roofs creates a combustible concealed space, the entire existing surface shall be covered with gypsum board, mineral fiber, glass fiber or other approved materials securely fastened in place.

❖ A new roof covering can be applied over the top of an existing roof of wood shakes or shingles only if the existing roof surface of shakes or shingles is protected to address the concern of concealed combustible spaces. The application of a new roof over wood shakes or shingles creates an extensive amount of concealed area, all with a high degree of combustible materials. By using gypsum board, mineral fiber, glass fiber or other similar materials in the reroofing installation, the combustible shakes or shingles will be protected by materials suitable for fireblocking.

R907.5 Reinstallation of materials. Existing slate, clay or cement tile shall be permitted for reinstallation, except that damaged, cracked or broken slate or tile shall not be reinstalled. Existing vent flashing, metal edgings, drain outlets, collars and metal counterflashings shall not be reinstalled where rusted, damaged or deteriorated. Aggregate surfacing materials shall not be reinstalled.

❖ Because all roofing materials should make a building weathertight, their reuse is strictly limited. Unless damaged, existing tile of cement, slate or clay may be reused. The reuse of all other roofing materials is prohibited. Roof accessories, such as metal edgings, flashing, drain outlets and collars, may be reinstalled only if they are in a suitable condition.

R907.6 Flashings. Flashings shall be reconstructed in accordance with approved manufacturer's installation instructions. Metal flashing to which bituminous materials are to be adhered shall be primed prior to installation.

❖ During reroofing operations, all flashings that are to remain must be reconstructed in a manner consistent with the manufacturer's instructions. In those cases where bituminous materials are applied to any existing metal flashing, a primer must be applied to the flashing to increase adhesion to the surface.

Bibliography

The following resource materials are referenced in this chapter or are relevant to the subject matter addressed in this chapter.

ASTM A 653M-04a, *Specification for Steel Sheet, Zinc-Coated (Galvanized) or Zinc-Iron Alloy-Coated (Galvannealed) by the Hot-Dip Process.* West Conshohocken, PA: ASTM International, 2004.

ASTM A 755M-01 (2003), *Specification for Steel Sheet, Metallic-Coated by the Hot-Dip Process and Prepainted by the Coil Coating Process for Exterior Exposed Building Products.* West Conshohocken, PA: ASTM International, 2003.

ASTM A 792M-03, *Specification for Steel Sheet, 55% Aluminum-Zinc Alloy-Coated by the Hot-Dip Process.* West Conshohocken, PA: ASTM International, 2003.

ASTM B 101-02, *Specification for Copper Sheets and Strip for Building Construction.* West Conshohocken, PA: ASTM International, 2002.

ASTM B 209-04, *Specification for Aluminum and Aluminum-alloy Sheet and Plate.* West Conshohocken, PA: ASTM International, 2004.

ASTM B 370-03, *Specification for Lead-Coated Copper Sheet and Strip for Building Construction.* West Conshohocken, PA: ASTM International, 2003.

ASTM C 836-03, *Specification for High Solids Content, Cold Liquid-Applied Elastomeric Waterproofing Membrane for Use with Separate Wearing Course.* West Conshohocken, PA: ASTM International, 2003.

ASTM C 957-04, *Specification for High Solids Content, Cold Liquid-Applied Elastomeric Waterproofing Membrane for Use with Integral Wearing Surface.* West Conshohocken, PA: ASTM International, 2004.

ASTM C 1029-02, *Specification for Spray-Applied Rigid Cellular Polyurethane Thermal Insulation.* West Conshohocken, PA: ASTM International, 2002.

ASTM C 1167-03, *Specification for Clay Roof Tiles.* West Conshohocken, PA: ASTM International, 2003.

ASTM C 1492-03, *Specification for Concrete Roof Tile.* West Conshohocken, PA: ASTM International, 2003.

ASTM D 41-e01, *Specification for Asphalt Primer Used in Roofing, Dampproofing, and Waterproofing.* West Conshohocken, PA: ASTM International, 2001.

ASTM D 43-00, *Specification for Coal Tar Primer Used in Roofing, dampproofing and Waterproofing.* West Conshohocken, PA: ASTM International, 2000.

ASTM D 225-04, *Specification for Asphalt Shingles (Organic Felt) Surfaced with Mineral Granules.* West Conshohocken, PA: ASTM International, 2004.

ASTM D 226-97a, *Specification for Asphalt-Saturated (Organic Felt) Used in Roofing and Waterproofing.* West Conshohocken, PA: ASTM International, 1997.

ASTM D 227-03, *Specification for Coal-Tar Saturated (Organic Felt) Used in Roofing and Waterproofing.* West Conshohocken, PA: ASTM International, 2003.

ASTM D 312-00, *Specification for Asphalt Used in Roofing.* West Conshohocken, PA: ASTM International, 2000.

ASTM D 450-(00)e01, *Specification for Coal-Tar Pitch Used in Roofing, Dampproofing and Waterproofing.* West Conshohocken, PA: ASTM International, 2000.

ASTM D 1227-00, *Specification for Emulsified Asphalt Used as a Protective Coating for Roofing.* West Conshohocken, PA: ASTM International, 2000.

ASTM D 1863-03, *Specification for Mineral Aggregate Used in Built-Up Roofs.* West Conshohocken, PA: ASTM International, 2003.

ASTM D 1970-01, *Specification for Self-Adhering Polymer Modified Bitumen Sheet Materials Used as Steep Roofing Underlayment for Ice-Dam Protection.* West Conshohocken, PA: ASTM International, 2001.

ASTM D 2178-97a, *Specification for Asphalt Glass Felt Used in Roofing and Waterproofing.* West Conshohocken, PA: ASTM International, 1997.

ASTM D 2626-04, *Specification for Asphalt-Saturated and Coated Organic Felt Base Sheet Used in Roofing.* West Conshohocken, PA: ASTM International, 2004.

ASTM D 2822-91(1997)e01, *Specification for Asphalt Roof Cement.* West Conshohocken, PA: ASTM International, 1997.

ASTM D 2823-90(1997)e01, *Specification for Asphalt Roof Coatings.* West Conshohocken, PA: ASTM International, 1997.

ASTM D 2824-04, *Specification for Aluminum-Pigmented Asphalt Roof Coatings, Non-Fibered Asbestos Fibered, and Fibered without Asbestos.* West Conshohocken, PA: ASTM International, 2004.

ASTM D 3019-e01, *Specification for Lap Cement Used with Asphalt Roll Roofing, Nonfibered, Asbestos Fibered, and Nonasbestos Fibered.* West Conshohocken, PA: ASTM International, 2001.

ASTM D 3161-03b, *Test Method for Wind Resistance of Asphalt Shingles (Fan Induced Method).* West Conshohocken, PA: ASTM International, 2003.

ASTM D 3462-04, *Specification for Asphalt Shingles Made From Glass Felt and Surfaced with Mineral Granules.* West Conshohocken, PA: ASTM International, 2004.

ASTM D 3468-99, *Specification for Liquid-Applied Neoprene and Chlorosulfanated Polyethylene Used in Roofing and Waterproofings.* West Conshohocken, PA: ASTM International, 1999.

ASTM D 3747-79 (2000)e01, *Specification for Emulsified Asphalt Adhesive for Adhering Roof Insulation.* West Conshohocken, PA: ASTM International, 2001.

ASTM D 3909-97b, *Specification for Asphalt Roll Roofing (Glass Felt) Surfaced with Mineral Granules.* West Conshohocken, PA: ASTM International, 1997.

ASTM D 4022-94(2000)e01, *Specification for Coal Tar Roof Cement, Asbestos Containing.* West Conshohocken, PA: ASTM International, 2000.

ASTM D 4434-04, *Specification for Poly (Vinyl Chloride) Sheet Roofing.* West Conshohocken, PA: ASTM International, 2004.

ASTM D 4479-00, *Specification for Asphalt Roof Coatings, Asbestos-Free.* West Conshohocken, PA: ASTM International, 2000.

ASTM D 4586-00, *Specification for Asphalt Roof Cement, Asbestos-Free.* West Conshohocken, PA: ASTM International, 2000.

ASTM D 4601-98, *Specification for Asphalt-Coated Glass Fiber Base Sheet Used in Roofing.* West Conshohocken, PA: ASTM International, 1998.

ASTM D 4637-04, *Specification for EPDM Sheet Used in Single-Ply Roof Membrane.* West Conshohocken, PA: ASTM International, 2004.

ASTM D 4869-04, *Specification for Asphalt-Saturated (Organic Felt) Shingle Underlayment Used in Roofing.* West Conshohocken, PA: ASTM International, 2004.

ASTM D 4897-01, *Specification for Asphalt Coated Glass-Fiber Venting Base Sheet Used in Roofing.* West Conshohocken, PA: ASTM International, 2001.

ASTM D 4990-97a, *Specification for Coal Tar Glass Felt Used in Roofing and Waterproofing.* West Conshohocken, PA: ASTM International, 1997.

ASTM D 5019-96e01, *Specification for Reinforced Non-vulcanized Polymeric Sheet Used in Roofing Membrane.* West Conshohocken, PA: ASTM International, 1996.

ASTM D 5643-94(2000)e01, *Specification for Coal Tar Cement Used in Roofing.* West Conshohocken, PA: ASTM International, 2000.

ASTM D 5665-99a, *Specification for Thermoplastic Fabrics Used in Cold-Applied Roofing and Waterproofing.* West Conshohocken, PA: ASTM International, 1999.

ASTM D 5726-98, *Specification for Thermoplastic Fabrics Used in Hot-Applied Roofing and Waterproofing.* West Conshohocken, PA: ASTM International, 1998.

ASTM D 6083-97a, *Specification for Liquid Applied Acrylic Coating Used in Roofing.* West Conshohocken, PA: ASTM International, 1997.

ASTM D 6162-00a, *Specification for Styrene Butadiene Styrene (SBS) Modified Bituminous Sheet Materials Using a Combination of Polyester and Glass Fiber Reinforcements.* West Conshohocken, PA: ASTM International, 2000.

ASTM D 6163-00e01, *Specification for Styrene Butadiene Styrene (SBS) Modified Bituminous Sheet Materials Using Glass Fiber Reinforcements.* West Conshohocken, PA: ASTM International, 2000.

ASTM D 6164-00, *Specification for Styrene Butadiene Styrene (SBS) Modified Bituminous Sheet Materials Using Polyester Reinforcements.* West Conshohocken, PA: ASTM International, 2000.

ASTM D 6298-00, *Specification for Fiberglass Reinforced Styrene-Butadiene-Styrene (SBS) Modified Bituminous Sheet with a Factory Applied Metal Surface.* West Conshohocken, PA: ASTM International, 2000.

ASTM D 6380-01[E1], *Standard Specification for Asphalt Roll Roofing (Organic Felt).* West Conshohocken, PA: ASTM International, 2001.

ASTM D 6694-01, *Standard Specification Liquid-Applied Silicone Coating Used in Spray Polurethane Foam Roofing.* West Conshohocken, PA: ASTM International, 2001.

ASTM D 6705-01, *Standard Guide for the Repair and Recoat of Spray Polyurethane Foam Roofing Systems.* West Conshohocken, PA: ASTM International, 2001.

ASTM D 6757-02, *Standard Specification for Inorganic Underlayment for Use with Steep Slope Roofing Products.* West Conshohocken, PA: ASTM International, 2002.

ASTM D 6878-03, *Standard Specification for Thermoplastic Polyolefin Based Sheet Roofing.* West Conshohocken, PA: ASTM International, 2003.

ASTM E 108-04, *Test Method for Fire Tests of Roof Coverings.* West Conshohocken, PA: ASTM International, 2004.

ASTM F 1667-03, *Specification for Driven Fasteners, Nails, Spikes, and Staples.* West Conshohocken, PA: ASTM International, 2003.

AWPA C1-00, *All Timber Products—Preservative Treatment by Pressure Processes.* Granbury, TX: American Wood–Preservers' Association, 2000.

CDA 4115-1929, *Copper in Architecture—Design Handbook.* New York: The Copper Development Association, 1929.

CGSB 37-GP-52M-84, *Roofing and Waterproofing Membrane, Sheet Applied, Elastomeric.* Ottawa, Ontario: Canadian General Standards Board, 1984.

CGSB CAN/CGSB 37.54-95, *Polyvinyl Chloride Roofing and Waterproofing Membrane.* Ottawa, Ontario: Canadian General Standards Board, 1995.

CGSB 37-GP-56M-80, Membrane, *Modified Bituminous, Prefabricated and Reinforced for Roofing—with December 1985 Amendment.* Ottawa, Ontario: Canadian General Standards Board, 1980.

CSSB-97, *Grading and Packing Rules for Western Red Cedar Shakes and Western Red Shingles of the Cedar Shake and Shingle Bureau.* Bellevue, WA: Cedar Shake & Shingle Bureau, 1997.

FM 4450-(1989), *Approved Standard for Class I Insulated Steel Deck Roofs—Supplements through 7/92.* Norwood, MA: Factory Mutual, 1989.

IPC-2006, *International Plumbing Code.* Falls Church, VA: International Code Council, 2006.

UL 790-04, *Tests for Fire Resistance of Roof Covering Materials.* Northbrook, IL: Underwriters Laboratories, 2004.

UL 1256-02, *Fire Test of Roof Deck Construction.* Northbrook, IL: Underwriters Laboratories, 2002.

Chapter 10:
Chimneys and Fireplaces

General Comments

Chapter 10 regulates two basic types of chimneys and fireplaces: factory-built and those constructed on site of masonry and other approved materials. Chimneys and fireplaces constructed of masonry rely on prescriptive requirements for the details of their construction; the factory-built type rely on the listing and labeling method of approval. This chapter also contains provisions for unvented gas log heaters. Seismic issues related to fireplaces and chimneys, including reinforcing requirements, are also addressed in Chapter 10.

Section R1001 regulates the construction of masonry fireplaces. Section R1002 establishes the standards for the use and installation of masonry heaters. Section R1003 regulates the construction of masonry chimneys.

Section R1004 establishes the standards for the use and installation of factory-built fireplaces including the use of unvented gas log heaters. Section R1005 establishes the standards for the use and installation of factory-built chimneys. Section R1006 requires the installation of an exterior air supply for use with both factory-built and masonry fireplaces.

Purpose

Chapter 10 contains requirements for the safe construction of masonry chimneys and fireplaces and establishes the standards for the use and installation of factory-built chimneys, fireplaces and masonry heaters.

SECTION R1001
MASONRY FIREPLACES

R1001.1 General. Masonry fireplaces shall be constructed in accordance with this section and the applicable provisions of Chapters 3 and 4.

❖ This section covers details for masonry fireplaces. Table R1001.1 and Figure R1001.1 show details for masonry fireplaces and chimneys.

R1001.2 Footings and foundations. Footings for masonry fireplaces and their chimneys shall be constructed of concrete or solid masonry at least 12 inches (305 mm) thick and shall extend at least 6 inches (152 mm) beyond the face of the fireplace or foundation wall on all sides. Footings shall be founded on natural, undisturbed earth or engineered fill below frost depth. In areas not subjected to freezing, footings shall be at least 12 inches (305 mm) below finished grade.

❖ The dead-load bearing pressure for a building structure, particularly if it is a light wood-frame building, is usually quite low, even though the bearing pressure on the foundation for a masonry or concrete fireplace and chimney can be several times higher. It is good practice to proportion the foundation of a masonry fireplace and chimney to have approximately the same bearing pressure as is present under the building structure itself, although the codes do not require such a design. Where the soil is compressible, differential settlements between the fireplace and chimney and its surrounding structure can cause cracking of the finish materials in the vicinity of the fireplace and chimney. Also, firestopping may be displaced so that a draft opening is created.

R1001.2.1 Ash dump cleanout. Cleanout openings located within foundation walls below fireboxes, when provided, shall be equipped with ferrous metal or masonry doors and frames constructed to remain tightly closed except when in use.

Cleanouts shall be accessible and located so that ash removal will not create a hazard to combustible materials.

❖ Noncombustible, tightly sealed cleanout doors must reduce the danger of fire spread through the cleanout openings. Cleanout openings must be easily accessible to allow ash removal. The ashes and other materials left in the fireplace may remain hot for some time. Cleanout doors must be tight closing to prevent air from entering the firebox and causing the unburned materials to re-ignite. The cleanout door must be located so that ashes and other unburned materials that may still be hot can be removed without creating a hazard to combustible materials.

R1001.3 Seismic reinforcing. Masonry or concrete chimneys in Seismic Design Category D_0, D_1 or D_2 shall be reinforced. Reinforcing shall conform to the requirements set forth in Table R1001.1 and Section R609, Grouted Masonry.

❖ Masonry and concrete have inherently brittle natures, and they are therefore weak when subjected to tensile forces. Using steel, an inherently ductile material, provides ductility in masonry and concrete construction. Chimneys in general and unreinforced masonry and concrete chimneys in particular do not perform well in earthquakes because of their tall and slender geometry. Observation of buildings after earthquakes has shown that chimneys suffered severe damage and sometimes completely collapsed. To improve the performance of masonry and concrete chimneys under severe earthquake loading, steel reinforcing bars are required as shown in Table R1001.1. This reinforcement is required in Seismic Design Categories D_0, D_1 and D_2 only. Reinforcement is not required in Seismic Design Categories A, B, and C, and buildings in seismic design categories higher than D must be designed in accordance with the *International Building Code®* (IBC®). See Section R301.2.2.

R1001.3.1 Vertical reinforcing. For chimneys up to 40 inches (1016 mm) wide, four No. 4 continuous vertical bars shall be placed between wythes of solid masonry or within the cells of hollow unit masonry and grouted in accordance with Section R609. Grout shall be prevented from bonding with the flue liner so that the flue liner is free to move with thermal expansion. For chimneys more than 40 inches (1016 mm) wide, two additional No. 4 vertical bars shall be provided for each additional flue incorporated into the chimney or for each additional 40 inches (1016 mm) in width or fraction thereof.

❖ To keep the chimney structure together and prevent severe damage or fracture, four No. 4 bars, continuous from bottom to top, are required. The bars are typically placed at the four corners, similar to the placement of vertical bars in concrete columns. The hollow cells in hollow unit masonry are used for placement of vertical bars, and if solid masonry is used, the reinforcing steel must be placed between the wythes. For wider chimneys or chimneys with more than one flue, four No. 4 bars are not adequate, and reinforcing bars in increments of two are added for each additional flue or each additional 40 inches (1016 mm) of chimney width.

R1001.3.2 Horizontal reinforcing. Vertical reinforcement shall be placed within $1/4$-inch (6 mm) ties, or other reinforcing of equivalent net cross-sectional area, placed in the bed joints according to Section R607 at a minimum of every 18 inches (457 mm) of vertical height. Two such ties shall be installed at each bend in the vertical bars.

❖ Vertical reinforcement under high seismic loading tends to buckle, then burst out of the concrete or masonry. To prevent this action, the vertical reinforcing required under Section R1001.3.1 must be placed in $1/4$-inch (6.4 mm) ties that are placed in the bed joints in accordance with Section R607. The ties are required at no more than 18-inch (457 mm) intervals vertically to form an effective cage for the vertical reinforcement. Where the vertical reinforcing bars have bends and abrupt changes in direction as a result of changes in sections of the chimney, two horizontal ties are required at each bend to handle the additional effects of stress concentration.

R1001.4 Seismic anchorage. Masonry or concrete chimneys in Seismic Design Categories D_0, D_1 or D_2 shall be anchored at each floor, ceiling or roof line more than 6 feet (1829 mm) above grade, except where constructed completely within the exterior walls. Anchorage shall conform to the requirements of Section R1001.4.1.

❖ The sway of a chimney under earthquake lateral movement may cause it to impact the roof and floor framing and ultimately pull away from the framing. To control or prevent this action in Seismic Design Categories D_0, D_1 and D_2, masonry and concrete chimneys must be anchored at each floor, ceiling or roof line to provide lateral bracing. Such anchorage is not required in three situations: 1) in Seismic Design Cate-

gories A, B, and C; 2) at floor, ceiling or roof lines 6 feet (1829 mm) or less above grade; and 3) where the chimney is constructed completely within the exterior walls and is therefore braced by the building elements surrounding it.

In Seismic Design Categories E and F, the reinforcing and anchorage must be designed in accordance with the IBC.

R1001.4.1 Anchorage. Two $3/16$-inch by 1-inch (5 mm by 25 mm) straps shall be embedded a minimum of 12 inches (305 mm) into the chimney. Straps shall be hooked around the outer bars and extend 6 inches (152 mm) beyond the bend. Each strap shall be fastened to a minimum of four floor ceiling or floor joists or rafters with two $1/2$-inch (13 mm) bolts.

❖ The anchors required under Section R1001.4 consist of a minimum of two straps, each $3/16$ inch by 1 inch (4.8 mm by 25.4 mm). The straps must be firmly connected to the building framing and chimney to control or prevent the chimney from pulling away or buckling. To accomplish this firm connection, the straps must be bolted to at least four joists or rafters using $1/2$-inch (12.7 mm) bolts. At the other end they must be embedded at least 12 inches (305 mm) into the chimney and hooked around the outer vertical reinforcing bars. Short hooks might pull out in earthquakes, and for this reason the minimum length of the hook around the vertical bar is 6 inches (152 mm).

R1001.5 Firebox walls. Masonry fireboxes shall be constructed of solid masonry units, hollow masonry units grouted solid, stone or concrete. When a lining of firebrick at least 2 inches (51 mm) thick or other approved lining is provided, the minimum thickness of back and side walls shall each be 8 inches (203 mm) of solid masonry, including the lining. The width of joints between firebricks shall not be greater than $1/4$ inch (6 mm). When no lining is provided, the total minimum thickness of back and side walls shall be 10 inches (254 mm) of solid masonry. Firebrick shall conform to ASTM C 27 or C 1261 and shall be laid with medium duty refractory mortar conforming to ASTM C 199.

❖ This section specifies the minimum thicknesses of refractory brick or solid masonry necessary to contain the generated heat.

Solid masonry walls forming the firebox must have a minimum total thickness of 8 inches (204 mm), including the refractory lining.

The refractory lining is to consist of a low-duty, fireclay refractory brick with a minimum thickness of 2 inches (51 mm), laid with medium-duty refractory mortar. Mortar joints are generally $1/16$ to $3/16$ inch (1.6 to 4.8 mm) thick, but not thicker than $1/4$ inch (6.4 mm), to reduce thermal movements and prevent joint deterioration.

Where a firebrick lining is not used in firebox construction, the wall thickness is not to be less than 10 inches (254 mm) of solid masonry. Firebrick must conform to ASTM C 27 or ASTM C 1261 and must be laid with medium-duty refractory mortar conforming to ASTM C 199.

TABLE R1001.1
SUMMARY OF REQUIREMENTS FOR MASONRY FIREPLACES AND CHIMNEYS

ITEM	LETTER[a]	REQUIREMENTS
Hearth slab thickness	A	4″
Hearth extension (each side of opening)	B	8″ fireplace opening < 6 square foot. 12″ fireplace opening ≥ 6 square foot.
Hearth extension (front of opening)	C	16″ fireplace opening < 6 square foot. 20″ fireplace opening ≥ 6 square foot.
Hearth slab reinforcing	D	Reinforced to carry its own weight and all imposed loads.
Thickness of wall of firebox	E	10″ solid brick or 8″ where a firebrick lining is used. Joints in firebrick $^1/_4$″ maximum.
Distance from top of opening to throat	F	8″
Smoke chamber wall thickness Unlined walls	G	6″ 8″
Chimney Vertical reinforcing[b]	H	Four No. 4 full-length bars for chimney up to 40″ wide. Add two No. 4 bars for each additional 40″ or fraction of width or each additional flue.
Horizontal reinforcing	J	$^1/_4$″ ties at 18″ and two ties at each bend in vertical steel.
Bond beams	K	No specified requirements.
Fireplace lintel	L	Noncombustible material.
Chimney walls with flue lining	M	Solid masonry units or hollow masonry units grouted solid with at least 4 inch nominal thickness.
Distances between adjacent flues	—	See Section R1003.13.
Effective flue area (based on area of fireplace opening)	P	See Section R1003.15.
Clearances: Combustible material Mantel and trim Above roof	R	See Sections R1001.11 and R1003.18. See Section R1001.11, Exception 4. 3′ at roofline and 2′ at 10′.
Anchorage[b] Strap Number Embedment into chimney Fasten to Bolts	S	$^3/_{16}$″ × 1″ Two 12″ hooked around outer bar with 6″ extension. 4 joists Two $^1/_2$″ diameter.
Footing Thickness Width	T	12″ min. 6″ each side of fireplace wall.

For SI: 1 inch = 25.4 mm, 1 foot = 304.8 mm, 1 square foot = 0.0929 m².

NOTE: This table provides a summary of major requirements for the construction of masonry chimneys and fireplaces. Letter references are to Figure R1001.1, which shows examples of typical construction. This table does not cover all requirements, nor does it cover all aspects of the indicated requirements. For the actual mandatory requirements of the code, see the indicated section of text.

a. The letters refer to Figure R1001.1.

b. Not required in Seismic Design Category A, B or C.

❖ This table provides a summary of the requirements for masonry fireplaces and chimneys directly correlated to Figure R1001.1. The first column lists various parts of the fireplace. The second column shows a letter designation corresponding to the item in the figure with the same letter. The third column contains the requirements for that particular fireplace or chimney.

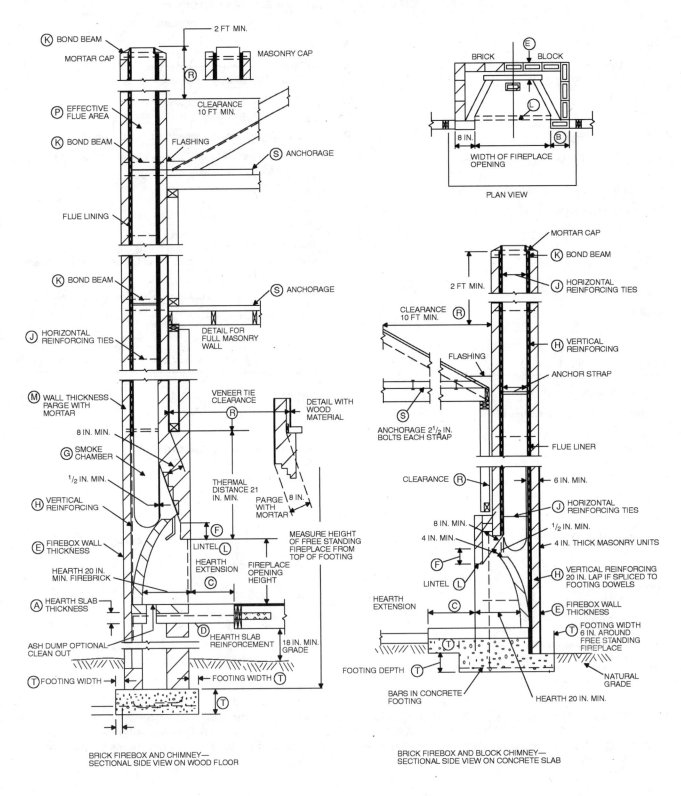

For SI: 1 inch = 25.4 mm, 1 foot = 304.8 mm.

**FIGURE R1001.1
FIREPLACE AND CHIMNEY DETAILS**

❖ See the commentary for Table R1001.1.

R1001.5.1 Steel fireplace units. Installation of steel fireplace units with solid masonry to form a masonry fireplace is permitted when installed either according to the requirements of their listing or according to the requirements of this section. Steel fireplace units incorporating a steel firebox lining, shall be constructed with steel not less than $1/4$ inch (6 mm) thick, and an air circulating chamber which is ducted to the interior of the building. The firebox lining shall be encased with solid masonry to provide a total thickness at the back and sides of not less than 8 inches (203 mm), of which not less than 4 inches (102 mm) shall be of solid masonry or concrete. Circulating air ducts used with steel fireplace units shall be constructed of metal or masonry.

❖ Steel fireplace units and an air chamber may be used as part of the required thickness of masonry fireplaces so that the combined thickness of the masonry and steel is 8 inches (203 mm). The steel may not be less than $1/4$ inch (6.4 mm) thick, and the masonry must be at least 4 inches thick. Where ducts are used as part of a steel fireplace unit to circulate warm air, they must be constructed of metal or masonry.

R1001.6 Firebox dimensions. The firebox of a concrete or masonry fireplace shall have a minimum depth of 20 inches (508 mm). The throat shall not be less than 8 inches (203 mm) above the fireplace opening. The throat opening shall not be less than 4 inches (102 mm) deep. The cross-sectional area of the passageway above the firebox, including the throat, damper and smoke chamber, shall not be less than the cross-sectional area of the flue.

Exception: Rumford fireplaces shall be permitted provided that the depth of the fireplace is at least 12 inches (305 mm) and at least one-third of the width of the fireplace opening, that the throat is at least 12 inches (305 mm) above the lintel and is at least $1/20$ the cross-sectional area of the fireplace opening.

❖ The proper functioning of the fireplace depends on the size of the face opening and the chimney dimensions, which in turn are related to the room size [see Commentary Figure R1001.6(1)]. This section specifies a minimum depth of 20 inches (508 mm) for the combustion chamber because that depth influences the draft requirement. The dimensions of the firebox (depth, opening size and shape) are usually based on two considerations: aesthetics and the need to prevent the room from overheating. Suggested dimensions for single-opening fireboxes are given in technical publications of the Brick Institute of America (BIA) and the National Concrete Masonry Association (NCMA).

This section also contains additional criteria for the throat's location and minimum cross-sectional area. Those criteria are based on many years of construction of successfully functioning fireplaces. The exception permits the use of Rumford fireplaces, which are tall, shallow fireplaces that can radiate a large amount of heat into a room.

The code reference to "depth of fireplace" is interpreted as the depth of the firebox [see Commentary Figure R1001.6(2)]. The throat must be made at least 12 inches (305 mm) above the lintel and at least 5 percent ($1/20$) of the cross-sectional area of the fireplace opening. Smoke chambers and flues for Rumford fireplaces should be sized and built like those of other masonry fireplaces. Even though those who build Rumford fireplaces do not totally agree about how they work, many books and guides address their construction.

R1001.7 Lintel and throat. Masonry over a fireplace opening shall be supported by a lintel of noncombustible material. The minimum required bearing length on each end of the fireplace opening shall be 4 inches (102 mm). The fireplace throat or damper shall be located a minimum of 8 inches (203 mm) above the lintel.

❖ A noncombustible lintel with a minimum bearing length of 4 inches (102 mm) on·each end must be provided to support masonry above the fireplace opening. Provisions should be made to allow for expansion of the steel lintels when they are heated.

R1001.7.1 Damper. Masonry fireplaces shall be equipped with a ferrous metal damper located at least 8 inches (203 mm) above the top of the fireplace opening. Dampers shall be installed in the fireplace or the chimney venting the fireplace, and shall be operable from the room containing the fireplace.

❖ A damper is used to close the chimney flue when the fireplace is not in use. This section provides guidance on its location and construction.

R1001.8 Smoke chamber. Smoke chamber walls shall be constructed of solid masonry units, hollow masonry units grouted solid, stone or concrete. Corbelling of masonry units shall not leave unit cores exposed to the inside of the smoke chamber. When a lining of firebrick at least 2 inches (51 mm) thick, or a lining of vitrified clay at least $5/8$ inch (16 mm) thick, is provided, the total minimum thickness of front, back and side walls shall be 6 inches (152 mm) of solid masonry, including the lining. Firebrick shall conform to ASTM C 27 or C 1261 and shall be laid with medium duty refractory mortar conforming to ASTM C 199. Where no lining is provided, the total minimum thickness of front, back and side walls shall be 8 inches (203 mm) of solid masonry. When the inside surface of the smoke chamber is formed by corbeled masonry, the inside surface shall be parged smooth.

❖ The smoke chamber is located directly above the firebox and must be constructed of solid masonry. A smoke shelf located behind the damper deflects downdrafts. As the downdraft hits the smoke shelf, it is turned upward by the damper assembly. Curved smoke shelves perform better; however, flat smoke shelves are acceptable.

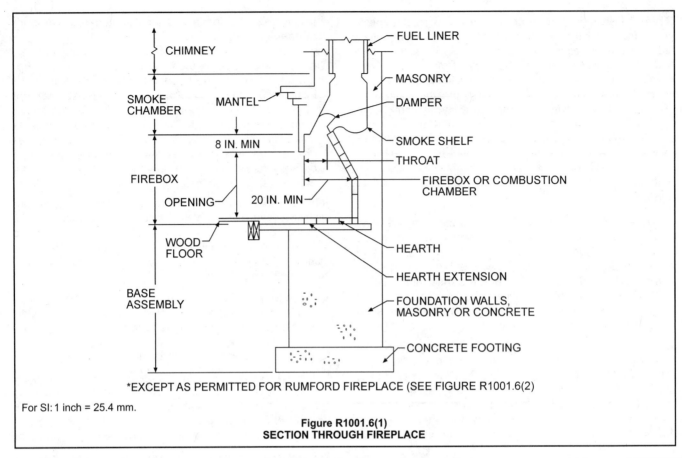

*EXCEPT AS PERMITTED FOR RUMFORD FIREPLACE (SEE FIGURE R1001.6(2)

For SI: 1 inch = 25.4 mm.

Figure R1001.6(1)
SECTION THROUGH FIREPLACE

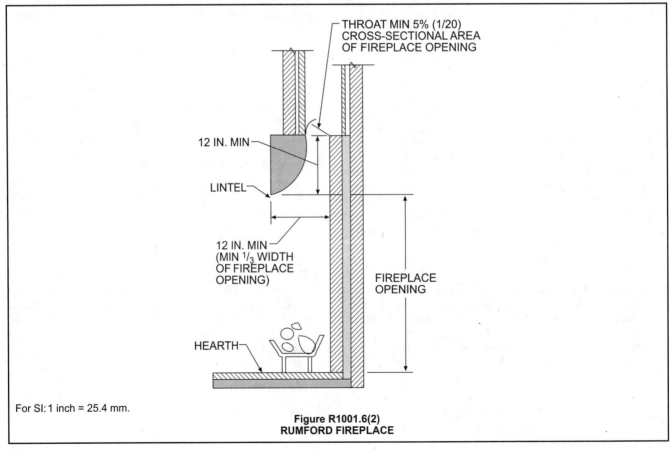

For SI: 1 inch = 25.4 mm.

Figure R1001.6(2)
RUMFORD FIREPLACE

R1001.8.1 Smoke chamber dimensions. The inside height of the smoke chamber from the fireplace throat to the beginning of the flue shall not be greater than the inside width of the fireplace opening. The inside surface of the smoke chamber shall not be inclined more than 45 degrees (0.79 rad) from vertical when prefabricated smoke chamber linings are used or when the smoke chamber walls are rolled or sloped rather than corbeled. When the inside surface of the smoke chamber is formed by corbeled masonry, the walls shall not be corbeled more than 30 degrees (0.52 rad) from vertical.

❖ This section provides specific dimensions for the smoke chamber of masonry fireplaces, which are necessary to provide proper flow of the smoke and products of combustion out of the fireplace and through the chimney. The height of the smoke chamber cannot exceed the inside width of the fireplace opening. The inside surface of the smoke chamber may be sloped 45 degrees (0.79 rad) from the vertical when constructed of a prefabricated smoke chamber lining or the smoke chamber walls are rolled or sloped. If the smoke chamber is sloped using corbeled masonry, the slope is limited to be no more than 30 degrees (0.52 rad) from the vertical. The slope is restricted more when using corbeled masonry because the surface is not smooth and will tend to create turbulence within the smoke chambers, thus affecting the flow of smoke to the flue.

R1001.9 Hearth and hearth extension. Masonry fireplace hearths and hearth extensions shall be constructed of concrete or masonry, supported by noncombustible materials, and reinforced to carry their own weight and all imposed loads. No combustible material shall remain against the underside of hearths and hearth extensions after construction.

❖ The hearth includes both the floor of the fire box and the projection in front of it. The hearth extension protects wood, carpet and combustible materials from being ignited by sparks, hot embers or ashes that may fall from the fire box.

R1001.9.1 Hearth thickness. The minimum thickness of fireplace hearths shall be 4 inches (102 mm).

❖ The minimum thickness of fireplace hearths is 4 inches (102 mm).

R1001.9.2 Hearth extension thickness. The minimum thickness of hearth extensions shall be 2 inches (51 mm).

Exception: When the bottom of the firebox opening is raised at least 8 inches (203 mm) above the top of the hearth extension, a hearth extension of not less than $^3/_8$-inch-thick (10 mm) brick, concrete, stone, tile or other approved noncombustible material is permitted.

❖ The hearth extension must be 2 inches (51 mm) thick unless the bottom of the fireplace opening is raised at least 8 inches (203 mm) above the top of the hearth extension. Where the bottom of the fireplace opening is raised at least 8 inches (203 mm) above the hearth extension, the hearth extension can be reduced to a thickness of $^3/_8$-inch-thick (9.5 mm) brick, concrete, stone, tile or other approved noncombustible material.

R1001.10 Hearth extension dimensions. Hearth extensions shall extend at least 16 inches (406 mm) in front of and at least 8 inches (203 mm) beyond each side of the fireplace opening. Where the fireplace opening is 6 square feet (0.6 m^2) or larger, the hearth extension shall extend at least 20 inches (508 mm) in front of and at least 12 inches (305 mm) beyond each side of the fireplace opening.

❖ The hearth must extend in front of and to the side of the firebox opening a sufficient dimension to prevent burning embers from landing on combustible surfaces.

R1001.11 Fireplace clearance. All wood beams, joists, studs and other combustible material shall have a clearance of not less than 2 inches (51 mm) from the front faces and sides of masonry fireplaces and not less than 4 inches (102 mm) from the back faces of masonry fireplaces. The air space shall not be filled, except to provide fire blocking in accordance with Section R1001.12.

Exceptions:

1. Masonry fireplaces listed and labeled for use in contact with combustibles in accordance with UL 127 and installed in accordance with the manufacturer's installation instructions are permitted to have combustible material in contact with their exterior surfaces.

2. When masonry fireplaces are part of masonry or concrete walls, combustible materials shall not be in contact with the masonry or concrete walls less than 12 inches (305 mm) from the inside surface of the nearest firebox lining.

3. Exposed combustible trim and the edges of sheathing materials such as wood siding, flooring and drywall shall be permitted to abut the masonry fireplace side walls and hearth extension in accordance with Figure R1001.11, provided such combustible trim or sheathing is a minimum of 12 inches (305 mm) from the inside surface of the nearest firebox lining.

4. Exposed combustible mantels or trim may be placed directly on the masonry fireplace front surrounding the fireplace opening providing such combustible materials are not placed within 6 inches (152 mm) of a fireplace opening. Combustible material within 12 inches (306 mm) of the fireplace opening shall not project more than $^1/_8$ inch (3 mm) for each 1-inch (25 mm) distance from such an opening.

❖ A 2-inch (51 mm) clearance is required from the front and side surfaces of the fireplace. A 4-inch (102 mm) clearance is required from the back of the fireplace. This air space created around the fireplace cannot be filled, except for approved fire blocking. See Figure R1001.11.

There are four exceptions to this section that allow combustible materials to come in contact with the fireplace. The first allows factory-built fireplaces constructed in accordance with UL 127 to be in contact with combustibles. The installation must carefully follow the manufacturer's installation instructions.

The second exception applies to masonry fireplaces constructed in concrete or masonry walls. In this case combustible materials must keep at least 12 inches (305 mm) from the inside surface of the firebox lining because of the conductive nature of the concrete or masonry walls.

Exception 3 permits combustible trim and edges of certain sheathing materials to abut the fireplace in accordance with Figure R1001.11. This exception permits the combustible trim to cover openings that would otherwise occur between the fireplace and other construction.

Exception 4 provides specific dimensions that protect combustible trim from radiant heat

R1001.12 Fireplace fireblocking. Fireplace fireblocking shall comply with the provisions of Section R602.8.

❖ Refer to the commentary for Section R602.8.

SECTION R1002
MASONRY HEATERS

R1002.1 Definition. A masonry heater is a heating appliance constructed of concrete or solid masonry, hereinafter referred to as masonry, which is designed to absorb and store heat from a solid-fuel fire built in the firebox by routing the exhaust gases through internal heat exchange channels in which the flow path downstream of the firebox may include flow in a horizontal or downward direction before entering the chimney and which delivers heat by radiation from the masonry surface of the heater.

❖ Masonry heaters are appliances designed to absorb and store heat from a relatively small fire and to radiate that heat into a building interior. They are thermally more efficient than traditional fireplaces because of their design. Interior passageways through the heater allow hot exhaust gases from the fire to transfer heat into the masonry, which then radiates into the building.

R1002.2 Installation. Masonry heaters shall be installed in accordance with this section and comply with one of the following:

1. Masonry heaters shall comply with the requirements of ASTM E 1602; or

2. Masonry heaters shall be listed and labeled in accordance with UL 1482 and installed in accordance with the manufacturer's installation instructions.

❖ ASTM E 1602 and UL 1482 contain guidelines for the installation of masonry heaters.

R1002.3 Footings and foundation. The firebox floor of a masonry heater shall be a minimum thickness of 4 inches (102 mm) of noncombustible material and be supported on a noncombustible footing and foundation in accordance with Section R1003.2.

❖ This section prescribes the minimum foundation requirements that will adequately support a masonry heating appliance. Section R1003.2 goes into specific detail on footings and foundations for masonry chimneys.

R1002.4 Seismic reinforcing. In Seismic Design Categories D_0, D_1 and D_2, masonry heaters shall be anchored to the masonry foundation in accordance with Section R1003.3. Seismic reinforcing shall not be required within the body of a masonry heater whose height is equal to or less than 3.5 times it's body width and where the masonry chimney serving the heater is not supported by the body of the heater. Where the masonry chimney shares a common wall with the facing of the masonry heater, the chimney portion of the structure shall be reinforced in accordance with Section R1003.

❖ Because of the large bulk and squat geometry of these heaters, seismic reinforcement is not typically required. Flexural tensile stresses, which typically cause damage to unreinforced masonry, rarely occur. Where

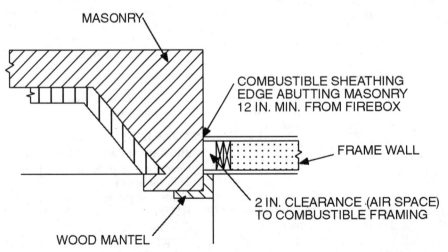

For SI: 1 inch = 25.4 mm.

Figure R1001.11
CLEARANCE FROM COMBUSTIBLES

chimneys extend above these heaters, however, seismic reinforcement is required by Section R1003.3. See the commentary to Sections R1003.3 and R1003.4 for seismic reinforcements and anchorage for chimneys.

R1002.5 Masonry heater clearance. Combustible materials shall not be placed within 36 inches (914 mm) of the outside surface of a masonry heater in accordance with NFPA 211 Section 8-7 (clearances for solid-fuel-burning appliances), and the required space between the heater and combustible material shall be fully vented to permit the free flow of air around all heater surfaces.

Exceptions:

1. When the masonry heater wall is at least 8 inches (203 mm) thick of solid masonry and the wall of the heat exchange channels is at least 5 inches (127 mm) thick of solid masonry, combustible materials shall not be placed within 4 inches (102 mm) of the outside surface of a masonry heater. A clearance of at least 8 inches (203 mm) shall be provided between the gas-tight capping slab of the heater and a combustible ceiling.

2. Masonry heaters tested and listed by an American National Standards Association (ANSI)-accredited laboratory to the requirements of UL1482 may be installed in accordance with the listing specifications and the manufacturer's written instructions.

❖ Heat conducted through masonry heater walls can ignite combustible structural materials in contact with these walls. For this reason, a minimum required clearance to combustibles from masonry heaters has been established. Because masonry heaters typically generate more heat for a longer period of time than traditional fireplaces, greater clearances to combustible materials are needed to reduce the risk of fire.

SECTION R1003
MASONRY CHIMNEYS

R1003.1 Definition. A masonry chimney is a chimney constructed of concrete or masonry, hereinafter referred to as masonry. Masonry chimneys shall be constructed, anchored, supported and reinforced as required in this chapter.

❖ A masonry chimney is a field-constructed assembly that can consist of masonry units, grout, reinforced concrete, rubble stone, fire-clay liners and mortars. A masonry chimney is permitted to serve residential (low-heat), medium- and high-heat appliances. This section outlines the general code requirements regarding construction details for all masonry chimneys, including those serving masonry fireplaces regulated by Section R1001.

R1003.2 Footings and foundations. Footings for masonry chimneys shall be constructed of concrete or solid masonry at least 12 inches (305 mm) thick and shall extend at least 6 inches (152 mm) beyond the face of the foundation or support wall on all sides. Footings shall be founded on natural undisturbed earth or engineered fill below frost depth. In areas not subjected to freezing, footings shall be at least 12 inches (305 mm) below finished grade.

❖ Masonry fireplaces and chimneys must be supported on adequate foundations because of their weight and the forces imposed on them by wind, earthquakes and other effects. This section prescribes minimum foundation requirements that are typically adequate to support a standard chimney (see Commentary Figure R1003.2).

R1003.3 Seismic reinforcing. Masonry or concrete chimneys shall be constructed, anchored, supported and reinforced as required in this chapter. In Seismic Design Category D_0, D_1 or D_2 masonry and concrete chimneys shall be reinforced and anchored as detailed in Section R1003.3.1, R1003.3.2 and R1003.4. In Seismic Design Category A, B or C, reinforcement and seismic anchorage is not required.

❖ Unreinforced fireplaces and chimneys subjected to strong ground motion have been severely damaged in past earthquakes. The requirements in this section provide minimum reinforcement in an effort to keep these structures together during such events. More substantial reinforcement, however, may be required in areas of high seismicity or for atypical chimneys.

R1003.3.1 Vertical reinforcing. For chimneys up to 40 inches (1016 mm) wide, four No. 4 continuous vertical bars, anchored in the foundation, shall be placed in the concrete, or between wythes of solid masonry, or within the cells of hollow unit masonry, and grouted in accordance with Section R609.1.1. Grout shall be prevented from bonding with the flue liner so that the flue liner is free to move with thermal expansion. For chimneys more than 40 inches (1016 mm) wide, two additional No. 4 vertical bars shall be installed for each additional 40 inches (1016 mm) in width or fraction thereof.

❖ These requirements are traditional minimum prescriptive provisions to help maintain the structural integrity of fireplaces and chimneys during earthquakes. More reinforcement may be required in areas of high seismicity or for atypical chimneys.

R1003.3.2 Horizontal reinforcing. Vertical reinforcement shall be placed enclosed within $^1/_4$-inch (6 mm) ties, or other reinforcing of equivalent net cross-sectional area, spaced not to exceed 18 inches (457 mm) on center in concrete, or placed in the bed joints of unit masonry, at a minimum of every 18 inches (457 mm) of vertical height. Two such ties shall be installed at each bend in the vertical bars.

❖ These requirements are traditional minimum prescriptive provisions to help maintain the structural integrity of fireplaces and chimneys during earthquakes. The vertical reinforcement required by Section R1003.3.1 must be enclosed within the horizontal reinforcement required by this section. More reinforcement may be required in areas of high seismicity or for atypical chimneys.

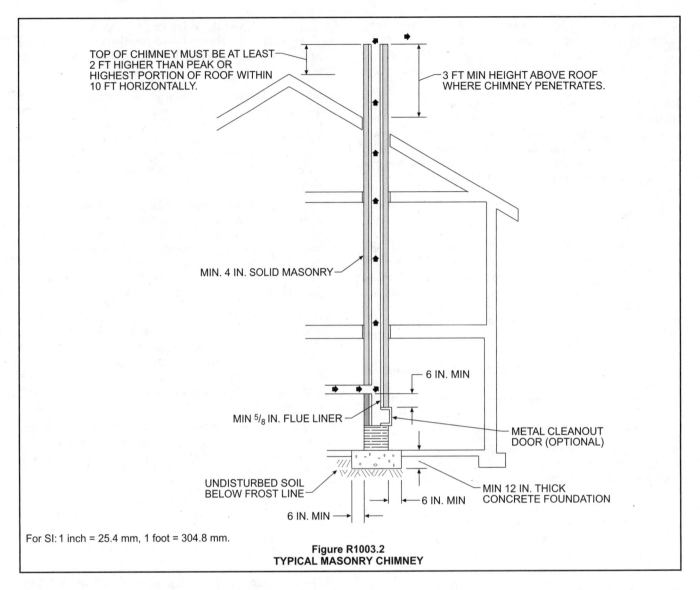

For SI: 1 inch = 25.4 mm, 1 foot = 304.8 mm.

**Figure R1003.2
TYPICAL MASONRY CHIMNEY**

R1003.4 Seismic anchorage. Masonry and concrete chimneys and foundations in Seismic Design Category D_0, D_1 or D_2 shall be anchored at each floor, ceiling or roof line more than 6 feet (1829 mm) above grade, except where constructed completely within the exterior walls. Anchorage shall conform to the requirements in Section R1003.4.1.

❖ Fireplaces and chimneys must be connected to floor and roof diaphragms to prevent overturning during earthquakes. Chimneys must be anchored at the ceiling line of roof and ceiling assemblies and at floor levels below the roof. Such anchorage is of lesser importance where the floor assembly is 6 feet (1829 mm) or less above grade.

R1003.4.1 Anchorage. Two $^3/_{16}$-inch by 1-inch (5 mm by 25 mm) straps shall be embedded a minimum of 12 inches (305 mm) into the chimney. Straps shall be hooked around the outer bars and extend 6 inches (152 mm) beyond the bend. Each strap shall be fastened to a minimum of four floor joists with two $^1/_2$-inch (13 mm) bolts.

❖ The prescriptive requirements in this section are traditional for typical fireplaces and chimneys. More substantial anchorage may be required in areas of high seismicity, for large fireplaces or where the distance between floor and roof diaphragms is large.

R1003.5 Corbeling. Masonry chimneys shall not be corbeled more than one-half of the chimney's wall thickness from a wall or foundation, nor shall a chimney be corbeled from a wall or foundation that is less than 12 inches (305 mm) thick unless it projects equally on each side of the wall, except that on the second story of a two-story dwelling, corbeling of chimneys on the exterior of the enclosing walls may equal the wall thickness. The projection of a single course shall not exceed one-half the unit height or one-third of the unit bed depth, whichever is less.

❖ Corbeling is the projection of masonry from the surface of the wall or fireplace in small increments for each course of masonry. The chimney should not be corbeled more than one-half of its wall thickness from the wall or foundation. A single course is not permitted to project more than one-half of the individual unit height or more than one-third of the individual unit depth. Exceeding these amounts could cause the unit to fail because it might not be able to carry the intended load [see Commentary Figure R1003.5(1)].

Additionally, masonry walls are not to be corbeled more than one-half of the chimney's wall thickness from the wall or foundation unless it projects equally on each side of the wall. If the corbeling is not equal on each side, the chimney might overturn or crack [see Commentary Figure R1003.5(2)].

R1003.6 Changes in dimension. The chimney wall or chimney flue lining shall not change in size or shape within 6 inches (152 mm) above or below where the chimney passes through floor components, ceiling components or roof components.

❖ Changes in the size or shape of a chimney change the stiffness of the chimney, and these areas have more potential for leaks or cracks in the chimney. The code prohibits these changes within 6 inches (152 mm) of where the chimney passes through the floor, ceiling or roof components.

R1003.7 Offsets. Where a masonry chimney is constructed with a fireclay flue liner surrounded by one wythe of masonry, the maximum offset shall be such that the centerline of the flue above the offset does not extend beyond the center of the chimney wall below the offset. Where the chimney offset is supported by masonry below the offset in an approved manner, the maximum offset limitations shall not apply. Each individual

corbeled masonry course of the offset shall not exceed the projection limitations specified in Section R1003.5.

❖ Limitations on chimney offsets provide a gradual transition to prevent critical stress concentrations at the bottom of the offset. The maximum offset of the center-

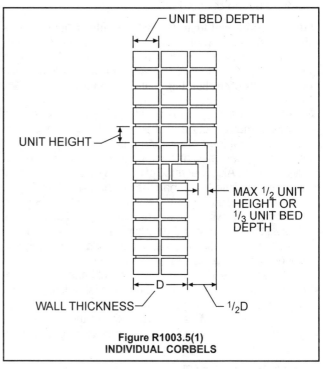

Figure R1003.5(1)
INDIVIDUAL CORBELS

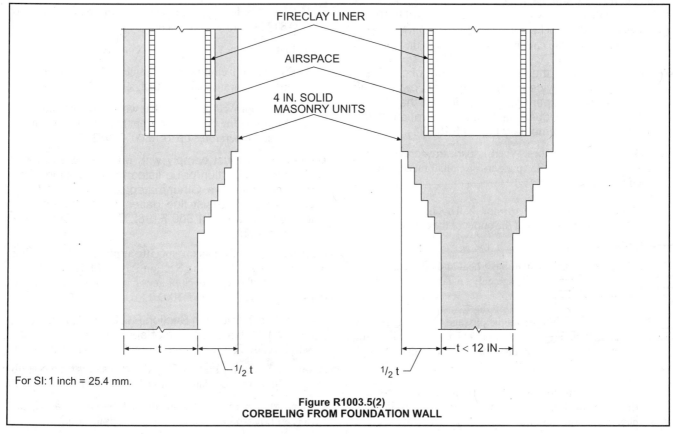

For SI: 1 inch = 25.4 mm.

Figure R1003.5(2)
CORBELING FROM FOUNDATION WALL

line of the flue to the centerline of the wall of the chimney below maintains structural stability for vertical loads. The slope of the offset is limited to that permitted for corbels in Section R1003.5. The offset limitation does not apply in cases where the offset is supported by masonry below the offset and in a manner approved by the building official.

R1003.8 Additional load. Chimneys shall not support loads other than their own weight unless they are designed and constructed to support the additional load. Construction of masonry chimneys as part of the masonry walls or reinforced concrete walls of the building shall be permitted.

❖ Chimneys are subject to considerable stresses resulting from thermal effects and therefore should not support any structural load other than their own weight, unless specifically designed as a supporting member for the additional load. Also, because of its heavy mass, a chimney will tend to settle more than the building structures. As a consequence, the chimney and any part of the building that it supports will settle at a greater rate and to a greater degree than the rest of the building, resulting in damage. If a chimney is subject to structural loads for which it is not designed, the additional stresses created could lead to cracks in the chimney. Thus, the chimney would become a hazard because of the potential of flames from the firebox penetrating the cracks and igniting combustible construction. Moreover, buildups of the products of combustion that leak through the cracks create hazardous conditions inside the building.

R1003.9 Termination. Chimneys shall extend at least 2 feet (610 mm) higher than any portion of a building within 10 feet (3048 mm), but shall not be less than 3 feet (914 mm) above the highest point where the chimney passes through the roof.

❖ The provisions for the termination height of chimneys above the roof provide for the necessary upward draft in the chimney. Experience has indicated that the required heights produce satisfactory operation of the chimney (see Commentary Figure R1003.2).

R1003.9.1 Spark arrestors. Where a spark arrestor is installed on a masonry chimney, the spark arrestor shall meet all of the following requirements:

1. The net free area of the arrestor shall not be less than four times the net free area of the outlet of the chimney flue it serves.

2. The arrestor screen shall have heat and corrosion resistance equivalent to 19-gage galvanized steel or 24-gage stainless steel.

3. Openings shall not permit the passage of spheres having a diameter greater than $^1/_2$ inch (13 mm) nor block the passage of spheres having a diameter less than $^3/_8$ inch (10 mm).

4. The spark arrestor shall be accessible for cleaning and the screen or chimney cap shall be removable to allow for cleaning of the chimney flue.

❖ This section contains specifications for spark arrestors, if they are provided. Their use is not mandated by the code, but owners and builders often install them.

R1003.10 Wall thickness. Masonry chimney walls shall be constructed of solid masonry units or hollow masonry units grouted solid with not less than a 4-inch (102 mm) nominal thickness.

❖ Masonry chimney walls must have a minimum nominal thickness of 4 inches (102 mm) of either solid masonry units or hollow masonry grouted solid for structural stability.

R1003.10.1 Masonry veneer chimneys. Where masonry is used to veneer a frame chimney, through-flashing and weep holes shall be installed as required by Section R703.

❖ Masonry veneer is a nonstructural facing material which provides ornamentation, protection or insulation. To be considered a veneer, the material cannot act structurally with the backing insofar as the structural strength of the assembly is concerned.

R1003.11 Flue lining (material). Masonry chimneys shall be lined. The lining material shall be appropriate for the type of appliance connected, according to the terms of the appliance listing and manufacturer's instructions.

❖ Masonry chimneys must be properly lined to allow for the smooth flow of products of combustion. This section regulates specific flue linings.

R1003.11.1 Residential-type appliances (general). Flue lining systems shall comply with one of the following:

1. Clay flue lining complying with the requirements of ASTM C 315 or equivalent.

2. Listed chimney lining systems complying with UL 1777.

3. Factory-built chimneys or chimney units listed for installation within masonry chimneys.

4. Other approved materials that will resist corrosion, erosion, softening or cracking from flue gases and condensate at temperatures up to 1,800°F (982°C).

❖ Flue linings must comply with one of the standards listed in this section or be listed for installation within masonry chimneys. Other materials may be used, provided they will resist flue gases and condensate at temperatures up to 1,800°F (982°C) and will not crack, soften or corrode.

R1003.11.2 Flue linings for specific appliances. Flue linings other than these covered in Section R1003.11.1, intended for use with specific types of appliances, shall comply with Sections R1003.11.3 through R1003.11.6.

❖ Except as provided in Section R1003.11.1, flue linings for use with specific types of appliances must comply with Section R1003.11.3 through Section R1003.11.6.

R1003.11.3 Gas appliances. Flue lining systems for gas appliances shall be in accordance with Chapter 24.

❖ Chapter 24 of the *International Residential Code*® (IRC®) covers flue linings for gas appliances.

R1003.11.4 Pellet fuel-burning appliances. Flue lining and vent systems for use in masonry chimneys with pellet fuel-burning appliances shall be limited to the following:

1. Flue lining systems complying with Section R1003.11.1.

2. Pellet vents listed for installation within masonry chimneys. (See Section R1003.11.6 for marking.)

❖ This section's provisions limit flue lining and vent systems.

R1003.11.5 Oil-fired appliances approved for use with Type L vent. Flue lining and vent systems for use in masonry chimneys with oil-fired appliances approved for use with Type L vent shall be limited to the following:

1. Flue lining systems complying with Section R1003.11.1.

2. Listed chimney liners complying with UL 641. (See Section R1003.11.6 for marking.)

❖ Flue lining and vent systems for use in masonry chimneys with oil-fired appliances approved for use with a Type L vent are limited to either flue lining systems complying with Section R1003.11.1 or listed chimney liners complying with UL 641.

R1003.11.6 Notice of usage. When a flue is relined with a material not complying with Section R1003.11.1, the chimney shall be plainly and permanently identified by a label attached to a wall, ceiling or other conspicuous location adjacent to where the connector enters the chimney. The label shall include the following message or equivalent language:

THIS CHIMNEY FLUE IS FOR USE ONLY WITH [TYPE OR CATEGORY OF APPLIANCE] APPLIANCES THAT BURN [TYPE OF FUEL]. DO NOT CONNECT OTHER TYPES OF APPLIANCES.

❖ Flues relined with materials not conforming to Section R1003.11.1 must be permanently identified in a conspicuous location adjacent to the connector's entrance into the building. Thus, current and future occupants of the building will know the type of appliance that may be attached to the flue.

R1003.12 Clay flue lining (installation). Clay flue liners shall be installed in accordance with ASTM C 1283 and extend from a point not less than 8 inches (203 mm) below the lowest inlet or, in the case of fireplaces, from the top of the smoke chamber to a point above the enclosing walls. The lining shall be carried up vertically, with a maximum slope no greater than 30 degrees (0.52 rad) from the vertical.

Clay flue liners shall be laid in medium-duty refractory mortar conforming to ASTM C 199 with tight mortar joints left smooth on the inside and installed to maintain an air space or insulation not to exceed the thickness of the flue liner separating the flue liners from the interior face of the chimney masonry walls. Flue liners shall be supported on all sides. Only enough mortar shall be placed to make the joint and hold the liners in position.

❖ Clay flue linings must be installed so there is no leakage through the liner to adjacent spaces, which would result in contaminated air. Flue linings must be installed in accordance with ASTM C 1283, beginning at

least 8 inches (203 mm) below the lowest inlet and extending to a point above the top of the chimney walls. Masonry chimneys atop masonry fireplaces should have the lining start at the top of the smoke chamber of the fireplace. If the chimney is corbeled, the lining may not slope more than 30 degrees (0.52 rad) from the vertical. A vent space is required around the lining and is not to be obstructed by the mortar used to set the liner. Only the amount of mortar needed to set the liners and hold them in place should be used. This air space will help limit the transfer of the heat build-up in the chimney to the masonry wall. This air space cannot be used for venting other appliances (see Commentary Figure R1003.12).

R1003.12.1 Listed materials. Listed materials used as flue linings shall be installed in accordance with the terms of their listings and manufacturer's instructions.

❖ Materials used as flue liners must be installed per the manufacturer's listing and instructions so the flue gases will be properly removed from the structure and not leak back in.

R1003.12.2 Space around lining. The space surrounding a chimney lining system or vent installed within a masonry chimney shall not be used to vent any other appliance.

Exception: This shall not prevent the installation of a separate flue lining in accordance with the manufacturer's installation instructions.

❖ The open space around a chimney lining system or vent that is installed within a masonry chimney is not to be used to vent any other appliance. This unused space is otherwise not designed or constructed to vent flue gases.

R1003.13 Multiple flues. When two or more flues are located in the same chimney, masonry wythes shall be built between adjacent flue linings. The masonry wythes shall be at least 4 inches (102 mm) thick and bonded into the walls of the chimney.

Exception: When venting only one appliance, two flues may adjoin each other in the same chimney with only the flue lining separation between them. The joints of the adjacent flue linings shall be staggered at least 4 inches (102 mm).

❖ Each separate fireplace should have its own flue. Even outdoor barbecues connected to a chimney should have their own separate flue. Flues in the same chimney must be separated by a 4-inch (102 mm) masonry wythe bonded into the walls of the chimney. Failure to properly separate flues within a common chimney could result in transfer of smoke or other products of combustion from one flue to the other as a result of the down-draft created by interior suction of the inactive flue. The draft could also be created by external wind effects forcing the smoke down the inactive flues as it exhausts the active flue. Two flues in the same chimney venting only one appliance need not be separated from each other; however, the flue joints must be staggered at least 4 inches (102 mm) (see Commentary Figure R1003.12).

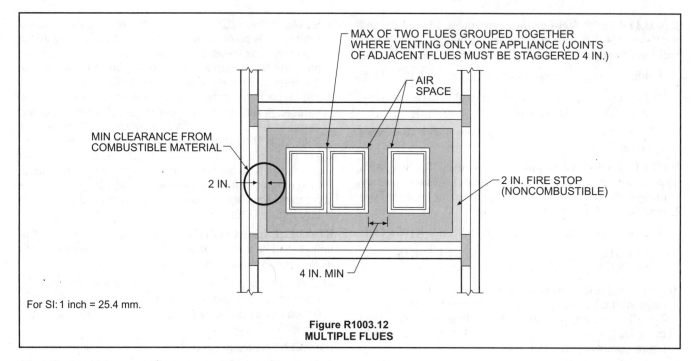

MAX OF TWO FLUES GROUPED TOGETHER
WHERE VENTING ONLY ONE APPLIANCE (JOINTS
OF ADJACENT FLUES MUST BE STAGGERED 4 IN.)

AIR
SPACE

MIN CLEARANCE FROM
COMBUSTIBLE MATERIAL

2 IN.

2 IN. FIRE STOP
(NONCOMBUSTIBLE)

4 IN. MIN

For SI: 1 inch = 25.4 mm.

Figure R1003.12
MULTIPLE FLUES

R1003.14 Flue area (appliance). Chimney flues shall not be smaller in area than that of the area of the connector from the appliance [see Tables R1003.14(1) and R1003.14(2)]. The sizing of a chimney flue to which multiple appliance venting systems are connected shall be in accordance with Section M1805.3.

❖ Tables R1001.14(1) and R1001.14(2) determine the sizes of chimney flues for appliances. In no case may the flue of the chimney have a smaller area than the area of the connector from the appliance. If the appliance is connected to a flue smaller than the appliance's flue, a back draft can occur, which will result in the appliance not operating properly and the products of combustion not being vented to the outside.

R1003.15 Flue area (masonry fireplace). Flue sizing for chimneys serving fireplaces shall be in accordance with Section R1003.15.1 or Section R1003.15.2.

❖ There are two methods for determining the flue size for masonry fireplaces. Option 1 uses a fraction of the area of the fireplace opening. Option 2 allows the use of a chart in determining the flue size.

R1003.15.1 Option 1. Round chimney flues shall have a minimum net cross-sectional area of at least $\frac{1}{12}$ of the fireplace opening. Square chimney flues shall have a minimum net cross-sectional area of $\frac{1}{10}$ of the fireplace opening. Rectangular chimney flues with an aspect ratio less than 2 to 1 shall have a minimum net cross-sectional area of $\frac{1}{10}$ of the fireplace opening. Rectangular chimney flues with an aspect ratio of 2 to 1 or more shall have a minimum net cross-sectional area of $\frac{1}{8}$ of the fireplace opening. Cross-sectional areas of clay flue linings are shown in Tables R1001.14(1) and R1001.14(2) or as provided by the manufacturer or as measured in the field.

❖ Option 1 indicates the use of different fractional amounts of the fireplace opening depending on the shape of the flue. Round chimney flues must have a net cross-sec-

tional area of $\frac{1}{12}$ of the fireplace opening, and square flues must have a net cross-sectional area of $\frac{1}{10}$ of the fireplace opening. Rectangular flues need either $\frac{1}{8}$ or $\frac{1}{10}$ of the net cross-sectional area of the fireplace opening depending on the aspect ratio of the flue. These fractional amounts vary because the effective flue area (EFA) of a square, smooth flue is equal to that of a smooth, round flue where the diameter of the round flue is the same as one side of the square flue (see Commentary Figure R1003.15.1). The crosshatched areas within the flue of the square or rectangular vent provide no draft.

TABLE R1003.14(1)
NET CROSS-SECTIONAL AREA OF ROUND FLUE SIZES[a]

FLUE SIZE, INSIDE DIAMETER (inches)	CROSS–SECTIONAL AREA (square inches)
6	28
7	38
8	50
10	78
$10^{3}/_{4}$	90
12	113
15	176
18	254

For SI: 1 inch = 25.4 mm, 1 square inch = 645.16 mm².
a. Flue sizes are based on ASTM C 315.

❖ Table R1003.14(1) lists the net cross-sectional area of round flues. The first column provides the interior flue diameter; the second column provides the cross-sectional area. For example, an 8-inch (203 mm) round flue requires a cross-sectional area of at least 50 square inches (32 258 mm²). Flue sizes are based on ASTM C 315.

TABLE R1003.14(2)
NET CROSS-SECTIONAL AREA OF SQUARE AND RECTANGULAR FLUE SIZES

FLUE SIZE, OUTSIDE NOMINAL DIMENSIONS (inches)	CROSS–SECTIONAL AREA (square inches)
4.5 × 8.5	23
4.5 × 13	34
8 × 8	42
8.5 × 8.5	49
8 × 12	67
8.5 × 13	76
12 × 12	102
8.5 × 18	101
13 × 13	127
12 × 16	131
13 × 18	173
16 × 16	181
16 × 20	222
18 × 18	233
20 × 20	298
20 × 24	335
24 × 24	431

For SI: 1 inch = 25.4 mm, 1 square inch = 645.16 mm².

❖ Table R1003.14(2) lists the net cross-sectional area of rectangular flues. The first column provides the outside dimensions of the flue. The second column provides the cross sectional area for the flue. For example, a rectangular flue with an exterior dimension of 11½ inches by 15½ inches (292 mm by 394 mm) requires a net cross-sectional area of at least 124 square inches (80 000 mm²). Flue sizes are based on ASTM C 315.

R1003.15.2 Option 2. The minimum net cross-sectional area of the chimney flue shall be determined in accordance with Figure R1003.15.2. A flue size providing at least the equivalent net cross-sectional area shall be used. Cross-sectional areas of clay flue linings are shown in Tables R1003.14(1) and R1003.14(2) or as provided by the manufacturer or as measured in the field. The height of the chimney shall be measured from the firebox floor to the top of the chimney flue.

❖ Option 2 entails using the chart in Figure R1003.15.2 to determine the required flue area. The flue must have the equivalent net cross-sectional area as determined by the chart. The cross-sectional area of clay flue linings is given in Tables R1003.14(1) and R1003.14(2) or as provided by the manufacturer. Alternately, the cross-sectional area of the flue may be field measured.

R1003.16 Inlet. Inlets to masonry chimneys shall enter from the side. Inlets shall have a thimble of fireclay, rigid refractory material or metal that will prevent the connector from pulling out of the inlet or from extending beyond the wall of the liner.

❖ Chimney inlet thimbles must not project into the flue. Inlets must have a thimble of fire clay, rigid refractory material or other metal. Thimbles must be installed so they will not dislodge from the inlet opening (see Commentary Figure R1003.16).

R1003.17 Masonry chimney cleanout openings. Cleanout openings shall be provided within 6 inches (152 mm) of the base of each flue within every masonry chimney. The upper edge of the cleanout shall be located at least 6 inches (152 mm) below the lowest chimney inlet opening. The height of the opening shall be at least 6 inches (152 mm). The cleanout shall be provided with a noncombustible cover.

Exception: Chimney flues serving masonry fireplaces where cleaning is possible through the fireplace opening.

❖ The code requires that cleanout openings be provided within 6 inches (152 mm) of the base of the flue. Addi-

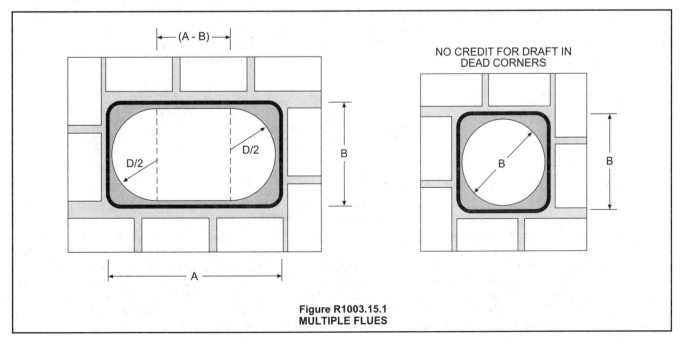

Figure R1003.15.1
MULTIPLE FLUES

tionally, the upper edge of the cleanout opening must be at least 6 inches (152 mm) below the lowest chimney outlet. The cleanout openings should be at least 6 inches (152 mm) high and have a noncombustible cover. The interior of the chimney may become coated with carbonaceous deposits and acids, and as a result the brick in the mortar will be chemically attacked, causing a gradual deterioration. The deposits will fall to the bottom of the chimney, and a cleanout will be necessary to remove them. Additionally, chimney-cleaning operations, which should be done periodically, will result in the accumulation of deposits at the base of the chimney, necessitating a cleanout opening. In the case of a fireplace chimney, the fireplace opening itself provides the cleanout, and no further opening is required.

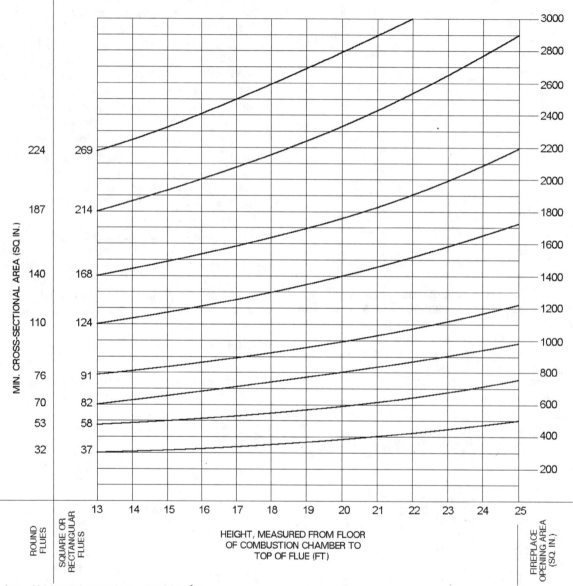

For SI: 1 foot = 304.8 mm, 1 square inch = 645.16 mm².

FIGURE R1003.15.2
FLUE SIZES FOR MASONRY CHIMNEYS

❖ Figure R1003.15.2 is used to determine the minimum flue size required for masonry chimneys. Two factors contribute to the determination of the flue size: 1) the height of the flue measured from the floor of the combustion chamber to the top of the flue, and 2) the fireplace opening dimension. In using the chart, one would go to the intersection of the x-y axis of the flue height and the fireplace opening size and then to the curved line above. Follow the curved line to the left to determine the minimum flue size. For example, for a flue with a height of 20 feet (6096 mm) and a fireplace opening of 1200 square inches (0.77 m²), the required net cross-sectional area would be 110 square inches (70 968 mm²) for a round flue or 124 square inches (80 000 mm²) for a rectangular flue.

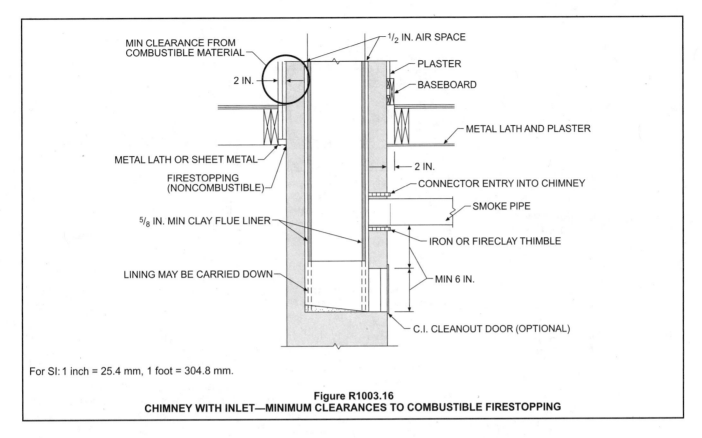

For SI: 1 inch = 25.4 mm, 1 foot = 304.8 mm.

Figure R1003.16
CHIMNEY WITH INLET—MINIMUM CLEARANCES TO COMBUSTIBLE FIRESTOPPING

R1003.18 Chimney clearances. Any portion of a masonry chimney located in the interior of the building or within the exterior wall of the building shall have a minimum air space clearance to combustibles of 2 inches (51 mm). Chimneys located entirely outside the exterior walls of the building, including chimneys that pass through the soffit or cornice, shall have a minimum air space clearance of 1 inch (25 mm). The air space shall not be filled, except to provide fire blocking in accordance with Section R1003.19.

Exceptions:

1. Masonry chimneys equipped with a chimney lining system listed and labeled for use in chimneys in contact with combustibles in accordance with UL 1777 and installed in accordance with the manufacturer's installation instructions are permitted to have combustible material in contact with their exterior surfaces.

2. When masonry chimneys are constructed as part of masonry or concrete walls, combustible materials shall not be in contact with the masonry or concrete wall less than 12 inches (305 mm) from the inside surface of the nearest flue lining.

3. Exposed combustible trim and the edges of sheathing materials, such as wood siding and flooring, shall be permitted to abut the masonry chimney side walls, in accordance with Figure R1003.18, provided such combustible trim or sheathing is a minimum of 12 inches (305 mm) from the inside surface of the nearest flue lining. Combustible material and trim shall not overlap the corners of the chimney by more than 1 inch (25 mm).

❖ Because of the passage of hot gases and products of combustion, the chimney becomes heated. The code requires that combustibles not be placed directly against the chimney to avoid potential heat transfer and eventual combustion of the combustibles. Masonry chimneys located on the interior of a structure or within an exterior wall require a 2-inch (51 mm) air-space clearance to combustibles. Masonry chimneys located entirely outside of the exterior walls are permitted to have a 1-inch (25.4 mm) clearance to combustibles. See Figure R1003.18. Three exceptions permit reduced clearances.

R1003.19 Chimney fireblocking. All spaces between chimneys and floors and ceilings through which chimneys pass shall be fireblocked with noncombustible material securely fastened in place. The fireblocking of spaces between chimneys and wood joists, beams or headers shall be self-supporting or be placed on strips of metal or metal lath laid across the spaces between combustible material and the chimney.

❖ As in concealed spaces in wood frame construction, spaces such as floors and ceilings that chimneys pass through must be fireblocked. Section R1003.18 prohibits chimneys from abutting combustible material. Therefore, the 1-inch (25.4 mm) to 2-inch (51 mm) air space required to separate the chimney from combustible material must be fireblocked. The fireblocking must be of noncombustible material to a depth of 1 inch (25.4 mm) and be supported by noncombustible material such as metal lath.

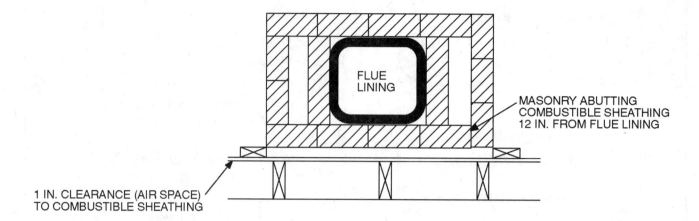

For SI: 1 inch = 25.4 mm.

FIGURE R1003.18
CLEARANCE FROM COMBUSTIBLES

R1003.20 Chimney crickets. Chimneys shall be provided with crickets when the dimension parallel to the ridgeline is greater than 30 inches (762 mm) and does not intersect the ridgeline. The intersection of the cricket and the chimney shall be flashed and counterflashed in the same manner as normal roof-chimney intersections. Crickets shall be constructed in compliance with Figure R1003.20 and Table R1003.20.

❖ Chimney crickets are required when the chimney width parallel to the ridgeline is more than 30 inches (762 mm), provided the chimney does not intersect the ridgeline. The cricket diverts the flow of water from the roof around the chimney. Flashing is required between the cricket and the chimney, as it is in other roof-chimney intersections.

TABLE R1003.20
CRICKET DIMENSIONS

ROOF SLOPE	H
12 - 12	$\frac{1}{2}$ of W
8 - 12	$\frac{1}{3}$ of W
6 - 12	$\frac{1}{4}$ of W
4 - 12	$\frac{1}{6}$ of W
3 - 12	$\frac{1}{8}$ of W

❖ Table R1003.20 establishes the height of chimney crickets as depicted in Figure R1003.20. For example, a cricket installed at the face of a chimney that is 4 feet wide (1219 mm) on a 6:12 roof slope, would need to be 1 foot high (305 mm) ($\frac{1}{4}$ of 4 feet).

SECTION R1004
FACTORY-BUILT FIREPLACES

R1004.1 General. Factory-built fireplaces shall be listed and labeled and shall be installed in accordance with the conditions of the listing. Factory-built fireplaces shall be tested in accordance with UL 127.

❖ The code requires that factory-built fireplaces be listed and labeled. In addition, factory-built fireplaces must be tested by an approved testing lab in accordance with UL 127 (see commentary, Section R1005).

R1004.2 Hearth extensions. Hearth extensions of approved factory-built fireplaces shall be installed in accordance with the listing of the fireplace. The hearth extension shall be readily distinguishable from the surrounding floor area.

❖ The primary requirements here are that the hearth extension must be installed in accordance with the listing of the fireplace, and it must be easily distinguished from the surrounding floor areas.

R1004.3 Decorative shrouds. Decorative shrouds shall not be installed at the termination of chimneys for factory-built fireplaces except where the shrouds are listed and labeled for use with the specific factory-built fireplace system and installed in accordance with the manufacturer's installation instructions.

❖ These provisions are the same as found in Section R1005.2 (see commentary, Section R1005.2).

R1004.4 Unvented gas log heaters. An unvented gas log heater shall not be installed in a factory-built fireplace unless the fireplace system has been specifically tested, listed and labeled for such use in accordance with UL 127.

❖ Unvented gas log heaters must not be installed in factory-built fireplaces if the fireplace has not been specifically tested, listed, and labeled for such use. Without such testing, there is no way to know how the unvented gas log heater will function in the fireplace, and it may cause harm to both the appliance and the occupants of the building.

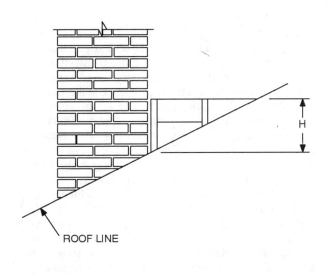
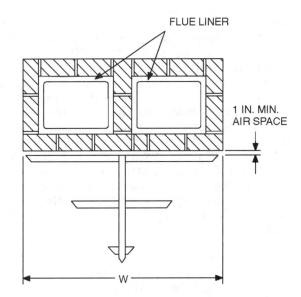

FLUE LINER

H

ROOF LINE

1 IN. MIN.
AIR SPACE

W

For SI: 1 inch = 25.4 mm.

FIGURE R1003.20
CHIMNEY CRICKET

❖ Figure R1003.20 depicts the height of chimney crickets and the clearance of combustibles to the chimney.

SECTION R1005
FACTORY-BUILT CHIMNEYS

R1005.1 Listing. Factory-built chimneys shall be listed and labeled and shall be installed and terminated in accordance with the manufacturer's installation instructions.

❖ Factory-built fireplaces can be used if they are listed and labeled and are installed following the manufacturer's installation instructions. "Listed" and "Labeled" are defined in Chapter 2.

R1005.2 Decorative shrouds. Decorative shrouds shall not be installed at the termination of factory-built chimneys except where the shrouds are listed and labeled for use with the specific factory-built chimney system and installed in accordance with the manufacturer's installation instructions.

❖ In general, decorative shrouds are not permitted on top of factory-built chimneys. Installing a shroud on top of a chimney that is not designed for one may result in improper venting of the appliance attached to the chimney. Decorative shrouds are permitted if they are listed and labeled for use with the specific chimney and are installed in accordance with the manufacturers' installation instructions.

R1005.3 Solid-fuel appliances. Factory-built chimneys installed in dwelling units with solid-fuel-burning appliances shall comply with the Type HT requirements of UL 103 and shall be marked "Type HT and "Residential Type and Building Heating Appliance Chimney."

Exception: Chimneys for use with open combustion chamber fireplaces shall comply with the requirements of UL 103 and shall be marked "Residential Type and Building Heating Appliance Chimney."

Chimneys for use with open combustion chamber appliances installed in buildings other than dwelling units shall comply with the requirements of UL 103 and shall be marked "Building Heating Appliance Chimney" or "Residential Type and Building Heating Appliance Chimney."

❖ Factory-built chimneys must withstand the high temperatures associated with solid fuel burning. For that reason the chimney must be clearly marked as Type HT (High Temperature) with the language "Residential Type and Building Heating Appliance Chimney." Factory-built chimneys for use with solid-fuel-burning appliances must comply with Type HT requirements of UL 103. The exception allows these chimneys to be used with open combustion chamber fireplace stoves that comply with UL 103.

R1005.4 Factory-built fireplaces. Chimneys for use with factory-built fireplaces shall comply with the requirements of UL 127.

❖ Chimneys installed and used for factory-built fireplaces must comply with UL 127.

R1005.5 Support. Where factory-built chimneys are supported by structural members, such as joists and rafters, those members shall be designed to support the additional load.

❖ Joists and rafters supporting factory-built chimneys must be designed to carry the load of the chimney in addition to other loads that the members carry.

R1005.6 Medium-heat appliances. Factory-built chimneys for medium-heat appliances producing flue gases having a temperature above 1,000°F (538°C), measured at the entrance to the chimney shall comply with UL 959.

❖ Factory-built chimneys for medium-heat appliances that produce flue gases with a temperature above 1,000°F (538°C) at the entry to the chimney must comply with UL 959.

Factory-built fireplaces are permitted when they are listed and labeled and are installed in accordance with the manufacturer's installation instructions. The word "approved" is defined in Chapter 2. Both factory-built fireplaces and chimneys for solid-fuel-burning appliances must meet the applicable UL standard as listed in the code. Agencies such as UL provide listing services for factory-built fireplaces and chimneys. The listing includes a description of the manufactured unit and requirements for clearance from combustible materials, hearth details if necessary, and other limitations placed on the unit to result in a safe installation. Factory-built fireplaces are installed in conjunction with factory-built chimneys; thus, the complete installation of fireplace and chimney is factory built (see Commentary Figure R1005.6).

SECTION R1006
EXTERIOR AIR SUPPLY

R1006.1 Exterior air. Factory-built or masonry fireplaces covered in this chapter shall be equipped with an exterior air supply to assure proper fuel combustion unless the room is mechanically ventilated and controlled so that the indoor pressure is neutral or positive.

❖ An adequate supply of combustion air must be provided for the fireplace. The code requires that both masonry and factory-built fireplaces have an exterior air supply. The air ducts used in either type of fireplace must be listed.

R1006.1.1 Factory-built fireplaces. Exterior combustion air ducts for factory-built fireplaces shall be a listed component of the fireplace and shall be installed according to the fireplace manufacturer's instructions.

❖ An exterior combustion air supply for factory-built fireplaces must be listed components, and the ducts must be installed in accordance with the manufacturer's instructions.

R1006.1.2 Masonry fireplaces. Listed combustion air ducts for masonry fireplaces shall be installed according to the terms of their listing and the manufacturer's instructions.

❖ If a masonry fireplace uses a listed combustion air duct, the air duct must be installed per its listing.

R1006.2 Exterior air intake. The exterior air intake shall be capable of supplying all combustion air from the exterior of the dwelling or from spaces within the dwelling ventilated with outside air such as non-mechanically ventilated crawl or attic spaces. The exterior air intake shall not be located within the garage or basement of the dwelling nor shall the air intake be located at an elevation higher than the firebox. The exterior air intake shall be covered with a corrosion-resistant screen of $^1/_4$-inch (6 mm) mesh.

❖ The exterior air supply must be capable of providing all of the necessary combustion air. This air must be taken from the exterior of the building or from spaces such as crawl or attic spaces that are adequately ventilated. The crawl or attic spaces cannot be mechanically ventilated. Attic and/or crawl space mechanical ventilating systems are primarily used to remove air from those areas.

They do this by exhausting the unwanted air or creating a negative pressure in those areas. If an air intake for a fireplace terminates in a crawl or attic space that has a mechanical ventilation system there is a potential for the air intake to perform exactly opposite of its designed intent. The exterior air intake cannot be in the garage or basement of the dwelling unit. The air intake must be lower than the fire box so that the firebox will properly draw in the combustion air. Where combustion air openings are located inside the firebox, the air intake opening on the outside of the dwelling cannot be located higher than the firebox. Such an installation could create a chimney effect, drawing the products of combustion up through the combustion air ducts. These ducts are not generally constructed of materials which can withstand the heat and sparks that could be drawn through them. Interior combustion air openings located outside the firebox are not subject to this requirement. The exterior air intake must be covered with a corrosion-resistant screen of $^1/_4$-inch (6.4 mm) mesh.

R1006.3 Clearance. Unlisted combustion air ducts shall be installed with a minimum 1-inch (25 mm) clearance to combustibles for all parts of the duct within 5 feet (1524 mm) of the duct outlet.

❖ Where masonry fireplaces use unlisted combustion air ducts, the ducts must be installed with a minimum clearance of 1 inch (25.4 mm) to combustibles for all parts of the ducts within 5 feet (1524 mm) of the outlets.

R1006.4 Passageway. The combustion air passageway shall be a minimum of 6 square inches (3870 mm²) and not more than 55 square inches (0.035 m²), except that combustion air systems for listed fireplaces shall be constructed according to the fireplace manufacturer's instructions.

❖ The combustion air passageway must be a minimum of 6 square inches (3870 mm²) and no more than 55 square inches (0.035 m²). Combustion air systems for listed fireplaces must have combustion air passageways in a size that is established by the manufacturer, which may be different from the dimensions indicated in this section. Combustion passageways supply combustion air to the firebox without relying on the air within the structure. Without this combustion air, the oxygen in the structure could be depleted to a level unacceptable for the occupants of the building.

R1006.5 Outlet. Locating the exterior air outlet in the back or sides of the firebox chamber or within 24 inches (610 mm) of the firebox opening on or near the floor is permitted. The outlet shall be closable and designed to prevent burning material from dropping into concealed combustible spaces.

❖ The combustion air opening may be located in the back or sides of the firebox chamber, or it may be located within 24 inches (610 mm) of the fire box opening on or near the floor. If the opening is on or near the floor, the outlet must be closeable and designed to prevent burning material from dropping into concealed combustible spaces. A noncombustible screen could be used for this purpose, if the mesh is fine enough to prevent burning embers from penetrating the screen.

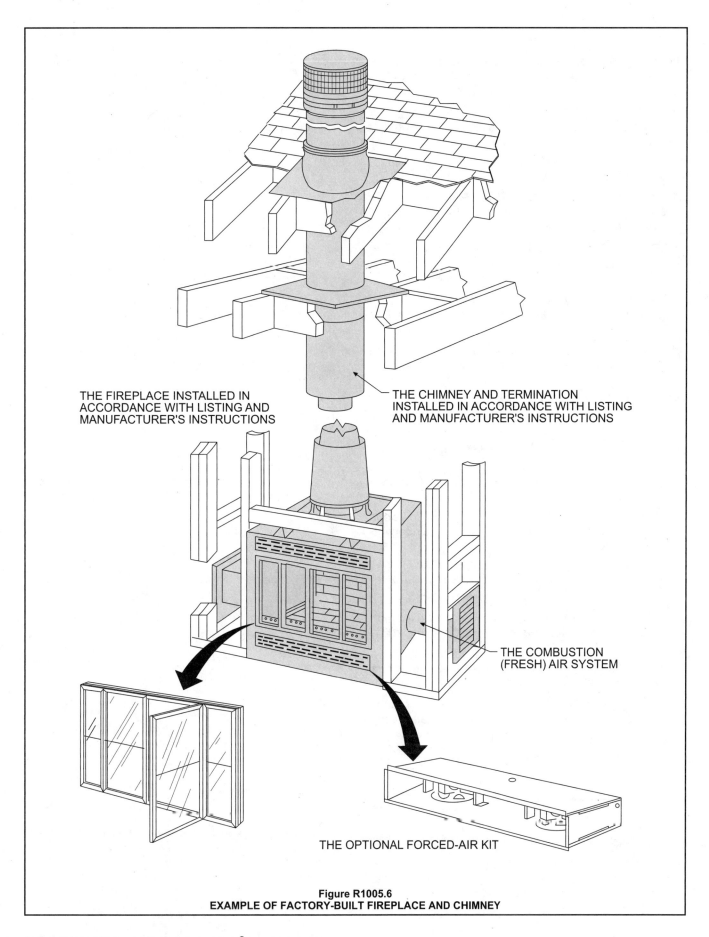

THE FIREPLACE INSTALLED IN ACCORDANCE WITH LISTING AND MANUFACTURER'S INSTRUCTIONS

THE CHIMNEY AND TERMINATION INSTALLED IN ACCORDANCE WITH LISTING AND MANUFACTURER'S INSTRUCTIONS

THE COMBUSTION (FRESH) AIR SYSTEM

THE OPTIONAL FORCED-AIR KIT

Figure R1005.6
EXAMPLE OF FACTORY-BUILT FIREPLACE AND CHIMNEY

Bibliography

The following resource materials are referenced in this chapter or are relevant to the subject matter addressed in this chapter.

ASTM C 27-98 (2002), *Classification of Fireclay and High-Alumina Refractory Brick.* West Conshohocken, PA: ASTM International, 2002.

ASTM C 315-02, *Specification for Clay Flue Linings.* West Conshohocken, PA: ASTM International, 2002.

ASTM C 1261-04, *Specification for Firebox Brick for Residential Fireplaces.* West Conshohocken, PA: ASTM International, 2004.

ASTM C 1283-03e01, *Practice for Installing Clay Flue Lining.* West Conshohocken, PA: ASTM International, 2003.

ASTM E 1602-03, *Guide for Construction of Solid Fuel Burning Masonry Heaters.* West Conshohocken, PA: ASTM International, 2003.

Amrhein, S. E. *Residential Masonry and Chimney Handbook.* Los Angeles, CA: Masonry Institute of America, 1989.

Olin, Howard B. *Construction Principles, Materials and Methods.* Chicago, IL: The Institute of Financial Education, 1983.

UL 103-01, *Factory-Built Chimneys for Residential Type and Building Heating Appliances—with Revisions through March 1999.* Northbrook, IL: Underwriters Laboratory, 2001.

UL 127-99, *Factory-Built Fireplaces—with Revisions through November 1999.* Northbrook, IL: Underwriters Laboratory, 1999.

UL 641-95, *Type L, Low-Temperature Venting Systems-with Revisions through April 1999*, Type L. Northbrook, IL: Underwriters Laboratory, 1995.

UL 959-01, *Medium-Heat Appliance Factory-Built Chimneys.* Northbrook, IL: Underwriters Laboratory, 2001.

UL 1482-98, *Solid-fuel Type Room Heaters with Revisions through July 2000.* Northbrook, IL: Underwriters Laboratory 2000.

UL 1777-04, *Chimney Liners—with Revisions through July 1998.* Northbrook, IL: Underwriters Laboratory, 2004.

Chapter 11: Energy Efficiency

General Comments

Chapter 11 contains the energy-efficiency-related requirements for the design and construction of buildings regulated under this code. The applicable portions of the building must comply with the provisions within this chapter for energy efficiency.

Section N1101 contains the scope and application of the chapter and also regulates material identification and labeling. Section N1102 contains the insulation *R*-value requirements and the window *U*-factor requirements for the building envelope, which includes the roof/ceiling assembly, wall assembly and floor assembly. Section 1103 contains the systems requirements for heating and cooling systems and includes requirements for equipment efficiency, duct installation, piping insulation and the requirements for service water heating performance.

Purpose

This chapter defines requirements for the portions of the building and building systems that impact energy use in new construction and promotes the effective use of energy. The provisions within the chapter promote energy efficiency in the building envelope, the heating and cooling system and the service water heating system of the building. Compliance with the chapter will provide a minimum level of energy efficiency for new construction. Greater levels of efficiency can be installed to decrease the energy use of new construction.

SECTION N1101
GENERAL

N1101.1 Scope. This chapter regulates the energy efficiency for the design and construction of buildings regulated by this code.

> **Exception:** Portions of the building envelope that do not enclose conditioned space.

❖ Chapter 11 applies to portions of the building thermal envelope that enclose conditioned space as shown in Commentary Figure N1101.1(1). Conditioned space is the area provided with heating and/or cooling either directly, through a positive heating/cooling supply system such as registers located in the space, or indirectly through an opening that allows heated or cooled air to communicate directly with the space. For example, a walk-in closet connected to a master bedroom suite may not contain a positive heating supply through a register, but it would be conditioned indirectly by the free passage of heated or cooled air into the space from the bedroom.

The exception exempts areas that are separated from the conditioned spaces of the building by the building envelope from the building envelope requirements such as insulation and fenestration. A good example of this would be an unconditioned garage or attic space. In the case of a garage, if the unconditioned garage area is separated from the conditioned portions of the residence by an assembly that meets the "building envelope" criteria (meaning that the wall between them is insulated), the exterior walls of the garage would not need to be insulated to separate the garage from the exterior climate.

The building thermal envelope consists of the wall, roof/ceiling and floor assemblies that surround the conditioned space. Raised floors over a crawlspace or garage or directly exposed to the outside air are considered to be part of the floor assembly. Walls surrounding a conditioned basement (in addition to surrounding conditioned spaces above grade) are part of the building envelope. The code defines above-grade walls surrounding conditioned spaces as exterior walls. This definition includes walls between the conditioned space and unconditioned garage, roof and basement knee walls, dormer walls, gable end walls, walls enclosing a mansard roof and basement walls with an average below-grade area that is less than 50 percent of the total basement gross wall area. This definition would not include walls separating an unconditioned garage from the outdoors. The roof/ceiling assembly is the surface where insulation will be installed, typically on top of the gypsum board [see commentary, Figure N1101.1(2)].

N1101.2 Compliance. Compliance shall be demonstrated by either meeting the requirements of the *International Energy Conservation Code* or meeting the requirements of this chapter. Climate zones from Figure N1101.2 or Table N1101.2 shall be used in determining the applicable requirements from this chapter.

❖ The *International Residential Code®* (IRC®) allows compliance with either Chapter 11 of the IRC or the provisions of the *International Energy Conservation Code®* (IECC®) to be used. A code user may evaluate both options and use the one that fits the project best, as these two differing methods can result in different requirements. Most requirements are given prescriptively. Two alternative "tradeoffs" are specified for many requirements, especially for the building thermal envelope requirements. For requirements specified by *U*-factors, an overall UA (*U*-factor times area) can be used to show equivalence. A performance-based annual energy calculation can also be met by showing overall energy equivalence.

Figure N1101.2 of the IRC shows the various Climate Zones in the country. The 2003 code's original 19 Climate Zones based on Heating Degree Days (HDD) have been combined, reorganized and changed into eight climate zones based on multiple climate variables. This is so that both heating and cooling considerations are accommodated. Zones 1 through 7 apply to various parts of the United States and are defined by county lines. Zones 7 and 8 apply to various parts of Alaska, and Hawaii is classified as Zone 1.

The climate zones have been divided into marine, dry and moist to deal with levels of humidity. For more details and background on the development of the new Climate Zones, code users can refer to the white paper at the following site: http://www.energycodes.gov/implement/pdfs/cliate_paper_reiew_draft_rev.pdf.

N1101.2.1 Warm humid counties. Warm humid counties are listed in Table N1101.2.1.

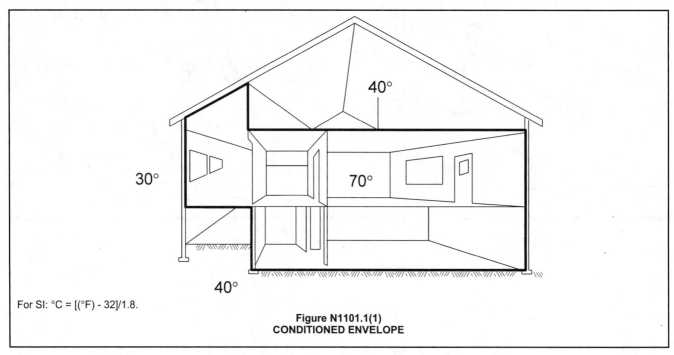

For SI: °C = [(°F) - 32]/1.8.

**Figure N1101.1(1)
CONDITIONED ENVELOPE**

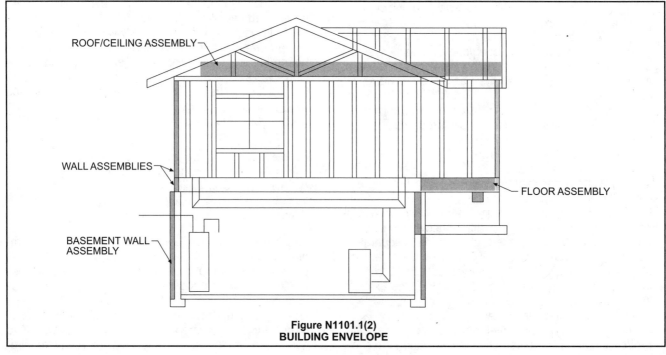

**Figure N1101.1(2)
BUILDING ENVELOPE**

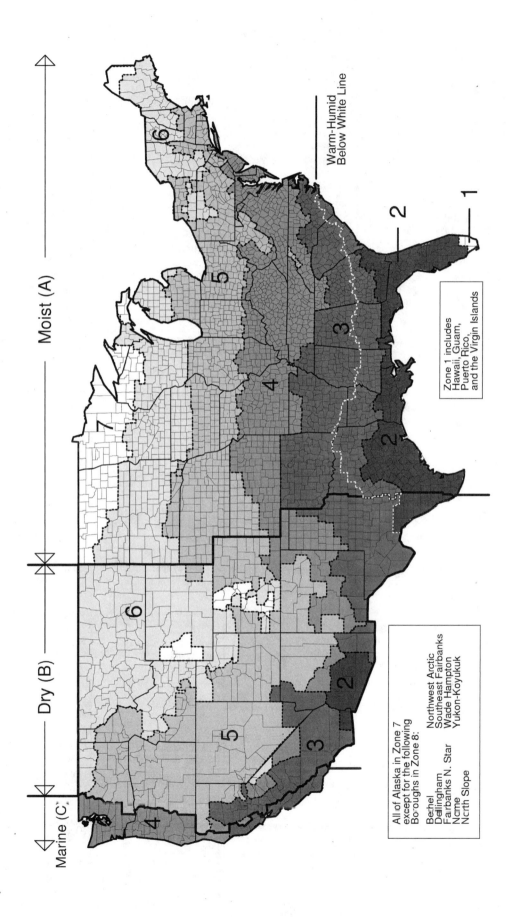

**FIGURE N1101.2
CLIMATE ZONES**

Warm-Humid
Below White Line

Moist (A)

Dry (B)

Marine (C)

Zone 1 includes
Hawaii, Guam,
Puerto Rico,
and the Virgin Islands

All of Alaska in Zone 7
except for the following
Boroughs in Zone 8:

Bethel   Northwest Arctic
Dellingham   Southeast Fairbanks
Farbanks N. Star Wade Hampton
Nome   Yukon-Koyukuk
North Slope

TABLE N1101.2
CLIMATE ZONES BY STATES AND COUNTIES

Alabama
Zone 3 except
Zone 2
Baldwin
Mobile

Alaska
Zone 7 except
Zone 8
Bethel
Dellingham
Fairbanks North Star
Nome
North Slope
Northwest Arctic
Southeast Fairbanks
Wade Hampton
Yukon-Koyukuk

Arizona
Zone 3 except
Zone 2
La Paz
Maricopa
Pima
Pinal
Yuma
Zone 4
Gila
Yavapai
Zone 5
Apache
Coconino
Navajo

Arkansas
Zone 3 except
Zone 4
Baxter
Benton
Boone
Carroll
Fulton
Izard
Madison
Marion
Newton
Searcy
Stone
Washington

California
Zone 3 Dry except
Zone 2
Imperial
Zone 3 Marine
Alameda
Marin

Mendocino
Monterey
Napa
San Benito
San Francisco
San Luis Obispo
San Mateo
Santa Barbara
Santa Clara
Santa Cruz
Sonoma
Ventura
Zone 4 Dry
Amador
Calaveras
El Dorado
Inyo
Lake
Mariposa
Trinity
Tuolumne
Zone 4 Marine
Del Norte
Humboldt
Zone 5
Lassen
Modoc
Nevada
Plumas
Sierra
Siskiyou
Zone 6
Alpine
Mono

Colorado
Zone 5 except
Zone 4
Baca
Las Animas
Otero
Zone 6
Alamosa
Archuleta
Chaffee
Conejos
Costilla
Custer
Dolores
Eagle
Moffat
Ouray
Rio Blanco
Saguache
San Miguel
Zone 7
Clear Creek
Grand

Gunnison
Hinsdale
Jackson
Lake
Mineral
Park
Pitkin
Rio Grande
Routt
San Juan
Summit

Connecticut
Zone 5

Delaware
Zone 4

Dist. of Columbia
Zone 4

Florida
Zone 2 except
Zone1
Broward
Dade
Monroe

Georgia
Zone 3 except
Zone 2
Appling
Atkinson
Bacon
Baker
Berrien
Brantley
Brooks
Bryan
Camden
Charlton
Chatham
Clinch
Colquitt
Cook
Decatur
Echols
Effingham
Evans
Glynn
Grady
Jeff Davis
Lanier
Liberty
Long
Lowndes
Mcintosh
Miller

Mitchell
Pierce
Seminole
Tattnall
Thomas
Toombs
Ware
Wayne
Zone 4
Banks
Catoosa
Dade
Dawson
Fannin
Floyd
Franklin
Gilmer
Gordon
Habersham
Hall
Lumpkin
Murray
Pickens
Rabun
Stephens
Towns
Union
Walker
White
Whitfield

Hawaii
Zone 1 Moist

Idaho
Zone 6 except
Zone 5
Ada
Benewah
Canyon
Cassia
Clearwater
Elmore
Gem
Gooding
Idaho
Jerome
Kootenai
Latah
Lewis
Lincoln
Minidoka
Nez Perce
Owyhee
Payette
Power
Shoshone
Twin Falls

Washington

Illinois
Zone 5 except
Zone 4
Alexander
Bond
Christian
Clay
Clinton
Crawford
Edwards
Effingham
Fayette
Franklin
Gallatin
Hamilton
Hardin
Jackson
Jasper
Jefferson
Johnson
Lawrence
Macoupin
Madison
Marion
Massac
Monroe
Montgomery
Perry
Pope
Pulaski
Randolph
Richland
Saline
Shelby
St clair
Union
Wabash
Washington
Wayne
White
Williamson

Indiana
Zone 5 except
Zone 4
Brown
Clark
Crawford
Daviess
Dearborn
Dubois
Floyd
Gibson
Greene
Harrison
Jackson

(continued)

TABLE N1101.2—continued
CLIMATE ZONES BY STATES AND COUNTIES

Jefferson
Jennings
Knox
Lawrence
Martin
Monroe
Ohio
Orange
Perry
Pike
Posey
Ripley
Scott
Spencer
Sullivan
Switzerland
Vanderburgh
Warrick
Washington

Iowa
Zone 5 except
Zone 6
Allamakee
Black Hawk
Bremer
Buchanan
Buena Vista
Butler
Calhoun
Cerro Gordo
Cherokee
Chickasaw
Clay
Clayton
Delaware
Dickinson
Emmet
Fayette
Floyd
Franklin
Grundy
Hamilton
Hancock
Hardin
Howard
Humboldt
Ida
Kossuth
Lyon
Mitchell
O'Brien
Osceola
Palo Alto
Plymouth
Pocahontas
Sac
Sioux

Webster
Winnebago
Winneshiek
Worth
Wright

Kansas
Zone 4 except
Zone 5
Cheyenne
Cloud
Decatur
Ellis
Gove
Graham
Greeley
Hamilton
Jewell
Lane
Logan
Mitchell
Ness
Norton
Osborne
Phillips
Rawlins
Republic
Rooks
Scott
Sheridan
Sherman
Smith
Thomas
Trego
Wallace
Wichita

Kentucky
Zone 4

Louisiana
Zone 2 except
Zone 3
Bienville
Bossier
Caddo
Caldwell
Catahoula
Claiborne
Concordia
De Soto
East Carroll
Franklin
Grant
Jackson
La Salle
Lincoln
Madison

Morehouse
Natchitoches
Ouachita
Red River
Richland
Sabine
Tensas
Union
Vernon
Webster
West Carroll
Winn

Maine
Zone 6 except
Zone 7
Aroostook

Maryland
Zone 4 except
Zone 5
Garrett

Massachusetts
Zone 5

Michigan
Zone 5 except
Zone 6
Alcona
Alger
Alpena
Antrim
Arenac
Benzie
Charlevoix
Cheboygan
Clare
Crawford
Delta
Dickinson
Emmet
Gladwin
Grand Traverse
Huron
Iosco
Isabella
Kalkaska
Lake
Leelanau
Manistee
Marquette
Mason
Mecosta
Menominee
Missaukee
Montmorency
Newaygo

Oceana
Ogemaw
Osceola
Oscoda
Otsego
Presque Isle
Roscommon
Sanilac
Wexford
Zone 7
Baraga
Chippewa
Gogebic
Houghton
Iron
Keweenaw
Luce
Mackinac
Ontonagon
Schoolcraft

Minnesota
Zone 6 except
Zone 7
Aitkin
Becker
Beltrami
Carlton
Cass
Clay
Clearwater
Cook
Crow Wing
Grant
Hubbard
Itasca
Kanabec
Kittson
Koochiching
Lake of the Wood
Mahnomen
Marshall
Mille Lacs
Norman
Otter Tail
Pennington
Pine
Polk
Red Lake
Roseau
St Louis
Wadena
Wilkin

Mississippi
Zone 3 except
Zone 2
Hancock

Harrison
Jackson
Pearl River
Stone

Missouri
Zone 4 except
Zone 5
Adair
Andrew
Atchison
Buchanan
Caldwell
Chariton
Clark
Clinton
Daviess
De Kalb
Gentry
Grundy
Harrison
Holt
Knox
Lewis
Linn
Livingston
Macon
Marion
Mercer
Nodaway
Pike
Putnam
Ralls
Schuyler
Scotland
Shelby
Sullivan
Worth

Montana
Zone 6

Nebraska
Zone 5

Nevada
Zone 5 except
Zone 3
Clark

New Hampshire
Zone 6 except
Zone 5
Cheshire
Hillsborough
Rockingham
Stafford

(continued)

TABLE N1101.2—continued
CLIMATE ZONES BY STATES AND COUNTIES

New Jersey
Zone 4 except
Zone 5
Bergen
Hunterdon
Mercer
Morris
Passaic
Somerset
Sussex
Warren

New Mexico
Zone 4 except
Zone 3
Chaves
Dona Ana
Eddy
Hidalgo
Lea
Luna
Otero
Zone 5
Catron
Colfax
Harding
Los Alamos
McKinley
Mora
Rio Arriba
San Juan
San Miguel
Sandoval
Santa Fe
Taos
Torrance

New York
Zone 5 except
Zone 4
Bronx
Kings
Nassau
New York
Queens
Richmond
Suffolk
Westchester
Zone 6
Allegany
Broome
Cattaraugus
Chenango
Clinton
Delaware
Essex
Franklin
Fulton

Hamilton
Herkimer
Jefferson
Lewis
Madison
Montgomery
Oneida
Otsego
Schoharie
Schuyler
St Lawrence
Steuben
Sullivan
Tompkins
Ulster
Warren
Wyoming

North Carolina
Zone 3 except
Zone 4
Alamance
Alexander
Bertie
Buncombe
Burke
Calwell
Caswell
Catawba
Chatham
Cherokee
Clay
Cleveland
Davie
Durham
Forsyth
Franklin
Gates
Graham
Granville
Guilford
Halifax
Harnett
Haywood
Henderson
Hertford
Iredell
Jackson
Lee
Lincoln
Macon
Madison
McDowell
Nash
Northampton
Orange
Person
Polk

Rockingham
Rutherford
Stokes
Surry
Swain
Transylvania
Vance
Wake
Warren
Wilkes
Yadkin
Zone 5
Alleghany
Ashe
Avery
Mitchell
Watauga
Yancey

North Dakota
Zone 7 except
Zone 6
Adams
Billings
Bowman
Burleigh
Dickey
Dunn
Emmons
Golden Valley
Grant
Hettinger
La Moure
Logan
McIntosh
McKenzie
Mercer
Morton
Oliver
Ransom
Richland
Sargent
Sioux
Slope
Stark

Ohio
Zone 5 except
Zone 4
Adams
Brown
Clermont
Gallia
Hamilton
Lawrence
Pike
Scioto
Washington

Oklahoma
Zone 3 Moist except
Zone 4 Dry
Beaver
Cimarron
Texas

Oregon
Zone 4 Marine except
Zone 5 Dry
Baker
Crook
Deschutes
Gilliam
Grant
Harney
Hood River
Jefferson
Klamath
Lake
Malheur
Morrow
Sherman
Umatilla
Union
Wallowa
Wasco
Wheeler

Pennsylvania
Zone 5 except
Zone 4
Bucks
Chester
Delaware
Montgomery
Philadelphia
York
Zone 6
Cameron
Clearfield
Elk
McKean
Potter
Susquehanna
Tioga
Wayne

Rhode Island
Zone 5

South Carolina
Zone 3

South Dakota
Zone 6 except
Zone 5
Bennett

Bon Homme
Charles Mix
Clay
Douglas
Gregory
Hutchinson
Jackson
Mellette
Todd
Tripp
Union
Yankton

Tennessee
Zone 4 except
Zone 3
Chester
Crockett
Dyer
Fayette
Hardeman
Hardin
Haywood
Henderson
Lake
Lauderdale
Madison
McNairy
Shelby
Tipton

Texas
Zone 2 Moist except
Zone 2 Dry
Bandera
Dimmit
Edwards
Frio
Kinney
La Salle
Maverick
Medina
Real
Uvalde
Val Verde
Webb
Zapata
Zavala
Zone 3 Dry
Andrews
Baylor
Borden
Brewster
Callahan
Childress
Coke
Coleman
Collingsworth

(continued)

TABLE N1101.2—continued
CLIMATE ZONES BY STATES AND COUNTIES

			Virginia	Wyoming
Concho	Wheeler	Stephens	**Zone 4**	
Cottle	Wilbarger	Tarrant		**Wisconsin**
Crane	Winkler	Titus	**Washington**	**Zone 6 except**
Crockett	**Zone 3 Moist**	Upshur	**Zone 4 Marine except**	**Zone 7**
Crosby	Archer	Van Zandt	**Zone 5 Dry**	Ashland
Culberson	Bianco	Wichita	Adams	Bayfield
Dawson	Bowie	Wise	Asotin	Burnett
Dickens	Brown	Wood	Benton	Douglass
Ector	Burnet	Young	Chelan	Florence
El Paso	Camp	**Zone 4**	Columbia	Forest
Fisher	Cass	Armstrong	Douglas	Iron
Foard	Clay	Bailey	Franklin	Langlade
Gaines	Collin	Briscoe	Garfield	Lincoln
Garza	Comanche	Carson	Grant	Oneida
Glasscock	Cooke	Castro	Kittitas	Price
Hall	Dallas	Cochran	Klickitat	Sawyer
Hardeman	Delta	Dallam	Lincoln	Taylor
Haskell	Denton	Deaf Smith	Skamania	Vilas
Hemphill	Eastland	Donley	Spokane	Washburn
Howard	Ellis	Floyd	Walla Walla	
Hudspeth	Erath	Gray	Whitman	**Wyoming**
Irion	Fannin	Hale	Yakima	**Zone 6 except**
Jeff Davis	Franklin	Hansford	**Zone 6 Dry**	**Zone 5**
Jones	Gillespie	Hartley	Ferry	Goshen
Kent	Grayson	Hockley	Okanogan	Platte
Kerr	Gregg	Hutchinson	Pend Oreille	**Zone 7**
Kimble	Hamilton	Lamb	Stevens	Lincoln
King	Harrison	Lipscomb		Sublette
Knox	Henderson	Moore	**West Virginia**	Teto
Loving	Hood	Ochiltree	**Zone 5 except**	
Lubbock	Hopkins	Oldham	**Zone 4**	**American Samoa**
Lynn	Hunt	Parmer	Berkely	**Zone 1 Moist**
Martin	Jack	Potter	Boone	
Mason	Johnson	Randall	Braxton	**Guam**
McCulloch	Kaufman	Roberts	Cabell	**Zone 1 Moist**
Menard	Kendall	Sherman	Calhoun	
Midland	Lamar	Swisher	Clay	**Northern Marianas**
Mitchell	Lampasas	Yoakum	Gilmer	**Zone 1 Moist**
Motley	Llano		Jackson	
Nolan	Marion	**Utah**	Jefferson	**Puerto Rico**
Pecos	Mills	**Zone 5 except**	Kanawha	**Zone 1 Moist**
Presidio	Montague	**Zone 3**	Lincoln	
Reagan	Morris	Washington	Logan	**U.S. Virgin Islands**
Reeves	Nacogdoches	**Zone 6**	Mason	**Zone 1 Moist**
Runnels	Navarro	Box Elder	McDowell	
Schleicher	Palo Pinto	Cache	Mercer	
Scurry	Panola	Carbon	Mingo	
Shackelford	Parker	Daggett	Monroe	
Sterling	Rains	Duchesne	Morgan	
Stonewall	Red River	Morgan	Pleasants	
Sutton	Rockwall	Rich	Putnam	
Taylor	Rusk	Summit	Ritchie	
Terrell	Sabine	Uintah	Roane	
Terry	San Augustine	Wasatch	Tyler	
Throckmorton	San Saba		Wayne	
Tom Green	Shelby	**Vermont**	Wirt	
Upton	Smith	**Zone 6**	Wood	
Ward	Somervell			

TABLE N1101.2.1
WARM HUMID COUNTIES

Alabama	Houston	Mississippi	Cass
Autauga	Irwin	All in Zone 2	Collin
Baldwin	Jenkins	Plus	Comanche
Barbour	Johnson	Adams	Dallas
Bullock	Laurens	Amite	Delta
Butler	Lee	Claiborne	Denton
Choctaw	Macon	Copiah	Ellis
Clarke	Marion	Covington	Erath
Coffee	Montgomery	Forrest	Franklin
Conecuh	Peach	Franklin	Gillespie
Covington	Pulaski	George	Gregg
Crenshaw	Quitman	Greene	Hamilton
Dale	Randolph	Hinds	Harrison
Dallas	Schley	Jefferson	Henderson
Elmore	Screven	Jefferson Davis	Hood
Escambia	Stewart	Jones	Hopkins
Geneva	Sumter	Lamar	Hunt
Henry	Taylor	Lawrence	Johnson
Houston	Telfair	Lincoln	Haufman
Lowndes	Terrell	Marion	Kendall
Macon	Tift	Perry	Lamar
Marengo	Treutlen	Pike	Lampasas
Mobile	Turner	Rankin	Llano
Monroe	Twiggs	Simpson	Marion
Montgomery	Webster	Smith	Mills
Perry	Wheeler	Walthall	Morris
Pike	Wilcox	Warren	Nacogdoches
Russell	Worth	Wayne	Navarro
Washington	**Hawaii**	Wilkinson	Palo Pinto
Wilcox	All	**North Carolina**	Panola
Arkansas	**Louisiana**	Brunswick	Parker
Columbia	All in Zone 2	Carteret	Rains
Hempstead	Plus	Columbus	Red River
Lafayette	Bienville	New Hanover	Rockwall
Little River	Bossier	Onslow	Rusk
Miller	Caddo	Pender	Sabine
Sevier	Caldwell	**South Carolina**	San Augustine
Union	Catahoula	Allendale	San Saba
Florida	Claiborne	Bamberg	Shelby
All	De Soto	Barnwell	Smith
Georgia	Franklin	Beaufort	Somervell
All in Zone 2	Grant	Berkeley	Tarrant
Plus	Jackson	Charleston	Titus
Ben Hill	La Salle	Colleton	Upshur
Bleckley	Lincoln	Dorchester	Van Zandt
Bulloch	Madison	Georgetown	Wood
Calhoun	Natchitoches	Hampton	**American Samoa**
Candler	Ouachita	Horry	All
Chattahoochee	Red River	Jasper	**Guam**
Clay	Richland	**Texas**	All
Coffee	Sabine	All in Zone 2	**Northern Marianas**
Crisp	Tensas	Plus	All
Dodge	Union	Blanco	**Puerto Rico**
Dooly	Vernon	Bowie	All
Dougherty	Webster	Brown	**U.S. Virgin Islands**
Early	Winn	Burnet	All
Emanuel		Camp	

N1101.3 Identification. Materials, systems and equipment shall be identified in a manner that will allow a determination of compliance with the applicable provisions of this chapter.

❖ The IRC requires that materials be labeled on site with the rated *R*-value [See Commentary Figures N1101.3(1) and N1101.3(2)] or, as in the case of blown in insulation products, have insulation certificates with the rated *R*-value.

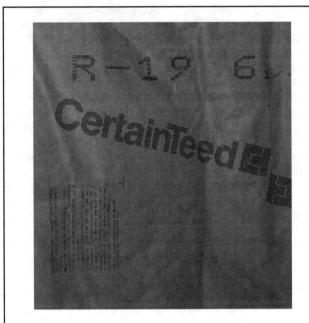

Figure N1101.3(1)
INSULATION *R*-VALUE STAMP

N1101.4 Building thermal envelope insulation. An *R*-value identification mark shall be applied by the manufacturer to each piece of building thermal envelope insulation 12 inches (305 mm) or more wide. Alternately, the insulation installers shall provide a certification listing the type, manufacturer and *R*-value of insulation installed in each element of the building thermal envelope. For blown or sprayed insulation (fiberglass and cellulose), the initial installed thickness, settled thickness, settled *R*-value, installed density, coverage area and number of bags installed shall be listed on the certification. For sprayed polyurethane foam (SPF) insulation, the installed thickness of the area covered and *R*-value of installed thickness shall be listed on the certificate. The insulation installer shall sign, date and post the certificate in a conspicuous location on the job site.

❖ Measuring blown-in insulation such as fiberglass or cellulose by attaching markers to the joist or truss provides a measure of the settling characteristics of blown-in insulation and its settled *R*-value.

A separate identification method is used for SPF. This is because spray polyurethane foam (SPF) has different distinctive characteristics as it is applied on the job site from the insulation that is traditionally blown in. SPF does not settle like traditional blown in

insulation so the initial thickness, density and coverage area remain the same. SPF actually expands upon initial application and could in fact cover the required thickness marker. Physical properties of cured SPF are dependent on the formula, not the number of drums of material used. The physical properties for the SPF including density and *R*-value are listed on the manufacturer's material data sheet.

SPF has tremendous adhesive properties, and when it is applied in uniform thickness over joists/trusses or roof/ceilings, it produces a wavelike pattern of high and low coverage in attics, that is reflective of the cavity between the joists/trusses. This wavelike pattern does not lend itself easily to visual inspection from a single vantage point within the attic. Therefore, SPF is most efficiently inspected by randomly probing the foam with a thin metal rod at the apparent low spots. After installation, SPF is typically covered with an ignition barrier. This ignition barrier could also prevent seeing the markers or their accuracy.

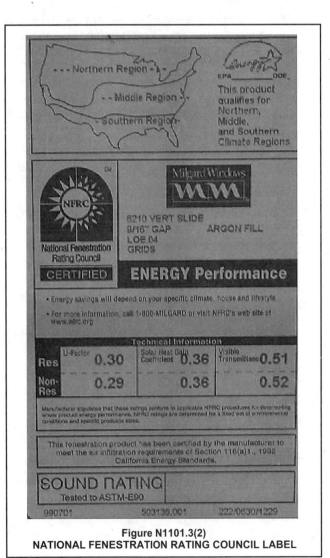

Figure N1101.3(2)
NATIONAL FENESTRATION RATING COUNCIL LABEL

N1101.4.1 Blown or sprayed roof/ceiling insulation. The thickness of blown in or sprayed roof/ceiling insulation (fiberglass or cellulose) shall be written in inches (mm) on markers that are installed at least one for every 300 ft^2 (28 m^2) throughout the attic space. The markers shall be affixed to the trusses or joists and marked with the minimum initial installed thickness with numbers a minimum of 1 inch (25 mm) high. Each marker shall face the attic access opening. Spray polyurethane foam thickness and installed R-value shall be listed on the certificate provided by the insulation installer.

❖ The amount of cellulose or fiberglass blown-in insulation is measured by attaching markers to the ceiling joists or trusses. These markers also provide a convenient method to inspect an entire attic from a single vantage point such as the attic access opening. This allows an inspector to accurately see how much settling occurs after the material has been applied and establish the correct R-value.

N1101.4.2 Insulation mark installation. Insulating materials shall be installed such that the manufacturer's R-value mark is readily observable upon inspection.

❖ To comply with this charging statement the insulation has to be installed with the markings face up and readable. In this way an inspector can readily identify the type and R-value of the insulation.

N1101.5 Fenestration product rating. U-factors of fenestration products (windows, doors and skylights) shall be determined in accordance with NFRC 100 by an accredited, independent laboratory, and labeled and certified by the manu-

facturer. Products lacking such a labeled U-factor shall be assigned a default U-factor from Tables N1101.5(1) and N1101.5(2). The solar heat gain coefficient (SHGC) of glazed fenestration products (windows, glazed doors and skylights) shall be determined in accordance with NFRC 200 by an accredited, independent laboratory, and labeled and certified by the manufacturer. Products lacking such a labeled SHGC shall be assigned a default SHGC from Table N1101.5(3).

❖ Not all manufacturers rate the U-factor of their products in accordance with NFRC 100, evident by the absence of a temporary NFRC label on some fenestration units. A default U-factor from Tables N1102.1 and N1102.1.2 must be assigned to the unit if it is not NFRC 100 rated. These tables assign U-factors based on the frame material, product type and number of glazing layers (single or double). When a composite of materials from two different product types is used, the product must be assigned the higher U-factor. The product cannot receive credit for a feature that cannot be seen. Because features such as argon-fill and low-emissivity glass are not visible, they do not receive credit in the default tables.

As with fenestration U- factors, default SHGC values must be assumed for all products not rated in accordance with NFRC 100. SHGC values from Table N1101.5(3) must be used for all units that do not have labels or are not listed in the NFRC Directory. The table assigns a SHGC value based on the frame type, window operation, layers of glazing, and the type of tint on the window.

TABLE N1101.5(1)
DEFAULT GLAZED FENESTRATION U-FACTORS

FRAME TYPE	SINGLE PANE	DOUBLE PANE	SKYLIGHT	
			Single	Double
Metal	1.2	0.8	2	1.3
Metal with thermal break	1.1	0.65	1.9	1.1
Nonmetal or metal clad	0.95	0.55	1.75	1.05
Glazed block	0.6			

❖ See the commentary to Section N1101.5.

TABLE N1101.5(2)
DEFAULT DOOR U-FACTORS

DOOR TYPE	U-FACTOR
Uninsulated metal	1.2
Insulated metal	0.6
Wood	0.5
Insulated, nonmetal edge, max 45% glazing, any glazing double pane	0.35

❖ See the commentary to Section N1101.5.

TABLE N1101.5(3)
DEFAULT GLAZED FENESTRATION SHGC

SINGLE GLAZED		DOUBLE GLAZED		GLAZED BLOCK
Clear	Tinted	Clear	Tinted	
0.8	0.7	0.7	0.6	0.6

❖ See the commentary to Section N1101.5.

N1101.6 Installation. All materials, systems and equipment shall be installed in accordance with the manufacturer's installation instructions and the provisions of this code.

❖ Manufacturers' installation instructions are evaluated by the listing agency verifying compliance with the applicable standard. The listing agency can require that the manufacturer alter, delete or add information to the instructions as necessary to achieve compliance with applicable standards and code requirements. Manufacturers' installation instructions are an enforceable extension of the code and must be in the hands of the code official when an inspection takes place.

N1101.6.1 Protection of exposed foundation insulation. Insulation applied to the exterior of basement walls, crawl space walls, and the perimeter of slab-on-grade floors shall have a rigid, opaque and weather-resistant protective covering to prevent the degradation of the insulation's thermal performance. The protective covering shall cover the exposed exterior insulation and extend a minimum of 6 inches (152 mm) below grade.

❖ The ultimate performance of insulation materials is directly proportional to the workmanship involved in the materials' initial installation as well as the materials' integrity over the life of the structure. Accordingly, foundation wall and slab edge insulation materials installed in the vicinity of the exterior grade line require protection from damage that could occur from contact by lawn mowing and maintenance equipment, garden hoses, garden tools, perimeter landscape materials, etc. In addition, the long-term thermal performance of foam plastic insulation materials is adversely affected by direct exposure to the sun. To protect the insulation from sunlight and physical damage, it must have a protective covering that is inflexible, puncture-resistant, opaque and weather-resistant.

N1101.7 Above code programs. The building official or other authority having jurisdiction shall be permitted to deem a national, state or local energy efficiency program to exceed the energy efficiency required by this chapter. Buildings approved in writing by such an energy efficiency program shall be considered in compliance with this chapter.

❖ Section N1101.7 offers the building official the flexibility of using national, state or local energy efficiency programs only if they exceed the requirements of Chapter 11. It is up to the building official to decide to accept a alternate energy efficiency program in place of the requirements of Chapter 11.

N1101.8 Certificate. A permanent certificate shall be posted on or in the electrical distribution panel. The certificate shall be completed by the builder or registered design professional. The certificate shall list the predominant *R*-values of insulation installed in or on ceiling/roof, walls, foundation (slab, basement wall, crawlspace wall and/or floor) and ducts outside conditioned spaces; *U*-factors for fenestration; and the solar heat gain coefficient (SHGC) of fenestration. Where there is more than one value for each component, the certificate shall list the value covering the largest area. The certificate shall list the type and efficiency of heating, cooling and service water heating equipment.

❖ This section is intended to increase the consumer's awareness of the energy-efficiency ratings for the various building elements in their home. The builder or registered design professional has to complete the certificate and place it inside the electrical panel (see Commentary Figure N1101.8).

The certificate must disclose the building's *R*-values, fenestration *U*-factors and fenestration SHGC, HVAC equipment types and efficiencies. The energy efficiency of a building as a system is a function of many elements considered as separate parts of the whole. It is difficult to have a proper identification and analysis of a building's energy efficiency once the building is completed because many of the elements may not be readily accessible. This information is also valuable for existing structures undergoing alterations and additions to help determine the appropriate sizing for the mechanical systems. This is meant to be a simple certificate that is easy to read. The certificate does not contain all the information required for compliance and cannot be substituted for information on the required construction documents. Instead the certificate is meant to provide the housing owner, occupant or buyer with a simple-to-understand overview of the home's energy efficiency. Where there is a mixture of insulation and/or fenestration values, the value applying to the largest area is specified. For example, if most of the wall insulation was *R*-19, but a limited area bordering the garage was *R*-13, the certificate would specify *R*-19 for the walls. (In contrast, plans and overall compliance would need to account for both *R*-values.)

The code specifies the minimum information on the certificate, but does not prohibit additional information being added so long as the required information is clearly visible. For example a builder might choose to list energy efficiency features beyond those required by the code.

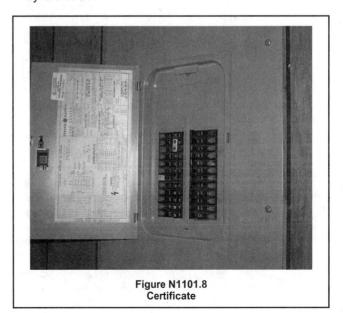

Figure N1101.8
Certificate

SECTION N1102
BUILDING THERMAL ENVELOPE

❖ The provisions of Section N1102 are the detailed requirements of the levels of insulation, the performance of openings (fenestrations) and air-leakage and moisture-control provisions which serve to establish the building's energy efficiency. When combined with the "systems" requirements (Section N1103), these two sections will provide the total package of energy conservation which the code requires.

The term Building Thermal Envelope is defined in Chapter 2 of the IRC as being "the basement walls, exterior walls, floor, roof and any other building element that enclose conditioned spaces." Therefore, when combined with the definition of "conditioned space," the code has defined the boundaries of the building which will be regulated by this section.

The building thermal envelope is a key term and resounding theme used throughout the energy requirements. It defines what portions of the building structure bound conditioned space and are thereby covered by the insulation and infiltration (air leakage) requirements of the code. The building thermal envelope includes all building components separating conditioned spaces (see commentary to Conditioned space) from unconditioned spaces or outside ambient conditions and through which heat is transferred. For example, the walls and doors separating an unheated garage (unconditioned space) from a living area (conditioned space) are part of the building envelope. The walls and doors separating an unheated garage from the outdoors are not part of the building thermal envelope. Walls, floors and other building components separating two conditioned spaces are not part of the building envelope. For example, interior partition walls, the common or party walls separating dwelling units in multifamily buildings, and the wall between a new conditioned addition and the existing conditioned space are not considered part of the building envelope.

Unconditioned spaces (areas having no heating or cooling sources) are considered outside the building thermal envelope. A space is conditioned if it is heated or cooled directly; communicates directly with a conditioned space; or where a space is indirectly supplied with heating, cooling/or both through uninsulated walls, floors or uninsulated ducts or HVAC piping. Boundaries that define the building envelope include the following:

- Building assemblies separating a conditioned space from outdoor ambient weather conditions.

- Building assemblies separating a conditioned space from the ground under or around that space, such as the ground around the perimeter of a slab or the soil at the exterior of a conditioned basement wall. Note that the code does not specify requirements for insulating basement floors or underneath slab floors (except at the perimeter edges).

- Building assemblies separating a conditioned space from an unconditioned garage, unconditioned sunroom or similar unheated/cooled area.

The code specifies requirements for ceiling, wall, floor, basement wall, slab-edge and crawlspace wall components of the building envelope. In some cases, it may be unclear how to classify a particular part of a building. For example, skylight shafts have properties of a wall assembly but are located in the ceiling assembly. In these situations, a determination needs to be made and approved by the code official prior to construction so that the proper level of insulation can be installed to complete the building thermal envelope.

N1102.1 Insulation and fenestration criteria. The building thermal envelope shall meet the requirements of Table N1102.1 based on the climate zone specified in Table N1101.2.

❖ This section serves as the basis for the code's general insulation and fenestration requirements. Therefore, this is the first place to determine what the requirements for the building thermal envelope will be. There are specific requirements for certain assemblies and locations which are addressed in Sections N1102.2 and N1102.3. Those requirements should be checked and would be considered the applicable requirements for those items based on the provisions of Section R102.1 which specify that "the specific requirement shall be applicable."

This section begins by establishing the requirements for the building thermal envelope by requiring compliance with the proper component insulation and fenestration requirements of Table N1102.1. However, once that general requirement is established, Sections N1102.1.1, N1102.1.2 and N1102.1.3 will provide three possible means of showing that the building thermal envelope will comply. Any of the three methods may be used at the discretion of the designer. The three options and their advantages are discussed in the commentary with the subsections. In general the later subsections will provide the designer with more options and flexibility but they will also require greater effort and documentation than using Table N1102.1 on an individual component basis.

Table N1102.1 lists the minimum R-value and maximum U-factor and SHGC requirements for different portions of the building thermal envelope, including basement and exterior walls, floor, ceiling and any other building elements that enclose conditioned space. Using the table begins with determining the Climate Zone for the proposed location from Table N1101.2 or Figure N1101.2. Once the Climate Zone has been determined, each of the R-value, U-factor or SHGC requirements must be met for the applicable component (e.g., ceilings, walls, floors, etc.).

Maximum glazing U-factor is the first column in Table N1102.1 that must be complied with. Each fenestration product in the proposed building must not exceed the maximum U-factor requirement presented in the table for a particular Climate Zone. For example, a single-family residence located in a Climate Zone 5 would require installation of glazed fenestration products with a maximum U-factor of 0.35. This would include all glazing in the walls of the building thermal en-

TABLE N1102.1
INSULATION AND FENESTRATION REQUIREMENTS BY COMPONENT[a]

CLIMATE ZONE	FENESTRATION U-FACTOR	SKYLIGHT[b] U-FACTOR	GLAZED FENESTRATION SHGC	CEILING R-VALUE	WOOD FRAME WALL R-VALUE	MASS WALL R-VALUE	FLOOR R-VALUE	BASEMENT[c] WALL R-VALUE	SLAB[d] R-VALUE AND DEPTH	CRAWL SPACE WALL R-VALUE
1	1.2	0.75	0.40	30	13	3	13	0	0	0
2	0.75	0.75	0.40	30	13	4	13	0	0	0
3	0.65	0.65	0.40[e]	30	13	5	19	0	0	5/13
4 except Marine	0.40	0.60	NR	38	13	5	19	10/13	10, 2 ft	10/13
5 and Marine 4	0.35	0.60	NR	38	19 or 13 + 5[g]	13	30[f]	10/13	10, 2 ft	10/13
6	0.35	0.60	NR	49	19 or 13 + 5[g]	15	30[f]	10/13	10, 4 ft	10/13
7 and 8	0.35	0.60	NR	49	21	19	30[f]	10/13	10, 4 ft	10/13

a. R-values are minimums. U-factors and SHGC are maximums. R-19 insulation shall be permitted to be compressed into a 2 × 6 cavity.

b. The fenestration U-factor column excludes skylights. The solar heat gain coefficient (SHGC) column applies to all glazed fenestration.

c. The first R-value applies to continuous insulation, the second to framing cavity insulation; either insulation meets the requirement.

d. R-5 shall be added to the required slab edge R-values for heated slabs.

e. There are no solar heat gain coefficient (SHGC) requirements in the Marine Zone.

f. Or insulation sufficient to fill the framing cavity, R-19 minimum.

g. "13+5" means R-13 cavity insulation plus R-5 insulated sheathing. If structural sheathing covers 25% or less of the exterior, R-5 sheathing is not required where structural sheathing is used. If structural sheathing covers more than 25% of exterior, structural sheathing shall be supplemented with insulated sheathing of at least R-2.

velope (e.g. vertical windows) and skylights in the roof would be limited to a maximum U-factor of 0.60. The proposed glazing U-factor should be called out in the building plans either on the floor plan or within a window schedule. This will provide the necessary information to the field inspector, who will then need to verify that what is on the plans is installed in the field. Fenestration products that do not have NFRC labels on them must use the default U-factors contained in Table N1101.5(1) or N1101.5(2). See Commentary Section N1101.5 and Table N1101.5(1). The lowest default U-factor included in the table for glazed fenestration is listed at 0.55 for a "nonmetal or metal-clad double-pane window." This U-factor will not meet the requirements of the IRC in Climate Zones 4 and higher.

Table N1102.1. See above.

❖ Table N1102.1 serves as the basis for establishing the building thermal envelope requirements based on the text of Section N1102.1 and sets the performance level for each of the individual components listed. See the commentary for Sections N1102.1, N1102.2 and N1102.3 for additional discussion related to the components in the table. The simplest compliance approach is to meet these requirements directly. Note that the requirements do not change based on the area of the components of the residence. These same requirements apply to changes in existing buildings; for example, additions.

A few specifics of Table N1102.1 may benefit from clarification.

When applying the fenestration requirements of this table, it is important to remember the definition of fen-

estration and that it does include items such as doors, glass block and other items as well as windows. Therefore, any door located in the building thermal envelope would still be subject to these limitations. Although vertical fenestration (vertical windows and doors) and skylights have a separate column for U-factor, the SHGC applies to both. This is reinforced by the provisions of Note b.

The ceiling R-value requirements are pre-calculated for insulation only and already assume a credible R-value for other building materials such as air films, interior sheathing and exterior sheathing. The only R-value for insulation that may be used to meet the requirements is that installed between the conditioned space and the vented air space in the roof/ceiling assembly. This is typically not an issue because most insulation is installed directly on top of the gypsum board ceiling and the ceiling location represents the building thermal envelope. Insulation installed in the ceiling must meet or exceed the required insulation level. These minimum ceiling R-values would still be applicable where the provisions of Section R806.4 were used to create a "conditioned attic assembly." In those cases the location of the insulation and air barrier (building thermal envelope) are simply located at the roof instead of at the ceiling line. See the commentary with Sections N1102.2.1 and N1102.2.2 for additional information regarding ceiling insulation requirements.

The R-values presented under the "Walls" column represent the sum of the insulation materials installed between the framing cavity and, if used, the insulating sheathing. See Section N1102.1.1 regarding how to compute the R-value. Insulating sheathing must have an *R*-value of at least R-2 to be considered. The R-2 limitation comes from the definition of "insulating sheathing." The *R*-value of non-insulative interior finishes, such as sheet rock, or exterior coverings like wood structural panel siding, is not considered when determining whether the proposed wall assembly meets the requirements. For example, in Commentary Figure N1102.1.1 the *R*-value of the cavity insulation installed between framing (R-13) is added to the insulating sheathing installed on the outside of the studs (R-6) resulting in an R-19 wall. The R-19 total insulation value can then be compared to the *R*-value requirement for the specific Climate Zone in Table N1102.1 to determine compliance.

The insulation *R*-value requirement for exterior walls assumes wood framing. Walls framed using steel studs or constructed of materials such as CMU are addressed in Sections N1102.2.4 and N1102.2.3. Whenever a residence has more than one type of wall (frame or mass) or more than one type of below grad wall (conditioned basement, or crawlspace) the requirement for each component is taken from the appropriate column in Table N1102.1.1.

Mass walls are defined within Chapter 2 and have additional requirements within Section N1102.2.3. Mass walls are intended to be "exterior walls" and do not include basement walls, which have a separate entry within the table.

Note a reminds the code user which level of performance is required. Therefore when dealing with *R*-values, a higher number would be better. When dealing with *U*-factors, the lower the number, the better the performance.

Per Note c, for "basement walls" and crawlspace walls the two numbers separated by a "/" represent the values for continuous and cavity insulation; either will meet the code. For example, in Climate Zone 6, the wall can either be covered with continuous insulation to a minimum level of R-10, or if some type of framing is used (such as a wood frame wall used to finish out a basement), R-13 insulation must be installed within the cavity. This higher level of cavity insulation adjusts for the bridging or reduction in energy efficiency that the framing elements would create.

Per Note d, heated slabs require R-5 insulation in Zones 1, 2 and 3; and R-15 slab edge insulation in Zones 4 and above. This R-15 insulation is the result of R-5 being "added" to the R-10 insulation level specified in the table for Climate Zones 4 through 8.

Per Note f, where R-30 under-floor insulation is required, less insulation may be used if the framing cavity is filled, down to a minimum of R-19. This recognizes that extending the framing solely to hold more insulation can cost more than it is worth.

Per Note g, R-13 cavity insulation and R-5 sheathing is deemed to meet the requirement for R-19 insulation. Also per Note g, structural sheathing can be substituted for insulated sheathing for up to 25 percent of the wall. Where structural sheathing is more than 25 percent of the wall, it must be supplemented with insu-

lated sheathing having at least a value of R-2 (see definition of Insulating sheathing).

N1102.1.1 *R*-value computation. Insulation material used in layers, such as framing cavity insulation and insulating sheathing, shall be summed to compute the component *R*-value. The manufacturer's settled *R*-value shall be used for blown insulation. Computed *R*-values shall not include an *R*-value for other building materials or air films.

❖ This section indicates how the *R*-value in Table N1102.1 is to be determined. Table N1102.1 specifies the required *R*-values for the insulation products, the nominal *R*-value. This is the *R*-value of the insulation products only. Although other products and features such as finish materials, air films and air spaces may contribute to overall energy efficiency, when determining the *R*-value in the code, these additional items are not considered and do not contribute to the nominal *R*-value. For example, if a wall had R-13 cavity insulation, gypsum board with an *R*-value of almost R-1, and exterior siding that has an *R*-value of R-1; the overall wall *R*-value is simply R-13 because the gypsum board and the exterior siding do not contribute to the *R*-value for purposes of determining code compliance. Where there is more than one layer of insulation the *R*-values for the layers are summed. For example a wall with R-13 batts within the framing cavity and R-4 insulated sheathing would be treated as an R-17 wall (13 plus 4 is 17). It is only insulation materials which may be summed to determine the component's *R*-value.

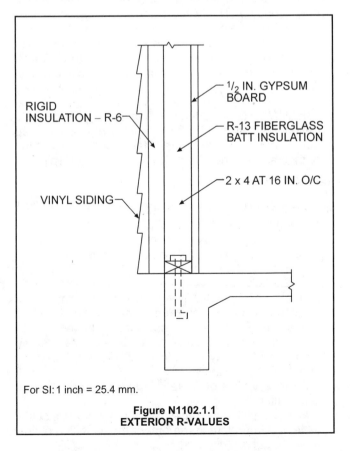

For SI: 1 inch = 25.4 mm.

Figure N1102.1.1
EXTERIOR R-VALUES

N1102.1.2 *U*-factor alternative. An assembly with a *U*-factor equal to or less than that specified in Table N1102.1.2 shall be permitted as an alternative to the *R*-value in Table N1102.1.

Exception: For mass walls not meeting the criterion for insulation location in Section N1102.2.3, the *U*-factor shall be permitted to be:

1. *U*-factor of 0.17 in Climate Zone 1

2. *U*-factor of 0.14 in Climate Zone 2

3. *U*-factor of 0.12 in Climate Zone 3

4. *U*-factor of 0.10 in Climate Zone 4 except Marine

5. *U*-factor of 0.082 in Climate Zone 5 and Marine 4

❖ For residences built with common insulation products, the most direct method of compliance is often the *R*-values in Table N1102.1. As an alternative, compliance can be demonstrated by calculating the *U*-factor for a component. Table N1102.1.2 gives *U*-factors that are deemed to be equivalent to the *R*-values in the prescriptive tables. Unlike the *R*-values in Table N1102.1, which consider only the insulation, *U*-factors consider all the parts of the construction. *U*-factors for a wall might include exterior siding, gyp-board and air films; all of which would be excluded from the *R*-value computation by Section N1102.1.1; for example, whether wall framing is 16 or 24 inches on center matters in computing the *U*-factor. Whether framing is metal or wood can have a significant impact on *U*-factor.

U-factors are well suited to several applications. Construction types that limit the amount of framing or include thermal breaks as part of their design may benefit from *U*-factor calculations. Components with complex or non-uniform geometries can use testing to establish *U*-factors. Compliance with the "total UA alternative" or trade-off approach in Section N1102.1.3 requires the use of the *U*-factor Tables.

The exception to this section addresses mass walls which do not meet the criteria established in Section N1102.2.3. The mass wall *U*-factors in Table N1102.1.2 apply to mass walls where the insulation is at least 50 percent on the outside or integral to the mass wall, which is the insulation locations that makes mass walls more effective. Mass walls are less effective if the insulation is located on the interior of the mass. Mass walls where more than 50 percent of the insulation is on the inside of the mass and therefore do not comply with Section 1102.2.3 would use the exception and have a lower *U*-factor requirement for Climate Zones 1 through 5. In Climate Zones 6 through 8, mass walls with insulation more than 50 percent on the interior would be required to meet the wood frame wall *R*-value because of the requirements found in Section N1102.2.3.

To correctly apply the exception that is found in Section N1102.1.2 it is important to understand the way the code is typically written. Generally when exceptions are placed into the code, they are less restrictive than the general requirement. Therefore either an exception may be used or compliance with the more restrictive base provision is still permitted. To properly apply the exception in this section, it is important to look at the mass wall requirements of Section N1102.2.3. Although items 1 through 4 of the exceptions in Section N1102.1.2 would appear to be more restrictive and require a lower *U*-factor than found in Table N1102.1.2, the code user is not really given the option to ignore the exception even though the exception states the *U*-factor "shall be permitted to be" as listed in the various items. To properly apply this exception it is important to note the requirement found in the last sentence of the base paragraph of Section N1102.2.3 dealing with mass walls. In general, mass walls which do not have the majority of their insulation located on the exterior or integral to the wall must comply with the "wood frame wall" insulation requirements which would mean a more restrictive *R*-value or *U*-factor than what is typically required for a mass wall. This exception, which is found in Section N1102.1.2, will permit the use of a more restrictive *U*-factor than what is typically required by Table N1102.1.2 for mass walls, but it will be less restrictive than the requirement would have been if the mass wall was forced to use the "frame wall" values

TABLE N1102.1.2
EQUIVALENT *U*-FACTORS[a]

CLIMATE ZONE	FENESTRATION *U*-FACTOR	SKYLIGHT *U*-FACTOR	CEILING *U*-FACTOR	FRAME WALL *U*-FACTOR	MASS WALL *U*-FACTOR	FLOOR *U*-FACTOR	BASEMENT WALL *U*-FACTOR	CRAWL SPACE WALL *U*-FACTOR
1	1.20	0.75	0.035	0.082	0.197	0.064	0.360	0.477
2	0.75	0.75	0.035	0.082	0.165	0.064	0.360	0.477
3	0.65	0.65	0.035	0.082	0.141	0.047	0.360	0.136
4 except Marine	0.40	0.60	0.030	0.082	0.141	0.047	0.059	0.065
5 and Marine 4	0.35	0.60	0.030	0.060	0.082	0.033	0.059	0.065
6	0.35	0.60	0.026	0.060	0.06	0.033	0.059	0.065
7 and 8	0.35	0.60	0.026	0.057	0.057	0.033	0.059	0.065

a. Nonfenestration *U*-factors shall be obtained from measurement, calculation or an approved source.

❖ This table provides the equivalent U-factors which may be used under Section N1102.1.2. See the commentary for Section N1102.1.2 for discussion related to this table.

from the table. Therefore when deciding whether to use the exception of Section N1102.1.2, code users would need to decide whether they wanted to use the more restrictive wood "frame wall" provisions as specified in Section N1102.2.3 or the exception in Section N1102.12 which lists specific *U*-factors that depend on the Climate Zone.

N1102.1.3 Total UA alternative. If the total building thermal envelope UA (sum of *U*-factor times assembly area) is less than or equal to the total UA resulting from using the *U*-factors in Table N1102.1.2, (multiplied by the same assembly area as in the proposed building), the building shall be considered in compliance with Table N1102.1. The UA calculation shall be done using a method consistent with the ASHRAE *Handbook of Fundamentals* and shall include the thermal bridging effects of framing materials. The SHGC requirements shall be met in addition to UA compliance.

❖ This alternative allows one portion of the building to make up for another. It recognizes that there may be reasons for less insulation in some parts of the residence, which can be compensated for by more insulation in other parts of the residence. The key concept is that the overall building thermal flow (UA) meets the code. This concept could allow a ceiling to make up for a wall or vice versa. As a practical matter, whether a building will comply by this method can sometimes be estimated quickly. A large area that is significantly over the required *R*-value will make up for a small area only mildly under the required *R*-value. Likewise it will sometimes be obvious that a small area that is mildly exceeds the requirement will not make up for a large area well below the requirement. When in doubt, look at the calculation.

This section will allow for such trade offs but only if the total UA for the proposed building is below the aggregate UA calculation using the required values in Table N1102.1.2 and the same assembly areas as the actual building. In other words, under this alternative, components with varying insulating values can be "traded-off" with one another as the builder sees fit just as long as the total UA calculation for the entire building complies with a calculation for that same house that uses the same assembly areas and the maximum UA values from Table N1102.1.2.

The UA is the sum of the component *U*-factors times each assembly area. The maximum allowable UA is the UA for a proposed design as if it was insulated to meet exactly the individual component *U*-factor requirements. This trade-off provision allows the type of insulation and installed fenestration to vary, which permits significant design flexibility. The desire for trade-offs in construction is common because of unexpected problems or design conflicts, and a UA trade-off analysis is usually calculated with the assistance of electronic compliance tools, depending on the jurisdiction. [For example, the Department of Energy has online compliance software for earlier versions of the IECC called REScheck, which can be downloaded from the DOE website at www.energycodes.gov. DOE

is expected to produce a new version of REScheck for the 2006 IECC. This new version, if approved by the jurisdiction as compliant with the code, can be used to perform a UA trade-off analysis under the 2006 IECC.]

This section explicitly prohibits the trade-off of SHGC requirements, requiring that the "SHGC requirements be met in addition to UA compliance." As a result, glazed fenestration must comply with the SHGC values shown in Table N1102.1 even if the *U*-factor is modified by trading off against some other component.

The requirements of this section establish specific additional requirements for any trade-off. First, the baseline house must have the same assembly areas as the proposed house (e.g., the same area of each assembly—fenestration, skylights, ceiling, wall and floor). Second, the calculation should be done consistent with the ASHRAE *Handbook of Fundamentals*. Third, the calculation must include the thermal bridging effects of framing materials. To meet these requirements, the calculation method must either specifically combine the actual framing and insulation paths (with their specific areas and *U*-values) or use framing factors such as those found in the ASHRAE Handbook for all framed building components (NOTE: this is not necessary for fenestration, which is a whole product value). To illustrate this approach, assume 1000 square feet (93 m²) of wall, of which 250 square feet (23 m²) is framing (assuming 0.81 *U*-factor) and 750 square feet (70 m²) is cavity (R-13 insulation). The baseline (general code requirement) and proposed opaque wall UAs are computed as follows:

- Baseline Opaque Wall for Climate Zone 3 = (0.082 × 1000) = 82. *[The 0.082 value used in the calculation is taken from Table N1102.1.2]*
- Proposed Opaque Wall UA = (0.81 × 250) + (0.077 × 750) = 78. *[The 0.81 value used in the calculation was given above as an assumption. The 0.077 value is determined based on the R-value of 13; that is, 1 ÷ 13 = 0.077.]*

Similar computations would be done for each assembly (such as fenestration or ceilings) and the baseline and proposed values then totaled and compared. If the baseline is greater than or equal to the proposed values, the house satisfies the UA alternative. The home will still need to meet all other prescriptive requirements including the fenestration SHGC in Table N1102.1, the air leakage of Section N1102.4 and the moisture control requirements of Section N1102.5 as examples.

It also should be noted that this alternative is limited to UA trade-offs for the building's thermal envelope. Unlike earlier versions of REScheck, the 2006 IRC does not authorize or establish any basis for HVAC trade-offs associated with its UA trade-off option. HVAC performance is simply not addressed under the 2006 IRC prescriptive or UA trade-off paths. It is addressed only in Section 404 of the IECC, under the Simulated Performance Alternative. As a result, the 2006 IRC elected to limit these "simplified trade-offs" to UA envelope trade-offs and defer any more complex trade-offs exclusively to the Simulated Performance Alternative in IECC Section 404. If the builder wishes to factor in HVAC performance for trade-offs, IECC Section 404 can be used and is permitted based on IRC Section N1101.2 accepting the IECC as a compliance option.

Documentation acceptable to the local building department generally must be submitted to the appropriate authority to certify acceptable component UA trade-offs.

N1102.2 Specific insulation requirements.

❖ This section contains specific requirements to be followed for the individual items listed within the subsections. Although Section N1102.1 and Tables N1102.1 and N1102.1.2 provide the general basis for complying with the energy requirements of this chapter, Section N1102.2 provides additional details regarding the actual construction of the assemblies or modifications that may affect the general requirements. Relying on the provisions of Section R102.1 will help assure that these specific provisions are properly followed.

N1102.2.1 Ceilings with attic spaces. When Section N1102.1 would require R-38 in the ceiling, R-30 shall be deemed to satisfy the requirement for R-38 wherever the full height of uncompressed R-30 insulation extends over the wall top plate at the eaves. Similarly R-38 shall be deemed to satisfy the requirement for R-49 wherever the full height of uncompressed R-38 insulation extends over the wall top plate at the eaves.

❖ Insulation installed in a typical roof assembly will be full height throughout the center portions of the assembly and will taper at the edges as the roof nears the top plate of the exterior wall system [See Commentary Figure N1102.2.1(1)]. The slope of the roof causes this tapering, which is further amplified by any baffling installed to direct ventilation air from the eave vents up and over the insulation. Because of this tapering, the installed R-value near the plate lines will be less than the rated R-value for the insulation. This is caused by compression (compressed insulation has a lower R-value than insulation installed to its full thickness) and the limited space between the floor of the attic and the roof sheathing near the exterior plate line. Thus a typical installation, on average, will have a lower R-value than that of the rated insulation. Because of this, the code will allow installation of a lower insulation value if it can be installed full thickness, to its rated R-value, over the plate line of the exterior wall. This allowance recognizes that a partial thermal "bypass" has been made more efficient by using insulation with the full R-value at the eaves. The full insulation R-value is sometimes achieved by what is termed an "energy truss" or "advanced framing." This can be achieved by using an oversized truss or raised-heel truss as shown in Commentary Figure N1102.2.1(2). Another way to achieve the full R-value would be by use of insulation with a higher R-value per inch at the eaves. The use of this option allows substituting R-30 for R-38 insulation, and R-38 may be substituted for R-49 insulation to meet the requirements of the code.

Note that the same option of using a reduced R-value would apply if the insulation was installed directly under the roof deck, rather than on the attic floor. (See Section R806.4 and the commentary for that section.) The insulation would be required to meet the lesser R-value, presuming the full R-value was met over the eaves. Of course this situation would also presume that the attic space beneath the insulation was not vented.

Note that this text applies only to the R-value portion of the code; there is no reduction in requirements if the

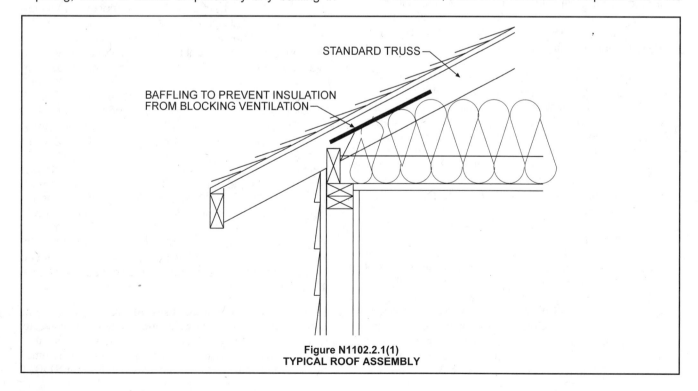

Figure N1102.2.1(1)
TYPICAL ROOF ASSEMBLY

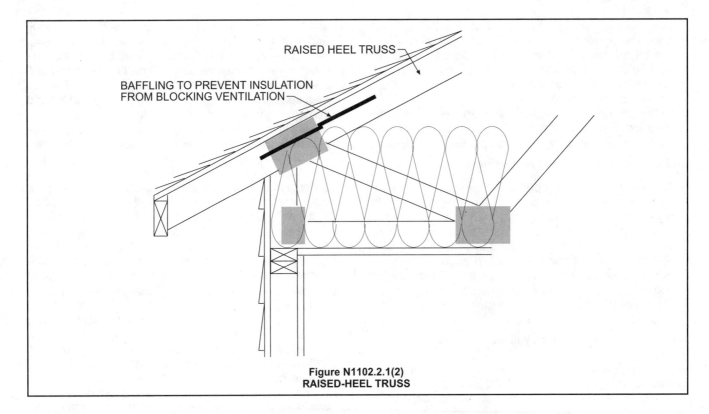

Figure N1102.2.1(2)
RAISED-HEEL TRUSS

U-factor alternative is used. In addition, if the residence had more than one separate attic space, it is possible this section could apply to one attic space but not another.

N1102.2.2 Ceilings without attic spaces. Where Section N1102.1 would require insulation levels above R-30 and the design of the roof/ceiling assembly does not allow sufficient space for the required insulation, the minimum required insulation for such roof/ceiling assemblies shall be R-30. This reduction of insulation from the requirements of Section N1102.1 shall be limited to 500 ft² (46 m²) of ceiling area.

❖ In situations where the ceiling is installed directly onto the roof rafters and no attic space is created, this section will allow a reduced level of ceiling insulation for a limited area. This section addresses the construction of what is typically called cathedral or vaulted ceilings which therefore result in that portion of the home not having an attic area above the ceiling. See definition of "attic" in Chapter 2. Based on the use of solid sawn lumber (2 × 8, 2 × 10 or 2 × 12) in conventional construction, as the ceiling *R*-value requirement increases it may be impossible to install the required ceiling insulation from Table N1102.1 within the available cavity depth. In addition the ventilation requirements of Section R806.3 will further reduce the available space by requiring a minimum space of 1-inch (25 mm) between the insulation and the roof sheathing. Therefore, when the depth of the cavity will not permit the required insulation level, this section permits a reduction to R-30 ceiling insulation instead of the normally required R-38 or R-49 requirement from Table N1102.1. This will generally result in reducing the required insulation level instead of having to increase the

depth of the framing members. This section in the code recognizes that increases in framing size done only to accommodate higher *R*-values are an expensive way to achieve a limited increase in *R*-value

It is important to notice that this section applies only to areas which have a required insulation level "above R-30" (Climate Zones 4 through 8) and to portions of ceiling assemblies that do not exceed 500 square feet (46 m²). The intent is that the 500 square feet (46 m²) limitation be the total aggregate exempted amount of the building's thermal envelope (ceiling) that can use this reduction. It is not the intent that a home could have multiple areas which were each under the 500 square feet (46 m²) limit but would aggregate to more than that amount. In situations which cannot meet these limitations (homes in Climate Zones 1 through 3 or homes needing more than 500 square feet (46 m²) of reduced ceiling insulation) the depth of the rafters would have to be increased to meet the table's required insulation level or other design changes or means of compliance would be necessary. For example, If more than 500 square feet (46 m²) of ceiling had an R-30 where a R-38 or R-49 was required, then some kind of tradeoff would be required to make up for the R-30 beyond 500 square feet (46 m²) One possibility would be that the area greater than 500 square feet (46 m²) could be traded off with an increased insulation or decreased window *U*-factor based on a total UA calculation under Section N1102.1.3.

N1102.2.3 Mass walls. Mass walls, for the purposes of this chapter, shall be considered walls of concrete block, concrete, insulated concrete form (ICF), masonry cavity, brick (other than brick veneer), earth (adobe, compressed earth block, rammed earth) and solid timber/logs. The provisions of Section N1102.1

for mass walls shall be applicable when at least 50 percent of the required insulation R-value is on the exterior of, or integral to, the wall. Walls that do not meet this criterion for insulation placement shall meet the wood frame wall insulation requirements of Section N1102.1.

Exception: For walls that do not meet this criterion for insulation placement, the minimum added insulation R-value shall be permitted to be:

1. R-value of 4 in Climate Zone 1

2. R-value of 6 in Climate Zone 2

3. R-value of 8 in Climate Zone 3

4. R-value of 10 in Climate Zone 4 except Marine

5. R-value of 13 in climate Zone 5 and Marine 4

❖ The residential code uses a simple definition for mass walls. Walls made of the specified materials and mass are mass walls. Mass walls may meet the lower mass wall R-value (as compared to frame wall values) specified in their respective Climate Zones because of the energy conserving characteristics of mass walls. Note that the difference between the wood-frame R-value and the mass wall R-value is greatest in southern climates. This recognizes that the thermal "averaging" provided by mass walls is most effective in warmer climates. In the very northern climates where there is almost continual heating during parts of the year, the thermal mass is of limited value.

The code defines a mass wall as a masonry or concrete wall having a mass equal to or greater than 30 pounds per square foot (146 kg/m²); a solid wood wall (e.g., in a log house) having a mass greater than 20 pounds per square foot (98 kg/m²); or any other walls having a heat capacity equal to or greater than 6 Btu/ft² · °F (266 J/m² · K). In general terms, the heat capacity is a measure of how well a material stores heat. The higher the heat capacity, the greater the amount of heat stored in the material. For example, a 6-inch (152 mm) heavyweight concrete wall has a heat capacity of 14 Btu/ft² · °F (620 J/m² · K) compared to a conventional 2 by 4 wood-framed wall with a heat capacity of approximately 3 Btu/ft² · °F (138 J/m² · K). Tables N1102.1 and N1102.1.2 are used to determine the insulation or equivalent U-factor requirements for mass walls. To use the tables, first consult either Table N1101.2 or Figure N1102.1 to determine the Climate Zone of the proposed project. The R-value for the assembly is then given for the mass wall in Table N1102.1 or the U-factor is given in Table N1102.2.

When dealing with the mass walls, the insulation location is important and the provisions of Section N1102.2.3 should be reviewed. The values listed in Tables N1102.1 and N1102.2 assume that the majority of the insulation ("at least 50 percent") is located on the exterior side of the wall or that it is integral to the wall. Note that the code offers two different R-values depending on the location of the insulation. If the insulation is 50 percent or more external to the mass or integral to the mass, it specifies the R-value in the Table N1102.1. Therefore in Climate Zone 3, a mass wall would require a minimum of R-5 insula-

tion. Mass walls with insulation more than 50 percent on the interior must have an R-value higher than specified for mass walls and are required to use the "wood frame wall" insulation requirement in Table N1102.1. So again, a mass wall in Climate Zone 3 with the majority of the insulation on the interior, the insulation level of R-13 from the wood frame wall column would be used.

Instead of using the higher "wood frame wall" values from Table N1102.1 for walls which have the major portion of the insulation located on the interior side of the wall, the exception in Section N1102.2.2.3 will permit the installation of specific R-values based on the Climate zone. For Climate Zones 1 through 5 the R-value is lower than required for wood frame walls. For Climate Zones 6 through 8 the exception is not applicable and the required insulation value is the wood-frame wall R-value in Table N1102.1.

For an example of a wall assembly that has insulation on the exterior of the wall [between the mass wall and the exterior] [see Commentary Figure N1102.2.3(1)]. Concrete masonry units with insulated cores or masonry cavity walls are examples of integral insulation. For mass walls with the insulation installed on the interior [an insulated furred wall located between the conditioned space and the mass wall] [see Commentary Figure N1102.2.3(2)]. As shown, these two figures would be examples of mass walls that meet or exceed the general requirements of Section N1102.2.3 and Table N1102.1 for climates zones 1 through 4, except for those in "Marine 4." If the exception in Section N1102.2.3 is used, the R-value of the insulation shown in Commentary Figure N1102.2.3(2) could be reduced to the level specified for each of the Climate Zones. As an example, the insulation in this wall could be reduced to a level of R-8 in Climate Zone 3 or the mass wall could be built as shown and be acceptable in Climate Zone 5 and Marine 4.

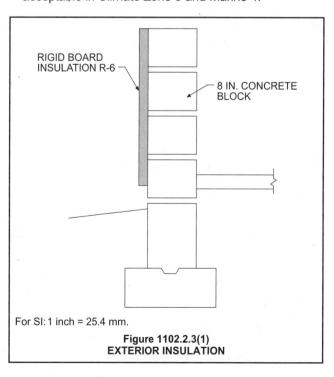

For SI: 1 inch = 25.4 mm.

**Figure 1102.2.3(1)
EXTERIOR INSULATION**

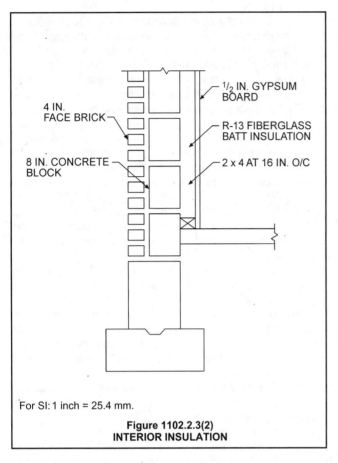

4 IN. FACE BRICK

8 IN. CONCRETE BLOCK

¹/₂ IN. GYPSUM BOARD

R-13 FIBERGLASS BATT INSULATION

2 x 4 AT 16 IN. O/C

For SI: 1 inch = 25.4 mm.

Figure 1102.2.3(2)
INTERIOR INSULATION

N1102.2.4 Steel-frame ceilings, walls and floors. Steel-frame ceilings, walls and floors shall meet the insulation requirements of Table N1102.2.4 or shall meet the U-factor requirements in Table N1102.1.2. The calculation of the U-factor for a steel-frame envelope assembly shall use a series-parallel path calculation method.

❖ The insulation requirements of Table N1102.1 are based on conventional wood-frame construction methods. Because the IRC also includes provisions applicable to steel framing methods, the energy provisions of Chapter 11 have been written to include this material. Table N1102.2.4 specifies combinations of cavity and continuous insulation for steel framing that are equivalent to the specified wood-frame component R-values in Table N1102.1.

Table N1102.1 cannot be used directly for steel-frame components. When using Table N1102.4, all listed options comply; the code user can choose any option on the list that corresponds to the correct R-value in the left-hand column. Steel has a much higher thermal conductivity (ability to transfer heat) than wood. Therefore, steel-frame cavities require a higher R-value and/or include a requirement for insulated sheathing that acts as a thermal break.

Instead of using this table, the code user could choose to calculate or measure a U-factor for a building component and show compliance based on meeting the U-factor requirement in Table N1102.1.2. The code user could use the total UA trade off in Section

TABLE N1102.2.4
STEEL-FRAME CEILING, WALL AND FLOOR INSULATION (R-VALUE)

WOOD FRAME R-VALUE REQUIREMENT	COLD-FORMED STEEL EQUIVALENT R-VALUE[a]
Steel Truss Ceilings[a]	
R-30	R-38 or R-30 + 3 or R-26 + 5
R-38	R-49 or R-38 + 3
R-49	R-38 + 5
Steel Joist Ceilings[b]	
R-30	R-38 in 2 × 4 or 2 × 6 or 2 × 8 R-49 in any framing
R-38	R-49 in 2 × 4 or 2 × 6 or 2 × 8 or 2 × 10
Steel Framed Wall	
R-13	R-13 + 5 or R-15 + 4 or R-21 + 3
R-19	R-13 + 9 or R-19 + 8 or R-25 + 7
R-21	R-13 +10 or R-19 + 9 or R-25 + 8
Steel Joist Floor	
R-13	R-19 in 2 × 6 R-19 + R-6 in 2 × 8 or 2 ×10
R-19	R-19 + R-6 in 2 × 6 R-19 + R-12 in 2 × 8 or 2 × 10

For SI: 1 inch = 25.4 mm.
a. Cavity insulation R-value is listed first, followed by continuous insulation R-value.
b. Insulation exceeding the height of the framing shall cover the framing.

❖ See the commentary for Section N1102.2.4.

N1102.1.3 or even the performance-based approach in Section 404 of the IECC to show compliance based on the overall building, even if the steel frame wall did not meet the code directly. Other combinations of cavity insulation and continuous insulation not shown in Table N1102.2.4 would also be allowed; however, if a combination from the Table is used, no additional calculation is required.

N1102.2.5 Floors. Floor insulation shall be installed to maintain permanent contact with the underside of the subfloor decking.

❖ Floors which are a part of the building thermal envelope, such as those over a crawlspace or an unconditioned garage are required to meet or exceed the floor *R*-value requirements listed in Table N1102.1. The insulation *R*-value requirements range from R-13 in warm climates R-30 in extremely cold climates. Insulation must be installed between the floor joists and must be well supported with netting, wire, wood strips, or another method of support so that the insulation does not droop or fall out of the joist cavities over time.

Some floor insulation has a tendency to sag or drop with time. This sag or drop exposes the sub floor directly to the temperature beneath the floor. Sagging also has a tendency to open airflow paths to parts of the floor producing cold spots and negating the value of the floor insulation in the affected section of the floor. Even small areas that lack insulation or allow air circulation between the floor insulation and the subfloor can have a marked effect on the energy efficiency of the floor. This section specifies that floor insulation must be installed so it will maintain "permanent contact" with the subfloor (meaning over the useful life of the residence).

Earlier editions of the IRC and other energy codes included a provision requiring that floor assemblies over outside air (not including a vented crawlspace or an unconditioned garage) be insulated to the higher *R*-value requirements for ceilings. This typically applied only if over 25 percent of the floor assembly area was directly exposed to the outside air. With the changes made in the 2006 IRC, this old requirement no longer exists. Therefore if a floor of a building extends over outside air, such as over an open car port, the floor is still insulated to the "floor" requirement of Table N1102.1 and the direct contact requirements of this section become important.

See additional code provisions which relate to the insulation of the floor in Section N1102.2.6 when dealing with basements or Section N1102.2.8 when dealing with crawlspaces.

N1102.2.6 Basement walls. Exterior walls associated with conditioned basements shall be insulated from the top of the basement wall down to 10 feet (3048 mm) below grade or to the basement floor, whichever is less. Walls associated with unconditioned basements shall meet this requirement unless the floor overhead is insulated in accordance with Sections N1102.1 and N1102.2.5.

❖ The walls of conditioned basements must be insulated to meet the requirements of Table N1102.1. Each wall of a basement must be considered separately to determine whether it is a basement wall or an exterior wall. It is a basement wall if it has an average below-grade wall area of 50 percent or more. A wall that is less than 50 percent below grade is an exterior wall and must meet the insulation requirements for walls. Most walls associated with basements will be at least 50 percent below grade and therefore must meet basement wall requirements. Walk-out basements offer a challenge in determining compliance with the code. A walk-out basement [see Commentary Figures N1102.2.6(1) and (2)] may have a back wall that is entirely below grade, a front wall or the walk-out portion that is entirely above grade, and two side walls with a grade line running diagonally. In this case the back wall must meet the requirements for basement walls, the front wall would need to meet the requirements for walls (either framed wall or mass wall), and the side walls would need to be evaluated to determine whether they were 50 percent or more below grade and therefore basement walls.

Basement insulation must extend up to 10 feet (3048 mm) under the ground, or at least as far as the basement wall extends under the ground. Heat flow into the ground occurs all along the buried portion of the wall, as well as along the above-ground portion of the wall. Heat flow below 10 feet (3048 mm) is greatly diminished, so the code requires basement insulation only down to 10 feet (3048 mm), or the depth of the basement wall.

The code does not specify whether the insulation is to be placed on the inside or outside of a basement wall. In some localities moisture considerations may suggest the type and location for insulation.

The last part of this section allows insulating unconditioned basement walls as an alternative to insulating the floor above the unconditioned basement. Therefore, it essentially shifts the location of the building thermal envelope from the floor to the basement walls. Although not required, insulating the unconditioned basement walls makes a good deal of sense if a basement is likely to be conditioned at some time after construction.

Because the rim joist between floors is a part of the building envelope, this must be insulated also if the basement is conditioned.

N1102.2.7 Slab-on-grade floors. Slab-on-grade floors with a floor surface less than 12 inches below grade shall be insulated in accordance with Table N1102.1. The insulation shall extend downward from the top of the slab on the outside or inside of the foundation wall. Insulation located below grade shall be extended the distance provided in Table N1102.1 by any combination of vertical insulation, insulation extending under the slab or insulation extending out from the building. Insulation extending away from the building shall be protected by pavement or by a minimum of 10 inches (254 mm) of soil. The top edge of the insulation installed between the exterior wall and the edge of the interior slab shall be permitted to be cut at a

45-degree (0.79 rad) angle away from the exterior wall. Slab-edge insulation is not required in jurisdictions designated by the code official as having a very heavy termite infestation.

❖ The perimeter edges of slab-on-grade floors must be insulated to the *R*-value listed in Table N1102.1. These requirements apply only to slabs 12 inches (305 mm) or less below grade. The listed *R*-value requirements in the table are for unheated slabs. A heated slab must add another R-5 to the required insulation levels based on Note d of Table N1102.1.

The insulation must extend downward from the top of the slab or downward to the bottom of the slab and then horizontally in either direction until the distance listed in Table N1102.1 is reached. See Commentary Figure N1102.2.7 for examples or how the distance is measured. Most of the heat loss from a slab will occur in the edge that is exposed directly to the outside air. The insulation must be installed to the top of the slab edge to prevent this heat loss. Slab insulation may be installed on the exterior of the slab edge or between the interior wall and the edge of the interior slab as in a non-monolithic slab. In this type of installation, the exposed insulation could cause problems with tack strips for carpeting. Therefore, the insulation is allowed to be cut at a 45-degree angle away from the exterior or the wall. If a monolithic slab and foundation is being used, the required insulation would obviously need to be installed on the exterior and then ei-

ther extended to the required depth or turned out to the exterior and be protected by either some type of pavement or a minimum of 10 inches (254 mm) of soil. Insulation which is exposed on or near the surface is easily damaged. This protection method assures that the insulation remains in place and provides the intended energy savings.

In areas with very heavy termite infestation, slab perimeter insulation need not be installed per Table N1102.1. These areas are identified in Figure R301.2(6) or the jurisdiction may base their determination on the local history and situation. It is important to understand that the revisions of the 2006 IRC energy requirements provide this exemption from the slab insulation provisions for any area with heavy termite infestations. The fact that this is an exemption and does not contain any requirement for a compensating increase of insulation at other locations is important. The 2000 and 2003 editions of the IRC only permitted the elimination of this slab edge insulation if one of the trade-off approaches from the IECC was used to make up for this lack of insulation at the slab edge. The requirements of IRC Section N1102.1.3 could still be used in areas which do not have a heavy termite infestation to eliminate the slab edge insulation if desired. Typically, slab perimeter insulation can be traded off entirely in these climates by increasing the ceiling or wall insulation *R*-values or by using glazing with a lower *U*-factor.

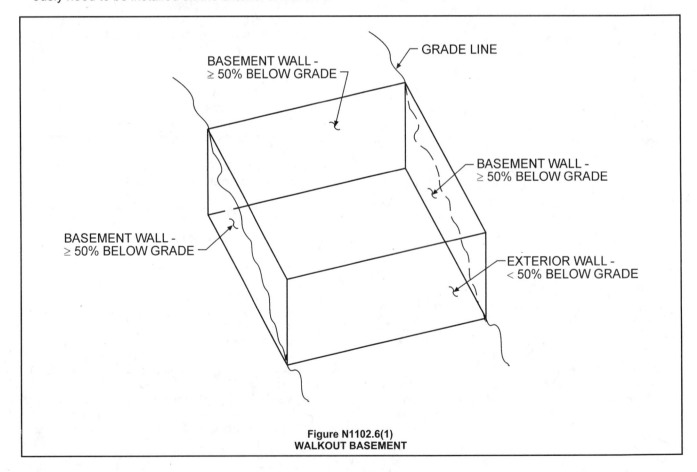

Figure N1102.6(1)
WALKOUT BASEMENT

Figure N1102.6(2)
WALKOUT BASEMENT

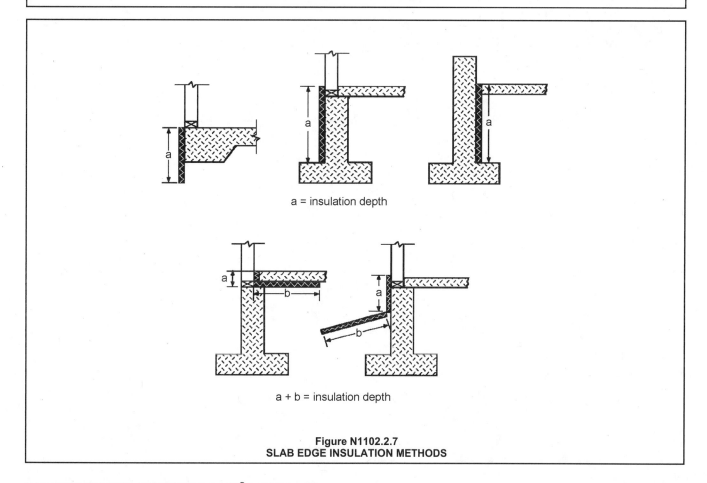

a = insulation depth

a + b = insulation depth

Figure N1102.2.7
SLAB EDGE INSULATION METHODS

N1102.2.8 Crawl space walls. As an alternative to insulating floors over crawl spaces, insulation of crawl space walls when the crawl space is not vented to the outside is permitted. Crawl space wall insulation shall be permanently fastened to the wall and extend downward from the floor to the finished grade level and then vertically and/or horizontally for at least an additional 24 inches (610 mm). Exposed earth in unvented crawl space foundations shall be covered with a continuous vapor retarder. All joints of the vapor retarder shall overlap by 6 inches (152 mm) and be sealed or taped. The edges of the vapor retarder shall extend at least 6 inches (152 mm) up the stem wall and shall be attached to the stem wall.

❖ The IRC allows for the insulation of crawlspace walls in crawlspaces instead of insulating the floor between the crawlspace and the conditioned space. In essence the code user is defining the thermal boundary as either the floor or the crawlspace wall. Because the ground under the crawlspace is tempered by the thermal mass of the dirt, the temperature of the crawlspace is usually more favorable than the outside temperature. This is a popular practice for freeze protection in colder climates because it is common to install plumbing in the crawlspace. The heat transferred through the uninsulated floor to the crawlspace helps keep the crawlspace temperature above freezing when the outside air temperature drops below freezing. To comply with this provision, the crawlspace must be mechanically vented or supplied with conditioned air from the living space. The IRC provides an exception from the general underfloor ventilation requirement, listed under Section R408.2. The provisions of Section R408.3 require that either mechanical ventilation or conditioned air be provided to the space.

The code also requires installation of insulation from the sill plate downward to the exterior finished grade level and then an additional 24 inches (610 mm) either vertically or horizontally. See Commentary Figure N1102.2.8. Under this insulation scenario, the rim joist is considered part of the conditioned envelope and must be insulated to the same level as the exterior wall. The insulation must be attached securely to the crawlspace wall so that it does not fall off. The code also requires installing a continuous vapor retarder on the floor of the crawlspace to prevent ground water vapor from entering the crawlspace. The vapor retarder must have a maximum permeance rating of 1 perm (see definition of "Vapor retarder"). Typically 6 mil (0.15 mm) polyethylene vapor retarder is used for this purpose. The vapor retarder is to be installed with all joints overlapped and sealed or taped to provide continuity. Also, the vapor retarder must extend up the crawlspace wall and be secured to the wall with an appropriate attachment such as either tape or a treated wood nailer.

N1102.2.9 Masonry veneer. Insulation shall not be required on the horizontal portion of the foundation that supports a masonry veneer.

❖ For exterior foundation insulation, the horizontal portion of the foundation that supports a masonry veneer need not be insulated. For slab edge insulation installed on the exterior of the slab, the code allows the insulation to start at the bottom of the masonry veneer and extend downward. This is essentially a matter of practicality and accommodates the construction of a "brick ledge" without the need for insulating the foundation at the point where the masonry would bear upon it.

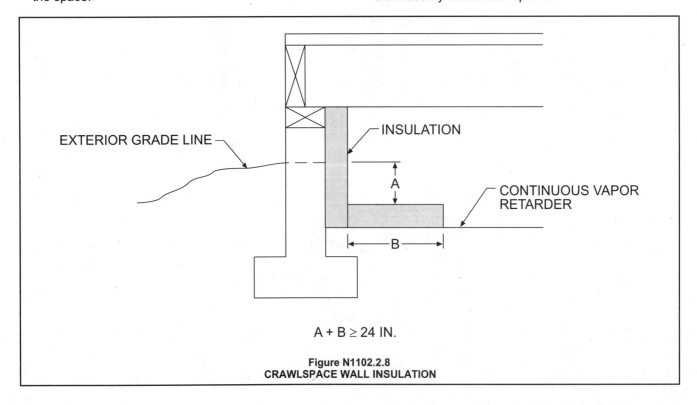

EXTERIOR GRADE LINE

INSULATION

A

CONTINUOUS VAPOR RETARDER

B

A + B ≥ 24 IN.

Figure N1102.2.8
CRAWLSPACE WALL INSULATION

N1102.2.10 Thermally isolated sunroom insulation. The minimum ceiling insulation *R*-values shall be R-19 in zones 1 through 4 and R-24 in zones 5 though 8. The minimum wall *R*-value shall be R-13 in all zones. New wall(s) separating the sunroom from conditioned space shall meet the building thermal envelope requirements.

❖ This section allows reductions in the insulation levels from those in Table N1102.1 and specifies the specific amounts of insulation required for a "thermally isolated sunroom." To apply these provisions correctly, review the definitions for both "sunroom" and "thermal isolation." To qualify as "thermally isolated" the sunroom must have not only a barrier (physical separation) to separate it from the conditioned spaces of the home but also have separate heating and cooling equipment serving only the sunroom or at a minimum controlled as a separate zone from the home's HVAC equipment. An adjustable duct into the sunroom does not constitute separate HVAC controls.

As this section specifies, when the wall separating the sunroom from the conditioned space of the home is new construction, it must meet the general building thermal envelope requirements from Table N1102.1. As an example, in Climate Zone 5 a wood frame wall would require a minimum of R-19 insulation and any doors or windows (fenestration) between the home and the sunroom would require a maximum *U*-factor of 0.35. (See Section N1102.3.5 regarding *U*-factor requirements.) If the sunroom was being added onto an existing home, the code would not apply the building thermal envelope requirements to the wall between the dwelling and the sunroom but would still require the "physical separation" which is required by the definition. Therefore it would not be permissible to simply have a cased opening between the sunroom and the home when using the provisions of this section.

It is also important to understand these provisions so that the code user is aware of what is not regulated by this section. It is possible to build sunrooms that are not "thermally isolated" but are considered as being within or a part of the building thermal envelope. In such situations, the general requirements of Section N1102.1 would be applied and the appropriate provisions of Table N1102.1 would also apply to the sunroom. Another option would be that the provisions of Section N1102.1.3 could be used to trade off better performance in other areas of the home against the poorer performing sunroom. If the sunroom is constructed either in compliance with the general provisions or using Section N1102.1.3, thermal isolation is not required. One additional option is that based on the scope of this section an unconditioned sunroom would not have to meet these requirements, presuming it was separated from the rest of the conditioned space by code-compliant components.

N1102.3 Fenestration.

❖ This section contains specific requirements that affect the requirements for the individual items listed within the subsections. Although Section N1102.1 and Tables N1102.1 and N1102.1.2 provide the general basis for complying with the energy requirements of this chapter, Section N1102.3 provides additional details regarding the application of the provisions or modifications which may affect the general fenestration requirements. Relying on the provisions of Section R102.1 will help assure that these specific provisions are properly applied.

The term "fenestration" in this section refers to opaque doors and the light-transmitting areas of a residential building's wall, floor or roof, generally window, skylight and non-opaque door products. (See "fenestration" definition in Chapter 2.) Prior to the 2006 edition, the IRC energy provisions applied only to buildings with 15 percent or less of glazing areas (fenestration). These earlier versions of the code established whole building performance requirements, with fenestration performance requirements as a derivative value dependent on window area, overall envelope area and the performance of other assemblies (e.g., walls, ceilings, floors). The 2006 IRC and IECC largely disconnect fenestration performance from area and other assemblies by establishing specific simplified prescriptive requirements (without area considerations) for these products in Table N1102.1.1 – specifically fenestration *U*-factors, skylight *U*-factors and glazed fenestration SHGCs.

The fenestration requirements of the code are critical to the overall energy efficiency of the residence. First, unlike opaque assemblies, glazed fenestration can transmit a substantial amount of heat gain through the glazing into the living space in both the summer and winter, resulting in a unique concern about solar heat gain (and, as a result, necessitating SHGC requirements). Second, the insulating value (*U*-factor) of typical fenestration is much higher than that of a typical wall. For example, a good low-e, insulated glass wood or vinyl fenestration product will have less than $1/4$ the insulating value of an equivalent area of R-13 insulated opaque wall. These issues have an affect not only on energy use, but overall occupant comfort, condensation and other issues.

In accordance with Section N1101.5, the *U*-factor and SHGC for each fenestration product must be obtained from a label attached to the product certifying that the values were determined in accordance with NFRC procedures by an accredited, independent lab or from a limited default table.

N1102.3.1 *U*-factor. An area-weighted average of fenestration products shall be permitted to satisfy the *U*-factor requirements.

❖ Section N1102.3.1 permits using the calculated "area weighted average" *U*-factor of all fenestration products in the building to satisfy the fenestration *U*-factor requirements set by Tables N1102.1 or N1102.1.2. As a result, if all fenestration products (window, door or skylight) do not meet the specific value, the user can still achieve compliance if the weighted average of all products is equal to or less than the specified value.

This option permits the use of some windows that have values worse than the prescriptive general requirement, so long as these more poorly-performing windows are offset by windows with values better than the requirement.

When applying this "area weighted" option it is important to remember that the term fenestration includes windows, skylights, doors with glazing and opaque doors; all of which would be included in the average calculation.

Using the *U*-factor requirement of 0.35 for Climate Zone 6 as an illustration, this section provides two options for compliance. The simplest option is to ensure that all windows and doors have labeled NFRC values of 0.35 or less. This approach is also more likely to ensure adequate performance and comfort throughout the home. Alternately, a weighted average may be taken of the values from all windows and doors to see if the "weighted average" is less than or equal to 0.35. As a simple example, assume 100 square feet of 0.32 windows, 100 square feet of 0.36 windows and one 20 square foot 0.40 U-factor door. [(100 × 0.32) + (100 × 0.36) + (20 × 0.40)] / (100 + 100 + 20) = 0.345 (weighted average *U*-factor). Therefore, because the weighted average *U*-factor is less than the required 0.35, the fenestration in this example would be in compliance with the code.

In accordance with Section N1101.5, the *U*-factor for each fenestration product must be obtained from a label attached to the product certifying that the *U*-factor was determined in accordance with NFRC procedures by an accredited, independent lab. In the absence of a NFRC-labeled *U*-factor, a value from the limited default tables [Tables N1101.5(1) and N1101.5(2)] must be used.

N1102.3.2 Glazed fenestration SHGC. An area-weighted average of fenestration products more than 50 percent glazed shall be permitted to satisfy the solar heat gain coefficient (SHGC) requirements.

❖ Under Table N1102.1, all glazed fenestration products in Climate Zones 1 through 3 must have an SHGC equal to or less than 0.40 (there is no requirement in Climate Zones 4 through 8). This requirement is intended to control unwanted solar gain in cooling-dominated climates to increase comfort, reduce air conditioning energy and peaks, reduce HVAC sizing and reduce energy costs.

Similar to Section N1102.3.1, Section N1102.3.2 allows some latitude for individual product variability by permitting this performance requirement to be met using an area-weighted average of all of the fenestration products that are more than 50 percent glazed. The 50 percent glazing threshold is established to exclude from the equation doors or other fenestration products that are either completely or largely opaque. The reason for this exclusion is that opaque elements do not allow solar heat gain as glazing does. The area-weighting calculation approach is explained above in an example with Section N1102.3.1. An additional example for SHGC is as follows:

Example:

Window 1	SHGC - 0.24	200 ft²
Window 2	SHGC - 0.42	100 ft²
Window 3	SHGC - 0.40	100 ft²
Sliding glass door	SHGC - 0.50	40 ft²

$$\frac{\left[\left(200\,ft^2 \times 0.24\right) + \left(100\,ft^2 \times 0.42\right) + \left(100\,ft^2 \times 0.40\right) + \left(40\,ft^2 \times 0.50\right)\right]}{440\,ft^2}$$

= SHGC - 0.34 Average

Using the figures in the example, even though window 2 and the sliding glass door both have SHGC values which exceed the 0.40 limitation of Table N1102.1 for Climate Zones 1 through 3, this design will comply because the weighted average is 0.34.

In accordance with Section N1101.5, the SHGC for each glazed fenestration product must be obtained from a label attached to the product certifying that the SHGC was determined in accordance with NFRC procedures by an accredited, independent lab. In the absence of a NFRC-labeled *U*-factor, a value from the limited default table (Table N1101.5(3)) must be used.

It is important to note that the SHGC requirement must be met by the fenestration product on a stand-alone basis. The code does not permit the alternative of a "permanent solar shading device" such as eave overhangs or awnings as was permitted by other energy codes and previous versions of the IECC to assist in code compliance.

N1102.3.3 Glazed fenestration exemption. Up to 15 square feet (1.4 m²) of glazed fenestration per dwelling unit shall be permitted to be exempt from *U*-factor and solar heat gain coefficient (SHGC) requirements in Section N1102.1.

❖ In addition to using the area-weighted average approach (Sections N1102.3.1 and N1102.3.2) to allow maximum compliance flexibility for builders, the code allows up to 15 square feet (1.4 m²) of the building's total glazed fenestration area to be exempt from the *U*-factor and SHGC requirements listed in Table N1102.1. All other glazing must meet or exceed the designated *U*-factor and SHGC requirements. The exempted glazing area should be designated on the building plan, either on the floor plan or within a window schedule. This will give the necessary information to the field inspector who will then need to verify that what is on the plans is installed in the field. This exemption allows the use of ornate or unique window, skylight or glazed door assemblies in a building without going to another compliance approach. The area, the *U*-factor and SHGC of the exempt product(s) should be excluded from the area weighting calculations that may be performed under Sections N1102.3.1 and N1102.3.2. In addition the exception provided by this section would also allow this 15 square feet (1.4 m²) of glazing to be excluded from the limits of Section N1102.5.1.

N1102.3.4 Opaque door exemption. One opaque door assembly is exempted from the *U*-factor requirement in Section N1102.1.

❖ Similar to the exemption provided in Section N1102.3.3, to enhance design flexibility, the IRC allows one opaque door to be exempt from fenestration *U*-factor requirements as contained in Table N1102.1 as well as the limitations of Section N1102.5.1. Although the IRC does not define it, an opaque door is generally considered to be a fenestration product with an overall glazing area of less than 50 percent. IECC Section 502.2.7 provides this limitation of 50 percent glazing when it is dealing with opaque doors. The opaque door exemption allows builders to use an ornate or otherwise *U*-factor noncompliant entrance door assembly in a building without going to another compliance approach. The area and the *U*-factor of the exempt product should be excluded from the area weighting calculations under Section N1102.3.1.

N1102.3.5 Thermally isolated sunroom *U*-factor. For zones 4 through 8 the maximum fenestration *U*-factor shall be 0.50 and the maximum skylight *U*-factor shall be 0.75. New windows and doors separating the sunroom from conditioned space shall meet the building thermal envelope requirements.

❖ Sunrooms that are thermally isolated have higher window and skylight *U*-factor requirements. This section sets the maximum fenestration *U*-factor at 0.50 for thermally isolated sunrooms constructed in Climate Zones 4 through 8, and the maximum skylight *U*-factor at 0.75. In other Climate Zones, the maximum fenestration and skylight *U*-factors for thermally isolated sunrooms are the maximums contained in Table N1102.1.

These reduced requirements apply only to sunrooms that are thermally isolated. Section 202 contains definitions for Sunroom and Thermal Isolation. There is additional discussion in the commentary to Section N1102.2.10 regarding the application of these requirements which is also applicable to this section. Based on these definitions, to satisfy this exception, a "thermally isolated sunroom" must:

1. Be a one-story structure attached to a dwelling,

2. Have a glazing area in excess of 40 percent of the gross area of the sunroom's exterior walls and roof,

3. Be thermally isolated by:

   3.1. Having physical and space conditioning separate from the other conditioned space in the home or

   3.2. Be controlled as a separate zone or have separate conditioning equipment.

In general, when using this section, walls (wood-frame, mass, or steel-frame) separating sunrooms from conditioned space must meet the appropriate requirement as if the wall was separating conditioned space from the exterior. If a new door or window opening is added to permit passage between or to provide a connection with the dwelling and the thermally isolated sunroom, the new door or window must comply with the fenestration *U*-factor as established in Table N1102.1. Where a sunroom is added to an existing home, existing doors and windows between the home and the sunroom are not affected by this section.

This section properly recognizes the different uses for and performance capabilities of thermally-isolated sunrooms, as compared with other additions or sunrooms that are not thermally isolated.

N1102.3.6 Replacement fenestration. Where some or all of an existing fenestration unit is replaced with a new fenestration product, including sash and glazing, the replacement fenestration unit shall meet the applicable requirements for *U*-factor and solar heat gain coefficient (SHGC) in Table N1102.1.

❖ Replacing only a glass pane in an existing sash and frame would not fall under this provision if the *U*-factor and SHGC will be equal to or lower than the values prior to the replacement. In situations where the existing values are not known or where the replacement is more than just replacing the glass pane, the provisions may be applicable. It is often common practice when fenestration is replaced to remove only the sash and glazing of an existing window and replace them with an entirely new fenestration product. Sometimes during the process the existing frame is also removed, but many times, the new fenestration product is custom made to fit in the existing space left after the sash and glazed portions are removed. In essence the new fenestration is installed in or over the existing frame. Whether the existing frame is removed or not, these types of replacements are regulated by this section.

Section N1102.3.6 requires that each fenestration unit replaced in a residence not exceed the maximum fenestration *U*-factor and SHGC for the applicable Climate Zone. This requirement applies to all replacement windows, even if the existing frame is not removed (e.g., the new window is placed inside the old frame), so long as the sash and glazing is replaced. In addition, remember that the definition of fenestration includes doors, which must meet the same *U*-factor requirements as windows. Therefore, the replacement of a door would also have to meet these requirements.

When dealing with replacement fenestration, the building official should be consulted to explain how this requirement will be applied. For simple ease of application for both the building official and the installer, the *U*-factor and SHGC requirements could simply be applied to each fenestration unit. Therefore, the *U*-factor and SHGC required for each unit would be the values listed directly in Table N1102.1. However, if acceptable to the building official and additional information is available regarding the performance of the remaining existing windows in the home, it may be reasonable to permit the use of the area-weighted values of Sections N1102.3.1 and N1102.3.2 or even the exemptions of Sections N1102.3.3 and N1102.3.4 to the replaced fenestration unit.

In accordance with Section N1101.5, the *U*-factor and SHGC for each replacement fenestration product must be obtained from a label attached to the product certifying that the values were determined in accordance with NFRC procedures by an accredited, independent lab. In the absence of a NFRC-labeled *U*-factor or SHGC, a value from the limited default tables [Table N1101.5(1) through N1101.5.(3)] must be used. The NFRC procedures do include applicable methods to test various replacement products.

N1102.4 Air leakage.

❖ Sealing the building envelope is critical to good thermal performance for the building. The seal will prevent warm, conditioned air from leaking out around doors, windows and other cracks during the heating season, thereby reducing the cost of heating the residence. During the hot summer months, a proper seal will stop hot air from entering the residence, helping to reduce the air-conditioning load on the building. Any penetration in the building envelope must be thoroughly sealed during the construction process, including holes made for the installation of plumbing, electrical and heating and cooling systems (see Commentary Figure N1102.4). The code lists several areas that must be caulked, gasketed, weatherstripped, wrapped or otherwise sealed to limit uncontrolled air movement. Most of the air sealing will be done prior to the installation of an interior wall covering because any penetra-

tion will be noticeable and accessible at this time. The code allows the use of airflow retarders (house wraps) or other solid materials as an acceptable method to meet this requirement. To be effective, the building thermal envelope seal must be:

- Impermeable to air flow.
- Continuous over the entire building envelope.
- Able to withstand the forces that may act on it during and after construction.
- Durable over the expected lifetime of the building.

It is unlikely that the same type of barrier will be used on all portions of the building's thermal envelope. Therefore, joints between the various elements, as well as joints or splices within products (such as the overlap in separate pieces of house wrap), must be effectively addressed to provide the continuity needed to perform as desired.

N1102.4.1 Building thermal envelope. The building thermal envelope shall be durably sealed to limit infiltration. The sealing methods between dissimilar materials shall allow for differential expansion and contraction. The following shall be caulked, gasketed, weatherstripped or otherwise sealed with an air barrier material, suitable film or solid material.

1. All joints, seams and penetrations.

2. Site-built windows, doors and skylights.

3. Openings between window and door assemblies and their respective jambs and framing.

4. Utility penetrations.

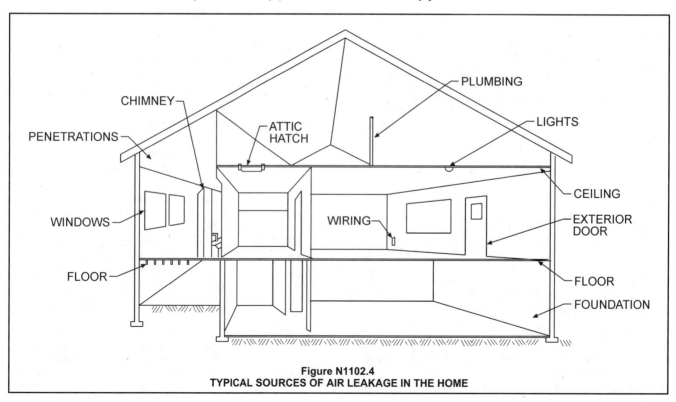

Figure N1102.4
TYPICAL SOURCES OF AIR LEAKAGE IN THE HOME

5. Dropped ceilings or chases adjacent to the thermal envelope.

6. Knee walls.

7. Walls and ceilings separating the garage from conditioned spaces.

8. Behind tubs and showers on exterior walls.

9. Common walls between dwelling units.

10. Other sources of infiltration.

❖ Air infiltration is a major source of energy use because the incoming air usually requires conditioning. The uncontrolled introduction of outside air (infiltration) creates a load that varies with time. Ventilation, the controlled introduction of fresh air, is more manageable and provides a more controlled air quality. Uncontrolled infiltration also has a tendency to create or aggravate moisture problems, providing an additional reason to limit infiltration.

The elements in this list include some of the largest sources of infiltration. Construction details not on this list must still be sealed by virtue of item 10. As stated above, it is the continuity of all of the barrier items which assure the building gets adequately sealed.

N1102.4.2 Fenestration air leakage. Windows, skylights and sliding glass doors shall have an air infiltration rate of no more than 0.3 cubic foot per minute per square foot [1.5(L/s)/m²], and swinging doors no more than 0.5 cubic foot per minute per square foot [2.5(L/s)/m²], when tested according to NFRC 400 or AAMA/WDMA/CSA 101/I.S.2/ A440 by an accredited, independent laboratory, and listed and labeled by the manufacturer.

Exception: Site-built windows, skylights and doors.

❖ Windows, skylights and doors should be tested and labeled by the manufacturer as meeting the air infiltration requirements. The intent of this section is to effectively complete the sealing of the building's thermal envelope by providing specific testing and performance criteria for windows, skylights and doors. This testing and labeling requirement provides an easy method for both the builder and the inspector to demonstrate compliance with the code and ease acceptance. While "site built" fenestration is exempted from these requirements, units would have to be "durably sealed" to limit infiltration according to the requirements in Section N1102.4.1.

N1102.4.3 Recessed lighting. Recessed luminaires installed in the building thermal envelope shall be sealed to limit air leakage between conditioned and unconditioned spaces by being:

1. IC-rated and labeled with enclosures that are sealed or gasketed to prevent air leakage to the ceiling cavity or unconditioned space; or

2. IC-rated and labeled as meeting ASTM E 283 when tested at 1.57 pounds per square foot (75 Pa) pressure differential with no more than 2.0 cubic feet per minute (0.944 L/s) of air movement from the conditioned space to the ceiling cavity; or

3. Located inside an airtight sealed box with clearances of at least 0.5 inch (13 mm) from combustible material and 3 inches (76 mm) from insulation.

❖ To correctly apply this provision, it is important to realize that it deals only with recessed lights that are "installed in the building thermal envelope." Therefore, lights which are located so that all sides of the luminaire are surrounded by conditioned space would not fall under this section's requirements. For example, a light located in a soffit would not be regulated if the soffit was below a ceiling which served as the building's thermal envelope. Additionally a light installed in the floor/ceiling assembly between a first floor living room and a bedroom above it would also be exempted.

Because of their typical location of installation, recessed lighting fixtures pose a potential fire hazard if incorrectly covered with insulation. In addition, the ceiling or a barrier directly above it often serve as the air leakage or moisture barrier for the home. Therefore, holes cut through the ceiling to install these fixtures also act as chimneys that transfer heat loss and moisture through the building envelope into attic spaces. The heat loss resulting from improperly insulated recessed lighting fixtures can be significant.

Recessed lighting fixtures must either be IC-(insulation contact) rated or installed in a sealed box that has the insulation at least 3 inches (76 mm) from the light fixture. IC-rated lights are typically double-can fixtures, with one can inside another (see Commentary Figure N1102.4.3). The outer can (in contact with insulation) is tested to make sure it remains cool enough to avoid a fire hazard. An IC-rated fixture should have the IC rating stamped on the fixture or printed on an attached label.

Recessed lights must be tightly sealed or gasketed to prevent air leakage through the fixture into the ceiling cavity by one of the three methods listed.

N1102.5 Moisture control. The building design shall not create conditions of accelerated deterioration from moisture condensation. Above-grade frame walls, floors and ceilings not ventilated to allow moisture to escape shall be provided with an approved vapor retarder. The vapor retarder shall be installed on the warm-in-winter side of the thermal insulation.

Exceptions:

1. In construction where moisture or its freezing will not damage the materials.

2. Frame walls, floors and ceilings in jurisdictions in Zones 1, 2, 3, 4A, and 4B. (Crawl space floor vapor retarders are not exempted.)

3. Where other approved means to avoid condensation are provided.

❖ In cold climates, warm, moist air inside the building can migrate through the building envelope and condense on building surfaces or within building component cavities as the migrating air is cooled. In warm climates, moist outside air can migrate into a cooled building and

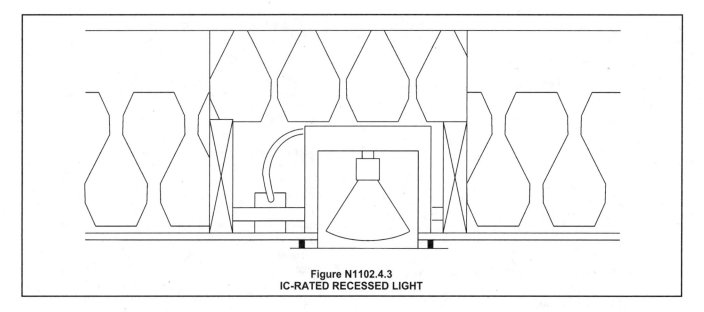

Figure N1102.4.3
IC-RATED RECESSED LIGHT

can also condense on building surfaces. Vapor retarders can help protect insulation and building materials from moisture damage, degradation and decay by preventing the water vapor from entering the external building envelope component cavities. In moderate and cold climates, the vapor retarder is installed on the warm-in-winter side of the wall (the interior) to prevent migration of water vapor from the inside to the exterior. Although not always required by the code, vapor retarders in warm, moist climates are sometimes installed on the warm-in-summer side (the exterior) to prevent inward migration from the exterior.

The warm-in-winter side is the interior side of the thermal insulation in the Climate Zones where this applies.

Sheet polyethylene is often used as a vapor retarder. The perm rating for sheet polyethylene is significantly below the 1 perm ($5.7 \cdot 10^{-11}$ kg/Pa $\cdot$ s $\cdot$ m^2) required by the code.

The code requires installation of vapor retarders in all unventilated framed ceilings, walls and floors. Unventilated areas are framed cavities without vents or other openings that allow significant amounts of air to move freely through the cavity and insulation. For example, a ceiling vapor retarder is not required in a ventilated attic where the insulation is exposed to the ventilated space. Per the definition in Chapter 2, the vapor retarder must have a perm rating of 1.0 or less when tested with the desiccant method using Procedure A of ASTM E 96.

The code also allows other approved alternatives to meet the vapor retarder requirements. A variety of materials act as a vapor retarder, even if not installed specifically for that purpose. For example, any material with an unperforated foil facing, some kinds of foam, some paints specifically formulated as vapor retarders and most kraft-faced fiberglass batt insulation would meet the 1 perm ($5.7 \cdot 10^{-11}$ kg/Pa $\cdot$ s $\cdot$ m^2) requirement in the code. For unfaced materials, the permeability varies with material thickness. Most product manufacturers list a materials perm based on a thickness of 1 inch (25

mm). Increasing the thickness of the material will decrease the perm. For example, 1 inch (25 mm) of extruded polystyrene, sheet foam product used for insulation, is about 1.1 perm, not low enough to be a vapor retarder by itself; however, increasing the thickness to 2 inches (51 mm) decreases the perm to about 0.55 perm, meeting the code requirement for a vapor retarder. Note that it is possible to get the required perm rating based on several materials in series. For example, most floors are constructed so that the floor achieves the required perm rating without the addition of materials specifically designated as a vapor retarder. The code official or other authority having jurisdiction must approve this alternative based on reason of accepted principles or tests by nationally recognized organizations.

The use of the exceptions needs to be evaluated on a case by case basis. Although the use of exception 3 will require review and approval of the proposed system by the code official, exception 2 will be a decision left to the designer. Although exception 2 does permit the elimination of the vapor retarder, the designer could still elect to include one or could elect to locate it at other than the warm-in-winter side. Such decisions must be made based on the historical experience in that location and consider the specific materials of construction. Exception 1 allows exempting exterior envelopes constructed from materials that are not damaged by moisture or freezing from the vapor retarder requirement in any Climate Zone. This exception should probably be reviewed and determined applicable by both the designer and the building official before deciding that it is applicable. Even when using the exceptions, the key principle which is stated in the first sentence of the section should always be considered.

N1102.5.1 Maximum fenestration U-factor. The area weighted average maximum fenestration U-factor permitted using tradeoffs from Section N1102.1.3 in Zones 6 through 8 shall be 0.55.

To comply with this section, the maximum *U*-factor for skylights shall be 0.75 in zones 6 through 8.

❖ This section is intended to clarify the application of the fenestration performance maximums and to set reasonable performance levels for these products when using the trade-offs of Section N1102.1.3.

This section does not define the minimum code requirements for fenestration *U*-factors which are set in Table N1102.1. Rather, this section sets limits on the trade-offs allowed based on the Total UA calculations (Section N1102.1.3). This section is in contrast to the basic principle used in Section N1102.1.3. Although that section will allow a *U*-factor that does not meet the general code requirements offset by the increased efficiency of another part of the residence, this section establishes a limitation on the level of efficiency that can be compensated for.

Note that this section does not set a limit on individual products because Sections N1102.3.1 and N1102.3.2 are not affected. Rather, this section sets an overall limit on the weighted average of those products. These limits are often called "hard limits" or "maximum trade-off" limits.

In situations where the code is applied to only a portion of the residence, such as only to the addition but not the existing residence, the weighted average could be calculated for just the addition.

The thermal properties (*U*-factor) of skylights are uniquely different from window products. After installation they perform differently than windows and are therefore rated differently. As a result, even the highest rated skylights cannot achieve the same level of *U*-factor performance as windows and glass doors. Therefore, the code provides different values for these "maximum" limits in the trade-off.

SECTION N1103
SYSTEMS

N1103.1 Controls. At least one thermostat shall be installed for each separate heating and cooling system.

❖ This provision assures that a separate thermostat is installed for each system. As an example, if separate systems are installed so that one serves the downstairs and one serves the upstairs of a two-story residence, two separate thermostats would be required, one regulating each level. This allows for greater flexibility, control and energy savings than would be possible if both systems were controlled by a single thermostat.

N1103.1.1 Heat pump supplementary heat. Heat pumps having supplementary electric-resistance heat shall have controls that, except during defrost, prevent supplemental heat operation when the heat pump compressor can meet the heating load.

❖ Heat pump systems must have controls that prevent supplementary electric resistance heater operation when the heating load can be met by the heat pump alone. Typically these controls will be thermostats that will have a "Heat Pump" designation on them. The make and model of the thermostat should be called out on the building plans so the inspector can verify that what is installed in the field matches the plans. Because change-outs in the field are common, the instructions that come with the thermostat can also be checked to verify that the thermostat is designed for use with a particular heat pump. To limit the hours of use of the electric resistance heating unit and provide the most cost efficient operation of the equipment, the specific control language is included in this section.

N1103.2 Ducts.

N1103.2.1 Insulation. Supply and return ducts shall be insulated to a minimum of R-8. Ducts in floor trusses shall be insulated to a minimum of R-6.

Exception: Ducts or portions thereof located completely inside the building thermal envelope.

❖ HVAC duct work located outside of the conditioned space must be insulated with insulation of minimum *R*-values. This includes both supply and return ducts.

The exception addresses duct work that is located in part or completely within a conditioned space. This duct work need not be insulated because by definition it is already within, and protected by, the insulation envelope.

N1103.2.2 Sealing. Ducts, air handlers, filter boxes and building cavities used as ducts shall be sealed. Joints and seams shall comply with Section M1601.3.1.

❖ Ducts must be sealed in accordance with the duct sealing requirements of Section M1601.3.1. Joints and seams that fail in a duct system can result in increased energy use because the conditioned air will be delivered to an unconditioned space, such as the crawlspace or attic instead of inside the conditioned envelope. Return-air duct work installed in basements or concealed building spaces may conduct chemicals or other products that produce potentially harmful fumes. Because the return air operates under negative pressure, any leaks could draw fumes, moisture, soil gases or odors from the surrounding area and direct them into the house. Therefore, sealing of return air duct work is also a requirement.

N1103.2.3 Building cavities. Building framing cavities shall not be used as supply ducts.

❖ Stud bays and other building cavities that are exposed to the differing outside temperatures cannot be used as supply air ducts. In addition, these spaces are limited to use for return air only because the negative pressures within the return air plenum with respect to surrounding spaces will decrease the likelihood of spreading smoke to other spaces via the plenum.

N1103.3 Mechanical system piping insulation. Mechanical system piping capable of carrying fluids above 105°F (40°C) or below 55°F (13°C) shall be insulated to a minimum of R-2.

❖ Heat losses during mechanical fluid distribution impact building energy use both in the energy required to make up for the lost heat and in the additional load that can be placed on the space cooling system if the heat is released to air-conditioned space. These losses can be effectively limited by insulating the mechanical system piping that conveys fluids at extreme temperatures.

N1103.4 Circulating hot water systems. All circulating service hot water piping shall be insulated to at least R-2. Circulating hot water systems shall include an automatic or readily accessible manual switch that can turn off the hot water circulating pump when the system is not in use.

❖ When the distribution piping is heated to maintain usage temperatures, such as in circulating hot water systems or systems using pipe heating cable, the system pump or heat trace cable must have conveniently located manual or automatic switches or other controls that can be set to optimize system operation or turn off the system during periods of reduced demand. The simplest of these devices is an automatic time clock.

N1103.5 Mechanical ventilation. Outdoor air intakes and exhausts shall have automatic or gravity dampers that close when the ventilation system is not operating.

❖ Mechanical ventilation is the alternative to having natural ventilation. Both natural and mechanical ventilation can be provided to a space. Unlike natural ventilation, mechanical ventilation does not depend on unpredictable air pressure differentials between the indoors and outdoors to create airflow. The volume of air supplied to a space must be approximately equal to the volume of the air removed from the space. Otherwise, the space will be either positively or negatively pressurized and the actual ventilation flow rate will be equivalent to the lower rate of either the air supply or air exhaust.

The requirements of this section are intended to reduce infiltration into the building when ventilation systems are off. Infiltration speeds up natural cooling or warming of the space during off-hours and can increase the energy use required to maintain normal temperatures.

Fans not equipped with an on/off switch that either introduces outside air into a building or exhaust air outside of a building must have dampers that automatically close when the fan is shut off. These dampers may be either gravity type or motorized, regardless of whether the fan is supplying or exhausting air.

N1103.6 Equipment sizing. Heating and cooling equipment shall be sized as specified in Section M1401.3.

❖ Section M1401.3 stipulates that heating and cooling equipment must be sized based on the building loads calculated in accordance with the Air Conditioning Contractors of America's (ACCA) Manual J. This manual contains a simplified method of calculating heating and cooling loads. It includes a room-by-room calculation method that allows the designer to determine the required capacity of the heating and cooling equipment. In addition, it provides a means to estimate the air-flow requirements for each of the areas in the house. This estimate can be used in sizing the duct system for the types of heating and cooling units that use air as the medium for heat transfer. Other approved methods may be used with the code official's approval.

INDEX

Note: *This is taken from the index for the 2006 International Residential Code. Volume I of the IRC Commentary includes only Chapters 1 through 11.*

N

NONCOMBUSTIBLE MATERIAL

Definition R202

NONMETALLIC BOXES E3805.3, E3808.16

NONMETALLIC RIGID CONDUIT E3701.4

**NONMETALLIC-SHEATHED
CABLE** E3605.4.4, E3701.4,
E3805.3.1, E3808.8.3

NOTCHING

Electrical E3302.1

For plumbing piping P2603.2

For mechanical systems M1308

Steel joists R505.3.5

Steel studs R603.3.4

Wood joists R502.8, R802.7.1

Wood studs R602.6

Wood top plates R602.6.1

O

OCCUPIED SPACE

Definition R202

OIL

Piping and connections M2202

Supply pumps and valves M2204

Tanks M2201

OPEN-TOP GAS BROILER UNITS M1505.1

OPENING

Requirements, combustion air M1702.2,
M1703.2.1

Waterproofing of piping penetrations P2606.1

OUTLET

Definition . . . . Chapter 24, E3401

OVEN, WALL-MOUNTED

Definition applied to
electrical provisions E3401

OVERCURRENT

Definition E3401

OVERCURRENT PROTECTION E3605

OVERCURRENT-PROTECTION RATING E3605.6

OVERLOAD

Definition E3401

P

PARAPETS R317.2.2, R606.2.4

PANELBOARD

Definition E3401

PANELBOARDS E3606

Clearance and dedicated space E3305.2,
E3305.3

Headroom E3305.7

PARTICLEBOARD

Floor R503.3

Walls R605

PERMITS R105

PIERS R606.6

PIPE

Insulation N1103.3, N1103.4

Materials listing . . . . Table M2101.1, P2608

Protection P2603

Standards, drain,
waste and vent . . . . Table P3002.1(1),
P3002.1(2)

Standards, sewer . . . . Table P3002.2

Standards, water supply P2904

PIPING

Connections and materials . . Table M2101.1, P2904

Drain, waste and vent P3002.1

Fuel-gas size determination . . . . Chapter 24

Protection P2603

Sizing methods, water distribution P2903

Support . . . . M2101.9, Chapter 24, P2605

PLANNING

Building . . . . Chapter 3

PLANS R106

PLASTER

Exterior R703.6

Interior R702.2

PLENUM

Definition R202

Definition, electrical installations E3401

PLUMBING

Fixture clearances R307

Fixtures . . . . Chapter 27

Inspection R109.1.2

Materials P2702.2, P2904, P3002

Requirements and definitions R202

System, definition R202

Traps . . . . Chapter 32

Vents . . . . Chapter 31

PLYWOOD

Application R703.3

Materials, walls R604

POTABLE WATER

Definition R202

PRESSURE, WATER SUPPLY

Maximum P2903.3.1

Minimum P2903.3

PRIVATE

Sewage disposal system . . . . Appendix I

PROHIBITED

Receptors P2706.3